3.2 Relation, Domain, Range — A set of ordered pairs is called a rela… all first elements in the ordered pairs, and the range is the set of all second el…

3.3 Function — A function is a relation that assigns to each element of a set X exactly one element of a set Y.

3.4 Slope — The slope m of the line through (x_1, y_1) and (x_2, y_2), with $x_2 - x_1 \neq 0$, is $\dfrac{y_2 - y_1}{x_2 - x_1}$. The slope of a vertical line is undefined.

3.5 Point-Slope Form — The line with slope m passing through the point (x_1, y_1) has equation $y - y_1 = m(x - x_1)$.

3.5 Slope-Intercept Form — The line with slope m and y-intercept b has equation $y = mx + b$.

3.7 Composition of Functions — $(g \circ f)(x) = g[f(x)]$

3.8 One-to-One Function — A function f is one-to-one if $f(a) = f(b)$ implies that $a = b$.

3.8 Inverse Functions — Let f be a one-to-one function. Then g is the inverse function of f if

$(f \circ g)(x) = x$ for every x in the domain of g,

and $(g \circ f)(x) = x$ for every x in the domain of f.

5.2 Logarithm — For $a > 0$, $a \neq 1$, and $x > 0$, $\log_a x$ is the power to which a must be raised to get x.

5.2 Properties of Logarithms — For $a > 0$, $a \neq 1$, $x > 0$, and $y > 0$,

$\log_a xy = \log_a x + \log_a y$; $\quad \log_a \dfrac{x}{y} = \log_a x - \log_a y$;

$\log_a x^r = r \cdot \log_a x$; $\quad \log_a a = 1$;

$\log_a 1 = 0$.

8.2 Arithmetic Sequences — nth term: $a_n = a_1 + (n - 1)d$ $\quad S_n = \dfrac{n}{2}(a_1 + a_n)$ or $S_n = \dfrac{n}{2}[2a_1 + (n - 1)d]$

8.3 Geometric Sequences — nth term: $a_n = a_1 r^{n-1}$ $\quad S_n = \dfrac{a_1(1 - r^n)}{1 - r}$ $(r \neq 1)$ $\quad S_\infty = \dfrac{a_1}{1 - r}$ $(-1 < r < 1)$

8.6 Permutations — The number of arrangements of n things taken r at a time is $P(n, r) = \dfrac{n!}{(n - r)!}$.

8.6 Combinations — The number of ways to choose r things from a group of n things is $\dbinom{n}{r} = \dfrac{n!}{(n - r)!\,r!}$.

8.7 Definition and Properties of Probability — In a sample space S with equally likely outcomes, $P(E) = \dfrac{n(E)}{n(S)}$.

For any events E and F:

$P(\text{a certain event}) = 1$; $\quad 0 \leq P(E) \leq 1$;

$P(\text{an impossible event}) = 0$; $\quad P(E') = 1 - P(E)$;

$P(E \text{ or } F) = P(E \cup F) = P(E) + P(F) - P(E \cap F)$.

ANNOTATED INSTRUCTOR'S EDITION

Fundamentals of College Algebra

ANNOTATED INSTRUCTOR'S EDITION

Fundamentals of College Algebra

CHARLES D. MILLER

MARGARET L. LIAL
American River College

DAVID I. SCHNEIDER
University of Maryland

FOURTH EDITION

HarperCollinsCollegePublishers

Sponsoring Editor: Anne Kelly
Managing Developmental Editor: Elaine Silverstein
Project Editor: Cathy Wacaser
Design Administrator: Jess Schaal
Text Design: David Lansdon
Cover Design: Lesiak/Crampton Design Inc.: Cynthia Crampton
Cover Photo: The Chicago Photographic Company
Production Administrator: Randee Wire
Compositor: CRWaldman Graphic Communications
Printer and Binder: R.R. Donnelley & Sons Company
Cover Printer: R.R. Donnelley & Sons Company

Ceiling detail on the cover: In 1921 the Apollo Theatre opened as a legitimate theatre presenting musical comedies. In 1927 it was remodeled for motion pictures. The new design, Moorish in inspiration, was accompanied by a new name. The Apollo became the United Artists Theatre. After years of neglect, it was razed in the late 1980s.

Library of Congress Cataloging-in-Publication Data

Miller, Charles David.
Fundamentals of college algebra / Charles D. Miller, Margaret L. Lial, David I. Schneider.—4th ed.
p. cm.
Includes index.
ISBN 0-673-46743-0
1. Algebra. I. Lial, Margaret L. II. Schneider, David I. III. Title.
QA154.2.M54 1993
512.9—dc20 93-5730
CIP

93 94 95 96 9 8 7 6 5 4 3 2 1

ANNOTATED INSTRUCTOR'S EDITION

Fundamentals of College Algebra

CHARLES D. MILLER

MARGARET L. LIAL
American River College

DAVID I. SCHNEIDER
University of Maryland

FOURTH EDITION

HarperCollinsCollegePublishers

Sponsoring Editor: Anne Kelly
Managing Developmental Editor: Elaine Silverstein
Project Editor: Cathy Wacaser
Design Administrator: Jess Schaal
Text Design: David Lansdon
Cover Design: Lesiak/Crampton Design Inc.: Cynthia Crampton
Cover Photo: The Chicago Photographic Company
Production Administrator: Randee Wire
Compositor: CRWaldman Graphic Communications
Printer and Binder: R.R. Donnelley & Sons Company
Cover Printer: R.R. Donnelley & Sons Company

Ceiling detail on the cover: In 1921 the Apollo Theatre opened as a legitimate theatre presenting musical comedies. In 1927 it was remodeled for motion pictures. The new design, Moorish in inspiration, was accompanied by a new name. The Apollo became the United Artists Theatre. After years of neglect, it was razed in the late 1980s.

Library of Congress Cataloging-in-Publication Data

Miller, Charles David.
Fundamentals of college algebra / Charles D. Miller, Margaret L. Lial, David I. Schneider.—4th ed.
p. cm.
Includes index.
ISBN 0-673-46743-0
1. Algebra. I. Lial, Margaret L. II. Schneider, David I. III. Title.
QA154.2.M54 1993
512.9—dc20 93-5730
CIP

93 94 95 96 9 8 7 6 5 4 3 2 1

Preface

Fundamentals of College Algebra, Fourth Edition, gives a mathematically sound development of the topics in algebra needed for success in later courses, particularly the study of calculus. The book is suitable for one-semester, two-semester, or two-quarter courses. We assume a prerequisite of intermediate algebra. For those students whose mathematical background is not solid or who have not studied mathematics recently, we have included a review of the basics of algebra in Chapter 1 and a review of equations and inequalities in Chapter 2.

In this new edition we have introduced a graphing calculator or computer (grapher) strand parallel to the regular text that can be integrated into the course if the instructor desires. References to graphers and discussions of graphing techniques are presented in the margin next to the applicable text. Where appropriate, exercise sets include clearly labeled exercises especially for graphers. Many of these exercises are exploratory. The grapher strand is not included in Chapter 1 because that material is often treated as review.

All of the successful features of the previous edition are carried over in this new edition. We have added many new features in response to recent instructional trends and users' comments. The text is organized so that the material flows logically and has been written in a clear, readable style for student comprehension. We have been careful to provide motivation and to relate new topics to those previously studied. The historical material, which has been very well received, provides background and a sense of continuity with the past. We have provided fully developed examples with comments printed at the side, and extensive section and review exercises (approximately 3,800 in all) that offer a variety of problems, including many applications that are current and interesting. Screened boxes set off important definitions, formulas, rules, and procedures to further aid students in learning and reviewing the course material. Cautionary remarks and notes highlight common student errors and difficulties as well as important comments.

We continue to provide an extensive supplemental package. For students, we offer a solution manual, interactive tutorial software, videotapes, and supplements containing further exploratory exercises. For instructors, we present an Annotated Instructor's Edition with all answers provided next to the exercises in a special section, overhead transparencies, a computerized test generator, a test manual, and solutions to even-numbered exercises.

New Features

Several new features, designed to assist students in the learning process, have been integrated into this edition. These features are illustrated on this page and the next. In addition, the use of full color and changes in format enhance the book's pedagogical features and increase its accessibility.

Grapher Comments

These permit the integration of graphing calculators and computers into courses using this text.

132 Chapter 3 Functions and Graphs

For Graphers

Some graphing utilities (for example, the Texas Instruments TI-81) will plot points and draw lines connecting two points. The graphs in Figure 10 can be drawn using a grapher with this capability. To graph relations with equations that can be put in the form $y =$ (an expression in x), however, it is easier to use the graph function of the graphing utility. The absolute value function is often a built-in function in a graphing utility. It is designated by abs(x) or $|x|$. If neither of these forms is available, use the fact that $|x| = \sqrt{x^2}$. Figure 12 shows the graph in Figure 11 drawn with a graphing utility.

EXAMPLE 3 Graph $y = |x|$.

Start with a table.

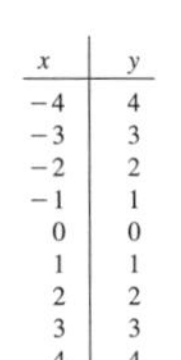

x	y
−4	4
−3	3
−2	2
−1	1
0	0
1	1
2	2
3	3
4	4

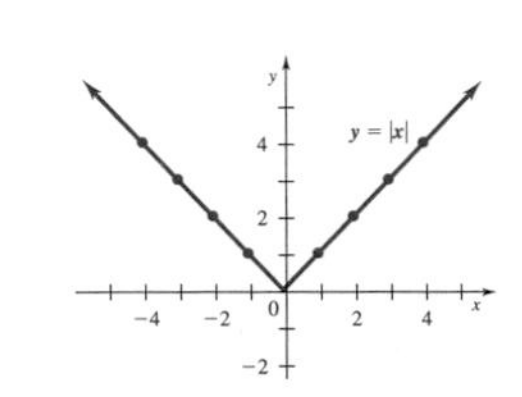

Figure 11

...this table to get the points in Figure 11. The graph drawn through these points ...de up of portions of two straight lines. The domain is $(-\infty, \infty)$ and the range ... $\infty)$.

...UTION There is a danger in the method used in Examples 2 and 3—we ...t choose a few values for x, find the corresponding values of y, begin to sketch ...ph through these few points, and then make a completely wrong guess as to ...hape of the graph. For example, choosing only −1, 0, and 1 as values of x in ...nple 3 above would produce only the three points $(-1, 1)$, $(0, 0)$, and $(1, 1)$. ...e three points alone would not give enough information to determine the proper ...n for $y = |x|$. However, this section involves only elementary graphs; when ...complicated graphs are presented later, we will develop more accurate methods ...orking with them. ■

...AMPLE 4 Graph $x = y^2 - 4$.

...ince $y^2 \geq 0$, the domain is $[-4, \infty)$. It is easier here to choose values of y, then ...he corresponding x-values. Choosing 1 for y, for example, gives $x = (1)^2 -$ ...−3. Choosing −1 for y gives the same result. The table shown with Figure 13 ...values of x corresponding to various values of y. The ordered pairs from this ...were used to get the points plotted in Figure 13. (Don't forget that x always ...first in the ordered pair.) A smooth curve was then drawn through the resulting ...s. Here, y can take on any value, so the range is $(-\infty, \infty)$. As mentioned earlier, ...omain is $[-4, \infty)$.

2.6 Inequalities 99

still a true statement. On the other hand, starting with $-3 < 5$ and multiplying both sides by the *negative* number −2 gives a true result only if the direction of the inequality symbol is reversed:

$$-3(-2) > 5(-2)$$
$$6 > -10. \blacksquare$$

For Graphers

A grapher can be used to solve the inequality in Example 1 as follows. Write the inequality with 0 on one side: $-3x + 12 > 0$. Graph $y = -3x + 12$. See Figure 9. Since $y = -3x + 12$, the y-values of the points on the graph show where $-3x + 12$ is positive or negative. The x-value where $y = 0$ is the dividing point. The graph shows that $y = -3x + 12 > 0$ when $x < 4$. This is the same solution found algebraically. Note that the graph does not tell you whether or not the endpoint is included in the solution. You must decide whether the endpoint satisfies the given inequality—that is, whether the inequality involves $>$ or $\geq$. The algebraic method is simpler for linear inequalities, but the graphical method is useful for inequalities of higher degree.

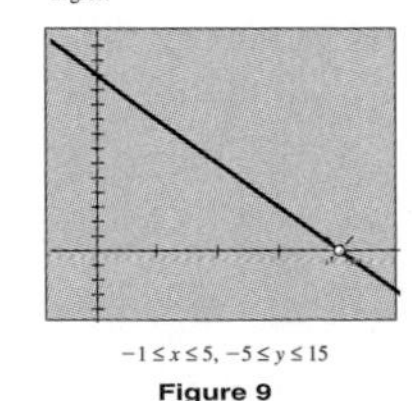

$-1 \leq x \leq 5$, $-5 \leq y \leq 15$

Figure 9

EXAMPLE 1 Solve the inequality $-3x + 5 > -7$.

Use the properties of inequalities. First, add −5 to both sides.

$$-3x + 5 + (-5) > -7 + (-5)$$
$$-3x > -12$$

Now multiply both sides by −1/3. Since $-1/3 < 0$, reverse the direction of the inequality symbol.

$$-\frac{1}{3}(-3x) < -\frac{1}{3}(-12)$$
$$x < 4$$

The original inequality is satisfied by any real number less than 4. The solution set can be written $\{x | x < 4\}$. A graph of the solution set is shown in Figure 8, where the parenthesis is used to show that 4 itself does not belong to the solution set.

0 4

Figure 8

The set $\{x | x < 4\}$, the solution set for the inequality in Example 1, is an example of an **interval**. A simplified notation, called **interval notation**, is used for writing intervals. With this notation, the interval in Example 1 can be written as just $(-\infty, 4)$. The symbol $-\infty$ is not a real number; the symbol is used as a convenience to show that the interval includes all real numbers less than 4. The interval $(-\infty, 4)$ is an example of an *open interval*, since the endpoint, 4, is not part of the interval. Examples of other sets written in interval notation are shown below. Square brackets are used to show that the given number *is* part of the graph. Whenever two real numbers a and b are used to write an interval, it is assumed that $a < b$.

Pedagogical Use of Color

Color is used to clarify the presentation of new concepts and to aid understanding in both examples and figures.

Connection Exercises

Exercises that require skills or techniques covered in previous sections or chapters are included in some section exercises and review exercise sets (about 150 in all). These provide review of earlier work and relate those topics to the current material.

New Exercises

Nearly 1,500 exercises are new to this edition. New exercises include drill-and-practice types as well as a variety of interesting applications.

Conceptual and Writing Exercises

To complement the drill and application exercises, several exercises especially requiring an understanding of the concepts introduced in a section are included in almost every exercise set (more than 400 in all). Most of these require the student to respond by writing a few sentences.

58. Temperature decreases with height above the earth's surface. We can use the following chart to find a function that describes this relationship.

Height (feet)	*Temperature (°F)*
1000	56
5000	41
10,000	23
15,000	5
20,000	−15
30,000	−47
36,100	−69

(a) Let x represent the height in thousands of feet and y represent the temperature in degrees Fahrenheit. Plot the ordered pairs (height, temperature) corresponding to the values in the table. The points should lie in an approximately linear pattern.
(b) Use the ordered pairs $(5, 41)$ and $(30, -47)$ to find the slope of the line through these two points. Then substitute the values from either ordered pair into the equation $y = mx + b$, which defines a linear function, to find b. Use the values of m and b to write the equation that defines the linear function describing the relationship between height and temperature.
(c) Test the function found in part (b) by substituting other heights from the chart to see if the predicted temperatures are close to the actual ones. What do you find?

59. Use the ordered pairs $(20, -15)$ and $(10, 23)$ to repeat Exercise 58(b). Compare th two equations. Are they similar? Do they give equally good predictions?

3.6 Symmetry, Translatio

One of the main objectives of this course i functions. Several graphing techniques are to graph functions that are defined by alte certain ways.

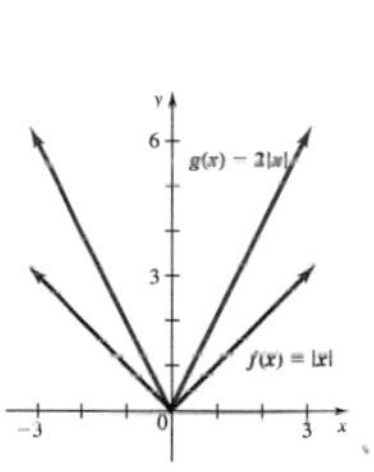

Figure 35

Stretching and Shrinking We begin by compares to the graph of $y = f(x)$.

EXAMPLE 1 Graph each of the fo

(a) $g(x) = 2|x|$.
Use a grapher or plot a few points in Figure 35. The graph of $f(x) = |x$ each y-value in $g(x)$ is twice the corre $g(x)$ is narrower than that of $f(x)$.

39. Explain why the graph of a polynomial having a as a zero of odd multiplicity crosses the x-axis at $x = a$.

40. Explain why the graph of a polynomial having a as a zero of even multiplicity touches, but does not cross, the x-axis at $x = a$.

In Exercises 41 and 42, find a cubic polynomial having the graph shown.

41.

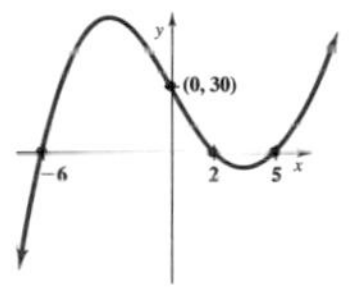

42.

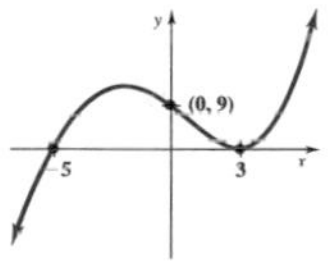

43. Give an example of a polynomial function that is never negative and has −4 and 1 as zeros.

44. Give an example of a polynomial function that has −3, 1, and 2 as zeros and is positive only between 1 and 2.

Determine the domain and range of the functions defined in Exercises 45 and 46.

45. $f(x) = \sqrt{x^3 - x^2 - 6x}$

46. $f(x) = \sqrt{x^3 - x}$

47. Summarize the steps used to graph a polynomial.

Graphing Utility Problems

48. The pressure of the oil in a reservoir tends to drop with time. By taking sample pressure readings for a particular oil reservoir, petroleum engineers have found that the change in pressure is given by $P(t) = t^3 - 25t^2 + 200t$, where t is time in years from the date of the first reading.
(a) Graph $P(t)$.
(b) Use the graph from part (a) to decide for what time periods the amount of change in pressure (drop) is increasing or decreasing.

49. During the early part of the twentieth century, the deer population of the Kaibab Plateau in Arizona experienced a rapid increase, because hunters had reduced the number of natural predators and because the deer were protected from hunters. The increase in population depleted the food resources and eventually caused the population to decline. For the period from 1905 to 1930, the deer population was approximated by $D(x) = -.125x^5 + 3.125x^4 + 4000$, where x is time in years from 1905.
(a) Graph $D(x)$.
(b) From the graph, over what period of time (from 1905 to 1930) was the population increasing? Relatively stable? Decreasing?

Graphing Calculator/Computer Exercises

A separate section in appropriate exercise sets is provided for those using the grapher strand. These exercises demonstrate how graphers can be used to clarify and illustrate concepts. There are approximately 150 of these new exercises.

New Content Highlights

• An intuitive introduction to the binomial theorem, using Pascal's triangle to give the coefficients, now follows the formulas for polynomial products in Chapter 1.

• The first two sections of Chapter 3 have been extensively rewritten to distribute the topics more evenly and improve the flow of the material. The section on quadratic functions has been deferred to Chapter 4, which treats polynomial functions. Section 3.6, on symmetry and translations, has been rewritten and now includes new material on stretching, shrinking, and reflections, so all types of graphing techniques are covered.

• We have increased the emphasis on graphs of polynomial functions in Chapter 4 by moving that section earlier in the chapter and broadening the discussion of general techniques of graphing that particularly apply to polynomial (and rational) functions.

• The first section of Chapter 6 now briefly reviews linear systems of two or three variables using elimination, and then presents methods of solution of nonlinear systems of equations.

• Chapter 7 has been rewritten to give more extensive coverage of ellipses and hyperbolas, including a thorough discussion of eccentricity.

• Chapter 8, ''Further Topics,'' has been rearranged for a more logical flow of ideas. The chapter now begins with the general concept of series, then presents the special cases of arithmetic and geometric sequences and series. This is followed by mathematical induction, which includes proofs of formulas for arithmetic and geometric series. Discussion of the binomial theorem, a special series, is next and includes a proof by induction. The last two sections on counting theory and probability are independent.

Supplements

Our extensive supplemental package includes an annotated instructor's edition that contains answers to all exercises in a special section at the back of the book, testing materials, solution manuals, software, and videotapes.

For the Instructor

Annotated Instructor's Edition With this volume, instructors have immediate access to the answers to every exercise in the text, excluding writing exercises and proofs. Each answer is printed in bold type next to or below the corresponding exercise in the annotated exercises section. In addition, challenging exercises, which will require most students to stretch beyond the concepts described in the text, are marked with the symbol ▲. The conceptual (◉), connection (◁▶), and writing (✎) exercises are also marked in the annotated exercises so that instructors may assign these problems at their discretion.

Instructor's Test Manual Included here are six versions of a chapter test for each chapter and a set of 130 to 250 test items per chapter, which can be used as an additional source of questions for tests, quizzes, or student review of difficult topics. Answers to all test forms and test items are provided.

Instructor's Solution Manual This manual includes complete, worked-out solutions to all even-numbered section exercises and chapter review exercises in the textbook (excluding most writing exercises).

HarperCollins Test Generator/Editor for Mathematics with QuizMaster Available in IBM and Macintosh versions, the test generator is fully networkable and allows instructors to select questions by objective, section, or chapter or to use a ready-made test for each chapter. The editor enables instructors to edit any preexisting data or to easily create their own questions. The software is algorithm driven, allowing the instructor to regenerate constants while maintaining problem type, providing a nearly unlimited number of available test or quiz items in multiple-choice or open-response format. The system features printed graphics and accurate mathematical symbols. QuizMaster enables instructors to create tests and quizzes using the Test Generator/Editor and save them on disk so that students can take them on a stand-alone computer or network. QuizMaster then grades the test or quiz, allowing the instructor to create reports on individual students or classes.

Transparencies Color overhead transparencies of figures, examples, definitions, procedures, properties, and problem-solving methods are available to assist instructors in presenting important points during their lectures.

For the Student

Student's Solution Manual Complete, worked-out solutions are given for odd-numbered section exercises and all chapter review exercises (excluding most writing exercises) in a volume available for purchase by students. In addition, a practice chapter test is provided for each chapter.

College Algebra with Trigonometry: Graphing Calculator Investigations This supplemental text, written by Dennis Ebersole of Northampton County Area Community College, provides investigations that help students visualize and explore key concepts, generalize and apply concepts, and identify patterns.

Videotapes A new videotape series has been developed to accompany *Fundamentals of College Algebra*, Fourth Edition. In a separate lesson for each section of the book, the series covers all objectives, topics, and problem-solving techniques within the text.

Interactive Tutorial Software with Management System This innovative package is also available in IBM and Macintosh versions and is fully networkable. As with the Test Generator/Editor, this software is algorithm driven, automatically regenerating constants so that students will not see the numbers repeat in a problem type if they return to any particular section. The tutorial is self-paced and provides unlimited opportunities to review lessons and to practice problem solving. When students give a wrong answer, they can request to see the problem worked out. The program is menu-driven for ease of use, and on-screen help can be obtained at any time with a single keystroke. Students' scores are automatically recorded and can be printed for a permanent record. The optional Management System lets instructors record student scores on disk and print diagnostic reports for individual students or classes.

Computer Software and Accompanying Manual: Visual Precalculus This package enhances the visualization and therefore the understanding of the ideas behind algebra. Although it has most of the computing power of standard mathematics packages, its focus is the teaching of concepts rather than calculations. Its combination of power and ease of use makes it the software of choice for students wanting to master algebra. Relevant examples and exercises from the text can be selected without having to be typed. The software is packaged with a manual containing documentation, walk-throughs, and exercises.

Acknowledgments

We want to thank the following professors for their contributions in reviewing portions of this text.

Emily Anne Battle, *University of Montevallo*
Charles D. Bedal, *Chandler Gilbert Community College*
Bobbie Parrino Cook, *Indian River Community College*
Roger E. Davis, *Pennsylvania College of Technology*
Lisa E. Downing, *Oakland Community College*
Constance Edwards, *Coastal Carolina College*
Sue Ehlers, *Anoka-Ramsey Community College*
Susan Farley, *Oklahoma University*
Susan S. Garstka, *Moraine Valley Community College*
Juan A. Gatica, *University of Iowa*
William L. Grimes, *Central Missouri State University*
Bob Harbison, *Blinn College*
Louise S. Hasty, *Austin Community College*
Brian Hayes, *Triton College*
Robert Hessel, *Rock Valley College*
Elizabeth Hodes, *Santa Barbara City College*
Norma F. James, *New Mexico State University*
Margret Kothmann, *University of Wisconsin—Stout*
William R. Livingston, *Missouri Southern State College*
James I. McCullough, *Arapahoe Community College*
LuAnn Malik, *Community College of Aurora*
Steve Martin, *Richard Bland College*
Robert Andrew Maynard, *Tidewater Community College*
Lawrence P. Merbach, *North Dakota State College of Science*
Peggy I. Miller, *University of Nebraska—Kearney*
Paul Riggs, *Jackson State University*
Ann Smith, *Hutchinson Community College*
Ben J. Sultenfuss, *Stephen F. Austin State University*
Jan Vandever, *South Dakota State University*

Paul Eldersveld, College of DuPage, deserves our gratitude for doing an excellent job coordinating all of the print ancillaries for us, an enormous and time-consuming task. Paul Van Erden, American River College, has created an accurate and complete index for us. Kitty Pellissier provided invaluable help in maintaining high standards of accuracy in the answer section. We especially thank the fine, professional staff at HarperCollins for their assistance and contributions to this project: Anne Kelly, Linda Youngman, Elaine Silverstein, and Janet Tilden.

Margaret L. Lial
David I. Schneider

Contents

1 Fundamentals of Algebra

Today, algebra is required in a great many fields, ranging from accounting to ecology. This is not surprising, since most topics in algebra were developed to help people solve applied problems. For example, algebra is used in this text to predict population growth, to determine the path of objects orbiting in space, and to investigate the costs versus the benefits of removing pollutants from a substance. To prepare for such problem-solving, the book begins with a review of the basics of algebra.

The name *algebra* is a Latin translation of the Arabic *al-jabr*. This word is part of the title of a book written by āl-Khwarizmī (ca. 800–847), a mathematician and astronomer who lived in Baghdad. The book is a collection of rules for solving equations. Since some techniques had been developed to solve geometric problems, negative numbers were not accepted as solutions. Part of the development of algebra to its current usefulness involves mathematicians' gradual recognition of negative and imaginary numbers. These enable us to write down *all* possible solutions of equations. We begin our study with a review of the real number system.

1.1 The Real Numbers

Numbers are the foundation of mathematics. The most common numbers in mathematics are the **real numbers**. These numbers can be written as decimals, either repeating: $\frac{1}{3} = .3333\overline{3}$, $\frac{3}{4} = .7500\overline{0}$, or $2\frac{4}{7} = 2.571428\overline{571428}$,

or nonrepeating: $\sqrt{2} = 1.4142135\ldots$ or $\pi = 3.14159\ldots.$

A repeating decimal such as $.75\overline{0}$ is also called a terminating decimal.

To use decimals, we may either *round off* or *truncate* them. For example, the number 2.56894. . . can be approximated to two decimal places as follows (the symbol $\approx$ means "is approximately equal to"):

$$2.56894\ldots \approx 2.57 \text{ by rounding;}$$
$$2.56894\ldots \approx 2.56 \text{ by truncating.}$$

In this text, we round decimals.

The real numbers are said to be **closed** under the operations of addition and multiplication. That is, for any two real numbers a and b, the **sum** $a + b$ and the **product** $a \cdot b$ are unique real numbers. Multiplication is written in a variety of ways. The symbols 2×8, $2 \cdot 8$, $2(8)$, and $(2)(8)$ all represent the product of 2 and 8, or 16. For writing products involving **variables** (letters used to represent numbers), no operation symbols may be necessary: $2x$ represents the product of 2 and x, while xy indicates the product of x and y.

The set of real numbers, together with the operations of addition and multiplication, form the **real number system**. (Informally, a **set** is a collection of objects.) The key properties of the real number system are given below, where a, b, and c are letters used to represent any real number.

Properties of the Real Numbers

For all real numbers a, b, and c:

Closure properties	$a + b$ is a real number. ab is a real number.
Commutative properties	$a + b = b + a$ $ab = ba$
Associative properties	$(a + b) + c = a + (b + c)$ $(ab)c = a(bc)$
Identity properties	There exists a unique real number 0 such that $a + 0 = a$ and $0 + a = a$. There exists a unqiue real number 1 such that $a \cdot 1 = a$ and $1 \cdot a = a$.
Inverse properties	There exists a unique real number $-a$ such that $a + (-a) = 0$ and $(-a) + a = 0$. If $a \neq 0$, there exists a unique real number $1/a$ such that $a \cdot \frac{1}{a} = 1$ and $\frac{1}{a} \cdot a = 1$.
Distributive property	$a(b + c) = ab + ac$

Let's consider some consequences of the last four properties.

The associative properties are used to add or multiply three or more numbers. For example, the associative property for addition says that the sum $a + b + c$ of the real numbers a, b, and c can be found either by first adding a and b, and then adding c to the result, indicated by the association

$$(a + b) + c,$$

or by first adding b and c, and then adding a to the result, indicated by

$$a + (b + c),$$

since, by the associative property, either method gives the same result.

The identity properties show that 0 and 1 are special numbers: the sum of 0 and any real number a is the number a, so that 0 preserves the identity of a real number under addition. For this reason, 0 is the **identity element for addition.** In the same way, 1 preserves the identity of a real number under multiplication, making 1 the **identity element for multiplication.**

According to the additive inverse property, for any real number a there is a real number, written $-a$, such that the sum of a and $-a$ is 0, or $a + (-a) = 0$. The number $-a$ is called the **additive inverse**, **opposite**, or **negative** of a. The additive inverse property also says that this number $-a$ is *unique*; that is, a given number has only one additive inverse.

CAUTION Don't confuse the *negative of a number* with a *negative number.* Since a is a variable, it can represent a positive or a negative number (as well as zero). The negative of a, written $-a$, can also be either a negative or a positive number (or zero). It is a common mistake to think that $-a$ *must* represent a negative number; however, if a is -3, for example, then $-a$ is $-(-3)$ or 3. ■

For each real number a except 0, there is a real number $1/a$ such that the product of a and $1/a$ is 1, or

$$a \cdot \frac{1}{a} = 1, \quad a \neq 0.$$

The symbol $1/a$ is often written a^{-1}.

Definition of a^{-1}

For every nonzero real number a,

$$a^{-1} = \frac{1}{a}.$$

For example, $2^{-1} = 1/2$, $5^{-1} = 1/5$, and $-3^{-1} = -1/3$.

The number $1/a$ or a^{-1} is called the **multiplicative inverse** or **reciprocal** of the number a. Every real number except 0 has a reciprocal. As with the additive inverse, the multiplicative inverse is unique—a given nonzero real number has only one multiplicative inverse.

The distributive property is particularly useful; we use it to rewrite certain sums as products or to rewrite some products as sums. Using the commutative property, we can rewrite the distributive property in an alternative form. Since $a(b + c) = (b + c)a$,

$$(b + c)a = ba + ca.$$

The distributive property can be extended to include more than two numbers in the sum, as follows:

$$a(b + c + d + e + \ldots + n) = ab + ac + ad + ae + \ldots + an.$$

This form is called the **extended distributive property**.

EXAMPLE 1 The following statements illustrate some properties of the real numbers.

(a) $-2 + 3 = 3 + (-2)$ Commutative property of addition

(b) $\sqrt{2}m = m\sqrt{2}$ Commutative property of multiplication

(The product of $\sqrt{2}$ and m is often written $m\sqrt{2}$, since $\sqrt{2}m$ is too easily confused with $\sqrt{2m}$.)

(c) $-\pi + (\pi + 3) = (-\pi + \pi) + 3$ Associative property of addition

(d) $3(9x) = (3 \cdot 9)x$ Associative property of multiplication

(e) $9 + (-9) = 0$ Inverse property of addition

(f) $6 \cdot 6^{-1} = 6 \cdot \dfrac{1}{6} = 1$ Inverse property of multiplication

(g) $6k^2 + 3k^3 = 2 \cdot 3k^2 + k \cdot 3k^2$
$= (2 + k)3k^2$ Distributive property

(This example shows that the distributive property can also be used in reverse to factor an expression.) ■

Other properties useful for solving equations are given at the top of the next page.

Further Properties of Real Numbers

For all real numbers a, b, and c:

Substitution property
If $a = b$, then a may replace b or b may replace a in any expression without affecting the truth or falsity of the statement.

Addition property If $a = b$, then $a + c = b + c$

Multiplication property If $a = b$, then $ac = bc$.

Properties of negatives $-(-a) = a$ $a(-b) = -(ab)$

$-a(b) = -(ab)$ $(-a)(-b) = ab$

Properties of zero $a \cdot 0 = 0$

$ab = 0$ **if and only if** $a = 0$ or $b = 0$.

The addition and multiplication properties say that the same number may be added or multiplied on both sides of a statement of equality. The second property of zero contains the phrase "if and only if." This phrase implies that "if $ab = 0$, then $a = 0$ or $b = 0$" and, conversely, "if $a = 0$ or $b = 0$, then $ab = 0$."

Definition of *If and Only If*

For any statements p and q,

p **if and only if** q

means

if p is true, then q is true and **if q is true, then p is true.**

The properties of real numbers given above apply to addition or multiplication. The two other common operations for the real numbers, subtraction and division, are defined in terms of the operations of addition and multiplication, respectively.

Subtraction is defined by saying that the difference of the numbers a and b, written $a - b$, is found by adding a and the *negative* of b.

Definition of Subtraction

For all real numbers a and b,

$$a - b = a + (-b).$$

Division of a real number a by a nonzero real number b is defined in terms of multiplication as follows.

Definition of Division

For all real numbers a and b, with $b \neq 0$,

$$\frac{a}{b} = a \cdot \frac{1}{b} = ab^{-1}.$$

That is, to divide a by b, multiply a by the reciprocal of b. The symbol $\div$ is also used to indicate division, as in $12 \div 3 = 4$.

Several useful properties of quotients are listed below.

Properties of Quotients

For all real numbers a, b, c, and d, with all denominators nonzero:

$$\frac{a}{b} = \frac{c}{d} \quad \textbf{if and only if} \quad ad = bc \qquad\qquad \frac{a}{b} + \frac{c}{d} = \frac{ad + bc}{bd}$$

$$\frac{ac}{bc} = \frac{a}{b} \qquad\qquad \frac{a}{b} - \frac{c}{d} = \frac{ad - bc}{bd}$$

$$\frac{a}{-b} = \frac{-a}{b} = -\frac{a}{b} \qquad\qquad \frac{a}{b} \cdot \frac{c}{d} = \frac{ac}{bd}$$

$$\frac{-a}{-b} = \frac{a}{b} \qquad\qquad \frac{a}{b} \div \frac{c}{d} = \frac{a}{b} \cdot \frac{d}{c}$$

EXAMPLE 2

(a) $6 - (-15) = 6 + (-(-15))$ Definition of subtraction

$= 6 + 15$ $\quad -(-a) = a$

$= 21$

(b) $\dfrac{-8}{4} = -8 \cdot \dfrac{1}{4} = -2$

(c) $\dfrac{-9}{0}$ is undefined since 0 has no multiplicative inverse; also, $\dfrac{0}{0}$ is undefined.

(d) $\dfrac{\frac{2}{3}}{\frac{5}{7}} = \dfrac{2}{3} \div \dfrac{5}{7} = \dfrac{2}{3} \cdot \dfrac{7}{5} = \dfrac{14}{15}$ ■

Order of Operations To avoid ambiguity when working problems, use the following **order of operations**, which has been generally agreed upon. (By the way, this order of operations is used by computers and many calculators.)

Order of Operations

1. Work separately above and below any fraction bar.
2. Use Steps 3–6 below within each set of parentheses or square brackets. Start with the innermost grouping and work outward.
3. Evaluate all exponents and roots.
4. Evaluate negations.
5. Do any multiplications or divisions in the order in which they occur, working from left to right.
6. Do any additions or subtractions in the order in which they occur, working from left to right.

For Graphers

Graphing calculators and computers use the order of operations shown here. You may want to use the problems given in Example 3 to experiment with your grapher. (We refer to graphing calculators and computers with graphing software generically as "graphing utilities" or "graphers" in this text.)

EXAMPLE 3 Use the order of operations given above to simplify each of the following expressions.

(a) $9 + 6 \div 3 = 9 + 2 = 11$

(b) $\dfrac{1 - 2(3 + 4)}{5 + 6 \cdot 2} = \dfrac{1 - 2(7)}{5 + 12}$

$$= \frac{1 - 14}{17} = \frac{-13}{17} = -\frac{13}{17}$$

(c) $4 \cdot 2^3 = 4 \cdot 8 = 32$

(d) $-2^2 = -(2^2) = -4$ ■

CAUTION A common error is evaluating negations before exponents. For instance, in Example 3(d) this type of error would give the incorrect calculation $-2^2 = (-2)^2 = 4$. ■

Subsets of the Real Numbers There are several subsets* of the set of real numbers that come up so often they are given special names, as listed below. Some of the subsets are written with **set-builder notation**; with this notation,

$$\{x | x \text{ has property } P\},$$

read "the set of all elements x such that x has property P," represents the set of all elements having some specified property P.

*Set A is a **subset** of set B if and only if every element of set A is also an element of set B.

Subsets of the Real Numbers	
Natural numbers	$\{1, 2, 3, 4, \ldots\}$
Whole numbers	$\{0, 1, 2, 3, 4, \ldots\}$
Integers	$\{\ldots, -3, -2, -1, 0, 1, 2, 3, \ldots\}$
Rational numbers	$\left\{\frac{p}{q} \,\middle|\, p \text{ and } q \text{ are integers}, \quad q \neq 0\right\}$
Irrational numbers	$\{x \mid x \text{ is a real number that is not rational}\}$

EXAMPLE 4 Let set $A = \{-8, -6, -3/4, 0, 3/8, 1/2, 1, \sqrt{2}, \sqrt{5}, 6, \sqrt{-1}\}$.

(a) The *natural numbers* in set A are 1 and 6.

(b) The *whole numbers* are 0, 1, and 6.

(c) The *integers* are -8, -6, 0, 1, and 6.

(d) The *rational numbers* are -8, -6, $-3/4$, 0, $3/8$, $1/2$, 1, and 6. (The number -8 is rational since -8 can be written as the quotient $-8/1$. Also, 6 is rational since $6 = 6/1$.)

(e) The *irrational numbers* are $\sqrt{2}$ and $\sqrt{5}$. (In further mathematics courses you will see that these numbers do not have repeating or terminating decimal expressions.)

(f) All elements of A are *real numbers* except $\sqrt{-1}$. Square roots of negative numbers are discussed later in this chapter. ■

The relationships among the various sets of numbers are shown in Figure 1. All the numbers shown are real numbers. (The number π shown in Figure 1 is the ratio

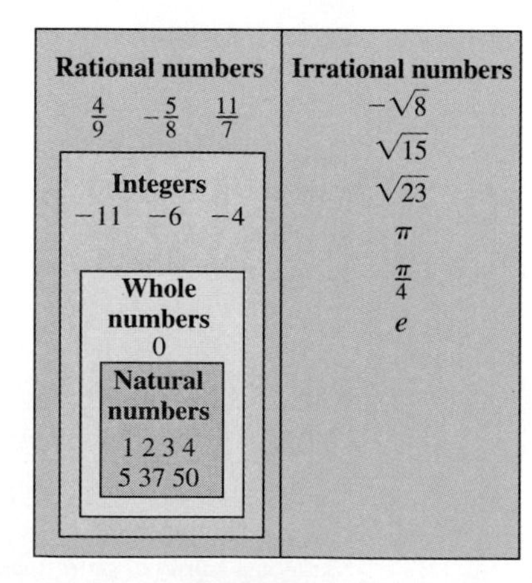

Figure 1

of the circumference of a circle to its diameter; π is approximately 3.14159. Also, e is an irrational number discussed later in this text in connection with exponential and logarithmic functions; e is approximately 2.7182818.)

Inequalities We often need to know which of two given real numbers is the smaller one. Deciding which is smaller is sometimes easier with a **number line**, a geometric representation of the set of real numbers. Figure 2 shows a number line with the points corresponding to several different numbers marked on it. A number that corresponds to a particular point on a line is called the **coordinate** of the point. For example, the leftmost marked point in Figure 2 has coordinate -4. The correspondence between points on a line and the real numbers is called a **coordinate system** for the line. (From now on, the phrase "the point on a number line with coordinate a" will be abbreviated as "the point with coordinate a," or simply "the point a.")

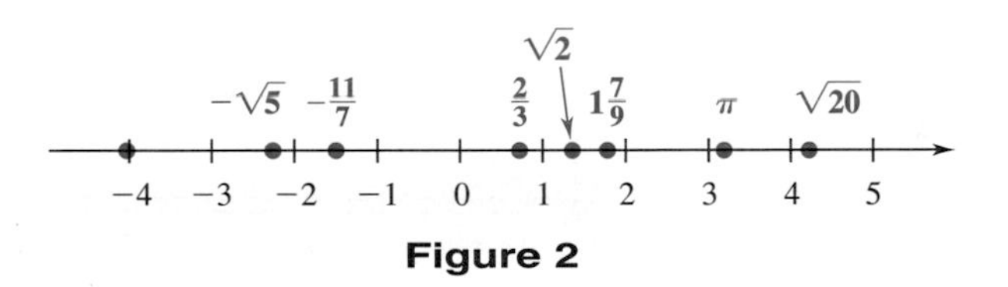

Figure 2

EXAMPLE 5 Locate the elements of the set $\{-2/3, 0, \sqrt{2}, \sqrt{5}, \pi, 4\}$ on a number line.

The numbers π, $\sqrt{2}$, and $\sqrt{5}$ are irrational. As mentioned above, $\pi \approx 3.14159$. From a calculator, $\sqrt{2} \approx 1.414$ and $\sqrt{5} \approx 2.236$. Using this information, place the given points on a number line as shown in Figure 3.

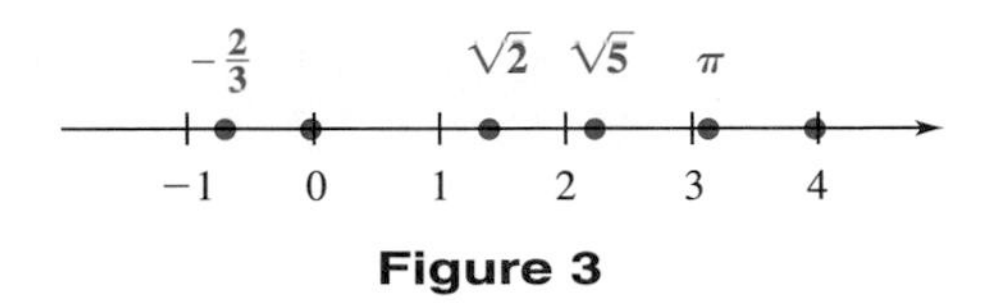

Figure 3

Suppose a and b are two real numbers. If the difference $a - b$ is positive, then a is greater than b, written $a > b$. If the difference $a - b$ is negative, then a is less than b, written $a < b$. The signs $<$ and $>$ were invented by the English mathematician Thomas Harriot (1560–1621), to simplify cumbersome notations used earlier.

These algebraic statements can be given a geometric interpretation. If $a - b$ is positive, so that $a > b$, then a would be to the *right* of b on a number line. Also, if $a < b$, then a would have to be to the *left* of b. Both the algebraic and geometric statements are summarized on the next page.

Inequality Statements

Statement	*Algebraic form*	*Geometric form*
$a > b$	$a - b$ is positive	a is to the right of b
$a < b$	$a - b$ is negative	a is to the left of b

EXAMPLE 6

(a) In Figure 3, $-2/3$ is the left of $\sqrt{2}$, so $-2/3 < \sqrt{2}$. Also, $\sqrt{2}$ is to the right of $-2/3$, giving $\sqrt{2} > -2/3$.

(b) The difference $-3 - (-8)$ is positive, showing that $-3 > -8$. The difference $-8 - (-3)$ is negative, so that $-8 < -3$. ■

The following variations on $<$ and $>$ are often used.

Definitions of Other Inequality Symbols

Symbol	*Meaning*
$\leq$	is less than or equal to
$\geq$	is greater than or equal to
$\nless$	is not less than
$\ngtr$	is not greater than

Statements involving these symbols, as well as $<$ and $>$, are called **inequalities**.

EXAMPLE 7

(a) $8 \leq 10$ (since $8 < 10$)

(b) $8 \leq 8$ (since $8 = 8$)

(c) $-9 \geq -14$ (since $-9 > -14$)

(d) $-8 \ngtr -2$ (since $-8 < -2$)

(e) $4 \nless 2$ (since $4 > 2$) ■

CAUTION The expression $a < b < c$ says that b is *between* a and c, since $a < b < c$ means $a < b$ and $b < c$. Also, $a \leq b \leq c$ means $a \leq b$ and $b \leq c$. When writing these ''between'' statements, make sure that both inequality symbols point in the same direction. For example, both $2 < 7 < 11$ and $5 > -1 > -6$ are true statements, but a statement such as $3 < 5 > -1$ is meaningless. ■

The following **properties of order** give the basic properties of $<$ and $>$.

Properties of Order

For all real numbers a, b, and c:

Transitive property	If $a < b$ and $b < c$, then $a < c$.
Addition property	If $a < b$, then $a + c < b + c$.
Multiplication property	If $a < b$, and $c > 0$, then $ac < bc$. If $a < b$, and $c < 0$, then $ac > bc$.
Trichotomy property	Given the real numbers a and b, either $a < b$, $a > b$, or $a = b$.

Absolute Value The distance on the number line between a number and 0 is called the **absolute value** of that number. The absolute value of the number a is written $|a|$. For example, the distance on the number line between 9 and 0 is 9, as is the distance between -9 and 0 (see Figure 4), so $|9| = 9$ and $|-9| = 9$.

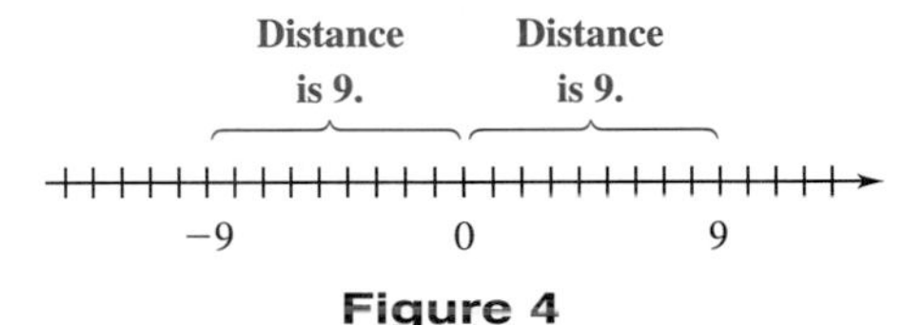

Figure 4

EXAMPLE 8

(a) $|-4| = 4$

(b) $|2\pi| = 2\pi$

(c) $-|8| = -(8) = -8$

(d) $-|-2| = -(2) = -2$ ■

The definition of absolute value can be stated as follows.

Definition of Absolute Value

For every real number a,

$$|a| = \begin{cases} a & \text{if } a \geq 0 \\ -a & \text{if } a < 0. \end{cases}$$

The second part of this definition requires some thought. If a is a negative number, that is, if $a < 0$, then $-a$ is positive. Thus, for a *negative a*,

$$|a| = -a.$$

For example, if $a = -5$, then $|a| = |-5| = -(-5) = 5$.

EXAMPLE 9 Write each of the following expressions without absolute value bars.

(a) $|\sqrt{5} - 2|$

For positive numbers we can reason as follows. Since $(\sqrt{5})^2 > 2^2$, $\sqrt{5} > 2$, making $\sqrt{5} - 2 > 0$, and $|\sqrt{5} - 2| = \sqrt{5} - 2$.

(b) $|\pi - 4|$

From Figure 3, $\pi < 4$, so $\pi - 4 < 0$, and $|\pi - 4| = -(\pi - 4) = -\pi + 4 = 4 - \pi$.

(c) $|m - 2|$ if $m < 2$

If $m < 2$, then $m - 2 < 0$, so $|m - 2| = -(m - 2) = 2 - m$. ■

The definition of absolute value can be used to prove the following properties of absolute value.

Properties of Absolute Value

For all real numbers a and b:

$|a| \geq 0$ $\qquad |a| \cdot |b| = |ab|$ $\qquad |-a| = |a|$

$\left|\dfrac{a}{b}\right| = \dfrac{|a|}{|b|}$, $b \neq 0$ $\qquad |a + b| \leq |a| + |b|$ (the triangle inequality)

The number line in Figure 5 shows the point A, with coordinate -3, and the point B, with coordinate 5. The distance between points A and B is 8 units, which can be found by subtracting the smaller coordinate from the larger. If $d(A, B)$ represents the distance between points A and B, then

$$d(A, B) = 5 - (-3) = 8.$$

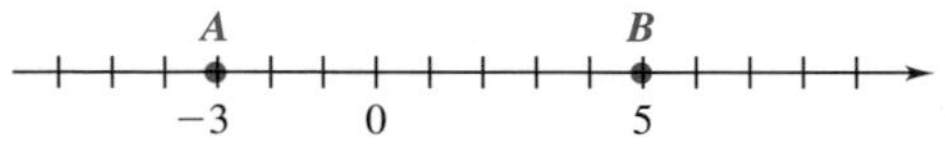

A B
−3 0 5

Figure 5

To avoid worrying about which coordinate is smaller, use the absolute value as in the definition below.

Definition of Distance

Suppose points A and B have coordinates a and b, respectively. The distance between A and B, written $d(A, B)$, is

$$d(A, B) = |a - b| = |b - a|.$$

EXAMPLE 10 Let points A, B, C, D, and E have coordinates as shown on the number line in Figure 6. Find the indicated distances.

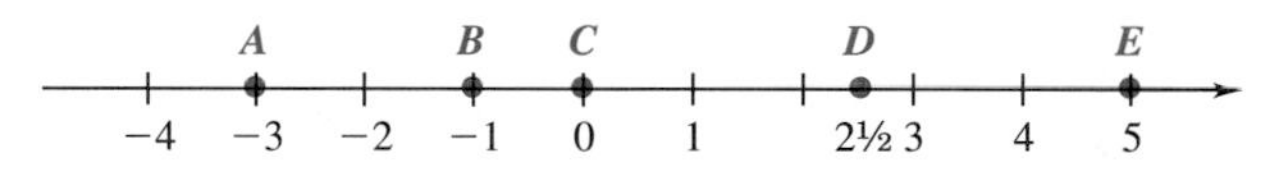

Figure 6

(a) $d(B, E)$

Since B has coordinate -1 and E has coordinate 5,

$$\begin{aligned} d(B, E) &= d(-1, 5) \\ &= |-1 - 5| = 6. \end{aligned}$$

(b) $d(D, A) = |2\frac{1}{2} - (-3)| = 5\frac{1}{2}$

(c) $d(B, C) = |-1 - 0| = 1$

(d) $d(E, E) = |5 - 5| = 0$

1.1 Exercises

Identify the properties illustrated in each of the following. Assume all variables are real numbers.

1. $17 \cdot 3 = 3 \cdot 17$

2. $-8 + 8 = 0$

3. $-8 + 8 = 8 + (-8)$

4. $12 + 0 = 0 + 12$

5. $-25 + 0 = -25$

6. $14 + (3 + 22) = (14 + 3) + 22$

7. $[13(-4)] \cdot 5 = 13[(-4) \cdot 5]$

8. $6(2 - k) = 12 - 6k$

Simplify each of the following expressions using the order of operations given in the text.

9. $24 \div 4 \cdot 5 \cdot 8$

10. $17 \cdot 6 \div 3 \div 4$

11. $(-6)(-4 + 12 \cdot 5)$

12. $-9 + 6 \cdot 5 + (-8)$

13. $1 + 2(-3)^2$

14. $-3^2 + 2 \cdot 3$

15. $\dfrac{15 \div 5 \cdot 4 \div 6 - 8}{-6 - (-5) - 8 \div 2}$

16. $\dfrac{-8 + (-4)(-6) \div 12}{4 - (-3)}$

For Exercises 17–26 choose all words from the following list that apply.

(a) natural number **(b)** whole number **(c)** integer **(d)** rational number
(e) irrational number **(f)** real number **(g)** undefined

17. 12 **18.** -9 **19.** $3/4$ **20.** $-5/9$

21. $\sqrt{8}$ **22.** $-\sqrt{2}$ **23.** $\sqrt{25}$ **24.** $-\sqrt{36}$

25. $8/0$ **26.** $0/(3 + 3)$

Use the distributive property to calculate the following values in your head.

27. $86 \cdot 19 + 14 \cdot 19$

28. $106\frac{5}{6} \cdot 1\frac{1}{2} - 6\frac{5}{6} \cdot 1\frac{1}{2}$

29. $15\frac{2}{5} \cdot 12\frac{3}{4} - 15\frac{2}{5} \cdot 2\frac{3}{4}$

30. $23 \cdot 80 + 23 \cdot 20$

Write the following numbers in numerical order, from smallest to largest. Use a calculator as necessary.

31. $\sqrt{6},\ -3,\ -\sqrt{5},\ -4,\ -6,\ 2$

32. $\sqrt{7},\ -4,\ 4,\ 2,\ \sqrt{3},\ -\sqrt{2},\ \sqrt{5}$

33. $3/4,\ \sqrt{2},\ \pi/2,\ 1.2,\ 8/5,\ 22/15$

34. $-9/8,\ -3,\ \pi/3,\ 1,\ -\sqrt{5},\ -9/5,\ -8/5$

Write an equivalent expression for each of the following without using absolute value bars.

35. $-|-3| + |-9|$ **36.** $5 - |-7|$ **37.** $|2 - \sqrt{3}|$ **38.** $|\sqrt{10} - 4|$

39. $|\sqrt{3} - 2|$ **40.** $|\pi - 3|$ **41.** $|x - 4|$, if $x > 4$ **42.** $|y - 3|$, if $y < 3$

43. $|7m - 56|$, if $m < 4$ **44.** $|2k - 7|$, if $k > 4$ **45.** $|3 + x^2|$ **46.** $|x^2 + 4|$

47. $|-1 - p^2|$ **48.** $|-r^4 - 16|$ **49.** $|\pi - 5| + 1$ **50.** $|3 - \sqrt{11}| + 2$

51. $|m - 3| + |m - 4|$, if $3 < m < 4$

52. $|z - 6| - |z - 5|$, if $5 < z < 6$

In Exercises 53–56, the coordinates of four points are given. Find **(a)** $d(A, B)$, **(b)** $d(B, C)$, **(c)** $d(D, A)$, **(d)** $d(A, C)$, **(e)** $d(A, B) + d(B, C)$.

53. $A, -4;\ B, -3;\ C, -2;\ D, 10$

54. $A, -8;\ B, -7;\ C, 11;\ D, 5$

55. $A, -3;\ B, -5;\ C, -12;\ D, -3$

56. $A, 0;\ B, 6;\ C, 9;\ D, -1$

57. Suppose that A, B, and C are three points on the number line with $d(A, B) = 3$ and $d(B, C) = 4$. What can you say about $d(A, C)$?

58. Under what conditions will the following statement be true? $a < b$ but $|b| > |a|$

Justify each of the following statements by giving the correct property from this section. Assume that all variables represent real numbers.

59. If $2k < 8$, then $k < 4$.

60. If $x + 8 < 15$, then $x < 7$.

61. If $-4x < 24$, then $x > -6$.

62. If $x < 5$ and $5 < m$, then $x < m$.

63. If $m > 0$, then $9m > 0$.

64. If $k > 0$, then $8 + k > 8$.

65. $|8 + m| \leq |8| + |m|$

66. $|k - m| \leq |k| + |-m|$

67. $|8| \cdot |-4| = |-32|$

68. $|12 + 11r| \geq 0$

69. $\left|\dfrac{-12}{5}\right| = \dfrac{|-12|}{|5|}$

70. $\left|\dfrac{6}{5}\right| = \dfrac{|6|}{|5|}$

71. If p is a real number, then $p < 5$, $p > 5$, or $p = 5$.

72. If z is a real number, then $z > -2$, $z < -2$, or $z = -2$.

For each of the following inequalities, what can be said about the signs of x and y?

73. $xy > 0$ **74.** $x^2y > 0$ **75.** $\frac{y^2}{x} < 0$ **76.** $\frac{x^3}{y} < 0$

Under what conditions are the following statements true?

77. $|x| = |y|$ **78.** $|x + y| = |x| + |y|$ **79.** $|x| \leq 0$ **80.** $|x - y| = |x| - |y|$

Evaluate the expressions in Exercises 81–84 for real numbers x and y, if no denominators are equal to 0.

81. $\frac{|x|}{x}$ **82.** $\left|\frac{x}{|x|}\right|$ **83.** $\left|\frac{x - y}{y - x}\right|$ **84.** $\left|\frac{x + y}{-x - y}\right|$

85. Suppose $x^2 \leq 81$. Must it then be true that $x \leq 9$?

86. Suppose $x^2 \geq 81$. Must it then be true that $x \geq 9$?

1.2 Exponents

Exponents are used to write the products of repeated factors. For example, the product $2 \cdot 2 \cdot 2$ can be written as 2^3, where the 3 shows that three factors of 2 appear in the product. The French philosopher and mathematician René Descartes (1596–1650) first used exponents extensively as an efficient way of writing $a \cdot a$ as a^2 and $a^2 \cdot a$ as a^3 and so on. Much progress in algebra has depended on such improved notations.

The symbol a^n is defined as follows.

Definition of a^n

If n is any positive integer and a is any real number,

$$a^n = a \cdot a \cdot a \cdots a,$$

where a appears as a factor n times.

The integer n is called the **exponent**, and a is the **base**. (Read a^n as "a to the nth power," or just "a to the nth.") For example,

$$(-6)^2 = (-6)(-6) = 36, \qquad 4^3 = 4 \cdot 4 \cdot 4 = 64,$$

$$\text{and} \quad \left(\frac{2}{3}\right)^4 = \frac{2}{3} \cdot \frac{2}{3} \cdot \frac{2}{3} \cdot \frac{2}{3} = \frac{16}{81}.$$

EXAMPLE 3

(a) $\left(\frac{1}{3}\right)^{-2} = \left(\frac{3}{1}\right)^2 = 9$

(b) $\left(\frac{2}{3}\right)^{-3} = \left(\frac{3}{2}\right)^3 = \frac{27}{8}$ ■

The meaning of a^n can be extended to rational values of the exponent. To start, let us define $a^{1/n}$ for positive integers n. Any definition of $a^{1/n}$ should be consistent with the rules of exponents. In particular, $(b^m)^n = b^{mn}$ should still be valid. Replacing b^m with $a^{1/n}$ gives

$$(a^{1/n})^n = a^{n/n} = a^1 = a.$$

This means that $a^{1/n}$ should be a real number whose nth power is a. Such a number is called an ***n*th root of *a***. When n is 2, an nth root is called a **square root** and when n is 3, it is called a **cube root**. Depending on the values of a and n, there are one, two, or no real nth roots. For instance, -5 has one cube root, and 5 has two square roots, one positive root written as $5^{1/2}$ and one negative root written as $-5^{1/2}$. However, -5 has no square root. Two real nth roots occur when a is positive and n is even. The value of $a^{1/n}$ is defined as follows.

Definition of $a^{1/n}$

Let n be a positive integer and a a real number. Then

$$a^{1/n} = \begin{cases} \textbf{the } n\textbf{th root of } a \textbf{ if } n \textbf{ is odd} \\ \textbf{the positive } n\textbf{th root of } a \textbf{ if } n \textbf{ is even and } a \geq 0. \end{cases}$$

EXAMPLE 4

(a) $(-32)^{1/5} = -2$ since $(-2)^5 = -32$.

(b) $\left(\frac{16}{9}\right)^{1/2} = \frac{4}{3}$ since $\left(\frac{4}{3}\right)^2 = \frac{16}{9}$.

(c) $(-9)^{1/2}$ is not defined. ■

For rational exponents in general, $a^{m/n}$ must be defined so that $(a^{1/n})^m = a^{m/n}$. (Isaac Newton, the greatest of the English mathematicians, first did this in 1676.) Thus $a^{m/n}$ is defined as follows.

Definition of $a^{m/n}$

For all integers m and all positive integers n such that m/n is in lowest terms, and all nonzero real numbers a for which $a^{1/n}$ is a real number,

$$a^{m/n} = (a^{1/n})^m.$$

EXAMPLE 5

(a) $8^{5/3} = (8^{1/3})^5 = 2^5 = 32$

(b) $25^{1.5} = 25^{3/2} = (25^{1/2})^3 = 5^3 = 125$ ■

We shall now show that $(a^{1/n})^m = (a^m)^{1/n}$, so that $a^{m/n}$ is also equal to $(a^m)^{1/n}$. To prove that $(a^{1/n})^m = (a^m)^{1/n}$, we shall use some of the properties of exponents. Start with $(a^{1/n})^m$, and raise it to the nth power, to get

$$[(a^{1/n})^m]^n.$$

Now use properties of exponents, getting

$$[(a^{1/n})^m]^n = (a^{1/n})^{mn} = [(a^{1/n})^n]^m = a^m.$$

Because of this result, $(a^{1/n})^m$ must be an nth root of a^m, so that

$$(a^{1/n})^m = (a^m)^{1/n}.$$

Since $a^{m/n} = (a^{1/n})^m$ by definition, then

$$a^{m/n} = (a^m)^{1/n}.$$

This result is summarized in the next theorem.

Theorem on $a^{m/n}$

For all integers m, and all positive integers n, such that m/n is in lowest terms, and all nonzero real numbers a for which all indicated powers exist,

$$a^{m/n} = (a^{1/n})^m \quad \text{or} \quad a^{m/n} = (a^m)^{1/n}.$$

For Graphers

When a grapher is used to evaluate fractional powers of negative numbers, the theorem on $a^{m/n}$ may not hold true. For example, the grapher will evaluate $[(-8)^{1/3}]^2$ or $[(-8)^2]^{1/3}$, but not $(-8)^{2/3}$. The simplest way to handle this is to evaluate $8^{2/3}$ and determine the appropriate sign mentally.

This theorem gives two ways to evaluate $a^{m/n}$. Find $a^{1/n}$ and raise the result to the mth power, or find the nth root of a^m. In practice, it is usually easier to find $(a^{1/n})^m$. For example, $27^{4/3}$ can be evaluated in either of two ways:

$$27^{4/3} = (27^{1/3})^4 = 3^4 = 81$$

or

$$27^{4/3} = (27^4)^{1/3} = 531{,}441^{1/3} = 81.$$

The form $(27^{1/3})^4$ is easier to evaluate.

We have shown that $a^m \cdot a^n = a^{m+n}$, $(a^m)^n = a^{mn}$, and $(a/b)^{-n} = (b/a)^n$ for all integers m and n. In a similar way, we could show that these properties and the additional definitions and properties shown below also hold true for rational exponents.

Definitions and Properties of Exponents

Let r and s be rational numbers. The results below are valid for all real numbers a and b for which all indicated expressions exist.

(a) $a^r \cdot a^s = a^{r+s}$ **(b)** $\dfrac{a^r}{a^s} = a^{r-s}$ **(c)** $(a^r)^s = a^{rs}$

(d) $(ab)^r = a^r b^r$ **(e)** $\left(\dfrac{a}{b}\right)^r = \dfrac{a^r}{b^r}$ **(f)** $a^0 = 1$

(g) $a^{-r} = \dfrac{1}{a^r}$ **(h)** $\left(\dfrac{a}{b}\right)^{-r} = \left(\dfrac{b}{a}\right)^r$

EXAMPLE 6 Rewrite each of the following using only positive exponents. Assume that all variables represent nonzero real numbers.

(a) $6y^{2/3} \cdot 2y^{1/2} = 12y^{2/3+1/2} = 12y^{7/6}$

(b)
$$3x^{-2}(4^{-1}x^{-5})^2 = 3x^{-2}(4^{-2}x^{-10})$$
$$= 3 \cdot 4^{-2} \cdot x^{-2+(-10)}$$
$$= 3 \cdot 4^{-2} \cdot x^{-12} = 3 \cdot \frac{1}{4^2} \cdot \frac{1}{x^{12}} = \frac{3}{16x^{12}}$$

(c)
$$(x+2)(x+2)^{-1/3}(x+2)^{-2} = (x+2)^{1+(-1/3)+(-2)}$$
$$= (x+2)^{-4/3} = \frac{1}{(x+2)^{4/3}}$$

(d) $$\frac{5k^{-3}}{10k^{-5}} = \frac{5}{10}k^{-3-(-5)} = \frac{1}{2}k^2, \text{ or } \frac{k^2}{2}$$

(e)
$$(8ab)^{-2/3}4a^{-1} = 8^{-2/3}a^{-2/3}b^{-2/3}4a^{-1} = \frac{1}{4} \cdot 4a^{-2/3+(-1)}b^{-2/3}$$
$$= a^{-5/3}b^{-2/3} = \frac{1}{a^{5/3}b^{2/3}}$$

(f)
$$\frac{(3m^3)^2(my)^{-1}}{(25y^{-4}m^{14})^{1/2}} = \frac{3^2m^6m^{-1}y^{-1}}{25^{1/2}y^{-2}m^7} = \frac{9}{5} \cdot \frac{m^{6+(-1)}}{m^7} \cdot \frac{y^{-1}}{y^{-2}}$$
$$= \frac{9}{5} \cdot m^{5-7}y^{-1-(-2)} = \frac{9}{5}m^{-2}y = \frac{9y}{5m^2}$$ ■

1.2 Exercises

Simplify each of the following. Assume all variables represent nonnegative real numbers. Write numerical answers without exponents.

1. $16^{1/2}$ **2.** $81^{1/2}$ **3.** $125^{2/3}$ **4.** $16^{3/4}$

5. -3^{-3} **6.** -5^{-4} **7.** $8^{-2/3}$ **8.** $32^{-4/5}$

9. $(-5)^{-2}$ **10.** $\left(\frac{2}{7}\right)^{-3}$ **11.** $\left(\frac{4}{9}\right)^{-3/2}$ **12.** $(121)^{-3/2}$

13. $\frac{12}{10^{-2}}$ **14.** $\frac{5}{2^{-4}}$ **15.** $(27x^6)^{2/3}$ **16.** $(64a^{12})^{5/6}$

17. $(9.864)^{-3}$ **18.** $(14.259)^{-2}$

19. Explain why a negative exponent is defined as a reciprocal: $a^{-n} = 1/a^n$.

20. Why is $a^{1/n}$ defined to be the nth root of a (with appropriate restrictions)?

21. Explain why a must be positive if n is even for $a^{1/n}$ to be defined.

22. Explain how you would evaluate $(-27)^{5/3}$.

Simplify the expressions in Exercises 23–46 by writing each with only positive exponents. Assume that all variables represent positive real numbers and that variables used as exponents represent rational numbers.

23. $(x^{2/3})(x^{5/3})$ **24.** $(a^{4/5})(a^{2/5})$ **25.** $(4^3)(4^{-3})(4^5)$

26. $5^6(5^{-3})(5^4)$ **27.** $(1 + n)^{1/2}(1 + n)^{3/4}$ **28.** $(m + 7)^{-1/6}(m + 7)^{-2/3}$

29. $(3p)^{-4}$ **30.** $(-5k)^{-2}$ **31.** $(2y^{3/4}z)(3y^{1/4}z^{-1/3})$

32. $(4a^{-1/2}b^{2/3})(2a^{3/2}b^{-1/3})$ **33.** $\frac{d^{-2}}{(d^8)(d^{-3})}$ **34.** $\frac{(t^5)(t^{-3})}{t^{-7}}$

35. $\frac{a^{4/3} \cdot b^{1/2}}{a^{2/3} \cdot b^{-3/2}}$ **36.** $\frac{x^{1/3}y^{2/3}}{x^{5/3}y^{-1/3}}$ **37.** $\left(\frac{a^{-1}}{b^2}\right)^{-3}$

38. $\left(\frac{2c^2}{d^3}\right)^{-2}$ **39.** $\frac{(5x)^{-2}(x^{-3})^{-4}}{(25^{-1}x^{-3})^{-1}}$ **40.** $\frac{(2k)^{-3}(k^{-5})^{-1}}{(6k^{-2})^{-1}(k^3)^{-6}}$

41. $\left(\frac{x^{13}(x^5)^{-7}}{(2x^7)^2}\right)^{-1/7}$ **42.** $\left(\frac{y^{2/5} \cdot y^{4/5}}{y^{1/5} \cdot y^{6/5}}\right)^{-5}$ **43.** $5p^r(6p^{3-2r}), \quad r < 3$

44. $(-z^{2r})(4z^{r+3})$ **45.** $\frac{(b^2)^{y+1}}{(2b^y)^3}, y > 2$ **46.** $\frac{(3x^n)^3}{(x^2)^{n-1}}$

47. Show that $(1 + 2^3)^{-1} + (1 + 2^{-3})^{-1} = 1$. **48.** Show that $(1 + x^m)^{-1} + (1 + x^{-m})^{-1} = 1$.

49. If $a^7 = 30$, what is a^{21}? **50.** If $a^{-3} = .2$, what is a^6?

51. If the lengths of the sides of a cube are tripled, by what factor will the volume change?

52. If the radius of a circle is doubled, by what factor will the area change?

One important application of mathematics to business and management concerns supply and demand. Usually, as the price of an item increases, the supply increases and the demand decreases. By studying past records of supply and demand at different prices, economists can construct an equation that describes (approximately) supply and demand for a given item. Such equations are provided in Exercises 53 and 54.

53. The price in dollars of a graphing calculator is approximated by p, where $p = 5x^{2/3} + 2x^{1/3}$ and x is the number (in hundreds) of units supplied. Find the price when the supply is 6400.

54. The demand for a certain commodity and the price in dollars are related by $p = 500 - 10x^{1/2}$, where x is the number (in hundreds) of units of the product demanded. Find the price when the demand is 2500.

In our system of government, the president is elected by the electoral college and not by individual voters. Because of this, smaller states have a greater voice in the selection of a president than they would otherwise have. Two political scientists have studied the problems of campaigning for president under the current system and have concluded that candidates should allot their money according to the formula

$$\text{Amount for large state} = \left(\frac{E_{large}}{E_{small}}\right)^{3/2} \times \text{Amount for small state}.$$

Here E_{large} represents the electoral vote of the large state, and E_{small} represents the electoral vote of the small state. Find the amount that should be spent in the larger states in Exercises 55–58 if \$1,000,000 *is spent in the small state and the following statements are true.*

55. The large state has 45 electoral votes and the small state has 5.

56. The large state has 40 electoral votes and the small state has 10.

57. There are six electoral votes in the small state; 28 in the large.

58. There are nine electoral votes in the small state; 32 in the large.

A Delta Airlines map gives a formula for calculating the visible distance from a jet plane to the horizon. On a clear day, this distance is approximated by $D = 1.22x^{1/2}$, where x is altitude in feet and D is distance to the horizon in miles. Find D for each of the following altitudes.

59. 5,000 ft **60.** 10,000 ft **61.** 30,000 ft **62.** 40,000 ft

The Galápagos Islands are a chain of islands ranging in size from 2 to 2249 sq mi. A biologist has shown that the number of different land-plant species on an island in this chain is related to the size of the island by $S = 28.6A^{.32}$, where A is the area of an island in square miles and S is the number of different plant species on that island. Estimate S (rounding to the nearest whole number) for islands with the following areas.

63. 100 sq mi **64.** 500 sq mi **65.** 300 sq mi **66.** 2000 sq mi

1.3 Polynomials; The Binomial Theorem

A variable is a letter used to represent an element from a given set. Unless otherwise specified, in this book variables will represent real numbers. An **algebraic expression** is the result of performing the basic operations of addition, subtraction, multiplication, division (except by 0), or extraction of roots on any collection of variables and numbers. The simplest algebraic expression, a polynomial, is discussed in this section.

A **term** is the product of a real number and one or more variables raised to powers. The real number is the **numerical coefficient**, or just the **coefficient**. For example, -3 is the coefficient in $-3m^4$, while -1 is the coefficient in $-p^2$.

A **polynomial** is defined as a term or a finite sum of terms, with only nonnegative integer exponents permitted on the variables. If the terms of a polynomial contain only the variable x, then the polynomial is called a *polynomial in x*. (Polynomials in other variables are defined similarly.)

Definition of a Polynomial in *x*

A **polynomial in *x*** is an expression of the form

$$a_n x^n + a_{n-1} x^{n-1} + \dots + a_1 x + a_0,$$

where a_n (read "*a*-sub-*n*"), $a_{n-1}, \dots, a_1$, and a_0 are real numbers and n is a nonnegative integer. If $a_n \neq 0$, then n is the **degree** of the polynomial, and a_n is called the **leading coefficient**.

By this definition,

$$3x^4 - 5x^2 + 2$$

is a polynomial of degree 4 with leading coefficient 3. A nonzero constant polynomial is said to have degree 0, but no degree is assigned to the polynomial 0. If all the coefficients of a polynomial are 0, the polynomial is called the **zero polynomial**. Polynomials are often denoted by names such as $f(x)$ and $g(x)$.

A polynomial can have more than one variable. A term containing more than one variable is said to have **degree** equal to the sum of all the exponents appearing on the variables in the term. For example, $-3x^4y^3z^5$ is of degree 12 because $4 + 3 + 5 = 12$. The **degree of a polynomial** in more than one variable is the highest degree of any term appearing in the polynomial. With this definition, the polynomial

$$2x^4y^3 - 3x^5y + x^6y^2$$

is of degree 8 because of the x^6y^2 term.

A polynomial containing exactly three terms is called a **trinomial**; one containing exactly two terms is a **binomial**; and a single-term polynomial is called a **monomial**. For example, $7x^9 - \sqrt{2}x^4 + 1$ is a trinomial of degree 9.

Since the variables used in polynomials represent real numbers, a polynomial represents a real number. This means that all the properties of the real numbers mentioned at the beginning of this chapter hold for polynomials. In particular, the distributive property holds, so that

$$3m^5 - 7m^5 = (3 - 7)m^5 = -4m^5.$$

Addition and Subtraction To *add* polynomials of one variable, use the distributive property to add coefficients of variables having the same powers; to *subtract* polynomials, subtract coefficients of variables having the same powers, again with the distributive property. That is, polynomials are added by adding like terms and subtracted by subtracting like terms.

EXAMPLE 1 Add or subtract, as indicated.

(a) $(2y^4 - 3y^2 + y) + (4y^4 + 7y^2 + 6y)$

$$= (2 + 4)y^4 + (-3 + 7)y^2 + (1 + 6)y$$
$$= 6y^4 + 4y^2 + 7y$$

(b) $(6r^4 + 2r^2) + (3r^3 + 9r) = 6r^4 + 3r^3 + 2r^2 + 9r$

(c) $(-3m^3 - 8m^2 + 4) - (m^3 + 7m^2 - 3)$

$$= (-3 - 1)m^3 + (-8 - 7)m^2 + [4 - (-3)]$$
$$= -4m^3 - 15m^2 + 7$$ ■

Multiplication The distributive property, together with some of the properties of exponents, is also used to find the product of two polynomials. For example, to find the product of $3x - 4$ and $2x^2 - 3x + 5$, treat $3x - 4$ as a single expression and use the distributive property as follows.

$$(3x - 4)(2x^2 - 3x + 5) = (3x - 4)(2x^2) - (3x - 4)(3x) + (3x - 4)(5)$$

Now use the distributive property three separate times on the right side of the equals sign to get

$$\begin{aligned}(3x - 4)(2x^2 - 3x + 5) &= (3x)(2x^2) - 4(2x^2) - (3x)(3x) - (-4)(3x) \\ &\quad + (3x)5 - 4(5) \\ &= 6x^3 - 8x^2 - 9x^2 + 12x + 15x - 20 \\ &= 6x^3 - 17x^2 + 27x - 20.\end{aligned}$$

It is sometimes more convenient to find such a product as follows.

$$\begin{array}{rrrrl} & 2x^2 & - \ 3x & + \ 5 & \\ & & 3x & - \ 4 & \\ \hline & -8x^2 & + \ 12x & - \ 20 & \longleftarrow -4(2x^2 - 3x + 5) \\ 6x^3 & - \ 9x^2 & + \ 15x & & \longleftarrow 3x(2x^2 - 3x + 5) \\ \hline 6x^3 & - \ 17x^2 & + \ 27x & - \ 20 & \longleftarrow \text{Add in columns.} \end{array}$$

EXAMPLE 2 Find each product.

(a) $(6m + 1)(4m - 3) = (6m)(4m) - (6m)(3) + 1(4m) - 1(3)$
$= 24m^2 - 14m - 3$

(b) $(2k^n - 5)(k^n + 3) = 2k^{2n} + 6k^n - 5k^n - 15$
$= 2k^{2n} + k^n - 15$

(c) $(2x + 7)(2x - 7) = 4x^2 - 14x + 14x - 49$
$= 4x^2 - 49$ ■

The products in Example 2 were found by multiplying the first terms, the outer terms, the inner terms, the last terms, and then adding the four products. This method, which is used to multiply any two binomials, is often called the **FOIL method** (for **F**irst, **O**uter, **I**nner, **L**ast).

Special Products Certain products occur so frequently that they should be memorized.

Special Products

Difference of two squares	$(x + y)(x - y) = x^2 - y^2$
Square of a binomial	$(x + y)^2 = x^2 + 2xy + y^2$
	$(x - y)^2 = x^2 - 2xy + y^2$
Cube of a binomial	$(x + y)^3 = x^3 + 3x^2y + 3xy^2 + y^3$
	$(x - y)^3 = x^3 - 3x^2y + 3xy^2 - y^3$

It is useful to memorize and be able to apply these special products.

EXAMPLE 3 Find each product.

(a) $(3p + 11)(3p - 11)$

Using the result above, replace x with $3p$ and y with 11. This gives

$$(3p + 11)(3p - 11) = (3p)^2 - 11^2 = 9p^2 - 121.$$

(b) $(2m + 5)^2 = (2m)^2 + 2(2m)(5) + (5)^2$
$= 4m^2 + 20m + 25$

(c) $(5k - 2z^5)^3 = (5k)^3 - 3(5k)^2(2z^5) + 3(5k)(2z^5)^2 - (2z^5)^3$
$= 125k^3 - 150k^2z^5 + 60kz^{10} - 8z^{15}$ ■

Binomial Theorem The square of a binomial and the cube of a binomial are special cases of the *binomial theorem*, which gives a pattern for finding any positive integer power of a binomial. The coefficients in the formula can be found from the following array of numbers, known as **Pascal's triangle**.

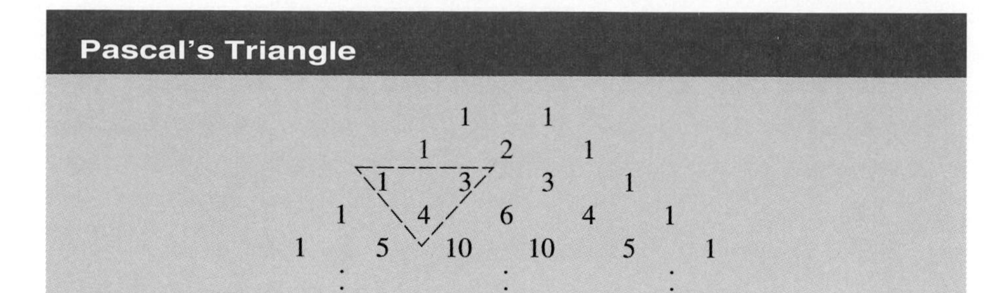

The triangle can be extended by observing the pattern. Each number is the sum of the two numbers above it, one to the right and one to the left. For example, in row 4, 1 is the sum of 1, 4 is the sum of 3 and 1, 6 is the sum of 3 and 3, and so on. Now notice the behavior of the exponents on x and y for the powers of the binomial we found with the special products.

$$(x + y)^2 = x^2 + 2xy + y^2$$
$$(x + y)^3 = x^3 + 3x^2y + 3xy^2 + y^3$$
$$(x + y)^4 = x^4 + 4x^3y + 6x^2y^2 + 4xy^3 + y^4$$

The last result can be verified by multiplying out $(x + y)^4$. These examples suggest that the variables in the expansion of $(x + y)^n$ should have the following pattern:

$$x^n, \quad x^{n-1}y, \quad x^{n-2}y^2, \quad x^{n-3}y^3, \quad \ldots, \quad xy^{n-1}, \quad y^n,$$

and the coefficients should come from the nth row of Pascal's triangle. Notice that the sum of the exponents in each term is n. The next example puts all this together.

EXAMPLE 4 Write out the terms of $(m - 2)^5$.

Let $n = 5$, $x = m$, and $y = -2$, since $(m - 2)^5 = [m + (-2)]^5$. Use the coefficients from row 5 of Pascal's triangle and the pattern shown above for the exponents.

$$\begin{aligned}(m - 2)^5 &= m^5 + 5m^4(-2) + 10m^3(-2)^2 + 10m^2(-2)^3 \\ &\quad + 5m(-2)^4 + (-2)^5 \\ &= m^5 - 10m^4 + 40m^3 - 80m^2 + 80m - 32\end{aligned}$$

An alternative method for finding the coefficients of $(x + y)^n$ is to multiply the exponent on x in any term by the coefficient of the term and divide the product by the number of the term to get the coefficient of the next term. For instance, in Example 4, the first term is $1m^5$, so the coefficient of the second term is $(5 \cdot 1)/1 = 5$; the second term is $5m^4(-2)$, so the coefficient of the third term is $(4 \cdot 5)/2 = 10$; and so on.

A complete discussion of the binomial theorem and a proof are given in Section 5 of the last chapter in this book.

Division To divide a polynomial by a monomial, divide each term of the polynomial by the monomial.

EXAMPLE 5 Divide.

(a) $$\frac{2m^5 - 6m^3}{2m^3} = \frac{2m^5}{2m^3} - \frac{6m^3}{2m^3} = m^2 - 3$$

The polynomial $m^2 - 3$ is the quotient of $2m^5 - 6m^3$ and $2m^3$.

(b) $$\frac{3y^6x^3 - 6y^3x^6 + 8y^5x}{3y^3x^3} = \frac{3y^6x^3}{3y^3x^3} - \frac{6y^3x^6}{3y^3x^3} + \frac{8y^5x}{3y^3x^3}$$

$$= y^3 - 2x^3 + \frac{8y^2}{3x^2}$$

This result is not a polynomial. ■

The quotient of two polynomials can be found with a **division algorithm** very similar to that used for dividing whole numbers. (An *algorithm* is a step-by-step procedure for working a problem.) This algorithm requires that both polynomials be written in descending order.

EXAMPLE 6 Divide $4m^3 - 8m^2 + 4m + 6$ by $2m - 1$.

Work as follows.

$4m^3$ divided by $2m$ is $2m^2$.
$-6m^2$ divided by $2m$ is $-3m$.
m divided by $2m$ is ½.

$$
\begin{array}{rll}
 & 2m^2 - 3m + \tfrac{1}{2} & \\
2m - 1 \,\big) & 4m^3 - 8m^2 + 4m + 6 & \\
 & \underline{4m^3 - 2m^2} & \leftarrow 2m^2(2m - 1) = 4m^3 - 2m^2 \\
 & -6m^2 + 4m & \leftarrow \text{Subtract; bring down the next term.} \\
 & \underline{-6m^2 + 3m} & \leftarrow -3m(2m - 1) = -6m^2 + 3m \\
 & m + 6 & \leftarrow \text{Subtract; bring down the next term.} \\
 & \underline{m - \tfrac{1}{2}} & \leftarrow (1/2)(2m - 1) = m - (1/2) \\
 & \tfrac{13}{2} & \leftarrow \text{Subtract. The remainder is } 13/2.
\end{array}
$$

In dividing these polynomials, $4m^3 - 2m^2$ is subtracted from $4m^3 - 8m^2 + 4m + 6$. The complete result, $-6m^2 + 4m + 6$, should be written under the line. However, it is customary to save work and "bring down" just the $4m$, the only term needed for the next step. By this work,

$$\frac{4m^3 - 8m^2 + 4m + 6}{2m - 1} = 2m^2 - 3m + \frac{1}{2} + \frac{13/2}{2m - 1}.$$ ■

The polynomial $3x^3 - 2x^2 - 150$ has a missing term, the term in which the power of x is 1. When a polynomial with a missing term is divided, it is useful to allow for that term by inserting a zero coefficient for the missing term, as shown in the next example.

EXAMPLE 7 Divide $3x^3 - 2x^2 - 150$ by $x^2 - 4$.

Both polynomials have missing terms. Insert each missing term with a 0 coefficient.

$$
\begin{array}{r}
3x - 2 \phantom{+ \mathbf{0x} - 150} \\
x^2 + \mathbf{0x} - 4\overline{)3x^3 - 2x^2 + \mathbf{0x} - 150} \\
\underline{3x^3 + 0x^2 - 12x } \\
-2x^2 + 12x - 150 \\
\underline{-2x^2 + 0x + 8} \\
12x - 158
\end{array}
$$

Since $12x - 158$ has lower degree than the divisor, it is the remainder, and the result of the division is written

$$\frac{3x^3 - 2x^2 - 150}{x^2 - 4} = 3x - 2 + \frac{12x - 158}{x^2 - 4}.$$

1.3 Exercises

Perform the indicated operations.

1. $(p^3 - 3p) + (2p^2 + 4p - 1) + (4p^3 - 3)$
2. $(5k^4 - 2k^2 + k) - (3k^3 + k^2 - 4) + 2k^4$
3. $(2m^2 - 6m + 7) + (2m^2 - 4m - 3) - (m^2 + 5)$
4. $(6x^3 + 2x^2 - 5x + 1) - (-x^3 - x^2 - 4) - (3x^2 + 2)$
5. $(5b^2 - 4b + 3) - [(2b^2 + b) - (3b + 4)]$
6. $-[(8x^3 + x - 3) + (2x^3 + x^2)] - (4x^2 + 3x - 1)$
7. $(y + 5)(2y - 1)$
8. $(3k - 4)(k + 2)$
9. $(2m + 1)(3m - 5)$
10. $(4p - 3)(2p + 7)$
11. $(6p + 5q)(3p - 7q)$
12. $(2z + y)(3z - 4y)$
13. $\left(\frac{2}{5}y + \frac{1}{8}z\right)\left(\frac{3}{5}y + \frac{1}{2}z\right)$
14. $\left(\frac{3}{4}r - \frac{2}{3}s\right)\left(\frac{5}{4}r + \frac{1}{3}s\right)$
15. $(5r + 2)(5r - 2)$
16. $(6z + 5)(6z - 5)$
17. $(3z + 2w)(3z - 2w)$
18. $(5r - 8t)(5r + 8t)$
19. $(3x + 8)^2$
20. $(5y - 3)^2$
21. $(4m + 2n)^2$
22. $(a - 6b)^2$
23. $(2z - 1)^3$
24. $(3m + 2)^3$
25. $(2x^3 + y^2)^2$
26. $(x^2 - y^2)^2$
27. $(x - 2y^2)^3$
28. $4x^2(3x^3 + 2x^2 - 5x + 1)$
29. $5m(3m^3 - 2m^2 + m - 1)$
30. $(2z - 1)(-z^2 + 3z - 4)$
31. $(3p - 1)(9p^2 + 3p + 1)$
32. $(2m + 1)(4m^2 - 2m + 1)$
33. $(k + 2)(12k^3 - 3k^2 + k + 1)$
34. $(m - n + k)(m + 2n - 3k)$
35. $(x + 2y - 3z)^2$
36. $(p - 2q + r)^2$
37. State the formula for the square of a binomial in words. Then state in words how you would use the formula to find $(a + b)^2$.

38. State the formula for the product of the sum and difference of two terms in words. Then state in words how you would use the product to find $(p + q)(p - q)$.

Find each of the following products. Assume all variables used as exponents represent integers.

39. $(x^y + 3)(x^y - 3)$
40. $(5 - z^a)(5 + z^a)$
41. $(4k^x - 3)(k^x + 1)$
42. $(2^a + 5)(2^a + 3)$
43. $(m^x - 2)^2$
44. $(z^r + 5)^2$
45. $(3k^a - 2)^3$
46. $(r^x - 4)^3$

Write out the terms of the binomial expansion for each of the following.

47. $(a + b)^6$
48. $(m + 4)^4$
49. $(3x - 2y)^4$
50. $(4m - 3p)^5$
51. $(2k + 1/2)^6$
52. $(3r + 1/3)^6$
53. $(y - 4z)^5$
54. $(3k - 2)^4$

The following are not products of polynomials. However, the same multiplication rules apply. Find each product. Assume that all variables represent positive real numbers.

55. $y^{4/3}(2y^{2/3} - 8y^{5/3})$
56. $z^{9/5}(5z^{4/5} + 3z^{7/5})$
57. $-2p(4p^{3/4} - 3p^{7/4})$
58. $-5y(3y^{9/10} + 4y^{3/10})$
59. $(x + x^{1/2})(x - x^{1/2})$
60. $(2z^{1/2} + z)(z^{1/2} - z)$
61. $(r^{1/2} - r^{-1/2})^2$
62. $(p^{1/2} - p^{-1/2})(p^{1/2} + p^{-1/2})$

63. Explain why the expressions in Exercises 55–62 are not polynomials.

Perform each of the following divisions.

64. $\dfrac{-4x^7 - 14x^6 + 10x^4 - 14x^2}{-2x^2}$
65. $\dfrac{-8r^3s - 12r^2s^2 + 20rs^3}{4rs}$
66. $\dfrac{10x^8 - 16x^6 - 4x^4}{-2x^6}$
67. $\dfrac{6m^3 + 7m^2 - 4m + 2}{3m + 2}$
68. $\dfrac{2x^3 + 6x^2 - 8x + 10}{2x - 1}$
69. $\dfrac{3x^4 - 6x^2 + 9x - 5}{3x + 3}$
70. $\dfrac{3x^4 + 2x^2 + 6x - 1}{3x^2 - x}$
71. $\dfrac{k^4 - 4k^2 + 2k + 5}{k^2 + 1}$
72. $\dfrac{5y^5 + 10y^4 - 5y^2 + 15y}{5y^3 - 2y + 1}$
73. $\dfrac{8z^5 + 4z^4 + 2z^2 - 5z + 16}{4z^2 - z + 2}$

In Exercises 74–79 find the coefficient of x^3 without finding the entire product.

74. $(2x^2 - 3x)(-5x^2 + 3x + 1)$
75. $(3x^3 - 4x^2 + 2)(x^3 - 1)$
76. $(1 + x^2)(1 + x)$
77. $(3 - x)(2 - x^2)$
78. $x^2(4 - 3x)^2$
79. $-4x^2(2 - x)(2 + x)$

80. Show that $(y - x)^3 = -(x - y)^3$.
81. Show that $(y - x)^2 = (x - y)^2$.

82. Show that $\left(x + \dfrac{1}{2}\right)^2 = x(x + 1) + \dfrac{1}{4}$.

83. Use the result of Exercise 82 to calculate the square of 9½ in your head.

84. Suppose one polynomial has degree 3 and another also has degree 3. Find all possible values for the degree of their **(a)** sum, **(b)** difference, **(c)** product.

85. If one polynomial has degree 3 and another has degree 4, find all possible values for the degree of their **(a)** sum, **(b)** difference, **(c)** product.

86. Generalize the results of Exercises 84 and 85: Suppose one polynomial has degree m and another has degree n, where m and n are natural numbers with $n < m$. Find all possible values for the degree of their **(a)** sum, **(b)** difference, **(c)** product.

1.4 Factoring

The process of finding polynomials whose product equals a given polynomial is called **factoring**. For example, since $x^2 + 3x = x(x + 3)$, both x and $x + 3$ are **factors** of $x^2 + 3x$, and $x(x + 3)$ is the **factored form** of $x^2 + 3x$. A polynomial is **factored completely** when it is written as a product of polynomials, none of which can be written as the product of polynomials of positive degree. A polynomial that cannot be written as a product of two polynomials of positive degree is a **prime** or **irreducible** polynomial.

Common Factors The first step in factoring a polynomial is to look for **common factors**; that is, expressions that are factors of each term of the given polynomial.

EXAMPLE 1 Factor a common factor from the polynomial $6x^2y^3 + 9xy^4 + 18y^5$.

Each term of this polynomial can be written with a factor of $3y^3$, so $3y^3$ is a common factor. Use the reverse of the distributive property to get

$$\begin{aligned} 6x^2y^3 + 9xy^4 + 18y^5 &= (3y^3)(2x^2) + (3y^3)(3xy) + (3y^3)(6y^2) \\ &= 3y^3(2x^2 + 3xy + 6y^2). \end{aligned}$$

Special Patterns Each of the special patterns of multiplication from the previous section can be used in reverse to get a pattern for factoring. One of the most common of these is the difference of two squares.

Difference of Two Squares

$$x^2 - y^2 = (x + y)(x - y)$$

EXAMPLE 2 Factor each of the following polynomials.

(a) $4m^2 - 9$

First, recognize that $4m^2 - 9$ is the difference of two squares, since $4m^2 = (2m)^2$ and $9 = 3^2$. Use the pattern for the difference of two squares with $2m$ replacing x and 3 replacing y.

$$\begin{aligned} 4m^2 - 9 &= (2m)^2 - 3^2 \\ &= (2m + 3)(2m - 3) \end{aligned}$$

(b) $144r^2 - 25s^2 = (12r + 5s)(12r - 5s)$

(c) $256k^4 - 625m^4$

Use the difference of two squares pattern twice, as follows.

$$\begin{aligned} 256k^4 - 625m^4 &= (\mathbf{16k^2})^2 - (\mathbf{25m^2})^2 \\ &= (\mathbf{16k^2} + \mathbf{25m^2})(\mathbf{16k^2} - \mathbf{25m^2}) \\ &= (16k^2 + 25m^2)(4k + 5m)(4k - 5m) \end{aligned}$$

(d) $$\begin{aligned} (a + 2b)^2 - 4c^2 &= (a + 2b)^2 - (2c)^2 \\ &= [(a + 2b) + 2c][(a + 2b) - 2c] \\ &= (a + 2b + 2c)(a + 2b - 2c) \end{aligned}$$ ■

In this chapter, a polynomial with only integer coefficients will be factored so that all factors have only integer coefficients. This assumption is sometimes summarized by saying that only **factoring over the integers** is permitted. With factoring over the integers, the polynomial $x^2 - 5$, for example, cannot be factored. While it is true that

$$x^2 - 5 = (x + \sqrt{5})(x - \sqrt{5}),$$

the two factors $x + \sqrt{5}$ and $x - \sqrt{5}$ have noninteger coefficients. When factored over the integers, $x^2 - 5$ is prime; when factored over the real numbers, it is not.

Factoring Trinomials Now we consider factoring trinomials of degree 2, such as $kx^2 + mx + n$, where k, m, and n are integers. Any factorization will be of the form $(ax + b)(cx + d)$ with a, b, c and d integers. Multiplying out the product $(ax + b)(cx + d)$ gives

$$(ax + b)(cx + d) = acx^2 + (ad + bc)x + bd,$$

which equals $kx^2 + mx + n$ if

$$ac = k, \qquad ad + bc = m, \qquad \text{and} \qquad bd = n. \tag{$*$}$$

In summary, to factor a trinomial $kx^2 + mx + n$, look for four integers a, b, c, and d satisfying the conditions given in (*). This is done by using FOIL backwards. If no such integers exist, the trinomial is prime.

EXAMPLE 3 Factor each of the following polynomials.

(a) $6p^2 - 7p - 5$

Use FOIL backwards. We must find integers so that

Product is −5.

$$6p^2 - 7p - 5 = (\quad p - \quad)(\quad p + \quad).$$

Product is 6.

We know that one integer must be positive and one must be negative to give a product of -5. We must choose the factors of 6 and -5 so that the sum of the inner and outer products will be $-7p$. Try various possibilities. Start with 2 and 3 as factors of 6, and -5 and 1 as factors of -5.

$$(2p - 5)(3p + 1) = 6p^2 - 13p - 5 \qquad \textbf{Incorrect}$$

To make another attempt, try

$$(3p - 5)(2p + 1) = 6p^2 - 7p - 5. \qquad \textbf{Correct}$$

Finally, $6p^2 - 7p - 5$ is factored as $(3p - 5)(2p + 1)$.

(b) $4x^3 + 6x^2r - 10xr^2$

A common factor is $2x$.

$$4x^3 + 6x^2r - 10xr^2 = 2x(2x^2 + 3xr - 5r^2)$$

Factoring $2x^2 + 3xr - 5r^2$ requires factors of $2x^2$ and of $-5r^2$ that will yield the correct middle term, $3xr$. By inspection,

$$4x^3 + 6x^2r - 10xr^2 = 2x(2x + 5r)(x - r).$$

(c) $r^2 + 6r + 7$

The factors of r^2 must be r and r, or $-r$ and $-r$. Factors of 7 are 1 and 7, or -1 and -7. With these factors it is not possible to get $6r$ for the middle term. Therefore, $r^2 + 6r + 7$ is prime. ■

Two other special types of factoring are listed below.

Difference and Sum of Cubes

Difference of two cubes	$x^3 - y^3 = (x - y)(x^2 + xy + y^2)$
Sum of two cubes	$x^3 + y^3 = (x + y)(x^2 - xy + y^2)$

EXAMPLE 4 Factor each polynomial.

(a) $m^3 - 64n^3$

Since $64n^3 = (4n)^3$, the given binomial is a difference of two cubes. To factor, use the first pattern above, replacing x with m and y with $4n$, to get

$$\begin{aligned} m^3 - 64n^3 &= m^3 - (4n)^3 \\ &= (m - 4n)[m^2 + m(4n) + (4n)^2] \\ &= (m - 4n)(m^2 + 4mn + 16n^2). \end{aligned}$$

(b) $8q^6 + 125p^9$

Write $8q^6$ as $(2q^2)^3$ and $125p^9$ as $(5p^3)^3$, showing that the given polynomial is a sum of two cubes. Factor as

$$\begin{aligned} 8q^6 + 125p^9 &= (2q^2)^3 + (5p^3)^3 \\ &= (2q^2 + 5p^3)[(2q^2)^2 - (2q^2)(5p^3) + (5p^3)^2] \\ &= (2q^2 + 5p^3)(4q^4 - 10q^2p^3 + 25p^6). \end{aligned}$$

(c) $(2a - 1)^3 + 8$

Use the pattern for the sum of two cubes, with $x = 2a - 1$ and $y = 2$. Doing so gives

$$\begin{aligned}(2a - 1)^3 + 8 &= [(2a - 1) + 2][(2a - 1)^2 - (2a - 1)2 + 2^2] \\ &= (2a - 1 + 2)(4a^2 - 4a + 1 - 4a + 2 + 4) \\ &= (2a + 1)(4a^2 - 8a + 7).\end{aligned}$$

■

Factoring by Grouping When a polynomial has more than three terms, it can sometimes be factored by grouping. For example, to factor

$$ax + ay + 6x + 6y,$$

collect the terms into two groups,

$$ax + ay + 6x + 6y = (ax + ay) + (6x + 6y),$$

and then factor each group, getting

$$ax + ay + 6x + 6y = a(x + y) + 6(x + y).$$

The quantity $(x + y)$ is now a common factor, which can be factored out, producing

$$ax + ay + 6x + 6y = (x + y)(a + 6).$$

It is not always obvious which terms should be grouped. Experience is the best teacher for techniques of factoring.

EXAMPLE 5 Factor by grouping.

(a) $mp^2 + 7m + 3p^2 + 21$

Group the terms as follows.

$$mp^2 + 7m + 3p^2 + 21 = (mp^2 + 7m) + (3p^2 + 21)$$

Find the common factor for each part.

$$\begin{aligned}(mp^2 + 7m) + (3p^2 + 21) &= m(p^2 + 7) + 3(p^2 + 7) \\ &= (p^2 + 7)(m + 3)\end{aligned}$$

(b) $2y^2 - 2z - ay^2 + az$

Grouping terms as above gives

$$\begin{aligned}2y^2 - 2z - ay^2 + az &= (2y^2 - 2z) + (-ay^2 + az) \\ &= 2(y^2 - z) + a(-y^2 + z).\end{aligned}$$

The expression $-y^2 + z$ is the negative of $y^2 - z$, so the terms should be grouped as follows:

$$\begin{aligned}2y^2 - 2z - ay^2 + az &= (2y^2 - 2z) - (ay^2 - az) \\ &= 2(y^2 - z) - a(y^2 - z) \\ &= (y^2 - z)(2 - a).\end{aligned}$$

(c) $x^2 - 6x + 9 - y^2 = (x^2 - 6x + 9) - y^2$
$= (x - 3)^2 - y^2$
$= (x - 3 + y)(x - 3 - y)$ ■

A more general method of factoring polynomials of degree higher than 2 is given in Section 4.3.

Expressions that are not polynomials can also be factored.

EXAMPLE 6 Factor out the specified common factor from the following expressions. Assume all variables represent positive numbers.

(a) $5x^{4/3} + 2x^{1/3}$; $x^{1/3}$

$$5x^{4/3} + 2x^{1/3} = x^{1/3}(5x + 2)$$

(b) $4y^{-1/2} - 3y^{1/2}$; $y^{-1/2}$

$$4y^{-1/2} - 3y^{1/2} = y^{-1/2}(4 - 3y)$$ ■

An important application of factoring is in simplifying fractions.

EXAMPLE 7 Simplify $\dfrac{3(m - 1)^{1/2} + (m - 1)^{-1/2}}{m - 1}$.

Factor the numerator and then simplify.

$$\frac{3(m - 1)^{1/2} + (m - 1)^{-1/2}}{m - 1} = \frac{(m - 1)^{-1/2}[3(m - 1) + 1]}{m - 1}$$

$$= \frac{3m - 2}{(m - 1)^{3/2}}$$ ■

1.4 Exercises

Factor as completely as possible. Assume all variables appearing as exponents represent integers.

1. $32m^8 + 28m^5 - 16m^3$

2. $-4x^3 + 16x^5 - 36x^6$

3. $-7jh^2 - 21j^2h - 35jh$

4. $16p^3q^2 - 24p^2q^3 + 40p^2q^4$

5. $2(3m + n) + 6m(3m + n)$

6. $3(z - 5)^2 + 6(z - 5)$

7. $6p^2 + 5p - 6$

8. $20x^2 - 7x - 6$

9. $6y^2 + 7y - 5$

10. $24a^2 - 38ab + 15b^2$

11. $12m^2 + 16mn - 35n^2$

12. $9x^2 - 6x^3 + x^4$

13. $10x^4 + 23x^2 + 12$

14. $49y^2 - 100$

15. $25a^2 - 16$

16. $144r^2 - 81s^2$

17. $81m^2 - 16n^2$

18. $121p^4 - 9q^4$

19. $81q^4 - 256m^4$

20. $p^8 - 1$

21. $y^{16} - 1$

22. $8m^3 - 27n^3$

23. $125x^3 - 1$

24. $x^2 + xy + 5x + 5y$

25. $x^3 - y^3 + x^2 - y^2$

26. $8x^2 - 14x^3 + 49x^2 - 16$

27. $a^2 + 2ab + b^2 - x^2 - 2xy - y^2$

28. $d^2 - 10d + 25 - c^2 + 4c - 4$

29. $(z - w)^2 - 4z^2$

30. $9(p - q)^2 - 25q^2$

31. $36(a + 5)^2 - 16a^2$

32. $49(m - 3)^2 - 4a^2$

33. $(x + y)^2 + 2(x + y)z - 15z^2$

34. $(m + n)^2 + 3(m + n)p - 10p^2$

35. $(p + q)^2 - (p - q)^2$

36. $(p - q)^2 - (p + q)^2$

37. $(r + 6)^3 - 216$

38. $(b + 3)^3 - 27$

39. $27 - (m + 2n)^3$

40. $125 - (4a - b)^3$

41. $8u^3 + 27$

42. $(3z - 1)^3 + 125$

43. $2rp - 6rq - tp + 3tq$

44. $3mx - 6my - nx + 2ny$

45. $n^3 - 3n^2p + 3np^2 - p^3$

46. $z^3 + 3z^2w + 3zw^2 + w^3$

47. $m^{2n} - 16$

48. $p^{4n} - 49$

49. $x^{3n} - y^{6n}$

50. $a^{4p} - b^{12p}$

51. $2x^{2n} - 23x^ny^n - 39y^{2n}$

52. $3a^{2x} + 7a^xb^x + 2b^{2x}$

53. $25q^{2r} - 30q^rt^p + 9t^{2p}$

54. $16m^{2p} + 56m^pn^q + 49n^{2q}$

55. $6(m + p)^{2k} + (m + p)^k - 15$

56. When asked to factor $6x^4 - 3x^2 - 3$ completely, a student gave the following result: $6x^4 - 3x^2 - 3 = (2x^2 + 1)(3x^2 - 3)$. Is this answer correct? Explain why or why not.

Factor the variable having the smallest exponent, together with any numerical common factor, from each of the following expressions. For example, factor $9x^{-2} - 6x^{-3}$ *as* $3x^{-3}(3x - 2)$.

57. $m^{-3} + m^{-1}$

58. $2p^{-2} + p^{-4}$

59. $9k^{-3} + 9k^{-2} + 18k^{-1}$

60. $16b^{-5} + 20b^{-4} - 6b^{-2}$

61. $15p^{-5} - 10p^{-1} + 30p^{-3}$

62. $48y^{-3} + 32y - 80y^2$

63. $4k^{7/4} + k^{3/4}$

64. $y^{9/2} - 3y^{5/2}$

65. $9z^{-1/2} + 2z^{1/2}$

66. $3m^{2/3} - 4m^{-1/3}$

67. $-(2a - 5)^{-3/2} + 3(2a - 5)^{-1/2}$

68. $(3k - 2)^{-1/2} + 4(3k - 2)^{-3/2}$

69. In Exercises 57–68, why do you think the variable with the smallest exponent was chosen in the common factor? What happens if the variable to one of the other powers is factored out?

Factor each expression. (The expressions in Exercises 70–77 arise in calculus from techniques called the product *and* quotient rules.*)*

70. $2(3x - 4)^2 + (x - 5)(2)(3x - 4)(3)$

71. $(5 - 2x)(3)(7x - 8)^2(7) + (7x - 8)^3(-2)$

72. $\dfrac{(p + 1)^{1/2} - p(\frac{1}{2})(p + 1)^{-1/2}}{p + 1}$

73. $\dfrac{(r - 2)^{2/3} - r(\frac{2}{3})(r - 2)^{-1/3}}{(r - 2)^{4/3}}$

74. $\dfrac{3(2x^2 + 5)^{1/3} - x(2x^2 + 5)^{-2/3}(4x)}{(2x^2 + 5)^{2/3}}$

75. $\dfrac{-(m^3 + m)^{2/3} + m(\frac{2}{3})(m^3 + m)^{-1/3}(3m^2 + 1)}{(m^3 + m)^{4/3}}$

76. $(x^{-1} - 5)^3(2)(2 - x^{-2})(2x^{-3}) + (2 - x^{-2})^2(3)(x^{-1} - 5)^2(-x^{-2})$

77. $(6 + x^{-4})^2(3)(3x - 2x^{-1})^2(3 + 2x^{-2}) + (3x - 2x^{-1})^3(2)(6 + x^{-4})(-4x^{-5})$

78. Factor $4(x^2 + 3)^{-3}(4x + 7)^{-4}$ from $(x^2 + 3)^{-2}(-3)(4x + 7)^{-4}(4) + (4x + 7)^{-3}(-2)(x^2 + 3)^{-3}(2x)$.

79. Factor $(7x - 8)^{-6}(3x + 2)^{-3}$ from $(7x - 8)^{-5}(-2)(3x + 2)^{-3}(3) + (3x + 2)^{-2}(-5)(7x - 8)^{-6}(7)$.

We can factor $x^4 + 4x^2 + 16$ by grouping as follows.

$$\begin{aligned} x^4 + 4x^2 + 16 &= (x^4 + 8x^2 + 16) - 4x^2 && \textbf{Group; replace } 4x^2 \textbf{ with } 8x^2 - 4x^2. \\ &= (x^2 + 4)^2 - (2x)^2 && \textbf{Factor the perfect squares.} \\ &= (x^2 + 4 + 2x)(x^2 + 4 - 2x) && \textbf{Difference of squares.} \end{aligned}$$

Use this procedure to factor each of the following polynomials.

80. $x^4 + 64$ **81.** $r^4 - 6r^2 + 1$ **82.** $p^4 + 9p^2 + 81$

83. $x^4 - 18x^2 + 1$ **84.** $z^4 - 11z^2 + 25$ **85.** $m^4 - 22m^2 + 9$

1.5 Rational Expressions

The quotient of two algebraic expressions (with denominator not 0) is a **fractional expression**. The most common fractional expressions are the quotients of two polynomials; these quotients are called **rational expressions**. Since fractional expressions involve quotients, it is important to note the values of the variables that would make the denominator zero. For example, -2 cannot replace x in the rational expression

$$\frac{x + 6}{x + 2}$$

since -2 (when substituted for x) makes the denominator equal to 0. This restriction on x can be written $x \neq -2$. In a similar way, the restriction on x for

$$\frac{(x + 6)(x + 4)}{(x + 2)(x + 4)}$$

can be written $x \neq -2$ and $x \neq -4$.

Just as the fraction $6/8$ is written in lowest terms as $3/4$, rational expressions can also be written in lowest terms. We do this with the *fundamental principle.*

Fundamental Principle

For any rational number a/b and nonzero real number c,

$$\frac{ac}{bc} = \frac{a}{b}.$$

EXAMPLE 1 Write each expression in lowest terms.

(a) $\dfrac{2p^2 + 7p - 4}{5p^2 + 20p}$

Factor the numerator and denominator to get

$$\frac{2p^2 + 7p - 4}{5p^2 + 20p} = \frac{(2p - 1)(p + 4)}{5p(p + 4)}.$$

By the fundamental principle,

$$\frac{2p^2 + 7p - 4}{5p^2 + 20p} = \frac{2p - 1}{5p}.$$

The restriction on p for the original expression is $p \neq 0$ and $p \neq -4$, so this result is valid only for values of p other than 0 and -4. From now on, we shall always assume such restrictions when writing rational expressions in lowest terms.

(b) $\dfrac{6 - 3k}{k^2 - 4}$

Factor to get

$$\frac{6 - 3k}{k^2 - 4} = \frac{3(2 - k)}{(k + 2)(k - 2)}.$$

The factors $2 - k$ and $k - 2$ have exactly opposite signs. Because of this, we multiply numerator and denominator by -1, as follows.

$$\frac{6 - 3k}{k^2 - 4} = \frac{3(2 - k)(-1)}{(k + 2)(k - 2)(-1)}$$

Since $(k - 2)(-1) = -k + 2$, or $2 - k$, the fraction becomes

$$\frac{6 - 3k}{k^2 - 4} = \frac{3(2 - k)(-1)}{(k + 2)(2 - k)},$$

finally giving

$$\frac{6 - 3k}{k^2 - 4} = \frac{-3}{k + 2}.$$

Working with the same expression in an alternate way would give the equivalent result $3/(-k - 2)$. ■

Multiplication and Division To multiply or divide rational expressions, again use properties and definitions from Section 1.1.

Multiplication and Division

For rational numbers a/b and c/d,

$$\frac{a}{b} \cdot \frac{c}{d} = \frac{ac}{bd}$$

$$\frac{a}{b} \div \frac{c}{d} = \frac{a}{b} \cdot \frac{d}{c}, \quad c \neq 0.$$

EXAMPLE 2 Multiply or divide, as indicated.

(a)
$$\frac{3m^2 - 2m - 8}{3m^2 + 14m + 8} \cdot \frac{3m + 2}{3m + 4} = \frac{(m - 2)(3m + 4)}{(m + 4)(3m + 2)} \cdot \frac{3m + 2}{3m + 4}$$
$$= \frac{(m - 2)(3m + 4)(3m + 2)}{(m + 4)(3m + 2)(3m + 4)} = \frac{m - 2}{m + 4}$$

(b)
$$\frac{5}{8m + 16} \div \frac{7}{12m + 24} = \frac{5}{8(m + 2)} \div \frac{7}{12(m + 2)}$$
$$= \frac{5}{8(m + 2)} \cdot \frac{12(m + 2)}{7}$$
$$= \frac{5 \cdot 12(m + 2)}{8 \cdot 7(m + 2)} = \frac{15}{14}$$

(c)
$$\frac{3p^2 + 11p - 4}{24p^3 - 8p^2} \div \frac{9p + 36}{24p^4 - 36p^3} = \frac{(p + 4)(3p - 1)}{8p^2(3p - 1)} \div \frac{9(p + 4)}{12p^3(2p - 3)}$$
$$= \frac{(p + 4)(3p - 1)(12p^3)(2p - 3)}{8p^2(3p - 1)(9)(p + 4)}$$
$$= \frac{12p^3(2p - 3)}{9 \cdot 8p^2} = \frac{p(2p - 3)}{6}$$

Addition and Subtraction Add or subtract rational expressions with properties given earlier.

Addition and Subtraction

For rational numbers a/b, c/b, and c/d,

$$\frac{a}{b} + \frac{c}{b} = \frac{a + c}{b} \quad \text{or} \quad \frac{a}{b} + \frac{c}{d} = \frac{ad + bc}{bd}$$
$$\frac{a}{b} - \frac{c}{b} = \frac{a - c}{b} \quad \text{or} \quad \frac{a}{b} - \frac{c}{d} = \frac{ad - bc}{bd}.$$

The results on the left side are for rational expressions with the same denominators. The results on the right side come from the fundamental principle. For example,

$$\frac{a}{b} + \frac{c}{d} = \frac{a \cdot d}{b \cdot d} + \frac{c \cdot b}{d \cdot b} = \frac{ad}{bd} + \frac{bc}{bd} = \frac{ad + bc}{bd}.$$

In practice, rational expressions are normally added or subtracted after rewriting all the rational expressions with the same denominator. The **least common denominator** of a set of rational expressions is the smallest expression that can be divided evenly (without remainder) by each denominator in the set. To find the least common denominator, first factor each denominator, then choose each prime factor to the highest power to which it occurs in any of the denominators. The product of these factors is the least common denominator. The next example illustrates this process.

EXAMPLE 3 Add or subtract, as indicated.

(a) $\dfrac{5}{9x^2} + \dfrac{1}{6x} - \dfrac{1}{x^2}$

The first denominator is factored as $3^2 \cdot x^2$, the second as $2 \cdot 3 \cdot x$, and the third as $1 \cdot x^2$. Choose each factor to its highest power: $3^2 \cdot x^2 \cdot 2$. The product, $18x^2$, is the least common denominator. Now write each of the fractions with this denominator by multiplying by a form of the number 1.

$$\frac{5}{9x^2} + \frac{1}{6x} - \frac{1}{x^2} = \frac{5 \cdot \mathbf{2}}{9x^2 \cdot \mathbf{2}} + \frac{1 \cdot \mathbf{3x}}{6x \cdot \mathbf{3x}} - \frac{1 \cdot \mathbf{18}}{x^2 \cdot \mathbf{18}}$$

$$= \frac{10 + 3x - 18}{18x^2} \qquad \textbf{Combine numerators.}$$

$$= \frac{3x - 8}{18x^2}$$

(b) $\dfrac{y + 2}{y^2 - y} - \dfrac{3y}{2y^2 - 4y + 2}$

To find the common denominator, first factor each denominator.

$$\frac{y + 2}{y^2 - y} - \frac{3y}{2y^2 - 4y + 2} = \frac{y + 2}{y(y - 1)} - \frac{3y}{2(y - 1)^2}$$

The common denominator is $2y(y - 1)^2$. Write each rational expression with this denominator, as follows.

$$\frac{y + 2}{y(y - 1)} - \frac{3y}{2(y - 1)^2} = \frac{2(y - 1)(y + 2)}{2y(y - 1)^2} - \frac{y \cdot 3y}{2y(y - 1)^2}$$

$$= \frac{2(y^2 + y - 2)}{2y(y - 1)^2} - \frac{3y^2}{2y(y - 1)^2}$$

$$= \frac{2y^2 + 2y - 4 - 3y^2}{2y(y - 1)^2} = \frac{-y^2 + 2y - 4}{2y(y - 1)^2}$$

Complex Fractions Any quotient of two rational expressions is called a **complex fraction**. Complex fractions can often be simplified by the methods shown in the following examples.

EXAMPLE 4 Simplify each complex fraction.

(a) $\dfrac{6 - \dfrac{5}{k}}{1 + \dfrac{5}{k}}$

Multiply both numerator and denominator by the common denominator k

$$\frac{k\left(6 - \frac{5}{k}\right)}{k\left(1 + \frac{5}{k}\right)} = \frac{6k - k\left(\frac{5}{k}\right)}{k + k\left(\frac{5}{k}\right)} = \frac{6k - 5}{k + 5}$$

(b) $\dfrac{\frac{a}{a+1} + \frac{1}{a}}{\frac{1}{a} + \frac{1}{a+1}}$

Multiply the numerator and denominator by the common denominator of all the fractions, in this case $a(a + 1)$. Doing so gives

$$\frac{\frac{a}{a+1} + \frac{1}{a}}{\frac{1}{a} + \frac{1}{a+1}} = \frac{\left(\frac{a}{a+1} + \frac{1}{a}\right)a(a+1)}{\left(\frac{1}{a} + \frac{1}{a+1}\right)a(a+1)} = \frac{a^2 + (a+1)}{(a+1) + a} = \frac{a^2 + a + 1}{2a + 1}.$$

As an alternative method of solution, first perform the indicated additions in the numerator and denominator, and then divide.

$$\frac{\frac{a}{a+1} + \frac{1}{a}}{\frac{1}{a} + \frac{1}{a+1}} = \frac{\frac{a^2 + 1(a+1)}{a(a+1)}}{\frac{1(a+1) + 1(a)}{a(a+1)}} = \frac{\frac{a^2 + a + 1}{a(a+1)}}{\frac{2a+1}{a(a+1)}}$$

$$= \frac{a^2 + a + 1}{a(a+1)} \cdot \frac{a(a+1)}{2a+1} = \frac{a^2 + a + 1}{2a + 1}$$ ■

The next example shows how negative exponents can lead to rational expressions.

EXAMPLE 5 Simplify $\dfrac{(x+y)^{-1}}{x^{-1} + y^{-1}}$. Write the result so that it has only positive exponents.

Use the definition of a negative integer exponent, then simplify the resulting complex fraction.

$$\frac{(x+y)^{-1}}{x^{-1} + y^{-1}} = \frac{\frac{1}{x+y}}{\frac{1}{x} + \frac{1}{y}} = \frac{\frac{1}{x+y}}{\frac{y+x}{xy}} = \frac{1}{x+y} \cdot \frac{xy}{x+y} = \frac{xy}{(x+y)^2}$$ ■

1.5 Exercises

Give the restrictions on x for each of the following.

1. $\frac{x-1}{2x+3}$

2. $\frac{x+3}{4x-1}$

3. $\frac{3x^2}{2x^2+11x+12}$

4. $\frac{x-10}{1+4x^2}$

5. $\frac{8x+1}{5x^2+2}$

6. $\frac{5x-1}{6x^2+5x-4}$

Write each of the following in lowest terms.

7. $\frac{16h^3}{2h^2}$

8. $\frac{27r^5}{9r}$

9. $\frac{5t+25}{3t+15}$

10. $\frac{32y+16}{18y+9}$

11. $\frac{-4(x-1)}{(x-1)(x+2)}$

12. $\frac{2(4m+3)}{(4m+3)(m-4)}$

13. $\frac{4r^2+24}{8r^2}$

14. $\frac{12m^2-60m}{4m^2}$

15. $\frac{3y^2-10y-8}{3y^2+5y+2}$

16. $\frac{p^2+2p-3}{p^2+4p-5}$

17. $\frac{2m^2+5m-25}{4m^2-25}$

18. $\frac{6z^2+7z+2}{3z^2+14z+8}$

Perform each operation.

19. $\frac{k(h-1)}{5} \div \frac{k}{h-1}$

20. $\frac{mn}{m+n} \div \frac{p}{m+n}$

21. $\frac{15p^3}{9p^2} \div \frac{6p}{10p^2}$

22. $\frac{5x^2}{15x^3} \div \frac{25x^4}{10x^2}$

23. $\frac{2k+8}{6} \div \frac{3k+12}{2}$

24. $\frac{5m+25}{10} \cdot \frac{12}{6m+30}$

25. $\frac{3p-18}{8p+48} \cdot \frac{4p+24}{9p-54}$

26. $\frac{12r+24}{36r-36} \div \frac{6r+12}{8r-8}$

27. $\frac{x^2+x}{5} \cdot \frac{25}{xy+y}$

28. $\frac{a^2-9}{a^2-a-20} \div \frac{4a+12}{2a-10}$

29. $\frac{m^2-10m+25}{12m-60} \cdot \frac{3m-15}{4m-20}$

30. $\frac{6r-18}{9r^2+6r-24} \cdot \frac{12r-16}{4r-12}$

31. $\frac{x^2+7x+10}{x^2+x-2} \div \frac{x^2+2x-15}{x^2+3x+2}$

32. $\frac{x^2+2x-15}{x^2+11x+30} \cdot \frac{x^2+2x-24}{x^2-8x+15}$

33. $\frac{3y^2-7y-20}{y^2+4y-32} \div \frac{9y^2-25}{2y^2+13y-24}$

34. $\frac{n^2-n-6}{n^2-2n-8} \div \frac{n^2-9}{n^2+7n+12}$

35. $\left(1+\frac{1}{x}\right)\left(1-\frac{1}{x}\right)$

36. $\left(3+\frac{2}{y}\right)\left(3-\frac{2}{y}\right)$

37. $\frac{x^3+y^3}{x^2-y^2} \cdot \frac{x+y}{x^2-xy+y^2}$

38. $\frac{8y^3-125}{4y^2-20y+25} \cdot \frac{2y-5}{y}$

39. $\frac{x^3+y^3}{x^3-y^3} \cdot \frac{x^2-y^2}{x^2+2xy+y^2}$

40. $\frac{8}{r}+\frac{6}{r}+\frac{r}{2}$

41. $\frac{3}{y}+\frac{4}{y}-\frac{y}{6}$

42. $\frac{8}{5p}+\frac{3}{4p}+\frac{2}{5}$

43. $\frac{2}{3y}-\frac{1}{4y}+\frac{y}{3}$

44. $\frac{6}{11z}-\frac{5}{2z}$

45. $\frac{a+1}{2}-\frac{a-1}{2}$

46. $\frac{y+6}{5}-\frac{y-6}{5}$

47. $\frac{2}{y}-\frac{1}{4}$

48. $\frac{6}{11}+\frac{3}{a}$

49. $\frac{1}{6m}+\frac{2}{5m}+\frac{4}{m}$

50. $\frac{8}{3p} + \frac{5}{4p} + \frac{9}{2p}$

51. $\frac{1}{y} + \frac{1}{y + 1}$

52. $\frac{5}{2(x + 3)} + \frac{7}{3(x + 3)}$

53. $\frac{2}{a + b} - \frac{1}{2(a + b)}$

54. $\frac{3}{m} - \frac{1}{m - 1}$

55. $\frac{1}{a + 1} - \frac{1}{a - 1}$

56. $\frac{1}{x + z} + \frac{1}{x - z}$

57. $\frac{m + 1}{m - 1} + \frac{m - 1}{m + 1}$

58. $\frac{3}{y - 2} + \frac{4}{2 - y}$

59. $\frac{5}{p - 4} - \frac{2}{4 - p}$

60. $\frac{m - 4}{3m - 4} + \frac{3m + 2}{4 - 3m}$

61. $\frac{1}{a^2 - 5a + 6} - \frac{1}{a^2 - 4}$

62. $\frac{5}{k^2 + 5k + 6} - \frac{-2}{k^2 + 2k - 3}$

63. $\frac{5x + 2}{x^2 - x - 20} + \frac{2x - 1}{x^2 - 4x - 5}$

64. $\frac{3y + 5}{y^2 - 9y + 20} + \frac{2y - 7}{y^2 - 2y - 8}$

65. $\left(\frac{3}{p - 1} - \frac{2}{p + 1}\right)\left(\frac{p - 1}{p}\right)$

66. $\left(\frac{y}{y^2 - 1} - \frac{y}{y^2 - 2y + 1}\right)\left(\frac{y - 1}{y + 1}\right)$

67. $\frac{\frac{1}{x + h} - \frac{1}{x}}{h}$

68. $\frac{1}{h}\left(\frac{1}{(x + h)^2 + 9} - \frac{1}{x^2 + 9}\right)$

69. $\frac{1 + \frac{1}{x}}{1 - \frac{1}{x}}$

70. $\frac{4 - \frac{1}{z}}{4 + \frac{1}{z}}$

71. $\frac{\frac{1}{k - 1} - \frac{1}{k}}{\frac{1}{k}}$

72. $\frac{\frac{1}{y + 3} - \frac{1}{y}}{\frac{1}{y}}$

73. $\frac{1 + \frac{1}{1 - b}}{1 - \frac{1}{1 + b}}$

74. $m - \frac{m}{m + \frac{1}{2}}$

75. In your own words, explain how to find the least common denominator for two fractions.

76. Describe the steps required to add three rational expressions. You may use an example to illustrate.

Perform all indicated operations and write all answers with positive integer exponents. (See Example 5.)

77. $\frac{3^{-1} - 4^{-1}}{4^{-1}}$

78. $\frac{6^{-1} + 5^{-1}}{6^{-1}}$

79. $\frac{a^{-1} + b^{-1}}{(ab)^{-1}}$

80. $\frac{p^{-1} - q^{-1}}{(pq)^{-1}}$

81. $\frac{r^{-1} + q^{-1}}{r^{-1} - q^{-1}} \cdot \frac{r - q}{r + q}$

82. $\frac{xy^{-1} + yx^{-1}}{x^2 + y^2}$

83. $(a + b)^{-1}(a^{-1} + b^{-1})$

84. $(m^{-1} + n^{-1})^{-1}$

Simplify each of the following expressions. (The expressions in Exercises 85–88 arise in calculus from techniques called the chain rule *and* quotient rule *that are used to determine the shape of a curve.)*

85. $\left(\frac{x^3 + 2}{x - 5}\right)^3\left(\frac{(x - 5)(3x^2) - (x^3 + 2)}{(x - 5)^2}\right)$

86. $\left(\frac{2x^2 - 9}{x^2 + 1}\right)^2\left(\frac{(x^2 + 1)(4x) - (2x^2 - 9)}{(x^2 + 1)^2}\right)$

87. $5\left(\frac{x + 1}{3x^2 - 4}\right)^4\left(\frac{(3x^2 - 4) - (x + 1)(6x)}{(3x^2 - 4)^2}\right)$

88. $7\left(\frac{x^2 + 2}{5x^2 + 3}\right)^6\left(\frac{(5x^2 + 3)(2x) - (x^2 + 2)(10x)}{(5x^2 + 3)^2}\right)$

1.6 Radicals

Earlier, we defined $a^{1/n}$ as the positive nth root of a. Radicals provide a convenient notation for nth roots. In this section we explore this notation and develop techniques for simplifying expressions involving radicals.

Definition of $\sqrt[n]{a}$

If a is a real number, n is a positive integer, and $a^{1/n}$ is a real number, then

$$\sqrt[n]{a} = a^{1/n}.$$

As we saw in Section 1.2, $a^{1/n}$ is defined only for nonnegative values of a when n is even, but is defined for all values of a when n is odd. Combining the definition of $\sqrt[n]{a}$ with the definition of $a^{1/n}$ from Section 1.2, we obtain the following characterization of $\sqrt[n]{a}$.

$\sqrt[n]{a}$

If a and b are nonnegative real numbers and n is a positive integer, or if both a and b are negative and n is an odd positive integer, then

$$\sqrt[n]{a} = b \quad \textbf{if and only if} \quad a = b^n.$$

The expression $\sqrt[n]{a}$ is called the **principal nth root** of a (abbreviated as the **nth root of a**); n is the **index** of the radical expression $\sqrt[n]{a}$. The number a is the **radicand**, and $\sqrt[n]{}$ is a **radical**. We abbreviate $\sqrt[2]{a}$ as just $\sqrt{a}$.

If n is a positive integer and a is a real number, then the following chart summarizes the conditions necessary for $\sqrt[n]{a}$ to exist.

Conditions on $\sqrt[n]{a}$

n is	$a > 0$	$a < 0$	$a = 0$
even	$\sqrt[n]{a}$ is the positive real number such that $(\sqrt[n]{a})^n = a$.	$\sqrt[n]{a}$ is undefined.	$\sqrt[n]{a}$ is 0.
odd	$\sqrt[n]{a}$ is the real number such that $(\sqrt[n]{a})^n = a$.		$\sqrt[n]{a}$ is 0.

If n is even, there are two nth roots of any positive number a; one root is positive and the other root is negative. In radical notation, $\sqrt[n]{a}$ denotes the positive nth root of a and $-\sqrt[n]{a}$ denotes the negative nth root of a.

EXAMPLE 1

(a) $\sqrt{\frac{4}{9}} = \frac{2}{3}$ **(b)** $-\sqrt[4]{16} = -2$ **(c)** $\sqrt{-4}$ is undefined.

(d) $\sqrt[5]{-32} = -2$ **(e)** $\sqrt[3]{1000} = 10$ ■

By the definition of $\sqrt[n]{a}$, for any positive integer n, if $\sqrt[n]{a}$ exists, then

$$(\sqrt[n]{a})^n = a.$$

If a is positive, or if a is negative and n is an odd positive integer, then

$$\sqrt[n]{a^n} = a.$$

Because of the conditions just given, it is *not* necessarily true that $\sqrt{x^2} = x$. For example, if $x = -5$,

$$\sqrt{x^2} = \sqrt{(-5)^2} = \sqrt{25} = 5 \neq x.$$

To account for the fact that a negative value of x can produce a positive result for the square root, we use the following rule, which involves absolute value.

For any real number a,

$$\sqrt{a^2} = |a|.$$

Also, if n is any *even* positive integer, then

$$\sqrt[n]{a^n} = |a|.$$

We shall prove only the first part of this statement, when $n = 2$. The statement is certainly true if a is positive. If a is negative, then $-a$ is positive. Since $(-a)^2 = a^2$,

$$\sqrt{(-a)^2} = -a.$$

Since $|a| = -a$ if a is negative, $\sqrt{a^2} = |a|$ for *all* real numbers a.

EXAMPLE 2

(a) $\sqrt{(-9)^2} = |-9| = 9$ **(b)** $\sqrt{13^2} = |13| = 13$

(c) $\sqrt{x^6} = |x^3|$ **(d)** $\sqrt[3]{x^6} = x^2$

(e) $\sqrt[3]{x^9} = x^3$ **(f)** $\sqrt[4]{x^4} = |x|$

(g) $\sqrt[4]{x^8} = x^2$ ■

Rational exponents can be expressed in terms of radicals in two ways.

Radical Definition of Rational Exponents

For all integers m and all positive integers n such that m/n is in lowest terms, and all nonzero numbers a for which $\sqrt[n]{a}$ is a real number,

$$a^{m/n} = (\sqrt[n]{a})^m = \sqrt[n]{a^m}.$$

To avoid difficulties when we are working with variable radicands, we will usually assume that all variables in radicands represent only nonnegative real numbers.

EXAMPLE 3 Write each of the following exponential expressions as radicals.

(a) $m^{1/3} = \sqrt[3]{m}$ **(b)** $2^{2/5} = \sqrt[5]{2^2} = \sqrt[5]{4}$ **(c)** $k^{3/2} = \sqrt{k^3}$ ■

We can convert radicals with different indexes to the same index by first writing them with rational exponents, and then using the rules for exponents.

EXAMPLE 4 Write each of the following with a single radical. Assume all variables represent nonnegative real numbers.

(a) $\sqrt[5]{y^4} \cdot \sqrt[3]{y} = y^{4/5}y^{1/3} = y^{17/15} = y^{1+2/15} = y \cdot y^{2/15} = y\sqrt[15]{y^2}$

(b) $\sqrt[4]{p^3q} \cdot \sqrt[5]{p^2q^7} = (p^3q)^{1/4}(p^2q^7)^{1/5} = p^{3/4}\,p^{2/5}\,q^{1/4}\,q^{7/5}$
$= p^{23/20}\,q^{33/20} = pq\sqrt[20]{p^3q^{13}}$ ■

Three key rules for working with radicals are given below. These rules are simply restatements in radical form of properties of exponents.

Rules for Radicals

For all real numbers a and b and positive integers m and n for which the indicated radicals exist,

$$\sqrt[n]{a} \cdot \sqrt[n]{b} = \sqrt[n]{ab}$$

$$\sqrt[n]{\frac{a}{b}} = \frac{\sqrt[n]{a}}{\sqrt[n]{b}}, \quad b \neq 0$$

$$\sqrt[m]{\sqrt[n]{a}} = \sqrt[mn]{a}.$$

EXAMPLE 5 Use the rules of radicals to simplify each of the following. Assume that all variables represent nonegative real numbers.

(a) $\sqrt{64 \cdot 5} = \sqrt{64} \cdot \sqrt{5} = 8\sqrt{5}$

(b) $\sqrt{7x^5yz^2} = \sqrt{x^4z^2} \cdot \sqrt{7xy} = x^2z\sqrt{7xy}$

(c) $\sqrt[3]{\dfrac{-11x^6}{64}} = \dfrac{\sqrt[3]{-11x^6}}{\sqrt[3]{64}} = \dfrac{x^2\sqrt[3]{-11}}{4} = \dfrac{-x^2\sqrt[3]{11}}{4}$

(d) $\sqrt[7]{\sqrt[3]{2}} = \sqrt[21]{2}$

(e) $\sqrt[4]{\sqrt{3}} = \sqrt[8]{3}$ ■

Simplifying Radicals By definition, an expression containing a radical is **simplified** when the following four conditions are satisfied.

Simplified Radicals

1. All possible factors have been removed from under the radical sign.
2. The index on the radical is as small as possible.
3. All radicals are removed from any denominators (a process called **rationalizing the denominator**).
4. All indicated operations have been performed (if possible).

EXAMPLE 6 Simplify each of the following. Assume that all variables represent nonnegative real numbers.

(a) $\sqrt{98x^3y} + 3x\sqrt{32xy}$

First remove all perfect square factors from under the radical. Then use the distributive property, as follows:

$$\begin{aligned}\sqrt{98x^3y} + 3x\sqrt{32xy} &= \sqrt{\mathbf{49} \cdot 2 \cdot \mathbf{x^2} \cdot x \cdot y} + 3x\sqrt{\mathbf{16} \cdot 2 \cdot x \cdot y} \\ &= \mathbf{7x}\sqrt{2xy} + (3x)(\mathbf{4})\sqrt{2xy} \\ &= 7x\sqrt{2xy} + 12x\sqrt{2xy} \\ &= 19x\sqrt{2xy}.\end{aligned}$$

(b)
$$\begin{aligned}\sqrt[3]{64m^4p^5} - \sqrt[3]{-27m^{10}p^{14}} &= \sqrt[3]{(64m^3p^3)(mp^2)} - \sqrt[3]{(-27m^9p^{12})(mp^2)} \\ &= 4mp\sqrt[3]{mp^2} - (-3m^3p^4)\sqrt[3]{mp^2} \\ &= 4mp\sqrt[3]{mp^2} + 3m^3p^4\sqrt[3]{mp^2} \\ &= (4 + 3m^2p^3)mp\sqrt[3]{mp^2}\end{aligned}$$
■

Multiplying radical expressions is much like multiplying polynomials.

EXAMPLE 7

$$(\sqrt{2} + 3)(\sqrt{8} - 5) = \sqrt{2}(\sqrt{8}) - \sqrt{2}(5) + 3\sqrt{8} - 3(5)$$
$$= \sqrt{16} - 5\sqrt{2} + 3(2\sqrt{2}) - 15$$
$$= 4 - 5\sqrt{2} + 6\sqrt{2} - 15$$
$$= -11 + \sqrt{2}$$

Rationalizing the Denominator The next example shows how to rationalize the denominator (remove any radicals from the denominator) in an expression containing radicals.

EXAMPLE 8 Simplify each of the following expressions.

(a) $\dfrac{4}{\sqrt{3}}$

To rationalize the denominator, multiply by $\sqrt{3}/\sqrt{3}$ (which equals 1) so that the denominator of the product is a rational number. Work as follows:

$$\frac{4}{\sqrt{3}} \cdot \frac{\sqrt{3}}{\sqrt{3}} = \frac{4\sqrt{3}}{3}.$$

(b) $\sqrt[4]{\dfrac{3}{5}}$

Start by using the fact that the radical of a quotient can be written as the quotient of radicals. To rationalize the denominator, multiply numerator and denominator by $\sqrt[4]{5^3}$. Use this number so that the denominator will be a rational number.

$$\sqrt[4]{\frac{3}{5}} = \frac{\sqrt[4]{3}}{\sqrt[4]{5}} = \frac{\sqrt[4]{3} \cdot \sqrt[4]{5^3}}{\sqrt[4]{5} \cdot \sqrt[4]{5^3}}$$
$$= \frac{\sqrt[4]{3 \cdot 5^3}}{\sqrt[4]{5^4}} = \frac{\sqrt[4]{375}}{5}$$

(c) $\sqrt[3]{\dfrac{9m^3n^5}{k^2p}}$

Write the radical as the quotient of two radicals. Then multiply by $\sqrt[3]{kp^2}/\sqrt[3]{kp^2}$, which will make the denominator $\sqrt[3]{k^3p^3} = kp$, so the denominator has no radical.

$$\sqrt[3]{\frac{9m^3n^5}{k^2p}} = \frac{\sqrt[3]{9m^3n^5}}{\sqrt[3]{k^2p}} \cdot \frac{\sqrt[3]{kp^2}}{\sqrt[3]{kp^2}}$$
$$= \frac{\sqrt[3]{9m^3n^5kp^2}}{\sqrt[3]{k^3p^3}} = \frac{mn\sqrt[3]{9n^2kp^2}}{kp}$$

In Example 9 below, the denominator is $1 - \sqrt{2}$. To rationalize this denominator, multiply the numerator and denominator by $1 + \sqrt{2}$. This number is chosen because if a is rational and b is a nonnegative rational number, then

$$(a - \sqrt{b})(a + \sqrt{b}) = a^2 - (\sqrt{b})^2 = a^2 - b,$$

which is a rational number. More generally, if a and b are nonnegative rational numbers,

$$(\sqrt{a} + \sqrt{b})(\sqrt{a} - \sqrt{b}) = a - b.$$

EXAMPLE 9 Rationalize the denominator of $\dfrac{1}{1 - \sqrt{2}}$.

As just mentioned, multiply the numerator and denominator by $1 + \sqrt{2}$.

$$\frac{1}{1 - \sqrt{2}} = \frac{1(1 + \sqrt{2})}{(1 - \sqrt{2})(1 + \sqrt{2})} = \frac{1 + \sqrt{2}}{1 - 2} = \frac{1 + \sqrt{2}}{-1} = -1 - \sqrt{2}$$ ■

1.6 Exercises

Write each of the following as a radical.

1. $6^{2/3}$ **2.** $5^{1/4}$ **3.** $(3z)^{1/5}$ **4.** $(4m)^{5/6}$ **5.** $2k^{2/3}$ **6.** $7p^{4/5}$

Simplify each of the following. Assume that all variables represent nonnegative real numbers and that no denominators are zero.

7. $\sqrt{98}$ **8.** $\sqrt{27}$ **9.** $\sqrt{375}$

10. $\sqrt{64}$ **11.** $-\sqrt{5/4}$ **12.** $-\sqrt{3/25}$

13. $4\sqrt{3} - 5\sqrt{12} + 3\sqrt{75}$ **14.** $2\sqrt{5} - 3\sqrt{20} + 2\sqrt{45}$ **15.** $\sqrt[4]{\dfrac{3}{2}}$

16. $\sqrt[4]{\dfrac{32}{81}}$ **17.** $\sqrt[4]{5\dfrac{1}{16}}$ **18.** $\sqrt[3]{3\dfrac{3}{8}}$

19. $\sqrt[3]{2} - \sqrt[3]{16} + 2\sqrt[3]{54}$ **20.** $3\sqrt[3]{4} - 5\sqrt[3]{256} + \sqrt[3]{32}$ **21.** $4/\sqrt{5} - 2/\sqrt{20} + 3\sqrt{5}$

22. $3/\sqrt{3} + 2/\sqrt{27} + 1/\sqrt{12}$ **23.** $-1/\sqrt[3]{5} + 3/\sqrt[3]{40} - 2/\sqrt[3]{135}$ **24.** $3/\sqrt[3]{2} + 5/\sqrt[3]{16} - 1/\sqrt[3]{128}$

25. $\sqrt{72p^5q^4r^6}$ **26.** $\sqrt[3]{16z^5x^8y^4}$ **27.** $\sqrt[4]{x^8y^6z^{10}}$

28. $\sqrt{a^3b^5} - 2\sqrt{a^7b^3} + \sqrt{a^3b^9}$ **29.** $\sqrt{p^7q^3} - \sqrt{p^5q^9} + \sqrt{p^9q}$ **30.** $(\sqrt{2} + 3)(\sqrt{2} - 3)$

31. $(\sqrt{5} + \sqrt{2})(\sqrt{5} - \sqrt{2})$ **32.** $(\sqrt{2} + \sqrt{7})^2$ **33.** $(\sqrt{5} - 1)^2$

34. $(3\sqrt[3]{2} - 4)(3\sqrt[3]{2} + 1)$ **35.** $(\sqrt[3]{6} + 2)(5\sqrt[3]{6} + 3)$ **36.** $\sqrt{\dfrac{x^5y^3}{z^2}}$

37. $\sqrt{\dfrac{g^3h^5}{r^3}}$ **38.** $-\sqrt[3]{\dfrac{k^5m^3r^2}{r^8}}$ **39.** $-\sqrt[3]{\dfrac{9x^5y^6}{z^5w^2}}$

40. $\sqrt[4]{\dfrac{g^3h^5}{9r^6}}$

41. $\sqrt[4]{\dfrac{32x^6}{y^5}}$

42. $\dfrac{\sqrt[3]{mn}\cdot\sqrt[3]{m^2}}{\sqrt[3]{n^2}}$

43. $\dfrac{\sqrt[3]{8m^2n^3}\cdot\sqrt[3]{2m^2}}{\sqrt[3]{32m^4n^3}}$

44. $\sqrt[3]{\sqrt{2}}$

45. $\sqrt[3]{\sqrt[4]{5}}$

46. $\sqrt[4]{\sqrt[5]{p}}$

47. $\sqrt[6]{\sqrt[3]{y}}$

48. $\sqrt{2(m+n)^2+2(m-n)^2}$

49. $\sqrt{2(x+y)^2-2(x-y)^2}$

50. $\sqrt{4x^4+x^4y^4}$

51. $\sqrt{-z^2+(z-9)^2}$

52. $\dfrac{6}{\sqrt{2}-1}$

53. $\dfrac{2}{1+\sqrt{5}}$

54. $\dfrac{\sqrt{3}}{4+\sqrt{3}}$

55. $\dfrac{2\sqrt{7}}{3-\sqrt{7}}$

56. $\dfrac{1}{\sqrt{m}-\sqrt{p}}$

57. $\dfrac{\sqrt{z}}{-\sqrt{z}+1}$

58. $\dfrac{-4}{\sqrt{1+x+3}}$

59. $\dfrac{-5}{1-\sqrt{3-p}}$

60. $\dfrac{\sqrt{x}+\sqrt{x+1}}{\sqrt{x}-\sqrt{x+1}}$

61. $\dfrac{\sqrt{p}+\sqrt{p^2-1}}{\sqrt{p}-\sqrt{p^2-1}}$

62. $\dfrac{5}{\sqrt[3]{a}+\sqrt[3]{b}}$

63. $\dfrac{1}{\sqrt[3]{m}-\sqrt[3]{n}}$

64. $\dfrac{6-\sqrt{3}}{(5-\sqrt{2})(3+\sqrt{5})}$

65. $\dfrac{1-\sqrt{7}}{(2+\sqrt{10})(1-\sqrt{5})}$

66. Explain how to rationalize the denominator of $\sqrt[3]{3/2}$.

67. Describe the steps required to multiply $\sqrt[3]{2}$ and $\sqrt{2}$.

Take the squared terms outside of the following radicals.

68. $\sqrt{5(3-\sqrt{10})^2}$

69. $\sqrt{(4-\sqrt{17})^2}$

Simplify Exercises 70 and 71, given $x \le 5$.

70. $\sqrt{(x-5)^2}$

71. $\sqrt{(5-x)^2(x-3)^4}$

72. Use a calculator to find an approximate value for $\sqrt{5+2\sqrt{6}}$.

73. Show that $\sqrt{5+2\sqrt{6}}=\sqrt{2}+\sqrt{3}$.

1.7 Complex Numbers

So far, we have worked only with real numbers in this book. The set of real numbers, however, does not include all the numbers needed in algebra. For example, with real numbers alone, it is not possible to find a number whose square is -1. Although as early as 50 B.C. square roots of negative numbers were known, they were not incorporated into an integrated number system until much later. Italian mathematician Girolamo Cardano (1501–1576) and others computed with them. As algebra became necessary in many applications, mathematicians distinguished between two types of solutions. Eventually these were named ''real'' and ''imaginary'' by Descartes (1596–1650).

The German mathematician Gottfried Leibniz, one of the founders of calculus, wrote to his Dutch colleague Christian Huygens in 1679 of the need for an expanded number system: ''I have no hope that we can get very far in physics until we have found some method of abridgment.'' The void was filled by the complex numbers, which combine real and imaginary numbers. It was the renowned Leonhard Euler in 1748 who first wrote $\sqrt{-1} = i$, and in 1832 Gauss stated that numbers of the form $a + bi$ are ''complex.'' The first application of the imaginary numbers to geometry came in 1796 when the Norwegian Caspar Wessel used them in surveying. Charles Steinmetz (1865–1923), an electrical engineer, is said to have ''generated electricity with the square root of minus one'' when he used complex numbers to develop a theory of alternating currents. Today complex numbers are used extensively in science and engineering.

To extend the real number system to include numbers such as $\sqrt{-1}$, the new number i is defined to have the property

$$i^2 = -1.$$

Thus, $i =$ the square root of -1. The number i is called the **imaginary unit**.

Numbers of the form $a + bi$, where a and b are real numbers, are called **complex numbers**. The real number a is called the **real part** of $a + bi$ and the real number b is called the **imaginary part**. Each real number is a complex number, since a real number a may be thought of as the complex number $a + 0i$. A complex number of the form $0 + bi$, where b is nonzero, is called an **imaginary number** (sometimes a *pure* imaginary number). Both the set of real numbers and the set of imaginary numbers are subsets of the set of complex numbers. (See Figure 7, which is an extension of Figure 1 in Section 1.1.) A complex number that is written in the form $a + bi$ or $a + ib$ is in **standard form**. (The form $a + ib$ is used to simplify certain symbols such as $i\sqrt{5}$, since $\sqrt{5}i$ could be too easily mistaken for $\sqrt{5i}$.)

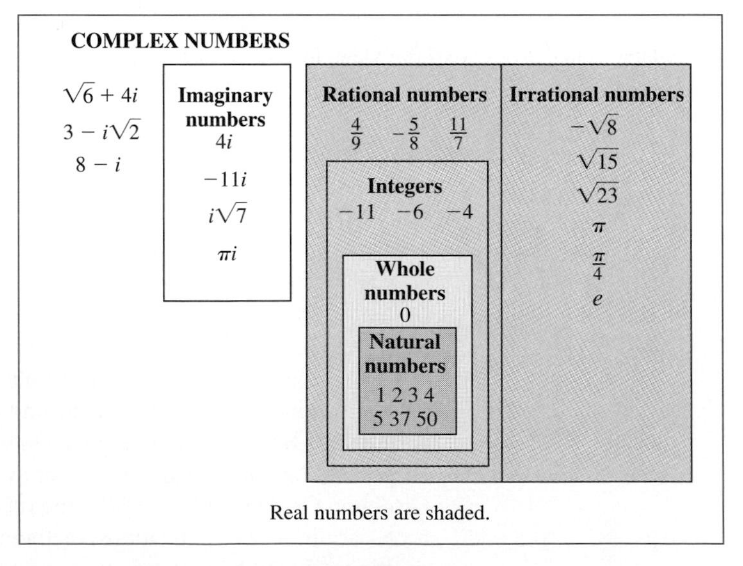

Figure 7

EXAMPLE 1 The following statements identify different kinds of complex numbers.

(a) -8 and $\sqrt{7}$ and π are real numbers and complex numbers.

(b) $3i$ and $-11i$ and $i\sqrt{14}$ are imaginary numbers and complex numbers.

(c) $1 - 2i$ and $8 - 8i\sqrt{3}$ are complex numbers. ■

Equality for complex numbers is defined as follows.

Definition of Equality

For real numbers a, b, c, and d,

$$a + bi = c + di \quad \textbf{if and only if} \quad a = c \textbf{ and } b = d.$$

EXAMPLE 2 Solve $2 + mi = k + 3i$ for real numbers m and k.

By the definition of equality, $2 + mi = k + 3i$ if and only if $2 = k$ and $m = 3$. ■

The imaginary unit is used to define the square root of a negative number.

Definition of $\sqrt{-a}$

If $a > 0$, then

$$\sqrt{-a} = i\sqrt{a}.$$

EXAMPLE 3

(a) $\sqrt{-16} = i\sqrt{16} = 4i$

(b) $\sqrt{-70} = i\sqrt{70}$ ■

Products or quotients with square roots of negative numbers may be simplified using the definitions $\sqrt{-a} = i\sqrt{a}$ for positive numbers a and $i^2 = -1$. The next example shows how to do this.

EXAMPLE 4

(a)
$$\begin{aligned} \sqrt{-7} \cdot \sqrt{-7} &= i\sqrt{7} \cdot i\sqrt{7} \\ &= i^2 \cdot (\sqrt{7})^2 \\ &= (-1) \cdot 7 \\ &= -7 \end{aligned}$$

(b)
$$\begin{aligned} \sqrt{-6} \cdot \sqrt{-10} &= i\sqrt{6} \cdot i\sqrt{10} \\ &= i^2 \cdot \sqrt{6 \cdot 10} \\ &= -1 \cdot 2\sqrt{15} \\ &= -2\sqrt{15} \end{aligned}$$

(c) $\dfrac{\sqrt{-20}}{\sqrt{-2}} = \dfrac{i\sqrt{20}}{i\sqrt{2}} = \sqrt{10}$ **(d)** $\dfrac{\sqrt{-48}}{\sqrt{24}} = \dfrac{i\sqrt{48}}{\sqrt{24}} = i\sqrt{2}$ ■

CAUTION When working with negative radicands, *use the definition* $\sqrt{-a} = i\sqrt{a}$ *before using any of the other rules for radicals.* In particular, the rule $\sqrt{c} \cdot \sqrt{d} = \sqrt{cd}$ is valid only when c and d are *not* both negative. For example,

$$\sqrt{(-4)(-9)} = \sqrt{36} = 6,$$

while

$$\sqrt{-4} \cdot \sqrt{-9} = 2i(3i) = 6i^2 = -6,$$

so

$$\sqrt{-4} \cdot \sqrt{-9} \neq \sqrt{(-4)(-9)}. ■$$

Operations on Complex Numbers With the definitions of $i^2 = -1$ and $\sqrt{-a} = i\sqrt{a}$, if $a > 0$, all the properties of real numbers can be extended to the complex numbers. As a result, complex numbers are added, subtracted, multiplied, and divided as shown by the following definitions and examples.

The *sum* and *difference* of two complex numbers $a + bi$ and $c + di$ are defined as follows.

Definition of Addition and Subtraction

For complex numbers $a + bi$ and $c + di$,

$$(a + bi) + (c + di) = (a + c) + (b + d)i$$

and

$$(a + bi) - (c + di) = (a - c) + (b - d)i.$$

EXAMPLE 5

(a) $(3 - 4i) + (-2 + 6i) = [3 + (-2)] + [-4 + 6]i = 1 + 2i$

(b) $(-9 + 7i) + (3 - 15i) = -6 - 8i$

(c) $(-4 + 3i) - (6 - 7i) = (-4 - 6) + [3 - (-7)]i = -10 + 10i$

(d) $(12 - 5i) - (8 - 3i) = 4 - 2i$ ■

The product of two complex numbers can be found by multiplying as if the numbers were binomials and using the fact that $i^2 = -1$, as follows.

$$\begin{aligned}(a + bi)(c + di) &= ac + adi + bic + bidi \\ &= ac + adi + bci + bdi^2 \\ &= ac + (ad + bc)i + bd(-1) \\ (a + bi)(c + di) &= (ac - bd) + (ad + bc)i\end{aligned}$$

Based on this result, the *product* of the complex numbers $a + bi$ and $c + di$ is defined in the following way.

Definition of Product

$$(a + bi)(c + di) = (ac - bd) + (ad + bc)i$$

This definition is hard to remember. To find a given product, it is usually better to multiply as with binomials and replace i^2 with -1. The next example shows this.

EXAMPLE 6 Find each of the following products.

(a) $(2 - 3i)(3 + 4i) = 2(3) + 2(4i) - 3i(3) - 3i(4i)$
$= 6 + 8i - 9i - 12i^2$
$= 6 - i - 12(-1)$
$= 18 - i$

(b) $(6 + 5i)(6 - 5i) = 6 \cdot 6 - 6 \cdot 5i + 6 \cdot 5i - (5i)^2$
$= 36 - 25i^2$
$= 36 - 25(-1)$
$= 36 + 25$
$= 61$

(c) i^{15}

Since $i^2 = -1$, the value of a power of i is found by writing the given power as a product involving i^2. For example, $i^3 = i^2 \cdot i = (-1) \cdot i = -i$. Also, $i^4 = i^2 \cdot i^2 = (-1)(-1) = 1$. Using i^4 to rewrite i^{15} gives

$$i^{15} = i^{12} \cdot i^3 = (i^4)^3 \cdot i^3 = (1)^3(-i) = -i.$$ ■

Methods similar to those in Example 6(c) give the following list of *powers of i*. Values of higher (or lower) powers continue in the same fashion.

Powers of *i*

$i^1 = i$	$i^5 = i$	$i^9 = i$
$i^2 = -1$	$i^6 = -1$	$i^{10} = -1$
$i^3 = -i$	$i^7 = -i$	$i^{11} = -i$
$i^4 = 1$	$i^8 = 1$	$i^{12} = 1$

Example 6(b) showed that $(6 + 5i)(6 - 5i) = 61$. The numbers $6 + 5i$ and $6 - 5i$ differ only in their middle signs; these numbers are **conjugates** of each other. Conjugates are useful because the product of a complex number and its conjugate is always a real number (see Exercise 68).

EXAMPLE 7 The following list shows several pairs of conjugates, along with the product of the conjugates.

Number	*Conjugate*	*Product*
$3 - i$	$3 + i$	$(3 - i)(3 + i) = 10$
$2 + 7i$	$2 - 7i$	$(2 + 7i)(2 - 7i) = 53$
$-6i$	$6i$	$(-6i)(6i) = 36$

The fact that the product of a complex number and its conjugate is always a real number is used to write the *quotient* of two complex numbers in standard form, as shown in the next example.

EXAMPLE 8

(a) Find $\dfrac{3 + 2i}{5 - i}$.

Multiply the numerator and denominator by the conjugate of $5 - i$.

$$\frac{3 + 2i}{5 - i} = \frac{(3 + 2i)(5 + i)}{(5 - i)(5 + i)}$$
$$= \frac{15 + 3i + 10i + 2i^2}{25 - i^2}$$
$$= \frac{13 + 13i}{26} = \frac{1}{2} + \frac{1}{2}i$$

To check this answer, show that

$$(5 - i)\left(\frac{1}{2} + \frac{1}{2}i\right) = 3 + 2i.$$

(b) Find $\dfrac{7 - 3i}{1 + 2i} \cdot \dfrac{1 - i}{4 - 3i}$.

$$\frac{7 - 3i}{1 + 2i} \cdot \frac{1 - i}{4 - 3i} = \frac{7 - 7i - 3i + 3i^2}{4 - 3i + 8i - 6i^2} = \frac{4 - 10i}{10 + 5i}$$

Multiply the numerator and denominator by the conjugate of $10 + 5i$.

$$= \frac{4 - 10i}{10 + 5i} \cdot \frac{10 - 5i}{10 - 5i} = \frac{40 - 20i - 100i + 50i^2}{100 - 25i^2}$$
$$= \frac{-10 - 120i}{125} = \frac{-10}{125} - \frac{120}{125}i$$
$$= \frac{-2}{25} - \frac{24}{25}i$$

1.7 Exercises

Identify each complex number as real or imaginary.

1. -21 **2.** -5 **3.** $\sqrt{6}$ **4.** π

5. $i\sqrt{7}$ **6.** $-4i$ **7.** $0 - 3i$ **8.** $0 + 6i$

Write each of the following in standard form.

9. $\sqrt{-81}$ **10.** $\sqrt{-14}$ **11.** $-\sqrt{-256}$ **12.** $-\sqrt{-64}$

13. $5 + \sqrt{-4}$ **14.** $-7 + \sqrt{-100}$ **15.** $-6 - \sqrt{-196}$ **16.** $13 + \sqrt{-16}$

17. $\sqrt{-5} \cdot \sqrt{-5}$ **18.** $\sqrt{-20} \cdot \sqrt{-20}$ **19.** $\sqrt{-8} \cdot \sqrt{-2}$ **20.** $\sqrt{-27} \cdot \sqrt{-3}$

21. $\dfrac{\sqrt{-40}}{\sqrt{-10}}$ **22.** $\dfrac{\sqrt{-190}}{\sqrt{-19}}$

Add or subract. Write each result in standard form.

23. $(5 - i) + (6 + 2i)$ **24.** $(7 + 3i) + (5 + 3i)$

25. $(-1 + 5i) - (-3 + 2i)$ **26.** $(-3 + 5i) - (-4 + 3i)$

27. $(2 - 5i) - (3 + 4i) - (-2 + i)$ **28.** $(-4 - i) - (2 + 3i) + (-4 + 5i)$

Multiply. Write each result in standard form.

29. $(2 + 4i)(-1 + 3i)$ **30.** $(1 + 3i)(2 - 5i)$ **31.** $(-3 + 2i)^2$

32. $(2 + i)^2$ **33.** $(2 + 3i)(2 - 3i)$ **34.** $(6 - 4i)(6 + 4i)$

35. $(\sqrt{2} + 2i)(\sqrt{2} - 2i)$ **36.** $(\sqrt{3} - 5i)(\sqrt{3} + 5i)$

37. $(2 - 7i)(2 + 7i)$ **38.** $i(4 + 9i)(4 - 9i)$

Divide. Write each result in standard form.

39. $\dfrac{4 - 3i}{4 + 3i}$ **40.** $(3 - 4i) \div (2 - 5i)$ **41.** $(1 - 3i) \div (1 + i)$ **42.** $\dfrac{5 + 6i}{5 - 6i}$

43. $\dfrac{2}{i}$ **44.** $\dfrac{-7}{3i}$ **45.** $\dfrac{1 - \sqrt{-5}}{3 + \sqrt{-4}}$ **46.** $\dfrac{2 + \sqrt{-3}}{1 - \sqrt{-9}}$

47. Explain why the method of dividing complex numbers (that is, multiplying both the numerator and the denominator by the conjugate of the denominator) works. That is, what property justifies this process?

48. Suppose that your friend, Susan Katz, tells you that she has discovered a method of simplifying a positive power of i. ''Just divide the exponent by 4,'' she says, ''and then look at the remainder. Then refer to the table of powers of i in this section. The large power of i is equal to i to the power indicated by the remainder. If the remainder is 0, the result is $i^0 = 1$.'' Explain why Susan's method works.

Find each of the following powers of i.

49. i^{25} **50.** $1/i^9$ **51.** $1/i^{12}$ **52.** i^{-6} **53.** i^{-15} **54.** i^{-49}

Perform the indicated operations, and write your answers in standard form.

55. $\frac{1-i}{2+i}\cdot\frac{4+3i}{1+i}$

56. $\frac{6+2i}{5-i}\cdot\frac{1-3i}{2+6i}$

57. $\frac{5-3i}{1+2i}\cdot\frac{2-4i}{1+i}$

58. $\frac{4-3i}{2+5i}+\frac{8-i}{2+5i}$

59. $\frac{6+2i}{1+3i}+\frac{2-i}{1-3i}$

60. $\frac{4-i}{3+4i}-\frac{3+2i}{3-4i}$

Use the definition of equality for complex numbers to solve the following equations for real numbers a and b.

61. $a - 3i = 5 + 3bi + 2a$

62. $4a - 2bi + 7 = 3i + 3a + 5$

63. $i(2b + 6) - 3 = 4(bi + a)$

64. $3i + 2(a - 1) = 4 + 2i(b + 3)$

Let $z = a + bi$ for real numbers a and b, and let $\bar{z} = a - bi$, the conjugate of z. For example, if $z = 8 - 9i$, then $\bar{z} = 8 + 9i$. Prove each of the following properties of conjugates.

65. $\bar{\bar{z}} = z$

66. $\bar{z} = z$ if and only if $b = 0$.

67. $\overline{-z} = -\bar{z}$

68. $z \cdot \bar{z}$ is a real number.

Evaluate $8z - z^2$ by replacing z with the indicated complex number in Exercises 69 and 70.

69. $2 + i$

70. $4 - 3i$

71. Find any restrictions on a and b so that the square $(a + bi)^2$ is real.

72. Find any restrictions on a and b so that the square $(a + bi)^2$ is imaginary.

73. Show that $\frac{\sqrt{2}}{2}+\frac{\sqrt{2}}{2}i$ is a square root of i.

74. Show that $\frac{\sqrt{3}}{2}-\frac{1}{2}i$ is a cube root of $-i$.

Chapter 1 Review Exercises

Simplify each of the expressions in Exercises 1–8.

1. $14 \div [5 - 4 \cdot (-3)]$

2. $-3 + (-5)(8) + 4$

3. $\frac{2-3(-7)-5}{7(-1)+(-5-2)}$

4. $\frac{-6-(-2)(-3)+4}{(-9)(-2)+5(-3)}$

5. $|3 - \pi|$

6. $|\sqrt{7} - 2|$

7. $|p - 5|$ if $p < 5$

8. $|-2x^2 - 1|$

In each of the following exercises, the coordinates of three points on a number line are given. Find **(a)** $d(A, B)$ *and* **(b)** $d(A, B) + d(B, C)$.

9. $A, -3; B, -2; C, -5$

10. $A, -8; B, 2; C, -4$

Under what conditions are the following statements true?

11. $|x| = x$

12. $|x| \le 0$

13. $d(A, B) = 0$

14. $d(A, B) + d(B, C) = d(A, C)$

Simplify each of the following. Assume all variables represent nonzero real numbers and variables used as exponents represent rational numbers.

15. $(3x^3 - 9x^2 - 5) - (-4x^3 + 6x^2) + (2x^3 - 9)$

16. $(2m - 6)(4m - 8)$

17. $(3r - 2)(r^2 + 4r - 8)$

18. $(a - 4b + c)^2$

19. $(9 - x^y)(9 + x^y)$

20. $\dfrac{2x^5t^2 \cdot 4x^3t}{16x^2t^3}$

21. $\dfrac{(2k^2)^2(3k^3)}{(4k^4)^2}$

22. Write out the terms of $(3x - y)^4$.

Factor as completely as possible.

23. $12k^2 - k - 20$

24. $12p^5 - 8p^4 + 20p^3$

25. $6m^2 - 13m - 5$

26. $10x^2 - 29xy - 21y^2$

27. $30m^5 - 35m^4n - 25m^3n^2$

28. $2x^2p^3 - 8xp^4 + 6p^5$

29. $16 - 81y^4$

30. $49m^8 - 9n^2$

31. $(x - 1)^2 - 4$

32. $8z^3 + 27w^3$

33. $r^9 - 8(r^3 - 1)^3$

34. $(2p - 1)^3 + 27p^3$

35. $7(3p - q) + m(3p - q)$

36. $6(z - 4)^2 + 9(z - 4)^3$

37. $2bx - b + 6x - 3$

38. $64 - y^{6p}$

39. Describe the steps needed to find the following sum:

$$\frac{2a + b}{4a^2 - b^2} + \frac{5a}{2a - b}.$$

Perform the indicated operations.

40. $\dfrac{5x^2y}{x + y} \cdot \dfrac{3x + 3y}{30xy^2}$

41. $\dfrac{2k - 10}{7k} \cdot \dfrac{14(3k - 1)}{8}$

42. $\dfrac{27m^3 - n^3}{3m - n} \div \dfrac{9m^2 + 3mn + n^2}{9m^2 - n^2}$

43. $\dfrac{x^2 + 2x - 15}{x^2 + 7x + 12} \div \dfrac{x^2 - 4x + 3}{x^2 + 3x - 4}$

44. $\dfrac{m}{4 - m} + \dfrac{3m}{m - 4}$

45. $\dfrac{3}{8z} - \dfrac{2}{6z} + \dfrac{1}{12}$

46. $\left(\dfrac{1}{(x + h)^2 + 16} - \dfrac{1}{x^2 + 16}\right) \div h$

47. $\dfrac{4a}{a^2 - a - 2} - \dfrac{1}{a^2 - 4}$

48. $\dfrac{x^{-1} - 2y^{-1}}{y^{-1} - x^{-1}}$

49. $\dfrac{\dfrac{6}{y + 5}}{\dfrac{3}{y^2 - 25} - 2}$

Simplify each of the following. Assume all variables represent positive real numbers.

50. $\dfrac{(p^4)(p^{-2})}{p^{-6}}$

51. $(-6x^2y^{-3}z^2)^{-2}$

52. $\dfrac{6^{-1}r^3s^{-2}}{6r^4s^{-3}}$

53. $\dfrac{(3m^{-2})^{-2}(m^2n^{-4})^3}{9m^{-3}n^{-5}}$

54. Give some examples of corresponding rules for exponents and radicals, and explain how they are related.

Simplify. Assume that all variables represent positive real numbers.

55. $\sqrt{128}$

56. $\sqrt[4]{405}$

57. $\sqrt{\dfrac{2^7y^8}{m^3}}$

58. $-\sqrt[3]{\dfrac{r^6m^5}{z^2}}$

59. $\sqrt[5]{\sqrt[3]{k}}$

60. $(\sqrt[3]{2} + 4)(\sqrt[3]{2^2} - 4\sqrt[3]{2} + 16)$

61. $\dfrac{\sqrt[4]{8p^2q^5} \cdot \sqrt[4]{2p^3q}}{\sqrt[4]{p^5q^2}}$

62. $-\sqrt{50x^5} + 3x^2\sqrt{72x} - 2x\sqrt{18x^3}$

63. $\sqrt{75y^5} - y^2\sqrt{108y} + 2y\sqrt{27y^3}$

64. $\dfrac{-12}{\sqrt[3]{4}}$

65. $\dfrac{3m}{\sqrt{m+2}}$

66. $\dfrac{15}{1 - \sqrt{3}}$

67. $\dfrac{4}{\sqrt{7} + 3}$

68. $\dfrac{\sqrt{x} - \sqrt{x-2}}{\sqrt{x} + \sqrt{x-2}}$

69. Give two ways to evaluate $125^{2/3}$ and then compare them. Which do you prefer? Why?

Simplify each of the following. Assume that all variables represent positive real numbers, and variables used as exponents are rational numbers.

70. $36^{-3/2}$

71. $(125m^6)^{-2/3}$

72. $(8r^{3/4}s^{2/3})(2r^{3/2}s^{5/3})$

73. $(7r^{1/2})(2r^{3/4})(-r^{1/6})$

74. $\dfrac{p^{-3/4} \cdot p^{5/4} \cdot p^{-1/4}}{p \cdot p^{3/4}}$

75. $\left(\dfrac{y^6x^3z^{-2}}{16x^5z^4}\right)^{-1/2}$

76. $\dfrac{m^{2+p} \cdot m^{-2}}{m^{3p}}$

77. $\dfrac{z^{-p+1} \cdot z^{-8p}}{z^{-9p}}$

Write in standard form.

78. $(-2 + i) - (3 + 6i)$

79. $(1 + 3i) - (2 - 5i) - (1 - 3i)$

80. $(7 - 2i)(3 + 4i)$

81. $(3 + 5i)(3 - 5i)$

82. $(4 + 7i)^2$

83. $(5 - 4i)^2$

84. $(2 - i)^3$

85. $(4 + i)^4$

86. i^{15}

87. i^{48}

88. $\dfrac{2 + 7i}{4 - 3i}$

89. $\dfrac{3 - 8i}{2 - i}$

90. $\dfrac{1 - 4i}{1 - i} \cdot \dfrac{3 + i}{1 + i}$

91. $\dfrac{2 + 5i}{1 - 3i} \cdot \dfrac{2 - 5i}{3 - i}$

92. $\sqrt{-27}$

93. $\sqrt{-72}$

Correct each incorrect statement in Exercises 94–105 by changing the right side of the equation.

94. $x(x^2 + 5) = x^3 + 5$

95. $-3^2 = 9$

96. $(m^2)^3 = m^5$

97. $(3x)(3y) = 3xy$

98. $\dfrac{\frac{a}{b}}{2} = \dfrac{2a}{b}$

99. $\dfrac{m}{r} \cdot \dfrac{n}{r} = \dfrac{mn}{r}$

100. $\dfrac{1}{\sqrt{a} + \sqrt{b}} = \dfrac{1}{\sqrt{a}} + \dfrac{1}{\sqrt{b}}$

101. $\dfrac{(2x)^3}{2y} = \dfrac{x^3}{y}$

102. $4 - (t + 1) = 4 - t + 1$

103. $\dfrac{1}{(-2)^3} = 2^{-3}$

104. $(-5)^2 = -5^2$

105. $\left(\dfrac{8}{7} + \dfrac{a}{b}\right)^{-1} = \dfrac{7}{8} + \dfrac{b}{a}$

106. For which of the following cases does $\sqrt{ab} = \sqrt{a} \cdot \sqrt{b}$?
(a) a and b positive **(b)** a positive, b negative **(c)** a and b negative

107. For what integer values of n does $\sqrt[n]{a^n} = a$?

108. For what values of x does $\sqrt{9ax^2} = 3x\sqrt{a}$?

2 Equations and Inequalities

In many cases, the study of algebra is really the study of equations. Applications of mathematics frequently require the solution of one or more equations. The study of inequalities has also become important as more and more applications—especially in fields such as business—utilize inequalities.

An **equation** is a statement that two expressions are equal, such as $14x^2 + 2 = 8x - 4$. An **inequality** expresses an order relationship between two expressions such as $11y > 5y + 3$. In this chapter we discuss the solution of several different kinds of equations and inequalities.

To *solve* an equation means to find the number or numbers that make the equation a true statement. For example, the number 7 makes $x + 2 = 9$ a true statement, so 7 is the **solution** or the **root** of the equation and is said to **satisfy** the equation. The set of all solutions for an equation is the **solution set** of the equation. The solution set of $x^2 - 1 = 0$ is $\{-1, 1\}$.

Any two equations with the same solution set are **equivalent equations**. For example, $x + 1 = 5$ and $6x + 3 = 27$ are equivalent equations since they both have the solution set $\{4\}$. On the other hand, the equations

$$x + 1 = 5 \quad \text{and} \quad (x - 4)(x + 2) = 0$$

are not equivalent. The number 4 is a solution of both equations, but the equation $(x - 4)(x + 2) = 0$ also has -2 as a solution.

One way to solve an equation is to rewrite it as successively simpler equivalent equations. These simpler equivalent equations are derived using the properties given in Chapter 1.

2.1 Linear Equations

In this section we discuss methods of solving equations that are equivalent to linear equations.

Linear Equation

An equation that can be written in the form

$$ax + b = 0,$$

where a and b are real numbers, with $a \neq 0$, is a **linear equation.**

Linear equations are solved using the properties that allow the same number to be added to both sides of an equation or to multiply both sides of an equation. The numbers are chosen so that the final form of the equation is $x = c$ for some real number c. Often it is necessary to begin by using other properties to first get the equation in the linear form $ax + b = 0$, or $ax = -b$.

For Graphers

A grapher can be used to solve equations in two ways. One way is to rewrite the equation with one side equal to 0 and then to let y equal the expression on the other side. For the equation in Example 1, we would have $y = 3(2x - 4) - 7 + (x + 5)$. To solve $y = 0$, we graph $y = 3(2x - 4) - 7 + (x + 5)$ and look for any value of x where the graph intersects the x-axis. As Figure 1(a) shows, $y = 0$ at $x = 2$. The range key is used to determine the viewing rectangle, the portion of the graph that will show on the screen. To get a reasonable graph, it is important to choose an appropriate viewing rectangle. This may require some trial and error. *(continued on page 61)*

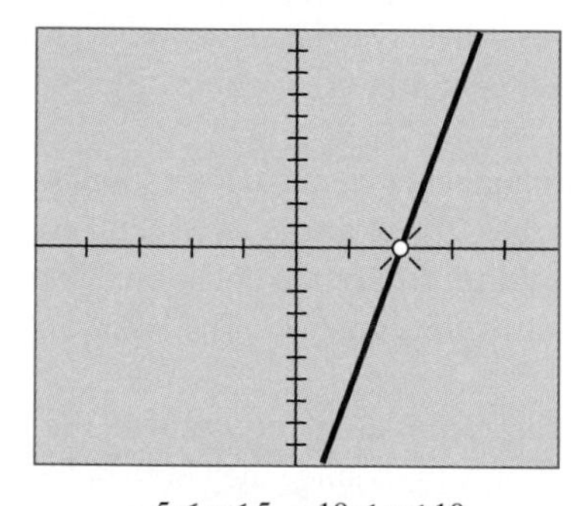

$-5 \le x \le 5$, $-10 \le y \le 10$

Figure 1(a)

EXAMPLE 1 Solve $3(2x - 4) + 3 = 10 - (x + 5)$.

Use the properties to get the following sequence of simpler equivalent equations.

$$3(2x - 4) + 3 = 10 - (x + 5)$$

$$3(2x - 4) = 7 - (x + 5) \qquad \text{Add } -3 \text{ to each side.}$$

$$6x - 12 = 7 - x - 5 \qquad \text{Distributive property}$$

$$6x - 12 = 2 - x \qquad \text{Collect like terms.}$$

Now add the same expressions to each side of the equation.

$$7x - 12 = 2 \qquad \text{Add } x \text{ to each side.}$$

$$7x = 14 \qquad \text{Add 12 to each side.}$$

Finally, multiply each side by the same number, 1/7 (or divide each side by 7).

$$\frac{1}{7} \cdot 7x = \frac{1}{7} \cdot 14$$

$$x = 2$$

For Graphers

(continued from page 60)

Another way to solve an equation is to graph the expressions on each side of the equation simultaneously, letting one equal y_1 and the other equal y_2 [see Figure 1(b)]. Next, use the trace key to find the x-value where the two graphs intersect. The zoom key may also be used to get a more accurate answer.

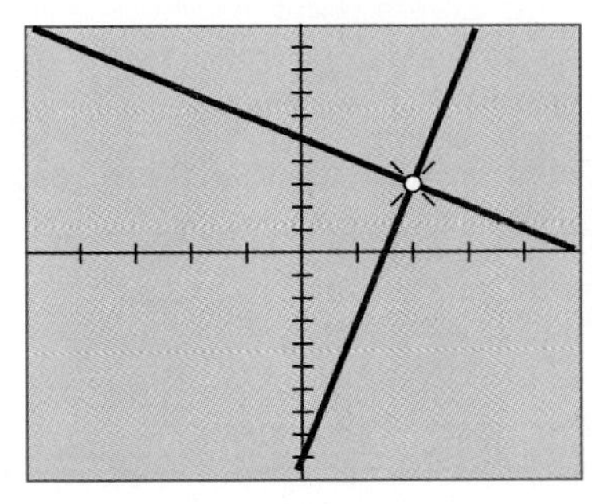

$-5 \le x \le 5, -10 \le y \le 10$

Figure 1(b)

Sometimes an *exact* answer cannot be obtained. In such cases, look for the answer that remains unchanged to the desired number of decimal places after repeated zooming.

To check this proposed solution, replace x with 2 in the original equation.

$$\begin{aligned} 3(2x - 4) + 3 &= 10 - (x + 5) && \text{Original equation} \\ 3(2 \cdot \mathbf{2} - 4) + 3 &= 10 - (\mathbf{2} + 5) && \text{Let } x = 2. \\ 3(4 - 4) + 3 &= 10 - (7) \\ 3 &= 3 && \text{True} \end{aligned}$$

Since replacing x with 2 results in a true statement, 2 is the solution of the given equation. The solution set is therefore $\{2\}$. ■

EXAMPLE 2 Solve $\dfrac{3p - 1}{3} - \dfrac{2p}{p - 1} = p.$

At first glance, this equation does not satisfy the definition of a linear equation given above. However, the equation does appear in proper form after algebraic simplification. To obtain a simpler equivalent equation, first multiply both sides by the common denominator, $3(p - 1)$, where we must assume $p \neq 1$.

$$\begin{aligned} \mathbf{3(p - 1)}\left(\frac{3p - 1}{3}\right) - \mathbf{3(p - 1)}\left(\frac{2p}{p - 1}\right) &= \mathbf{3(p - 1)}p \\ (p - 1)(3p - 1) - 3(2p) &= 3p(p - 1) \\ 3p^2 - 4p + 1 - 6p &= 3p^2 - 3p \end{aligned}$$

An even simpler equivalent equation comes from combining terms and adding $-3p^2$ to both sides.

$$\begin{aligned} -10p + 1 &= -3p \\ 1 &= 7p && \text{Add } 10p \text{ to each side.} \\ \frac{1}{7} &= p && \text{Multiply each side by } \frac{1}{7}. \end{aligned}$$

Substitute $1/7$ for p in the given equation to verify that $1/7$ is the root of the equation. The restriction $p \neq 1$ does not affect the solution set here, since $1/7 \neq 1$. ■

EXAMPLE 3 Solve $\dfrac{x}{x - 2} = \dfrac{2}{x - 2} + 2.$

Multiply both sides of the equation by $x - 2$, assuming that $x - 2 \neq 0$ (or $x \neq 2$).

$$\begin{aligned} x &= 2 + 2(x - 2) \\ x &= 2 + 2x - 4 \\ x &= 2 \end{aligned}$$

It was necessary to assume $x - 2 \neq 0$ in order to multiply both sides of the equation by $x - 2$. The proposed solution of 2, however, makes $x - 2 = 0$, meaning that the given equation has no solution. (To see that 2 is not a solution, substitute 2 for x in the given equation.) ■

EXAMPLE 4 Solve $5(2x + 1) - 2x = 8x + 5$.

Use the distributive property on the left side to get

$$\begin{aligned} 10x + 5 - 2x &= 8x + 5 \\ 8x + 5 &= 8x + 5 \\ 0 &= 0. \end{aligned}$$

This last equation is true for every real number, which indicates that the given equation is an *identity*. ■

In general, an equation is an **identity** if it is true for all values of its variables that are defined. For instance, $3x + 4x = 7x$ and $x^2 - 3x + 2 = (x - 2)(x - 1)$ are identities. They are defined and true for all real numbers. The identity

$$\frac{2x^2 + 3x}{x} = 2x + 3$$

is defined and true for all real numbers except 0.

An equation that is not an identity is called a **conditional equation**. An equation is conditional if there are numbers for which the equation is defined but which do not satisfy the equation. Thus, the equations solved in Examples 1–3 are conditional equations.

EXAMPLE 5 Decide whether the following equations are identities or conditional equations.

(a) $9p^2 - 25 = (3p + 5)(3p - 5)$

Since the product of $3p + 5$ and $3p - 5$ is $9p^2 - 25$, the equation is meaningful and true for *every* value of p and is an identity.

(b) $\dfrac{(x - 3)(x + 2)}{x - 3} = x + 2$

The equation is an identity since it is meaningful for all real numbers except 3 and is true for all such numbers.

(c) $5y - 4 = 11$

Choosing the value 3 as a replacement for y gives

$$\begin{aligned} 5 \cdot 3 - 4 &= 11 \\ 11 &= 11, \end{aligned}$$

a true statement. On the other hand, $y = 4$ gives

$$5 \cdot 4 - 4 = 11$$
$$16 = 11,$$

a false statement. Since the equation is meaningful for all real numbers and true for $y = 3$, but not true for $y = 4$, the equation is conditional. ■

Sometimes an equation has more than one letter. To solve for a specified variable, treat the other letters as constants. (As a general rule, letters from the beginning of the alphabet, such as a, b, and c, are used to represent constants, while letters such as x, y, and z represent variables.)

EXAMPLE 6 Solve the equation $3(2x - 5a) + 4b = 4x - 2$ for x in terms of the other variables.

Using the distributive property gives

$$6x - 15a + 4b = 4x - 2.$$

Treat x as the variable and the other letters as constants. Get all terms with x on one side and all terms without x on the other side.

$$6x - 4x = 15a - 4b - 2$$
$$2x = 15a - 4b - 2$$
$$x = \frac{15a - 4b - 2}{2}$$ ■

2.1 Exercises

Decide whether each of the following equations is an identity or a conditional equation.

1. $2y - y^2 = y(2 - y)$

2. $m^2 - 4 = (m + 2)(m - 2)$

3. $2x + 5 = 2(x + 5)$

4. $3x + 8 - x = 3(x + 8) - x$

5. $\frac{z - 2}{z} = 1 - \frac{2}{z}$

6. $\frac{y - 1}{y + 3} = -\frac{1}{3}$

7. $2(x - 1) = x - 1 + x - 1$

8. $4p + 16 = 5(p + 4) - (p + 4)$

Decide which of the following pairs of equations are equivalent.

9. $4x - 1 = 10$
$12x - 3 = 30$

10. $5 = 8 - 2x$
$2x = 3$

11. $\frac{y + 2}{y + 3} = \frac{4}{y + 3}$
$y + 2 = 4$

12. $\frac{2x + 5}{9} = \frac{4x}{9}$
$5 = 2x$

13. $\frac{x}{x - 2} = \frac{2}{x - 2}$
$x = 2$

14. $\frac{x + 3}{x + 1} = \frac{2}{x + 1}$
$x = -1$

15. $x = 4$
$x^2 = 16$

16. $z^2 = 9$
$z = 3$

17. Explain the difference between an identity and a conditional equation.

18. Make a complete list of the steps needed to solve any linear equation. (Some equations will not require every step.)

Solve each of the following equations. Check each solution.

19. $.3x - .7 = .3 + .2x$

20. $.04x - 2.01 = 3.18x + 4.72$

21. $(3/4)x - 5 + (2/3) = (5/3) - x$

22. $(-1/2) + (1/4)y + 2 = (3/4)y$

23. $3r + 2 - 5(r + 1) = 6r + 4$

24. $5(a + 3) + 4a - 5 = -(2a - 4)$

25. $2[m - (4 + 2m) + 3] = 2m + 2$

26. $4[2p - (3 - p) + 5] = -7p - 2$

27. $\frac{3x - 2}{7} = \frac{x + 2}{5}$

28. $\frac{2p + 5}{5} = \frac{p + 2}{3}$

29. $\frac{3k - 1}{4} = \frac{5k + 2}{8}$

30. $\frac{9x - 1}{6} = \frac{2x + 7}{3}$

31. $\frac{x}{3} - 7 = 6 - \frac{3x}{4}$

32. $\frac{y}{3} + 1 = \frac{2y}{5} - 4$

33. $\frac{1}{4p} + \frac{2}{p} = 3$

34. $\frac{2}{t} + 6 = \frac{5}{2t}$

35. $\frac{m}{2} - \frac{1}{m} = \frac{6m + 5}{12}$

36. $\frac{-3k}{2} + \frac{9k - 5}{6} = \frac{11k + 8}{k}$

37. $\frac{2r}{r - 1} = 5 + \frac{2}{r - 1}$

38. $\frac{3x}{x + 2} = \frac{1}{x + 2} - 4$

39. $\frac{5}{2a + 3} + \frac{1}{a - 6} = 0$

40. $\frac{2}{x + 1} = \frac{3}{2x - 5}$

41. $\frac{4}{x - 3} - \frac{8}{2x + 5} + \frac{3}{x - 3} = 0$

42. $\frac{8}{3x + 1} + \frac{2}{x - 1} = \frac{5}{3x + 1}$

43. $\frac{2}{4 - 3x} - 5 = \frac{2}{4 - 3x}$

44. $\frac{5}{2y - 3} = 1 - \frac{2}{2y - 3}$

45. $\frac{3a}{a + 1} - 2 = \frac{6}{a + 1}$

46. $\frac{5p}{2p - 1} = \frac{15}{2p - 1} + 4$

47. $2(m + 1)(m - 1) = (2m + 3)(m - 2)$

48. $(2y - 1)(3y + 2) = 6(y + 2)^2$

49. $(3x - 4)^2 - 5 = 3(x + 5)(3x + 2)$

50. $(2x + 5)^2 = 3x^2 + (x + 3)^2$

Solve each equation in Exercises 51–58 for x.

51. $2(x - a) + b = 3x + a$

52. $5x - (2a + c) = a(x + 1)$

53. $ax + b = 3(x - a)$

54. $4a - ax = 3b + bx$

55. $\frac{4x}{2a + 1} = ax - 1$

56. $\frac{a}{3x + 2} + b = 2a$

57. $a^2(2x - 3) = 4x$

58. $a(x + a) = b(x + b)$

Graphing Utility Problems

Use a graphing utility to solve the equations in Exercises 59 and 60.

59. **(a)** Solve $\frac{2x}{x - 1} = 5 + \frac{2}{x - 1}$.

(b) Solve $2x = 5(x - 1) + 2$.

(c) Are the equations in (a) and (b) equivalent, identities, or neither? Explain how the graphs illustrate your answer.

60. **(a)** Solve $\frac{4x - 3}{5} = \frac{2x}{5}$.

(b) Solve $4x - 3 = 2x$.

(c) Are the equations in (a) and (b) equivalent, identities, or neither? Explain how the graphs illustrate your answer.

2.2 Formulas and Applications

Mathematics is an important problem-solving tool. Many times the solution of a problem depends on the use of a formula which expresses a relationship among several variables. For example, the formula

$$A = \frac{24f}{b(p + 1)} \qquad (*)$$

gives the approximate annual interest rate for a consumer loan paid off with monthly payments. Here f is the finance charge on the loan, p is the number of payments, and b is the original amount of the loan.

Suppose the number of payments, p, must be found when the other quantities are known. To do this, solve the equation for p by treating p as the variable and the other letters as constants. Begin by multiplying both sides of formula (*) by $p + 1$.

$$(p + 1)A = \frac{24f}{b}$$

Multiplying both sides by $1/A$ gives

$$p + 1 = \frac{24f}{Ab}.$$

(Here we must assume $A \neq 0$. Why is this a very safe assumption?)

Finally, add -1 to both sides.

$$p = \frac{24f}{Ab} - 1$$

This process is called **solving for a specified variable.**

EXAMPLE 1

Solve for x in the equation $p\left(q - \frac{x}{B}\right) = x.$

To get all terms with x on one side of the equation and all terms without x on the other, first use the distributive property.

$$pq - p\left(\frac{x}{B}\right) = x$$

$$pq - \frac{px}{B} = x$$

Eliminate the denominator, B, by assuming $B \neq 0$ and multiplying both sides by B.

$$\mathbf{B}pq - \mathbf{B}\left(\frac{px}{B}\right) = \mathbf{B}x$$

$$Bpq - px = Bx$$

Then add px to both sides to get the two terms with x together.

$$Bpq = Bx + px$$

$$Bpq = x(B + p) \qquad \text{Factor on the right side.}$$

Assuming $B \neq -p$ permits multiplying both sides by $1/(B + p)$ to get

$$x = \frac{Bpq}{B + p}.$$

One of the main reasons for learning mathematics is to be able to use it in solving practical problems. However, for most students, learning how to apply mathematical skills to real situations is the most difficult task they face. In the rest of this section we give a few hints that may help you with applications.

A common difficulty with "word problems" is trying to do everything at once. It is usually best to attack the problem in stages.

Solving Word Problems

1. Read the problem carefully.
2. Decide on an unknown, usually the quantity you are asked to find, and name it with a variable that you *write down.* Many students, eager to get on with writing an equation, try to skip this step. But it is important. If you don't know what "x" represents, how can you write a meaningful equation or interpret a result?
3. Decide on variable expressions to represent any other unknowns in the problem. For example, if x represents the width of a rectangle and you know that the length is one unit more than twice the width, *write down* "$1 + 2x =$ the length of the rectangle."
4. If possible, draw a sketch or prepare a table showing the variables appearing in the problem and the relationships among them.
5. Use the information given in the problem to write a verbal equation. For example, if the problem states that the perimeter of a rectangle is 48 inches, *write down* "$2 \cdot$ length $+ 2 \cdot$ width $= 48$."
6. Use the results of Steps 3 and 4 to convert the verbal equation into an algebraic equation in the variable from Step 2.
7. Solve the equation from Step 6.
8. Make sure you have answered the question in the problem.
9. Check your answer *in the words of the original problem.*

Notice how each of the preceding steps is carried out in the examples on the following pages.

EXAMPLE 2 If the length of a side of a square is increased by 3 centimeters, the perimeter of the new square is 40 centimeters more than twice the length of a side of the original square. Find the length of a side of the original square.

First, what should the variable represent? Since the length of a side of the original square is needed, let the variable represent that length. Write this down:

Let x = Length of side of original square.

The length of the side of the new square can be expressed in terms of x:

$x + 3$ = Length of side of new square.

Now draw a figure using the given information, as in Figure 2.

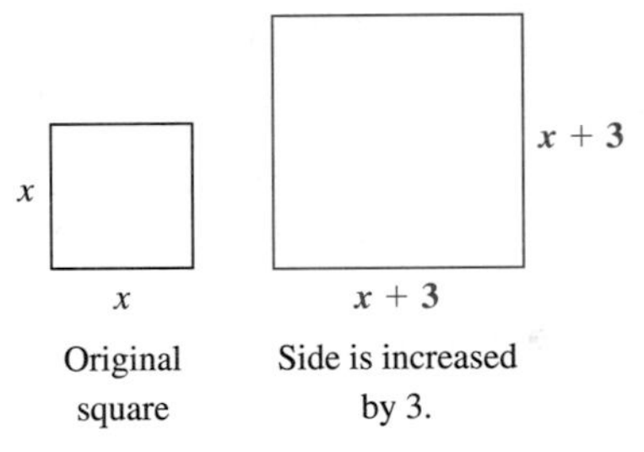

Figure 2

Use the information given in the problem to write a verbal equation.

Perimeter of the new square = 40 + Twice the length of a side of the original square

Transform the verbal equation into an algebraic equation.

$$4(x + 3) = 40 + 2x$$

Solve the equation for the variable x.

$$\begin{aligned} 4x + 12 &= 40 + 2x \\ 2x &= 28 \\ x &= 14 \end{aligned}$$

Therefore, the length of a side of the original square is 14 centimeters.

Check the solution using the words of the problem.

Length of side of new square:	$14 + 3 = 17$ cm	
Perimeter of new square:	$4(17) = 68$ cm	Same
Twice length of side of original square:	$2(14) = 28$ cm	
40 + Twice length of side of original square:	$40 + 28 = 68$ cm	Same

The next example is a constant velocity problem. The components *distance*, *rate*, and *time* are denoted by the letters d, r, and t, respectively. (The *rate* is also called the *speed* or *velocity*.) These variables are related by the equation

$$d = rt.$$

This equation is easily solved for r and t; $r = d/t$ and $t = d/r$.

EXAMPLE 3 Marge and Ed are traveling to a business conference. Marge travels 110 miles in the same time that Ed travels 140 miles. Ed travels 15 miles per hour faster than Marge. Find the average rate of each person.

$$\begin{aligned} \text{Let } x &= \text{Marge's rate,} \\ x + 15 &= \text{Ed's rate.} \end{aligned}$$

Summarize the given information in a table.

	d	r	t
Marge	110	x	$\frac{110}{x}$
Ed	140	$x + 15$	$\frac{140}{x + 15}$

← Use $t = d/r$.

The unused information from the problem is ''in the same time that.''

$$\begin{aligned} \text{Marge's time} &= \text{Ed's time} \\ \frac{110}{x} &= \frac{140}{x + 15} \\ x(x + 15)\frac{110}{x} &= x(x + 15)\frac{140}{x + 15} && \text{Multiply both sides by } x(x + 15). \\ 110(x + 15) &= 140x \\ 110x + 1650 &= 140x \\ 1650 &= 30x \\ 55 &= x && \text{Marge's rate} \\ x + 15 = 55 + 15 &= 70 && \text{Ed's rate} \end{aligned}$$

Therefore, Marge's rate is 55 miles per hour and Ed's rate is 70 miles per hour.

Check:

Time traveled by Marge: $110/55 = 2$ hr
Time traveled by Ed: $140/70 = 2$ hr

Same ■

EXAMPLE 4 For a chemistry class, the instructor needs a 20% solution of potassium permanganate. She has a 15% solution on hand, as well as a 30% solution. How many liters of the 15% solution should she add to 3 liters of the 30% solution to get the 20% solution? (See Figure 3).

Let $x =$ Number of liters of 15% solution to be added.

Strength	*Liters of solution*	*Liters of potassium permanganate*
15%	x	$.15x$
30%	3	$.30(3)$
20%	$3 + x$	$.20(3 + x)$

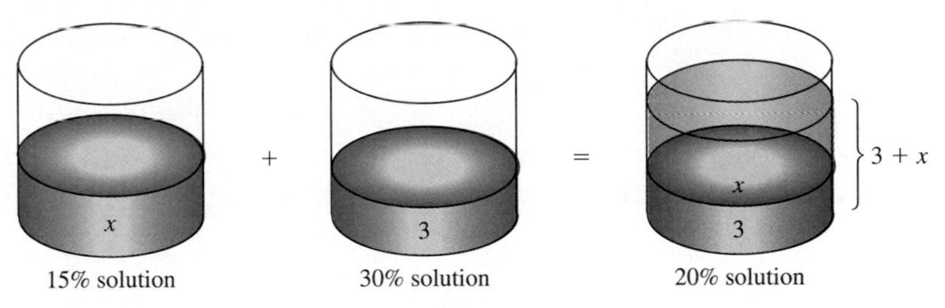

Figure 3

Liters of pot. permang. in 15% solution	+	Liters of pot. permang. in 30% solution	=	Liters of pot. permang. in 20% solution
$.15x$	+	$.30(3)$	=	$.20(3 + x)$

$$.15x + .90 = .60 + .20x$$
$$.30 = .05x$$
$$6 = x$$

Therefore, 6 liters of the 15% solution should be added. (Note that this assumes that none is spilled.) Now check the results.

Pot. permang. contributed by 15% solution:	$.15(6) = .90$ liters
Pot. permang. contributed by 30% solution:	$.30(3) = .90$ liters
Pot. permang. in mixture:	$.90 + .90 = 1.8$ liters
Volume of mixture:	$6 + 3 = 9$ liters
Concentration of pot. permang. in mixture:	$1.8/9 = .20$ or 20%

The next example involves rates of work. The letters r, t, and A represent the rate at which work is done, the time, and the amount of work accomplished, respectively. These variables are related by the equation

$$A = rt.$$

Amounts of work are often measured in terms of the number of jobs accomplished. For instance, if one job is accomplished in t hours, then $A = 1$ and $r = 1/t$.

EXAMPLE 5 One computer can do a job twice as fast as another. Working together, both computers can do the job in 8/3 hours. How long would it take the faster computer, working alone, to do the job?

Let t = Numbers of hours for faster computer working alone to complete the job,

$2t$ = Numbers of hours for slower computer working alone to complete the job.

	r	t	A
Faster computer	$\frac{1}{t}$	$\frac{8}{3}$	$\frac{1}{t} \cdot \frac{8}{3}$ ← Use $A = rt$.
Slower computer	$\frac{1}{2t}$	$\frac{8}{3}$	$\frac{1}{2t} \cdot \frac{8}{3}$

Since both computers contribute toward completion of the job, we have the following equation.

Amount of work done by faster computer + Amount of work done by slower computer = Amount of work done by both

$$\frac{1}{t} \cdot \frac{8}{3} + \frac{1}{2t} \cdot \frac{8}{3} = 1$$

$$\frac{8}{3t} + \frac{8}{6t} = 1$$

$$16 + 8 = 6t \qquad \text{Multiply both sides by } 6t.$$

$$4 = t$$

The faster computer could do the entire job, working alone, in 4 hours.

Check:

Rate of faster computer: 1/4 job per hour

Rate of slower computer: 1/8 job per hour

Amount of job done by faster computer in 8/3 hours: $\frac{1}{4} \cdot \frac{8}{3} = \frac{2}{3}$

Amount of job done by slower computer in 8/3 hours: $\frac{1}{8} \cdot \frac{8}{3} = \frac{1}{3}$

Amount of job done by both computers working together: $\frac{2}{3} + \frac{1}{3} = 1$ ■

NOTE In problems involving rates of work, the formula given above and used in Example 5 assumes a uniform rate. In other words, the work does not speed up or slow down as the job is carried out. ■

When P dollars is invested at the annual interest rate r for t years, the amount of simple interest earned, I, is given by

$$I = Prt.$$

EXAMPLE 6 At the end of the year, Chuck Hickman receives a $10,000 bonus from his company. He wants to earn annual interest of $610 on the money. He plans to invest part of it in a certificate of deposit earning 5% annual interest and the rest in a stock fund earning 7% annual interest. How much should he invest in the certificate of deposit?

Let x = Amount invested at 5%,

$10{,}000 - x$ = Amount invested at 7%.

Make a chart showing the information from the problem.

r	P	I
.05	x	$.05x(1)$ ← $I = Prt$
.07	$10{,}000 - x$	$.07(10{,}000 - x)(1)$ ←

The total interest earned is $610.

$$\begin{aligned} .05x + .07(10{,}000) - .07x &= 610 \\ -.02x + 700 &= 610 \\ -.02x &= -90 \\ x &= 4500 \qquad \text{Amount invested at 5\%} \\ 10{,}000 - x = 10{,}000 - 4500 &= 5500 \qquad \text{Amount invested at 7\%} \end{aligned}$$

Thus, Hickman should invest $4500 at 5% and $5500 at 7%.

Check:

Interest earned from the 5% investment: $(.05)(4500) = \$225$

Interest earned from the 7% investment: $(.07)(5500) = \$385$

Total interest earned: $\$225 + \$385 = \$610$ ■

2.2 Exercises

Solve each of the following equations for y.

1. $3x - 4y = 12$

2. $6x + 2y = 18$

3. $-5x + 3y + 27 = 0$

4. $-8x - 5y - 35 = 0$

5. $x = \dfrac{y}{3} + \dfrac{3}{4}$

6. $x = \dfrac{4y}{5} - \dfrac{2}{3}$

Solve each of the following for the variable indicated.

7. $x = m - ax$ for x

8. $kr - p = br + c$ for r

9. $s = \frac{1}{2}gt^2$ for g

10. $A = \frac{1}{2}(B + b)h$ for h

11. $A = \frac{1}{2}(B + b)h$ for B

12. $C = \frac{5}{9}(F - 32)$ for F

13. $S = 2\pi(r_1 + r_2)h$ for r_1

14. $A = P\left(1 + \frac{i}{m}\right)$ for m

15. $P = \frac{E^2R}{r + R}$ for R

16. $\frac{1}{R} = \frac{1}{r_1} + \frac{1}{r_2}$ for R

17. $m = \frac{Ft}{v_1 - v_2}$ for v_2

18. Refer to Example 1. Suppose someone tells you that there is no reason to solve for x, since the right side of the formula is already x. Is this correct? Explain.

19. Suppose two acid solutions are mixed. One is 26% acid and the other is 32% acid. Which of the following concentrations cannot possibly be the concentration of the mixture? Explain.
(a) 36% **(b)** 28% **(c)** 30% **(d)** 31%

20. Suppose that a computer that originally sells for x dollars has been discounted 30%. Which of the following expressions does not represent its sale price?
(a) $x - .30x$ **(b)** $.70x$ **(c)** $\frac{7}{10}x$ **(d)** $x - .30$

Solve each of the following problems.

21. The length of a rectangular label is 3 cm less than twice the width. The perimeter is 54 cm. Find the width.

22. A puzzle piece in the shape of a triangle has a perimeter of 30 cm. Two sides of the triangle are each twice as long as the shortest side. Find the length of the shortest side.

23. A pharmacist wishes to strengthen a mixture that is 10% alcohol to one that is 30% alcohol. How much pure alcohol should be added to 7 liters of the 10% mixture?

24. A student needs 10% hydrochloric acid for a chemistry experiment. How much 5% acid should be mixed with 60 ml of 20% acid to get a 10% solution?

25. An automobile radiator contains a 10-qt mixture of water and antifreeze that is 40% antifreeze. How much should the owner drain from the radiator and replace with pure antifreeze so that the liquid in the radiator will be 80% antifreeze?

26. In Exercise 24, suppose the student has only pure acid and 5% acid. How much pure acid should be added to the 5% acid to get 12 ml of 10% acid?

27. A recycling bin is in the shape of a closed rectangular box. Find the height of the bin if its length is 18 ft, its width is 8 ft, and its surface area is 496 sq ft.

28. A right circular cylinder has radius 6 inches and volume 144π cu inches. What is its height?

29. On a vacation trip, Le Hong averaged 50 mph traveling from Denver to Minneapolis. Returning by a different route that covered the same number of miles, he averaged 55 mph. What is the distance between the two cities if his total traveling time was 32 hr?

30. Lindsay left by plane to visit her mother in Hartford, 420 km away. Fifteen minutes later, her mother left to meet her at the airport. She drove the 20 km to the airport at 40 km per hr, arriving just as the plane taxied in. What was the speed of the plane?

31. Russ and Janet are running in the Apple Hill Fun Run. Russ runs at 7 mph, Janet at 5 mph. If they leave the starting point at the same time, how long will it be before they are $1/2$ mi apart?

32. If the run in Exercise 31 has a staggered start, and Janet starts first, with Russ starting 10 min later, how long will it be before he catches up with her?

33. Jon gets to work in 20 min when he drives his car. Riding his bike (by the same route) takes him 45 min. His average driving speed is 14.5 mph greater than his average speed on the bike. How far does he travel to work?

34. In the morning, Karen drove to a business appointment at 50 mph. Her average speed on the return trip in the afternoon was 40 mph. The return trip took $1/4$ hr longer because of heavy traffic. How far did she travel to the appointment?

35. Jim Macias drove from Philadelphia to New York, a distance of 100 miles, at an average speed of 45 mph and made the return trip at an average speed of 36 mph.

(a) What was his average speed for the entire trip?

(b) Why was the average speed closer to the lower speed?

(c) Show that the average speed will be the same no matter what the distance between the two cities.

36. The distance between New York and London is 3469 miles. If an airplane has a cruising speed of 350 mph and a tail wind of 50 mph, how many miles out will it reach the point of no return? (A tail wind blows in the same direction as the plane. The point of no return is the point on the flight where it will take the same amount of time to fly on to the destination as to fly back to the starting point.)

37. A town gardener can mow the lawn in a small park in 2 hr. Another gardener can mow the same lawn in 3 hr. How long would it take them to mow the park lawn if they work together?

38. Two painters working together can paint an average-sized room in 2.5 hr. One of the painters can do the same job alone in 4 hr. How long would it take the other painter working alone to do the job?

39. A sewage treatment plant has two inlet pipes to its settling pond. One can fill the pond in 10 hr, the other in 12 hr. If the first pipe is open for 5 hr and then the second pipe is opened, how long will it take to fill the pond after the second pipe is opened?

40. Two chemical plants are polluting a river. If plant A produces a predetermined maximum amount of pollution in half the time as plant B, and together they produce the maximum pollution in 26 hr, how long will it take plant B alone?

41. With both taps open, Mark can fill his kitchen sink in 5 min. When full, the sink drains in 10 min. How long will it take to fill the sink if Mark forgets to put in the stopper?

42. If Mark (see Exercise 41) remembers to put in the stopper after 1 min, how much longer will it take to fill the sink?

43. A VCR is on sale for \$245. If the sale price is 30% less than the regular price, what was the regular price?

44. A jeweler prices his items 60% over their wholesale price. If a watch sells for \$152, what is its wholesale price?

45. Anne Kelly received \$52,000 profit from the sale of some land. She invested part at 7.5% interest and the rest at 5.5% interest. She earned a total of \$3280 interest during the first year. How much did she invest at each rate?

46. Jim Marshall invests \$20,000 received from an insurance settlement in two ways, some at 5% and some at 6%. Altogether, he makes \$1080 the first year in interest. How much is invested at each rate?

47. Mary Ellen Heise earned \$48,000 from royalties on her book. She paid a 28% income tax on the royalties. She invested \$15,000 at one rate and the balance at a rate that was 1% lower, earning \$1878 annual interest on the two investments. What was the higher interest rate?

48. Kevin Connors won \$100,000 in a state lottery. He paid income taxes of 33% on the winnings. He invested \$40,000 of the balance at one interest rate and the remaining \$27,000 at a rate that was 2% lower. The two investments earned \$5080 per year in interest. What was the lower rate?

49. Diane Gray bought two plots of land for a total of \$120,000. On the first plot, she made a profit of 15%. On the second, she lost 10%. Her total profit was \$5500. How much did she pay for each piece of land?

50. Suppose \$10,000 is invested at 6%. How much additional money must be invested at 8% to produce a yield on the entire amount invested of 7.2%?

51. Cathy Wacaser earns take-home pay of \$592 a week. If her deductions for taxes, retirement, union dues, and medical plan amount to 26% of her wages, what is her weekly pay before deductions?

52. Barbara Burnett gives 10% of her net income to her church. This amounts to \$80 a month. In addition, her paycheck deductions are 24% of her gross monthly income. What is her gross monthly income?

53. Adam Bryer wishes to sell a piece of property for \$125,000. He wants the money to be paid off in two ways: a short-term note at 10.5% and a long-term note at 9%. Find the amount of each note if the total annual interest on the two notes is \$12,600.

54. A bank pays 4% interest on passbook accounts and 6% on long-term deposits. Suppose a depositor divides \$20,000 among the two types of deposits. Find the amount deposited at each rate if the total annual income from interest is \$1060.

55. If x represents the number of pennies in a jar, which of the following equations cannot be correct for finding x? (*Hint:* Solve each equation and consider the solution.)
(a) $5x + 3 = 9$ **(b)** $12x + 3 = -4$
(c) $100x = 50(x + 3)$ **(d)** $6(x + 4) = x + 24$

56. Which of the following cannot be a correct equation to solve a geometry problem, if x represents the length of a rectangle? (*Hint:* Solve each equation and consider the solution.)
(a) $2x + 2(x - 1) = 14$ **(b)** $-2x + 7(5 - x) = 62$
(c) $4(x + 2) + 4x = 8$ **(d)** $2x + 2(x - 3) = 22$

57. Biologists can estimate the number of individual members of a species in an area. Suppose, for example, that 100 animals of the species are caught and marked. A period of time is permitted to elapse, and then b animals are caught. If c of these ($c \leq b$) are marked, find an expression involving b and c that estimates the total number of individuals in the area.

58. Suppose you invest B dollars, some at m% and the rest at n%. If a total of I dollars in interest is earned per year, find the amount invested at each rate.

2.3 Quadratic Equations

An equation of the form $ax + b = 0$ is a first-degree equation because x has the exponent 1. In $ax^2 + bx + c = 0$ the highest exponent is 2. Hence, it is a second-degree equation. If we consider a square having sides of length x, then $x \cdot x$ or x^2 is the area of the square. Consequently, we often say "x squared" for x^2. The Latin word for square is *quadratum* so we refer to $ax^2 + bx + c$ as a *quadratic* and to second-degree equations as *quadratic equations.*

Quadratic Equation

An equation that can be written in the form

$$ax^2 + bx + c = 0,$$

where a, b, and c are real numbers with $a \neq 0$, is a **quadratic equation.**

(Why is the restriction $a \neq 0$ necessary?) A quadratic equation written in the form $ax^2 + bx + c = 0$ is in **standard form.**

The simplest method of solving a quadratic equation, but one that is not always easily applied, is factoring. This method depends on the **zero-factor property.**

Zero-Factor Property

If a and b are complex numbers, with $ab = 0$, then $a = 0$ or $b = 0$ or both. (Also, if $a = 0$ or $b = 0$, then $ab = 0$.)

For Graphers

Use a grapher to solve the equation in Example 1(a). The equation is in standard form, so let $y = 6x^2 + 7x - 3$. (Graphers use only the variable x for equations, so we use x here instead of r.) Graph the polynomial and look for the x-values where the graph crosses the x-axis. See Figure 4.

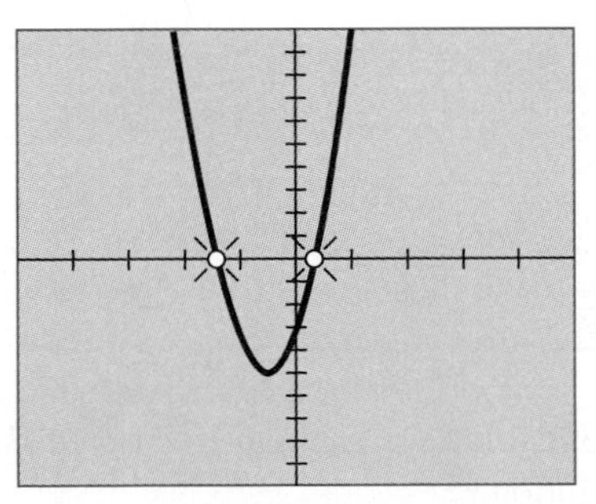

$-5 \le x \le 5$, $-10 \le y \le 10$

Figure 4

In Example 1, the zero-factor property is used to solve a quadratic equation.

EXAMPLE 1 Solve each equation.

(a) $6r^2 + 7r = 3$

First write the equation in standard form as

$$6r^2 + 7r - 3 = 0.$$

Now factor $6r^2 + 7r - 3$ to get

$$(3r - 1)(2r + 3) = 0.$$

By the zero-factor property, the product $(3r - 1)(2r + 3)$ can equal 0 only if

$$3r - 1 = 0 \quad \text{or} \quad 2r + 3 = 0.$$

Solve each of these linear equations separately to find that the solutions of the original equation are $1/3$ and $-3/2$. Check these solutions by substituting back in the original equation.

(b) $x^2 = 5x$

To use the zero-factor property here, first get zero on one side.

$$x^2 = 5x$$

$$x^2 - 5x = 0 \qquad \text{Add } -5x \text{ to both sides.}$$

$$x(x - 5) = 0 \qquad \text{Factor.}$$

$$x = 0 \quad \text{or} \quad x - 5 = 0 \qquad \text{Set each factor equal to 0.}$$

$$x = 0 \quad \text{or} \quad x = 5$$

Check that the solutions are 0 and 5 by substituting them in the original equation. ■

CAUTION An error that students often make when solving equations like Example 1(b) is to begin by dividing both sides of the equation by x to obtain $x = 5$. Notice that if this is done, one of the two solutions is lost. This happens because when $x = 0$, division by x is not defined. ■

A quadratic equation of the form $x^2 = k$ can be solved by factoring over the real numbers using the following sequence of equivalent equations.

$$x^2 = k$$

$$x^2 - k = 0$$

$$(x - \sqrt{k})(x + \sqrt{k}) = 0$$

$$x - \sqrt{k} = 0 \quad \text{or} \quad x + \sqrt{k} = 0$$

$$x = \sqrt{k} \quad \text{or} \quad x = -\sqrt{k}$$

This proves the following statement, sometimes called the **square root property**, which gives a direct way to solve equations of the form $x^2 = k$.

Square Root Property

The solution set of $x^2 = k$ is

$$\{\sqrt{k}, -\sqrt{k}\}.$$

The solutions are often abbreviated as $\pm\sqrt{k}$. Both solutions are real if $k > 0$ and imaginary if $k < 0$. If $k = 0$, there is only one solution.

EXAMPLE 2 Solve each equation.

(a) $z^2 = 17$

The solution set is $\{\pm\sqrt{17}\}$.

(b) $m^2 = -25$

Since $\sqrt{-25} = 5i$, the solution set of $m^2 = -25$ is $\{\pm 5i\}$.

(c) $(y - 4)^2 = 12$

Use a generalization of the square root property, working as follows.

$$(y - 4)^2 = 12$$
$$y - 4 = \pm\sqrt{12}$$
$$y = 4 \pm \sqrt{12}$$
$$y = 4 \pm 2\sqrt{3}$$

(d) $(2x - 5)^2 = 4$

$$2x - 5 = \pm 2 \qquad \text{Generalized square root property}$$
$$2x = 5 \pm 2$$
$$x = \frac{5 \pm 2}{2}$$
$$x = \frac{7}{2} \text{ or } x = \frac{3}{2}$$

Completing the Square As suggested by the equation in Example 2(c), a quadratic equation can be solved with the square root property by first writing the given equation in the form $(x + n)^2 = k$ for suitable numbers n and k. This has been known since Babylonian times (2000 B.C.). Euclid (ca. 330–270 B.C.) described methods for solving quadratics by geometric figures. Hence we say ''complete the square.''

The next two examples show how to write a quadratic equation in the form $(x + n)^2 = k$ by *completing the square*. (We will use this process again in later chapters when drawing the graphs of equations.)

EXAMPLE 3 Solve $x^2 - 2x = 15$.

To rewrite this equation in the form $(x + n)^2 = k$, the left side must be rewritten as a perfect square trinomial. Expanding $(x + n)^2$ gives $x^2 + 2xn + n^2$. Get $x^2 - 2x$ in this form by first looking at the terms of first degree, $-2x$ and $2xn$. These terms are equal if

$$2xn = -2x$$

or

$$n = -1.$$

The value of n is always half the coefficient of the first-degree term. Here, the first-degree term is $-2x$, so $n = -2/2 = -1$. If $n = -1$, then $n^2 = (-1)^2 = 1$. Thus, $x^2 - 2x$ can be converted into a perfect square by adding 1, since

$$x^2 - 2x + 1 = (x - 1)^2.$$

If 1 is added to $x^2 - 2x$, the left side of the given equation, then 1 must also be added to the right side, with $x^2 - 2x = 15$ becoming

$$x^2 - 2x + 1 = 15 + 1,$$

or, after factoring on the left,

$$(x - 1)^2 = 16.$$

This equation, $(x - 1)^2 = 16$, is nothing more than the original equation, $x^2 - 2x = 15$, rewritten in an alternate form. In this new form, the equation can be solved by the square root property.

$$x - 1 = 4 \quad \text{or} \quad x - 1 = -4$$

There are two solutions:

$$x = 1 + 4 = 5 \quad \text{and} \quad x = 1 - 4 = -3.$$

In Example 3 we rewrote the equation $x^2 - 2x = 15$ as $(x - 1)^2 = 16$ by completing the square, to get it in a form suitable for solution by the square root property. A summary of the steps involved in completing the square is given below.

Completing the Square

To solve $ax^2 + bx + c = 0$, $a \neq 0$, by **completing the square**:

1. If $a \neq 1$, multiply both sides by $1/a$. Then rewrite the equation so that the constant is alone on one side of the equals sign.
2. Square half the coefficient of x and add the square to both sides.
3. Factor, and use the square root property.

EXAMPLE 4 Solve $9z^2 - 12z - 1 = 0$.

To rewrite this equation in the form $(z + n)^2 = k$, it is necessary that the coefficient of z^2 be 1. To get this coefficient, multiply both sides by $1/9$.

$$z^2 - \frac{4}{3}z - \frac{1}{9} = 0$$

Now add $1/9$ on both sides.

$$z^2 - \frac{4}{3}z = \frac{1}{9}$$

The coefficient of the first-degree term is $-4/3$. Half of $-4/3$ is $-2/3$, and $(-2/3)^2 = 4/9$, which should be added to both sides, giving

$$z^2 - \frac{4}{3}z + \frac{4}{9} = \frac{1}{9} + \frac{4}{9}.$$

Factoring on the left yields

$$\left(z - \frac{2}{3}\right)^2 = \frac{5}{9}.$$

Now use the square root property to find z.

$$z - \frac{2}{3} = \pm\sqrt{\frac{5}{9}}$$

$$z - \frac{2}{3} = \pm\frac{\sqrt{5}}{3}$$

$$z = \frac{2}{3} \pm \frac{\sqrt{5}}{3}$$

These two solutions can be written as

$$z = \frac{2 \pm \sqrt{5}}{3}.$$

Quadratic Formula The method of completing the square can be used to solve any quadratic equation. However, in the long run it is better to start with the general quadratic equation $ax^2 + bx + c = 0$, and, using the method of completing the square, solve this equation for x in terms of the constants a, b, and c. The result will be a general formula for solving any quadratic equation. To find this general formula, start with the fact that in a quadratic equation $a \neq 0$, and multiply both sides by $1/a$ to get

$$x^2 + \frac{b}{a}x + \frac{c}{a} = 0.$$

Add $-c/a$ to both sides.

$$x^2 + \frac{b}{a}x = -\frac{c}{a}$$

Now take half of b/a, and square the result.

$$\frac{1}{2} \cdot \frac{b}{a} = \frac{b}{2a} \quad \text{and} \quad \left(\frac{b}{2a}\right)^2 = \frac{b^2}{4a^2}$$

Add the square to both sides.

$$x^2 + \frac{b}{a}x + \frac{b^2}{4a^2} = \frac{b^2}{4a^2} - \frac{c}{a}$$

The expression on the left side of the equals sign can be written as the square of a binomial, while the expression on the right can be simplified. Doing all this yields

$$\left(x + \frac{b}{2a}\right)^2 = \frac{b^2 - 4ac}{4a^2}.$$

By the square root property, this last statement leads to

$$x + \frac{b}{2a} = \sqrt{\frac{b^2 - 4ac}{4a^2}} \quad \text{or} \quad x + \frac{b}{2a} = -\sqrt{\frac{b^2 - 4ac}{4a^2}}.$$

Since $4a^2 = (2a)^2$,

$$x + \frac{b}{2a} = \frac{\sqrt{b^2 - 4ac}}{|2a|} \quad \text{or} \quad x + \frac{b}{2a} = \frac{-\sqrt{b^2 - 4ac}}{|2a|}.$$

If $a > 0$, then $|2a| = 2a$, giving

$$x = \frac{-b + \sqrt{b^2 - 4ac}}{2a} \quad \text{or} \quad x = \frac{-b - \sqrt{b^2 - 4ac}}{2a}. \qquad (*)$$

If $a < 0$, then $|2a| = -2a$, giving the same two solutions as in (*), except in reversed order. In either case, the solutions can be written as

$$x = \frac{-b + \sqrt{b^2 - 4ac}}{2a} \quad \text{or} \quad x = \frac{-b - \sqrt{b^2 - 4ac}}{2a}.$$

A more compact form of this result, called the **quadratic formula**, is given below.

Quadratic Formula

The solutions of the quadratic equation $ax^2 + bx + c = 0$, where $a \neq 0$, are

$$x = \frac{-b \pm \sqrt{b^2 - 4ac}}{2a}.$$

For Graphers

An important feature of a grapher is that it can be programmed to evaluate formulas such as the one in the quadratic theorem. Since this formula is used frequently throughout mathematics and its applications, it is worth programming. Refer to your manual for details.

When a solution is an irrational number, as in Example 5, the grapher gives an approximation that can be accurate to as many places as desired. For practical purposes, this is more useful than an exact answer, such as $2 + \sqrt{3}$.

EXAMPLE 5 Use the quadratic formula to solve $x^2 - 4x + 1 = 0$.

Here $a = 1$, $b = -4$, and $c = 1$. Substitute these values into the quadratic formula.

$$x = \frac{-b \pm \sqrt{b^2 - 4ac}}{2a}$$

$$= \frac{-(-4) \pm \sqrt{(-4)^2 - 4(1)(1)}}{2(1)}$$

$$= \frac{4 \pm \sqrt{16 - 4}}{2} = \frac{4 \pm 2\sqrt{3}}{2}$$

$$= \frac{2(2 \pm \sqrt{3})}{2}$$

$$x = 2 \pm \sqrt{3}$$

The solutions are $x = 2 + \sqrt{3}$ and $x = 2 - \sqrt{3}$. ■

EXAMPLE 6 Solve $2y^2 = y - 4$.

To find the values of a, b and c, first rewrite the equation as $2y^2 - y + 4 = 0$. Then $a = 2$, $b = -1$, and $c = 4$.

$$y = \frac{-(-1) \pm \sqrt{(-1)^2 - 4(2)(4)}}{2(2)} \quad \text{By the quadratic formula}$$

$$= \frac{1 \pm \sqrt{1 - 32}}{4}$$

$$= \frac{1 \pm \sqrt{-31}}{4} = \frac{1 \pm i\sqrt{31}}{4}$$

$$y = \frac{1}{4} \pm \frac{\sqrt{31}}{4} i \quad \text{in standard form.}$$

The Discriminant The quantity under the radical in the quadratic formula, $b^2 - 4ac$, is called the **discriminant**. The value of the discriminant determines whether the solutions of quadratic equations with real coefficients are real or complex (but not real) and whether there are one or two solutions, as follows.

Discriminant (Real Coefficients)

Discriminant	*Number of solutions*	*Kind of solutions*
Positive	Two	Real
Zero	One	Real
Negative	Two	Complex (not real)

EXAMPLE 7 Find a value of k so that there is exactly one solution to the equation

$$16p^2 + kp + 25 = 0.$$

A quadratic equation with real coefficients will have exactly one solution if the discriminant is zero. Here, $a = 16$, $b = k$, and $c = 25$, giving the discriminant

$$b^2 - 4ac = k^2 - 4(16)(25) = k^2 - 1600.$$

$$\text{The discriminant is 0 if } k^2 - 1600 = 0$$

$$\text{or if } k^2 = 1600,$$

$$\text{from which } k = \pm 40.$$

When the numbers a, b, and c are *integers*, the value of the discriminant can be used to determine whether the solution will be rational, irrational, or nonreal complex, as follows.

Discriminant (Integer Coefficients)

Discriminant	*Number of solutions*	*Kind of solutions*
Positive, perfect square	Two	Rational
Positive, but not a perfect square	Two	Irrational
Zero	One	Rational
Negative	Two	Complex (not real)

CAUTION The restriction that a, b, and c be integers is important. For example, for the equation

$$x^2 - \sqrt{5}x - 1 = 0$$

the discriminant is $b^2 - 4ac = 5 + 4 = 9$, which would indicate two rational solutions. By the quadratic formula, however, the two solutions

$$x = \frac{\sqrt{5} \pm 3}{2}$$

are *irrational* numbers. ■

EXAMPLE 8 Use the discriminant to determine whether the solutions of $5x^2 + 2x - 4 = 0$ are rational, irrational, or nonreal complex.

Since $a = 5$, $b = 2$, and $c = -4$, the discriminant is

$$b^2 - 4ac = (2)^2 - (4)(5)(-4) = 84.$$

The discriminant is positive, so there are two real number solutions. Since 84 is not a perfect square, the solutions will be irrational numbers. ■

Applied problems often lead to quadratic equations, as in the next example.

EXAMPLE 9 Michael wants to make an exposed gravel border of uniform width around a rectangular gazebo in his garden. The gazebo is 10 feet long and 6 feet wide. He has enough material to cover 36 square feet. How wide will the border be?

A sketch of the gazebo with border is shown in Figure 5.

Let x = the width of the border,

$6 + 2x$ = the width of the larger rectangle,

$10 + 2x$ = the length of the larger rectangle.

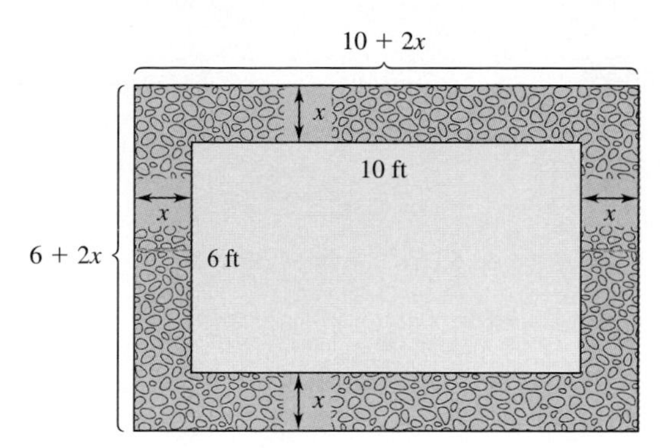

Figure 5

Since the area of the border is the difference of the areas of the larger rectangle and the gazebo,

$$\begin{aligned}
\text{Area of larger rectangle} - \text{Area of gazebo} &= \text{Amount of material} \\
(6 + 2x)(10 + 2x) - 10 \cdot 6 &= 36 \\
60 + 32x + 4x^2 - 60 &= 36 \\
4x^2 + 32x - 36 &= 0 \\
x^2 + 8x - 9 &= 0 \qquad \text{Divide by 4.} \\
(x + 9)(x - 1) &= 0.
\end{aligned}$$

The solutions are -9 and 1. Since -9 cannot be the width of the border, the border must be 1 foot wide.

Check. Figure 6 shows the areas of the different parts of the border. The sum of the areas checks out to be 36, as required.

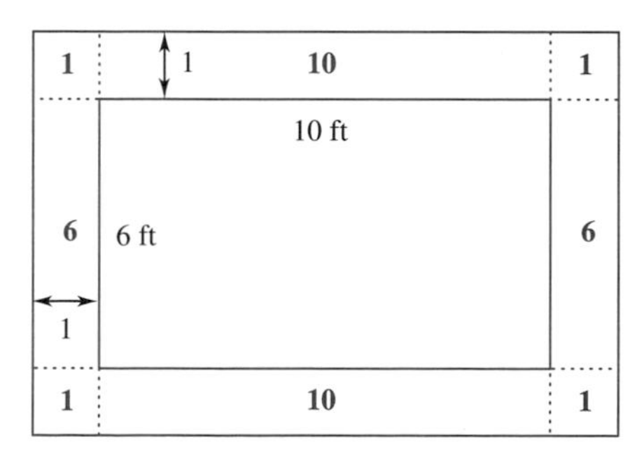

Figure 6

2.3 Exercises

Solve the following equations by factoring.

1. $p^2 = 5p - 6$ **2.** $6x + 9 = -x^2$ **3.** $3q + 4 = q^2$ **4.** $2x^2 - 8x = 0$

5. $-5x^2 + 30x = 0$ **6.** $6y^2 - 5y - 50 = 0$ **7.** $21p^2 = 10 - 29p$ **8.** $6r^2 + 7r = 3$

Solve the following equations by the square root property.

9. $x^2 = 20$ **10.** $y^2 = 48$ **11.** $(t - 2)^2 = 7$

12. $(4z + 3)^2 = 12$ **13.** $(5r + 3)^2 = 3$ **14.** $(-2w + 5)^2 = 8$

Solve the following equations by completing the square.

15. $p^2 - 8p + 15 = 0$ **16.** $m^2 + 4m = 1/3$ **17.** $9z^2 = 12z - 8$

18. $3k^2 - 12k + 12 = 75$ **19.** $2p^2 = -(2p + 1)$ **20.** $r^2 + 8r = -13$

Solve the following equations by the quadratic formula.

21. $m^2 = m + 1$ **22.** $x^2 + 7 = 6x$ **23.** $y^2 - 3y = 2$

24. $11p^2 = 7p - 1$ **25.** $m^2 - \sqrt{2}m - 1 = 0$ **26.** $z^2 - \sqrt{3}z - 2 = 0$

27. $\sqrt{2}p^2 - 3p + \sqrt{2} = 0$ **28.** $-\sqrt{6}k^2 - 2k + \sqrt{6} = 0$

Solve the following equations by any method.

29. $x^2 - x = 6$ **30.** $2s^2 = 10$ **31.** $6n^2 = 3n$

32. $9p^2 = 25 + 30p$ **33.** $4 - \frac{11}{x} - \frac{3}{x^2} = 0$ **34.** $3 = \frac{4}{p} + \frac{2}{p^2}$

35. $\frac{2x - 3}{x + 1} = \frac{x + 1}{x + 2}$ **36.** $\frac{x}{x - 1} = \frac{3x}{x - 2}$

37. Solve $8x^2 + 2x = 5$ by factoring, completing the square, and using the quadratic formula. Compare the three methods, noting advantages and disadvantages of each.

38. In your own words, write a brief explanation of the process of completing the square as you might explain it to another student.

Solve each of the following equations by factoring first and then using the quadratic formula.

39. $x^3 - 64 = 0$ **40.** $8p^3 + 125 = 0$ **41.** $64r^3 - 343 = 0$

Evaluate the discriminant $b^2 - 4ac$ and use it to predict the type of solutions in each of the following. Do not solve the equations.

42. $8y^2 = 14y - 3$ **43.** $3m^2 - 5m + 2 = 0$ **44.** $9k^2 + 11k + 4 = 0$ **45.** $4p^2 = 6p + 3$

Find all values of k for which the following equations have exactly one solution.

46. $25m^2 - 10m + k = 0$ **47.** $y^2 + 11y + k = 0$

48. $kr^2 + (2k + 6)r + 16 = 0$ **49.** $ky^2 + 2(k + 4)y + 25 = 0$

Solve Exercises 50–55 for the indicated variables. Assume that all variables represent positive real numbers.

50. $L = \dfrac{d^4k}{h^2}$ for h

51. $F = \dfrac{kMv^2}{r}$ for v

52. $s = s_0 + gt^2 + k$ for t

53. $P = \dfrac{E^2R}{(r + R)^2}$ for R

54. $S = 2\pi rh + 2\pi r^2$ for r

55. $pm^2 - 8qm + \dfrac{1}{r} = 0$ for m

56. Is it possible for the solution of a quadratic equation with real coefficients to consist of a single irrational number? Explain.

57. Is it possible for the solution of a quadratic equation with real coefficients to consist of one real and one nonreal root? Explain.

For each of the following, find a quadratic equation that has the given numbers as solutions.

58. 4, 5

59. -3, 2

60. $1 + \sqrt{2}$, $1 - \sqrt{2}$

61. i, $-i$

Solve the following problems.

62. A shopping center has a rectangular area of 40,000 sq yd enclosed on three sides for a parking lot. The length is 200 yd more than twice the width. What are the dimensions of the lot?

63. An ecology center wants to set up an experimental garden. It has 300 m of fencing to enclose a rectangular area of 5000 sq m. Find the dimensions of the rectangle.

64. A rectangular poster is to have an 18-inch-by-23-inch illustration in the center with equal margins on all four sides. How wide should the margins be if the poster has an area of 594 sq inches?

65. A landscape architect has included a rectangular flower bed measuring 9 ft by 5 ft in her plans for a new building. She wants to use two colors of flowers in the bed, one in the center and the other for a border of the same width on all four sides. If she can get just enough plants to cover 24 sq ft for the border, how wide can the border be?

66. Alfredo went into a frame it yourself shop. He wanted a frame 3 cm longer than its width. The frame he chose extends 1.5 cm beyond the picture on each side. Find the outside dimensions of the frame if the area of the unframed picture is 70 sq cm.

67. Juanita wants to buy a rug for a room that is 12 ft wide and 15 ft long. She wants to leave a uniform strip of floor around the rug. She can afford 108 sq ft of carpeting. What dimensions should the rug have?

68. An experienced roofer can do a complete roof in a housing development in half the time that it takes an inexperienced roofer to do the job. If the two work together on a roof, they complete the job in $2\frac{2}{3}$ hr. How long would it take the experienced roofer to do a roof working alone?

69. Two typists are working on a special project. The experienced typist could complete the project in 2 hr less time than the new typist. Together they complete the project in 2.4 hr. How long would it have taken the experienced typist to complete the project working alone?

70. It takes two copy machines 5/6 hr to make the copies for a company newsletter. One copy machine takes 1 hr longer than the other to do the job alone. The slower machine is out of order. How long will it take the faster machine to complete the job alone?

71. In 1991 Rick Mears won the (500-mi) Indianapolis 500 race. His speed (rate) was 100 mph (to the nearest mph) faster than that of the 1911 winner, Ray Harroun. Mears completed the race in 3.74 hr less time than Harroun. Find Mears' rate to the nearest whole number.

72. The Branson family traveled 100 mi to a lake for their vacation. On the return trip their average speed was 10 mph faster. The total time for the round trip was $3\frac{2}{3}$ hr. What was the family's average speed on their trip to the lake?

Use the Pythagorean theorem, $c^2 = a^2 + b^2$, to solve each of the following.

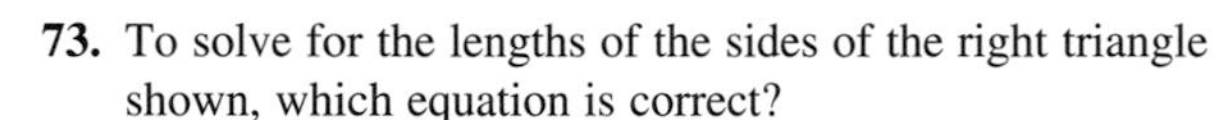

73. To solve for the lengths of the sides of the right triangle shown, which equation is correct?

(a) $x^2 = (2x - 2)^2 + (x + 4)^2$

(b) $x^2 + (x + 4)^2 = (2x - 2)^2$

(c) $x^2 = (2x - 2)^2 - (x + 4)^2$

(d) $x^2 + (2x - 2)^2 = (x + 4)^2$

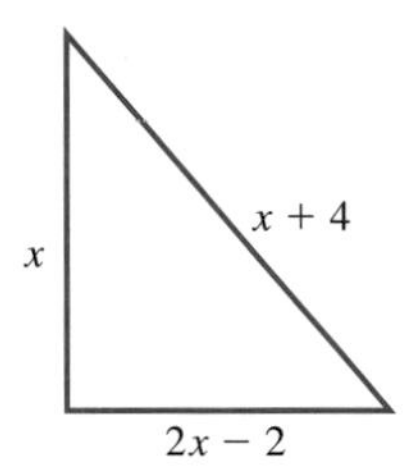

74. If a rectangle is r ft long and s ft wide, which one of the following is the length of its diagonal in terms of r and s?
(a) rs **(b)** $r + s$ **(c)** $\sqrt{r^2 + s^2}$ **(d)** $r^2 + s^2$

75. A boat is being pulled into a dock with a rope attached at water level. When the boat is 12 ft from the dock, the length of the rope from the boat to the dock is 3 ft longer than twice the height of the dock above the water. Find the height of the dock above the water.

76. Chris and Josh have received walkie-talkies for Christmas. If they leave from the same point at the same time, Chris walking north at 2.5 mph and Josh walking east at 3 mph, how long will they be able to talk to each other if the range of the walkie-talkies is 4 mi? Round your answer to the nearest minute.

Let r_1 and r_2 be the solutions of the quadratic equation $ax^2 + bx + c = 0$. Show that the equations in Exercises 77 and 78 are true.

77. $r_1 + r_2 = -\dfrac{b}{a}$

78. $r_1 r_2 = \dfrac{c}{a}$

79. Suppose one solution of the equation $km^2 + 10m = 8$ is -4. Find the value of k, and the other solution.

For the equations in Exercises 80 and 81, **(a)** *solve for x in terms of y,* **(b)** *solve for y in terms of x.*

80. $4x^2 - 2xy + 3y^2 = 2$

81. $3y^2 + 4xy - 9x^2 = -1$

Graphing Utility Problems

For each of the following equations, use a grapher to solve the equation. Then solve the equation algebraically and compare the results.

82. $2x^2 = x - 1$

83. $3x^2 + 2 = x$

84. What limitation on solving equations with a grapher is suggested by Exercises 82 and 83?

2.4 Other Types of Equations

The equation $12m^4 - 11m^2 + 2 = 0$ is not a quadratic equation because of the m^4 term. However, it can be written as a quadratic equation by making the substitutions

$$x = m^2 \quad \text{and} \quad x^2 = m^4$$

which change the equation into

$$12x^2 - 11x + 2 = 0.$$

This quadratic equation can be solved for x, and then from $x = m^2$, the values of m, the solutions of the original equation, can be found.

An equation such as $12m^4 - 11m^2 + 2 = 0$ is said to be **quadratic in form** if it can be written as

$$au^2 + bu + c = 0,$$

where $a \neq 0$ and u is some algebraic expression.

For Graphers

Graph the equation in Example 1 using a grapher. Since the four solutions are very close together, the x-min and x-max also should be close. The values found in Example 1 suggest using $[-1, 1]$. It is a good idea to start with the standard rectangle (most graphers have one that is easily set), and then adjust the x-min, x-max, y-min, and y-max until the graph looks reasonable. See Figure 7.

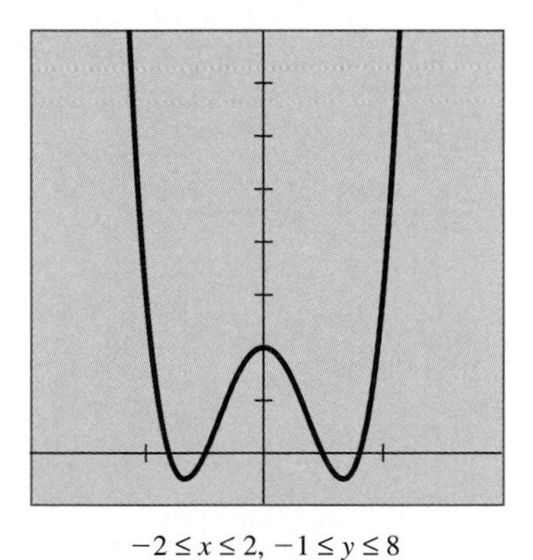

$-2 \leq x \leq 2, \ -1 \leq y \leq 8$

Figure 7

EXAMPLE 1 Solve $12m^4 - 11m^2 + 2 = 0$.

As mentioned above, this equation is quadratic in form. By making the substitution $u = m^2$, the equation becomes

$$12u^2 - 11u + 2 = 0,$$

which can be solved by factoring, as follows.

$$(3u - 2)(4u - 1) = 0$$

$$u = 2/3 \quad \text{or} \quad u = 1/4$$

To find m, use the fact that $u = m^2$ and replace u with $2/3$ and then with $1/4$.

$$m^2 = \frac{2}{3} \qquad \text{or} \qquad m^2 = \frac{1}{4}$$

$$m = \pm\sqrt{\frac{2}{3}} \qquad\qquad m = \pm\sqrt{\frac{1}{4}}$$

$$m = \pm\frac{\sqrt{2}}{\sqrt{3}}$$

$$m = \frac{\pm\sqrt{6}}{3} \qquad \text{or} \qquad m = \pm\frac{1}{2}$$

The given equation $12m^4 - 11m^2 + 2 = 0$ has the four solutions $-\sqrt{6}/3$, $\sqrt{6}/3$, $-1/2$, and $1/2$. As before, check these solutions by substituting back into the *original* equation. ■

EXAMPLE 2 Solve $6p^{-2} + p^{-1} = 2$.

Let $u = p^{-1}$ and rearrange terms to get

$$6u^2 + u - 2 = 0.$$

Factor on the left, and then place each factor equal to 0.

$$(3u + 2)(2u - 1) = 0$$

$$3u + 2 = 0 \quad \text{or} \quad 2u - 1 = 0$$

$$u = -\frac{2}{3} \qquad\qquad u = \frac{1}{2}$$

Since $u = p^{-1} = \frac{1}{p}$, $\quad \frac{1}{p} = -\frac{2}{3} \quad$ or $\quad \frac{1}{p} = \frac{1}{2}$,

from which $\quad p = -\frac{3}{2} \quad$ or $\quad p = 2.$ ■

EXAMPLE 3 Solve $4p^4 - 16p^2 + 13 = 0$.

Let $u = p^2$ to get

$$4u^2 - 16u + 13 = 0.$$

To solve this equation, use the quadratic formula with $a = 4$, $b = -16$, and $c = 13$. Substitute these values into the quadratic formula.

$$u = \frac{-(-16) \pm \sqrt{(-16)^2 - 4(4)(13)}}{2(4)}$$

$$= \frac{16 \pm \sqrt{48}}{8} = \frac{16 \pm 4\sqrt{3}}{8} = \frac{4 \pm \sqrt{3}}{2}$$

Since $p^2 = u$, $\quad p^2 = \frac{4 + \sqrt{3}}{2} \quad$ or $\quad p^2 = \frac{4 - \sqrt{3}}{2}$.

Finally, $\quad p = \pm\sqrt{\frac{4 + \sqrt{3}}{2}} \quad$ or $\quad p = \pm\sqrt{\frac{4 - \sqrt{3}}{2}}$.

Rationalizing the denominators gives the four solutions

$$\frac{\pm\sqrt{8 + 2\sqrt{3}}}{2}, \quad \frac{\pm\sqrt{8 - 2\sqrt{3}}}{2}.$$ ■

To solve equations containing radicals or rational exponents, such as $x = \sqrt{15 - 2x}$, or $(x + 1)^{1/2} = x$, use the following result.

If P and Q are algebraic expressions, then every solution of the equation $P = Q$ is also a solution of the equation $(P)^n = (Q)^n$, for any positive integer n.

CAUTION Be very careful when using this result. It does not say that the equations $P = Q$ and $(P)^n = (Q)^n$ are equivalent; it says only that each solution of the original equation $P = Q$ is also a solution of the new equation $(P)^n = (Q)^n$. However, *the new equation may have more solutions than the original equation.* For example, the solution set of the equation $x = -2$ is $\{-2\}$. Squaring both sides of the equation $x = -2$ gives the new equation $x^2 = 4$, which has solution set $\{-2, 2\}$. Since the solution sets are not equal, the equations are not equivalent. As this example shows, it is essential to check all proposed solutions back in the original equation. ■

EXAMPLE 4 Solve $x = \sqrt{15 - 2x}$.

The equation $x = \sqrt{15 - 2x}$ can be solved by squaring both sides as follows.

$$x^2 = (\sqrt{15 - 2x})^2$$
$$x^2 = 15 - 2x$$
$$x^2 + 2x - 15 = 0$$
$$(x + 5)(x - 3) = 0$$
$$x = -5 \quad \text{or} \quad x = 3$$

Now it is necessary to check the proposed solutions in the *original* equation,

$$x = \sqrt{15 - 2x}.$$

If $x = -5$, does $x = \sqrt{15 - 2x}$?	If $x = 3$, does $x = \sqrt{15 - 2x}$?
$-5 = \sqrt{15 + 10}$	$3 = \sqrt{15 - 6}$
$-5 = 5$ False	$3 = 3$ True

As this check shows, only 3 is a solution of the given equation. ■

EXAMPLE 5 Solve $\sqrt{2x + 3} - \sqrt{x + 1} = 1$.

Separate the radicals by writing the equation as

$$\sqrt{2x + 3} = 1 + \sqrt{x + 1}.$$

Now square both sides. Be very careful when squaring on the right side of this equation. Recall that $(a + b)^2 = a^2 + 2ab + b^2$; replace a with 1 and b with

$\sqrt{x+1}$ to get the next equation, the result of squaring both sides of $\sqrt{2x+3} = 1 + \sqrt{x+1}$.

$$2x + 3 = 1 + 2\sqrt{x+1} + x + 1$$
$$x + 1 = 2\sqrt{x+1}$$

One side of the equation still contains a radical; to eliminate it, square both sides again.

$$x^2 + 2x + 1 = 4(x + 1)$$
$$x^2 - 2x - 3 = 0$$
$$(x - 3)(x + 1) = 0$$
$$x = 3 \quad \text{or} \quad x = -1$$

Check these proposed solutions in the original equation.

$$\text{If } x = 3, \text{ does } \sqrt{2x+3} - \sqrt{x+1} = 1?$$
$$\sqrt{9} - \sqrt{4} = 1$$
$$3 - 2 = 1 \qquad \text{True}$$
$$\text{If } x = -1, \text{ does } \sqrt{2x+3} - \sqrt{x+1} = 1?$$
$$\sqrt{1} - \sqrt{0} = 1$$
$$1 - 0 = 1 \qquad \text{True}$$

Both proposed solutions 3 and -1 are solutions of the original equation. ■

EXAMPLE 6 Solve $(5x^2 - 6)^{1/4} = x$.

Since the equation involves a fourth root, begin by raising both sides to the fourth power.

$$[(5x^2 - 6)^{1/4}]^4 = x^4$$
$$5x^2 - 6 = x^4$$
$$x^4 - 5x^2 + 6 = 0$$

Now substitute y for x^2.

$$y^2 - 5y + 6 = 0$$
$$(y - 3)(y - 2) = 0$$
$$y = 3 \quad \text{or} \quad y = 2$$

Since $y = x^2$,

$$x^2 = 3 \quad \text{or} \quad x^2 = 2$$
$$x = \pm\sqrt{3} \quad \text{or} \quad x = \pm\sqrt{2}.$$

Checking the four proposed solutions, $\sqrt{3}$, $-\sqrt{3}$, $\sqrt{2}$, and $-\sqrt{2}$, in the original equation shows that only $\sqrt{3}$ and $\sqrt{2}$ are solutions. ■

2.4 Exercises

Find all real solutions for each of the following equations.

1. $m^4 + 2m^2 - 15 = 0$
2. $2x^4 + 5x^2 = 3$
3. $5 = 7r^2 - 2r^4$
4. $3 = 8x^2 - 4x^4$
5. $(g - 2)^2 + 8 = 6(g - 2)$
6. $(p + 2)^2 - 2(p + 2) = 15$
7. $4p^{-2} - 9p^{-1} = 5$
8. $6 + 11x^{-1} - 3x^{-2} = 0$
9. $(y + 3)^{2/3} - 2(y + 3)^{1/3} = 3$
10. $(r - 1)^{2/3} = 12 - (r - 1)^{1/3}$
11. $3 + \dfrac{5}{p^2 + 1} = \dfrac{2}{(p^2 + 1)^2}$
12. $\dfrac{7}{2y - 3} + \dfrac{3}{(2y - 3)^2} = 6$
13. $a^3 - 8a^{3/2} + 7 = 0$
14. $r^3 - 13r^{3/2} + 40 = 0$
15. $5(m^2 + 1)^{-2} = 4(m^2 + 1)^{-1} + 1$
16. $1 + 3(r^2 - 1)^{-1} = 28(r^2 - 1)^{-2}$
17. $20(2 - \sqrt{m})^2 + 11(2 - \sqrt{m}) = 3$
18. $2(1 + 2\sqrt{x})^2 - (1 + 2\sqrt{x}) = 21$

What is wrong with each solution in Exercises 19 and 20?

19. Solve $4x^4 - 11x^2 - 3 = 0$.
Let $t = x^2$.

$$4t^2 - 11t - 3 = 0$$
$$(4t + 1)(t - 3) = 0$$
$$4t + 1 = 0 \quad \text{or} \quad t - 3 = 0$$
$$t = -1/4 \quad \text{or} \quad t = 3$$

The solutions are $-1/4$ and 3.

20. Solve $x = \sqrt{3x + 4}$.
Square both sides to get

$$x^2 = 3x + 4.$$
$$x^2 - 3x - 4 = 0$$
$$(x - 4)(x + 1) = 0$$
$$x - 4 = 0 \quad \text{or} \quad x + 1 = 0$$
$$x = 4 \quad \text{or} \quad x = -1$$

The solutions are 4 and -1.

Solve the following equations.

21. $\sqrt{6m + 7} - 1 = m + 1$
22. $\sqrt{3z + 7} = 3z + 5$
23. $\sqrt{4x} - x + 3 = 0$
24. $\sqrt{2t} + 4 = t$
25. $\sqrt{4k + 5} - 2 = 2k - 7$
26. $\sqrt[3]{2z} = \sqrt[3]{5z + 2}$
27. $\sqrt[3]{4n + 3} = \sqrt[3]{2n - 1}$
28. $\sqrt[3]{2x^2 - 5x + 4} = \sqrt[3]{2x^2}$
29. $\sqrt[3]{t^2 + 2t - 1} = \sqrt[3]{t^2 + 3}$
30. $(3m + 7)^{1/3} - (4m + 2)^{1/3} = 0$
31. $(2r + 5)^{1/3} - (6r - 1)^{1/3} = 0$
32. $\sqrt[4]{q - 15} = 2$
33. $\sqrt[4]{3x + 1} = 1$
34. $(3t^2 + 52t)^{1/4} - 4 = 0$
35. $(z^2 + 24z)^{1/4} - 3 = 0$
36. $\sqrt{2m} = \sqrt{m + 7} - 1$
37. $\sqrt{y} = \sqrt{y - 5} + 1$
38. $\sqrt{r + 2} - 1 = \sqrt{3r + 7}$
39. $\sqrt{2p - 5} - 2 = \sqrt{p - 2}$
40. $\sqrt{x + 4} - \sqrt{x + 3} = \sqrt{3x + 10}$
41. $\sqrt{5x - 1} + \sqrt{2 - x} = \sqrt{8x + 1}$
42. $\sqrt{3\sqrt{2m + 3}} = \sqrt{5m - 6}$
43. $\sqrt{2\sqrt{7x + 2}} = \sqrt{3x + 2}$
44. $3 - \sqrt{x} = \sqrt{2\sqrt{x} - 3}$
45. $\sqrt{x} + 2 = \sqrt{4 + 7\sqrt{x}}$
46. $(z - 3)^{2/5} = (4z)^{1/5}$
47. $(2r - 1)^{2/3} = r^{1/3}$
48. $(2k - 9)^{-2/3} + 4(2k - 9)^{1/3} = 0$
49. $p(2 + p)^{-1/2} + (2 + p)^{1/2} = 0$

50. How can we tell that the equation $x^{1/4} = -2$ has no real solution without actually going through a solution process?

Solve each equation for the indicated variable.

51. $d = k\sqrt{h}$ for h

52. $v = \dfrac{k}{\sqrt{d}}$ for d

53. $P = 2\sqrt{\dfrac{L}{g}}$ for L

54. $c = \sqrt{a^2 + b^2}$ for a

55. $x^{2/3} + y^{2/3} = a^{2/3}$ for y

56. $m^{3/4} + n^{3/4} = 1$ for m

Graphing Utility Problems

Use a grapher to solve each of the following equations. Be careful to use parentheses around the radicand. Check your answers with an algebraical solution.

57. $\sqrt{2x + 1} = 2\sqrt{x}$

58. $\sqrt{5x + 3} = 3\sqrt{x - 1}$

59. $\sqrt{x + 5} - 2 = \sqrt{x - 1}$

60. $\sqrt{3x - 2} + 1 = \sqrt{3x + 1}$

2.5 Variation

In many applications of mathematics, it is necessary to write relationships between variables. For example, in chemistry the ideal gas law shows how temperature, pressure, and volume are related. In physics, various formulas in optics show how the focal length of a lens and the size of an image are related.

Two related quantities that vary together, either both increasing or both decreasing proportionally, are said to *vary directly*. For example, if you work more hours, you earn more money (assuming an hourly wage). This is stated more precisely as follows.

Directly Proportional

y **varies directly** as x, or y **is directly proportional** to x, means that a nonzero real number k (called the **constant of variation**) exists such that

$$y = kx.$$

The phrase "directly proportional" is sometimes abbreviated as just "proportional."

EXAMPLE 1 Suppose the value of y varies directly as the value of x, and that $y = 12$ when $x = 5$. Find y when $x = 21$.

If y varies directly as x, then a real number k exists such that

$$y = kx.$$

To find the value of k, use the given information: $y = 12$ when $x = 5$. Replacing y with 12 and x with 5 gives

$$12 = k \cdot 5 \quad \text{or} \quad k = \frac{12}{5}.$$

In this example, then, the relationship between y and x is given by

$$y = \frac{12}{5}x.$$

Now find y when x is 21.

$$y = \frac{12}{5} \cdot 21 = \frac{252}{5}$$ ■

The x in the direct variation equation may be replaced with any power of x, as long as x is not in a denominator, without changing the type of variation involved. For example, the area of a square of side x is given by the formula $A = x^2$, so that the area *varies directly as the square* of the length of a side. Here $k = 1$.

The case where y increases as x decreases is an example of *inverse variation.*

Inverse Variation

Let n be a positive real number. Then y **varies inversely as the nth power** of x means that there exists a real number k such that

$$y = \frac{k}{x^n}.$$

Here, too, x may be replaced by any appropriate expression in x.

EXAMPLE 2 In a certain manufacturing process, the cost of producing a single item varies inversely as the square of the number of items produced. If 100 items are produced, each costs \$2. Find the cost per item if 400 items are produced.

We can let x represent the number of items produced and y the cost per item, and write

$$y = \frac{k}{x^2}$$

for some nonzero constant k. Since $y = 2$ when $x = 100$,

$$2 = \frac{k}{100^2} \quad \text{or} \quad k = 20{,}000.$$

Thus, the relationship between x and y is given by

$$y = \frac{20{,}000}{x^2}.$$

When 400 items are produced, the cost per item is given by

$$\begin{aligned} y &= \frac{20{,}000}{400^2} \\ &= .125, \text{ or } 12.5¢. \end{aligned}$$

One variable may depend on more than one other variable. Such variation is called *combined variation.* If a variable depends on the product of two or more other variables, we refer to that as *joint variation.*

Joint Variation

y **varies jointly** as x and z if there exists a real number k such that

$$y = kxz.$$

As before, x and z may be replaced by variable expressions.

The steps involved in solving a problem in variation can be summarized as follows.

Solving Variation Problems

1. Write, in algebraic form, the general relationship among the variables. Use the constant k.
2. Substitute given values of the variables and find the value of k.
3. Substitute this value of k into the formula from Step 1, thus obtaining a specific formula.
4. Solve for the required unknown.

Notice how these steps are followed in Examples 1 through 3.

EXAMPLE 3 The number of vibrations per second (the pitch) of a steel guitar string varies directly as the square root of the tension and inversely as the length of the string. If the number of vibrations per second is 5 when the tension is 225 kilograms and the length is .60 meter, find the number of vibrations per second when the tension is 196 kilograms and the length is .65 meter.

Let n represent the number of vibrations per second, T represent the tension, and L represent the length of the string. Then, from the information in the problem,

$$n = \frac{k\sqrt{T}}{L}.$$

Substitute the given values for n, T, and L to find k.

$$5 = \frac{k\sqrt{225}}{.60} \qquad \text{Let } n = 5,\ T = 225,\ L = .60.$$

$$3 = k\sqrt{225} \qquad \text{Multiply by .60.}$$

$$3 = 15k \qquad \sqrt{225} = 15$$

$$k = \frac{1}{5} = .2 \qquad \text{Divide by 15.}$$

Now substitute for k and use the second set of values for T and L to find n.

$$n = \frac{.2\sqrt{196}}{.65} \qquad \text{Let } k = .2,\ T = 196,\ L = .65.$$

$$n = 4.3$$

The number of vibrations per second is 4.3. ■

2.5 Exercises

Express each of the following as an equation. Use k as the constant of proportionality if none is given.

1. y is proportional to x with a constant of proportionality of 16.
2. y varies inversely as x with a constant of variation of 2.6.
3. x is inversely proportional to y.
4. p varies inversely as y.
5. r varies jointly as s and t.
6. R is proportional to m and p.
7. w is proportional to x^2 and inversely proportional to y.
8. c varies directly as d and inversely as f^2 and g.

Write each of the formulas in Exercises 9–14 as an English phrase using the words varies *or* proportional.

9. $c = 2\pi r$, where c is the circumference of a circle of radius r
10. $d = s/5$, where d is the approximate distance (in miles) from a storm and s is the number of seconds between seeing lightning and hearing thunder
11. $v = d/t$, where v is the average speed when traveling d mi in t hr
12. $d = 1/(4\pi nr^2)$, where d is the average distance a gas atom of radius r travels between collisions and n is the number of atoms per unit volume
13. $s = kx^3$, where s is the strength of a muscle of length x

14. $f = mv^2/r$, where f is the centripetal force of an object of mass m moving along a circle of radius r at velocity v

15. What happens to y if y varies inversely as x, and x is doubled?

16. If y varies directly as x, and x is halved, how is y changed?

17. Suppose y is directly proportional to x, and x is replaced by $(1/3)x$. What happens to y?

18. What happens to y if y is inversely proportional to x, and x is tripled?

19. Suppose p varies directly as r^3 and inversely as t^2. If r is halved and t is doubled, what happens to p?

20. If m varies directly as p^2 and q^4, and p doubles while q triples, what happens to m?

21. Simple interest varies jointly as principal, rate, and time. If \$1000 left at interest for 2 yr earned \$110, find the amount of interest earned by \$5000 in 5 yr at the same rate.

22. In electric current flow, the resistance (measured in units called ohms) offered by a fixed length of wire of a given material varies inversely as the square of the diameter of the wire. If a wire .01 inch in diameter has a resistance of .4 ohm, what is the resistance of a wire of the same length and material but .03 inch in diameter?

Photographers use the fact that the amount of light required to take a picture varies directly as the square of the F-stop setting and inversely as the ASA number of the film and the shutter speed. For an F-stop of 8, 200 ASA film, and a shutter speed of 1/100 sec, 800 footcandles of light is required. Use this information to solve Exercises 23–25.

23. What F-stop should be used with 200 ASA film and a shutter speed of 1/250 when 500 footcandles of light is available?

24. What illumination is needed when a photographer is using 400 ASA film, a shutter speed of 1/60 sec, and an F-stop of 5.6?

25. If 125 footcandles of light is available and an F-stop of 2 is used with 200 ASA film, what shutter speed should be used?

26. In a certain manufacturing process, the cost of producing a single item varies inversely as the square of the number of items produced. If 60 items are produced, each costs \$2. Find the cost per item if 400 items are produced. How many items must be produced to reduce the cost to \$1?

27. Hooke's law for an elastic spring states that the distance a spring stretches varies directly as the force applied. If a force of 15 lb stretches a certain spring 8 inches, how much will a force of 30 lb stretch the spring? See the figure.

28. The roof of a new sports arena rests on round concrete pillars. The maximum load a cylindrical column of circular cross-section can hold varies directly as the fourth power of the diameter and inversely as the square of the height. The arena has columns that are 9 m tall, 1 m in diameter, and will support a load of 8 metric tons. How many metric tons will be supported by a column 12 m high and 2/3 m in diameter?

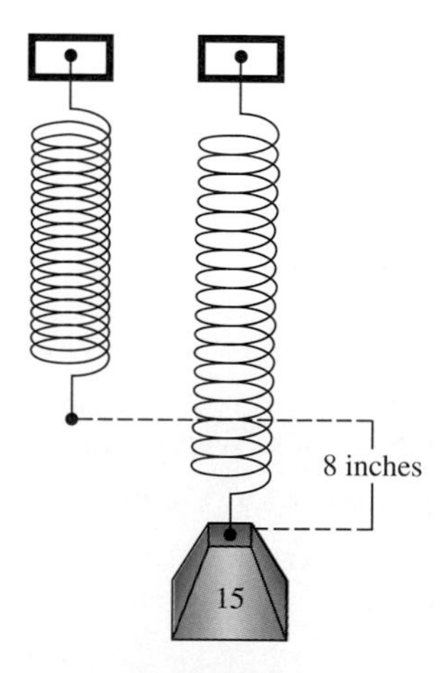

Exercise 27

29. The sports arena in Exercise 28 requires a beam 16 m long, 4 cm wide, and 8 cm high. The maximum load of a horizontal beam that is supported at both ends varies directly as the width and the square of the height and inversely as the length between supports. If a beam of the same material 8 m long, 12 cm wide, and 15 cm high can support a maximum of 400 kg, what is the maximum load the beam in the arena will support?

30. The area of a triangle varies jointly as the lengths of its base and its height. A triangle with a base of 10 ft and a height of 4 ft has an area of 20 square ft. Find the area of a triangle with a base of 3 cm and a height of 8 cm.

31. The number of long-distance phone calls between two cities in a certain time period varies directly as the populations p_1 and p_2 of the cities, and inversely as the distance between them. If 10,000 calls are made between two cities 500 mi apart, having populations of 50,000 and 125,000, find the number of calls between two cities 800 mi apart having populations of 20,000 and 80,000.

32. The distance that a person can see to the horizon from a point above the surface of the earth varies directly as the square root of the height. A person on a hill 121 m high can see 15 km to the horizon. How far can a pilot see from a plane flying at 8100 m?

33. The maximum speed possible on a length of railroad track is directly proportional to the cube root of the amount of money spent on maintaining the track. Suppose that a maximum speed of 25 km per hr is possible on a stretch of track for which $450,000 was spent on maintenance. Find the maximum speed if the amount spent on maintenance is increased to $1,750,000.

34. The force needed to keep a car from skidding on a curve varies inversely as the radius of the curve and jointly as the weight of the car and the square of the speed. It takes 3000 lb of force to keep a 2000-lb car from skidding on a curve of radius 500 ft at 30 mph. What force is needed to keep the same car from skidding on a curve of radius 800 ft at 60 mph?

35. A measure of malnutrition, called the *pelidisi*, varies directly as the cube root of a person's weight in grams and inversely as the person's sitting height in cm. A person with a pelidisi below 100 is considered to be undernourished, while a pelidisi greater than 100 indicates overfeeding. A person who weighs 48,820 g and has a sitting height of 78.7 cm has a pelidisi of 100. Find the pelidisi (to the nearest whole number) of a person whose weight is 54,430 g and whose sitting height is 88.9 cm. Is this individual undernourished or overfed?

36. When the brakes of a car are applied, the speed that the car was traveling is proportional to the square root of the distance that the car travels before coming to a stop. Under certain conditions, a car moving at 60 mph will travel 18 m after the brakes are applied. Determine the formula giving speed in terms of the stopping distance.

37. In Exercise 36, how much stopping distance does the car have at 55 mph?

38. In Exercise 36, what speed would produce a stopping distance of 20 m?

39. Suppose a nuclear bomb is detonated at a certain site. The effects of the bomb will be felt over a distance from the point of detonation that is directly proportional to the cube root of the yield of the bomb. Suppose a 100-kiloton bomb has certain effects to a radius of 3 km from the point of detonation. Find the distance that the effects would be felt for a 1500-kiloton bomb.

40. Under certain conditions, the length of time that it takes for fruit to ripen during the growing season varies inversely as the average maximum temperature during the season. If it takes 25 days for fruit to ripen with an average maximum temperature of 80°F, find the number of days it would take at 75°F.

Graphing Utility Problems

41. Graph $y = k/x$ for various values of k. Explain how the graph illustrates the idea that as x increases, y decreases.

42. Graph $y = kx$ for various values of k. Explain how the graph illustrates the idea that as x increases, y increases.

43. In Exercises 41 and 42, are negative values of k meaningful? Explain.

2.6 Inequalities

An equation says that two expressions are equal, while an **inequality** says that one expression is greater than, greater than or equal to, less than, or less than or equal to, another. As with equations, a value of the variable for which the inequality is true is a *solution* of the inequality; the set of all such solutions makes up the solution set of the inequality. Two inequalities with the same solution set are equivalent.

Inequalities are solved with the following properties.

Properties of Inequalities

For real numbers a, b, and c,

(a) if $a < b$, then $a + c < b + c$,

(b) if $a < b$, and if $c > 0$, then $ac < bc$,

(c) if $a < b$, and if $c < 0$, then $ac > bc$.

Similar properties are valid if $<$ is replaced with $>$, $\leq$, or $\geq$.

CAUTION Pay careful attention to Property (c): if both sides of an inequality are multiplied by a negative number, the direction of the inequality symbol must be reversed. For example, starting with the true statement $-3 < 5$, multiplying both sides by the positive number 2 gives

$$-3 \cdot 2 < 5 \cdot 2$$
$$-6 < 10,$$

still a true statement. On the other hand, starting with $-3 < 5$ and multiplying both sides by the *negative* number -2 gives a true result only if the direction of the inequality symbol is reversed:

$$-3(-2) > 5(-2)$$
$$6 > -10. \blacksquare$$

For Graphers

A grapher can be used to solve the inequality in Example 1 as follows. Write the inequality with 0 on one side: $-3x + 12 > 0$. Graph $y = -3x + 12$. See Figure 9. Since $y = -3x + 12$, the y-values of the points on the graph show where $-3x + 12$ is positive or negative. The x-value where $y = 0$ is the dividing point. The graph shows that $y = -3x + 12 > 0$ when $x < 4$. This is the same solution found algebraically. Note that the graph does not tell you whether or not the endpoint is included in the solution. You must decide whether the endpoint satisfies the given inequality—that is, whether the inequality involves $>$ or $\geq$. The algebraic method is simpler for linear inequalities, but the graphical method is useful for inequalities of higher degree.

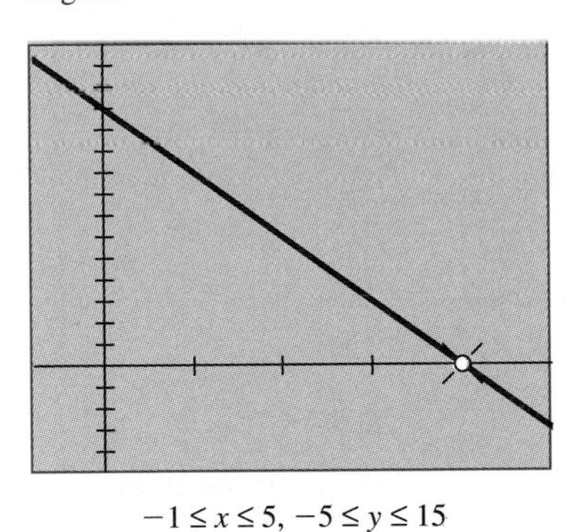

$-1 \leq x \leq 5, -5 \leq y \leq 15$

Figure 9

EXAMPLE 1 Solve the inequality $-3x + 5 > -7$.

Use the properties of inequalities. First, add -5 to both sides.

$$-3x + 5 + (-5) > -7 + (-5)$$
$$-3x > -12$$

Now multiply both sides by $-1/3$. Since $-1/3 < 0$, reverse the direction of the inequality symbol.

$$-\frac{1}{3}(-3x) < -\frac{1}{3}(-12)$$
$$x < 4$$

The original inequality is satisfied by any real number less than 4. The solution set can be written $\{x|x < 4\}$. A graph of the solution set is shown in Figure 8, where the parenthesis is used to show that 4 itself does not belong to the solution set.

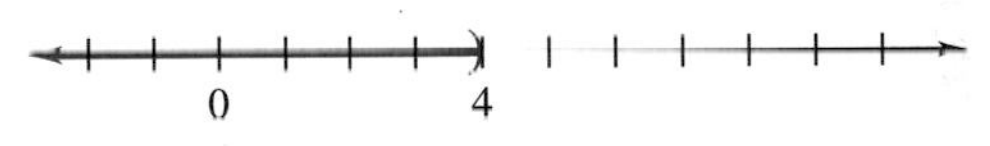

Figure 8

The set $\{x|x < 4\}$, the solution set for the inequality in Example 1, is an example of an **interval**. A simplified notation, called **interval notation**, is used for writing intervals. With this notation, the interval in Example 1 can be written as just $(-\infty, 4)$. The symbol $-\infty$ is not a real number; the symbol is used as a convenience to show that the interval includes all real numbers less than 4. The interval $(-\infty, 4)$ is an example of an *open interval*, since the endpoint, 4, is not part of the interval. Examples of other sets written in interval notation are shown below. Square brackets are used to show that the given number *is* part of the graph. Whenever two real numbers a and b are used to write an interval, it is assumed that $a < b$.

Type of interval	*Set*	*Interval notation*	*Graph*
Open interval	$\{x \mid x > a\}$	(a, ∞)	a
	$\{x \mid a < x < b\}$	(a, b)	a b
	$\{x \mid x < b\}$	$(-\infty, b)$	b
Half-open interval	$\{x \mid a < x \le b\}$	$(a, b]$	a b
	$\{x \mid a \le x < b\}$	$[a, b)$	a b
Closed interval	$\{x \mid a \le x \le b\}$	$[a, b]$	a b
	$\{x \mid \ge a\}$	$[a, \infty)$	a
	$\{x \mid x \le b\}$	$(-\infty, b]$	b

EXAMPLE 2 Solve $4 - 3y \le 7 + 2y$. Write the solution in interval notation and graph the solution on a number line.

Write the following series of equivalent inequalities.

$$4 - 3y \le 7 + 2y$$
$$\mathbf{-4 - 2y} + 4 - 3y \le \mathbf{-4 - 2y} + 7 + 2y$$
$$-5y \le 3$$
$$(\mathbf{-1/5})(-5y) \ge (\mathbf{-1/5})(3)$$
$$y \ge -3/5$$

In set notation, the solution set is $\{y \mid y \ge -3/5\}$, while in interval notation the solution set is $[-3/5, \infty)$. See Figure 10 for the graph of the solution set.

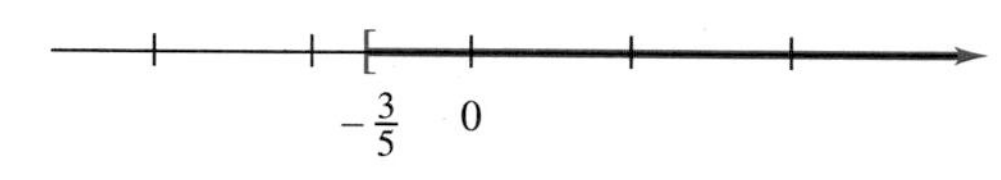

Figure 10

From now on, we shall use interval notation to write the solutions of all inequalities.

The inequality $-2 < 5 + 3m < 20$ says that $5 + 3m$ is between -2 and 20. Solve this inequality using an extension of the properties of inequalities given above.

For Graphers

An inequality like $-2 < 5 + 3m < 20$ can be solved with a grapher by slightly altering the graphical method given earlier. (Since graphers use only the variable x, we change $5 + 3m$ to $5 + 3x$.) Here, it is not possible to get 0 on one side, since there are three "sides." Instead, we must graph three expressions, -2, $5 + 3x$, and 20, simultaneously. Use the trace key to approximate the x-values where the graph of $5 + 3x$ intersects the graphs for -2 and 20. Use the zoom key to get more accurate values. Then decide on what interval the graph of $5 + 3x$ lies between the graphs of -2 and 20. See Figure 12. (Don't forget to decide about the endpoints separately.)

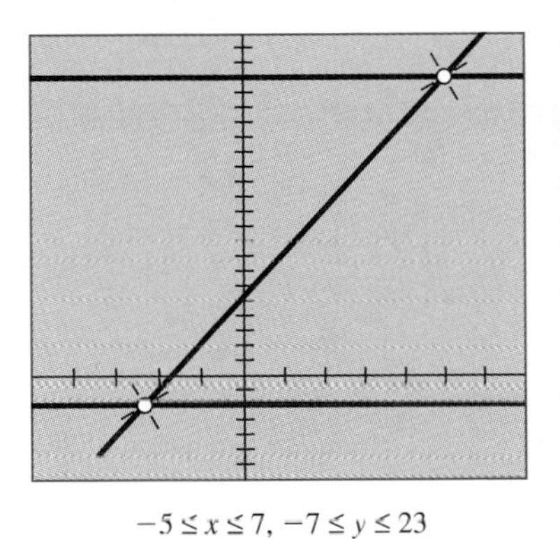

$-5 \leq x \leq 7, -7 \leq y \leq 23$

Figure 12

EXAMPLE 3 Solve $-2 < 5 + 3m < 20$.

Write equivalent inequalities as follows.

$$-2 < 5 + 3m < 20$$
$$-7 < 3m < 15$$
$$-7/3 < m < 5$$

The solution, graphed in Figure 11, is the interval $(-7/3, 5)$.

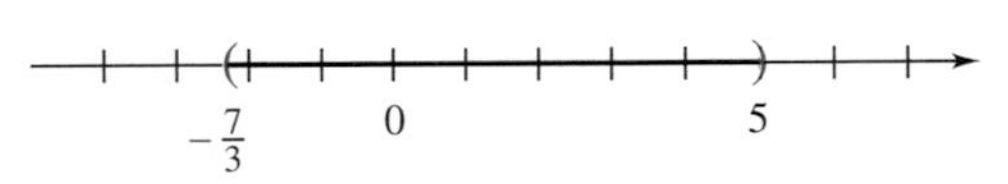

Figure 11

Quadratic Inequalities In Section 2.3 we discussed quadratic equations. Now we will look at *quadratic inequalities.*

Quadratic Inequality

A **quadratic inequality** is an inequality that can be written in the form

$$ax^2 + bx + c < 0,$$

for real numbers $a \neq 0$, b, and c. (The symbol $<$ can be replaced with $>$, $\leq$, or $\geq$.)

Since quadratic equations usually have two solutions, while linear equations have just one, solving quadratic inequalities requires a little more work than solving linear inequalities.

EXAMPLE 4 Solve the quadratic inequality $x^2 - x - 12 < 0$.

Begin by finding the values of x that satisfy $x^2 - x - 12 = 0$.

$$x^2 - x - 12 = 0$$
$$(x + 3)(x - 4) = 0$$
$$x = -3 \quad \text{or} \quad x = 4$$

These two points, -3 and 4, divide a number line into the three regions shown in Figure 13(a). If a point in region A, for example, leads to negative values for the polynomial $x^2 - x - 12$, then all points in region A will lead to negative values.

The regions that make $x^2 - x - 12$ negative can be found by selecting a test point from each region and substituting it into the inequality. For example,

in **region A**, $(-\infty, -3)$, choose -4: $(-4)^2 - (-4) - 12 = \mathbf{8 > 0}$;
in **region B**, $(-3, 4)$, choose 0: $0^2 - 0 - 12 = \mathbf{-12 < 0}$;
in **region C**, $(4, \infty)$, choose 5: $5^2 - 5 - 12 = \mathbf{8 > 0}$.

Only the points in region B, the interval $(-3, 4)$, make the expression $x^2 - x - 12$ negative. The graph of this solution is shown in Figure 13(b).

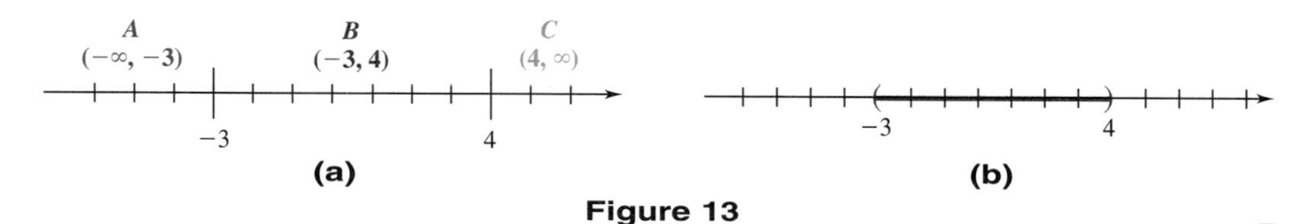

Figure 13

EXAMPLE 5 Solve the inequality $2x^2 + 5x - 12 \geq 0$.

Find the values of x that satisfy $2x^2 + 5x - 12 = 0$.

$$2x^2 + 5x - 12 = 0$$
$$(2x - 3)(x + 4) = 0$$
$$x = 3/2 \quad \text{or} \quad x = -4$$

These two points divide the number line into the three regions shown in Figure 14. We choose -5 in $(-\infty, -4)$, 0 in $(-4, 3/2)$, and 2 in $(3/2, \infty)$ as test points.

If $x = -5$, $2x^2 + 5x - 12 = 2(-5)^2 + 5(-5) - 12 = 13 > 0$.
If $x = 0$, $2x^2 + 5x - 12 = 2(0)^2 + 5(0) - 12 = -12 < 0$.
If $x = 2$, $2x^2 + 5x - 12 = 2(2)^2 + 5(2) - 12 = 6 > 0$.

The tests show that $2x^2 + 5x - 12 \geq 0$ in the interval $(-\infty, -4]$ and also in the interval $[3/2, \infty)$. Since both of these intervals belong to the solution, the result can be written as the *union** of the two intervals, or as

$$(-\infty, -4] \cup [3/2, \infty).$$

The graph of the solution is shown in Figure 14.

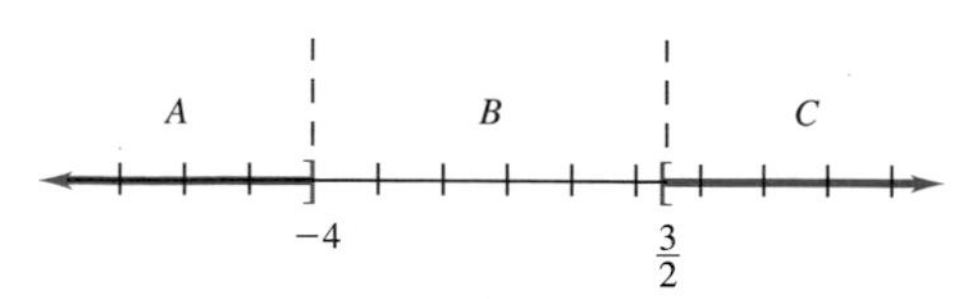

Figure 14

For Graphers

Try the graphical method to solve the inequality in Example 5. The resulting graph should look like the one in Figure 15.

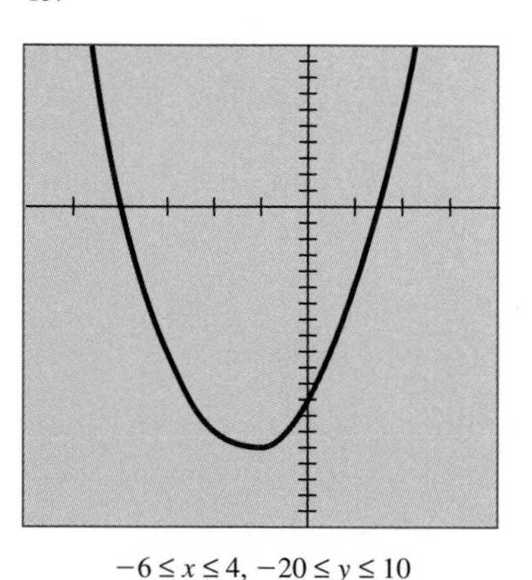

$-6 \leq x \leq 4, -20 \leq y \leq 10$

Figure 15

*The union of sets A and B, written $A \cup B$, is defined as $A \cup B = \{x | x \text{ is an element of } A \text{ or } x \text{ is an element of } B\}$.

Rational Inequalities The inequalities discussed in the remainder of this section involve quotients of algebraic expressions, and for this reason they are called **rational inequalities**. These inequalities can be solved in much the same way that quadratic inequalities are solved.

EXAMPLE 6

Solve the inequality $\dfrac{5}{x + 4} \geq 1$.

It is tempting to begin the solution by multiplying both sides of the inequality by $x + 4$, but to do this it would be necessary to consider whether $x + 4$ is positive or negative. Instead, subtract 1 from both sides of the inequality, getting

$$\frac{5}{x + 4} - 1 \geq 0.$$

Writing the left side as a single fraction gives

$$\frac{5 - (x + 4)}{x + 4} \geq 0 \quad \text{or} \quad \frac{1 - x}{x + 4} \geq 0.$$

Since the sign (positive or negative) depends on the signs of the numerator and denominator, the quotient (like the product) will change sign only when the denominator is 0 or when the numerator is 0. This happens when

$$\begin{aligned} 1 - x &= 0 \quad \text{or} \quad & x + 4 &= 0 \\ x &= 1 \quad \text{or} \quad & x &= -4. \end{aligned}$$

Now test a point in each of the intervals $(-\infty, -4)$, $(-4, 1)$, and $(1, \infty)$. Values in the interval $(-4, 1)$ give a positive quotient and are part of the solution. With a quotient, the endpoints must be considered separately to make sure that no denominator is 0. With this inequality, -4 gives a 0 denominator while 1 satisfies the given inequality. In interval notation, the solution is $(-4, 1]$, as shown in Figure 16.

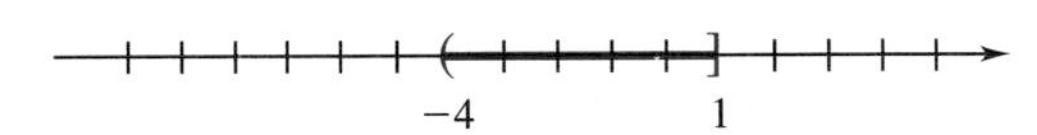

Figure 16

CAUTION As suggested by Example 6, we need to be very careful with endpoints of the intervals in the solution of rational inequalities. ■

For Graphers

There are two ways to use a grapher for Example 7: getting one side equal to 0, as we did for Example 1, or graphing the two expressions simultaneously, as we did for Example 3. To solve the inequality in Example 7, decide on which intervals the graph of the rational expression is less than the graph of $y = 5$. Using the graphical method with this inequality may involve some difficulty. First, it is necessary to use a small enough set of x-values to clearly see where the graph crosses the x-axis or the two graphs intersect. Second, since the intervals are determined by fractions, not integers, it is difficult to decide exactly where the boundaries of the intervals are. The trace key is used to locate these x-values. (See your manual for more information.) To determine the interval $(-4/3, \infty)$, use the vertical line that the graph approaches. If your calculator does not show such a line, use the trace key to approximate $-4/3$.

EXAMPLE 7 Solve $\dfrac{2x - 1}{3x + 4} < 5$.

Begin by subtracting 5 on both sides and combining the terms on the left into a single fraction.

$$\frac{2x - 1}{3x + 4} < 5$$

$$\frac{2x - 1}{3x + 4} - 5 < 0 \qquad \text{Subtract 5 on each side.}$$

$$\frac{2x - 1 - 5(3x + 4)}{3x + 4} < 0 \qquad \text{Write as a single fraction.}$$

$$\frac{-13x - 21}{3x + 4} < 0$$

Solve the equations

$$-13x - 21 = 0 \quad \text{and} \quad 3x + 4 = 0,$$

getting the solutions

$$x = -\frac{21}{13} \quad \text{and} \quad x = -\frac{4}{3}.$$

Use the values $-21/13$ and $-4/3$ to divide the number line into three intervals, and test a point in each region to see that values of x in the two intervals $(-\infty, -21/13)$ and $(-4/3, \infty)$ make the quotient negative, as required. Neither endpoint satisfies the given inequality, so the solution set is written $(-\infty, -21/13) \cup (-4/3, \infty)$. See Figure 17.

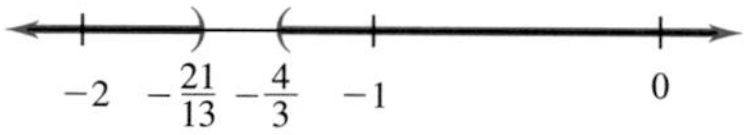

Figure 17

2.6 Exercises

Write each of the following in interval notation. Graph each interval.

1. $-3 < x < 2$ **2.** $-9 > x$ **3.** $x \leq -4$ **4.** $-4 \geq x \geq -5$

Using the variable x, write each of the following intervals as an inequality.

5. $[-2, 5)$ **6.** $(-\infty, 3]$

7.
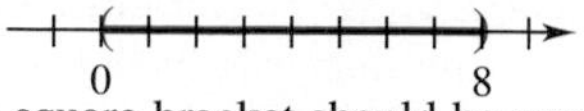

8.
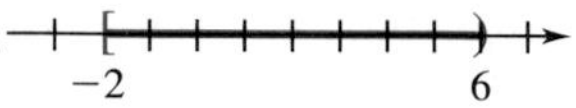

9. Explain how to determine whether a parenthesis or a square bracket should be used when graphing the solution of a linear inequality.

10. The three-part inequality $a < x < b$ means "a is less than x and x is less than b." Which one of the following inequalities is not satisfied by some real number x?
(a) $-3 < x < 5$ **(b)** $0 < x < 4$ **(c)** $-3 < x < -2$ **(d)** $-7 < x < -10$

Solve the following inequalities. Write the solutions in interval notation.

11. $4 - 3x \leq 10$

12. $2y + 7 \geq 12$

13. $2(m + 5) - 3m + 1 \geq 5$

14. $6m - (2m + 3) \geq 4m - 5$

15. $\frac{2x + 4}{5} \leq -2x - 1$

16. $\frac{3z + 7}{-5} \leq 4 - z$

17. $-2 \leq y - 3 \leq 4$

18. $-3 \leq 2t \leq 6$

19. $-8 > 2r + 9 > -12$

20. $5 > 4a - 5 > -2$

21. $-5 < \frac{x + 3}{-2} < 4$

22. $-1 < \frac{3m - 2}{-3} < 5$

23. $4 \geq 3 - \frac{5}{4}m > -4$

24. $6 > 2 - \frac{3}{2}k > -3$

25. $x^2 - 6x + 9 < 9$

26. $m^2 + 16m + 64 > 64$

27. $r^2 + 4r > -3$

28. $z^2 + 6z < -8$

29. $4x^2 + 3x + 1 \leq 0$

30. $x^2 + 5x - 2 < 0$

31. $4m^3 + 7m^2 - 2m > 0$

32. $6p^3 - 11p^2 + 3p > 0$

33. $\frac{m + 2}{m - 4} \leq 0$

34. $\frac{r - 3}{r - 1} > 0$

35. $\frac{k + 6}{k + 3} > 1$

36. $\frac{a - 2}{a - 5} < -1$

37. $\frac{1}{3k - 5} < \frac{1}{3}$

38. $\frac{1}{2m - 7} < \frac{5}{4}$

39. $\frac{5}{x - 6} \geq \frac{2}{x - 6}$

40. $\frac{6}{5 - x} > \frac{3}{5 - x}$

41. $\frac{-2}{3r + 2} > \frac{1}{r}$

42. $\frac{6}{2 - 3h} \geq -\frac{2}{h}$

43. $\frac{5}{y - 1} < \frac{2}{y - 2}$

44. $\frac{8}{n + 3} < \frac{2}{n + 1}$

In each of the following inequalities, the intervals on the number line are $(-\infty, 2)$, $(2, 5)$, and $(5, \infty)$. Without actually solving the inequality, state whether the numbers 2 and 5 will be included in or excluded from the solution of the inequality.

45. $(x - 2)(x - 5) \geq 0$

46. $(x - 5)(x - 2) < 0$

47. $\frac{x - 2}{x - 5} < 0$

48. $\frac{x - 5}{x - 2} \geq 0$

Solve the following rational inequalities using methods similar to those used above.

49. $\frac{9x - 8}{4x^2 + 25} < 0$

50. $\frac{(5x - 3)^3}{(8x - 25)^2} \leq 0$

51. $\frac{(9x - 11)(2x + 7)}{(3x - 8)^3} > 0$

52. Which of the following inequalities have solution set $(-\infty, \infty)$? Explain.

(a) $(x + 3)^2 \geq 0$ **(b)** $(5x - 6)^2 \leq 0$ **(c)** $(6y + 4)^2 > 0$

(d) $(8p - 7)^2 < 0$ **(e)** $\frac{x^2 + 7}{2x^2 + 4} \geq 0$ **(f)** $\frac{2x^2 + 8}{x^2 + 9} < 0$

53. Which of the inequalities in Exercise 52 has no real solution? Explain.

Use the discriminant to find the values of k for which the following equations have real solutions.

54. $x^2 - kx + 8 = 0$

55. $x^2 + kx - 5 = 0$

56. $x^2 + kx + 2k = 0$

57. $kx^2 + 4x + k = 0$

In Exercises 58–61, find a quadratic inequality having the given solution.

58. $(-\infty, 2) \cup (5, \infty)$

59. $(2, 5)$

60. $[-4, 3]$

61. $(-\infty, -3] \cup [4, \infty)$

In Exercises 62–65, find a rational inequality having the given solution.

62. $(-\infty, -3] \cup (0, \infty)$ **63.** $[-1, 5)$ **64.** $(4, 9]$ **65.** $(-\infty, 4) \cup [9, \infty)$

66. A student attempted to solve the inequality

$$\frac{2x - 1}{x + 2} \leq 0$$

by multiplying both sides by $x + 2$ to get

$$2x - 1 \leq 0$$
$$x \leq \frac{1}{2}.$$

He wrote the solution as $(-\infty, 1/2)$. Is his solution correct? Explain.

67. A student solved the inequality $p^2 \leq 16$ by taking the square root of both sides to get $p \leq 4$. She wrote the solution as $(-\infty, 4]$. Is her solution correct? Explain.

A product will break even or produce a profit only if the revenue from selling the product at least equals the cost of producing it. Find all x-intervals in Exercises 68 and 69 for which the product will at least break even.

68. The cost to produce x underwater cameras is $C = 80x + 10{,}000$; the revenue is $R = 50x$.

69. The cost to produce x chocolate bars is $C = 125x + 5000$; the revenue is $R = 150x$.

70. An analyst has found that his company's profits, in hundreds of thousands of dollars, are given by $P = 3x^2 - 35x + 50$, where x is the amount (in hundreds of dollars) spent on advertising. For what values of x does the company make a profit?

71. The commodities market is very unstable; money can be made or lost quickly on investments in soybeans, wheat, and so on. Suppose that an investor kept track of her total profit, P, at time t, in months, after she began investing, and found that $P = 4t^2 - 29t + 30$. Find the time intervals in which she has made a profit. (*Hint:* $t > 0$ in this case.)

72. The manager of a large apartment complex has found that the profit is given by $P = -x^2 + 250x - 15{,}000$, where x is the number of apartments rented. For what values of x does the complex produce a profit?

73. A projectile is fired from ground level. After t sec its height above the ground is $220t - 16t^2$ ft. For what time period is the projectile at least 624 ft above the ground?

74. A physicist has found that the velocity (in feet per second) of a moving particle is given by $2t^2 - 5t - 12$, where t is time in seconds since he began his observations. (Here t can be positive or negative; think of t sec before his observations began.) Find the time intervals in which the velocity has been negative.

75. Oliver's video club charges an annual fee of \$30 and rents videos for \$2 per day. Stan's video club has no annual fee, but charges \$3 per day to rent videos. Let x be the number of days of rentals during the year.
(a) Express the cost of renting the videos from Oliver in terms of x.
(b) Express the cost of renting the videos from Stan in terms of x.
(c) Find all x-intervals for which renting from Stan is cheaper.

76. Two companies, A and B, offer you a sales position. Both jobs are essentially the same, but Company A pays a straight 7% commission on sales and Company B pays \$100 per week plus 5% commission. Let x be the weekly sales.
(a) Express a week's earnings from Company A in terms of x.
(b) Express a week's earnings from Company B in terms of x.
(c) Find all x-intervals for which Company A pays a better salary.

77. The formula for converting from Celsius to Fahrenheit temperature is $F = 9C/5 + 32$. What temperature range in degrees Fahrenheit corresponds to 0° to 30°C?

Graphing Utility Problems

Use a grapher to solve the following inequalities.

78. $x^3 - 2x + 3 > 0$ **79.** $2x^4 - 5x^3 - 5x - 1 < 0$ **80.** $\dfrac{x}{x^2 - 9} < 0$

2.7 Absolute Value Equations and Inequalities

In this section we discuss methods of solving equations and inequalities involving absolute value. Recall from Chapter 1 that the absolute value of a number a, written $|a|$, gives the distance between a and 0 on a number line. By this definition, the absolute value equation $|x| = 3$ can be solved by finding all real numbers at a distance of 3 units from 0. As shown in the graph of Figure 18, there are two numbers satisfying this condition, 3 and -3, making $\{3, -3\}$ the solution set of the equation $|x| = 3$. This idea leads to the following properties, which are useful for solving absolute value problems.

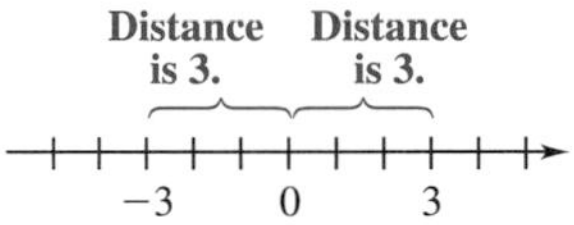

Distance is 3. Distance is 3.

-3 0 3

Figure 18

Properties of Absolute Value

If b is positive, then

$$|a| = b \quad \textbf{if and only if} \quad a = b \text{ or } a = -b.$$

For any values of a and b,

$$|a| = |b| \quad \textbf{if and only if} \quad a = b \text{ or } a = -b.$$

For Graphers

The best method for solving absolute value equations with a grapher is to graph each side of the equation simultaneously. For instance, in Example 1, graph $y_1 = |x - 4|$ and $y_2 = 3$. Then locate the x-values of the points where these two graphs intersect. See Figure 19.

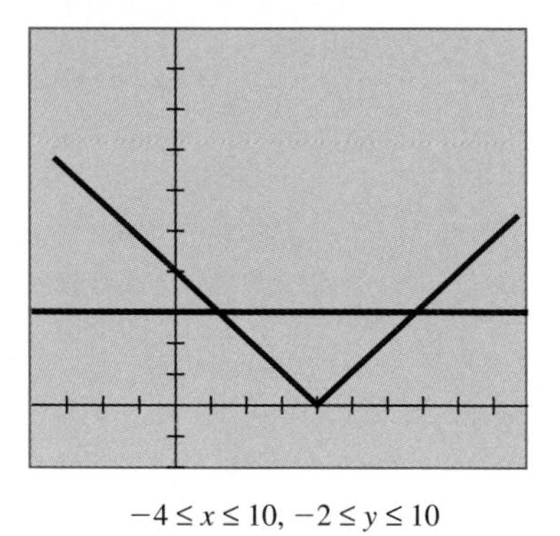

$-4 \le x \le 10,\ -2 \le y \le 10$

Figure 19

EXAMPLE 1 Solve $|p - 4| = 3$.

Let $a = p - 4$ and $b = 3$ in the first property above. (Note that $b = 3$ is positive, as required.)

$$p - 4 = \mathbf{3} \quad \text{or} \quad p - 4 = \mathbf{-3}$$
$$p = 7 \quad \text{or} \quad p = 1$$

Check that 7 and 1 are solutions of $|p - 4| = 3$. ■

EXAMPLE 2 Solve $|4m - 3| = |m + 6|$.

Let $a = 4m - 3$ and $b = m + 6$ in the second property above.

$$4m - 3 = \boldsymbol{m + 6} \quad \text{or} \quad 4m - 3 = \boldsymbol{-(m + 6)}$$

Solve each of these equations separately. Starting with $4m - 3 = m + 6$ gives

$$4m - 3 = m + 6$$
$$3m = 9$$
$$m = 3.$$

If $4m - 3 = -(m + 6)$, then

$$4m - 3 = -m - 6$$
$$5m = -3$$
$$m = -\frac{3}{5}.$$

The solutions of $|4m - 3| = |m + 6|$ are $-3/5$ and 3. ■

Absolute Value Inequalities The method used to solve absolute value equations can be generalized to solve inequalities involving absolute value.

EXAMPLE 3 Solve each of the following.

(a) $|x| < 5$

Since absolute value gives the distance from a number to 0, the inequality $|x| < 5$ will be satisfied by all real numbers whose distance from 0 is less than 5. As shown in Figure 20, the solution includes all numbers from -5 to 5, or $-5 < x < 5$. In interval notation, the solution is written as the open interval $(-5, 5)$.

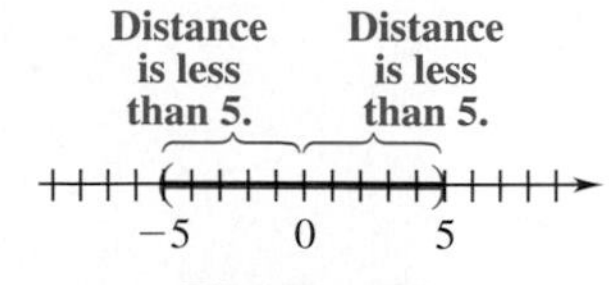

Figure 20

(b) $|x| > 5$

In a similar way, the solution of $|x| > 5$ is made up of all real numbers whose distance from 0 is greater than 5. This includes those numbers greater than 5 or those less than -5:

$$x < -5 \quad \text{or} \quad x > 5.$$

In interval notation, the solution is written $(-\infty, -5) \cup (5, \infty)$. A graph of the solution set is shown in Figure 21.

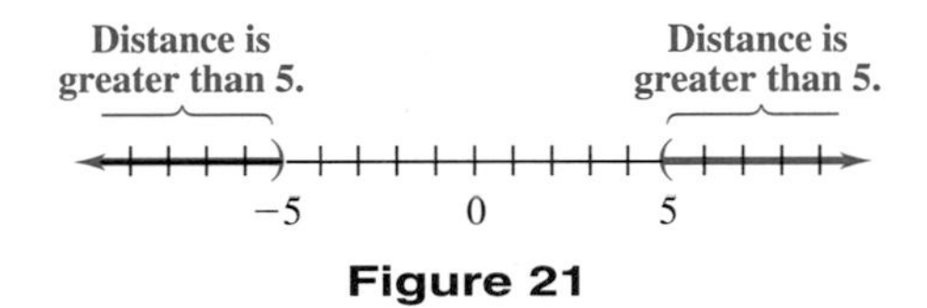

Figure 21

The following properties of absolute value can be obtained from the definitions of absolute value and inequalities.

Additional Properties of Absolute Value

If b is a positive number,

(a) $|a| < b$ **if and only if** $-b < a < b$;

(b) $|a| > b$ **if and only if** $a < -b$ **or** $a > b$.

NOTE The $<$ symbol in part (a) leads to a single interval, $-b < a < b$, while the $>$ symbol in part (b) indicates two separate intervals. ■

EXAMPLE 4 Solve $|x - 2| < 5$.

To solve this inequality on the number line, find all real numbers whose distance from 2 is less than 5. As shown in Figure 22, the solution set is the interval $(-3, 7)$.

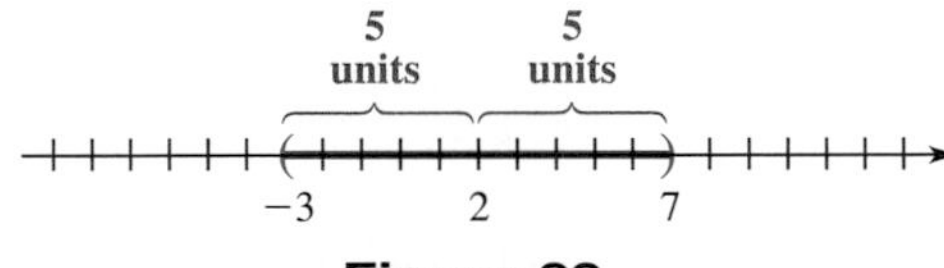

Figure 22

An algebraic solution of this inequality can be found using Property (a) above. Let $a = x - 2$ and $b = 5$, so that $|x - 2| < 5$ if and only if

$$-5 < x - 2 < 5.$$

For Graphers

To solve the inequality in Example 4 with a grapher, graph the expressions in each part of the inequality and locate the x-values of the intersection points. Then decide on which intervals the graph of $y = |x - 2|$ is below (and therefore less than) the graph of $y = 5$. You could also solve this inequality by graphing $y = |x - 2| - 5$ and locating the intersection points with the x-axis. Remember to use parentheses around the quantity in absolute value bars.

Adding 2 to each portion of this inequality produces

$$-3 < x < 7,$$

again giving the interval solution $(-3, 7)$. ■

EXAMPLE 5 Solve $|x - 8| \geq 1$.

We need to find all numbers whose distance from 8 is greater than or equal to 1. As shown in Figure 23 the solution set is $(-\infty, 7] \cup [9, \infty)$. To find the solution using Property (b) above, let $a = x - 8$ and $b = 1$ so that $|x - 8| \geq 1$ if and only if

$$x - 8 \leq -1 \quad \text{or} \quad x - 8 \geq 1.$$

Solve each inequality separately to get the same solution set, $(-\infty, 7] \cup [9, \infty)$, as mentioned above.

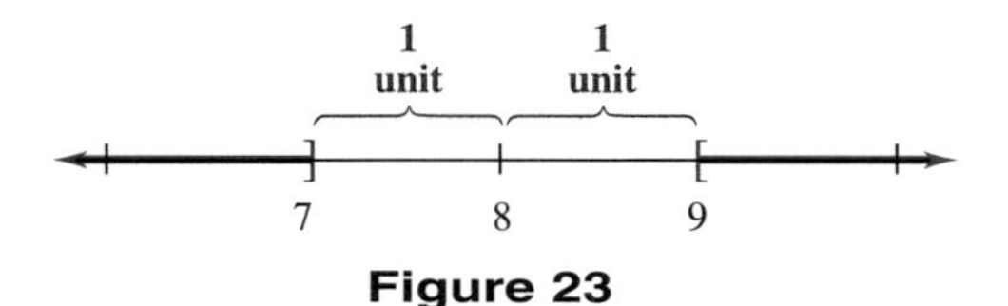

Figure 23

■

EXAMPLE 6 Solve $|2 - 7m| - 1 > 4$.

In order to use the properties of absolute value given above, first add 1 to both sides; this gives

$$|2 - 7m| > 5.$$

Now use Property (b). By this property, $|2 - 7m| > 5$ if and only if

$$2 - 7m < -5 \quad \text{or} \quad 2 - 7m > 5.$$

Solve each of these inequalities separately to get the solution set $(-\infty, -3/7) \cup (1, \infty)$. ■

EXAMPLE 7 Solve $|2 - 5x| \geq -4$.

As stated in Property (b), b must be a positive number, so property (b) does not apply here. However, since the absolute value of a number is always nonnegative, $|2 - 5x| \geq -4$ is always true. The solution set includes all real numbers. In interval notation, the solution set is $(-\infty, \infty)$. ■

EXAMPLE 8 Solve $\left|\frac{y+1}{2-y}\right| < 3.$

Use Property (a) to eliminate the absolute value signs.

$$-3 < \frac{y+1}{2-y} < 3$$

Because of the variable in the denominator, the inequality must be written as two inequalities connected by *and*, each of which is solved separately.

$$-3 < \frac{y+1}{2-y} \quad \textbf{and} \quad \frac{y+1}{2-y} < 3$$

Use the method shown in Section 2.6 to solve these inequalities. Get 0 on one side, then simplify the other side.

$$0 < \frac{y+1}{2-y} + 3 \quad \text{and} \quad \frac{y+1}{2-y} - 3 < 0$$

$$0 < \frac{y+1+6-3y}{2-y} \quad \text{and} \quad \frac{y+1-6+3y}{2-y} < 0$$

$$0 < \frac{7-2y}{2-y} \quad \text{and} \quad \frac{4y-5}{2-y} < 0$$

Setting the numerator and denominator in each inequality equal to 0 and solving the four equations gives the three values $7/2$, 2, and $5/4$. These values determine the four regions to be tested: $(-\infty, 5/4)$, $(5/4, 2)$, $(2, 7/2)$, and $(7/2, \infty)$. Testing a number from each region in the *original* inequality shows that the solution is $(-\infty, 5/4) \cup (7/2, \infty)$. ■

EXAMPLE 9 Write each statement using absolute value.

(a) k is not less than 5 units from 8.

Since the distance from k to 8, written $|k - 8|$ or $|8 - k|$, is not less than 5, the distance is greater than or equal to 5. Write this as

$$|k - 8| \geq 5.$$

(b) n is within .001 of 6.

This statement indicates that n may be .001 more than 6 or .001 less than 6. That is, the distance of n from 6 is no more than .001, written

$$|n - 6| \leq .001.$$ ■

2.7 Exercises

Solve each of the following equations.

1. $|x + 3| = 5$
2. $|x - 8| = 1$
3. $|2m + 6| = 4$
4. $|3p - 5| = 7$
5. $|6 - 2x| + 4 = -3$
6. $|-5a + 1| - 2 = -4$
7. $\left|\frac{z - 5}{3}\right| = 12$
8. $\left|\frac{m + 1}{5}\right| = 9$
9. $\left|\frac{4}{r - 2}\right| = 7$
10. $\left|\frac{1}{2h + 3}\right| = 5$
11. $\left|\frac{4y + 3}{y + 2}\right| = 5$
12. $\left|\frac{3a - 4}{2a + 3}\right| = 1$
13. $|2k - 3| = |5k + 4|$
14. $|p + 1| = |3p - 1|$
15. $|4 - 3y| = |7 + 2y|$
16. $|2 + 5a| = |4 - 6a|$

Solve each of the following inequalities. Give the solution in interval notation.

17. $|m| > 1$
18. $|z| > 4$
19. $|a| < -2$
20. $|b| > -5$
21. $|x| - 3 \le 7$
22. $|r| + 3 \le 10$
23. $|4x - 1| < 9$
24. $|4 - 3x| < 3$
25. $|2m - 5| > 10$
26. $|6x - 3| > 9$
27. $|4z + 3| \ge 7$
28. $|6b + 1| \ge 19$
29. $\left|2x + \frac{1}{4}\right| - 1 < 7$
30. $\left|3x + \frac{2}{3}\right| + 2 < 4$
31. $\left|\frac{3x - 2}{x}\right| < 2$
32. $\left|\frac{6 - 2y}{y}\right| < 4$
33. $\left|\frac{4 + y}{y - 2}\right| > 3$
34. $\left|\frac{5 - 5p}{p + 2}\right| > 0$
35. $\left|\frac{4}{q - 1}\right| \le 0$
36. $\left|\frac{3}{t - 4}\right| \le 0$
37. Explain why it is incorrect to write the absolute value inequality $|x| > 6$ in any of the following ways: $-6 > x > 6$, $-6 > x < 6$, or $-6 < x > 6$.
38. Without actually going through the solution process, we can say that the equation $|5x - 6| = 6x$ cannot have a negative solution. Explain why this is true.

Solve the equations in Exercises 39–46.

39. $|x + 1| = 2x$
40. $|m - 3| = 4m$
41. $|2k - 1| = k + 2$
42. $|3r + 5| = r + 3$
43. $|p| = |p|^2$
44. $5|m| = |m|^2$
45. $|1 - 6q|^2 - 4|1 - 6q| - 45 = 0$
46. $|6a - 5|^2 + 4|6a - 5| - 12 = 0$
47. Is $|a - b|^2$ always equal to $(b - a)^2$?
48. The temperatures on the surface of Mars in degrees Celsius approximately satisfy the inequality $|C + 84| \le 56$. What range of temperatures corresponds to this inequality?
49. Dr. Tydings has found that, over the years, 95% of the babies that she has delivered have weighed y lb, where $|y - 8.0| \le 1.5$. What range of weights corresponds to this inequality?

50. The industrial process that is used to convert methanol to gasoline is carried out at a temperature range of 680° F to 780° F. Using F as the variable, write an absolute value inequality that corresponds to this range.

51. When a model kite was flown in crosswinds in tests to determine its limits of power extraction, it attained speeds of 98 to 148 ft per sec in winds of 16 to 26 ft per sec. Using x as the variable in each case, write absolute value inequalities that correspond to these ranges.

Solve each inequality in Exercises 52–55.

52. $|m^2 + 1| < 3$

53. $|z^2 - 5| \geq 4$

54. $|m + 6| \leq |2m - 3|$
(*Hint:* First divide each side by $|2m - 3|$.)

55. $|1 - 3x| < 2|x + 5|$

Write each statement in Exercises 56–62 using absolute value notation. For example, write "k is at least 4 units from 1" as $|k - 1| \geq 4$.

56. x is within 4 units of 2.

57. z is no less than 2 units from 12.

58. p is at least 5 units from 9.

59. k is 6 units from 1.

60. r is 5 units from 3.

61. If x is within .0004 units of 2, then y is within .00001 units of 7.

62. y is within 10^{-6} units of 10 whenever x is within 2×10^{-4} units of 5.

63. If $|x - 2| < 3$, find the values of m and n such that $m < 3x + 5 < n$.

64. If $|x + 8| < 16$, find the values of p and q such that $p < 2x - 1 < q$.

65. If $|x - 1| < 10^{-6}$, show that $|7x - 7| < 10^{-5}$.

Graphing Utility Problems

Use a grapher to solve the following inequalities.

66. $|1 - x| > x + 1$

67. $|2x + 3| < x - 1$

68. $|3x + 5| < |x - 1|$

Chapter 2 Review Exercises

Solve the following equations and check your answers.

1. $2m + 7 = 3m + 1$

2. $4k - 2(k - 1) = 12$

3. $\frac{y}{3} - \frac{2y}{5} = 6 + \frac{y}{2}$

4. $\frac{x + 1}{2} = \frac{2x - 5}{3}$

5. $\frac{10}{4z - 4} = \frac{1}{1 - z}$

6. $(m + 2)(3m - 1) = 3m^2 + 5m$

Solve each of the following for the indicated variable.

7. $2x - 5k = 2(kx + 3)$ for x

8. $F = \frac{9}{5}C + 32$ for C

9. $A = P + Pi$ for P

10. $A = I\left(1 - \frac{j}{n}\right)$ for j

11. $\frac{1}{k} = \frac{1}{r_1} + \frac{1}{r_2}$ for r_1

12. $P(r + R)^2 = PE^2R$ for E

13. $\frac{xy^2 - 5xy + 4}{3x} = 2p$ for x

Solve each of the following problems.

14. A computer printer is on sale for 15% off. The sale price is \$425. What was the original price?

15. A realtor borrowed \$90,000 to develop some property. He was able to borrow part of the money at 10.5% interest and the rest at 9%. The annual interest on the two loans amounts to \$8925. How much was borrowed at each rate?

16. An excursion boat travels upriver to a landing and then returns to its starting point. The trip upriver takes 1.2 hr and the trip back takes .9 hr. If the average speed on the return trip is 5 mph faster than on the trip upriver, what is the boat's speed upriver?

17. Wei-jen and Alan Luan are canvassing their neighborhood for their candidate for the school board. Alan can canvass the entire neighborhood alone in 6 hr. Working together, they complete the job in 4 hr. How long would it take Wei-jen to canvass the neighborhood, working alone?

Solve each equation.

18. $(b + 7)^2 = 5$

19. $(3y - 2)^2 = 8$

20. $2a^2 + a - 15 = 0$

21. $12x^2 = 8x - 1$

22. $3x^2 + 2x = -5$

23. $\sqrt{2}x^2 - 4x + \sqrt{2} = 0$

24. $2 - \frac{4}{y} = \frac{21}{y^2}$

25. $2 + \frac{4}{a} + \frac{3}{a^2} = 0$

26. Discuss the method you chose (or would choose) to solve Exercises 19, 21, and 23 and explain why you made that choice.

Evaluate the discriminant for each of the following, and use it to predict the type of solutions for the equation.

27. $4p^2 + 8 = 3p$

28. $5x^2 - 2x = 7$

29. $2k^2 + k - 1 = 0$

30. $7m^2 - 2m = 5$

31. $5x^2 = 8x + 1$

32. $4x^2 + 4x + 5 = 0$

Solve the following problems.

33. Tony Romero has a rectangular-shaped flower box that measures 4 ft by 6 ft. He wants to double the available area by increasing the length and width by the same amount. What should the new dimensions be?

34. Paula Cunningham plans to replace the vinyl floor covering in her 10-by-12-ft kitchen. She wants to have a border of a special material with the same width on each side. She can afford only 21 sq ft of this material. How wide a border can she have?

35. It takes two gardeners 3 hr (working together) to mow the lawns in a city park. One gardener could do the entire job in 1 hr less time than the other. How long would it take the slower gardener to complete the work alone? Give the answer to the nearest tenth of an hour.

36. In a marathon (a 26-mi run), the winner finished 2/5 hr before a friend. If the difference in their rates was 4/3 mph, what was the winner's average speed in the race?

37. Suppose that one solution of the equation $km^2 - 11m = 3$ is 3. Find the value of k and the other solution.

Solve each equation. Check your answers.

38. $3y^4 + 2y^2 = 16$

39. $2(z - 1)^2 - 11(z - 1) = 21$

40. $(2n + 3)^{2/3} + (2n + 3)^{1/3} = 6$

41. $\sqrt{2m - 1} = 3\sqrt{m}$

42. $\sqrt{4y + 3} = \sqrt{2y - 5}$

43. $\sqrt{3x + 4} = 2x + 1$

44. $x - 2 = \sqrt{3x - 6}$

45. $\sqrt{k} = \sqrt{k + 3} - 1$

46. $\sqrt{x^2 + 3x} - 2 = 0$

47. $\sqrt[3]{2y - 3} = \sqrt[3]{y + 1}$

48. $\sqrt[3]{5z} = \sqrt[3]{8z + 4}$

49. $\sqrt{3 + x} = \sqrt{3x + 7} - 2$

50. $\sqrt{4 + 3y} = \sqrt{y + 5} + 1$

51. $(x - 2)^{2/3} = x^{1/3}$

Write each of the statements below as an equation.

52. Y varies jointly as M and the square of N and inversely as the cube of X.

53. A varies jointly as the third power of t and the fourth power of s, and inversely as p and the square of h.

Solve each problem below.

54. The power a windmill obtains from the wind varies directly as the cube of the wind velocity. If a wind blowing at 10 km per hr produces 10,000 units of power, how much power is produced by a wind of 15 km per hr?

55. Hooke's law for an elastic spring states that the distance a spring stretches varies directly as the force applied. If a force of 32 lb stretches a certain spring 48 inches, how much will a force of 24 lb stretch the spring?

56. The weight of an object varies inversely as the square of the distance between the object and the center of the earth. If a man weighs 90 kg on the surface of the earth, how much would he weigh 800 km above the surface? (The radius of the earth is about 6400 km.)

57. The force of the wind on a sail varies jointly as the area of the sail and the square of the wind velocity. If the force is 8 lb when the velocity is 15 mph and the area is 3 sq ft, find the force when the area is 6 sq ft and the velocity is 22.5 mph.

58. Without actually solving the inequality, explain why 3 cannot be in the solution set of

$$\frac{2x + 5}{x - 3} < 0.$$

59. What is wrong with the following "solution" of $\dfrac{1}{x - 3} \geq 2$?

$$(x - 3)\left(\frac{1}{x - 3}\right) \geq (x - 3)(2) \qquad \text{Multiply by } x - 3.$$

$$1 \geq 2x - 6 \qquad \text{Distributive property}$$

$$7 \geq 2x$$

$$\frac{7}{2} \geq x$$

The solution is $(-\infty, 7/2)$.

Solve each inequality. Write solutions in interval notation.

60. $-a + 3 \geq 2(5a + 1)$
61. $-(2k + 4) < k - 5$
62. $2r - 5 > -r + 3(r + 2)$
63. $4m - 3(2m + 2) \leq 6(1 - m)$
64. $5 \leq 2x - 3 \leq 7$
65. $-8 < 3a - 5 < -1$
66. $-5 < \frac{2p - 1}{-3} \leq 2$
67. $3 < \frac{6z + 5}{-2} < 7$
68. $x^2 + 3x - 4 \leq 0$
69. $p^2 + 4p > 21$
70. $z^3 - 16z \leq 0$
71. $2r^3 - 3r^2 - 5r < 0$
72. $\frac{3a - 2}{a} > 4$
73. $\frac{5p + 2}{p} < -1$
74. $\frac{3}{r - 1} \leq \frac{5}{r + 3}$
75. $\frac{3}{x + 2} > \frac{2}{x - 4}$

76. If $a < b$, on what x-interval is $(x - a)(x - b)$ positive? Negative? Where is the product zero?

77. On what x-interval is $(x - a)^2$ positive? Negative? Where is it zero?

Work the following problems.

78. A projectile is thrown upward. Its height in feet above the ground after t sec is $320t - 16t^2$.
(a) After how many seconds in the air will it hit the ground?
(b) During what time interval is the projectile more than 576 ft above the ground?

79. A company that produces videotapes has found that the revenue from the sale of x units of tapes is $R = 8x$. The cost to produce x units of tapes is $C = 3x + 1500$. In what interval will the company at least break even?

Solve each equation.

80. $|a + 4| = 7$
81. $|-y + 2| = -4$
82. $\left|\frac{7}{2 - 3a}\right| = 9$
83. $\left|\frac{8r - 1}{3r + 2}\right| = 7$
84. $|5r - 1| = |2r + 3|$
85. $|k + 7| = |k - 8|$

Solve each inequality. Write solutions with interval notation.

86. $|m - 3| \leq 7$
87. $|3 - r| < 2$
88. $|z - 2| \geq -1$
89. $|2x + 9| \leq 3$
90. $|5m - 8| \leq 2$
91. $|7k - 3| > 5$
92. $|2p - 1| > 2$
93. $|3r + 7| - 5 > 0$

94. Solve the following equations and inequalities. Compare the solutions.
(a) $\frac{x - 1}{2x + 5} = 4$
(b) $\frac{x - 1}{2x + 5} < 4$
(c) $\frac{x - 1}{2x + 5} > 4$
(d) $\left|\frac{x - 1}{2x + 5}\right| = 4$
(e) $\left|\frac{x - 1}{2x + 5}\right| < 4$
(f) $\left|\frac{x - 1}{2x + 5}\right| > 4$

95. Homing pigeons avoid flying over large bodies of water, preferring to fly around them instead. (One possible explanation is the fact that extra energy is required to fly over water because air pressure drops over water in the daytime.) Assume that a pigeon released from a boat 1 mi from the shore of a lake (point B in the figure) flies first to point P on the shore and then along the straight edge of the lake to reach its home at L. If L is 2 mi from point A, the point on the shore closest to the boat, and if a pigeon needs $4/3$ as much energy to fly over water as over land, find an expression for the total energy expended, assuming 1 unit of energy per mi over land.

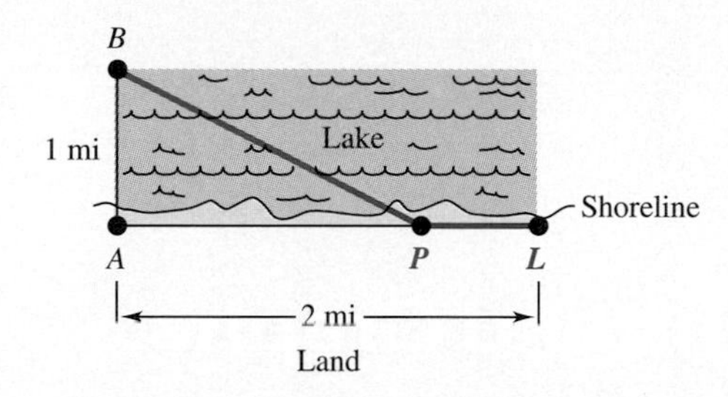

96. A hiker is at a point on a riverbank. He wants to get to his cabin, located 3 mi north and 8 mi west (see the figure). He can travel 5 mph along the river but only 2 mph on the very rocky ground away from the river. If he travels $8 - x$ mi along the river and then walks in a straight line to the cabin, find an expression for the total time that he travels.

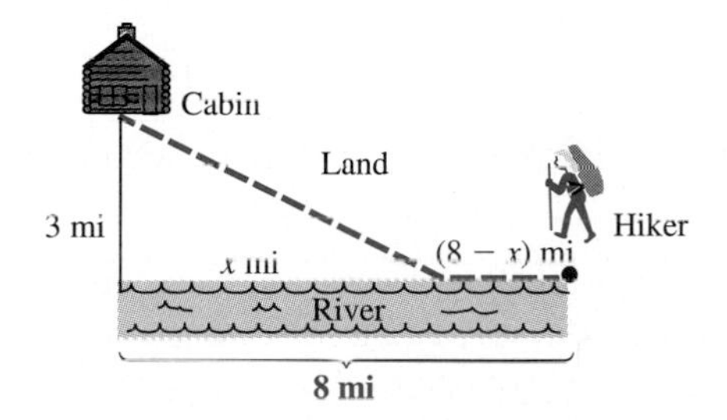

*In Exercises 97 and 98, solve for x and express the solution in terms of intervals.**

97. $\dfrac{1}{|x - 4|} < \dfrac{1}{|x + 7|}$

98. $\dfrac{1}{|x - 3|} - \dfrac{1}{|x + 4|} \geq 0$

99. Find the smallest value of M such that $|1/x| \leq M$ for all x in the interval $[2, 7]$.

100. Find the smallest value of M such that $|1/(x + 7)| \leq M$ for all x in the interval $(-4, 2)$.

*Exercises 97–100 from *Calculus with Analytic Geometry*, Fourth Edition, by Howard Anton, pp. 26, 27.

3 Functions and Graphs

The equations and inequalities in the previous chapter involved only *one* variable. However, it is very common for a practical application to use *two* variables, with the value of one variable dependent on the value of the other. As an example, Figure 1(a)* shows how the speed in miles per hour of a Porsche 928 depends on the time t in seconds after the car has started from a dead stop. Figure 1(b)† shows how a person's blood pressure depends on time. (Systolic and diastolic pressures are the upper and lower limits in the periodic changes in pressure that produce the pulse. The length of time between peaks is called the *period* of the pulse.)

Both Figures 1(a) and 1(b) show a *function*, a rule or procedure giving just one value of a variable from a given value of another variable. In each of these examples, a given value of time can be used to find just one value of the other variable, speed or blood pressure, respectively. The study of functions is a major theme of this text. Before getting to a more complete discussion of functions later in the chapter, we need to look at the topic of *graphing* with two variables, since functions are often studied by looking at their *graphs*.

*Figure 1(a): From "Road Tests," *Road & Track*, April 1985. Copyright © 1985, CBS Consumer Publishing Division. Reprinted by permission.

†Figure 1(b): From *Calculus for the Life Sciences* by Rodolfo De Sapio. W. H. Freeman and Company. Copyright © 1978.

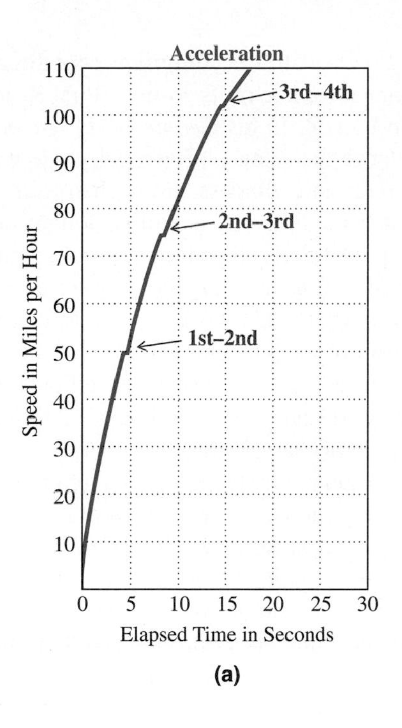

(a)

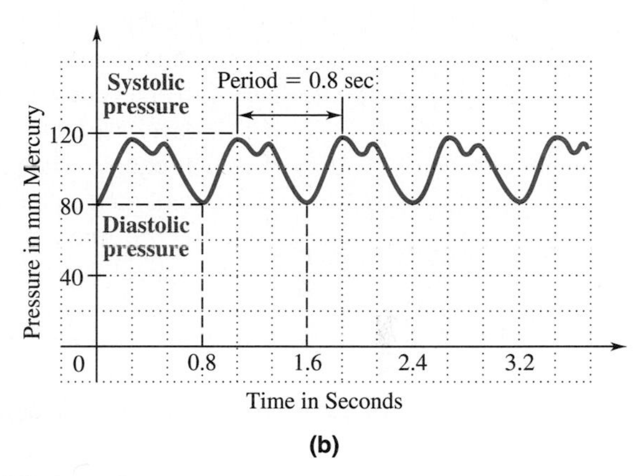

(b)

Figure 1

3.1 Cartesian Coordinate Systems

In Chapter 1 we saw how to set up a correspondence between real numbers and points on a number line. This idea can be extended to two dimensions: in two dimensions the correspondence is between *pairs* of numbers and points on a *plane*. One way to get such a correspondence is with perpendicular number lines that intersect at 0, with one horizontal, called the ***x*-axis**, and one vertical, called the ***y*-axis**.

Starting at the **origin** (the point at which the axes intersect), the x-axis is made into a number line by placing positive numbers to the right and negative numbers to the left. The y-axis is made into a number line with positive numbers going up and negative numbers down. The plane into which the coordinate system is introduced is the **coordinate plane**, or ***xy*-plane**. The x-axis and y-axis divide the plane into four regions, or **quadrants**, labeled as shown in Figure 2. The points on the x-axis and y-axis themselves belong to no quadrant.

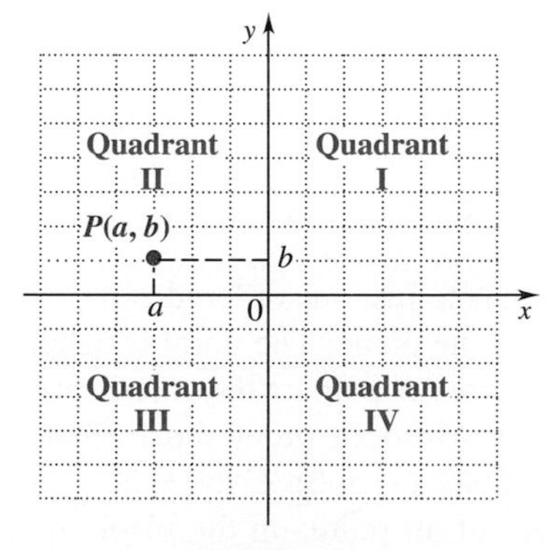

Figure 2

The x-axis and y-axis set up a **rectangular coordinate system**, or **Cartesian coordinate system**, named for one of its co-inventors, René Descartes; the other co-inventor was Pierre de Fermat. In his *Geometry* (*La géométrie*) of 1637, Descartes attempted to bring order to the science of geometry. He was constructing problems in the classic way by ruler and compass and wished to include curves that have an algebraic equation, but if he did so he would no longer be doing geometry. In the course of his investigations he developed the two-axis system and *ordered pairs* of numbers, which enabled him to have algebraic descriptions of geometrical curves. Other mathematicians translated his work and wrote commentaries, so that in less than a century the subject of analytic geometry with Cartesian coordinates was well developed. Descartes' (false) claim that one can tell everything about a curve from its equation helped motivate Isaac Newton (1642–1724) to develop results that eventually led to the invention of the calculus.

To find a pair of numbers corresponding to a given point P in Figure 2, draw a vertical line from P cutting the x-axis at a. Draw a horizontal line cutting the y-axis at b. Then point P has **coordinates** (a, b), where (a, b) is an **ordered pair** of numbers, two numbers written in parentheses in which the sequence of the numbers is important. For example, (3, 4) and (4, 3) in Figure 3 are not the same ordered pair, since the sequence of the numbers is different. When an ordered pair represents values of x and y, the value for x is written first. The x-value of the ordered pair is called the **abscissa**, and the y-value is called the **ordinate**.

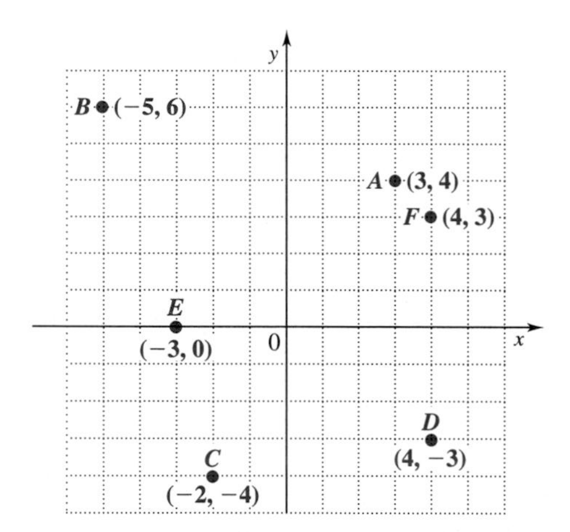

Figure 3

By this method, each point of the plane corresponds to just one ordered pair, and each ordered pair corresponds to exactly one point of the plane. The point is called the **graph** of the ordered pair, and the numbers in the ordered pair are the coordinates of the point. For example, point A in Figure 3 corresponds to the ordered pair (3, 4). Also in Figure 3, point B corresponds to the ordered pair $(-5, 6)$, C to $(-2, -4)$, D to $(4, -3)$, E to $(-3, 0)$, and F to (4, 3). The set of all points in the plane that correspond to a set of ordered pairs is the **graph** of the set of ordered pairs.

EXAMPLE 1 Graph the set of all points (x, y) satisfying the inequality $|x| \leq 2$.

As shown in the last chapter, $|x| \leq 2$ means

$$-2 \leq x \leq 2.$$

The graph of the set of all ordered pairs (x, y) whose x-coordinates satisfy the inequality $-2 \leq x \leq 2$ is bounded by vertical lines through $(2, 0)$ and $(-2, 0)$. The region satisfying $|x| \leq 2$ is shaded in Figure 4.

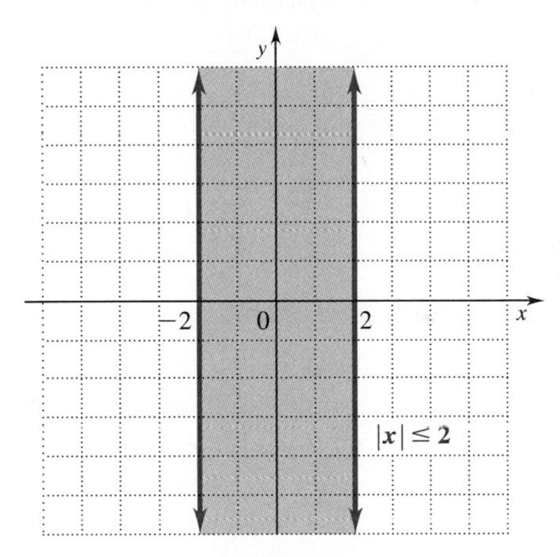

Figure 4

The distance formula, derived next, is quite useful for determining the equations of sets of points from a description of the set. We will use this approach in this section to derive the general equation of a circle.

Distance Formula We frequently need to find the distance between two given points in a plane. Recall the Pythagorean theorem: in a right triangle with legs of lengths a and b and hypotenuse of length c, $a^2 + b^2 = c^2$. This theorem may be used to obtain a formula for the distance between any two points in the plane. To get this distance formula, we start with two points on a horizontal line, as in Figure 5(a) on the next page. We use the symbol $P(x_1, y_1)$ to represent point P having coordinates (x_1, y_1) (read "x-sub-one, y-sub-one"). The distance between points $P(x_1, y_1)$ and $Q(x_2, y_1)$ can be found by subtracting the x-coordinates. (Absolute value is used to make sure that the distance is not negative—recall the work with distance in Chapter 1.) From this, the distance between points P and Q is $|x_1 - x_2|$. If $d(P, Q)$ represents the distance between P and Q, then

$$d(P, Q) = |x_1 - x_2|.$$

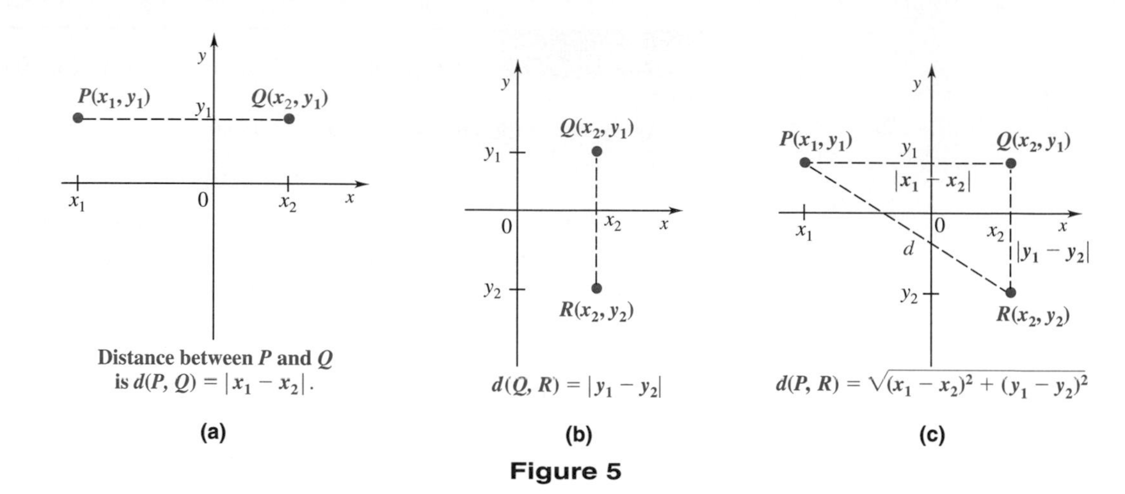

Figure 5

Figure 5(b) shows points $Q(x_2, y_1)$ and $R(x_2, y_2)$ on a vertical line. To find the distance between Q and R, subtract the y-coordinates.

$$d(Q, R) = |y_1 - y_2|$$

Finally, Figure 5(c) shows two points, $P(x_1, y_1)$ and $R(x_2, y_2)$, which are *not* on a horizontal or vertical line. To find $d(P, R)$, construct the right triangle shown in the figure. One side of this triangle is horizontal and has length $|x_1 - x_2|$. The other side is vertical and has length $|y_1 - y_2|$. By the Pythagorean theorem,

$$[d(P, R)]^2 = |x_1 - x_2|^2 + |y_1 - y_2|^2.$$

Since $|x_1 - x_2|^2$ equals the expression $(x_1 - x_2)^2$ and $|y_1 - y_2|^2$ equals $(y_1 - y_2)^2$, the **distance formula** can be written as follows.

Distance Formula

Suppose $P(x_1, y_1)$ and $R(x_2, y_2)$ are two points in a coordinate plane. Then the distance between P and R, written $d(P, R)$, is

$$d(P, R) = \sqrt{(x_1 - x_2)^2 + (y_1 - y_2)^2}.$$

EXAMPLE 2 Find the distance between $P(-8\sqrt{3}, 4)$ and $Q(\sqrt{3}, -2)$.

According to the distance formula,

$$\begin{aligned} d(P, Q) &= \sqrt{(-8\sqrt{3} - \sqrt{3})^2 + [4 - (-2)]^2} \\ &= \sqrt{(-9\sqrt{3})^2 + 6^2} \\ &= \sqrt{243 + 36} = \sqrt{279} = 3\sqrt{31}. \end{aligned}$$

A calculator gives 16.703 as an approximate value for $3\sqrt{31}$. ■

EXAMPLE 3 Are the points $M(-2, 5)$, $N(12, 3)$, and $Q(10, -11)$ the vertices of a right triangle?

To decide whether or not the triangle determined by these three points is a right triangle, use the converse of the Pythagorean theorem: if the sides a, b, and c of a triangle satisfy $a^2 + b^2 = c^2$, then the triangle is a right triangle. A triangle with the three given points as vertices is shown in Figure 6. This triangle is a right triangle if the square of the length of the hypotenuse equals the sum of the squares of the lengths of the legs. Use the distance formula to find the length of each side of the triangle.

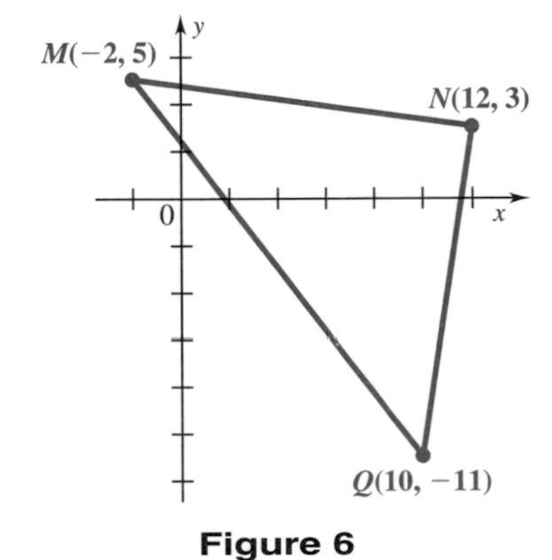

Figure 6

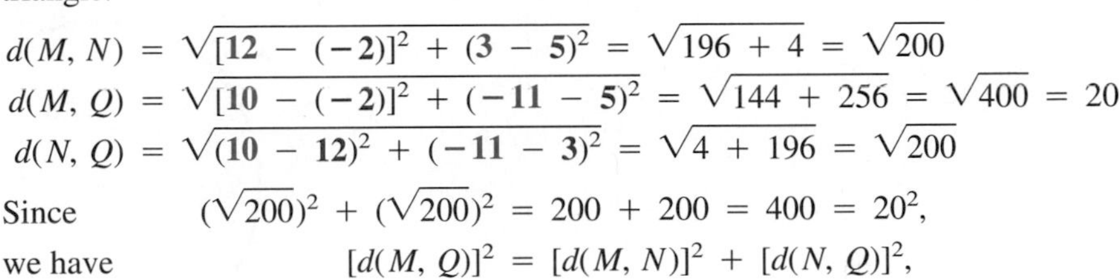

$$d(M, N) = \sqrt{[12 - (-2)]^2 + (3 - 5)^2} = \sqrt{196 + 4} = \sqrt{200}$$
$$d(M, Q) = \sqrt{[10 - (-2)]^2 + (-11 - 5)^2} = \sqrt{144 + 256} = \sqrt{400} = 20$$
$$d(N, Q) = \sqrt{(10 - 12)^2 + (-11 - 3)^2} = \sqrt{4 + 196} = \sqrt{200}$$

Since $$(\sqrt{200})^2 + (\sqrt{200})^2 = 200 + 200 = 400 = 20^2,$$

we have $$[d(M, Q)]^2 = [d(M, N)]^2 + [d(N, Q)]^2,$$

proving that the triangle is a right triangle with hypotenuse connecting M and Q. ■

Circles A **circle** is the set of all points in a plane that lie a given distance from a given point. The given distance is the **radius** of the circle and the given point is the **center**. The equation of a circle can be found by using the distance formula.

For example, Figure 7 shows a circle of radius 3 with center at the origin. To find the equation of this circle, let (x, y) be any point on the circle. The distance between (x, y) and the center of the circle, $(0, 0)$, is given by

$$\sqrt{(x - 0)^2 + (y - 0)^2}.$$

For Graphers

In order to use a grapher to draw the graph of this circle, you must first solve the equation $x^2 + y^2 = 9$, for y, then draw the graphs of the two resulting equations, $y = \sqrt{9 - x^2}$ and $y = -\sqrt{9 - x^2}$, on the same coordinate system. You may find that the graph appears elliptical rather than circular. Some graphers have a command to correct this.

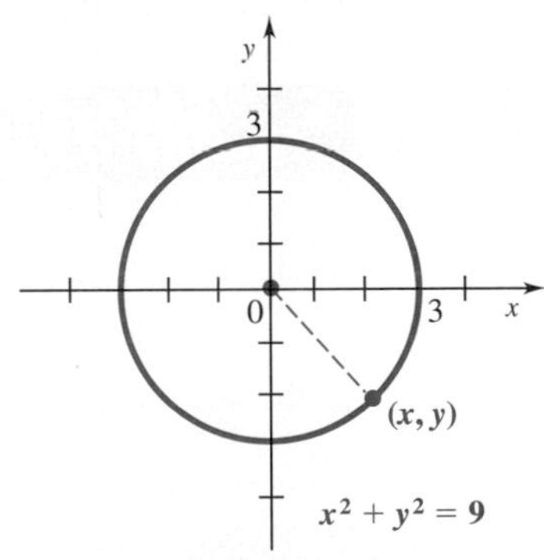

Figure 7

Since this distance equals the radius, 3,

$$\sqrt{(x - 0)^2 + (y - 0)^2} = 3$$
$$\sqrt{x^2 + y^2} = 3$$
$$x^2 + y^2 = 9.$$

EXAMPLE 4 Find an equation for the circle having radius 6 and center at $(-3, 4)$.

This circle is shown in Figure 8. Its equation can be found by using the distance formula. Start by letting (x, y) be any point on the circle. The distance from (x, y) to $(-3, 4)$ is given by

$$\sqrt{[x - (-3)]^2 + (y - 4)^2} = \sqrt{(x + 3)^2 + (y - 4)^2}.$$

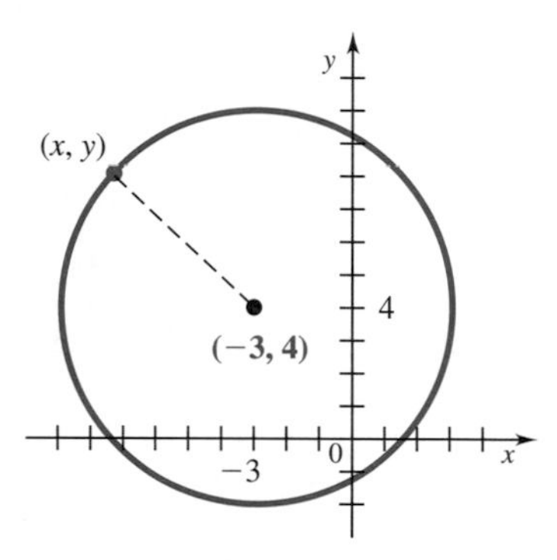

Figure 8

This distance equals the radius, 6. Therefore,

$$\sqrt{(x + 3)^2 + (y - 4)^2} = 6$$

or

$$(x + 3)^2 + (y - 4)^2 = 36.$$

Generalizing the work of Example 4 to a circle of radius r and center (h, k) would give the following result.

Standard Form of the Equation of a Circle

The circle with center (h, k) and radius r has equation

$$(x - h)^2 + (y - k)^2 = r^2,$$

the **standard form** of the equation of a circle. As a special case,

$$x^2 + y^2 = r^2$$

is the equation of a circle with radius r and center at the origin.

Starting with the standard form of the equation of a circle, $(x - h)^2 + (y - k)^2 = r^2$, and squaring $x - h$ and $y - k$ gives

$$x^2 - 2hx + h^2 + y^2 - 2ky + k^2 = r^2$$

$$x^2 + y^2 - 2hx - 2ky + (h^2 + k^2 - r^2) = 0.$$

Letting $-2h = D$, $-2k = E$, and $h^2 + k^2 - r^2 = F$, the equation becomes

$$x^2 + y^2 + Dx + Ey + F = 0, \qquad (*)$$

where D, E, and F are real numbers, the **general form** of the equation of a circle. Also, starting with an equation similar to (*), the process of *completing the square* discussed in Section 2.3 can be used to get an equation of the form

$$(x - h)^2 + (y - k)^2 = m$$

for some number m. If $m > 0$, then $r^2 = m$, and the graph is that of a circle with radius $\sqrt{m}$. If $m = 0$, the graph is the single point (h, k), and if $m < 0$, the graph has no points.

EXAMPLE 5 Decide whether or not each equation has a circle as its graph.

(a) $x^2 - 6x + y^2 + 10y + 25 = 0$

Since this equation has the form of equation (*) above, it either represents a circle, a single point, or no points at all. To decide which, complete the square on x and y separately, as explained in Section 2.3. Start with

$$(x^2 - 6x \qquad) + (y^2 + 10y \qquad) = -25.$$

Half of -6 is -3, and $(-3)^2 = 9$. Also, half of 10 is 5, and $5^2 = 25$. Add 9 and 25 on the left, and to compensate, add 9 and 25 on the right.

$$(x^2 - 6x + \mathbf{9}) + (y^2 + 10y + \mathbf{25}) = -25 + \mathbf{9} + \mathbf{25}$$
$$(x - 3)^2 + (y + 5)^2 = 9$$

Since $9 > 0$, the equation represents a circle which has its center at $(3, -5)$ and radius 3.

(b) $x^2 + 10x + y^2 - 4y + 33 = 0$

Complete the square as above.

$$(x^2 + 10x + 25) + (y^2 - 4y + 4) = -33 + 25 + 4$$
$$(x + 5)^2 + (y - 2)^2 = -4$$

Since $-4 < 0$, there are no ordered pairs (x, y), with x and y both real numbers, satisfying the equation. The graph of the given equation contains no points. ■

EXAMPLE 6 Find the equation of a circle with center $(-2, 3)$ that has a diameter with one endpoint at $(1, 0)$.

The diameter of a circle goes through the center of the circle and has endpoints on the circle, so the point $(1, 0)$ is on the circle and satisfies its equation. Thus, the radius is the distance between $(-2, 3)$ and $(1, 0)$.

$$r = \sqrt{(-3)^2 + 3^2} = \sqrt{18} = 3\sqrt{2}$$

Since the center is at $(-2, 3)$, the equation is

$$(x + 2)^2 + (y - 3)^2 = 18.$$

Midpoint Formula The distance formula is used to find the length of a line segment in a plane; the **midpoint formula** is used to find the coordinates of the midpoint of a line segment.

To develop this formula, let $A(x_1, y_1)$ and $B(x_2, y_2)$ be two different points in a plane (see Figure 9). Assume that A and B are not on a horizontal or vertical line. Let C be the intersection of the horizontal line through A and the vertical line through B. Let B' (read "B-prime") be the midpoint of segment AB. Draw a line through B' and parallel to segment BC. Let C' be the point where this line cuts segment AC. If the coordinates of B' are (x', y'), then C' has coordinates (x', y_1). Since B' is the midpoint of AB, point C' must be the midpoint of segment AC (why?), and

$$d(C, C') = d(C', A),$$

or

$$|x_2 - x'| = |x' - x_1|.$$

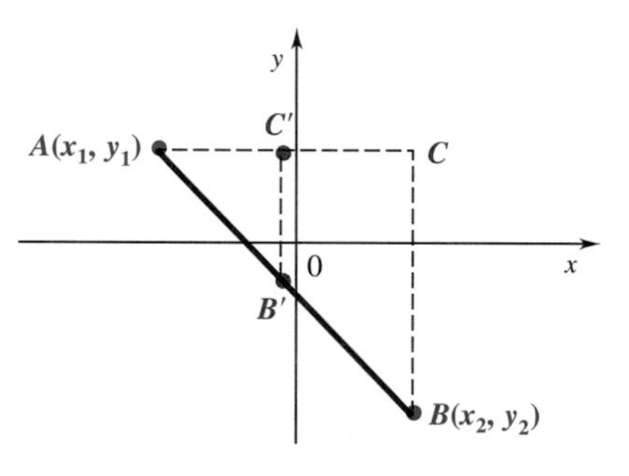

Figure 9

Because $d(C, C')$ and $d(C', A)$ must be positive, the only solutions for this equation are found if

$$x_2 - x' = x' - x_1,$$

or

$$x_2 + x_1 = 2x'.$$

Finally,

$$x' = \frac{x_1 + x_2}{2},$$

so that the x-coordinate of the midpoint is the average of the x-coordinates of the endpoints of the segment. In a similar manner, the y-coordinate of the midpoint is $(y_1 + y_2)/2$, proving the following result.

Midpoint Formula

The coordinates of the midpoint of the line segment with endpoints having coordinates (x_1, y_1) and (x_2, y_2) is

$$\left(\frac{x_1 + x_2}{2}, \frac{y_1 + y_2}{2}\right).$$

In other words, the coordinates of the midpoint of a segment are the *average* of the x-coordinates and the *average* of the y-coordinates of the endpoints of the segment.

EXAMPLE 7 Find the coordinates of the midpoint M of the segment with endpoints having coordinates $(8, -4)$ and $(-9, 6)$.

Use the midpoint formula to find that the coordinates of M are

$$\left(\frac{8 + (-9)}{2}, \frac{-4 + 6}{2}\right) = \left(-\frac{1}{2}, 1\right).$$

EXAMPLE 8 A line segment has an endpoint with coordinates $(2, -8)$ and a midpoint with coordinates $(-1, -3)$. Find the coordinates of the other endpoint of the segment.

The x-coordinate of the midpoint is found from $(x_1 + x_2)/2$. Here, the x-coordinate of the midpoint is -1. To find x_2, use this formula, with $x_1 = 2$.

$$-1 = \frac{2 + x_2}{2}$$

$$-2 = 2 + x_2$$

$$-4 = x_2$$

In the same way, $y_2 = 2$; the endpoint has coordinates $(-4, 2)$.

3.1 Exercises

If the point (a, b) is in the third quadrant, in what quadrant is each of the following points?

1. $(-a, b)$ **2.** $(a, -b)$ **3.** (b, a) **4.** $(-a, -b)$

In Exercises 5–10, graph the set of all points satisfying the conditions for ordered pairs (x, y).

5. $y = 0$ **6.** $y \leq 0$ **7.** $x > 0$

8. $x = 0$ **9.** $x/y < 0$ **10.** $xy > 0$

Find the distance $d(P, Q)$ and the coordinates of the midpoint of segment PQ. If necessary, round the distance to the nearest thousandth.

11. $P(4, 8)$, $Q(-1, 3)$
12. $P(5, -2)$, $Q(3, -4)$
13. $P(-7, -5)$, $Q(5, 6)$
14. $P(-2, -6)$, $Q(-4, -3)$
15. $P(\sqrt{2}, -\sqrt{5})$, $Q(3\sqrt{2}, 4\sqrt{5})$
16. $P(5\sqrt{7}, -\sqrt{3})$, $Q(-\sqrt{7}, 8\sqrt{3})$
17. $P(3, -7)$, $Q(-5, 19)$
18. $P(-9, -2)$, $Q(-1, -15)$
19. The distance formula equates the distance between two points in a plane with the square root of a quantity. Explain why this formula always produces a nonnegative result.

Find the coordinates of the other endpoint of the segments with endpoints and midpoints having the given coordinates.

20. Endpoint $(-3, 6)$, midpoint $(5, 8)$
21. Endpoint $(2, -8)$, midpoint $(3, -5)$
22. Endpoint $(6, -1)$, midpoint $(-2, 5)$
23. Endpoint $(-5, 3)$, midpoint $(-7, 6)$

Decide whether or not the following points are the vertices of a right triangle.

24. $(-2, 5)$, $(1, 5)$, $(1, 9)$
25. $(-9, -2)$, $(-1, -2)$, $(-9, 11)$
26. $(-4, 0)$, $(1, 3)$, $(-6, -2)$
27. $(-8, 2)$, $(5, -7)$, $(3, -9)$

Use the distance formula to decide whether or not the following points lie on a straight line. (Hint: *The points lie on a straight line if the sum of the two smallest distances equals the largest distance.*)

28. $(0, 7)$, $(3, -5)$, $(-2, 15)$
29. $(1, -4)$, $(2, 1)$, $(-1, -14)$
30. $(0, -9)$, $(3, 7)$, $(-2, -19)$
31. $(1, 3)$, $(5, -12)$, $(-1, 11)$

Find all values of x or y such that the distance between the given points is as indicated.

32. $(3, y)$ and $(-2, 9)$ is 12
33. $(x, 11)$ and $(5, -4)$ is 17
34. (x, x) and $(2x, 0)$ is 4
35. (y, y) and $(0, 4y)$ is 6
36. What does the graph of the equation $(x - h)^2 + (y - k)^2 = A$ look like if $A < 0$? If $A = 0$? Explain why.

Graph each of the following.

37. $x^2 + y^2 = 36$
38. $x^2 + y^2 = 81$
39. $x^2 - 4x + y^2 + 12y + 4 = 0$
40. $x^2 + 6x + y^2 + 8y = -9$
41. $x^2 - 2x + y^2 = -1$
42. $x^2 + 8x + y^2 - 14y + 65 = 0$
43. $3x^2 + 12x + 3y^2 = 63$
44. $2x^2 + 2y^2 = 4y + 96$

Find equations for each of the following circles.

45. Center $(1, 4)$, radius 3
46. Center $(-2, 5)$, radius 4
47. Center $(-8, 6)$, radius 5
48. Center $(3, -2)$, radius 2
49. Center $(-1, 2)$, passing through $(2, 6)$
50. Center $(2, -7)$, passing through $(-2, -4)$
51. Center $(-3, -2)$, tangent to the x-axis
52. Center $(5, -1)$, tangent to the y-axis
53. Endpoints of a diameter at $(1, 5)$ and $(7, 8)$
54. Endpoints of a diameter at $(3, -5)$ and $(-7, 2)$

Use the appropriate formulas to find the circumference and area of each of the circles with equations as follows.

55. $(x - 2)^2 + (y + 4)^2 = 25$

56. $(x + 3)^2 + (y - 1)^2 = 9$

The unit circle *is a circle centered at the origin with a radius of 1. In Exercises 57 and 58, show that each point lies on the unit circle.*

57. $(-\sqrt{2}/2, -\sqrt{2}/2)$

58. $(-1/2, \sqrt{3}/2)$

59. Find all points (x, y) with $x = y$ that are 4 units from $(1, 3)$.

60. Find all points satisfying $x + y = 0$ that are 8 units from $(-2, 3)$.

61. Decide whether each of the following points is *inside*, *on*, or *outside* the circle with center $(1, -4)$ and radius 6: **(a)** $(3, -2)$ **(b)** $(9, 1)$ **(c)** $(7, -4)$ **(d)** $(0, 9)$.

62. A circle is tangent to both axes, has its center in the third quadrant, and has a radius of $\sqrt{2}$. Find an equation for the circle.

63. Find the coordinates of a point whose distance from $(1, 0)$ is $\sqrt{10}$ and whose distance from $(5, 4)$ is $\sqrt{10}$.

64. Find the equation of the circle of smallest radius that contains the points $(1, 4)$ and $(-3, 2)$ within or on its boundary.

65. One circle has center at $(3, 4)$ and radius 5. A second circle has center at $(-1, 3)$ and radius 4. Do the circles cross?

66. Does the circle with radius 6 and center at $(0, 5)$ cross the circle with radius 4 and center at $(-5, -4)$?

67. One circle has center (a, b) and radius 4, while another has center (c, d) and radius s. State an inequality that is true if and only if the two circles intersect.

68. Find the coordinates of the points that divide the line segment joining $(4, 5)$ and $(10, 14)$ into three equal parts.

69. Show that the points $(-2, 2)$, $(13, 10)$, $(21, -5)$, and $(6, -13)$ are the vertices of a square.

70. Are the points $A(1, 1)$, $B(5, 2)$, $C(3, 4)$, $D(-1, 3)$ the vertices of a parallelogram (opposite sides equal in length)? Of a rhombus (all sides equal in length)?

71. Find the equation of the circle shown in the figure.

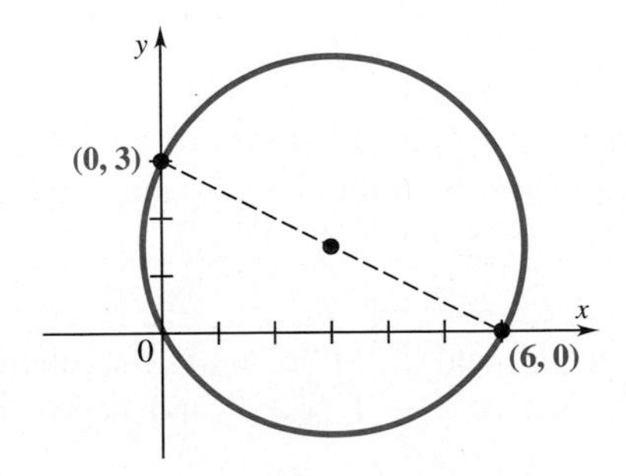

72. A rhombus is a parallelogram with all sides equal in length. The figure shows a rhombus with one vertex at the origin. Find the values of a and b.

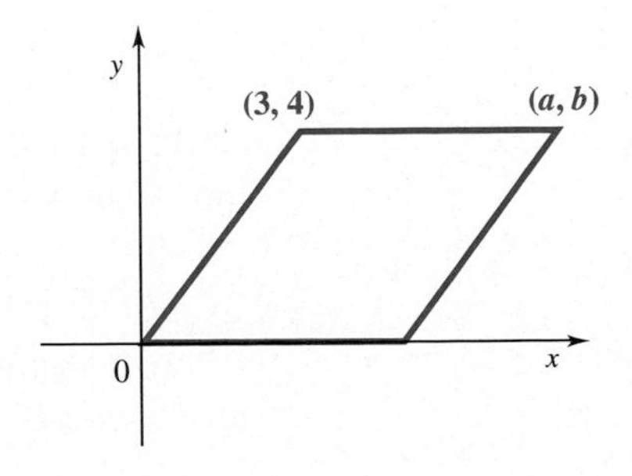

3.2 Graphing Relations

For Graphers

Now we can see the reason for the name of the RANGE key on a graphing calculator. The name is somewhat misleading, because this key is used to determine *both* the domain and range intervals. Determining suitable domain and range intervals is the most important (and sometimes the most difficult) step in producing a graph with a graphing utility. It may be necessary to experiment with different intervals to get a satisfactory graph.

A set of ordered pairs is called a **relation**. The **domain** of the relation is the set of all first elements in the ordered pairs, and the **range** is the set of all second elements.

Although any set of ordered pairs is a relation, in mathematics we are most interested in those relations that are solution sets of equations. We may say that an equation *defines a relation*, or that it is the *equation of the relation*. For simplicity, we often refer to equations such as

$$y = 3x + 5 \qquad \text{or} \qquad x^2 + y^2 = 16$$

as relations, although technically the solution set of the equation is the relation.

EXAMPLE 1 For each relation defined below, give three ordered pairs that belong to the relation, and state the domain and range of the relation.

(a) $\{(2, 5), (7, -1), (10, 3), (-4, 0), (0, 5)\}$

Three ordered pairs from the relation are any three of the five ordered pairs in the set. The domain is the set of first elements,

$$\{2, 7, 10, -4, 0\},$$

and the range is the set of second elements,

$$\{5, -1, 3, 0\}.$$

(b) $y = 4x - 1$

To find an ordered pair of the relation, choose any number for x or y and substitute it in the equation to get the corresponding value. For example, let $x = -2$. Then

$$y = 4(-2) - 1 = -9,$$

giving the ordered pair $(-2, -9)$. If $y = 3$, then

$$\begin{aligned} 3 &= 4x - 1 \\ 4 &= 4x \\ 1 &= x, \end{aligned}$$

and the ordered pair is $(1, 3)$. Verify that $(0, -1)$ also belongs to the relation. Since x and y can take any real-number values, both the domain and range are $(-\infty, \infty)$.

(c) $x = \sqrt{y - 1}$

Verify that the ordered pairs $(1, 2)$, $(0, 1)$, and $(2, 5)$ belong to the relation. Since x equals the nonnegative square root of $y - 1$, $x \geq 0$ and hence, the domain is restricted to $[0, \infty)$. Also, only nonnegative numbers have a square root, so the range is determined by the inequality

$$y - 1 \geq 0$$
$$y \geq 1,$$

giving $[1, \infty)$ as the range. ■

The set of all points in the plane corresponding to the ordered pairs of a relation is the **graph of the relation**. For now we will find graphs of relations by determining a reasonable number of ordered pairs, locating the corresponding points, and then connecting the points, guessing at the shape of the entire graph. A graphing utility provides another option that is becoming more widely available. Later in this text, we will develop methods for quickly identifying the graphs of many relations.

EXAMPLE 2 Graph $y = -4x + 3$.

Find several ordered pairs by selecting values for x (or y) and then finding the corresponding values of y (or x). For example, if $x = -3$, then $y = -4(-3) + 3 = 15$, producing the ordered pair $(-3, 15)$. Additional ordered pairs found in this way are given in the table with Figure 10. The ordered pairs from this table lead to the points that have been plotted in Figure 10(a). These points suggest that the entire graph is a straight line, as drawn in Figure 10(b). Notice that both the domain and the range are $(-\infty, \infty)$.

x	y	*Ordered Pair*
-3	15	$(-3, 15)$
-2	11	$(\ 2, 11)$
-1	7	$(-1, 7)$
0	3	$(0, 3)$
1	-1	$(1, -1)$
2	-5	$(2, -5)$
3	-9	$(3, -9)$

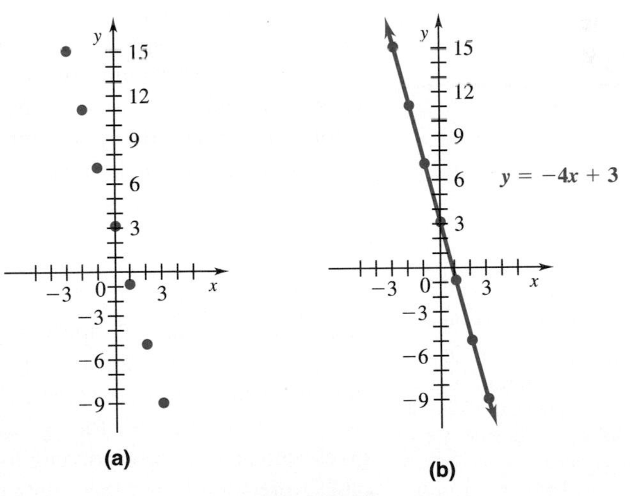

Figure 10 ■

For Graphers

Some graphing utilities (for example, the Texas Instruments TI-81) will plot points and draw lines connecting two points. The graphs in Figure 10 can be drawn using a grapher with this capability. To graph relations with equations that can be put in the form $y =$ (an expression in x), however, it is easier to use the graph function of the graphing utility. The absolute value function is often a built-in function in a graphing utility. It is designated by abs(x) or $|x|$. If neither of these forms is available, use the fact that $|x| = \sqrt{x^2}$. Figure 12 shows the graph in Figure 11 drawn with a graphing utility.

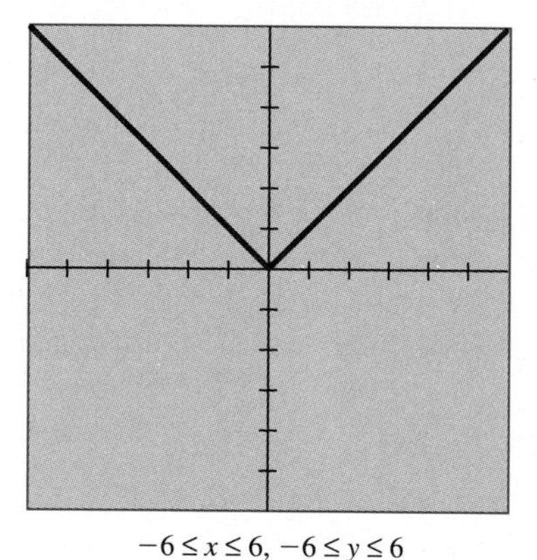

$-6 \le x \le 6, -6 \le y \le 6$

Figure 12

EXAMPLE 3 Graph $y = |x|$.

Start with a table.

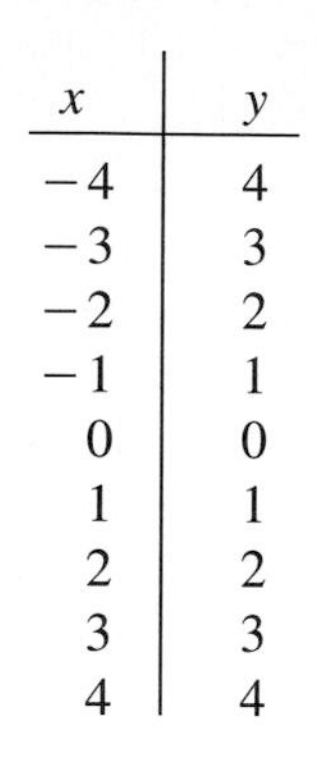

x	y
-4	4
-3	3
-2	2
-1	1
0	0
1	1
2	2
3	3
4	4

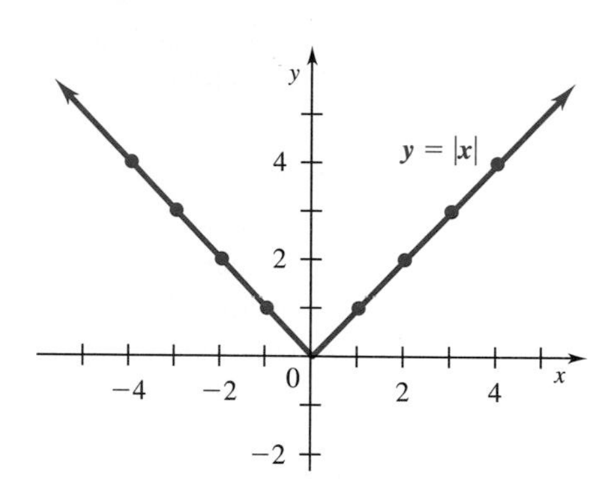

Figure 11

Use this table to get the points in Figure 11. The graph drawn through these points is made up of portions of two straight lines. The domain is $(-\infty, \infty)$ and the range is $[0, \infty)$. ■

CAUTION There is a danger in the method used in Examples 2 and 3—we might choose a few values for x, find the corresponding values of y, begin to sketch a graph through these few points, and then make a completely wrong guess as to the shape of the graph. For example, choosing only -1, 0, and 1 as values of x in Example 3 above would produce only the three points $(-1, 1)$, $(0, 0)$, and $(1, 1)$. These three points alone would not give enough information to determine the proper graph for $y = |x|$. However, this section involves only elementary graphs; when more complicated graphs are presented later, we will develop more accurate methods of working with them. ■

For Graphers

A graphing utility requires the relation to have the form $y =$ (an expression in x). To use a grapher to graph $x = y^2 - 4$, we must first solve the equation for y. This gives two equations, $y = \sqrt{x + 4}$ and $y = -\sqrt{x + 4}$. Then we can use the grapher to draw both graphs on the same coordinate system.

EXAMPLE 4 Graph $x = y^2 - 4$.

Since $y^2 \ge 0$, the domain is $[-4, \infty)$. It is easier here to choose values of y, then find the corresponding x-values. Choosing 1 for y, for example, gives $x = (1)^2 - 4 = -3$. Choosing -1 for y gives the same result. The table shown with Figure 13 gives values of x corresponding to various values of y. The ordered pairs from this table were used to get the points plotted in Figure 13. (Don't forget that x always goes first in the ordered pair.) A smooth curve was then drawn through the resulting points. Here, y can take on any value, so the range is $(-\infty, \infty)$. As mentioned earlier, the domain is $[-4, \infty)$.

x	y
5	3
0	2
-3	1
-4	0
-3	-1
0	-2
5	-3

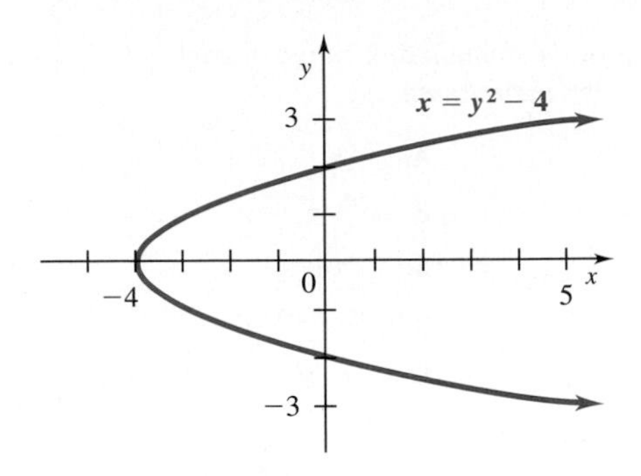

Figure 13

For Grapherṡ

If you use a graphing utility to graph this relation, you will need to place parentheses around the quantity $x + 4$.

EXAMPLE 5 Graph $y = \sqrt{x + 4}$.

The domain is determined by the fact that $x + 4 \geq 0$ or $x \geq -4$, giving $[-4, \infty)$. Again, the range is $[0, \infty)$. Selecting some values of x in the domain and calculating the corresponding y-values leads to the ordered pairs shown in the table.

x	-4	-3	0	5
y	0	1	2	3

Plotting these points and drawing a curve through them gives the graph in Figure 14. This graph has an endpoint at $(-4, 0)$, as indicated by the domain.

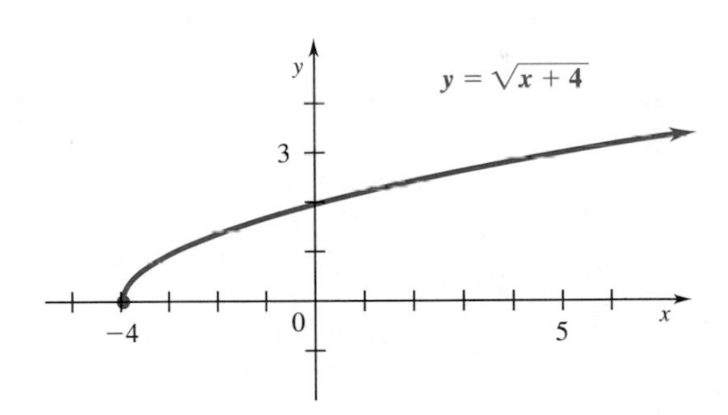

Figure 14

3.2 Exercises

1. Discuss the distinctions among a relation, the equation of a relation, and the graph of a relation.

2. In your own words, write a few sentences describing a relation, its domain, and its range.

In Exercises 3–22, give the domain and range of each relation, then use point plotting or a graphing utility to graph the relation.

3. $y = 3x - 2$
4. $y = -2x + 4$
5. $3x = y^2$
6. $4x = y^2$
7. $16x^2 = -y$
8. $4x^2 = -y$
9. $y = |x| + 4$
10. $y = |x| - 3$
11. $x = |y|$
12. $x = |y| - 1$
13. $y = -|x + 1|$
14. $y = -|x - 2|$
15. $x = \sqrt{y} - 2$
16. $x = -\sqrt{y} + 1$
17. $x = -\sqrt{y - 2}$
18. $x = \sqrt{y - 4}$
19. $y = \sqrt{2x + 4}$
20. $y = \sqrt{3x + 9}$
21. $y = -2\sqrt{x}$
22. $y = -\sqrt{x}$

Decide whether the given ordered pair is in the relation defined by the given equation.

23. $(3, -1)$, $2x^2 - y^2 = 17$
24. $(-2, 5)$, $3y - x^2 = 2x + 9$

25. A tool box has a square end with sides of length s and a length 2 inches more than twice the measure of the sides of the square end. Write a relation giving the surface area of the closed box.

26. Write a relation giving the volume of the box in Exercise 25.

27. The volume of a cone is found with the formula $V = (1/3)Bh$, where B is the area of the base of the cone and h is the height. Write a relation expressing the volume of a circular cone whose height is three times the radius of the base.

28. The formula $A = (1/2)bh$ gives the area of a triangle with base b and height h. Write a relation expressing the area of a triangle in terms of its base if its base is 2/3 its height.

Graphing Utility Problems

Use a graphing utility to graph each of the following relations. You will have to rewrite each relation as two equations first. Use the graph to determine the domain and range of each relation.

29. $9x^2 - 4y^2 = 36$
30. $9x^2 + 4y^2 = 36$
31. $x^2 + y^2 - 4x = 32$
32. $x^2 + y^2 + 6x - 4y = 3$

3.3 Functions

In business, the price of an item often is directly related to the cost of producing the item. The relationship may be expressed as an equation. In such a situation it would be undesirable to have a particular cost produce more than one price. A special type of relation, which assigns exactly one range value to each value in the domain, is most suitable for applications and is so useful it is given a special name.

Definition of Function

A **function** is a relation that assigns to each element of a set X exactly one element of a set Y.

Set X is the domain and set Y is the range of the function.

We can think of a function as an input-output machine. If we input an element from the domain, the function (machine) outputs an element belonging to the range. See Figure 15.

For Graphers

Now we can see why, when using a graphing utility, we had to express some relations by using two equations: $x = y^2$ as $y = \sqrt{x}$ and $y = -\sqrt{x}$, for example. Graphing utilities will graph only functions of x.

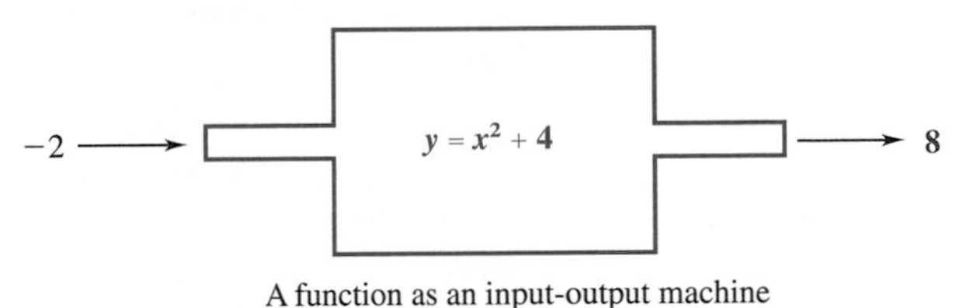

A function as an input-output machine

Figure 15

EXAMPLE 1 Decide whether the following relations are functions.

(a) The x^2 key on a calculator

Each real number input for x produces exactly one square, so this is an example of a function.

(b) The optical reader at the checkout counter in many stores that converts codes to prices

Since each code is assigned just one price, the set of ordered pairs (code, price) is a function.

(c) $\{(x, y) \mid x = |y|\}$

Most values of x are assigned two y-values by the equation $x = |y|$. For example, both (3, 3) and (3, −3) belong to this relation, so it is not a function.

(d) $\{(x, y) \mid y = x\}$

Typical ordered pairs are (1, 1), (−2, −2), (3, 3), and so on. Each x-value is assigned just one y-value (which equals that x-value), so this relation is a function, sometimes called the **identity function.** ■

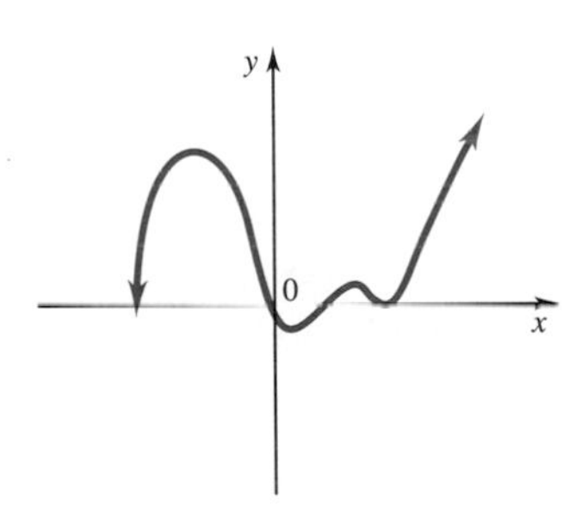

Figure 16

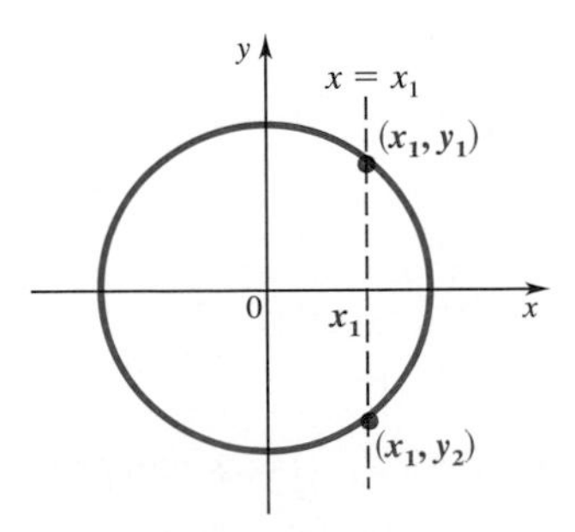

Figure 17

There is a quick way to tell from its graph if a relation is a function. In the graph of Figure 16, for any x that might be chosen, exactly one value of y can be found, showing that this is the graph of a function. On the other hand, the graph in Figure 17 is not the graph of a function. For example, the vertical line through x_1 leads to two different values of y, namely y_1 and y_2. This example suggests the **vertical line test** for a function.

Vertical Line Test

If each vertical line intersects a graph in no more than one point, the graph is the graph of a function.

EXAMPLE 2 Use graphs to decide whether the following relations are functions.

(a) $\{(x, y) \mid y = |x|\}$

This relation was graphed in Figure 11 in Section 3.2. By the vertical line test, the graph shows that $y = |x|$ defines a function, called the **absolute value function**.

(b) $\{(x, y) \mid y = \sqrt{x + 4}\}$

The vertical line test shows that this relation, whose graph is shown in Figure 14 in Section 3.2, is a function. This function is an example of a **root function**. ■

***f*(*x*) Notation** It is common to use the letters f, g, and h to name functions. If f is a function and x is an element in the domain X, then $f(x)$ is used to represent the element in the range Y that corresponds to x, and we say $y = f(x)$. (The notation $f(x)$ is read "f of x" or "f at x.") For example, if f is used to name the function in Example 2(a), then $f(x) = |x|$, and

$$f(1) = 1, \qquad f(-2) = 2, \qquad \text{and} \qquad f(3) = 3.$$

For a given element x in set X, the corresponding element $f(x)$ in set Y is called the **value** or **image** of f at x. The set of all possible values of $f(x)$ makes up the range of the function. Throughout this book, if the domain for a function is not given, it will be assumed to be the largest possible set of real numbers for which $f(x)$ is a real number. For example, suppose function f is defined by

$$f(x) = \frac{-4x}{2x - 3}.$$

With this equation any real number can be used for x except $x = 3/2$, which makes the denominator equal to 0. Assuming that the domain is the largest possible set of real numbers for which $f(x)$ is a real number makes the domain $(-\infty, 3/2) \cup (3/2, \infty)$.

EXAMPLE 3 Find the domain and range for the functions defined by the following rules.

(a) $f(x) = x^2$

Any number may be squared, so the domain is $(-\infty, \infty)$. To determine the range of y, solve the equation for x.

$$f(x) = x^2$$
$$y = x^2$$
$$x = \pm\sqrt{y}$$

This shows that y can take any value in the interval $[0, \infty)$ as x takes values in the interval $(-\infty, \infty)$.

(b) $f(x) = \sqrt{x^2 + x - 6}$

The domain includes those values of x that make

$$x^2 + x - 6 \geq 0.$$

Factor. $$(x + 3)(x - 2) \geq 0$$

Solve this quadratic inequality to get $(-\infty, -3] \cup [2, \infty)$ for the domain. The radical indicates the nonnegative square root, so the range is $[0, \infty)$. ■

For most of the functions in this book, the domain can be found with algebraic methods already discussed. The range, however, at this level of mathematics must often be found from the graph, as shown in Figure 18.

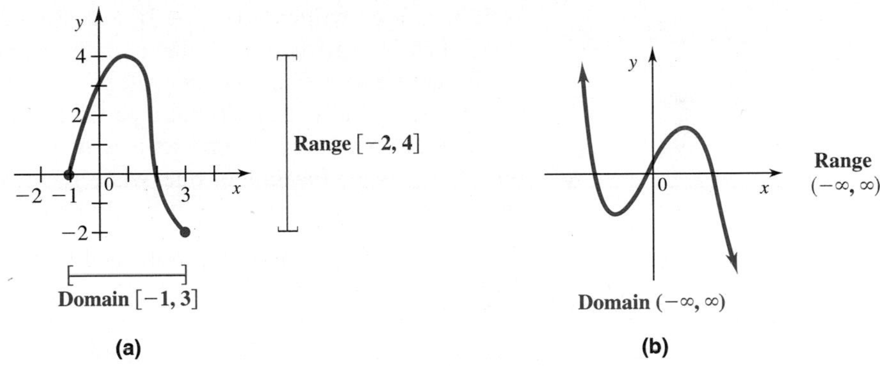

Figure 18

NOTE Based on the definition of a function, functions f and g are **equal** if and only if f and g have exactly the same domains and $f(x) = g(x)$ for every value of x in the domain. ■

In Figure 18(a) the range is a closed interval $[-2, 4]$, but in Figure 18(b), the range is the set of real numbers $(-\infty, \infty)$. The largest number in the range (if there is one) is called the **maximum**, and the smallest number in the range (if there is one) is called the **minimum**. The function graphed in Figure 18(a) has a maximum of 4 and a minimum of -2. These values are important in many applications. One of the major applications of calculus is finding maximum and minimum function values.

A function f is said to be increasing on an interval if, for any two numbers x_1 and x_2 in the interval,

$$f(x_1) < f(x_2) \quad \text{whenever} \quad x_1 < x_2;$$

and f is decreasing on the interval if

$$f(x_1) > f(x_2) \quad \text{whenever} \quad x_1 < x_2.$$

This means that the graph of f goes *up* from left to right on x-intervals where f is increasing and *down* from left to right on x-intervals where f is decreasing. For example, in Figure 18(a), the function is increasing on $[-1, 1]$ and decreasing on $[1, 3]$.

Suppose a function is defined by $f(x) = -3x + 2$. To emphasize that this statement is used to find values in the range of f, it is common to write

$$y = -3x + 2.$$

When a function is written in the form $y = f(x)$, x is called the **independent variable**, and y the **dependent variable**.

There is no reason to restrict the variables to x or y; different areas of study use different variables. For example, it is common to use t for time in physics, or p for price in economics. The function defined by $f(x) = -3x + 2$ above could just as well have been written $g(t) = -3t + 2$. Both functions have the real numbers as domain and assign the same value to each real number.

By definition, a function f is a rule that assigns to each element of one set exactly one element of a second set. For a particular value of x in the first set, the corresponding element in the second set is written $f(x)$. There is a distinction between f and $f(x)$: f is the function or rule, while $f(x)$ is the value obtained by applying the rule to an element x. However, it is very common to abbreviate

"the function defined by the rule $y = f(x)$"

as simply

"the function $y = f(x)$."

EXAMPLE 4 Let $g(x) = 3\sqrt{x}$ and $f(x) = 1 + 4x$. Find each of the following and give the corresponding ordered pair.

(a) $g(16)$

To find $g(16)$, replace x in $g(x) = 3\sqrt{x}$ with 16.

$$g(16) = 3\sqrt{16} = 3 \cdot 4 = 12$$

The ordered pair is $(16, 12)$.

(b) $f(-3) = 1 + 4(-3) = -11$, and the ordered pair is $(-3, -11)$.

(c) $g(-4)$ does not exist; -4 is not in the domain of g since $\sqrt{-4}$ is not a real number. There is no ordered pair here.

(d) $g(m) = 3\sqrt{m}$, if m represents a nonnegative real number; the ordered pair is $(m, 3\sqrt{m})$.

(e) $g(f(3))$

First find $f(3)$, as follows.

$$f(3) = 1 + 4 \cdot 3 = 1 + 12 = 13$$

Now, $g(f(3)) = g(13) = 3\sqrt{13}$, and the corresponding ordered pair is $(3, 3\sqrt{13})$.

(f) $g(x + h)$

Substitute $x + h$ for x in $g(x) = 3\sqrt{x}$.

$$g(x + h) = 3\sqrt{x + h}$$

The corresponding ordered pair is $(x + h, 3\sqrt{x + h})$. ■

EXAMPLE 5 Let $f(x) = 2x^2 - 3x$. If h represents any nonzero number, then the quotient

$$\frac{f(x + h) - f(x)}{h}, \quad h \neq 0,$$

represents the slope of the line through $(x, f(x))$ and $(x + h, f(x + h))$. This expression is called a *difference quotient* and is used in calculus to determine the steepness of a curve at a point. Find and simplify the quotient.

To find $f(x + h)$, replace x in $f(x)$ with $x + h$.

$$f(x) = 2x^2 - 3x$$

$$f(x + h) = 2(x + h)^2 - 3(x + h) \qquad \text{Substitute } x + h \text{ for } x \text{ in } f(x).$$

Now, subtract $f(x)$, which equals $2x^2 - 3x$, to get $f(x + h) - f(x)$. Then

$$\begin{aligned}\frac{f(x + h) - f(x)}{h} &= \frac{[2(x + h)^2 - 3(x + h)] - (2x^2 - 3x)}{h} \\ &= \frac{2(x^2 + 2xh + h^2) - 3x - 3h - 2x^2 + 3x}{h} \\ &= \frac{2x^2 + 4xh + 2h^2 - 3x - 3h - 2x^2 + 3x}{h} \\ &= \frac{4xh + 2h^2 - 3h}{h} \\ &= \frac{h(4x + 2h - 3)}{h} \\ &= 4x + 2h - 3.\end{aligned}$$

■

CAUTION In Example 5, note that $f(x + h) \neq f(x) + f(h)$, since

$$f(x) + f(h) = 2x^2 - 3x + 2h^2 - 3h$$

and, as shown above,

$$f(x + h) = 2x^2 + 4xh + 2h^2 - 3x - 3h. \blacksquare$$

EXAMPLE 6 A rectangle has the dimensions shown in Figure 19. Give the diagonal of the rectangle, d, as a function of the width, x.

By the Pythagorean theorem,

$$\begin{aligned} d^2 &= x^2 + (x + 3)^2 \\ &= x^2 + x^2 + 6x + 9 \\ &= 2x^2 + 6x + 9 \\ d &= \sqrt{2x^2 + 6x + 9}. \end{aligned}$$

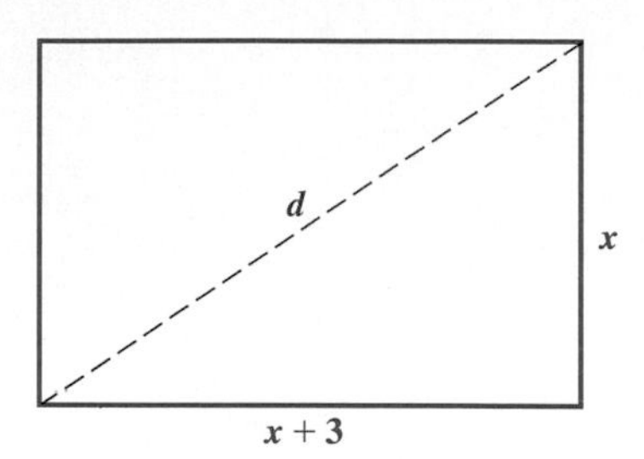

Figure 19

Only the positive square root is meaningful here, since d represents the diagonal of a rectangle. ■

The concept of *function*, as with all of mathematics, required many years to be developed and clarified. Leibniz used the term *function* in 1694 to refer to any part of a curve, a point on it, the extent of its curvature, or its slope at a point. By the eighteenth century, mathematicians said a function was "an analytic expression composed in whatever way of a variable and numbers." In the second half of the eighteenth century a function was considered to be some sort of an equation, possibly with infinitely many terms.

The conception of a function as a correspondence, as in our text, required a new geometric, rather than algebraic, point of view. It was Lejeune Dirichlet (1805–1859), a German mathematician, who wrote "$f(x)$ is a real function of a real variable x if, to every real number x, there corresponds a real number $f(x)$."

3.3 Exercises

Decide whether each of the following relations is a function.

1. The $\sqrt{x}$ key on a calculator

2. The $1/x$ key on a calculator

3. $\{(x, y) \mid 2x^2 = y + 3\}$

4. $\{(x, y) \mid y^2 = 3x + 1\}$

5. $\{(x, y) \mid 2x - y = 1\}$

6. $\{(x, y) \mid y - x = 4\}$

7. $\{(x, y) \mid x^2 + y^2 = 4\}$

8. $\{(x, y) \mid x^2 - y^2 = 4\}$

9.

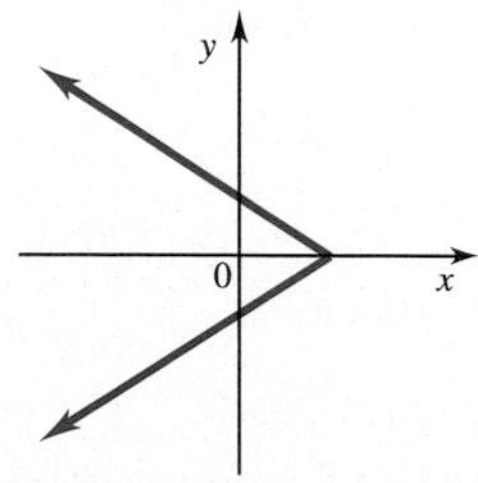

10.

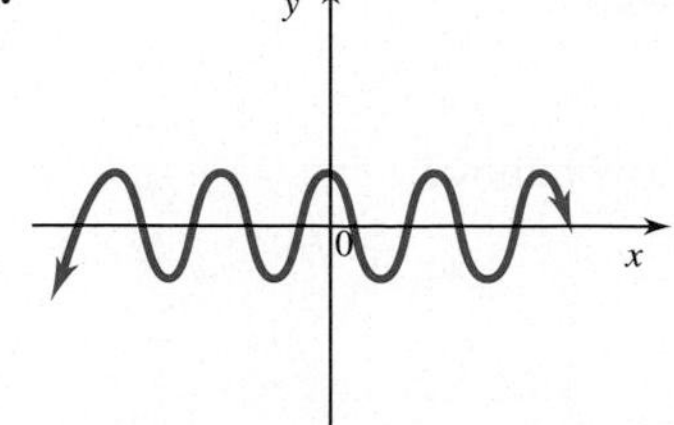

11.

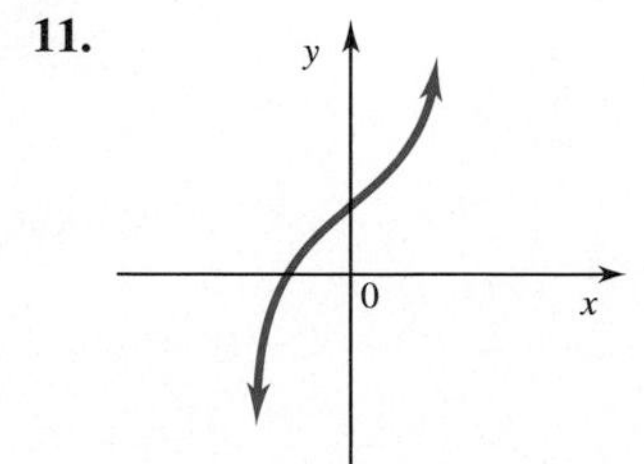

12. Compare relations and functions. How are they alike? How are they different? Be specific; use examples.

For each of the following, **(a)** *find* $f(-2)$, $f(0)$, $f(4)$; **(b)** *find the x-value(s) that correspond(s) to the following function values:* 3, 0, 2.

13.

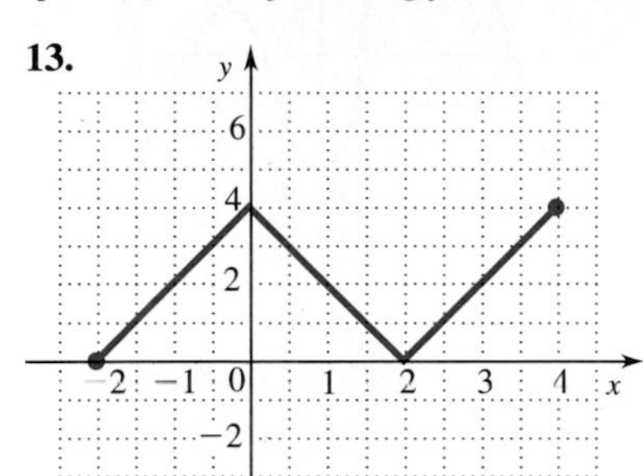

14.

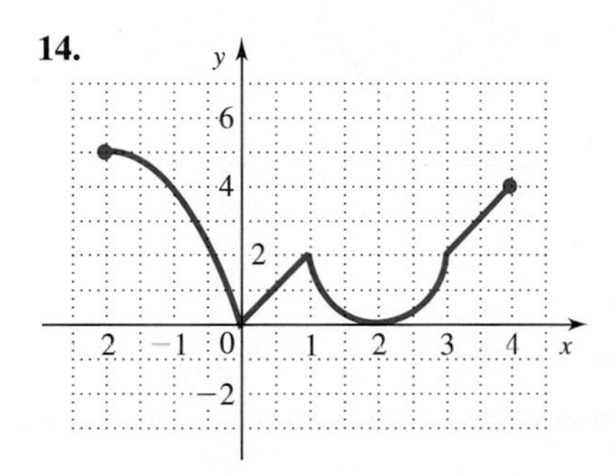

15.

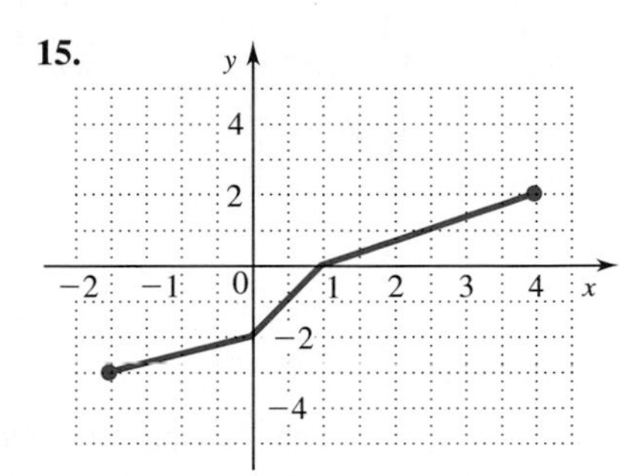

Let $f(x) = 3 - 2x^2$, $g(x) = \sqrt{4x - 2}$, *and* $h(x) = x/(2 - x)$. *Find each of the following.*

16. $f(-2)$ **17.** $h(3)$ **18.** $g(4)$ **19.** $g(5)$

20. $f(-3) + 2$ **21.** $f(4) - 5$ **22.** $f(1) + g(1)$ **23.** $[g(1)]^2$

24. $[f(-1)]^2$ **25.** $f\left(\frac{1}{a}\right)$ **26.** $g(3a - 1)$ **27.** $\frac{1}{f(a)}$

28. $f(\sqrt{a})$ **29.** $\frac{f(5a)}{g(a)}$ **30.** $\sqrt{h(a)}$

Let $f(x) = \begin{cases} 2x^2 & \text{if } x < 0 \\ 3x & \text{if } 0 \le x \le 1 \\ |6 - x| & \text{if } x > 1. \end{cases}$ *Find each of the following.*

31. $f(-3)$ **32.** $f(1)$ **33.** $f(10)$ **34.** $f\left(\frac{1}{3}\right) + f(3)$

35. Figure 18 shows how the range of a function can be determined by looking at the graph of the function. Describe in words how one can tell the range from the graph.

Give the domain and range of the functions defined or graphed in Exercises 36–51. Give only the domain in Exercises 40 and 41.

36. $g(x) = x^2 - 3$ **37.** $h(x) = (x - 2)^4$ **38.** $f(x) = \sqrt{8 + x}$ **39.** $f(x) = (3x + 2)^{1/2}$

40. $g(x) = \frac{2}{x^2 - 3x + 2}$ **41.** $h(x) = \frac{-4}{x^2 + 5x + 4}$ **42.** $r(x) = -\sqrt{x^2 - 4x - 5}$

43. $r(x) = -\sqrt{x^2 + 7x + 10}$ **44.** $f(x) = |3x - 4| + 2$ **45.** $k(x) = -|2x - 7| + 5$

46.

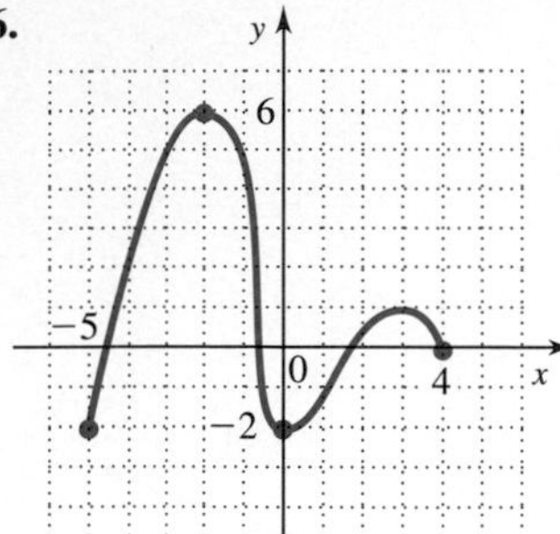

47.

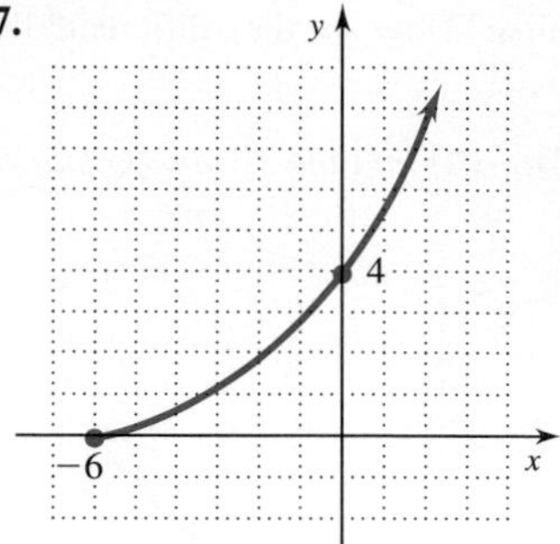

48.

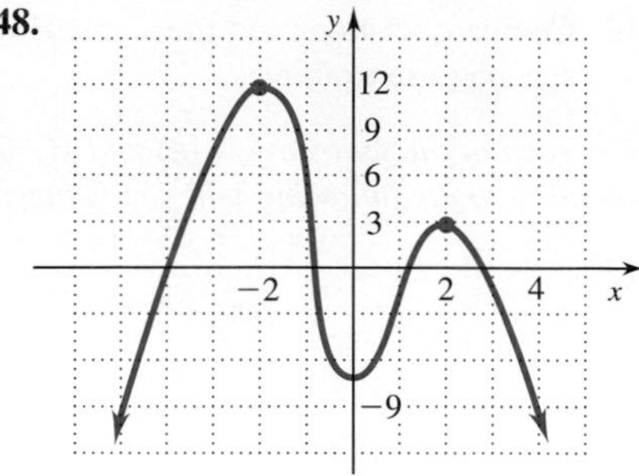

49. Exercise 11

50. Exercise 14

51. Exercise 13

52. Identify the maximum and minimum function values of the functions graphed in each of the indicated exercises.
(a) Exercise 46 **(b)** Exercise 47 **(c)** Exercise 48

53. Give the intervals where the function graphed in each of the indicated exercises is increasing and decreasing.
(a) Exercise 46 **(b)** Exercise 47 **(c)** Exercise 48

54. The figure shows the typical relationship between temperature and time of day in April for a town.
(a) Is it the graph of a function?
(b) What is the temperature at 8 A.M.? At 12 noon? At 6 P.M.?
(c) At what time(s) is the temperature 40°? 65°?
(d) What are the maximum and minimum temperatures shown?

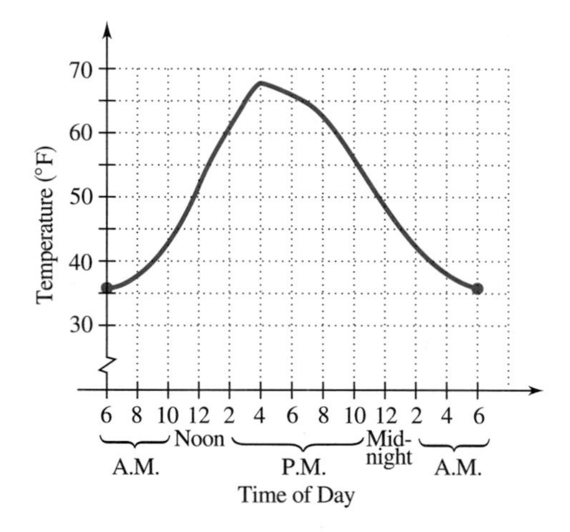

55. The cost of an automobile trip depends on the distance traveled, as shown in the figure.
(a) Is this the graph of a function?
(b) What is the cost to travel 100 miles? 250 miles? 500 miles?
(c) How many miles can be traveled for \$100? For \$25?
(d) Assume the graph continues in the same way. Does this relation have a maximum or minimum value?

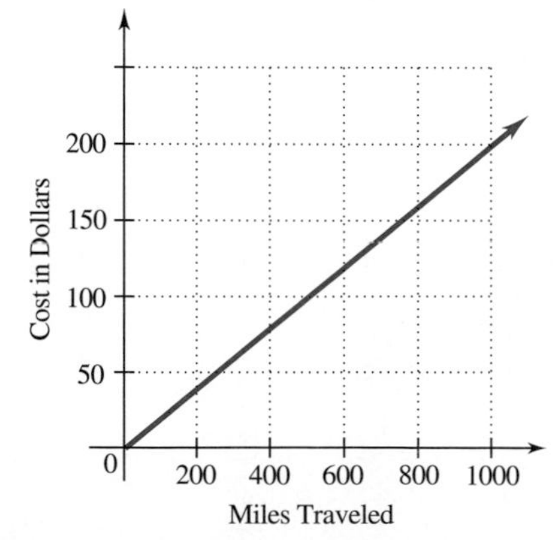

56. If $f(x) = x^2/x$ and $g(x) = x$, does $f(x) = g(x)$? Explain why.

For the functions defined in Exercises 57–60, find **(a)** $f(x + h)$, **(b)** $f(x + h) - f(x)$, *and* **(c)** $\frac{f(x + h) - f(x)}{h}$. *(Assume* $h \neq 0$*).*

57. $f(x) = x^2 - 4$ **58.** $f(x) = 8 - 3x^2$ **59.** $f(x) = 6x + 2$ **60.** $f(x) = 4x + 11$

61. A storage container in the shape of a cube with an edge of length e yd is to be insulated at the cost of \$2 per square yard. Find the total cost $C(e)$ for the insulation job as a function of e in yards.

62. A cylindrical water tank has a radius of 6 ft and a height of 9 ft. (See the figure.) The tank is filled with water to a depth of h ft. Express the volume of the water as a function of h.

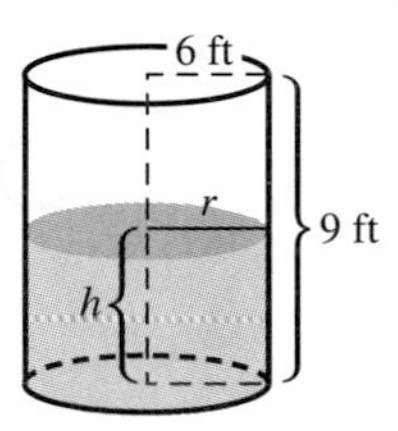

Exercise 62

When a function models a real-life situation, the domain is usually restricted by practical considerations. For instance, in Exercise 62, the value of h must be in the interval [0, 9]. *Determine a meaningful domain for the functions in Exercises 63–65.*

63. A ball is dropped from a 144-ft building. Its height in feet at time t (in seconds) after being dropped is $s(t) = -16t^2 + 144$.

64. According to the economic law of demand, the price that should be charged in order to sell x units of a certain commodity is $p(x) = 100 - .05x$ dollars.

65. If a worker earning ten dollars an hour works at least 40 hr a week, with time and a half for overtime, then her weekly salary for h hr of work is $s(h) = 400 + 15(h - 40)$ dollars.

66. An architect is designing a building shaped like a rectangle with semicircles of radius r meters attached to each end. (See the figure.) The perimeter of the building is to be 440 m, and the radius must be at least 5 m.
(a) Show that each straight portion of the building must have length $220 - \pi r$ meters.
(b) Express the area of the entire building as a function of r.
(c) Determine a meaningful domain for the function in part (b).

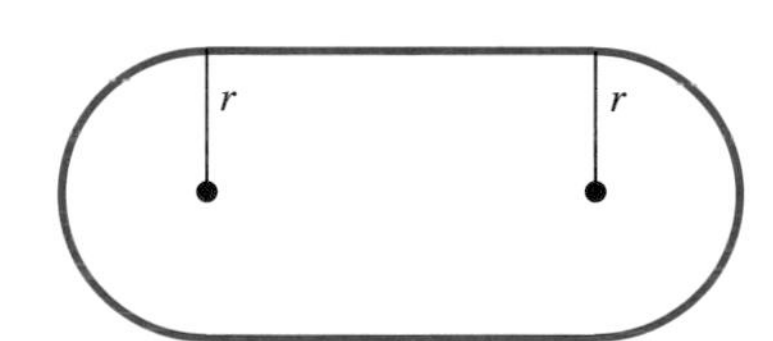

Exercise 66

67. Sixty feet of fencing is used to enclose a rectangular garden in which each side of the garden is at least 12 ft long and one side, of length x, is an extension of a 10-ft stone wall. See the figure.
(a) Show that the length of each side perpendicular to the side with the stone wall is $25 - x$ ft.
(b) Express the area of the garden as a function of x.
(c) Determine a meaningful domain for the function in part (b).

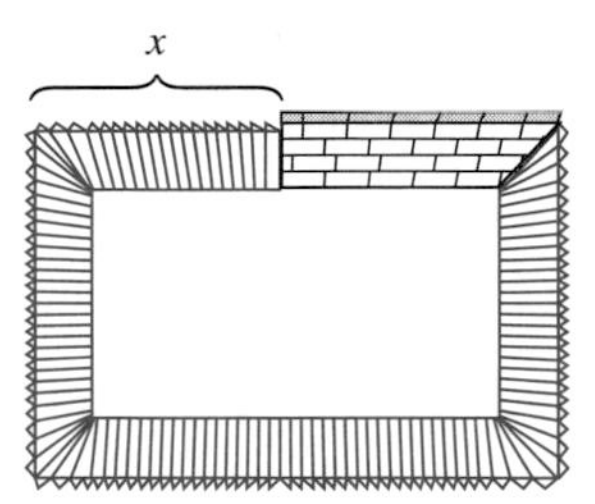

Exercise 67

68. Give the area of a circle as a function of its radius r; also, give the circumference as a function of the radius.

69. The number of BTUs (British Thermal Units) required to cool a room is 3000 more than 15 times the area of the room in square feet. Let B represent the number of BTUs and A represent the area in square feet. Give A as a function of B.

70. A rectangle is inscribed in a circle of radius r. Let x represent the length of one side of the rectangle. Give the area of the rectangle as a function of r.

71. The height of a circular cone is half the radius of the base. Give the volume of the cone as a function of the radius of the base.

27. Explain why the equation $y = 5$ defines a linear function, but $x = 5$ does not. What must be true about the slope of a line in order for the line to be the graph of a function?

Use slopes to decide whether or not the following points lie on a straight line. (Hint: *Find the slope of the line through the first two points, and then the slope of the line through the last two points.*)

28. $M(4, 3)$, $N(-1, -5)$, $P(2, 0)$

29. $A(1, 10/3)$, $B(0, 5)$, $C(-3, 10)$

Graph the functions defined in Exercises 30–33.

30. $f(x) = \begin{cases} x - 1 & \text{if } x \le 3 \\ 2 & \text{if } x > 3 \end{cases}$

31. $h(x) = \begin{cases} -2 & \text{if } x \ge 1 \\ 2 & \text{if } x < 1 \end{cases}$

32. $p(x) = \begin{cases} |x| & \text{if } x > -2 \\ x & \text{if } x \le -2 \end{cases}$

33. $r(x) = \begin{cases} |x| - 1 & \text{if } x > -1 \\ x - 1 & \text{if } x \le -1 \end{cases}$

Write an expression similar to those in Exercises 30–33 to define the piecewise functions graphed in Exercises 34–37. Give the domain and range of each function.

34.

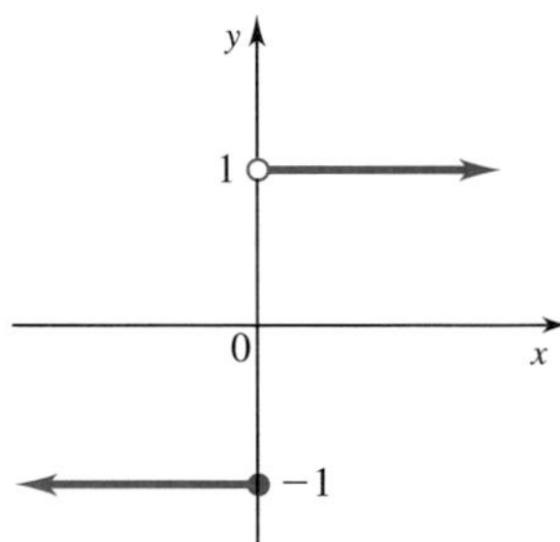

35.

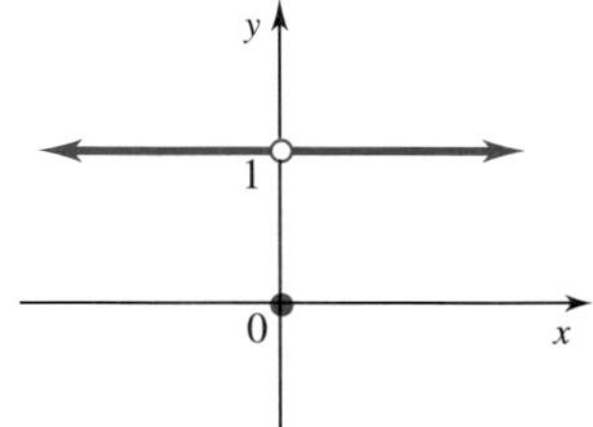

36.

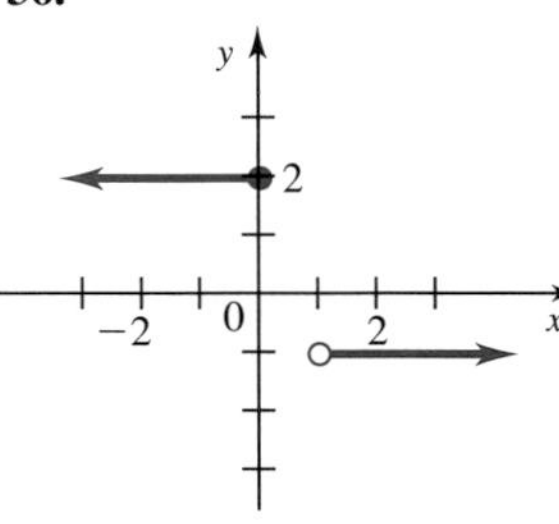

37.

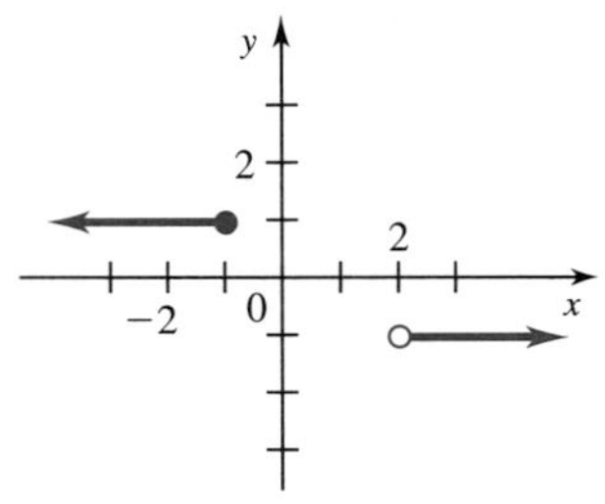

38. When a diabetic takes long-acting insulin, the insulin reaches its peak effect on the blood sugar level in about 3 hr. This effect remains fairly constant for 5 hr, then declines, and is very low until the next injection. In a typical patient, the level of insulin might be given by the following function.

$$i(t) = \begin{cases} 40t + 100 & \text{if } 0 \le t \le 3 \\ 220 & \text{if } 3 < t \le 8 \\ -80t + 860 & \text{if } 8 < t \le 10 \\ 60 & \text{if } 10 < t \le 24 \end{cases}$$

Here $i(t)$ is the blood sugar level, in appropriate units, at time t measured in hours from the time of the injection. Suppose a patient takes insulin at 6 A.M. Find the blood sugar level at each of the following times: **(a)** 7 A.M. **(b)** 9 A.M. **(c)** 10 A.M. **(d)** noon **(e)** 2 P.M. **(f)** 5 P.M. **(g)** midnight. **(h)** Graph $y = i(t)$.

39. The snow depth in Michigan's Isle Royale National Park varies throughout the winter. In a typical winter, the snow depth in inches is approximated by the following function.

$$f(x) = \begin{cases} 6.5x & \text{if } 0 \le x \le 4 \\ -5.5x + 48 & \text{if } 4 < x \le 6 \\ -30x + 195 & \text{if } 6 < x \le 6.5 \end{cases}$$

Here, x represents the time in months with $x = 0$ representing the beginning of October, $x = 1$ representing the beginning of November, and so on.
(a) Graph $f(x)$
(b) In what month is the snow deepest? What is the deepest snow depth?
(c) In what months does the snow begin and end?

Graph the functions defined as follows.

40. $f(x) = [\![-x]\!]$ **41.** $f(x) = [\![2x]\!]$ **42.** $g(x) = [\![2x - 1]\!]$

43. $h(x) = [\![3x + 1]\!]$ **44.** $k(x) = [\![3x]\!]$ **45.** $r(x) = [\![3x]\!] + 1$

46. Describe how the y-values of the greatest-integer function are determined for negative x-values.

47. At Rick's Rentals, a lift truck can be rented for \$74 per day or fraction of a day plus a fixed charge of \$55. Let $T(x)$ represent the cost to rent a lift truck from Rick for x days.
Find the following: **(a)** $T(1)$, **(b)** $T(1.25)$ **(c)** $T(3.5)$
(d) Graph $y = T(x)$. **(e)** Give the domain and range of T.

48. A mail-order firm charges 30¢ to mail a package weighing 1 oz or less, and then 27¢ for each additional ounce or fraction of an ounce. Let $M(x)$ be the cost of mailing a package weighing x oz.
Find **(a)** $M(.75)$ **(b)** $M(1.6)$ **(c)** $M(4)$.
(d) Graph $y = M(x)$.
(e) Give the domain and range of M.

49. Use the greatest-integer function and write an expression for the number of ounces for which postage will be charged on a package weighing x oz (see Exercise 48).

50. A car rental cost \$37 for one day, which includes 50 free miles. Each additional 25 miles or portion costs \$10. Graph the ordered pairs (miles, cost) for a one-day rental. Use the greatest-integer function to write an expression for the cost of a one-day rental for x miles.

51. Suppose that the demand and price for a certain model of electric motor are related by $p - 16 - (5/4)q$, where p is the price in hundreds of dollars and q is the demand in appropriate units. That is, for q units to be sold, the price must be set at $16 - (5/4)q$ hundred dollars.
(a) Find the price at the following demand levels: 0 units, 4 units, 8 units.
(b) Find the demand for the motors at the following prices: \$600, \$1100, \$1600.
(c) According to this function, what is the maximum price that can be charged? What is the maximum demand for the motors?
(d) Graph $p = 16 - (5/4)q$.
(e) Suppose the price (in hundreds of dollars) and the supply are related by $p = (3/4)q$, where q represents the supply and p the price. Find the supply at the following prices: \$0, \$1000, \$2000.
(f) Graph $p = (3/4)q$ on the same axes used for part (d).
(g) Find the equilibrium supply (the supply at the point where the supply and demand graphs cross).
(h) Find the equilibrium price (the price at the equilibrium supply).

52. Let the supply and demand equations for a new textbook in advanced mathematics be as follows, where x is in thousands and p is in dollars.

$$\text{Supply: } p = (3/2)x \quad \text{and} \quad \text{Demand: } p = 81 - (3/4)x$$

(a) Graph these on the same axes.
(b) Find the equilibrium demand. (See Exercise 51.)
(c) Find the equilibrium price.

In Exercises 53–56, decide which of the following numbers gives the slope of the linear function f: **(a)** 3, **(b)** 1, **(c)** 1/3, **(d)** $-1/3$, **(e)** -1, **(f)** -3. *(*Hint: *Use the slope formula and consider the sign of the differences in the numerator and denominator as well as their comparative sizes.)*

53. Number line: $f(b)$, a, b, $f(a)$

54. Number line: a, $f(a)$, $f(b)$, b

55. Number line: a, b, $f(a)$, $f(b)$

56. Number line: a, $f(b)$, b, $f(a)$

Graphing Utility Problems

Use a graphing utility to graph each of the following lines. You may need to rewrite the equation before graphing it.

57. $5y - 4x = 10$

58. $(2y - 2)/(x - 3) = 4$

59. Graph $y = x + 4$ and $y = (x^2 + 3x - 4)/(x - 1)$. Do you see a difference between the graphs? Should there be a difference? If so, describe the difference.

60. Graph $y = |x|$. Then graph $y = |x - 2|$, $y = |x| + 2$, and $y = 2|x|$. How are these four graphs related? How do they differ?

3.5 Equations of a Line

In the previous section we assumed that the graph of a linear function is a straight line. In this section we develop various forms for the equation of a line.

The first equation given is for a **vertical line**, the only line that is not the graph of a function. The vertical line through the point $(k, 0)$ goes through all points of the form (k, y); this fact is used to determine the equation of a vertical line.

Equation of a Vertical Line

For any value of y, an equation of the vertical line through (k, y) is

$$x = k.$$

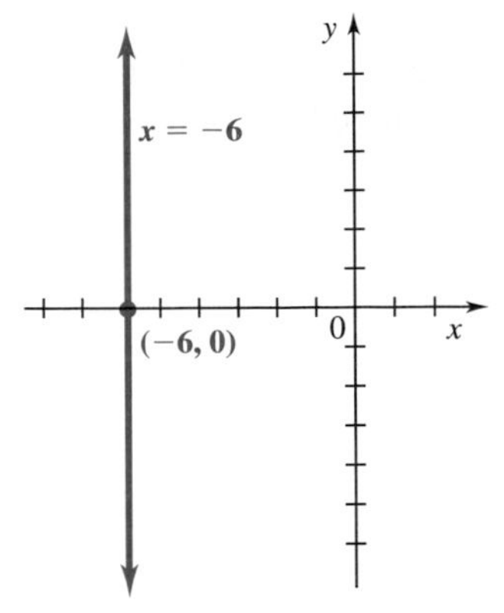

Figure 31

For example, the vertical line through $(-6, 0)$ in Figure 31 has equation $x = -6$.

Point-Slope Form Suppose now that a line has slope m and goes through the fixed point (x_1, y_1), as in Figure 32. Let (x, y) be any other point on this line. By the definition of slope, the slope of this line is

$$\frac{y - y_1}{x - x_1}.$$

Since the slope of the line is m,

$$\frac{y - y_1}{x - x_1} = m.$$

Multiplying both sides by $x - x_1$ gives

$$y - y_1 = m(x - x_1). \qquad (*)$$

Since (x, y) represents any point on the line except (x_1, y_1), and since (x_1, y_1) also satisfies equation $(*)$, all points on the line satisfy equation $(*)$. Thus, equation $(*)$ is the equation of the given line.

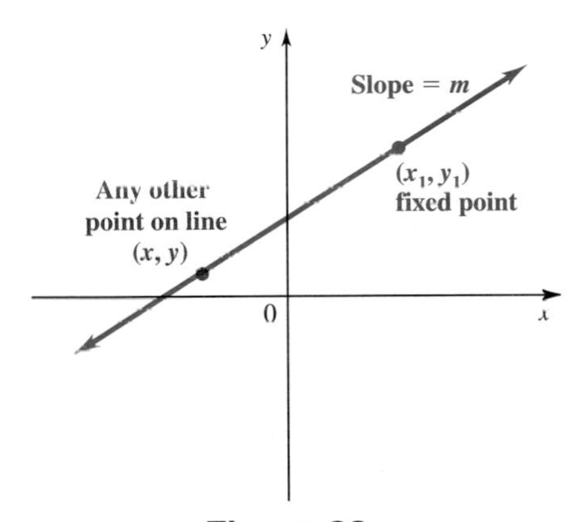

Figure 32

Point-Slope Form

The line with slope m passing through the point (x_1, y_1) has equation

$$\mathbf{y - y_1 = m(x - x_1)},$$

the **point-slope form** of the equation of a line.

The various forms of the equation of a line can be summarized as follows.

Equations of a Line

General equation	*Description of line*
$ax + by + c = 0$	**Standard form** x-intercept $-c/a$, y-intercept $-c/b$, slope $-a/b$ (if $a \neq 0$ and $b \neq 0$)
$x = k$	**Vertical line** x-intercept k, no y-intercept, line has undefined slope
$y = k$	**Horizontal line** No x-intercept, y-intercept k, line has slope 0
$y = mx + b$	**Slope-intercept form** Slope is m, y-intercept is b
$y - y_1 = m(x - x_1)$	**Point-slope form** Slope is m, line passes through (x_1, y_1)

3.5 Exercises

Write an equation in standard form for each of the following lines.

1. Through (2, 1), $m = 5$

2. Through $(-3, 2)$, $m = -4$

3. Through $(-5, 3)$, $m = -1$

4. Through (0, 0), $m = 2$

5. $f(3) = 1$, $m = 2/3$

6. $f(4) = -1$, $m = .4$

7. $f(.6) = 2$, $m = 0$

8. $f(0) = 1/2$, $m = -5$

9. Through (2, 3), vertical

10. Through $(3, -1)$, horizontal

11. Through the origin, horizontal

12. Through $(-5, 1/2)$, vertical

13. x-intercept 3, vertical

14. $f(-1) = 2$, horizontal

15. Through the origin, bisects second and fourth quadrants

16. Through the origin, bisects first and third quadrants

17. Through (3, 4) and (5, 6)

18. Through $(-1, 2)$ and (7, 3)

19. Through $(-4, 2)$ and (4, 18)

20. Through (.1, .3) and (.3, .1)

21. Through (2, 7) and (3, 7)

22. Through (3, 4) and (3, 5)

23. Through $(3, -1)$, parallel to $x - 2y = 4$

24. Through (0, 4), parallel to $-2x + 4y = 7$

25. Through $(-3, 4)$, perpendicular to $y = 4x + 6$

26. Through (5, 6), perpendicular to x-axis

27. $f(-2) = 8$, perpendicular to y-axis

28. $f(4) = 3$, perpendicular to y-axis

29. $f(9) = 0$, parallel to $x = 3y + 4$

30. $f(-2) = 5$, parallel to $y = .2x + 3$

31. x-intercept 3, y-intercept 6

32. x-intercept -4, y-intercept 2

For each linear equation in Exercises 33–38, identify its graph in (a)–(f) below.

33. $y = -x - 1$

34. $y = 1 - x$

35. $y = \frac{1}{2}x$

36. $y = 2$

37. $y = 1 - 2x$

38. $y = 2x - 1$

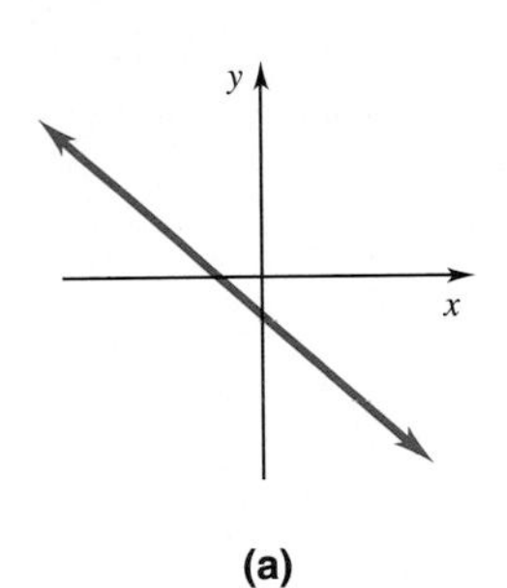

(a)

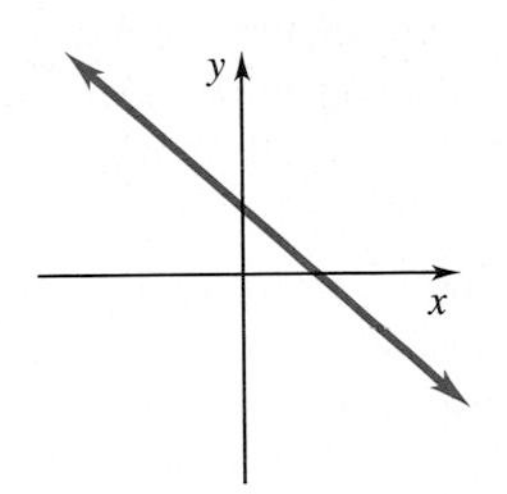

(b)

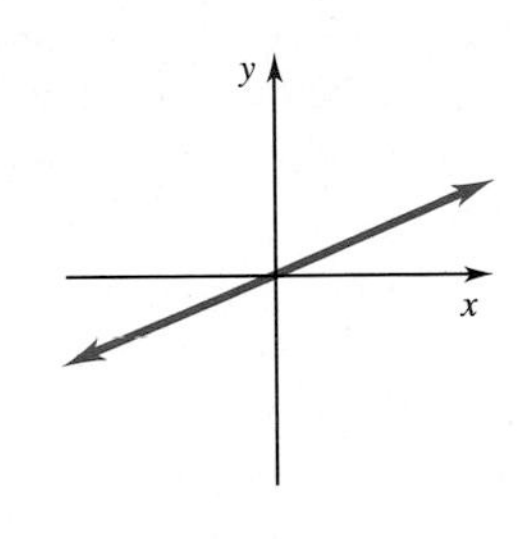

(c)

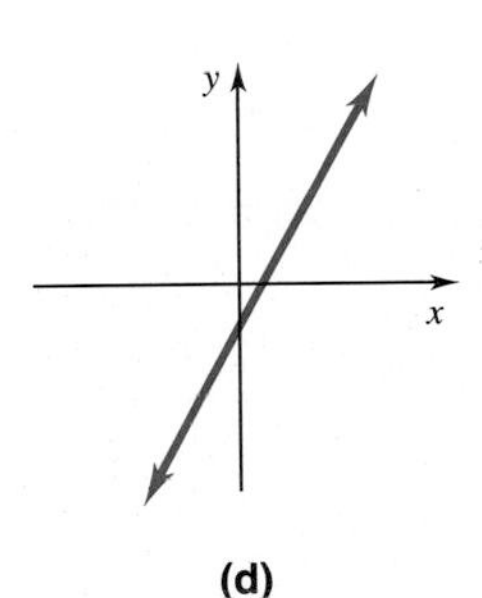

(d)

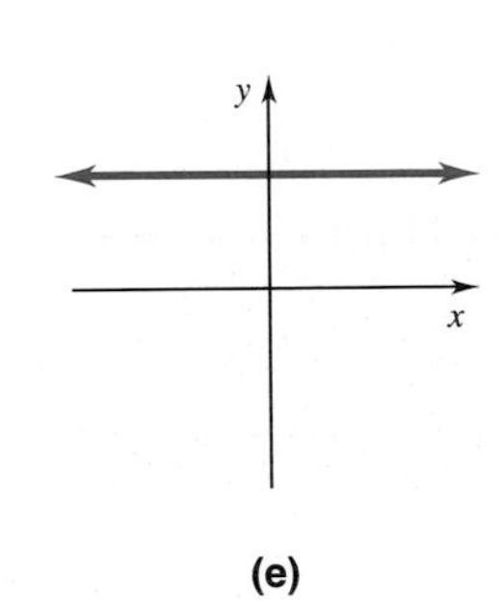

(e)

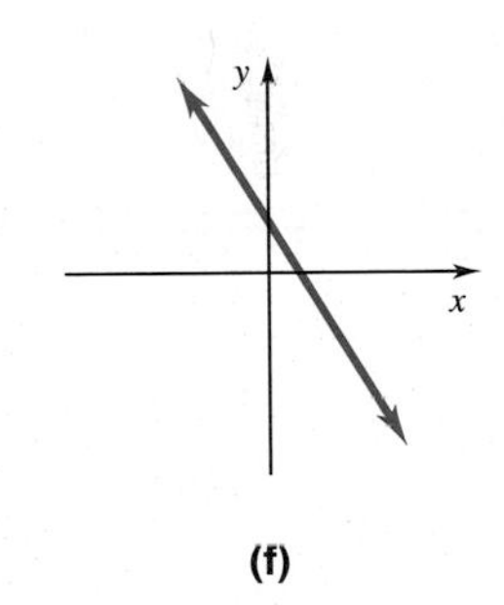

(f)

39. If $f(4) = -1$, $f(2) = 0$, and f is a linear function, find $f(x)$ and $f(-4)$.

40. A linear function has $f(3) = 4$ and $f(4) = 5$. Find $f(x)$ and $f(8)$.

41. Find r so that the line through $(2, 6)$ and $(-4, r)$ is as follows.
(a) Parallel to $2x - 3y = 4$
(b) Perpendicular to $x + 2y = 1$

42. Find k so that the line through $(4, -1)$ and $(k, 2)$ is as follows.
(a) Parallel to $3y + 2x = 6$
(b) Perpendicular to $2y - 5x = 1$

43. Two lines intersect at the point $(5, -4)$ and are each perpendicular to one of the axes. What are the equations of the lines?

44. The following table gives some points on the line $y = mx + b$. Find m and b.

x	2.6	2.8	3	3.2	3.4
y	6.8	6.9	7	7.1	7.2

45. Use slopes to show that the quadrilateral with vertices at $(3/2, 6)$, $(-3/2, 3)$, $(-5/2, 5)$, and $(5/2, 2)$ is not a parallelogram (opposite sides parallel).

46. Use slopes to show that the square with vertices at $(-1, 4)$, $(5, 4)$, $(5, -2)$, and $(-1, -2)$ has diagonals that are perpendicular.

*Many real-life situations can be described approximately by a straight-line graph. One way to find the equation of such a straight line is to use two typical data points from the graph and the point-slope form of the equation of a line. In each of Exercises 47–52, assume that the data can be approximated fairly closely by a straight line. Use the given information to find the equation of the line.** *Then answer the question asked in the problem.*

47. Temperatures of 32° and 212° Fahrenheit correspond to temperatures of 0° and 100° Centigrade. Let y be the Centigrade temperature corresponding to the Fahrenheit temperature x. An increase in temperature of 1° Fahrenheit corresponds to what change in degrees Centigrade?

48. Suppose a baseball is thrown at 85 mph. The ball will travel 320 ft when hit by a bat swung at 50 mph and will travel 440 ft when hit by a bat swung at 80 mph. Let y be the number of feet traveled by the ball when hit by a bat swung at x mph. (*Note:* This function is valid for $50 \leq x \leq 90$, where the bat is 35 inches long, weighs 32 oz, and is swung slightly upward to drive the ball at an angle of 35°.†) How much further will a ball travel for each one-mile-per-hour increase in the speed of the bat?

49. The number of farms in the United States declined from 6 million in 1920 to 2 million in 1980. Let y be the number of farms (in millions) x years after 1900. How many farms were there in 1960?

50. The average size of a farm in the United States increased from 100 acres in 1920 to 700 acres in 1980. Let y be the average size x years after 1900. In what year was the average size 400 acres?

51. The worldwide consumption of cigarettes increased from 2.5 trillion in 1960 to 4 trillion in 1980. Let y be the consumption of cigarettes (in trillions) x years after 1940. In what year will the consumption reach 5.5 trillion?

52. The amount of tropical rain forests in Central America decreased from 130,000 sq mi in 1969 to about 80,000 sq mi in 1985. Let y be the amount (in ten-thousands of square miles) x years after 1965. How large will the rain forests be in the year 1997?

53. Use your own words to describe how to find the equation of a line through two given points.

54. Discuss the advantages and disadvantages of each of the three forms of the equation of a line. When would you use each of them and why? Which form is written in function notation?

55. The product of the slopes of two perpendicular lines is -1. Is this true for *any* two perpendicular lines? Explain. (*Hint:* Is slope defined for every line?)

56. Show that the line $y = x$ is the perpendicular bisector of the segment containing (a, b) and (b, a), where $a \neq b$. (*Hint:* Use the midpoint formula and the slope formula.)

57. Let $y = m_1x + b_1$ and $y = m_2x + b_2$ be the equations of two nonvertical lines.

(a) Set $m_1x + b_1$ equal to $m_2x + b_2$ and solve for x to find the x-coordinate of the point of intersection of the lines.

(b) If $m_1 = m_2$ and $b_1 \neq b_2$, the lines must be parallel. Why?

*The information for Exercises 49–52 was taken from *Mathematics and Global Survival*, Second Edition, by Richard H. Schwartz (Ginn Press, 1991).

†Adair, Robert K., *The Physics of Baseball*; (New York: Harper & Row, 1990).

58. Temperature decreases with height above the earth's surface. We can use the following chart to find a function that describes this relationship.

Height (feet)	*Temperature (°F)*
1000	56
5000	41
10,000	23
15,000	5
20,000	−15
30,000	−47
36,100	−69

(a) Let x represent the height in thousands of feet and y represent the temperature in degrees Fahrenheit. Plot the ordered pairs (height, temperature) corresponding to the values in the table. The points should lie in an approximately linear pattern.
(b) Use the ordered pairs $(5, 41)$ and $(30, -47)$ to find the slope of the line through these two points. Then substitute the values from either ordered pair into the equation $y = mx + b$, which defines a linear function, to find b. Use the values of m and b to write the equation that defines the linear function describing the relationship between height and temperature.
(c) Test the function found in part (b) by substituting other heights from the chart to see if the predicted temperatures are close to the actual ones. What do you find?

59. Use the ordered pairs $(20, -15)$ and $(10, 23)$ to repeat Exercise 58(b). Compare the two equations. Are they similar? Do they give equally good predictions?

3.6 Symmetry, Translation, and Reflection

One of the main objectives of this course is to recognize and learn to graph various functions. Several graphing techniques are presented in this section that show how to graph functions that are defined by altering the equation of another function in certain ways.

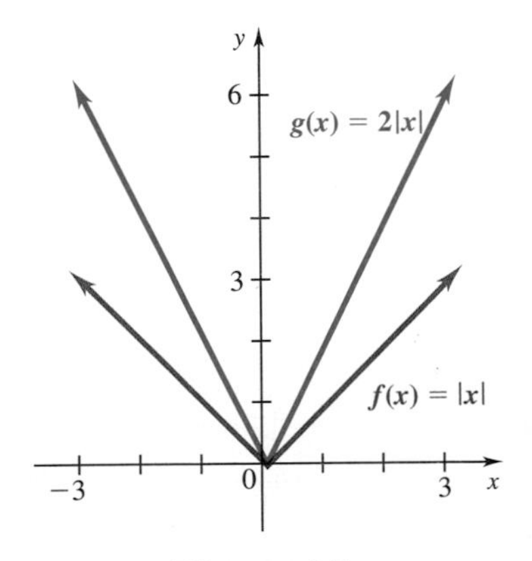

Figure 35

Stretching and Shrinking We begin by considering how the graph of $y = af(x)$ compares to the graph of $y = f(x)$.

EXAMPLE 1 Graph each of the following.

(a) $g(x) = 2|x|$.

Use a grapher or plot a few points to get the graph of $g(x)$, shown in blue in Figure 35. The graph of $f(x) = |x|$ is shown in red for comparison. Since each y-value in $g(x)$ is twice the corresponding y-values in $f(x)$, the graph of $g(x)$ is narrower than that of $f(x)$.

(b) $h(x) = (1/2)|x|$

The graph of $h(x)$ is again the same general shape as that of $f(x)$, but here, the coefficient $1/2$ causes the graph of $h(x)$ to be broader than the graph of $f(x)$. See Figure 36.

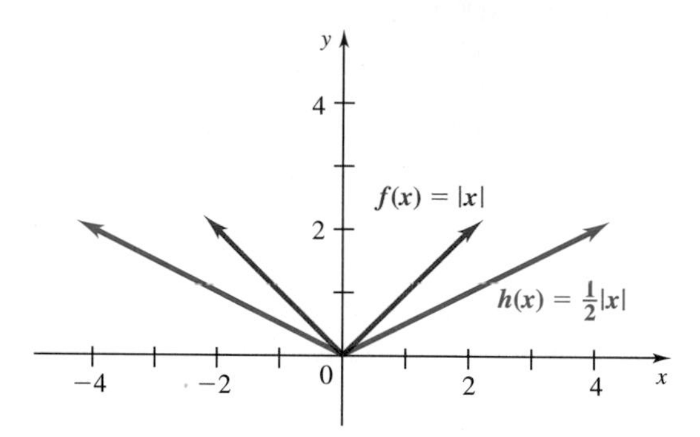

Figure 36

The graphs in Figures 35 and 36 suggest the following generalizations.

Stretching and Shrinking

The graph of $g(x) = af(x)$ has the same shape as the graph of $f(x)$, and it is

narrower than the graph of $f(x)$ if $|a| > 1$;
broader than the graph of $f(x)$ if $0 < |a| < 1$.

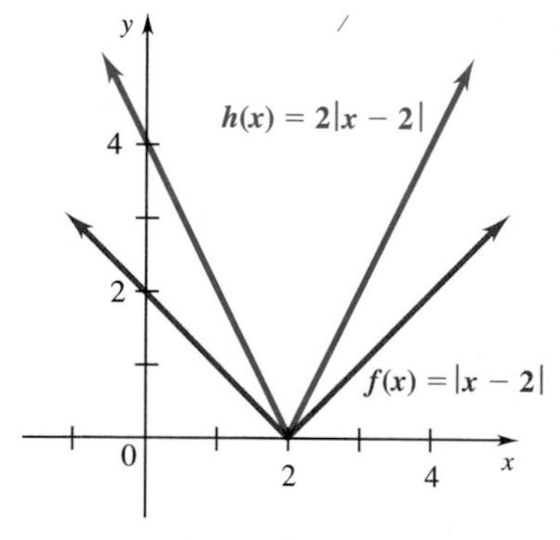

Figure 37

EXAMPLE 2 Graph $h(x) = |2x - 4|$.

Factor out 2 as follows.

$$\begin{aligned} h(x) &= |2x - 4| \\ &= |2(x - 2)| \\ &= |2| \cdot |x - 2| \\ &= 2|x - 2| \end{aligned}$$

The graph of h will be narrower than the graph of $f(x) = |x - 2|$, because $2 > 1$. See Figure 37.

EXAMPLE 3 Graph $g(x) = -|x|$.

Use a grapher or find enough points to sketch the graph of $g(x) = -|x|$. The result is shown in blue in Figure 38. The graph of $f(x) = |x|$ is shown in red for comparison. As the equation suggests, every y-value of the graph of $g(x) = -|x|$ is the negative of the corresponding y-value of $f(x) = |x|$. This has the effect of reflecting the graph about the x-axis.

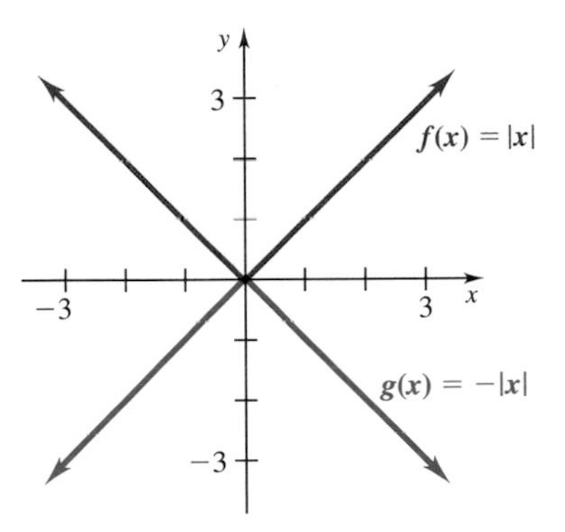

Figure 38

The graph of the function defined by $f(x) = \sqrt{x}$ $(x \geq 0)$ is shown in red in Figure 39. The figure also shows the graph of $g(x) = \sqrt{-x}$ $(x \leq 0)$. The graphs in Figures 38 and 39 suggest the following generalization.

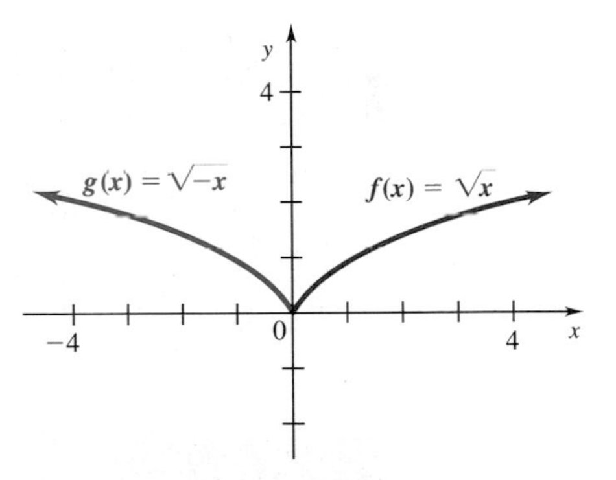

Figure 39

Reflection About an Axis

The graph of $y = -f(x)$ is the same as the graph of $y = f(x)$ reflected about the x-axis.

The graph of $y = f(-x)$ is the same as the graph of $y = f(x)$ reflected about the y-axis.

EXAMPLE 8 Graph $f(x) = -|x + 3| + 1$.

The "vertex" of this graph is translated 3 units to the left and 1 unit up, as shown in Figure 48. The graph opens downward because of the negative sign in front of the absolute value symbol, making the "vertex" the *highest* point on the graph. The graph is symmetric with respect to the line $x = -3$.

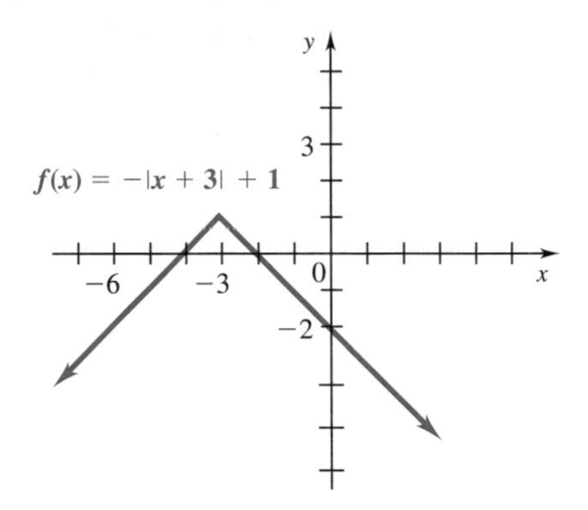

Figure 48

In general, the graph of a function g, defined by $g(x) = f(x) + c$, where c is a real number, can be found from the graph of the function f as follows. For every point (x, y) on the graph of f, there will be a corresponding point $(x, y + c)$ on the graph of g. The new graph will be the same as the graph of f, but translated c units upward if c is positive, or $|c|$ units downward if c is negative. The graph of g is called a **vertical translation** of the graph of f. Figure 49 shows a graph of a function f and two different vertical translations of f.

If a function g is defined by $g(x) = f(x - c)$, for each ordered pair (x, y) of f, there will be a corresponding ordered pair $(x - c, y)$ on the graph of g. This has the effect of translating the graph of f horizontally; c units to the right if c is positive and $|c|$ units to the left if c is negative. Figure 50 shows the graph of $y = f(x)$ along with the graphs of $y = f(x - 3)$ and $y = f(x + 2)$; each of these graphs is obtained from that of $y = f(x)$ by a **horizontal translation**.

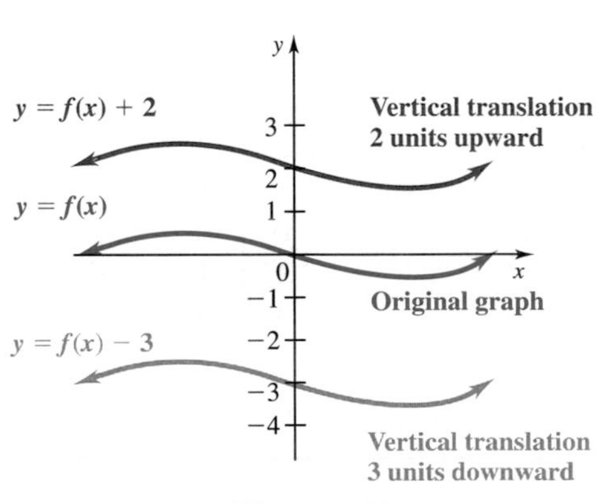

Figure 49

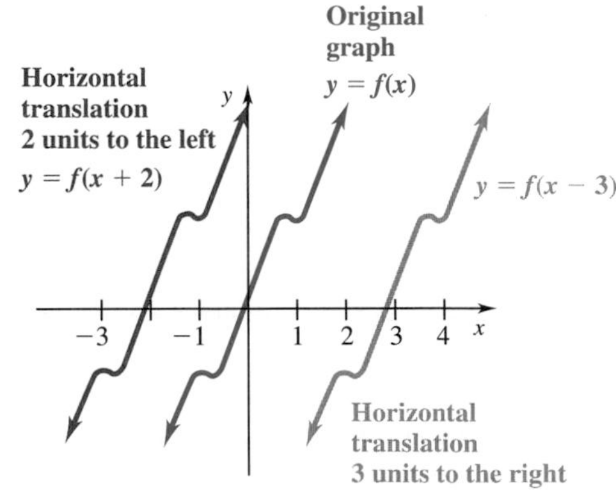

Figure 50

Translations of the Graph of a Function

Let f be a function, and let c be a positive number.

To graph:	*Shift the graph of* $y = f(x)$ *by c units:*
$y = f(x) + c$	upward
$y = f(x) - c$	downward
$y = f(x + c)$	left
$y = f(x - c)$	right

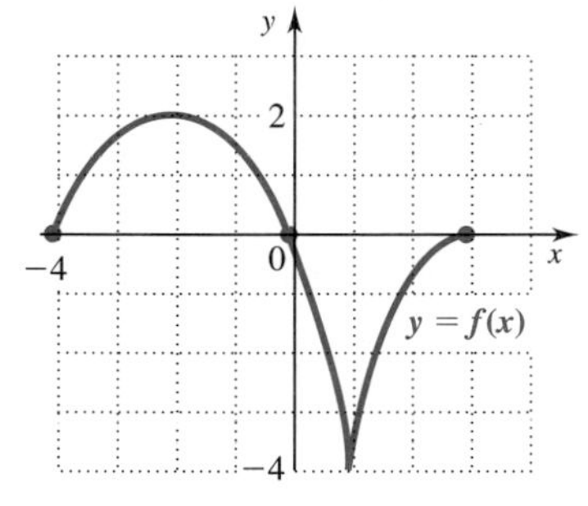

Figure 51

EXAMPLE 9 A graph of a function defined by $y = f(x)$ is shown in Figure 51. Use this graph to find each of the following graphs.

(a) $g(x) = f(x) + 3$

This graph is the same as the graph in Figure 51, translated 3 units upward. See Figure 52(a).

(b) $h(x) = f(x + 3)$

To get the graph of $y = f(x + 3)$, translate the graph of $y = f(x)$ three units to the left. See Figure 52(b).

(c) $k(x) = f(x - 2) + 3$

This graph will look like the graph of $f(x)$ translated 2 units to the right and 3 units up, as shown in Figure 52(c).

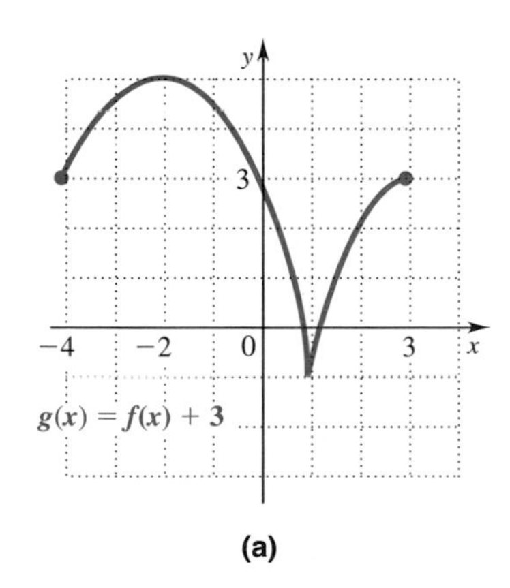

(a)

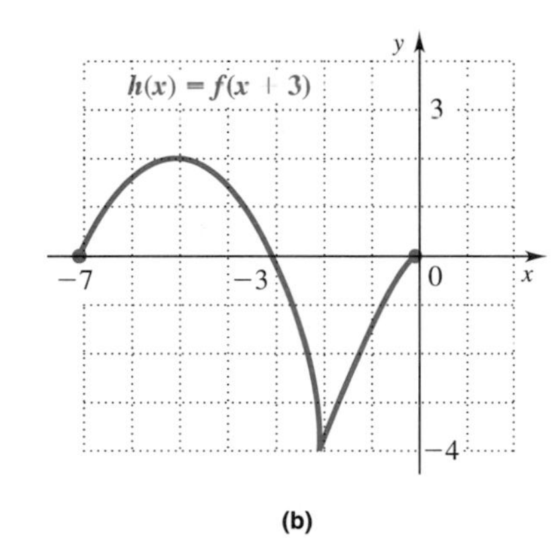

(b)

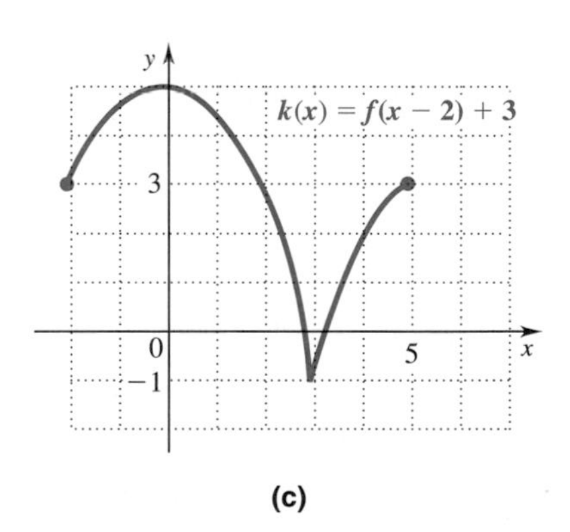

(c)

Figure 52

3.6 Exercises

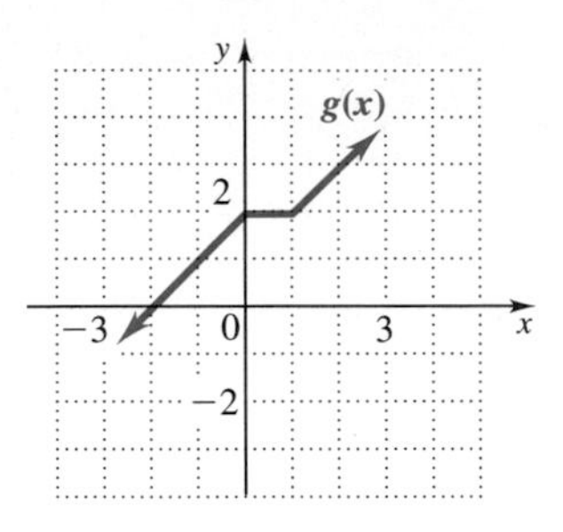

1. Given the graph of $g(x)$ in the figure, sketch the graph of each of the following and explain how it is obtained from the graph of $y = g(x)$.
(a) $y = g(-x) + 1$ **(b)** $y = g(x - 2)$
(c) $y = g(x + 1) - 2$ **(d)** $y = -g(x) + 2$

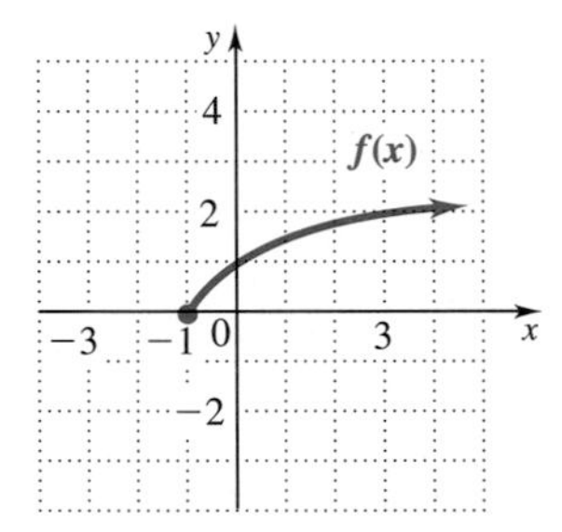

2. Use the graph of $f(x)$ in the figure to obtain the graph of each of the following. Explain how each graph is related to the graph of $f(x)$.
(a) $y = -f(x)$ **(b)** $y = 2f(x)$
(c) $y = f(x - 1) + 3$ **(d)** $y = f(-x)$

Plot the following points, and then plot the points that are symmetric to the given point with respect to the **(a)** *x-axis,* **(b)** *y-axis,* **(c)** *origin.*

3. $(5, -3)$ **4.** $(-6, 1)$ **5.** $(-4, -2)$ **6.** $(-8, 0)$

Without graphing, determine whether each equation has a graph that is symmetric with respect to the x-axis, the y-axis, or the origin.

7. $y = (x + 3)(x - 5)$ **8.** $x^2 + y^2 = 6$ **9.** $xy = -3$ **10.** $3x/y = 2$

11. $y = -x^3$ **12.** $y + 1 = (x - 3)^3$ **13.** $x^2 + 4y^2 = 3$ **14.** $y = 1/(2 - x^2)$

Use the techniques presented in this section to help graph each of the following.

15. $|x| = y + 1$ **16.** $y - 2 = |x + 3|$ **17.** $x/y = 2$ **18.** $xy = 1$

19. $y = -(x + 1)^3$ **20.** $y = (-x + 1)^3$ **21.** $y = 2x^2 - 1$ **22.** $y = (2/3)(x - 2)^2$

23. $y + 2 = (1 - x)^2$ **24.** $x^2 = y^2$ **25.** $(x - 1)^4 = y - 3$ **26.** $x^2y = 2$

Suppose that $f(2) = 3$*. For each assumption in Exercises 27–29, find another value of the function.*

27. $f(x)$ is symmetric with respect to the origin.

28. $f(x)$ is symmetric with respect to the y-axis.

29. $f(x)$ is symmetric with respect to the line $x = 6$.

30. Find the function whose graph can be obtained by translating the graph of $f(x) = 2x + 5$ up 2 units and left 3 units.

31. Find the function whose graph can be obtained by translating the graph of $f(x) = 3 - x$ down 2 units and right 3 units.

Explain why each of the following statements is true.

32. The function $f(x) = 0$ is both even and odd.

33. A nonzero function cannot be both even and odd.

34. If (a, b) is on the graph of an even function, then so is $(-a, b)$.

35. If (a, b) is on the graph of an odd function, then so is $(-a, -b)$.

36. Complete the left half of the graph of $f(x)$ in the figure for each of the following conditions:
(a) $f(-x) = f(x)$ **(b)** $f(-x) = -f(x)$.

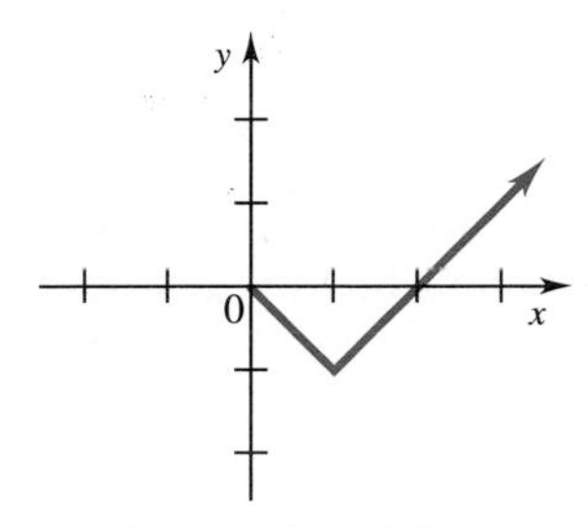

Exercise 36

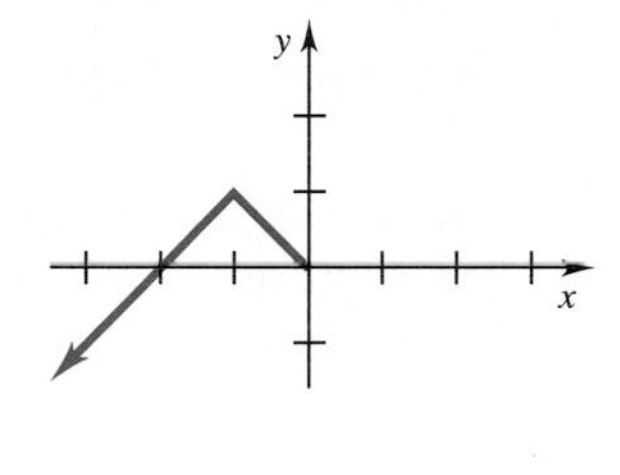

Exercise 37

37. Complete the right half of the graph of $f(x)$ in the figure for each of the following conditions: **(a)** $f(x)$ is odd, **(b)** $f(x)$ is even.

In Exercises 38 and 39, let F be some algebraic expression involving x as the only variable.

38. Suppose the equation $y = F$ is changed to $y = c \cdot F$, for some constant c. What is the effect on the graph of $y = F$? Discuss the effect depending on whether $c > 0$ or $c < 0$, and $|c| > 1$ or $|c| < 1$.

39. Suppose $y = F(x)$ is changed to $y = F(x + h)$. How are the graphs of these equations related? Is the graph of $y = F(x) + h$ the same as the graph of $y = F(x + h)$? If not, how do they differ?

Sketch examples of graphs having the given characteristics.

40. Symmetric with respect to the x-axis but not to the y-axis

41. Symmetric with respect to the origin but to neither the x-axis nor the y-axis

Prove each statement below.

42. A graph symmetric with respect to both the x-axis and the y-axis is also symmetric with respect to the origin.

43. A graph possessing two of the three types of symmetry, with respect to the x-axis, y-axis, and origin, must possess the third type of symmetry also.

Answer true *or* false *in Exercises 44–46.*

44. The graph of a nonzero function cannot be symmetric with respect to the x-axis.

45. The graph of an even function is symmetric with respect to the y-axis.

46. The graph of an odd function is symmetric with respect to the x-axis.

Graphing Utility Problems

Let $f(x) = 8x^3 - 12x^2 + 2x + 1$. *Sketch* $f(x)$ *as y1 and each function defined as follows as y2. Start with x- and y-ranges of* $[-1, 2]$ *and* $[-2, 2]$ *and adjust as necessary. In each case, describe the translation or reflection involved.*

47. $y = f(x - 1)$ **48.** $y = f(x) - 1$ **49.** $y = -f(x)$ **50.** $y = f(-x)$

3.7 Algebra of Functions; Composite Functions

Economists frequently use the equation "profit equals revenue minus cost," or $P = R - C$. That is, the profit function is found by subtracting the cost function from the revenue function. New functions can be formed by using other operations as well.

The various operations on functions are defined below.

Definition of Operations on Functions

Given two functions f and g, then for all values of x for which both $f(x)$ and $g(x)$ exist, the functions $f + g$, $f - g$, fg, and f/g are defined as follows.

Sum	$(f + g)(x) = f(x) + g(x)$
Difference	$(f - g)(x) = f(x) - g(x)$
Product	$(fg)(x) = f(x) \cdot g(x)$
Quotient	$\left(\dfrac{f}{g}\right)(x) = \dfrac{f(x)}{g(x)}, \quad g(x) \neq 0$

NOTE The condition $g(x) \neq 0$ in the definition of the quotient means that the domain of $\left(\dfrac{f}{g}\right)(x)$ consists of all values of x for which $g(x)$ is not zero. The condition does not mean that $g(x)$ is a function that is never zero. ■

EXAMPLE 1 Let $f(x) = x^2 + 1$, and $g(x) = 3x + 5$. Find each of the following.

(a) $(f + g)(1)$

Since $f(1) = 2$ and $g(1) = 8$, use the definition above to get

$$(f + g)(1) = f(1) + g(1) = 2 + 8 = 10.$$

(b) $(f - g)(-3) = f(-3) - g(-3) = 10 - (-4) = 14$

(c) $(fg)(5) = f(5) \cdot g(5) = 26 \cdot 20 = 520$

(d) $\left(\dfrac{f}{g}\right)(0) = \dfrac{f(0)}{g(0)} = \dfrac{1}{5}$ ■

EXAMPLE 2 Let $f(x) = 8x - 9$ and $g(x) = \sqrt{2x - 1}$.

(a) $(f + g)(x) = f(x) + g(x) = 8x - 9 + \sqrt{2x - 1}$

(b) $(f - g)(x) = f(x) - g(x) = 8x - 9 - \sqrt{2x - 1}$

(c) $(fg)(x) = f(x) \cdot g(x) = (8x - 9)\sqrt{2x - 1}$

(d) $\left(\dfrac{f}{g}\right)(x) = \dfrac{f(x)}{g(x)} = \dfrac{8x - 9}{\sqrt{2x - 1}}$ ■

In Example 2, the domain of f is the set of all real numbers, while the domain of g, where $g(x) = \sqrt{2x - 1}$, includes just those real numbers that make $2x - 1 \geq 0$; the domain of g is the interval $[1/2, \infty)$. The domains of $f + g$, $f - g$, and fg are thus $[1/2, \infty)$. With f/g, the denominator cannot be zero, so the value $1/2$ is excluded from the domain. The domain of f/g is $(1/2, \infty)$.

The domains of $f + g$, $f - g$, fg, and f/g are summarized below. (Recall that the intersection of two sets is the set of all elements belonging to *both sets*.)

Domains

For functions f and g, the domains of $f + g$, $f - g$, and fg include all real numbers in the intersection of the domains of f and g, while the domain of f/g includes those real numbers in the intersection of the domains of f and g for which $g(x) \neq 0$.

Composition of Functions The sketch in Figure 53 shows a function f that assigns to each element x of set X some element y of set Y. Suppose also that a function g takes each element of set Y and assigns a value z of set Z. Using both f and g, then, an element x in X is assigned an element z in Z. The result of this process is a new function h, which takes an element x in X and assigns an element z in Z.

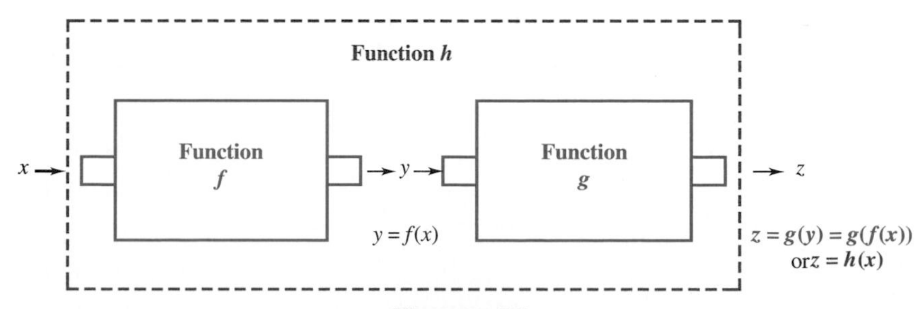

Figure 53

This function h is called the **composition** of functions g and f, written $g \circ f$, (read "g of f"), and defined as follows.

Definition of Composition of Functions

Let f and g be functions. The **composite function**, or **composition**, of g and f, written $g \circ f$, is defined by

$$(g \circ f)(x) = g(f(x)),$$

for all x in the domain of f such that $f(x)$ is in the domain of g.

The notation $g(f(x))$ is read "g of f of x".

EXAMPLE 3 Given $f(x) = 2x - 1$ and $g(x) = 4/(x - 1)$, find each of the following.

(a) $f(g(2))$

First find $g(2)$. Since $g(x) = 4/(x - 1)$,

$$g(2) = \frac{4}{2 - 1} = \frac{4}{1} = 4.$$

Now find $f(g(2)) = f(4)$:

$$f(x) = 2x - 1$$
$$f(g(2)) = f(4) = 2(4) - 1 = 7.$$

(b) $g(f(-3))$

$$f(-3) = 2(-3) - 1 = -7$$
$$g(f(-3)) = g(-7) = \frac{4}{-7 - 1} = \frac{4}{-8} = -\frac{1}{2}$$ ■

EXAMPLE 4 Let $f(x) = 4x + 1$ and $g(x) = 2x^2 + 5x$. Find each of the following.

(a) $(g \circ f)(x)$

By definiton, $(g \circ f)(x) = g(f(x))$. Use the given functions as follows.

$$\begin{aligned}(g \circ f)(x) &= g(f(x)) \\ &= g(\mathbf{4x + 1}) \\ &= 2(\mathbf{4x + 1})^2 + 5(\mathbf{4x + 1}) \\ &= 2(16x^2 + 8x + 1) + 20x + 5 \\ &= 32x^2 + 16x + 2 + 20x + 5 \\ &= 32x^2 + 36x + 7\end{aligned}$$

(b) $(f \circ g)(x)$

Use the definition above with f and g interchanged, so that $(f \circ g)(x)$ becomes $f(g(x))$.

$$\begin{aligned}(f \circ g)(x) &= f(g(x)) \\ &= f(2x^2 + 5x) \\ &= 4(2x^2 + 5x) + 1 \\ &= 8x^2 + 20x + 1\end{aligned}$$

As this example shows, it is not always true that $f \circ g = g \circ f$. In fact, $f \circ g$ is very rarely equal to $g \circ f$. In Example 4, the domain of both composite functions is the set of all real numbers.

CAUTION In general, the composite function $f \circ g$ is not the same as the product $f \cdot g$. For example, with f and g defined as in Example 4,

$$f \circ g = 8x^2 + 20x + 1$$

but

$$f \cdot g = (4x + 1)(2x^2 + 5x) = 8x^3 + 22x^2 + 5x.$$

EXAMPLE 5 Let $f(x) = 1/x$ and $g(x) = \sqrt{3 - x}$. Find $f \circ g$ and $g \circ f$. Give the domain of each.

First find $f \circ g$.

$$\begin{aligned}(f \circ g)(x) &= f(g(x)) = f(\sqrt{3 - x}) \\ &= \frac{1}{\sqrt{3 - x}}\end{aligned}$$

The denominator $\sqrt{3 - x}$ is a nonzero real number only when $3 - x > 0$, or $x < 3$, so that the domain of $f \circ g$ is the interval $(-\infty, 3)$.

Use the same functions to find $g \circ f$, as follows.

$$\begin{aligned}(g \circ f)(x) &= g(f(x)) = g\left(\frac{1}{x}\right) \\ &= \sqrt{3 - \frac{1}{x}} = \sqrt{\frac{3x - 1}{x}}\end{aligned}$$

The domain of $g \circ f$ is the set of all real numbers x such that $x \neq 0$ and $3 - f(x) \geq 0$. Since $3 - f(x) = 3 - (1/x) = (3x - 1)/x$, this is the same as $(3x - 1)/x \geq 0$ and $x \neq 0$. By the methods in Section 2.6, the set $(-\infty, 0) \cup [1/3, \infty)$ is the domain of $g \circ f$.

In Example 5 we found $(f \circ g)(x)$ given $f(x)$ and $g(x)$. Sometimes it is necessary to work backwards and find $f(x)$ and $g(x)$ given $(f \circ g)(x)$. This skill is useful in calculus.

EXAMPLE 6 Find functions f and g such that

$$(f \circ g)(x) = (x^2 - 5)^3 - 4(x^2 - 5) + 3.$$

One pair of functions that will work is

$$f(x) = x^3 - 4x + 3 \quad \text{and} \quad g(x) = x^2 - 5.$$

Then

$$\begin{aligned}(f \circ g)(x) &= f(g(x)) \\ &= f(x^2 - 5) \\ &= (x^2 - 5)^3 - 4(x^2 - 5) + 3.\end{aligned}$$

Other pairs of functions f and g might also work. For instance,

$$f(x) = (x - 5)^3 - 4(x - 5) + 3$$

and

$$g(x) = x^2.$$

3.7 Exercises

For each of the pairs of functions defined as follows, find $f + g$, $f - g$, fg, and f/g. Give the domain of each.

1. $f(x) = 3x + 4, \quad g(x) = 2x - 5$
2. $f(x) = 6 - 3x, \quad g(x) = -4x + 1$
3. $f(x) = 2x^2 - 3x, \quad g(x) = x^2 - x + 3$
4. $f(x) = 4x^2 + 2x - 3, \quad g(x) = x^2 - 3x + 2$
5. $f(x) = \sqrt{4x - 1}, \quad g(x) = \sqrt{x + 3}$
6. $f(x) = \sqrt{5x - 4}, \quad g(x) = \sqrt{3x - 1}$

Let $f(x) = 5x^2 - 2x$ and let $g(x) = 6x + 4$. Find each of the following.

7. $(f + g)(3)$
8. $(f - g)(-5)$
9. $(fg)(4)$
10. $(fg)(-3)$
11. $\left(\dfrac{f}{g}\right)(-1)$
12. $\left(\dfrac{f}{g}\right)(4)$
13. $(f - g)(m)$
14. $(f + g)(2k)$
15. $(f \circ g)(2)$
16. $(f \circ g)(-5)$
17. $(g \circ f)(2)$
18. $(g \circ f)(-5)$

Find $f \circ g$ and $g \circ f$ for each of the pairs of functions defined as follows.

19. $f(x) = -6x + 9, \quad g(x) = 5x + 7$
20. $f(x) = 8x + 12, \quad g(x) = 3x - 1$
21. $f(x) = 4x^2 + 2x + 8, \quad g(x) = x + 5$
22. $f(x) = 5x + 3, \quad g(x) = -x^2 + 4x + 3$
23. $f(x) = \dfrac{2}{x^4}, \quad g(x) = 2 - x$
24. $f(x) = \dfrac{1}{x}, \quad g(x) = x^2$
25. $f(x) = 9x^2 - 11x, \quad g(x) = 2\sqrt{x + 2}$
26. $f(x) = \sqrt{x + 2}, \quad g(x) = 8x^2 - 6$
27. $f(x) = \begin{cases} 2x & \text{if } x < 0 \\ 3x - 5 & \text{if } x \geq 0 \end{cases}$

    $g(x) = \begin{cases} x + 3 & \text{if } x \geq 1 \\ 4x & \text{if } x < 1 \end{cases}$
28. $f(x) = \begin{cases} 3x - 1 & \text{if } x \leq -1 \\ x - 1 & \text{if } x > -1 \end{cases}$

    $g(x) = \begin{cases} x^2 + 1 & \text{if } x < 0 \\ x - 1 & \text{if } x \geq 0 \end{cases}$

29. Describe the steps required to find the composite function $f \circ g$, given $f(x) = 2x - 5$ and $g(x) = x^2 + 3$.

30. Composition is an operation that is unique to functions. Is composition of functions commutative? That is, does $f \circ g = g \circ f$ for all functions f and g? Explain.

The graphs of functions f and g are shown. Use these graphs to find the values in Exercises 31–38.

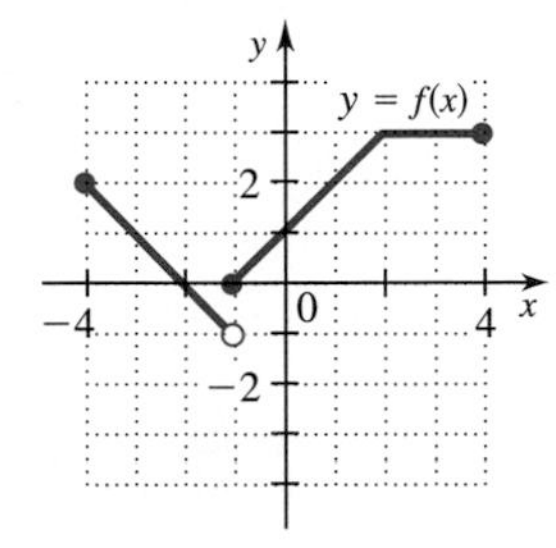

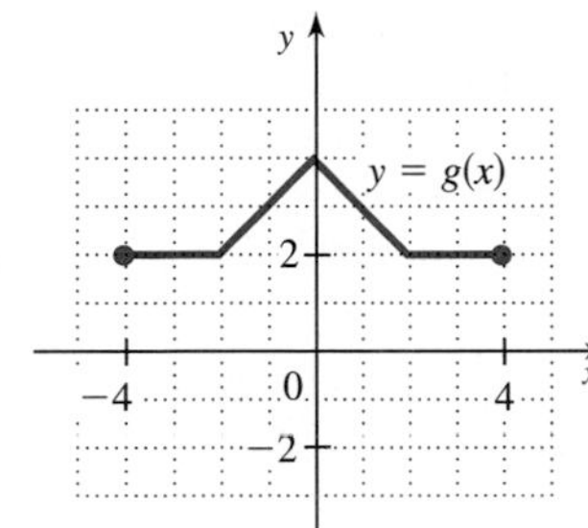

31. $f(1) + g(1)$

32. $f(4) - g(3)$

33. $f(-2) \cdot g(4)$

34. $\dfrac{f(4)}{g(2)}$

35. $(f \circ g)(2)$

36. $(g \circ f)(2)$

37. $(g \circ f)(-4)$

38. $(f \circ g)(-2)$

39. The graphs of functions f and g are shown. Draw the graph of $f \circ g$.

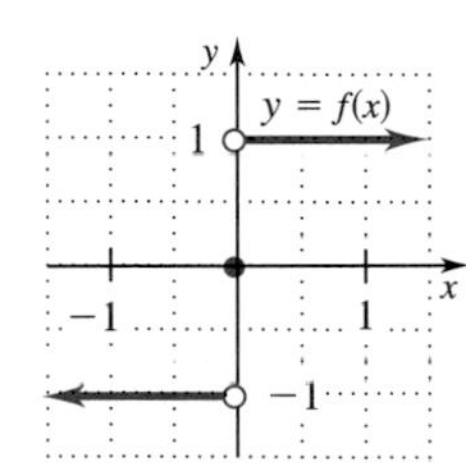

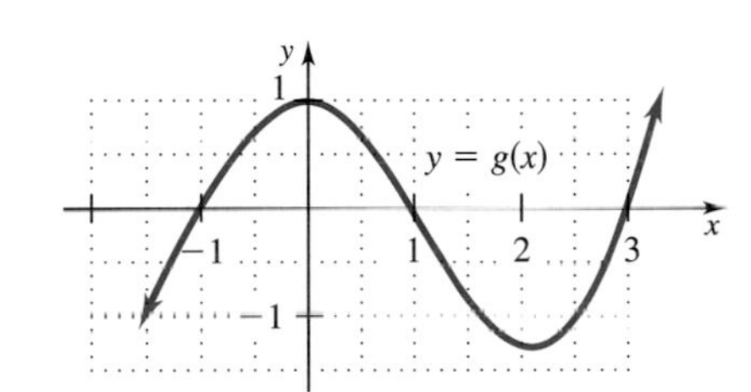

For each of the pairs of functions defined as follows, show that $(f \circ g)(x) = x$ and $(g \circ f)(x) = x$.

40. $f(x) = 2x + 3, \quad g(x) = (x - 3)/2$

41. $f(x) = (x + 5)/3, \quad g(x) = 3x - 5$

42. $f(x) = 2x^3 - 1, \quad g(x) = \sqrt[3]{(x + 1)/2}$

43. $f(x) = x^3 + 4, \quad g(x) = \sqrt[3]{x - 4}$

In Exercises 44–49, a function h is defined. Find functions f and g such that $h(x) = (f \circ g)(x)$. Many such pairs of functions exist.

44. $h(x) = (2x - 5)^3$

45. $h(x) = (8x^2 - 3x)^4$

46. $h(x) = \sqrt{x^2 - 1}$

47. $h(x) = \dfrac{1}{x^2 + 2}$

48. $h(x) = \dfrac{(x - 2)^2 + 1}{5 - (x - 2)^2}$

49. $h(x) = (x + 2)^3 - 3(x + 2)^2$

50. The demand for a new camera is given by $D(p) = (-p^2/100) + 500$, where p is the price in dollars. The price, in terms of the cost c to make the camera, is expressed as $p(c) = 2c - 10$. Find the demand in terms of the cost c.

51. The population P of a certain mammal depends on the number x (in hundreds) of a smaller mammal that serves as its primary food supply. The number x (in hundreds) of the smaller mammal depends upon the amount (in appropriate units) of its food supply, a type of plant. Suppose $P(x) = 2x^2 + 1$ and $x = f(a) = 3a + 2$. Find $(P \circ f)(a)$, the relationship between the population P of the larger mammal and the amount a of plants available to serve as food for the smaller mammal.

52. When a thermal inversion layer is over a city (as happens often in Los Angeles), pollutants cannot rise vertically but are trapped below the layer and must disperse horizontally. Assume that a factory smokestack begins emitting a pollutant at 8 A.M. Assume that the pollutant disperses horizontally, forming a circle. If t represents the time, in hours, since the factory began emitting pollutants ($t = 0$ represents 8 A.M.), assume that the radius of the circle of pollution is $r(t) = 2t$ mi. Let $A(r) = \pi r^2$ represent the area of a circle of radius r. Find and interpret $(A \circ r)(t)$.

53. An oil well off the Gulf Coast is leaking, with the leak spreading oil over the surface of the gulf as a circle. At any time t, in minutes, after the beginning of the leak, the radius of the circular oil slick on the water surface is $r(t) = 4t$ ft. Let $A(r) = \pi r^2$ represent the area of a circle of radius r. Find and interpret $(A \circ r)(t)$.

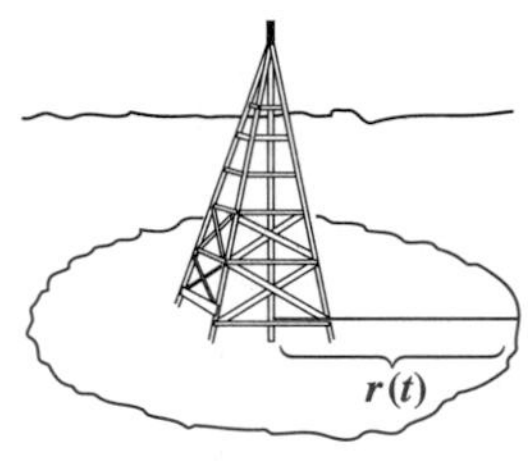

Exercise 53

54. The area of a square is x^2 square inches. If 3 inches is added to one dimension and 1 inch is subtracted from the other dimension, express the area $A(x)$ of the resulting rectangle as a product of two functions.

55. A charter flight charges a fare of \$300 per person plus \$20 per person for each unsold seat on the plane. The plane holds 200 passengers. Let x represent the number of unsold seats.

(a) Find an expression in x for the number of people flying.
(b) Find an expression in x for the price per ticket.
(c) Write an expression for the total revenue $R(x)$.

56. The revenue from sales of the camera in Exercise 50 is given by $R(x) = xp$, where x, the number sold, is given by $1.5c$. Find $R(c)$.

57. Suppose $g(x) = x - 5$.

(a) For any function f, the graph of $f \circ g$ is a translation of the graph of f. Describe the translation.
(b) For any function f, the graph of $g \circ f$ is a translation of the graph of f. Describe the translation.

58. Suppose $g(x) = -x$. How are the graphs of $f \circ g$ and $g \circ f$ related to the graph of f? (*Hint:* See Exercise 57.)

59. Let $f(x) = x/(x - 1)$ for $x \neq 1$. Show that $f \circ f = x$.

60. If $f(x) = 1 + x$ and $g(x) = 1/(1 + x)$, what is the domain of $f(x) \cdot g(x)$?

Graphing Utility Problems

61. Graph $f(x) = 2x + 5$ and $g(x) = (x - 5)/2$ on the same coordinate system. Add the line $y = x$. What do you notice about the graphs of f and g?

62. Repeat Exercise 61 for $f(x) = .5x^3 - 1$ and $g(x) = \sqrt[3]{2(x + 1)}$.

63. Graph $f(x) = |2x - 1|/(2x - 1)$ using x- and y-ranges of $[-3, 3]$. Separately, graph $g(x) = (2x - 1)/(2x - 1)$. Are the graphs the same? Explain. What is the domain of each function? Do the graphs show the domains correctly?

3.8 Inverse Functions

The operations of addition and subtraction "undo" one another. That is, adding 5 to a number and then subtracting 5 from the sum gives back the original number. Certain pairs of functions also "undo" one another. To have this property the functions must both be *one-to-one functions.*

One-to-One Functions Given the function $y = x^2$, it is possible for two different values of x to lead to the same value of y. For example, the value $y = 4$ is obtained from either of two values of x: both $2^2 = 4$ and $(-2)^2 = 4$. On the other hand, for the function $y = 6x$, a given value of y can be found from exactly one value of x. For this function, if $y = 30$, then $x = 5$; there is no other value of x that will produce a value of 30 for y.

This second function, $y = 6x$, is an example of a one-to-one function. A function f is **one-to-one** if $f(a) = f(b)$ implies that $a = b$. In a one-to-one function, each y-value corresponds to exactly one x-value. In the function defined by $y = 6x$, each y-value is assigned to exactly one x-value, so the function is one-to-one. The function defined by $y = x^2$, however, has one y-value assigned to *two* x-values, which indicates that it is not a one-to-one function.

EXAMPLE 1 Decide whether the functions defined as follows are one-to-one.

(a) $f(x) = -4x + 12$

For this function, any value of y, say $y = b$, corresponds to exactly one x-value, since if $y = b$,

$$\begin{aligned} b &= -4x + 12 \\ 4x &= 12 - b \\ x &= \frac{12 - b}{4}. \end{aligned}$$

Thus, the function is one-to-one.

(b) $g(x) = \sqrt{25 - x^2}$

Suppose $y = k$. Then $g(x) = k$, and

$$\begin{aligned} k &= \sqrt{25 - x^2} \\ k^2 &= 25 - x^2 \\ x^2 &= 25 - k^2 \\ x &= \sqrt{25 - k^2} \quad \text{or} \quad -\sqrt{25 - k^2}. \end{aligned}$$

Here $y = k$ corresponds to more than one x-value, so the function g is not one-to-one. ■

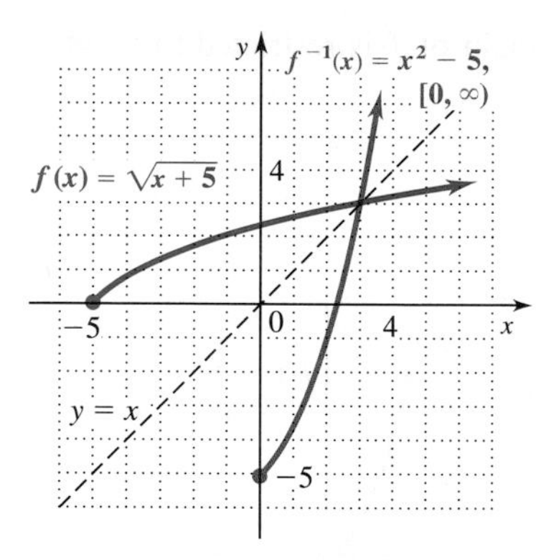

Figure 60

EXAMPLE 4 Let $f(x) = \sqrt{x + 5}$ with domain $[-5, \infty)$. Find $f^{-1}(x)$.

As shown by the graph in Figure 60, the function f is one-to-one and has an inverse function. To find this inverse function, start with $y = \sqrt{x + 5}$ and solve for x.

$$y = \sqrt{x + 5}$$
$$y^2 = x + 5$$
$$y^2 - 5 = x$$

Exchanging x and y gives

$$x^2 - 5 = y.$$

We cannot give just $x^2 - 5$ as $f^{-1}(x)$. In the definition of f above, the domain was given as $[-5, \infty)$. The range of f is $[0, \infty)$. Since the range of f equals the domain of f^{-1}, the definition of the function f^{-1} must be given as

$$f^{-1}(x) = x^2 - 5, \quad \text{domain } [0, \infty).$$

As a check, the range of f^{-1}, $[-5, \infty)$, equals the domain of f. Verify that $(f \circ f^{-1})(x) = x$ and $(f^{-1} \circ f)(x) = x$. ■

3.8 Exercises

Which of the functions graphed or defined as follows are one-to-one?

1.

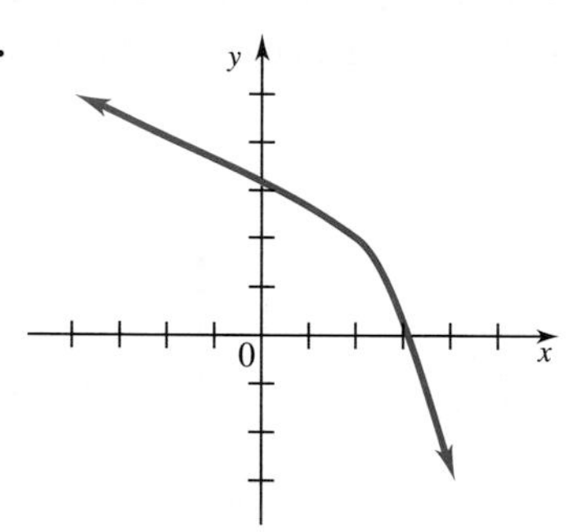

2.

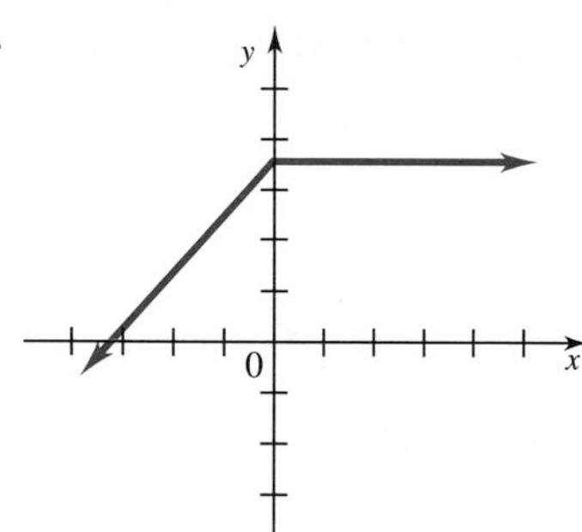

3.

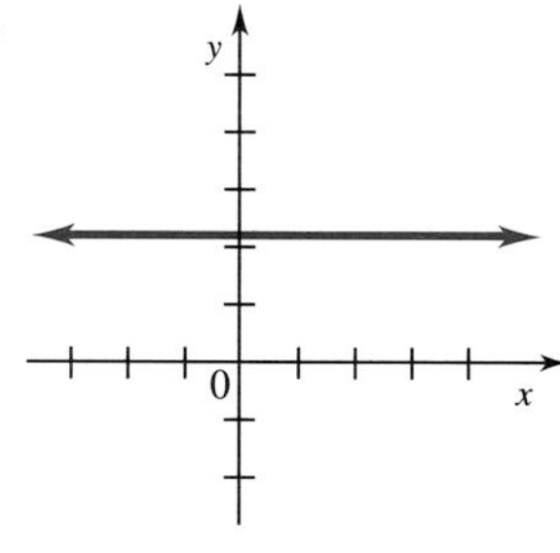

4.

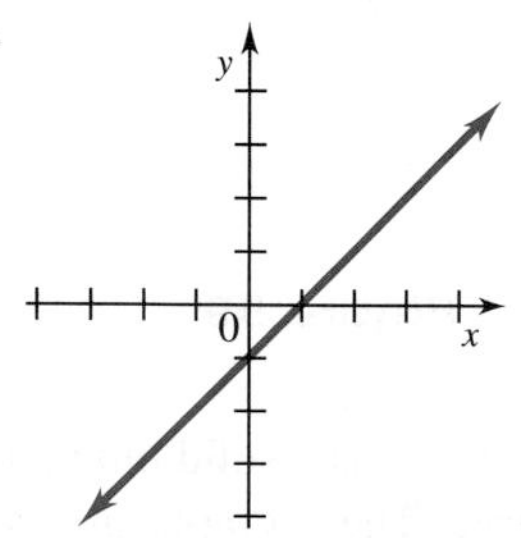

5.

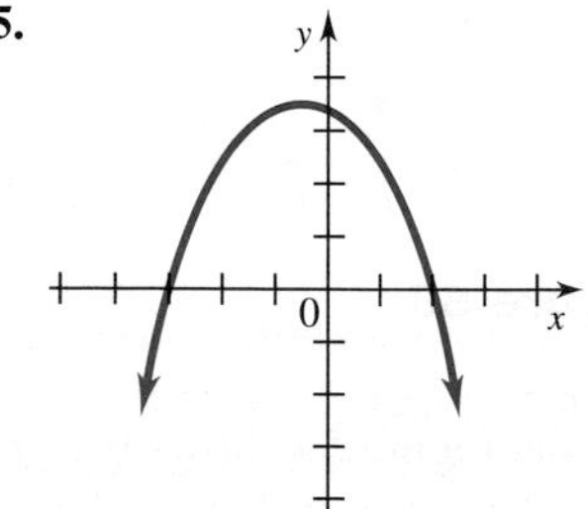

6.

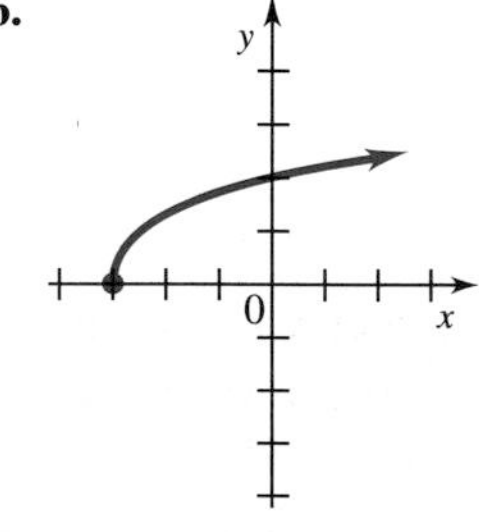

7. $y = 5x - 6$
8. $y = 1 - x$
9. $y = (1 - x)^2$
10. $y = (x - 5)^3$
11. $y = |25 - x^2|$
12. $y = -|16 - x^2|$
13. $y = \sqrt{36 - x^2}$
14. $y = -\sqrt{100 - x^2}$
15. $y = \dfrac{1}{x + 2}$
16. $y = \dfrac{-4}{x - 8}$
17. $y = x^3 - 1$
18. $y = -\sqrt[3]{x + 5}$

Which of the pairs of functions graphed or defined as follows are inverses of each other?

19.
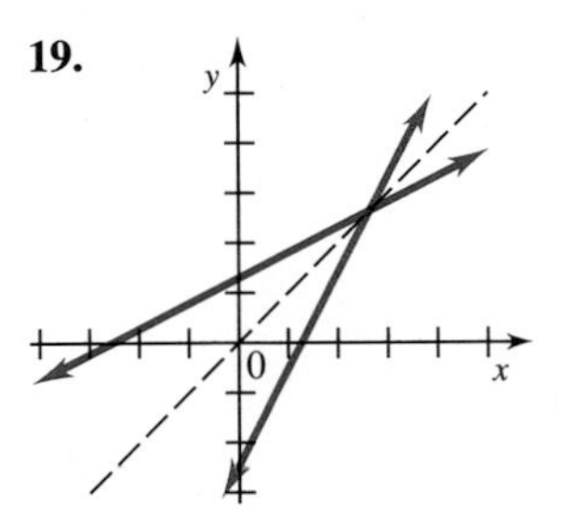

20.
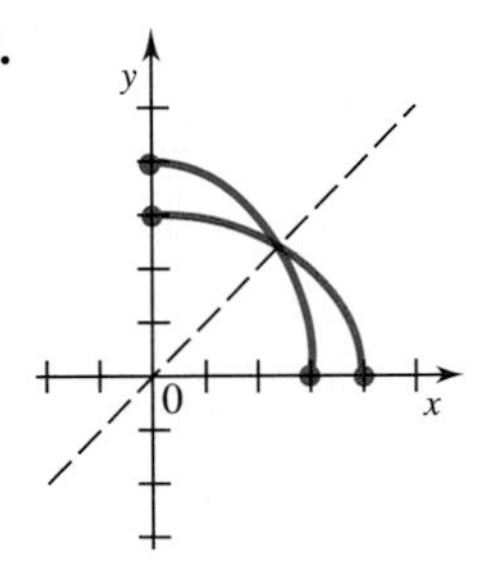

21.
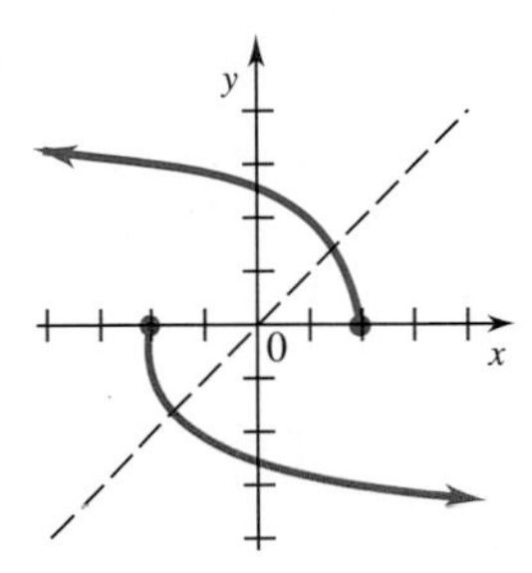

22.
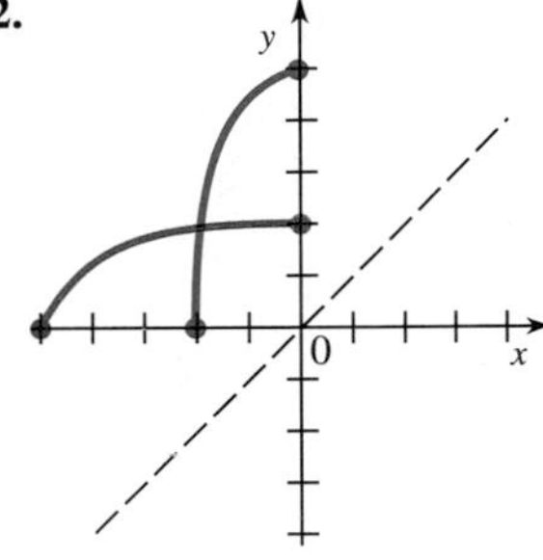

23. $f(x) = 2x + 4, \quad g(x) = \dfrac{1}{2}x - 2$
24. $f(x) = 5x - 5, \quad g(x) = \dfrac{1}{5}x + 1$
25. $f(x) = \dfrac{1}{x + 1}, \quad g(x) = \dfrac{1 - x}{x}$
26. $f(x) = \dfrac{2}{x + 6}, \quad g(x) = \dfrac{6x + 2}{x}$
27. $f(x) = x^2 + 3$, domain $[0, \infty)$, and $g(x) = \sqrt{x - 3}$, domain $[3, \infty)$
28. $f(x) = \sqrt{x + 8}$, domain $[-8, \infty)$ and $g(x) = x^2 - 8$, domain $[0, \infty)$
29. $f(x) = -|x + 5|$, domain $[-5, \infty)$, and $g(x) = |x - 5|$, domain $[5, \infty)$
30. $f(x) = |x - 1|$, domain $[-1, \infty)$, and $g(x) = |x + 1|$, domain $[1, \infty)$

Graph the inverse of each one-to-one function.

31.
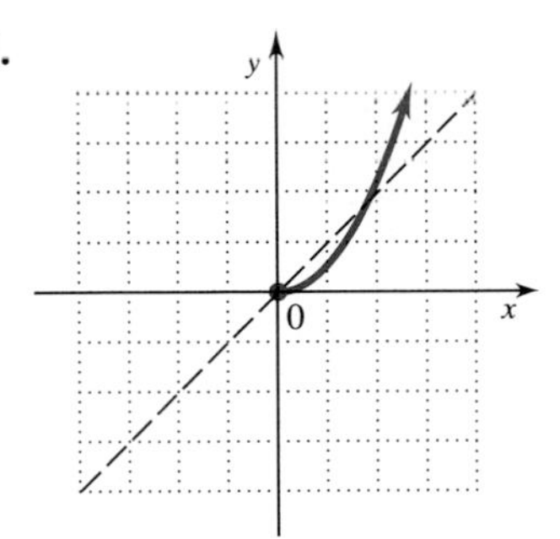

32.
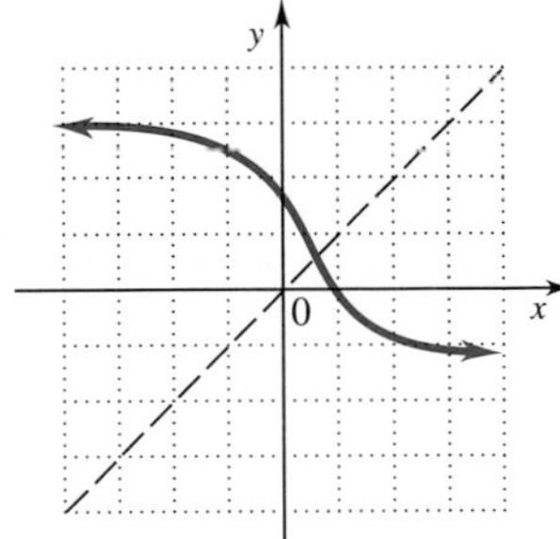

33.
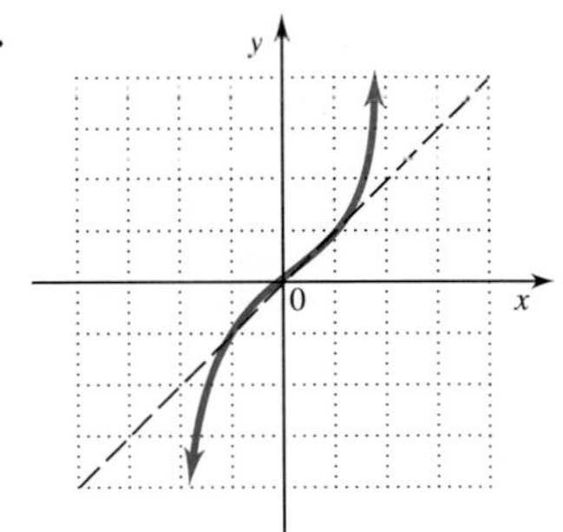

34.
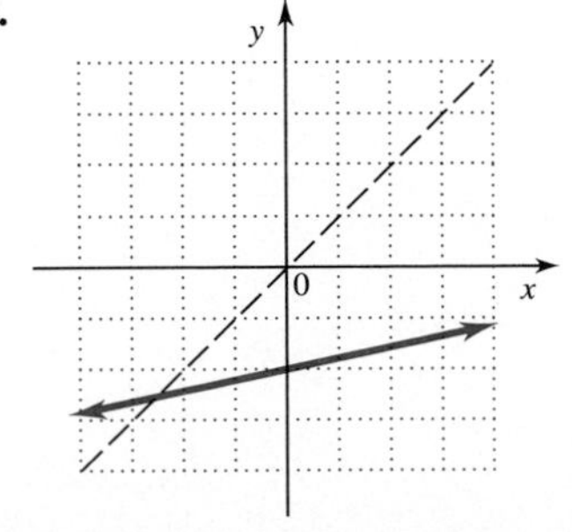

35.
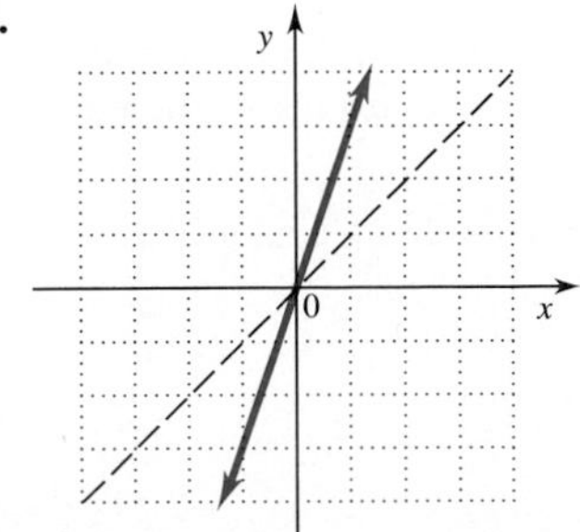

36.
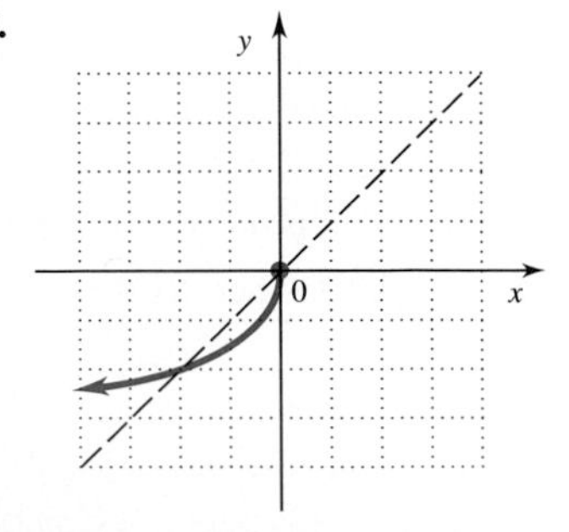

37. As if explaining to another student, describe the steps needed to find the inverse of $f(x) = (x - 3)/x$.

38. As if you were writing a note to a classmate, describe how a function and its inverse are related.

39. Explain why the function defined by $f(x) = x^4$ does not have an inverse. Give examples of ordered pairs to illustrate your explanation.

40. Explain why the function defined by $f(x) = x^5$ is one-to-one.

For each function defined as follows that is one-to-one, write an equation for the inverse function in the form $y = f^{-1}(x)$.

41. $y = 8x - 3$ **42.** $y = 2x - 7$ **43.** $y = -x^3 - 2$ **44.** $y = -x^3 + 1$

45. $y = -x^2 + 2$ **46.** $y = 2x^2 - 3$ **47.** $y = \frac{4}{x}$ **48.** $y = \frac{1}{x}$

49. $y = (x - 3)^2$, domain $(-\infty, 3]$ **50.** $y = \sqrt{2 - x}$

51. $f(x) = \sqrt{6 + x}$ **52.** $f(x) = (x - 2)^2 + 1$, domain $[2, \infty)$

The graph of a function f is shown in the figure. Use the graph to find each of the following values.

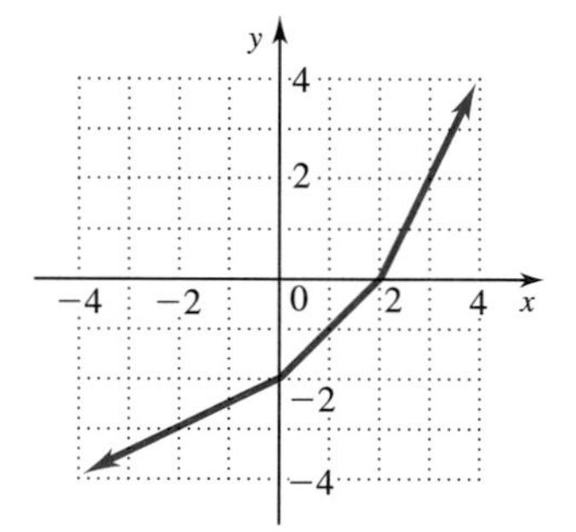

53. $f^{-1}(4)$ **54.** $f^{-1}(2)$ **55.** $f^{-1}(0)$

56. $f^{-1}(-2)$ **57.** $f^{-1}(-3)$ **58.** $f^{-1}(-4)$

Let $f(x) = x^2 + 5x$ for $x \geq -5/2$. Find the value of the expression in Exercises 59 and 60, rounding to the nearest hundredth.

59. $f^{-1}(7)$ **60.** $f^{-1}(-3)$

61. Suppose that $f(x)$ is the number of cars that can be built for x dollars. What does $f^{-1}(1000)$ represent?

62. Suppose that $f(r)$ is the volume (in cubic inches) of a sphere of radius r inches. What does $f^{-1}(5)$ represent?

63. The brightness B of a star is related to its distance from Earth, d, by the equation $B = k/d$, where k is a constant. Write an equation that expresses d as a function of B.

64. Young's rule for determining the correct medicine dosage c for a five-year-old child from the adult dosage d is $c = f(d) = (5/17)d$. Find an expression in terms of c for f^{-1}.

65. If a line has slope a, what is the slope of its reflection in the line $y = x$?

66. Find $f^{-1}(f(2))$, where $f(2) = 3$.

Let f be a function having an inverse. Prove the statements in Exercises 67 and 68.

67. Every x-intercept of f is a y-intercept of f^{-1}.

68. Every y-intercept of f is an x-intercept of f^{-1}.

Graphing Utility Problems

Graph each of the following using the given x- and y-ranges. Use the graph to decide which functions are one-to-one. If a function is one-to-one, give the equation of its inverse function and graph the inverse function on the same coordinate system.

69. $f(x) = 6x^3 + 11x^2 - x - 6$; $[-3, 2], [-10, 10]$

70. $f(x) = x^4 - 5x^2 + 6$; $[-3, 3], [-1, 8]$

71. $f(x) = (x - 5)/(x + 3)$; $[-8, 8], [-6, 8]$

72. $f(x) = -x/(x - 4)$; $[-4, 4], [-4, 4]$

Chapter 3 Review Exercises

In Exercises 1 and 2, find $d(P, Q)$ and the midpoint of segment PQ.

1. $P(3, -1)$ and $Q(-4, 5)$

2. $P(-8, 2)$ and $Q(3, -7)$

3. Find the other endpoint of a line segment having one end at $(-5, 7)$ and having its midpoint at $(1, -3)$.

4. Are the points $(5, 7)$, $(3, 9)$, $(6, 8)$ the vertices of a right triangle?

5. Find all possible values of k so that $(-1, 2)$, $(-10, 5)$, and $(-4, k)$ are the vertices of a right triangle.

6. Find all possible values of x so that the distance between $(x, -9)$ and $(3, -5)$ is 6.

7. Find all points (x, y) with $x = 6$ so that (x, y) is 4 units from $(1, 3)$.

8. Find all points (x, y) with $x + y = 0$ so that (x, y) is 6 units from $(-2, 3)$.

Prove each of the following.

9. The medians to two equal sides of an isosceles triangle are equal in length. (A median is a line segment from a vertex of a triangle to the midpoint of the opposite side.)

10. The line segment connecting midpoints of two sides of a triangle is half as long as the third side.

Find equations for each of the circles below.

11. Center $(5, -2)$, radius 4

12. Center $(\sqrt{5}, -\sqrt{7})$, radius $\sqrt{3}$

13. Center $(-8, 1)$, passing through $(0, 16)$

14. Center $(-3, 5)$, passing through $(1, 8)$

Find the center and radius of each of the following that are circles.

15. $x^2 - 6x + y^2 - 10y + 30 = 0$

16. $x^2 + y^2 - 6x + 8y = 11$

17. $x^2 + 11x + y^2 - 5y + 46 = 0$

18. $x^2 + y^2 + x + 5y = 7/2$

Tell whether each of the following is true or false.

19. The number a is in the domain of a relation if and only if the vertical line $x = a$ intersects the graph of the relation.

20. The number b is in the range of a relation if and only if the horizontal line $y = b$ intersects the graph of the relation.

Decide whether the following curves are graphs of functions of x. Give the domain and range of each relation.

21.

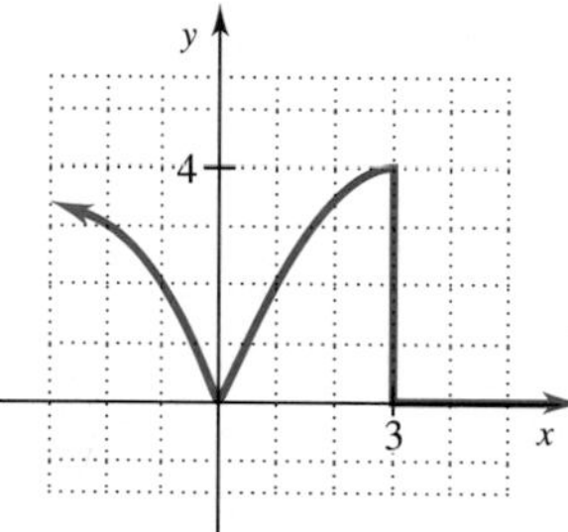

22.

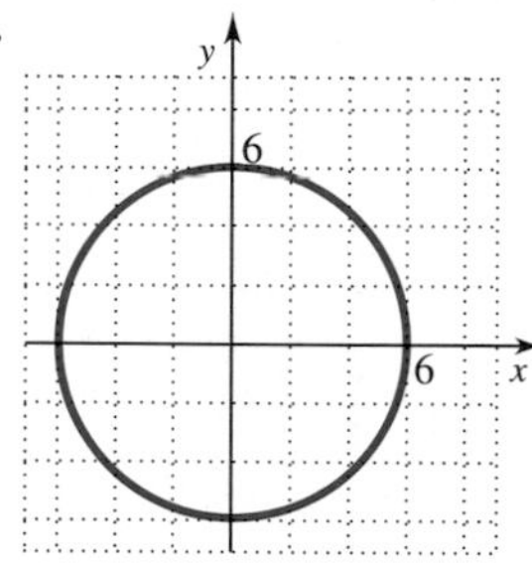

23.

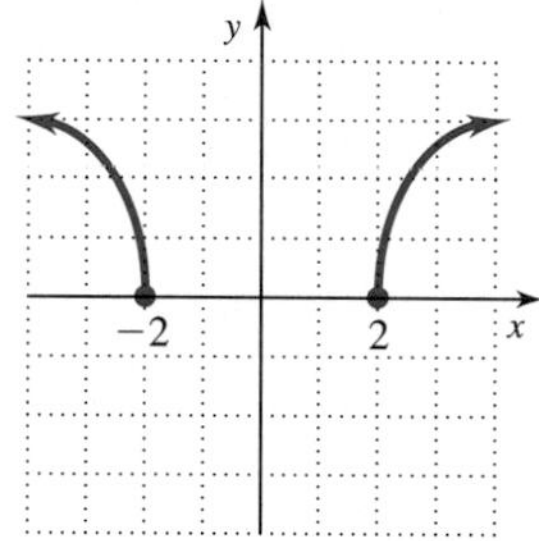

24.

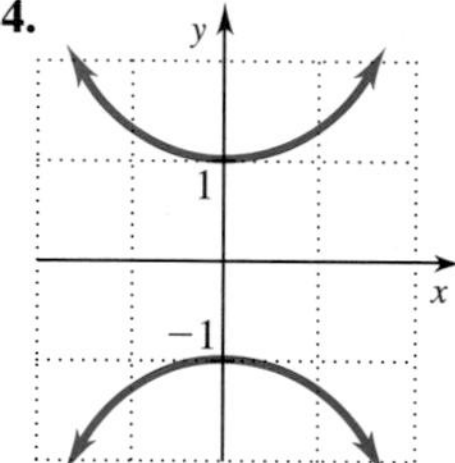

25.

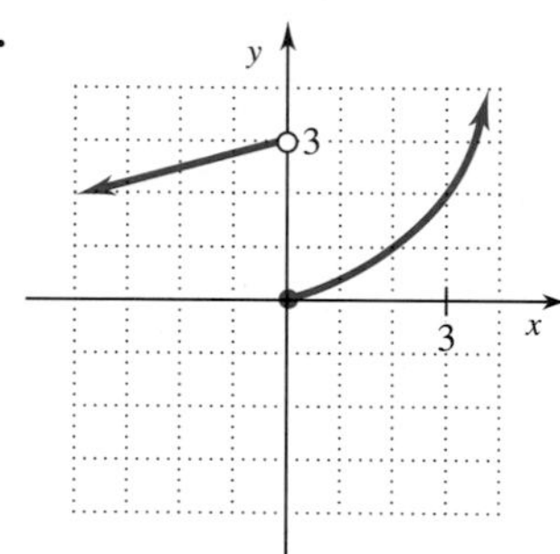

26.

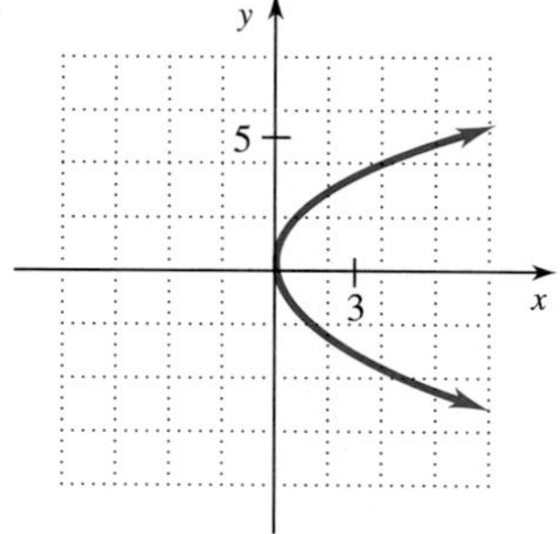

Identify any equations that define y as a function of x.

27. $x = \frac{1}{2}y^2$

28. $y = 3 - x^2$

29. $y = \frac{-8}{x}$

30. $y = \sqrt{x - 7}$

Give the domain of the functions defined as follows.

31. $y = -4 + |x|$

32. $y = \frac{8 + x}{8 - x}$

33. $y = -\sqrt{\frac{5}{x^2 + 9}}$

34. $y = \sqrt{49 - x^2}$

Find the slope for each of the following lines.

35. Through (8, 7) and (1/2, −2)

36. Through (2, −2), and (3, −4)

37. $11x + 2y = 3$

38. $y + 6 = 0$

Write an equation in standard form for each of the following lines.

39. Through (5, −23) and (−3, 1)

40. Through (1/4, 5/2) and (−3, −4)

41. Through (2, 6), with slope 7/3

42. Through (−4, 8), with slope −4/5

43. No x-intercept, y-intercept -1

44. x-intercept -2, y-intercept 3

45. Through $(-4, 2)$, perpendicular to $2x - y = 3$

46. Through $(3, -1)$, parallel to $3x + 5y = 15$

Graph each function defined as follows.

47. $3y = x$

48. $f(x) = [\![x - 1]\!]$

49. $f(x) = \begin{cases} -4x + 2 & \text{if } x \le 1 \\ 3x - 5 & \text{if } x > 1 \end{cases}$

50. $f(x) = x^2 - 4$

51. The line through $(-3, -2)$, with $m = -1$

52. $f(x) = \begin{cases} 3x + 1 & \text{if } x < 2 \\ -x + 4 & \text{if } x \ge 2 \end{cases}$

53. $y = 6 - x^2$

54. The college fieldhouse used for graduation ceremonies has 5000 seats available for family and friends of the graduates. Use the greatest-integer function to write an expression for the number of tickets that can be allocated to each of x graduates.

Decide whether the equations in Exercises 55–62 have graphs that are symmetric with respect to the x-axis, the y-axis, the origin, or none of these.

55. $3y^2 - 5x^2 = 15$

56. $x + y^2 = 8$

57. $y^3 = x + 1$

58. $x^2 = y^3$

59. $|y| = -x$

60. $|x + 2| = |y - 3|$

61. $|x| = |y|$

62. $xy = 8$

Describe how the graphs of the following functions can be obtained from the graph of $f(x) = |x|$.

63. $g(x) = -|x|$

64. $h(x) = |x| - 2$

65. $k(x) = 2|x - 4|$

Let $f(x) = 3x - 4$. *Find an equation for each of the following reflections of the graph of* $f(x)$.

66. About the x-axis

67. About the y-axis

68. About the origin

69. About the line $y = x$

70. The graph of a function f is shown in the figure. Sketch the graph of each function defined as follows.

(a) $y = f(x) + 3$

(b) $y = f(x - 2)$

(c) $y = f(x + 3) - 2$

(d) $y = |f(x)|$

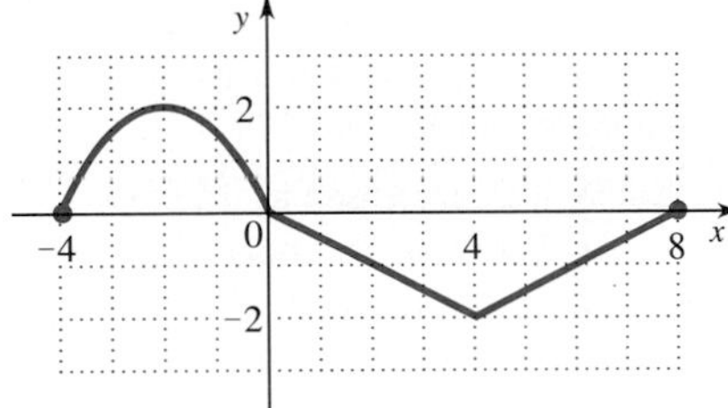

Let $f(x) = 2x + 5$ *and* $g(x) = x^2 - 3x - 4$. *Find each of the following.*

71. $(f + g)(x)$

72. $(fg)(x)$

73. $(f - g)(4)$

74. $(f + g)(-4)$

75. $(f + g)(2k)$

76. $(fg)(1 + r)$

77. $(f/g)(3)$

78. $(f/g)(-1)$

79. The domain of $(fg)(x)$

80. The domain of $(f/g)(x)$

Let $f(x) = \sqrt{x - 2}$ *and* $g(x) = x^2$. *Find each of the following.*

81. $(f \circ g)(x)$

82. $(g \circ f)(x)$

83. $(g \circ f)(3)$

84. $(f \circ g)(-6)$

Which of the functions graphed or defined as follows are one-to-one?

85.

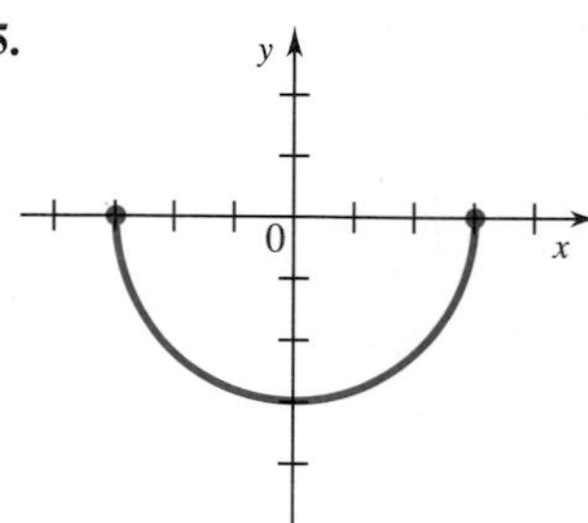

86.

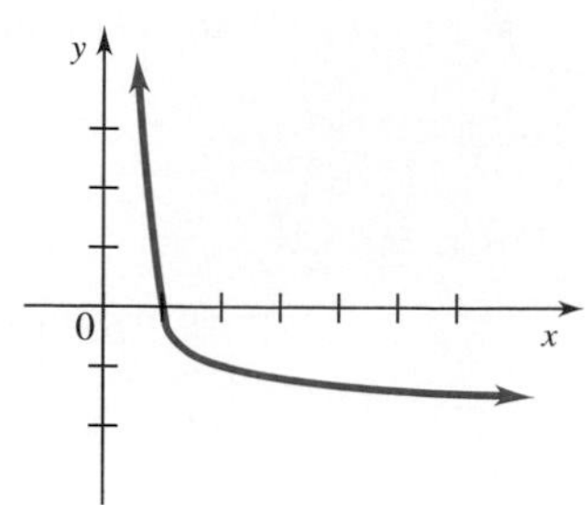

87.

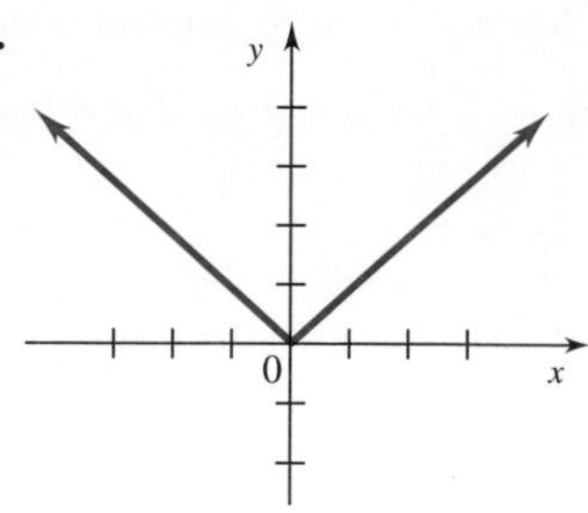

88. $y = \dfrac{8x - 9}{5}$

89. $y = -x^2 + 11$

90. $y = \sqrt{100 - x^2}$

For each of the functions defined in Exercises 91–94, write an equation for the inverse function in the form $y = f^{-1}(x)$ and then graph f and f^{-1}.

91. $f(x) = 12x + 3$

92. $f(x) = x^3 - 3$

93. $f(x) = x^2 - 6$

94. $f(x) = \sqrt{25 - x^2}$

95. Suppose $f(t)$ is the amount an investment will grow to become t years after 1992. What does $f^{-1}(\$50{,}000)$ represent?

96. Find the point of intersection of $y = f(x)$ and $y = f^{-1}(x)$ where $f(x) = 3x - 5$.

97. Cylindrical cans make the most efficient use of materials when their height is the same as the diameter of their top.

(a) Express the volume V of such a can as a function of the diameter d of its top.

(b) Express the surface area S of such a can as a function of the diameter d of its top. (*Hint:* The curved side is made from a rectangle whose length is the circumference of the top of the can.)

98. A baseball diamond is a square 90 ft long on each side. Casey runs a constant 30 ft per sec whether he hits a ground ball or a home run. Today in his first time at bat, he hit a home run. Write an expression that measures his line-of-sight distance from second base as a function of the time t, in seconds, after he left home plate.*

99. Alice, on vacation in Canada, found that her U.S. dollars were increased by 16%. On her return, she expected a 16% decrease when converting her Canadian money back into U.S. dollars. Write an equation for each of these conversion functions. Show that one is not the inverse of the other. What should the conversion factor be for Canadian to U.S. dollars?

*Exercise 98: From *Calculus*, 4th Edition by Stanley I. Grossman. Copyright © 1988 by Harcourt Brace Jovanovich, Inc. Reprinted with permission of Harcourt Brace Jovanovich, Inc. and the author.

100. The figure shows average prices for domestic crude oil from mid-1986 to mid-1992.

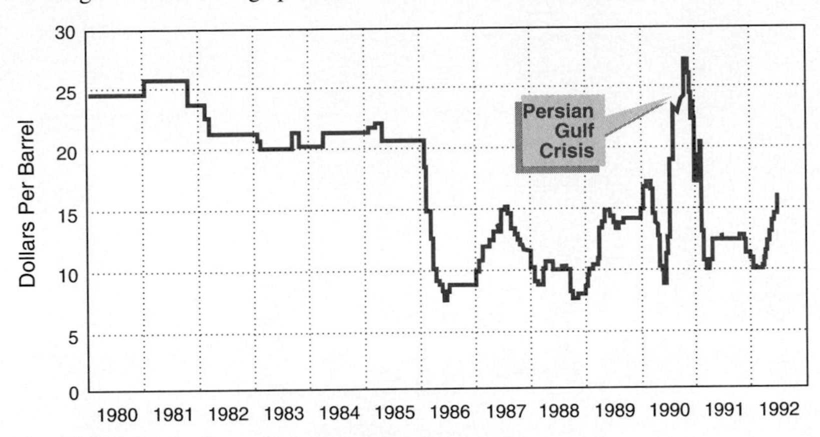

(a) Is this the graph of a function? Of a one-to-one function?
(b) In what year were oil prices lowest? Highest?
(c) What was the lowest price? The highest price?
(d) What is the general trend of prices over the given period?
(e) What do the horizontal portions of the graph indicate?

101. The figure shows the number of jobs gained or lost in the Sacramento area from September 1991 to May 1992.

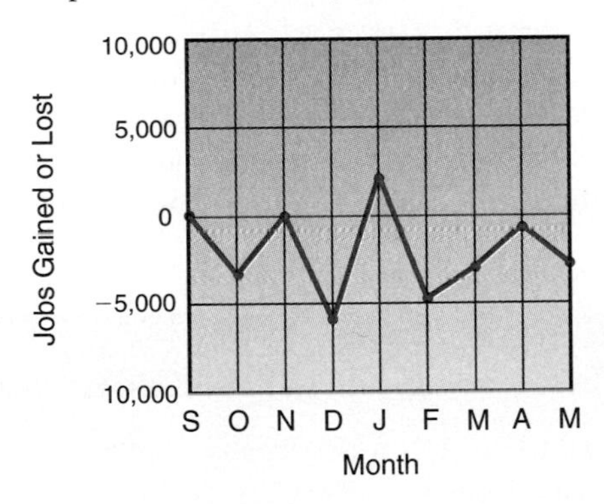

(a) Is this the graph of a function? A one-to-one function?
(b) In what month were the most jobs lost? The most gained?
(c) What was the largest number of lost jobs? The most gained?
(d) Do these data show an upward or downward trend? If so, what is it?

102. Explain the similarities and differences between relations and functions.

103. Discuss what is meant by inverse functions. Give the conditions that are necessary for two functions to be inverses. Describe how a pair of inverse functions are related.

104. Discuss the connections between the following concepts: the distance formula, the equation of a circle, and functions of the form $y = \sqrt{a - x^2}$ or $y = -\sqrt{a - x^2}$.

4 Polynomial and Rational Functions

Functions such as $f(x) = 2x^3 - \frac{1}{2}x + 5$, $f(x) = x^4 + \sqrt{2}x^3 - 4x^2$, and $f(x) = x^2 - ix + (2 + 3i)$ are examples of *polynomial functions.* Polynomial functions are the simplest type of function, because a polynomial involves only the operations of addition, subtraction, and multiplication. In calculus it is shown how polynomial functions can be used to approximate more complicated functions.

Definition of Polynomial Function

A **polynomial function of degree *n***, where n is a nonnegative integer, is a function defined by an expression of the form

$$f(x) = a_n x^n + a_{n-1}x^{n-1} + \ldots + a_1 x + a_0,$$

where $a_n, a_{n-1}, \ldots, a_1$, and a_0 are complex numbers, with $a_n \neq 0$.*

For the polynomial $f(x) = 2x^3 - \frac{1}{2}x + 5$, n is 3 and the polynomial has the form $a_3x^3 + a_2x^2 + a_1x + a_0$, where a_3 is 2, a_2 is 0, a_1 is $-\frac{1}{2}$, and a_0 is 5. The polynomial functions defined by $f(x) = x^4 + \sqrt{2}x^3 - 4x^2$ and $f(x) = x^2 - ix +$

*Remember, *complex numbers* include imaginary numbers such as $5i$ or $-i\sqrt{3}$, real numbers, and numbers such as $4 - 2i$ or $6 + i\sqrt{7}$.

$(2 + 3i)$ have degrees 4 and 2, respectively. The number a_n is the **leading coefficient** of $f(x)$. The function defined by $f(x) = 0$ is called the **zero polynomial**. The zero polynomial has no degree. However, a polynomial $f(x) = a_0$ for a nonzero number a_0 has degree 0.

In this chapter we discuss the graphs of polynomial functions of degree 2 or higher and methods of finding, or at least approximating, the values of x that satisfy $f(x) = 0$, called the **zeros** of $f(x)$. The chapter ends with a section on *rational functions*, which are defined as quotients of polynomials.

4.1 Quadratic Functions

In Section 3.4, we discussed several first-degree functions, where the highest power of the variable is 1. In this section we look at functions of degree 2, called *quadratic functions.*

Definition of Quadratic Function

A function f is a **quadratic function** if

$$f(x) = ax^2 + bx + c,$$

where a, b, and c are real numbers with $a \neq 0$.

The simplest quadratic function is given by $f(x) = x^2$ with $a = 1$, $b = 0$, and $c = 0$. To find some points on the graph of this function, choose some values for x and find the corresponding values for $f(x)$, as in the chart with Figure 1. Then plot these points, and draw a smooth curve through them. (The reason for drawing a smooth curve depends on ideas from calculus). This graph is called a **parabola**. Every quadratic function has a graph that is a parabola.

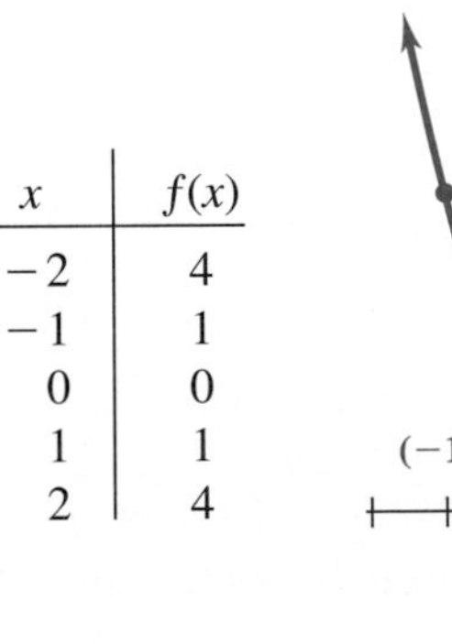

x	$f(x)$
−2	4
−1	1
0	0
1	1
2	4

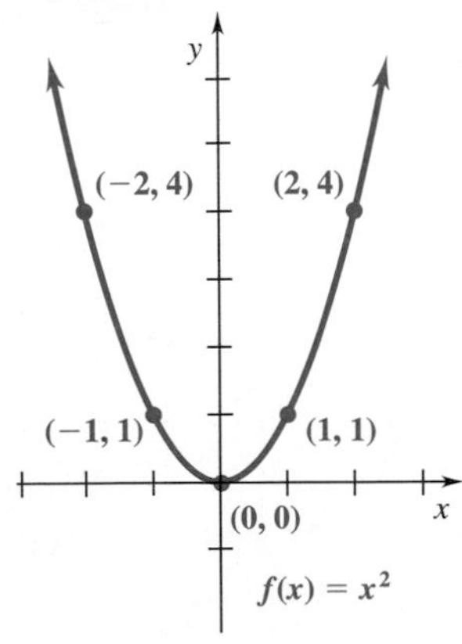

Figure 1

Parabolas are symmetric with respect to a line (the y-axis in Figure 1). The line of symmetry for a parabola is called the **axis** of the parabola. The point where the axis intersects the parabola is the **vertex** of the parabola. As Figure 2 shows, the vertex of a parabola that opens downward is the highest point of the graph and the vertex of a parabola that opens upward is the lowest point of the graph.

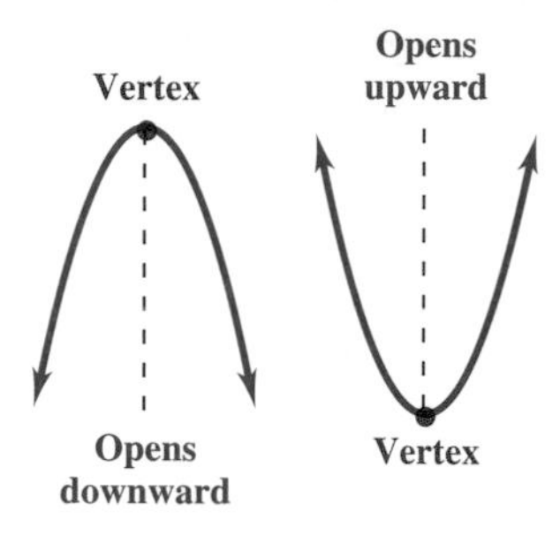

Figure 2

The methods of Section 3.6 can be applied to the graph of $f(x) = x^2$ to obtain the graph of *any* quadratic function. The graph of $g(x) = ax^2$ is a parabola with vertex at the origin that opens upward if a is positive and downward if a is negative. The width of $g(x)$ is determined by the magnitude of a. That is, $g(x)$ is narrower than $f(x) = x^2$ if $|a| > 1$ and is broader than $f(x) = x^2$ if $|a| < 1$. By completing the square, a technique discussed in Chapter 2, any quadratic function can be written in the form

$$h(x) = a(x - h)^2 + k.$$

The graph of $h(x)$ is the same as the graph of $g(x) = ax^2$ translated h units horizontally (to the right if h is positive and to the left if h is negative) and translated k units vertically (up if k is positive and down if k is negative).

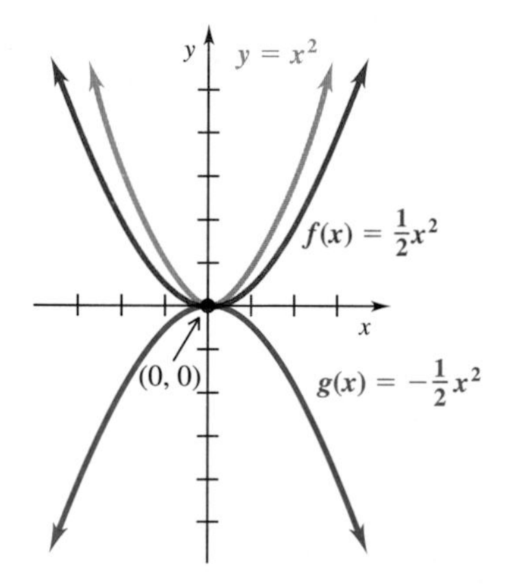

Figure 3

EXAMPLE 1 Graph the functions defined as follows.

(a) $g(x) = -\frac{1}{2}x^2$

The function $g(x)$ can be thought of as $-(\frac{1}{2}x^2)$. The graph of $\frac{1}{2}x^2$ is a broad version of the graph of x^2 and the graph of $g(x) = -(\frac{1}{2}x^2)$ is a reflection of the graph of $\frac{1}{2}x^2$ in the x-axis. See Figure 3. The vertex is (0, 0) and the axis of the parabola is the line $x = 0$ (the y-axis).

(b) $h(x) = -\frac{1}{2}(x - 4)^2 + 3$

The function $h(x)$ is $g(x - h) + k$, where $g(x)$ is the function of part (a), h is 4, and k is 3. Therefore, $h(x)$ is obtained by translating the graph of $g(x)$ 4 units to the right and 3 units up. See Figure 4. The vertex is (4, 3) and the axis of the parabola is the line $x = 4$.

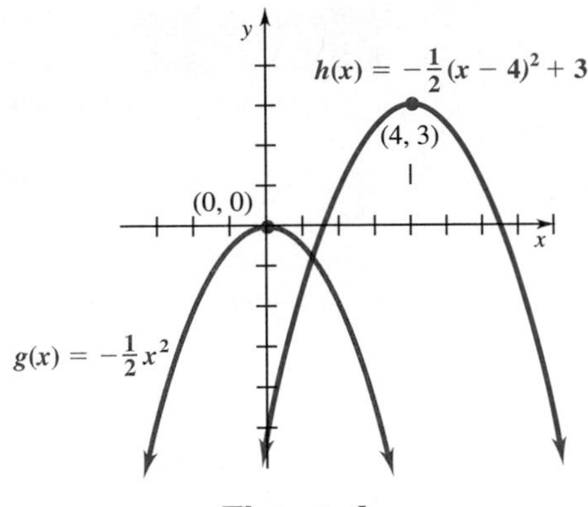

Figure 4

In general, the graph of the quadratic function

$$f(x) = a(x - h)^2 + k$$

is a parabola with vertex (h, k) and axis $x = h$. The parabola opens upward if a is positive and opens downward if a is negative. With these facts in mind, we will apply *completing the square* to the graphing of a quadratic function.

EXAMPLE 2 Graph $f(x) = x^2 - 6x + 7$.

To graph this parabola, $x^2 - 6x + 7$ must be rewritten in the form $(x - h)^2 + k$. Start as follows.

$$f(x) = (x^2 - 6x \quad) + 7$$

As shown earlier, a number must be added inside the parentheses to get a perfect square trinomial. To find this number, take half the coefficient of x and then square the result. Half of -6 is -3, and $(-3)^2$ is 9. Now add and subtract 9 inside the parentheses. (This is the same as adding 0.)

$$f(x) = (x^2 - 6x + 9 - 9) + 7$$

Group the terms and factor as follows.

$$f(x) = (x^2 - 6x + 9) - 9 + 7$$
$$f(x) = (x - 3)^2 - 2$$

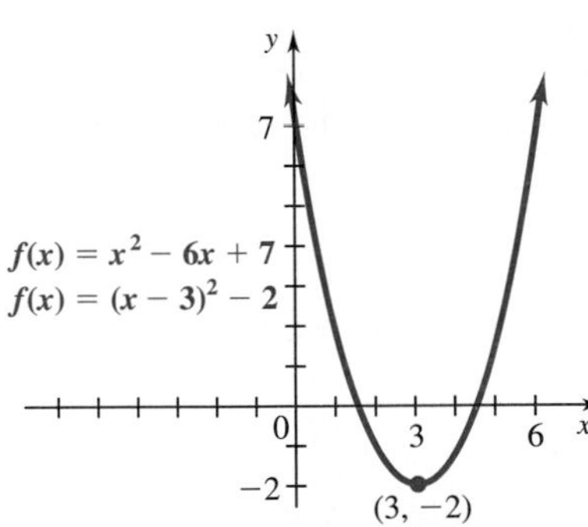

Figure 5

This result shows that the vertex of the parabola is $(3, -2)$ and the axis is the line $x = 3$. The graph is shown in Figure 5. ■

NOTE In Example 2 we added and subtracted 9 *on the same side* of the equation to complete the square. This differs from adding the same number to *each side of the equation*, as when we completed the square in Chapter 2. Here, since we want just y on one side of the equation, we had to change slightly that step in the process of completing the square. ■

EXAMPLE 3 Graph $f(x) = -3x^2 - 2x + 1$.

To complete the square, first factor out -3 from the x-terms.

$$f(x) = -3\left(x^2 + \frac{2}{3}x \qquad \right) + 1$$

(This is necessary to make the coefficient of x^2 equal to 1.) Half the coefficient of x is $1/3$, and $(1/3)^2 = 1/9$. Add and subtract $1/9$ inside the parentheses as follows.

$$f(x) = -3\left(x^2 + \frac{2}{3}x + \frac{1}{9} - \frac{1}{9}\right) + 1$$

Use the distributive property and simplify.

$$f(x) = -3\left(x^2 + \frac{2}{3}x + \frac{1}{9}\right) - 3\left(-\frac{1}{9}\right) + 1$$

$$f(x) = -3\left(x^2 + \frac{2}{3}x + \frac{1}{9}\right) + \frac{1}{3} + 1$$

$$f(x) = -3\left(x^2 + \frac{2}{3}x + \frac{1}{9}\right) + \frac{4}{3}$$

$$f(x) = -3\left(x + \frac{1}{3}\right)^2 + \frac{4}{3} \qquad \text{Factor.}$$

Now the equation of the parabola is written in the form $f(x) = a(x - h)^2 + k$. In this form, the equation shows that the axis of the parabola is the vertical line

$$x + \frac{1}{3} = 0 \quad \text{or} \quad x = -\frac{1}{3}$$

and that the vertex is $(-1/3, 4/3)$. Additional points can be found by substituting x-values near the vertex into the original equation. For example, $(1/2, -3/4)$ is on the graph, which is shown in Figure 6. The intercepts are often good additional points to find. Here, the y-intercept is

$$y = -3(0)^2 - 2(0) + 1 = 1,$$

giving the point $(0, 1)$. The x-intercepts are found by setting $f(x)$ equal to zero in the original equation.

$$0 = -3x^2 - 2x + 1$$

$$3x^2 + 2x - 1 = 0 \qquad \text{Multiply by } -1.$$

$$(3x - 1)(x + 1) = 0 \qquad \text{Factor.}$$

Therefore, the x-intercepts are $1/3$ and -1.

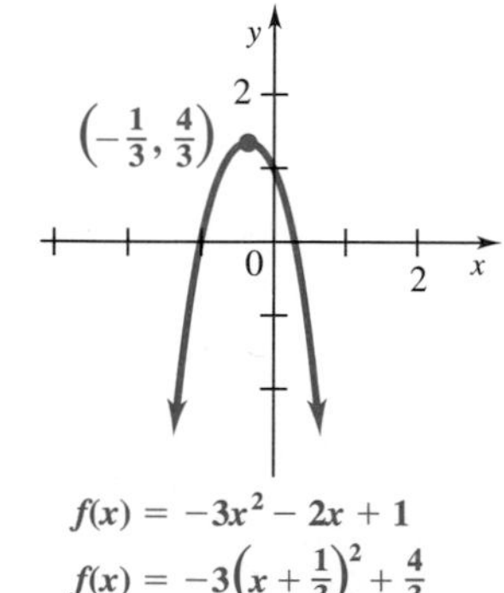

$f(x) = -3x^2 - 2x + 1$

$f(x) = -3\left(x + \frac{1}{3}\right)^2 + \frac{4}{3}$

Figure 6

We can now generalize the work above to get a formula for the vertex of a parabola. Starting with the general quadratic function $f(x) = ax^2 + bx + c$ and completing the square will change the function to the form $f(x) = a(x - h)^2 + k$. Begin by factoring a from the first two terms.

$$f(x) = ax^2 + bx + c$$
$$= a\left(x^2 + \frac{b}{a}x \qquad\right) + c$$

Now add $\left(\frac{1}{2} \cdot \frac{b}{a}\right)^2 = \frac{b^2}{4a^2}$ in the parentheses and subtract $a\left(\frac{b^2}{4a^2}\right)$ from c.

$$f(x) = a\left(x^2 + \frac{b}{a}x + \frac{b^2}{4a^2}\right) + c - a\left(\frac{b^2}{4a^2}\right)$$
$$= a\left(x + \frac{b}{2a}\right)^2 + c - \frac{b^2}{4a}$$

Comparing the last result with $f(x) = a(x - h)^2 + k$ shows that

$$h = -\frac{b}{2a} \quad \text{and} \quad k = c - \frac{b^2}{4a}.$$

Letting $x = h$ in $f(x) = a(x - h)^2 + k$ gives

$$f(h) = a(h - h)^2 + k = k,$$

so $k = f(h)$, or $k = f(-b/(2a))$.

The following statement summarizes this discussion.

Graph of a Quadratic Function

The quadratic function defined by $f(x) = ax^2 + bx + c$ can be written in the form

$$\mathbf{y = f(x) = a(x - h)^2 + k}, \quad a \neq 0,$$

where $$h = -\frac{b}{2a} \quad \text{and} \quad k = f(h).$$

The graph of f has the following characteristics:

1. It is a parabola with vertex (h, k), and the vertical line $x = h$ as axis.

2. It opens upward if $a > 0$ and downward if $a < 0$.

3. It is broader than $y = x^2$ if $|a| < 1$ and narrower than $y = x^2$ if $|a| > 1$.

The vertex and axis of a parabola can be found from its equation either by completing the square or by memorizing the formula $h = -b/2a$ and letting $k = f(h)$.

EXAMPLE 4 Find the axis and the vertex of the parabola having equation $f(x) = 2x^2 + 4x + 5$ using the formula given above.

Here $a = 2$, $b = 4$, and $c = 5$. The axis of the parabola is the vertical line

$$x = h = -\frac{b}{2a} = -\frac{4}{2(2)} = -1.$$

The vertex is the point

$$(-1, f(-1)) = (-1, 3).$$

The fact that the vertex of a parabola of the form $f(x) = ax^2 + bx + c$ is the highest or lowest point on the graph can be used in applications to find a maximum or a minimum value.

EXAMPLE 5 Ms. Whitney owns and operates Aunt Emma's Pie Shop. She has hired a consultant to analyze her business operations. The consultant tells her that her profit $P(x)$ is given by

$$P(x) = 120x - x^2,$$

where x is the number of units of pies that she makes. How many units of pies should be made to maximize profit? What is the maximum possible profit?

The profit function can be rewritten as $P(x) = -x^2 + 120x + 0$, a quadratic function with $a = -1$, $b = 120$, and $c = 0$. Complete the square to find that the vertex of the parabola is (60, 3600). Since $a < 0$, the parabola opens downward and the vertex is the highest point on the graph, producing a *maximum* rather than a minimum. Figure 7 shows the portion of the profit function that lies in Quadrant I. (Why is Quadrant I the only one of interest here?) The maximum profit of \$3600 is achieved when 60 units of pies are made. In this case, profit increases as more and more pies are made up to 60 units and then decreases as more and more pies are made past this point.

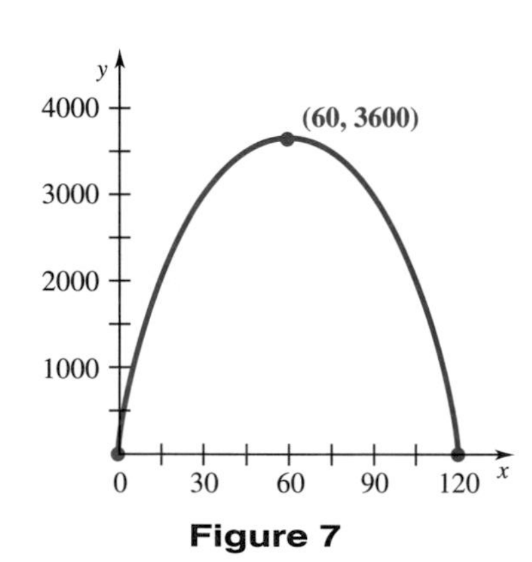

Figure 7

4.1 Exercises

1. Graph the functions defined as follows on the same coordinate system.

(a) $f(x) = 2x^2$ **(b)** $f(x) = -3x^2$ **(c)** $f(x) = -\frac{2}{3}x^2$ **(d)** $f(x) = \frac{1}{4}x^2$

(e) How does the coefficient affect the shape of the graph?

2. Graph the functions defined as follows on the same coordinate system.

(a) $f(x) = x^2 + 3$ **(b)** $f(x) = x^2 - 1$ **(c)** $f(x) = x^2 + \frac{1}{2}$ **(d)** $f(x) = x^2 - \frac{3}{2}$

(e) How do these graphs differ from the graph of $f(x) = x^2$?

3. Graph the functions defined as follows on the same coordinate system.
(a) $f(x) = (x - 1)^2$ **(b)** $f(x) = (x + 1)^2$ **(c)** $f(x) = (x + 3)^2$ **(d)** $f(x) = (x - 2)^2$
(e) How do these graphs differ from the graph of $f(x) = x^2$?

Give the range of each of the functions defined as follows.

4. $f(x) = (x - 3)^2 + 4$

5. $f(x) = -(x + 3)^2 - 1$

6. $g(x) = 2(x + 4)^2 - \frac{1}{2}$

Graph the functions defined as follows. Give the vertex, axis, x-intercepts, and y-intercepts of each graph.

7. $f(x) = (x + 2)^2$

8. $f(x) = (x - 3)^2$

9. $g(x) = 2(x + 3)^2 - 4$

10. $h(x) = (x - 5)^2 - 4$

11. $F(x) = -3(x - 2)^2 + 1$

12. $k(x) = -2(x + 3)^2 + 2$

13. $H(x) = \frac{1}{2}(x + 2)^2 - 3$

14. $G(x) = -\frac{2}{3}(x - 2)^2 - 1$

15. $f(x) = x^2 + 4x + 3$

16. $f(x) = x^2 + 6x + 5$

17. $k(x) = -x^2 + 6x - 6$

18. $g(x) = -x^2 - 4x + 2$

19. $g(x) = -3x^2 + 24x - 46$

20. $f(x) = 2x^2 - 4x + 5$

The figure shows the graph of a quadratic function $f(x)$.

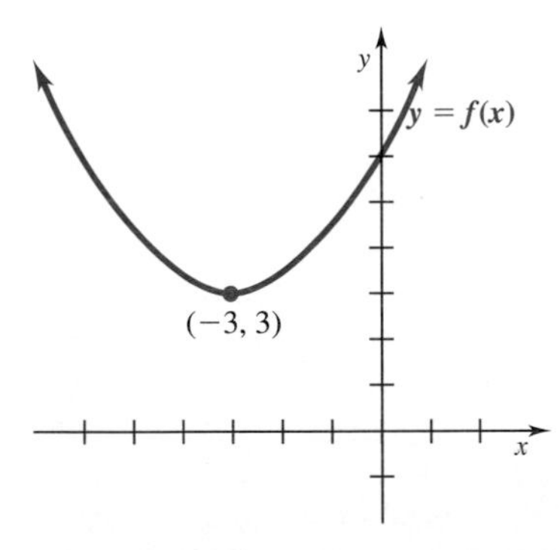

21. What is the minimum value of $f(x)$?

22. For what value of x is $f(x)$ as small as possible?

23. How many solutions are there to the equation $f(x) = 1$?

24. How many solutions are there to the equation $f(x) = 4$?

25. Glenview College wants to construct a rectangular parking lot on land bordered on one side by a highway. It has 320 ft of fencing which it will use to fence off the other three sides. What should be the dimensions of the lot if the enclosed area is to be a maximum? See the figure. (*Hint:* Let x represent the width of the lot and let $320 - 2x$ represent the length. Graph the area parabola, $A = x(320 - 2x)$, and investigate the vertex.)

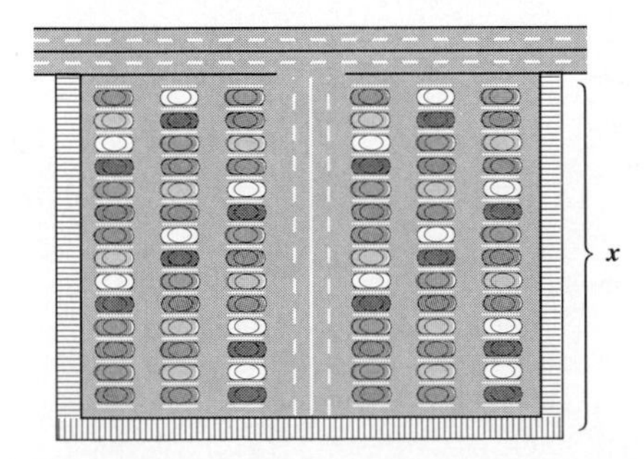

26. The number of mosquitoes, $M(X)$, in millions, in a certain area of Kentucky depends on the June rainfall, x, in inches, approximately as $M(x) = 10x - x^2$. Find the rainfall that will produce the maximum number of mosquitoes.

27. The revenue and cost of a bus trip depend on the number of unsold seats. If the revenue $R(x)$ is given by $R(x) = 5000 + 40x - x^2$, and the cost $C(x)$ is given by $C(x) = 2000 - 10x$, find the maximum profit. (*Hint:* Revenue minus cost equals profit.)

28. Find the dimensions of a rectangle of maximum area whose perimeter is 8 ft. (*Hint:* Let x be the length and h be the height. Then $2x + 2h = 8$ or $h = 4 - x$.)

29. If an object is thrown upward with an initial velocity of 32 ft per sec, then its height after t sec is given by $h(t) = 32t - 16t^2$. Find the maximum height attained by the object. Find the number of seconds it takes the object to hit the ground.

30. Find two numbers whose sum is 20 and whose product is a maximum. (*Hint:* Let x and $20 - x$ be the two numbers, and write an equation for the product.)

31. Find two numbers whose difference is 30 and whose product is a minimum.

32. A charter flight charges a fare of $200 per person, with a surcharge of $4 per person for each unsold seat on the plane. If the plane holds 100 passengers, and if x represents the number of unsold seats, find the following.

(a) An expression for the total revenue received for the flight (*Hint:* Multiply the number of people flying, $100 - x$, by the price per ticket.)
(b) The graph for the expression in part (a)
(c) The number of unsold seats that will produce the maximum revenue
(d) The maximum revenue

33. The daily measurement (in particles) of a certain type of pollen during the first 10 days of June is approximated by the function $G(x) = 15 + 24x - 2x^2$, where x is the day in June, with $x = 1$ representing June 1.

(a) Sketch the graph of $G(x)$.
(b) Find the maximum pollen measurement, and determine when it occurs.

34. The demand for a certain type of cosmetic is given by $p = 500 - x$, where p is the price per unit when x units are demanded.

(a) Find the revenue, $R(x)$, obtained when x units are demanded. (*Hint:* Revenue = Number of units demanded $\times$ Price per unit.)
(b) Graph the revenue function defined by $y = R(x)$.
(c) From the graph of the revenue function, estimate the price that will produce the maximum revenue.
(d) What is the maximum revenue?

35. During the course of a year, the number of volunteers available to run a food bank each month is approximated by $V(x)$, where $V(x) = 2x^2 - 32x + 150$ between the months of January and August. Here x is time in months, with $x = 1$ representing January. From August to December, $V(x)$ is approximated by $V(x) = 31x - 226$. Find the number of volunteers in each of the following months:

(a) January **(b)** May **(c)** August **(d)** October **(e)** December.
(f) Sketch a graph of $y = V(x)$ for January through December. In what month are the fewest volunteers available?

36. Between the months of June and October, the percent of maximum possible chlorophyll production in a leaf is approximated by $C(x) = 10x + 50$. Here x is time in months, with $x = 1$ representing June. From October through December, $C(x)$ is approximated by $C(x) = -20(x - 5)^2 + 100$.
Find the percent of maximum possible chlorophyll production in each of the following months:
(a) June **(b)** July **(c)** September **(d)** October **(e)** November **(f)** December.
(g) Sketch a graph of $C(x)$ from June through December. In what month is chlorophyll production a maximum?

37. The cross section of an irrigation ditch is shaped like a parabola measuring 18 m across the top and 12 m deep. How wide is the ditch 8 m from the top?

38. An arch is shaped like a parabola. It is 30 m wide at the base and 15 m high. How wide is the arch 10 m from the ground?

39. Suppose that a quadratic function with $a > 0$ is written in the form $f(x) = a(x - h)^2 + k$. Match each of the items **(a)**, **(b)**, and **(c)** with one of the items **(A)**, **(B)**, or **(C)**.

(a) k is positive. **(A)** $f(x)$ intersects the x-axis at only one point.
(b) k is negative. **(B)** $f(x)$ does not intersect the x-axis.
(c) k is zero. **(C)** $f(x)$ intersects the x-axis twice.

The figures below show several possible graphs of $f(x) = ax^2 + bx + c$. For the restrictions on a, b, and c given in Exercises 40–45, select the corresponding graph from (a) through (f) below.

40. $a < 0, \quad b^2 - 4ac = 0$ **41.** $a > 0, \quad b^2 - 4ac < 0$ **42.** $a < 0, \quad b^2 - 4ac < 0$

43. $a < 0, \quad b^2 - 4ac > 0$ **44.** $a > 0, \quad b^2 - 4ac > 0$ **45.** $a > 0, \quad b^2 - 4ac = 0$

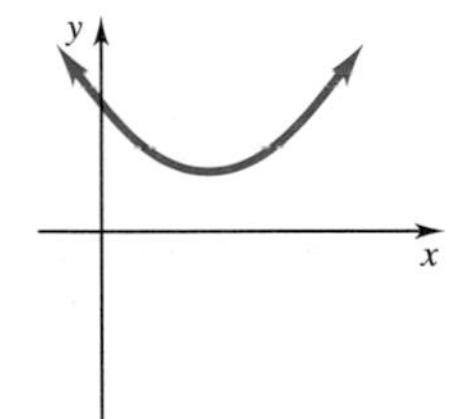

(a)

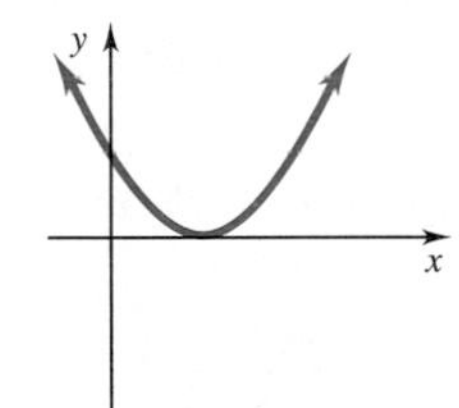

(b)

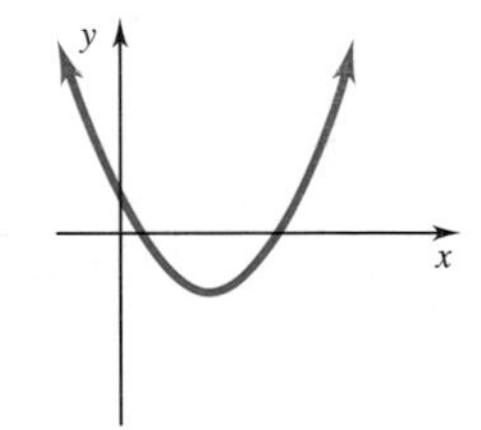

(c)

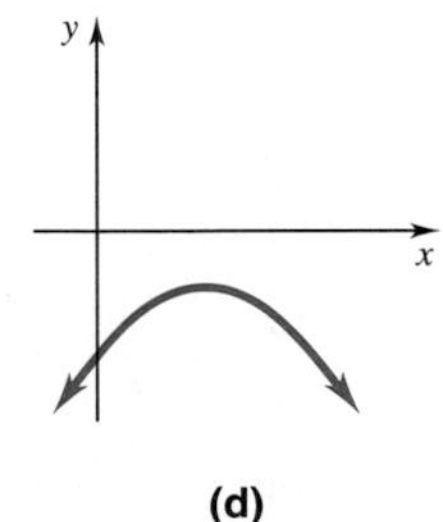

(d)

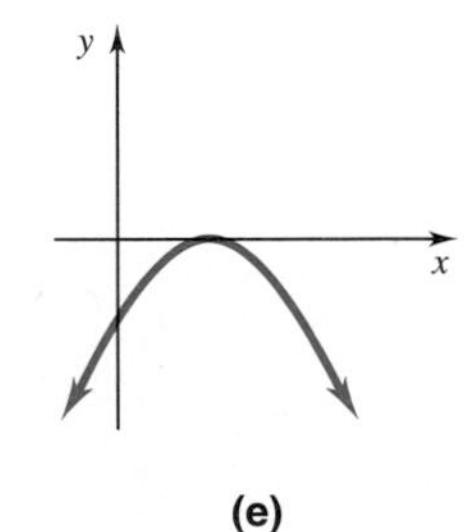

(e)

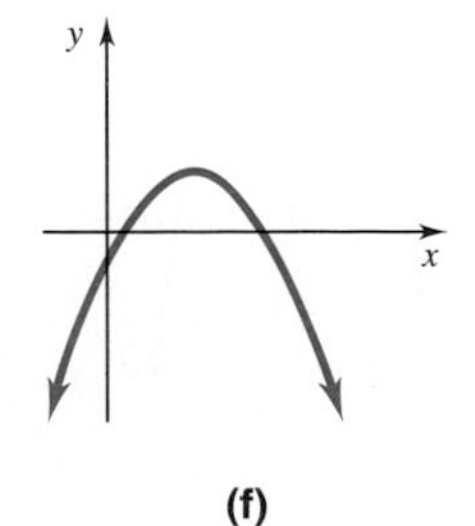

(f)

46. Find a value of c so that $y = x^2 - 10x + c$ has exactly one x-intercept.

47. For what values of a does $y = ax^2 - 8x + 4$ have no x-intercepts?

48. Find b so that $y = x^2 + bx + 9$ has exactly one x-intercept.

49. Find the quadratic function having x-intercepts 2 and 5, and y-intercept 5.

50. Find the quadratic function having x-intercepts 1 and -2, and y-intercept 4.

51. Find the largest possible value of y if $y = -(x - 2)^2 + 9$. Then find the following.

(a) The largest possible value of $\sqrt{-(x - 2)^2 + 9}$
(b) The smallest possible value of $1/[-(x - 2)^2 + 9]$

52. Find the smallest possible value of y if $y = 3 + (x + 5)^2$. Then find the following.

(a) The smallest possible value of $\sqrt{3 + (x + 5)^2}$
(b) The largest possible value of $1/[3 + (x + 5)^2]$

53. From the distance formula in Section 3.1, the distance between the two points $P(x_1, y_1)$ and $R(x_2, y_2)$ is $d(P, R) = \sqrt{(x_1 - x_2)^2 + (y_1 - y_2)^2}$. Using the results of Exercises 51 and 52, find the closest point on the line $y = 2x$ to the point (1, 7). (*Hint:* Every point on $y = 2x$ has the form $(x, 2x)$, and the closest point has the minimum distance.)

54. Let $f(x) = a(x - h)^2 + k$, and show that $f(h + x) = f(h - x)$. Why does this show that the parabola is symmetric with respect to its axis?

Graphing Utility Problems

55. In Section 3.6 we saw how certain changes to an equation cause the graph of the equation to be stretched, shrunk, reflected about an axis, or translated vertically or horizontally. It is important to notice that the order in which these changes are done affects the final graph. For example, stretching and then shifting vertically produces a graph that differs from the one produced by shifting vertically, then stretching. To see this, use your grapher to graph $y = 3x^2 - 2$ and $y = 3(x^2 - 2)$, and then compare the results. Are the two expressions equivalent algebraically?

In Exercises 56 and 57, find a polynomial function whose graph matches the one in the figure. Then use your grapher to graph the function and verify your result.

56.

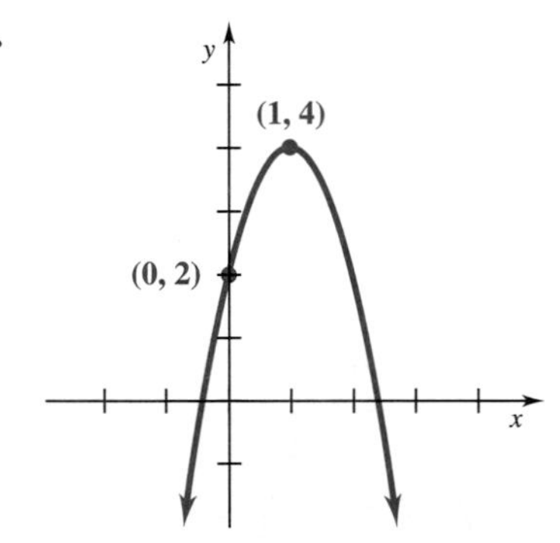

57.

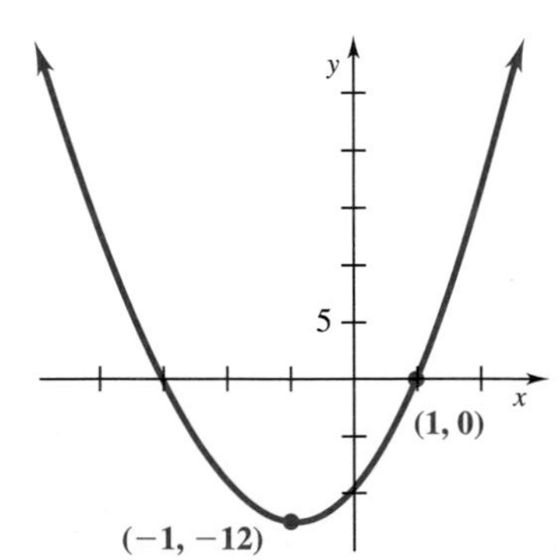

4.2 Synthetic Division

The quotient of two polynomials was found in Chapter 1 with a division algorithm similar to that used to divide with whole numbers.

Division Algorithm

Let $f(x)$ and $g(x)$ be polynomials with $g(x)$ of lower degree than $f(x)$ and $g(x)$ of degree one or more. There exist unique polynomials $q(x)$ and $r(x)$ such that

$$\mathbf{f(x) = g(x) \cdot q(x) + r(x),}$$

where either $r(x) = 0$ or the degree of $r(x)$ is less than the degree of $g(x)$.

Recall that $q(x)$ is the quotient polynomial and $r(x)$ is the remainder polynomial or the remainder.

A shortcut method of performing long division with certain polynomials, called **synthetic division**, will be useful in applying the theorems presented in the next few sections. The method is used only when a polynomial is divided by a first-degree binomial of the form $x - k$, where the coefficient of x is 1. To illustrate, notice the example worked on the left below. On the right the division process is simplified by omitting all variables and writing only coefficients, with 0 used to represent the coefficient of any missing terms. Since the coefficient of x in the divisor is always 1 in these divisions, it too can be omitted. These omissions simplify the problem as shown on the right below.

$$\begin{array}{r} 3x^2 + 10x \quad + 40 \qquad\quad \\ x - 4\overline{)3x^3 - \quad 2x^2 \qquad\qquad - 150} \\ \underline{3x^3 - 12x^2} \qquad\qquad\qquad \\ 10x^2 \qquad\qquad\qquad \\ \underline{10x^2 - 40x} \qquad\qquad \\ 40x - 150 \\ \underline{40x - 160} \\ 10 \end{array} \qquad\qquad \begin{array}{r} \mathbf{3} \quad \mathbf{10} \quad \mathbf{40} \qquad \\ -4\overline{)3 - \quad 2 + \quad 0 - 150} \\ \underline{\mathbf{3} - 12} \qquad\qquad\quad \\ 10 \qquad\qquad\quad \\ \underline{\mathbf{10} - 40} \qquad\quad \\ 40 - 150 \\ \underline{\mathbf{40} - 160} \\ 10 \end{array}$$

The numbers in color that are repetitions of the numbers directly above them can also be omitted.

$$\begin{array}{r} 3 \quad 10 \quad 40 \qquad \\ -4\overline{)3 - \quad 2 + \quad 0 - 150} \\ \underline{- 12} \qquad\qquad\quad \\ 10 \qquad\qquad\quad \\ \underline{- 40} \qquad\quad \\ 40 - 150 \\ \underline{- 160} \\ 10 \end{array}$$

The entire problem can now be condensed vertically, and the top row of numbers can be omitted since it duplicates the bottom row if the 3 is brought down.

$$\begin{array}{r|rrrr} -4 & 3 & -2 & 0 & -150 \\ & & -12 & -40 & -160 \\ \hline & 3 & 10 & 40 & 10 \end{array}$$

The rest of the bottom row is obtained by subtracting -12, -40, and -160 from the corresponding terms above.

With synthetic division it is useful to change the sign of the divisor, so the -4 at the left is changed to 4, which also changes the sign of the numbers in the second row. To compensate for this change, subtraction is changed to addition. Doing this gives the following result.

$$\begin{array}{r|rrrr} 4 & 3 & -2 & 0 & -150 \\ & & 12 & 40 & 160 \\ \hline & 3 & 10 & 40 & 10 \end{array}$$

In summary, to use synthetic division to divide a polynomial by a binomial of the form $x - k$, begin by writing the coefficients of the polynomial in decreasing powers of the variable, using 0 as the coefficient of any missing powers. The number k is written to the left in the same row. In the example above, $x - k$ is $x - 4$, so k is 4. Next bring down the leading coefficient of the polynomial, 3 in the example above, as the first number in the last row. Multiply the 3 by 4 to get the first number in the second row, 12. Add 12 to -2; this gives 10, the second number in the third row. Multiply 10 by 4 to get 40, the next number in the second row. Add 40 to 0 to get the third number in the third row, and so on. This process of multiplying each result in the third row by k and adding the product to the number in the next column is repeated until there is a number in the last row for each coefficient in the first row.

CAUTION To avoid incorrect results, it is essential to use a 0 for any missing terms, including a missing constant, when you set up the division. ■

EXAMPLE 1 Use synthetic division to divide $5x^3 - 6x^2 - 28x - 2$ by $x + 2$.

Begin by writing

$$\begin{array}{r|rrrr} -2 & 5 & -6 & -28 & -2. \end{array}$$

The value of k is -2, since k is found by writing $x + 2$ as $x - (-2)$. Next, bring down the 5.

$$\begin{array}{r|rrrr} -2 & 5 & -6 & -28 & -2 \\ & & & & \\ \hline & \mathbf{5} & & & \end{array}$$

Now, multiply -2 by 5 to get -10, and add it to the -6 in the first row. The result is -16.

$$\begin{array}{r|rrrr} -2 & 5 & -6 & -28 & -2 \\ & & \mathbf{-10} & & \\ \hline & 5 & \mathbf{-16} & & \end{array}$$

Next, $(-2)(-16) = 32$. Add this to the -28 in the first row.

$$\begin{array}{r|rrrr} -2 & 5 & -6 & -28 & -2 \\ & & -10 & \mathbf{32} & \\ \hline & 5 & -16 & \mathbf{4} & \end{array}$$

Finally, $(-2)(4) = -8$, which is added to the -2 to get -10.

$$\begin{array}{r|rrrr} -2 & 5 & -6 & -28 & -2 \\ & & -10 & 32 & \mathbf{-8} \\ \hline & 5 & -16 & 4 & \mathbf{-10} \end{array}$$

The coefficients of the quotient polynomial and the remainder are read directly from the bottom row. Since the degree of the quotient will always be one less than the degree of the polynomial to be divided,

$$\frac{5x^3 - 6x^2 - 28x - 2}{x + 2} = 5x^2 - 16x + 4 + \frac{-10}{x + 2}.$$ ■

The result of the division in Example 1 can be written as

$$5x^3 - 6x^2 - 28x - 2 = (x + 2)(5x^2 - 16x + 4) + (-10)$$

by multiplying both sides by the denominator $x + 2$. The following theorem is a generalization of the division process illustrated above.

> For any polynomial $f(x)$ and any complex number k, there exists a unique polynomial $q(x)$ and number r such that
>
> $$\boldsymbol{f(x) = (x - k)q(x) + r.}$$

For example, in the synthetic division above,

$$\underbrace{5x^3 - 6x^2 - 28x - 2}_{f(x)} = \underbrace{(x + 2)}_{(x - k)} \cdot \underbrace{(5x^2 - 16x + 4)}_{q(x)} + \underbrace{(-10)}_{r}.$$

This theorem is a special case of the division algorithm given earlier. Here $g(x)$ is the first-degree polynomial $x - k$.

Evaluating $f(k)$ By the division algorithm, $f(x) = (x - k)q(x) + r$. This equality is true for all complex values of x, so it is true for $x = k$. Replacing x with k gives

$$f(k) = (k - k)q(k) + r$$
$$f(k) = r.$$

This proves the following **remainder theorem**, which gives a new method of evaluating polynomial functions.

Remainder Theorem

If the polynomial $f(x)$ is divided by $x - k$, the remainder is $f(k)$.

As an illustration of this theorem, we have seen that when the polynomial $f(x) = 5x^3 - 6x^2 - 28x - 2$ is divided by $x + 2$ or $x - (-2)$, the remainder is -10. Now substituting -2 for x in $f(x)$ gives

$$\begin{aligned} f(-2) &= 5(-2)^3 - 6(-2)^2 - 28(-2) - 2 \\ &= -40 - 24 + 56 - 2 \\ &= -10. \end{aligned}$$

As shown here, the simpler way to find the value of a polynomial is often by using synthetic division. By the remainder theorem, instead of replacing x by -2 to find $f(-2)$, divide $f(x)$ by $x + 2$ using synthetic division as in Example 1. Then $f(-2)$ is the remainder, -10.

$$\begin{array}{r|rrrr} -2 & 5 & -6 & -28 & -2 \\ & & -10 & 32 & -8 \\ \hline & 5 & -16 & 4 & \mathbf{-10} \end{array} \quad \leftarrow f(-2)$$

EXAMPLE 2 Let $f(x) = -x^4 + 3x^2 - 4x - 5$. Find $f(-3)$.

Use the remainder theorem and synthetic division.

$$\begin{array}{r|rrrrr} -3 & -1 & 0 & 3 & -4 & -5 \\ & & 3 & -9 & 18 & -42 \\ \hline & -1 & 3 & -6 & 14 & -47 \end{array}$$

The remainder when $f(x)$ is divided by $x - (-3) = x + 3$ is -47, so $f(-3) = -47$. ■

The remainder theorem gives a quick way to decide if a number k is a zero of a polynomial $f(x)$. Use synthetic division to find $f(k)$; if the remainder is zero, then $f(k) = 0$ and k is a zero of $f(x)$. A zero of $f(x)$ is also called a **root** or **solution** of the equation $f(x) = 0$.

EXAMPLE 3 Decide whether or not the given number is a zero of the given polynomial.

(a) 2; $f(x) = x^3 - 4x^2 + 9x - 10$

Use synthetic division.

$$\begin{array}{r|rrrr} 2 & 1 & -4 & 9 & -10 \\ & & 2 & -4 & 10 \\ \hline & 1 & -2 & 5 & 0 \end{array}$$

Since the remainder is 0, $f(2) = 0$, and 2 is a zero of the polynomial $f(x) = x^3 - 4x^2 + 9x - 10$.

(b) -4; $f(x) = x^4 + x^2 - 3x + 1$

Remember to use a coefficient of 0 for the missing x^3 term in the synthetic division.

$$\begin{array}{r|rrrrr} -4 & 1 & 0 & 1 & -3 & 1 \\ & & -4 & 16 & -68 & 284 \\ \hline & 1 & -4 & 17 & -71 & 285 \end{array}$$

The remainder is not 0, so -4 is not a zero of $f(x) = x^4 + x^2 - 3x + 1$. In fact, $f(-4) = 285$.

(c) $1 + 2i$; $f(x) = x^4 - 2x^3 + 4x^2 + 2x - 5$

Use synthetic division and operations with complex numbers.

$$\begin{array}{r|rrrrr} 1+2i & 1 & -2 & 4 & 2 & -5 \\ & & 1+2i & -5 & -1-2i & 5 \\ \hline & 1 & -1+2i & -1 & 1-2i & 0 \end{array}$$

Since the remainder is zero, $1 + 2i$ is a zero of the given polynomial. ■

4.2 Exercises

Use synthetic division to perform each of the following divisions.

1. $\dfrac{x^3 + 4x^2 - 5x + 42}{x + 6}$

2. $\dfrac{x^3 + 2x^2 - 8x - 17}{x - 3}$

3. $\dfrac{4x^3 - 3x - 2}{x + 1}$

4. $\dfrac{3x^3 - 4x + 2}{x - 1}$

5. $\dfrac{x^4 - 3x^3 - 4x^2 + 12x}{x - 3}$

6. $\dfrac{x^4 - 3x^3 - 5x^2 + 2x - 16}{x + 2}$

7. $\dfrac{x^5 + 3x^4 + 2x^3 + 2x^2 + 3x + 1}{x + 2}$

8. $\dfrac{\frac{1}{3}x^3 - \frac{2}{9}x^2 + \frac{1}{27}x + 1}{x - \frac{1}{3}}$

9. $\dfrac{x^3 - 1}{x - 1}$

10. $\dfrac{x^5 - 1}{x - 1}$

Express each polynomial in the form $f(x) = (x - k)q(x) + r$ for the given value of k.

11. $f(x) = 2x^3 + x^2 + x - 8$; $k = -1$

12. $f(x) = 2x^3 + 3x^2 - 16x + 10$; $k = -4$

13. $f(x) = -x^3 + 2x^2 + 4$; $k = -2$

14. $f(x) = -4x^3 + 2x^2 - 3x - 10$; $k = 2$

15. $f(x) = 4x^4 - 3x^3 - 20x^2 - x$; $k = 3$

16. $f(x) = 2x^4 + x^3 - 15x^2 + 3x$; $k = -3$

For each of the following polynomials, use the remainder theorem and synthetic division to find $f(k)$.

17. $k = 5;\quad f(x) = -x^2 + 2x + 7$

18. $k = -3;\quad f(x) = 3x^2 + 8x + 5$

19. $k = 3;\quad f(x) = x^2 - 4x + 5$

20. $k = -2;\quad f(x) = x^2 + 5x + 6$

21. $k = 2;\quad f(x) = 2x^2 - 3x - 3$

22. $k = 4;\quad f(x) = -x^3 + 8x^2 + 63$

23. $k = -1;\quad f(x) = x^3 - 4x^2 + 2x + 1$

24. $k = 2;\quad f(x) = 2x^3 - 3x^2 - 5x + 4$

25. $k = 3;\quad f(x) = 2x^5 - 10x^3 - 19x^2 - 45$

26. $k = 4;\quad f(x) = x^4 + 6x^3 + 9x^2 + 3x - 3$

27. $k = -8;\quad f(x) = x^6 + 7x^5 - 5x^4 + 22x^3 - 16x^2 + x + 19$

28. $k = -\frac{1}{2};\quad f(x) = 6x^3 - 31x^2 - 15x$

29. $k = 2 + i;\quad f(x) = x^2 - 5x + 1$

30. $k = 3 - 2i;\quad f(x) = x^2 - x + 3$

Use synthetic division to decide whether or not the given number is a zero of the given polynomial.

31. $3;\quad f(x) = 2x^3 - 6x^2 - 9x + 4$

32. $-6;\quad f(x) = 2x^3 + 9x^2 - 16x + 12$

33. $-5;\quad f(x) = x^3 + 7x^2 + 10x$

34. $-2;\quad f(x) = 2x^3 - 3x^2 - 5x$

35. $\frac{2}{5};\quad f(x) = 5x^4 + 2x^3 - x + 15$

36. $\frac{1}{2};\quad f(x) = 2x^4 - 3x^2 + 4$

37. $2 - i;\quad f(x) = x^2 + 3x + 4$

38. $1 - 2i;\quad f(x) = x^2 - 3x + 5$

39. $i;\quad f(x) = x^3 + 2ix^2 + 2x + i$

40. $-i;\quad f(x) = x^3 - ix^2 + 3x + 5i$

41. Why must the function $r(x)$ in the division algorithm either be zero or have degree less than the degree of $g(x)$? (*Hint:* Think about the division algorithm for whole numbers.)

42. Find the remainder when the polynomial $x^{99} - 2x^{52} + x^2$ is divided by $x + 1$.

43. Find the remainder when the polynomial $x^{101} + 3x^{20} + x^3$ is divided by $x - i$.

44. Find the value of k so that $(x^2 + 4x + 8) \div (x - k)$ has a remainder of 4.

4.3 Zeros of Polynomial Functions

By the remainder theorem, if $f(k) = 0$, then the remainder when $f(x)$ is divided by $x - k$ is zero. This means that $x - k$ is a factor of $f(x)$. Conversely, if $x - k$ is a factor of $f(x)$, then $f(k)$ must equal 0. This is summarized in the following **factor theorem**.

Factor Theorem

The polynomial $x - k$ is a factor of the polynomial $f(x)$ if and only if $f(k) = 0$.

EXAMPLE 1 Is $x - 1$ a factor of $f(x) = 2x^4 + 3x^2 - 5x + 7$?

By the factor theorem, $x - 1$ will be a factor of $f(x)$ only if $f(1) = 0$. Use synthetic division and the remainder theorem to decide.

$$\begin{array}{r|rrrrr} 1 & 2 & 0 & 3 & -5 & 7 \\ & & 2 & 2 & 5 & 0 \\ \hline & 2 & 2 & 5 & 0 & 7 \end{array}$$

Since the remainder is 7, $f(1) = 7$ and not 0, so $x - 1$ is not a factor of the polynomial $f(x)$. ■

EXAMPLE 2 Is $x - i$ a factor of the following polynomial?

$$f(x) = 3x^3 + (-4 - 3i)x^2 + (5 + 4i)x - 5i$$

The only way $x - i$ can be a factor of $f(x)$ is for $f(i)$ to be 0. To see if this is the case, use synthetic division.

$$\begin{array}{r|rrrr} i & 3 & -4 - 3i & 5 + 4i & -5i \\ & & 3i & -4i & 5i \\ \hline & 3 & -4 & 5 & \mathbf{0} \end{array}$$

The remainder is 0, so $f(i) = 0$, and $x - i$ is a factor of $f(x)$. ■

The factor theorem can be used to factor a polynomial of higher degree into linear factors. Linear factors are factors of the form $ax - b$ for complex numbers a and b.

EXAMPLE 3 Factor $f(x)$ into linear factors given that k is a zero of $f(x)$.

(a) $f(x) = 6x^3 + 19x^2 + 2x - 3; \quad k = -3$

Since $k = -3$ is a zero of $f(x)$, $x - (-3) = x + 3$ is a factor. Use synthetic division to divide $f(x)$ by $x + 3$.

$$\begin{array}{r|rrrr} -3 & 6 & 19 & 2 & -3 \\ & & -18 & -3 & 3 \\ \hline & 6 & 1 & -1 & 0 \end{array}$$

The quotient is $6x^2 + x - 1$, so

$$f(x) = (x + 3)(6x^2 + x - 1).$$

Factor $6x^2 + x - 1$ as $(2x + 1)(3x - 1)$ to get

$$f(x) = (x + 3)(2x + 1)(3x - 1),$$

where all factors are linear.

(b) $f(x) = 3x^3 + (-1 + 3i)x^2 + (-12 + 5i)x + 4 - 2i; \quad k = 2 - i$

One factor is $x - (2 - i)$ or $x - 2 + i$. Divide $f(x)$ by $x - (2 - i)$.

$$\begin{array}{r|rrrr} 2 - i & 3 & -1 + 3i & -12 + 5i & 4 - 2i \\ & & 6 - 3i & 10 - 5i & -4 + 2i \\ \hline & 3 & 5 & -2 & 0 \end{array}$$

By the division algorithm,

$$f(x) = (x - 2 + i)(3x^2 + 5x - 2).$$

Factor $3x^2 + 5x - 2$ as $(3x - 1)(x + 2)$; then a linear factored form of $f(x)$ is

$$f(x) = (x - 2 + i)(3x - 1)(x + 2).$$

The next theorem says that every polynomial of degree 1 or more has a zero, which means that every such polynomial can be factored. The theorem was first proved by Carl Friedrich Gauss (1777–1855) as part of his doctoral dissertation completed in 1799. This theorem, which Gauss named the "fundamental theorem of algebra," had challenged the world's finest mathematicians for at least 200 years. Peter Rothe had stated it in 1608, followed by Albert Girard in 1629 and René Descartes in 1637. Jean LeRond D'Alembert (1717–1783) thought he had a proof in 1746; consequently it is known in France today as D'Alembert's theorem. Two of the world's greatest mathematicians, Euler and Lagrange, had attempted unsuccessfully to solve it in 1749 and 1772, respectively. Their errors were noted in Gauss's dissertation. Gauss's proof uses advanced mathematical concepts outside of the field of algebra. To this day, no purely algebraic proof has been discovered.

Fundamental Theorem of Algebra

Every polynomial of degree 1 or more has at least one complex zero.

From the fundamental theorem, if $f(x)$ is of degree 1 or more then there is some number k_1 such that $f(k_1) = 0$. By the factor theorem, then

$$f(x) = (x - k_1) \cdot q_1(x)$$

for some polynomial $q_1(x)$. If $q_1(x)$ is of degree 1 or more, the fundamental theorem and the factor theorem can be used to factor $q_1(x)$ in the same way. There is some number k_2 such that $q_1(k_2) = 0$, so that

$$q_1(x) = (x - k_2)q_2(x)$$

and

$$f(x) = (x - k_1)(x - k_2)q_2(x).$$

Assuming that $f(x)$ has degree n and repeating this process n times gives

$$f(x) = a(x - k_1)(x - k_2) \ldots (x - k_n),$$

where a is the leading coefficient of $f(x)$. Each of these factors leads to a zero of $f(x)$, so $f(x)$ has the n zeros $k_1, k_2, k_3, \ldots, k_n$. This result suggests the next theorem.

Number of Zeros Theorem

A polynomial of degree n has at most n distinct zeros.

NOTE The theorem says that there exist *at most n* distinct zeros. For example, the polynomial $f(x) = x^3 + 3x^2 + 3x + 1 = (x + 1)^3$ is of degree 3 but has only one zero, -1. Actually, the zero -1 occurs three times, since there are three factors of $x + 1$; this zero is called a **zero of multiplicity 3.** ■

For Graphers

A graphing utility is useful for finding the number of distinct real zeros of a polynomial function. Since the zeros are x-intercepts, they can be found by zooming in on the intercepts of the graph and using TRACE to locate them to the required degree of accuracy. It is important, however, to be sure you have an x- range large enough to show all real zeros. In the next section we show how to decide whether the x-range is large enough.

Although it cannot identify imaginary zeros, a graph of a polynomial function can indicate their existence. The fundamental theorem says that every polynomial has at least one complex zero, and the number of zeros theorem tells us that a polynomial of degree n has at most n zeros. If a graph showing all real zeros shows fewer than n real zeros, either the missing zeros are imaginary or there are one or more zeros of multiplicity greater than 1. (See Exercises 53–58.)

EXAMPLE 4 Find a polynomial $f(x)$ of degree 3 that satisfies the following conditions.

(a) Zeros of -1, 2, and 4; $f(1) = 3$

These three zeros give $x - (-1) = x + 1$, $x - 2$, and $x - 4$ as factors of $f(x)$. Since $f(x)$ is to be of degree 3, these are the only possible factors by the theorem just above. Therefore, $f(x)$ has the form

$$f(x) = a(x + 1)(x - 2)(x - 4)$$

for some real number a. To find a, use the fact that $f(1) = 3$.

$$\begin{aligned} f(1) = a(1 + 1)(1 - 2)(1 - 4) &= \mathbf{3} \\ a(2)(-1)(-3) &= 3 \\ 6a &= 3 \\ a &= \frac{1}{2} \end{aligned}$$

Thus,
$$f(x) = \frac{1}{2}(x + 1)(x - 2)(x - 4),$$

or, by multiplication,

$$f(x) = \frac{1}{2}x^3 - \frac{5}{2}x^2 + x + 4.$$

(b) -2 is a zero of multiplicity 3; $f(-1) = 4$

The polynomial $f(x)$ has the form

$$\begin{aligned} f(x) &= a(x + 2)(x + 2)(x + 2) \\ &= a(x + 2)^3. \end{aligned}$$

Since $f(-1) = 4$,

$$\begin{aligned} f(-1) = a(-1 + 2)^3 &= 4, \\ a(1)^3 &= 4 \end{aligned}$$

or
$$a = 4,$$

and $f(x) = 4(x + 2)^3 = 4x^3 + 24x^2 + 48x + 32.$ ■

For Graphers

In Example 6, we saw how to find additional zeros when at least one zero is known. A graphing utility can be used to locate a real zero by zooming and tracing. If the exact value of the zero can be determined, then the polynomial can be factored, and the techniques of this section might be used to find other zeros. (See Exercises 59–64.)

Use synthetic division to find $q(x)$.

$$\begin{array}{r|rrrr} 1+i & 1 & -6-i & 11+5i & -6-6i \\ & & 1+i & -5-5i & 6+6i \\ \hline & 1 & -5 & 6 & 0 \end{array}$$

Since $q(x) = x^2 - 5x + 6$, $f(x)$ can be written as

$$f(x) = [x - (1 - i)][x - (1 + i)](x^2 - 5x + 6).$$

Now find the zeros of the quadratic polynomial $x^2 - 5x + 6$. Factoring the polynomial shows that the zeros are 2 and 3, so the four zeros of $f(x)$ are $1 - i$, $1 + i$, 2, and 3. ■

4.3 Exercises

Use the factor theorem to decide whether or not the second polynomial is a factor of the first.

1. $4x^2 + 2x + 54;\quad x - 4$

2. $5x^2 - 14x + 10;\quad x + 2$

3. $x^3 + 2x^2 - 3;\quad x - 1$

4. $2x^3 + x + 2;\quad x + 1$

5. $2x^4 + 5x^3 - 2x^2 + 5x + 6;\quad x + 3$

6. $5x^4 + 16x^3 - 15x^2 + 8x + 16;\quad x + 4$

For each of the following, find a polynomial of degree 3 with only real coefficients that satisfies the given conditions.

7. Zeros of -3, 1, and 4; $f(2) = 30$

8. Zeros of 1, -1, and 0; $f(2) = 3$

9. Zeros of -2, 1, and 0; $f(-1) = -1$

10. Zeros of 2, -3, and 5; $f(3) = 6$

11. Zeros of 5, i, and $-i$; $f(2) = 5$

12. Zeros of -2, i, and $-i$; $f(-3) = 30$

For each of the following polynomial functions, find all zeros and their multiplicities.

13. $f(x) = 7x^3 + x$

14. $f(x) = (x + 1)^2(x - 1)^3(x^2 - 10)$

15. $f(x) = 3(x - 2)(x + 3)(x - 1 + i)$

16. $f(x) = 5x^2(x + 1 - \sqrt{2})(2x + 5)$

17. $f(x) = (x^2 + x - 2)^5(x - 1 + \sqrt{3})^2$

18. $f(x) = (7x - 2)^3(x^2 + 9)^2$

For each of the following, find a polynomial of lowest degree with only real coefficients and having the given zeros.

19. $3 + i$ and $3 - i$

20. $7 - 2i$ and $7 + 2i$

21. $1 + \sqrt{2}$, $1 - \sqrt{2}$, and 3

22. $1 - \sqrt{3}$, $1 + \sqrt{3}$, and 1

23. $-2 + i$, $-2 - i$, 3, and -3

24. $3 + 2i$, -1, and 2

25. 2 and $3i$

26. -1 and $6 - 3i$

27. $1 + 2i$, 2 (multiplicity 2)

28. $2 + i$, -3 (multiplicity 2)

For each of the following polynomials, one zero is given. Find all others.

29. $f(x) = x^3 - x^2 - 4x - 6;\ 3$

30. $f(x) = x^3 + 4x^2 - 5;\ 1$

31. $f(x) = 4x^3 + 6x^2 - 2x - 1;\ 1/2$

32. $f(x) = x^3 - 7x^2 + 17x - 15;\ 2 - i$

33. $f(x) = x^4 + 5x^2 + 4;\ -i$

34. $f(x) = x^4 + 10x^3 + 27x^2 + 10x + 26;\ i$

Factor $f(x)$ into linear factors given that k is a zero of $f(x)$.

35. $f(x) = 2x^3 - 3x^2 - 17x + 30;\ k = 2$

36. $f(x) = 2x^3 - 3x^2 - 5x + 6;\ k = 1$

37. $f(x) = 6x^3 + 13x^2 - 14x + 3;\ k = -3$

38. $f(x) = 6x^3 + 17x^2 - 63x + 10;\ k = -5$

39. $f(x) = x^3 + (7 - 3i)x^2 + (12 - 21i)x - 36i;\ k = 3i$

40. $f(x) = 2x^3 + (11 - 4i)x^2 + (12 - 22i)x - 24i;\ k = 2i$

41. Show that -2 is a zero of multiplicity 2 of $f(x) = x^4 + 2x^3 - 7x^2 - 20x - 12$ and find all other complex zeros. Then write $f(x)$ in factored form.

42. Show that -1 is a zero of multiplicity 3 of $f(x) = x^5 - 4x^3 - 2x^2 + 3x + 2$ and find all other complex zeros. Then write $f(x)$ in factored form.

43. How many real zeros (counting multiplicities) are possible for a polynomial with real coefficients of degree 5? Explain your reasoning.

44. Explain why a polynomial of degree 3 with real coefficients has at least one real zero.

45. Explain why it is not possible for a polynomial of degree 3 with real coefficients to have zeros of 1, 2, and $1 + i$.

46. Show that the zeros of $f(x) = x^3 + ix^2 - (7 - i)x + (6 - 6i)$ are $1 - i$, 2, and -3. Does the conjugate zeros theorem apply? Why?

47. The displacement after time t of a particle moving along a straight line is given by $s(t) = 2t^4 - t^3 + t^2 - 14t$, where t is in seconds and s is in meters. The displacement is 0 after 2 sec. At what other times is the displacement 0?

48. The cost function (in thousands of dollars) for producing squeeze bottles is given by $C(x) = x^3 - 7x^2 + 20x - 12$ and the revenue function (in thousands of dollars) is given by $R(x) = 4x$, where x is the number of bottles produced (in hundred-thousands). Cost equals revenue if 200,000 bottles are produced ($x = 2$). Find all other break-even points.

If c and d are complex numbers, prove each of the following statements. (Hint: *Let $c = a + bi$ and $d = m + ni$ and form all the conjugates, the sums, and the products.*)

49. $\overline{c + d} = \overline{c} + \overline{d}$

50. $\overline{cd} = \overline{c} \cdot \overline{d}$

51. $\overline{a} = a$ for any real number a.

52. $\overline{c^n} = (\overline{c})^n$

Graphing Utility Problems

53. Graph $f(x) = x^3 - 9.8x^2 - 2x$ first with the ranges $-1 \le x \le 1$, $-3 \le y \le 1$, and then with the ranges $-20 \le x \le 20$, $-1000 \le y \le 1000$. Use TRACE to estimate the three real zeros of $f(x)$. Explain why both sets of ranges were needed to find the zeros.

54. Graph $f(x) = x^4 - 1.9x^2 + .9$ with the ranges $-10 \le x \le 10$, $-1 \le y \le 1000$, and use TRACE to estimate the zeros of $f(x)$. Repeat with the ranges $-2 \le x \le 2$, $-1 \le y \le 10$ and $.9 \le x \le 1.1$ and $-.01 \le y \le .01$. Use TRACE to estimate the three real zeros of $f(x)$. Explain why all three sets of ranges were needed to find the zeros.

In Exercises 55–58, determine the number of distinct real zeros of the polynomial.

55. $f(x) = x^4 + 2x^3 - 4x^2 + 2x - 7$

56. $f(x) = x^3 - 3x^2 + 4$

57. $f(x) = 3x^4 - x^2 - 5x - 2$

58. $f(x) = 10x^4 - 109x^3 + 92x^2 - 20x$

In Exercises 59–64, each polynomial has one or more rational zeros. Use ZOOM *and* TRACE *to find the rational zero (or zeros) and then factor the polynomial to find the other zeros.*

59. $f(x) = x^3 - x^2 - x - 2$

60. $f(x) = 2x^3 - 3x^2 - 12x + 18$

61. $f(x) = 6x^3 + x^2 - 5x - 2$

62. $f(x) = 4x^3 - 7x^2 + 6x + 2$

63. $f(x) = x^4 - x^3 + 2x^2 + x - 3$

64. $f(x) = 2x^4 - 7x^3 + 10x^2 - x - 4$

4.4 Graphs of Polynomial Functions

We have already discussed the graphs of polynomial functions of degree 0 to 2. In this section we show how to graph polynomial functions of degree 3 or more. The domains will be restricted to real numbers, since we will be graphing on the real-number plane. Without calculus, it is necessary to plot many points to determine the graph of a polynomial function. The theorems in this chapter will be helpful in deciding which points to plot.

The domain of every polynomial function is the set of all real numbers. The range of a polynomial function of odd degree is also the set of all real numbers. Some typical graphs of polynomial functions of odd degree are shown in Figure 8. These graphs suggest that for every polynomial function f of odd degree there is at least one real value of x that makes $f(x) = 0$. The zeros are the x-intercepts of the graph.

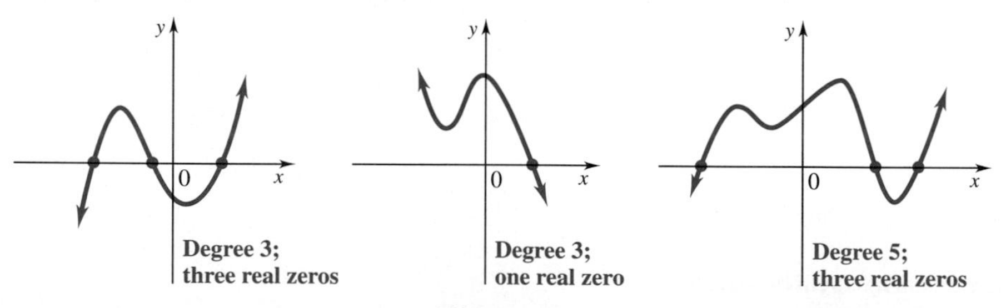

Figure 8

A polynomial function of even degree will have a range of the form $(-\infty, k]$ or else $[k, \infty)$ for some real number k. Figure 9 shows two typical graphs of polynomial functions of even degree.

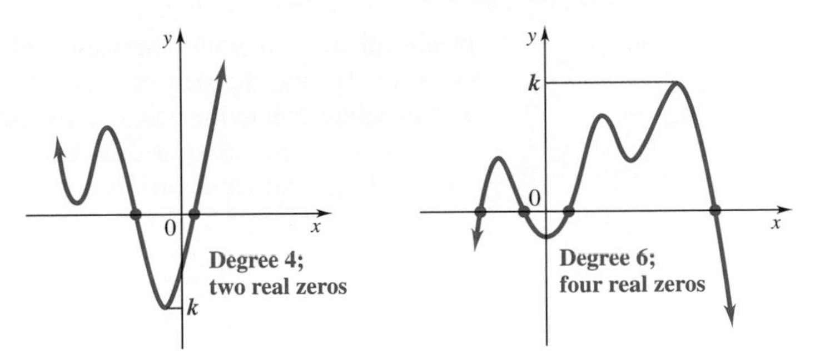

Figure 9

The graphs in Figures 8 and 9 show that polynomial functions often have *turning points* where the function changes from increasing to decreasing or from decreasing to increasing. A polynomial function of degree n has at most $n - 1$ turning points with at least one turning point between each pair of successive zeros. The graphs shown above illustrate this.

The end behavior of a polynomial graph is determined by the term of highest degree. That is, a polynomial of the form $f(x) = a_nx^n + a_{n-1}x^{n-1} + \cdots + a_0$ has the same end behavior as the polynomial $f(x) = a_nx^n$. For instance, the polynomial $f(x) = 2x^3 - 8x^2 + 9$ has the same end behavior as $f(x) = 2x^3$. It is large and positive for large positive values of x and large and negative for negative values of x of large absolute value. The arrows at the ends of the graph look like those of the first graph in Figure 8; the right arrow points up and the left arrow points down. Table 1 summarizes the end behavior of polynomials.

End Behavior of $f(x) = a_nx^n + a_{n-1}x^{n-1} + \cdots + a_0$

Degree	*Sign of a_n*	*Left arrow*	*Right arrow*	*Example*
Odd	Positive	Down	Up	First graph of Figure 8
Odd	Negative	Up	Down	Second graph of Figure 8
Even	Positivc	Up	Up	First graph of Figure 9
Even	Negative	Down	Down	Second graph of Figure 9

If the zeros of a polynomial function are known, its graph can be approximated without plotting very many points; this method is shown in the next example.

EXAMPLE 1 Graph $f(x) = (x - 1)(2x + 3)(x + 2)$.

The three zeros of $f(x)$ are $x = 1$, $x = -3/2$ and $x = -2$. These three zeros divide the x-axis into four regions, shown in Figure 10.

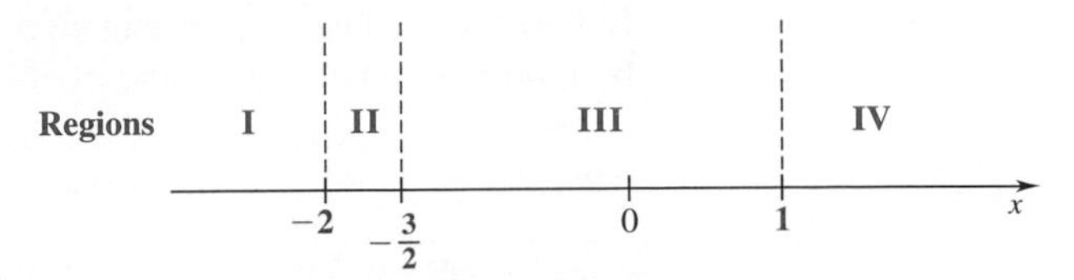

Figure 10

In any of these regions, the values of $f(x)$ are either always positive or always negative. To find the sign of $f(x)$ in each region, select an x-value in each region and substitute it into the equation for $f(x)$ to determine if the values of the function are positive or negative in that region. A typical selection of test points and the results of the tests are shown below.

Region	*Test point*	*Value of $f(x)$*	*Sign of $f(x)$*
I $(-\infty, -2)$	-3	-12	Negative
II $(-2, -3/2)$	$-7/4$	$11/32$	Positive
III $(-3/2, 1)$	0	-6	Negative
IV $(1, \infty)$	2	28	Positive

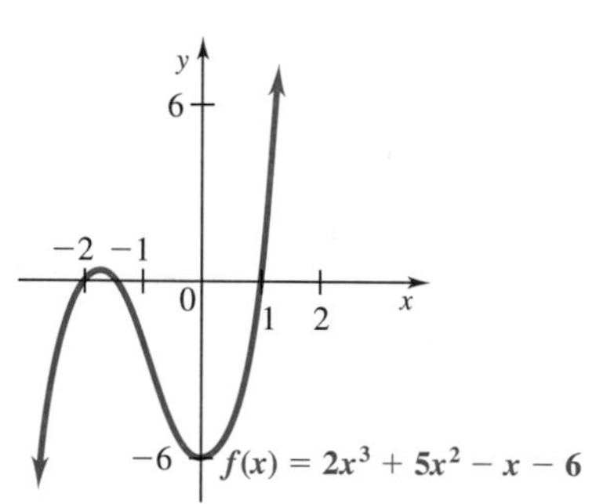

Figure 11

Plot the three zeros and the test points and connect them with a smooth curve to get the graph. When the values of $f(x)$ are negative, the graph is below the x-axis, and when $f(x)$ takes on positive values, the graph is above the x-axis, as shown in Figure 11. The sketch could be improved by plotting additional points in each region. Notice that the left arrow points down and the right arrrow points up. This end behavior is correct since when the linear factors are multiplied out, the highest term of the polynomial is $2x^3$. ■

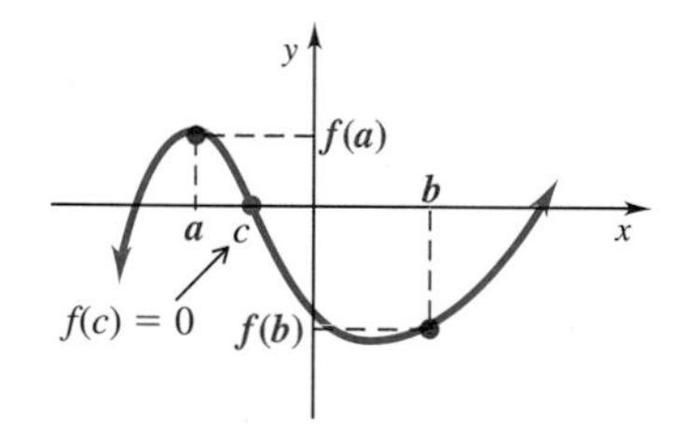

Figure 12

As Example 1 shows, the key to graphing a polynomial function is locating its zeros. In the special case where the zeros are rational numbers, the zeros can be found by a technique presented in the next section. Occasionally, irrational zeros can be found by inspection. For instance, $f(x) = x^3 - 2$, has the irrational zero $\sqrt[3]{2}$. Two theorems presented in this section apply to the zeros of every polynomial function with real coefficients. The first theorem uses the fact that graphs of polynomial functions are unbroken curves, with no gaps or sudden jumps. The proof requires advanced methods, so it is not given here. Figure 12 illustrates the theorem.

Intermediate Value Theorem for Polynomials

If $f(x)$ is a polynomial with only real coefficients, and if for real numbers a and b, the values $f(a)$ and $f(b)$ are opposite in sign, then there exists at least one real zero between a and b.

This theorem helps to identify intervals where zeros of polynomials are located. If $f(a)$ and $f(b)$ are opposite in sign, then 0 is between $f(a)$ and $f(b)$, and there must be a number c between a and b where $f(c) = 0$.

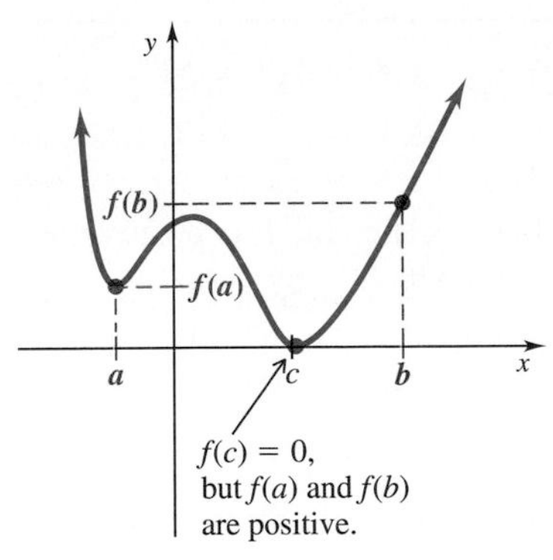

Figure 13

CAUTION The converse of the theorem is false. The fact that $f(c) = 0$, for c between a and b, does not imply that 0 is between $f(a)$ and $f(b)$. See Figure 13. ■

EXAMPLE 2 Show that $f(x) = x^3 - 2x^2 - x + 1$ has a real zero between 2 and 3.

Use synthetic division to find $f(2)$ and $f(3)$.

$$\begin{array}{r|rrrr} 2 & 1 & -2 & -1 & 1 \\ & & 2 & 0 & -2 \\ \hline & 1 & 0 & -1 & -1 \end{array} \qquad \begin{array}{r|rrrr} 3 & 1 & -2 & -1 & 1 \\ & & 3 & 3 & 6 \\ \hline & 1 & 1 & 2 & 7 \end{array}$$

Since $f(2)$ is negative but $f(3)$ is positive, there must be a real zero between 2 and 3. ■

The intermediate value theorem for polynomials is helpful in limiting the search for real zeros to smaller and smaller intervals. In Example 2 the theorem was used to verify that there is a real zero between 2 and 3. The theorem could then be used repeatedly to locate the zero more accurately. The next theorem, the *boundedness theorem,* shows how the bottom row of a synthetic division can be used to place upper and lower bounds on the possible real zeros of a polynomial.

Boundedness Theorem

Let $f(x)$ be a polynomial of degree $n \geq 1$ with real coefficients and with a positive leading coefficient. If $f(x)$ is divided synthetically by $x - c$ and

(a) if $c > 0$ and all numbers in the bottom row of the synthetic division are nonnegative, then $f(x)$ has no zero greater than c;

(b) if $c < 0$ and the numbers in the bottom row of the synthetic division alternate in sign (with 0 considered positive or negative, as needed), then $f(x)$ has no zero less than c.

An outline of the proof of part (a) is given here. The proof for part (b) is similar. By the division algorithm, if $f(x)$ is divided by $x - c$, then

$$f(x) = (x - c)q(x) + r,$$

where all coefficients of $q(x)$ are nonnegative, $r \geq 0$, and $c > 0$. If $x > c$, then $x - c > 0$. Since $q(x) > 0$, and $r \geq 0$,

$$f(x) = (x - c)q(x) + r > 0.$$

This means that $f(x)$ will never be 0 for $x > c$.

EXAMPLE 3 Show that the real zeros of $f(x) = 2x^4 - 5x^3 + 3x + 1$ satisfy the following conditions.

(a) No real zero is greater than 3.

Since $f(x)$ has real coefficients and the leading coefficient, 2, is positive, the boundedness theorem can be used. Divide $f(x)$ synthetically by $x - 3$.

$$\begin{array}{r|rrrrr} 3 & 2 & -5 & 0 & 3 & 1 \\ & & 6 & 3 & 9 & 36 \\ \hline & 2 & 1 & 3 & 12 & 37 \end{array}$$

Since $3 > 0$ and all numbers in the last row of the synthetic division are nonnegative, $f(x)$ has no real zero greater than 3.

(b) No real zero is less than -1.

Divide $f(x)$ by $x + 1$.

$$\begin{array}{r|rrrrr} -1 & 2 & -5 & 0 & 3 & 1 \\ & & -2 & 7 & -7 & 4 \\ \hline & 2 & -7 & 7 & -4 & 5 \end{array}$$

Here $-1 < 0$ and the numbers in the last row alternate in sign, so $f(x)$ has no zero less than -1. ■

EXAMPLE 4 Approximate the real zeros of $f(x) = x^4 - 6x^3 + 8x^2 + 2x - 1$.

The term of highest degree is x^4. Therefore, for negative values of x large in absolute value, the graph of $f(x)$ assumes large positive values. By substitution, $f(0)$ is easily seen to be the constant term, -1. Therefore, by the intermediate value theorem, $f(x)$ has a zero on the left half of the x-axis. We will determine $f(-1)$, $f(-2)$, $f(-3)$, and so on, until there is a positive value. Success occurs on the first try.

$$\begin{array}{r|rrrrr} -1 & 1 & -6 & 8 & 2 & -1 \\ & & -1 & 7 & -15 & 13 \\ \hline & 1 & -7 & 15 & -13 & \mathbf{12 > 0} \end{array}$$

Therefore, since $f(-1) = 12 > 0$ and $f(0) = -1 < 0$, $f(x)$ has a zero between -1 and 0. The leading coefficient of $f(x)$ is positive and the numbers in the last row of the synthetic division alternate in sign. Since $-1 < 0$, by the boundedness theorem $f(x)$ has no zeros to the left of -1.

To locate the zero of $f(x)$ in the interval $(-1, 0)$, try $c = -.5$ in the synthetic division. Divide $f(x)$ by $x + .5$.

$$\begin{array}{r|rrrrr} -.5 & 1 & -6 & 8 & 2 & -1 \\ & & -.5 & 3.25 & -5.625 & 1.8125 \\ \hline & 1 & -6.5 & 11.25 & -3.625 & .8125 \end{array}$$

Since $f(-.5) > 0$ and $f(0) < 0$, there is a real zero between $-.5$ and 0. Try $c = -.4$.

$$\begin{array}{r|rrrrr} -.4 & 1 & -6 & 8 & 2 & -1 \\ & & -.4 & 2.56 & -4.224 & .8896 \\ \hline & 1 & -6.4 & 10.56 & -2.224 & -.1104 \end{array}$$

Since $f(-.5)$ is positive, but $f(-.4)$ is negative, there is a zero between $-.5$ and $-.4$. The value of $f(-.4)$ is closer to zero than $f(-.5)$, so it is probably safe to say that, to one decimal place of accuracy, $-.4$ is an approximation to a real zero of $f(x)$. A more accurate result can be found, if desired, by continuing this process.

To locate the remaining real zeros of $f(x)$, continue in the same way. Use synthetic division to find $f(1)$, $f(2)$, $f(3)$, and so on, until there is a change in sign. It is helpful to use the shortened form of synthetic division shown below. Only the last row of the synthetic division is shown for each division. The first row of the chart is used for each division and the work in the second row of the division is done mentally.

x					$f(x)$	
	1	−6	8	2	−1	
−1	1	−7	15	−13	12	←Zero between −1 and 0
0	1	−6	8	2	−1	←Zero between 0 and 1
1	1	−5	3	5	4	
2	1	−4	0	2	3	←Zero between 2 and 3
3	1	−3	−1	−1	−4	←Zero between 3 and 4
4	1	−2	0	2	7	

Since the polynomial is of degree 4, there are no more than 4 zeros. Expand the table to approximate the real zeros to the nearest tenth. For example, for the zero between 0 and 1, the work might go as follows. Start halfway between 0 and 1 with $x = .5$. Since $f(.5) > 0$ and $f(0) < 0$, try $x = .4$ next.

x					$f(x)$	
	1	−6	8	2	−1	
.5	1	−5.5	5.25	4.63	1.31	
.4	1	−5.6	5.76	4.30	.72	
.3	1	−5.7	6.29	3.89	.17	←Zero between .3 and .2
.2	1	−5.8	6.84	3.37	−.33	

The value $f(.3) = .17$ is closer to 0 than $f(.2) = -.33$, so to the nearest tenth, the zero is .3. Use synthetic division to verify that the remaining two zeros are approximately 2.4 and 3.7. ■

For Graphers

As an alternative, a graphing utility can be used to search for real zeros of a polynomial. First choose x- and y-ranges that show the complete graph (the boundedness theorem can be used to determine suitable values for the x- and y-ranges). Then use the ZOOM and TRACE features to approximate the zeros.

In *Geometry*, published as an appendix to *A Discourse on the Method of Rightly Conducting the Reason and Seeking Truth in the Sciences* (1637), René Descartes stated for the first time a general rule, which had been useful earlier in specific cases. The rule gives a method to determine limits to the number of positive and the number of negative roots of a polynomial. This labor-shortening rule became known as *Descartes' rule of signs* because of his exposition. His *Geometry* had a profound effect on mathematicians. It is the oldest mathematical text that we can read without having difficulty with the notation.

Descartes' Rule of Signs

Let $f(x)$ be a polynomial with real coefficients and terms in descending order.

(a) The number of positive real zeros of $f(x)$ either is equal to the number of variations in sign occurring in the coefficients of $f(x)$, or else is less than the number of variations by a positive even integer.

(b) The number of negative real zeros of $f(x)$ either is equal to the number of variations in sign of $f(-x)$, or else is less than the number of variations by a positive even integer.

In the theorem, the number of variations in sign of the coefficients of $f(x)$ or $f(-x)$ refers to changes from positive to negative in successive terms of the polynomial. Missing terms (those with 0 coefficients) are counted as no change in sign and can be ignored.

For the purposes of this theorem, zeros of multiplicity k count as k zeros. For example,

$$\begin{aligned} f(x) &= (x - 1)^4 \\ &= \underbrace{+\, x^4 \;-}_{1} \underbrace{\; 4x^3 \;+}_{2} \underbrace{\; 6x^2 \;-}_{3} \underbrace{\; 4x \;+}_{4} 1 \end{aligned}$$

has four changes of sign. By Descartes' rule of signs, $f(x)$ has four, two, or no positive real zeros. In this case, there are four; each of the four positive real zeros is 1. Also,

$$f(-x) = x^4 + 4x^3 + 6x^2 + 4x + 1$$

which has no changes in sign, indicating there are no negative real zeros, as expected.

The polynomial in Example 4,

$$f(x) = x^4 - 6x^3 + 8x^2 + 2x - 1,$$

has three variations in sign:

$$\underbrace{+\, x^4 \;-}_{1} \underbrace{\; 6x^3 \;+}_{2} 8x^2 \underbrace{+\; 2x \;-}_{3} 1.$$

By Descartes' rule of signs, $f(x)$ has either three or one $(3 - 2 = 1)$ positive real zeros. Example 4 showed that $f(x)$ has three positive real zeros. Since $f(x)$ is of degree 4 and has three positive real zeros, it must have one negative real zero, which corresponds to the result in Example 4. This could be verified with part (b) of Descartes' rule of signs. Since

$$\begin{aligned} f(-x) &= (-x)^4 - 6(-x)^3 + 8(-x)^2 + 2(-x) - 1 \\ &= x^4 + 6x^3 + 8x^2 - 2x - 1 \end{aligned}$$

has only one variation in sign, $f(x)$ has only one negative real zero.

EXAMPLE 5 Apply Descartes' rule of signs to

$$f(x) = x^5 + 5x^4 + 3x^2 + 2x + 1.$$

The polynomial $f(x)$ has no variations in sign and so has no positive real zeros. Here

$$f(-x) = -x^5 + 5x^4 + 3x^2 - 2x + 1,$$

with three variations in sign, so $f(x)$ has either three or one negative real zeros. The other zeros are nonreal complex numbers. ■

EXAMPLE 6 Graph the function defined by $f(x) = 8x^3 - 12x^2 + 2x + 1$.

By Descartes' rule of signs, $f(x)$ has two or no positive real zeros and one negative real zero, so the graph of $f(x)$ will have two or no positive x-intercepts and one negative x-intercept. To find some ordered pairs belonging to the graph, use synthetic division to evaluate, say $f(3)$, in hopes of finding a number greater than all zeros of $f(x)$.

$$\begin{array}{r|rrrr} 3 & 8 & -12 & 2 & 1 \\ & & 24 & 36 & 114 \\ \hline & 8 & 12 & 38 & 115 \end{array}$$

Since $f(3) = 115$, the point $(3, 115)$ belongs to the graph. Also, since the bottom row is all positive, there are no zeros greater than 3. Now find $f(-1)$.

$$\begin{array}{r|rrrr} -1 & 8 & -12 & 2 & 1 \\ & & -8 & 20 & -22 \\ \hline & 8 & -20 & 22 & -21 \end{array}$$

From this result, the point $(-1, -21)$ belongs to the graph. Since the signs in the last row alternate, -1 is less than any zero of $f(x)$. This shows that the x-intercepts are between -1 and 3. Using the shortened form of synthetic division shown in the chart on the next page, find several points between -1 and 3 to help in sketching the graph.

x				$f(x)$	*Ordered pair*	
	8	−12	2	1		
3	8	12	38	115	(3, 115)	
2	8	4	10	21	(2, 21)	
1	8	−4	−2	−1	(1, −1)	←Zero
0	8	−12	2	1	(0, 1)	←Zero
−1	8	−20	22	−21	(−1, −21)	←Zero

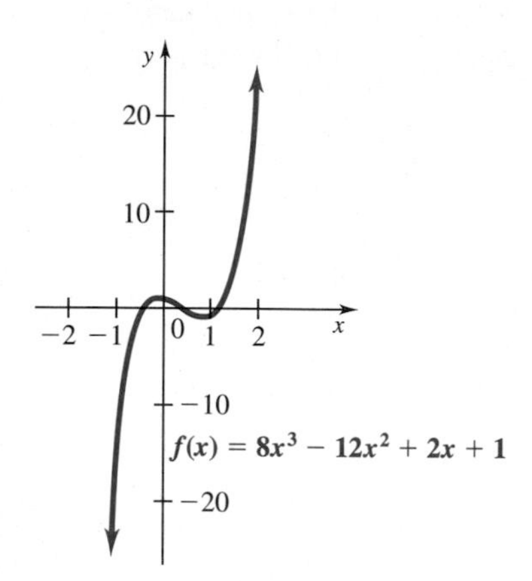

Figure 14

All three possible real zeros have been located. Since the numbers in the row for $x = 2$ are all positive and $2 > 0$, 2 is greater than or equal to any zero of $f(x)$ and the top row of the chart is not needed.

By the intermediate value theorem, there is a zero between 0 and 1 and between −1 and 0, as well as between 1 and 2. To get the graph, plot the points from the chart and then draw a continuous curve through them, as in Figure 14. ■

EXAMPLE 7 Graph the function defined by $f(x) = 3x^4 - 14x^3 + 24x - 3$.

To draw this graph, shown in Figure 15, first use Descartes' rule of signs to see that there are three or one positive real zeros and one negative real zero. Thus, there are three or one positive x-intercepts and one negative x-intercept. To find points to plot, use synthetic division to make a table like the one shown below. Start with $x = 0$ and work up through the positive integers until a row with all positive numbers is found. Then work down through the negative integers until a row with alternating signs is found.

x					$f(x)$	*Ordered pair*	
	3	−14	0	24	−3		
5	3	1	5	49	242	(5, 242)	
4	3	−2	−8	−8	−35	(4, −35)	←Zero
3	3	−5	−15	−21	−66	(3, −66)	
2	3	−8	−16	−8	−19	(2, −19)	
1	3	−11	−11	13	10	(1, 10)	←Zero
0	3	−14	0	54	−3	(0, −3)	←Zero
−1	3	−17	17	7	−10	(−1, −10)	
−2	3	−20	40	−56	109	(−2, 109)	←Zero

Since the row in the chart for $x = 5$ contains all positive numbers, the polynomial has no zero greater than 5. Also, since the row for $x = -2$ has numbers that alternate in sign, there is no zero less than −2. By the changes in sign of $f(x)$, the polynomial has zeros between -2 and -1, 0 and 1, 1 and 2, and 4 and 5. Three are positive and one is negative, as predicted. Each of these zeros is an x-intercept on the graph. Plotting the points found above and drawing a continuous curve through them gives the graph shown in Figure 15.

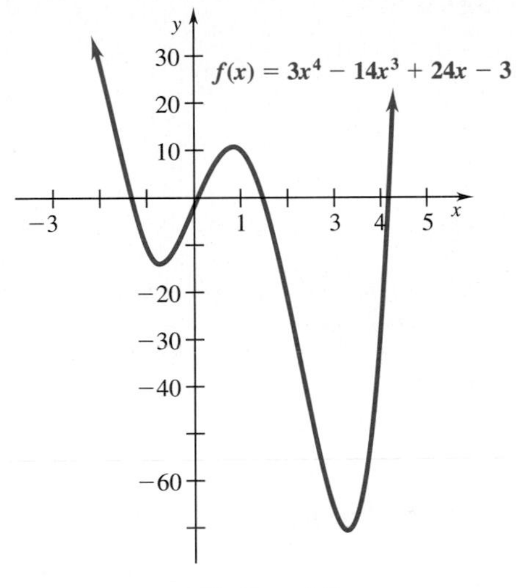

Figure 15

CAUTION As the graph in Figure 16 suggests, it is possible to have several zeros between a pair of consecutive integers or two or more zeros without a sign change. Predicting the number of positive and negative zeros ahead of time will help to avoid missing any zeros in such cases. ■

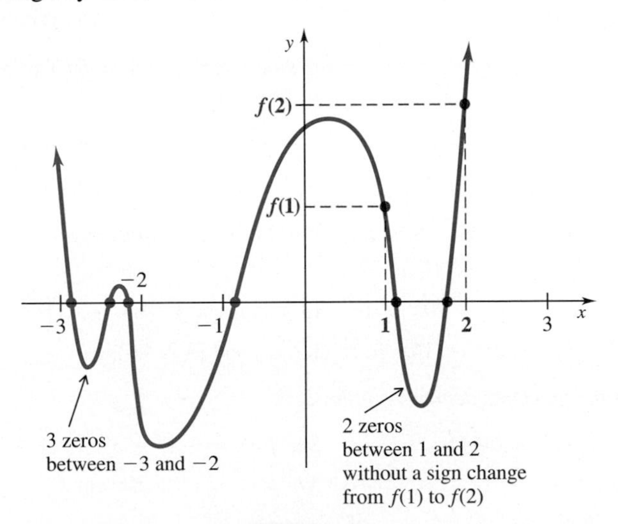

Figure 16

4.4 Exercises

Graph each of the following polynomials.

1. $f(x) = (x - 1)(x - 3)(x - 5)$
2. $f(x) = -2x(x + 4)(x - 2)$
3. $f(x) = -(x + 1)(2x - 3)(2x - 1)$
4. $f(x) = x^3 + x^2 - 6x$
5. $f(x) = x^3 - 4x^2 + 4x$
6. $f(x) = \frac{1}{2}(x + 1)(x - 1)^2$
7. $f(x) = -3x(x + 1)^2(x - 1)$
8. $f(x) = 2x^4 - 5x^3$
9. $f(x) = x^5 - 2x^3$
10. $f(x) = x^3 - 3x^2 + 3x - 1$
11. $f(x) = (x^2 - 2)(x - 4)^2$
12. $f(x) = -\frac{1}{2}(x + 4)x(x^2 + 5)$

Use the intermediate value theorem for polynomials to show that the following polynomials have a real zero between the numbers given.

13. $f(x) = 2x^2 - 7x + 4$; 2 and 3
14. $f(x) = 3x^2 - x - 4$; 1 and 2
15. $f(x) = 2x^3 - 5x^2 - 5x + 7$; 0 and 1
16. $f(x) = 2x^3 - 9x^2 + x + 20$; 2 and 2.5
17. $f(x) = 2x^4 - 4x^2 + 4x - 8$; 1 and 2
18. $f(x) = x^4 - 4x^3 - x + 3$; 1 and .5

Show that the real zeros of each of the following polynomials satisfy the given conditions.

19. $f(x) = 4x^3 - 3x^2 + 4x + 7$; no real zero greater than 1
20. $f(x) = x^4 - x^3 + 2x^2 - 3x - 5$; no real zero greater than 2
21. $f(x) = x^4 + x^3 - x^2 + 3$; no real zero less than -2
22. $f(x) = x^5 + 2x^3 - 2x^2 + 5x + 5$; no real zero less than -1

Apply Descartes' rule of signs to determine the possible number of positive and negative real zeros of the following polynomials.

23. $f(x) = 5x^3 + 2x^2 - x + 2$
24. $f(x) = x^3 + 6x^2 - 4$
25. $f(x) = x^4 + 6x^3 - 3x^2 + x - 17$
26. $f(x) = x^4 - 2x^3 + x^2 - 1$

For each of the following polynomials, approximate each zero as a decimal to the nearest tenth.

27. $f(x) = x^3 + 3x^2 - 2x - 6$
28. $f(x) = x^3 - 3x + 3$
29. $f(x) = -2x^4 - x^2 + x + 5$
30. $f(x) = -x^4 + 2x^3 + 3x^2 + 6$

The following polynomials have zeros in the given intervals. Approximate these zeros to the nearest hundredth.

31. $f(x) = x^4 + x^3 - 6x^2 - 20x - 16$; $[3.2, 3.3]$ and $[-1.4, -1.1]$
32. $f(x) = x^4 - 3x^3 - 2x^2 - 16x + 5$; $[.2, .3]$ and $[4.2, 4.3]$

Graph each of the functions defined as follows.

33. $f(x) = x^3 - 2x^2 - x + 1$
34. $f(x) = -x^3 - x^2 + 2x + 1$
35. $f(x) = -4x^3 + 7x^2 - 2$
36. $f(x) = 5x^3 - 9x^2 + 1$
37. $f(x) = x^4 - 5x^2 + 2$
38. $f(x) = 2x^4 - 6x^3 + 7x - 2$

39. Explain why the graph of a polynomial having a as a zero of odd multiplicity crosses the x-axis at $x = a$.

40. Explain why the graph of a polynomial having a as a zero of even multiplicity touches, but does not cross, the x-axis at $x = a$.

In Exercises 41 and 42, find a cubic polynomial having the graph shown.

41.

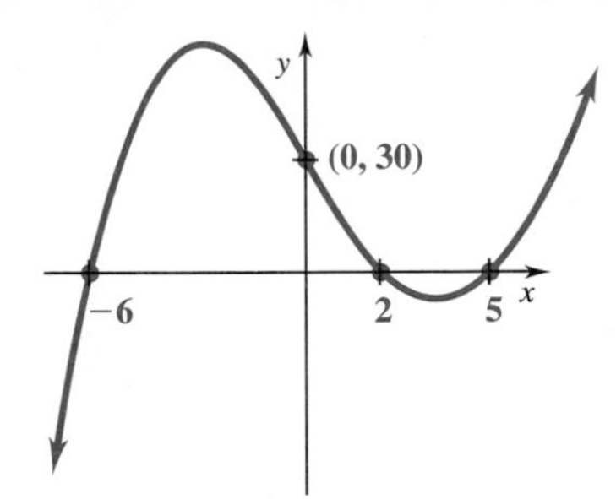

42.

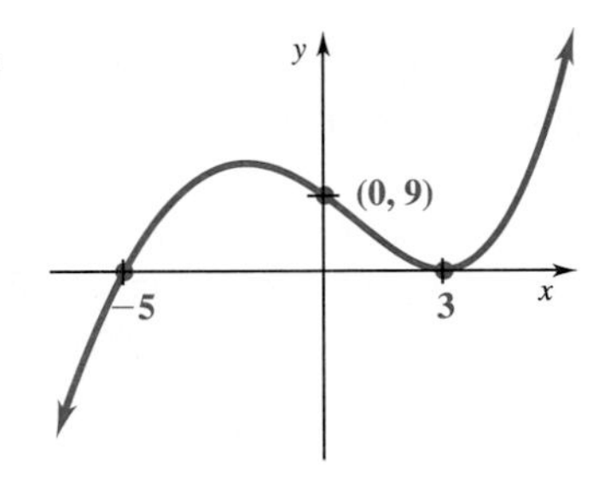

43. Give an example of a polynomial function that is never negative and has -4 and 1 as zeros.

44. Give an example of a polynomial function that has -3, 1, and 2 as zeros and is positive only between 1 and 2.

Determine the domain and range of the functions defined in Exercises 45 and 46.

45. $f(x) = \sqrt{x^3 - x^2 - 6x}$

46. $f(x) = \sqrt{x^3 - x}$

47. Summarize the steps used to graph a polynomial.

Graphing Utility Problems

48. The pressure of the oil in a reservoir tends to drop with time. By taking sample pressure readings for a particular oil reservoir, petroleum engineers have found that the change in pressure is given by $P(t) = t^3 - 25t^2 + 200t$, where t is time in years from the date of the first reading.

(a) Graph $P(t)$.

(b) Use the graph from part (a) to decide for what time periods the amount of change in pressure (drop) is increasing or decreasing.

49. During the early part of the twentieth century, the deer population of the Kaibab Plateau in Arizona experienced a rapid increase, because hunters had reduced the number of natural predators and because the deer were protected from hunters. The increase in population depleted the food resources and eventually caused the population to decline. For the period from 1905 to 1930, the deer population was approximated by $D(x) = -.125x^5 + 3.125x^4 + 4000$, where x is time in years from 1905.

(a) Graph $D(x)$.

(b) From the graph, over what period of time (from 1905 to 1930) was the population increasing? Relatively stable? Decreasing?

50. The polynomial function A defined by $A(x) = -.015x^3 + 1.06x$ gives the approximate alcohol concentration (in tenths of a percent) in an average person's bloodstream x hours after drinking about 8 oz of 100-proof whiskey. This function is approximately valid for x in the interval $[0, 8.2]$.

(a) Graph $y = A(x)$.

(b) Using the graph from part (a), estimate the time of maximum alcohol concentration.

(c) In California a person is legally drunk if the blood alcohol concentration exceeds .08%. Using the graph from part (a), estimate the period in which this average person is legally drunk.

51. A survey team measures the concentration (in parts per million) of a particular toxin in a local river. On a normal day, the concentration of the toxin at time x (in hours) after the factory upstream dumps its waste is given by $g(x) = -.006x^4 + .14x^3 - .05x^2 + .02x$, where $0 \le x \le 24$.

(a) Graph $g(x)$.

(b) Estimate the time at which the concentration is greatest.

(c) A concentration greater than 100 parts per million is considered pollution. Using the graph from part (a), estimate the period during which the river is polluted.

4.5 Rational Zeros of Polynomial Functions

By the fundamental theorem of algebra, every polynomial of degree 1 or more has a zero. However, the fundamental theorem merely says that a zero exists. It gives no help at all in identifying zeros. Other theorems can be used to find any rational zeros of polynomials with rational coefficients or to find decimal approximations of any irrational zeros.

The **rational zeros theorem** gives a useful method for finding a set of possible zeros of a polynomial with integer coefficients.

Rational Zeros Theorem

Let

$$f(x) = a_nx^n + a_{n-1}x^{n-1} + \cdots + a_1x + a_0, \quad a_n \neq 0,$$

be a polynomial with only integer coefficients. If p/q is a rational number written in lowest terms and if p/q is a zero of $f(x)$, then p is a factor of the constant term a_0 and q is a factor of the leading coefficient a_n.

This theorem can be proven as follows:

$f(p/q) = 0$ since p/q is a zero of $f(x)$, so

$$a_n(p/q)^n + a_{n-1}(p/q)^{n-1} + \cdots + a_1(p/q) + a_0 = 0.$$

This also can be written as

$$a_n(p^n/q^n) + a_{n-1}(p^{n-1}/q^{n-1}) + \cdots + a_1(p/q) + a_0 = 0.$$

Multiply both sides of this last result by q^n and add $-a_0q^n$ to both sides.

$$a_np^n + a_{n-1}p^{n-1}q + \cdots + a_1pq^{n-1} = -a_0q^n$$

Factoring out p gives

$$p(a_np^{n-1} + a_{n-1}p^{n-2}q + \cdots + a_1q^{n-1}) = -a_0q^n.$$

This result shows that $-a_0q^n$ equals the product of the two factors, p and $(a_np^{n-1} + \cdots + a_1q^{n-1})$. For this reason, p must be a factor of $-a_0q^n$. Since it was assumed that p/q is written in lowest terms, p and q have no common factor other than 1, so p is not a factor of q^n. Thus p must be a factor of a_0. In a similar way it can be shown that q is a factor of a_n.

EXAMPLE 1 Find all rational zeros of $f(x) = 2x^3 - x^2 - 2x + 1$.

Since $f(x)$ has only integer coefficients, the rational zeros theorem applies. For this polynomial, $a_0 = 1$ and $a_n = a_3 = 2$. By the rational zeros theorem, p must be a factor of $a_0 = 1$ so that p is either 1 or -1. Also q must be a factor of $a_3 = 2$, so q is ± 1 or ± 2. Any rational zeros are in the form p/q, and so must be either ± 1 or $\pm 1/2$. Descartes' rule of signs indicates that the polynomial has two positive zeros or none, and one negative zero. Since a negative zero is guaranteed, use synthetic division to check out -1.

$$\begin{array}{r|rrrr} -1 & 2 & -1 & -2 & 1 \\ & & -2 & 3 & -1 \\ \hline & 2 & -3 & 1 & 0 \end{array}$$

One zero is -1. To find any other zeros, set the quotient polynomial, $2x^2 - 3x + 1$, equal to zero.

$$2x^2 - 3x + 1 = 0$$

$$(2x - 1)(x - 1) = 0$$

$$x = \frac{1}{2} \quad \text{or} \quad x = 1$$

The rational zeros of $f(x)$ are -1, $1/2$, and 1. ∎

NOTE When the quotient is a second-degree polynomial, either factoring or the quadratic formula can be used to get any remaining zeros. ∎

EXAMPLE 2 Find all rational zeros of $f(x) = 2x^4 - 11x^3 + 14x^2 - 11x + 12$.

All coefficients are integers, so the rational zeros theorem can be used. If p/q is to be a rational zero of $p(x)$, by the rational zeros theorem p must be a factor of $a_0 = 12$ and q must be a factor of $a_4 = 2$. The possible values of p are ± 1, ± 2, ± 3, ± 4, ± 6, or ± 12, while q must be ± 1 or ± 2. The possible rational zeros are found by forming all possible quotients of the form p/q; then any rational zero of $f(x)$ will come from the list

$$\pm 1, \quad \pm 1/2, \quad \pm 2, \quad \pm 3, \quad \pm 3/2, \quad \pm 4, \quad \pm 6, \quad \pm 12.$$

Though it is not certain whether any of these numbers are zeros, if $f(x)$ has any rational zeros, they will be in the list above. (Descartes' rule of signs indicates four, two, or no positive zeros and two or no negative zeros.) The proposed rational zeros can be checked by synthetic division. A search of the possibilities leads to the discovery that 4 is a rational zero of $f(x)$.

$$\begin{array}{r|rrrrr} 4 & 2 & -11 & 14 & -11 & 12 \\ & & 8 & -12 & 8 & -12 \\ \hline & 2 & -3 & 2 & -3 & 0 \end{array}$$

As a fringe benefit of this calculation, zeros of the simpler polynomial $q(x) = 2x^3 - 3x^2 + 2x - 3$ can now be sought. Any rational zero of $q(x)$ will have a numerator of ± 3 or ± 1, with a denominator of ± 1 or ± 2. Thus, any rational zeros of $q(x)$ will come from the list

$$\pm 3, \quad \pm 3/2, \quad \pm 1, \quad \pm 1/2.$$

Descartes' rule of signs indicates three or one positive zeros. Again use synthetic division and trial and error to find that $3/2$ is a zero.

$$\begin{array}{r|rrrr} 3/2 & 2 & -3 & 2 & -3 \\ & & 3 & 0 & 3 \\ \hline & 2 & 0 & 2 & 0 \end{array}$$

The quotient is $2x^2 + 2$, which, by the quadratic formula, has i and $-i$ as zeros. They are, however, complex zeros. The rational zeros of

$$f(x) = 2x^4 - 11x^3 + 14x^2 - 11x + 12$$

and 4 and $3/2$. ■

EXAMPLE 3 Find all rational zeros of $f(x) = 6x^4 + 7x^3 - 12x^2 - 3x + 2$, and factor the polynomial.

For a rational number p/q to be a zero of $f(x)$, p must be a factor of $a_0 = 2$ and q must be a factor of $a_4 = 6$. Thus, p can be ± 1 or ± 2 and q can be ± 1, ± 2, ± 3, or ± 6. The rational zeros, p/q, must be from the following list.

$$\pm 1, \quad \pm 2, \quad \pm 1/2, \quad \pm 1/3, \quad \pm 1/6, \quad \pm 2/3$$

Descartes' rule of signs indicates two positive zeros or none, and two negative zeros or none. Check 1 first because it is easy.

$$\begin{array}{r|rrrrr} 1 & 6 & 7 & -12 & -3 & 2 \\ & & 6 & 13 & 1 & -2 \\ \hline & 6 & 13 & 1 & -2 & 0 \end{array}$$

The 0 remainder shows that 1 is a zero. Now use the quotient polynomial $6x^3 + 13x^2 + x - 2$ and synthetic division to find that -2 is also a zero.

$$\begin{array}{r|rrrr} -2 & 6 & 13 & 1 & -2 \\ & & -12 & -2 & 2 \\ \hline & 6 & 1 & -1 & 0 \end{array}$$

The new quotient polynomial is $6x^2 + x - 1$. Use the quadratic formula or factor to solve the equation $6x^2 + x - 1 = 0$. The remaining two zeros are $1/3$ and $-1/2$.

Factor the polynomial $f(x)$ in the following way. Since the four zeros of $f(x) = 6x^4 + 7x^3 - 12x^2 - 3x + 2$ are $1, -2, 1/3$, and $-1/2$, the factors are $x - 1, x + 2, x - 1/3$, and $x + 1/2$, and

$$f(x) = a(x - 1)(x + 2)\left(x - \frac{1}{3}\right)\left(x + \frac{1}{2}\right).$$

Since the leading coefficient of $f(x)$ is 6, let $a = 6$. Then

$$\begin{aligned} f(x) &= \mathbf{6}(x - 1)(x + 2)\left(x - \frac{1}{3}\right)\left(x + \frac{1}{2}\right) \\ &= (x - 1)(x + 2)(\mathbf{3})\left(x - \frac{1}{3}\right)(\mathbf{2})\left(x + \frac{1}{2}\right) \\ &= (x - 1)(x + 2)(3x - 1)(2x + 1). \end{aligned}$$

To find any rational zeros of a polynomial with fractional coefficients, first multiply the polynomial by a number that will clear it of all fractional coefficients. Then use the rational zeros theorem, which requires only integer coefficients.

EXAMPLE 4 Find all rational zeros of

$$f(x) = x^4 - \frac{1}{6}x^3 + \frac{2}{3}x^2 - \frac{1}{6}x - \frac{1}{3}.$$

To find the values of x that make $f(x) = 0$, or

$$x^4 - \frac{1}{6}x^3 + \frac{2}{3}x^2 - \frac{1}{6}x - \frac{1}{3} = 0,$$

first multiply both sides by 6 to eliminate all fractions. This gives

$$6x^4 - x^3 + 4x^2 - x - 2 = 0.$$

This polynomial will have the same zeros as $f(x)$. The possible rational zeros are of the form p/q, where p is ± 1 or ± 2 and q is $\pm 1, \pm 2, \pm 3$, or ± 6. Then p/q may be

$$\pm 1, \quad \pm 2, \quad \pm\frac{1}{2}, \quad \pm\frac{1}{3}, \quad \pm\frac{1}{6}, \quad \text{or} \quad \pm\frac{2}{3}.$$

Descartes' rule of signs indicates three or one positive zeros and one negative zero. Use synthetic division to find that $-1/2$ and $2/3$ are zeros.

$$\begin{array}{r|rrrrr} -1/2 & 6 & -1 & 4 & -1 & -2 \\ & & -3 & 2 & -3 & 2 \\ \hline & 6 & -4 & 6 & -4 & 0 \end{array} \qquad \begin{array}{r|rrrr} 2/3 & 6 & -4 & 6 & -4 \\ & & 4 & 0 & 4 \\ \hline & 6 & 0 & 6 & 0 \end{array}$$

The final quotient is $q(x) = 6x^2 + 6 = 6(x^2 + 1)$. The zeros of this polynomial are i and $-i$. Since these zeros are not rational numbers, there are just two rational zeros: $-1/2$ and $2/3$. ■

CAUTION Remember, the rational zeros theorem can be used only if the coefficients of $f(x)$ are integers. Polynomial functions with rational coefficients can be rewritten with integer coefficients in order to use the theorem, but the theorem cannot be used with polynomial functions with irrational or imaginary coefficients. ■

4.5 Exercises

List all possible rational zeros for the following polynomials.

1. $f(x) = 2x^3 + 7x^2 + 12x - 1$

2. $f(x) = 2x^3 + 20x^2 + 68x - 40$

3. $f(x) = 8x^3 + 19x^2 - 32x - 2$

4. $f(x) = 15x^3 + 59x^2 + 4x - 1$

5. $f(x) = x^4 + 4x^3 + 3x^2 - 10x + 75$

6. $f(x) = x^4 - 2x^3 + x^2 + 18$

Find all rational zeros of the following polynomials.

7. $f(x) = x^3 - 3x^2 - 25x - 21$

8. $f(x) = x^3 - 2x^2 - 4x + 8$

9. $f(x) = x^3 + 9x^2 - 14x - 24$

10. $f(x) = x^3 + 3x^2 - 4x - 12$

11. $f(x) = x^3 + 6x^2 - x - 30$

12. $f(x) = x^3 - x^2 - 10x - 8$

13. $f(x) = x^4 + 9x^3 + 21x^2 - x - 30$

14. $f(x) = x^4 + 4x^3 - 7x^2 - 34x - 24$

Find the rational zeros of the following polynomials; then write each polynomial in factored form with each factor having only integer coefficients.

15. $f(x) = 6x^3 + x^2 - 5x - 2$

16. $f(x) = 15x^3 + 61x^2 + 2x - 8$

17. $f(x) = 2x^3 + 7x^2 + 12x - 8$

18. $f(x) = 2x^3 + 20x^2 + 68x - 40$

19. $f(x) = x^4 + 2x^3 - 13x^2 - 38x - 24$

20. $f(x) = 6x^4 + x^3 - 7x^2 - x + 1$

21. $f(x) = x^4 - 4x^3 - 7x^2 + 34x - 24$

22. $f(x) = x^4 - 2x^3 + x^2 + 18$

23. $f(x) = x^5 + 3x^4 - 5x^3 - 11x^2 + 12$

24. $f(x) = 4x^5 + 4x^4 - 37x^3 - 37x^2 + 9x + 9$

Find all rational zeros of the following polynomials.

25. $f(x) = x^3 + \frac{1}{2}x^2 - \frac{11}{2}x - 5$

26. $f(x) = x^3 - \frac{4}{3}x^2 - \frac{13}{3}x - 2$

27. $f(x) = x^4 + \frac{1}{4}x^3 + \frac{11}{4}x^2 + x - 5$

28. $f(x) = \frac{10}{7}x^4 - x^3 - 7x^2 + 5x - \frac{5}{7}$

29. $f(x) = \frac{1}{3}x^5 + x^4 - \frac{5}{3}x^3 - \frac{11}{3}x^2 + 4$

30. $f(x) = x^5 + x^4 - \frac{37}{4}x^2 + \frac{9}{4}x + \frac{9}{4}$

Find all integer solutions of the following equations.

31. $6x^3 - 31x^2 + 3x + 10 = 0$

32. $12x^3 - 5x^2 - 9x + 1 = 0$

33. After a 2-inch slice is cut off the top of a cube, the resulting solid has a volume of 32 cubic inches. Find the dimensions of the original cube.

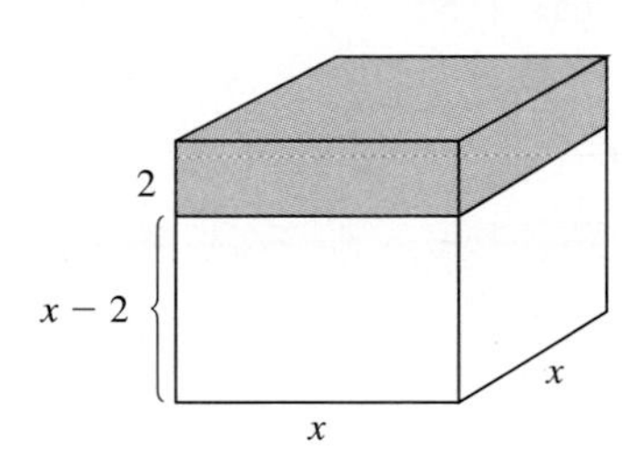

Exercise 33

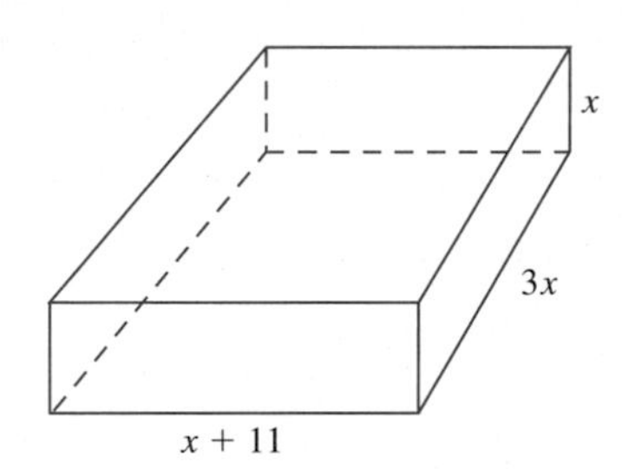

Exercise 34

34. The width of a rectangular box is three times its height, and its length is 11 inches more than its height. Find the dimensions of the box if its volume is 720 cubic inches.

35. A rectangle of area 42 square inches has its base on the x-axis and its upper corners on the parabola $y = 16 - x^2$. Find the dimensions of the rectangle. (*Hint:* What are the coordinates of P?)

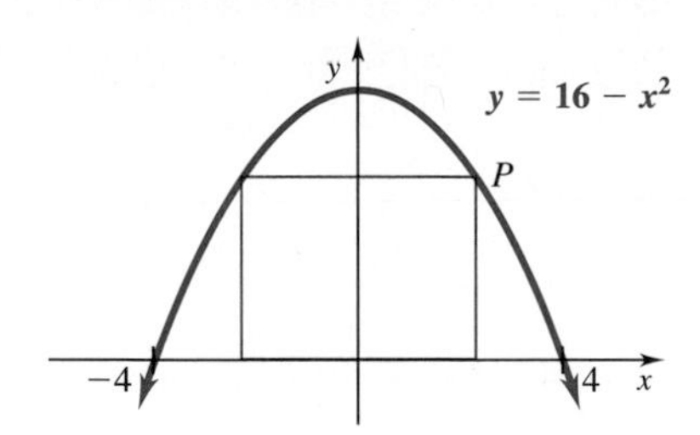

Exercise 35

36. Show that $f(x) = x^2 - 7$ has no rational zeros, so $\sqrt{7}$ must be irrational.

37. Show that for any prime p, $\sqrt{p}$ is irrational. (*Hint:* Look at $x^2 - p$.)

38. Show that $f(x) = x^4 + 4x^2 - 13$ has no rational zeros.

39. Show that any integer zeros of a polynomial with integer coefficients must be factors of the constant term a_0.

After factoring the polynomial and locating its zeros, sketch the graph of each function.

40. $f(x) = 2x^3 - 7x^2 + 4x + 4$

41. $f(x) = 3x^3 + 8x^2 + 3x - 2$

42. $f(x) = 2x^3 + x^2 - x$

43. $f(x) = 3x^3 - 8x^2 - 5x + 6$

44. $f(x) = x^4 - 3x^2 + 2$

45. $f(x) = 2x^4 + 5x^3 - 5x - 2$

46. Describe some situations in which Descartes' rule of signs would be helpful in searching for rational zeros.

47. Describe some situations in which the boundedness theorem would be helpful in searching for rational zeros.

Graphing Utility Problems

In Exercises 48–53, use the rational zeros theorem to determine the possible rational zeros of the polynomial. Then use the graph of the polynomial to find the actual rational zeros.

48. $f(x) = 2x^3 - 30x^2 + 142x - 210$

49. $f(x) = 6x^3 - 19x^2 - 65x + 50$

50. $f(x) = -6x^3 + 13x^2 + 99x - 70$

51. $f(x) = 20x^3 + 12x^2 - 128x + 48$

52. $f(x) = -9x^4 - 33x^3 + 48x^2 + 132x - 48$

53. $f(x) = x^4 - 17x^3 + 69x^2 + 17x - 70$

4.6 Rational Functions

A rational expression is a fraction that is the quotient of two polynomials. A function defined by a rational expression is called a *rational function.*

Definition of Rational Function

If $p(x)$ and $q(x)$ are polynomials with $q(x) \neq 0$, then

$$f(x) = \frac{p(x)}{q(x)}$$

is a **rational function.**

Since any values of x such that $q(x) = 0$ are excluded from the domain, a rational function usually has a graph that has one or more breaks in it.

The simplest rational function with a variable denominator is defined by

$$f(x) = \frac{1}{x}.$$

The domain of this function is the set of all real numbers except 0. The number 0 cannot be used as a value of x, but it is helpful to find the values of $f(x)$ for several values of x close to 0. The following table shows what happens to $f(x)$ as x gets closer and closer to 0 from either side.

x approaches 0.

x	-1	$-.1$	$-.01$	$-.001$	$.001$	$.01$	$.1$	1
$f(x)$	-1	-10	-100	-1000	1000	100	10	1

$|f(x)|$ gets larger and larger.

The table suggests that $|f(x)|$ gets larger and larger as x gets closer and closer to 0, written in symbols as

$$|f(x)| \to \infty \quad \text{as} \quad x \to 0.$$

(The symbol $x \to 0$ means that x approaches as close as desired to 0, without ever being equal to 0.) Since x cannot equal 0, the graph of $f(x) = 1/x$ will never intersect the vertical line $x = 0$. The line $x = 0$, the y-axis, is called a *vertical asymptote* for the graph. The graph gets closer and closer to the y-axis as x gets closer and closer to 0.

On the other hand, as $|x|$ gets larger and larger, the values of $f(x) = 1/x$ get closer and closer to 0. (See the table.)

x	$-10{,}000$	-1000	-100	-10	10	100	1000	$10{,}000$
$f(x)$	$-.0001$	$-.001$	$-.01$	$-.1$	$.1$	$.01$	$.001$	$.0001$

Letting $|x|$ get larger and larger without bound (written $|x| \to \infty$) causes the graph of $y = 1/x$ to move closer and closer to the horizontal line $y = 0$, the x-axis. That is,

$$\lim_{|x| \to \infty} f(x) = 0,$$

read as "the limit of $f(x)$ as $|x|$ becomes infinitely large is zero." The line $y = 0$ is called a *horizontal asymptote*.

To graph $f(x)$, first replace x with $-x$, getting $f(-x) = 1/(-x) = -1/x = -f(x)$, showing that $f(x) = 1/x$ is an odd function, symmetric with respect to the origin. Choosing some positive values of x and finding the corresponding values of $f(x)$ gives the first-quadrant part of the graph shown in Figure 17. The other part of the graph (in the third quadrant) can be found by symmetry.

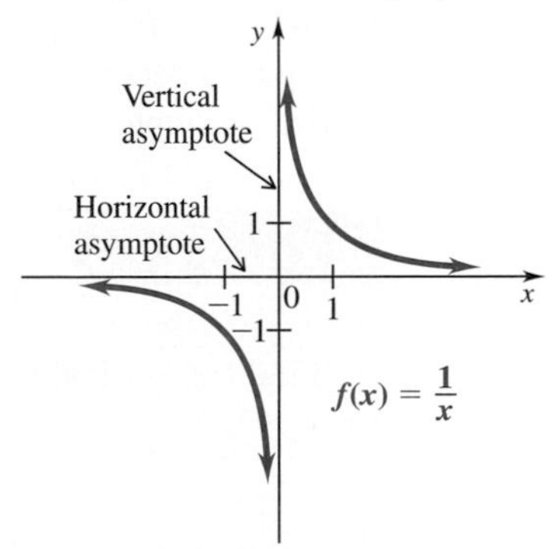

Figure 17

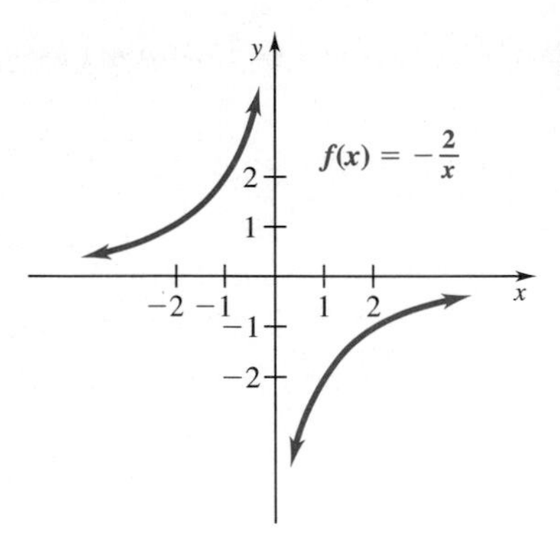

Figure 18

EXAMPLE 1 Graph $f(x) = -\dfrac{2}{x}$.

Rewrite $f(x)$ as

$$f(x) = -2 \cdot \frac{1}{x}.$$

Compared to $f(x) = 1/x$, the graph will be reflected about the x-axis (because of the negative sign) and each point will be twice as far from the x-axis. See the graph in Figure 18. The y-axis is the vertical asymptote and the horizontal asymptote is $y = 0$, or the x-axis. ■

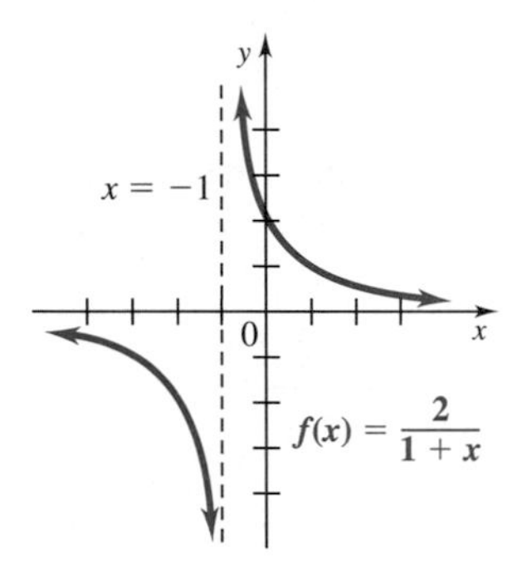

Figure 19

EXAMPLE 2 Graph $f(x) = \dfrac{2}{1 + x}$.

The domain of this function is the set of all real numbers except -1. Since $x \neq -1$, the graph cannot cross the line $x = -1$, which is thus a vertical asymptote. The horizontal asymptote is $y = 0$. As shown in Figure 19, the graph is similar to that of $f(x) = 1/x$, translated 1 unit to the left. The y-intercept is 2. ■

The examples above suggest the following definitions of vertical and horizontal asymptotes.

Definition of Asymptotes

For the rational function defined by $y = f(x)$, and for real numbers a and b,

if $\lim_{x \to a} |f(x)| = \infty$, then the line $x = a$ is a **vertical asymptote**;

if $\lim_{|x| \to \infty} f(x) = b$, then the line $y = b$ is a **horizontal asymptote**.

When the rational function $f(x) = p(x)/q(x)$ is written in lowest terms, vertical asymptotes occur at values of x that make the denominator 0. Therefore, to find all vertical asymptotes, find all solutions to $q(x) = 0$.

EXAMPLE 3 Graph $f(x) = \dfrac{1}{(x - 1)(x + 4)}$.

Find the vertical asymptotes by setting the denominator equal to 0 and solving for x.

$$(x - 1)(x + 4) = 0$$
$$x = 1 \quad \text{or} \quad x = -4$$

As $|x|$ gets larger and larger, $|(x - 1)(x + 4)|$ also gets larger and larger, bringing y closer and closer to 0. This means that the x-axis is a horizontal asymptote. The vertical asymptotes divide the x-axis into three regions. Find the sign of $f(x)$ in each region by using a test point, as was done with polynomials earlier in this chapter.

Region	*Test point*	*Value of $f(x)$*	*Sign of $f(x)$*
$x < -4$	-5	$1/6$	+
$-4 < x < 1$	-1	$-1/6$	−
$1 < x$	2	$1/6$	+

Finding the y-intercept by letting $x = 0$ and then plotting it and a few additional points (shown with the graph) gives the result shown in Figure 20. ■

x	y
-5	$1/6$
-2	$-1/6$
-1	$-1/6$
0	$-1/4$
2	$1/6$

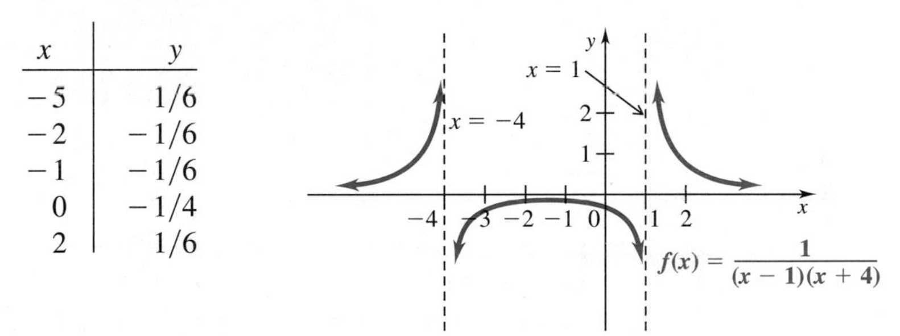

Figure 20

EXAMPLE 4 Graph $f(x) = \dfrac{x + 1}{(2x - 1)(x + 3)}$.

The graph has as vertical asymptotes the lines $x = 1/2$ and $x = -3$. To find any horizontal asymptote, multiply the factors in the denominator.

$$f(x) = \frac{x + 1}{(2x - 1)(x + 3)} = \frac{x + 1}{2x^2 + 5x - 3}$$

Now divide each term in the numerator and denominator by x^2, since 2 is the highest exponent on x.

$$f(x) = \frac{\dfrac{x}{x^2} + \dfrac{1}{x^2}}{\dfrac{2x^2}{x^2} + \dfrac{5x}{x^2} - \dfrac{3}{x^2}} = \frac{\dfrac{1}{x} + \dfrac{1}{x^2}}{2 + \dfrac{5}{x} - \dfrac{3}{x^2}}$$

As $|x|$ gets larger and larger, the quotients $1/x$, $1/x^2$, $5/x$, and $3/x^2$ all approach 0. For example,

$$\text{if } x = 10, 100, 1000, \ldots,$$

$$\text{then } \frac{1}{x} = \frac{1}{10}, \frac{1}{100}, \frac{1}{1000}, \ldots,$$

$$\text{and } \frac{1}{x^2} = \frac{1}{100}, \frac{1}{10000}, \frac{1}{1000000}, \ldots.$$

Thus, as $|x| \to \infty$, the value of $f(x)$ approaches

$$\frac{0 + 0}{2 + 0 - 0} = \frac{0}{2} = 0.$$

The line $y = 0$ is therefore a horizontal asymptote.

Replacing x with 0 gives $-1/3$ as the y-intercept. For the x-intercept, the value of y must be 0. The only way that y can equal 0 is if the numerator, $x + 1$, is 0. This happens when $x = -1$, making -1 the x-intercept.

Since this graph has an x-intercept, consider the sign of $f(x)$ in each region determined by a vertical asymptote or an x-intercept. Here there are four regions to be considered, as shown in the chart below.

Region	*Test point*	*Value of f(x)*	*Sign of f(x)*
$(-\infty, -3)$	-4	$-1/3$	$-$
$(-3, -1)$	-2	$1/5$	$+$
$(-1, 1/2)$	0	$-1/3$	$-$
$(1/2, \infty)$	2	$1/5$	$+$

Using the asymptotes and intercepts and plotting a few points (shown in the table of values) gives the graph in Figure 21.

x	y
-4	$-1/3$
-2	$1/5$
-1	0
0	$-1/3$
1	$1/2$
2	$1/5$

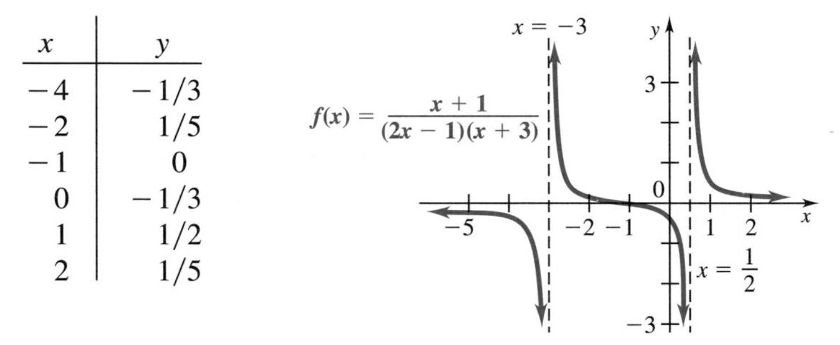

Figure 21

EXAMPLE 5 Graph $f(x) = \dfrac{2x - 5}{x - 3}$.

The vertical asymptote is the line $x = 3$. Work as in the previous example to find any horizontal asymptote. Since the highest exponent on x is 1, divide each term in the numerator and denominator by x.

$$f(x) = \frac{2x - 5}{x - 3} = \frac{\frac{2x}{x} - \frac{5}{x}}{\frac{x}{x} - \frac{3}{x}} = \frac{2 - \frac{5}{x}}{1 - \frac{3}{x}}$$

As $|x|$ gets larger and larger, both $3/x$ and $5/x$ approach 0:

$$\frac{2-0}{1-0} = \frac{2}{1} = 2,$$

showing that the line $y = 2$ is a horizontal asymptote.

Letting $x = 0$ gives the y-intercept $5/3$. The x-intercept is found when the numerator, $2x - 5$, is 0, so the x-intercept is $5/2$. The following chart shows the sign of $f(x)$ in the regions determined by the x-intercept and the vertical asymptote.

Region	*Test point*	*Value of f(x)*	*Sign of f(x)*
$(-\infty, 5/2)$	0	$5/3$	+
$(5/2, 3)$	$11/4$	-2	−
$(3, \infty)$	4	3	+

Using this information and plotting the ordered pairs shown with the graph produces the graph shown in Figure 22.

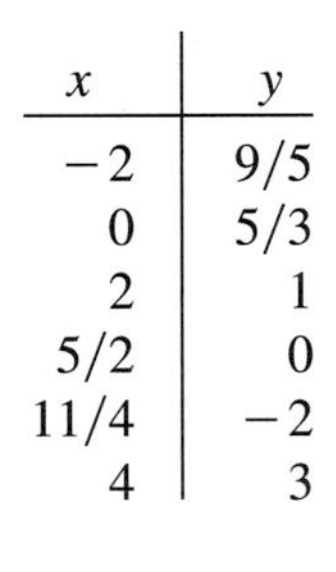

x	y
-2	$9/5$
0	$5/3$
2	1
$5/2$	0
$11/4$	-2
4	3

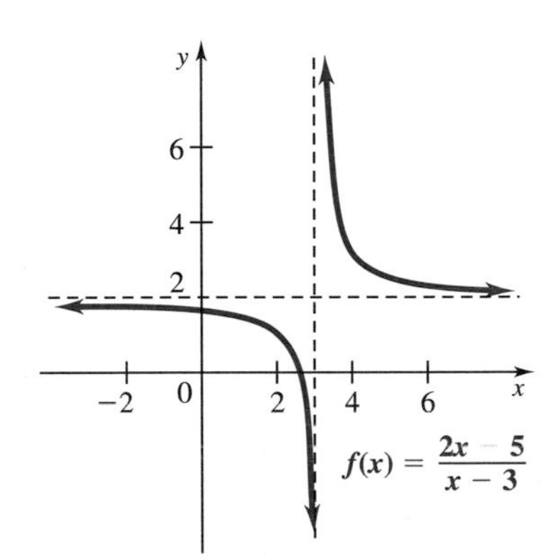

Figure 22

There is an alternate way to sketch the graph in Example 5. Use synthetic division to divide $2x - 5$ by $x - 3$, getting

$$\frac{2x-5}{x-3} = 2 + \frac{1}{x-3}.$$

This result shows that the graph of $f(x) = (2x - 5)/(x - 3)$ is the same as that of $1/(x - 3)$, translated 2 units upward. Since the graph of $1/(x - 3)$ is the same as that of $1/x$ translated 3 units to the right, the final graph is that of $1/x$ translated 3 units to the right and 2 units upward.

EXAMPLE 6

Graph $f(x) = \dfrac{3(x + 1)(x - 2)}{(x + 4)^2}$.

The only vertical asymptote is the line $x = -4$. Find any horizontal asymptotes by multiplying the factors in the numerator and denominator.

$$f(x) = \frac{3x^2 - 3x - 6}{x^2 + 8x + 16}$$

Now divide the numerator and denominator by x^2.

$$f(x) = \frac{\dfrac{3x^2}{x^2} - \dfrac{3x}{x^2} - \dfrac{6}{x^2}}{\dfrac{x^2}{x^2} + \dfrac{8x}{x^2} + \dfrac{16}{x^2}} = \frac{3 - \dfrac{3}{x} - \dfrac{6}{x^2}}{1 + \dfrac{8}{x} + \dfrac{16}{x^2}}$$

Letting $|x|$ get larger and larger shows that $y = 3/1$, or $y = 3$, is a horizontal asymptote. The y-intercept is $-3/8$, and the x-intercepts are -1 and 2. The sign of $f(x)$ in each region is shown below.

Region	*Test point*	*Value of f(x)*	*Sign of f(x)*
$(-\infty, -4)$	-8	13.125	+
$(-4, -1)$	-2	3	+
$(-1, 2)$	0	$-3/8$	−
$(2, \infty)$	4	.46875	+

See the final graph in Figure 23.

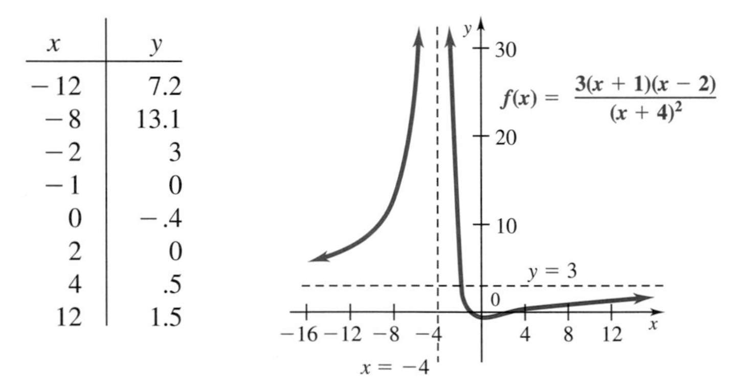

x	y
−12	7.2
−8	13.1
−2	3
−1	0
0	−.4
2	0
4	.5
12	1.5

Figure 23

In Figure 23, the graph of the function actually crosses the horizontal asymptote. While it is possible for the graph of a rational function to cross a horizontal asymptote, such a graph can never cross a vertical asymptote. The reason that a graph can cross a horizontal asymptote comes from the definition: only values of x where $|x|$ is relatively large are used in finding horizontal asymptotes. The value of x where the graph crosses the horizontal asymptote of Figure 23 is found by letting $f(x) = 3$, since the horizontal asymptote is the line $y = 3$. Solving $f(x) = 3$ gives $x = -2$, so the graph crosses the horizontal asymptote at $(-2, 3)$.

The next example shows a rational function defined by an expression having a numerator with higher degree than the denominator. Polynomial division must be used for these functions.

EXAMPLE 7 Graph $f(x) = \dfrac{x^2 + 1}{x - 2}$.

The vertical asymptote is the line $x = 2$. Dividing the numerator and denominator by x^2 as in the examples above would show that the graph has no horizontal asymptote. Since the numerator has higher degree than the denominator, divide to rewrite the equation for the function in another form.

$$\begin{array}{r|rrr} 2 & 1 & 0 & 1 \\ & & 2 & 4 \\ \hline & 1 & 2 & 5 \end{array}$$

By this result,

$$f(x) = \frac{x^2 + 1}{x - 2} = x + 2 + \frac{5}{x - 2}.$$

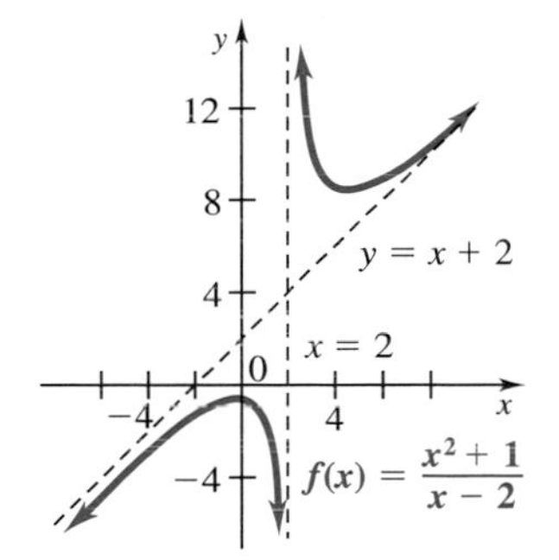

Figure 24

For very large values of $|x|$, $5/(x - 2)$ is close to 0, and the graph approaches the line $y = x + 2$. The y-intercept is $-1/2$. There is no x-intercept. Using the asymptotes, the y-intercept, and additional points as needed leads to the graph in Figure 24. ■

The line $y = x + 2$ in Example 7 is an **oblique asymptote** (neither vertical nor horizontal). In general, if the degree of the numerator is exactly one more than the degree of the denominator, the rational function will have an oblique asymptote. The equation of this asymptote is found by dividing the numerator by the denominator and dropping the remainder, since the remainder becomes close to 0 for large values of $|x|$.

To summarize: when graphing a rational function, use the following procedure. (Assume that the rational function is defined by an expression written in lowest terms.)

The table suggests that as n increases, the value of $(1 + 1/n)^n$ gets closer and closer to some fixed number. It turns out that this is indeed the case. This fixed number is called e.

Value of *e*

To ten decimal places,

$$e \approx \mathbf{2.7182818284.}$$

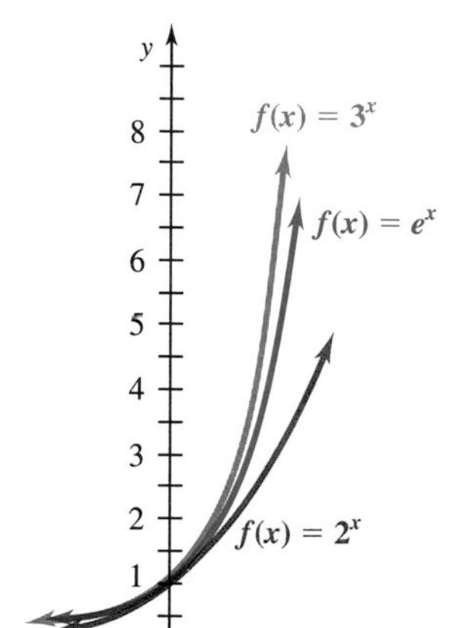

Figure 6

Scientific and business calculators give values of e^x. To obtain a specific value, key in the number x and then press the key labeled e^x. (If your calculator does not have an e^x key, press the INV key and then the $\ln x$ key. The reason that this method works will be apparent in Section 2 of this chapter.)

In Figure 6, the functions defined by $f(x) = 2^x$, $f(x) = e^x$, and $f(x) = 3^x$ are graphed for comparison.

It can be shown that in many situations involving growth or decay of a population, the amount or number present at time t can be closely approximated by an exponential function with base e. The next example illustrates exponential growth.

EXAMPLE 8 Suppose the population of a midwestern city is approximated by

$$P(t) = 10{,}000e^{.04t},$$

where t represents time measured in years. Find the population of the city at the following times: **(a)** $t = 0$, **(b)** $t = 5$.

(a) The population at time $t = 0$ is

$$\begin{aligned} P(0) &= 10{,}000e^{(.04)0} \\ &= 10{,}000e^{0} \\ &= 10{,}000(1) \\ &= 10{,}000, \end{aligned}$$

written $P_0 = 10{,}000$.

(b) The population of the city at year $t = 5$ is

$$\begin{aligned} P(5) &= 10{,}000e^{(.04)5} \\ &= 10{,}000e^{.2} \\ &= 10{,}000(1.2214) = 12{,}214. \end{aligned}$$

In five years the population of the city will be about 12,200. ■

5.1 Exercises

1. Graph each of the functions defined as follows. Compare the graphs to that of $f(x) = 2^x$.
(a) $f(x) = 2^x - 1/2$ **(b)** $f(x) = 2^x + 3$ **(c)** $f(x) = 2^{x+2}$ **(d)** $f(x) = -2^{x-4}$

2. Graph each of the functions defined as follows. Compare the graphs to that of $f(x) = 3^{-x}$.
(a) $f(x) = 3^{-x} - 2$ **(b)** $f(x) = 3^{-x} + 4$ **(c)** $f(x) = 3^{-x-2}$ **(d)** $f(x) = 3^{-x+4}$

Graph each of the functions defined as follows.

3. $f(x) = 3^x$ **4.** $f(x) = (.5)^x$ **5.** $f(x) = (2/3)^x$ **6.** $f(x) = 4^{-x}$

7. $f(x) = e^x$ **8.** $f(x) = e^{1-x}$ **9.** $f(x) = 4^x$ **10.** $f(x) = 9^x$

11. $f(x) = e^{|x|}$ **12.** $f(x) = 3^{-|x|}$

Solve each equation in Exercises 13–30. (Hint: *In Exercise 26, note that* $16/81 = (2/3)^4$.)

13. $5^r = 625$ **14.** $4^x = 2$ **15.** $\left(\frac{1}{2}\right)^k = 8$ **16.** $\left(\frac{3}{5}\right)^x = \frac{125}{27}$

17. $3^{2x-5} = \frac{1}{9}$ **18.** $5^{2p+1} = 25$ **19.** $\frac{1}{27} = b^{-3}$ **20.** $\frac{1}{27} = k^{-1/2}$

21. $4 = r^{2/3}$ **22.** $z^{5/2} = 32$ **23.** $8^{2x} = 2^{x+3}$ **24.** $4^t = 16^{1-t}$

25. $\left(\frac{1}{2}\right)^{-x} = \left(\frac{1}{4}\right)^{x+1}$ **26.** $\left(\frac{2}{3}\right)^{k-1} = \left(\frac{16}{81}\right)^{k+1}$ **27.** $2^{|x|} = 128$

28. $3^{-|x|} = \frac{1}{81}$ **29.** $e^{-5x} = (e^2)^x$ **30.** $e^{3(1+x)} = e^{-8x}$

31. \$5000 is invested for 4 yr at 8% compound interest. Find the final amount on deposit if the interest is compounded as follows:
(a) annually **(b)** quarterly **(c)** daily (365 days).

32. Find the final amount on deposit if \$5800 is left at interest for 6 yr at 13% and interest is compounded as follows:
(a) annually **(b)** quarterly **(c)** daily (365 days).

33. Suppose the termite population of a house is given by $P(t)$, where $P(t) = 10{,}000e^{.1t}$, with t representing time in days after some initial day.
Find the following to the nearest thousand.
(a) $P(0)$ **(b)** $P(2)$ **(c)** $P(8)$ **(d)** $P(10)$
(e) Graph $y = P(t)$.

34. Suppose the quantity in grams of a radioactive substance present at time t is $Q(t) = 500e^{-.05t}$. Let t be time measured in days from some initial day.
Find the quantity (to the nearest 10 g) present at each of the following times.
(a) $t = 0$ **(b)** $t = 4$ **(c)** $t = 8$ **(d)** $t = 20$
(e) Graph $y = Q(t)$.

Give an equation of the form $f(x) = a^x$ to define the exponential function whose graph contains the given point.

35. $(3, 8)$ **36.** $(-3, 64)$ **37.** $(-.5, .4)$ **38.** $(2/3, 4)$

Use properties of exponents to write each of the following in the form $f(t) = ka^t$, where k is a constant. (Hint: *Recall* $4^{x+y} = 4^x \cdot 4^y$*.)*

39. $f(t) = 3^{2t+3}$

40. $f(t) = 2^{7-t}$

41. $f(t) = (1/3)^{1-2t}$

42. $f(t) = 7^{2t-1}$

43. Consider a function of the form $y = Pa^x$. Show that when the value of x is increased by 1, the value of y is multiplied by a.

Suppose f is an exponential function of the form $f(x) = a^x$ and $f(3) = 4$. Determine the function values in Exercises 44–46.

44. $f(-3)$

45. $f(0)$

46. $f(6)$

47. What two points on the graph of $f(x) = a^x$ can be found without any computation?

The pressure of the atmosphere p(h) in pounds per square inch is given by $p(h) = p_0e^{-kh}$, where h is the height above sea level and p_0 and k are constants. The pressure at sea level is 15 lb/sq in and the pressure is 9 lb/sq in at a height of 12,000 ft. Use this information in Exercises 48 and 49. (The exercises can be answered without solving for k.)

48. Find the pressure at an altitude of 3000 ft.

49. What would be the pressure encountered by a satellite at an altitude of 150,000 ft?

50. When defining an exponential function, explain why we require $a > 0$.

51. A function of the form x^r, where r is a constant, is called a *power function*. Discuss the difference between an exponential function and a power function.

52. Explain in your own words what the number e is.

53. The techniques of this section can be used to solve $a^x = b$ only in special situations. Describe such a situation.

54. Explain why the graph of $y = a^x$ (where $a > 0$) has a y-intercept but no x-intercept.

Let $f(x) = a^x$ define an exponential function of base a.

55. Is f odd, even, or neither?

56. Prove that $f(m + n) = f(m) \cdot f(n)$ for any real numbers m and n.

Graphing Utility Problems

Any points where the graphs of functions f and g intersect give solutions of the form $f(x) = g(x)$. Use this idea to estimate the solutions of the following equations.

57. $x = 2^x$

58. $5e^{3x} = 75$

59. $6^{-x} = 1 - x$

60. $3x + 2 = 4^x$

61. Graph the function $f(x) = (1 + (1/x))^x$ and the horizontal line $y = 2.71828$ with $1 \le x \le 10{,}000$ and $0 \le y \le 3$. Observe that $f(x)$ gets closer and closer to the line as x gets large.

62. The function e^x grows faster than any power function. Graph the function x^2/e^x for $0 \le x \le 10$ and the function x^{10}/e^x for $0 \le x \le 25$ and observe that the values approach 0 as x gets large. (*Note:* For any n, x^n/e^x approaches 0 as x gets large.)

5.2 Logarithmic Functions

Exponential functions defined by $f(x) = a^x$ for all positive values of a, where $a \neq 1$, were discussed in the previous section. As mentioned there, exponential functions are one-to-one, and so have inverse functions. In this section we discuss the inverses of exponential functions. The inverse of the function a^x is the function *logarithm to the base a*, written $\log_a x$. *Log* is an abbreviation for *logarithm*. Read $\log_a x$ as "the logarithm of x to the base a."

Definition of $\log_a x$

For $a > 0$, $a \neq 1$, and $x > 0$, $\log_a x$ is the power to which a must be raised to get x.

Consider the following simple fill-in-the-box problems.

$$4^3 = \square \qquad 5^{\square} = 25$$

The answers, of course, are

$$4^3 = \boxed{64} \qquad 5^{\boxed{2}} = 25.$$

When we solve the problem on the left, we are "doing" exponents. When we solve the right problem, we are "doing" logarithms. That is, we are finding the power to which 5 must be raised in order to get 25. Therefore, $2 = \log_5 25$. In a certain sense, logarithms are just exponents.

If $s = \log_a r$, then the power to which a *must* be raised to get r is s, or $r = a^s$.

$$s = \log_a r \quad \textbf{if and only if} \quad r = a^s.$$

This key statement should be memorized. It is important to remember the location of the base and exponent in each part.

$$\text{Logarithmic form: } \underset{\substack{\uparrow\\ \text{Exponent}}}{s} = \log_{\underset{\text{Base}}{a}} r$$

Exponent ↓ Logarithmic form: $s = \log_a r$ ↑ Base

Exponent ↓ Exponential form: $a^s = r$ ↑ **Base**

EXAMPLE 1 The chart below shows several pairs of equivalent statements. The same statement is written in both exponential and logarithmic forms.

Exponential form	*Logarithmic form*
$2^3 = 8$	$\log_2 8 = 3$
$(1/2)^{-4} = 16$	$\log_{1/2} 16 = -4$
$10^5 = 100{,}000$	$\log_{10} 100{,}000 = 5$
$3^{-4} = 1/81$	$\log_3 (1/81) = -4$
$5^1 = 5$	$\log_5 5 = 1$
$(3/4)^0 = 1$	$\log_{3/4} 1 = 0$

We can often solve equations with logarithms by rewriting them in exponential form, as shown in the next example.

EXAMPLE 2 Solve each of the following equations.

(a) $\log_x \frac{8}{27} = 3$

First, write the expression in exponential form; then solve.

$$x^3 = \frac{8}{27} = \left(\frac{2}{3}\right)^3$$

$$x = \frac{2}{3}$$

The solution is 2/3.

(b) $\log_4 x = 5/2$

In exponential form, the given statement becomes

$$4^{5/2} = x$$

$$(4^{1/2})^5 = x$$

$$2^5 = x$$

$$32 = x.$$

The solution is 32.

The logarithmic function with base a is defined as follows.

Definition of Logarithmic Function

If $a > 0$, $a \neq 1$, and $x > 0$, then the function f defined by

$$\mathbf{f(x) = \log_a x}$$

is the **logarithmic function with base *a*.**

Exponential and logarithmic functions are inverses of each other. Since the domain of an exponential function is the set of all real numbers, the range of a logarithmic function will also be the set of all real numbers. In the same way, both the range of an exponential function and the domain of a logarithmic function are the set of all positive real numbers, so logarithms can be found for positive numbers only.

EXAMPLE 3 Graph the logarithmic function $f(x) = \log_2 x$.

One way to graph a logarithmic function is to begin with its inverse function. Here, the inverse has equation $y = 2^x$. The graph of the equation $y = 2^x$ is shown with a blue curve in Figure 7. To get the graph of $y = \log_2 x$, reverse the ordered pairs used to graph $y = 2^x$. For example, the pairs $(-1, 1/2)$, $(0, 1)$, $(1, 2)$, and $(2, 4)$ for $y = 2^x$ become $(1/2, -1)$, $(1, 0)$, $(2, 1)$, and $(4, 2)$ for $x = 2^y$ or $y = \log_2 x$. The graph of the equation $y = \log_2 x$ is shown as a red curve. As the graph suggests, the function defined by $f(x) = \log_2 x$ is increasing for all its domain, is one-to-one, and has the y-axis as a vertical asymptote. ■

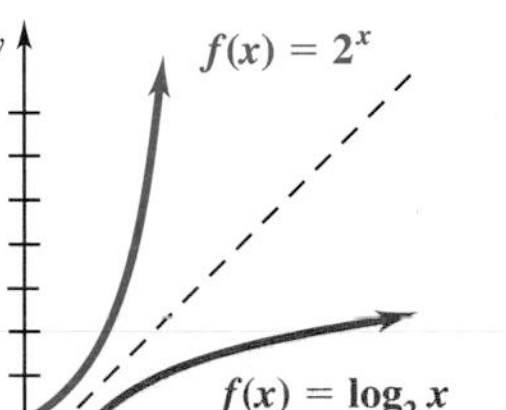

Figure 7

The graph of a logarithmic function of the form $f(x) = \log_a x$ has the following features.

Graph of $f(x) = \log_a x$

1. The point $(1, 0)$ is on the graph.
2. If $a > 1$, $f(x)$ is an increasing function; if $0 < a < 1$, $f(x)$ is a decreasing function.
3. The y-axis is a vertical asymptote.
4. The domain is $(0, \infty)$, and the range is $(-\infty, \infty)$.

EXAMPLE 4 Graph $f(x) = \log_2 (x - 1)$.

The graph of $f(x) = \log_2 (x - 1)$ will be the graph of $f(x) = \log_2 x$, translated one unit to the right. The vertical asymptote is $x = 1$. The domain of the function defined by $f(x) = \log_2 (x - 1)$ is $(1, \infty)$, since logarithms can be found only for positive numbers. To find some ordered pairs to plot, use the equivalent equation

$$x - 1 = 2^y \quad \text{or} \quad x = 2^y + 1,$$

choosing values for y and then calculating each of the corresponding x-values. See Figure 8. ■

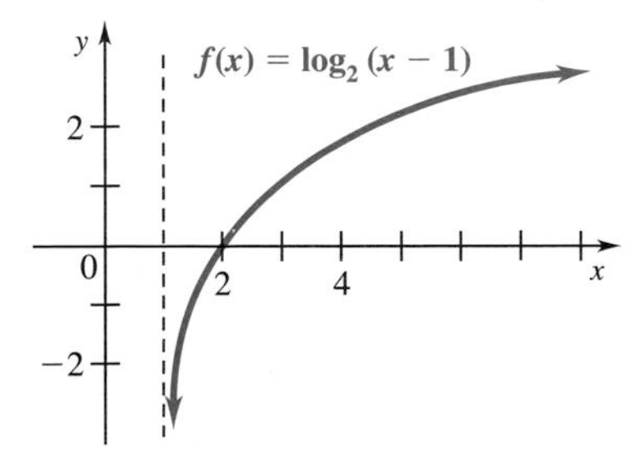

Figure 8

Common and Natural Logarithms The bases 10 and e are so important for logarithms that scientific calculators have $\log_{10}$ and $\log_e$ keys. Base 10 logarithms are called **common logarithms** and $\log_{10} x$ is often abbreviated as $\log x$. Base e logarithms are called **natural logarithms** and $\log_e x$ is often abbreviated as $\ln x$. The graphs of $\log x$ and $\ln x$ are the reflections of their inverses 10^x and e^x in the line $y = x$. See Figures 9 and 10.

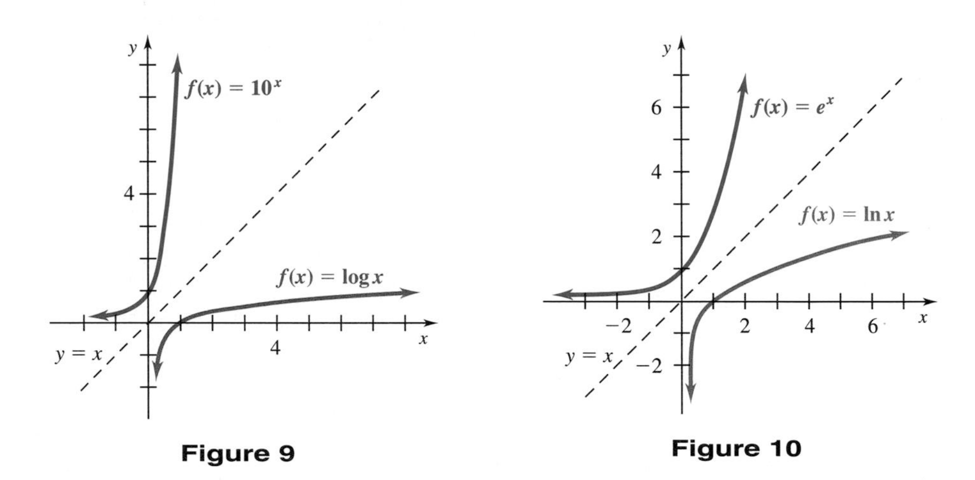

Figure 9

Figure 10

In the next section we will look at applications of natural logarithms. The following example gives an application of common logarithms. Other applications of common logarithms appear in the exercises.

EXAMPLE 5 In chemistry, the pH of a solution is defined as

$$\text{pH} = -\log_{10}\,[\text{H}^+],$$

where $[\text{H}^+]$ is the hydrogen ion concentration in moles per liter. The number pH is a measure of the acidity or alkalinity of solutions. Pure water has a pH of 7.0, with values greater than 7.0, indicating alkalinity and values less than 7.0 indicating acidity. Find the following.

(a) The pH of milk with $[\text{H}^+] = 4 \times 10^{-7}$

$$\begin{aligned} \text{pH} &= -\log_{10} .0000004 \qquad 4 \times 10^{-7} = .0000004 \\ &= -(-6.39794) \\ &\approx 6.4 \end{aligned}$$

It is customary to round pH values to two significant digits.

(b) The hydrogen ion concentration of shampoo with pH = 5.5

$$\begin{aligned} \text{pH} &= -\log_{10}[\text{H}^+] \\ 5.5 &= -\log_{10}[\text{H}^+] \\ -5.5 &= \log_{10}[\text{H}^+] \end{aligned}$$

Write this expression as

$$\begin{aligned} [\text{H}^+] &= 10^{-5.5} \\ &\approx 3.2 \times 10^{-6} \end{aligned}$$

rounded to two significant digits. ■

Compositions of the exponential and logarithmic functions can be used to get two useful properties. If $f(x) = a^x$ and $g(x) = \log_a x$, then

$$f[g(x)] = a^{\log_a x}$$

and

$$g[f(x)] = \log_a a^x.$$

Since $\log_a x$ is the power to which a must be raised to get x, raising a to the power $\log_a x$ gives x. That is $a^{\log_a x} = x$. Similarly, since the power to which a must be raised to get a^x is x, $\log_a a^x = x$. These results give the following useful theorem.

Theorem

If $a > 0$ and $a \neq 1$, then

$$\mathbf{\log_a a^x = x.}$$

Also, if $x > 0$, then

$$\mathbf{a^{\log_a x} = x.}$$

EXAMPLE 6 Simplify each of the following.

(a) $\log_5 5^3 = 3$

(b) $10^{\log 7} = 10^{\log_{10} 7} = 7$

(c) $\ln e^{x^2} = \log_e e^{x^2} = x^2$ ■

Since a logarithmic statement can be written as an exponential statement, it is not surprising that there are properties of logarithms based on the properties of exponents. The properties of logarithms allow us to change the form of logarithmic statements so that products can be converted to sums, quotients can be converted to differences, and powers can be converted to products. These properties will be used to solve logarithmic and exponential equations later in this chapter.

Properties of Logarithms

If x and y are any positive real numbers, r is any real number, and a is any positive real number, $a \neq 1$, then the following properties are true.

1. $\log_a xy = \log_a x + \log_a y$

2. $\log_a \frac{x}{y} = \log_a x - \log_a y$

3. $\log_a x^r = r \cdot \log_a x$

4. $\log_a a = 1$

5. $\log_a 1 = 0$

To prove Property 1 of logarithms, let $m = \log_a x$ and $n = \log_a y$. Then, changing to exponential form,

$$a^m = x \quad \text{and} \quad a^n = y.$$

Multiplication gives

$$a^m \cdot a^n = xy.$$

By a property of exponents,

$$a^{m+n} = xy.$$

Now write this statement in logarithmic form.

$$\log_a xy = m + n$$

Since $m = \log_a x$ and $n = \log_a y$,

$$\log_a xy = \log_a x + \log_a y.$$

To prove Property 2, use m and n as defined above.

$$\frac{a^m}{a^n} = \frac{x}{y}$$

$$a^{m-n} = \frac{x}{y} \qquad \frac{a^m}{a^n} = a^{m-n}$$

In logarithmic form, this statement can be written

$$\log_a \frac{x}{y} = m - n$$

or

$$\log_a \frac{x}{y} = \log_a x - \log_a y.$$

For Property 3,

$$(a^m)^r = x^r \qquad \text{or} \qquad a^{mr} = x^r.$$

Again using logarithmic form,

$$\log_a x^r = mr$$

or

$$\log_a x^r = r \cdot \log_a x.$$

Finally, Properties 4 and 5 follow directly from the exponential form since $a^1 = a$ and $a^0 = 1$.

The properties of logarithms are useful for rewriting expressions with logarithms in different forms, as shown in the next examples.

EXAMPLE 7 Assuming all variables represent positive real numbers, use the properties of logarithms to write each of the following in a different form.

(a) $\log_6 7 \cdot 9 = \log_6 7 + \log_6 9$

(b) $\log \frac{15}{7} = \log 15 - \log 7$

(c) $\ln \sqrt{8} = \ln 8^{1/2} = \frac{1}{2} \ln 8$

(d) $\log_a \frac{mnq}{p^2} = \log_a mnq - \log_a p^2$

$$= \log_a m + \log_a n + \log_a q - 2 \log_a p$$

(e) $\log_a \sqrt[3]{m^2} = \log_a m^{2/3} = \frac{2}{3} \log_a m$

(f) $\log_b \sqrt[n]{\frac{x^3 y^5}{z^m}} = \frac{1}{n} \log_b \frac{x^3 y^5}{z^m}$

$$= \frac{1}{n} (\log_b x^3 + \log_b y^5 - \log_b z^m)$$

$$= \frac{3}{n} \log_b x + \frac{5}{n} \log_b y - \frac{m}{n} \log_b z$$ ■

EXAMPLE 8 Use the properties of logarithms to write each of the following as a single logarithm with a coefficient of 1. Assume all variables represent positive real numbers.

(a) $\log_3 (x + 2) + \log_3 x - \log_3 2 = \log_3 \frac{(x + 2)x}{2}$

(b) $2 \log_a m - 3 \log_a n = \log_a m^2 - \log_a n^3 = \log_a \frac{m^2}{n^3}$

(c) $\frac{1}{2}\log_b m + \frac{3}{2}\log_b 2n - \log_b m^2 n$

$$= \log_b m^{1/2} + \log_b (2n)^{3/2} - \log_b m^2 n$$

$$= \log_b \frac{m^{1/2}(2n)^{3/2}}{m^2 n} = \log_b \frac{2^{3/2} n^{1/2}}{m^{3/2}}$$ ■

EXAMPLE 9 Assume $\log_2 5 = 2.322$. Find the base 2 logarithms of 25 and 5/2.

By the properties of logarithms,

$$\log_2 25 = \log_2 5^2 = 2\log_2 5 = 2(2.322) = 4.644$$

$$\log_2 \frac{5}{2} = \log_2 5 - \log_2 2 = 2.322 - 1 = 1.322.$$ ■

Logarithms to Other Bases A calculator will give the values of either natural logarithms (base e) or common logarithms (base 10). Sometimes, however, it is convenient to use logarithms to other bases. For example, base 2 logarithms are important in computer science. The **change-of-base theorem** is used to convert logarithms from one base to another.

Change-of-Base Theorem

If x is any positive number and if a and b are positive real numbers $a \neq 1$, $b \neq 1$, then

$$\log_a x = \frac{\log_b x}{\log_b a}.$$

For Graphers

The change-of-base theorem is needed to graph logarithmic functions having bases other than 10 and e. For instance, to obtain the graph of $f(x) = \log_2 x$, graph the function $f(x) = \log x / \log 2$.

To prove this result, use the definition of logarithm to write $y = \log_a x$ as $x = a^y$ or $x = a^{\log_a x}$ (for positive x and positive a, $a \neq 1$). Now take base b logarithms of both sides of this last equation.

$$\log_b x = \log_b a^{\log_a x}$$

or $$\log_b x = (\log_a x)(\log_b a), \qquad \log_b a^r = r \log_b a$$

from which $$\log_a x = \frac{\log_b x}{\log_b a}.$$

NOTE As an aid in remembering the change-of-base theorem, notice that x is above a on both sides of the equation. ■

EXAMPLE 10 Use the change-of-base theorem to approximate each of the following. Round to the nearest hundredth.

(a) $\log_5 27$

With natural logarithms as the base b logarithms, $x = 27$, $a = 5$, and $b = e$. Substituting into the change-of-base theorem gives

$$\log_5 27 = \frac{\ln 27}{\ln 5} \approx \frac{3.2958}{1.6094} \approx 2.05.$$

To check, use a calculator with a y^x key to verify that $5^{2.05} \approx 27$.

(b) $\log_2 .1$

Use common logarithms, with $x = .1$ and $a = 2$.

$$\log_2 .1 = \frac{\log .1}{\log 2} = \frac{-1}{.30103} \approx -3.322$$

EXAMPLE 11 One measure of the diversity of the species in an ecological community is given by

$$H = -[P_1 \log_2 P_1 + P_2 \log_2 P_2 + \ldots + P_n \log_2 P_n],$$

where $P_1, P_2, \ldots, P_n$ are the proportions of a sample belonging to each of n species in the sample. For example, in a community with two species, where there are 90 of one species and 10 of the other, $P_1 = 90/100 = .9$ and $P_2 = 10/100 = .1$, so

$$H = -[.9 \log_2 .9 + .1 \log_2 .1].$$

The value of $\log_2 .1$ was found in Example 10(b) above. Now find $\log_2 .9$.

$$\log_2 .9 = \frac{\ln .9}{\ln 2} \approx \frac{-.1054}{.6931} \approx -.152$$

Therefore, $$H \approx -[(.9)(-.152) + (.1)(-3.32)] \approx .47.$$

The properties of logarithms were developed over a twenty-five-year period by the Scottish mathematician John Napier of Merchiston (1550–1617). In 1614 Napier published *Mirifici logarithmorum canonis descriptio* (A Description of the Wonderful Law of Logarithms), a work that is second only to Newton's *Principia* in the history of British mathematics. Napier's work, which astonished the scientific world, contains the first description of logarithms, the first table of logarithms, and the first use of the word "logarithm" (meaning "number of the ratio"). Napier had been studying sequences of numbers obtained by repeated multiplication (exponentiation) and an ancient trigonometric formula that reduces multiplication to addition. Professor Henry Briggs of the University of London was so impressed by Napier's achievement that he traveled to Edinburgh to meet him. Together they determined that a log table would be more useful if the base was 10 rather than Napier's base of $(1 - 1/10^7)^{10^7}$, which is approximately $1/e$.

5.2 Exercises

For each of the following statements, write an equivalent statement in logarithmic form.

1. $10^3 = 1000$

2. $7^2 = 49$

3. $\left(\frac{1}{2}\right)^{-4} = 16$

4. $e^0 = 1$

For each of the following statements, write an equivalent statement in exponential form.

5. $\log_6 36 = 2$

6. $\log .0001 = -4$

7. $\ln 1 = 0$

8. $\log_8 \left(\frac{1}{64}\right) = -2$

Find the value of each of the following. Assume all variables represent positive real numbers.

9. $\log_5 25$

10. $\log_3 243$

11. $\log_8 8$

12. $\log_5 125^2$

13. $\ln e^5$

14. $\ln \frac{1}{e}$

15. $\ln \sqrt[3]{e}$

16. $\log \sqrt{10}$

17. $e^{\ln 5}$

18. $10^{\log 2}$

19. $5^{\log_5 (x+1)}$

20. $8^{\log_8 2x}$

21. $e^{\ln 3 + \ln 5}$

22. $e^{3(\ln 2)}$

23. $(e^5)^{\ln 2}$

24. $\ln (6 + 2) - \ln 2$

Solve each of the following equations.

25. $\log_x 256 = 8$

26. $\log_x \frac{1}{16} = -2$

27. $\log_y 12 = \frac{1}{2}$

28. $\log_r 4 = \frac{1}{3}$

Write each of the following as a sum or difference of logarithms (or constants times logarithms). Simplify the result if possible. Assume all variables represent positive real numbers.

29. $\log_2 \frac{6x}{y}$

30. $\log \frac{p^2}{3q}$

31. $\log_7 \frac{\sqrt{5}}{9}$

32. $\log_2 \frac{2\sqrt{3}}{5}$

33. $\log_5 (x + y)$

34. $\log_6 (7m + 3q)$

35. $\log_a \frac{p^2q^3}{r^2}$

36. $\log_z \frac{x^5y^3}{3}$

Write each of the expressions in Exercises 37–42 as a single logarithm with a coefficient of 1. Assume that all variables represent positive real numbers.

37. $\log_a x + \log_a y - \log_a m$

38. $\log_h (4m + 1) + \log_h 2m - \log_h 3$

39. $2 \log_m a - 3 \log_m b^2$

40. $\frac{1}{3}(\log_y p^3 + \log_y q^2) - \frac{1}{2}(\log_y q^{4/3})$

41. $-\frac{3}{4} \log_x a^6b^8 + \frac{2}{3} \log_x a^9b^3$

42. $\log_n (pqr) + 2 \log_n \left(\frac{p}{q}\right)$

43. Show that $\log \left(\frac{1}{2}\sqrt{2}\right)$ simplifies to $-\frac{1}{2} \log 2$.

44. What two points on the graph of $y = \log_a x$, $a > 1$, can be found without computation?

45. Explain why the natural logarithm function is defined only for $x > 0$.

46. The function $f(x) = \ln|x|$ plays a prominent role in calculus. Give its domain, range, and symmetries.

Use the properties of logarithms to relate the graphs of each of the following functions to the graph of $f(x) = \ln x$.

47. $f(x) = \ln(ex)$

48. $f(x) = \ln(x/e)$

For each of the following functions, identify the corresponding graph below.

49. $f(x) = \log_2 x$

50. $f(x) = \log_2 2x$

51. $f(x) = \log_2 \dfrac{1}{x}$

52. $f(x) = \log_2 \dfrac{x}{2}$

53. $f(x) = \log_2 (x - 1)$

54. $f(x) = \log_2 (-x)$

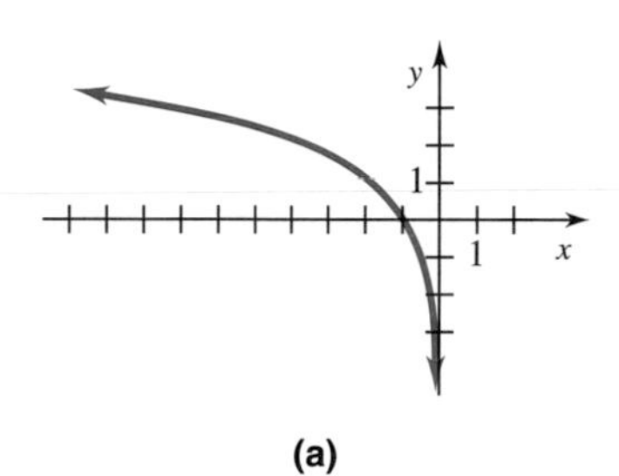

(a)

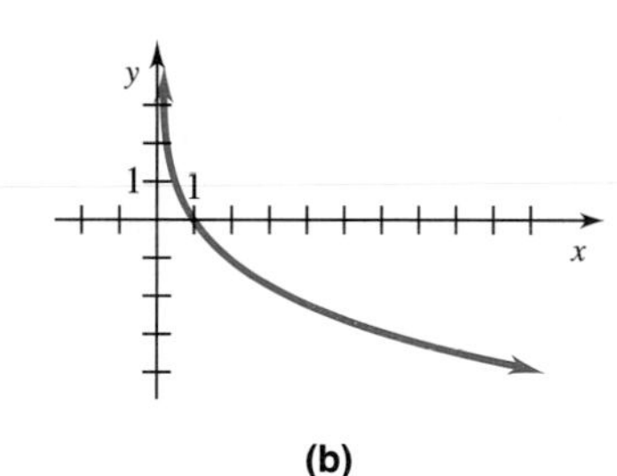

(b)

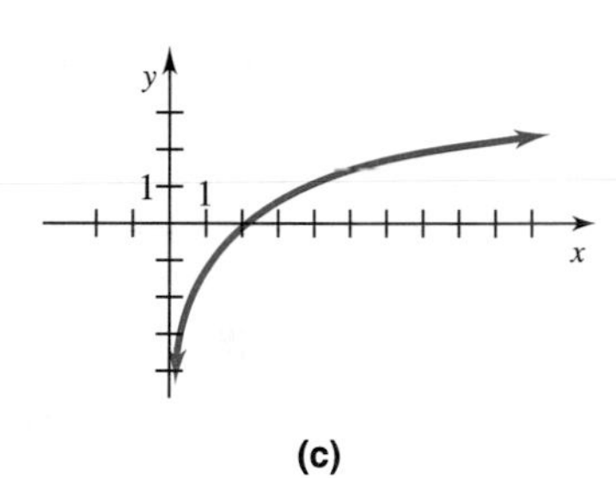

(c)

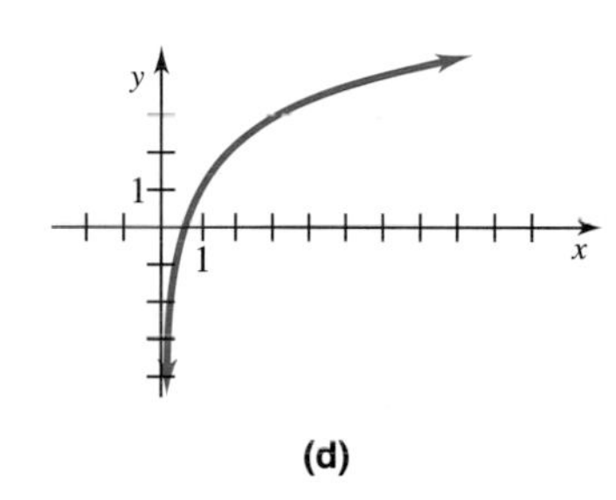

(d)

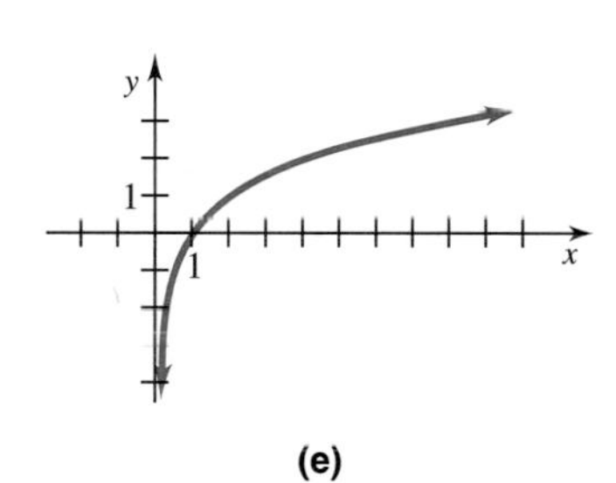

(e)

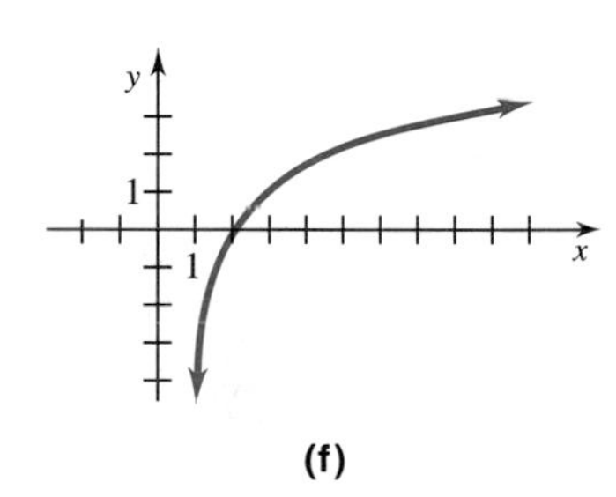

(f)

Graph each of the functions defined as follows. Give any x- and y-intercepts for each graph.

55. $f(x) = \log_5 x$

56. $f(x) = \log_3 (2 - x)$

57. $f(x) = \log_3 (x + 3) - 2$

58. $f(x) = \log_5 (x - 1) + 2$

59. $f(x) = \ln(x - 1)$

60. $f(x) = 1 - \log x$

Given $\log 2 = .3010$ *and* $\log 3 = .4771$, *evaluate the logarithms in Exercises 61–68 without using a calculator.*

61. $\log 6$

62. $\log \dfrac{2}{3}$

63. $\log 30$

64. $\log 8$

65. $\log 1.5$

66. $\log 12$

67. $\log 9$

68. $\log 20$

69. Which is larger, $\log_7 2$ or $\log_6 2$?

70. Which is larger, $\log_{1/2} 3$ or $\log_{1/3} 3$?

Suppose f *is a logarithmic function and* $f(3) = 2$. *Determine the function values in Exercises 71–73.*

71. $f(1/9)$

72. $f(1)$

73. $f(27)$

74. The population of an animal species that is introduced into a certain area may grow rapidly at first but then grow more slowly as time goes on. A logarithmic function can provide an excellent description of such growth. Suppose that the population of foxes in an area t months after the foxes were introduced there is given by $F(t) = 500 \log(2t + 3)$.
Find the population of foxes at the following times:
(a) When first released into the area (that is, when $t = 0$)
(b) After 3 months **(c)** After 15 months
(d) Graph $y = F(t)$.

75. A company determines that its monthly sales S are related to its advertising budget x by the function $S(x) = 11{,}000 \log_{10}(3x + 1) + 10{,}000$, where x is in thousands of dollars.
Find the monthly sales for the following advertising budgets:
(a) \$0 **(b)** \$3000 **(c)** \$10,000.
(d) Graph $y = S(x)$.

76. The loudness of sounds is measured in a unit called a *decibel.* To measure with this unit we first assign an intensity of I_0 to a very faint sound, called the *threshold sound.* If a particular sound has an intensity I, then the decibel rating d of this louder sound is given by

$$d = 10 \log_{10} \frac{I}{I_0}.$$

Find the decibel ratings of the following sounds, having intensities as given. Round answers to the nearest whole number.
(a) Whisper, 115 I_0 **(b)** Rock music, 895,000,000,000 I_0
(c) Jetliner at takeoff, 109,000,000,000,000 I_0
(d) If the intensity of a sound is doubled, by how much is the decibel rating increased? What happens to the decibel rating if the sound intensity is tripled?

77. The *Richter scale rating* of an earthquake of intensity I is given by $\log_{10}(I/I_0)$, where I_0 is the intensity of an earthquake of a certain (small) size. Find the Richter scale ratings of earthquakes having the following intensities:
(a) 1000 I_0 **(b)** 1,000,000 I_0 **(c)** 100,000,000 I_0.

78. The San Francisco earthquake of 1906 had a Richter scale rating of 8.6. Express the intensity of this earthquake as a multiple of I_0 (see Exercise 77).

79. The San Francisco earthquake of 1989 had a Richter scale rating of 7.1. How much more powerful was the 1906 earthquake than the 1989 earthquake?

80. The number of years, n, since two independently evolving languages split off from a common ancestral language is approximated by $n \approx -7600 \log r$, where r is the ratio of words from the ancestral language common to both languages.
Find n if **(a)** $r = .9$ **(b)** $r = .3$.
(c) How many years have elapsed since the split if half of the words of the ancestral language are common to both languages?

81. Suppose the number of paramecia in a colony is given by $y = y_0 e^{.5t}$, where t is time in hours and y_0 is the population at $t = 0$.
(a) If $y_0 = 1000$, find y when $t = 4$.
(b) Find y_0 if there are 5400 paramecia when $t = 2$.

82. In the central Sierra Nevada Mountains of California, the percent of moisture that falls as snow rather than rain is approximated reasonably well by $p = 86.3 \ln h - 680$, where p is the percent of snow at an altitude h (in feet). (Assume $h \geq 3000$.)
Find the percent of moisture that falls as snow at the following altitudes.
(a) 3000 ft **(b)** 4000 ft **(c)** 7000 ft
(d) Graph p.

Find each of the following logarithms to the nearest hundredth.

83. $\log_2 10$ **84.** $\log_8 12$ **85.** $\log_{1/2} 3$ **86.** $\log_{12} 62$
87. $\log_{1/10} 9.2$ **88.** $\log_{200} 175$ **89.** $\log_{2.9} 7.5$ **90.** $\log_{6.7} .84$

Use the change-of-base theorem to evaluate the following expressions.

91. $\dfrac{\log_3 25}{\log_3 5}$ **92.** $\dfrac{\log_8 32}{\log_8 2}$

In Exercises 93 and 94, assume that a and b represent positive numbers other than 1, and x is any real number.

93. Show that $1/\log_a b = \log_b a$. **94.** Show that $a^x = e^{x \ln a}$.

95. Explain the error in the following proof that $2 < 1$.

$$\frac{1}{9} < \frac{1}{3}$$

$$\left(\frac{1}{3}\right)^2 < \frac{1}{3} \qquad \text{Rewrite left side.}$$

$$\log \left(\frac{1}{3}\right)^2 < \log \frac{1}{3} \qquad \text{Take log of both sides.}$$

$$2 \cdot \log \frac{1}{3} < 1 \cdot \log \frac{1}{3} \qquad \text{Use the third property of logarithms.}$$

$$2 < 1 \qquad \text{Divide both sides by } \log \frac{1}{3}.$$

96. The first logarithm property can be verbalized as ''The logarithm of the product of two numbers is the sum of their logarithms.'' State the second and third properties in words.

Graphing Utility Problems

97. Graph $f(x) = 10^x$ and $y = 5$ on the same axes and determine the x-coordinate of their point of intersection. Express this number in terms of a logarithm.

98. Use the technique of Exercise 97 to estimate ln 3.

99. Graph the function $f(x) = \log_5 x$. **100.** Graph the function $f(x) = \log_x 5$.

Estimate the solutions of the following equations.

101. $\ln x = x - 2$ **102.** $e^{-x} = \ln x$ **103.** $\ln x = e^{.3x} - 2$ **104.** $\log x = 2 - x/2$

105. The function $\ln x$ grows so slowly that for any r with $0 < r < 1$, $(\ln x)/x^r$ approaches 0 as x gets large. Test this statement by graphing $(\ln x)/\sqrt{x}$ and $(\ln x)/\sqrt[5]{x}$.

5.3 Exponential and Logarithmic Equations

In Section 1 of this chapter we solved exponential equations such as $(1/3)^x = 81$ by writing each side of the equation as a power of 3. That method cannot be used to solve an equation such as $7^x = 12$, since 12 cannot easily be written as a power of 7. However, the equation $7^x = 12$ can be solved by taking the logarithm of each side. This approach is justified by the following property.

Property of Logarithms

If $x > 0$, $y > 0$, $a > 0$, and $a \neq 1$, then

$$x = y \quad \text{if and only if} \quad \log_a x = \log_a y.$$

The fact that $\log_a x = \log_b x$ implies that $x = y$ because a logarithmic function is one-to-one.

EXAMPLE 1 Solve the equation $7^x = 12$.

While the result above is valid for any appropriate base a, the best practical base to use is either base e or base 10, since these are the logarithms on calculators. Taking base e (natural) logarithms of both sides gives

$$\ln 7^x = \ln 12$$

$$x \cdot \ln 7 = \ln 12$$

$$x = \frac{\ln 12}{\ln 7} = \frac{2.4849}{1.9459} \approx 1.2770.$$

A check using the y^x key shows that $7^{1.277} \approx 12$. ■

In Example 1, had we used common logarithms instead of natural logarithms to solve for x, the final result would have been

$$x = \frac{\log 12}{\log 7} = \frac{1.0792}{.8451} \approx 1.2770.$$

CAUTION A common mistake is to confuse $(\ln 12)/(\ln 7)$ with $\ln (12/7)$. However,

$$\frac{\ln 12}{\ln 7} \neq \ln \frac{12}{7} = \ln 12 - \ln 7. \blacksquare$$

EXAMPLE 2 Solve $3^{2x-1} = 4^{x+2}$.

Taking natural logarithms on both sides gives

$$\ln 3^{2x-1} = \ln 4^{x+2}.$$

Now use a property of logarithms.

$$\begin{aligned}
(2x - 1) \ln 3 &= (x + 2) \ln 4 \\
2x \ln 3 - \ln 3 &= x \ln 4 + 2 \ln 4 && \text{Distributive property} \\
2x \ln 3 - x \ln 4 &= 2 \ln 4 + \ln 3 \\
x(2 \ln 3 - \ln 4) &= 2 \ln 4 + \ln 3 && \text{Factor out } x. \\
x &= \frac{2 \ln 4 + \ln 3}{2 \ln 3 - \ln 4} \\
x &= \frac{\ln 16 + \ln 3}{\ln 9 - \ln 4} && \text{Properties of logarithms} \\
x &= \frac{\ln 48}{\ln \frac{9}{4}}
\end{aligned}$$

This quotient could be approximated by a decimal if desired.

$$x = \frac{\ln 48}{\ln 2.25} \approx \frac{3.8712}{.8109} \approx 4.774$$

To the nearest thousandth, the solution is 4.774. ■

EXAMPLE 3 Solve $e^{x^2} = 200$.

Take natural logarithms on both sides; then use properties of logarithms.

$$\begin{aligned}
e^{x^2} &= 200 \\
\ln e^{x^2} &= \ln 200 \\
x^2 &= \ln 200 && \ln e^{x^2} = x^2 \\
x &= \pm\sqrt{\ln 200} \\
x &\approx \pm 2.302
\end{aligned}$$

to the nearest thousandth. ■

EXAMPLE 4 Solve $3 = 5(1 - e^x)$.

First solve for e^x.

$$\begin{aligned}
3 &= 5(1 - e^x) \\
\frac{3}{5} &= 1 - e^x \\
e^x &= 1 - \frac{3}{5} = \frac{2}{5}
\end{aligned}$$

Now take the natural logarithm on each side.

$$\ln e^x = \ln \frac{2}{5}$$

Since $\ln e^x = x$, $\quad x = \ln \frac{2}{5} = \ln .4 \approx -.916$

to the nearest thousandth. ■

The properties of logarithms given earlier are useful in solving logarithmic equations, as shown in the next examples.

EXAMPLE 5 Solve $\log_a (x + 6) - \log_a (x + 2) = \log_a x$.

Using a property of logarithms, rewrite the equation as

$$\log_a \frac{x + 6}{x + 2} = \log_a x.$$

Then, since logarithmic functions are one-to-one,

$$\frac{x + 6}{x + 2} = x$$

$$x + 6 = x(x + 2)$$

$$x + 6 = x^2 + 2x$$

$$x^2 + x - 6 = 0$$

$$(x - 2)(x + 3) = 0 \qquad \text{Factor.}$$

$$x = 2 \quad \text{or} \quad x = -3.$$

$\text{Log}_a\, x$ cannot be evaluated for $x = -3$, since the domain of $\log_a x$ includes only positive numbers. Verify by substitution that $x = 2$ is the only solution. ■

EXAMPLE 6 Solve $\log_{10} (3x + 2) + \log_{10} (x - 1) = 1$.

By a property of logarithms, $\log_{10} (3x + 2)(x - 1) = 1$. Write this expression in exponential form.

$$(3x + 2)(x - 1) = 10^1$$

$$3x^2 - x - 2 = 10$$

$$3x^2 - x - 12 = 0$$

Now use the quadratic formula to arrive at

$$x = \frac{1 \pm \sqrt{1 + 144}}{6}.$$

If $x = (1 - \sqrt{145})/6$, then $x - 1 < 0$ and $\log (x - 1)$ does not exist. For this reason, $(1 - \sqrt{145})/6$ must be discarded as a solution. A calculator can help to show that $(1 + \sqrt{145})/6$ is the only solution. ■

EXAMPLE 7 Solve $\log_3 (3m^2)^{1/4} - 1 = 2$

Solve for the logarithm first.

$$\log_3 (3m^2)^{1/4} - 1 = 2$$
$$\log_3 (3m^2)^{1/4} = 3$$

Now write the expression in exponential form.

$$(3m^2)^{1/4} = 3^3$$
$$3^{1/4}m^{1/2} = 3^3$$
$$m^{1/2} = 3^{11/4}$$

Square both sides to get $m = (3^{11/4})^2 = 3^{11/2}$

or $m \approx 420.888.$ ■

EXAMPLE 8 Suppose

$$P(t) = 10{,}000e^{.4t}$$

gives the population of a city at time t (in years). In how many years will the population double?

If the population is doubled, it will be 20,000. To find the time t when that occurs, let $P(t) = 20{,}000$ and substitute this value into the given function. Then divide by 10,000 on both sides of the equation.

$$P(t) = 10{,}000e^{.4t}$$
$$20{,}000 = 10{,}000e^{.4t}$$
$$2 = e^{.4t}$$

Now take natural logarithms on both sides.

$$\ln 2 = \ln e^{.4t}$$

Since $\ln e^{.4t} = .4t$, $\ln 2 = .4t$

with $t = \dfrac{\ln 2}{.4} \approx 1.733.$

The population of the city will double in about 1¾ years. ■

For Graphers

Graphing utilities can be helpful in determining the number of solutions to equations and estimating the solutions. For instance, to examine the equation from Example 9, we could graph the functions $f(x) = \dfrac{e^x - 1}{e^{-x} - 1}$ and $g(x) = -3$ and find their intersection points.

EXAMPLE 9 Solve the equation $\dfrac{e^x - 1}{e^{-x} - 1} = -3.$

Begin by multiplying both sides by $e^{-x} - 1$.

$$e^x - 1 = -3(e^{-x} - 1)$$
$$e^x - 1 = -3e^{-x} + 3$$
$$e^x - 4 + 3e^{-x} = 0 \qquad \text{Get 0 alone on one side.}$$
$$e^{2x} - 4e^x + 3 = 0 \qquad \text{Multiply both sides by } e^x.$$

Rewrite this equation as

$$(e^x)^2 - 4e^x + 3 = 0,$$

a quadratic equation in e^x. Proceed as in Section 2.4. Let $u = e^x$.

$$\begin{aligned}
u^2 - 4u + 3 &= 0 \\
(u - 1)(u - 3) &= 0 \\
u - 1 = 0 \quad &\text{or} \quad u - 3 = 0 \\
u = 1 \quad & \quad\quad u = 3 \\
e^x = 1 \quad & \quad\quad e^x = 3 \qquad \text{Replace } u \text{ with } e^x. \\
x = 0 \quad & \quad\quad x = \ln 3
\end{aligned}$$

Check the solutions by substitution into the original equation. When $x = 0$, the denominator of the original equation is 0, and therefore $x = 0$ is not a valid solution. When $x = \ln 3$, $e^x = 3$ and $e^{-x} = 1/3$. Then

$$\frac{e^x - 1}{e^{-x} - 1} = \frac{3 - 1}{\frac{1}{3} - 1} = \frac{2}{-\frac{2}{3}} = 2 \cdot -\frac{3}{2} = -3.$$

Therefore, the solution is $x = \ln 3$. ■

CAUTION In solving equations like the one in Example 9 that involve a^x (for any value of a), always solve first for a^x. Then use logarithms to get x. ■

Methods for solving exponential or logarithmic equations are summarized below.

Solving Exponential or Logarithmic Equations

In summary, to solve an exponential or logarithmic equation, first use the properties of algebra to change the given equation into one of the following forms, where a and b are real numbers with appropriate restrictions.

1. $a^{f(x)} = b$

   To solve, take logarithms on both sides.
2. $\log_a f(x) = b$

   Solve by changing to the exponential form $a^b = f(x)$.
3. $\log_a f(x) = \log_a g(x)$

   From the given equation, obtain the equation $f(x) = g(x)$, then solve algebraically.
4. In a more complicated equation, such as the one in Example 9, it is necessary to first solve for $e^{f(x)}$ or $\log_a f(x)$ and then solve the resulting equation using one of the methods given above.

5.3 Exercises

Solve the following equations. Give answers as decimals rounded to the nearest hundredth.

1. $5^x = 7$
2. $4^x = 12$
3. $6^{1-2k} = 8$
4. $3^{k-3} = 11$
5. $4^{3m-1} = 12^{m+2}$
6. $3^{2m-5} = 13^{m-1}$
7. $e^x = 4$
8. $e^{1-y} = 10$
9. $3e^{4a+1} = 9$
10. $10e^{3z-7} = 5$
11. $2^x = -3$
12. $\left(\frac{1}{5}\right)^q = -3$
13. $100(1 + .02)^{3+n} = 150$
14. $500(1 + .05)^{p/4} = 200$
15. $2^{x^2-9} = 36$
16. $5^{3-x^2} = 6$
17. $4(e^x - 1) = 20$
18. $\frac{1}{5}(e^{-x} + 1) = 3$
19. $\log(t - 1) = 1$
20. $\log q^{-1} = 1$
21. $\log(x + 2) = 2 - \log x$
22. $\log(z - 6) = 2 - \log(z + 15)$
23. $\ln(y + 1) = \ln(y - 2) + \ln 2$
24. $\ln p^2 - \ln(p + 2) = \ln 6$
25. $\ln(3 + 2y) - \ln(1 + y) = \ln 10$
26. $\log_3(a - 3) = 1 + \log_3(a + 1)$
27. $\log_4(z + 3) + \log_4(z - 3) = 1$
28. $\log 2w + \log(3w - 7) = \log 3$
29. $5^{\log_5(x+1)} = 9$
30. $\log_2 \sqrt{2y^2} - 1 = \frac{1}{2}$
31. $\log_x(5x - 6) = 2$
32. $\ln e^{x^3-2} = 6$
33. $7^{2x \log_7 4} = 256$
34. $\log_3(\log_3 x) = 1$
35. $\log z = \sqrt{\log z}$
36. $\log x^2 = (\log x)^2$

37. The amount of a radioactive specimen present at time t (measured in seconds) is $A(t) = 1000(10)^{-.04t}$, where $A(t)$ is measured in grams. Find the half-life of the specimen—that is, the time it will take until exactly half the specimen remains.

A large cloud of radioactive debris from a nuclear explosion has floated over the Pacific Northwest, contaminating much of the hay supply. Consequently, farmers in the area are concerned that the cows who eat the hay will give contaminated milk. The percent of the initial amount of radioactive iodine still present in the hay after t days is approximated by $P(t) = 100e^{-.1t}$, *where t is time measured in days.*

38. Find the half-life of the radioactive iodine in the hay (the time until only 50 percent of the iodine is left).

39. Some scientists feel that the hay is safe after the level of radioactive iodine has declined to 10 percent of the original amount. Based on this assumption, find the number of days before the hay could be used. Other scientists believe that the hay is not safe until the level of radioactive iodine has declined to 1 percent of the original amount. Find the number of days this would take.

Solve the following equations for x. (Hint: *In Exercises 42–45 multiply by* e^x.)

40. $2^{2x} - 2^x - 2 = 0$
41. $5^{2x} + 2 \cdot 5^x - 3 = 0$
42. $2e^x + 1 - 3e^{-x} = 0$
43. $e^x - 5 + 6e^{-x} = 0$
44. $\frac{e^x - e^{-x}}{2} = 4$
45. $\frac{e^x + e^{-x}}{2} = 3$

Solve each of the following equations for the indicated variables. Use logarithms to the appropriate bases.

46. $P = 1000e^{t/1000}$ for t

47. $I = \dfrac{E}{R}(1 - e^{-Rt/2})$ for t

48. $T = T_0 + (T_1 - T_0)10^{-kt}$ for t

49. $A = \dfrac{Pr}{1 - (1 + r)^{-t}}$ for t

50. $G = \dfrac{mH}{H + (m - H)e^{-kmt}}$ for t

51. $\log_5 (x + y) = \log_5 (2x - 1)$ for x

52. $\dfrac{10^x - 10^{-x}}{2} = y$ for x

53. $\dfrac{10^x + 10^{-x}}{2} = y$ for x

Find the formula, domain, and range of $f^{-1}(x)$.

54. $f(x) = \ln (x - 1)$

55. $f(x) = \log (x + 5)$

56. $f(x) = e^{3x+1}$

57. $f(x) = 3 \cdot 10^x$

Solve the inequalities in Exercises 58–63 for x. The inequalities in Exercises 60 and 61 are studied in calculus to determine where certain functions are increasing.

58. $\log_2 x < -1$

59. $\log_x 64 < 3$

60. $\log_3 x > 3$

61. $\log_x .2 < -.1$

62. $x^2e^x - e^x > 0$

63. $\dfrac{e^x - xe^x}{e^{2x}} > 0$

64. Recall (from Exercise 77 of Section 5.2) the formula for the Richter scale of earthquake intensity, $R = \log_{10} (I/I_0)$. Solve this formula for I.

65. Solve the formula for compound interest, $A = P\left(1 + \dfrac{r}{n}\right)^{nt}$, for t, and describe the meaning of t.

66. Use natural logarithms to solve $A = Pe^{rt}$ for t.

67. Explain why an equation of the form $2^x = a$ does not always have a solution.

68. Without solving, explain why an equation of the form $\log_2 x = a$ must always have a solution for x.

Graphing Utility Problems

69. For each of the functions in Exercises 54–57, graph the function, its inverse, and the line $y = x$. Convince yourself that the two functions are indeed inverses by observing that the two functions are reflections of each other in the line. Also, find the coordinates of a point on one graph, call it (a, b), and check that the point (b, a) is on the other graph.

Solve the following equations by finding the intersection point(s) of the graph of a function and a horizontal line.

70. $e^x + \ln x = 5$

71. $e^x - \ln(x + 1) = 3$

5.4 Exponential Growth and Decay

In many situations that occur in biology, economics, and the social sciences, a quantity changes at a rate proportional to the amount present. In such cases the amount present at time t is a function of t called the **exponential growth and decay function**.

Exponential Growth and Decay Function

Let y_0 be the amount or number present at time $t = 0$. Then, under certain conditions, the amount present at any time t is given by

$$y = y_0 e^{kt},$$

where k is a constant.

In this section we see how to determine the constant k of the function from given data. When $k > 0$, the function describes growth; when $k < 0$, the function describes decay. Radioactive decay is an important application; it has been shown that radioactive substances decay exponentially; that is, in the function

$$y = y_0 e^{kt},$$

k is a negative number.

EXAMPLE 1 If 600 grams of a radioactive substance are present initially and three years later only 300 grams remain, how much of the substance will be present after six years?

To express the situation as an exponential equation,

$$y = y_0 e^{kt},$$

we first find y_0 and then find k. From the statement of the problem, $y = 600$ when $t = 0$ (that is, initially), so

$$600 = y_0 e^{k(0)}$$
$$600 = y_0,$$

giving the exponential decay equation

$$y = 600\, e^{kt}.$$

Since 300 grams of the substance remains after three years, use the fact that $y = 300$ when $t = 3$ to find k.

$$300 = 600\, e^{3k}$$
$$\frac{1}{2} = e^{3k}$$

Take natural logarithms on both sides, then solve for k.

$$\ln \frac{1}{2} = \ln e^{3k}$$

$$\ln .5 = 3k \qquad \ln e^x = x$$

$$\frac{\ln .5}{3} = k$$

$$k \approx -.231,$$

giving
$$y = 600e^{-.231t}$$

as the exponential decay equation. To find the amount present after six years, let $t = 6$.

$$y = 600\ e^{-.231(6)} \approx 600\ e^{-1.386} \approx 150$$

After six years, about 150 grams of the substance remain. ■

As mentioned in the exercises for the previous section, the *half-life* of a radioactive substance is the time it takes for half of a given amount of the substance to decay.

EXAMPLE 2 The amount in grams of a certain radioactive substance at time t is given by

$$y = y_0 e^{-.1t},$$

where t is in days. Find the half-life of the substance.

Find the time t when y will equal $y_0/2$. That is, solve the equation

$$\frac{y_0}{2} = y_0 e^{-.1t}$$

for t. Divide both sides by y_0.

$$\frac{1}{2} = e^{-.1t}$$

Now take natural logarithms on both sides and solve for t.

$$\ln \frac{1}{2} = \ln e^{-.1t}$$

$$\ln \frac{1}{2} = -.1t \qquad \ln e^x = x$$

$$t = \frac{-\ln 1/2}{.1} \approx \frac{.6931}{.1} \approx 6.9$$

Therefore, the half-life is 6.9 days. ■

EXAMPLE 3 Carbon 14, also known as radiocarbon, is a radioactive form of carbon that is found in all living plants and animals. After a plant or animal dies, the radiocarbon disintegrates, with a half-life of 5600 years. Scientists can determine the age of the remains by comparing the amount of radiocarbon with the amounts present in living plants and animals. This technique is called *carbon dating*. The amount of radiocarbon present after t years is given by

$$y = y_0 e^{-(\ln 2)(1/5600)t},$$

where y_0 is the amount present in living plants and animals.

(a) Verify the formula for $t = 5600$.
Substitute 5600 for t. Then

$$y = y_0 e^{-(\ln 2)(1/5600)5600} = y_0 e^{-\ln 2} = y_0 (e^{\ln 2})^{-1} = y_0 2^{-1} = \frac{1}{2} y_0.$$

This result is correct. Since 5600 years is the half-life of carbon 14, half of the initial amount should remain after 5600 years.

(b) A round table hanging in Winchester Castle (in England) was alleged to have belonged to King Arthur, who lived in the fifth century. A chemical analysis recently showed that the table had 91% of the amount of radiocarbon present in living wood. How old is the table?

The amount of radiocarbon present in the round table after t years is $.91y_0$. Therefore, in the equation

$$y - y_0 e^{-(\ln 2)(1/5600)t}$$

replace y with $.91y_0$ and solve for t.

$$\begin{aligned}
.91y_0 &= y_0 e^{-(\ln 2)(1/5600)t} \\
.91 &= e^{-(\ln 2)(1/5600)t} \\
\ln .91 &= \ln e^{-(\ln 2)(1/5600)t} \\
\ln .91 &= -(\ln 2)(1/5600)t \\
t &= \frac{(5600)\ln .91}{-\ln 2} \approx 760
\end{aligned}$$

The table is about 760 years old and therefore could not have belonged to King Arthur. ■

The compound interest formula

$$A = P\left(1 + \frac{r}{n}\right)^{nt}$$

was discussed in Section 1 of this chapter. The table presented there shows that increasing the frequency of compounding makes smaller and smaller differences in the amount of interest earned. In fact, it can be shown that even if interest is compounded at intervals of time as small as one chooses (such as each hour, each minute, or each second), the total amount of interest earned will be only slightly more than the interest earned with daily compounding. This is true even for a process called **continuous compounding**, which can be described loosely as compounding every instant. As suggested in Section 1, the value of the expression $(1 + 1/n)^n$ approaches e as n gets larger. Because of this, the formula for continuous compounding involves the number e.

Continuous Compounding

If P dollars is deposited at a rate of interest r compounded continuously for t years, the final amount on deposit is

$$A = Pe^{rt}$$

dollars.

EXAMPLE 4 Suppose \$5000 is deposited in an account paying 8% interest compounded continuously for five years. Find the total amount on deposit at the end of five years.

Let $P = 5000$, $t = 5$, and $r = .08$. Then

$$A = 5000e^{.08(5)} = 5000e^{.4} \approx 5000(1.49182) = 7459.1$$

or \$7459.10. Check that daily compounding would have produced a compound amount about 30¢ less. ■

EXAMPLE 5 How long will it take for the money in an account that is compounded continuously at 8% interest to double?

Use the formula for continuous compounding, $A = Pe^{rt}$, to find the time t that makes $A = 2P$. Substitute $2P$ for A and .08 for r; then solve for t.

$$A = Pe^{rt}$$
$$2P = Pe^{.08t}$$
$$2 = e^{.08t}$$

Taking natural logarithms on both sides gives

$$\ln 2 = \ln e^{.08t}$$
$$\ln 2 = .08t \qquad \ln e^x = x$$
$$\frac{\ln 2}{.08} = t$$
$$8.664 = t.$$

It will take about 8⅔ years for the amount to double. ■

5.4 Exercises

1. A population that initially contains 40,000 people is growing exponentially with a growth constant $k = .03$.
(a) Give a formula for the size of the population after t yr.
(b) How large will the population be after 20 yr?
(c) After how many years will the population reach 60,000?

2. Ten grams of radioactive material with decay constant $k = -.025$ is buried in the ground.
(a) Give a formula for the amount after t yr.
(b) How much will remain after 4 yr?
(c) When will 60 percent of the material have distintegrated?

3. After a bactericide is introduced into a culture of 50,000 bacteria, the number of bacteria decreases exponentially. After 9 hr there are only 20,000 bacteria.
(a) Find the value of the decay constant k.
(b) In how many hours will the original population of 50,000 bacteria be reduced by half?

4. The average tuition for public colleges increased exponentially from $1533 in 1970 to $8174 in 1990.
(a) Determine the annual rate of increase in tuition from 1970 to 1990.
(b) If tuition continues to increase at the rate found in (a), what will be the average tuition in the year 2000?

5. A population of bacteria in a culture is increasing exponentially. The original culture of 25,000 bacteria contains 40,000 bacteria after 10 hr. How long will it be until there are 60,000 bacteria in the culture? (*Hint:* This problem must be solved in two steps: first find the growth constant, and then find the time requested.)

6. A radioactive substance is decaying exponentially. The amount of substance is reduced from 800 g to 400 g after 4 days. How much remains after 10 days? (*Hint:* This problem must be solved in two steps: first find the decay constant, and then find the amount requested.)

7. The amount of a certain radioactive specimen present at time t (in days) decreases exponentially. If 5000 g decreased to 4000 g in 5 days, find the half-life of the specimen.

8. A population of insects is growing exponentially. After 2 months the population has increased from 100 to 150. How many months will it take for the population to reach 500?

9. The amount (in grams) of a substance involved in a chemical reaction increases exponentially as temperature increases (for $0°\text{C} \leq t \leq 50°\text{C}$). At 0°C, 25 g reacts, and at 20°C, 75 g reacts. At what temperature will 150 g react?

Exercises 10–12 refer to the carbon dating process of Example 3 in the text.

10. Suppose an Egyptian mummy is discovered in which the amount of carbon 14 is only half of the original amount. About how long ago did the Egyptian die?

11. If an object contains $1/4$ of the amount of carbon 14 that it originally had, how old is the object? How old if the amount is $1/8$?

12. The Lascaux caves of France contain prehistoric paintings of animals. Charcoal found in these caves contains only about 15 percent of the amount of carbon 14 in living trees. Estimate the age of the paintings.

Nuclear energy derived from radioactive isotopes can be used to supply power to space vehicles. Suppose that the output of the radioactive power supply for a certain satellite is $P(t) = 30\,e^{-t/250}$ *watts, where t is the time in days.**

13. How much power will be available at the end of one year? (Assume a 365-day year.)

14. How long will it take for the power to drop to half its original strength?

15. The equipment aboard the satellite requires 10 watts of power to operate properly. What is the operational life of the satellite?

Find each of the amounts in Exercises 16–21, assuming continuous compounding.

16. $1,000 at 9% for 1 yr

17. $2,000 at 8% for 5 yr

18. $15,000 at 10% for 6 yr

19. $135 at 4% for 3 yr

20. $580 at 15% for 6 yr

21. $10,000 at 12% for 3 yr

22. Assuming an inflation rate of 5% compounded continuously, how long will it take for prices to double?

23. For Exercise 22, how long would it take for prices to double if the inflation rate is 10%?

24. Solve $A = Pe^{rt}$ for r, and describe the meaning of t.

25. Solve $2P = Pe^{rt}$ for t, and describe the meaning of t.

Newton's law of cooling says that the rate at which a body cools is proportional to the difference in temperature between the body and the environment into which it is introduced. The temperature $f(t)$ *of the body at time t in appropriate units after being introduced into an environment having constant temperature* T_0 *is* $f(t) = T_0 + Ce^{-kt}$, *where C and k are constants. Use this result in Exercises 26 and 27.*

26. Boiling water, at 100°C, is placed in a freezer at 0°C. The temperature of the water is 50°C after 24 min. Find the temperature of the water after 96 min.

27. A piece of metal is heated to 300°C and then placed in a cooling liquid at 50°C. After 4 min, the metal has cooled to 175°C. Find its temperature after 12 min.

28. Explain why the half-life of a radioactive material does not depend on the amount of material present initially.

29. The number of bacteria in a jar doubles every minute. If the jar is full at 2:00 P.M., when was it half full?

30. If interest is compounded continuously and the interest rate is tripled, what effect will this have on the time required for an investment to double?

Graphing Utility Problems

31. Suppose an 80-g sample of radioactive material has decay constant $-.02$. On the same axes, graph the three functions $y = 80e^{-.02t}$, $y = 40$, and $y = 20$. Determine the t-coordinates of the intersection points of the decay curve and the horizontal lines and observe that the time required for the material to decay from 80 g to 40 g is the same as the time required to decay from 40 g to 20 g. Explain why this is true.

*Bernice Kastner, Phd., *Spacemathematics.* (NASA, 1985).

32. On the same axes, graph $y = 5\left(1 + \frac{.08}{2}\right)^{2x}$ and $y = 5e^{.08x}$, for $0 \le x \le 25$, $0 \le y \le 75$. Compare the graphs and determine the two y-values when $x = 25$. Repeat with the 2 in the first function replaced by 12.

Many environmental situations place effective limits on the growth of the number of an organism in an area. Many such limited growth situations are described by the logistic function, *defined by*

$$G(t) = \frac{MG_0}{G_0 + (M - G_0)e^{-kMt}}$$

where G_0 is the initial number present, M is the maximum possible size of the population, and k is a positive constant. Assume $G_0 = 100$, $M = 2500$, $k = .0004$, and t is time in decades (10-yr periods).

33. Graph the S-shaped limited growth function using $0 \le t \le 8$, $0 \le y \le 2500$.

34. Estimate the value of $G(2)$ from the graph. Then evaluate $G(2)$ to find the population after 20 yr.

35. Find the x-coordinate of the intersection of the curve with the horizontal line $y = 1000$ to estimate the number of decades required for the population to reach 1000. Then solve $G(t) = 1000$ algebraically to obtain the exact value of t.

Chapter 5 Review Exercises

Graph each of the functions defined as follows.

1. $f(x) = 2^x$

2. $f(x) = \left(\frac{1}{2}\right)^x$

3. $f(x) = \left(\frac{1}{3}\right)^{x-2}$

4. $f(x) = e^x + e^{-x}$

5. $f(x) = (x - 1)e^{-x}$

6. $f(x) = \log_3 x$

7. $f(x) = \log_2 (x - 1)$

8. $f(x) = \ln \frac{1}{x}$

Solve each of the following equations.

9. $16^p = 32$

10. $9^{y-1} = 27^{2y}$

11. $\frac{-8}{125} = b^{-3}$

12. $\frac{1}{2} = \left(\frac{b}{4}\right)^{1/4}$

The amount (in grams) of a certain radioactive material present after t days is given by $A(t) = 800e^{-.04t}$. Find $A(t)$ for the following values of t.

13. $t = 0$

14. $t = 5$

15. $t = 10$

How much would \$1200 *amount to at* 10% *interest compounded continuously for the following number of years?*

16. 1 yr

17. 2 yr

18. 10 yr

19. Historically, the consumption of electricity has increased at a continuous rate of 6% per year. If it continued to increase at this rate, find the number of years before exactly twice as much electricity would be needed.

Multiplying both sides of equation (4) by $-1/17$ gives the equivalent system

$$\begin{aligned} 3x - 4y &= 1 \\ y &= 2. \end{aligned}$$

Now substitute 2 for y in equation (1).

$$\begin{aligned} 3x - 4(2) &= 1 \\ 3x - 8 &= 1 \\ 3x &= 9 \\ x &= 3 \end{aligned}$$

The solution of the original system is (3, 2). The graphs of the equations of the system in Figure 1 confirm that (3, 2) satisfies both equations of the system.

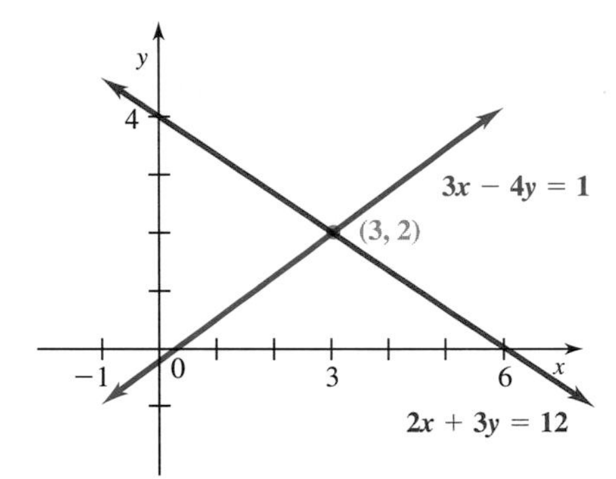

Figure 1

EXAMPLE 3 Solve the system

$$\begin{aligned} 3x - 2y &= 4 &\quad (1) \\ -6x + 4y &= 7. &\quad (2) \end{aligned}$$

Multiply both sides of equation (1) by 2 and add it to equation (2).

$$\begin{aligned} 6x - 4y &= 8 \\ -6x + 4y &= 7 \\ \hline 0 &= 15 \end{aligned}$$

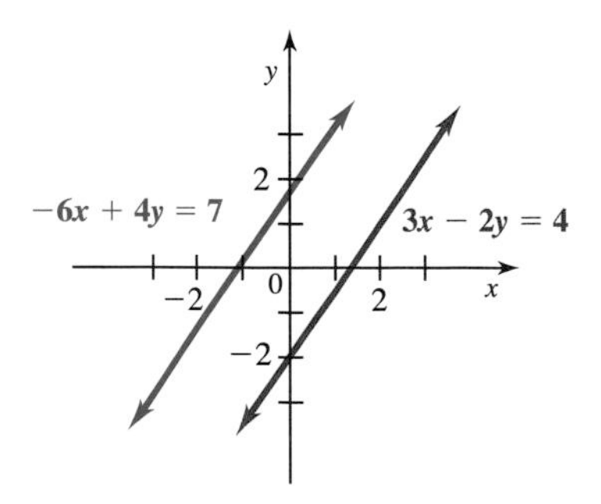

Figure 2

The new equivalent system is

$$\begin{aligned} 3x - 2y &= 4 \\ 0 &= 15. \end{aligned}$$

Since $0 = 15$ is never true, the system has no solution. As suggested by Figure 2, this means that the graphs of the equations of the system never intersect (the lines are parallel). The system has no solution.

A system of equations with no solution is **inconsistent**, and the graphs of an inconsistent linear system of two equations in two variables are parallel lines. If the two equations of a system of two variables are multiples of each other, the equations are said to be **dependent equations**, and their graphs will coincide as in Figure 3. In this case the system has infinitely many solutions.

EXAMPLE 4 Solve the system

$$8x - 2y = -4 \quad (1)$$
$$-4x + y = 2 \quad (2)$$

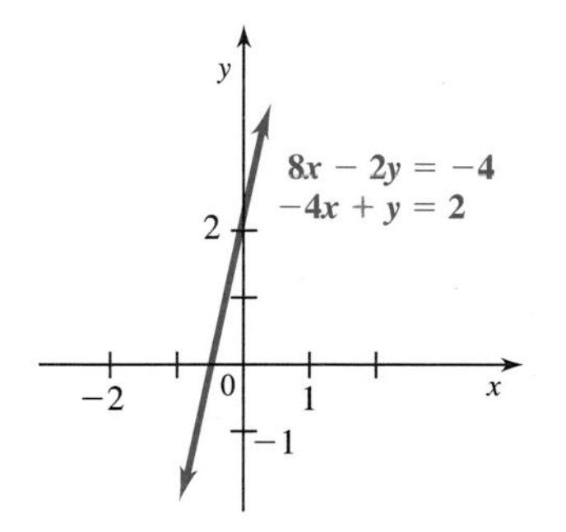

Figure 3

(graphed in Figure 3) algebraically.

Multiply both sides of equation (1) by $1/2$, and add the result to equation (2), to get the equivalent system

$$8x - 2y = -4$$
$$0 = 0.$$

The second equation, $0 = 0$, is always true, which indicates that the equations of the original system are equivalent. (With this system, the second transformation can be used to change either equation into the other.) Any ordered pair (x, y) that satisfies either equation will satisfy the system. From equation (2),

$$-4x + y = 2,$$

or

$$y = 2 + 4x.$$

The solution of the system can be written in the form of a set of ordered pairs $(x, 2 + 4x)$, for any real number x. Typical ordered pairs in the solution set are $(0, 2 + 4 \cdot 0) = (0, 2)$, $(-4, 2 + 4(-4)) = (-4, -14)$, $(3, 14)$, and $(7, 30)$. As shown in Figure 3, both equations of the original system lead to the same straight-line graph. ■

Applications of mathematics often require the solution of a system of equations. A typical application is shown in the next example.

EXAMPLE 5 An animal feed is made from two ingredients: corn and soybeans. One unit of each ingredient provides units of protein and fat as shown in the table below. How many units of each ingredient should be used to make a feed that contains 7 units of protein and 9 units of fat?

	Protein	*Fat*
Corn	.25	.4
Soybeans	.4	.2

Let x represent the number of units of corn and y the number of units of soybeans required. Since the total amount of protein is to be 7 units,

$$.25x + .4y = 7.$$

For the 9 units of fat, $.4x + .2y = 9.$

Multiply the first equation on both sides by 100 and the second equation by 10 to get the system

$$\begin{aligned} 25x + 40y &= 700 \\ 4x + 2y &= 90. \end{aligned}$$

Now use the elimination method to solve this system. Replace the first equation by the sum of the first equation and -20 times the second equation to eliminate y from the first equation.

$$\begin{array}{rcrcr} 25x & + & 40y & = & 700 \\ -80x & - & 40y & = & -1800 \\ \hline -55x & & & = & -1100 \end{array}$$

The result is the system

$$\begin{aligned} -55x &= -1100 \\ 4x + 2y &= 90. \end{aligned}$$

Multiplying both sides of the first equation by $-1/55$ gives the equivalent system

$$\begin{aligned} x &= 20 \\ 4x + 2y &= 90. \end{aligned}$$

Now substitute 20 for x in the second equation.

$$\begin{aligned} 4(20) + 2y &= 90 \\ 2y &= 10 \\ y &= 5 \end{aligned}$$

The solution of the original system is (20, 5). The feed should contain 20 units of corn and 5 units of soybeans to meet the given requirements. ■

Transformations can also be used to solve a system of three linear equations in three variables. In that case, first eliminate a variable from any two of the equations. Then eliminate the *same variable* from a different pair of equations. Eliminate a second variable using the resulting two equations in two variables to get an equation with just one variable whose value can now be determined. Find the values of the remaining variables by substitution.

EXAMPLE 6 Solve the system

$$\begin{aligned} 3x + 9y + 6z &= 3 && \textbf{(1)} \\ 2x + y - z &= 2 && \textbf{(2)} \\ x + y + z &= 2. && \textbf{(3)} \end{aligned}$$

To begin, eliminate z by simply adding equations (2) and (3) to get

$$3x + 2y = 4. \qquad \textbf{(4)}$$

To eliminate z from another pair of equations, multiply both sides of equation (2) by 6 and add the result to equation (1).

$$\begin{array}{rl} 3x + 9y + 6z &= 3 \\ 12x + 6y - 6z &= 12 \\ \hline 15x + 15y &= 15 \end{array} \qquad \textbf{(5)}$$

To eliminate x from equations (4) and (5), multiply both sides of equation (4) by -5 and add the result to equation (5). Solve the new equation for y.

$$\begin{array}{rl} -15x - 10y &= -20 \\ 15x + 15y &= 15 \\ \hline 5y &= -5 \\ y &= -1 \end{array}$$

Using $y = -1$, find x from equation (4) by substitution.

$$\begin{aligned} 3x + 2(-1) &= 4 \\ x &= 2 \end{aligned}$$

Substitute 2 for x and -1 for y in equation (3) to find z.

$$\begin{aligned} 2 + (-1) + z &= 2 \\ z &= 1 \end{aligned}$$

Verify that the **ordered triple** $(2, -1, 1)$ satisfies all three equations. ■

Nonlinear Systems A system of equations in which at least one equation is *not* linear is called a **nonlinear system**. The substitution method works well for solving many such systems, particularly when one of the equations is linear, as in the next example.

EXAMPLE 7 Solve the system

$$x^2 - y = 4 \qquad \textbf{(1)}$$
$$x + y = -2. \qquad \textbf{(2)}$$

When one of the equations in a nonlinear system is linear, it is usually best to begin by solving the linear equation for either variable. With this system, begin by solving equation (2) for y, giving

$$y = -2 - x.$$

Now substitute this result for y in equation (1) to get

$$\begin{aligned} x^2 - (\mathbf{-2 - x}) &= 4 \\ x^2 + 2 + x &= 4 \\ x^2 + x - 2 &= 0 \\ (x + 2)(x - 1) &= 0 \\ x = -2 \quad \text{or} \quad x &= 1. \end{aligned}$$

Substituting -2 for x in equation (2) gives $y = 0$. Also, if $x = 1$, then $y = -3$. The solutions of the given system are $(-2, 0)$ and $(1, -3)$. A graph of the system is shown in Figure 4.

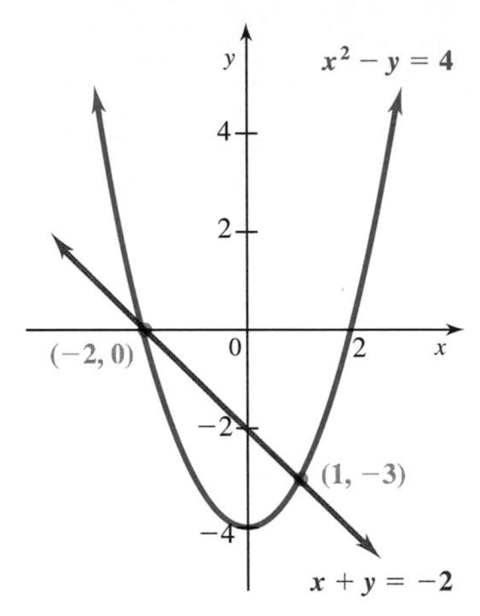

Figure 4

CAUTION If we had solved for x in equation (2) to begin the solution, we would find $y = 0$ or $y = -3$. Substituting $y = 0$ into equation (1) gives $x^2 = 4$, so $x = 2$ or $x = -2$, leading to the ordered pairs $(2, 0)$ and $(-2, 0)$. The ordered pair $(2, 0)$ does not satisfy equation (2), however. This shows the *necessity* of checking by substituting all potential solutions into each equation of the system. ■

Nonlinear systems where both variables are squared in both equations are best solved by elimination, as shown in the next example.

For Graphers

To graph this system, you must split both equations into two functions, so four functions will have to be graphed.

EXAMPLE 8 Solve the system

$$x^2 + y^2 = 4 \quad (1)$$
$$2x^2 - y^2 = 8. \quad (2)$$

Adding equation (1) to equation (2) (to eliminate y) gives the new system

$$x^2 + y^2 = 4$$
$$3x^2 = 12. \quad (3)$$

Solve equation (3) for x.

$$x^2 = 4$$
$$x = 2 \quad \text{or} \quad x = -2$$

Find y by substituting back into equation (1). If $x = 2$, then $y = 0$, and if $x = -2$, then $y = 0$. The solutions of the given system are $(2, 0)$ and $(-2, 0)$. See Figure 5.

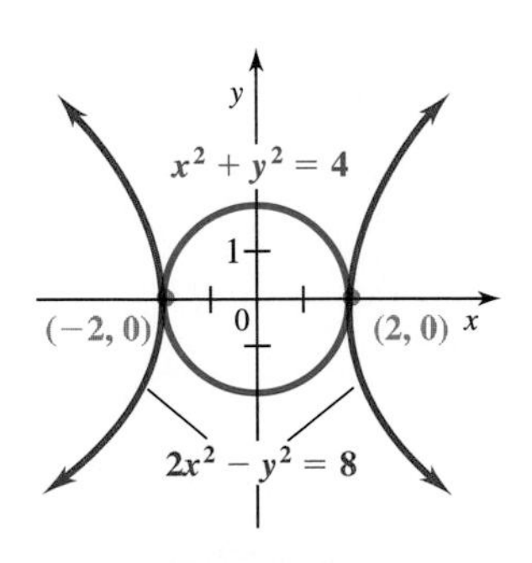

Figure 5

NOTE The elimination method works with the system in Example 8 since the system can be thought of as a system of linear equations where the variables are x^2 and y^2. In other words, the system is **linear in x^2 and y^2**. To see this, substitute u for x^2 and v for y^2. The resulting system is linear in u and v. ■

6.1 Exercises

Solve each of the following systems by the substitution method.

1. $x - 5y = 8$
$x = 6y$

2. $8x - 10y = -22$
$3x + y = 6$

3. $6x - y = 5$
$y = 11x$

4. $4x - 5y = -11$
$2x + y = 5$

5. $7x - y = -10$
$3y - x = 10$

6. $4x + 5y = 7$
$9y = 31 + 2x$

7. $-2x = 6y + 18$
$-29 = 5y - 3x$

8. $3x - 7y = 15$
$3x + 7y = 15$

9. $3y = 5x + 6$
$x + y = 2$

Solve each system by elimination.

10. $4x + 2y = 6$
$5x - 2y = 12$

11. $2x - y = 8$
$3x + 2y = 5$

12. $3x - y = -4$
$x + 3y = 12$

13. $2x - 3y = -7$
$5x + 4y = 17$

14. $4x + 3y = -1$
$2x + 5y = 3$

15. $5x + 7y = 6$
$10x - 3y = 46$

16. $12x - 5y = 9$
$3x - 8y = -18$

17. $6x + 7y = -2$
$7x - 6y = 26$

18. $\frac{x}{2} + \frac{y}{3} = 4$
$\frac{3x}{2} + \frac{3y}{2} = 15$

19. $\frac{x}{5} + 3y = 46$
$2x - \frac{y}{5} = 7$

20. $\frac{x}{3} + \frac{2y}{5} = 4$
$x + y = 11$

21. $\frac{3x}{2} - \frac{y}{3} = 5$
$\frac{5x}{2} + \frac{2y}{3} = 12$

22. $\frac{3x}{2} + \frac{y}{2} = -2$
$\frac{x}{2} + \frac{y}{2} = 0$

23. $\frac{2x - 1}{3} + \frac{y + 2}{4} = 4$
$\frac{x + 3}{2} - \frac{x - y}{3} = 3$

24. $\frac{x + 6}{5} + \frac{2y - x}{10} = 1$
$\frac{x + 2}{4} + \frac{3y + 2}{5} = -3$

25. $-2x + 5y + 3z = -6$
$4x - y - 2z = -6$
$3x + 4y - 2z = -14$

26. $-3x + y - z = -13$
$x - y - z = 3$
$x + 2y - z = -3$

27. $x + y + z = 2$
$2x - y - z = 1$
$3x - y + z = 0$

28. $x + y + z = 2$
$2x + y - z = 5$
$x - y + z = -2$

29. $2x + y + z = 5$
$-x + z = 2$
$3x - y + z = 6$

30. $x + 3y + 4z = 11$
$2x - 3y + 2z = 13$
$3x - z = 3$

31. Consider the linear equation $2x + 3y = 6$. Find a second linear equation for which the system of two linear equations will have the following: **(a)** exactly one solution; **(b)** no solution; **(c)** infinitely many solutions.

32. Consider the linear equation $ax + by = c$, where $a \neq 0$ and $b \neq 0$. Find a second linear equation for which the system of two linear equations will have **(a)** no solution; **(b)** infinitely many solutions.

33. Suppose that one of the three allowable transformations of a systm of equations has been performed. Is there always an allowable transformation that will convert the new system back to the original system? Explain.

Write a system of linear equations for each of the following, and then use the system to solve the problem.

34. At the local drugstore, 2 candy bars and 5 lollipops cost \$1.92, while 3 candy bars and 7 lollipops cost \$2.78. Find the cost of a candy bar and the cost of a lollipop.

35. At the Sharp Ranch, 6 goats and 5 sheep cost \$305, while 2 goats and 9 sheep cost \$285. Find the cost of a goat and the cost of a sheep.

36. The perimeter of a rectangle is 42 cm. The longer side is 7 cm longer than the shorter side. Find the length of the longer side.

37. During summer vacation Hector and Ann earned a total of \$1088. Hector worked 8 days fewer than Ann and earned \$2 per day less. Find the number of days he worked and the daily wage he made if the total number of days worked by both was 72.

38. Chuck Sullivan won \$100,000 in a lottery. He invested part of the money at 5% and part at 6%. His total annual income from the two investments is \$5,500. How much does he have invested at each rate?

39. Ms. Caminiti has some money invested at 4% and three times as much invested at 6%. Her total annual income from the two investments is \$1100. How much is invested at each rate?

40. A cash drawer contains only fives and twenties. There are eight more fives than twenties. The total value of the money is \$215. How many of each type of bill are there?

41. Thirty liters of a 50% alcohol solution are to be made by mixing a 70% solution and a 20% solution. How many liters of each solution should be used?

42. A merchant wishes to make 100 lb of a coffee blend that can be sold for \$4 per pound. This blend is to be made by mixing coffee worth \$6 per pound with coffee worth \$3 per pound. How many pounds of each will be needed?

43. How many gallons of full-fat milk (4.5% butterfat) must be mixed with 250 gal of skim milk (0% butterfat) to get lowfat milk (2% butterfat)?

44. Two trains leave towns 192 km apart, traveling toward one another. One train travels 40 km/hr faster than the other. They pass one another 2 hr later. What is the speed of each train?

45. The perimeter of a triangle is 33 cm. The longest side is 3 cm longer than the medium side. The medium side is twice as long as the shortest side. Find the length of each side of the triangle.

46. Carrie O'Day invests $30,000 in lottery winnings in three ways. With part of the money, she buys a mutual fund paying 9% per year. She uses the second part, $2000 more than the first, to buy utility bonds paying 10% per year. She invests the rest in a tax-free 5% bond. The first year her investments bring a return of $2500. How much is invested at each rate?

The position of a particle moving in a straight line is given by $s = at^2 + bt + c$, where t is time in seconds and a, b, and c are real numbers.

47. If $s(0) = 5$, $s(1) = 23$, and $s(2) = 37$, find $s(8)$.

48. If $s(0) = -10$, $s(1) = 6$, and $s(2) = 30$, find $s(10)$.

49. The graph of each equation in a system of three linear equations in two variables corresponds to a line. Describe geometrically what happens when the system has a unique solution, no solution, or infinitely many solutions.

Solve each system by the substitution method. In Exercises 50–55, graph the system.

50. $x^2 = y - 1$, $y = 3x + 5$

51. $2x^2 = 3y + 23$, $y = 2x - 5$

52. $x^2 + y^2 = 5$, $-3x + 4y = 2$

53. $y = x^2$, $x + y = 2$

54. $x^2 + y^2 = 45$, $x + y = -3$

55. $y = x^2 + 6x + 9$, $x + 2y = -2$

56. $x^2 - y = -1$, $3x = y - 11$

57. $x = y - 2$, $y^2 = 3x^2 + 4$

58. $y = -x^2 + 2$, $x - y = 0$

59. $y = x^2 - 2x + 1$, $x - 3y = -1$

60. $x^2 - 3y^2 = 22$, $x + 3y = 2$

61. $3x^2 + 2y^2 = 5$, $x - y = 2$

Solve the following systems by any method.

62. $x^2 + y^2 = 32$, $x^2 - y^2 = 0$

63. $x^2 + y^2 = 25$, $2x^2 - y^2 = 2$

64. $x^2 + y^2 = 4$, $2x^2 - 3y^2 = -12$

65. $5x^2 - y^2 = 0$, $3x^2 + 4y^2 = 0$

66. $3x^2 + 2y^2 = 5$, $-4x^2 + 3y^2 = -1$

67. $2x^2 + 2y^2 = 20$, $3x^2 + 3y^2 = 30$

68. $3x^2 + 5y^2 = 17$, $2x^2 - 3y^2 = 5$

69. $x^2 + y^2 = 4$, $4x^2 + 4y^2 = 15$

70. $xy = -4$, $2x + y = -7$

71. $xy = 6$, $x + y = 5$

72. $xy = 8$, $3x + 2y = -16$

73. $x + 3y = -6$, $xy = 3$

74. $xy = -15$, $4x + 3y = 3$

75. $2xy + 1 = 0$, $x + 16y = 2$

76. $-5xy + 2 = 0$, $x - 15y = 5$

Use any method to solve the following problems.

77. Find two numbers whose sum is 17 and whose product is 42.

78. Find two numbers whose sum is 10 and whose squares differ by 20.

79. Find two numbers whose squares have a sum of 100 and a difference of 28.

80. The longest side of a right triangle is 13 m in length. One of the other sides is 7 m longer than the shortest side. Find the length of the two shorter sides of the triangle.

Linear Programming An important application of mathematics to business and social science is called linear programming. **Linear programming** is used to find such things as minimum cost and maximum profit. It was developed to solve problems in allocating supplies for the U.S. Air Force during the Second World War. The basic ideas of this technique will be explained with an example.

EXAMPLE 4 The Smith Company makes two products, tape decks and amplifiers. Each tape deck gives a profit of $30, while each amplifier produces $70. The company must manufacture at least ten tape decks per day to satisfy one of its customers, but no more than fifty because of production problems. Also, the number of amplifiers produced cannot exceed sixty per day. As a further requirement, the number of tape decks cannot exceed the number of amplifiers. How many of each should the company manufacture in order to obtain the maximum profit?

To begin, translate the statement of the problem into symbols.

Let $x =$ number of tape decks to be produced daily,

$y =$ number of amplifiers to be produced daily.

According to the statement of the problem given above, the company must produce at least ten tape decks (ten or more), so that

$$x \geq 10.$$

The requirement that no more than 50 tape decks may be produced means that

$$x \leq 50.$$

Since no more than 60 amplifiers may be made in one day,

$$y \leq 60.$$

The fact that the number of tape decks may not exceed the number of amplifiers translates as

$$x \leq y.$$

The number of tape decks and of amplifiers cannot be negative, so

$$x \geq 0 \quad \text{and} \quad y \geq 0.$$

Listing all the restrictions, or **constraints**, that are placed on production gives

$$x \geq 10, \quad x \leq 50, \quad y \leq 60, \quad x \leq y, \quad x \geq 0, \quad y \geq 0.$$

To find the maximum possible profit that the company can make, subject to these constraints, begin by sketching the graph of each constraint. The only feasible values of x and y are those that satisfy all constraints—that is, the values that lie in the intersection of the graphs of the constraints. The intersection is shown in Figure 12. Any point lying inside the shaded region or on the boundary in Figure 12 satisfies the restrictions as to the number of tape decks and amplifiers that may be produced. (For practical purposes, however, only points with integer coefficients are useful.) This region is called the **region of feasible solutions**.

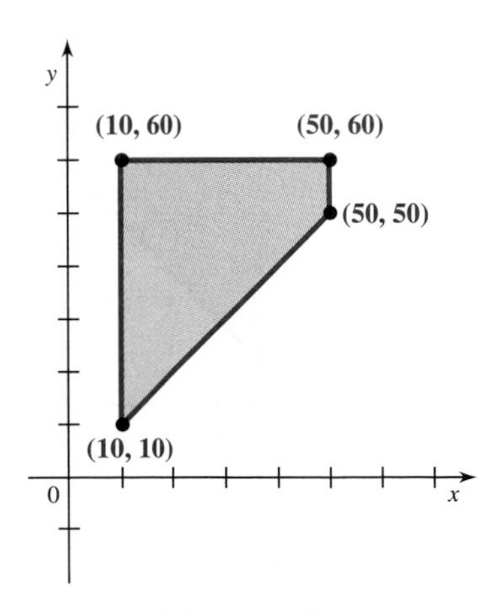

Figure 12

Since each tape deck gives a profit of \$30, the daily profit from the production of x tape decks is $30x$ dollars. Also, the profit from the production of y amplifiers will be $70y$ dollars per day. The total daily profit is thus

$$\text{Profit} = 30x + 70y.$$

The problem of the Smith Company may now be stated as follows: Find values of x and y in the shaded region of Figure 12 that will produce the maximum possible value of $30x + 70y$. To locate the point (x, y) that gives the maximum profit, add to the graph of Figure 12 lines corresponding to profits of \$0, \$1000, \$3000, and \$7000.

$$30x + 70y = 0$$
$$30x + 70y = 1000$$
$$30x + 70y = 3000$$
$$30x + 70y = 7000$$

For instance, each point on the line $30x + 70y = 3000$ corresponds to production values that yield a profit of \$3000. Figure 13 shows the region of feasible solutions together with these lines. The lines are parallel, and the higher the line, the higher the profit. The line $30x + 70y = 7000$ has the highest profit but does not contain any points of the region of feasible solutions. To find the feasible solution of greatest profit, lower the line $30x + 70y = 7000$ until it contains a feasible solution—that is, until it just touches the region of feasible solutions. This occurs at point A, a **vertex** (or corner point) of the region. See Figure 14. Since the coordinates of this point are (50, 60), the maximum profit is obtained when fifty tape decks and sixty amplifiers are produced each day. The maximum profit will be $30(50) + 70(60) = 5700$ dollars per day.

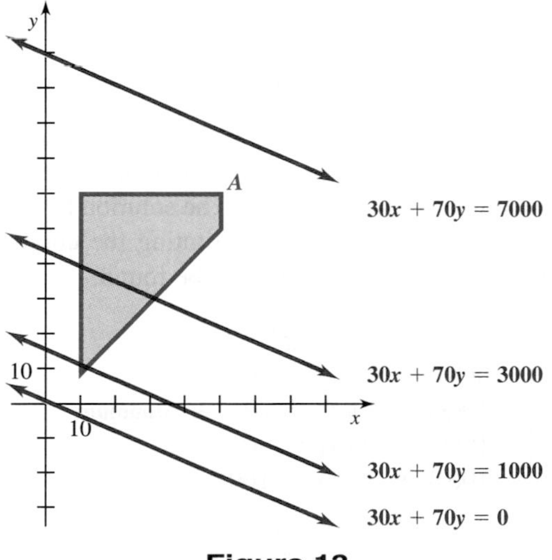

Figure 13

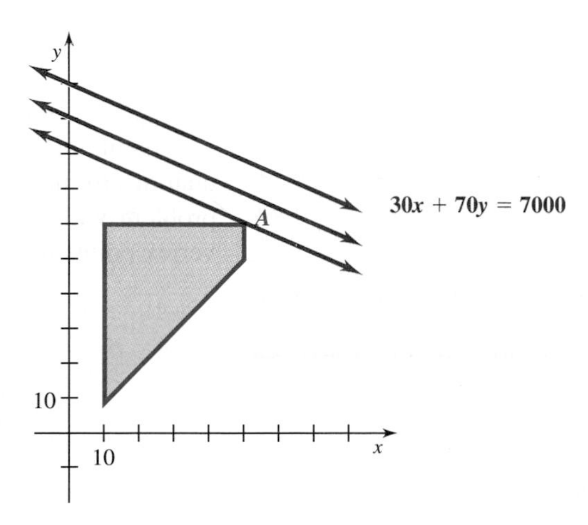

Figure 14

29. Find a system of linear inequalities for which the graph is the region in the first quadrant between the pair of lines $x + 2y - 8 = 0$ and $x + 2y = 12$.

30. Find a linear inequality in two variables whose graph does not intersect the graph of $y \geq 3x + 5$.

31. Find an inequality in two variables whose graph does not intersect the graph of $y \leq |x|$.

The graphs in Exercises 32–35 show regions of feasible solutions. Find the maximum and minimum values of the given expressions.

32. $3x + 5y$

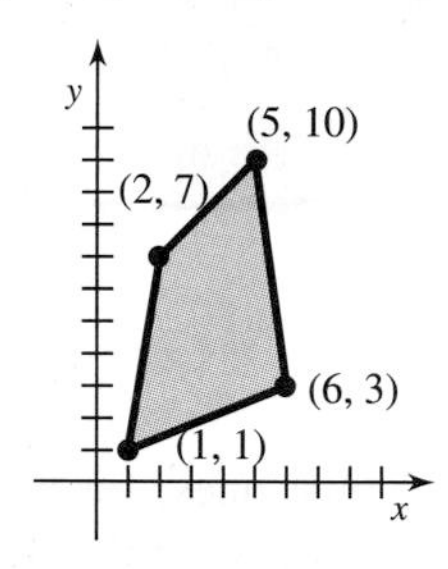

33. $6x + y$

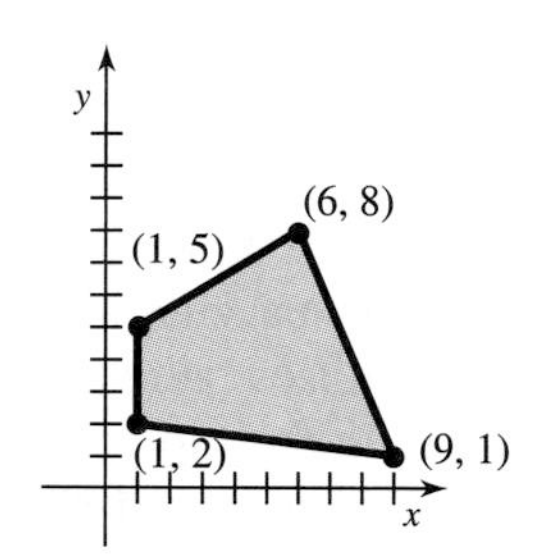

34. $40x + 75y$

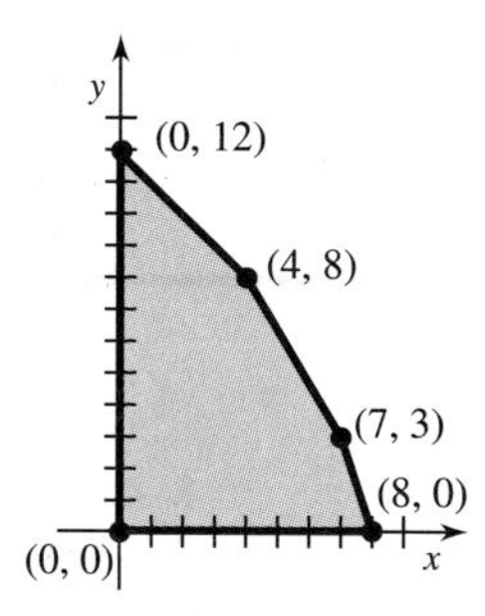

35. $35x + 125y$

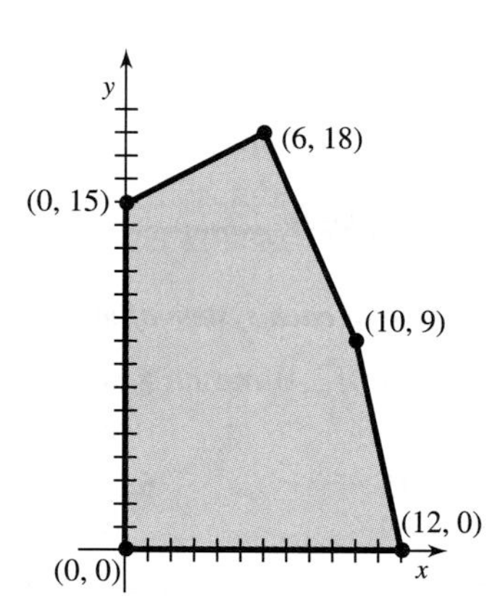

For Exercises 36–39, find the maximum and minimum values of the given expressions over the feasible set shown below.

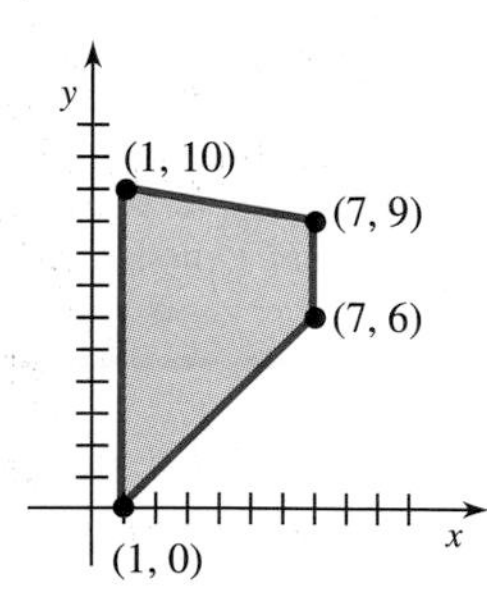

36. $3x + 5y$ **37.** $5x + 5y$ **38.** $10y$ **39.** $3x - y$

In Exercises 40–43, use graphical methods to solve each problem.

40. Find $x \geq 0$ and $y \geq 0$ such that

$$2x + 3y \leq 6$$
$$4x + y \leq 6,$$

and $5x + 2y$ is maximized.

41. Find $x \geq 0$ and $y \geq 0$ such that

$$x + y \leq 10$$
$$5x + 2y \geq 20$$
$$2y \geq x,$$

and $x + 3y$ is minimized.

42. Find $x \geq 2$ and $y \geq 5$ such that

$$3x - y \geq 12$$
$$x + y \leq 15,$$

and $2x + y$ is minimized.

43. Find $x \geq 10$ and $y \geq 20$ such that

$$2x + 3y \leq 100$$
$$5x + 4y \leq 200,$$

and $x + 3y$ is maximized.

44. Why is there no maximum value for the cost function in Example 5? In general, what sort of feasible set has no maximum value for a linear expression $ax + by$ where $a \geq 0$ and $b \geq 0$?

Write a system of inequalities for each of the following problems and then graph the region of feasible solutions of the system.

45. Ms. Oliveras was given the following advice. She should supplement her daily diet with at least 6000 USP units of vitamin A, at least 195 mg of vitamin C, and at least 600 USP units of vitamin D. Ms. Oliveras finds that Mason's Pharmacy carries Brand X and Brand Y vitamins. Each Brand X pill contains 3000 USP units of A, 45 mg of C, and 75 USP units of D, while the Brand Y pills contain 1000 USP units of A, 50 mg of C, and 200 USP units of D.

46. The California Almond Growers have 2400 boxes of almonds to be shipped from their plant in Sacramento to Des Moines and San Antonio. The Des Moines market needs at least 1000 boxes, while the San Antonio market must have at least 800 boxes.

Solve each of the following linear programming problems.

47. The manufacturing process requires that oil refineries must manufacture at least two barrels of gasoline for every barrel of fuel oil. To meet the winter demand for fuel oil, at least 3 million barrels per day must be produced. The demand for gasoline is no more than 6.4 million barrels per day. If the price of gasoline is \$1.90 and the price of fuel oil is \$1.50 per gallon, how much of each should be produced to maximize revenue?

48. Theo, who is dieting, requires two food supplements, I and II. He can get these supplements from two different products, A and B, as shown in the following table.

Supplement (g/serving)		*I*	*II*
Product	*A*	3	2
	B	2	4

Theo's physician has recommended that he include at least 15 g of each supplement in his daily diet. If product A costs 25¢ per serving and product B costs 40¢ per serving, how can he satisfy his requirements most economically?

49. Use your own words to describe what a linear programming problem is and how it can be solved graphically.

51. At rush hours, substantial traffic congestion is encountered at the traffic intersections shown in the figure. (All streets are one-way.) The city wishes to improve the signals at these corners so as to speed the flow of traffic. The traffic engineers first gather data. As the figure shows, 700 cars per hour come down M Street to intersection A, and 300 cars per hour come to intersection A on 10th Street. A total of x_1 of these cars leave A on M Street, while x_4 cars leave A on 10th Street. The number of cars entering A must equal the number leaving, so that

$$x_1 + x_4 = 700 + 300$$

or

$$x_1 + x_4 = 1000.$$

For intersection B, x_1 cars enter B on M Street, and x_2 cars enter B on 11th Street. The figure shows that 900 cars leave B on 11th, while 200 leave on M. We have

$$x_1 + x_2 = 900 + 200$$

$$x_1 + x_2 = 1100.$$

At intersection C, 400 cars enter on N Street and 300 on 11th Street, while x_2 leave on 11th Street and x_3 leave on N street. This gives

$$x_2 + x_3 = 400 + 300$$

$$x_2 + x_3 = 700.$$

Finally, intersection D has x_3 cars entering on N and x_4 entering on 10th. There are 400 cars leaving D on 10th and 200 leaving on N.

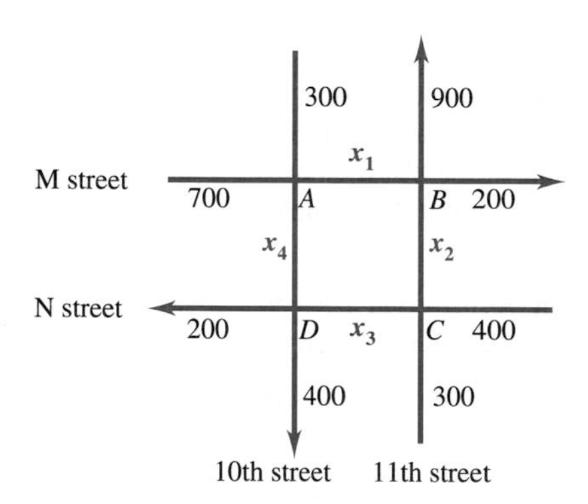

(a) Set up an equation for intersection D.

(b) Use the four equations to set up an augmented matrix, and then use the Gaussian method to reduce it to triangular form.

(c) Since you got a row of all zeros, the system of equations does not have a unique solution. Write three equations, corresponding to the three nonzero rows of the matrix. Solve each of the equations for x_4.

(d) One of your equations should have been $x_4 = 1000 - x_1$. What is the largest possible value of x_1 so that x_4 is not negative?

(e) Your second equation should have been $x_4 = x_2 - 100$. Find the smallest possible value of x_2 so that x_4 is not negative.

(f) Find the largest possible values of x_3 and x_4 so that neither variable is negative.

(g) Use the results of (a)–(f) to give a solution for the problem in which all the equations are satisfied and all variables are nonnegative. Is the solution unique?

For each of the following equations, determine the constants A and B that make the equation an identity. (Hint: Combine terms on the right and set coefficients of corresponding terms in the numerators equal.)

52. $\dfrac{1}{(x-1)(x+1)} = \dfrac{A}{x-1} + \dfrac{B}{x+1}$

53. $\dfrac{x+4}{x^2} = \dfrac{A}{x} + \dfrac{B}{x^2}$

54. $\dfrac{x}{(x-a)(x+a)} = \dfrac{A}{x-a} + \dfrac{B}{x+a}$

55. $\dfrac{2x}{(x+2)(x-1)} = \dfrac{A}{x+2} + \dfrac{B}{x-1}$

56. Describe the Gaussian reduction method in your own words.

Graphing Utility Problems

If your graphing utility is capable of performing row operations, use the utility to solve the following systems of linear equations by the Gaussian reduction method.

57. $\begin{aligned} \sqrt{2}x - 5y &= .3 \\ -.7x + 3y &= 3/2 \end{aligned}$

58. $\begin{aligned} 5.1x + \sqrt{3}y &= \sqrt{2} \\ 9x - 2.2y &= 5 \end{aligned}$

59. $\begin{aligned} .3x + 2.7y - \sqrt{2}z &= 3 \\ \sqrt{7}x - 20y + 12z &= -2 \\ 4x + \sqrt{3}y - 1.2z &= 3/4 \end{aligned}$

60. $\begin{aligned} \sqrt{5}x - 1.2y + z &= -3 \\ (1/2)x - 3y + 4z &= 4/3 \\ 4x + 7y - 9z &= \sqrt{2} \end{aligned}$

6.4 Properties of Matrices

The use of matrix notation in solving a system of linear equations was shown in the previous section. In this section, the algebraic properties of matrices are discussed.

It is customary to use capital letters to name matrices. Also, subscript notation is often used to name the elements of a matrix, as in the following matrix A.

$$A = \begin{bmatrix} a_{11} & a_{12} & a_{13} & \cdots & a_{1n} \\ a_{21} & a_{22} & a_{23} & \cdots & a_{2n} \\ a_{31} & a_{32} & a_{33} & \cdots & a_{3n} \\ \vdots & \vdots & \vdots & & \vdots \\ a_{m1} & a_{m2} & a_{m3} & \cdots & a_{mn} \end{bmatrix}$$

With this notation, the first-row, first-column element is a_{11} (read "a-sub-one-one"); the second-row, third-column element is a_{23}; and in general, the ith-row, jth-column element is a_{ij}.

Matrices are classified by their size—that is, by the number of rows and columns that they contain. For example, the matrix

$$\begin{bmatrix} 2 & 7 & -5 \\ 3 & -6 & 0 \end{bmatrix}$$

NOTE Examples 5 and 6 showed that the order in which two matrices are to be multiplied may determine whether their product can be found. Example 7 shows that even when both AB and BA can be found, they may not be equal. In general, for matrices A and B, $AB \neq BA$, so *matrix multiplication is not commutative.* ■

Matrix multiplication does, however, satisfy the associative and distributive properties.

Properties of Matrix Multiplication

If A, B, and C are matrices such that all the following products and sums exist, then

$$(AB)C = A(BC)$$
$$A(B + C) = AB + AC$$
$$(B + C)A = BA + CA.$$

For proofs of these results for the special cases when A, B, and C are square matrices, see Exercises 42 and 43. The identity and inverse properties for matrix multiplication are discussed in a later section of this chapter.

6.4 Exercises

Find the values of the variables in each of the following.

1. $\begin{bmatrix} w & x \\ y & z \end{bmatrix} = \begin{bmatrix} 3 & 2 \\ -1 & 4 \end{bmatrix}$

2. $\begin{bmatrix} 0 & 5 & x \\ -1 & 3 & y+2 \\ 4 & 1 & z \end{bmatrix} = \begin{bmatrix} 0 & w+3 & 6 \\ -1 & 3 & 0 \\ 4 & 1 & 8 \end{bmatrix}$

3. $\begin{bmatrix} 2 & 5 & 6 \\ 1 & m & n \end{bmatrix} = \begin{bmatrix} z & y & w \\ 1 & 8 & -2 \end{bmatrix}$

4. $\begin{bmatrix} -7+z & 4r & 8s \\ 6p & 2 & 5 \end{bmatrix} + \begin{bmatrix} -9 & 8r & 3 \\ 2 & 5 & 4 \end{bmatrix} = \begin{bmatrix} 2 & 36 & 27 \\ 20 & 7 & 12a \end{bmatrix}$

5. $\begin{bmatrix} a+2 & 3z+1 & 5m \\ 8k & 0 & 3 \end{bmatrix} + \begin{bmatrix} 3a & 2z & 5m \\ 2k & 5 & 6 \end{bmatrix} = \begin{bmatrix} 10 & -14 & 80 \\ 10 & 5 & 9 \end{bmatrix}$

In each of the following, determine whether the two matrices are equal.

6. $[1 \quad 2]$, $\begin{bmatrix} 1 \\ 2 \end{bmatrix}$

7. $\begin{bmatrix} 1 & 2 \\ 3 & 4 \end{bmatrix}$, $\begin{bmatrix} 1 & 2 & 0 \\ 3 & 4 & 0 \end{bmatrix}$

Perform each of the operations in Exercises 8–13, whenever possible.

8. $\begin{bmatrix} 6 & -9 & 2 \\ 4 & 1 & 3 \end{bmatrix} - \begin{bmatrix} -8 & 2 & 5 \\ 6 & -3 & 4 \end{bmatrix}$

9. $\begin{bmatrix} 9 & 4 \\ -8 & 2 \end{bmatrix} + \begin{bmatrix} -3 & 2 \\ -4 & 7 \end{bmatrix}$

10. $\begin{bmatrix} -6 & 8 \\ 0 & 0 \end{bmatrix} - \begin{bmatrix} 0 & 0 \\ -4 & -2 \end{bmatrix}$

11. $\begin{bmatrix} 1 & -4 \\ 2 & -3 \\ -8 & 4 \end{bmatrix} - \begin{bmatrix} -6 & 9 \\ -2 & 5 \\ -7 & -12 \end{bmatrix}$

12. $\begin{bmatrix} 3x + y & x - 2y & 2x \\ 5x & 3y & x + y \end{bmatrix} + \begin{bmatrix} 2x & 3y & 5x + y & x - y \\ 3x + 2y & x & 2x & 4y \end{bmatrix}$

13. $\begin{bmatrix} 4k - 8y \\ 6z - 3x \\ 2k + 5a \\ -4m + 2n \end{bmatrix} - \begin{bmatrix} 5k + 6y \\ 2z + 5x \\ 4k + 6a \\ 4m - 2n \end{bmatrix}$

14. A hardware chain does an inventory of a particular size of screw and finds that its Adelphi store has 100 flat-head and 150 round-head screws, its Beltsville store has 125 flat and 50 round, and its College Park store has 175 flat and 200 round. Write this information first as a 3×2 matrix and then as a 2×3 matrix.

15. At the grocery store, Miguel bought 4 quarts of milk, 2 loaves of bread, 4 potatoes, and an apple. Mary bought 2 quarts of milk, a loaf of bread, 5 potatoes, and 4 apples. Write this information first as a 2×4 matrix and then as a 4×2 matrix.

16. For any size matrix A, the zero matrix O of the same size is the *additive identity* such that $A + 0 = A$. For any 2×2 matrix A, find the 2×2 matrix I that is the *multiplicative identity*, that is, for which $A \cdot I = A$.

Let $A = \begin{bmatrix} -2 & 4 \\ 0 & 3 \end{bmatrix}$ *and* $B = \begin{bmatrix} -6 & 2 \\ 4 & 0 \end{bmatrix}$. *Find each of the following.*

17. $2A$

18. $-3B$

19. $2A - B$

20. $-2A + 4B$

21. $-A + \frac{1}{2}B$

22. $\frac{3}{4}A - B$

Find the matrix products in Exercises 23–34, whenever possible.

23. $\begin{bmatrix} 1 & 2 \\ 3 & 4 \end{bmatrix}\begin{bmatrix} -1 \\ 7 \end{bmatrix}$

24. $\begin{bmatrix} 3 & -4 & 1 \\ 5 & 0 & 2 \end{bmatrix}\begin{bmatrix} -1 \\ 4 \\ 2 \end{bmatrix}$

25. $\begin{bmatrix} 5 & 6 \\ 3 & 4 \end{bmatrix}\begin{bmatrix} 0 & 3 \\ -1 & 2 \end{bmatrix}$

26. $\begin{bmatrix} -2 & 1 & 3 \\ 7 & 0 & -1 \\ 0 & 2 & 1 \end{bmatrix}\begin{bmatrix} 0 & 1 & 1 \\ -1 & 2 & 0 \\ 3 & 1 & 4 \end{bmatrix}$

27. $\begin{bmatrix} -2 & 1 & 4 \\ 0 & 1 & 2 \end{bmatrix}\begin{bmatrix} -2 & 1 & 0 \\ 0 & -2 & 0 \\ 4 & 1 & 2 \end{bmatrix}$

28. $\begin{bmatrix} -1 & 0 & 0 \\ 2 & 1 & 4 \end{bmatrix}\begin{bmatrix} 4 & -2 & 5 \\ 0 & 1 & 4 \\ 2 & -9 & 0 \end{bmatrix}$

29. $\begin{bmatrix} -3 & 0 & 2 & 1 \\ 4 & 0 & 2 & 6 \end{bmatrix}\begin{bmatrix} -4 & 2 \\ 0 & 1 \end{bmatrix}$

30. $\begin{bmatrix} -1 & 2 & 4 & 1 \\ 0 & 2 & -3 & 5 \end{bmatrix}\begin{bmatrix} 1 & 2 & 4 \\ -2 & 5 & 1 \end{bmatrix}$

31. $[5 \quad -1 \quad 2]\begin{bmatrix} 2 \\ 1 \\ -1 \end{bmatrix}$

32. $[7 \quad 5 \quad 4 \quad -6]\begin{bmatrix} 1 \\ 0 \\ 1 \\ 0 \end{bmatrix}$

33. $\begin{bmatrix} 2 & 1 & -3 \\ 1 & 0 & 4 \end{bmatrix}\begin{bmatrix} 1 \\ 0 \\ 0 \end{bmatrix}$

34. $\begin{bmatrix} -6 \\ -1 \\ -2 \end{bmatrix}[3 \quad 0 \quad 1]$

35. The Bread Box, a small neighborhood bakery, sells four main items: sweet rolls, bread, cake, and pie. The amount of certain major ingredients (measured in cups except for eggs) required to make these items is given in matrix A.

$$A = \begin{array}{c} \\ \end{array}\begin{bmatrix} 1 & 4 & 1/4 & 1/4 & 1 \\ 0 & 3 & 0 & 1/4 & 0 \\ 4 & 3 & 2 & 1 & 1 \\ 0 & 1 & 0 & 1/3 & 0 \end{bmatrix}$$

	Eggs	Flour	Sugar	Shortening	Milk
Dozen rolls	1	4	1/4	1/4	1
Loaf of bread	0	3	0	1/4	0
Cake (1)	4	3	2	1	1
Pie (1)	0	1	0	1/3	0

The cost per cup or per egg (in cents) for each ingredient when purchased in large lots and in small lots is given by matrix B.

Cost

$$B = \begin{bmatrix} 5 & 5 \\ 8 & 10 \\ 10 & 12 \\ 12 & 15 \\ 5 & 6 \end{bmatrix}$$

Large lot	Small lot
5	5
8	10
10	12
12	15
5	6

(a) Use matrix multiplication to find a matrix representing the comparative costs per item under the two purchase options.

Suppose a day's orders consist of 20 dozen sweet rolls, 200 loaves of bread, 50 cakes, and 60 pies.

(b) Represent these orders as a 1×4 matrix and use matrix multiplication to write as a matrix the amount of each ingredient required to fill the day's orders.

(c) Use matrix multiplication to find a matrix representing the costs under the two purchase options to fill the day's orders.

36. In what ways is the matrix I like the real number 1? (See Exercise 16.)

For Exercises 37–51, let

$$A = \begin{bmatrix} a_{11} & a_{12} \\ a_{21} & a_{22} \end{bmatrix}, \quad B = \begin{bmatrix} b_{11} & b_{12} \\ b_{21} & b_{22} \end{bmatrix}, \quad \text{and} \quad C = \begin{bmatrix} c_{11} & c_{12} \\ c_{21} & c_{22} \end{bmatrix}$$

where all the elements are real numbers. Decide which of the following statements are true for these three matrices. If a statement is true, prove that it is true. If it is false, give a numerical example to show it is false.

37. $A + B = B + A$ (commutative property)

38. $A + (B + C) = (A + B) + C$ (associative property)

39. $A + B$ is a 2×2 matrix. (closure property)

40. There exists a matrix O such that $A + O = A$ and $O + A = A$. (identity property)

41. There exists a matrix $-A$ such that $A + (-A) = O$ and $-A + A = O$. (inverse property)

42. $(AB)C = A(BC)$ (associative property)

43. $A(B + C) = AB + AC$ (distributive property)

44. AB is a 2×2 matrix. (closure property)

45. $c(A + B) = cA + cB$ for any real number c.

46. $(c + d)A = cA + dA$ for any real numbers c and d.

47. $c(A)d = cd(A)$

48. $(cd)A = c(dA)$

49. $(A + B)(A - B) = A^2 - B^2$ (where $A^2 = AA$)

50. $(A + B)^2 = A^2 + 2AB + B^2$

51. If $AB = O$, then $A = O$ or $B = O$.

The **transpose**, A^T, *of a matrix A is found by exchanging the rows and columns of A. That is if* $A = \begin{bmatrix} a & b \\ c & d \end{bmatrix}$, *then* $A^T = \begin{bmatrix} a & c \\ b & d \end{bmatrix}$. *Show that each of the following equations are true for matrices A and B, where* $B = \begin{bmatrix} m & n \\ p & q \end{bmatrix}$.

52. $(A^T)^T = A$

53. $(A + B)^T = A^T + B^T$

54. $(AB)^T = B^T A^T$

Graphing Utility Problems

In Exercises 55–58, calculate the given expression, where

$$A = \begin{bmatrix} 1.3 & 17 & -.8 \\ 12 & 3/2 & 5.4 \\ .24 & -5 & 9 \end{bmatrix} \quad \text{and} \quad B = \begin{bmatrix} 45 & 7.2 & 4/5 \\ 9/8 & -43 & 2.2 \\ -11 & .33 & 3/4 \end{bmatrix}.$$

55. $A + B$

56. $3A$

57. $2A - 3B$

58. AB

A stochastic matrix is a square matrix for which the sum of the entries in each row is 1. *An example is* $A = \begin{bmatrix} .8 & .2 \\ .4 & .6 \end{bmatrix}$. *Use A in Exercises 59 and 60.*

59. Show that A^2 is also a stochastic matrix. (*Note:* $A^2 = AA$.)

60. Compute A^3, A^4, A^5, and A^6 by successively multiplying by A. (*Note:* The rows of these powers should get closer and closer to $[2/3 \quad 1/3]$.)

61. Repeat Exercises 59 and 60 for the matrix $\begin{bmatrix} .1 & .6 & .3 \\ .3 & .4 & .3 \\ .7 & .2 & .1 \end{bmatrix}$. What matrix do the rows get closer and closer to?

6.5 Determinants

Given any square matrix A, there is a unique real number associated with A—the *determinant* of A. In this section determinants are defined, and in the next section determinants are used to solve systems of linear equations. This technique was first published by the Swiss mathematician Gabriel Cramer (1704–1752) in 1750. Although Leibniz knew the technique in 1693 and Seki Kowa of Japan knew it in 1683, their work did not become known to others. By 1812, Cauchy (1789–1857), who introduced the word *determinant*, had proved that the determinant of the product of two matrices is the product of their determinants. Carl Gustave Jacobi (1804–1851) gave the convincing arguments that made determinants acceptable.

The determinant of a matrix A is written $|A|$. The determinant of a 2×2 matrix is defined as follows.

Definition of Determinant of a 2 × 2 Matrix

If $A = \begin{bmatrix} a_{11} & a_{12} \\ a_{21} & a_{22} \end{bmatrix}$, then $|A| = \begin{vmatrix} a_{11} & a_{12} \\ a_{21} & a_{22} \end{vmatrix} = a_{11}a_{22} - a_{21}a_{12}$.

NOTE Notice that matrices are enclosed with square brackets, while determinants are denoted with vertical bars. ■

EXAMPLE 1 Let $A = \begin{bmatrix} -3 & 4 \\ 6 & 8 \end{bmatrix}$. Find $|A|$.

Use the definition above.

$$|A| = \begin{vmatrix} -3 & 4 \\ 6 & 8 \end{vmatrix} = -3(8) - 6(4) = -48$$ ■

The determinant of a 3×3 matrix A is defined as follows.

Definition of Determinant of a 3 × 3 Matrix

If $A = \begin{bmatrix} a_{11} & a_{12} & a_{13} \\ a_{21} & a_{22} & a_{23} \\ a_{31} & a_{32} & a_{33} \end{bmatrix}$, then

$$|A| = \begin{vmatrix} a_{11} & a_{12} & a_{13} \\ a_{21} & a_{22} & a_{23} \\ a_{31} & a_{32} & a_{33} \end{vmatrix} = (a_{11}a_{22}a_{33} + a_{12}a_{23}a_{31} + a_{13}a_{21}a_{32}) \\ -(a_{31}a_{22}a_{13} + a_{32}a_{23}a_{11} + a_{33}a_{21}a_{12}).$$

The terms on the right side of the equation in the definition of $|A|$ can be rearranged to get

$$\begin{vmatrix} \mathbf{a_{11}} & a_{12} & a_{13} \\ \mathbf{a_{21}} & a_{22} & a_{23} \\ \mathbf{a_{31}} & a_{32} & a_{33} \end{vmatrix} = \mathbf{a_{11}}(a_{22}a_{33} - a_{32}a_{23}) - \mathbf{a_{21}}(a_{12}a_{33} - a_{32}a_{13}) \\ + \mathbf{a_{31}}(a_{12}a_{23} - a_{22}a_{13}).$$

Each of the quantities in parentheses above represents a determinant of a 2 × 2 matrix which is the part of the 3 × 3 matrix left when the row and column of the multiplier are eliminated as shown below.

$$a_{11}(a_{22}a_{33} - a_{32}a_{23}) \qquad \begin{bmatrix} a_{11} & a_{12} & a_{13} \\ a_{21} & a_{22} & a_{23} \\ a_{31} & a_{32} & a_{33} \end{bmatrix}$$

$$a_{21}(a_{12}a_{33} - a_{32}a_{13}) \qquad \begin{bmatrix} a_{11} & a_{12} & a_{13} \\ a_{21} & a_{22} & a_{23} \\ a_{31} & a_{32} & a_{33} \end{bmatrix}$$

$$a_{31}(a_{12}a_{23} - a_{22}a_{13}) \qquad \begin{bmatrix} a_{11} & a_{12} & a_{13} \\ a_{21} & a_{22} & a_{23} \\ a_{31} & a_{32} & a_{33} \end{bmatrix}$$

These determinants of 2 × 2 matrices are called **minors** of an element in the 3 × 3 matrix. The symbol M_{ij} represents the determinant of the matrix that results when row i and column j are eliminated. The following list gives some of the minors from the matrix above.

Element	*Minor*	*Element*	*Minor*
a_{11}	$M_{11} = \begin{vmatrix} a_{22} & a_{23} \\ a_{32} & a_{33} \end{vmatrix}$	a_{22}	$M_{22} = \begin{vmatrix} a_{11} & a_{13} \\ a_{31} & a_{33} \end{vmatrix}$
a_{21}	$M_{21} = \begin{vmatrix} a_{12} & a_{13} \\ a_{32} & a_{33} \end{vmatrix}$	a_{23}	$M_{23} = \begin{vmatrix} a_{11} & a_{12} \\ a_{31} & a_{32} \end{vmatrix}$
a_{31}	$M_{31} = \begin{vmatrix} a_{12} & a_{13} \\ a_{22} & a_{23} \end{vmatrix}$	a_{33}	$M_{33} = \begin{vmatrix} a_{11} & a_{12} \\ a_{21} & a_{22} \end{vmatrix}$

In a 4 × 4 matrix, the minors are determinants of 3 × 3 matrices, and an $n \times n$ matrix has minors that are determinants of $(n - 1) \times (n - 1)$ matrices.

To find the determinant of a 3 × 3 or larger matrix, first choose any row or column. Then the minor of each element in that row or column must be multiplied by 1 or −1, depending on whether the sum of the row numbers and column numbers is even or odd. The product of a minor and the number 1 or −1 is called a *cofactor.*

Definition of Cofactor

Let M_{ij} be the minor for element a_{ij} in an $n \times n$ matrix. The **cofactor** of a_{ij}, written A_{ij}, is

$$A_{ij} = (-1)^{i+j} \cdot M_{ij}.$$

Finally, the determinant of a 3×3 or larger matrix is found as follows.

Finding the Determinant of a Matrix

Multiply each element in any row or column of the matrix by its cofactor. The sum of these products gives the value of the determinant.

The process of forming this sum of products is called **expansion by a given row or column**. (See Exercises 55 and 56.)

For Graphers

Consult the manual to see how to find the determinant of a matrix with your graphing utility.

EXAMPLE 2 Evaluate $\begin{vmatrix} 2 & -3 & -2 \\ -1 & -4 & -3 \\ -1 & 0 & 2 \end{vmatrix}$. Expand by the second column.

To find this determinant, first get the minors of each element in the second column.

$$M_{12} = \begin{vmatrix} -1 & -3 \\ -1 & 2 \end{vmatrix} = -1(2) - (-1)(-3) = \mathbf{-5}$$

$$M_{22} = \begin{vmatrix} 2 & -2 \\ -1 & 2 \end{vmatrix} = 2(2) - (-1)(-2) = \mathbf{2}$$

$$M_{32} = \begin{vmatrix} 2 & -2 \\ -1 & -3 \end{vmatrix} = 2(-3) - (-1)(-2) = \mathbf{-8}$$

Now find the cofactor of each of these minors.

$$A_{12} = (-1)^{1+2} \cdot M_{12} = (-1)^3 \cdot (-5) = (-1)(-5) = 5$$

$$A_{22} = (-1)^{2+2} \cdot M_{22} = (-1)^4 \cdot (2) = 1 \cdot 2 = 2$$

$$A_{32} = (-1)^{3+2} \cdot M_{32} = (-1)^5 \cdot (-8) = (-1)(-8) = 8$$

The determinant is found by multiplying each cofactor by its corresponding element in the matrix and finding the sum of these products.

$$\begin{vmatrix} 2 & \mathbf{-3} & -2 \\ -1 & \mathbf{-4} & -3 \\ -1 & \mathbf{0} & 2 \end{vmatrix} = a_{12} \cdot A_{12} + a_{22} \cdot A_{22} + a_{32} \cdot A_{32}$$

$$= \mathbf{-3}(5) + (\mathbf{-4})(2) + (\mathbf{0})(8)$$

$$= -15 + (-8) + 0 = -23$$ ■

Exactly the same answer would be found using any row or column of the matrix. One reason that column 2 was used here is that it contains a 0 element, so that it was not really necessary to calculate M_{32} and A_{32} above. One learns quickly that 0's are friends in work with determinants.

Instead of calculating $(-1)^{i+j}$ for a given element, the following sign checkerboards can be used.

Array of Signs

For 3 × 3 matrices

$$\begin{matrix} + & - & + \\ - & + & - \\ + & - & + \end{matrix}$$

For 4 × 4 matrices

$$\begin{matrix} + & - & + & - \\ - & + & - & + \\ + & - & + & - \\ - & + & - & + \end{matrix}$$

The signs alternate for each row and column, beginning with + in the first row, first column position. Thus, these arrays of signs can be reproduced as needed. If we expand a 3 × 3 matrix about row 3, for example, the first minor would have a + sign associated with it, the second minor a − sign, and the third minor a + sign. These arrays of signs can be extended in this way for determinants of 5 × 5, 6 × 6, and larger matrices.

EXAMPLE 3

Evaluate $\begin{vmatrix} -1 & -2 & 3 & 2 \\ 0 & 1 & 4 & -2 \\ 3 & -1 & 4 & 0 \\ 2 & 1 & 0 & 3 \end{vmatrix}$.

Expand about the fourth row, and do the arithmetic that has been left out.

$$-2\begin{vmatrix} -2 & 3 & 2 \\ 1 & 4 & -2 \\ -1 & 4 & 0 \end{vmatrix} + 1\begin{vmatrix} -1 & 3 & 2 \\ 0 & 4 & -2 \\ 3 & 4 & 0 \end{vmatrix} - 0\begin{vmatrix} -1 & -2 & 2 \\ 0 & 1 & -2 \\ 3 & -1 & 0 \end{vmatrix} + 3\begin{vmatrix} -1 & -2 & 3 \\ 0 & 1 & 4 \\ 3 & -1 & 4 \end{vmatrix}$$

$$= -2(6) + 1(-50) - 0 + 3(-41) = -185$$

There are several theorems that make it easier to calculate determinants. The theorems are true for square matrices of any order, but they are proved here only for determinants of 3 × 3 matrices.

Determinant Theorem 1

If every element in a row (or column) of matrix A is 0, then $|A| = 0$.

Determinant Theorem 5

If two rows (or columns) of a matrix A are identical, then $|A| = 0$.

To prove this theorem, note that if two rows or columns of a matrix A are interchanged to form matrix B, then $|A| = -|B|$; while if two rows of matrix A are identical and are interchanged, we still have matrix A. But then $|A| = -|A|$, which can only happen if $|A| = 0$.

EXAMPLE 8

Since two rows are identical, $\begin{vmatrix} -4 & 2 & 3 \\ 0 & 1 & 6 \\ -4 & 2 & 3 \end{vmatrix} = 0.$ ■

The last theorem of this section is perhaps the most useful of all.

Determinant Theorem 6

Changing a row (or column) of a matrix by adding to it a constant times another row (or column) does not change the determinant of the matrix.

This theorem is proved in much the same way as the others in this section. (see Exercises 59 and 60 below.) It provides a powerful method for simplifying the work of finding the determinant of a 3×3 or larger matrix, as shown in the next example.

EXAMPLE 9

Let $A = \begin{bmatrix} -2 & 4 & 1 \\ 2 & 1 & 5 \\ 4 & 0 & 2 \end{bmatrix}$. Find $|A|$.

First obtain a new matrix B (using Determinant Theorem 6) by adding row 1 to row 2 and then (using Theorem 6 again) adding 2 times row 1 to row 3.

$$B = \begin{bmatrix} -2 & 4 & 1 \\ 0 & 5 & 6 \\ 0 & 8 & 4 \end{bmatrix} \qquad \begin{matrix} \\ R_1 + R_2 \\ 2R_1 + R_3 \end{matrix}$$

Now find $|B|$ by expanding about the first column.

$$|B| = -2\begin{vmatrix} 5 & 6 \\ 8 & 4 \end{vmatrix} = -2(20 - 48) = 56$$

By the theorem above, $|B| = |A|$, so $|A| = 56$. ■

The following examples show how the properties of determinants are used to simplify the calculation of determinants.

EXAMPLE 10 Without expanding, show that the value of the following determinant is 0.

$$\begin{vmatrix} 2 & 5 & -1 \\ 1 & -15 & 3 \\ -2 & 10 & -2 \end{vmatrix}$$

Examining the columns of the array shows that each element in the second column is -5 times the corresponding element in the third column. By Determinant Theorem 6, add to the elements of the second column the results of multiplying the elements of the third column by 5 (abbreviated below as $5C_3 + C_2$), to get the determinant

$$\begin{vmatrix} 2 & 0 & -1 \\ 1 & 0 & 3 \\ -2 & 0 & -2 \end{vmatrix}. \qquad 5C_3 + C_2$$

The value of this determinant is 0, by Theorem 1. ■

EXAMPLE 11 Find $|A| = \begin{vmatrix} 4 & 2 & 1 & 0 \\ -2 & 4 & -1 & 7 \\ -5 & 2 & 3 & 1 \\ 6 & 4 & -3 & 2 \end{vmatrix}$.

The goal is to change row 1 of the matrix (any row or column could be selected) to a row in which every element but one is 0. To begin, multiply the elements of column 2 of the matrix by -2 and add the results to the elements of column 1.

$$\begin{vmatrix} 0 & 2 & 1 & 0 \\ -10 & 4 & -1 & 7 \\ -9 & 2 & 3 & 1 \\ -2 & 4 & -3 & 2 \end{vmatrix} \qquad -2C_2 + C_1$$

Add to the elements of column 2 of the matrix the results of multiplying the elements of column 3 by -2.

$$\begin{vmatrix} 0 & 0 & 1 & 0 \\ -10 & 6 & -1 & 7 \\ -9 & -4 & 3 & 1 \\ -2 & 10 & -3 & 2 \end{vmatrix} \qquad -2C_3 + C_2$$

Row 1 of the matrix has only one nonzero number, so expand about the first row.

$$|A| = +1 \begin{vmatrix} -10 & 6 & 7 \\ -9 & -4 & 1 \\ -2 & 10 & 2 \end{vmatrix}$$

Now change column 3 of the matrix to a column with two zeros.

$$\begin{vmatrix} 53 & 34 & 0 \\ -9 & -4 & 1 \\ -2 & 10 & 2 \end{vmatrix} \quad -7R_2 + R_1$$

$$\begin{vmatrix} 53 & 34 & 0 \\ -9 & -4 & 1 \\ 16 & 18 & 0 \end{vmatrix} \quad -2R_2 + R_3$$

Finally, expand about column 3 of the matrix to find the value of $|A|$.

$$|A| = -1\begin{vmatrix} 53 & 34 \\ 16 & 18 \end{vmatrix} = -1(954 - 544) = -410$$ ■

In Example 11, working with *rows* of the matrix led to a *column* with only one nonzero number and working with *columns* of the matrix led to a *row* with one nonzero number.

6.5 Exercises

Find the value of each determinant. All variables represent real numbers.

1. $\begin{vmatrix} 2 & 5 \\ 4 & -7 \end{vmatrix}$
2. $\begin{vmatrix} 3 & 4 \\ 5 & -2 \end{vmatrix}$
3. $\begin{vmatrix} -9 & 7 \\ 2 & 6 \end{vmatrix}$
4. $\begin{vmatrix} 0 & 4 \\ 4 & 0 \end{vmatrix}$
5. $\begin{vmatrix} y & 3 \\ -2 & x \end{vmatrix}$
6. $\begin{vmatrix} y & 2 \\ 8 & y \end{vmatrix}$
7. $\begin{vmatrix} 3 & 8 \\ m & n \end{vmatrix}$
8. $\begin{vmatrix} 2m & 8n \\ 8n & 2m \end{vmatrix}$

Find the cofactor of each element in the second row for the following determinants.

9. $\begin{vmatrix} -2 & 0 & 1 \\ 1 & 2 & 0 \\ 4 & 2 & 1 \end{vmatrix}$
10. $\begin{vmatrix} 1 & -1 & 2 \\ 1 & 0 & 2 \\ 0 & -3 & 1 \end{vmatrix}$
11. $\begin{vmatrix} 1 & 2 & -1 \\ 2 & 3 & -2 \\ -1 & 4 & 1 \end{vmatrix}$
12. $\begin{vmatrix} 2 & -1 & 4 \\ 3 & 0 & 1 \\ -2 & 1 & 4 \end{vmatrix}$

Find the value of each determinant. All variables represent real numbers.

13. $\begin{vmatrix} 1 & 0 & 0 \\ 0 & 1 & 0 \\ 0 & 0 & 1 \end{vmatrix}$
14. $\begin{vmatrix} 1 & 0 & 0 \\ 0 & -1 & 0 \\ 1 & 0 & 1 \end{vmatrix}$
15. $\begin{vmatrix} -2 & 0 & 1 \\ 0 & 1 & 0 \\ 0 & 0 & -1 \end{vmatrix}$
16. $\begin{vmatrix} 1 & -2 & 3 \\ 0 & 0 & 0 \\ 1 & 10 & -12 \end{vmatrix}$
17. $\begin{vmatrix} 0 & 5 & 2 \\ 0 & 3 & -1 \\ 0 & -4 & 7 \end{vmatrix}$
18. $\begin{vmatrix} 3 & 3 & -1 \\ 2 & 6 & 0 \\ -6 & -6 & 2 \end{vmatrix}$
19. $\begin{vmatrix} 0 & 3 & y \\ 0 & 4 & 2 \\ 1 & 0 & 1 \end{vmatrix}$
20. $\begin{vmatrix} 3 & 2 & 0 \\ 0 & 1 & x \\ 2 & 0 & 0 \end{vmatrix}$
21. $\begin{vmatrix} i & j & k \\ 0 & -4 & 2 \\ -1 & 3 & 1 \end{vmatrix}$
22. $\begin{vmatrix} i & j & k \\ -1 & 2 & 4 \\ 3 & 0 & 5 \end{vmatrix}$
23. $\begin{vmatrix} 2 & 0 & 0 & 1 \\ -2 & 0 & 6 & 0 \\ 2 & 4 & 0 & 1 \\ 2 & 4 & 1 & 2 \end{vmatrix}$

24. $\begin{vmatrix} .4 & -.8 & .6 \\ .3 & .9 & .7 \\ 3.1 & 4.1 & -2.8 \end{vmatrix}$ **25.** $\begin{vmatrix} -.3 & -.1 & .9 \\ 2.5 & 4.9 & -3.2 \\ -.1 & .4 & .8 \end{vmatrix}$ **26.** $\begin{vmatrix} -.5 & -.7 & .9 \\ 1.4 & 3.6 & -.2 \\ 1.5 & 2.1 & -2.7 \end{vmatrix}$

Tell why each determinant has a value of 0. *All variables represent real numbers.*

27. $\begin{vmatrix} 2 & 3 \\ 2 & 3 \end{vmatrix}$ **28.** $\begin{vmatrix} -5 & -5 \\ 6 & 6 \end{vmatrix}$ **29.** $\begin{vmatrix} 2 & 0 \\ 3 & 0 \end{vmatrix}$ **30.** $\begin{vmatrix} 6 & -8 \\ -3 & 4 \end{vmatrix}$

31. $\begin{vmatrix} 1 & 0 & 0 \\ 1 & 0 & 1 \\ 3 & 0 & 0 \end{vmatrix}$ **32.** $\begin{vmatrix} -1 & 2 & 4 \\ 4 & -8 & -16 \\ 3 & 0 & 5 \end{vmatrix}$ **33.** $\begin{vmatrix} 7z & 8x & 2y \\ z & x & y \\ 7z & 7x & 7y \end{vmatrix}$ **34.** $\begin{vmatrix} m & 2 & 2m \\ 3n & 1 & 6n \\ 5p & 6 & 10p \end{vmatrix}$

Use the appropriate theorems from this section to tell why each statement is true. Do not evaluate the determinants. All variables represent real numbers.

35. $\begin{vmatrix} 4 & -2 \\ 3 & 8 \end{vmatrix} = \begin{vmatrix} 4 & 3 \\ -2 & 8 \end{vmatrix}$

36. $\begin{vmatrix} 2 & 1 & 6 \\ 3 & 0 & 2 \\ 4 & 1 & 8 \end{vmatrix} = \begin{vmatrix} 2 & 3 & 4 \\ 1 & 0 & 1 \\ 6 & 2 & 8 \end{vmatrix}$

37. $\begin{vmatrix} -1 & 8 & 9 \\ 0 & 2 & 1 \\ 3 & 2 & 0 \end{vmatrix} = -\begin{vmatrix} 8 & -1 & 9 \\ 2 & 0 & 1 \\ 2 & 3 & 0 \end{vmatrix}$

38. $\begin{vmatrix} 2 & 6 \\ 3 & 5 \end{vmatrix} = -\begin{vmatrix} 3 & 5 \\ 2 & 6 \end{vmatrix}$

39. $-\frac{1}{2}\begin{vmatrix} 5 & -8 & 2 \\ 3 & -6 & 9 \\ 2 & 4 & 4 \end{vmatrix} = \begin{vmatrix} 5 & 4 & 2 \\ 3 & 3 & 9 \\ 2 & -2 & 4 \end{vmatrix}$

40. $3\begin{vmatrix} 6 & 0 & 2 \\ 4 & 1 & 3 \\ 2 & 8 & 6 \end{vmatrix} = \begin{vmatrix} 6 & 0 & 2 \\ 4 & 3 & 3 \\ 2 & 24 & 6 \end{vmatrix}$

41. $\begin{vmatrix} 3 & -4 \\ 2 & 5 \end{vmatrix} = \begin{vmatrix} 3 & -4 \\ 5 & 1 \end{vmatrix}$

42. $\begin{vmatrix} -1 & 6 \\ 3 & -5 \end{vmatrix} = \begin{vmatrix} -1 & 5 \\ 3 & -2 \end{vmatrix}$

43. $\begin{vmatrix} -4 & 2 & 1 \\ 3 & 0 & 3 \\ -1 & 4 & -2 \end{vmatrix} = \begin{vmatrix} -4 & 2 & 1 + (-4)k \\ 3 & 0 & 3 + 3k \\ -1 & 4 & -2 + (-1)k \end{vmatrix}$

44. $2\begin{vmatrix} 4 & 2 & -1 \\ m & 2n & 3p \\ 5 & 1 & 0 \end{vmatrix} = \begin{vmatrix} 4 & 2 & -1 \\ 2m & 4n & 6p \\ 5 & 1 & 0 \end{vmatrix}$

Use the method of Examples 10 and 11 to find the value of each determinant.

45. $\begin{vmatrix} -5 & 10 \\ 6 & -12 \end{vmatrix}$ **46.** $\begin{vmatrix} 2 & 4 \\ 3 & 6 \end{vmatrix}$ **47.** $\begin{vmatrix} 6 & 8 & -12 \\ -1 & 0 & 2 \\ 4 & 0 & -8 \end{vmatrix}$ **48.** $\begin{vmatrix} 4 & 8 & 0 \\ -1 & -2 & 1 \\ 2 & 4 & 3 \end{vmatrix}$

49. $\begin{vmatrix} -2 & 2 & 3 \\ 0 & 2 & 1 \\ -1 & 4 & 0 \end{vmatrix}$ **50.** $\begin{vmatrix} 3 & 1 & 2 \\ 2 & 0 & 1 \\ 1 & 0 & -2 \end{vmatrix}$ **51.** $\begin{vmatrix} -4 & 1 & 4 \\ 2 & 0 & 1 \\ 0 & 2 & 4 \end{vmatrix}$ **52.** $\begin{vmatrix} 6 & 3 & 2 \\ 1 & 0 & 2 \\ 5 & 7 & 3 \end{vmatrix}$

53. $\begin{vmatrix} 2 & -1 & 1 & 0 \\ 1 & 1 & 0 & 1 \\ 0 & -1 & 1 & 1 \\ 1 & 2 & 1 & 2 \end{vmatrix}$ **54.** $\begin{vmatrix} 1 & 0 & 2 & 2 \\ 2 & 4 & 1 & -1 \\ 1 & -3 & 1 & 0 \\ 1 & 1 & 0 & 1 \end{vmatrix}$

Let $A = \begin{bmatrix} a_{11} & a_{12} & a_{13} \\ a_{21} & a_{22} & a_{23} \\ a_{31} & a_{32} & a_{33} \end{bmatrix}$ *for Exercises 55–60.*

55. Find $|A|$ by expansion about row 3 of the matrix. Show that your result is really equal to $|A|$ as given in the definition of the determinant of a 3×3 matrix.

56. Repeat Exercise 55 for column 3.

57. Obtain matrix B by exchanging columns 1 and 3 of matrix A. Show that $|B| = -|A|$.

58. Obtain matrix B by multiplying each element of row 3 of matrix A by the real number k. Show that $|B| = k \cdot |A|$.

59. Obtain matrix B by adding to column 1 of matrix A the result of multiplying each element of column 2 of A by the real number k. Show that $|B| = |A|$.

60. Obtain matrix B by adding to row 1 of matrix A the result of multiplying each element of row 3 of A by the real number k. Show that $|B| = |A|$.

61. Let A and B be any 2×2 matrices. Show that $|AB| = |A| \cdot |B|$, where $|AB|$ is the determinant of matrix AB.

62. Show that $\begin{vmatrix} a_{11}+a & a_{12} & a_{13} \\ a_{21}+b & a_{22} & a_{23} \\ a_{31}+c & a_{32} & a_{33} \end{vmatrix} = \begin{vmatrix} a_{11} & a_{12} & a_{13} \\ a_{21} & a_{22} & a_{23} \\ a_{31} & a_{32} & a_{33} \end{vmatrix} + \begin{vmatrix} a & a_{12} & a_{13} \\ b & a_{22} & a_{23} \\ c & a_{32} & a_{33} \end{vmatrix}$.

Use this fact and Determinant Theorems 4 and 5 to prove Determinant Theorem 6.

Determinants can be used to find the area of a triangle, given the coordinates of its vertices. Given a triangle PQR with vertices (x_1, y_1), (x_2, y_2), *and* (x_3, y_3), *as in the figure, it can be shown that the area of the triangle is given by A, where*

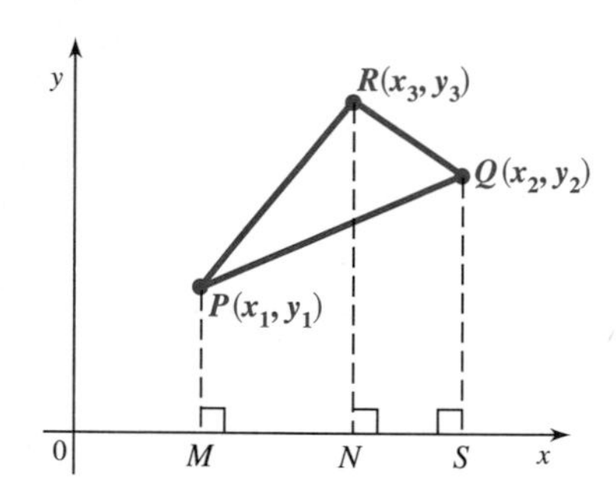

$$A = \frac{1}{2}\begin{vmatrix} x_1 & y_1 & 1 \\ x_2 & y_2 & 1 \\ x_3 & y_3 & 1 \end{vmatrix}.$$

The points (x_1, y_1), (x_2, y_2), (x_3, y_3) *must be taken in counterclockwise order; if this is not done, than A may have the wrong sign. Alternatively, we could define A as the absolute value of 1/2 the determinant shown above. Use the formula given to find the area of the triangles in Exercises 63 and 64.*

63. (a) $P(0, 0)$, $Q(0, 2)$, $R(1, 1)$
(b) $P(0, 1)$, $Q(2, 0)$, $R(1, 3)$
(c) $P(2, 5)$, $Q(-1, 3)$, $R(4, 0)$

64. (a) $P(2, -2)$, $Q(0, 0)$, $R(-3, -4)$
(b) $P(4, 7)$, $Q(5, -2)$, $R(1, 1)$
(c) $P(1, 2)$, $Q(4, 3)$, $R(3, 5)$

65. Prove that the straight line through the distinct points (x_1, y_1) and (x_2, y_2) has equation

$$\begin{vmatrix} x & y & 1 \\ x_1 & y_1 & 1 \\ x_2 & y_2 & 1 \end{vmatrix} = 0.$$

66. Use the result of Exercise 65 to show that three distinct points (x_1, y_1), (x_2, y_2), and (x_3, y_3) lie on a straight line if

$$\begin{vmatrix} x_1 & y_1 & 1 \\ x_2 & y_2 & 1 \\ x_3 & y_3 & 1 \end{vmatrix} = 0.$$

67. Show that the lines $a_1x + b_1y = c_1$ and $a_2x + b_2y = c_2$, when $c_1 \neq c_2$, are parallel if

$$\begin{vmatrix} a_1 & b_1 \\ a_2 & b_2 \end{vmatrix} = 0.$$

68. Prove that $\begin{vmatrix} 1 & 1 & 1 \\ a & b & c \\ a^2 & b^2 & c^2 \end{vmatrix} = (a - b)(b - c)(c - a)$.

69. Find $\begin{vmatrix} 1 & 1 & 1 \\ 1 + x & 1 + y & 1 \\ 1 & 1 & 1 \end{vmatrix}$.

70. Prove that $\begin{vmatrix} 1 & 1 & 1 \\ a & b & c \\ bc & ca & ab \end{vmatrix} = (a - b)(a - c)(c - b)$.

71. In your own words, describe the method for obtaining the determinant of a matrix.

72. Give a method using determinants to decide whether two lines intersect at a single point.

Graphing Utility Problems

For the pairs of matrices in Exercises 73 and 74, show that $|AB| = |A| \cdot |B|$.

73. $A = \begin{bmatrix} 2 & 5 \\ 1 & 7 \end{bmatrix}$, $B = \begin{bmatrix} 9 & 4 \\ 2 & 6 \end{bmatrix}$

74. $A = \begin{bmatrix} 4 & -2 & 11 \\ 5 & 7 & 9 \\ 8 & 6 & 3 \end{bmatrix}$, $B = \begin{bmatrix} 6 & 8 & 7 \\ 3 & 12 & -5 \\ 7 & 1 & 6 \end{bmatrix}$

6.6 Cramer's Rule

We have now seen how to solve a system of n linear equations with n variables using the following methods: elimination, substitution, and row transformations of matrices. Most of these systems can also be solved with determinants, as shown here.

To see how determinants arise in solving a system, write the linear system

$$\begin{aligned} a_{11}x + a_{12}y &= b_1 \\ a_{21}x + a_{22}y &= b_2 \end{aligned}$$

where each equation has at least one nonzero coefficient. Solve this system using matrix methods; begin by writing the augmented matrix

$$\left[\begin{array}{cc|c} a_{11} & a_{12} & b_1 \\ a_{21} & a_{22} & b_2 \end{array}\right].$$

Multiply each element of row 1 by $1/a_{11}$. (Here we assume $a_{11} \neq 0$.) This gives the matrix of an equivalent system:

$$\left[\begin{array}{cc|c} 1 & a_{12}/a_{11} & b_1/a_{11} \\ a_{21} & a_{22} & b_2 \end{array}\right]. \qquad (1/a_{11})\mathrm{R}_1$$

Multiply each element of row 1 by $-a_{21}$, and add the results to the corresponding element of row 2.

$$\left[\begin{array}{cc|c} 1 & a_{12}/a_{11} & b_1/a_{11} \\ 0 & a_{22} - a_{21}a_{12}/a_{11} & b_2 - a_{21}b_1/a_{11} \end{array}\right] \qquad -a_{21}\mathrm{R}_1 + \mathrm{R}_2$$

Multiply each element of row 2 by a_{11}.

$$\left[\begin{array}{cc|c} 1 & a_{12}/a_{11} & b_1/a_{11} \\ 0 & a_{11}a_{22} - a_{21}a_{12} & a_{11}b_2 - a_{21}b_1 \end{array}\right] \qquad a_{11}\mathrm{R}_2$$

This matrix leads to the system of equations

$$x + \frac{a_{12}}{a_{11}} y = \frac{b_1}{a_{11}}$$
$$(a_{11}a_{22} - a_{21}a_{12})y = a_{11}b_2 - a_{21}b_1.$$

From the second equation of this system,

$$y = \frac{a_{11}b_2 - a_{21}b_1}{a_{11}a_{22} - a_{21}a_{12}}.$$

Both the numerator and denominator here may be written as determinants:

$$y = \frac{\begin{vmatrix} a_{11} & b_1 \\ a_{21} & b_2 \end{vmatrix}}{\begin{vmatrix} a_{11} & a_{12} \\ a_{21} & a_{22} \end{vmatrix}}. \qquad \mathbf{(1)}$$

Inserting this value of y into the first equation above shows that x can also be written with determinants as

$$x = \frac{\begin{vmatrix} b_1 & a_{12} \\ b_2 & a_{22} \end{vmatrix}}{\begin{vmatrix} a_{11} & a_{12} \\ a_{21} & a_{22} \end{vmatrix}}. \qquad \mathbf{(2)}$$

The denominator for finding both x and y is just the determinant of the matrix of coefficients of the original system. This determinant is often denoted D, so that

$$D = \begin{vmatrix} a_{11} & a_{12} \\ a_{21} & a_{22} \end{vmatrix}.$$

In equation (1), the numerator is the determinant of a matrix obtained by replacing the coefficients of y in D with the respective constants: D_y is defined as

$$D_y = \begin{vmatrix} a_{11} & b_1 \\ a_{21} & b_2 \end{vmatrix}.$$

In the same way, from equation (2), D_x is defined as

$$D_x = \begin{vmatrix} b_1 & a_{12} \\ b_2 & a_{22} \end{vmatrix}.$$

With this notation, the solution of the given system is

$$x = \frac{D_x}{D} \quad \text{and} \quad y = \frac{D_y}{D}.$$

The system has a single solution as long as $D \neq 0$. We have now proved much of the next theorem, called **Cramer's rule.**

Cramer's Rule for 2 Equations in 2 Variables

Given the system

$$\begin{aligned} a_{11}x + a_{12}y &= b_1 \\ a_{21}x + a_{22}y &= b_2, \end{aligned}$$

if $D \neq 0$, the system has the unique solution

$$\boldsymbol{x = \frac{D_x}{D}} \quad \text{and} \quad \boldsymbol{y = \frac{D_y}{D}}.$$

Cramer was looking for a method to determine the equation of a curve when he knew several points on the curve. In 1750 he wrote down the general equation for a curve and then substituted each point for which he had two coordinates into the equation. For this system of equations he gave "a rule very convenient and general to solve any number of equations and unknowns which are of no more than first degree." This is the rule that now bears his name.

EXAMPLE 1 Use Cramer's rule to solve the system

$$5x + 7y = -1$$
$$6x + 8y = 1.$$

To use Cramer's rule, first evaluate D, D_x, and D_y.

$$D = \begin{vmatrix} 5 & 7 \\ 6 & 8 \end{vmatrix} = 5(8) - 6(7) = \mathbf{-2}$$

$$D_x = \begin{vmatrix} -1 & 7 \\ 1 & 8 \end{vmatrix} = (-1)(8) - (1)(7) = \mathbf{-15}$$

$$D_y = \begin{vmatrix} 5 & -1 \\ 6 & 1 \end{vmatrix} = 5(1) - (6)(-1) = \mathbf{11}$$

By Cramer's rule, $x = \mathbf{-15/(-2)} = 15/2$ and $y = \mathbf{11/(-2)} = -11/2$. The solution is $(15/2, -11/2)$, as can be verified by substituting within the given system.

By much the same method as used above, Cramer's rule can be generalized to a system of n linear equations with n variables.

General Form of Cramer's Rule

Let an $n \times n$ system have linear equations of the form

$$a_{11}x_1 + a_{12}x_2 + a_{13}x_3 + \cdots + a_{1n}x_n = b_1.$$

Define D as the determinant of the $n \times n$ matrix of all coefficients of the variables. Define D_{x1} as the determinant obtained from D by replacing the entries in column 1 of D with the constants of the system. Define D_{xi} as the determinant obtained from D by replacing the entries in column i with the constants of the system. If $D \neq 0$, the unique solution of the system is

$$x_1 = \frac{D_{x1}}{D}, x_2 = \frac{D_{x2}}{D}, x_3 = \frac{D_{x3}}{D}, \ldots, x_n = \frac{D_{xn}}{D}.$$

EXAMPLE 2 Use Cramer's rule to solve the system

$$x + y - z + 2 = 0$$
$$2x - y + z + 5 = 0$$
$$x - 2y + 3z - 4 = 0.$$

To use Cramer's rule, the system must be rewritten in the form

$$\begin{aligned} x + y - z &= -2 \\ 2x - y + z &= -5 \\ x - 2y + 3z &= 4. \end{aligned}$$

The determinant of coefficients, D, is

$$D = \begin{vmatrix} 1 & 1 & -1 \\ 2 & -1 & 1 \\ 1 & -2 & 3 \end{vmatrix}.$$

To find D_x, replace the elements of the first column of D with the constants of the system. Find D_y and D_z in a similar way.

$$D_x = \begin{vmatrix} -2 & 1 & -1 \\ -5 & -1 & 1 \\ 4 & -2 & 3 \end{vmatrix}, \quad D_y = \begin{vmatrix} 1 & -2 & -1 \\ 2 & -5 & 1 \\ 1 & 4 & 3 \end{vmatrix}, \quad D_z = \begin{vmatrix} 1 & 1 & -2 \\ 2 & -1 & -5 \\ 1 & -2 & 4 \end{vmatrix}$$

Verify that $D = -3$, $D_x = 7$, $D_y = -22$, and $D_z = -21$. Then, by Cramer's rule,

$$x = \frac{D_x}{D} = \frac{7}{-3} = -\frac{7}{3},$$

$$y = \frac{D_y}{D} = \frac{-22}{-3} = \frac{22}{3},$$

$$\text{and} \quad z = \frac{D_z}{D} = \frac{-21}{-3} = 7.$$

The solution of the system is $(-7/3, 22/3, 7)$. ■

CAUTION As shown in Example 2, each equation in the system must be written in the form $a_1x_1 + a_2x_2 + \ldots + a_nx_n = k$ before using Cramer's rule. ■

EXAMPLE 3 Use Cramer's rule to solve the system

$$\begin{aligned} 2x - 3y + 4z &= 10 \\ 6x - 9y + 12z &= 24 \\ x + 2y - 3z &= 5. \end{aligned}$$

Verify that $D = 0$, so Cramer's rule does not apply. Use another method to determine that this system is inconsistent and thus has no solution. ■

Several different methods for solving systems of equations have now been shown. In general, if a small system of linear equations must be solved by pencil and paper, substitution is the best method if the various variables can easily be found in terms of each other. This happens rarely. The next choice, perhaps the best choice of all, is the elimination method. Some people like the Gaussian reduction method, which is really just a systematic way of doing the elimination method. The Gaussian reduction method is probably superior where four or more equations are involved. Cramer's rule is seldom the method of choice simply because it involves more calculations than any other method.

6.6 Exercises

Use Cramer's rule to solve each of the following systems of linear equations.

1. $\begin{aligned} 2x + 5y &= 12 \\ x + 3y &= 7 \end{aligned}$

2. $\begin{aligned} 3x + 2y &= 8 \\ 5x + 4y &= 14 \end{aligned}$

3. $\begin{aligned} -x + 2y &= 1 \\ 4x - 7y &= -2 \end{aligned}$

4. $\begin{aligned} 2x - 5y &= 17 \\ x + 3y &= -8 \end{aligned}$

5. $\begin{aligned} 4x + 2y &= 11 \\ 3x - y &= 2 \end{aligned}$

6. $\begin{aligned} 6x - y &= -1 \\ 3x + y &= 3 \end{aligned}$

7. $\begin{aligned} 5x + 12y &= 2 \\ -3x + y &= 7 \end{aligned}$

8. $\begin{aligned} 3x + 5y &= 5 \\ -4x + 6y &= 6 \end{aligned}$

9. $\begin{aligned} 6x - 15y &= 4 \\ -2x + 5y &= 9 \end{aligned}$

10. $\begin{aligned} -8x + 6y &= -7 \\ 4x - 3y &= 0 \end{aligned}$

11. $\begin{aligned} 3x - z &= -10 \\ y + 4z &= 8 \\ x + 2z &= -1 \end{aligned}$

12. $\begin{aligned} 4x - y + 3z &= -3 \\ 3x + y + z &= 0 \\ 2x - y + 4z &= 0 \end{aligned}$

13. $\begin{aligned} 5x + 2y + z &= 15 \\ 2x - y + z &= 9 \\ 4x + 3y + 2z &= 13 \end{aligned}$

14. $\begin{aligned} 2x - y + 4z &= -2 \\ 3x + 2y - z &= -3 \\ x + 4y + 2z &= 17 \end{aligned}$

15. $\begin{aligned} x + y + z &= 4 \\ 2x - y + 3z &= 4 \\ 4x + 2y - z &= -15 \end{aligned}$

16. $\begin{aligned} 4x - 3y + z &= -1 \\ 5x + 7y + 2z &= -2 \\ 3x - 5y - z &= 1 \end{aligned}$

17. $\begin{aligned} 2x - 3y + z &= 8 \\ -x - 5y + z &= -4 \\ 3x - 5y + 2z &= 12 \end{aligned}$

18. $\begin{aligned} x + 2y + 3z &= 4 \\ 4x + 3y + 2z &= 1 \\ -x - 2y - 3z &= 0 \end{aligned}$

19. $\begin{aligned} 2x - y + 3z &= 1 \\ -2x + y - 3z &= 2 \\ 5x - y + z &= 2 \end{aligned}$

20. $\begin{aligned} x - 2y + 3z &= 4 \\ 5x + 7y - z &= 2 \\ 2x + 2y - 5z &= 3 \end{aligned}$

21. $\begin{aligned} -3x - 2y - z &= 4 \\ 4x + y + z &= 5 \\ 3x - 2y + 2z &= 1 \end{aligned}$

22. $\begin{aligned} 2x + 3y &= 13 \\ 2y - z &= 5 \\ x + 2z &= 4 \end{aligned}$

23. $\begin{aligned} 3x + 5y &= -7 \\ 2x + 7z &= 2 \\ 4y + 3z &= -8 \end{aligned}$

24. $\begin{aligned} 5x - y &= -4 \\ 3x + 2z &= 4 \\ 4y + 3z &= 22 \end{aligned}$

25. $\begin{aligned} 5x - 2y &= 3 \\ 4y + z &= 8 \\ x + 2z &= 4 \end{aligned}$

26. $\begin{aligned} x + 2y &= 10 \\ 3x + 4z &= 7 \\ -y - z &= 1 \end{aligned}$

27. $\begin{aligned} x - y + z + w &= 6 \\ 2y - w &= -7 \\ x - z &= -1 \\ y + w &= 1 \end{aligned}$

28. $\begin{aligned} x + z &= 0 \\ y + 2z + w &= 0 \\ 2x - w &= 0 \\ x + 2y + 3z &= -2 \end{aligned}$

For the following two exercises, use the system of equations

$$\begin{aligned} a_1x + b_1y &= c_1 \\ a_2x + b_2y &= c_2. \end{aligned}$$

29. Assume $D_x = 0$ and $D_y = 0$. Show that if $c_1c_2 \neq 0$, then $D = 0$, and the equations are dependent.

30. Assume $D = 0$, $D_x = 0$, and $b_1b_2 \neq 0$. Show that $D_y = 0$.

31. State Cramer's rule in your own words.

Graphing Utility Problems

Solve each of the following systems of linear equations by Cramer's rule.

32. $17x + .5y = \sqrt{3}$
$9x + .3y = \frac{3}{5}$

33. $\sqrt{3}x + 6y = 5$
$4.5x - 19y = \sqrt{2}$

34. $3x + 1.4y - z = \sqrt{3}$
$\sqrt{7}x - 6y + .6z = -1$
$x + \sqrt{3}y - 5.1z = 5$

35. $9x - 2y + \sqrt{2}z = 1$
$-7x + 3y + \sqrt{3}z = \frac{4}{3}$
$.24x + 17y - z = \sqrt{2}$

6.7 Matrix Inverses

As shown in the exercises for an earlier section, the commutative, associative, closure, identity, and inverse properties hold for *addition* of matrices of the same size. Also, the associative and closure properties hold for *multiplication* of square matrices of the same size. There is no commutative property for multiplication, but the distributive property is valid for matrices of the proper size.

In this section two additional properties are discussed, the identity and inverse properties for multiplication of certain matrices. For the identity property to hold, there must be a matrix I such that

$$AI = A \quad \text{and} \quad IA = A$$

for any square matrix A. (Compare these products to the statement of the identity property for real numbers: $a \cdot 1 = a$ and $1 \cdot a = a$ for any real number a.)

2 × 2 Identity

If I_2 represents the 2×2 identity, then

$$I_2 = \begin{bmatrix} 1 & 0 \\ 0 & 1 \end{bmatrix}.$$

To verify that I_2 is the 2×2 identity matrix, show that $AI = A$ and $IA = A$ for any 2×2 matrix. Let

$$A = \begin{bmatrix} x & y \\ z & w \end{bmatrix}.$$

Then
$$AI = \begin{bmatrix} x & y \\ z & w \end{bmatrix}\begin{bmatrix} 1 & 0 \\ 0 & 1 \end{bmatrix} = \begin{bmatrix} x \cdot 1 + y \cdot 0 & x \cdot 0 + y \cdot 1 \\ z \cdot 1 + w \cdot 0 & z \cdot 0 + w \cdot 1 \end{bmatrix} = \begin{bmatrix} x & y \\ z & w \end{bmatrix} = A$$

and
$$IA = \begin{bmatrix} 1 & 0 \\ 0 & 1 \end{bmatrix}\begin{bmatrix} x & y \\ z & w \end{bmatrix} = \begin{bmatrix} 1 \cdot x + 0 \cdot z & 1 \cdot y + 0 \cdot w \\ 0 \cdot x + 1 \cdot z & 0 \cdot y + 1 \cdot w \end{bmatrix} = \begin{bmatrix} x & y \\ z & w \end{bmatrix} = A.$$

Generalizing from this example, there is an $n \times n$ identity matrix having 1's on the main diagonal and 0's elsewhere.

$n \times n$ Identity Matrix

The $n \times n$ identity matrix is given by I_n where

$$I_n = \begin{bmatrix} \mathbf{1} & \mathbf{0} & \cdots & \mathbf{0} \\ \mathbf{0} & \mathbf{1} & \cdots & \mathbf{0} \\ \vdots & \vdots & a_{ij} & \vdots \\ \mathbf{0} & \mathbf{0} & \cdots & \mathbf{1} \end{bmatrix}.$$

The element $a_{ij} = 1$ when $i = j$ (the diagonal elements) and $a_{ij} = 0$ otherwise.

For every nonzero real number a, there is a multiplicative inverse $1/a$ such that

$$a \cdot \frac{1}{a} = 1 \quad \text{and} \quad \frac{1}{a} \cdot a = 1.$$

(Recall: $1/a$ is also written a^{-1}.) In a similar way, if A is an $n \times n$ matrix, then its **multiplicative inverse**, written A^{-1}, must satisfy both

$$AA^{-1} = I_n \quad \text{and} \quad A^{-1}A = I_n.$$

This means that only a square matrix can have a multiplicative inverse.

CAUTION Although $a^{-1} = 1/a$ for any nonzero real number a, if A is a matrix,

$$A^{-1} \neq \frac{1}{A}.$$

In fact, $1/A$ has no meaning, since 1 is a *number* and A is a *matrix*. ■

The matrix A^{-1} can be found by using the row operations introduced earlier in this chapter. As an example, let us find the inverse of

$$A = \begin{bmatrix} 2 & 4 \\ 1 & -1 \end{bmatrix}.$$

Let the unknown inverse matrix be

$$A^{-1} = \begin{bmatrix} x & y \\ z & w \end{bmatrix}.$$

By the definition of matrix inverse, $AA^{-1} = I_2$, or

$$AA^{-1} = \begin{bmatrix} 2 & 4 \\ 1 & -1 \end{bmatrix}\begin{bmatrix} x & y \\ z & w \end{bmatrix} = \begin{bmatrix} 1 & 0 \\ 0 & 1 \end{bmatrix}.$$

By matrix multiplication,

$$\begin{bmatrix} 2x + 4z & 2y + 4w \\ x - z & y - w \end{bmatrix} = \begin{bmatrix} 1 & 0 \\ 0 & 1 \end{bmatrix}.$$

Setting corresponding elements equal gives the system of equations

$$2x + 4z = 1 \qquad \textbf{(1)}$$
$$2y + 4w = 0 \qquad \textbf{(2)}$$
$$x - z = 0 \qquad \textbf{(3)}$$
$$y - w = 1. \qquad \textbf{(4)}$$

Since equations (1) and (3) involve only x and z, while equations (2) and (4) involve only y and w, these four equations lead to two systems of equations,

$$\begin{aligned} 2x + 4z &= 1 \\ x - z &= 0 \end{aligned} \quad \text{and} \quad \begin{aligned} 2y + 4w &= 0 \\ y - w &= 1. \end{aligned}$$

Writing the two systems as augmented matrices gives

$$\left[\begin{array}{cc|c} 2 & 4 & 1 \\ 1 & -1 & 0 \end{array}\right] \quad \text{and} \quad \left[\begin{array}{cc|c} 2 & 4 & 0 \\ 1 & -1 & 1 \end{array}\right].$$

Each of these systems can be solved by the Gaussian method. However, since the elements to the left of the vertical bar are identical, the two systems can be combined into one matrix,

$$\left[\begin{array}{cc|cc} 2 & 4 & 1 & 0 \\ 1 & -1 & 0 & 1 \end{array}\right],$$

and solved simultaneously using matrix row transformations. We need to change the numbers on the left of the vertical bar to the 2×2 identity matrix.

Start by exchanging the two rows to get a 1 in the upper left corner.

$$\left[\begin{array}{rr|rr} 1 & -1 & 0 & 1 \\ 2 & 4 & 1 & 0 \end{array}\right]$$

Multiply row 1 by -2 and add the results to row 2 to get

$$\left[\begin{array}{rr|rr} 1 & -1 & 0 & 1 \\ 0 & 6 & 1 & -2 \end{array}\right]. \qquad -2R_1 + R_2$$

Now, to get a 1 in the second-row, second-column position, multiply row 2 by $1/6$.

$$\left[\begin{array}{rr|rr} 1 & -1 & 0 & 1 \\ 0 & 1 & 1/6 & -1/3 \end{array}\right] \qquad (1/6)R_2$$

Finally, add row 2 to row 1 to get a 0 in the first-row, second-column position.

$$\left[\begin{array}{rr|rr} 1 & 0 & 1/6 & 2/3 \\ 0 & 1 & 1/6 & -1/3 \end{array}\right] \qquad R_2 + R_1$$

The numbers in the first column to the right of the vertical bar give the values of x and z. The second column gives the value of y and w. That is,

$$\left[\begin{array}{rr|rr} 1 & 0 & x & y \\ 0 & 1 & z & w \end{array}\right] = \left[\begin{array}{rr|rr} 1 & 0 & \mathbf{1/6} & \mathbf{2/3} \\ 0 & 1 & \mathbf{1/6} & \mathbf{-1/3} \end{array}\right],$$

so that

$$A^{-1} = \begin{bmatrix} x & y \\ z & w \end{bmatrix} = \begin{bmatrix} 1/6 & 2/3 \\ 1/6 & -1/3 \end{bmatrix}.$$

To check, multiply A by A^{-1}. The result should be I_2.

$$AA^{-1} = \begin{bmatrix} 2 & 4 \\ 1 & -1 \end{bmatrix}\begin{bmatrix} 1/6 & 2/3 \\ 1/6 & -1/3 \end{bmatrix} = \begin{bmatrix} 1/3 + 2/3 & 4/3 - 4/3 \\ 1/6 - 1/6 & 2/3 + 1/3 \end{bmatrix} = \begin{bmatrix} 1 & 0 \\ 0 & 1 \end{bmatrix} = I_2$$

Verify that $A^{-1}A = I_2$, also. Finally,

$$A^{-1} = \begin{bmatrix} 1/6 & 2/3 \\ 1/6 & -1/3 \end{bmatrix}.$$

For Grraphers

Consult your manual to see how to find the inverse of a matrix with your graphing utility.

In summary, the following method is used to find the inverse of the $n \times n$ matrix A.

1. Form the augmented matrix $[A|I_n]$, where I_n is the $n \times n$ identity matrix.
2. Perform row transformations on $[A|I_n]$ until a matrix of the form $[I_n|B]$ is obtained. (If this is not possible, then A does not have an inverse.)
3. Matrix B is the desired matrix A^{-1}.

NOTE To confirm that two $n \times n$ matrices A and B are inverses of each other, it is sufficient to show that $AB = I_n$. It is not necessary to show also that $BA = I_n$. ■

EXAMPLE 1

Find A^{-1} if $A = \begin{bmatrix} 1 & 0 & 1 \\ 2 & -2 & -1 \\ 3 & 0 & 0 \end{bmatrix}$.

Use row transformations, going through as many steps as needed.

Step 1 Write the augmented matrix $[A|I_3]$.

$$\left[\begin{array}{ccc|ccc} 1 & 0 & 1 & 1 & 0 & 0 \\ 2 & -2 & -1 & 0 & 1 & 0 \\ 3 & 0 & 0 & 0 & 0 & 1 \end{array}\right]$$

Step 2 Since 1 is already in the upper left-hand corner as required, begin by selecting the row operation which will result in a 0 for the first element in the second row. Add to each element in the second row the result of multiplying the first row by -2.

$$\left[\begin{array}{ccc|ccc} 1 & 0 & 1 & 1 & 0 & 0 \\ 0 & -2 & -3 & -2 & 1 & 0 \\ 3 & 0 & 0 & 0 & 0 & 1 \end{array}\right] \quad -2R_1 + R_2$$

Step 3 To get 0 for the first element in the third row, add to the third row the results of multiplying each element of the first row by -3.

$$\left[\begin{array}{ccc|ccc} 1 & 0 & 1 & 1 & 0 & 0 \\ 0 & -2 & -3 & -2 & 1 & 0 \\ 0 & 0 & -3 & -3 & 0 & 1 \end{array}\right] \quad -3R_1 + R_3$$

Step 4 To get 1 for the second element in the second row, multiply the second row by $-1/2$.

$$\left[\begin{array}{ccc|ccc} 1 & 0 & 1 & 1 & 0 & 0 \\ 0 & 1 & 3/2 & 1 & -1/2 & 0 \\ 0 & 0 & -3 & -3 & 0 & 1 \end{array}\right] \quad (-1/2)R_2$$

Step 5 To get 1 for the third element in the third row, multiply the third row by $-1/3$.

$$\left[\begin{array}{ccc|ccc} 1 & 0 & 1 & 1 & 0 & 0 \\ 0 & 1 & 3/2 & 1 & -1/2 & 0 \\ 0 & 0 & 1 & 1 & 0 & -1/3 \end{array}\right] \quad (-1/3)R_3$$

Step 6 To get 0 for the third element in the first row, add to the first row the results of multiplying each element in row 3 by -1.

$$\left[\begin{array}{ccc|ccc} 1 & 0 & 0 & 0 & 0 & 1/3 \\ 0 & 1 & 3/2 & 1 & -1/2 & 0 \\ 0 & 0 & 1 & 1 & 0 & -1/3 \end{array}\right] \quad -1R_3 + R_1$$

Step 7 To get 0 for the third element in the second row, add to the second row the results of multiplying each element of row 3 by $-3/2$.

$$\left[\begin{array}{ccc|ccc} 1 & 0 & 0 & 0 & 0 & 1/3 \\ 0 & 1 & 0 & -1/2 & -1/2 & 1/2 \\ 0 & 0 & 1 & 1 & 0 & -1/3 \end{array}\right] \qquad (-3/2)\text{R}_3 + \text{R}_2$$

From the last transformation, we get the desired inverse.

$$A^{-1} = \begin{bmatrix} 0 & 0 & 1/3 \\ -1/2 & -1/2 & 1/2 \\ 1 & 0 & -1/3 \end{bmatrix}$$

Confirm this by forming the product $A^{-1}A$, or AA^{-1}, each of which should be equal to I_3. ■

EXAMPLE 2

Find A^{-1} given $A = \begin{bmatrix} 2 & -4 \\ 1 & -2 \end{bmatrix}$.

Using row operations to transform the first column of the augmented matrix

$$\left[\begin{array}{cc|cc} 2 & -4 & 1 & 0 \\ 1 & -2 & 0 & 1 \end{array}\right]$$

results in the following matrices:

$$\left[\begin{array}{cc|cc} 1 & -2 & 1/2 & 0 \\ 1 & -2 & 0 & 1 \end{array}\right] \qquad (1/2)\text{R}_1$$

$$\left[\begin{array}{cc|cc} 1 & -2 & 1/2 & 0 \\ 0 & 0 & -1/2 & 1 \end{array}\right]. \qquad (-1)\text{R}_1 + \text{R}_2$$

At this point, the matrix should be changed so that the second-row, second-column element will be 1. Since that element is now 0, there is no way to complete the desired transformation, so matrix A^{-1} does not exist. ■

If the inverse of a matrix exists, it is unique. That is, any given square matrix has no more than one inverse. The proof of this is left to Exercise 45 of this section.

Solving Systems by Inverses Matrix inverses can be used to solve square linear systems of equations. (A square system has the same number of equations as variables.) For example, given the linear system

$$\begin{aligned} a_{11}x + a_{12}y + a_{13}z &= b_1 \\ a_{21}x + a_{22}y + a_{23}z &= b_2 \\ a_{31}x + a_{32}y + a_{33}z &= b_3, \end{aligned}$$

the definition of matrix multiplication can be used to rewrite the system as

$$\begin{bmatrix} a_{11} & a_{12} & a_{13} \\ a_{21} & a_{22} & a_{23} \\ a_{31} & a_{32} & a_{33} \end{bmatrix} \cdot \begin{bmatrix} x \\ y \\ z \end{bmatrix} = \begin{bmatrix} b_1 \\ b_2 \\ b_3 \end{bmatrix}. \quad \textbf{(1)}$$

(To see this, multiply the matrices on the left.)

$$\text{If} \quad A = \begin{bmatrix} a_{11} & a_{12} & a_{13} \\ a_{21} & a_{22} & a_{23} \\ a_{31} & a_{32} & a_{33} \end{bmatrix}, \quad X = \begin{bmatrix} x \\ y \\ z \end{bmatrix}, \quad \text{and} \quad B = \begin{bmatrix} b_1 \\ b_2 \\ b_3 \end{bmatrix},$$

the system given in (1) becomes

$$AX = B.$$

If A^{-1} exists, then both sides of $AX = B$ can be multiplied on the left to get

$$\begin{aligned} A^{-1}(AX) &= A^{-1}B && \\ (A^{-1}A)X &= A^{-1}B && \text{Associative property} \\ I_3X &= A^{-1}B && \text{Inverse property} \\ X &= A^{-1}B. && \text{Identity property} \end{aligned}$$

Matrix $A^{-1}B$ gives the solution of the system.

Solution of the Matrix Equation $AX = B$

If A is an $n \times n$ matrix with inverse A^{-1}, X is an $n \times 1$ matrix of variables, and B is an $n \times 1$ matrix, then the matrix equation

$$\mathbf{AX = B}$$

has the solution

$$\mathbf{X = A^{-1}B.}$$

For Graphers

This method of solving systems of linear equations can be carried out with a graphing utility. Find the inverse matrix, then multiply the inverse and matrix B.

This method of using matrix inverses to solve systems of equations is useful when the inverse is already known or when many systems of the form $AX = B$ must be solved and only B changes.

EXAMPLE 3 Use the method of matrix inverses to solve the following systems.

(a) $2x - 3y = 4$
$x + 5y = 2$

To represent the system as a matrix equation, use one matrix for the coefficients, one for the variables, and one for the constants, as follows.

$$A = \begin{bmatrix} 2 & -3 \\ 1 & 5 \end{bmatrix}, \quad X = \begin{bmatrix} x \\ y \end{bmatrix}, \quad \text{and} \quad B = \begin{bmatrix} 4 \\ 2 \end{bmatrix}$$

As shown above, the solution is given by $X = A^{-1}B$.
To solve the system, first find A^{-1}. Verify that

$$A^{-1} = \begin{bmatrix} 5/13 & 3/13 \\ -1/13 & 2/13 \end{bmatrix}.$$

Next, find the product $A^{-1}B$.

$$A^{-1}B = \begin{bmatrix} 5/13 & 3/13 \\ -1/13 & 2/13 \end{bmatrix}\begin{bmatrix} 4 \\ 2 \end{bmatrix} = \begin{bmatrix} 2 \\ 0 \end{bmatrix}$$

Since $X = A^{-1}B$,

$$X = \begin{bmatrix} x \\ y \end{bmatrix} = \begin{bmatrix} 2 \\ 0 \end{bmatrix}.$$

The final matrix shows that the solution of the system is (2, 0).

(b) $2x - 3y = 1$
$x + 5y = 20$

This system has the same matrix of coefficients. Only matrix B is different. Use A^{-1} from part (a) and multiply by B to get

$$X = A^{-1}B = \begin{bmatrix} 5/13 & 3/13 \\ -1/13 & 2/13 \end{bmatrix}\begin{bmatrix} 1 \\ 20 \end{bmatrix} = \begin{bmatrix} 5 \\ 3 \end{bmatrix},$$

giving the solution (5, 3). ■

6.7 Exercises

Decide whether or not the given matrices are inverses of each other. (Check to see if their product is the identity matrix I_n.)

1. $\begin{bmatrix} 5 & 7 \\ 2 & 3 \end{bmatrix}$ and $\begin{bmatrix} 3 & -7 \\ -2 & 5 \end{bmatrix}$

2. $\begin{bmatrix} 2 & 3 \\ 1 & 1 \end{bmatrix}$ and $\begin{bmatrix} -1 & 3 \\ 1 & -2 \end{bmatrix}$

3. $\begin{bmatrix} -1 & 2 \\ 3 & -5 \end{bmatrix}$ and $\begin{bmatrix} -5 & -2 \\ -3 & -1 \end{bmatrix}$

4. $\begin{bmatrix} 2 & 1 \\ 3 & 2 \end{bmatrix}$ and $\begin{bmatrix} 2 & 1 \\ -3 & 2 \end{bmatrix}$

5. $\begin{bmatrix} 0 & 1 & 0 \\ 0 & 0 & -2 \\ 1 & -1 & 0 \end{bmatrix}$ and $\begin{bmatrix} 1 & 0 & 1 \\ 1 & 0 & 0 \\ 0 & -1 & 0 \end{bmatrix}$

6. $\begin{bmatrix} 1 & 2 & 0 \\ 0 & 1 & 0 \\ 0 & 1 & 0 \end{bmatrix}$ and $\begin{bmatrix} 1 & -2 & 0 \\ 0 & 1 & 0 \\ 0 & -1 & 1 \end{bmatrix}$

7. $\begin{bmatrix} -1 & -1 & -1 \\ 4 & 5 & 0 \\ 0 & 1 & -3 \end{bmatrix}$ and $\begin{bmatrix} 15 & 4 & -5 \\ -12 & -3 & 4 \\ -4 & -1 & 1 \end{bmatrix}$

8. $\begin{bmatrix} 1 & 3 & 3 \\ 1 & 4 & 3 \\ 1 & 3 & 4 \end{bmatrix}$ and $\begin{bmatrix} 7 & -3 & -3 \\ -1 & 1 & 0 \\ -1 & 0 & 1 \end{bmatrix}$

Find the inverse, if it exists, for each matrix.

9. $\begin{bmatrix} -1 & 2 \\ -2 & -1 \end{bmatrix}$

10. $\begin{bmatrix} 1 & -1 \\ 2 & 0 \end{bmatrix}$

11. $\begin{bmatrix} -1 & -2 \\ 3 & 4 \end{bmatrix}$

12. $\begin{bmatrix} 3 & -1 \\ -5 & 2 \end{bmatrix}$

13. $\begin{bmatrix} 5 & 10 \\ -3 & -6 \end{bmatrix}$

14. $\begin{bmatrix} -6 & 4 \\ -3 & 2 \end{bmatrix}$

15. $\begin{bmatrix} 1 & 0 & 1 \\ 0 & -1 & 0 \\ 2 & 1 & 1 \end{bmatrix}$

16. $\begin{bmatrix} 1 & 0 & 0 \\ 0 & -1 & 0 \\ 1 & 0 & 1 \end{bmatrix}$

17. $\begin{bmatrix} 1 & 3 & 3 \\ 1 & 4 & 3 \\ 1 & 3 & 4 \end{bmatrix}$

18. $\begin{bmatrix} -2 & 2 & 4 \\ -3 & 4 & 5 \\ 1 & 0 & 2 \end{bmatrix}$

19. $\begin{bmatrix} 2 & 2 & -4 \\ 2 & 6 & 0 \\ -3 & -3 & 5 \end{bmatrix}$

20. $\begin{bmatrix} 2 & 4 & 6 \\ -1 & -4 & -3 \\ 0 & 1 & -1 \end{bmatrix}$

21. $\begin{bmatrix} 1 & 1 & 0 & 2 \\ 2 & -1 & 1 & -1 \\ 3 & 3 & 2 & -2 \\ 1 & 2 & 1 & 0 \end{bmatrix}$

22. $\begin{bmatrix} 1 & -2 & 3 & 0 \\ 0 & 1 & -1 & 1 \\ -2 & 2 & -2 & 4 \\ 0 & 2 & -3 & 1 \end{bmatrix}$

Solve each system of equations by using the inverse of the coefficient matrix.

23. $\begin{aligned} x - 2y &= 2 \\ 3x - 5y &= 6 \end{aligned}$

24. $\begin{aligned} 3x + 7y &= 10 \\ -x - 9y &= -20 \end{aligned}$

25. $\begin{aligned} -x - 2y &= 8 \\ 3x + 4y &= 24 \end{aligned}$

26. $\begin{aligned} -x + y &= 1 \\ 2x - y &= 1 \end{aligned}$

27. $\begin{aligned} 3x - 6y &= 1 \\ -5x + 9y &= -1 \end{aligned}$

28. $\begin{aligned} 3x - 6y &= 2 \\ -5x + 9y &= 1 \end{aligned}$

Solve each system of equations by using the inverse of the coefficient matrix. The inverses for the first four problems are found in Exercises 17–20 above. Assume b is a constant in Exercises 35–38.

29. $\begin{aligned} x + 3y + 3z &= 1 \\ x + 4y + 3z &= 0 \\ x + 3y + 4z &= -1 \end{aligned}$

30. $\begin{aligned} -2x + 2y + 4z &= 3 \\ -3x + 4y + 5z &= 1 \\ x + 2z &= 2 \end{aligned}$

31. $\begin{aligned} 2x + 2y - 4z &= 12 \\ 2x + 6y &= 16 \\ -3x - 3y + 5z &= -20 \end{aligned}$

32. $\begin{aligned} 2x + 4y + 6z &= 4 \\ -x - 4y - 3z &= 8 \\ y - z &= -4 \end{aligned}$

33. $\begin{aligned} x + y - 3z &= 4 \\ 2x + 4y - 4z &= 8 \\ -x + y + 4z &= -3 \end{aligned}$

34. $\begin{aligned} x + 2y + 3z &= 5 \\ 2x + 3y + 2z &= 2 \\ -x - 2y - 4z &= -1 \end{aligned}$

35. $\begin{aligned} 4x + 2y &= 7 \\ bx + 5y &= 8 \end{aligned}$

36. $\begin{aligned} 2x + 3y &= 4 \\ 5x + 6y &= b \end{aligned}$

37. $\begin{aligned} 2x - y &= b^2 \\ x + y &= b \end{aligned}$

38. $\begin{aligned} bx + 4y &= 1 \\ x + by &= b \end{aligned}$

Solve each system of equations by using the inverse of the coefficient matrix. The inverses were found in Exercises 21 and 22.

39.
$$\begin{aligned} x + y + 2w &= 3 \\ 2x - y + z - w &= 3 \\ 3x + 3y + 2z - 2w &= 5 \\ x + 2y + z &= 3 \end{aligned}$$

40.
$$\begin{aligned} x - 2y + 3z &= 1 \\ y - z + w &= -1 \\ -2x + 2y - 2z + 4w &= 2 \\ 2y - 3z + w &= -3 \end{aligned}$$

Let $A = \begin{bmatrix} a & b \\ c & d \end{bmatrix}$ *and let* O *be the* 2×2 *matrix of all zeros.*

Show that the statements in Exercises 41–44 are true.

41. $A \cdot O = O \cdot A = O$

42. For square matrices A and B of the same order, if $AB = O$ and if A^{-1} exists, then $B = O$.

43. $A \cdot A^{-1} = A^{-1} \cdot A = I_2$

44. $A^{-1} = \dfrac{1}{ad - bc} \cdot \begin{bmatrix} d & -b \\ -c & a \end{bmatrix}$

45. Prove that, if it exists, the inverse of a matrix is unique. (*Hint:* Assume there are two inverses B and C for some matrix A, so that $AB = BA = I$ and $AC = CA = I$. Multiply both sides of $AB = I$ on the left by C and then simplify.)

46. The Bread Box Bakery sells three types of cakes, each requiring the amounts of the basic ingredients shown in the following matrix.

	Types of cakes		
	I	**II**	**III**
Flour (in cups)	2	4	2
Sugar (in cups)	2	1	2
Eggs	2	1	3

To fill its daily orders for these three kinds of cake, the bakery uses 72 cups of flour, 48 cups of sugar, and 60 eggs.

(a) Write a 3×1 matrix for the amounts used daily.

(b) Let the number of daily orders for cakes be a 3×1 matrix X with entries x_1, x_2, and x_3. Write a matrix equation that you can solve for X, using the given matrix and the matrix from part (a).

(c) Solve the equation you wrote in part (b) to find the number of daily orders for each type of cake.

47. Let $A = \begin{bmatrix} a & b \\ c & d \end{bmatrix}$. Under what conditions on a, b, c, d does A^{-1} exist?

(*Hint:* See Exercise 44.)

48. Give an example of two matrices A and B, where $(AB)^{-1} \neq A^{-1}B^{-1}$.

49. Suppose A and B are matrices where A^{-1}, B^{-1}, and AB all exist. Show that $(AB)^{-1} = B^{-1}A^{-1}$.

50. Let $A = \begin{bmatrix} a & 0 & 0 \\ 0 & b & 0 \\ 0 & 0 & c \end{bmatrix}$, where a, b, and c are nonzero real numbers. Find A^{-1}.

51. Let $A = \begin{bmatrix} 1 & 0 & 0 \\ 0 & 0 & -1 \\ 0 & 1 & -1 \end{bmatrix}$. Show that $A^3 = I$ and use this result to find the inverse of A.

52. What are the inverses of I, $-A$ (in terms of A), and kA (k a scalar)?

53. Give two ways to use matrices to solve a system of linear equations. Will they both work in all situations? In which situtations does each method excel?

54. Discuss the similarities and differences between solving the linear equation $ax = b$ and solving the matrix equation $AX = B$.

Graphing Utility Problems

Find the inverses of the following matrices.

55. $\begin{bmatrix} \sqrt{2} & .5 \\ -17 & 1/2 \end{bmatrix}$

56. $\begin{bmatrix} 2/3 & .7 \\ 22 & \sqrt{3} \end{bmatrix}$

Use matrix inversion to solve the following systems of linear equations.

57. $$\begin{aligned} x - \sqrt{2}y &= 2.6 \\ \frac{3}{4}x + y &= 7 \end{aligned}$$

58. $$\begin{aligned} 2.1x + y &= \sqrt{5} \\ \sqrt{2}x - 2y &= 5 \end{aligned}$$

6.8 Partial Fractions

In Chapter 1 sums of rational expressions were found by combining two or more rational expressions into one rational expression. Here the reverse problem is considered: given one rational expression, express it as the sum of two or more rational expressions. A special type of sum of rational expressions is called the **partial fraction decomposition**; each term in the sum is a **partial fraction**. The technique of decomposing a rational expression into partial fractions is useful in calculus and other areas of mathematics.

To form a partial fraction decomposition of a rational expression, use the following steps.

Partial Fraction Decomposition of $\dfrac{f(x)}{g(x)}$

Step 1 If $f(x)/g(x)$ is not a proper fraction (a fraction with the numerator of lower degree than the denominator), divide $f(x)$ by $g(x)$. For example,

$$\frac{x^4 - 3x^3 + x^2 + 5x}{x^2 + 3} = x^2 - 3x - 2 + \frac{14x + 6}{x^2 + 3}.$$

Then apply the following steps to the remainder, which is a proper fraction.

Step 2 Factor $g(x)$ completely into factors of the form $(ax + b)^m$ or $(cx^2 + dx + e)^n$, where $cx^2 + dx + e$ is irreducible and m and n are integers.

Step 3

(a) For each distinct linear factor $(ax + b)$, the decomposition must include the term

$$\frac{A}{ax + b}.$$

(b) For each repeated linear factor $(ax + b)^m$, the decomposition must include the terms

$$\frac{A_1}{ax + b} + \frac{A_2}{(ax + b)^2} + \cdots + \frac{A_m}{(ax + b)^m}.$$

Step 4

(a) For each distinct quadratic factor $(cx^2 + dx + e)$, the decomposition must include the term

$$\frac{Bx + C}{cx^2 + dx + e}.$$

(b) For each repeated quadratic factor $(cx^2 + dx + e)^n$ the decomposition must include the terms

$$\frac{B_1x + C_1}{cx^2 + dx + e} + \frac{B_2x + C_2}{(cx^2 + dx + e)^2} + \cdots + \frac{B_nx + C_n}{(cx^2 + dx + e)^n}.$$

Step 5 Use algebraic techniques to solve for the constants in the numerators of the decomposition.

To find the constants in Step 5, the goal is to get a system of equations with as many equations as there are unknowns in the numerators. One method for getting these equations is to substitute values for x on both sides of the rational equation formed from Steps 3 or 4 on the preceding page.

EXAMPLE 1 Find the partial fraction decomposition of

$$\frac{2x^4 - 8x^2 + 5x - 2}{x^3 - 4x}.$$

The given fraction is not a proper fraction; the numerator has higher degree than the denominator. Perform the division.

$$\begin{array}{r|l} & 2x \\ \hline x^3 - 4x & 2x^4 - 8x^2 + 5x - 2 \\ & \underline{2x^4 - 8x^2} \\ & \qquad\qquad 5x - 2 \end{array}$$

The quotient is $$\frac{2x^4 - 8x^2 + 5x - 2}{x^3 - 4x} = 2x + \frac{5x - 2}{x^3 - 4x}.$$

Now work with the remainder fraction. Factor the denominator as $x^3 - 4x = x(x + 2)(x - 2)$. Since the factors are **distinct linear factors**, use Step 3(a) to write the decomposition as

$$\frac{5x - 2}{x^3 - 4x} = \frac{A}{x} + \frac{B}{x + 2} + \frac{C}{x - 2}, \qquad \textbf{(1)}$$

where A, B, and C are constants that need to be found. Multiply both sides of equation (1) by $x(x + 2)(x - 2)$, getting

$$5x - 2 = A(x + 2)(x - 2) + Bx(x - 2) + Cx(x + 2). \qquad \textbf{(2)}$$

Equation (1) is an identity, since both sides represent the same rational expression. Thus, equation (2) is also an identity. Equation (1) holds for all values of x except 0, -2, and 2. However, equation (2) holds for all values of x. In particular, substituting 0 for x in equation (2) gives

$$-2 = -4A, \qquad \text{so that} \qquad A = \frac{1}{2}.$$

Similarly, choosing $x = -2$ gives

$$-12 = 8B, \qquad \text{so that} \qquad B = -\frac{3}{2}.$$

Finally, choosing $x = 2$,

$$8 = 8C, \qquad \text{so that} \qquad C = 1.$$

The remainder rational expression can be written as the following sum of partial fractions:

$$\frac{5x - 2}{x^3 - 4x} = \frac{1}{2x} + \frac{-3}{2(x + 2)} + \frac{1}{x - 2},$$

and the given rational expression can be written as

$$\frac{2x^4 - 8x^2 + 5x - 2}{x^3 - 4x} = 2x + \frac{1}{2x} + \frac{-3}{2(x + 2)} + \frac{1}{x - 2}.$$

Check the work by combining the terms on the right. ■

EXAMPLE 2 Find the partial fraction decomposition of $\frac{2x}{(x - 1)^3}$.

This is a proper fraction. The denominator is already factored with **repeated linear factors**. Write the decomposition as shown, by using Step 3(b) above.

$$\frac{2x}{(x - 1)^3} = \frac{A}{x - 1} + \frac{B}{(x - 1)^2} + \frac{C}{(x - 1)^3}$$

Clear the denominators by multiplying both sides of this equation by $(x - 1)^3$.

$$2x = A(x - 1)^2 + B(x - 1) + C$$

Substituting 1 for x leads to $C = 2$, so that

$$2x = A(x - 1)^2 + B(x - 1) + 2. \quad \textbf{(1)}$$

The only root has been substituted and values for A and B still need to be found. However, *any* number can be substituted for x. For example, when we choose $x = -1$ (because it is easy to substitute), equation (1) becomes

$$-2 = 4A - 2B + 2$$

$$-4 = 4A - 2B$$

$$-2 = 2A - B. \quad \textbf{(2)}$$

Substituting 0 for x in equation (1) gives

$$0 = A - B + 2$$

$$2 = -A + B. \quad \textbf{(3)}$$

Now solve the system of equations (2) and (3) to get $A = 0$ and $B = 2$. The partial fraction decomposition is

$$\frac{2x}{(x - 1)^3} = \frac{2}{(x - 1)^2} + \frac{2}{(x - 1)^3}.$$

Three substitutions were needed because there were three constants to evaluate, A, B, and C.

To check this result, combine the terms on the right. ■

EXAMPLE 3 Find the partial fraction decomposition of $\dfrac{x^2 + 3x - 1}{(x + 1)(x^2 + 2)}$.

This denominator has **distinct linear and quadratic factors** where neither is repeated. Since $x^2 + 2$ cannot be factored, it is irreducible. The partial fraction decomposition is

$$\frac{x^2 + 3x - 1}{(x + 1)(x^2 + 2)} = \frac{A}{x + 1} + \frac{Bx + C}{x^2 + 2}.$$

Multiply both sides by $(x + 1)(x^2 + 2)$ to get

$$x^2 + 3x - 1 = A(x^2 + 2) + (Bx + C)(x + 1). \qquad \textbf{(1)}$$

First substitute -1 for x to get

$$\begin{aligned} (-1)^2 + 3(-1) - 1 &= A[(-1)^2 + 2] + 0 \\ -3 &= 3A \\ A &= -1. \end{aligned}$$

Replace A with -1 in equation (1) and substitute any value for x. For instance, if $x = 0$,

$$\begin{aligned} 0^2 + 3(0) - 1 &= -1(0^2 + 2) + (B \cdot 0 + C)(0 + 1) \\ -1 &= -2 + C \\ C &= 1. \end{aligned}$$

Now, letting $A = -1$ and $C = 1$, substitute again in equation (1) using another number for x. For $x = 1$,

$$\begin{aligned} 3 &= -3 + (B + 1)(2) \\ 6 &= 2B + 2 \\ B &= 2. \end{aligned}$$

Using $A = -1$, $B = 2$, and $C = 1$, the partial fraction decomposition is

$$\frac{x^2 + 3x - 1}{(x + 1)(x^2 + 2)} = \frac{-1}{x + 1} + \frac{2x + 1}{x^2 + 2}.$$

Again, this work can be checked by combining terms on the right. ■

For fractions with denominators that have quadratic factors, another method is often more convenient. The system of equations is formed by equating coefficients of like terms on both sides of the partial fraction decomposition. For instance, in Example 3, after both sides were multiplied by the common denominator, the equation was

$$x^2 + 3x - 1 = A(x^2 + 2) + (Bx + C)(x + 1).$$

Multiplying on the right and collecting like terms, we have

$$x^2 + 3x - 1 = Ax^2 + 2A + Bx^2 + Bx + Cx + C$$
$$x^2 + 3x - 1 = (A + B)x^2 + (B + C)x + (C + 2A).$$

Now, equating the coefficients of like powers of x gives the three equations

$$1 = A + B$$
$$3 = B + C$$
$$-1 = C + 2A.$$

Solving this system of equations for A, B, and C would give the partial fraction decomposition. The next example uses a combination of the two methods.

EXAMPLE 4

Find the partial decomposition of $\dfrac{2x}{(x^2 + 1)^2(x - 1)}$.

This expression has both a linear factor and a **repeated quadratic factor**. By Steps 3(a) and 4(b) from the beginning of this section,

$$\frac{2x}{(x^2 + 1)^2(x - 1)} = \frac{Ax + B}{x^2 + 1} + \frac{Cx + D}{(x^2 + 1)^2} + \frac{E}{x - 1}.$$

Multiplication of both sides by $(x^2 + 1)^2(x - 1)$ leads to

$$2x = (Ax + B)(x^2 + 1)(x - 1) + (Cx + D)(x - 1) + E(x^2 + 1)^2. \quad \textbf{(1)}$$

If $x = 1$, equation (1) reduces to

$$2 = 4E, \quad \text{or} \quad E = \frac{1}{2}.$$

Substituting $1/2$ for E in equation (1) and combining terms on the right gives

$$2x = (A + 1/2)x^4 + (-A + B)x^3 + (A - B + C + 1)x^2 +$$
$$(-A + B + D - C)x + (-B - D + 1/2). \quad \textbf{(2)}$$

To get additional equations involving the unknowns, equate the coefficients of like powers of x on the two sides of equation (2). Setting corresponding coefficients of x^4 equal,

$$0 = A + \frac{1}{2} \quad \text{or} \quad A = -\frac{1}{2}.$$

From the corresponding coefficients of x^3,

$$0 = -A + B.$$

Since $A = -1/2$,

$$B = -1/2.$$

Using the coefficients of x^2,

$$0 = A - B + C + 1.$$

Since $A = -1/2$ and $B = -1/2$,

$$C = -1.$$

Finally, from the coefficients of x,

$$2 = -A + B + D - C.$$

Substituting for A, B, and C gives

$$D = 1.$$

With $A = -1/2$, $B = -1/2$, $C = -1$, $D = 1$, and $E = 1/2$, the given fraction has the partial fraction decomposition

$$\frac{2x}{(x^2+1)^2(x-1)} = \frac{-\frac{1}{2}x - \frac{1}{2}}{x^2+1} + \frac{-x+1}{(x^2+1)^2} + \frac{\frac{1}{2}}{x-1}$$

or

$$\frac{2x}{(x^2+1)^2(x-1)} = \frac{-(x+1)}{2(x^2+1)} + \frac{-x+1}{(x^2+1)^2} + \frac{1}{2(x-1)}.$$

The two methods discussed in this section are summarized below.

Methods of Solving for Constants

To solve for the constants in the numerators of a partial fraction decomposition, use either of the following methods or a combination of the two.

Method 1 For Linear Factors

1. Multiply both sides of the rational expression by the common denominator.
2. Substitute the root of each factor in the resulting equation. For repeated linear factors, substitute as many other numbers as necessary to find all the constants in the numerators. The number of substitutions required will equal the number of constants $A, B, \ldots$.

Method 2 For Quadratic Factors

1. Multiply both sides of the rational expression by the common denominator.
2. Collect terms on the right side of the resulting equation.
3. Equate the coefficients of like terms to get a system of equations.
4. Solve the system to find the constants in the numerators.

6.8 Exercises

Find the partial fraction decomposition for the following rational expressions.

1. $\dfrac{5}{3x(2x+1)}$
2. $\dfrac{3x-1}{x(x+1)}$
3. $\dfrac{4x+2}{(x+2)(2x-1)}$
4. $\dfrac{x+2}{(x+1)(x-1)}$
5. $\dfrac{x}{x^2+4x-5}$
6. $\dfrac{5x-3}{(x+1)(x-3)}$
7. $\dfrac{2x}{(x+1)(x+2)^2}$
8. $\dfrac{2}{x^2(x+3)}$
9. $\dfrac{4}{x(1-x)}$
10. $\dfrac{4x^2-4x^3}{x^2(1-x)}$
11. $\dfrac{4x^2-x-15}{x(x+1)(x-1)}$
12. $\dfrac{2x+1}{(x+2)^3}$
13. $\dfrac{x^2}{x^2+2x+1}$
14. $\dfrac{3}{x^2+4x+3}$
15. $\dfrac{2x^5+3x^4-3x^3-2x^2+x}{2x^2+5x+2}$
16. $\dfrac{6x^5+7x^4-x^2+2x}{3x^2+2x-1}$
17. $\dfrac{x^3+4}{9x^3-4x}$
18. $\dfrac{x^3+2}{x^3-3x^2+2x}$
19. $\dfrac{-3}{x^2(x^2+5)}$
20. $\dfrac{2x+1}{(x+1)(x^2+2)}$
21. $\dfrac{3x-2}{(x+4)(3x^2+1)}$
22. $\dfrac{3}{x(x+1)(x^2+1)}$
23. $\dfrac{1}{x(2x+1)(3x^2+4)}$
24. $\dfrac{x^4+1}{x(x^2+1)^2}$
25. $\dfrac{3x-1}{x(2x^2+1)^2}$
26. $\dfrac{3x^4+x^3+5x^2-x+4}{(x-1)(x^2+1)^2}$
27. $\dfrac{-x^4-8x^2+3x-10}{(x+2)(x^2+4)^2}$
28. $\dfrac{x^2}{x^4-1}$
29. $\dfrac{5x^5+10x^4-15x^3+4x^2+13x-9}{x^3+2x^2-3x}$
30. $\dfrac{3x^6+3x^4+3x}{x^4+x^2}$

Graphing Utility Problems

Determine whether each of the following partial fraction decompositions are correct by graphing the left side and the right side of the equation on the same coordinate system and observing whether the graphs overlap.

31. $\dfrac{4x^2-3x-4}{x^3+x^2-2x} = \dfrac{2}{x} + \dfrac{-1}{x-1} + \dfrac{3}{x+2}$
32. $\dfrac{1}{(x-1)(x+2)} = \dfrac{1}{x-1} - \dfrac{1}{x+2}$
33. $\dfrac{x^3-2x}{(x^2+2x+2)^2} = \dfrac{x-2}{x^2+2x+2} + \dfrac{2}{(x^2+2x+2)^2}$
34. $\dfrac{2x+4}{x^2(x-2)} = \dfrac{-2}{x} + \dfrac{-2}{x^2} + \dfrac{2}{x-2}$

Chapter 6 Review Exercises

Use the elimination or substitution method to solve each of the following linear systems. Identify any inconsistent systems or systems with dependent equations.

1. $\begin{aligned} 3x - 5y &= 7 \\ 2x + 3y &= 30 \end{aligned}$
2. $\begin{aligned} -x + 4y &= 3 \\ x + 2y &= 9 \end{aligned}$
3. $\begin{aligned} 6x - 2y &= 4 \\ 4x + 5y &= 9 \end{aligned}$
4. $\begin{aligned} \frac{1}{6}x + \frac{1}{3}y &= 8 \\ \frac{1}{4}x + \frac{1}{2}y &= 12 \end{aligned}$

5. $\begin{aligned} .2x - .3y &= 1 \\ .3x + .5y &= 11 \end{aligned}$

6. $\begin{aligned} 3x - 2y &= 0 \\ 9x + 8y &= 7 \end{aligned}$

7. $\begin{aligned} 2x - 5y + 3z &= -1 \\ x + 4y - 2z &= 9 \\ -x + 2y + 4z &= 5 \end{aligned}$

8. $\begin{aligned} 5x - y &= 26 \\ 4y + 3z &= -4 \\ 3x + 3z &= 15 \end{aligned}$

Write a system of linear equations for each of the following, and then use the system to solve the problem.

9. Three hundred people attending a club's banquet paid a total of \$4060. Each member paid \$13 and each nonmember paid \$15. How many members and how many nonmembers attended the banquet?

10. A cup of uncooked rice contains 15 g of protein and 810 cal. A cup of uncooked soybeans contains 22.5 g of protein and 270 cal. How many cups of each should be used for a meal containing 9.5 g of protein and 324 cal?

11. A company sells 3½″ diskettes for 40¢ each and sells 5¼″ diskettes for 30¢ each. The company receives \$38 for an order of 100 diskettes. However, the customer neglected to specify how many of each size to send. Determine the number of each size of diskette that should be sent.

12. The Waputi Indians make woven blankets, rugs, and skirts. Each blanket requires 24 hr for spinning the yarn, 4 hr for dyeing the yarn, and 15 hr for weaving. Rugs require 30, 5, and 18 hr and skirts 12, 3, and 9 hr, respectively. If there are 306, 59, and 201 hr available for spinning, dyeing, and weaving, respectively, how many of each item can be made? (*Hint:* Simplify the equations you write, if possible, before solving the system.)

Find solutions for the following systems with the specified arbitrary variable.

13. $\begin{aligned} 2x - 6y + 4z &= 5 \\ 5x + y - 3z &= 1 \end{aligned}$; z

14. $\begin{aligned} 3x - 4y + z &= 2 \\ 2x + y &= 1 \end{aligned}$; x

Solve each system in Exercises 15–18.

15. $\begin{aligned} x^2 &= 2y - 3 \\ x + y &= 3 \end{aligned}$

16. $\begin{aligned} 2x^2 + 3y^2 &= 30 \\ x^2 + y^2 &= 13 \end{aligned}$

17. $\begin{aligned} xy &= -2 \\ y - x &= 3 \end{aligned}$

18. $\begin{aligned} x^2 + 2xy + y^2 &= 4 \\ x - 3y &= -2 \end{aligned}$

19. Find a value of b so that the straight line $3x - y = b$ touches the circle $x^2 + y^2 = 25$ at only one point.

20. Do the circle $x^2 + y^2 = 144$ and the line $x + 2y = 8$ have any points in common? If so, what are they?

Graph the solution of each system of inequalities in Exercises 21–23.

21. $\begin{aligned} x - 3y &\geq 6 \\ y^2 &\leq 16 - x^2 \end{aligned}$

22. $\begin{aligned} x + y &\leq 6 \\ 2x - y &\geq 3 \end{aligned}$

23. $\begin{aligned} x^2 + y^2 &\leq 144 \\ x^2 + y^2 &\geq 16 \end{aligned}$

24. A bakery makes both cakes and cookies. Each batch of cakes requires 2 hr in the oven and 3 hr in the decorating room. Each batch of cookies needs 1½ hr in the oven and ¾ hr in the decorating room. The oven is available no more than 16 hr a day, while the decorating room can be used no more than 12 hr per day. Set up a system of inequalities expressing this information, and then graph the system.

25. A candy company has 100 kg of chocolate-covered nuts and 125 kg of chocolate-covered raisins to be sold as two different mixtures. One mix will contain ½ nuts and ½ raisins, while the other mix will contain ⅓ nuts and ⅔ raisins. Set up a system of inequalities expressing this information, and then graph the system.

In Exercises 26–29, use graphical methods to find nonnegative values of x and y that meet the constraints.

26. Maximize $2x + 3y$ subject to

$$x + 2y \le 24$$
$$3x + 4y \le 60.$$

27. Maximize $5x + 7y$ subject to

$$2x + 4y \ge 40$$
$$3x + 2y \le 60.$$

28. Minimize $4x + 3y$ subject to

$$x + 2y \le 12$$
$$4x + 3y \ge 12$$
$$y \le 2x.$$

29. Minimize $x + 2y$ subject to

$$5x + 4y \ge 20$$
$$2x + 6y \ge 24$$
$$x + y \le 12.$$

30. In Exercise 24, a batch of cookies produces a profit of \$20; the profit on a batch of cakes is \$30. Find the number of batches of each item which will maximize profit.

31. In Exercise 25, how much of each mixture should be made to maximize revenue if the first mix sells for \$6.00 per kilogram and the second mix sells for \$4.80 per kilogram?

Use the Gaussian reduction method to solve each of the following.

32. $\begin{aligned} 5x + 2y &= -10 \\ 3x - 5y &= -6 \end{aligned}$

33. $\begin{aligned} 2x + 3y &= 10 \\ -3x + y &= 18 \end{aligned}$

34. $\begin{aligned} 5x - 8y + z &= 1 \\ 3x - 2y + 4z &= 3 \\ 10x - 16y + 2z &= 3 \end{aligned}$

35. $\begin{aligned} 2x - y + 4z &= -1 \\ -3x + 5y - z &= 5 \\ 2x + 3y + 2z &= 3 \end{aligned}$

Find the values of all variables in the following.

36. $\begin{bmatrix} 5 & x+2 \\ -6y & z \end{bmatrix} = \begin{bmatrix} a & 3x-1 \\ 5y & 9 \end{bmatrix}$

37. $\begin{bmatrix} -6+k & 2 & a+3 \\ -2+m & 3p & 2r \end{bmatrix} + \begin{bmatrix} 3-2k & 5 & 7 \\ 5 & 8p & 5r \end{bmatrix} = \begin{bmatrix} 5 & y & 6a \\ 2m & 11 & -35 \end{bmatrix}$

Perform each of the following operations whenever possible.

38. $\begin{bmatrix} 3 \\ 2 \\ 5 \end{bmatrix} - \begin{bmatrix} 8 \\ -4 \\ 6 \end{bmatrix} + \begin{bmatrix} 1 \\ 0 \\ 2 \end{bmatrix}$

39. $\begin{bmatrix} 3 & -4 & 2 \\ 5 & -1 & 6 \end{bmatrix} + \begin{bmatrix} -3 & 2 & 5 \\ 1 & 0 & 4 \end{bmatrix}$

40. $\begin{bmatrix} -3 & 4 \\ 2 & 8 \end{bmatrix}\begin{bmatrix} -1 & 0 \\ 2 & 5 \end{bmatrix}$

41. $\begin{bmatrix} 2 & 5 & 8 \\ 1 & 9 & 2 \end{bmatrix} - \begin{bmatrix} 3 & 4 \\ 7 & 1 \end{bmatrix}$

42. $\begin{bmatrix} 1 & -2 & 4 & 2 \\ 0 & 1 & -1 & 8 \end{bmatrix}\begin{bmatrix} -1 \\ 2 \\ 0 \\ 1 \end{bmatrix}$

43. $\begin{bmatrix} 3 & 2 & -1 \\ 4 & 0 & 6 \end{bmatrix}\begin{bmatrix} -2 & 0 \\ 0 & 2 \\ 3 & 1 \end{bmatrix}$

Find each of the following determinants.

44. $\begin{vmatrix} -2 & 4 \\ 0 & 3 \end{vmatrix}$

45. $\begin{vmatrix} -1 & 8 \\ 2 & 9 \end{vmatrix}$

46. $\begin{vmatrix} -2 & 4 & 1 \\ 3 & 0 & 2 \\ -1 & 0 & 3 \end{vmatrix}$

47. $\begin{vmatrix} -1 & 2 & 3 \\ 4 & 0 & 3 \\ 5 & -1 & 2 \end{vmatrix}$

48. $\begin{vmatrix} -1 & 0 & 2 & -3 \\ 0 & 4 & 4 & -1 \\ -6 & 0 & 3 & -5 \\ 0 & -2 & 1 & 0 \end{vmatrix}$

Explain why each of the following statements is true.

49. $\begin{vmatrix} 4 & 6 \\ 3 & 5 \end{vmatrix} = \begin{vmatrix} 4 & 3 \\ 6 & 5 \end{vmatrix}$

50. $\begin{vmatrix} 8 & 9 & 2 \\ 0 & 0 & 0 \\ 3 & 1 & 4 \end{vmatrix} = 0$

51. $\begin{vmatrix} 4 & 6 & 2 \\ -3 & 8 & -5 \\ 4 & 6 & 2 \end{vmatrix} = 0$

52. $\begin{vmatrix} 8 & 2 \\ 4 & 3 \end{vmatrix} = 2\begin{vmatrix} 4 & 1 \\ 4 & 3 \end{vmatrix}$

53. $\begin{vmatrix} 8 & 2 & -5 \\ -3 & 1 & 4 \\ 2 & 0 & 5 \end{vmatrix} = -\begin{vmatrix} 8 & -5 & 2 \\ -3 & 4 & 1 \\ 2 & 5 & 0 \end{vmatrix}$

54. $\begin{vmatrix} 5 & -1 & 2 \\ 3 & -2 & 0 \\ -4 & 1 & 2 \end{vmatrix} = \begin{vmatrix} 5 & -1 & 2 \\ 8 & -3 & 2 \\ -4 & 1 & 2 \end{vmatrix}$

Solve each of the following systems by Cramer's rule if possible. Identify any dependent equations or inconsistent systems.

55. $3x + 7y = 2$
$5x - y = -22$

56. $3x + y = -1$
$5x + 4y = 10$

57. $5x - 2y - z = 8$
$-5x + 2y + z = -8$
$x - 4y - 2z = 0$

58. $3x + 2y + z = 2$
$4x - y + 3z = -16$
$x + 3y - z = 12$

Find the inverse of each of the following matrices that has an inverse.

59. $\begin{bmatrix} -4 & 2 \\ 0 & 3 \end{bmatrix}$

60. $\begin{bmatrix} 2 & 1 \\ 5 & 3 \end{bmatrix}$

61. $\begin{bmatrix} 2 & 3 & 5 \\ -2 & -3 & -5 \\ 1 & 4 & 2 \end{bmatrix}$

62. $\begin{bmatrix} 2 & -1 & 0 \\ 1 & 0 & 1 \\ 1 & -2 & 0 \end{bmatrix}$

Use the method of matrix inverses to solve each of the following.

63. $2x + y = 5$
$3x - 2y = 4$

64. $x + y + z = 1$
$2x - y = -2$
$3y + z = 2$

65. $x = -3$
$y + z = 6$
$2x - 3z = -9$

66. $3x - 2y + 4z = 1$
$4x + y - 5z = 2$
$-6x + 4y - 8z = -2$

Let $f(x) = ax + b$ and $g(x) = cx + d$. Also, let $\begin{bmatrix} a & b \\ 0 & 1 \end{bmatrix}$ correspond to f and $\begin{bmatrix} c & d \\ 0 & 1 \end{bmatrix}$ correspond to g. *

67. Show that $\begin{bmatrix} a & b \\ 0 & 1 \end{bmatrix}\begin{bmatrix} c & d \\ 0 & 1 \end{bmatrix}$ corresponds to $f \circ g$.

68. Assume that $a \neq 0$ and show that f^{-1} corresponds to $\begin{bmatrix} a & b \\ 0 & 1 \end{bmatrix}^{-1}$.

Find the partial fraction decomposition of the following rational expressions.

69. $\dfrac{x + 2}{x^3 + 2x^3 + x}$

70. $\dfrac{x - 1}{x^3 - x^2 + 4x}$

71. $\dfrac{2x + 1}{(x + 1)(x^2 - 3x + 5)}$

72. $\dfrac{x + 3}{x^3 + 64}$

73. Discuss four ways of solving a system of two linear equations in two variables. Discuss the advantages and disadvantages of each method.

*Exercises 67 and 68 are from *College Algebra and Trigonometry* by John Schiller and Marie Wurster.

7 Analytic Geometry

In this chapter we discuss **conic sections**, which are curves that represent the intersection of a plane and a cone. See Figure 1. Two examples of conic sections, circles and parabolas, were studied in Chapters 3 and 4. In this chapter we look at parabolas in more detail and study two other types of conic sections, called *ellipses* and *hyperbolas*. In the last section of this chapter, we discuss the special characteristics of the equations and graphs of the four types of conic sections.

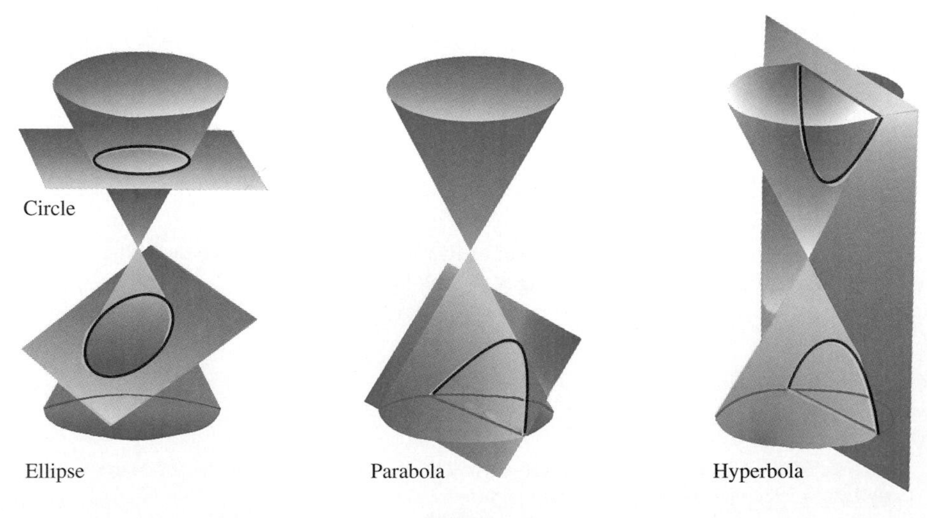

Figure 1

7.1 Parabolas

From Chapter 4, we know that the graph of the equation $y - h = a(x - k)^2$ is a parabola with vertex at (k, h) and the vertical line $x = k$ as axis. The equation $x - h = a(y - k)^2$ also has a parabola as its graph. The graph of this new equation is a reflection of the graph of $y - h = a(x - k)^2$ about the line $y = x$ since the new equation can be obtained by interchanging x and y in the original equation. Since the graph of $y - h = a(x - k)^2$ has a vertical axis and vertex at (k, h), the graph of $x - h = a(y - k)^2$ has a horizontal axis and vertex at (h, k).

Parabola with Horizontal Axis

The parabola with vertex at (h, k) and the horizontal line $y = k$ as axis has an equation of the form

$$x - h = a(y - k)^2.$$

The parabola opens to the right if $a > 0$ and to the left if $a < 0$.

For Graphers

Since the graph of $x - h = a(y - k)^2$ is not the graph of a function, it usually must be obtained with a graphing utility by graphing two functions. For instance, when $a > 0$, the graph can be obtained by graphing both $y = k + \sqrt{\frac{x - h}{a}}$ and $y = k - \sqrt{\frac{x - h}{a}}$ for $x \geq h$.

EXAMPLE 1 Graph $x = 2y^2 + 6y + 5$.

Write this equation in the form $x - h = a(y - k)^2$ by completing the square on y as follows.

$$\begin{aligned}
x &= 2(y^2 + 3y \qquad) + 5 \\
&= 2\left(y^2 + 3y + \frac{9}{4} - \frac{9}{4}\right) + 5 \\
&= 2\left(y^2 + 3y + \frac{9}{4}\right) + 2\left(-\frac{9}{4}\right) + 5 \\
&= 2\left(y + \frac{3}{2}\right)^2 + \frac{1}{2} \\
x - \frac{1}{2} &= 2\left(y - \left(-\frac{3}{2}\right)\right)^2
\end{aligned}$$

As this result shows, the vertex of the parabola is the point $(1/2, -3/2)$. The axis is the horizontal line $y = k$, or $y = -3/2$. Using the vertex and the axis and plotting a few additional points gives the graph in Figure 2. ■

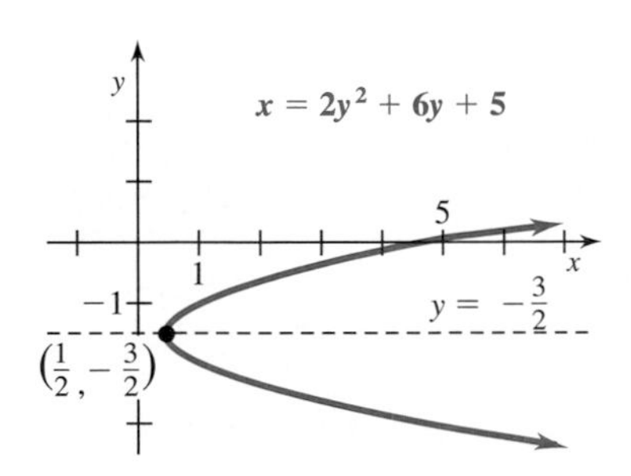

Figure 2

An equation of a parabola can be developed from the definition of a parabola as a set of points.

Definition of Parabola

A **parabola** is the set of points in a plane equidistant from a fixed point and a fixed line. The fixed point is called the **focus** and the fixed line is called the **directrix** of the parabola.

An equation of a parabola can be found from the definition as follows. Let the directrix be the line $y = -p$ and the focus be the point F with coordinates $(0, p)$, as shown in Figure 3. To get the equation of the set of points that are the same distance from the line $y = -p$ and the point $(0, p)$, choose one such point P and give it coordinates (x, y). Since $d(P, F)$ and $d(P, D)$ must have the same length, using the distance formula gives

$$\begin{aligned} d(P, F) &= d(P, D) \\ \sqrt{(x-0)^2 + (y-p)^2} &= \sqrt{(x-x)^2 + (y-(-p))^2} \\ \sqrt{x^2 + (y-p)^2} &= \sqrt{(y+p)^2} \\ x^2 + y^2 - 2yp + p^2 &= y^2 + 2yp + p^2 \\ x^2 &= 4py. \end{aligned}$$

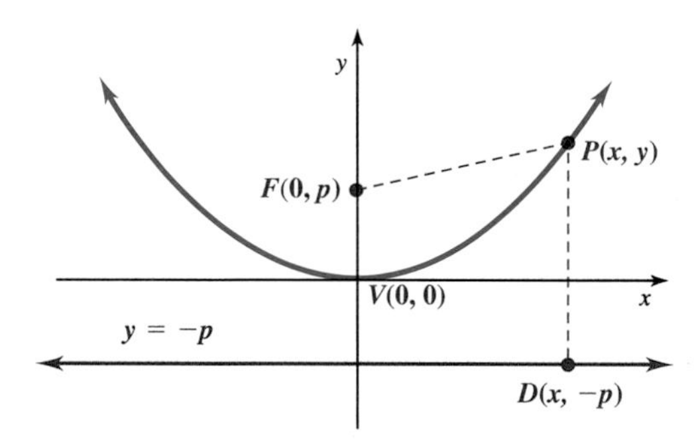

Figure 3

This discussion is summarized below.

Parabola with a Vertical Axis and Vertex (0, 0)

The parabola with focus at $(0, p)$ and directrix $y = -p$ has equation

$$\mathbf{x^2 = 4py.}$$

The parabola has a vertical axis, opens upward if $p > 0$, and opens downward if $p < 0$.

The definition of a parabola given above has led to another form of the equation $y = ax^2$, discussed in Chapter 4, with $a = 1/(4p)$.

If the directrix is the line $x = -p$, and the focus is at $(p, 0)$ using the definition of a parabola and the distance formula leads to the equation of a parabola with a horizontal axis. (See Exercise 50.)

Parabola with a Horizontal Axis and Vertex (0, 0)

The parabola with focus at $(p, 0)$ and directrix $x = -p$ has equation

$$y^2 = 4px.$$

The parabola opens to the right if $p > 0$ or to the left if $p < 0$, and it has a horizontal axis.

EXAMPLE 2 Find the focus, directrix, vertex, and axis of the following parabolas.

(a) $x^2 = 8y$

The equation has the form $x^2 = 4py$, so set $4p = 8$, from which $p = 2$. Since the x-term is squared, the parabola is vertical, with focus at $(0, p) = (0, 2)$ and directrix $y = -2$. The vertex is $(0, 0)$, and the axis of the parabola is the y-axis. (See Figure 4.)

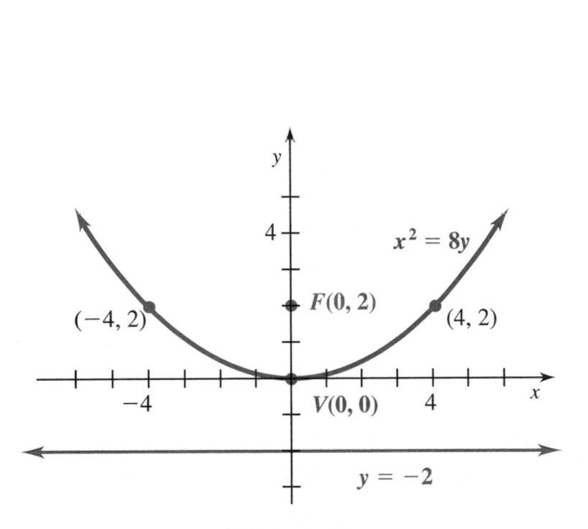

Figure 4

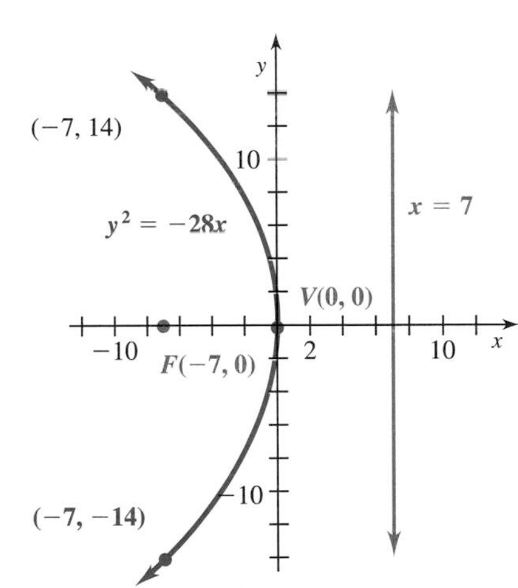

Figure 5

(b) $y^2 = -28x$

This equation has the form $y^2 = 4px$, with $4p = -28$, so $p = -7$. The parabola is horizontal, with focus $(-7, 0)$, directrix $x = 7$, vertex $(0, 0)$, and x-axis as axis of the parabola. Since p is negative, the graph opens to the left, as shown in Figure 5. ■

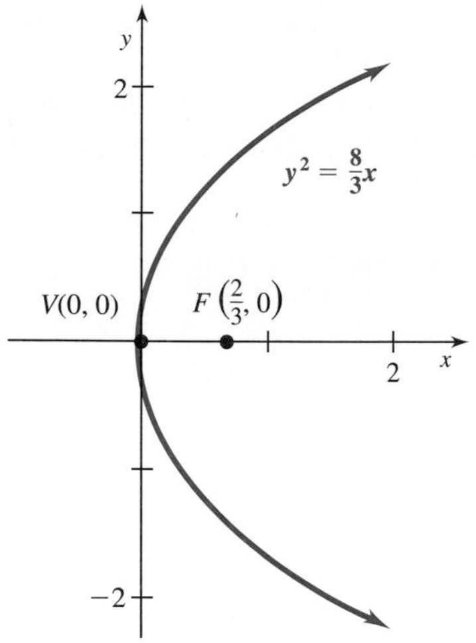

Figure 6

EXAMPLE 3 Write an equation for each of the following parabolas.

(a) Focus $(2/3, 0)$ and vertex at the origin

Since the focus $(2/3, 0)$ is on the x-axis, the parabola is horizontal and opens to the right because $p = 2/3$ is positive. (See Figure 6.) The equation, which will have the form $y^2 = 4px$, is

$$y^2 = 4\left(\frac{2}{3}\right)x \quad \text{or} \quad y^2 = \frac{8}{3}x.$$

(b) Vertical axis, vertex at the origin, through the point $(-2, 12)$

The parabola will have an equation of the form $x^2 = 4py$ because the axis is vertical. Since the point $(-2, 12)$ is on the graph, it must satisfy the equation. Substitute $x = -2$ and $y = 12$ into $x^2 = 4py$ to get

$$(-2)^2 = 4p(12)$$
$$4 = 48p$$
$$p = \frac{1}{12},$$

which gives

$$x^2 = \frac{1}{3}y \quad \text{or} \quad y = 3x^2$$

as an equation of the parabola. ■

The equations $x^2 = 4py$ and $y^2 = 4px$ can be extended to parabolas having vertex at (h, k) by replacing x and y by $x - h$ and $y - k$.

The parabola with vertex (h, k) has an equation of the form

$$(x - h)^2 = 4p(y - k) \qquad \textbf{Vertical axis}$$
$$\text{or} \quad (y - k)^2 = 4p(x - h), \qquad \textbf{Horizontal axis}$$

where the focus is distance p or $-p$ from the vertex.

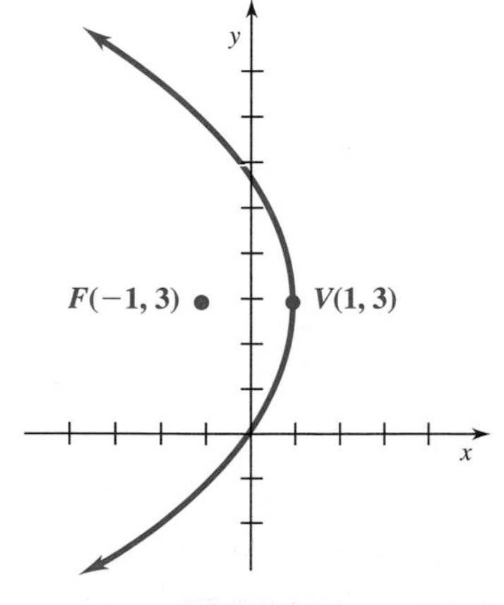

Figure 7

EXAMPLE 4 Write an equation for a parabola with vertex at $(1, 3)$ and focus at $(-1, 3)$.

Since the focus is to the left of the vertex, the axis is horizontal and the parabola opens to the left. (See Figure 7.) Since the distance between the vertex and the focus is $1 - (-1)$ or 2, the equation of the parabola is

$$(y - 3)^2 = 4(-2)(x - 1)$$
$$\text{or} \quad (y - 3)^2 = -8(x - 1).$$

The minus sign was chosen because the parabola opens to the left. ■

Parabolas have a special property, called the *reflecting property*, that make them useful in the design of telescopes, radar equipment, auto headlights, and solar furnaces. When a ray of light or a sound wave traveling parallel to the axis of a parabolic shape bounces off the parabola, it passes through the focus. For example, in the solar furnace shown in Figure 8, a parabolic mirror collects light at the focus and thereby generates intense heat at that point. The reflecting property can be used in reverse. If a light source is placed at the focus, then the reflected light rays will be directed straight ahead. This is the reason why the reflector in a car headlight is parabolic.

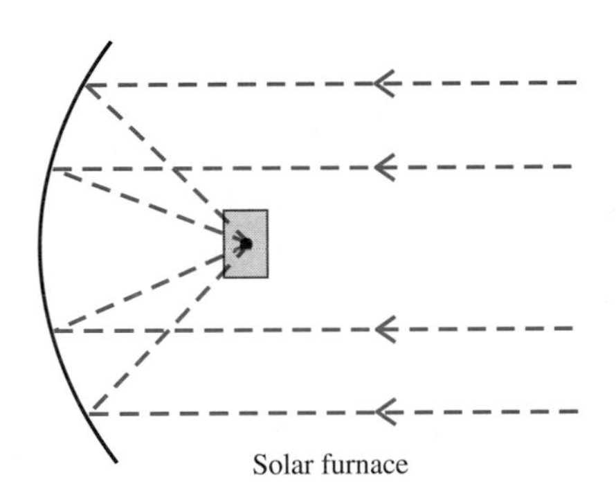

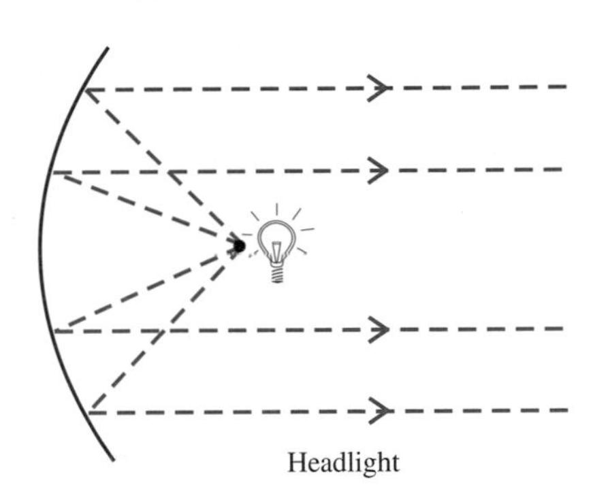

Figure 8

7.1 Exercises

Graph each horizontal parabola.

1. $x = -y^2$
2. $x = y^2 + 2$
3. $x = (y - 3)^2$
4. $x = (y + 1)^2$
5. $x = (y - 4)^2 + 2$
6. $x = (y + 2)^2 - 1$
7. $x = -3(y - 1)^2 + 2$
8. $x = -2(y + 3)^2$
9. $x = \frac{1}{2}(y - 1)^2 + 4$
10. $x = -\frac{1}{3}(y - 3)^2 + 3$
11. $x = y^2 + 4y + 2$
12. $x = 2y^2 - 4y + 6$
13. $x = -4y^2 - 4y + 3$
14. $x = -2y^2 + 2y - 3$
15. $2x = y^2 - 4y + 6$
16. $x + 3y^2 + 18y + 22 = 0$

Give the focus, directrix, and axis for each of the following parabolas.

17. $x^2 = 24y$
18. $y = 8x^2$
19. $y = -4x^2$
20. $9y = x^2$
21. $x = -32y^2$
22. $x = 16y^2$
23. $x = (-1/4)y^2$
24. $x = (-1/16)y^2$

25. $(y - 3)^2 = 12(x - 1)$ **26.** $(x + 2)^2 = 20y$ **27.** $(x - 7)^2 = 16(y + 5)$ **28.** $(y - 2)^2 = 24(x - 3)$

Write an equation for each of the following parabolas with vertex at the origin.

29. Focus $(5, 0)$ **30.** Focus $(-1/2, 0)$ **31.** Focus $(0, 1/4)$ **32.** Focus $(0, -1/3)$

33. Through $(\sqrt{3}, 3)$, opening upward **34.** Through $(2, -2\sqrt{2})$, opening to the right

35. Through $(3, 2)$, symmetric with respect to the x-axis **36.** Through $(2, -4)$, symmetric with respect to the y-axis.

Write an equation for each of the following parabolas.

37. Vertex $(4, 3)$, focus $(4, 5)$ **38.** Vertex $(-2, 1)$, focus $(-2, -3)$

39. Vertex $(-5, 6)$, focus $(2, 6)$ **40.** Vertex $(1, 2)$, focus $(4, 2)$

41. The cross-section of a parabolic mirror in a telescope is 8 ft across and 4 inches deep. Find the distance of the focus from the vertex.

42. The cable in the center portion of a bridge is supported as shown in the figure to form a parabola. The center vertical cable is 10 ft high, the supports are 210 ft high, and the distance between the two supports is 400 ft. Find the height of the remaining vertical cables, if the vertical cables are evenly spaced. (Ignore the width of the supports and cables.)

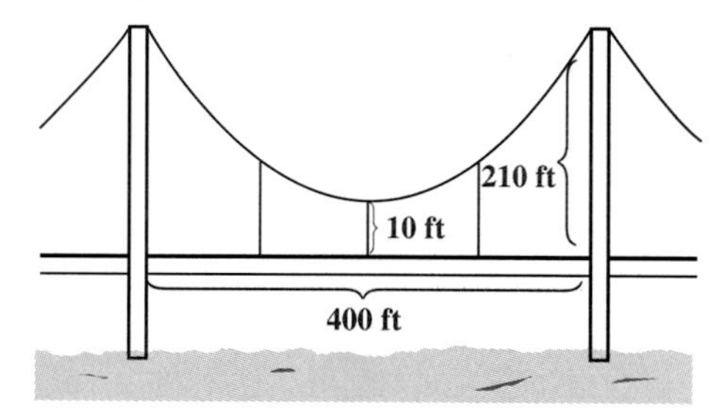

Exercise 42

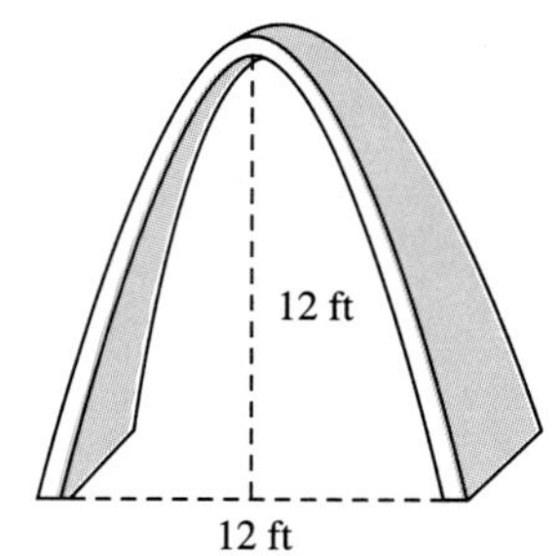

Exercise 43

43. An arch in the shape of a parabola has the dimensions shown in the figure. How wide is the arch 9 ft up?

44. Find the equation of the parabola that has vertex $(1, 2)$, has its axis parallel to the x-axis, and passes through the point $(13, 4)$.

45. Find the equation of the parabola having a horizontal axis and passing through the points $(-3/4, 0)$, $(0, 3)$, and $(0, -1)$.

46. A parabolic reflector for a car's headlight is 6 inches across and 2 inches deep. Find the distance of the bulb from the vertex. (*Note:* The bulb is located at the focus.)

47. Explain why the vertex is the point on a parabola that is closest to the focus.

48. Explain why parabolas with horizontal axes were not discussed in the section on quadratic functions.

49. How can you tell by inspecting the equation of a parabola whether the axis is horizontal or vertical?

50. Prove that the parabola with focus $(p, 0)$ and directrix $x = -p$ has equation $y^2 = 4px$.

51. How many (nonfocal) points are needed to determine a parabola? (*Hint:* Look at Exercises 44 and 45.)

Graphing Utility Problems

52. Graph the parabolas $x = \frac{1}{2}y^2$, $x = y^2$, and $x = 2y^2$ using the same scale. How does the coefficient a affect the shape of the graph of $x = ay^2$?

53. The parabola $x = 8y^2$ has $F = (2, 0)$ as focus and the line $x = -2$ as directrix. Graph the parabola and use TRACE to find the coordinates of several points on the graph. Verify that for each point P, the distance of P from the point F is the same as the distance of P from the line L.

Write an equation for each of the following parabolas. Then graph the equation and confirm each of the given properties.

54. Vertex (2, 3), passes through (−18, 1), opens to the left

55. Vertex (−1, 5), passes through (2, 4), opens to the right

7.2 Ellipses

An ellipse is defined as follows.

Definition of Ellipse

An **ellipse** is the set of all points in a plane the sum of whose distances from two fixed points is constant. Each fixed point is called a **focus** (plural, **foci**) of the ellipse.

For example, the ellipse in Figure 9 has foci at points F and F'. By the definition, the ellipse is made up of all points P such that the sum $d(P, F) + d(P, F')$ is constant. The ellipse in Figure 9 has its center at the origin. As the vertical line test shows, the graph of Figure 9 is not the graph of a function, since one value of x can lead to two values of y.

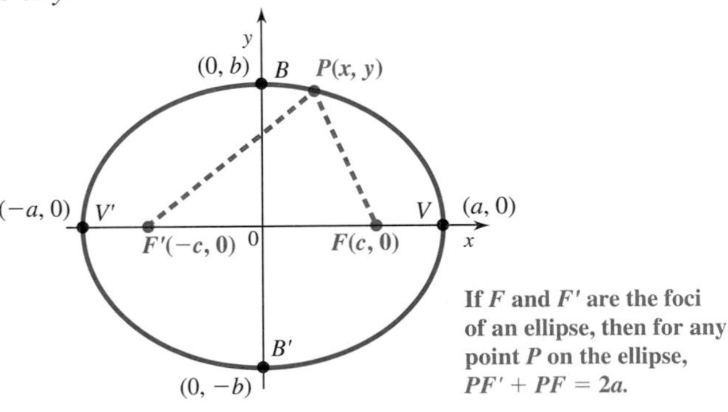

If F and F' are the foci of an ellipse, then for any point P on the ellipse, $PF' + PF = 2a$.

Figure 9

To obtain an equation for an ellipse centered at the origin, let the two foci have coordinates $(-c, 0)$ and $(c, 0)$, respectively. Let the sum of the distances from any point $P(x, y)$ on the ellipse to the two foci be $2a$. By the distance formula, segment PF has length

$$d(P, F) = \sqrt{(x - c)^2 + y^2},$$

while segment PF' has length

$$d(P, F') = \sqrt{[x - (-c)]^2 + y^2} = \sqrt{(x + c)^2 + y^2}.$$

The sum of the lengths $d(P, F)$ and $d(P, F')$ must be $2a$.

$$\sqrt{(x - c)^2 + y^2} + \sqrt{(x + c)^2 + y^2} = 2a$$

$$\sqrt{(x - c)^2 + y^2} = 2a - \sqrt{(x + c)^2 + y^2} \qquad \text{Isolate } \sqrt{(x - c)^2 + y^2}.$$

$$(x - c)^2 + y^2 = 4a^2 - 4a\sqrt{(x + c)^2 + y^2} + (x + c)^2 + y^2 \qquad \text{Square both sides.}$$

$$x^2 - 2cx + c^2 + y^2 = 4a^2 - 4a\sqrt{(x + c)^2 + y^2} + x^2 + 2cx + c^2 + y^2$$

$$4a\sqrt{(x + c)^2 + y^2} = 4a^2 + 4cx \qquad \text{Isolate } 4a\sqrt{(x + c)^2 + y^2}.$$

$$a\sqrt{(x + c)^2 + y^2} = a^2 + cx \qquad \text{Divide both sides by 4.}$$

$$a^2[x^2 + 2cx + c^2 + y^2] = a^4 + 2ca^2x + c^2x^2 \qquad \text{Square both sides.}$$

$$a^2x^2 + 2ca^2x + a^2c^2 + a^2y^2 = a^4 + 2ca^2x + c^2x^2 \qquad \text{Multiply.}$$

$$a^2x^2 + a^2c^2 + a^2y^2 = a^4 + c^2x^2 \qquad \text{Subtract } 2ca^2x \text{ from both sides.}$$

$$a^2x^2 - c^2x^2 + a^2y^2 = a^4 - a^2c^2 \qquad \text{Rearrange terms.}$$

$$(a^2 - c^2)x^2 + a^2y^2 = a^2(a^2 - c^2) \qquad \text{Factor.}$$

$$\frac{x^2}{a^2} + \frac{y^2}{a^2 - c^2} = 1 \qquad (*) \qquad \text{Divide both sides by } a^2(a^2 - c^2).$$

Since $B(0, b)$ is on the ellipse in Figure 9, we have

$$d(B, F) + d(B, F') = 2a$$

$$\sqrt{(-c)^2 + b^2} + \sqrt{c^2 + b^2} = 2a$$

$$2\sqrt{c^2 + b^2} = 2a$$

$$\sqrt{c^2 + b^2} = a$$

$$c^2 + b^2 = a^2$$

$$b^2 = a^2 - c^2.$$

Replacing $a^2 - c^2$ with b^2 in equation (*) gives

$$\frac{x^2}{a^2} + \frac{y^2}{b^2} = 1,$$

the **standard form** of the equation of an ellipse centered at the origin with foci on the x-axis.

Letting $y = 0$ in the standard form gives

$$\frac{x^2}{a^2} + \frac{0^2}{b^2} = 1$$

$$\frac{x^2}{a^2} = 1$$

$$x^2 = a^2$$

$$x = \pm a$$

as the x-intercepts of the ellipse. The points $V'(-a, 0)$ and $V(a, 0)$ are the **vertices** of the ellipse; the segment VV' is the **major axis**. In a similar manner, letting $x = 0$ shows that the y-intercepts are $\pm b$; the segment connecting $(0, b)$ and $(0, -b)$ is the **minor axis**. We assumed throughout the work above that the foci were on the x-axis. If the foci were on the y-axis, an almost identical proof could be used to get the standard form

$$\frac{x^2}{b^2} + \frac{y^2}{a^2} = 1.$$

CAUTION Do not be confused by the two standard forms—in one case a^2 is associated with x^2; in the other case a^2 is associated with y^2. However, in practice it is necessary only to find the intercepts of the graph—if the positive x-intercept is larger than the positive y-intercept, the major axis is horizontal; otherwise, it is vertical. When using the relationship $a^2 - c^2 = b^2$, or $a^2 - b^2 = c^2$, choose a^2 and b^2 so that $a^2 > b^2$. ■

A summary of this work with ellipses follows.

Standard Equations for Ellipses

The ellipse with center at the origin and equation

$$\frac{\mathbf{x}^2}{\mathbf{a}^2} + \frac{\mathbf{y}^2}{\mathbf{b}^2} = \mathbf{1} \quad (a > b)$$

has vertices $(\pm a, 0)$, endpoints of the minor axis $(0, \pm b)$, and foci $(\pm c, 0)$, where $c^2 = a^2 - b^2$.

The ellipse with center at the origin and equation

$$\frac{\mathbf{x}^2}{\mathbf{b}^2} + \frac{\mathbf{y}^2}{\mathbf{a}^2} = \mathbf{1} \quad (a > b)$$

has vertices $(0, \pm a)$, endpoints of the minor axis $(\pm b, 0)$, and foci $(0, \pm c)$, where $c^2 = a^2 - b^2$.

An ellipse is symmetric with respect to its major axis, its minor axis, and its center.

For Graphers

Since the graph of an ellipse is not the graph of a function, it usually must be obtained with a graphing utility by graphing two functions. For instance, the graph of $\frac{x^2}{a^2} + \frac{y^2}{b^2} = 1$ can be obtained by graphing both $y = \frac{b}{a}\sqrt{a^2 - x^2}$ and $y = -\frac{b}{a}\sqrt{a^2 - x^2}$ for $-a \leq x \leq a$.

EXAMPLE 1 Graph $4x^2 + 9y^2 = 36$ and find the coordinates of the foci.

To obtain the standard form for the equation of an ellipse, divide each side by 36 to get

$$\frac{x^2}{9} + \frac{y^2}{4} = 1.$$

The x-intercepts of this ellipse are ± 3, and the y-intercepts ± 2. Additional ordered pairs satisfying the equation of the ellipse may be found if desired by choosing x-values and using the equation to find the corresponding y-values. The graph of the ellipse is shown in Figure 10. Since $9 > 4$, find the foci by letting $c^2 = 9 - 4 = 5$ so that $c = \sqrt{5}$. The major axis is along the x-axis, so the foci are at $(-\sqrt{5}, 0)$ and $(\sqrt{5}, 0)$.

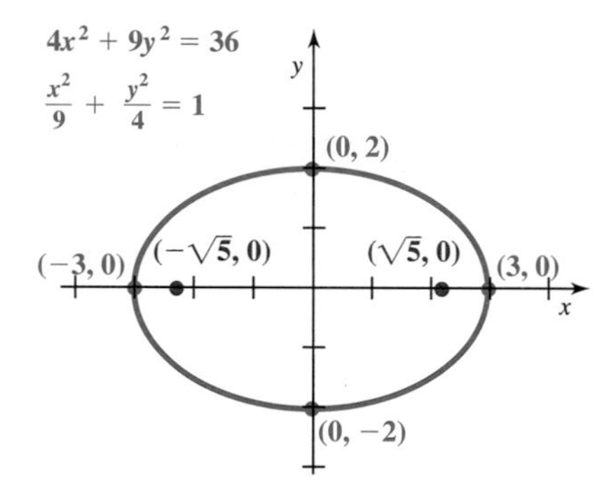

Figure 10

EXAMPLE 2 Find the equation of the ellipse having center at the origin, foci at $(0, 3)$ and $(0, -3)$, and major axis of length 8 units.

Since the major axis is 8 units long,

$$2a = 8$$

or

$$a = 4.$$

To find b^2, use the relationship $a^2 - b^2 = c^2$. Here $a = 4$ and $c = 3$. Substituting for a and c gives

$$\begin{aligned} a^2 - b^2 &= c^2 \\ 4^2 - b^2 &= 3^2 \\ 16 - b^2 &= 9 \\ b^2 &= 7. \end{aligned}$$

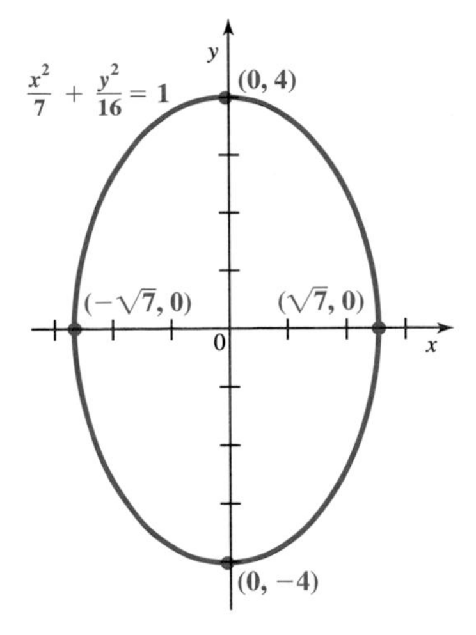

Figure 11

Since the foci are on the y-axis, the larger intercept, a, is used to find the denominator for y^2, giving the equation in standard form as

$$\frac{x^2}{7} + \frac{y^2}{16} = 1.$$

A graph of this ellipse is shown in Figure 11. ■

EXAMPLE 3 Graph $\frac{y}{4} = \sqrt{1 - \frac{x^2}{25}}$.

Square both sides to get

$$\frac{y^2}{16} = 1 - \frac{x^2}{25} \quad \text{or} \quad \frac{x^2}{25} + \frac{y^2}{16} = 1$$

as the equation of an ellipse with x-intercepts ± 5 and y-intercepts ± 4. Since $\sqrt{1 - x^2/25} \geq 0$, the only possible values of y are those making $y/4 \geq 0$, giving the half-ellipse shown in Figure 12. While the graph of the ellipse $x^2/25 + y^2/16 = 1$ is not the graph of a function, the half-ellipse in Figure 12 *is* the graph of a function. The domain of this function is the interval $[-5, 5]$, and the range is $[0, 4]$.

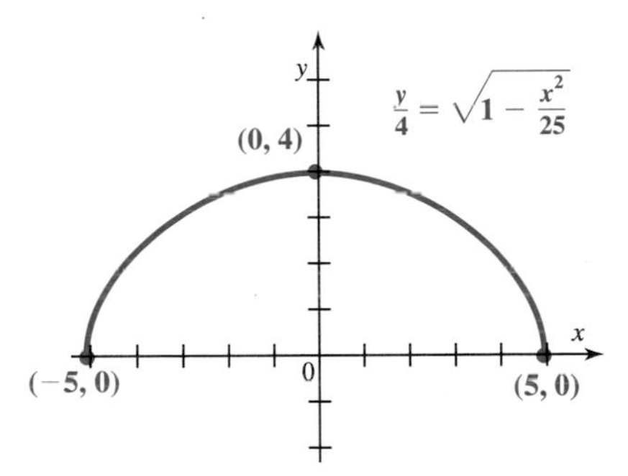

Figure 12

■

Just as a circle need not have its center at the origin, an ellipse may also have its center translated away from the origin.

Ellipse Centered at (h, k)

An ellipse centered at (h, k) with horizontal major axis of length $2a$ has equation

$$\frac{(x - h)^2}{a^2} + \frac{(y - k)^2}{b^2} = 1.$$

There is a similar result for ellipses having a vertical major axis.

21. Center at (5, 2); minor axis vertical, with length 8; $c = 3$

22. Center at $(-3, 6)$; major axis vertical, with length 10; $c = 2$

23. Vertices at (4, 9), (4, 1); minor axis with length 6

24. Foci at $(-3, -3)$, $(7, -3)$; (2, 1) on ellipse

25. Foci at $(0, -3)$, (0, 3); (8, 3) on ellipse

26. Foci at $(-4, 0)$, (4, 0); sum of distances from foci to point on ellipse is 9 (*Hint:* Consider one of the vertices.)

27. Foci at (0, 4), $(0, -4)$; sum of distances from foci to point on ellipse is 10

28. Eccentricity $\frac{1}{2}$; vertices at $(-4, 0)$, (4, 0)

29. Eccentricity $\frac{3}{4}$; foci at $(0, -2)$, (0, 2)

30. Eccentricity $\frac{2}{3}$; foci at $(0, -9)$, (0, 9)

31. Eccentricity $\frac{2}{5}$; vertices at $(-20, 0)$, (20, 0)

Sketch the graph of each of the following. Identify any that are the graphs of functions.

32. $\frac{y}{2} = \sqrt{1 - \frac{x^2}{25}}$

33. $\frac{x}{4} = \sqrt{1 - \frac{y^2}{9}}$

34. $x = -\sqrt{1 - \frac{y^2}{64}}$

35. $y = -\sqrt{1 - \frac{x^2}{100}}$

36. The figure shows an ellipse with its foci labeled. Answer the following questions with a method that does not use the equation of the ellipse.
(a) Is the point (2, 4) on the ellipse?
(b) What is the length of the major axis of the ellipse?

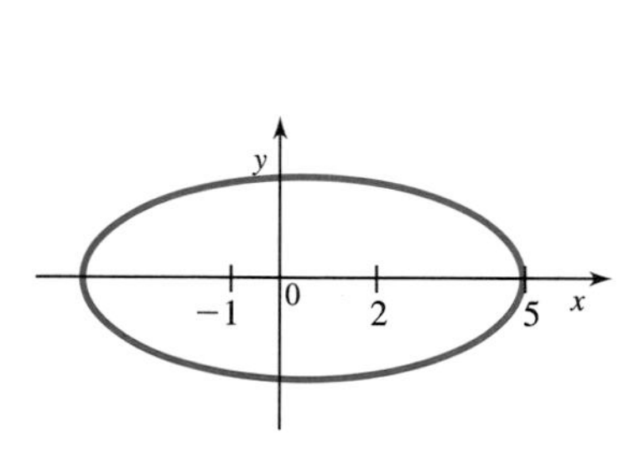

Exercise 36

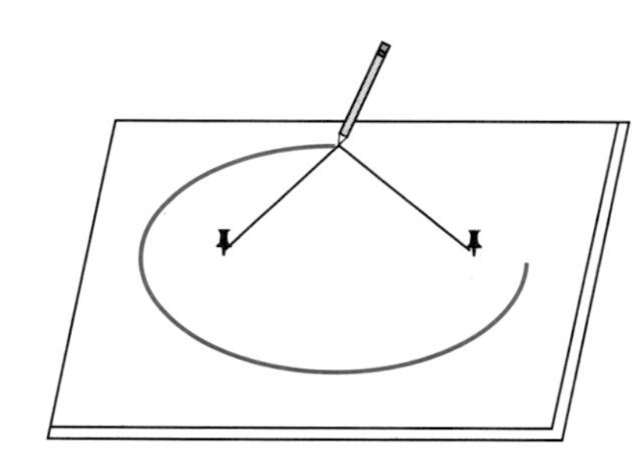

Exercise 37

37. Draftspeople often use the method shown in the sketch to draw an ellipse. Explain why the method works.

38. Suppose you know the foci of an ellipse. What additional piece of information would allow you to graph the ellipse? (Give several examples.) Explain how this information would be used.

39. Halley's Comet has an elliptical orbit of eccentricity .9673 with the sun at one of the foci. The greatest distance of the comet from the sun is 3281 million miles. Find the shortest distance between Halley's comet and the sun.

40. The orbit of planet Earth is an ellipse with the sun at one focus. The distance between Earth and the sun ranges from 91.4 to 94.6 million miles. Find the eccentricity of Earth's orbit.

41. An arch of a bridge has the shape of the top half of an ellipse. The arch is 40 ft wide and 12 ft high at the center. Find the equation of the ellipse. Find the height of the arch 10 ft from the center of the bottom.

Graphing Utility Problems

In Exercises 42–45, find the equation for the ellipse that has its center at the origin and satisfies the given conditions. Then graph the equation and confirm that the conditions hold.

42. Horizontal major axis with length 6; minor axis with length 4

43. Vertical major axis with length 10; minor axis with length 2

44. x-intercepts ± 1; y-intercepts $\pm \frac{2}{3}$

45. x-intercepts $\pm \frac{3}{4}$; y-intercepts ± 2

46. Graph the ellipse $\frac{x^2}{16} + \frac{y^2}{12} = 1$. The ellipse has foci $(-2, 0)$ and $(2, 0)$. Use TRACE to find the coordinates of several points on the ellipse. For each of these points P, verify that

$$[\text{Distance of } P \text{ from } (-2, 0)] + [\text{Distance of } P \text{ from } (2, 0)] = 8.$$

47. Find the equation of an ellipse consisting of all points P in the plane for which the sum of the distances of P from $(-4, 0)$ and $(4, 0)$ is 10. Then graph the ellipse and use TRACE to find the coordinates of several points on the graph of the ellipse. For each of these points, verify that the sum of the distances is indeed 10.

7.3 Hyperbolas

An ellipse was defined as the set of all points in a plane the sum of whose distances from two fixed points is a constant. A *hyperbola* is defined similarly.

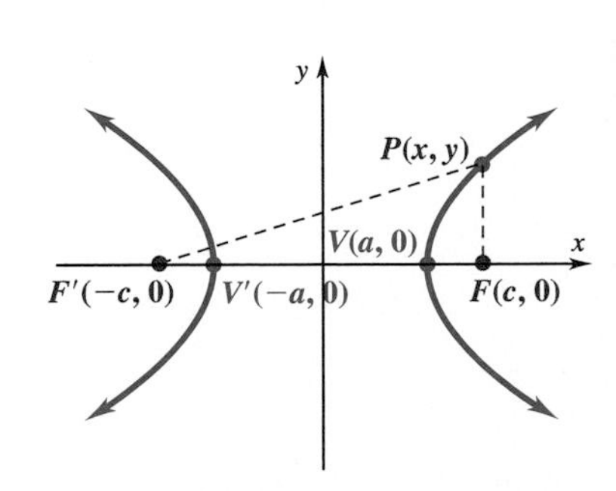

Figure 17

Definition of Hyperbola

A **hyperbola** is the set of all points in a plane the *difference* of whose distances from two fixed points is a constant. The two fixed points are called the **foci** of the hyperbola.

Suppose a hyperbola has center at the origin and foci at $F'(-c, 0)$ and $F(c, 0)$. The midpoint of the segment $F'F$ is the **center** of the hyperbola and the points $V'(-a, 0)$ and $V(a, 0)$ are the **vertices** of the hyperbola. The line segment $V'V$ is the **transverse axis** of the hyperbola. (See Figure 17.)

As with the ellipse,

$$d(V, F') - d(V, F) = (c + a) - (c - a) = 2a,$$

so the constant in the definition is $2a$, and

$$d(P, F') - d(P, F) = 2a.$$

The distance formula and algebraic manipulation similar to that used for finding an equation for an ellipse (see Exercise 47) produce the result

$$\frac{x^2}{a^2} - \frac{y^2}{c^2 - a^2} = 1.$$

Letting $b^2 = c^2 - a^2$ gives

$$\frac{x^2}{a^2} - \frac{y^2}{b^2} = 1$$

as an equation of the hyperbola in Figure 17.

Letting $y = 0$ shows that the x-intercepts are $\pm a$. If $x = 0$ the equation becomes

$$\frac{0^2}{a^2} - \frac{y^2}{b^2} = 1$$

$$-\frac{y^2}{b^2} = 1$$

$$y^2 = -b^2,$$

which has no real-number solutions, showing that this hyperbola has no y-intercepts.

EXAMPLE 1 Graph $\dfrac{x^2}{16} - \dfrac{y^2}{9} = 1.$

This hyperbola has x-intercepts 4 and -4 and no y-intercepts. To sketch the graph, we can find some other points that lie on the graph. For example, letting $x = 6$ gives

$$\frac{6^2}{16} - \frac{y^2}{9} = 1$$

$$-\frac{y^2}{9} = 1 - \frac{6^2}{16}$$

$$\frac{y^2}{9} = \frac{20}{16}$$

$$y^2 = \frac{180}{16} = \frac{45}{4}$$

$$y \approx \pm 3.4.$$

The graph includes the points (6, 3.4) and (6, -3.4). Also, letting $x = -6$ would still give $y \approx \pm 3.4$, with the points (-6, 3.4) and (-6, -3.4) also on the graph. These points, along with other points on the graph, were used to help sketch the final graph shown in Figure 18.

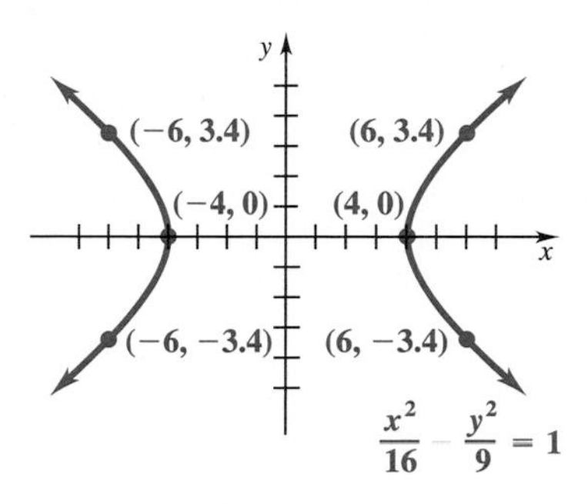

Figure 18

Starting with the equation for a hyperbola $(x^2/a^2) - (y^2/b^2) = 1$ and solving for y, we have

$$\frac{x^2}{a^2} - 1 = \frac{y^2}{b^2}$$

$$\frac{x^2 - a^2}{a^2} = \frac{y^2}{b^2}$$

or

$$y = \pm\frac{b}{a}\sqrt{x^2 - a^2}. \qquad (*)$$

If x^2 is very large in comparison to a^2, the difference $x^2 - a^2$ would be very close to x^2. If this happens, then the points satisfying equation (*) above would be very close to one of the lines

$$y = \pm\frac{b}{a}x.$$

Thus, as $|x|$ gets larger and larger, the points of the hyperbola $x^2/a^2 - y^2/b^2 = 1$ come closer and closer to the lines $y = (\pm b/a)x$. These lines, called the **asymptotes** of the hyperbola, are very helpful for graphing the hyperbola. They make it possible to avoid having to plot points.

For Graphers

Since the graph of a hyperbola is not the graph of a function, it usually must be obtained with a graphing utility by graphing two functions. For instance, the graph of $\frac{x^2}{a^2} - \frac{y^2}{b^2} = 1$ can be obtained by graphing both $y = \frac{b}{a}\sqrt{x^2 - a^2}$ and $y = -\frac{b}{a}\sqrt{x^2 - a^2}$ for $x \le -a$ and $x \ge a$.

EXAMPLE 2 Graph $\frac{x^2}{25} - \frac{y^2}{49} = 1$ and find the coordinates of the foci.

For this hyperbola, $a = 5$ and $b = 7$. With these values, $y = (\pm b/a)x$ becomes $y = (\pm 7/5)x$. If we choose $x = 5$, then $y = (\pm 7/5)(5) = \pm 7$. Choosing $x = -5$ also gives $y = \pm 7$. These four points, (5, 7), (5, -7), (-5, 7), and (-5, -7), are the corners of the rectangle shown in Figure 19. The extended diagonals of this rectangle are the asymptotes of the hyperbola. The hyperbola crosses the x-axis at 5 and -5, as shown in Figure 19. Find the foci by letting $c^2 = a^2 + b^2 = 25 + 49 = 74$ so that $c = \pm\sqrt{74}$. Therefore, the foci are $(\sqrt{74}, 0)$ and $(-\sqrt{74}, 0)$.

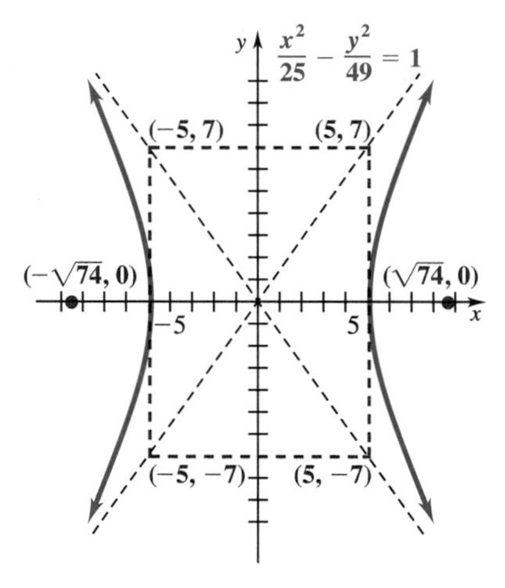

Figure 19

The rectangle used to graph the hyperbola in Example 2 is called the **fundamental rectangle**.

While $a > b$ for an ellipse, the examples above show that for hyperbolas, it is possible that $a > b$ or $a < b$; other examples would show that a might equal b, also. If the foci of a hyperbola are on the y-axis, the equation of the hyperbola is of the form

$$\frac{y^2}{a^2} - \frac{x^2}{b^2} = 1, \quad \text{with asymptotes} \quad y = \pm\frac{a}{b}x.$$

CAUTION If the foci of the hyperbola are on the x-axis, we found that the asymptotes have equations $y = \pm(b/a)x$, while foci on the y-axis lead to asymptotes $y = \pm(a/b)x$. There is an obvious chance for confusion here; to avoid mistakes write the equation of the hyperbola in either the form

$$\frac{x^2}{a^2} - \frac{y^2}{b^2} = 1 \quad \text{or} \quad \frac{y^2}{a^2} - \frac{x^2}{b^2} = 1,$$

and replace 1 with 0. Solving the resulting equation for y produces the proper equations for the asymptotes. (The reason why this process works is explained in more advanced courses.) ■

The basic information on hyperbolas is summarized as follows.

Standard Equations for Hyperbolas

The hyperbola with center at the origin and equation

$$\frac{x^2}{a^2} - \frac{y^2}{b^2} = 1$$

has vertices $(\pm a, 0)$, asymptotes $y = \pm\frac{b}{a}x$, and foci $(\pm c, 0)$, where $c^2 = a^2 + b^2$.

The hyperbola with center at the origin and equation

$$\frac{y^2}{a^2} - \frac{x^2}{b^2} = 1$$

has vertices $(0, \pm a)$, asymptotes $y = \pm\frac{a}{b}x$, and foci $(0, \pm c)$, where $c^2 = a^2 + b^2$.

EXAMPLE 3 Graph $25y^2 - 4x^2 = 100$.

Divide each side by 100 to get

$$\frac{y^2}{4} - \frac{x^2}{25} = 1.$$

This hyperbola is centered at the origin, has foci on the y axis, and has y-intercepts 2 and -2. To find the equations of the asymptotes, replace 1 with 0.

$$\frac{y^2}{4} - \frac{x^2}{25} = 0$$

$$\frac{y^2}{4} = \frac{x^2}{25}$$

$$y^2 = \frac{4x^2}{25}$$

$$y = \pm\frac{2}{5}x$$

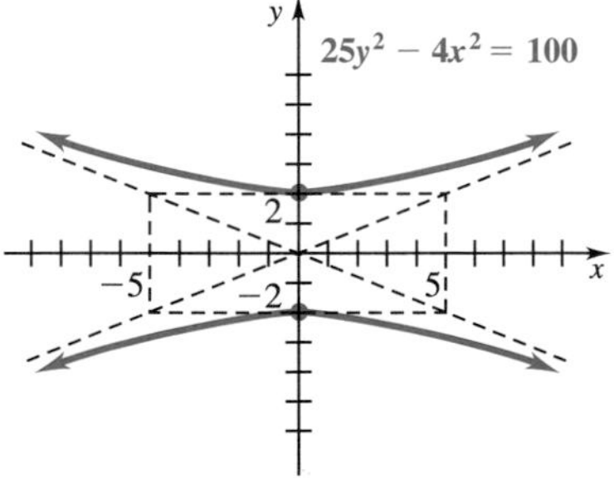

Figure 20

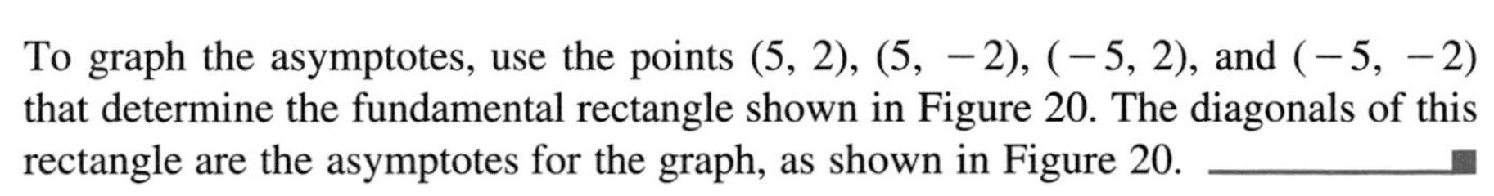

To graph the asymptotes, use the points $(5, 2)$, $(5, -2)$, $(-5, 2)$, and $(-5, -2)$ that determine the fundamental rectangle shown in Figure 20. The diagonals of this rectangle are the asymptotes for the graph, as shown in Figure 20. ■

In each of the graphs of hyperbolas considered so far, the center is the origin and the asymptotes pass through the origin. This feature holds in general; the asymptotes of *any* hyperbola pass through the center of the hyperbola. The next example involves a hyperbola with its center translated away from the origin.

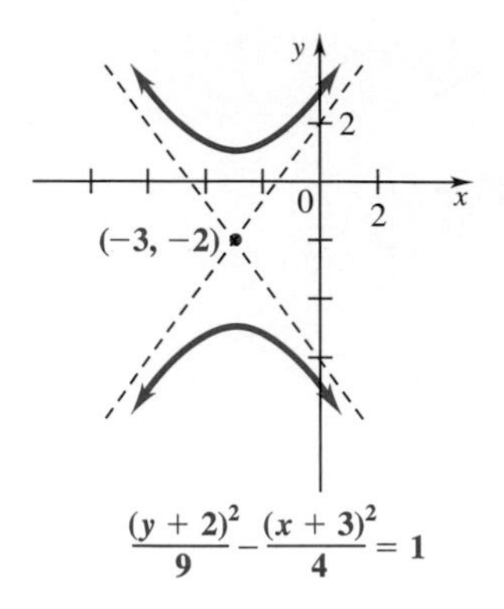

Figure 21

EXAMPLE 4 Graph $\dfrac{(y+2)^2}{9} - \dfrac{(x+3)^2}{4} = 1$.

This equation represents a hyperbola centered at $(-3, -2)$. For this vertical hyperbola, $a = 3$ and $b = 2$. Locate the y-values of the vertices by taking the y-value of the center, -2, and adding and subtracting 3. The x-values of the vertices are -3. Thus, the vertices are at $(-3, 1)$ and $(-3, -5)$. The asymptotes have slopes $\pm 3/2$ and pass through the center $(-3, -2)$. The equations of the asymptotes, $y + 2 = \pm(3/2)(x + 3)$, can be found either by using the point-slope form of the equation of a line or by replacing 1 by 0 in the equation of the hyperbola as was done in Example 3. The completed graph appears in Figure 21. ■

Our final example shows an equation whose graph is only half of a hyperbola.

EXAMPLE 5 Graph $x = -\sqrt{1 + 4y^2}$.

Squaring both sides gives

$$x^2 = 1 + 4y^2$$

or

$$x^2 - 4y^2 = 1.$$

To find the asymptotes, rewrite 4 as $1/(1/4)$ to change this equation into

$$x^2 - \frac{y^2}{1/4} = 0$$

or

$$\frac{1}{4}x^2 = y^2,$$

giving

$$y = \pm\frac{1}{2}x.$$

Since the given equation $x = -\sqrt{1 + 4y^2}$ restricts x to negative values, the graph is the left branch of the hyperbola, as shown in Figure 22.

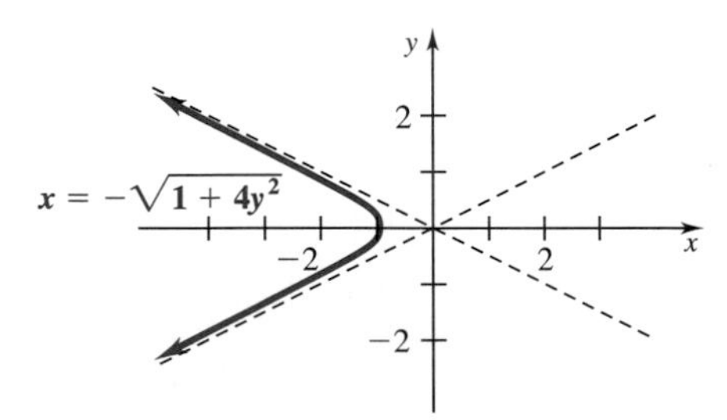

Figure 22

■

The **eccentricity**, e, of a hyperbola is defined by

$$e = \frac{\sqrt{a^2 + b^2}}{a} = \frac{c}{a}.$$

Since $c > a$, we have $e > 1$. Narrow hyperbolas have e near 1 and wide hyperbolas have large e. See Figure 23.

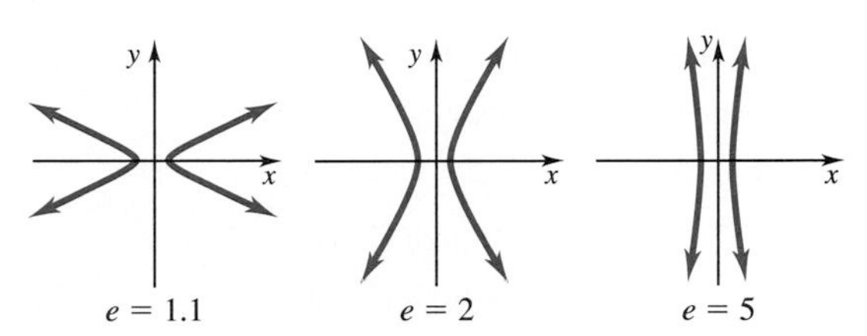

Figure 23

EXAMPLE 6 Find the equation of the hyperbola with eccentricity 2 and foci at $(-9, 5)$ and $(-3, 5)$.

Since the foci have the same y-coordinate, the line through them, and therefore the hyperbola, is horizontal. The center of the hyperbola is halfway between the two foci at $(-6, 5)$. The distance from each focus to the center is $c = 3$. Since $e = \frac{c}{a}$,

$$a = \frac{c}{e} = \frac{3}{2}$$

$$b^2 = c^2 - a^2 = 9 - \frac{9}{4} = \frac{27}{4}.$$

Therefore, the equation of the hyperbola is

$$\frac{x^2}{9/4} - \frac{y^2}{27/4} = 1.$$

7.3 Exercises

Sketch the graph of each of the following hyperbolas. Give the center, vertices, foci, and equations of the asymptotes for each figure.

1. $\frac{x^2}{16} - \frac{y^2}{9} = 1$

2. $\frac{x^2}{25} - \frac{y^2}{144} = 1$

3. $\frac{y^2}{25} - \frac{x^2}{49} = 1$

4. $\frac{y^2}{64} - \frac{x^2}{4} = 1$

5. $x^2 - y^2 = 9$

6. $x^2 - 4y^2 = 64$

7. $9x^2 - 25y^2 = 225$

8. $25x^2 - 4y^2 = -100$

9. $4x^2 - y^2 = -16$

10. $\frac{x^2}{4} - y^2 = 4$

11. $9x^2 - 4y^2 = 1$

12. $25y^2 - 9x^2 = 1$

13. $\dfrac{(y-7)^2}{36} - \dfrac{(x-4)^2}{64} = 1$

14. $\dfrac{(x+6)^2}{144} - \dfrac{(y+4)^2}{81} = 1$

15. $\dfrac{(x+3)^2}{16} - \dfrac{(y-2)^2}{9} = 1$

16. $\dfrac{(y+5)^2}{4} - \dfrac{(x-1)^2}{16} = 1$

17. $(x-8)^2 - 5(y+7)^2 = 25$

18. $2(x-2)^2 - (y-10)^2 = 8$

19. $16(x+5)^2 - (y-3)^2 = 1$

20. $4(x+9)^2 - 25(y+6)^2 = 100$

Find equations for each of the following hyperbolas.

21. x-intercepts ± 4; foci at $(-5, 0)$, $(5, 0)$

22. y-intercepts ± 9; foci at $(0, -15)$, $(0, 15)$

23. Vertices at $(0, 6)$, $(0, -6)$; asymptotes $y = \pm(1/2)x$

24. Vertices at $(-10, 0)$, $(10, 0)$; asymptotes $y = \pm 5x$

25. Vertices at $(-3, 0)$, $(3, 0)$; passing through $(6, 1)$

26. Vertices at $(0, 5)$, $(0, -5)$; passing through $(3, 10)$

27. Foci at $(0, \sqrt{13})$, $(0, -\sqrt{13})$; asymptotes $y = \pm 5x$

28. Foci at $(-\sqrt{45}, 0)$, $(\sqrt{45}, 0)$; asymptotes $y = \pm 2x$

29. Vertices at $(4, 5)$, $(4, 1)$; asymptotes $y - 3 = \pm 7(x - 4)$

30. Vertices at $(5, -2)$, $(1, -2)$; asymptotes $y + 2 = \pm\dfrac{3}{2}(x - 3)$

31. Center at $(1, -2)$; focus at $(4, -2)$; vertex at $(3, -2)$

32. Center at $(9, -7)$; focus at $(9, 3)$; vertex at $(9, -1)$

33. Eccentricity 3; center at $(0, 0)$; vertex at $(0, 7)$

34. Center at $(8, 7)$; focus at $(13, 7)$; eccentricity $5/3$

35. Vertices at $(-2, 10)$, $(-2, 2)$; eccentricity $5/4$

36. Foci at $(9, 2)$, $(-11, 2)$; eccentricity $25/9$

37. Vertices at $(5, 5)$, $(5, 11)$; fundamental rectangle of area 18

38. Vertices at $(0, 6)$, $(8, 6)$; fundamental rectangle of area 20

Sketch the graph of each of the following. Identify any that are the graphs of functions.

39. $\dfrac{y}{3} = \sqrt{1 + \dfrac{x^2}{16}}$

40. $\dfrac{x}{3} = -\sqrt{1 + \dfrac{y^2}{25}}$

41. $5x = -\sqrt{1 + 4y^2}$

42. $3y = \sqrt{4x^2 - 16}$

43. Describe the method you would use to determine (without graphing) whether a hyperbola opens left and right or up and down.

44. The figure shows a hyperbola with its foci labeled. Answer the following questions with a method that does not use the equation of the hyperbola.
(a) Is the point $(10, 2)$ on the hyperbola?
(b) What are the x-intercepts of the hyperbola?

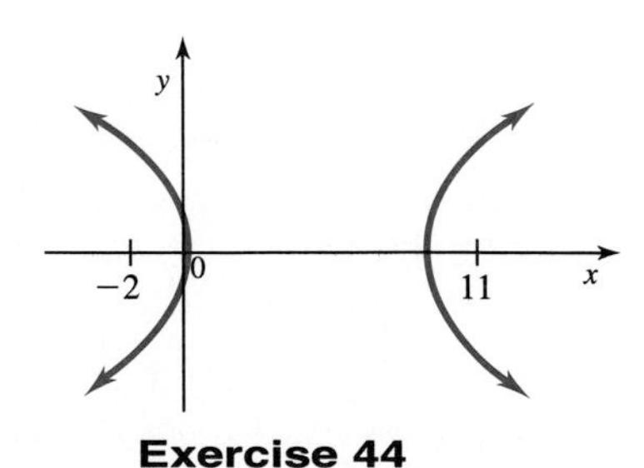

Exercise 44

45. Ships and planes often use a location finding system called LORAN. With this system, a radio transmitter at M on the figure sends out a series of pulses. When each pulse is received at transmitter S, it then sends out a pulse. A ship at P receives pulses from both M and S. A receiver on the ship measures the difference in the arrival times of the pulses. The navigator then consults a special map, showing certain curves according to the differences in arrival times. In this way, the ship can be located as lying on a portion of which curve? (This method requires three transmitters acting as two pairs.)

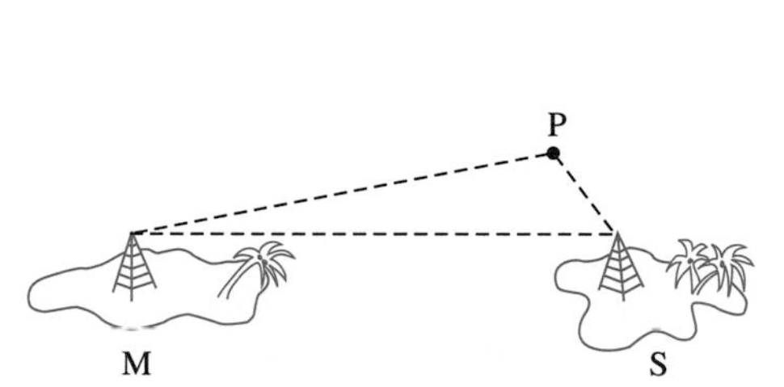

Exercise 45

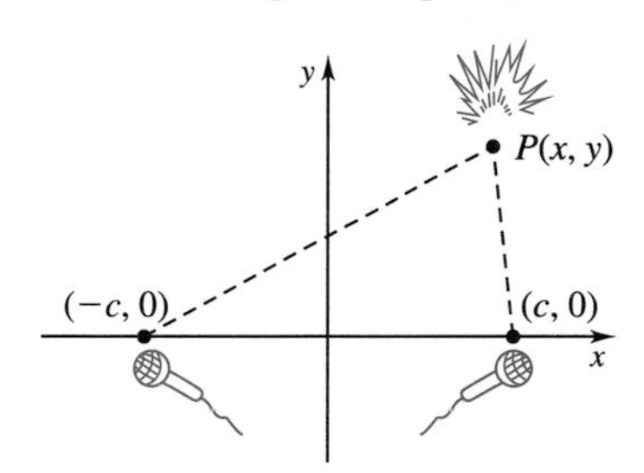

Exercise 46

46. Microphones are placed at points $(-c, 0)$ and $(c, 0)$. An explosion occurs at point $P(x, y)$ having positive x-coordinate. (See the figure.) The sound is detected at the closer microphone t sec before being detected at the farther microphone. Assume that sound travels at a speed of 330 m per sec, and show that P must be on the hyperbola

$$\frac{x^2}{330^2t^2} - \frac{y^2}{4c^2 - 330^2t^2} = \frac{1}{4}.$$

47. Suppose a hyperbola has center at the origin, foci at $F'(-c, 0)$ and $F(c, 0)$, and the value $d(P, F') - d(P, F) = 2a$. Let $b^2 = c^2 - a^2$, and show that an equation of the hyperbola is

$$\frac{x^2}{a^2} - \frac{y^2}{b^2} = 1.$$

Graphing Utility Problems

In Exercises 48–51, find the equations for the asymptotes of the hyperbola. Then graph the hyperbola and one or both (if possible) of the asymptotes to confirm that the equations of the asymptotes are correct.

48. $\dfrac{y^2}{3} - \dfrac{x^2}{4} = 1$

49. $\dfrac{x^2}{5} - \dfrac{y^2}{6} = 1$

50. $\dfrac{(x - 4)^2}{5} - \dfrac{(y - 5)^2}{4} = 1$

51. $\dfrac{(x + 7)^2}{8} - \dfrac{(y - 6)^2}{2} = 1$

52. Graph the hyperbola $\dfrac{x^2}{4} - \dfrac{y^2}{12} = 1$. The hyperbola has foci $(-4, 0)$ and $(4, 0)$. Use TRACE to find the coordinates of several points on the right half of the hyperbola. For each of these points P, verify that

$$[\text{Distance of } P \text{ from } (-4, 0)] - [\text{Distance of } P \text{ from } (4, 0)] = 4.$$

Use TRACE to find the coordinates of several points on the left half of the hyperbola. For each of these points P, verify that

$$[\text{Distance of } P \text{ from } (4, 0)] - [\text{Distance of } P \text{ from } (-4, 0)] = 4.$$

53. Find the equation of a hyperbola consisting of all points P in the plane for which the difference of the distances of P from $(-5, 0)$ and $(5, 0)$ is 8. Then graph the hyperbola and use TRACE to find the coordinates of several points on the graph of the hyperbola. For each of these points, verify that the differences of the distances is indeed 8.

54. Graph the upper half of the hyperbola $\frac{y^2}{4} - \frac{x^2}{25} = 1$ and the asymptote $y = \frac{2}{5}x$. Use ZOOM and TRACE to confirm that the curve approaches the asymptote as x gets larger and larger.

7.4 Conic Sections

The graphs of parabolas, circles, hyperbolas, and ellipses are called **conic sections** since each graph can be obtained by cutting a cone with a plane as suggested by Figure 1 at the beginning of the chapter. It turns out that all conic sections of the types presented in this chapter have equations of the form

$$Ax^2 + Cy^2 + Dx + Ey + F = 0,$$

where either A or C must be nonzero. The graphs of the conic sections are summarized in the following chart. Ellipses and hyperbolas having centers not at the origin can be shown in much the same way as we show circles and parabolas. Following the chart, the special characteristics of the equations of each of the conic sections are summarized.

Equation	*Graph*	*Description*	*Identification*
$y - k = a(x - h)^2$	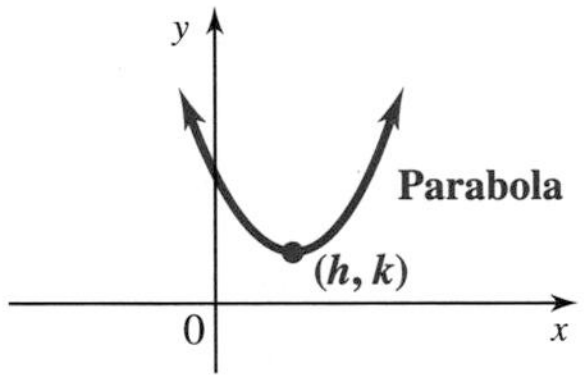 Parabola; (h, k); y; x; 0	Opens upward if $a > 0$, downward if $a < 0$. Vertex is at (h, k).	x^2 term. y is not squared.
$x - h = a(y - k)^2$	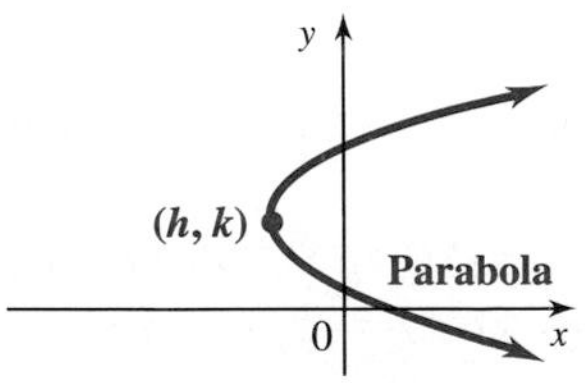 Parabola; (h, k); y; x; 0	Opens to right if $a > 0$, to left if $a < 0$. Vertex is at (h, k).	y^2 term. x is not squared.
$(x - h)^2 + (y - k)^2 = r^2$	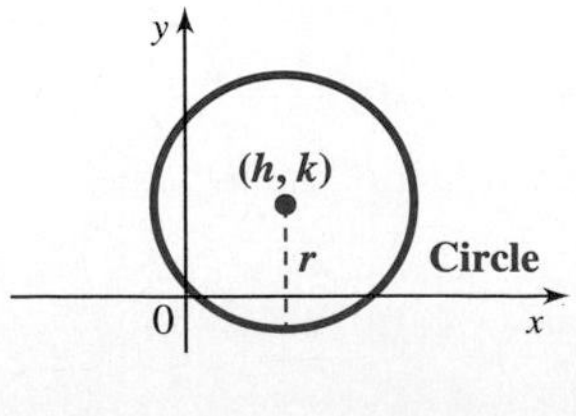 Circle; (h, k); r; y; x; 0	Center is at (h, k), radius is r	x^2 and y^2 terms have the same positive coefficient.

$$\frac{x^2}{a^2} + \frac{y^2}{b^2} = 1 \quad (a > b)$$

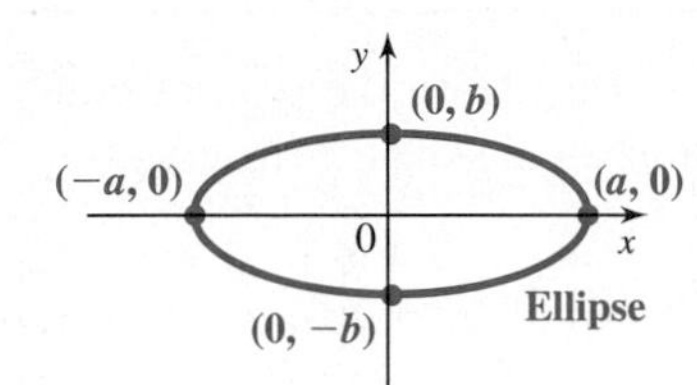

x-intercepts are a and $-a$.
y-intercepts are b and $-b$

x^2 and y^2 terms have different positive coefficients.

$$\frac{x^2}{b^2} + \frac{y^2}{a^2} = 1 \qquad (a > b)$$

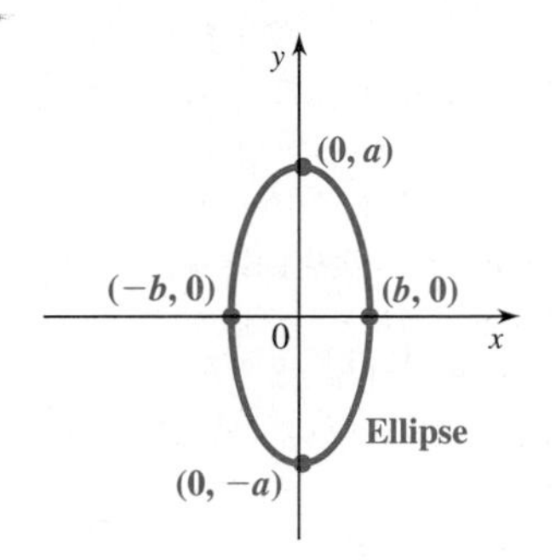

x-intercepts are b and $-b$.
y-intercepts are a and $-a$.

x^2 and y^2 terms have different positive coefficients.

$$\frac{x^2}{a^2} - \frac{y^2}{b^2} = 1$$

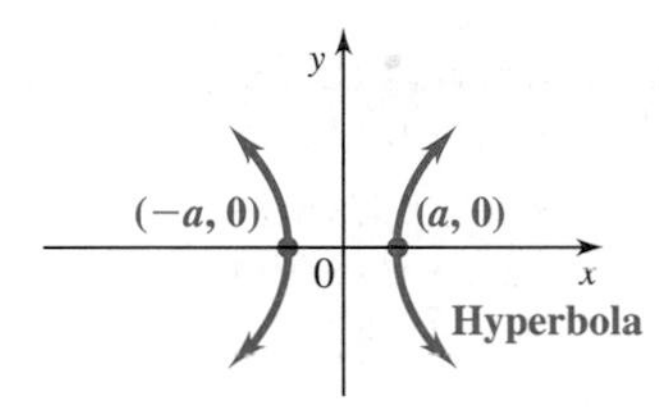

x-intercepts are a and $-a$.
Asymptotes found from (a, b), $(a, -b)$, $(-a, -b)$, and $(-a, b)$.

x^2 has a positive coefficient.
y^2 has a negative coefficient.

$$\frac{y^2}{a^2} - \frac{x^2}{b^2} = 1$$

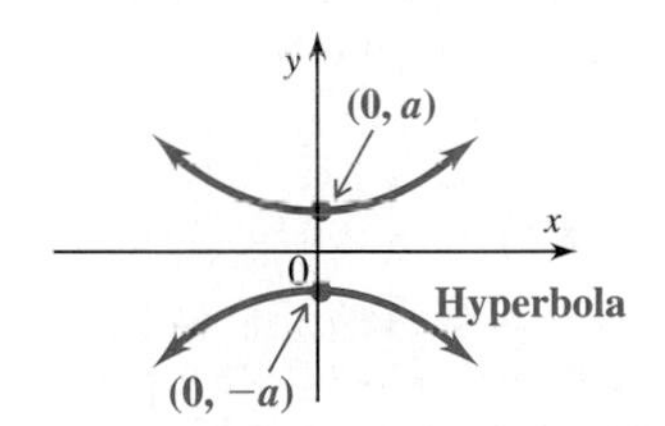

y-intercepts are a and $-a$
Asymptotes found from (b, a), $(b, -a)$, $(-b, a)$, and $(-b, -a)$.

y^2 has a positive coefficient.
x^2 has a negative coefficient.

Equations of Conic Sections

Conic section	*Characteristic*	*Example*
Parabola	Either $A = 0$ or $C = 0$, but not both.	$x^2 = y + 4$ $(y - 2)^2 = -(x + 3)$
Circle	$A = C \neq 0$	$x^2 + y^2 = 16$
Ellipse	$A \neq C, AC > 0$	$\frac{x^2}{16} + \frac{y^2}{25} = 1$
Hyperbola	$AC < 0$	$x^2 - y^2 = 1$

EXAMPLE 1 Decide on the type of conic section represented by each of the following equations, and sketch each graph.

(a) $x^2 = 25 + 5y^2$

Rewriting the equation as

$$x^2 - 5y^2 = 25$$

or

$$\frac{x^2}{25} - \frac{y^2}{5} = 1$$

shows that the equation represents a hyperbola centered at the origin, with asymptotes

$$\frac{x^2}{25} - \frac{y^2}{5} = 0,$$

or

$$y = \frac{\pm\sqrt{5}}{5}x.$$

The x-intercepts are ± 5; the graph is shown in Figure 24.

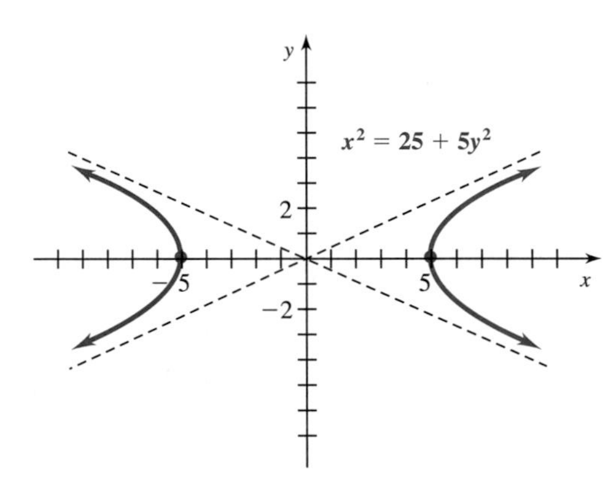

Figure 24

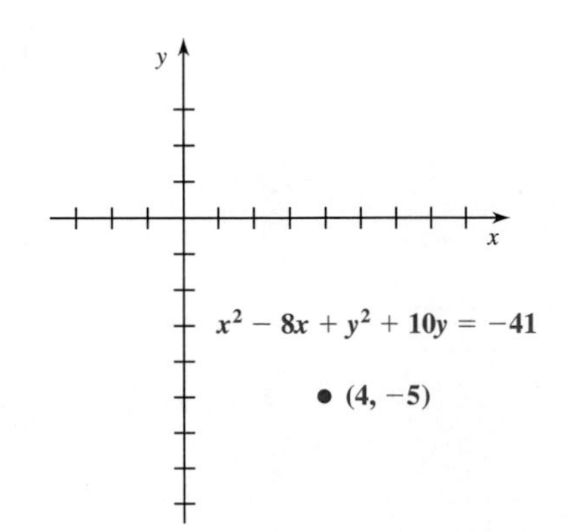

Figure 25

(b) $x^2 - 8x + y^2 + 10y = -41$

Complete the square on both x and y, as follows:

$$(x^2 - 8x + 16 - 16) + (y^2 + 10y + 25 - 25) = -41$$
$$(x - 4)^2 + (y + 5)^2 = 16 + 25 - 41$$
$$(x - 4)^2 + (y + 5)^2 = 0.$$

The result shows that the equation is that of a circle of radius 0; that is the point $(4, -5)$. See Figure 25. Had a negative number been obtained on the right (instead of 0), the equation would have no solution at all, and there would be no graph.

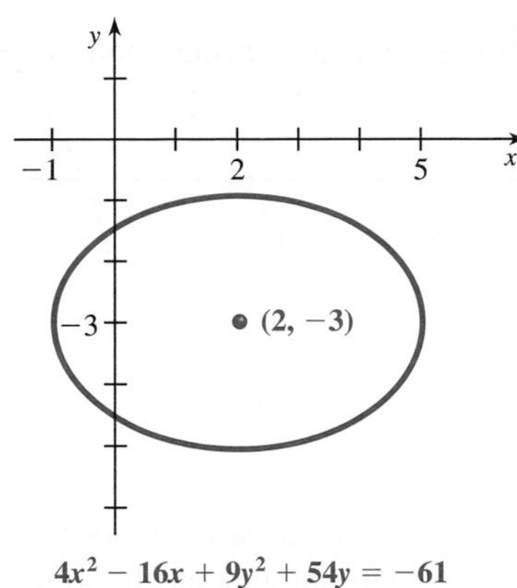

$4x^2 - 16x + 9y^2 + 54y = -61$

Figure 26

(c) $4x^2 - 16x + 9y^2 + 54y = -61$

Since the coefficients of the x^2 and y^2 terms are unequal and both positive, this equation might represent an ellipse. (It might also represent a single point or no points at all.) To find out, complete the square on x and y.

$$4(x^2 - 4x \quad\ ) + 9(y^2 + 6y \quad\ ) = -61$$
$$4(x^2 - 4x + 4 - 4) + 9(y^2 + 6y + 9 - 9) = -61$$
$$4(x^2 - 4x + 4) - 16 + 9(y^2 + 6y + 9) - 81 = -61$$
$$4(x - 2)^2 + 9(y + 3)^2 = 36$$
$$\frac{(x - 2)^2}{9} + \frac{(y + 3)^2}{4} = 1$$

This equation represents an ellipse having center at $(2, -3)$ and graph as shown in Figure 26.

(d) $x^2 - 6x + 8y - 7 = 0$

Since only one variable is squared (x, and not y), the equation represents a parabola. Rearrange the terms to get the term with y (the variable that is not squared) alone on one side. Then complete the square on the other side of the equation.

$$8y = -x^2 + 6x + 7$$
$$8y = -(x^2 - 6x \quad\ ) + 7$$
$$8y = -(x^2 - 6x + 9) + 7 + 9$$
$$8y = -(x - 3)^2 + 16$$
$$y = -\frac{1}{8}(x - 3)^2 + 2 \qquad \text{Multiply both sides by } 1/8.$$
$$y - 2 = -\frac{1}{8}(x - 3)^2 \qquad \text{Subtract 2 from both sides.}$$

The parabola has vertex at $(3, 2)$, and opens downward, as shown in Figure 27.

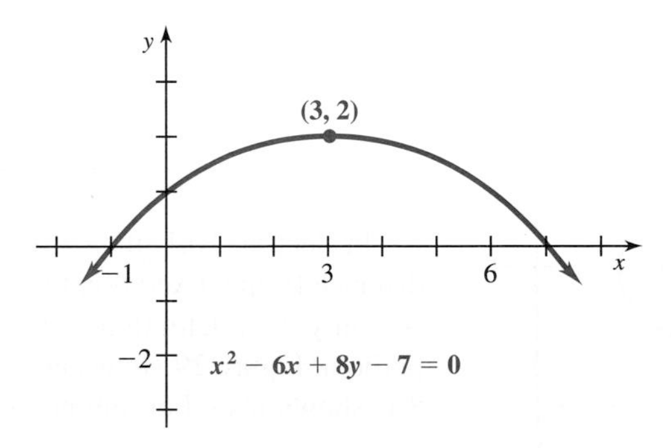

$x^2 - 6x + 8y - 7 = 0$

Figure 27

In Exercises 30–35, find the eccentricity of the conic section. The point shown on the x-axis is a focus and the line shown is a directrix.

30.

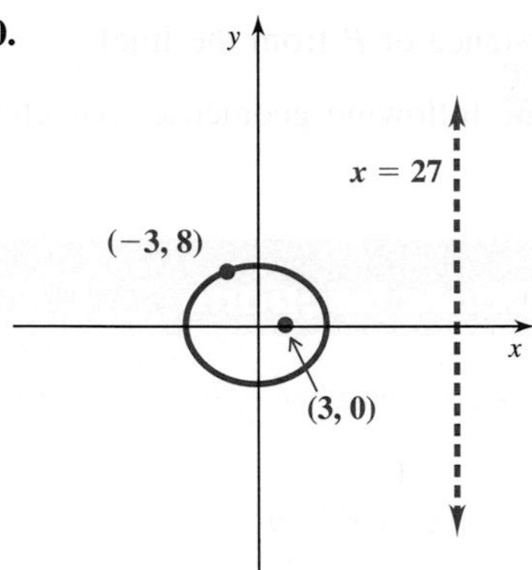

31.

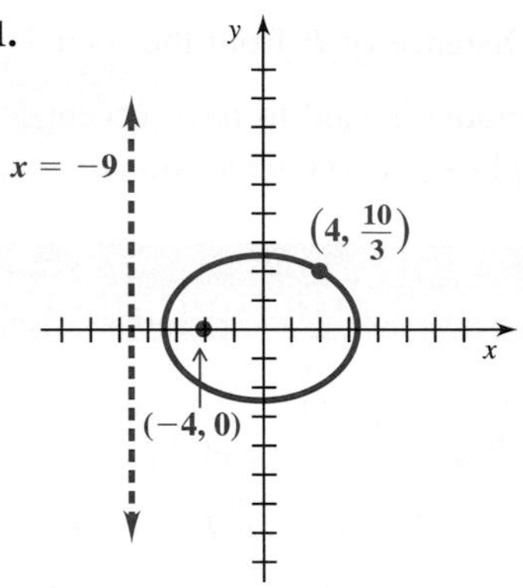

32.

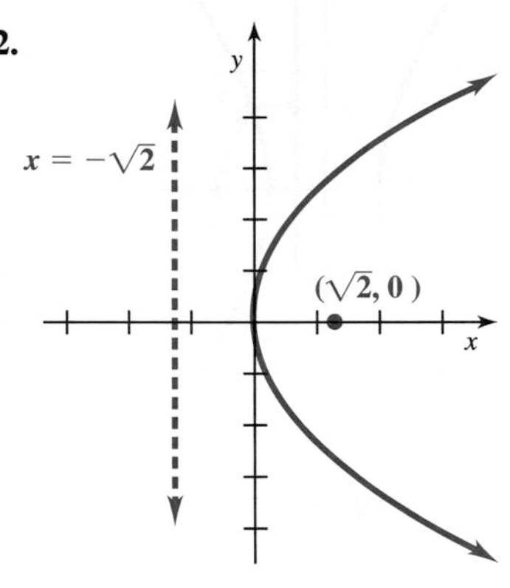

33.

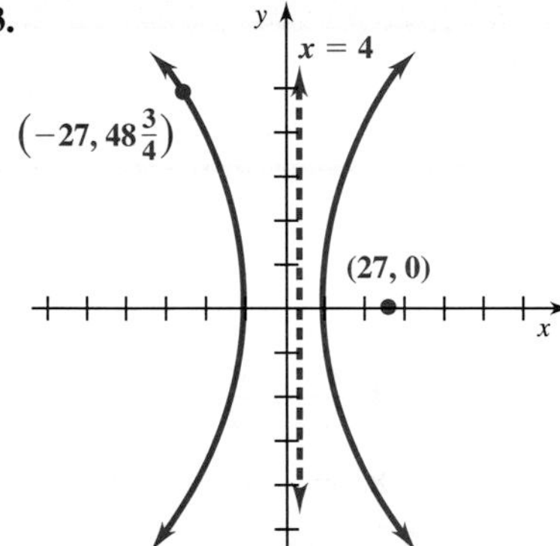

34.

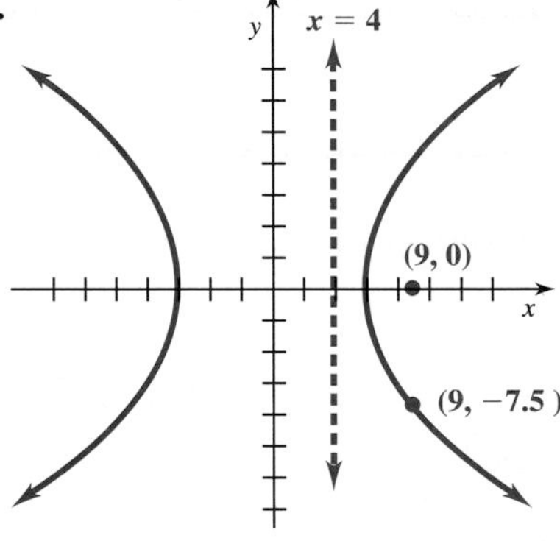

35.

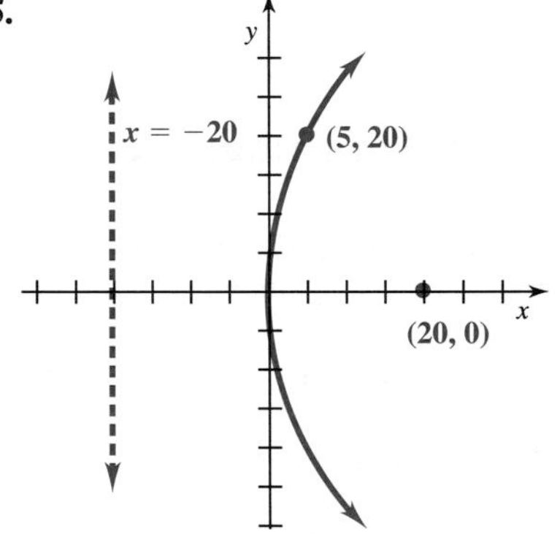

36. What is the eccentricity of a circle? (*Hint:* Think of a circle as an ellipse with $a = b$.)

37. If $Ax^2 + Cy^2 + Dx + Ey + F = 0$ is the general equation of an ellipse, find its center point by completing the square.

Graphing Utility Problems

38. Graph the ellipse $\dfrac{x^2}{16} + \dfrac{y^2}{12} = 1$. Use TRACE to find the coordinates of several points on the ellipse. For each of these points P, verify that

$$[\text{Distance of } P \text{ from } (2, 0)] = \frac{1}{2}\,[\text{Distance of } P \text{ from the line } x = 8].$$

39. Graph the hyperbola $\dfrac{x^2}{4} - \dfrac{y^2}{12} = 1$. Use TRACE to find the coordinates of several points on the hyperbola. For each of these points P, verify that

$$[\text{Distance of } P \text{ from } (4, 0)] = 2\,[\text{Distance of } P \text{ from the line } x = 1].$$

Chapter 7 Review Exercises

Graph each of the following. Give the vertex and axis of each figure.

1. $x = 4(y - 5)^2 + 2$
2. $x = -(y + 1)^2 - 7$
3. $x = 5y^2 - 5y + 3$
4. $x = 2y^2 - 4y + 1$

Graph each of the following. Give the focus, directrix, and axis of each figure.

5. $y^2 = -\frac{2}{3}x$
6. $y^2 = 2x$
7. $3x^2 = y$
8. $x^2 + 2y = 0$

Write an equation for each parabola with vertex at the origin.

9. Focus $(4, 0)$
10. Focus $(0, -3)$
11. Through $(2, 5)$, opening to the right
12. Through $(-3, 4)$, opening upward

Graph each of the following and identify each graph. Give the coordinates of the vertices for each ellipse or hyperbola, and give the equations of the asymptotes for each hyperbola.

13. $\frac{x^2}{5} + \frac{y^2}{9} = 1$
14. $\frac{x^2}{16} + \frac{y^2}{4} = 1$
15. $\frac{x^2}{64} - \frac{y^2}{36} = 1$
16. $\frac{y^2}{25} - \frac{x^2}{9} = 1$
17. $\frac{(x + 1)^2}{16} + \frac{(y - 1)^2}{16} = 1$
18. $(x - 3)^2 + (y + 2)^2 = 9$
19. $\frac{100x^2}{49} + \frac{9y^2}{16} = 1$
20. $\frac{25x^2}{9} + \frac{4y^2}{25} = 1$
21. $4x^2 + 9y^2 = 36$
22. $x^2 = 16 + y^2$
23. $\frac{(x - 3)^2}{4} + (y + 1)^2 = 1$
24. $\frac{(x - 2)^2}{9} + \frac{(y + 3)^2}{4} = 1$
25. $\frac{(y + 2)^2}{4} - \frac{(x + 3)^2}{9} = 1$
26. $\frac{x}{3} = -\sqrt{1 - \frac{y^2}{16}}$
27. $\frac{(x + 1)^2}{16} - \frac{(y - 2)^2}{4} = 1$
28. $x = -\sqrt{1 - \frac{y^2}{36}}$
29. $y = -\sqrt{1 + x^2}$
30. $y = -\sqrt{1 - \frac{x^2}{25}}$

Write an equation for each of the following conic sections (centers at the origin).

31. Ellipse; vertex at $(0, 4)$, focus at $(0, 2)$
32. Ellipse; x-intercept 6, focus at $(-2, 0)$
33. Hyperbola; focus at $(0, -5)$, transverse axis of length 8
34. Hyperbola; y-intercept -2, passing through $(2, 3)$

For the equations in Exercises 35–42, name the conic section and sketch the graph.

35. $y^2 + 9x^2 = 9$
36. $9x^2 - 16y^2 = 144$
37. $3y^2 - 5x^2 = 30$
38. $y^2 + x = 4$
39. $4x^2 - y = 0$
40. $x^2 + y^2 = 25$
41. $4x^2 - 8x + 9y^2 + 36y = -4$
42. $25x^2 + 50x + 4y^2 - 24y = 39$

43. Find the equation of the ellipse consisting of all points in the plane the sum of whose distances from (0, 0) and (4, 0) is 8.

44. Find the equation of the hyperbola consisting of all points in the plane for which the difference of the distances from (0, 0) and (0, 4) is 2.

45. The orbit of Venus is an ellipse with the sun at one of the foci. The eccentricity of the orbit is $e = .006775$ and the major axis has length 134.5 million miles. Find the smallest and greatest distances of Venus from the sun.

46. Comet Swift-Tuttle has an elliptical orbit of eccentricity $e = .964$, with the sun at one of the foci. Find the equation of the comet given that the closest it comes to the sun is 89 million miles.

Graphing Utility Problems

In Exercises 47–52, find the equation of the conic section that satisfies the given conditions. Then graph the equation and confirm that the conditions hold.

47. Parabola with focus at (3, 2) and directrix $x = -3$

48. Parabola with vertex at $(-3, 2)$ and y-intercepts 5 and -1

49. Ellipse with foci at $(-2, 0)$ and $(2, 0)$ and major axis of length 10

50. Ellipse with foci at (0, 3) and $(0, -3)$ and vertex at (0, 7)

51. Hyperbola with x-intercepts ± 3; foci at $(-5, 0)$, $(5, 0)$

52. Hyperbola with foci at (0, 12), $(0, -12)$; asymptotes $y = \pm x$

8 Further Topics in Algebra

In this chapter we explore topics related to sums of n terms, where n is a positive integer. First, we look at *sequences* (lists of numbers) and *series* (sums of sequences), paying special attention to two very useful types of sequence, *arithmetic* and *geometric*. Next, we discuss the binomial theorem, which gives a formula for writing out the terms of the series that is equivalent to $(x + y)^n$. Then we see how *mathematical induction* is used to prove theorems about sequences and series. The chapter ends with two related sections on counting theory and probability theory.

8.1 Sequences and Series

Sequences Defined informally, a sequence is a list of numbers. We are most interested in lists of numbers that satisfy some pattern. For example,

$$2, 4, 6, 8, 10, \ldots$$

is a list of the natural-number multiples of 2. This can be written as $2n$, where n is a natural number, so a sequence may be defined (as is a function) by a variable expression. More formally, a sequence is defined as follows.

Sequence

A **sequence** is a function that has a set of natural numbers as its domain.

Instead of using $f(x)$ notation to indicate a sequence, it is customary to use a_n, where n represents an element in the domain of a sequence. Thus, $a_n = f(n)$. The letter n is used instead of x as a reminder that n represents a *natural number*. The elements in the range of a sequence, called the **terms** of the sequence, are a_1, a_2, $a_3, \ldots$. The elements of both the domain and the range of a sequence are *ordered.* The first term (range element) is found by letting $n = 1$, the second term is found by letting $n = 2$, and so on. The **general term**, or ***n*th term**, of the sequence is a_n.

Figure 1 shows graphs of $f(x) = 2x$ and $a_n = 2n$. Notice that $f(x)$ defines a "continuous" function, while a_n is "discontinuous."

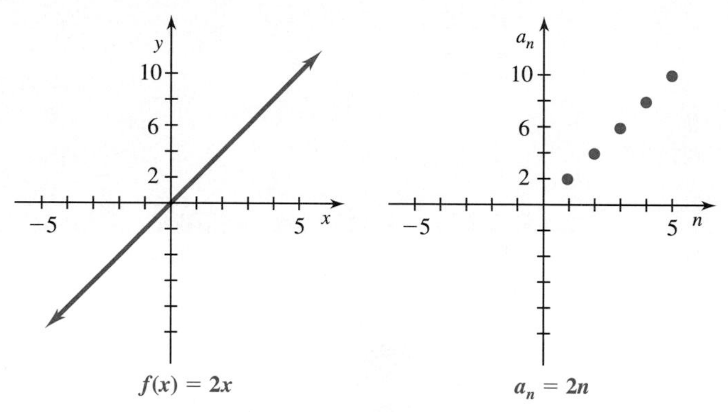

Figure 1

EXAMPLE 1 Write the first five terms for each of the following sequences.

(a) $a_n = \dfrac{n + 1}{n + 2}$

Replacing n, in turn, with 1, 2, 3, 4, and 5 gives

$$\frac{2}{3}, \frac{3}{4}, \frac{4}{5}, \frac{5}{6}, \frac{6}{7}.$$

(b) $a_n = (-1)^n \cdot n$

Replace n with 1, 2, 3, 4, and 5 to get

$$n = 1\colon\ a_1 = (-1)^1 \cdot 1 = -1$$
$$n = 2\colon\ a_2 = (-1)^2 \cdot 2 = 2$$
$$n = 3\colon\ a_3 = (-1)^3 \cdot 3 = -3$$
$$n = 4\colon\ a_4 = (-1)^4 \cdot 4 = 4$$
$$n = 5\colon\ a_5 = (-1)^5 \cdot 5 = -5.$$

(c) $b_n = \dfrac{(-1)^n}{2^n}$

Here, we have $b_1 = -1/2$, $b_2 = 1/4$, $b_3 = -1/8$, $b_4 = 1/16$, and $b_5 = -1/32$. ■

A sequence is a **finite sequence** if the domain is the set $\{1, 2, 3, 4, \ldots, n\}$, where n is a natural number. An **infinite sequence** has the set of all natural numbers as its domain.

EXAMPLE 2 The sequence of natural-number multiples of 2,

$$2, 4, 6, 8, 10, 12, 14, \ldots,$$

is infinite, but the sequence of days in June is finite:

$$1, 2, 3, 4, \ldots, 29, 30.$$

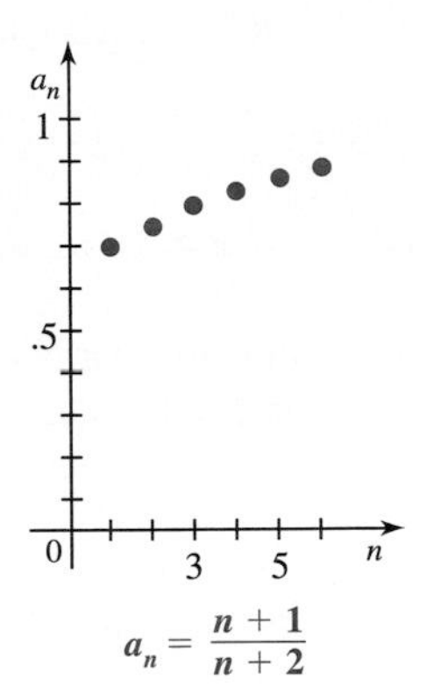

$a_n = \dfrac{n+1}{n+2}$

Figure 2

If the terms of an infinite sequence get closer and closer to some real number, the sequence is said to be **convergent** and to **converge** to that real number. Graphs of sequences illustrate this property. The sequence in Example 1(a) is graphed in Figure 2. What number do you think this sequence converges to? A sequence that does not converge to some number is **divergent**.

Some sequences are defined by a **recursive definition**, one in which each term is defined as an expression involving the previous term. On the other hand, the sequences in Example 1 were defined *explicitly*, with a formula for a_n that does not depend on a previous term.

EXAMPLE 3 Find the first four terms for the sequences defined as follows.

(a) $a_1 = 4$; for $n > 1$, $a_n = 2 \cdot a_{n-1} + 1$

This is an example of a recursive definition. We know $a_1 = 4$. Since $a_n = 2 \cdot a_{n-1} + 1$,

$$a_2 = 2 \cdot \mathbf{a_1} + 1 = 2 \cdot \mathbf{4} + 1 = \mathbf{9}$$
$$a_3 = 2 \cdot \mathbf{a_2} + 1 = 2 \cdot \mathbf{9} + 1 = \mathbf{19}$$
$$a_4 = 2 \cdot \mathbf{a_3} + 1 = 2 \cdot \mathbf{19} + 1 = 39.$$

(b) $a_1 = 2$; for $n > 1$, $a_n = a_{n-1} + n - 1$

$$a_1 = \mathbf{2}$$
$$a_2 = \mathbf{a_1} + 2 - 1 = \mathbf{2} + 1 = \mathbf{3}$$
$$a_3 = \mathbf{a_2} + 3 - 1 = \mathbf{3} + 2 = \mathbf{5}$$
$$a_4 = \mathbf{a_3} + 4 - 1 = \mathbf{5} + 3 = 8$$

Series Suppose a sequence has terms $a_1, a_2, a_3, \ldots$. Then S_n is defined as the sum of the first n terms. That is,

$$S_n = a_1 + a_2 + a_3 + \ldots + a_n.$$

The sum of the first n terms of a sequence is called a **series**. Special notation is used to represent a series. The symbol Σ, the Greek capital letter *sigma*, is used to indicate a sum.

Series

A **finite series** is an expression of the form

$$S_n = a_1 + a_2 + a_3 + \dots + a_n = \sum_{i=1}^{n} a_i,$$

and an **infinite series** is an expression of the form

$$S_n = a_1 + a_2 + a_3 + \dots + a_n + \dots = \sum_{i=1}^{\infty} a_i.$$

The letter i is called the **index of summation**. Do not confuse this use of i with the use of i to represent an imaginary number. Other letters can be used instead of i. Also, the initial value of the index can be a number other than 1.

EXAMPLE 4 Evaluate the series $\sum_{k=1}^{6} (2^k + 1)$.

Write out each of the 6 terms, then evaluate the sum.

$$\begin{aligned}\sum_{k=1}^{6} (2^k + 1) &= (2^1 + 1) + (2^2 + 1) + (2^3 + 1) + (2^4 + 1) + \\ &\quad (2^5 + 1) + (2^6 + 1) \\ &= (2 + 1) + (4 + 1) + (8 + 1) + (16 + 1) + \\ &\quad (32 + 1) + (64 + 1) \\ &= 3 + 5 + 9 + 17 + 33 + 65 = 132\end{aligned}$$

EXAMPLE 5 Write out the terms for each of the following series. Evaluate each sum if possible.

(a) $\sum_{j=3}^{6} a_j = a_3 + a_4 + a_5 + a_6$

(b) $\sum_{i=1}^{3} (6x_i - 2)$ if $x_1 = 2, x_2 = 4, x_3 = 6$

Let $i = 1, 2,$ and 3 respectively to get

$$\sum_{i=1}^{3} (6x_i - 2) = (6x_1 - 2) + (6x_2 - 2) + (6x_3 - 2).$$

Now substitute the given values for x_1, x_2, and x_3.

$$\begin{aligned}\sum_{i=1}^{3} (6x_i - 2) &= (6 \cdot 2 - 2) + (6 \cdot 4 - 2) + (6 \cdot 6 - 2) \\ &= 10 + 22 + 34 = 66\end{aligned}$$

(c) $\sum_{i=1}^{4} f(x_i)\Delta x$ if $f(x) = x^2$, $x_1 = 0$, $x_2 = 2$, $x_3 = 4$, $x_4 = 6$, and $\Delta x = 2$

$$\begin{aligned}\sum_{i=1}^{4} f(x_i)\Delta x &= f(x_1)\Delta x + f(x_2)\Delta x + f(x_3)\Delta x + f(x_4)\Delta x \\ &= {x_1}^2\Delta x + {x_2}^2\Delta x + {x_3}^2\Delta x + {x_4}^2\Delta x \\ &= 0^2(2) + 2^2(2) + 4^2(2) + 6^2(2) \\ &= 0 + 8 + 32 + 72 = 112\end{aligned}$$

A given series can be represented by summation notation in more than one way, as shown in the next example.

EXAMPLE 6 Use summation notation to rewrite each series with the index of summation starting at the indicated number.

(a) $\sum_{i=1}^{8} (3i - 4)$; 0

Let the new index be j. Since the new index is to start at 0, which is $1 - 1$, $j = i - 1$, or $i = j + 1$. Substitute $j + 1$ for i in the summation.

$$\begin{aligned}\sum_{i=1}^{8} (3i - 4) &= \sum_{j+1=1}^{j+1=8} [3(j + 1) - 4] \\ &= \sum_{j=0}^{j=7} (3j - 1) \quad \text{or} \quad \sum_{j=0}^{7} (3j - 1)\end{aligned}$$

(b) $\sum_{i=2}^{10} i^2$; -1

Here, if the new index is j, then $i = j + 3$ and

$$\begin{aligned}\sum_{i=2}^{10} i^2 &= \sum_{j+3=2}^{j+3=10} (j + 3)^2 \\ &= \sum_{j=-1}^{j=7} (j + 3)^2 \quad \text{or} \quad \sum_{j=-1}^{7} (j + 3)^2.\end{aligned}$$

Polynomial functions, defined by expressions of the form

$$f(x) = a_n x^n + a_{n-1}x^{n-1} + \cdots + a_1 x + a_0,$$

can be written in compact form, using summation notation, as

$$f(x) = \sum_{i=0}^{n} a_n x^n.$$

Several properties of summation are given below. These provide useful shortcuts for evaluating series. The proofs of some of these properties are given in the exercises for Section 5 of this chapter.

Summation Properties

If $a_1, a_2, a_3, \ldots, a_n$ and $b_1, b_2, b_3, \ldots, b_n$ are two sequences, and c is a constant, then for every positive integer n,

(a) $\sum_{i=1}^{n} c = nc$

(b) $\sum_{i=1}^{n} ca_i = c\sum_{i=1}^{n} a_i$

(c) $\sum_{i=1}^{n} (a_i + b_i) = \sum_{i=1}^{n} a_i + \sum_{i=1}^{n} b_i$

(d) $\sum_{i=1}^{n} (a_i - b_i) = \sum_{i=1}^{n} a_i - \sum_{i=1}^{n} b_i.$

To prove property (a), expand the series to get

$$c + c + c + c + \cdots + c,$$

where there are n terms of c, so the sum is nc.

Property (c) also can be proved by first expanding the series:

$$\sum_{i=1}^{n} (a_i + b_i) = (a_1 + b_1) + (a_2 + b_2) + \cdots + (a_n + b_n).$$

Now use the commutative and associative properties to rearrange the terms.

$$\begin{aligned}\sum_{i=1}^{n} (a_i + b_i) &= (a_1 + a_2 + \cdots + a_n) + (b_1 + b_2 + \cdots + b_n) \\ &= \sum_{i=1}^{n} a_i + \sum_{i=1}^{n} b_i\end{aligned}$$

Proofs of the other two properties are left for the exercises of Section 5.

The following results are proved in the text and exercises of Sections 2 and 5 of this chapter.

$$\sum_{i=1}^{n} i^2 = 1^2 + 2^2 + \cdots + n^2 = \frac{n(n+1)(2n+1)}{6}$$

and

$$\sum_{i=1}^{n} i = 1 + 2 + \cdots + n = \frac{n(n+1)}{2}$$

These summations are used in the next example.

EXAMPLE 7 Use the properties of series to evaluate $\sum_{i=1}^{6} (i^2 + 3i + 5)$.

$$\sum_{i=1}^{6} (i^2 + 3i + 5) = \sum_{i=1}^{6} i^2 + \sum_{i=1}^{6} 3i + \sum_{i=1}^{6} 5 \qquad \text{Property (c)}$$

$$= \sum_{i=1}^{6} i^2 + 3\sum_{i=1}^{6} i + \sum_{i=1}^{6} 5 \qquad \text{Property (b)}$$

$$= \sum_{i=1}^{6} i^2 + 3\sum_{i=1}^{6} i + 6(5) \qquad \text{Property (a)}$$

By substituting the results given just before this example, we get

$$= \frac{6(6+1)(2 \cdot 6 + 1)}{6} + 3\left[\frac{6(6+1)}{2}\right] + 6(5)$$

$$= 91 + 3(21) + 6(5)$$

$$= 184.$$

8.1 Exercises

Write the first five terms of each of the following sequences.

1. $a_n = 4n + 10$
2. $a_n = 6n - 3$
3. $a_n = 2^{n-1}$
4. $a_n = -3^n$
5. $a_n = (1/3)^n(n - 1)$
6. $a_n = (-2)^n(n)$
7. $a_n = (-1)^n(2n)$
8. $a_n = (-1)^{n-1}(n + 1)$
9. $a_n = \dfrac{4n - 1}{n^2 + 2}$
10. $a_n = \dfrac{n^2 - 1}{n^2 + 1}$

11. Your friend does not understand what is meant by the *n*th term or general term of a sequence. How would you explain this idea?

12. How are sequences related to functions? Discuss some similarities and some differences.

Find the first six terms for the sequences defined as follows.

13. $a_1 = -2, \quad a_n = a_{n-1} + 3, \quad$ for $n > 1$

14. $a_1 = -1, \quad a_n = a_{n-1} - 4, \quad$ for $n > 1$

15. $a_1 = 1, \quad a_2 = 1, \quad a_n = a_{n-1} + a_{n-2}, \quad$ for $n \geq 3$ (the Fibonacci sequence)

16. $a_1 = 2, \quad a_n = n \cdot a_{n-1}, \quad$ for $n > 1$

Evaluate the terms for each of the following sums where $x_1 = -2,\ x_2 = -1,\ x_3 = 0,\ x_4 = 1,\ x_5 = 2$.

17. $\sum_{i=1}^{5} (2x_i + 3)$
18. $\sum_{i=1}^{4} x_i^2$
19. $\sum_{i=1}^{3} (3x_i - x_i^2)$
20. $\sum_{i=1}^{3} (x_i^2 + 1)$
21. $\sum_{i=2}^{5} \dfrac{x_i + 1}{x_i + 2}$
22. $\sum_{i=1}^{5} \dfrac{x_i}{x_i + 3}$

Evaluate the terms of $\sum_{i=1}^{4} f(x_i)\Delta x$ *with* $x_1 = 0$, $x_2 = 2$, $x_3 = 4$, $x_4 = 6$, *and* $\Delta x = .5$ *for the functions defined as follows.*

23. $f(x) = 4x - 7$

24. $f(x) = 6 + 2x$

25. $f(x) = 2x^2$

26. $f(x) = x^2 - 1$

27. $f(x) = \dfrac{-2}{x + 1}$

28. $f(x) = \dfrac{5}{2x - 1}$

Use summation notation to rewrite each series with the index of summation starting at the indicated number.

29. $\sum_{i=1}^{5} (6 - 3i)$; 3

30. $\sum_{i=1}^{7} (5i + 2)$; -2

31. $\sum_{i=1}^{10} 2(3)^i$; 0

32. $\sum_{i=-1}^{6} 5(2)^i$; 3

33. $\sum_{i=-1}^{9} (i^2 - 2i)$; 0

34. $\sum_{i=3}^{11} (2i^2 + 1)$; 0

Use the summation properties to evaluate each series. The following sums may be needed.

$$\sum_{i=1}^{n} i = \frac{n(n + 1)}{2}$$

$$\sum_{i=1}^{n} i^2 = \frac{n(n + 1)(2n + 1)}{6}$$

$$\sum_{i=1}^{n} i^3 = \frac{n^2(n + 1)^2}{4}$$

35. $\sum_{i=1}^{5} (5i + 3)$

36. $\sum_{i=1}^{5} (8i - 1)$

37. $\sum_{i=1}^{5} (4i^2 - 2i + 6)$

38. $\sum_{i=1}^{6} (2 + i - i^2)$

39. $\sum_{i=1}^{4} (3i^3 + 2i - 4)$

40. $\sum_{i=1}^{6} (i^2 + 2i^3)$

Use the summation properties to write each of the following without Σ. (Hint: *Think of n as a constant.)*

41. $\sum_{i=1}^{n} \left[4 + \left(\frac{2}{n}\right)i\right]\frac{2}{n}$

42. $\sum_{i=1}^{n} \left[\left(\frac{2}{n}\right)i + 1\right]\frac{2}{n}$

43. $\sum_{i=1}^{n} \left[3 + \left(\frac{1}{n}\right)i\right]^2\frac{1}{n}$

44. $\sum_{i=1}^{n} \left[5 + \left(\frac{3}{n}\right)i\right]^2\frac{3}{n}$

Graphing Utility Problems

For each sequence defined by a_n, *graph the corresponding function defined by* $f(x)$. *Use the graph to decide whether the sequence converges, and if it does, determine the number to which it converges.*

45. $a_n = \dfrac{n + 2}{2n}$

46. $a_n = 2e^n$

47. $a_n = n(n + 1)$

48. $a_n = \left(1 + \dfrac{1}{n}\right)^n$

8.2 Arithmetic Sequences and Series

A sequence in which each term after the first is obtained by adding a fixed number to the previous term is an **arithmetic sequence** (or **arithmetic progression**). The fixed number that is added is the **common difference**. The sequence

$$5, 9, 13, 17, 21, \ldots$$

is an arithmetic sequence since each term after the first is obtained by adding 4 to the previous term. That is,

$$\begin{aligned} 9 &= 5 + 4 \\ 13 &= 9 + 4 \\ 17 &= 13 + 4 \\ 21 &= 17 + 4, \end{aligned}$$

and so on. The common difference is 4.

EXAMPLE 1 Find the common difference, d, for the arithmetic sequence

$$-9, -7, -5, -3, -1, \ldots.$$

Since this sequence is arithmetic, d can be found by choosing any two adjacent terms and subtracting the first from the second. If we choose -7 and -5,

$$d = -5 - (-7) = 2.$$

Choosing -9 and -7 would give $d = -7 - (-9) = 2$, the same result. ■

EXAMPLE 2 Write the first five terms for each arithmetic sequence.

(a) $a_1 = 7, d = -3$

The first term is 7, and each succeeding term is found by adding -3 to the preceding term.

$$\begin{aligned} a_2 &= 7 + d = 7 + (-3) = 4 \\ a_3 &= 4 + d = 4 + (-3) = 1 \\ a_4 &= 1 + d = 1 + (-3) = -2 \\ a_5 &= -2 + d = -2 + (-3) = -5 \end{aligned}$$

(b) $a_1 = -12, d = 5$

Use the definition of an arithmetic sequence.

$$\begin{aligned} a_1 &= -12 \\ a_2 &= -12 + d = -12 + 5 = -7 \\ a_3 &= -7 + d = -7 + 5 = -2 \\ a_4 &= -2 + d = -2 + 5 = 3 \\ a_5 &= 3 + d = 3 + 5 = 8 \end{aligned}$$

■

If a_1 is the first term of an arithmetic sequence and d is the common difference, then the terms of the sequence are given by

$$\begin{aligned}
a_1 &= a_1 \\
a_2 &= a_1 + d \\
a_3 &= a_2 + d = a_1 + d + d = a_1 + 2d \\
a_4 &= a_3 + d = a_1 + 2d + d = a_1 + 3d \\
a_5 &= a_1 + 4d \\
a_6 &= a_1 + 5d,
\end{aligned}$$

and, by this pattern,

$$a_n = a_1 + (n - 1)d.$$

This result can be proven by mathematical induction (see Section 5 of this chapter); a summary is given below.

*n*th Term of an Arithmetic Sequence

In an arithmetic sequence with first term a_1 and common difference d, the nth term, a_n, is given by

$$\mathbf{a_n = a_1 + (n - 1)d.}$$

EXAMPLE 3 Find a_{13} and a_n for the arithmetic sequence

$$-3, 1, 5, 9, \ldots.$$

Here $a_1 = -3$ and $d = 1 - (-3) = 4$. To find a_{13}, substitute 13 for n in the formula for the nth term of an arithmetic sequence.

$$\begin{aligned}
a_{13} &= a_1 + (\mathbf{13} - 1)d \\
a_{13} &= -3 + (12)4 \\
a_{13} &= -3 + 48 \\
a_{13} &= 45
\end{aligned}$$

To find a_n, substitute values for a_1 and d in the formula for a_n.

$$\begin{aligned}
a_n &= \mathbf{-3} + (n - 1)\mathbf{4} \\
a_n &= -3 + 4n - 4 \\
a_n &= 4n - 7
\end{aligned}$$

EXAMPLE 4 Find a_{18} and a_n for the arithmetic sequence having $a_2 = 9$ and $a_3 = 15$.

First find d: $d = a_3 - a_2 = 15 - 9 = 6$.

Since $\quad a_2 = a_1 + d, \quad 9 = a_1 + 6 \quad$ and $\quad a_1 = 3$.

Then $\quad a_{18} = 3 + (18 - 1) \cdot 6 = 105,$

and $\quad a_n = 3 + (n - 1) \cdot 6 = 3 + 6n - 6 = 6n - 3.$ ■

EXAMPLE 5 A child building a tower with blocks uses 15 for the first row. Each row has 2 blocks fewer than the previous row. If there are 8 rows in the tower, how many blocks are used for the top row?

The number of blocks in each row forms an arithmetic sequence with $a_1 = 15$ and $d = -2$. Since there are 8 rows in the tower, $n = 8$ and we want to find a_8. Using the formula

$$a_n = a_1 + (n - 1)d$$

gives

$$a_8 = 15 + (8 - 1)(-2) = 1.$$

There is just one block in the top row. ■

EXAMPLE 6 Suppose that an arithmetic sequence has $a_8 = -16$ and $a_{16} = -40$. Find a_1.

Since $a_8 = a_1 + (8 - 1)d$, replacing a_8 with -16 gives

$$-16 = a_1 + 7d \quad \text{or} \quad a_1 = -16 - 7d.$$

Similarly, $\quad -40 = a_1 + 15d \quad$ or $\quad a_1 = -40 - 15d.$

From these two equations, by the substitution method given earlier,

$$-16 - 7d = -40 - 15d$$
$$d = -3.$$

To find a_1, substitute -3 for d in $a_1 = -16 - 7d$. (We could have used $a_1 = -40 - 15d$.)

$$a_1 = -16 - 7(-3)$$
$$a_1 = 5$$ ■

Sum of the First *n* Terms It is often necessary to add the terms of an arithmetic sequence. For example, suppose that a person borrows \$3000 and agrees to pay \$100 per month plus interest of 1% per month on the unpaid balance until the loan is paid off. The first month \$100 is paid to reduce the loan, plus interest of $(.01)3000 = 30$ dollars. The second month another \$100 is paid toward the loan and $(.01)2900 = 29$ dollars) is paid for interest. Since the loan is reduced by \$100 each month, interest payments decrease by $(.01)100 = 1$ dollar each month, forming the arithmetic sequence

$$30, 29, 28, \ldots, 3, 2, 1.$$

The total amount of interest paid is given by the sum S_n of the terms of this sequence. A formula will be developed here to find this sum without adding all thirty numbers directly.

Since the sequence is arithmetic, the sum of the first n terms can be written as follows.

$$S_n = a_1 + [a_1 + d] + [a_1 + 2d] + \dots + [a_1 + (n - 1)d]$$

The formula for the general term was used in the last expression. Now write the same sum in reverse order, beginning with a_n and *subtracting d.*

$$S_n = a_n + [a_n - d] + [a_n - 2d] + \dots + [a_n - (n - 1)d]$$

Adding respective sides of these two equations term by term gives

$$S_n + S_n = (a_1 + a_n) + (a_1 + a_n) + \dots + (a_1 + a_n)$$

or

$$2S_n = n(a_1 + a_n).$$

since there are n terms of $a_1 + a_n$ on the right. Now solve for S_n to get

$$S_n = \frac{n}{2}(a_1 + a_n).$$

Using the formula $a_n = a_1 + (n - 1)d$, this result for S_n can also be written as

$$S_n = \frac{n}{2}[a_1 + a_1 + (n - 1)d]$$

or

$$S_n = \frac{n}{2}[2a_1 + (n - 1)d].$$

an alternative formula for the sum of the first n terms of an arithmetic sequence.

Sum of the First *n* Terms of an Arithmetic Sequence

If an arithmetic sequence has first term a_1 and common difference d, then the sum of the first n terms, S_n, is given by

$$S_n = \sum_{i=1}^{n} a_i = \frac{n}{2}(a_1 + a_n)$$

or

$$S_n = \sum_{i=1}^{n} a_i = \frac{n}{2}[2a_1 + (n - 1)d].$$

NOTE The first formula of the theorem is used when the first and last terms are known; otherwise the second formula is used. ■

Either one of these formulas can be used to find the amount of interest the person above will pay on the \$3000 loan. In the sequence of interest payments $a_1 = 30$, $d = -1$, $n = 30$, and $a_n = 1$.

Choosing the first formula, $$S_n = \frac{n}{2}(a_1 + a_n)$$

gives $$S_n = \frac{30}{2}(30 + 1)$$
$$= 15(31) = 465,$$

so a total of \$465 interest will be paid over 30 months.

EXAMPLE 7

(a) Find S_{12} for the arithmetic sequence $-9, -5, -1, 3, 7, \ldots$.

Here $a_1 = -9$, $d = 4$, and $n = 12$. Use the second formula above.

$$S_n = \frac{n}{2}[2a_1 + (n - 1)d]$$
$$S_{12} = \frac{12}{2}[2(-9) + 11(4)]$$
$$= 6(-18 + 44) = 156$$

(b) Use the formula for S_n to find the sum of the first 60 positive integers.

In this example, $n = 60$, $a_1 = 1$, and $a_{60} = 60$, so it is convenient to use the first of the two formulas:

$$S_n = \frac{n}{2}(a_1 + a_n)$$
$$S_{60} = \frac{60}{2}(1 + 60)$$
$$= 30 \cdot 61 = 1830.$$

EXAMPLE 8 The sum of the first 17 terms of an arithmetic sequence is 187. If $a_{17} = -13$, find a_1 and d.

Use the first formula for S_n, with $n = 17$, to find a_1.

$$187 = \frac{17}{2}(a_1 - 13)$$
$$374 = 17(a_1 - 13)$$
$$22 = a_1 - 13$$
$$a_1 = 35$$

From the formula for a_n,

$$a_{17} = a_1 + (17 - 1)d,$$

and

$$-13 = 35 + 16d$$
$$-48 = 16d$$
$$d = -3.$$

EXAMPLE 9 Find each of the following sums.

(a) $\sum_{i=1}^{10} (4i + 8)$

This sum can be written as

$$\sum_{i=1}^{10} (4i + 8)$$
$$= [4(1) + 8] + [4(2) + 8] + [4(3) + 8] + \ldots + [4(10) + 8]$$
$$= 12 + 16 + 20 + \ldots + 48,$$

the sum of the first ten terms of the arithmetic sequence having

$$a_1 = 4 \cdot 1 + 8 = 12,$$
$$n = 10,$$

and

$$a_n = a_{10} = 4 \cdot 10 + 8 = 48.$$

By a formula for S_n,

$$\sum_{i=1}^{10} (4i + 8) = S_{10} = \frac{10}{2}(12 + 48) = 5(60) = 300.$$

(b) $\sum_{k=3}^{9} (4 - 3k)$

The first few terms are

$$[4 - 3(3)] + [4 - 3(4)] + [4 - 3(5)] + \ldots$$
$$= -5 + (-8) + (-11) + \ldots.$$

Thus, $a_1 = -5$ and $d = -3$. If the sequence started with $k = 1$, there would be 9 terms. Since it starts at 3, 2 of those terms are missing, so there are 7 terms and $n = 7$.

$$\sum_{k=3}^{9} (4 - 3k) = \frac{7}{2}[2(-5) + (6)(-3)] = -98$$ ■

8.2 Exercises

Write the terms of the arithmetic sequences satisfying each of the following conditions.

1. $a_1 = 5,\ d = -2,\ n = 6$

2. $a_1 = 4,\ d = 3,\ n = 5$

3. $a_2 = 10,\ d = -3,\ n = 4$

4. $a_3 = 10,\ d = -2,\ n = 5$

5. $a_1 = 3 - \sqrt{2},\ a_2 = 3,\ n = 5$

6. $a_1 = -5,\ a_2 = -5 + \sqrt{3},\ n = 4$

For each of the following sequences, find d and a_n.

7. 18, 15, 12, 9, 6, 3, . . .

8. 5, 11, 17, 23, 29, 35, . . .

9. 6, 10, 14, 18, 22, . . .

10. 27, 22, 17, 12, . . .

11. $\sqrt{3} + 1, 2\sqrt{3} + 1, 3\sqrt{3} + 1, 4\sqrt{3} + 1, \ldots$

12. $5 - \sqrt{7}, 6 - \sqrt{7}, 7 - \sqrt{7}, 8 - \sqrt{7}, \ldots$

13. $x - 2, x - 1, x, x + 1, x + 2, \ldots$

14. $3x + y, 3x + 2y, 3x + 3y, 3x + 4y, \ldots$

Find a_{10} and a_n for each of the following sequences.

15. $a_1 = 3, d = 5$

16. $a_1 = -4, d = 8$

17. $a_1 = -1, d = 5$

18. $a_1 = 3, d = -4$

19. $a_1 = 5, a_3 = 12$

20. $a_2 = 4, a_4 = -2$

21. $a_1 = x, a_2 = x + 3$

22. $a_2 = y + 1, d = -5$

Find a_1 for each of the following arithmetic sequences.

23. $a_5 = 27, a_{15} = 87$

24. $a_{12} = 60, a_{20} = 84$

25. $S_{16} = -160, a_{16} = -25$

26. $S_{28} = 2926, a_{28} = 199$

27. Which of the following is not an arithmetic sequence?
(a) 4, 6, 8, 10, . . . **(b)** −2, 6, 14, 22, . . .
(c) 1/2, 1, 3/2, 2, . . . **(d)** 5, 10, 20, 40, . . .
For the sequence that is not arithmetic, explain how each term after the first is determined by using the previous term of the sequence.

Find the sum of the first ten terms for each of the following arithmetic sequences.

28. $a_4 = 16, a_5 = 19$

29. $a_2 = 12, a_5 = 24$

30. 4, 12, 20, . . .

31. 20, 15, 10, . . .

32. $a_1 = 9.428, d = -1.723$

33. $a_1 = -3.119, d = 2.422$

34. $a_4 = 2.556, a_5 = 3.004$

35. $a_7 = 11.192, a_9 = 4.812$

Evaluate each of the following sums.

36. $\sum_{i=1}^{8} (5i - 4)$

37. $\sum_{i=1}^{12} (8i + 17)$

38. $\sum_{i=1}^{15} (-4i - 1)$

39. $\sum_{i=1}^{20} (-2 - 6i)$

40. $\sum_{j=1}^{500} 4j$

41. $\sum_{j=1}^{1200} 2j$

42. $\sum_{i=6}^{17} (3i + 7)$

43. $\sum_{i=8}^{21} (2 - 5i)$

In Exercises 44 and 45, find the value of x for which the sequence is arithmetic.

44. $5, x, 19$

45. $2, x, -5$

46. The sum of the first n terms in an arithmetic sequence is given by the formula $S_n = n(3n + 2)$. Find the fourteenth term in the sequence.

47. An arithmetic sequence has $a_{24} - a_{12} = 48$. Find the common difference.

48. Find the sum of the first n positive integers.

49. Find the sum of the first n odd positive integers.

50. If a clock strikes the proper number of tones each hour on the hour, how many tones will it strike in a month of 30 days?

51. A display of stacked canned goods in a grocery store has 31 cans on the bottom, 25 on the next row, and 1 can on top. Assume the number of cans in the layers form an arithmetic sequence. How many cans are in the display?

52. A skydiver falls 5 m during the first second, 15 m during the second, 25 m during the third, and so on. How many meters will the diver fall during the tenth second? During the first ten seconds?

46. $\sum_{i=1}^{\infty} 3(1/4)^{i-1}$ **47.** $\sum_{i=1}^{\infty} 5(-1/4)^{i-1}$ **48.** $\sum_{k=1}^{\infty} (.3)^k$ **49.** $\sum_{k=1}^{\infty} 10^{-k}$

50. Explain the difference between an arithmetic sequence and a geometric sequence.

51. The final step in processing a black-and-white photographic print is to immerse the print in a chemical called ''fixer.'' The print is then washed in running water. Under certain conditions, 98% of the fixer in a print will be removed with 15 min of washing. How much of the original fixer would be left after 1 hr of washing?

52. A scientist has a vat containing 100 L of a pure chemical. Twenty liters is drained and replaced with water. After complete mixing, 20 L of the mixture is drained and replaced with water. What will be the strength of the mixture after 9 such drainings?

53. The half-life of a radioactive substance is the time it takes for half the substance to decay. Suppose the half-life of a substance is 3 yr, and 10^{15} molecules of the substance are present initially. How many molecules will be present after 15 yr?

54. Each year a machine loses 20% of the value it had at the beginning of the year. Find the value of the machine at the end of 6 yr if it cost $100,000 new.

55. A bicycle wheel rotates 400 times in one minute. If the rider removes his or her feet from the pedals, the wheel will start to slow down. Each minute, it will rotate only 3/4 as many times as in the preceding minute. How many times will the wheel rotate in the fifth minute after the rider's feet are removed from the pedals?

56. A piece of paper is .008 inch thick. Suppose the paper is folded in half, so that its thickness doubles, for 12 times in a row. How thick is the folded paper?

57. A sugar factory receives an order for 1000 units of sugar. The production manager thus orders production of 1000 units of sugar. He forgets, however, that the production of sugar requires some sugar (to prime the machines, for example), and so he ends up with only 900 units of sugar. He then orders an additional 100 units, and receives only 90 units. A further order for 10 units produces 9 units. Finally seeing he is wrong, the manager decides to try mathematics. He views the production process as an infinite geometric progression with $a_1 = 1000$ and $r = .1$. Using this, find the number of units of sugar that he should have ordered originally.

58. After a person pedaling a bicycle removes his or her feet from the pedals, the wheel rotates 400 times the first minute. As it continues to slow down, it rotates in each minute only 3/4 as many times as in the previous minute. How many times will the wheel rotate before coming to a complete stop?

59. A pendulum bob swings through an arc 40 cm long on its first swing. Each swing thereafter, it swings only 80% as far as on the previous swing. How far will it swing altogether before coming to a complete stop?

60. Mitzi drops a ball from a height of 10 m and notices that on each bounce the ball returns to about 3/4 of its previous height. About how far will the ball travel before it comes to rest? (*Hint:* Consider the sum of two sequences).

61. Each person has two parents, four grandparents, eight great-grandparents, and so on. What is the total number of ancestors a person has, going back five generations? Ten generations?

62. Certain medical conditions are treated with a fixed dose of a drug administered at regular intervals. Suppose that a person is given 2 mg of a drug each day and that during each 24-hr period, the body utilizes 40% of the amount of drug that was present at the beginning of the period.

(a) Show that the amount of the drug present in the body at the end of n days is

$$\sum_{i=1}^{n} 2(.6)^i.$$

(b) What will be the approximate quantity of the drug in the body at the end of each day after the treatment has been administered for a long period of time?

(c) What is the maximum daily dosage that will guarantee that the amount of the drug in the body never exceeds 2 mg?

63. A sequence of equilateral triangles is constructed. The first triangle has sides 2 m in length. To get the second triangle, midpionts of the sides of the original triangle are connected. What is the length of the side of the eighth such triangle? See the figure below.

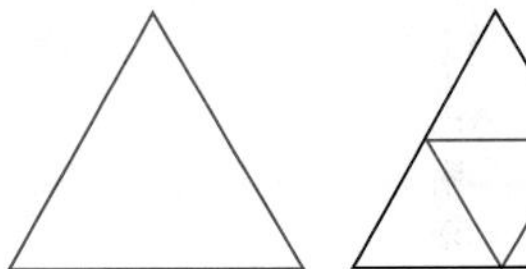
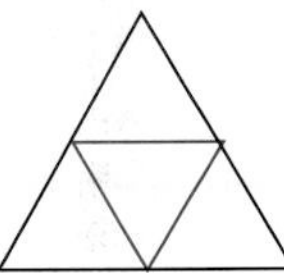
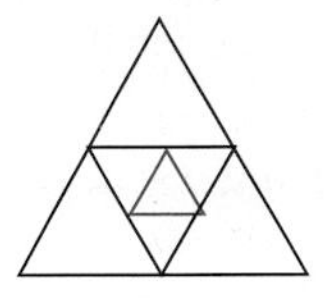
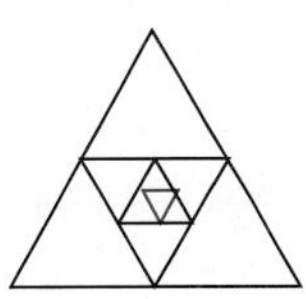

64. In Exercise 63, if the process could be continued indefinitely, what would be the total perimeter of all the triangles? What would be the total area of all the triangles, disregarding the overlapping?

65. Find three numbers x, y, and z that are consecutive terms of both an arithmetic sequence and a geometric sequence.

66. Let $a_1, a_2, a_3, \ldots$ and $b_1, b_2, b_3, \ldots$ be geometric sequences. Let $d_n = c\ a_n\ b_n$ for any real number c and every positive integer n. Show that $d_1, d_2, d_3, \ldots$ is a geometric sequence.

Suppose that $a_1, a_2, a_3, a_4, a_5, \ldots$ is a geometric sequence with common ratio r. Show that the following sequences are geometric and give their common ratios.

67. $a_1, a_4, a_7, a_{10}, \ldots$

68. $(a_1)^2, (a_2)^2, (a_3)^2, (a_4)^2, (a_5)^2, \ldots$

69. $\sqrt{a_1}, \sqrt{a_2}, \sqrt{a_3}, \sqrt{a_4}, \sqrt{a_5}, \ldots$

70. $\frac{3}{a_1}, \frac{3}{a_2}, \frac{3}{a_3}, \frac{3}{a_4}, \frac{3}{a_5}, \ldots$

Explain why the following sequences are geometric.

71. log 6, log 36, log 1296, log 1,679,616, . . .

72. log 2, log 4, log 16, log 256, . . .

8.4 The Binomial Theorem

In Chapter 1 we observed a parallel between the numbers in Pascal's triangle, shown below, and the coefficients of the terms in expansions of powers of binomials of the form $(x + y)^n$. As we saw earlier, the nth row of the triangle gives the coefficients of the terms of $(x + y)^n$. Also, the *variables* in the expansion have the pattern

$$x^n,\ x^{n-1}y,\ x^{n-2}y^2,\ x^{n-3}y^3,\ \ldots,\ xy^{n-1},\ y^n.$$

As the rows of Pascal's triangle show, there are $n + 1$ terms in the expansion of $(x + y)^n$. For example, in the fifth row, there are 6 coefficients, and therefore 6 terms, in the expansion of $(x + y)^5$.

Pascal's Triangle

$$
\begin{array}{ccccccccccc}
 & & & & 1 & & 1 & & & & \\
 & & & 1 & & 2 & & 1 & & & \\
 & & 1 & & 3 & & 3 & & 1 & & \\
 & 1 & & 4 & & 6 & & 4 & & 1 & \\
1 & & 5 & & 10 & & 10 & & 5 & & 1
\end{array}
$$

Although it is possible to use Pascal's triangle to find the coefficients of $(x + y)^n$ for any positive integer value of n, this becomes impractical for large values of n because of the need to write out all the preceding rows. A more efficient way of finding these coefficients uses factorial notation. The number $n!$ (read "n-factorial") is defined as follows.

Definition of *n*-factorial

For any positive integer n,

$$\mathbf{n! = n(n - 1)(n - 2) \ldots (3)(2)(1)}$$

and

$$\mathbf{0! = 1.}$$

For example, $5! = 5 \cdot 4 \cdot 3 \cdot 2 \cdot 1 = 120$, $7! = 7 \cdot 6 \cdot 5 \cdot 4 \cdot 3 \cdot 2 \cdot 1 = 5040$, $2! = 2 \cdot 1 = 2$, and so on.

Now look at the coefficients of the expression

$$(x + y)^5 = x^5 + 5x^4y + 10x^3y^2 + 10x^2y^3 + 5xy^4 + y^5.$$

The coefficient of the second term, $5x^4y$, is 5, and the exponents on the variables are 4 and 1. Note that

$$5 = \frac{5!}{4!1!}.$$

The coefficient of the third term is 10, with exponents of 3 and 2, and

$$10 = \frac{5!}{3!2!}.$$

The last term (the sixth term) can be written as $y^5 = 1x^0y^5$, with coefficient 1, and exponents of 0 and 5. Since $0! = 1$, check that

$$1 = \frac{5!}{0!5!}.$$

Generalizing from these examples, the coefficient for the term of the expansion of $(x + y)^n$ in which the variable part is $x^r y^{n-r}$ (where $r \leq n$) will be

$$\frac{n!}{r!(n - r)!}.$$

This number, called a **binomial coefficient**, is often symbolized $\binom{n}{r}$ (read "n above r").

Definition of Binomial Coefficient

For nonnegative integers n and r, with $r \leq n$, the symbol $\binom{n}{r}$ is defined as

$$\binom{n}{r} = \frac{n!}{r!(n - r)!}.$$

For Graphers

Most graphers have the ability to calculate binomial coefficients. This function is often found in the statistics mode.

These binomial coefficients are just numbers from Pascal's triangle. For example, $\binom{3}{0}$ is the first number in the third row, and $\binom{7}{4}$ is the fifth number in the seventh row. Another common notation for the binomial coefficient is ${}_nC_r$. Many calculators have a key to use for finding binomial coefficients. Others can be programmed to calculate them.

EXAMPLE 1

(a) $\binom{6}{2} = \frac{6!}{2!(6-2)!} = \frac{6!}{2!4!} = \frac{6 \cdot 5 \cdot 4 \cdot 3 \cdot 2 \cdot 1}{2 \cdot 1 \cdot 4 \cdot 3 \cdot 2 \cdot 1} = 15$

(b) $\binom{8}{0} = \frac{8!}{0!(8-0)!} = \frac{8!}{0!8!} = \frac{8!}{1 \cdot 8!} = 1$

(c) $\binom{10}{10} = \frac{10!}{10!(10-10)!} = \frac{10!}{10!0!} = 1$ ■

Refer again to Pascal's triangle. Notice the symmetry in each row. This suggests that the binomial coefficients should have the same property. That is,

$$\binom{n}{r} = \binom{n}{n-r}.$$

This is true, since

$$\binom{n}{r} = \frac{n!}{r!(n-r)!} \quad \text{and} \quad \binom{n}{n-r} = \frac{n!}{(n-r)!r!}.$$

Our conjectures about the expansion of $(x + y)^n$ may be summarized as follows.

1. There are $n + 1$ terms in the expansion.
2. The first term is x^n and the last term is y^n.
3. The exponent on x decreases by 1, and the exponent on y increases by 1 in each succeeding term.
4. The sum of the exponents on x and y in any term is n.
5. The coefficient of the term with $x^r y^{n-r}$ or $x^{n-r}y^r$ is $\binom{n}{r}$.

These observations about the expansion of $(x + y)^n$ for any positive integer value of n suggest the **binomial theorem**.

Binomial Theorem

For any positive integer n and any complex numbers x and y,

$$(x + y)^n = x^n + \binom{n}{1}x^{n-1}y + \binom{n}{2}x^{n-2}y^2 + \binom{n}{3}x^{n-3}y^3 + \dots + \binom{n}{r}x^{n-r}y^r + \dots + \binom{n}{n-1}xy^{n-1} + y^n.$$

As stated above, the binomial theorem is a conjecture, determined inductively by looking at $(x + y)^n$ for several values of n. A proof of the binomial theorem using *mathematical induction* is given in Section 5 of this chapter.

NOTE The binomial theorem also looks much more manageable written in summation notation. The theorem can be summarized as follows:

$$(x + y)^n = \sum_{r=0}^{n} \binom{n}{r} x^{n-r}y^r. \blacksquare$$

EXAMPLE 2 Write out the binomial expansion of $(x + y)^9$. Using the binomial theorem,

$$(x + y)^9 = x^9 + \binom{9}{1} x^8y + \binom{9}{2} x^7y^2 + \binom{9}{3} x^6y^3 + \binom{9}{4} x^5y^4 + \binom{9}{5} x^4y^5 + \binom{9}{6} x^3y^6 + \binom{9}{7} x^2y^7 + \binom{9}{8} xy^8 + y^9.$$

Now evaluate each of the binomial coefficients.

$$\begin{aligned}(x + y)^9 &= x^9 + \frac{9!}{1!8!} x^8y + \frac{9!}{2!7!} x^7y^2 + \frac{9!}{3!6!} x^6y^3 + \frac{9!}{4!5!} x^5y^4 \\ &\quad + \frac{9!}{5!4!} x^4y^5 + \frac{9!}{6!3!} x^3y^6 + \frac{9!}{7!2!} x^2y^7 + \frac{9!}{8!1!} xy^8 + y^9 \\ &= x^9 + 9x^8y + 36x^7y^2 + 84x^6y^3 + 126x^5y^4 + 126x^4y^5 \\ &\quad + 84x^3y^6 + 36x^2y^7 + 9xy^8 + y^9 \quad \blacksquare\end{aligned}$$

EXAMPLE 3 Expand $\left(a - \frac{b}{2}\right)^5$.

Write the binomial as follows.

$$\left(a - \frac{b}{2}\right)^5 = \left(a + \left(-\frac{b}{2}\right)\right)^5$$

Now use the binomial theorem with $x = a$, $y = -\frac{b}{2}$, and $n = 5$ to get

$$\begin{aligned}\left(a - \frac{b}{2}\right)^5 &= a^5 + \binom{5}{1}a^4\left(-\frac{b}{2}\right) + \binom{5}{2}a^3\left(-\frac{b}{2}\right)^2 + \binom{5}{3}a^2\left(-\frac{b}{2}\right)^3 + \binom{5}{4}a\left(-\frac{b}{2}\right)^4 + \left(-\frac{b}{2}\right)^5 \\ &= a^5 + 5a^4\left(-\frac{b}{2}\right) + 10a^3\left(-\frac{b}{2}\right)^2 + 10a^2\left(-\frac{b}{2}\right)^3 + 5a\left(-\frac{b}{2}\right)^4 + \left(-\frac{b}{2}\right)^5 \\ &= a^5 - \frac{5}{2}a^4b + \frac{5}{2}a^3b^2 - \frac{5}{4}a^2b^3 + \frac{5}{16}ab^4 - \frac{1}{32}b^5. \quad \blacksquare\end{aligned}$$

NOTE As Example 3 illustrates, any expansion of the *difference* of two terms has alternating signs. ■

EXAMPLE 4 Expand $\left(\frac{3}{m^2} - 2\sqrt{m}\right)^4$. (Assume $m > 0$.)

By the binomial theorem,

$$\left(\frac{3}{m^2} - 2\sqrt{m}\right)^4 = \left(\frac{3}{m^2}\right)^4 + \binom{4}{1}\left(\frac{3}{m^2}\right)^3(-2\sqrt{m})^1 + \binom{4}{2}\left(\frac{3}{m^2}\right)^2(-2\sqrt{m})^2$$
$$+ \binom{4}{3}\left(\frac{3}{m^2}\right)^1(-2\sqrt{m})^3 + (-2\sqrt{m})^4$$
$$= \frac{81}{m^8} + 4\left(\frac{27}{m^6}\right)(-2m^{1/2}) + 6\left(\frac{9}{m^4}\right)(4m)$$
$$+ 4\left(\frac{3}{m^2}\right)(-8m^{3/2}) + 16m^2.$$

Here, we used the fact that $\sqrt{m} = m^{1/2}$. Finally,

$$\left(\frac{3}{m^2} - 2\sqrt{m}\right)^4 = \frac{81}{m^8} - \frac{216}{m^{11/2}} + \frac{216}{m^3} - \frac{96}{m^{1/2}} + 16m^2.$$ ■

Earlier, we wrote the binomial theorem in summation notation as $\sum_{r=0}^{n}\binom{n}{r} x^{n-r}y^r$, which gives the form of each term. We can use this form to write any particular term of a binomial expansion without writing out the entire expansion. For example, to find the tenth term of $(x + y)^n$, where $n \geq 9$, first notice that in the tenth term y is raised to the ninth power (since y has the power 1 in the second term, the power 2 in the third term, and so on). Because the exponents on x and y in any term must have a sum of n, the exponent on x in the tenth term is $n - 9$. Thus, the tenth term of the expansion is

$$\binom{n}{9} x^{n-9}y^9 = \frac{n!}{9!(n - 9)!} x^{n-9}y^9.$$

This same idea can be used to obtain the result given in the following theorem.

*k*th Term of the Binomial Expansion

The kth term of the binomial expansion of $(x + y)^n$, where $n \geq k - 1$, is

$$\binom{n}{k - 1} x^{n-(k-1)}y^{k-1}.$$

To find the kth term of a binomial expansion, use the following steps.

1. Find $k - 1$. This is the exponent on the second term of the binomial.
2. Subtract the exponent on the second term from n to get the exponent on the first term of the binomial.
3. Determine the coefficient by using the exponents found in the first two steps and n.

EXAMPLE 5 Find the seventh term of $(a + 2b)^{10}$.

In the seventh term $2b$ has an exponent of 6, while a has an exponent of $10 - 6$, or 4. The seventh term is

$$\binom{10}{6}a^4(2b)^6 = 210a^4(64b^6)$$
$$= 13{,}440a^4b^6.$$

8.4 Exercises

Evaluate each of the following binomial coefficients.

1. $\binom{10}{6}$ 2. $\binom{8}{3}$ 3. $\binom{12}{0}$ 4. $\binom{20}{17}$ 5. $\binom{25}{2}$ 6. $\binom{28}{28}$

Write out the binomial expansion for each of the following.

7. $(m^2 + n)^6$ 8. $(a + b^2)^5$ 9. $(z - 3w)^5$ 10. $(2k - h)^6$
11. $(r/2 - 3)^5$ 12. $(5 - x/4)^5$ 13. $(x^{3/5} + 2y^{4/5})^5$ 14. $(3z^{1/3} + w^{2/3})^4$
15. What is true of the signs of the terms of the expansion of $(x - y)^n$ if $y > 0$? Explain why this is so.

Write only the first four terms in each of the following expansions.

16. $(a + 2b)^{15}$ 17. $(3m - n)^{20}$ 18. $(4m^{-1} + m^{-2})^{12}$ 19. $(k^{-2} + 3k^2)^9$

In Exercises 20–25 write the indicated term of the binomial expansion.

20. Sixth term of $(4h - j)^8$
21. Eighth term of $(2c - 3d)^{14}$
22. Fifteenth term of $(a^2 + b)^{22}$
23. Twelfth term of $(2x + y^2)^{16}$
24. Fifteenth term of $(x - y^3)^{20}$
25. Tenth term of $(a^3 + 3b)^{11}$
26. Find the middle term of $(3x^7 + 2y^3)^8$.
27. Find the two middle terms of $(-2m^{-1} + 3n^{-2})^{11}$.
28. Find the value of n for which the coefficients of the fifth and eighth terms in the expansion of $(x + y)^n$ are the same.
29. Find the term in the expansion of $(3 + \sqrt{x})^{11}$ that contains x^4.
30. Find the value of n for which the coefficient of x^2 in the expansion of $(1 + x)^n$ is 21.

In later courses, it is shown that

$$(1 + x)^n = 1 + nx + \frac{n(n - 1)}{2!}x^2 + \frac{n(n - 1)(n - 2)}{3!}x^3 + \ldots$$

for any real number n (not just positive integer values) and any real number x where $|x| < 1$. *This result, a generalized binomial theorem, may be used to find approximate values of powers and roots. For example,*

$$\begin{aligned}(1.008)^{1/4} &= (1 + .008)^{1/4} \\ &= 1 + \frac{1}{4}(.008) + \frac{1/4(-3/4)}{2!}(.008)^2 + \frac{1/4(-3/4)(-7/4)}{3!}(.008)^3 + \ldots \\ &\approx 1.002.\end{aligned}$$

Use this result to approximate the quantities in Exercises 31–34 to the nearest thousandth.

31. $(1.02)^{-3}$ **32.** $1/(1.04)^5$ **33.** $(1.01)^{3/2}$ **34.** $(1.03)^{.2}$

35. Let $n = -1$ and expand $(1 + x)^{-1}$.

36. Use polynomial division to find the first four terms when $1 + x$ is divided into 1. Compare the result with the result of Exercise 35. What do you find? Explain.

37. Find the sum of the first four terms in the expansion of $(1 + 3)^{1/2}$ using $x = 3$ and $n = 1/2$ in the formula above. Is the result close to $(1 + 3)^{1/2} = 4^{1/2} = 2$? Why not? Explain.

38. Use the result above to show that for small values of x, $\sqrt{1 + x} \approx 1 + \frac{1}{2}x$.

39. When $(4x - 5)^7$ is written in the form $a_7x^7 + a_6x^6 + \ldots + a_1x + a_0$, what is the sum of the numbers $a_7, a_6, \ldots, a_1, a_0$? (*Hint:* This question can be answered without determining the values of the coefficients.)

40. Show that $\binom{n}{0} + \binom{n}{1} + \binom{n}{2} + \ldots + \binom{n}{n} = 2^n$. (*Hint:* Set $x = 1$ in the binomial expansion of $(1 + x)^n$.)

8.5 Mathematical Induction

Many results in mathematics are claimed to be true for every positive integer. Any of these results could be checked for $n = 1$, $n = 2$, $n = 3$, and so on, but since the set of positive integers is infinite it would be impossible to check every possible case. For example, let S_n represent the statement that the sum of the first n positive integers is $n(n + 1)/2$.

$$S_n\!: 1 + 2 + 3 + \ldots + n = \frac{n(n + 1)}{2}$$

The truth of this statement is easily verified for the first few values of n:

If $n = 1$, then S_1 is $\quad 1 = \frac{1(1 + 1)}{2}$, $\quad$ which is true.

If $n = 2$, then S_2 is $\quad 1 + 2 = \frac{2(2 + 1)}{2}$, $\quad$ which is true.

If $n = 3$, then S_3 is $\quad 1 + 2 + 3 = \frac{3(3 + 1)}{2}$, $\quad$ which is true.

If $n = 4$, then S_4 is $\quad 1 + 2 + 3 + 4 = \frac{4(4 + 1)}{2}$, $\quad$ which is true.

Continuing in this way for any amount of time would still not prove that S_n is true for *every* positive integer value of n. To prove that such statements are true for every positive integer value of n, the following principle is often used.

Principle of Mathematical Induction

Let S_n be a statement concerning the positive integer n. Suppose that

(a) S_1 is true;

(b) for any positive integer k, S_k implies S_{k+1}.

Then S_n is true for every positive integer value of n

A proof by mathematical induction can be explained as follows. By (a) above, the statement is true when $n = 1$. By (b) above, the fact that the statement is true for $n = 1$ implies that it is true for $n = 1 + 1 = 2$. Using (b) again, it is thus true for $2 + 1 = 3$, for $3 + 1 = 4$, for $4 + 1 = 5$, and so on. By continuing in this way, the statement must be true for *every* positive integer, no matter how large.

The situation is similar to that of a number of dominoes lined up as shown in Figure 4. If the first domino is pushed over, it pushes the next, which pushes the next, and so on until all are down.

Figure 4

Another example of the principle of mathematical induction is an infinite ladder. Suppose the rungs are spaced so that, whenever you are on a rung, you know you can move to the next rung. Then *if* you can get to the first rung, you can go as high up the ladder as you wish.

As these comments show, two separate steps are required for a proof by mathematical induction.

Proof by Induction

1. Prove that the statement is true for $n = 1$.
2. Show that, for any positive integer k, S_k implies S_{k+1}.

In the next example, mathematical induction is used to prove the statement S_n mentioned at the beginning of this section.

EXAMPLE 1 Let S_n represent the statement

$$1 + 2 + 3 + \ldots + n = \frac{n(n + 1)}{2}.$$

Prove that S_n is true for every positive integer n.

The proof by mathematical induction is as follows.

Step 1 Show that the statement is true when $n = 1$. If $n = 1$, S_n becomes S_1, which is

$$1 = \frac{1(1 + 1)}{2},$$

and is true.

Step 2 Show that S_k implies S_{k+1}, where S_k is the statement

$$1 + 2 + 3 + \ldots + k = \frac{k(k + 1)}{2},$$

and S_{k+1} is the statement

$$1 + 2 + 3 + \ldots + k + (k + 1) = \frac{(k + 1)[(k + 1) + 1]}{2}.$$

Start with S_k.

$$1 + 2 + 3 + \ldots + k = \frac{k(k + 1)}{2}$$

How can S_k be changed algebraically to match S_{k+1}? Adding $k + 1$ to both sides of S_k gives

$$1 + 2 + 3 + \ldots + k + (k + 1) = \frac{k(k + 1)}{2} + (k + 1).$$

Then, factoring on the right gives

$$= (k + 1)\left(\frac{k}{2} + 1\right)$$

$$= (k + 1)\left(\frac{k + 2}{2}\right)$$

$$1 + 2 + 3 + \ldots + k + (k + 1) = \frac{(k + 1)[(k + 1) + 1]}{2}.$$

This final result is the statement for $n = k + 1$; it has been shown that S_k implies S_{k+1}. The two steps required for a proof by mathematical induction have now been completed, so that the statement S_n is true for every positive integer value of n. ■

EXAMPLE 2 Prove: if x is a real number between 0 and 1, then for every positive integer n, $0 < x^n < 1$.

Here S_1 is the statement

$$\text{if } 0 < x < 1, \text{ then } 0 < x^1 < 1,$$

which is true. S_k is the statement

$$\text{if } 0 < x < 1, \text{ then } 0 < x^k < 1.$$

Now show that this implies S_{k+1}. Multiply all members of $0 < x^k < 1$ by x to get

$$x \cdot 0 < x \cdot x^k < x \cdot 1.$$

(Here the fact that $0 < x$ is used.) Simplify to get

$$0 < x^{k+1} < x.$$

Since it is given that $x < 1$,

$$x^{k+1} < x < 1,$$

and

$$0 < x^{k+1} < 1.$$

This work shows that S_k implies S_{k+1}, and since S_1 is true, the given statement is true for every positive integer n. ■

Some statements S_n are not true for the first few values of n, but are true for all values of n that are at least equal to some fixed integer j. The following slightly generalized form of the principle of mathematical induction takes care of these cases.

Generalized Principle of Mathematical Induction

Let S_n be a statement concerning the positive integer n. Let j be a fixed positive integer. Suppose that

(a) S_j is true;

(b) for any positive integer k, $k \geq j$, S_k implies S_{k+1}.

Then S_n is true for all positive integers n, where $n \geq j$.

EXAMPLE 3 Let S_n represent the statement $2^n > 2n + 1$. Show that S_n is true for all values of n such that $n \geq 3$.

(Check that S_n is false for $n = 1$ and $n = 2$.) As before, the proof requires two steps.

Step 1 Show that S_n is true for $n = 3$. If $n = 3$, S_n is

$$2^3 > 2 \cdot 3 + 1,$$

or

$$8 > 7,$$

which is true.

Step 2 Now show that S_k implies S_{k+1}, where $k \geq 3$ and

$$S_k \text{ is } 2^k > 2k + 1,$$
$$S_{k+1} \text{ is } 2^{k+1} > 2(k + 1) + 1.$$

Multiply both sides of $2^k > 2k + 1$ by 2, obtaining

$$2 \cdot 2^k > 2(2k + 1),$$

or

$$2^{k+1} > 4k + 2.$$

Rewrite $4k + 2$ as $2(k + 1) + 2k$, giving

$$2^{k+1} > 2(k + 1) + 2k. \qquad \textbf{(1)}$$

Since k is a positive integer greater than 3,

$$2k > 1. \qquad \textbf{(2)}$$

Adding $2(k + 1)$ to both sides of inequality (2) gives

$$2(k + 1) + 2k > 2(k + 1) + 1. \qquad \textbf{(3)}$$

From inequalities (1) and (3),

$$2^{k+1} > 2(k + 1) + 2k > 2(k + 1) + 1,$$

or

$$2^{k+1} > 2(k + 1) + 1,$$

as required. Thus, S_k implies S_{k+1}, and this, together with the fact that S_3 is true, shows that S_n is true for every positive integer value of n greater than or equal to 3. ■

EXAMPLE 4 The binomial theorem can be proved by mathematical induction. That is, for any positive integer n and any complex numbers x and y,

$$(x + y)^n = x^n + \binom{n}{1} x^{n-1}y + \binom{n}{2} x^{n-2}y^2 + \binom{n}{3} x^{n-3}y^3 + \ldots + \binom{n}{r} x^{n-r}y^r + \ldots + \binom{n}{n-1} xy^{n-1} + y^n. \quad \textbf{(4)}$$

Let S_n be statement (4) above. Begin by verifying S_n for $n = 1$.

$$S_1\!: (x + y)^1 = x^1 + y^1,$$

which is true.

Now assume that S_n is true for the positive integer k. Statement S_k becomes (using the definition of the binomial coefficient)

$$S_k\!: \ (x + y)^k = x^k + \frac{k!}{1!(k-1)!} x^{k-1}y + \frac{k!}{2!(k-2)!} x^{k-2}y^2 + \ldots + \frac{k!}{(k-1)!1!} xy^{k-1} + y^k. \quad \textbf{(5)}$$

Multiply both sides of equation (5) by $x + y$.

$$\begin{aligned}(x + y)^k \cdot (x + y) &= x(x + y)^k + y(x + y)^k \\ &= \left[x \cdot x^k + \frac{k!}{1!(k-1)!} x^k y + \frac{k!}{2!(k-2)!} x^{k-1}y^2 + \ldots + \frac{k!}{(k-1)!1!} x^2y^{k-1} + xy^k\right] \\ &\quad + \left[x^k \cdot y + \frac{k!}{1!(k-1)!} x^{k-1}y^2 + \ldots + \frac{k!}{(k-1)!1!} xy^k + y \cdot y^k\right]\end{aligned}$$

Rearrange terms to get

$$\begin{aligned}(x + y)^{k+1} &= x^{k+1} + \left[\frac{k!}{1!(k-1)!} + 1\right] x^k y + \left[\frac{k!}{2!(k-2)!} + \frac{k!}{1!(k-1)!}\right] x^{k-1}y^2 + \ldots \\ &\quad + \left[1 + \frac{k!}{(k-1)!1!}\right] xy^k + y^{k+1}. \quad \textbf{(6)}\end{aligned}$$

The first expression in brackets in equation (6) simplifies to $\binom{k+1}{1}$. To see this, note that

$$\binom{k+1}{1} = \frac{(k+1)(k)(k-1)(k-2)\ldots 1}{1 \cdot (k)(k-1)(k-2)\ldots 1} = k + 1.$$

Also, $$\frac{k!}{1!(k-1)!} + 1 = \frac{k(k-1)!}{1(k-1)!} + 1 = k + 1.$$

The second expression becomes $\binom{k+1}{2}$, the last $\binom{k+1}{k}$, and so on. The result of equation (6) is just equation (5) with every k replaced by $k + 1$. Thus, the truth of S_n when $n = k$ implies the truth of S_n for $n = k + 1$, which completes the proof of the theorem by mathematical induction. ■

8.5 Exercises

Write out in full and verify each of the statements S_1, S_2, S_3, S_4, and S_5 for each of the following. Then use mathematical induction to prove that each of the given statements is true for every positive integer n.

1. $2 + 4 + 6 + \ldots + 2n = n(n+1)$

2. $1 + 3 + 5 + \ldots + (2n-1) = n^2$

Use the method of mathematical induction to prove that each of the following statements is true for every positive integer n.

3. $2 + 4 + 8 + \ldots + 2^n = 2^{n+1} - 2$

4. $1^2 + 2^2 + 3^2 + \ldots + n^2 = \frac{n(n+1)(2n+1)}{6}$

5. $1^3 + 2^3 + 3^3 + \ldots + n^3 = \frac{n^2(n+1)^2}{4}$

6. $3 + 3^2 + 3^3 + \ldots + 3^n = \frac{3(3^n - 1)}{2}$

7. $5 \cdot 6 + 5 \cdot 6^2 + 5 \cdot 6^3 + \ldots + 5 \cdot 6^n = 6(6^n - 1)$

8. $\frac{1}{1 \cdot 2} + \frac{1}{2 \cdot 3} + \frac{1}{3 \cdot 4} + \ldots + \frac{1}{n(n+1)} = \frac{n}{n+1}$

9. $\frac{1}{1 \cdot 4} + \frac{1}{4 \cdot 7} + \frac{1}{7 \cdot 10} + \ldots + \frac{1}{(3n-2)(3n+1)} = \frac{n}{3n+1}$

10. $\frac{1}{2} + \frac{1}{2^2} + \frac{1}{2^3} + \ldots + \frac{1}{2^n} = 1 - \frac{1}{2^n}$

In the following statements S_n, find a value of n for which S_n is not true or prove S_n by mathematical induction.

11. $2^n > 2n$

12. $3^n > 2n + 1$

13. $1 \cdot 4 + 2 \cdot 9 + 3 \cdot 16 + \ldots + n(n+1)^2 = \frac{n(n+1)(n+2)(3n+5)}{12}$

14. $2^n > n^2$

15. $n! > 2n$

16. $1 \cdot 2 + 2 \cdot 3 + 3 \cdot 4 + \ldots + n(n+1) = \frac{n(n+1)(n+2)}{3}$

Prove each result in Exercises 17–26 by mathematical induction.

17. $(a^m)^n = a^{mn}$ (Assume a and m are constant.)

18. $(ab)^n = a^n b^n$ (Assume a and b are constant.)

19. $2^n > 2n$, if $n \geq 3$

20. $3^n > 2n + 1$, if $n \geq 2$

21. If $a > 1$, then $a^n > 1$

22. If $a > 1$, then $a^n > a^{n-1}$

23. If $0 < a < 1$, then $a^n < a^{n-1}$

24. $2^n > n^2$, for $n > 4$

25. If $n \geq 4$, then $n! > 2^n$, where $n! = n(n-1)(n-2) \ldots (3)(2)(1)$.

26. $4^n > n^4$, for $n \geq 5$

27. Suppose that Step 2 in a proof by mathematical induction can be satisfied, but Step 1 cannot. May we conclude that the proof is complete? Explain.

28. What is wrong with the following "proof" by mathematical induction?
Prove: Any natural number equals the next natural number.

To begin, we assume the statement is true for some natural number k:

$$k = k + 1.$$

We must now show that the statement is true for $n = k + 1$. If we add 1 to both sides, we have

$$k + 1 = k + 1 + 1$$
$$k + 1 = k + 2.$$

Hence, if the statement is true for $n = k$, it is also true for $n = k + 1$. Thus, the theorem is proved.

29. Suppose that n straight lines (with $n \geq 2$) are drawn in a plane, where no two lines are parallel and no three lines pass through the same point. Show that the number of points of intersection of the lines is $(n^2 - n)/2$.

30. The series of sketches below starts with an equilateral triangle having sides of length 1. In the following steps, equilateral triangles are constructed on each side of the preceding figure. The lengths of the sides of these new triangles is 1/3 the length of the sides of the preceding triangles. Develop a formula for the number of sides of the nth figure. Use mathematical induction to prove your answer.

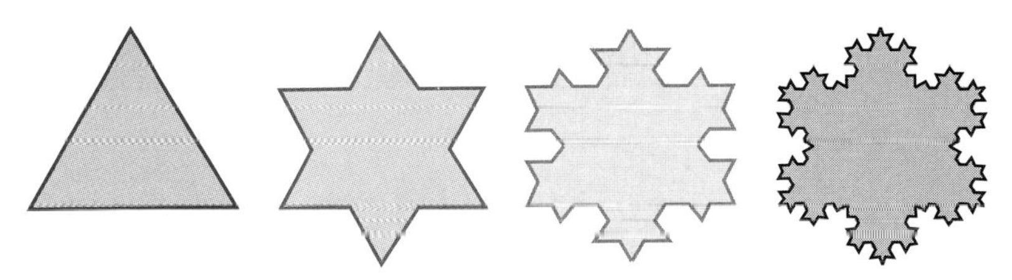

31. Find the perimeter of the nth figure in Exercise 30.

32. Show that the area of the nth figure in Exercise 30 is

$$\sqrt{3}\left[\frac{2}{5} - \frac{3}{20}\left(\frac{4}{9}\right)^{n-1}\right].$$

33. A pile of n rings, each smaller than the one below it, is on a peg. Two other pegs are attached to a board with this peg. In the game called the *Tower of Hanoi* puzzle, all the rings must be moved to a different peg, with only one ring moved at a time, and with no ring ever placed on top of a smaller ring. Find the least number of moves that would be required. Prove your result with mathematical induction.

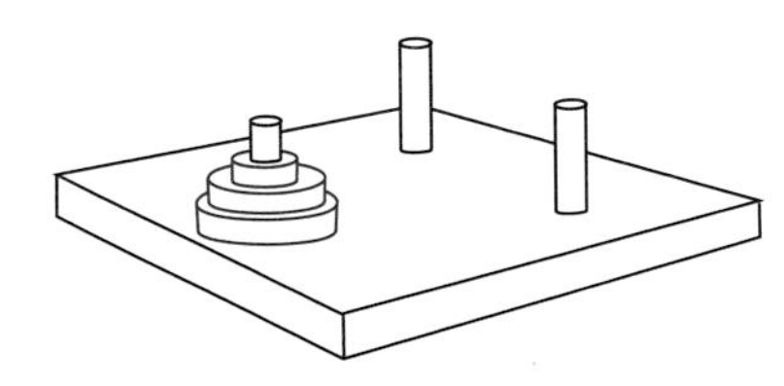

Exercise 33

(b) The sum of the points showing is at most 4.

"At most 4" can be written as "2 or 3 or 4." (A sum of 1 is meaningless here.) Then

$$\begin{aligned} P(\textbf{at most 4}) &= P(\textbf{2 or 3 or 4}) \\ &= P(2) + P(3) + P(4), \end{aligned} \qquad (*)$$

since the events represented by "2," "3," and "4" are mutually exclusive.

The sample space for this experiment includes the 36 possible pairs of numbers shown in Figure 9. The pair (1, 1) is the only one with a sum of 2, so $P(2) = 1/36$. Also $P(3) = 2/36$ since both (1, 2) and (2, 1) give a sum of 3. The pairs, (1, 3), (2, 2), and (3, 1) have a sum of 4, so $P(4) = 3/36$. Substituting into equation (*) above gives

$$\begin{aligned} P(\text{at most 4}) &= \frac{1}{36} + \frac{2}{36} + \frac{3}{36} \\ &= \frac{6}{36} = \frac{1}{6}. \end{aligned}$$

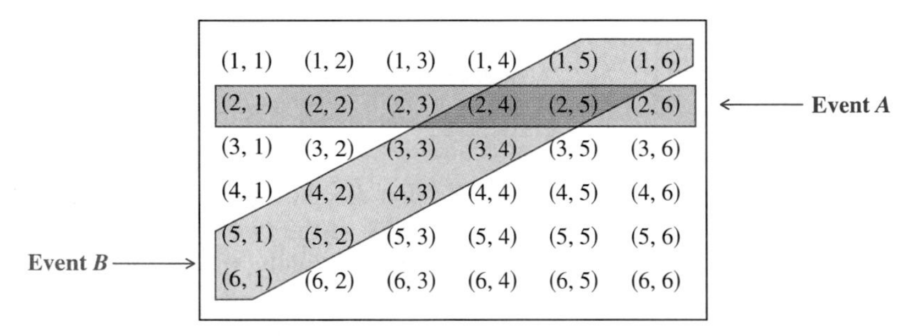

Figure 9

8.7 Exercises

Write a sample space with equally likely outcomes for each of the following experiments.

1. A two-headed coin is tossed once.

2. Two ordinary coins are tossed.

3. Three ordinary coins are tossed.

4. Five slips of paper marked with the numbers 1, 2, 3, 4, and 5 are placed in a box. After mixing well, two slips are drawn.

5. An unprepared student takes a three-question true/false quiz in which he guesses the answer to all three questions.

6. A die is rolled and then a coin is tossed.

Write the events in Exercises 7–10 in set notation and give the probability of each event.

7. In the experiment from Exercise 2:
(a) Both coins show the same face.
(b) At least one coin turns up heads.

8. In Exercise 1:
(a) The result of the toss is heads.
(b) The result of the toss is tails.

9. In Exercise 4:
(a) Both slips are marked with even numbers.
(b) Both slips are marked with odd numbers.
(c) Both slips are marked with the same number.
(d) One slip is marked with an odd number and the other with an even number.

10. In Exercise 5:
(a) The student gets all three answers correct.
(b) He gets all three answers wrong.
(c) He gets exactly two answers correct.
(d) He gets at least one answer correct.

11. A student gives the answer to a probability problem as $6/5$. Explain why this answer must be incorrect.

12. If the probability of an event is .857, what is the probability that the event will not occur?

13. A marble is drawn at random from a box containing 3 yellow, 4 white, and 8 blue marbles. Find each probability in (a) – (c).
(a) A yellow marble is drawn.
(b) A blue marble is drawn.
(c) A black marble is drawn.
(d) What are the odds in favor of drawing a yellow marble?
(e) What are the odds against drawing a blue marble?

14. A baseball player with a batting average of .300 comes to bat. What are the odds in favor of his getting a hit?

15. In Exercise 4, what are the odds that the sum of the numbers on the two slips of paper is 5?

16. If the odds that it will rain are 4 to 5, what is the probability of rain?

17. If the odds that a candidate will win an election are 3 to 2, what is the probability that the candidate will lose?

18. Ms. Bezzone invites 10 relatives to a party: her mother, two uncles, three brothers, and four cousins. If the chances of any one guest arriving first are equally likely, find the following probabilities.
(a) The first guest is an uncle or a cousin.
(b) The first guest is a brother or a cousin.
(c) The first guest is an uncle or her mother.

19. A card is drawn from a well-shuffled deck of 52 cards. Find the probability that the card is the following.
(a) A queen **(b)** Red **(c)** A black 3 **(d)** A club or red

20. In Exercise 19, find the probability of the following.
(a) A face card (K, Q, J of any suit) **(b)** Red or a 3
(c) Less than a four (consider aces as ones)

21. Two dice are rolled. Find the probability of the following events.
(a) The sum of the points is at least 10.
(b) The sum of the points is either 7 or at least 10.
(c) The sum of the points is 3 or the dice both show the same number.

22. If a marble is drawn from a bag containing 2 yellow, 5 red, and 3 blue marbles, what are the probabilities of the following results?
(a) The marble is yellow or blue.
(b) The marble is yellow or red.
(c) The marble is green.

23. The law firm of Alam, Bartolini, Chinn, Dickinson, and Ellsberg has two senior partners, Alam and Bartolini. Two of the attorneys are to be selected to attend a conference. Assuming that all are equally likely to be selected, find the following probabilities.
(a) Chinn is selected.
(b) Alam and Dickinson are selected.
(c) At least one senior partner is selected.

24. The management of a firm wants to survey its workers, who are classified as follows for the purpose of an interview: 30% have worked for the company more than 5 years; 28% are female; 65% contribute to a voluntary retirement plan; half of the female workers contribute to the retirement plan. Find the following probabilities.
(a) A male worker is selected.
(b) A worker is selected who has been employed by the company for 5 years or less.
(c) A worker is selected who contributes to the retirement plan or is female.

25. The table shows the probabilities of a person accumulating specific amounts of credit card charges over a 12-month period.

Charges	*Probability*
Under $100	.31
$100–$499	.18
$500–$999	.18
$1000–$1999	.13
$2000–$2999	.08
$3000–$4999	.05
$5000–$9999	.06
$10,000 or more	.01

Find the probabilities that a person's total charges during the period are the following.
(a) $500–$999 **(b)** $5000–$9999
(c) $500 to $2999 **(d)** $3000 or more

26. In Exercise 25, find the probabilities that a person charges the following amounts.
(a) $100–$499 **(b)** $2000–$2999
(c) Less than $2000 **(d)** More than $499

In most animals and plants, it is very unusual for the number of main parts of the organism (arms, legs, toes, flower petals, etc.) to vary from generation to generation. Some species, however, have meristic variability, *in which the number of certain body parts varies from generation to generation. One researcher studied the front feet of certain guinea pigs and produced the following probabilities.**

$$P(\text{only four toes, all perfect}) = .77$$
$$P(\text{one imperfect toe and four good ones}) = .13$$
$$P(\text{exactly five good toes}) = .10$$

Find the probability of each of the following events.

27. No more than four good toes

28. Five toes, whether perfect or not

The probabilities for the outcomes of an experiment having sample space $S = \{s_1, s_2, s_3, s_4, s_5, s_6\}$ *are shown here.*

Outcomes	s_1	s_2	s_3	s_4	s_5	s_6
Probability	.17	.03	.09	.46	.21	.04

Let $E = \{s_1, s_2, s_5\}$, *and let* $F = \{s_4, s_5\}$. *Find each probability in Exercises 29–34.*

29. $P(E)$

30. $P(F)$

31. $P(E \cap F)$

32. $P(E \cup F)$

33. $P(E' \cup F')$

34. $P(E' \cap F)$

Chapter 8 Review Exercises

Use summation notation to rewrite each sum with the index of summation starting at the indicated number.

1. $\sum_{i=1}^{8} (3 + 2i);\ -2$

2. $\sum_{i=2}^{9} (4 - 6i);\ 0$

Use the properties of summation, along with sums given earlier, to evaluate each summation.

3. $\sum_{i=1}^{4} (i^2 + 2i)$

4. $\sum_{i=1}^{6} (8 + i^3)$

5. How can you decide whether a given sequence or series is arithmetic or geometric?

Write the first five terms for each sequence in Exercises 6–15.

6. $a_1 = 8, a_2 = 4, a_n = a_{n-1} - a_{n-2}$ for $n > 2$

7. $b_1 = 5, b_2 = -1, b_n = -2 \cdot b_{n-2}$ if n is odd, and $b_n = 2 \cdot b_{n-2}$ if n is even

*Excerpt from ''Analysis of Variability in Number of Digits in an Inbred Strain of Guinea Pigs'' by S. Wright, in *Genetics*, v. 19 (1934), 506–36. Reprinted by permission of Genetics Society of America.

8. Arithmetic, $a_2 = 10, d = -2$

9. Arithmetic, $a_2 = 5, a_3 = 3$

10. Arithmetic, $a_3 = \pi, a_4 = 1$

11. Arithmetic, $a_1 = 4 - \sqrt{3}, a_2 = 3$

12. Geometric, $a_1 = 6, r = 2$

13. Geometric, $a_3 = 4, r = -1/2$

14. Geometric, $a_1 = -5, a_2 = -1$

15. Geometric, $a_2 = 3, a_5 = 12$

16. An arithmetic sequence has $a_5 = -3$ and $a_{15} = 17$. Find a_1 and a_n.

17. A geometric sequence has $a_1 = -8$ and $a_7 = -1/8$. Find a_4 and a_n.

Find a_8 for each of the following arithmetic sequences.

18. $a_1 = 6, \ d = 2$

19. $a_1 = -4, \ d = 3$

20. $a_1 = 6x - 9, \ a_2 = 5x + 1$

21. $a_3 = 11m, \ a_5 = 7m - 4$

Find S_{12} for each of the following arithmetic sequences.

22. $a_1 = 2, \ d = 3$

23. $a_2 = 6, \ d = 10$

24. $a_1 = -4k, \ d = 2k$

Find a_5 for each of the following geometric sequences.

25. $a_1 = -2, \ r = 3$

26. $a_3 = 4, \ r = 1/5$

27. $a_1 = 3y, \ a_2 = y^3$

28. $a_2 = \sqrt{2}, \ a_4 = 3\sqrt{2}$

Find S_4 for each of the following geometric sequences.

29. $a_1 = 3, \ r = 2$

30. $a_1 = -1, \ r = 3$

31. $a_1 = p, \ r = -2p$

Determine whether each of the following sequences is arithmetic, geometric, or neither. If the sequence is arithmetic, find the common difference. If the sequence is geometric, find the common ratio.

32. $8, \ -4, \ 2, \ -1, \ \frac{1}{2}, \ \ldots$

33. $-3, \ 0, \ 3, \ 6, \ 9, \ \ldots$

34. $\ln 1, \ \ln 2, \ \ln 3, \ \ln 4, \ \ldots$

35. $\ln 2, \ \ln 4, \ \ln 8 \ \ln 16, \ \ldots$

36. Explain the difference between a sequence and a series.

Evaluate each of the following sums that converge.

37. $24 + 8 + 8/3 + 8/9 + \ldots$

38. $-3/4 + 1/2 - 1/3 + 2/9 - \ldots$

39. $1/12 + 1/6 + 1/3 + 2/3 + \ldots$

40. $.9 + .09 + .009 + .0009 + \ldots$

Evaluate each of the following sums that exist.

41. $\sum_{i=1}^{7} (-1)^{i-1}$

42. $\sum_{i=1}^{5} (i^2 + i)$

43. $\sum_{i=1}^{4} \frac{i + 1}{i}$

44. $\sum_{j=1}^{10} (3j - 4)$

45. $\sum_{j=1}^{2500} j$

46. $\sum_{i=1}^{5} 4 \cdot 2^i$

47. $\sum_{i=1}^{\infty} \left(\frac{4}{7}\right)^i$

48. $\sum_{i=1}^{\infty} -2\left(\frac{6}{5}\right)i$

Evaluate each of the following sums, where $x_1 = 0, \ x_2 = 1, \ x_3 = 2, \ x_4 = 3, \ x_5 = 4, \ x_6 = 5$.

49. $\sum_{i=1}^{4} (x_i^2 - 6)$

50. $\sum_{i=1}^{6} f(x_i)\Delta x; \quad f(x) = (x - 2)^3, \ \Delta x = .1$

Write each of the following sums using summation notation.

51. $4 - 1 - 6 - \ldots - 66$

52. $10 + 14 + 18 + \ldots + 86$

53. $4 + 12 + 36 + \ldots + 972$

54. $\frac{5}{6} + \frac{6}{7} + \frac{7}{8} + \ldots + \frac{12}{13}$

55. What is the binomial theorem used for? Give examples.

Use the binomial theorem to expand each of the following.

56. $(x + 2y)^4$

57. $(3z - 5w)^3$

58. $\left(3\sqrt{x} - \frac{1}{\sqrt{x}}\right)^5$

59. $(m^3 - m^{-2})^4$

Find the indicated term or terms for each of the following expansions.

60. Sixth term of $(4x - y)^8$

61. Seventh term of $(m - 3n)^{14}$

62. First four terms of $(x + 2)^{12}$

63. Last three terms of $(2a + 5b)^{16}$

64. Describe a proof by mathematical induction.

65. What kinds of statements are proved by mathematical induction? Give examples.

Use mathematical induction to prove that each of the following is true for every positive integer n.

66. $1 + 3 + 5 + 7 + \ldots + (2n - 1) = n^2$

67. $2 + 6 + 10 + 14 + \ldots + (4n - 2) = 2n^2$

68. $2 + 2^2 + 2^3 + \ldots + 2^n = 2(2^n - 1)$

69. $1^3 + 3^3 + 5^3 + \ldots + (2n - 1)^3 = n^2(2n^2 - 1)$

70. How do permutations and combinations differ? How are they alike?

Find the value of each expression in Exercises 71–74.

71. $P(9, 2)$

72. $P(6, 0)$

73. $\binom{8}{3}$

74. $\binom{10}{5}$

75. Four students are to be assigned to 4 different summer jobs. Each student is qualified for all 4 jobs. In how many ways can the jobs be assigned?

76. Nine football teams are competing for first-, second-, and third-place titles in a statewide tournament. In how many ways can the winners be determined?

77. John Jacobs, who is furnishing his apartment, wants to buy a new sofa. He can select from 5 different styles, each available in 3 different fabrics, with 6 color choices. How many different sofas are available?

Write sample spaces for the following.

78. 2 coins are tossed.

79. A card is drawn from a deck containing only the twelve face cards.

80. A sample of 2 headsets from a box of 5 is tested for defects.

81. The age of one student is determined from a class of college freshmen whose ages range from 17 to 25.

A company sells typewriters and copiers. Let E be the event "a customer buys a typewriter," and let F be the event "a customer buys a copier." Write each of the following using ∩, ∪, or ′ as necessary.

82. A customer buys neither.

83. A customer buys at least one.

Find the odds in favor of a card drawn from an ordinary deck being as follows.

84. A club

85. A black jack

86. A red face card or a queen

87. A sample shipment of five transistors is chosen at random. The probability of exactly 0, 1, 2, 3, 4, or 5 transistors being defective is given in the following table.

Number defective	0	1	2	3	4	5
Probability	.31	.25	.18	.12	.08	.06

Find the probability that at most two are defective.

Answers to Selected Exercises

Chapter 1 Fundamentals of Algebra

Section 1.1 (page 13)

1. commutative **3.** commutative **5.** identity **7.** associative **9.** 240 **11.** -336 **13.** 19 **15.** $6/5$ **17.** natural, whole, integer, rational, real **19.** rational, real **21.** irrational, real **23.** ($\sqrt{25} = 5$) natural, whole, integer, rational, real **25.** undefined **27.** 1900 **29.** 154 **31.** $-6, -4, -3, -\sqrt{5}, 2, \sqrt{6}$ **33.** $3/4, 1.2, \sqrt{2}, 22/15, \pi/2, 8/5$ **35.** 6 **37.** $2 - \sqrt{3}$ **39.** $2 - \sqrt{3}$ **41.** $x - 4$ **43.** $56 - 7m$ **45.** $3 + x^2$ **47.** $1 + p^2$ **49.** $6 - \pi$ **51.** 1 **53.** **(a)** 1 **(b)** 1 **(c)** 14 **(d)** 2 **(e)** 2 **55.** **(a)** 2 **(b)** 7 **(c)** 0 **(d)** 9 **(e)** 9 **57.** $d(A, C)$ is 7 or 1. **59.** multiplication property of order **61.** multiplication property of order **63.** multiplication property of order **65.** triangle inequality, $|a + b| \le |a| + |b|$ **67.** property of absolute value, $|a| \cdot |b| = |ab|$ **69.** property of absolute value, $|a/b| = |a|/|b|$ **71.** trichotomy property **73.** $x > 0$ and $y > 0$, or $x < 0$ and $y < 0$. **75.** y may be positive or negative, but x must be negative. **77.** if $x = y$ or $x = -y$ **79.** if $x = 0$ **81.** -1 if $x < 0$ and 1 if $x > 0$ **83.** 1 **85.** Yes; x must satisfy $-9 \le x \le 9$.

Section 1.2 (page 21)

1. 4 **3.** 25 **5.** $-1/27$ **7.** $1/4$ **9.** $1/25$ **11.** $27/8$ **13.** 1200 **15.** $9x^4$ **17.** .001042 **23.** $x^{7/3}$ **25.** 4^5 **27.** $(1 + n)^{5/4}$ **29.** $1/(81p^4)$ **31.** $6yz^{2/3}$ **33.** $1/d^7$ **35.** $a^{2/3}b^2$ **37.** a^3b^6 **39.** $x^7/5^4$ **41.** $2^{2/7}x^{36/7}$ **43.** $30p^{3-r}$ **45.** $1/(2^3b^{y-2})$ **49.** 27,000 **51.** 27 **53.** \$88 **55.** \$27,000,000 **57.** about \$10,000,000 **59.** about 86.3 mi **61.** about 211 mi **63.** about 125 **65.** about 177

Section 1.3 (page 28)

1. $5p^3 + 2p^2 + p - 4$ **3.** $3m^2 - 10m - 1$ **5.** $3b^2 - 2b + 7$ **7.** $2y^2 + 9y - 5$ **9.** $6m^2 - 7m - 5$ **11.** $18p^2 - 27pq - 35q^2$ **13.** $\frac{6}{25}y^2 + \frac{11}{40}yz + \frac{1}{16}z^2$ **15.** $25r^2 - 4$ **17.** $9z^2 - 4w^2$ **19.** $9x^2 + 48x + 64$ **21.** $16m^2 + 16mn + 4n^2$ **23.** $8z^3 - 12z^2 + 6z - 1$ **25.** $4x^6 + 4x^3y^2 + y^4$ **27.** $x^3 - 6x^2y^2 + 12xy^4 - 8y^6$ **29.** $15m^4 - 10m^3 + 5m^2 - 5m$ **31.** $27p^3 - 1$ **33.** $12k^4 + 21k^3 - 5k^2 + 3k + 2$ **35.** $x^2 + 4xy - 6xz + 4y^2 - 12yz + 9z^2$ **39.** $x^{2y} - 9$ **41.** $4k^{2x} + k^x - 3$ **43.** $m^{2x} - 4m^x + 4$ **45.** $27k^{3a} - 54k^{2a} + 36k^a - 8$ **47.** $a^6 + 6a^5b + 15a^4b^2 + 20a^3b^3 + 15a^2b^4 + 6ab^5 + b^6$ **49.** $81x^4 - 216x^3y + 216x^2y^2 - 96xy^3 + 16y^4$ **51.** $64k^6 + 96k^5 + 60k^4 + 20k^3 + (15/4)k^2 + (3/8)k + (1/64)$ **53.** $y^5 - 20y^4z + 160y^3z^2 - 640y^2z^3 + 1280yz^4 - 1024z^5$ **55.** $2y^2 - 8y^3$ **57.** $-8p^{7/4} + 6p^{11/4}$ **59.** $x^2 - x$ **61.** $r - 2 + r^{-1}$ or $r - 2 + 1/r$ **65.** $-2r^2 - 3rs + 5s^2$ **67.** $2m^2 + m - 2 + \frac{6}{3m + 2}$ **69.** $x^3 - x^2 - x + 4 + \frac{-17}{3x + 3}$ **71.** $k^2 - 5 + \frac{2k + 10}{k^2 + 1}$ **73.** $2z^3 + \frac{3}{2}z^2 - \frac{5}{8}z - \frac{13}{32} + \frac{(-133/32)z + (269/16)}{4z^2 - z + 2}$ **75.** -1 **77.** 1 **79.** 0 **83.** $90\frac{1}{4}$ **85.** **(a)** 4 **(b)** 4 **(c)** 7

Section 1.4 (page 34)

1. $4m^3(8m^5 + 7m^2 - 4)$ **3.** $-7jh(h + 3j + 5)$ **5.** $2(3m + n)(1 + 3m)$ **7.** $(3p - 2)(2p + 3)$ **9.** $(3y + 5)(2y - 1)$
11. $(6m - 7n)(2m + 5n)$ **13.** $(5x^2 + 4)(2x^2 + 3)$ **15.** $(5a + 4)(5a - 4)$ **17.** $(9m + 4n)(9m - 4n)$
19. $(9q^2 + 16m^2)(3q + 4m)(3q - 4m)$ **21.** $(y^8 + 1)(y^4 + 1)(y^2 + 1)(y + 1)(y - 1)$ **23.** $(5x - 1)(25x^2 + 5x + 1)$
25. $(x - y)(x^2 + xy + y^2 + x + y)$ **27.** $(a + b + x + y)(a + b - x - y)$ **29.** $(-z - w)(3z - w)$ **31.** $20(a + 3)(a + 15)$
33. $(x + y + 5z)(x + y - 3z)$ **35.** $4pq$ **37.** $r(r^2 + 18r + 108)$ **39.** $(3 - m - 2n)(9 + 3m + 6n + m^2 + 4mn + 4n^2)$
41. $(2u + 3)(4u^2 - 6u + 9)$ **43.** $(p - 3q)(2r - t)$ **45.** $(n - p)^3$ **47.** $(m^n + 4)(m^n - 4)$ **49.** $(x^n - y^{2n})(x^{2n} + x^n y^{2n} + y^{4n})$
51. $(2x^n + 3y^n)(x^n - 13y^n)$ **53.** $(5q^r - 3t^p)^2$ **55.** $[3(m + p)^k + 5][2(m + p)^k - 3]$ **57.** $m^{-3}(1 + m^2)$ **59.** $9k^{-3}(1 + k + 2k^2)$
61. $5p^{-5}(3 - 2p^4 + 6p^2)$ **63.** $k^{3/4}(4k + 1)$ **65.** $z^{-1/2}(9 + 2z)$ **67.** $2(3a - 8)(2a - 5)^{-3/2}$ **71.** $(7x - 8)^2(121 - 56x)$
73. $(r - 6)/[3(r - 2)^{5/3}]$ **75.** $(3m^3 - m)/[3(m^3 + m)^{5/3}]$ **77.** $(6 + x^{-4})(3x - 2x^{-1})^2(54 + 36x^{-2} - 15x^{-4} + 22x^{-6})$
79. $(7x - 8)^{-6}(3x + 2)^{-3}(-147x - 22)$ **81.** $(r^2 - 1 + 2r)(r^2 - 1 - 2r)$ **83.** $(x^2 - 1 + 4x)(x^2 - 1 - 4x)$
85. $(m^2 - 3 + 4m)(m^2 - 3 - 4m)$

Section 1.5 (page 41)

1. $x \neq -3/2$ **3.** $x \neq -3/2, x \neq -4$ **5.** no restrictions **7.** $8h$ **9.** $5/3$ **11.** $-4/(x + 2)$ **13.** $(r^2 + 6)/(2r^2)$
15. $(y - 4)/(y + 1)$ **17.** $(m + 5)/(2m + 5)$ **19.** $(h - 1)^2/5$ **21.** $25p^2/9$ **23.** $2/9$ **25.** $1/6$ **27.** $5x/y$
29. $(m - 5)/16$ **31.** $(x + 2)(x + 1)/[(x - 1)(x - 3)]$ **33.** $(2y - 3)/(3y - 5)$ **35.** $(x^2 - 1)/x^2$ **37.** $(x + y)/(x - y)$
39. $(x^2 - xy + y^2)/(x^2 + xy + y^2)$ **41.** $(42 - y^2)/(6y)$ **43.** $(5 + 4y^2)/(12y)$ **45.** 1
47. $(8 - y)/(4y)$ **49.** $137/(30m)$ **51.** $(2y + 1)/[y(y + 1)]$ **53.** $3/[2(a + b)]$ **55.** $-2/[(a + 1)(a - 1)]$
57. $(2m^2 + 2)/[(m - 1)(m + 1)]$ **59.** $7/(p - 4)$ or $-7/(4 - p)$ **61.** $5/[(a - 2)(a - 3)(a + 2)]$
63. $(7x^2 + 14x - 2)/[(x - 5)(x + 4)(x + 1)]$ **65.** $(p + 5)/[p(p + 1)]$ **67.** $-1/[x(x + h)]$ **69.** $(x + 1)/(x - 1)$ **71.** $1/(k - 1)$
73. $(2 - b)(1 + b)/[b(1 - b)]$ **77.** $1/3$ **79.** $a + b$ **81.** -1 **83.** $1/(ab)$ **85.** $(x^3 + 2)^3(2x^3 - 15x^2 - 2)/(x - 5)^5$
87. $[-5(x + 1)^4(3x^2 + 6x + 4)]/(3x^2 - 4)^6$

Section 1.6 (page 48)

1. $\sqrt[3]{36}$ **3.** $\sqrt[5]{3z}$ **5.** $2\sqrt[3]{k^2}$ **7.** $7\sqrt{2}$ **9.** $5\sqrt{15}$ **11.** $-\sqrt{5}/2$ **13.** $9\sqrt{3}$ **15.** $\sqrt[4]{24}/2$ **17.** $3/2$ **19.** $5\sqrt[3]{2}$ **21.** $18\sqrt{5}/5$
23. $-\sqrt[3]{25}/30$ **25.** $6p^2q^2r^3\sqrt{2p}$ **27.** $x^2yz^2\sqrt[4]{y^2z^2}$ **29.** $(pq - q^4 + p^2)p^2\sqrt{pq}$ **31.** 3 **33.** $6 - 2\sqrt{5}$ **35.** $5\sqrt[3]{36} + 13\sqrt[3]{6} + 6$
37. $gh^2\sqrt{ghr}/r^2$ **39.** $-xy^2\sqrt[3]{9x^2zw}/(z^2w)$ **41.** $2x\sqrt[4]{2x^2y^3}/y^2$ **43.** $\sqrt[3]{4}/2$ **45.** $\sqrt[12]{5}$ **47.** $\sqrt[18]{y}$ **49.** $2\sqrt{2xy}$ **51.** $3\sqrt{9 - 2z}$
53. $-(1 - \sqrt{5})/2$ **55.** $3\sqrt{7} + 7$ **57.** $(-z - \sqrt{z})/(z - 1)$ or $(z + \sqrt{z})/(1 - z)$ **59.** $(-5 - 5\sqrt{3 - p})/(p - 2)$
61. $(p^2 + p + 2\sqrt{p(p^2 - 1)} - 1)/(-p^2 + p + 1)$ **63.** $(\sqrt[3]{m^2} + \sqrt[3]{mn} + \sqrt[3]{n^2})/(m - n)$ **65.** $(1 - \sqrt{7})(2 - \sqrt{10})(1 + \sqrt{5})/24$
69. $\sqrt{17} - 4$ **71.** $(5 - x)(x - 3)^2$

Section 1.7 (page 55)

1. real **3.** real **5.** imaginary **7.** imaginary **9.** $0 + 9i$ **11.** $0 - 16i$ **13.** $5 + 2i$ **15.** $-6 - 14i$ **17.** $-5 + 0i$
19. $-4 + 0i$ **21.** $2 + 0i$ **23.** $11 + i$ **25.** $2 + 3i$ **27.** $1 - 10i$ **29.** $-14 + 2i$ **31.** $5 - 12i$ **33.** $13 + 0i$
35. $6 + 0i$ **37.** $53 + 0i$ **39.** $7/25 - (24/25)i$ **41.** $-1 - 2i$ **43.** $0 - 2i$ **45.** $(3 - 2\sqrt{5})/13 + (-2 - 3\sqrt{5})i/13$ **49.** i
51. 1 **53.** i **55.** $2/5 - (11/5)i$ **57.** $-38/5 + (16/5)i$ **59.** $17/10 - (11/10)i$ **61.** $a = -5; b = 1$ **63.** $a = -3/4; b = 3$
69. $13 + 4i$ **71.** $a = 0$ or $b = 0$

Chapter 1 Review Exercises (page 56)

1. $14/17$ **2.** -39 **3.** $-9/7$ **4.** $-8/3$ **5.** $\pi - 3$ **6.** $\sqrt{7} - 2$ **7.** $5 - p$ **8.** $2x^2 + 1$ **9. (a)** 1 **(b)** 4 **10. (a)** 10
(b) 16 **11.** if $x \geq 0$ **12.** if $x = 0$ **13.** if A and B are the same point **14.** if B is between A and C on a line **15.** $9x^3 - 15x^2 - 14$
16. $8m^2 - 40m + 48$ **17.** $3r^3 + 10r^2 - 32r + 16$ **18.** $a^2 - 8ab + 2ac + 16b^2 - 8bc + c^2$ **19.** $81 - x^{2y}$ **20.** $x^6/2$
21. $3/(4k)$ **22.** $81x^4 - 108x^3y + 54x^2y^2 - 12xy^3 + y^4$ **23.** $(3k - 4)(4k + 5)$ **24.** $4p^3(3p^2 - 2p + 5)$ **25.** $(3m + 1)(2m - 5)$
26. $(2x - 7y)(5x + 3y)$ **27.** $5m^3(3m - 5n)(2m + n)$ **28.** $2p^3(x - 3p)(x - p)$ **29.** $(4 + 9y^2)(2 + 3y)(2 - 3y)$
30. $(7m^4 + 3n)(7m^4 - 3n)$ **31.** $(x + 1)(x - 3)$ **32.** $(2z + 3w)(4z^2 - 6zw + 9w^2)$ **33.** $(-r^3 + 2)(7r^6 - 10r^3 + 4)$
34. $(5p - 1)(7p^2 - p + 1)$ **35.** $(3p - q)(7 + m)$ **36.** $3(z - 4)^2(3z - 10)$ **37.** $(2x - 1)(b + 3)$
38. $(2 + y^p)(4 - 2y^p + y^{2p})(2 - y^p)(4 + 2y^p + y^{2p})$ **40.** $x/(2y)$ **41.** $(k - 5)(3k - 1)/(2k)$ **42.** $(3m + n)(3m - n)$
43. $(x + 5)/(x + 3)$ **44.** $2m/(m - 4)$ **45.** $(1 + 2z)/(24z)$ **46.** $(-2x - h)/\{[(x + h)^2 + 16](x^2 + 16)\}$
47. $(4a^2 + 7a - 1)/[(a - 2)(a + 1)(a + 2)]$ **48.** $(y - 2x)/(x - y)$ **49.** $6(y - 5)/(53 - 2y^2)$ **50.** p^8 **51.** $y^6/(36x^4z^4)$
52. $s/(36r)$ **53.** $m^{13}/(81n^7)$ **55.** $8\sqrt{2}$ **56.** $3\sqrt[4]{5}$ **57.** $2^3y^4\sqrt{2m}/m^2$ **58.** $-r^2m\sqrt[3]{m^2z}/z$ **59.** $\sqrt[15]{k}$ **60.** 66 **61.** $2q$
62. $7x^2\sqrt{2x}$ **63.** $5y^2\sqrt{3y}$ **64.** $-6\sqrt[3]{2}$ **65.** $3m\sqrt{m + 2}/(m + 2)$ **66.** $-15(1 + \sqrt{3})/2$ **67.** $-2(\sqrt{7} - 3)$
68. $x - \sqrt{x(x - 2)} - 1$ **70.** $1/216$ **71.** $1/(25m^4)$ **72.** $16r^{9/4}s^{7/3}$ **73.** $-14r^{17/12}$ **74.** $1/p^{3/2}$ **75.** $4xz^3/y^3$ **76.** $1/m^{2p}$ **77.** z
78. $-5 - 5i$ **79.** $-2 + 11i$ **80.** $29 + 22i$ **81.** $34 + 0i$ **82.** $-33 + 56i$ **83.** $9 - 40i$ **84.** $2 - 11i$ **85.** $161 + 240i$
86. $0 - i$ **87.** $1 + 0i$ **88.** $-13/25 + (34/25)i$ **89.** $14/5 - (13/5)i$ **90.** $7/2 - (11/2)i$ **91.** $0 + (29/10)i$ **92.** $0 + 3i\sqrt{3}$

93. $0 + 6i\sqrt{2}$ **94.** $x^3 + 5x$ **95.** -9 **96.** m^6 **97.** $9xy$ **98.** $a/(2b)$ **99.** mn/r^2 **100.** $\dfrac{\sqrt{a} - \sqrt{b}}{a - b}$ **101.** $4x^3/y$
102. $4 - t - 1$ **103.** $(-2)^{-3}$ **104.** 5^2 **105.** $7b/(8b + 7a)$ **106.** (a) and (b) **107.** n odd and positive **108.** $x \geq 0$

Chapter 2 Equations and Inequalities

Section 2.1 (page 63)

1. identity **3.** conditional **5.** identity **7.** identity **9.** equivalent **11.** equivalent **13.** not equivalent **15.** not equivalent **19.** 10 **21.** 24/7 **23.** $-7/8$ **25.** -1 **27.** 3 **29.** 4 **31.** 12 **33.** 3/4 **35.** $-12/5$ **37.** no solution **39.** 27/7 **41.** $-59/6$ **43.** no solution **45.** 8 **47.** -4 **49.** $-19/75$ **51.** $x = -3a + b$ **53.** $x = (3a + b)/(3 - a)$ **55.** $x = (2a + 1)/(2a^2 + a - 4)$ **57.** $x = 3a^2/(2a^2 - 4)$ **59.** **(a)** no solution **(b)** 1

Section 2.2 (page 71)

1. $y = (3x - 12)/4$ **3.** $y = (5x - 27)/3$ **5.** $y = (12x - 9)/4$ **7.** $x = m/(1 + a)$ **9.** $g = 2s/t^2$ **11.** $B = 2A/h - b$ **13.** $r_1 = S/(2\pi h) - r_2$ **15.** $R = Pr/(E^2 - P)$ **17.** $v_2 = (mv_1 - Ft)/m$ **19.** (a) **21.** 10 cm **23.** 2 liters **25.** $6\frac{2}{3}$ qt **27.** 4 ft **29.** about 838.1 mi **31.** 15 min **33.** 8.7 mi **35.** **(a)** 40 mph **(b)** He drove for a longer time at the lower speed. **37.** $1\frac{1}{5}$ hr **39.** $2\frac{8}{11}$ hr **41.** 10 min **43.** \$350 **45.** \$31,000 at 5.5%; \$21,000 at 7.5% **47.** 6% **49.** \$70,000 for land that makes a profit; \$50,000 for land that produces a loss **51.** \$800 **53.** \$90,000 at 10.5% and \$35,000 at 9% **55.** (a) and (b) **57.** $100b/c$

Section 2.3 (page 84)

1. 2, 3 **3.** $-1, 4$ **5.** 0, 6 **7.** $-5/3, 2/7$ **9.** $\pm 2\sqrt{5}$ **11.** $2 \pm \sqrt{7}$ **13.** $(-3 \pm \sqrt{3})/5$ **15.** 3, 5 **17.** $(2 \pm 2i)/3$ **19.** $(-1 \pm i)/2$ **21.** $(1 \pm \sqrt{5})/2$ **23.** $(3 \pm \sqrt{17})/2$ **25.** $(\sqrt{2} \pm \sqrt{6})/2$ **27.** $\sqrt{2}/2, \sqrt{2}$ **29.** $-2, 3$ **31.** 0, 1/2 **33.** $-1/4, 3$ **35.** $(1 \pm \sqrt{29})/2$ **39.** $4, -2 \pm 2i\sqrt{3}$ **41.** $7/4, (-7 \pm 7i\sqrt{3})/8$ **43.** 1; two rational solutions **45.** 84; two irrational solutions **47.** 121/4 **49.** 1, 16 **51.** $v = \sqrt{FrkM}/(kM)$ **53.** $R = [(E^2 - 2Pr) \pm E\sqrt{E^2 - 4Pr}]/(2P)$ **55.** $m = (4qr \pm \sqrt{16q^2r^2 - rp})/(pr)$ **59.** $x^2 + x - 6 = 0$ **61.** $x^2 + 1 = 0$ **63.** 50 m by 100 m **65.** 1 ft **67.** 9 ft by 12 ft **69.** 4 hr **71.** 176 mph **73.** (d) **75.** 5 ft **79.** $k = 3$; solution is 2/3 **81.** $x = (2y \pm \sqrt{31y^2 + 9})/9, y = (-2x \pm \sqrt{31x^2 - 3})/3$ **83.** $(1 \pm i\sqrt{23})/6$

Section 2.4 (page 91)

1. $\pm\sqrt{3}$ **3.** $\pm\sqrt{10}/2, \pm 1$ **5.** 4, 6 **7.** $(-9 \pm \sqrt{161})/10$ **9.** $-4, 24$ **11.** no real solutions **13.** $1, 7^{2/3}$ **15.** 0 **17.** 121/16, 81/25 **21.** $-1, 3$ **23.** 9 **25.** 5 **27.** -2 **29.** 2 **31.** 3/2 **33.** 0 **35.** $-27, 3$ **37.** 9 **39.** 27 **41.** 2/9, 1 **43.** $2, -2/9$ **45.** 0, 9 **47.** 1/4, 1 **49.** -1 **51.** $h = d^2/k^2$ **53.** $L = p^2g/4$ **55.** $y = \pm(a^{2/3} - x^{2/3})^{3/2}$ **57.** 1/2 **59.** 5/4

Section 2.5 (page 95)

1. $y = 16x$ **3.** $x = k/y$ **5.** $r = kst$ **7.** $w = kx^2/y$ **9.** The circumference of a circle varies directly as (or is proportional to) the radius. **11.** The average speed varies directly as the distance and inversely as the time. **13.** The strength of a muscle varies directly as (or is proportional to) the cube of its length. **15.** y is half as large as it was before. **17.** y is one-third as large as it was before. **19.** p is 1/32 as large as it was before. **21.** \$1375 **23.** 4 **25.** 1/250 sec **27.** 16 inches **29.** 18.96 kg **31.** 1600 **33.** about 39.3 km per hour **35.** 92; undernourished **37.** 15.125 m **39.** about 7.4 km

Section 2.6 (page 104)

1. $(-3, 2)$

−3 2

3. $(-\infty, -4]$

−4

5. $-2 \leq x < 5$ **7.** $0 < x < 8$ **11.** $[-2, \infty)$ **13.** $(-\infty, 6]$ **15.** $(-\infty, -3/4]$ **17.** $[1, 7]$

19. $(-21/2, -17/2)$ **21.** $(-11, 7)$ **23.** $[-4/5, 28/5)$ **25.** $(0, 6)$ **27.** $(-\infty, -3) \cup (-1, \infty)$ **29.** no solution **31.** $(-2, 0) \cup (1/4, \infty)$ **33.** $[-2, 4)$ **35.** $(-3, \infty)$ **37.** $(-\infty, 5/3) \cup (8/3, \infty)$ **39.** $(6, \infty)$ **41.** $(-\infty, -2/3) \cup (-2/5, 0)$ **43.** $(-\infty, 1) \cup (2, 8/3]$ **45.** included **47.** excluded **49.** $(-\infty, 8/9)$ **51.** $(-7/2, 11/9) \cup (8/3, \infty)$ **55.** all real numbers **57.** $[-2, 2]$ **59.** $x^2 - 7x + 10 < 0$ **61.** $x^2 - x - 12 \geq 0$ **63.** $\dfrac{x + 1}{x - 5} \leq 0$ **65.** $\dfrac{x - 9}{x - 4} \geq 0$ **69.** $[200, \infty)$ **71.** $(0, 5/4)$ or $(6, \infty)$ **73.** $(4, 9.75)$ **75.** **(a)** $C = 30 + 2x$ **(b)** $C = 3x$ **(c)** $(0, 30)$ **77.** 32°F to 86°F **79.** $(-1.95, 2.83)$

Section 2.7 (page 112)

1. $-8, 2$ **3.** $-5, -1$ **5.** no solution **7.** $-31, 41$ **9.** 10/7, 18/7 **11.** $-7, -13/9$ **13.** $-7/3, -1/7$ **15.** $-3/5, 11$ **17.** $(-\infty, -1) \cup (1, \infty)$ **19.** no solution **21.** $[-10, 10]$ **23.** $(-2, 5/2)$ **25.** $(-\infty, -5/2) \cup (15/2, \infty)$ **27.** $(-\infty, -5/2] \cup [1, \infty)$ **29.** $(-33/8, 31/8)$ **31.** $(2/5, 2)$ **33.** $(1/2, 2) \cup (2, 5)$ **35.** no solution **39.** 1 **41.** $3, -1/3$ **43.** $0, \pm 1$ **45.** $5/3, -4/3$ **47.** yes **49.** $[6.5, 9.5]$ **51.** $|x - 123| \leq 25, |x - 21| \leq 5$ **53.** $(-\infty, -3] \cup [-1, 1] \cup [3, \infty)$ **55.** $(-9/5, 11)$ **57.** $|z - 12| \geq 2$ **59.** $|k - 1| = 6$ **61.** If $|x - 2| \leq .0004$, then $|y - 7| \leq .00001$. **63.** $m = 2, n = 20$ **67.** no solution

Chapter 2 Review Exercises (page 113)

1. 6 **2.** 5 **3.** $-180/17$ **4.** 13 **5.** no solution **6.** no solution **7.** $x = (6 + 5k)/(2 - 2k)$ **8.** $C = (5/9)(F - 32)$ **9.** $P = A/(1 + i)$ **10.** $j = n - nA/I$ **11.** $r_1 = kr_2/(r_2 - k)$ **12.** $E = |r + R|\sqrt{R}/R$ **13.** $x = -4/(y^2 - 5y - 6p)$ **14.** \$500 **15.** \$55,000 at 10.5% and \$35,000 at 9% **16.** 15 mph **17.** 12 hr **18.** $-7 \pm \sqrt{5}$ **19.** $(2 \pm 2\sqrt{2})/3$ **20.** $5/2, -3$ **21.** $1/2, 1/6$ **22.** $(-1 \pm i\sqrt{14})/3$ **23.** $\sqrt{2} \pm 1$ **24.** $(2 \pm \sqrt{46})/2$ **25.** $(-2 \pm i\sqrt{2})/2$ **27.** -119; two complex (nonreal) solutions **28.** 144; two rational solutions **29.** 9; two rational solutions **30.** 144; two rational solutions **31.** 84; two irrational solutions **32.** -64; two complex (nonreal) solutions **33.** 6 ft by 8 ft **34.** 1/2 ft **35.** 6.5 hr **36.** 10 mph **37.** $k = 4, m = -1/4$ **38.** $\pm\sqrt{2}, \pm 2i\sqrt{6}/3$ **39.** $-1/2, 8$ **40.** $-15, 5/2$ **41.** no solution **42.** no solution **43.** $3/4$ **44.** 2, 5 **45.** 1 **46.** $1, -4$ **47.** 4 **48.** $-4/3$ **49.** 6 **50.** 4 **51.** 4, 1 **52.** $Y = (kMN^2)/X^3$ **53.** $A = kt^3s^4/(ph^2)$ **54.** 33,750 units **55.** 36 inches **56.** about 71.1 kg **57.** 36 lb **60.** $(-\infty, 1/11]$ **61.** $(1/3, \infty)$ **62.** no solution **63.** $(-\infty, 3]$ **64.** $[4, 5]$ **65.** $(-1, 4/3)$ **66.** $[-5/2, 8)$ **67.** $(-19/6, -11/6)$ **68.** $[-4, 1]$ **69.** $(-\infty, -7) \cup (3, \infty)$ **70.** $(-\infty, -4] \cup [0, 4]$ **71.** $(-\infty, -1) \cup (0, 5/2)$ **72.** $(-2, 0)$ **73.** $(-1/3, 0)$ **74.** $(-3, 1) \cup [7, \infty)$ **75.** $(-2, 4) \cup (16, \infty)$ **76.** $(-\infty, a) \cup (b, \infty)$; (a, b); at $x = a$ and $x = b$ **77.** $(-\infty, a) \cup (a, \infty)$; at $x = a$ **78.** **(a)** 20 sec **(b)** for values of t in $(2, 18)$ **79.** $[300, \infty)$ **80.** $-11, 3$ **81.** no solution **82.** $11/27, 25/27$ **83.** $-15/13, -13/29$ **84.** $4/3, -2/7$ **85.** $1/2$ **86.** $[-4, 10]$ **87.** $(1, 5)$ **88.** $(-\infty, \infty)$ **89.** $[-6, -3]$ **90.** $[6/5, 2]$ **91.** $(-\infty, -2/7) \cup (8/7, \infty)$ **92.** $(-\infty, -1/2) \cup (3/2, \infty)$ **93.** $(-\infty, -4) \cup (-2/3, \infty)$ **95.** $(4/3)\sqrt{1 + x^2} + 2 - x$ **96.** $(8 - x)/5 + \sqrt{9 + x^2}/2$ **97.** $(-\infty, -7) \cup (-7, -3/2)$ **98.** $[-1/2, 3) \cup (3, \infty)$ **99.** $M = 1/2$ **100.** $M = 1/3$

Chapter 3 Functions and Graphs

Section 3.1 (page 127)

1. IV **3.** III

5. **7.**

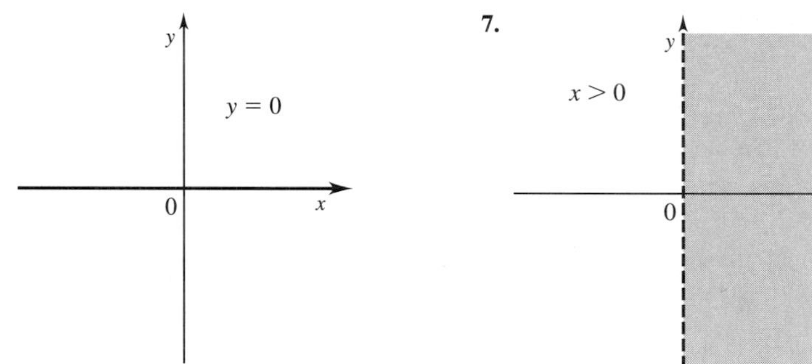

9.

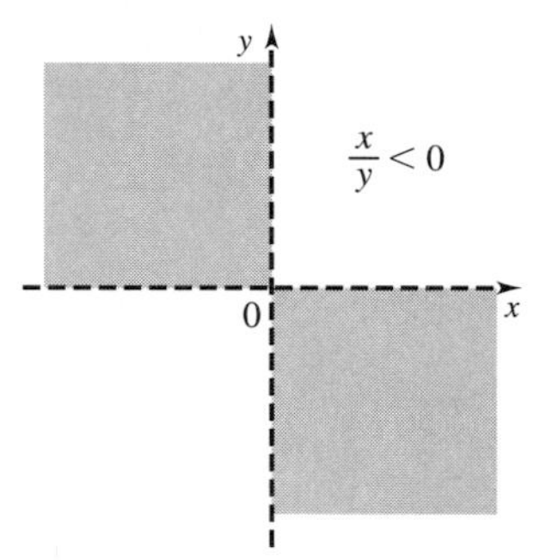

11. 7.071; (1.5, 5.5) **13.** 16.279; $(-1, 1/2)$ **15.** 11.533; $(2\sqrt{2}, 3\sqrt{5}/2)$ **17.** 27.203; $(-1, 6)$ **21.** $(4, -2)$ **23.** $(-9, 9)$ **25.** yes **27.** yes **29.** yes **31.** no **33.** $13, -3$ **35.** $3\sqrt{10}/5, -3\sqrt{10}/5$

37.

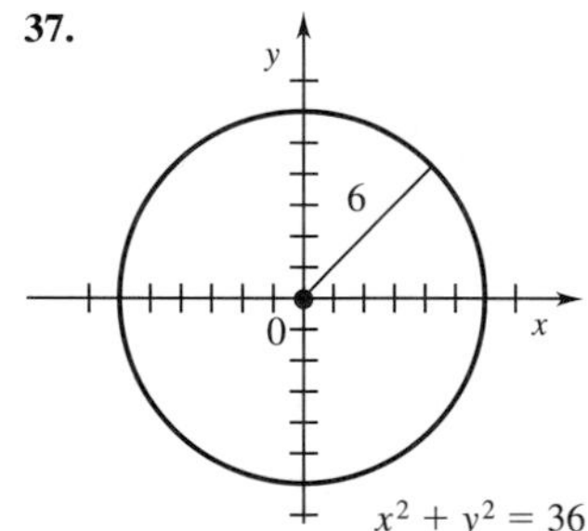

39.

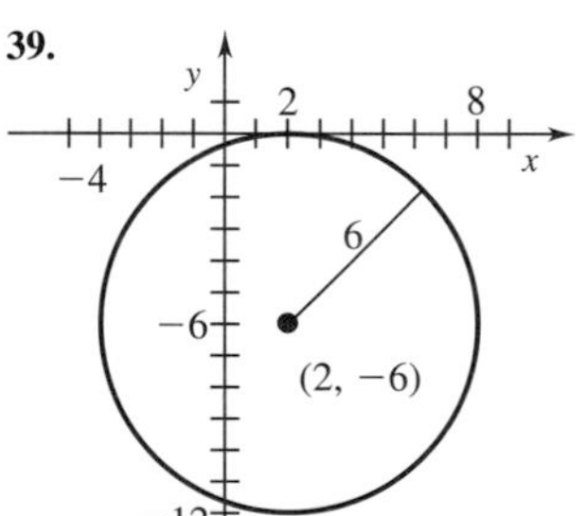

41.

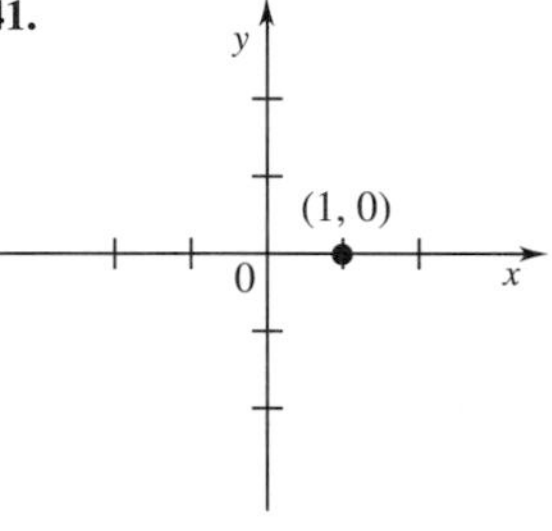

43.

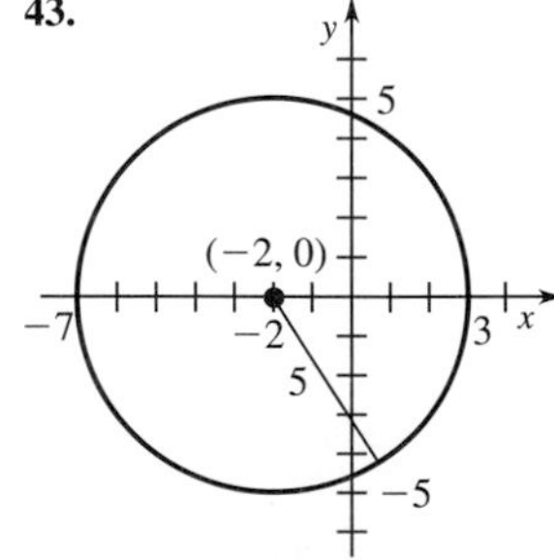

45. $(x - 1)^2 + (y - 4)^2 = 9$ **47.** $(x + 8)^2 + (y - 6)^2 = 25$ **49.** $(x + 1)^2 + (y - 2)^2 = 25$ **51.** $(x + 3)^2 + (y + 2)^2 = 4$ **53.** $(x - 4)^2 + (y - 13/2)^2 = 45/4$ **55.** $C = 10\pi, A = 25\pi$ **59.** $(2 + \sqrt{7}, 2 + \sqrt{7}), (2 - \sqrt{7}, 2 - \sqrt{7})$ **61.** **(a)** inside **(b)** outside **(c)** on **(d)** outside **63.** (2, 3) or (4, 1) **65.** yes **67.** $4 + s \geq \sqrt{(a - c)^2 + (b - d)^2} \geq |4 - s|$ **71.** $(x - 3)^2 + (y - 3/2)^2 = 45/4$

Section 3.2 (page 133)

3. domain: $(-\infty, \infty)$; range: $(-\infty, \infty)$

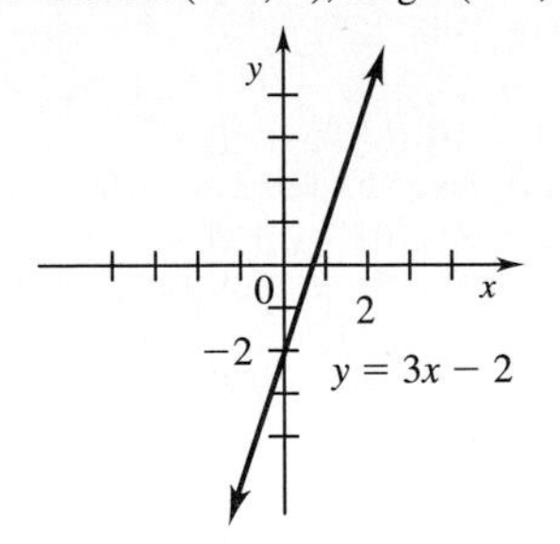

5. domain: $[0, \infty)$; range: $(-\infty, \infty)$

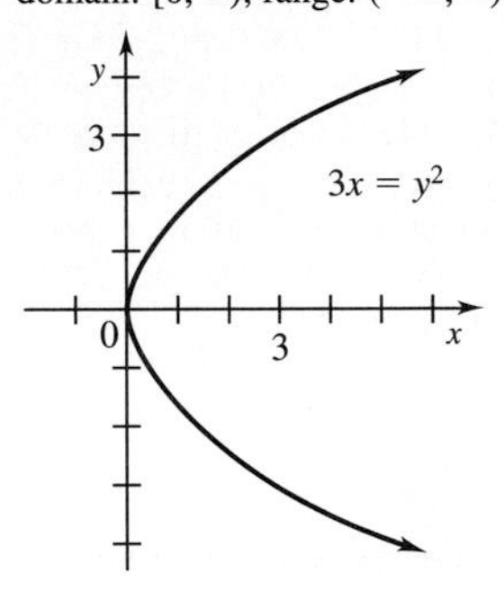

7. domain: $(-\infty, \infty)$; range: $(-\infty, 0]$

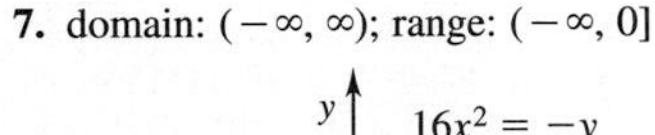
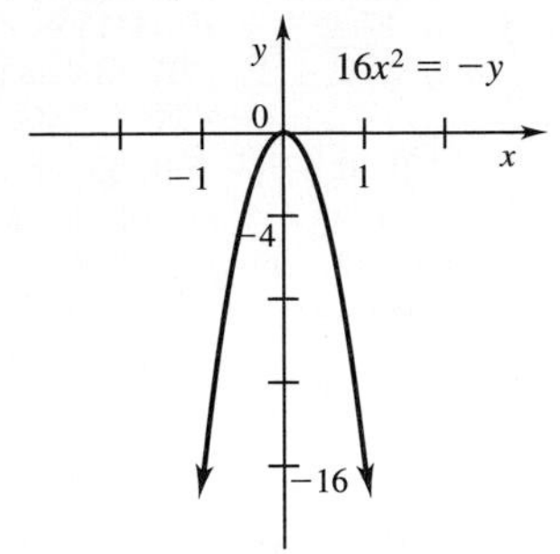

9. domain: $(-\infty, \infty)$; range: $[4, \infty)$

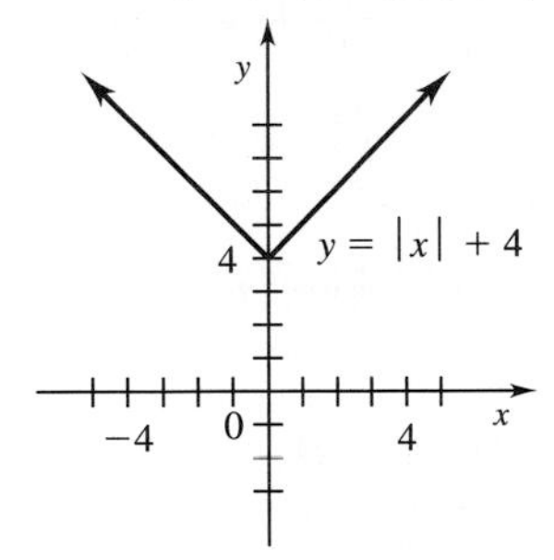

11. domain: $[0, \infty)$; range: $(-\infty, \infty)$

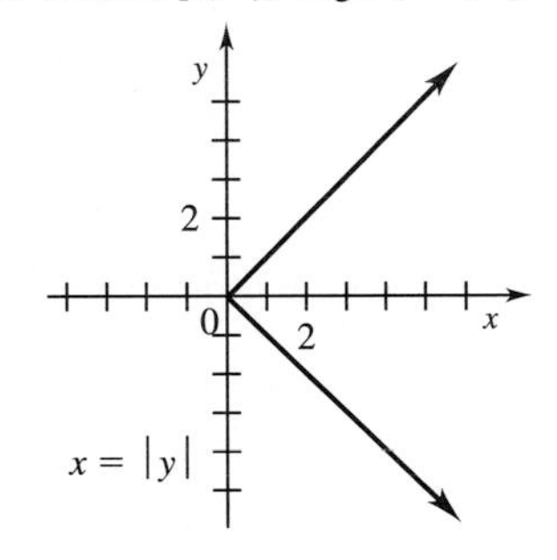

13. domain: $(-\infty, \infty)$; range: $(-\infty, 0]$

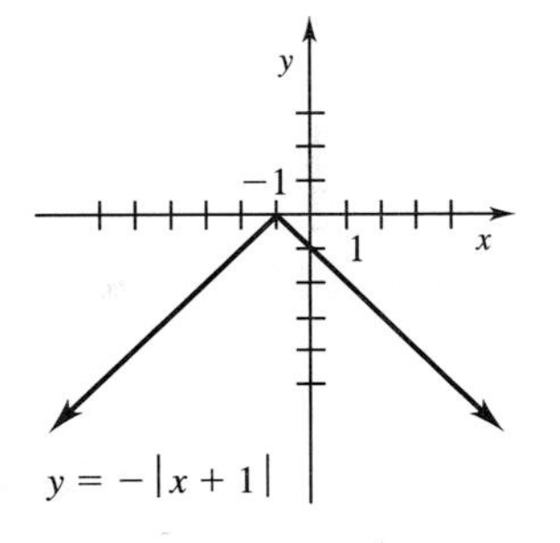

15. domain: $[-2, \infty)$; range: $[0, \infty)$

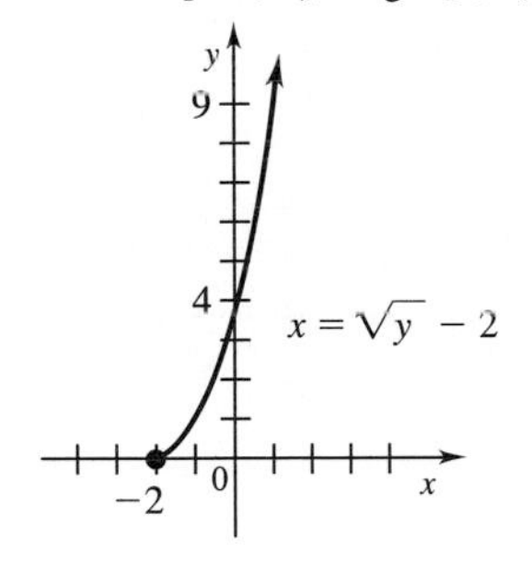

17. domain: $(-\infty, 0]$; range: $[2, \infty)$

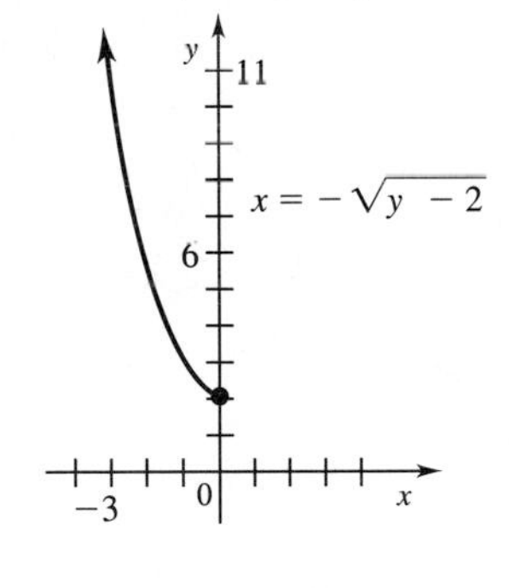

19. domain: $[-2, \infty)$; range: $[0, \infty)$

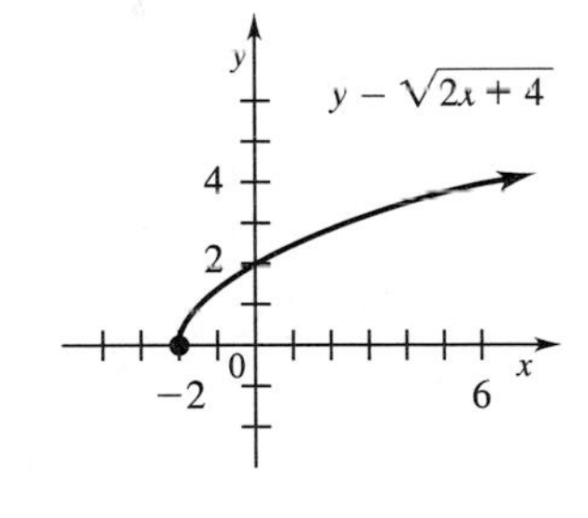

21. domain: $[0, \infty)$; range: $(-\infty, 0]$

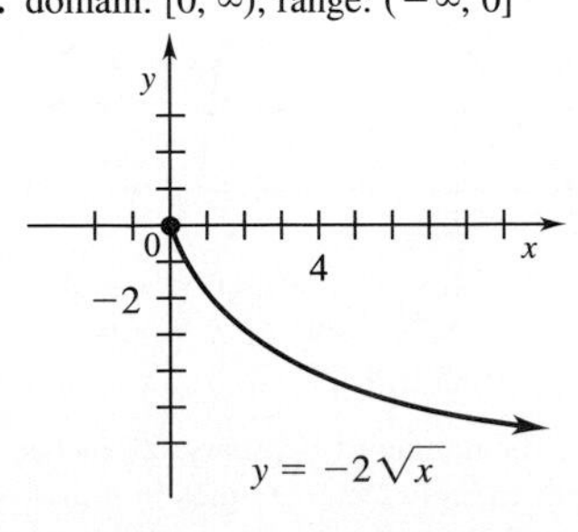

23. yes **25.** $A = 10s^2 + 8s$ **27.** $V = \pi r^3$
29. $(-\infty, -2] \cup [2, \infty)$; $(-\infty, \infty)$ **31.** $[-4, 8]$; $[-6, 6]$

Section 3.3 (page 140)

1. yes **3.** yes **5.** yes **7.** no **9.** no **11.** yes **13. (a)** 0, 4, 4 **(b)** $-.5$, .5, and 3.5; -2 and 2; -1, 1, and 3 **15. (a)** -3, -2, 2 **(b)** none; 1; 4 **17.** -3 **19.** $\sqrt{18}$ or $3\sqrt{2}$ **21.** -34 **23.** 2 **25.** $3 - 2/a^2$ or $(3a^2 - 2)/a^2$ **27.** $1/(3 - 2a^2)$ **29.** $(3 - 50a^2)/\sqrt{4a - 2}$ **31.** 18 **33.** 4 **37.** $(-\infty, \infty)$; $[0, \infty)$ **39.** $[-2/3, \infty)$; $[0, \infty)$ **41.** $(-\infty, -4) \cup (-4, -1) \cup (-1, \infty)$ **43.** $(-\infty, -5] \cup [-2, \infty)$; $(-\infty, 0]$ **45.** $(-\infty, \infty)$; $(-\infty, 5]$ **47.** $[-6, \infty)$; $[0, \infty)$ **49.** $(-\infty, \infty)$; $(-\infty, \infty)$ **51.** $[-2, 4]$; $[0, 4]$ **53. (a)** $[-5, -2]$, $[0, 3]$; $[-2, 0]$, $[3, 4]$ **(b)** $[-6, \infty)$; none **(c)** $(-\infty, -2]$, $[0, 2]$; $[-2, 0]$, $[2, \infty)$ **55. (a)** yes **(b)** \$20, \$50, \$100 **(c)** about 500 mi, about 125 mi **(d)** no maximum; a minimum of 0 **57. (a)** $x^2 + 2xh + h^2 - 4$ **(b)** $2xh + h^2$ **(c)** $2x + h$ **59. (a)** $6x + 6h + 2$ **(b)** $6h$ **(c)** 6 **61.** $C(e) = 12e^2$ **63.** $[0, 3]$ **65.** $[40, 168)$ **67. (b)** $A(x) = -x^2 + 15x + 250$ **(c)** $[2, 13]$ **69.** $A = (B - 3000)/15$ **71.** $V = \pi r^3/6$ **73.** $[1, 3]$ **75.** 306.25; 276

Section 3.4 (page 151)

1.

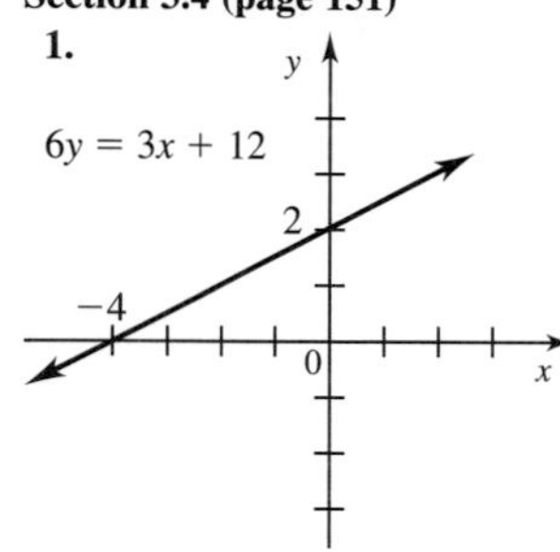

3.

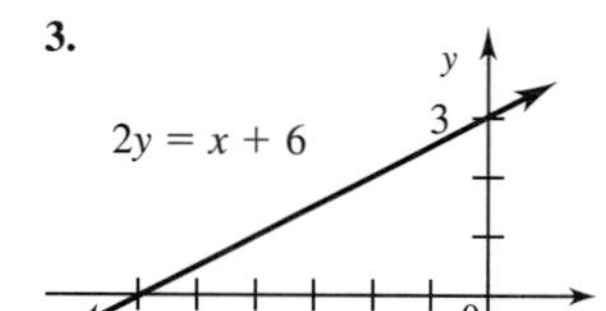

5.

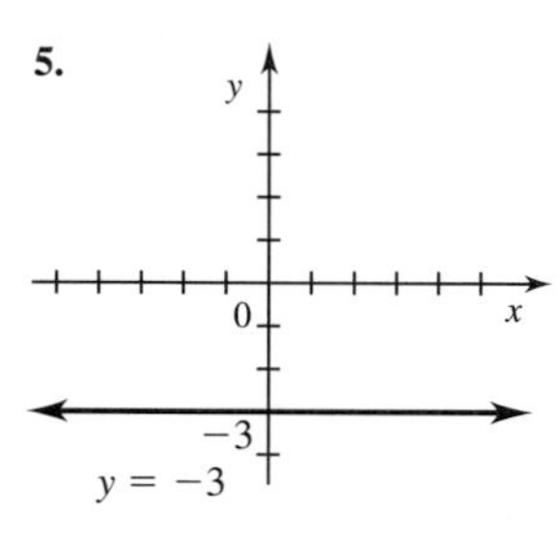

7.

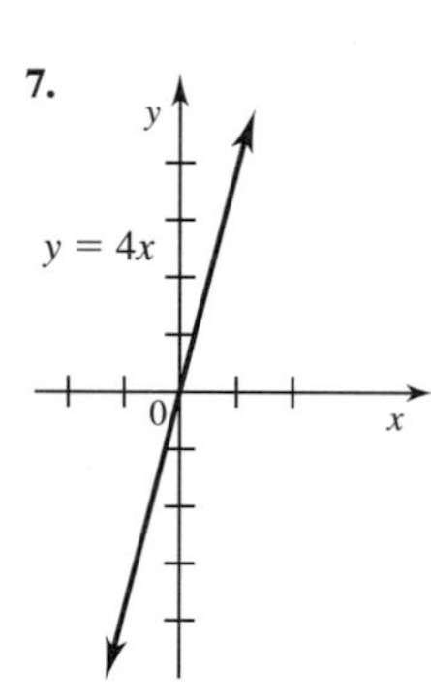

9.

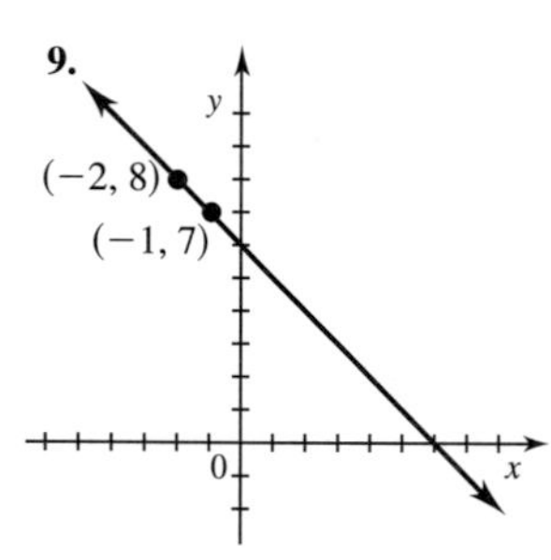

11.

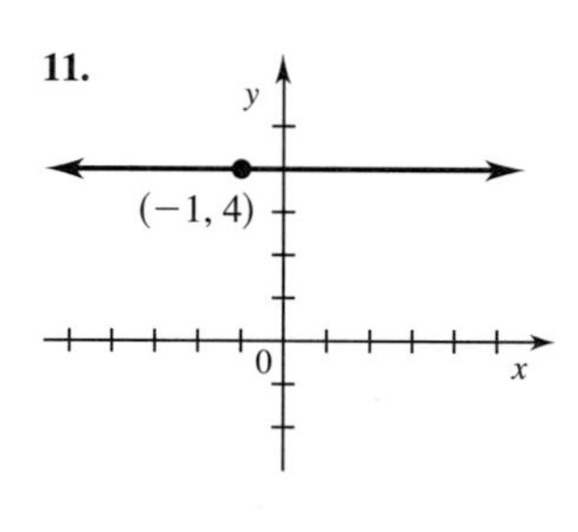

13.

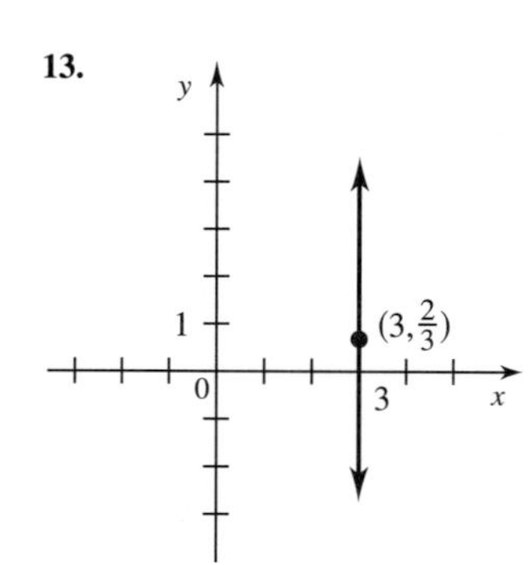

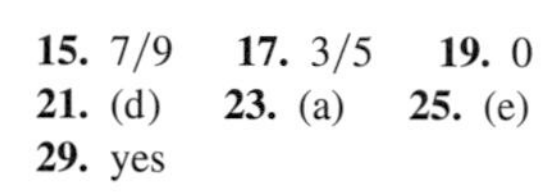

15. 7/9 **17.** 3/5 **19.** 0
21. (d) **23.** (a) **25.** (e)
29. yes

31.

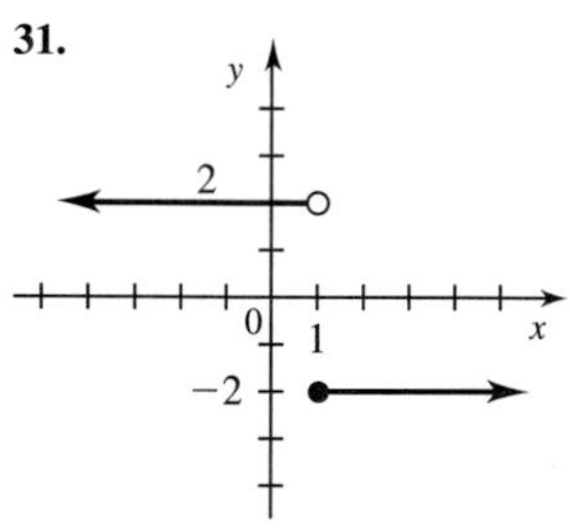

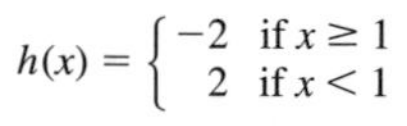

$$h(x) = \begin{cases} -2 & \text{if } x \geq 1 \\ 2 & \text{if } x < 1 \end{cases}$$

33.

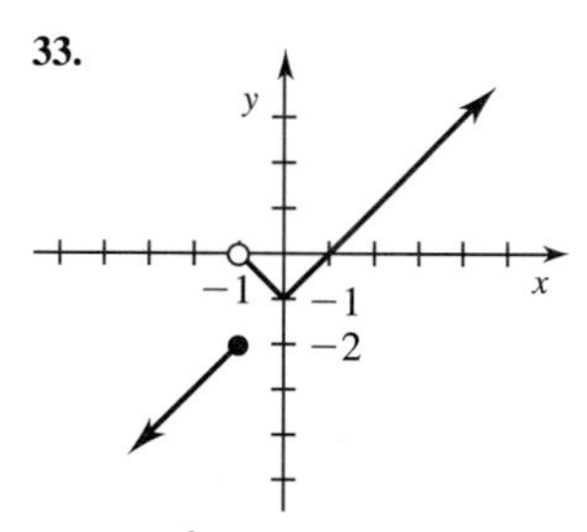

$$r(x) = \begin{cases} |x| - 1 & \text{if } x > -1 \\ x - 1 & \text{if } x \leq -1 \end{cases}$$

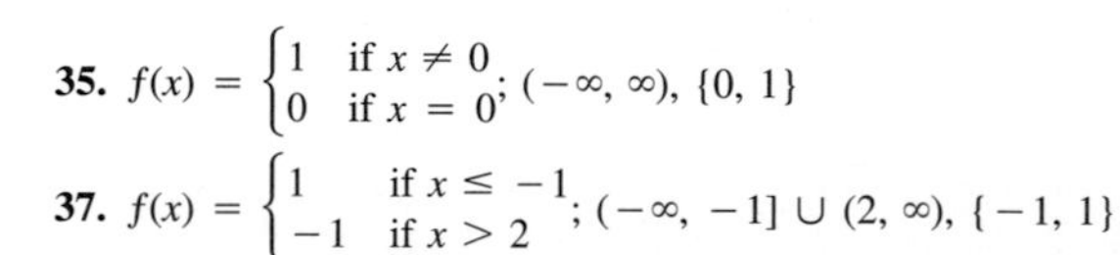

35. $f(x) = \begin{cases} 1 & \text{if } x \neq 0 \\ 0 & \text{if } x = 0 \end{cases}$; $(-\infty, \infty)$, $\{0, 1\}$

37. $f(x) = \begin{cases} 1 & \text{if } x \leq -1 \\ -1 & \text{if } x > 2 \end{cases}$; $(-\infty, -1] \cup (2, \infty)$, $\{-1, 1\}$

39. (a)

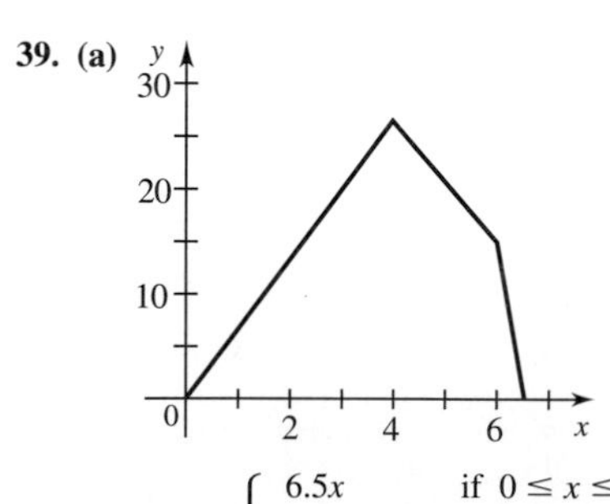

$$f(x) = \begin{cases} 6.5x & \text{if } 0 \leq x \leq 4 \\ -5.5x + 48 & \text{if } 4 < x \leq 6 \\ -30x + 195 & \text{if } 6 < x \leq 6.5 \end{cases}$$

(b) at the beginning of February; 26 inches
(c) begins in early October; ends in mid-April

41.

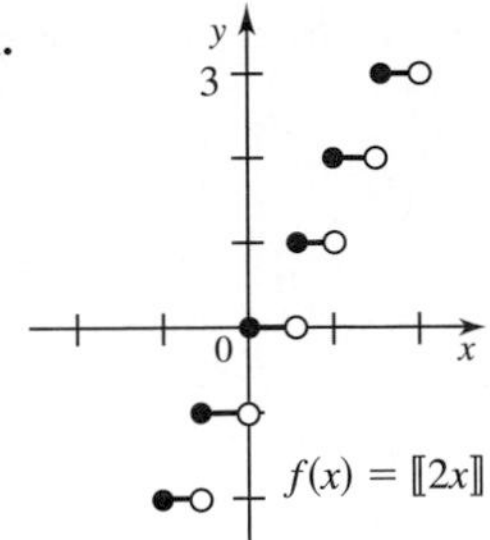

43.

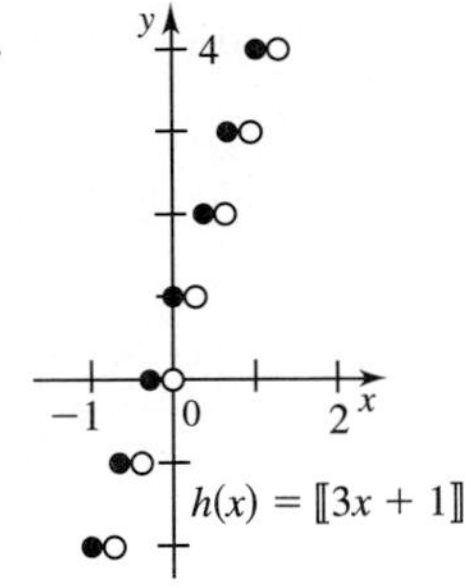

45.

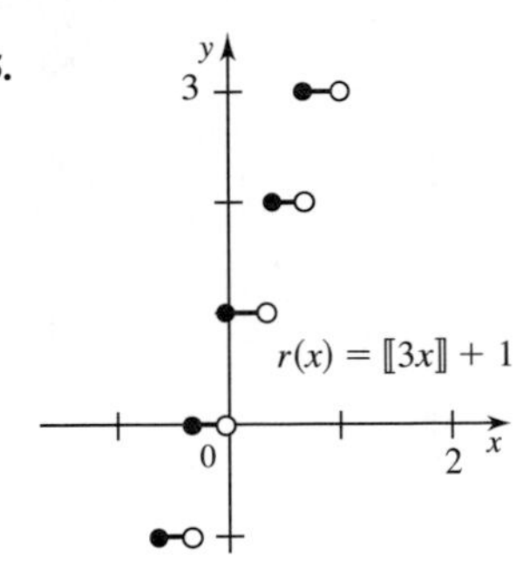

47. **(a)** \$129 **(b)** \$203 **(c)** \$351

(d)

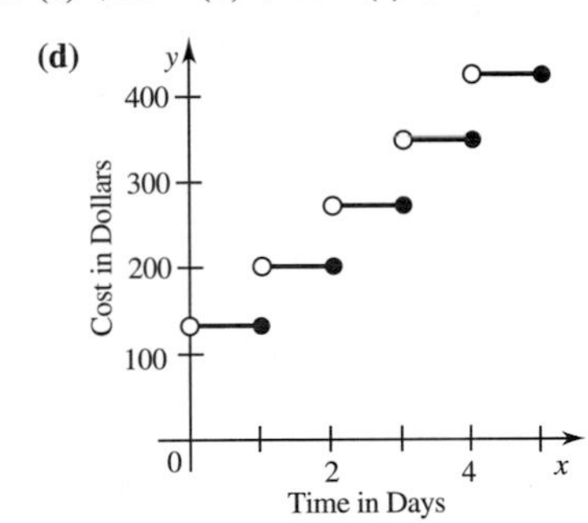

(e) $(0, \infty)$; $\{129, 203, 277, 351, \ldots\}$

49. $-[\![-x]\!]$

51. **(a)** \$1600, \$1100, \$600
(b) 8 units, 4 units, 0 units
(c) \$1600, 12.8 units
(d)

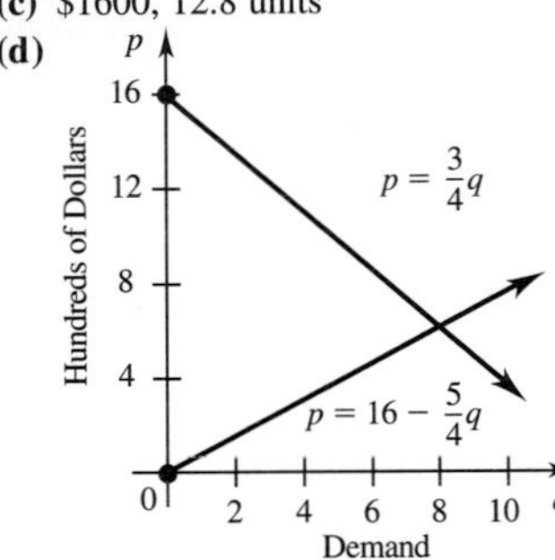

(e) 0, 13⅓, 26⅔ **(f)** See (d).
(g) 8 units **(h)** \$600

53. **(f)** -3 **55.** **(b)** 1

57.

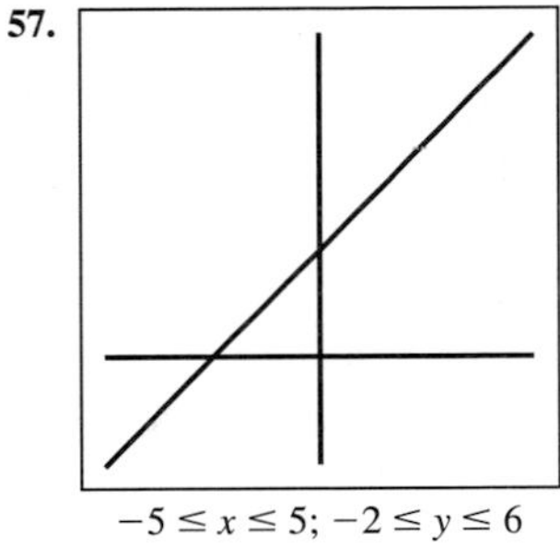

$-5 \le x \le 5$; $-2 \le y \le 6$

Section 3.5 (page 160)

1. $5x - y - 9 = 0$ **3.** $x + y + 2 = 0$ **5.** $2x - 3y - 3 = 0$ **7.** $y - 2 = 0$ **9.** $x - 2 = 0$ **11.** $y = 0$ **13.** $x - 3 = 0$ **15.** $x + y = 0$ **17.** $x - y + 1 = 0$ **19.** $2x - y + 10 = 0$ **21.** $y - 7 = 0$ **23.** $x - 2y - 5 = 0$ **25.** $x + 4y - 13 = 0$ **27.** $y - 8 = 0$ **29.** $x - 3y - 9 = 0$ **31.** $2x + y - 6 = 0$ **33.** (a) **35.** (c) **37.** (f) **39.** $f(x) = (-1/2)x + 1$; $f(-4) = 3$ **41.** **(a)** 2 **(b)** -6 **43.** $x = 5$ and $y = -4$ **47.** $y = (5/9)(x - 32)$; $5/9$ **49.** $y = (-1/15)x + 22/3$; about 3.3 million **51.** $y = (3/40)x + 1$; 2000 **57.** **(a)** $x = (b_2 - b_1)/(m_1 - m_2)$ **59.** $y = -3.8x + 61$

Section 3.6 (page 172)

1. **(a)**

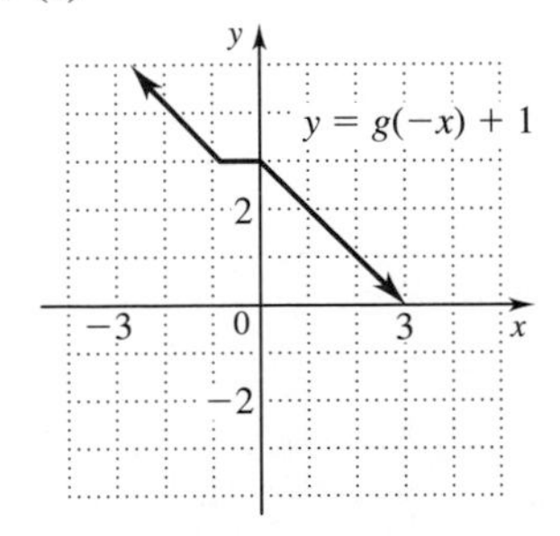

The graph of $g(x)$ is reflected about the y-axis and translated up 1 unit.

(b)

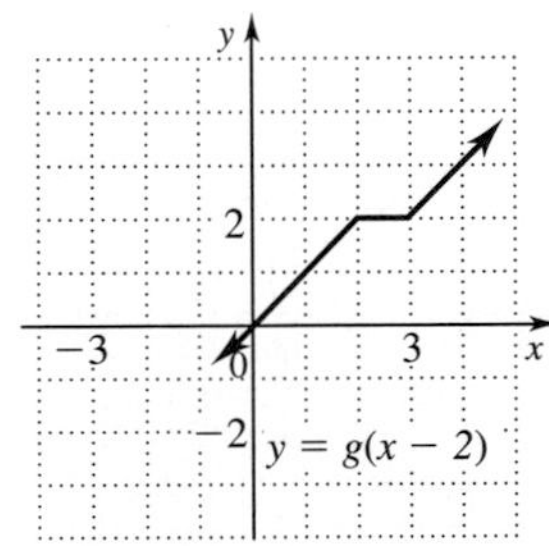

The graph of $g(x)$ is translated to the right 2 units.

(c)

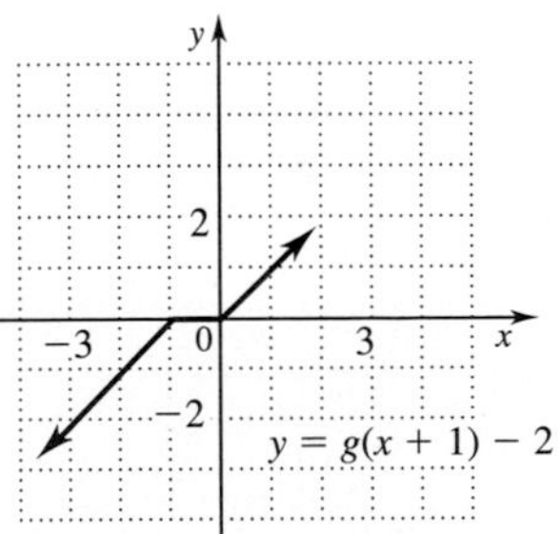

The graph of $g(x)$ is translated to the left 1 unit and down 2 units.

(d)

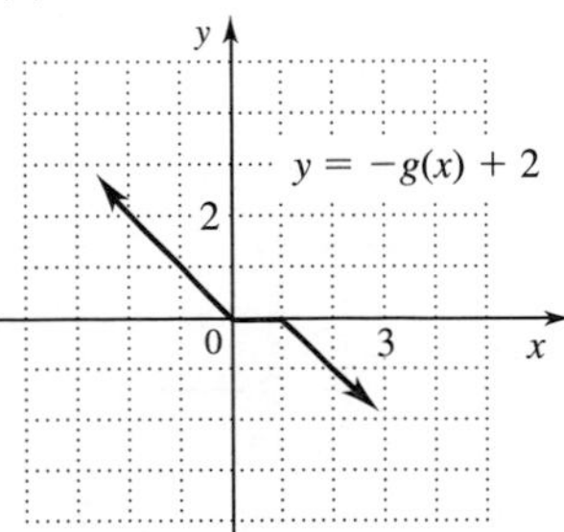

The graph of $g(x)$ is reflected about the x-axis and translated up 2 units.

3.

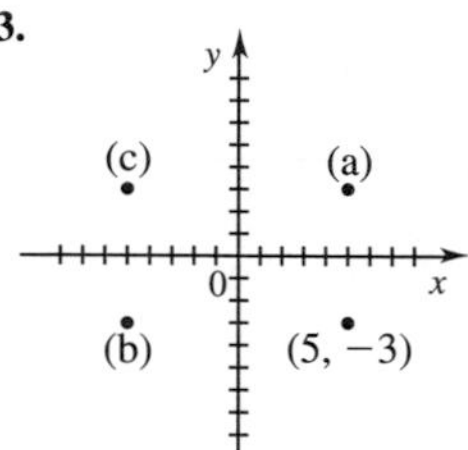

5.

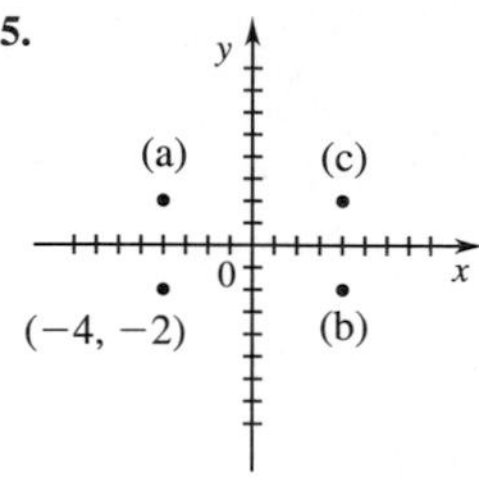

7. none of these **9.** origin **11.** origin **13.** x-axis, y-axis, origin

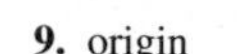

15.

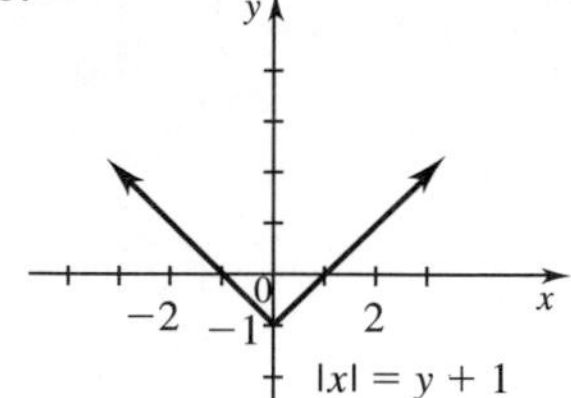

17.

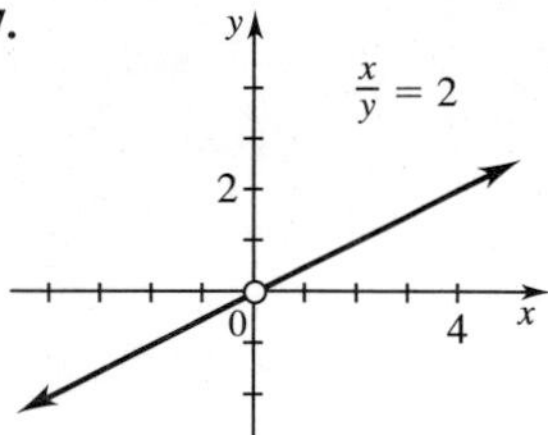

19.

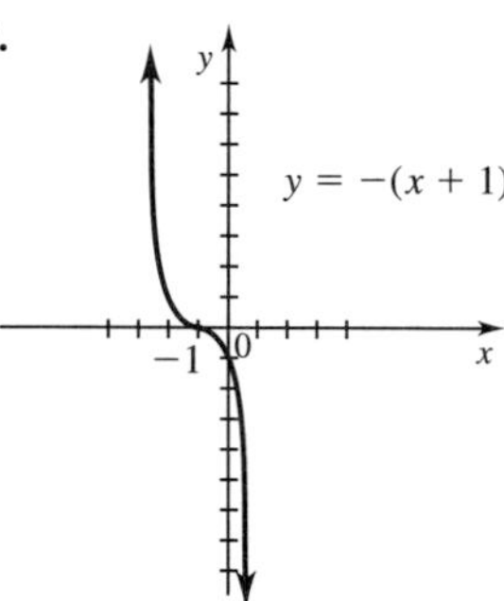

21.

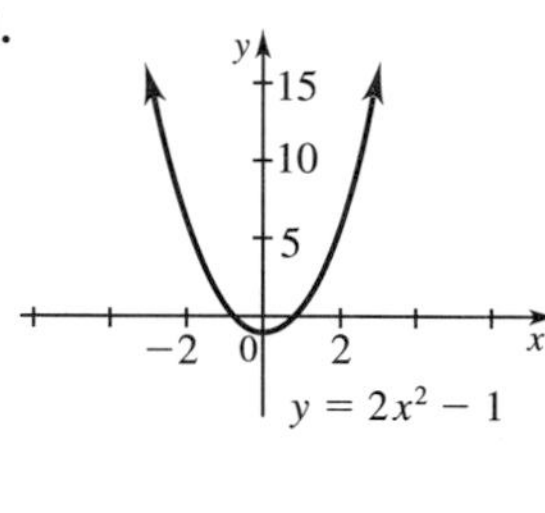

23.

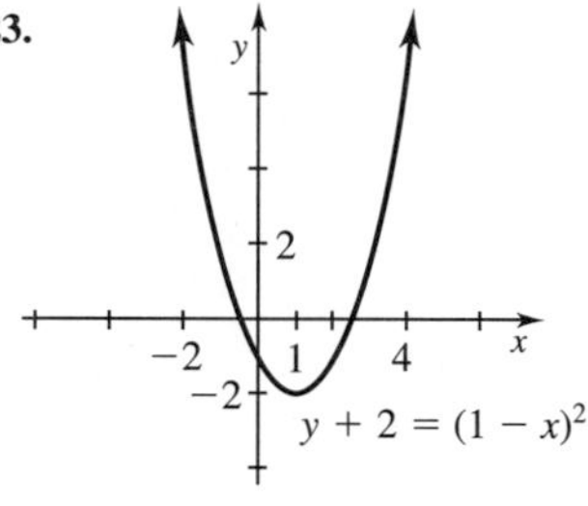

25.

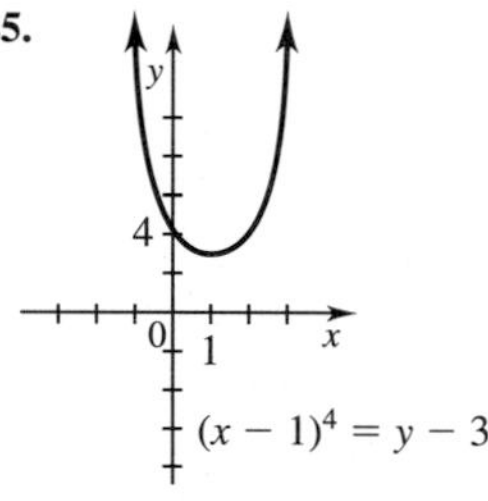

27. $f(-2) = -3$ **29.** $f(10) = 3$ **31.** $g(x) = -x + 4$ **37. (a)**

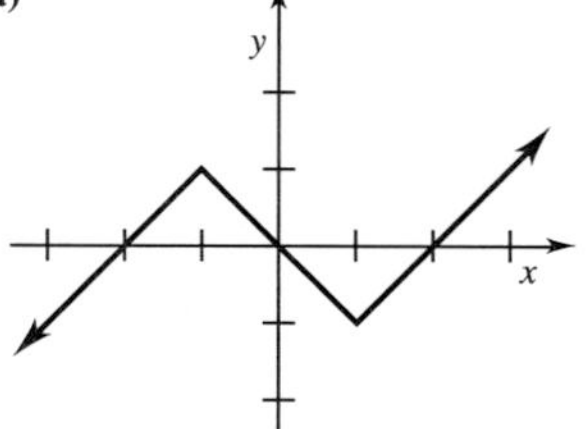

(b)

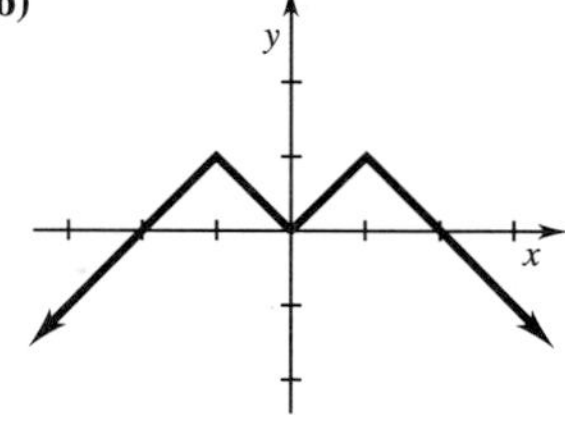

41. Many answers are possible. **45.** true

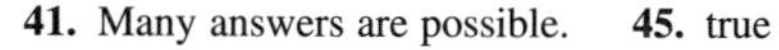

Section 3.7 (page 178)

1. $5x - 1$; $x + 9$; $6x^2 - 7x - 20$; $(3x + 4)/(2x - 5)$; all domains are $(-\infty, \infty)$ except for that of f/g, which is $(-\infty, 5/2) \cup (5/2, \infty)$
3. $3x^2 - 4x + 3$; $x^2 - 2x - 3$; $2x^4 - 5x^3 + 9x^2 - 9x$; $(2x^2 - 3x)/(x^2 - x + 3)$; all domains are $(-\infty, \infty)$ **5.** $\sqrt{4x - 1} + \sqrt{x + 3}$; $\sqrt{4x - 1} - \sqrt{x + 3}$; $\sqrt{(4x - 1)(x + 3)}$; $\sqrt{(4x - 1)/(x + 3)}$; all domains are $[1/4, \infty)$ except for that of f/g, which is $(-\infty, -3) \cup [1/4, \infty)$
7. 61 **9.** 2016 **11.** $-7/2 = -3.5$ **13.** $5m^2 - 8m - 4$ **15.** 1248 **17.** 100 **19.** $-30x - 33$; $-30x + 52$
21. $4x^2 + 42x + 118$; $4x^2 + 2x + 13$ **23.** $2/(2 - x)^4$; $2 - 2/x^4$ **25.** $36x + 72 - 22\sqrt{x + 2}$; $2\sqrt{9x^2 - 11x + 2}$
27. $(f \circ g)(x) = \begin{cases} 8x & \text{if } x < 0 \\ 12x - 5 & \text{if } 0 \le x \le 1 \\ 3x + 4 & \text{if } x > 1 \end{cases}$ $(g \circ f)(x) = \begin{cases} 8x & \text{if } x < 0 \\ 12x - 20 & \text{if } 0 \le x < 2 \\ 3x - 2 & \text{if } x \ge 2 \end{cases}$ **31.** 5 **33.** 0 **35.** 3 **37.** 2

39.

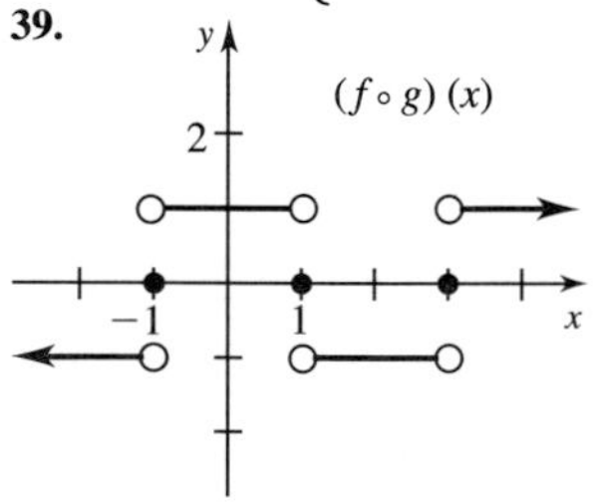

45. Many answers are possible. **47.** Many answers are possible.
49. Many answers are possible. **51.** $18a^2 + 24a + 9$ **53.** $16\pi t^2$ sq ft
55. (a) $200 - x$ **(b)** $300 + 20x$ **(c)** $(200 - x)(300 + 20x)$

Section 3.8 (page 188)

1. one-to-one **3.** not one-to-one **5.** not one-to-one **7.** one-to-one **9.** not one-to-one **11.** not one-to-one **13.** not one-to-one
15. one-to-one **17.** one-to-one **19.** inverses **21.** not inverses **23.** inverses **25.** inverses **27.** inverses **29.** not inverses

31.

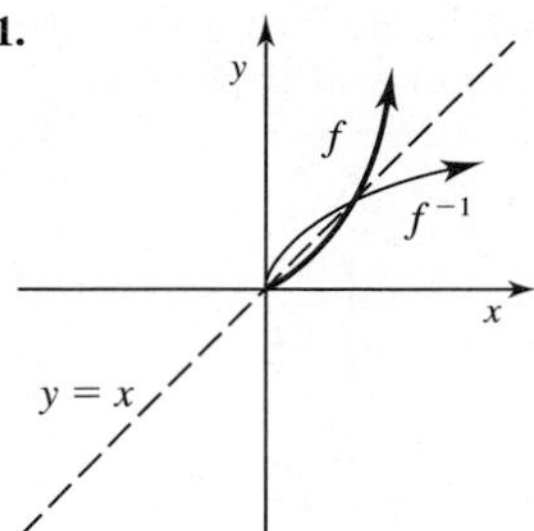

33.

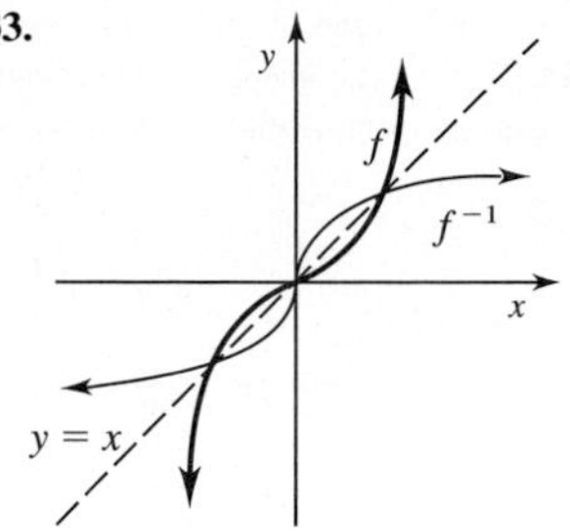

35.

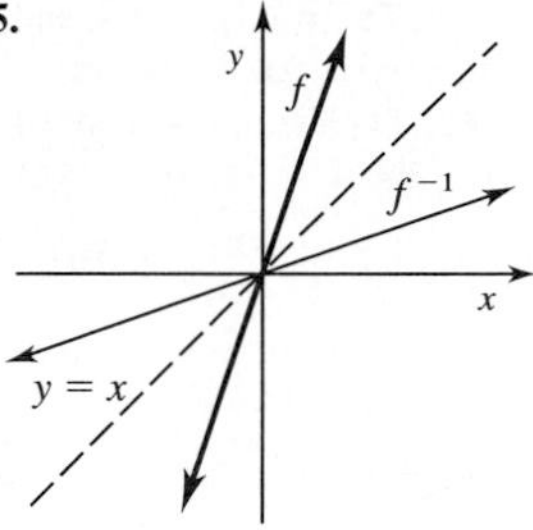

41. $f^{-1}(x) = (x + 3)/8$ **43.** $f^{-1}(x) = \sqrt[3]{-x - 2}$ **45.** not one-to-one **47.** $f^{-1}(x) = 4/x$ **49.** $f^{-1}(x) = 3 - \sqrt{x}$ **51.** $f^{-1}(x) = x^2 - 6$; domain $[0, \infty)$ **53.** 4 **55.** 2 **57.** -2 **59.** 1.14 **61.** the number of dollars required to build 1000 cars **63.** $d = k/B$ **65.** $1/a$ **69.** not one-to-one **71.** one-to-one; $f^{-1}(x) = (-5 - 3x)/(x - 1)$

Chapter 3 Review Exercises (page 191)

1. $d(P, Q) = \sqrt{85}$; $(-1/2, 2)$ **2.** $d(P, Q) = \sqrt{202}$; $(-5/2, -5/2)$ **3.** $(7, -13)$ **4.** yes **5.** -7, 1, 8, 23 **6.** $3 \pm 2\sqrt{5}$ **7.** No such points exist. **8.** $\left(\frac{-5 + \sqrt{71}}{2}, \frac{5 - \sqrt{71}}{2}\right), \left(\frac{-5 - \sqrt{71}}{2}, \frac{5 + \sqrt{71}}{2}\right)$ **11.** $(x - 5)^2 + (y + 2)^2 = 16$ **12.** $(x - \sqrt{5})^2 + (y + \sqrt{7})^2 = 3$ **13.** $(x + 8)^2 + (y - 1)^2 = 289$ **14.** $(x + 3)^2 + (y - 5)^2 = 25$ **15.** $(3, 5)$; 2 **16.** $(3, -4)$; 6 **17.** not a circle **18.** $(-1/2, -5/2)$; $\sqrt{10}$ **19.** true **20.** true **21.** no; $(-\infty, \infty)$; $[0, \infty)$ **22.** no; $[-6, 6]$; $[-6, 6]$ **23.** yes; $(-\infty, -2] \cup [2, \infty)$; $[0, \infty)$ **24.** no; $(-\infty, \infty)$; $(-\infty, -1] \cup [1, \infty)$ **25.** yes; $(-\infty, \infty)$; $(-\infty, \infty)$ **26.** no; $[0, \infty)$; $(-\infty, \infty)$ **27.** not a function **28.** function **29.** function **30.** function **31.** $(-\infty, \infty)$ **32.** $(-\infty, 8) \cup (8, \infty)$ **33.** $(-\infty, \infty)$ **34.** $[-7, 7]$ **35.** $6/5$ **36.** -2 **37.** $-11/2$ **38.** 0 **39.** $3x + y + 8 = 0$ **40.** $2x - y + 2 = 0$ **41.** $7x - 3y + 4 = 0$ **42.** $4x + 5y - 24 = 0$ **43.** $y + 1 = 0$ **44.** $3x - 2y + 6 = 0$ **45.** $x + 2y = 0$ **46.** $3x + 5y - 4 = 0$

47.

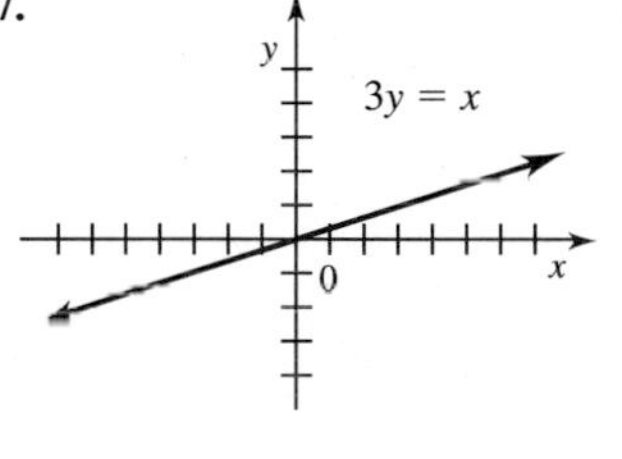

48.

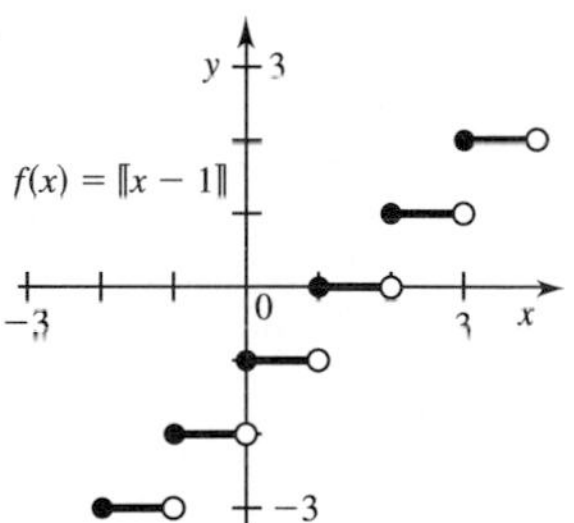

49.

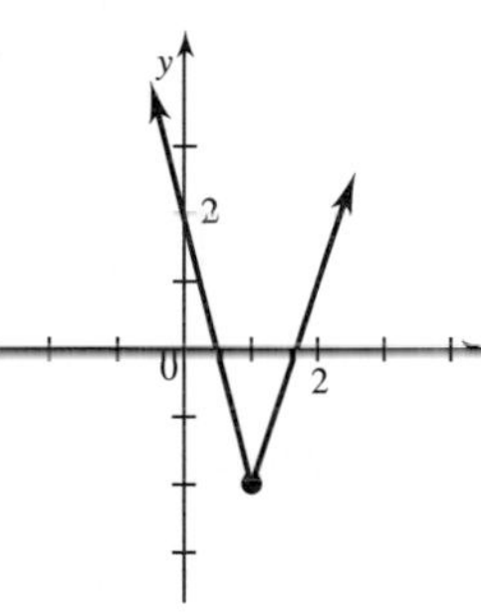

50.

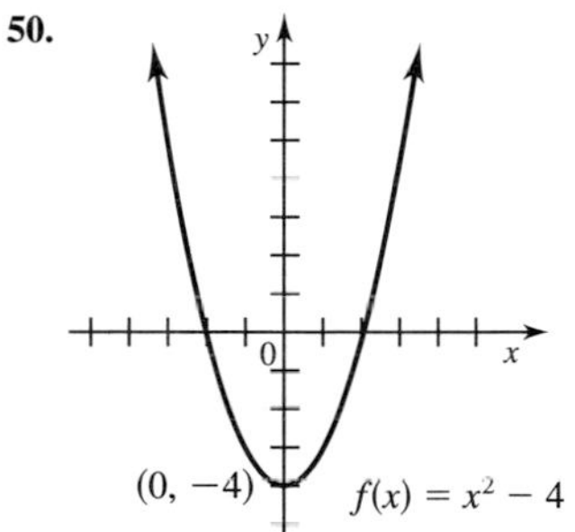

51.

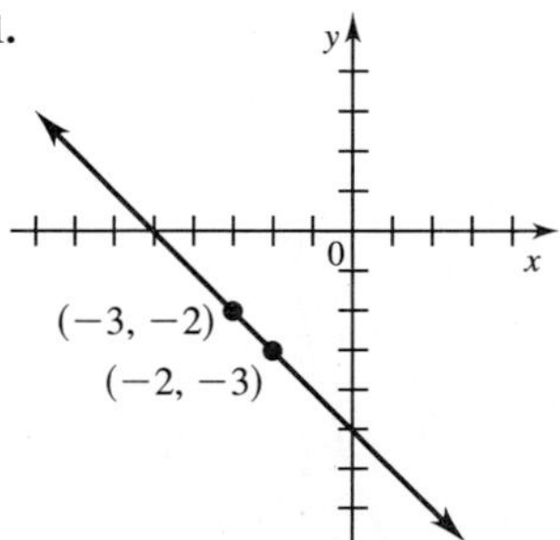

52.

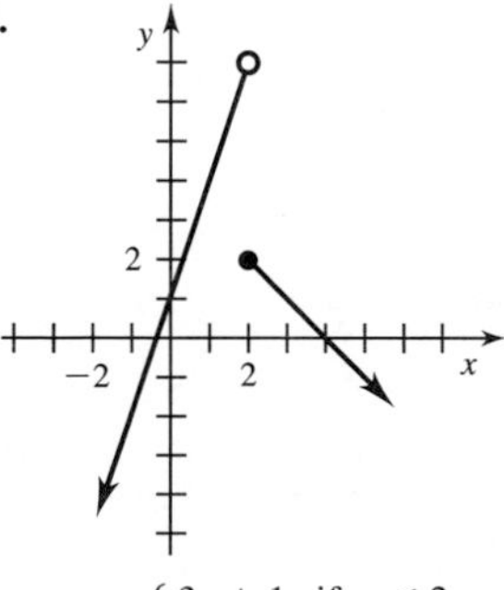

53.

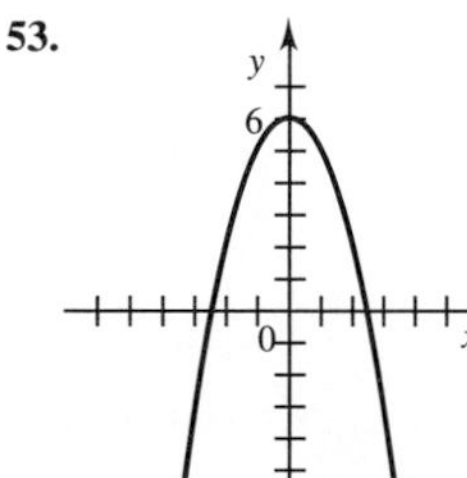

54. $y = [\![5000/x]\!]$ **55.** x-axis, y-axis, and origin **56.** x-axis **57.** none of these symmetries **58.** y-axis **59.** x-axis **60.** none of these symmetries **61.** x-axis, y-axis, and origin **62.** origin **63.** Reflect the graph of $f(x)$ about the x-axis. **64.** Translate the graph of $f(x)$ down 2 units. **65.** Double the y-values of $f(x)$ and translate 4 units to the right. **66.** $y = -3x + 4$ **67.** $y = -3x - 4$ **68.** $y = 3x + 4$ **69.** $y = (1/3)x + 4/3$

70. (a)

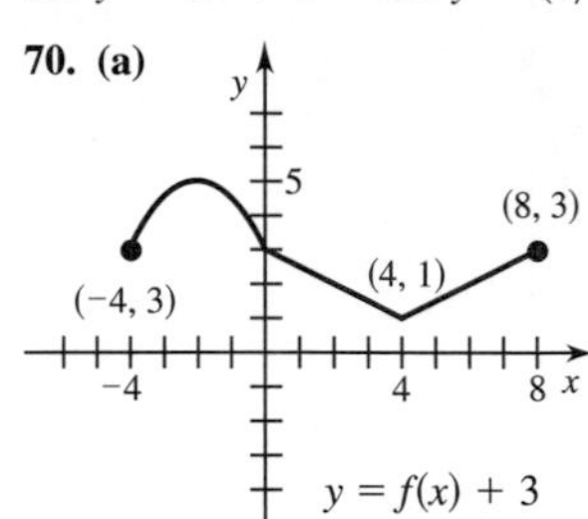

(b)

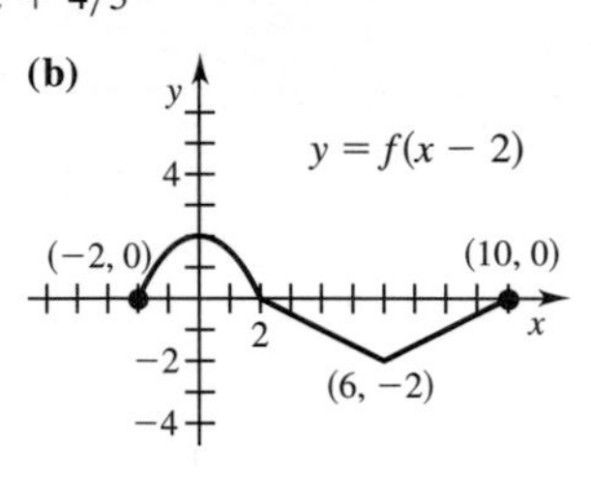

(c)

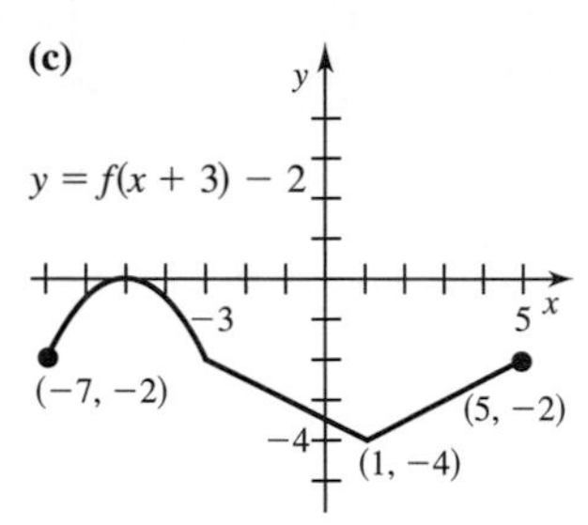

(d)

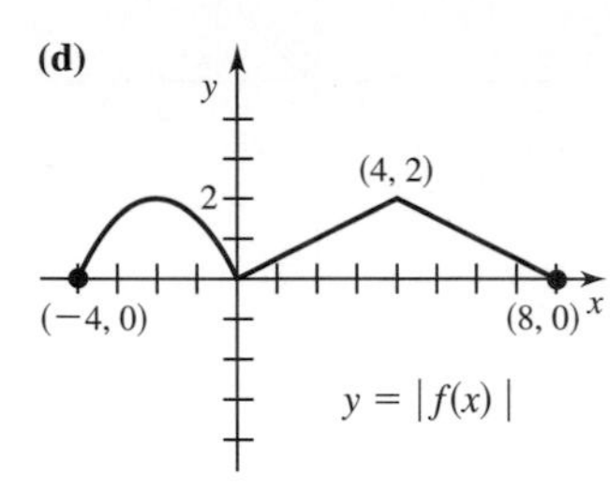

71. $x^2 - x + 1$ **72.** $2x^3 - x^2 - 23x - 20$ **73.** 13 **74.** 21 **75.** $4k^2 - 2k + 1$ **76.** $2r^3 + 5r^2 - 19r - 42$ **77.** $-11/4$ **78.** undefined **79.** $(-\infty, \infty)$ **80.** $(-\infty, -1) \cup (-1, 4) \cup (4, \infty)$ **81.** $\sqrt{x^2 - 2}$ **82.** $x - 2$ **83.** 1 **84.** $\sqrt{34}$ **85.** not one-to-one **86.** one-to-one **87.** not one-to-one **88.** one-to-one **89.** not one-to-one **90.** not one-to-one

91. $f^{-1}(x) = (x - 3)/12$

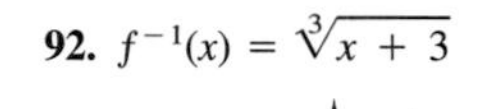

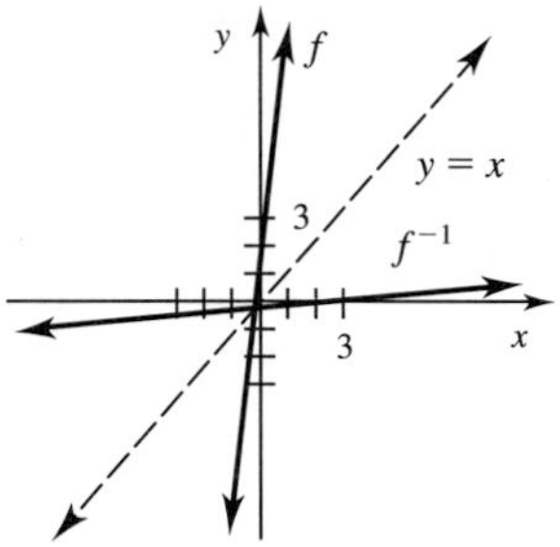

92. $f^{-1}(x) = \sqrt[3]{x + 3}$

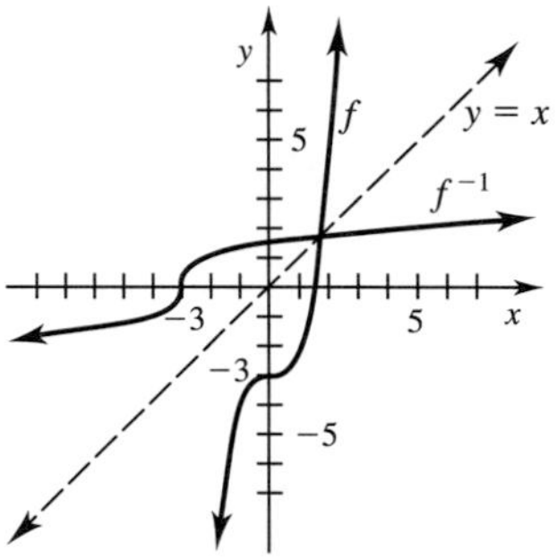

93. f has no inverse. **94.** f has no inverse. **95.** the number of years after 1992 required for the investment to reach \$50,000 **96.** $(5/2, 5/2)$ **97. (a)** $V = \pi d^3/4$ **(b)** $S = 3\pi d^2/2$ **98.** Let $f(t)$ represent the distance from second base at time t, where $0 \le t \le 12$. Then

$$f(t) = \begin{cases} 30\sqrt{t^2 - 6t + 18} & \text{if } 0 \le t \le 3 \\ 180 - 30t & \text{if } 3 < t \le 6 \\ 30t - 180 & \text{if } 6 < t \le 9 \\ 30\sqrt{t^2 - 18t + 90} & \text{if } 9 < t \le 12. \end{cases}$$

99. For the 16% increase, $y = 1.16x$; the inverse is $y = .8621x$. For the 16% decrease, $y = .84x$; the inverse is $y = 1.1905x$. The Canadian to U.S. dollars conversion factor should be .8621.

100. (a) no; no **(b)** 1986; 1990 **(c)** about \$7; about \$27.50 **(d)** down **(e)** a constant, stable price **101. (a)** yes; no **(b)** December; January **(c)** about 6,000; about 2,000 **(d)** a slight downward trend

Chapter 4 Polynomial and Rational Functions

Section 4.1 (page 202)

1.

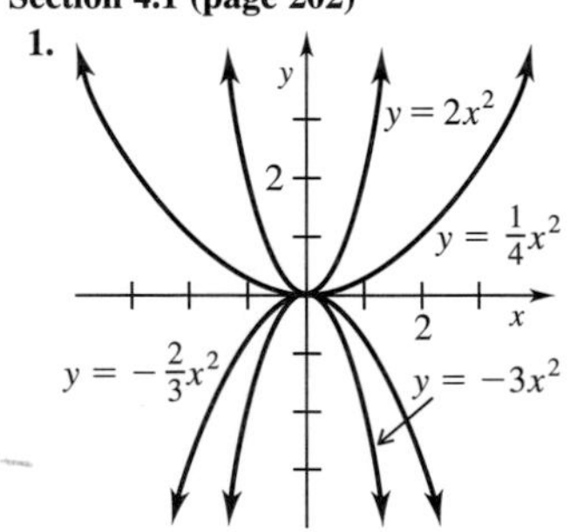

3.

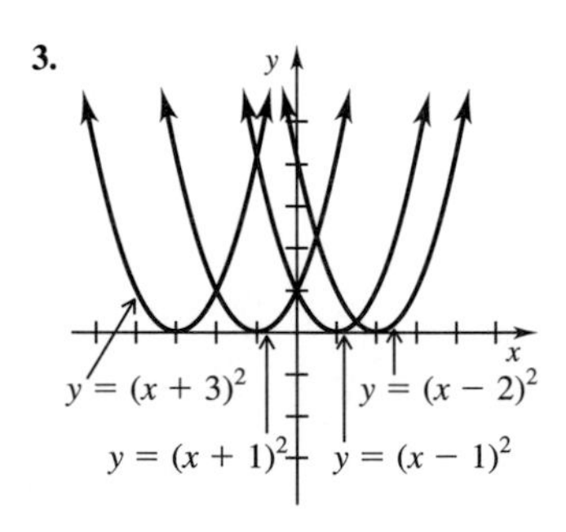

5. $(-\infty, -1]$
7. vertex: $(-2, 0)$; axis: $x = -2$; x-intercept: -2; y-intercept: 4

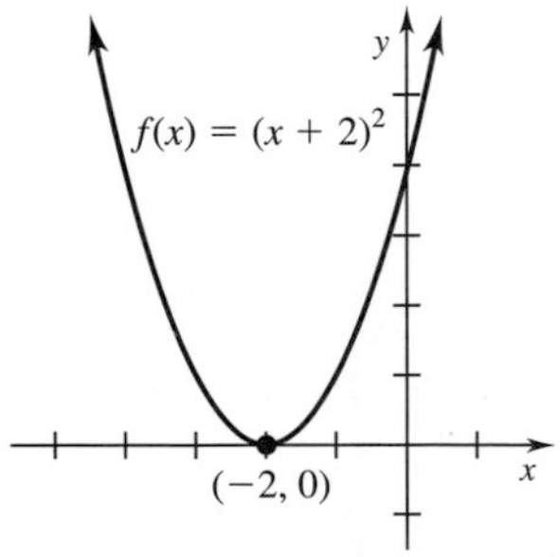

9. vertex: $(-3, -4)$; axis: $x = -3$; x-intercepts: $-3 \pm \sqrt{2}$; y-intercept: 14

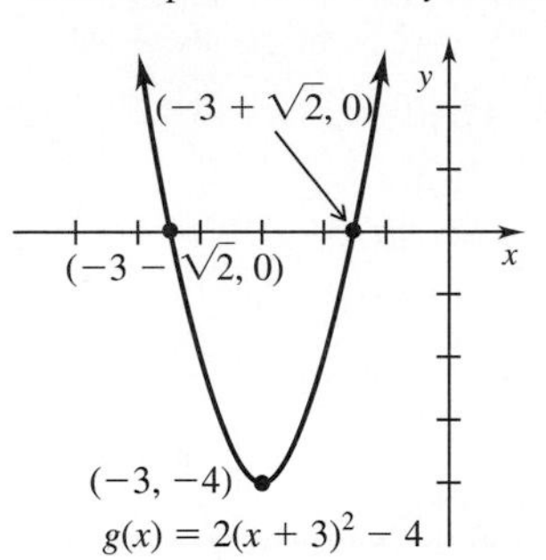

11. vertex: $(2, 1)$; axis: $x = 2$; x-intercepts: $(6 \pm \sqrt{3})/3$; y-intercept: -11

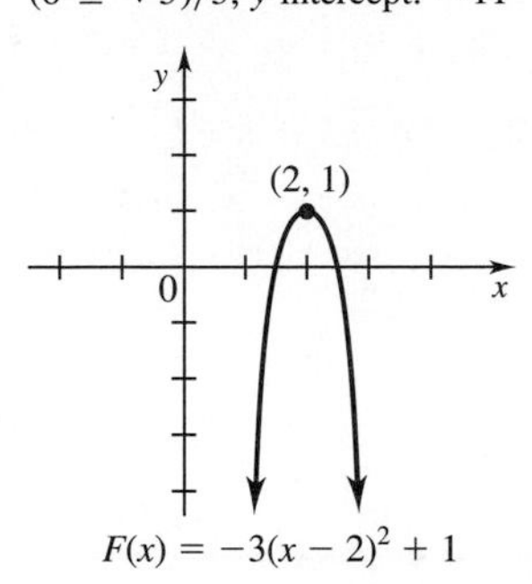

13. vertex: $(-2, -3)$; axis: $x = -2$; x-intercepts: $-2 \pm \sqrt{6}$; y-intercept: -1

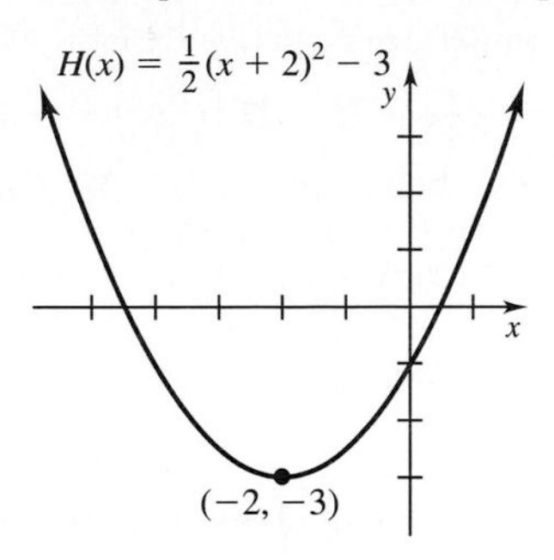

15. vertex: $(-2, -1)$; axis: $x = -2$; x-intercepts: $-1, -3$; y-intercept: 3

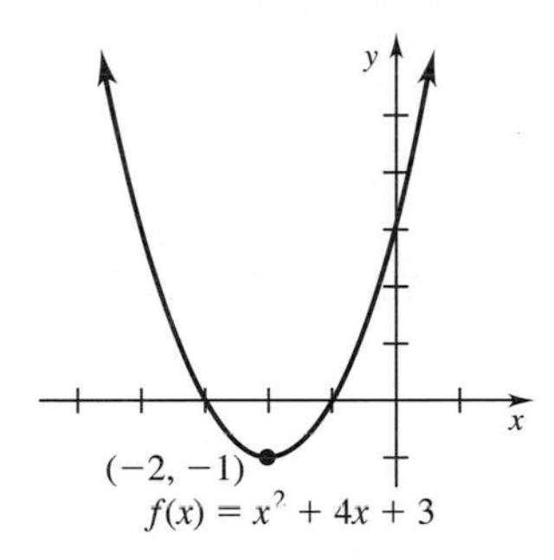

17. vertex: $(3, 3)$; axis: $x = 3$; x-intercepts: $3 \pm \sqrt{3}$; y-intercept: -6

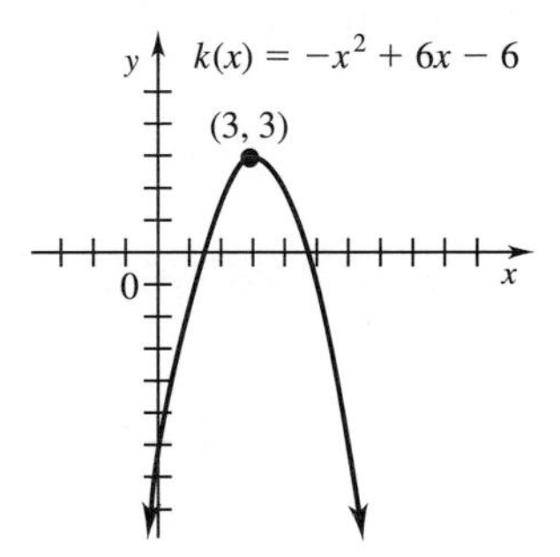

19. vertex: $(4, 2)$; axis: $x = 4$; x-intercepts: $(12 \pm \sqrt{6})/3$, y-intercept: -46

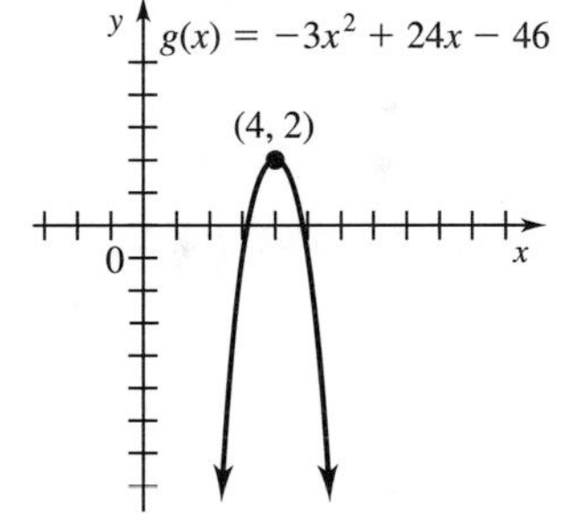

21. 3 **23.** 0 solutions **25.** 80 ft wide, 160 ft long **27.** Maximum profit is \$3625 with 25 unsold seats. **29.** max. height: 16 ft; 2 sec **31.** 15 and -15

33. **(a)**

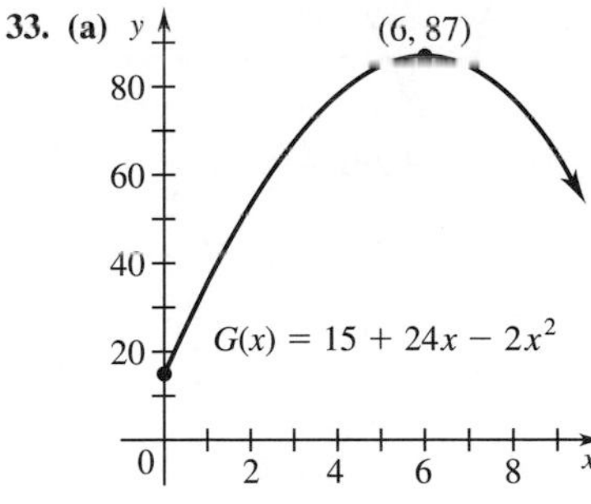

(b) 87 particles on June 6

35. **(a)** 120 **(b)** 40 **(c)** 22 **(d)** 84 **(e)** 140 **(f)** minimum at $x = 8$ (August)

37. $6\sqrt{3}$ m **39.** **(a)** B **(b)** C **(c)** A **41.** (a) **43.** (f) **45.** (b) **47.** $a > 4$

49. $f(x) = \frac{x^2}{2} - \frac{7}{2}x + 5$ **51.** $y = 9$ **(a)** $\sqrt{9} = 3$ **(b)** $\frac{1}{9}$ **53.** (3, 6) **55.** no

57. $f(x) = 3(x + 1)^2 - 12$

Section 4.2 (page 211)

1. $x^2 - 2x + 7$ **3.** $4x^2 - 4x + 1 + [-3/(x + 1)]$ **5.** $x^3 - 4x$ **7.** $x^4 + x^3 + 2x - 1 + 3/(x + 2)$ **9.** $x^2 + x + 1$ **11.** $f(x) = (x + 1)(2x^2 - x + 2) - 10$ **13.** $f(x) = (x + 2)(-x^2 + 4x - 8) + 20$ **15.** $f(x) = (x - 3)(4x^3 + 9x^2 + 7x + 20) + 60$ **17.** -8 **19.** 2 **21.** -1 **23.** -6 **25.** 0 **27.** 11 **29.** $-6 - i$ **31.** no **33.** yes **35.** no **37.** no **39.** yes **43.** 3

Section 4.3 (page 218)

1. no **3.** yes **5.** yes **7.** $f(x) = -3x^3 + 6x^2 + 33x - 36$ **9.** $f(x) = -\frac{1}{2}x^3 - \frac{1}{2}x^2 + x$ **11.** $f(x) = -\frac{1}{3}x^3 + \frac{5}{3}x^2 - \frac{1}{3}x + \frac{5}{3}$

13. 0, $\pm\frac{\sqrt{7}}{7}i$ **15.** 2, -3, $1 - i$ **17.** -2 (multiplicity 5), 1 (multiplicity 5), $1 - \sqrt{3}$ (multiplicity 2)

Note: Other answers are possible in Exercises 19–27. **19.** $x^2 - 6x + 10$ **21.** $x^3 - 5x^2 + 5x + 3$ **23.** $x^4 + 4x^3 - 4x^2 - 36x - 45$

25. $x^3 - 2x^2 + 9x - 18$ **27.** $x^4 - 6x^3 + 17x^2 - 28x + 20$ **29.** $-1 + i, -1 - i$ **31.** $-1 \pm \frac{\sqrt{2}}{2}$ **33.** $i, \pm 2i$

35. $f(x) = (x - 2)(2x - 5)(x + 3)$ **37.** $f(x) = (x + 3)(3x - 1)(2x - 1)$ **39.** $f(x) = (x - 3i)(x + 4)(x + 3)$

41. $-1, 3$; $f(x) = (x + 2)^2(x + 1)(x - 3)$ **43.** 1, 3, or 5 **47.** 0, 2 sec; all others are nonreal complex **55.** 2 **57.** 2

59. 2, $(-1 \pm i\sqrt{3})/2$ **61.** $-2/3, -1/2, 1$ **63.** $\pm 1, (1 \pm i\sqrt{11})/2$

Section 4.4 (page 230)

1.

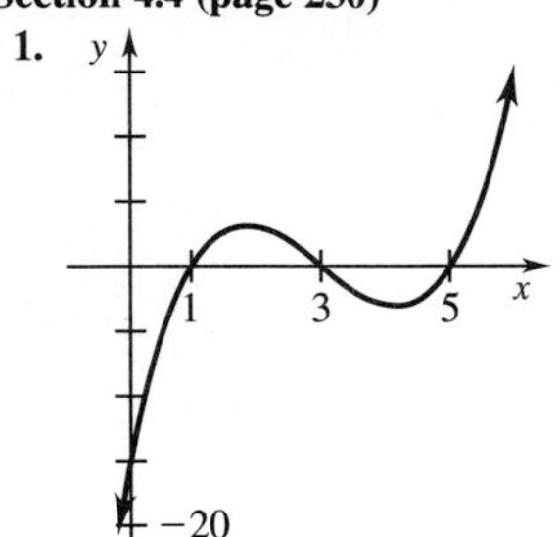

$f(x) = (x - 1)(x - 3)(x - 5)$

3.

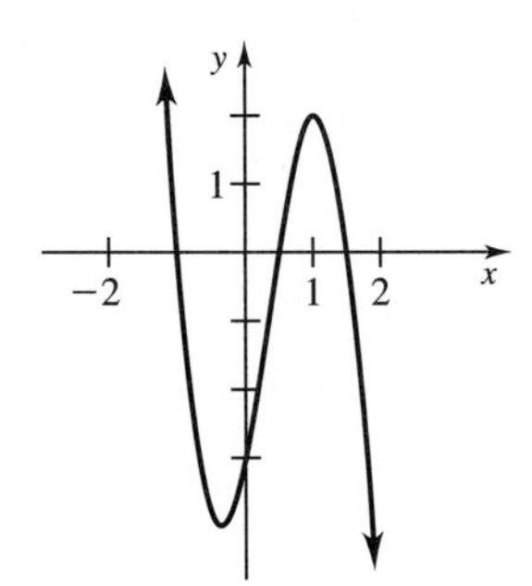

$f(x) = -(x + 1)(2x - 3)(2x - 1)$

5.

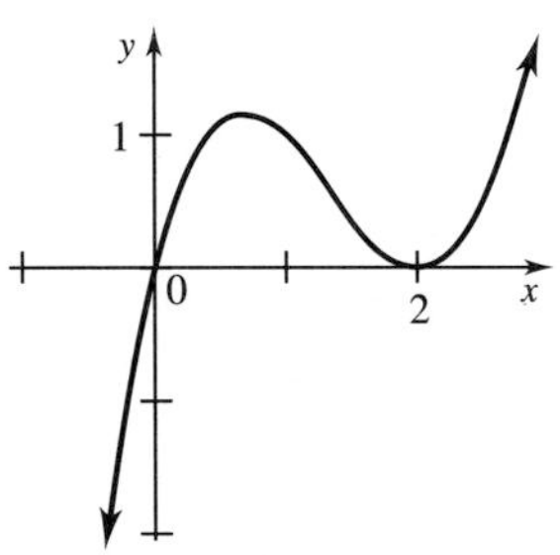

$f(x) = x^3 - 4x^2 + 4x$

7.

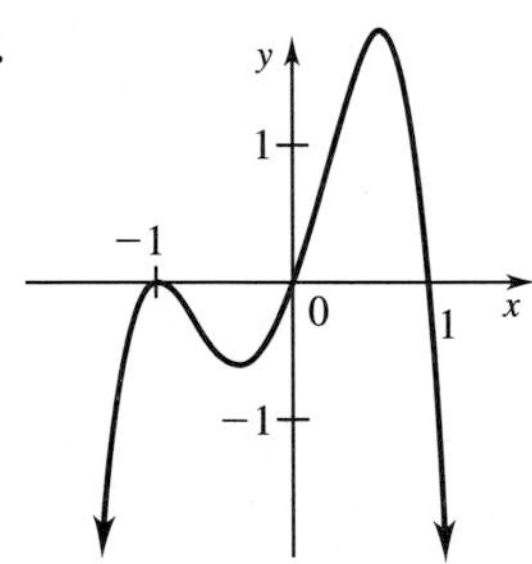

$f(x) = -3x(x + 1)^2(x - 1)$

9.

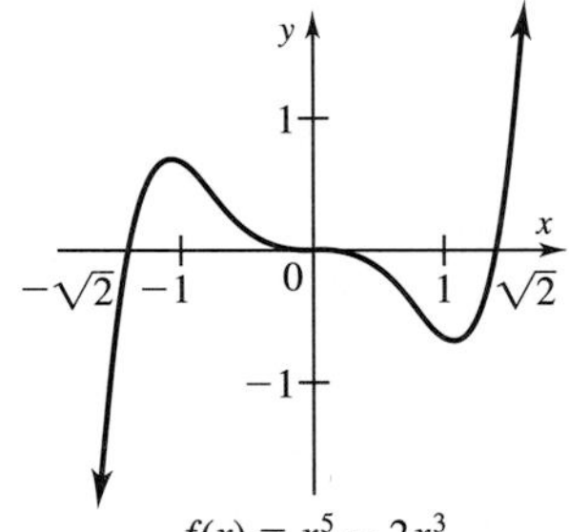

$f(x) = x^5 - 2x^3$

11.

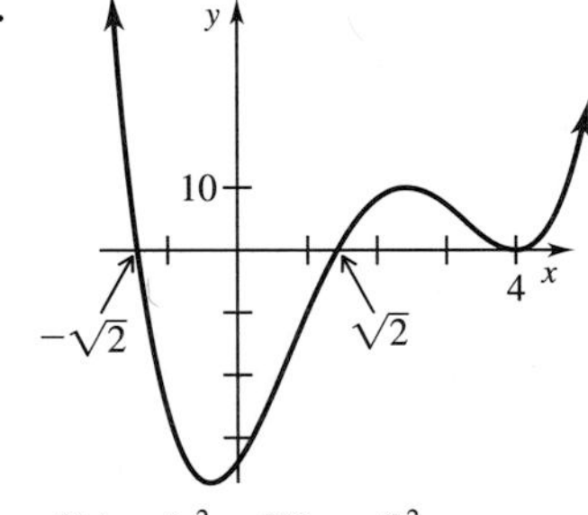

$f(x) = (x^2 - 2)(x - 4)^2$

23. 2 or 0; 1 **25.** 3 or 1; 1 **27.** $-3, -1.4, 1.4$ **29.** $-1.1, 1.2$ **31.** $3.24, -1.24$

33.

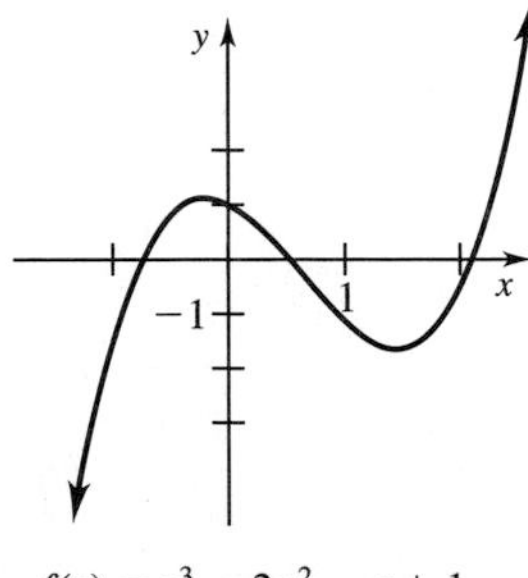

$f(x) = x^3 - 2x^2 - x + 1$

35.

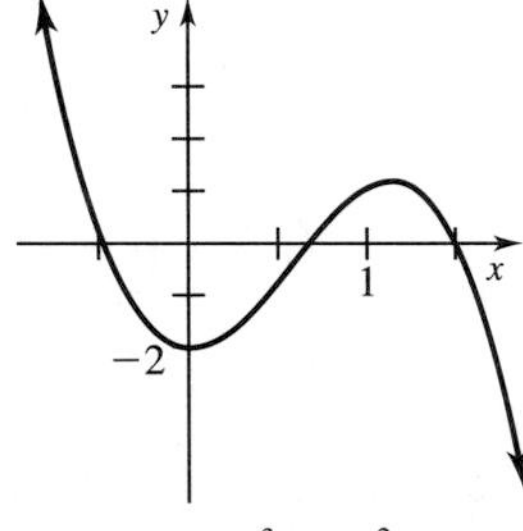

$f(x) = -4x^3 + 7x^2 - 2$

37.

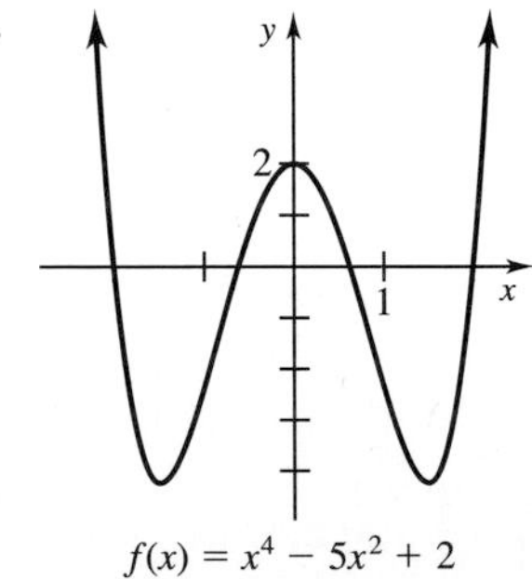

$f(x) = x^4 - 5x^2 + 2$

41. $f(x) = .5(x + 6)(x - 2)(x - 5)$ **43.** One possible answer is $f(x) = (x + 4)^2(x - 1)^2$. **45.** domain: $[-2, 0] \cup [3, \infty)$; range: $[0, \infty)$

49. **(b)** 1910 to 1925; 1905 to 1910; 1925 to 1930 **51.** **(b)** 17.3 hr **(c)** from 11.4 hr to 21.2 hr

Section 4.5 (page 236)

1. $\pm 1, \pm 1/2$ **3.** $\pm 1, \pm 2, \pm 1/2, \pm 1/4, \pm 1/8$ **5.** $\pm 1, \pm 3, \pm 5, \pm 15, \pm 25, \pm 75$ **7.** $-1, -3, 7$ **9.** no rational zeros

11. $2, -3, -5$ **13.** $1, -2, -3, -5$ **15.** $-1/2, -2/3, 1$; $f(x) = (2x + 1)(3x + 2)(x - 1)$ **17.** $1/2$; $f(x) = (2x - 1)(x^2 + 4x + 8)$

19. $-1, -2, -3, 4$; $f(x) = (x + 1)(x + 2)(x + 3)(x - 4)$ **21.** $1, 2, -3, 4$; $f(x) = (x - 1)(x - 2)(x + 3)(x - 4)$
23. 1; $f(x) = (x - 1)(x^4 + 4x^3 - x^2 - 12x - 12)$ **25.** $-2, -1, 5/2$ **27.** $1, -5/4$ **29.** 1 **31.** 5 **33.** $4 \times 4 \times 4$ inches
35. 5×7 inches or $(\sqrt{37} - 3) \times (3\sqrt{37} + 9)/2$ inches

41.

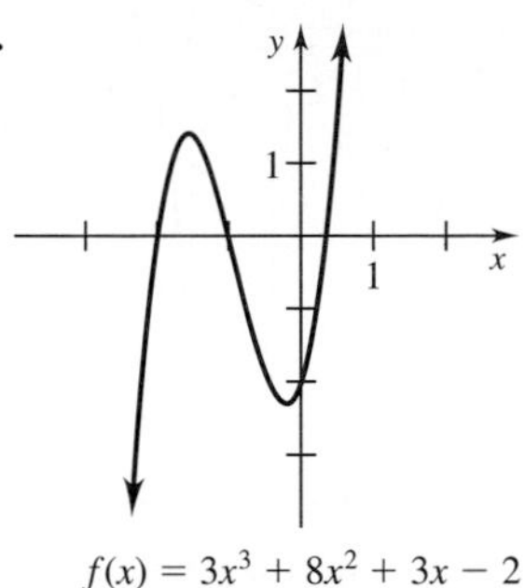

43.

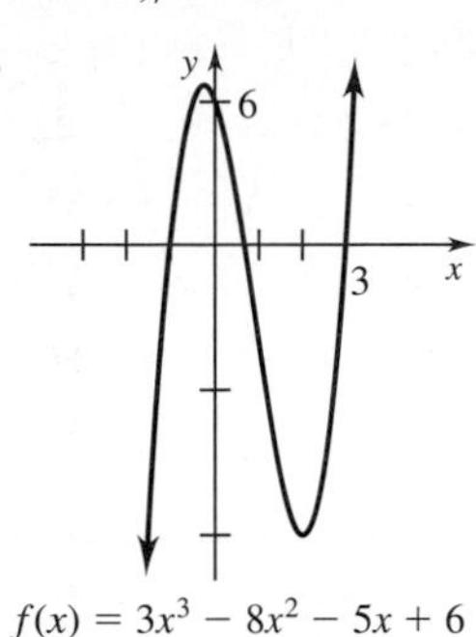

45.

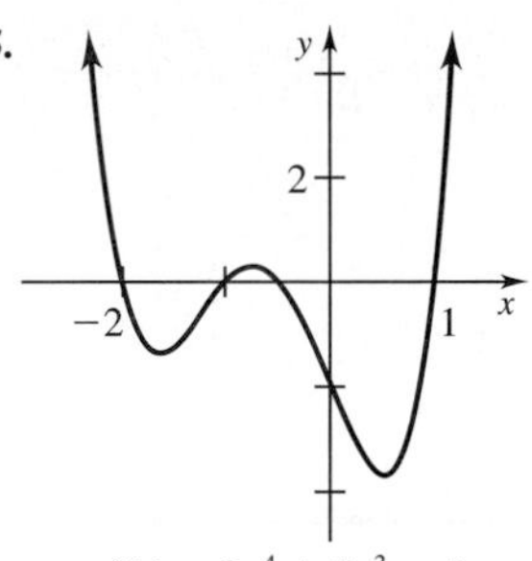

49. $-5/2, 2/3, 5$
51. $-3, 2/5, 2$
53. $-1, 1, 7, 10$

Section 4.6 (page 247)

1. vertical asymptote: $x = 3$; horizontal asymptote: $y = 0$ **3.** vertical asymptote: $x = 2$; horizontal asymptote: $y = 3$ **5.** vertical asymptote: $x = -3$; oblique asymptote: $y = x - 3$ **7.** vertical asymptotes: $x = 5, x = -2$; oblique asymptote: $y = 2x + 6$ **9.** vertical asymptotes: $x = -2, x = 5/2$; horizontal asymptote: $y = 1/2$

11. (a)

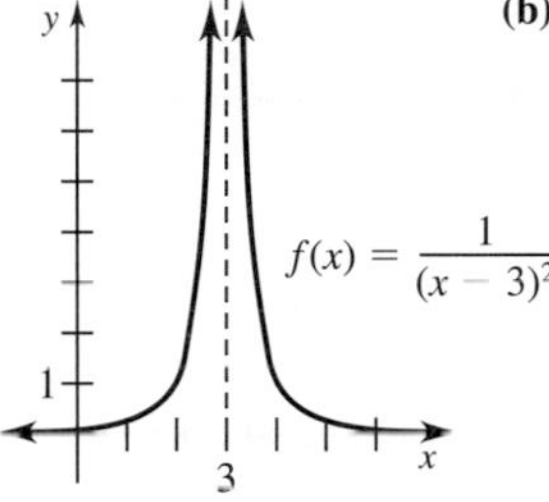

(b)

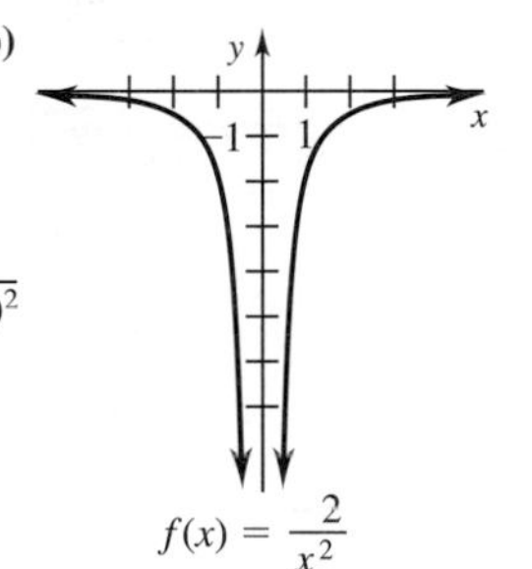

(c)

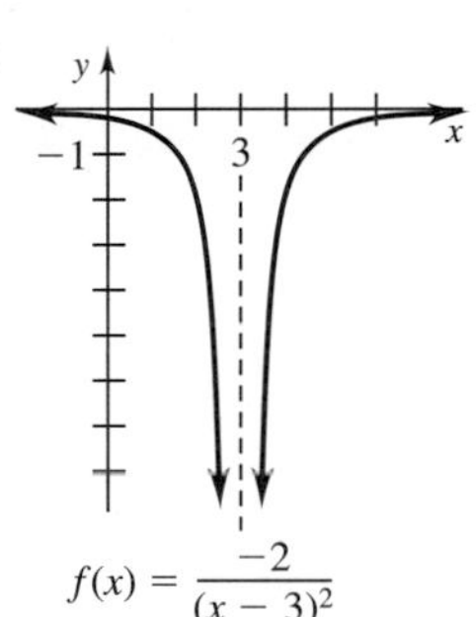

13. asymptote
15. missing point
17. missing point

19.

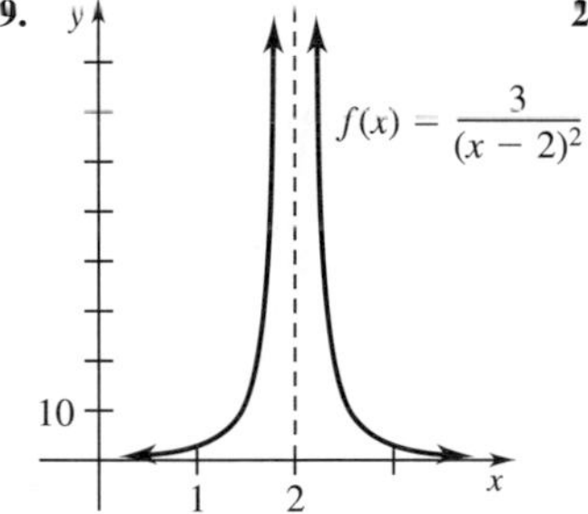

21.

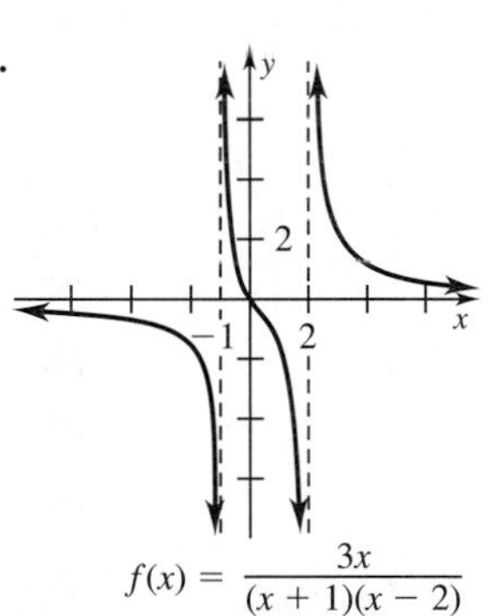

23.

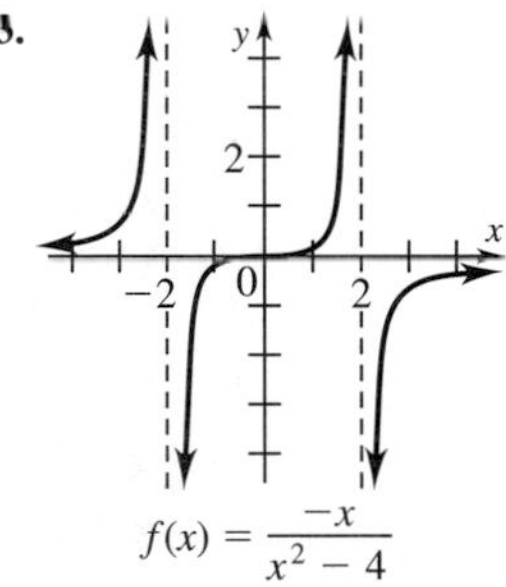

25.

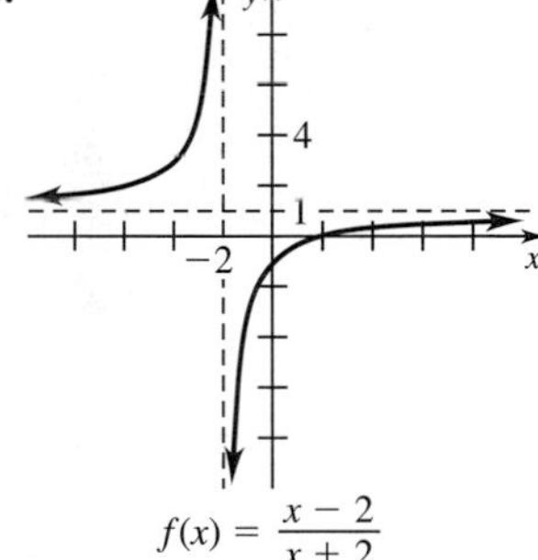

27.

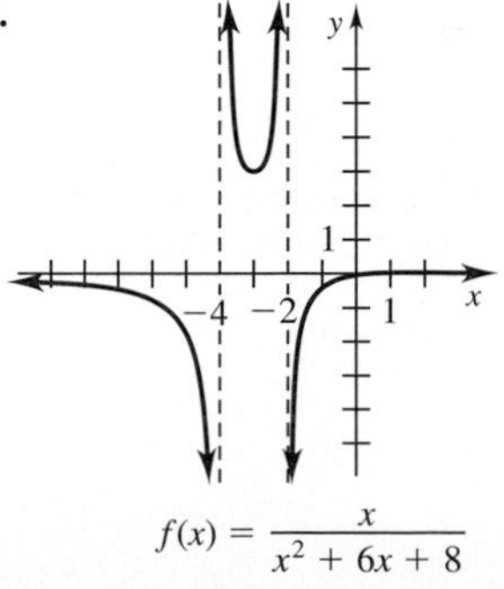

29.

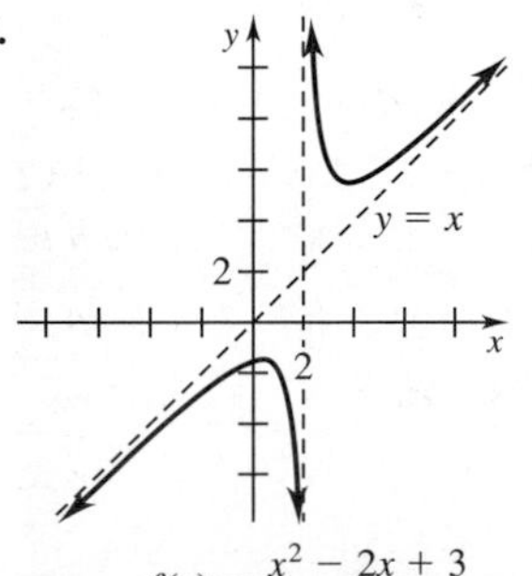

31.

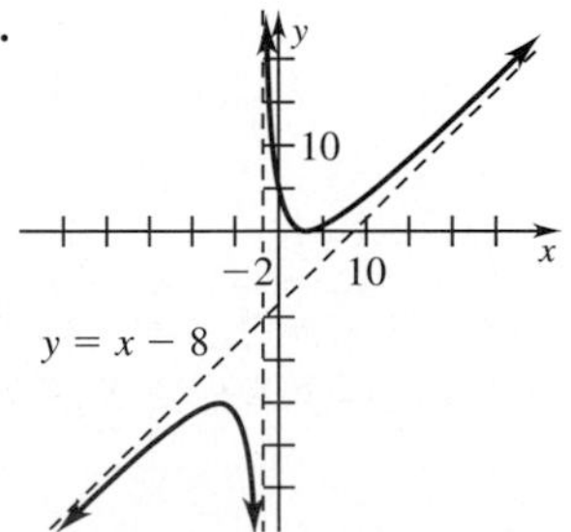

33.

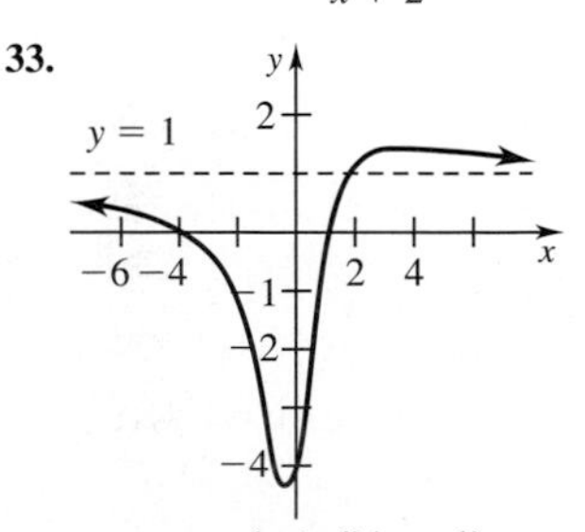

35.

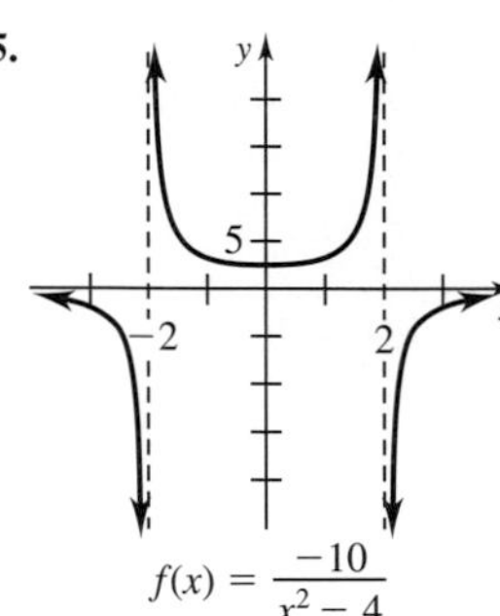

$f(x) = \dfrac{-10}{x^2 - 4}$

37.

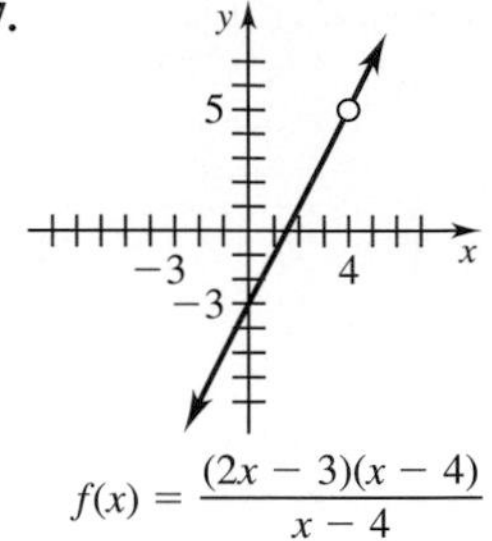

$f(x) = \dfrac{(2x - 3)(x - 4)}{x - 4}$

39.

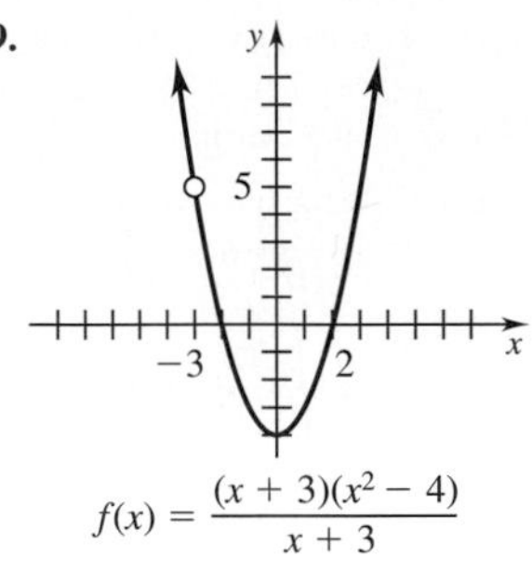

$f(x) = \dfrac{(x + 3)(x^2 - 4)}{x + 3}$

41.

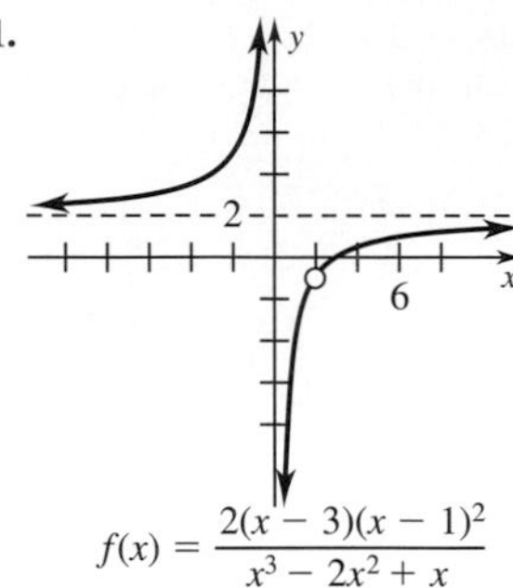

$f(x) = \dfrac{2(x - 3)(x - 1)^2}{x^3 - 2x^2 + x}$

Note: In Exercises 43–45, other answers are possible.

43. $f(x) = \dfrac{x - 2}{x(x - 4)}$ **45.** $f(x) = \dfrac{(x - 3)(x + 2)}{(x - 2)(x + 2)}$

47. (a)

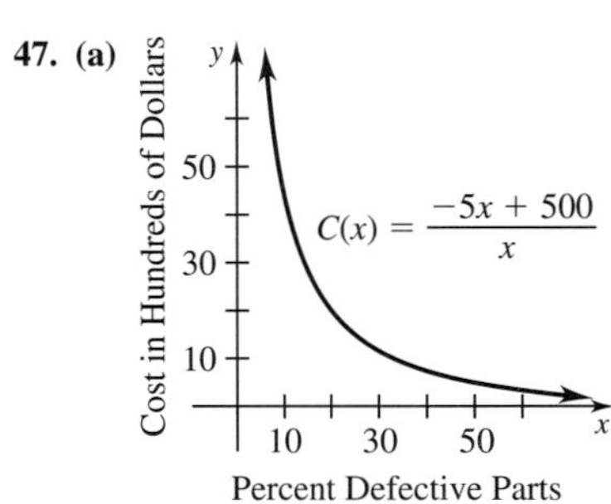

(b) no

51. 36.6%; \$268 million

Chapter 4 Review Exercises (page 250)

1. vertex: $(-4, -5)$; axis: $x = -4$; x-intercepts: $-4 \pm (1/3)\sqrt{15}$; y-intercept: 43

2. vertex: $(6, 7)$; axis: $x = 6$; x-intercepts: $6 \pm (1/2)\sqrt{42}$; y-intercept: -17

3. vertex: $(-2, 11)$; axis: $x = -2$; x-intercepts: $-2 \pm (1/3)\sqrt{33}$; y-intercept: -1

4. vertex: $(1/2, 2)$; axis: $x = 1/2$; x-intercepts: none; y-intercept: 3

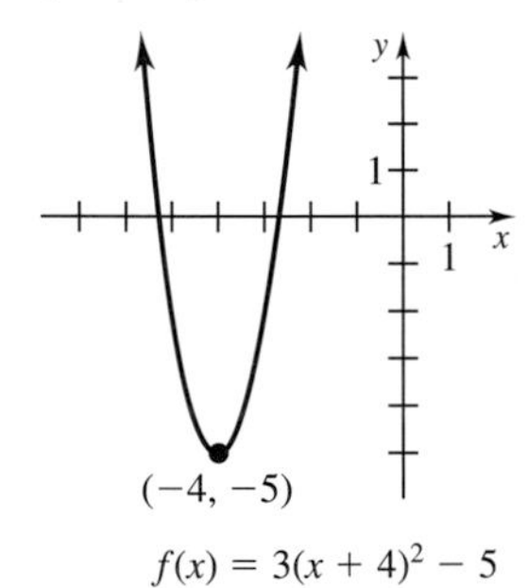

$f(x) = 3(x + 4)^2 - 5$

(6, 7)

$f(x) = -\frac{2}{3}(x - 6)^2 + 7$

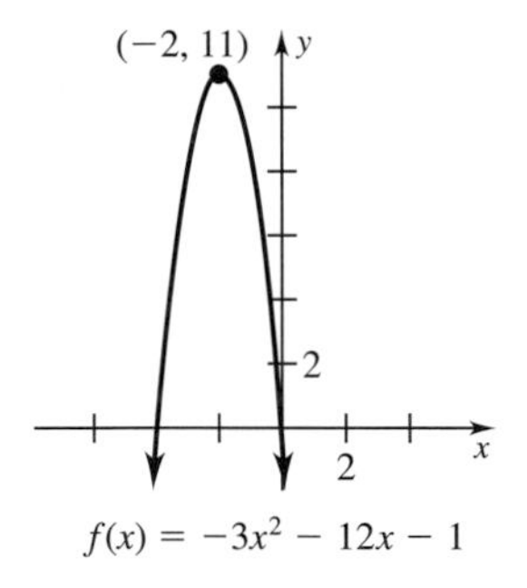

$f(x) = -3x^2 - 12x - 1$

$(\frac{1}{2}, 2)$

$f(x) = 4x^2 - 4x + 3$

5. k **6.** h **7.** $ah^2 + k$ **8.** $\dfrac{-k}{a} \geq 0$; $\left(h \pm \sqrt{\dfrac{-k}{a}}, 0\right)$ **9.** $c - \dfrac{b^2}{4a}$ **10.** 90 m × 45 m **11.** $2x^2 + 5x + 1$; 2

12. $4x^2 + 23x + 118$; 595 **13.** $x^2 + 4x + 1$; -7 **14.** $3x^2 + 2x + 1$; 8 **15.** -1 **16.** 6 **17.** 28 **18.** 40

Note: In Exercises 19–22, other answers are possible. **19.** $f(x) = x^3 - 10x^2 + 17x + 28$ **20.** $f(x) = x^3 - 13x^2 + 46x - 48$

21. $f(x) = x^4 - 5x^3 + 3x^2 + 15x - 18$ **22.** $f(x) = x^4 + 5x^3 + x^2 - 9x + 2$ **23.** no **24.** no **25.** no **26.** no **27.** yes

28. $f(x) = 2x^4 - 4x^3 + 10x^2 - 68x + 60$ **29.** $f(x) = -2x^3 + 6x^2 + 12x - 16$

Note: In Exercises 30 and 31, other answers are possible. **30.** $f(x) = x^4 - 3x^2 - 4$ **31.** $f(x) = x^4 - x^3 + 3x^2 - 9x - 54$

32. Any polynomial that can be factored into $a(x - b)^2(x - c)^2$ works.

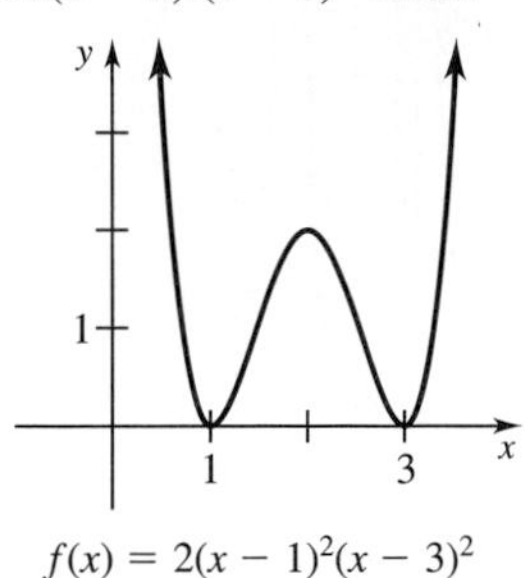

33. Any polynomial that can be factored into $a(x - b)^3$ works.

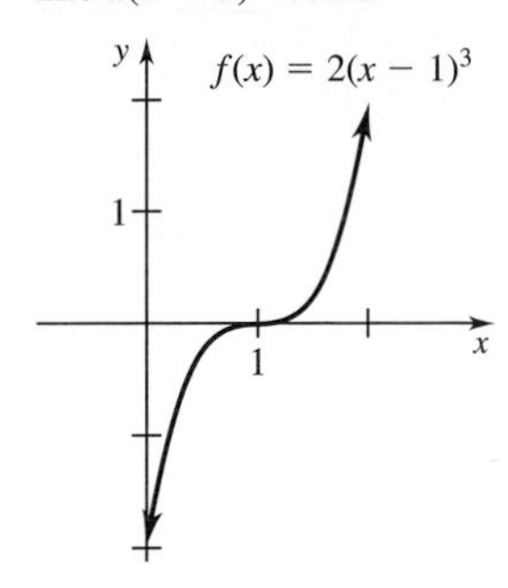

34. $1 - i, 1 + i, 4, -3$
35. $1, -1/2, \pm 2i$ **36.** $1/2, -1, 5$
37. $-2, 1/3, 5$ **38.** $-1/5, 1, 5/2$
41. $-6, 4, -16, -2$ **42.** -9
43. $13/2$

47. $-.7, 1, 7.7$

$f(x) = x^3 - 8x^2 + 2x + 5$

48. $-2.3, 4.6$

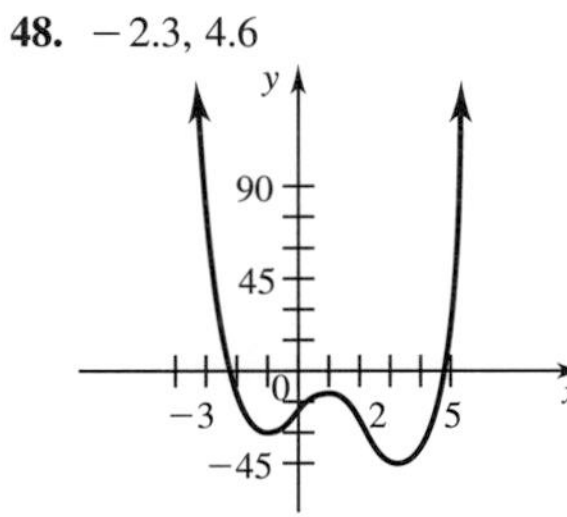

49.

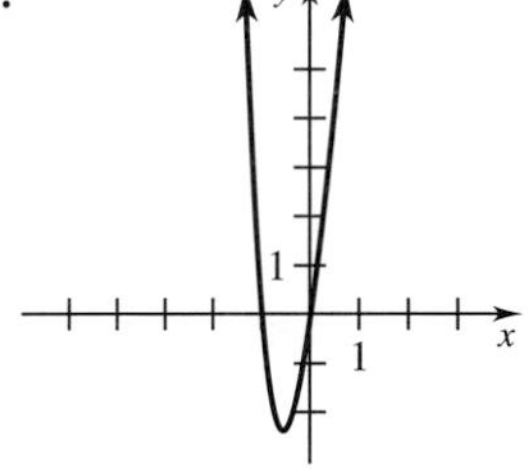

50.

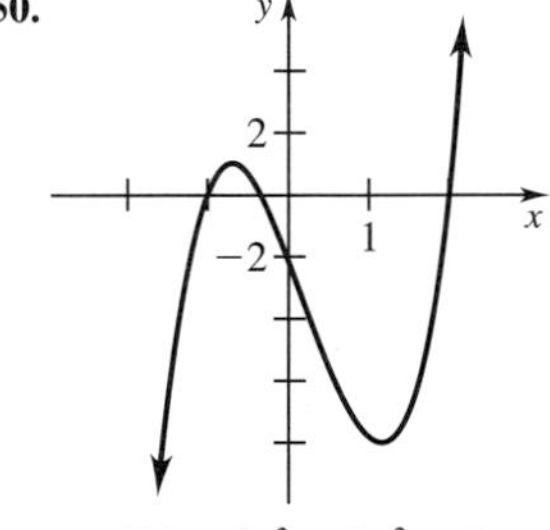

51. **(a)** 1 positive; 2 or 0 negative
(b) $x = -3.9$
(c)

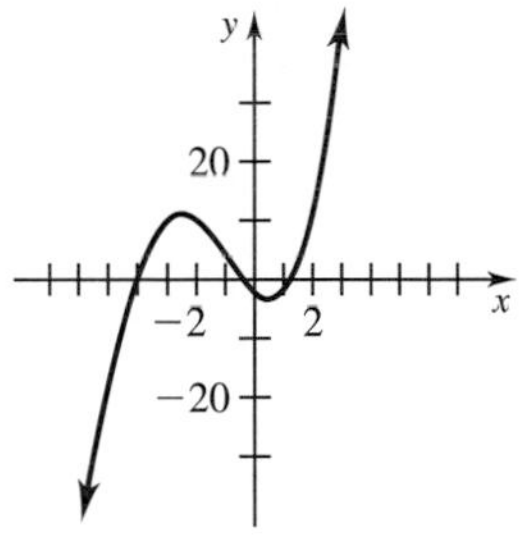

53.

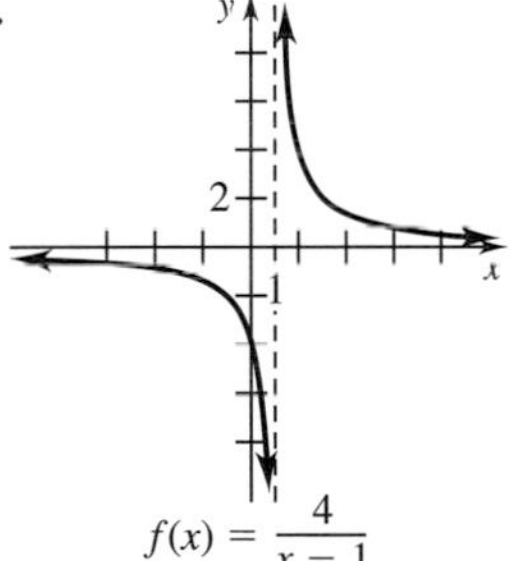

54.

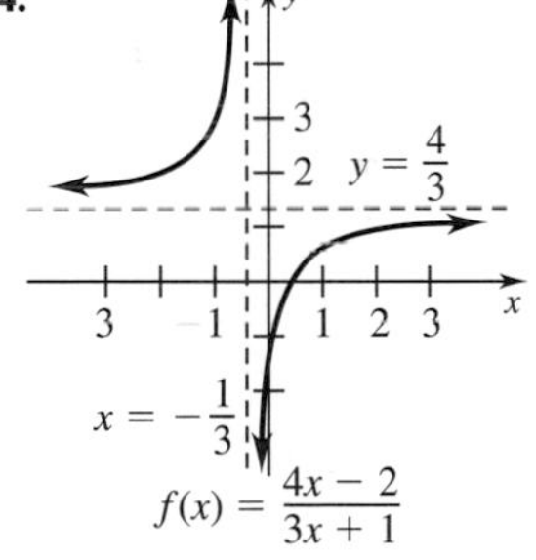

55.

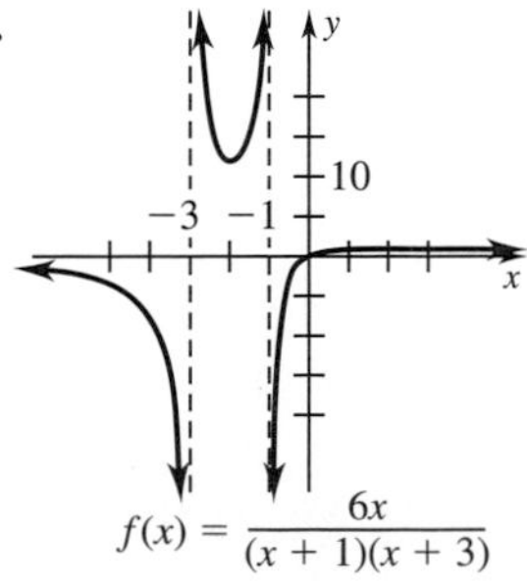

56.

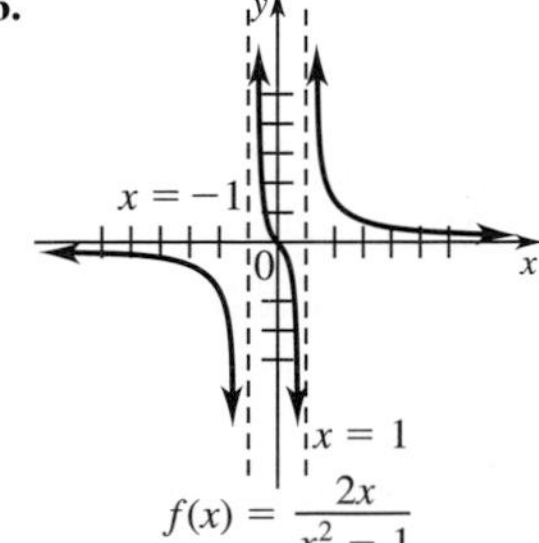

57.

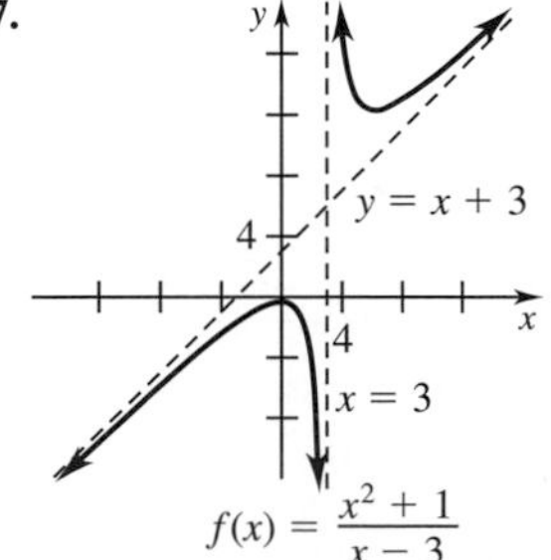

58.

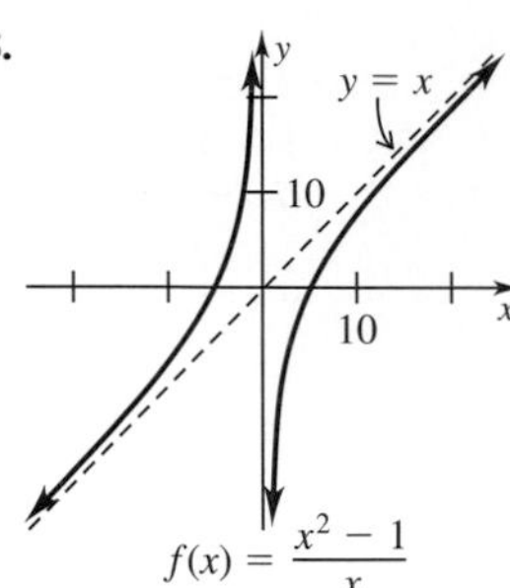

59. (a)

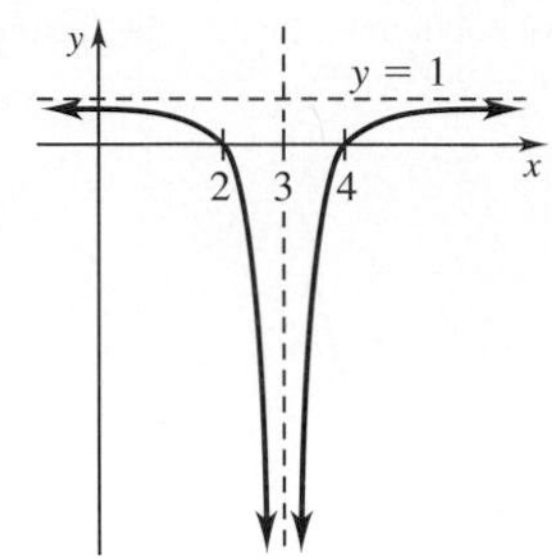

(b) One possibility is

$$f(x) = \frac{(x - 2)(x - 4)}{(x - 3)^2}.$$

60. (a)

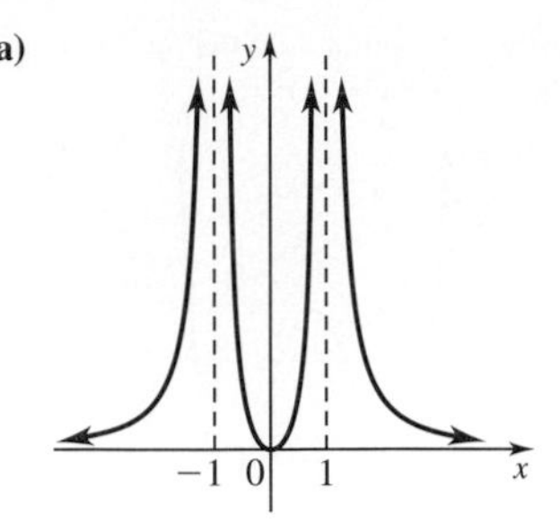

(b) One possibility is

$$f(x) = \frac{x^2}{(x^2 - 1)^2}.$$

Chapter 5 Exponential and Logarithmic Functions

Section 5.1 (page 263)

1. (a)

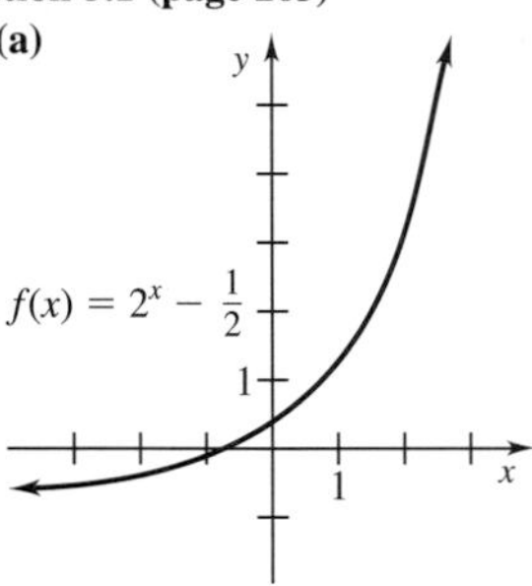

(b)

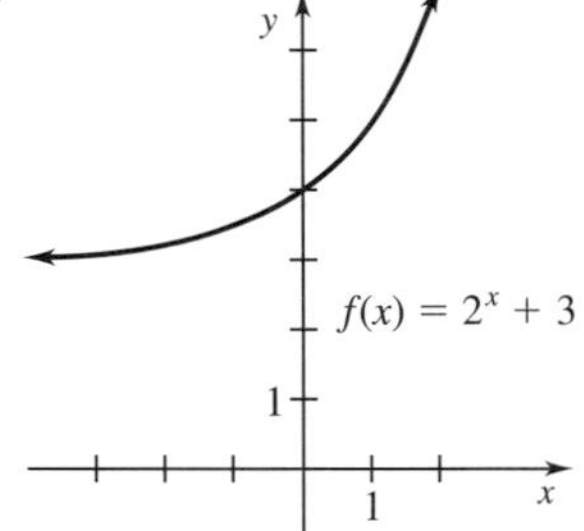

(c)

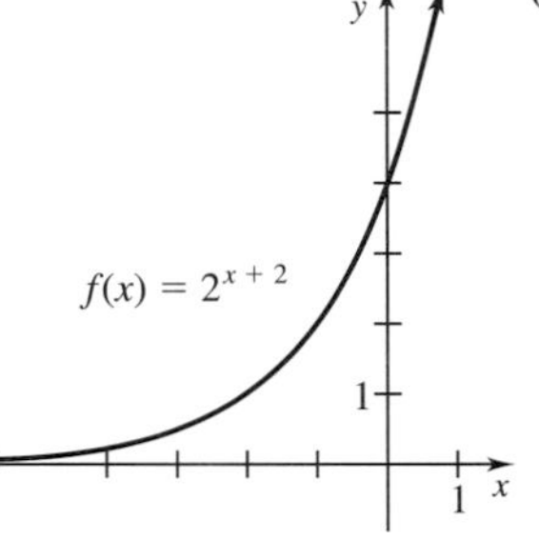

(d)

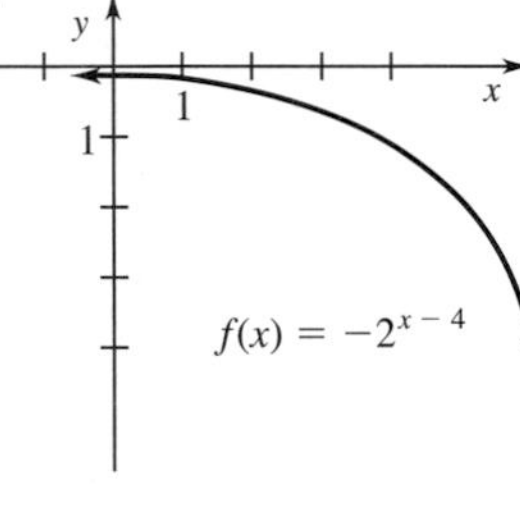

3.

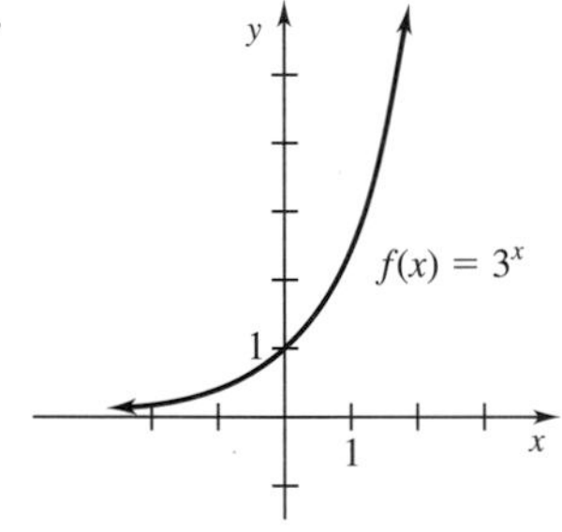

5.

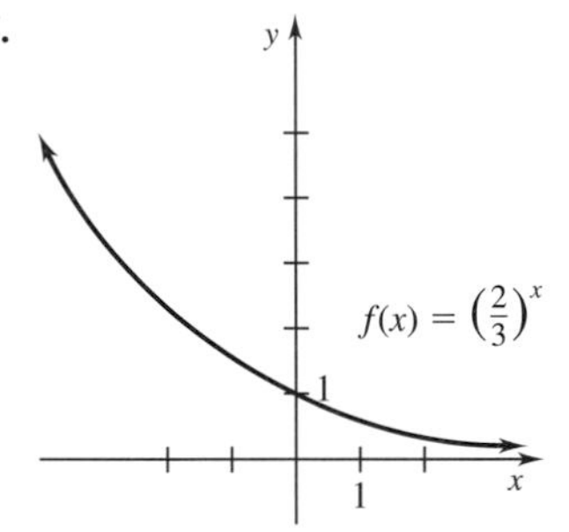

7.

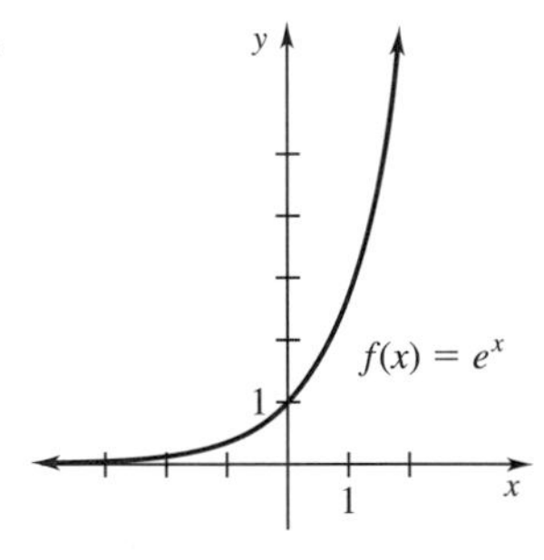

9.

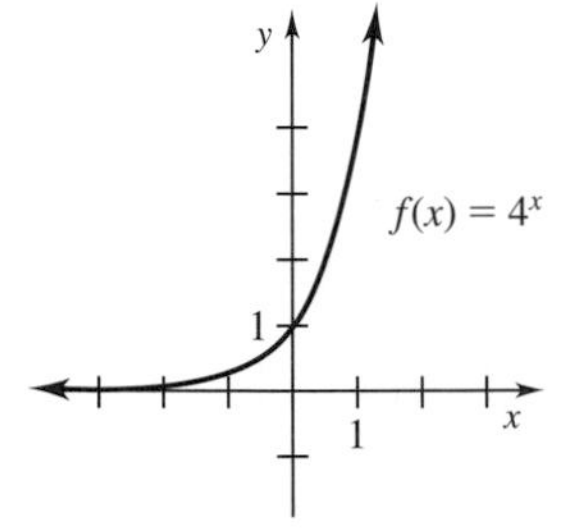

11.

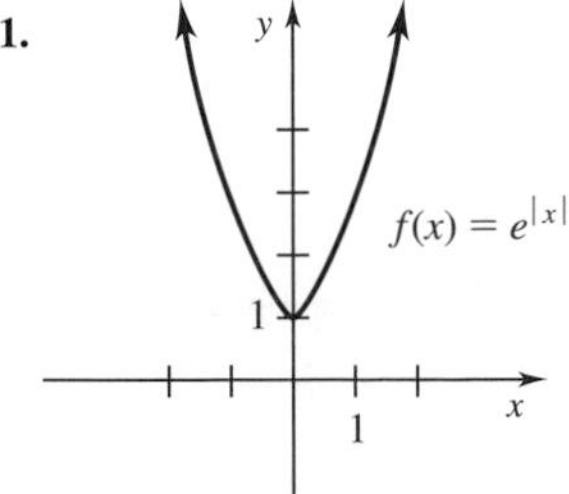

13. 4 **15.** −3 **17.** 3/2 **19.** 3 **21.** 8 **23.** 3/5 **25.** −2/3 **27.** ±7 **29.** 0 **31. (a)** \$6,802.44 **(b)** \$6,863.93 **(c)** \$6,885.40

33. **(a)** 10,000 **(b)** 12,000 **(c)** 22,000 **(d)** 27,000 **(e)**

35. $f(x) = 2^x$ **37.** $f(x) = (25/4)^x$ **39.** $f(t) = 27 \cdot 9^t$ **41.** $f(t) = (1/3) \cdot 9^t$ **45.** 1 **47.** $(0, 1), (1, a)$ **49.** .025 lb per sq inch **55.** neither **57.** no solution **59.** $x = 0; x \approx .73$

Section 5.2 (page 274)

1. $\log_{10} 1000 = 3$ **3.** $\log_{1/2} 16 = -4$ **5.** $6^2 = 36$ **7.** $e^0 = 1$ **9.** 2 **11.** 1 **13.** 5 **15.** 1/3 **17.** 5 **19.** $x + 1$ **21.** 15 **23.** 32 **25.** 2 **27.** 144 **29.** $\log_2 6 + \log_2 x - \log_2 y$ **31.** $(1/2) \log_7 5 - \log_7 9$ **33.** not possible **35.** $2 \log_a p + 3 \log_a q - 2 \log_a r$ **37.** $\log_a \frac{xy}{m}$ **39.** $\log_m \frac{a^2}{b^6}$ **41.** $\log_x \frac{a^{3/2}}{b^4}$ **47.** $f(x) = 1 + \ln x$ **49.** e **51.** b **53.** f

55. x-intercept: 1

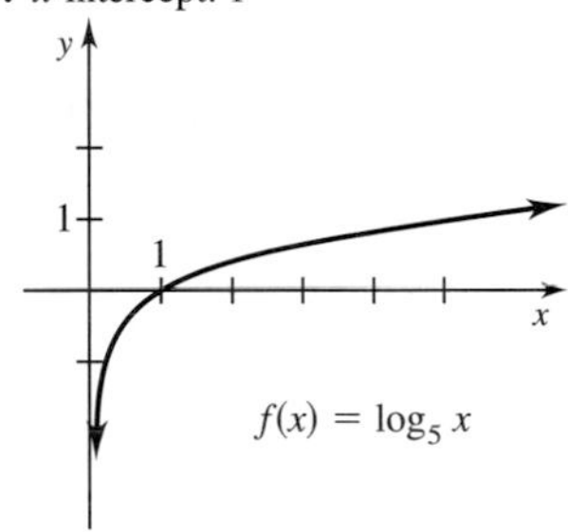

57. x-intercept: 6; y-intercept: -1

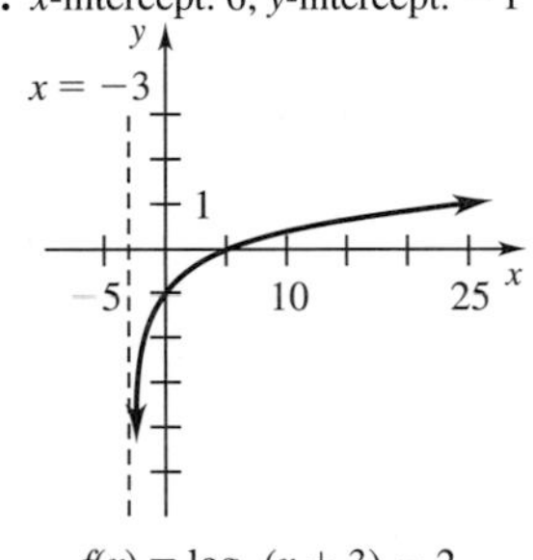

59. x-intercept: 2

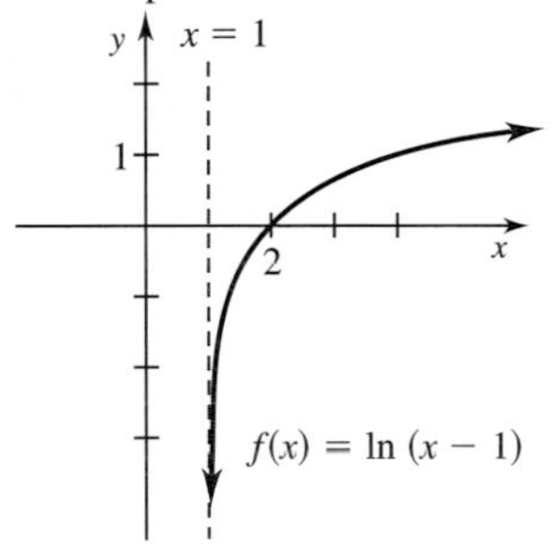

61. .7781 **63.** 1.4771 **65.** .1761 **67.** .9542 **69.** $\log_6 2$ **71.** -4 **73.** 6 **75.** **(a)** 10,000 **(b)** 21,000 **(c)** $\approx$ 26,000 **(d)**

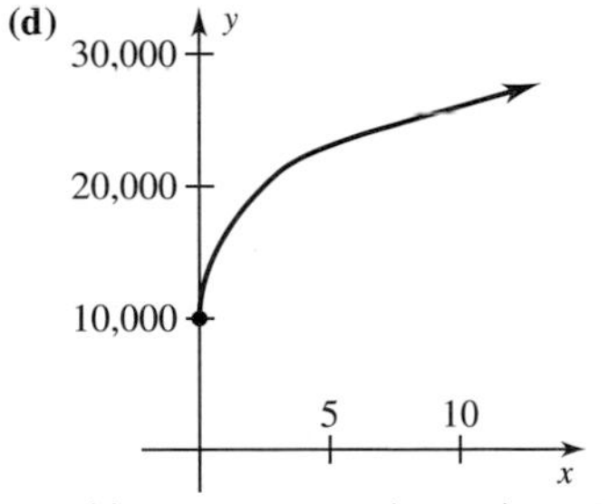

$S(x) = 11{,}000 \log_{10} (3x + 1) + 10{,}000$

77. **(a)** 3 **(b)** 6 **(c)** 8 **79.** 31.6 times more powerful **81.** **(a)** $\approx$ 7400 **(b)** $\approx$ 2000 **83.** 3.32 **85.** -1.58 **87.** $-.96$ **89.** 1.89 **91.** 2 **97.** $x \approx .70$, log 5 **99.**

101. $x \approx .16, x \approx 3.1$ **103.** $x \approx .42, x \approx 4.1$

Section 5.3 (page 283)

1. 1.21 **3.** $-.08$ **5.** 3.80 **7.** 1.39 **9.** .02 **11.** no solution **13.** 17.48 **15.** ± 3.76 **17.** 1.79 **19.** 11 **21.** 9.05 **23.** 5 **25.** $-7/8$ **27.** 3.61 **29.** 8 **31.** 2, 3 **33.** 2 **35.** 1, 10 **37.** 7.53 sec **39.** 23 days, 46 days **41.** 0 **43.** ln 2, ln 3 **45.** $\ln(3 \pm 2\sqrt{2})$ **47.** $t = (-2/R) \ln (1 - RI/E)$ **49.** $t = \frac{\ln [A/(A - Pr)]}{\ln (1 + r)}$ **51.** $x = y + 1$ **53.** $x = \log (y + \sqrt{y^2 - 1})$ **55.** $10^x - 5, (-\infty, \infty), (-5, \infty)$ **57.** $\log \left(\frac{x}{3}\right), (0, \infty), (-\infty, \infty)$ **59.** $(0, 1) \cup (4, \infty)$ **61.** $(1, 5^{10})$ **63.** $(-\infty, 1)$

65. $t = \dfrac{\ln (A/P)}{n \ln (1 + r/n)}$ **71.** $\approx -.93, \approx 1.35$

Section 5.4 (page 289)

1. **(a)** $40{,}000\, e^{.03t}$ **(b)** $\approx 73{,}000$ **(c)** ≈ 13.5 yr **3.** **(a)** $\approx -.1$ **(b)** ≈ 6.9 hr **5.** ≈ 19 hr **7.** about 16 days **9.** about 33°C **11.** about 11,200 yr, about 16,800 yr **13.** about 6.97 watts **15.** about 275 days **17.** \$2983.65 **19.** \$152.21 **21.** \$14,333.29 **23.** about 6.9 yr **25.** $t = \dfrac{\ln 2}{r}$ **27.** 81.25°C **29.** 1:59 P.M. **33.**

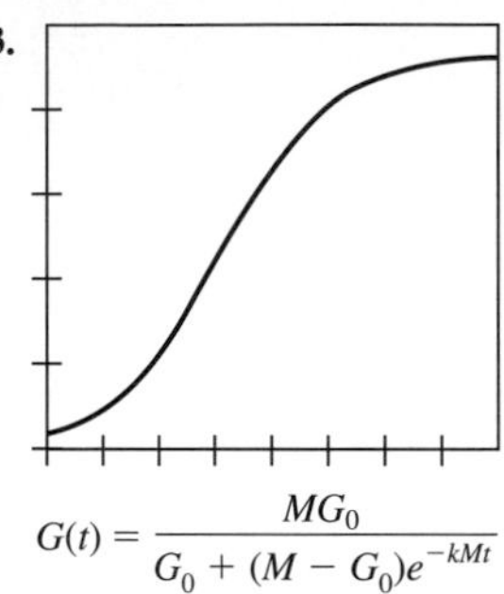

35. 2.8; 2.7726

Chapter 5 Review Exercises (page 291)

1.

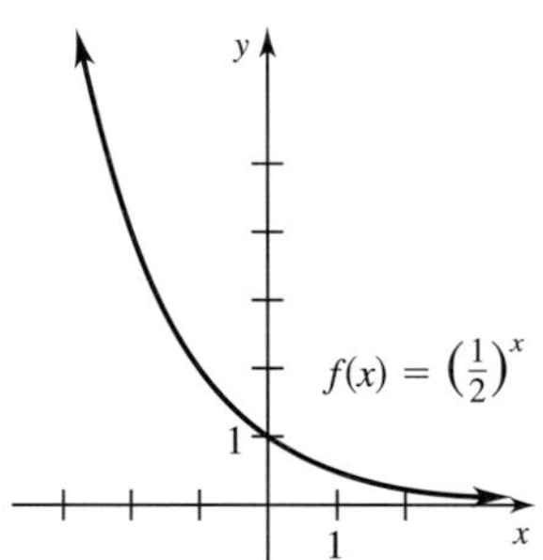

2.

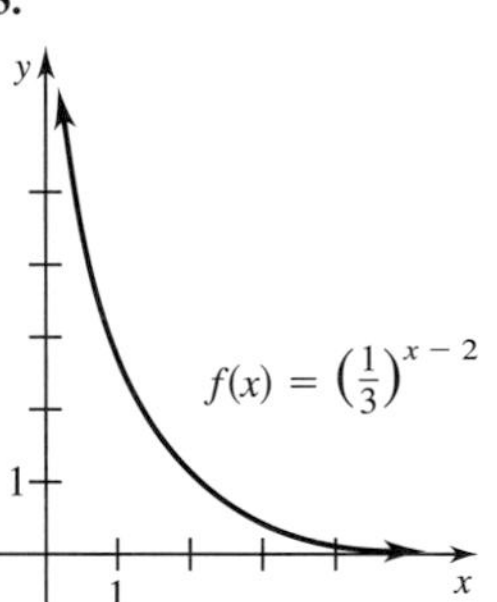

3.

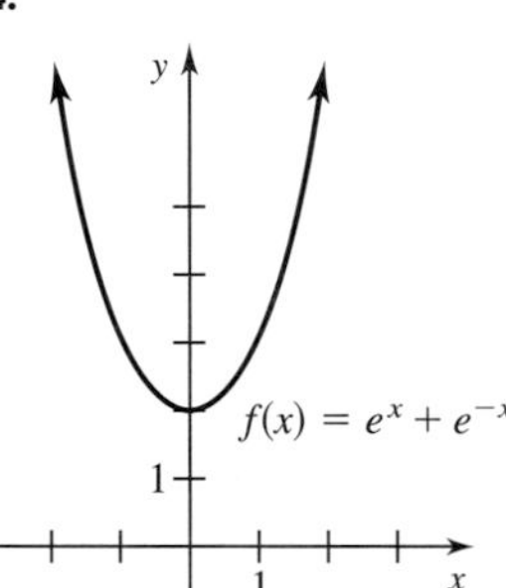

4.

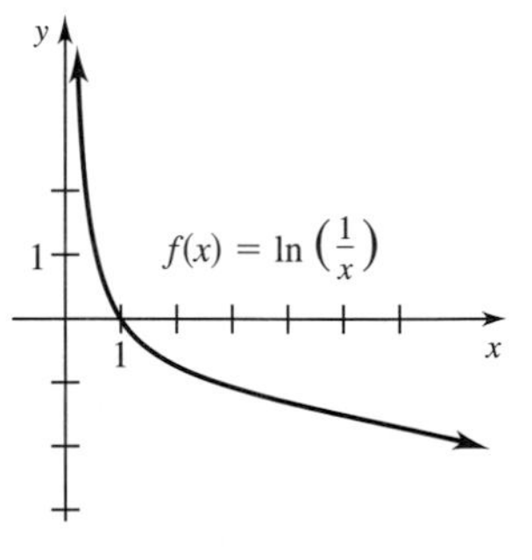

5.

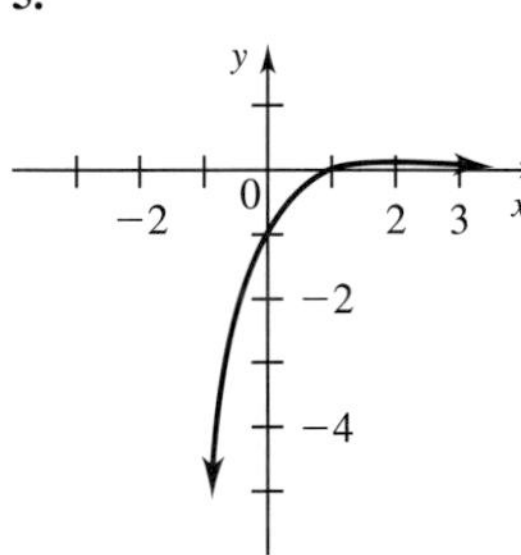

6.

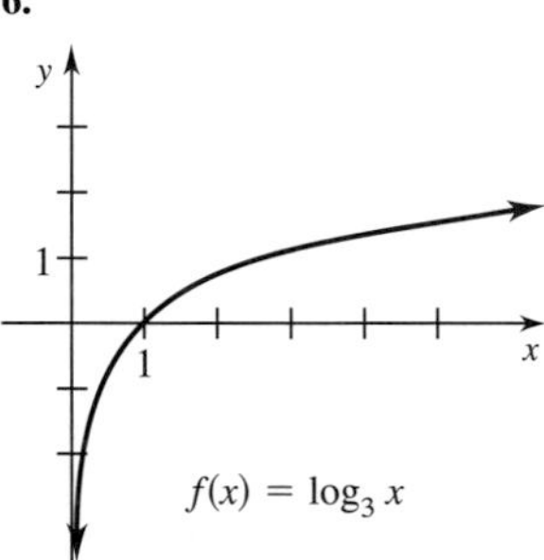

7.

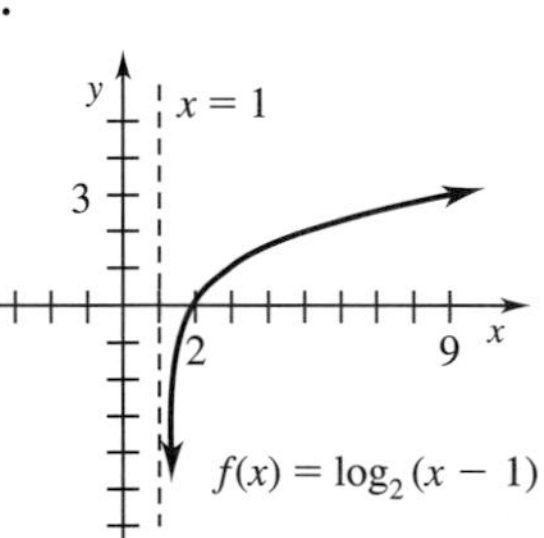

8.

9. 5/4 **10.** $-1/2$ **11.** $-5/2$ **12.** 1/4 **13.** 800 g **14.** ≈ 650 g **15.** ≈ 540 g **16.** \$1326.21 **17.** \$1465.68 **18.** \$3261.94 **19.** about 12 yr **20.** about 35 yr **21.** $\log_2 256 = 8$ **22.** $\log_{100} 10 = 1/2$ **23.** $\log_{1/16} (1/2) = 1/4$ **24.** $\log_{2/3} (9/4) = -2$ **25.** $\ln 1.1052 = .1$ **26.** $\log_2 5 = 2.322$ **27.** $\log 3 = .4771$ **28.** $\ln 12 = 2.4849$ **29.** $2^{5/2} = \sqrt{32}$ **30.** $e^{3.806662} = 45$ **31.** 3 **32.** 2 **33.** $\log_3 m + \log_3 n - \log_3 p$ **34.** $(1/3) \log_3 7 - \log_3 4$ **35.** $\log_3 49 + 2 \log_3 x + (2/3) \log_3 m + (1/3) \log_3 p$ **36.** cannot be rewritten using the properties of logarithms **37.** -2.7183 **38.** .0400 **39.** 1.7842 **40.** 2.2553 **41.** -3.2403 **42.** 2.6080 **43.** 1.490 **44.** -2.222 **45.** 1.386 **46.** $-.862$ **47.** .747 **48.** 1.609 **49.** 7.071 **50.** 8 **51.** 2 **52.** 81 **53.** 2

54. 17.520, −19.520 **55.** $x = [\ln(3 + \sqrt{10})]/\ln 5$ **56.** $x = -.485$ **57.** $x = \dfrac{1+a}{1-a}$ **58.** $t = \dfrac{1}{n}\ln\left|\ln\dfrac{r}{r_0}\right|$ **59.** $c = de^{(N-a)/b}$
60. $t = (\ln[P/(k-P)])/r$ **61.** $15 million **62.** $15.6 million **63.** $16.4 million **64.** $16.56 million **65.** **(a)** .0054 g/l **(b)** .00073 g/l **(c)** .000013 g/l **(d)** .75 ml
66. **(a)** 207 **(b)** 235 **(c)** 249 **(d)**

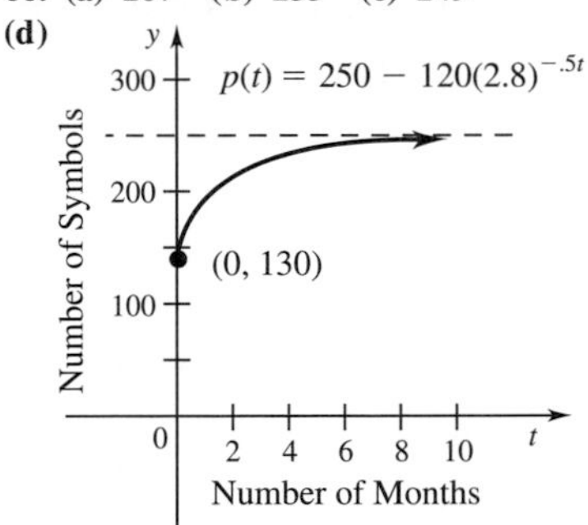

67. **(a)** B **(b)** D **(c)** C **(d)** A **68.** by a factor of $2^{1/2} \approx 1.4$ **69.** by a factor of $2^5 = 32$ **70.** domain: $(5, \infty)$; range: $(-\infty, \infty)$ **71.** domain: $(-\infty, -1) \cup (2, \infty)$; range: $(-\infty, \infty)$ **72.** 10 sec **74.** $14,038.99 **75.** $16,095.74 **76.** $580,792.63 **77.** .0719, .0725 **78.** **(a)** about 15,800 **(b)** about 7.6 yr **79.** The correct statement is $\log_5 125 - \log_5 25 = \log_5 \dfrac{125}{25} = \log_5 5 = 1$.

Chapter 6 Systems of Equations and Inequalities

Section 6.1 (page 303)

1. (48, 8) **3.** (−1, −11) **5.** (−1, 3) **7.** (3, −4) **9.** (0, 2) **11.** (3, −2) **13.** (1, 3) **15.** (4, −2) **17.** (2, −2) **19.** (5, 15) **21.** (4, 3) **23.** (5, 2) **25.** (−2, −2, 0) **27.** (1, 2, −1) **29.** (1, 0, 3) Each part of Exercise 31 has many correct answers. We give one example for each part. **31.** **(a)** $x + y = 1$ **(b)** $2x + 3y = 5$ **(c)** $4x + 6y = 12$ **33.** yes **35.** goats cost $30, sheep cost $25 **37.** 32 days at $14 **39.** $5000 at 4%, $15,000 at 6% **41.** 18 liters of 70%, 12 liters of 20% **43.** 200 gal **45.** 15 cm, 12 cm, 6 cm **47.** 37 **51.** (−1, −7), (4, 3)

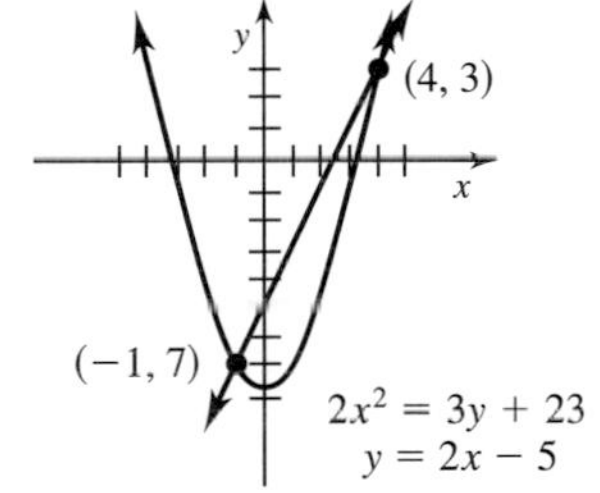

53. (1, 1), (−2, 4)

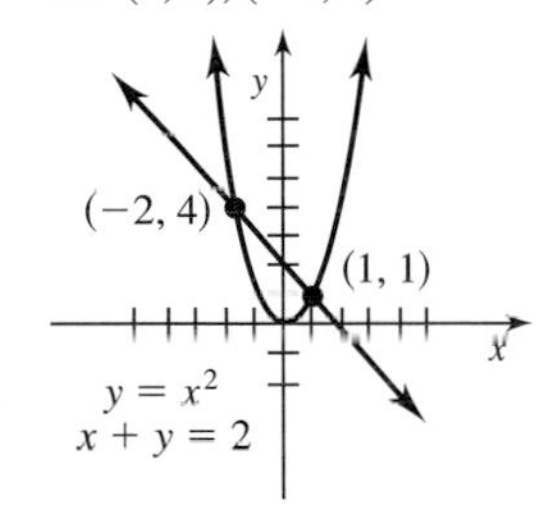

55. (−4, 1), (−5/2, 1/4)

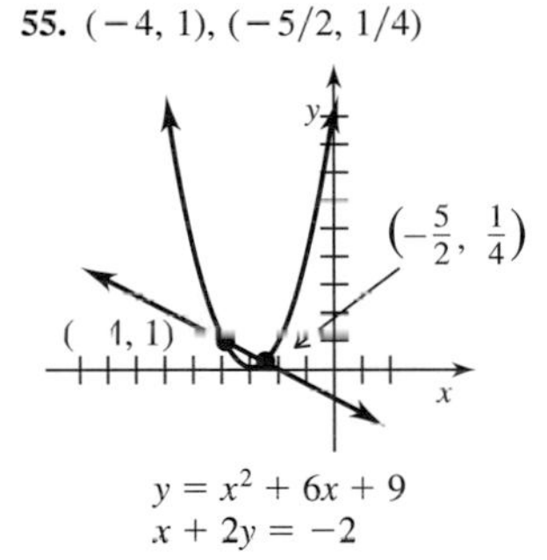

57. (0, 2), (2, 4) **59.** (2, 1), (1/3, 4/9) **61.** (−3/5, 7/5), (−1, 1) **63.** (3, 4), (3, −4), (−3, 4), (−3, −4) **65.** (0, 0) **67.** $\{(x, y) | x^2 + y^2 = 10\}$ **69.** no solution **71.** (2, 3), (3, 2) **73.** (−3, −1) **75.** (4, −1/8), (−2, 1/4) **77.** 14 and 3 **79.** 8 and 6, 8 and −6, −8 and 6, −8 and −6 **81.** $a < 0$ **83.** $\pm 3\sqrt{5}$ **87.** (.820, −2.51) **89.** (.892, .453) **91.** none **93.** (2, 3)

Section 6.2 (page 313)

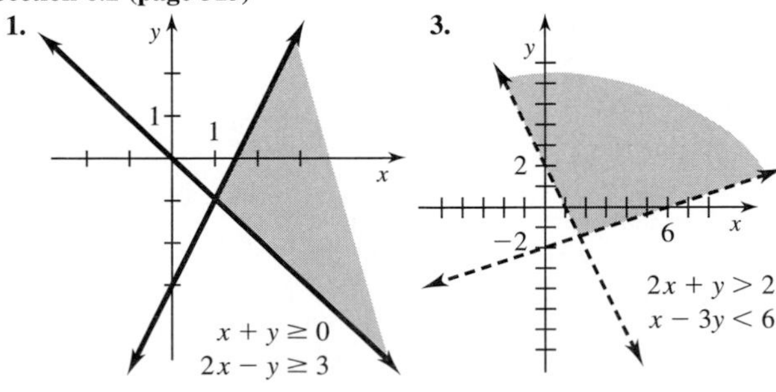

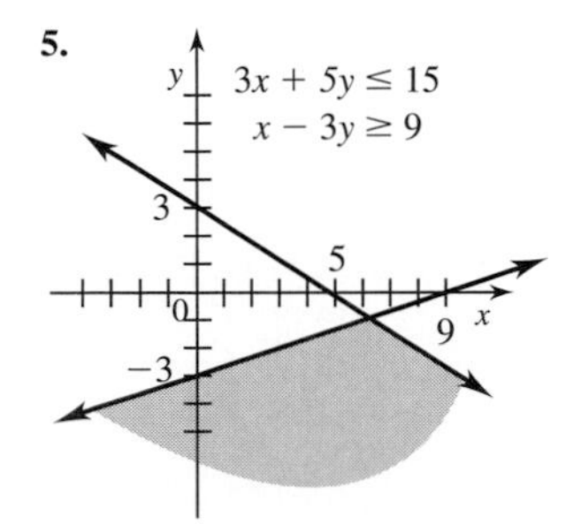

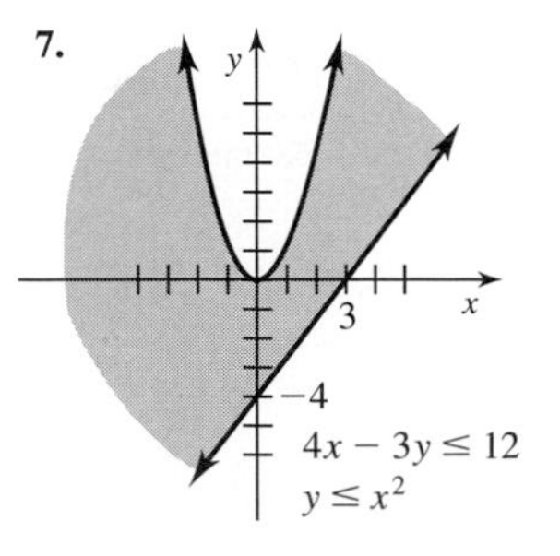

9.

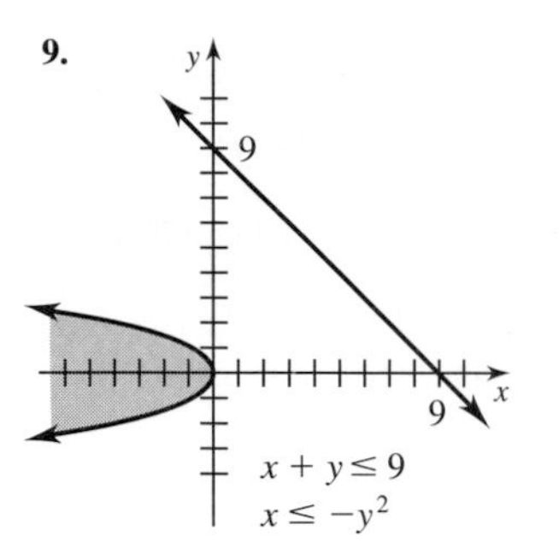

11.

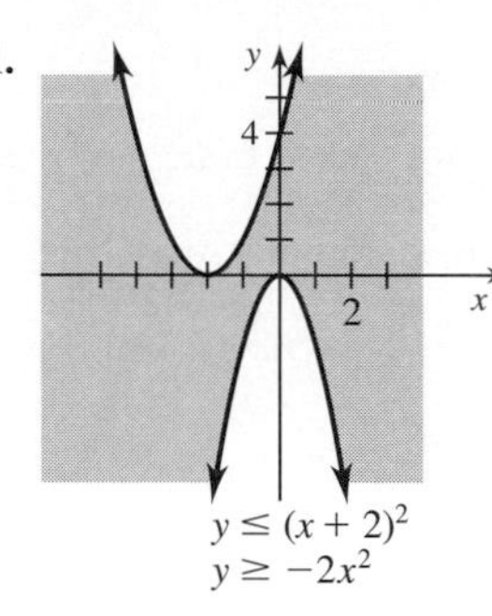

13.

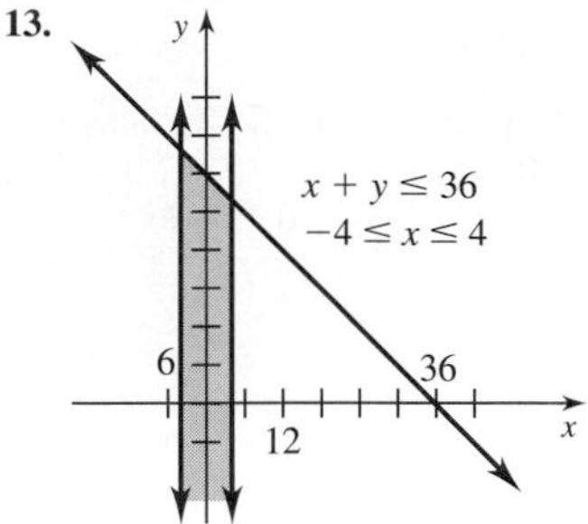

15.

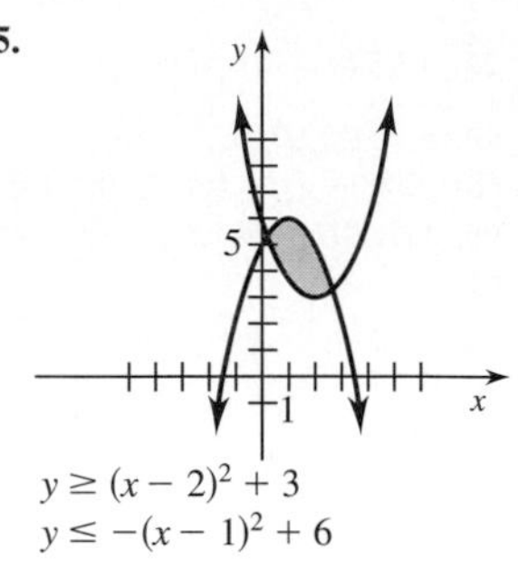

17.

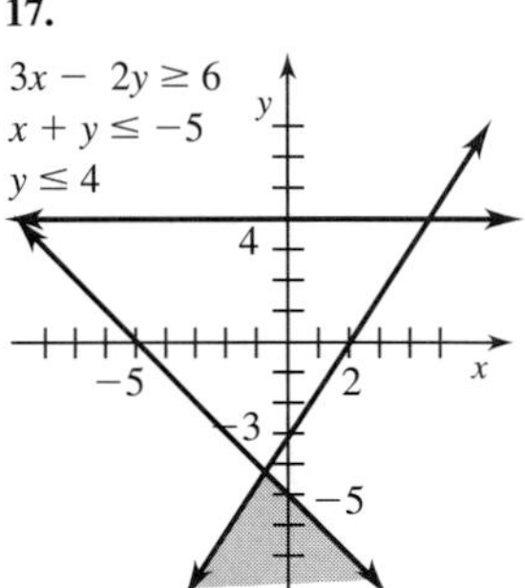

19.

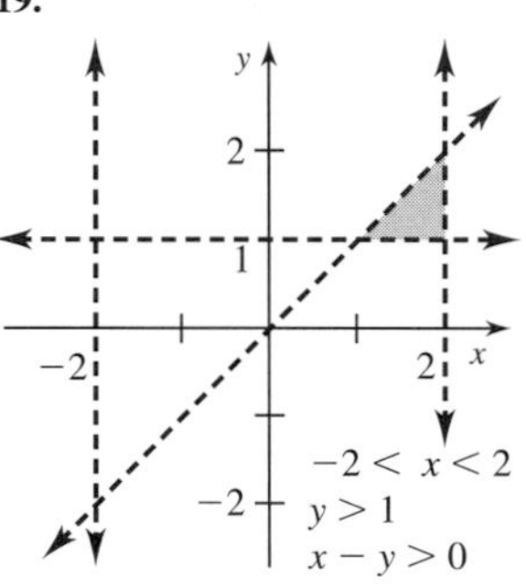

21.

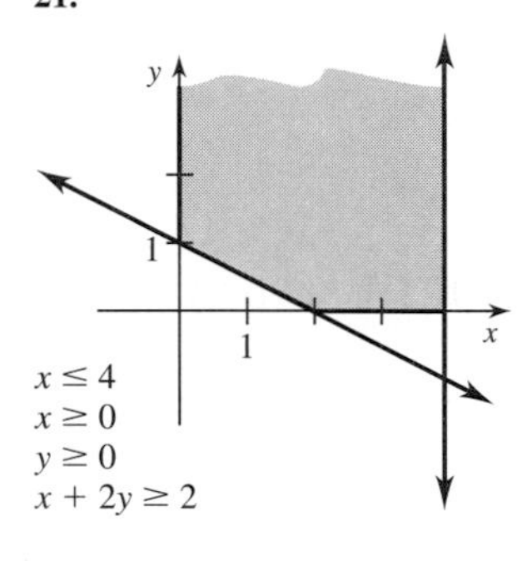

23.

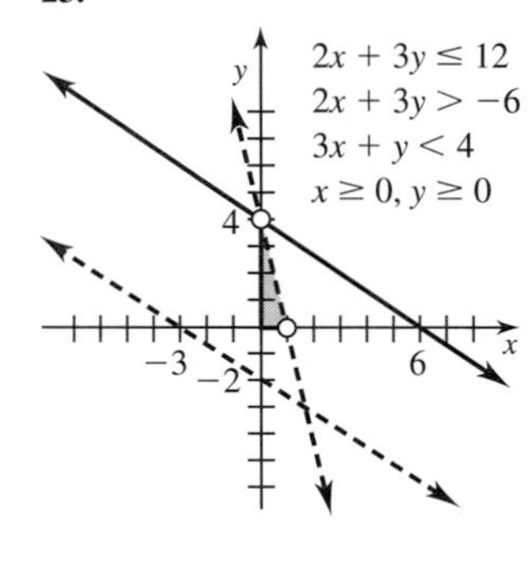

25.

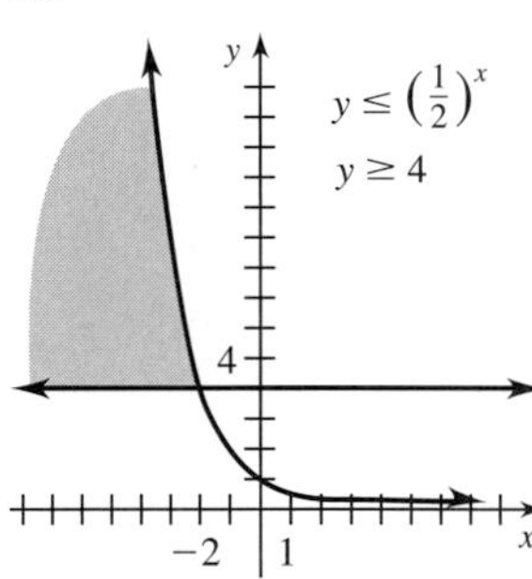

27.

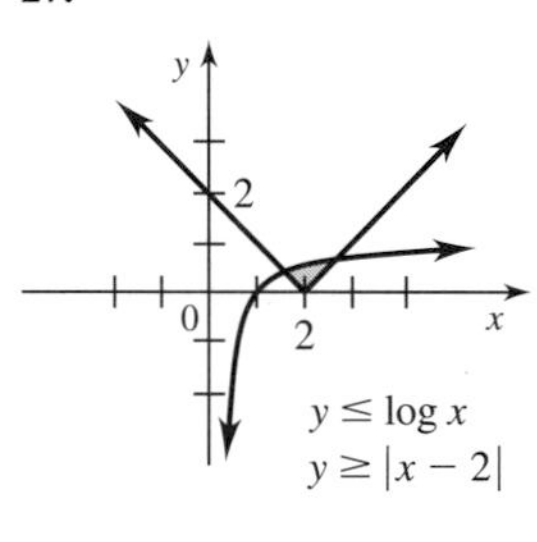

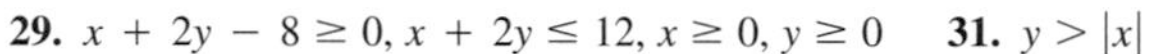

29. $x + 2y - 8 \ge 0$, $x + 2y \le 12$, $x \ge 0$, $y \ge 0$ **31.** $y > |x|$ **33.** maximum of 55 at (9, 1); minimum of 8 at (1, 2) **35.** maximum of 2460 at (6, 18); minimum of 0 at (0, 0) **37.** maximum of 80 at (7, 9); minimum of 5 at (1, 0) **39.** maximum of 15 at (7, 6); minimum of -7 at (1, 10) **41.** minimum of 25/3 at (10/3, 5/3) **43.** maximum of 90 at (10, 80/3) **45.** Let x = number of Brand X pills and y = number of Brand Y pills. Then $3000x + 1000y \ge 6000$, $45x + 50y \ge 195$, $75x + 200y \ge 600$, $x \ge 0$, $y \ge 0$.

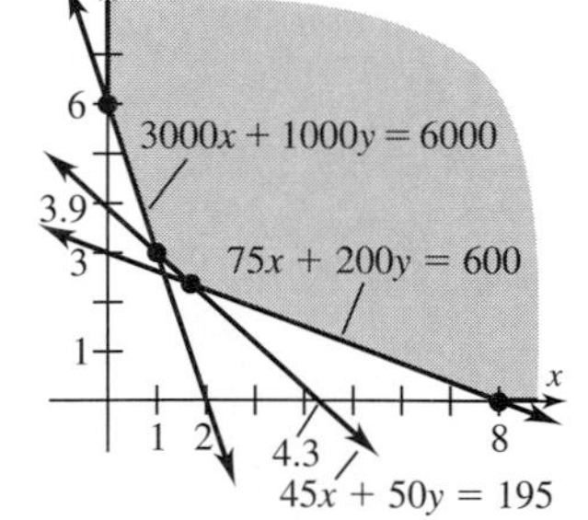

47. 6.4 million gal of gasoline and 3.2 million gal of fuel oil, for maximum revenue of \$16,960,000 **51.** maximum of 19.78 at (3.10, 5.24) **53.** maximum of 102 at (0, 8.5)

Section 6.3 (page 324)

1. $\left[\begin{array}{cc|c} 3 & 4 & 15 \\ -4 & 2 & 5 \end{array}\right]$ **3.** $\left[\begin{array}{cc|c} 3 & 5 & -13 \\ 2 & 3 & -9 \end{array}\right]$ **5.** $\left[\begin{array}{cc|c} 2 & 7 & 1 \\ 5 & 0 & -15 \end{array}\right]$ **7.** $\left[\begin{array}{ccc|c} 4 & -2 & 3 & 4 \\ 3 & 5 & 1 & 7 \\ 5 & -1 & 4 & 7 \end{array}\right]$ **9.** $\left[\begin{array}{ccc|c} 1 & 0 & 0 & 6 \\ 0 & 1 & 2 & 2 \\ 1 & 0 & -3 & 6 \end{array}\right]$
11. $x - 5y = -18, 6x + 2y = 20$ **13.** $x + y + z = 0, y = 6, z = 3$ **15.** $3x + y + 4z = 12, -2x + 3y = 5, 4y + 7z = 11$
17. $(-3, 4)$ **19.** $(x, 3 - 2x)$ **21.** $(2, -7)$ **23.** $(1, 1)$ **25.** $(-1, 23, 16)$ **27.** $(2, 4, 5)$ **29.** $(1/2, 1, -1/2)$ **31.** $(2, 1, -1)$
33. no solution **35.** no solution **37.** $\left(\frac{13}{11} + \frac{2}{11}z, \frac{7}{11}z + \frac{18}{11}, z\right)$ **39.** $\left(2 - \frac{5}{2}z, 3 - z, z\right)$ **41.** $(0, 2, -2, 1)$ **43.** wife, 40 days; husband, 32 days **45.** 9.6 cm^3 of 7%, 30.4 cm^3 of 2% **49.** 44.4 g of A, 133.3 g of B, 222.2 g of C **51. (a)** $x_3 + x_4 = 600$
(b) $\left[\begin{array}{cccc|c} 1 & 0 & 0 & 1 & 1000 \\ 1 & 1 & 0 & 0 & 1100 \\ 0 & 1 & 1 & 0 & 700 \\ 0 & 0 & 1 & 1 & 600 \end{array}\right]$, $\left[\begin{array}{cccc|c} 1 & 0 & 0 & 1 & 1000 \\ 0 & 1 & 0 & -1 & 100 \\ 0 & 0 & 1 & 1 & 600 \\ 0 & 0 & 0 & 0 & 0 \end{array}\right]$ **(c)** $x_4 = 1000 - x_1, x_4 = x_2 - 100, x_4 = 600 - x_3$ **(d)** 1000 **(e)** 100
(f) 600, 600 **(g)** $x_1 = 1000, x_2 = 100, x_3 = 600, x_4 = 0$; no **53.** $A = 1, B = 4$ **55.** $A = 4/3, B = 2/3$ **57.** $(11.311, 3.139)$
59. $(.571, 7.041, 11.442)$

Section 6.4 (page 334)

1. $w = 3, x = 2, y = -1, z = 4$ **3.** $m = 8, n = -2, z = 2, y = 5, w = 6$ **5.** $a = 2, z = -3, m = 8, k = 1$ **7.** not equal
9. $\begin{bmatrix} 6 & 6 \\ -12 & 9 \end{bmatrix}$ **11.** $\begin{bmatrix} 7 & -13 \\ 4 & -8 \\ -1 & 16 \end{bmatrix}$ **13.** $\begin{bmatrix} -k - 14y \\ 4z - 8x \\ -2k - a \\ -8m + 4n \end{bmatrix}$ **15.** $\begin{bmatrix} 4 & 2 & 4 & 1 \\ 2 & 1 & 5 & 4 \end{bmatrix}$; $\begin{bmatrix} 4 & 2 \\ 2 & 1 \\ 4 & 5 \\ 1 & 4 \end{bmatrix}$ **17.** $\begin{bmatrix} -4 & 8 \\ 0 & 6 \end{bmatrix}$ **19.** $\begin{bmatrix} 2 & 6 \\ -4 & 6 \end{bmatrix}$
21. $\begin{bmatrix} -1 & -3 \\ 2 & -3 \end{bmatrix}$ **23.** $\begin{bmatrix} 13 \\ 25 \end{bmatrix}$ **25.** $\begin{bmatrix} -6 & 27 \\ -4 & 17 \end{bmatrix}$ **27.** $\begin{bmatrix} 20 & 0 & 8 \\ 8 & 0 & 4 \end{bmatrix}$ **29.** cannot be multiplied **31.** [7] **33.** $\begin{bmatrix} 2 \\ 1 \end{bmatrix}$
35. (a) $\begin{bmatrix} 47.5 & 57.75 \\ 27 & 33.75 \\ 81 & 95 \\ 12 & 15 \end{bmatrix}$ **(b)** [20 200 50 60], [220 890 105 125 70] **(c)** [11,120 13,555] **37.** true **39.** true **41.** true
43. true **45.** true **47.** true **49.** false **51.** false **55.** $\begin{bmatrix} 46.3 & 24.2 & 0 \\ 13.125 & -41.5 & 7.6 \\ -10.76 & -4.67 & 9.75 \end{bmatrix}$ **57.** $\begin{bmatrix} -132.4 & 12.4 & -4 \\ 20.625 & 132 & 4.2 \\ 33.48 & -10.99 & 15.75 \end{bmatrix}$
61. [1/3 5/12 1/4]

Section 6.5 (page 346)

1. -34 **3.** -68 **5.** $yx + 6$ **7.** $3n - 8m$ **9.** $2, -6, 4$ **11.** $-6, 0, -6$ **13.** 1 **15.** 2 **17.** 0 **19.** $6 - 4y$
21. $-10i - 2j - 4k$ **23.** -40 **25.** $-.051$ **27.** Two rows are identical. **29.** One column is all zeros. **31.** One column is all zeros.
33. Multiply second row by 7, then two rows are identical. **35.** Rows and columns are interchanged. **37.** Two columns are interchanged.
39. Each element of second column is multiplied by $-1/2$. **41.** Elements of first row are multiplied by 1; products are added to elements of second row. **43.** Elements of first column are multiplied by constant k; products are added to third column. **45.** 0 **47.** 0 **49.** 12
51. 16 **53.** -6 **63. (a)** 1 **(b)** 5/2 **(c)** 19/2 **69.** 0

Section 6.6 (page 354)

1. $(1, 2)$ **3.** $(3, 2)$ **5.** $(3/2, 5/2)$ **7.** $(-2, 1)$ **9.** $D = 0$; no solution **11.** $(-3, 4, 1)$ **13.** $(3, -1, 2)$ **15.** $(-4, 3, 5)$
17. $(4, 0, 0)$ **19.** $D = 0$; no solution **21.** $(5, -4, -11)$ **23.** $(1, -2, 0)$ **25.** $(24/19, 63/38, 26/19)$ **27.** $(0, -2, 1, 3)$
33. $(1.727, .3347)$ **35.** $(.0311, .1229, .6828)$

Section 6.7 (page 372)

1. yes **3.** no **5.** no **7.** yes **9.** $\begin{bmatrix} -1/5 & -2/5 \\ 2/5 & -1/5 \end{bmatrix}$ **11.** $\begin{bmatrix} 2 & 1 \\ -3/2 & -1/2 \end{bmatrix}$ **13.** no inverse **15.** $\begin{bmatrix} -1 & 1 & 1 \\ 0 & -1 & 0 \\ 2 & -1 & -1 \end{bmatrix}$
17. $\begin{bmatrix} 7 & -3 & -3 \\ -1 & 1 & 0 \\ -1 & 0 & 1 \end{bmatrix}$ **19.** $\begin{bmatrix} -15/4 & -1/4 & -3 \\ 5/4 & 1/4 & 1 \\ -3/2 & 0 & -1 \end{bmatrix}$ **21.** $\begin{bmatrix} 1/2 & 0 & 1/2 & -1 \\ 1/10 & -2/5 & 3/10 & -1/5 \\ -7/10 & 4/5 & -11/10 & 12/5 \\ 1/5 & 1/5 & -2/5 & 3/5 \end{bmatrix}$ **23.** $(2, 0)$ **25.** $(40, -24)$
27. $(-1, -2/3)$ **29.** $(10, -1, -2)$ **31.** $(11, -1, 2)$ **33.** $(0, 1, -1)$ **35.** $\left(\frac{19}{20 - 2b}, \frac{-7b + 32}{20 - 2b}\right)$ **37.** $\left(\frac{b^2 + b}{3}, \frac{-b^2 + 2b}{3}\right)$

39. (1, 0, 2, 1) **47.** $ad - bc \neq 0$ **51.** $A^{-1} = A^2 = \begin{bmatrix} 1 & 0 & 0 \\ 0 & -1 & 1 \\ 0 & -1 & 0 \end{bmatrix}$ **55.** $\begin{bmatrix} .0543 & -.0543 \\ 1.8464 & .1536 \end{bmatrix}$ **57.** $(-3.542, -4.343)$

Section 6.8 (page 372)

1. $\frac{5}{3x} + \frac{-10}{3(2x+1)}$ **3.** $\frac{6}{5(x+2)} + \frac{8}{5(2x-1)}$ **5.** $\frac{5}{6(x+5)} + \frac{1}{6(x-1)}$ **7.** $\frac{-2}{x+1} + \frac{2}{x+2} + \frac{4}{(x+2)^2}$ **9.** $\frac{4}{x} + \frac{4}{1-x}$
11. $\frac{15}{x} + \frac{-5}{(x+1)} + \frac{-6}{(x-1)}$ **13.** $1 + \frac{-2}{x+1} + \frac{1}{(x+1)^2}$ **15.** $x^3 - x^2 + \frac{-1}{3(2x+1)} + \frac{2}{3(x+2)}$
17. $\frac{1}{9} + \frac{-1}{x} + \frac{25}{18(3x+2)} + \frac{29}{18(3x-2)}$ **19.** $\frac{-3}{5x^2} + \frac{3}{5(x^2+5)}$ **21.** $\frac{-2}{7(x+4)} + \frac{6x-3}{7(3x^2+1)}$
23. $\frac{1}{4x} + \frac{-8}{19(2x+1)} + \frac{-(9x+24)}{76(3x^2+4)}$ **25.** $\frac{-1}{x} + \frac{2x}{2x^2+1} + \frac{2x+3}{(2x^2+1)^2}$ **27.** $\frac{-1}{x+2} + \frac{3}{(x^2+4)^2}$
29. $5x^2 + \frac{3}{x} + \frac{-1}{x+3} + \frac{2}{x-1}$ **31.** correct **33.** incorrect

Chapter 6 Review Exercises (page 372)

1. (9, 4) **2.** (5, 2) **3.** (1, 1) **4.** $(48 - 2y, y)$, dependent equations **5.** (20, 10) **6.** (1/3, 1/2) **7.** (3, 2, 1) **8.** (5, −1, 0) **9.** 220 members, 80 nonmembers **10.** 1/3 cup of rice, 1/5 cup of soybeans **11.** 80 3 1/2″ diskettes, 20 5 1/4″ diskettes **12.** 5 blankets, 3 rugs, 8 skirts **13.** $(11/32 + (7/16)z, -23/32 + (13/16)z, z)$ **14.** $(x, 1 - 2x, 6 - 11x)$ **15.** (1, 2), (−3, 6) **16.** (3, 2), (3, −2), (−3, 2), (−3, −2) **17.** (−1, 2), (−2, 1) **18.** (−2, 0), (1, 1) **19.** $\pm 5\sqrt{10}$
20. yes; $\left(\frac{8 - 8\sqrt{41}}{5}, \frac{16 + 4\sqrt{41}}{5}\right), \left(\frac{8 + 8\sqrt{41}}{5}, \frac{16 - 4\sqrt{41}}{5}\right)$

21.

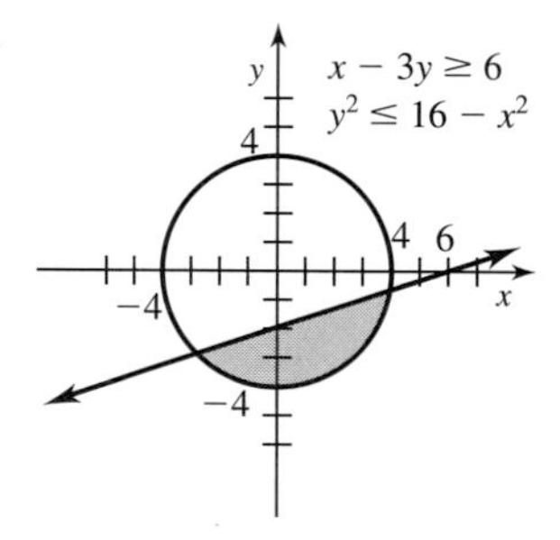

22.

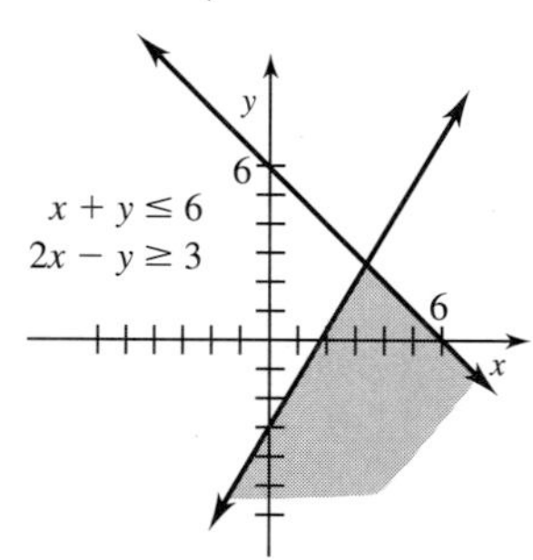

23.

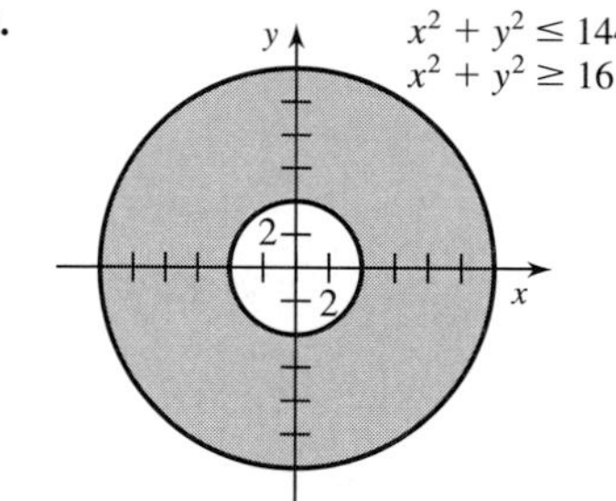

24. Let x = number of batches of cakes, y = number of batches of cookies. Then $2x + (3/2)y \leq 16$; $3x + (3/4)y \leq 12$; $x \geq 0$; $y \geq 0$.

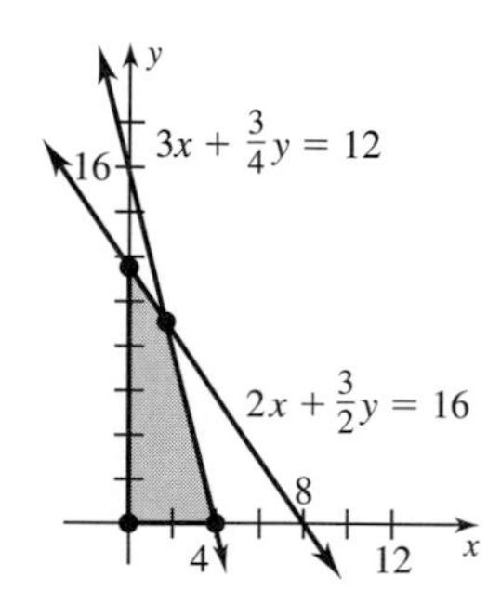

25. Let x = number of kg of first mix, y = number of kg of second mix. Then $(1/2)x + (1/3)y \leq 100$, $(1/2)x + (2/3)y \leq 125$, $x \geq 0$, $y \geq 0$.

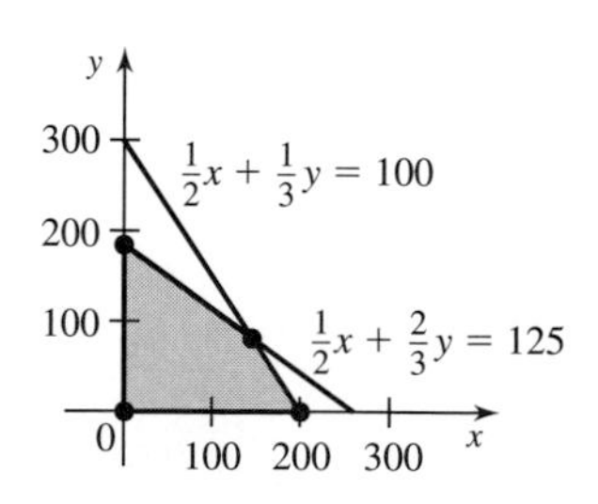

26. There is a maximum of 42 at (12, 6). **27.** There is a maximum of 210 at (0, 30). **28.** There is a minimum of 12 at (3, 0) or at (6/5, 12/5) or at any point on the line joining these points. **29.** There is a minimum of 92/11 at (12/11, 40/11). **30.** The maximum profit is $220, which occurs when 2 batches of cakes and 8 batches of cookies are made. **31.** 150 kg of the half-and-half, 75 kg of the other for a maximum revenue of $1260 **32.** (−2, 0) **33.** (−4, 6) **34.** no solution **35.** (0, 1, 0) **36.** $a = 5$, $x = 3/2$, $y = 0$, $z = 9$ **37.** $k = -8$, $y = 7$, $a = 2$, $m = 3$, $p = 1$, $r = -5$

38. $\begin{bmatrix} -4 \\ 6 \\ 1 \end{bmatrix}$ **39.** $\begin{bmatrix} 0 & -2 & 7 \\ 6 & -1 & 10 \end{bmatrix}$

40. $\begin{bmatrix} 11 & 20 \\ 14 & 40 \end{bmatrix}$ **41.** not possible **42.** $\begin{bmatrix} -3 \\ 10 \end{bmatrix}$

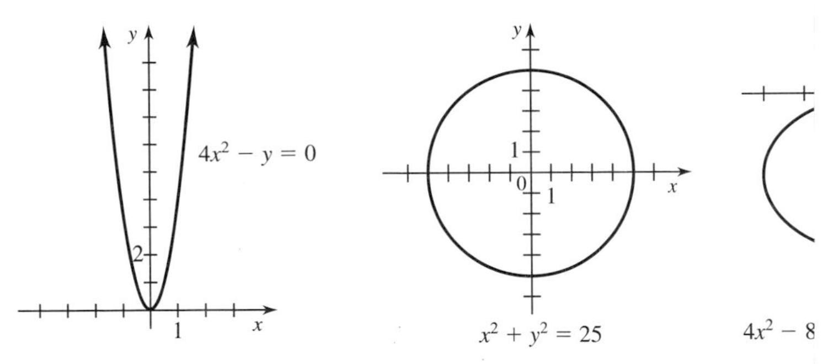

43. $(x - 2)^2/16 + y^2/12 = 1$ **44.** $(y - 2)^2 - x^2/3 = 1$ **45.** 66.8 and 67.
47. $x = (1/12)(y - 2)^2$ **48.** $x = (1/3)(y - 2)^2 - 3$ **49.** $x^2/25 + y^2/21 =$
52. $y^2/72 - x^2/72 = 1$

Chapter 8 Further Topics in Algebra

Section 8.1 (page 415)
1. 14, 18, 22, 26, 30 **3.** 1, 2, 4, 8, 16 **5.** 0, 1/9, 2/27, 1/27, 4/243 **7.** −
13. −2, 1, 4, 7, 10, 13 **15.** 1, 1, 2, 3, 5, 8 **17.** $-1 + 1 + 3 + 5 + 7$ **1**

23. $-3.5 + .5 + 4.5 + 8.5$ **25.** $0 + 4 + 16 + 36$ **27.** $-1 - 1/3 - 1/$

33. $\sum_{j=0}^{10} [(j - 1)^2 - 2(j - 1)]$ or $\sum_{j=0}^{10} (j^2 - 4j + 3)$ **35.** 90 **37.** 220 **3**
43. $9 + [3(n + 1)]/n + [(n + 1)(2n + 1)]/(6n^2)$ **45.** converges to 1/2 **47**

Section 8.2 (page 422)
1. 5, 3, 1, −1, −3, −5 **3.** 13, 10, 7, 4 **5.** 3 $\sqrt{2}$, 3, 3 + $\sqrt{2}$, 3 + $2\sqrt{}$
11. $\sqrt{3}$; $1 + \sqrt{3}n$ **13.** 1; $x - 3 + n$ **15.** 48; $-2 + 5n$ **17.** 44; $-6 +$
23. 3 **25.** 5 **29.** 260 **31.** −25 **33.** 77.8 **35.** 159.77 **37.** 828 **3**
47. 4 **49.** n^2 **51.** 96 **53.** $32,000 **55.** the sixth week; 700 mln **57.** −

Section 8.3 (page 431)
1. 5/3, 5, 15, 45 **3.** 5/8, 5/4, 5/2, 5, 10 **5.** 80; $5(-2)^{n-1}$ **7.** −108; $(-$
$-(-3)^{n+1}$ **11.** −324; $-4(3)^{n-1}$ **13.** 125/4; $(4/5)(5/2)^{n-1}$ or $5^{n-2}/2^{n-3}$
21. −2; 1/2 **23.** 682 **25.** 99/8 **27.** 860.95 **29.** 363 **31.** 189/4 **3**
43. 256 **45.** 3/20 **47.** 4 **49.** 1/9 **51.** .000016% **53.** $1/32 \times 10^{15}$ o
57. 10,000/9 units **59.** 200 cm **61.** 62; 2046 **63.** 1/64 m **65.** $x = y =$

Section 8.4 (page 439)
1. 210 **3.** 1 **5.** 300 **7.** $m^{12} + 6m^{10}n + 15m^8n^2 + 20m^6n^3 + 15m^4n^4 +$
$405zw^4 - 243w^5$ **11.** $r^5/32 - 15r^4/16 + 45r^3/8 - 135r^2/2 + 405r/2 - 24$
$80x^{3/5}y^{16/5} + 32y^4$ **17.** $3^{20}m^{20} - 20 \cdot 3^{19}m^{19}n + 190 \cdot 3^{18}m^{18}n^2 - 1140 \cdot 3^{1}$
21. $-3432 \cdot 6^7c^7d^7$ **23.** $139{,}776x^5y^{22}$ **25.** $55 \cdot 3^9a^6b^9$ **27.** $462(2^6)(3^5)m^-$
33. 1.015 **35.** $1 - x + x^2 - x^3 + x^4 - \ldots$ **39.** −1

Section 8.5 (page 446)
1. S_1: $2 = 1(1 + 1)$; S_2: $2 + 4 = 2(2 + 1)$; S_3: $2 + 4 + 6 = 3(3 + 1)$; S

$10 = 5(5 + 1)$ **11.** $n = 1$ **13.** true for all positive integers n **15.** $n = 1,$

Section 8.6 (page 454)
1. 19,958,400 **3.** 72 **5.** 5 **7.** 6 **9.** 1 **11.** 495 **13.** 720 **15. (a)**
17. $2.052371412 \times 10^{10}$ **19.** 120; 30,240 **21.** 720; 151,200 **23.** 10,626
(b) 150 **39.** 210; 5,040 **41.** 35 **43. (a)** 220 **(b)** 55 **(c)** 105

43. $\begin{bmatrix} -9 & 3 \\ 10 & 6 \end{bmatrix}$ **44.** −6 **45.** −25 **46.** −44 **47.** −1 **48.** 138 **49.** Rows and columns are interchanged. **50.** One row is all zeros. **51.** Row 1 and row 3 are identical. **52.** Elements of first row are multiplied by 2. **53.** Columns 2 and 3 are interchanged. **54.** New row 2 is the sum of old rows 1 and 2. **55.** (−4, 2) **56.** (−2, 5) **57.** $(16/9, (8 - 9z)/18, z)$, dependent equations **58.** (−4, 6, 2)

59. $\begin{bmatrix} -1/4 & 1/6 \\ 0 & 1/3 \end{bmatrix}$ **60.** $\begin{bmatrix} 3 & -1 \\ -5 & 2 \end{bmatrix}$ **61.** no inverse **62.** $\begin{bmatrix} 2/3 & 0 & -1/3 \\ 1/3 & 0 & -2/3 \\ -2/3 & 1 & 1/3 \end{bmatrix}$ **63.** (2, 1) **64.** (−1, 0, 2) **65.** (−3, 5, 1)

66. $((6z + 5)/11, (31z + 2)/11, z)$ **69.** $\dfrac{2}{x} + \dfrac{-2}{x + 1} + \dfrac{-1}{(x + 1)^2}$ **70.** $\dfrac{-1}{4x} + \dfrac{x + 3}{4(x^2 - x + 4)}$ **71.** $\dfrac{-1}{9(x + 1)} + \dfrac{(x + 14)}{9(x^2 - 3x + 5)}$

72. $\dfrac{-1}{48(x + 4)} + \dfrac{x + 40}{48(x^2 - 4x + 16)}$

Chapter 7 Analytic Geometry

Section 7.1 (page 381)

1.

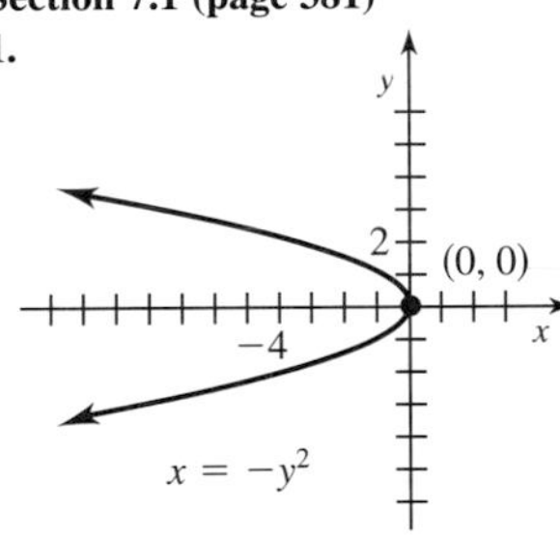

3.

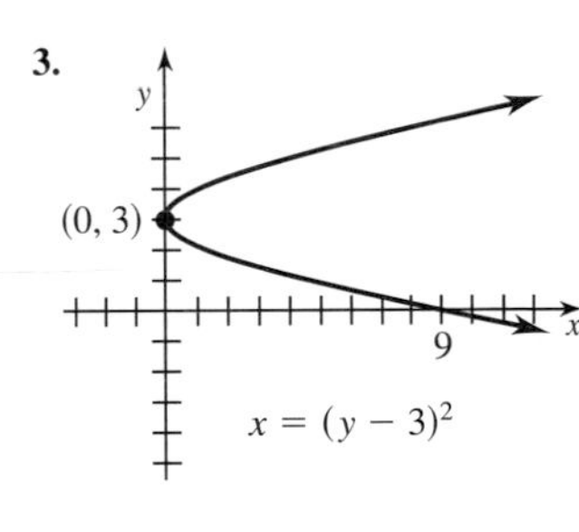

5.

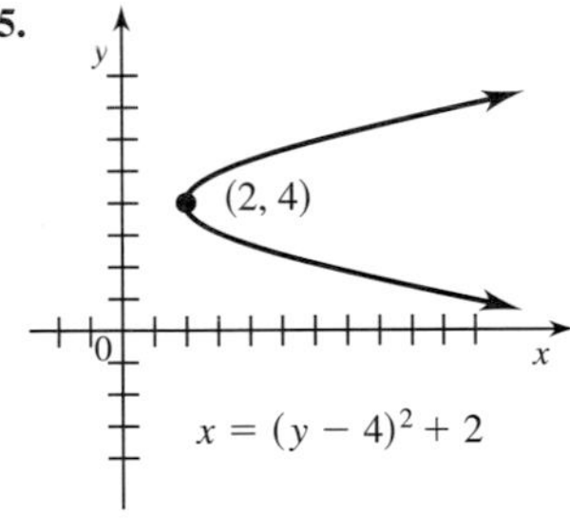

7.

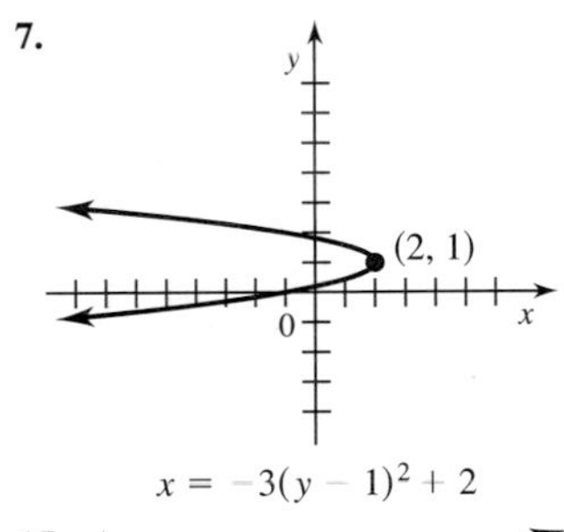

9.

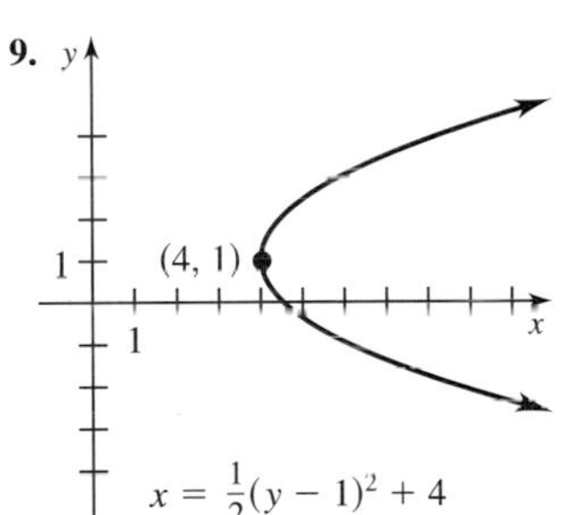

11.

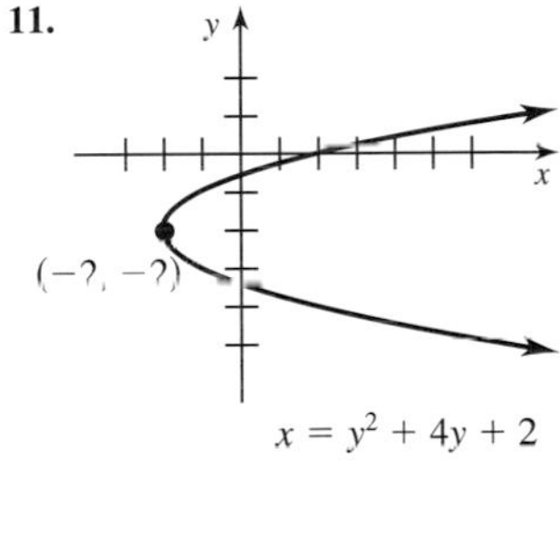

13.

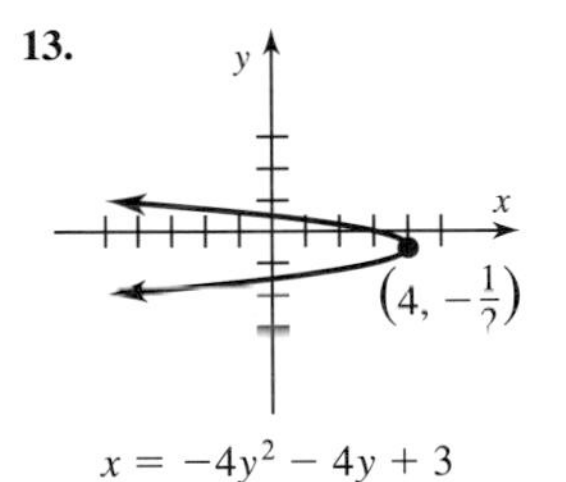

15.

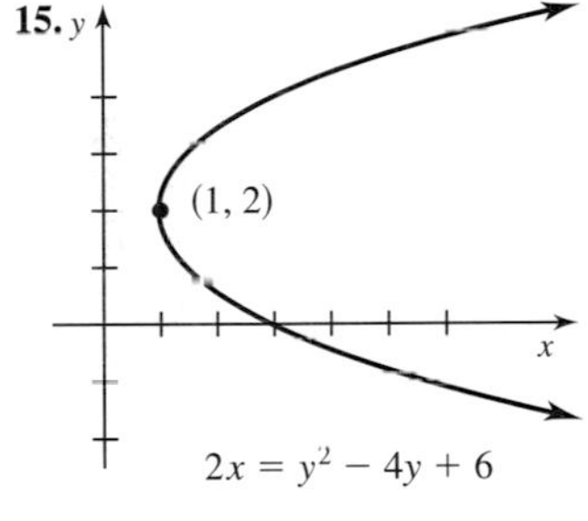

17. (0, 6), $y = -6$, y-axis **19.** (0, −1/16), $y = 1/16$, y-axis **21.** (−1/128, 0), $x = 1/128$, x-axis **23.** (−1, 0), $x = 1$, x-axis **25.** (4, 3), $x = -2$, $y = 3$ **27.** (7, −1), $y = -9$, $x = 7$ **29.** $y^2 = 20x$ **31.** $x^2 = y$ **33.** $x^2 = y$ **35.** $y^2 = (4/3)x$ **37.** $(x - 4)^2 = 8(y - 3)$ **39.** $(y - 6)^2 = 28(x + 5)$ **41.** 12 ft **43.** 6 ft **45.** $(y - 1)^2 = 4(x + 1)$ **55.** $(y - 5)^2 = (1/3)(x + 1)$

Section 7.2 (page 389)

1. (0, 0); (−5, 0), (5, 0); (0, −3), (0, 3); (−4, 0)(4, 0)

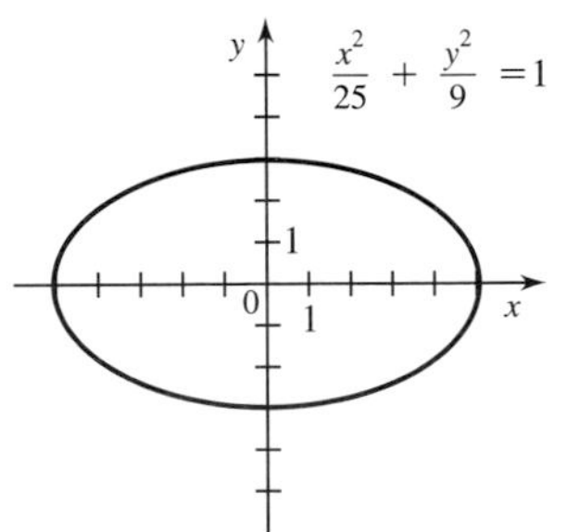

3. (0, 0); (3, 0), (−3, 0); (0, −1), (0, 1); $(-2\sqrt{2}, 0)$, $(2\sqrt{2}, 0)$

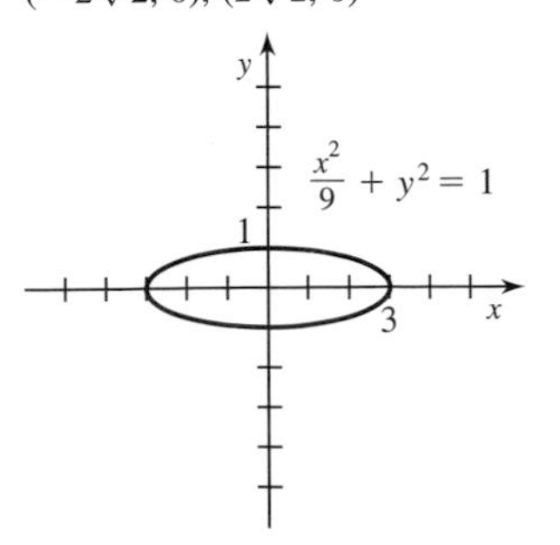

5. (0, 0); (0, −9)(0, 9); (−3, 0), (3, 0); $(0, -6\sqrt{2})$, $(0, 6\sqrt{2})$

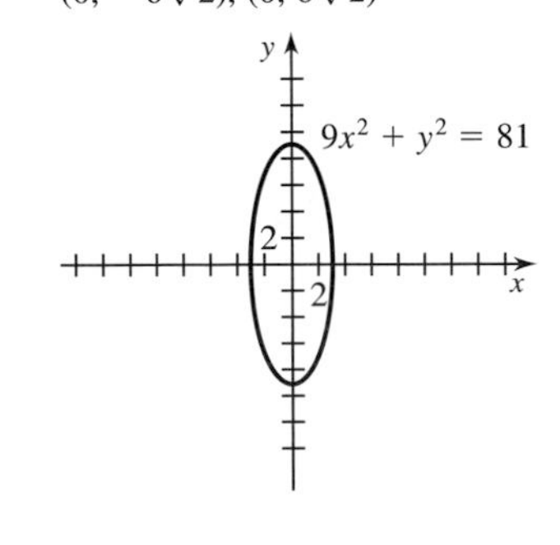

19. ellipse; $(0, -4/3)$, $(0, 4/3)$

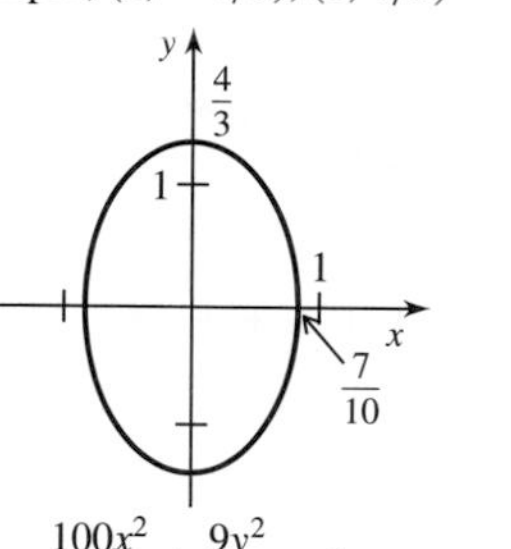

$$\frac{100x^2}{49} + \frac{9y^2}{16} = 1$$

20. ellipse; $(0, 5/2)$, $(0, -5/2)$

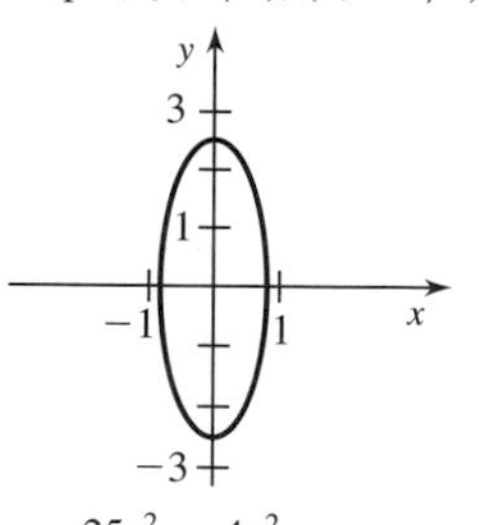

$$\frac{25x^2}{9} + \frac{4y^2}{25} = 1$$

21. ellipse;

$4x^2 +$

23. ellipse; $(1, -1)$, $(5, -1)$

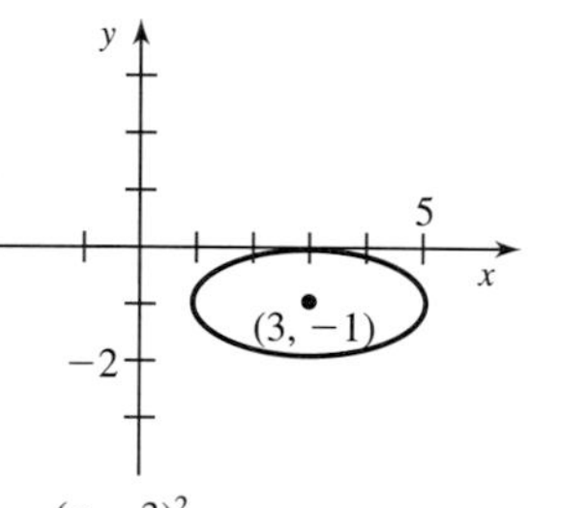

$$\frac{(x-3)^2}{4} + (y+1)^2 = 1$$

24. ellipse; $(5, -3)$, $(-1, -3)$

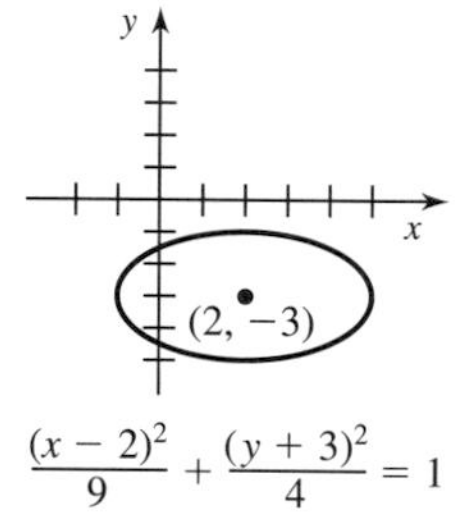

$$\frac{(x-2)^2}{9} + \frac{(y+3)^2}{4} = 1$$

25. hyperbo
$y = \pm(2$

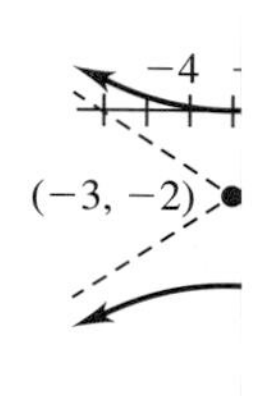

$\frac{(y+}{4}$

27. hyperbola; $(3, 2)$, $(-5, 2)$;
$y = \pm(1/2)(x+1) + 2$

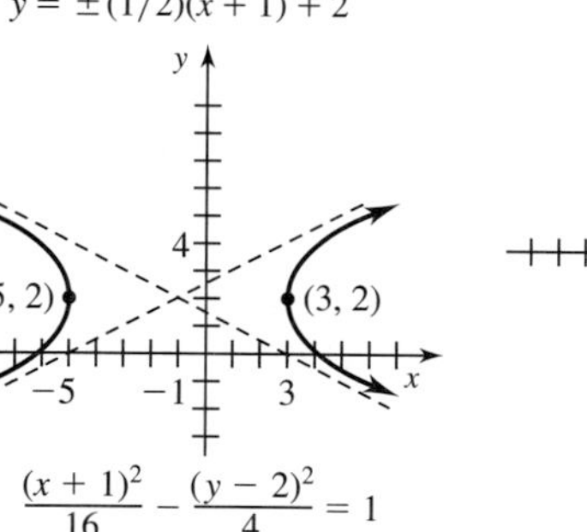

$$\frac{(x+1)^2}{16} - \frac{(y-2)^2}{4} = 1$$

28. semi-ellipse; $(0, -6)$, $(0, 6)$

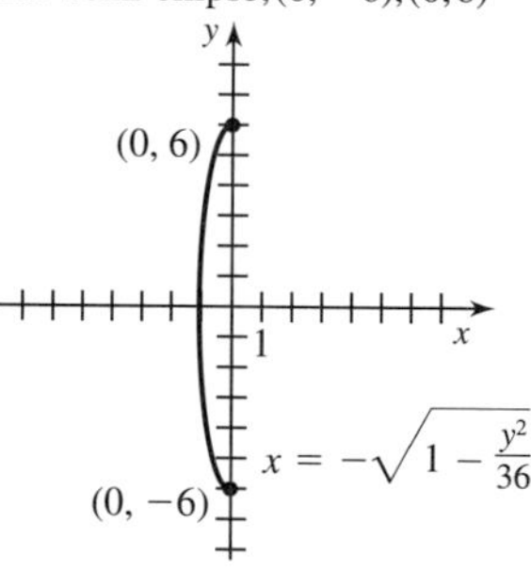
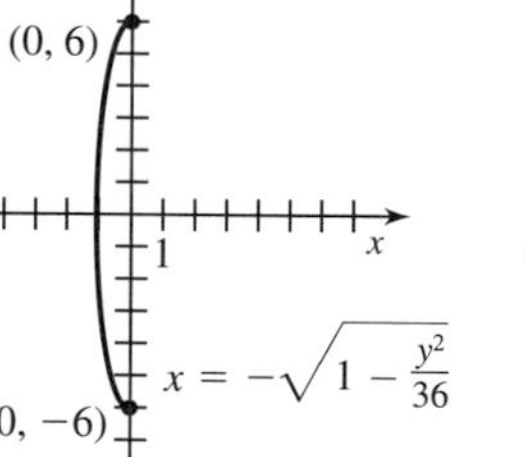

$$x = -\sqrt{1 - \frac{y^2}{36}}$$

29. semi-hy
$y = \pm x$

31. $x^2/12 + y^2/16 = 1$

32. $x^2/36 + y^2/32 = 1$

33. $y^2/16 -$

35. ellipse

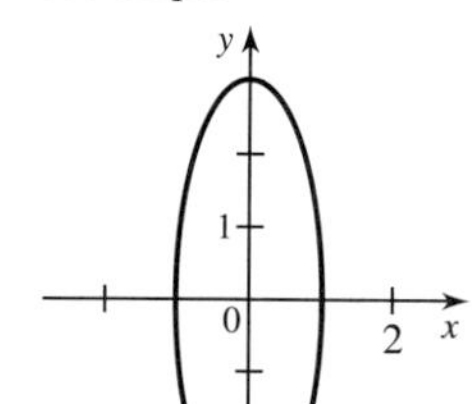

$y^2 + 9x^2 = 9$

36. hyperbola

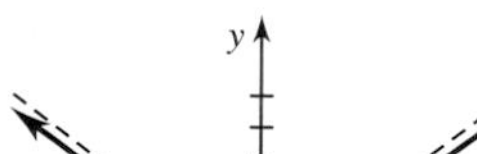
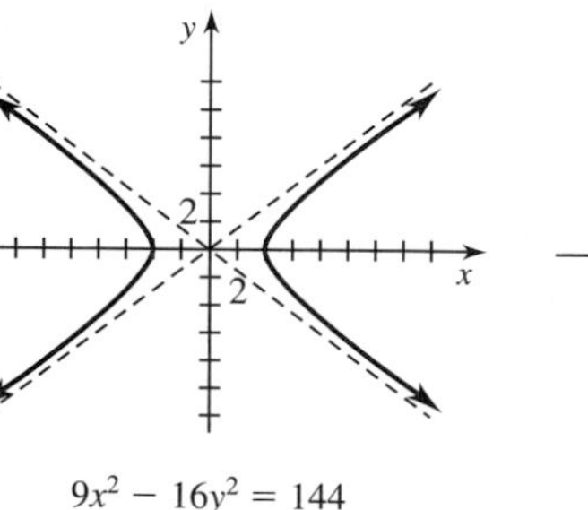

$9x^2 - 16y^2 = 144$

37. hyperbol

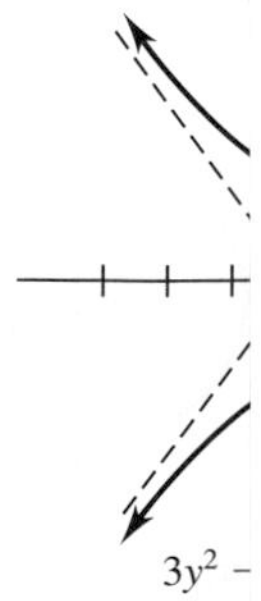

$3y^2 -$

Simplify the expressions in Exercises 23–46 by writing each with only positive exponents. Assume that all variables represent positive real numbers and that variables used as exponents represent rational numbers.

23. $(x^{2/3})(x^{5/3})$ $\mathbf{x^{7/3}}$

24. $(a^{4/5})(a^{2/5})$ $\mathbf{a^{6/5}}$

25. $(4^3)(4^{-3})(4^5)$ $\mathbf{4^5}$

26. $5^6(5^{-3})(5^4)$ $\mathbf{5^7}$

27. $(1+n)^{1/2}(1+n)^{3/4}$ $\mathbf{(1+n)^{5/4}}$

28. $(m+7)^{-1/6}(m+7)^{-2/3}$ $\mathbf{1/(m+7)^{5/6}}$

29. $(3p)^{-4}$ $\mathbf{1/(81p^4)}$

30. $(-5k)^{-2}$ $\mathbf{1/(25k^2)}$

31. $(2y^{3/4}z)(3y^{1/4}z^{-1/3})$ $\mathbf{6yz^{2/3}}$

32. $(4a^{-1/2}b^{2/3})(2a^{3/2}b^{-1/3})$ $\mathbf{8ab^{1/3}}$

33. $\dfrac{d^{-2}}{(d^8)(d^{-3})}$ $\mathbf{1/d^7}$

34. $\dfrac{(t^5)(t^{-3})}{t^{-7}}$ $\mathbf{t^9}$

35. $\dfrac{a^{4/3} \cdot b^{1/2}}{a^{2/3} \cdot b^{-3/2}}$ $\mathbf{a^{2/3}b^2}$

36. $\dfrac{x^{1/3}y^{2/3}}{x^{5/3}y^{-1/3}}$ $\mathbf{y/x^{4/3}}$

37. $\left(\dfrac{a^{-1}}{b^2}\right)^{-3}$ $\mathbf{a^3b^6}$

38. $\left(\dfrac{2c^2}{d^3}\right)^{-2}$ $\mathbf{d^6/(4c^4)}$

39. $\dfrac{(5x)^{-2}(x^{-3})^{-4}}{(25^{-1}x^{-3})^{-1}}$ $\mathbf{x^7/5^4}$

40. $\dfrac{(2k)^{-3}(k^{-5})^{-1}}{(6k^{-2})^{-1}(k^3)^{-6}}$ $\mathbf{3k^{18}/4}$

41. $\left(\dfrac{x^{13}(x^5)^{-7}}{(2x^7)^2}\right)^{-1/7}$ $\mathbf{2^{2/7}x^{36/7}}$

42. $\left(\dfrac{y^{2/5} \cdot y^{4/5}}{y^{1/5} \cdot y^{6/5}}\right)^{-5}$ $\mathbf{y}$

43. $5p^r(6p^{3-2r})$, $r < 3$ $\mathbf{30p^{3-r}}$

44. $(-z^{2r})(4z^{r+3})$ $\mathbf{-4z^{3r+3}}$

45. $\dfrac{(b^2)^{y+1}}{(2b^y)^3}$, $y > 2$ $\mathbf{1/(2^3b^{y-2})}$

46. $\dfrac{(3x^n)^3}{(x^2)^{n-1}}$ $\mathbf{27x^{n+2}}$

▲ **47.** Show that $(1+2^3)^{-1} + (1+2^{-3})^{-1} = 1$.

▲ **48.** Show that $(1+x^m)^{-1} + (1+x^{-m})^{-1} = 1$.

◉ **49.** If $a^7 = 30$, what is a^{21}? **27,000**

◉ **50.** If $a^{-3} = .2$, what is a^6? **25**

◉ **51.** If the lengths of the sides of a cube are tripled, by what factor will the volume change? **27**

◉ **52.** If the radius of a circle is doubled, by what factor will the area change? **4**

One important application of mathematics to business and management concerns supply and demand. Usually, as the price of an item increases, the supply increases and the demand decreases. By studying past records of supply and demand at different prices, economists can construct an equation that describes (approximately) supply and demand for a given item. Such equations are provided in Exercises 53 and 54.

53. The price in dollars of a graphing calculator is approximated by p, where $p = 5x^{2/3} + 2x^{1/3}$ and x is the number (in hundreds) of units supplied. Find the price when the supply is 6400. **$88**

54. The demand for a certain commodity and the price in dollars are related by $p = 500 - 10x^{1/2}$, where x is the number (in hundreds) of units of the product demanded. Find the price when the demand is 2500. **$450**

In our system of government, the president is elected by the electoral college and not by individual voters. Because of this, smaller states have a greater voice in the selection of a president than they would otherwise have. Two political scientists have studied the problems of campaigning for president under the current system and have concluded that candidates should allot their money according to the formula

$$\text{Amount for large state} = \left(\frac{E_{\text{large}}}{E_{\text{small}}}\right)^{3/2} \times \text{Amount for small state}.$$

Here E_{large} represents the electoral vote of the large state, and E_{small} represents the electoral vote of the small state. Find the amount that should be spent in the larger states in Exercises 55–58 if $1,000,000 is spent in the small state and the following statements are true.

55. The large state has 45 electoral votes and the small state has 5. **$27,000,000**

56. The large state has 40 electoral votes and the small state has 10. **$8,000,000**

57. There are six electoral votes in the small state; 28 in the large. **about $10,000,000**

58. There are nine electoral votes in the small state; 32 in the large. **about $6,700,000**

A Delta Airlines map gives a formula for calculating the visible distance from a jet plane to the horizon. On a clear day, this distance is approximated by $D = 1.22x^{1/2}$, where x is altitude in feet and D is distance to the horizon in miles. Find D for each of the following altitudes.

59. 5,000 ft **about 86.3 mi** **60.** 10,000 ft **about 122 mi** **61.** 30,000 ft **about 211 mi** **62.** 40,000 ft **about 244 mi**

The Galápagos Islands are a chain of islands ranging in size from 2 to 2249 sq mi. A biologist has shown that the number of different land-plant species on an island in this chain is related to the size of the island by $S = 28.6A^{.32}$, where A is the area of an island in square miles and S is the number of different plant species on that island. Estimate S (rounding to the nearest whole number) for islands with the following areas.

63. 100 sq mi **about 125** **64.** 500 sq mi **about 209** **65.** 300 sq mi **about 177** **66.** 2000 sq mi **326**

1.3 Exercises (text page 28)

Perform the indicated operations.

1. $(p^3 - 3p) + (2p^2 + 4p - 1) + (4p^3 - 3)$ $\mathbf{5p^3 + 2p^2 + p - 4}$

2. $(5k^4 - 2k^2 + k) - (3k^3 + k^2 - 4) + 2k^4$ $\mathbf{7k^4 - 3k^3 - 3k^2 + k + 4}$

3. $(2m^2 - 6m + 7) + (2m^2 - 4m - 3) - (m^2 + 5)$ $\mathbf{3m^2 - 10m - 1}$

4. $(6x^3 + 2x^2 - 5x + 1) - (-x^3 - x^2 - 4) - (3x^2 + 2)$ $\mathbf{7x^3 - 5x + 3}$

5. $(5b^2 - 4b + 3) - [(2b^2 + b) - (3b + 4)]$ $\mathbf{3b^2 - 2b + 7}$

6. $-[(8x^3 + x - 3) + (2x^3 + x^2)] - (4x^2 + 3x - 1)$ $\mathbf{-10x^3 - 5x^2 - 4x + 4}$

7. $(y + 5)(2y - 1)$ $\mathbf{2y^2 + 9y - 5}$

8. $(3k - 4)(k + 2)$ $\mathbf{3k^2 + 2k - 8}$

9. $(2m + 1)(3m - 5)$ $\mathbf{6m^2 - 7m - 5}$

10. $(4p - 3)(2p + 7)$ $\mathbf{8p^2 + 22p - 21}$

11. $(6p + 5q)(3p - 7q)$ $\mathbf{18p^2 - 27pq - 35q^2}$

12. $(2z + y)(3z - 4y)$ $\mathbf{6z^2 - 5zy - 4y^2}$

13. $\left(\frac{2}{5}y + \frac{1}{8}z\right)\left(\frac{3}{5}y + \frac{1}{2}z\right)$ $\mathbf{\frac{6}{25}y^2 + \frac{11}{40}yz + \frac{1}{16}z^2}$

14. $\left(\frac{3}{4}r - \frac{2}{3}s\right)\left(\frac{5}{4}r + \frac{1}{3}s\right)$ $\mathbf{\frac{15}{16}r^2 - \frac{7}{12}rs - \frac{2}{9}s^2}$

15. $(5r + 2)(5r - 2)$ $\mathbf{25r^2 - 4}$

16. $(6z + 5)(6z - 5)$ $\mathbf{36z^2 - 25}$

17. $(3z + 2w)(3z - 2w)$ $\mathbf{9z^2 - 4w^2}$

18. $(5r - 8t)(5r + 8t)$ $\mathbf{25r^2 - 64t^2}$

19. $(3x + 8)^2$ $\mathbf{9x^2 + 48x + 64}$

20. $(5y - 3)^2$ $\mathbf{25y^2 - 30y + 9}$

21. $(4m + 2n)^2$ $\mathbf{16m^2 + 16mn + 4n^2}$

22. $(a - 6b)^2$ $\mathbf{a^2 - 12ab + 36b^2}$

23. $(2z - 1)^3$ **$8z^3 - 12z^2 + 6z - 1$**

24. $(3m + 2)^3$ **$27m^3 + 54m^2 + 36m + 8$**

25. $(2x^3 + y^2)^2$ **$4x^6 + 4x^3y^2 + y^4$**

26. $(x^2 - y^2)^2$ **$x^4 - 2x^2y^2 + y^4$**

27. $(x - 2y^2)^3$ **$x^3 - 6x^2y^2 + 12xy^4 - 8y^6$**

28. $4x^2(3x^3 + 2x^2 - 5x + 1)$ **$12x^5 + 8x^4 - 20x^3 + 4x^2$**

29. $5m(3m^3 - 2m^2 + m - 1)$ **$15m^4 - 10m^3 + 5m^2 - 5m$**

30. $(2z - 1)(-z^2 + 3z - 4)$ **$-2z^3 + 7z^2 - 11z + 4$**

31. $(3p - 1)(9p^2 + 3p + 1)$ **$27p^3 - 1$**

32. $(2m + 1)(4m^2 - 2m + 1)$ **$8m^3 + 1$**

33. $(k + 2)(12k^3 - 3k^2 + k + 1)$ **$12k^4 + 21k^3 - 5k^2 + 3k + 2$**

34. $(m - n + k)(m + 2n - 3k)$ **$m^2 + mn - 2n^2 - 2km + 5kn - 3k^2$**

35. $(x + 2y - 3z)^2$ **$x^2 + 4xy - 6xz + 4y^2 - 12yz + 9z^2$**

36. $(p - 2q + r)^2$ **$p^2 - 4pq + 2pr + 4q^2 - 4qr + r^2$**

37. State the formula for the square of a binomial in words. Then state in words how you would use the formula to find $(a + b)^2$.

38. State the formula for the product of the sum and difference of two terms in words. Then state in words how you would use the product to find $(p + q)(p - q)$.

Find each of the following products. Assume all variables used as exponents represent integers.

39. $(x^y + 3)(x^y - 3)$ **$x^{2y} - 9$**

40. $(5 - z^a)(5 + z^a)$ **$25 - z^{2a}$**

41. $(4k^x - 3)(k^x + 1)$ **$4k^{2x} + k^x - 3$**

42. $(2^a + 5)(2^a + 3)$ **$2^{2a} + 8 \cdot 2^a + 15$**

43. $(m^x - 2)^2$ **$m^{2x} - 4m^x + 4$**

44. $(z^r + 5)^2$ **$z^{2r} + 10z^r + 25$**

45. $(3k^a - 2)^3$ **$27k^{3a} - 54k^{2a} + 36k^a - 8$**

46. $(r^x - 4)^3$ **$r^{3x} - 12r^{2x} + 48r^x - 64$**

Write out the terms of the binomial expansion for each of the following.

47. $(a + b)^6$ **$a^6 + 6a^5b + 15a^4b^2 + 20a^3b^3 + 15a^2b^4 + 6ab^5 + b^6$**

48. $(m + 4)^4$ **$m^4 + 16m^3 + 96m^2 + 256m + 256$**

49. $(3x - 2y)^4$ **$81x^4 - 216x^3y + 216x^2y^2 - 96xy^3 + 16y^4$**

50. $(4m - 3p)^5$ **$1024m^5 - 3840m^4p + 5760m^3p^2 - 4320m^2p^3 + 1620mp^4 - 243p^5$**

51. $(2k + 1/2)^6$ **$64k^6 + 96k^5 + 60k^4 + 20k^3 + (15/4)k^2 + (3/8)k + (1/64)$**

52. $(3r + 1/3)^6$ **$729r^6 + 486r^5 + 135r^4 + 20r^3 + (5/3)r^2 + (2/27)r + (1/729)$**

53. $(y - 4z)^5$ **$y^5 - 20y^4z + 160y^3z^2 - 640y^2z^3 + 1280yz^4 - 1024z^5$**

54. $(3k - 2)^4$ **$81k^4 - 216k^3 + 216k^2 - 96k + 16$**

The following are not products of polynomials. However, the same multiplication rules apply. Find each product. Assume that all variables represent positive real numbers.

55. $y^{4/3}(2y^{2/3} - 8y^{5/3})$ **$2y^2 - 8y^3$**

56. $z^{9/5}(5z^{4/5} + 3z^{7/5})$ **$5z^{13/5} + 3z^{16/5}$**

57. $-2p(4p^{3/4} - 3p^{7/4})$ **$-8p^{7/4} + 6p^{11/4}$**

58. $-5y(3y^{9/10} + 4y^{3/10})$ **$-15y^{19/10} - 20y^{13/10}$**

59. $(x + x^{1/2})(x - x^{1/2})$ **$x^2 - x$**

60. $(2z^{1/2} + z)(z^{1/2} - z)$ **$2z - z^{3/2} - z^2$**

61. $(r^{1/2} - r^{-1/2})^2$ **$r - 2 + r^{-1}$ or $r - 2 + 1/r$**

62. $(p^{1/2} - p^{-1/2})(p^{1/2} + p^{-1/2})$ **$p - p^{-1}$ or $p - 1/p$**

63. Explain why the expressions in Exercises 55–62 are not polynomials.

Perform each of the following divisions.

64. $\dfrac{-4x^7 - 14x^6 + 10x^4 - 14x^2}{-2x^2}$ **$2x^5 + 7x^4 - 5x^2 + 7$**

65. $\dfrac{-8r^3s - 12r^2s^2 + 20rs^3}{4rs}$ **$-2r^2 - 3rs + 5s^2$**

66. $\dfrac{10x^8 - 16x^6 - 4x^4}{-2x^6}$ **$-5x^2 + 8 + 2/x^2$**

67. $\dfrac{6m^3 + 7m^2 - 4m + 2}{3m + 2}$ **$2m^2 + m - 2 + \dfrac{6}{3m + 2}$**

68. $\dfrac{2x^3 + 6x^2 - 8x + 10}{2x - 1}$ **$x^2 + (7/2)x - (9/4) + \dfrac{31/4}{2x - 1}$**

69. $\dfrac{3x^4 - 6x^2 + 9x - 5}{3x + 3}$ **$x^3 - x^2 - x + 4 + \dfrac{-17}{3x + 3}$**

70. $\dfrac{3x^4 + 2x^2 + 6x - 1}{3x^2 - x}$ **$x^2 + \dfrac{1}{3}x + \dfrac{7}{9} + \dfrac{(61/9)x - 1}{3x^2 - x}$**

71. $\dfrac{k^4 - 4k^2 + 2k + 5}{k^2 + 1}$ **$k^2 - 5 + \dfrac{2k + 10}{k^2 + 1}$**

72. $\dfrac{5y^5 + 10y^4 - 5y^2 + 15y}{5y^3 - 2y + 1}$ **$y^2 + 2y + \dfrac{2}{5} + \dfrac{-2y^2 + (69/5)y - (2/5)}{5y^3 - 2y + 1}$**

73. $\dfrac{8z^5 + 4z^4 + 2z^2 - 5z + 16}{4z^2 - z + 2}$ **$2z^3 + \dfrac{3}{2}z^2 - \dfrac{5}{8}z - \dfrac{13}{32} + \dfrac{(-133/32)z + (269/16)}{4z^2 - z + 2}$**

◉ *In Exercises 74–79 find the coefficient of x^3 without finding the entire product.*

74. $(2x^2 - 3x)(-5x^2 + 3x + 1)$ **21**

75. $(3x^3 - 4x^2 + 2)(x^3 - 1)$ **−1**

76. $(1 + x^2)(1 + x)$ **1**

77. $(3 - x)(2 - x^2)$ **1**

78. $x^2(4 - 3x)^2$ **−24**

79. $-4x^2(2 - x)(2 + x)$ **0**

◉ 80. Show that $(y - x)^3 = -(x - y)^3$.

◉ 81. Show that $(y - x)^2 = (x - y)^2$.

◉ 82. Show that $\left(x + \dfrac{1}{2}\right)^2 = x(x + 1) + \dfrac{1}{4}$.

◉ 83. Use the result of Exercise 82 to calculate the square of 9½ in your head. **90¼**

◉ 84. Suppose one polynomial has degree 3 and another also has degree 3. Find all possible values for the degree of their **(a)** sum, **(b)** difference, **(c)** product.
(a) 0, 1, 2, or 3 (b) 0, 1, 2, or 3 (c) 6

◉ 85. If one polynomial has degree 3 and another has degree 4, find all possible values for the degree of their **(a)** sum, **(b)** difference, **(c)** product.
(a) 4 (b) 4 (c) 7

◉ 86. Generalize the results of Exercises 84 and 85: Suppose one polynomial has degree m and another has degree n, where m and n are natural numbers with $n < m$. Find all possible values for the degree of their **(a)** sum, **(b)** difference, **(c)** product.
(a) m (b) m (c) $m + n$

1.4 Exercises (text page 34)

Factor as completely as possible. Assume all variables appearing as exponents represent integers.

1. $32m^8 + 28m^5 - 16m^3$ **$4m^3(8m^5 + 7m^2 - 4)$**

2. $-4x^3 + 16x^5 - 36x^6$ **$-4x^3(1 - 4x^2 + 9x^3)$**

3. $-7jh^2 - 21j^2h - 35jh$ **$-7jh(h + 3j + 5)$**

4. $16p^3q^2 - 24p^2q^3 + 40p^2q^4$ **$8p^2q^2(2p - 3q + 5q^2)$**

5. $2(3m + n) + 6m(3m + n)$ **$2(3m + n)(1 + 3m)$**

6. $3(z - 5)^2 + 6(z - 5)$ **$3(z - 5)(z - 3)$**

7. $6p^2 + 5p - 6$ **$(3p - 2)(2p + 3)$**

8. $20x^2 - 7x - 6$ **$(4x - 3)(5x + 2)$**

9. $6y^2 + 7y - 5$ **$(3y + 5)(2y - 1)$**

10. $24a^2 - 38ab + 15b^2$ **$(4a - 3b)(6a - 5b)$**

11. $12m^2 + 16mn - 35n^2$ **$(6m - 7n)(2m + 5n)$**

12. $9x^2 - 6x^3 + x^4$ **$x^2(3 - x)^2$**

13. $10x^4 + 23x^2 + 12$ **$(5x^2 + 4)(2x^2 + 3)$**

14. $49y^2 - 100$ **$(7y + 10)(7y - 10)$**

15. $25a^2 - 16$ **$(5a + 4)(5a - 4)$**

16. $144r^2 - 81s^2$ **$9(4r + 3s)(4r - 3s)$**

17. $81m^2 - 16n^2$ **$(9m + 4n)(9m - 4n)$**

18. $121p^4 - 9q^4$ **$(11p^2 + 3q^2)(11p^2 - 3q^2)$**

19. $81q^4 - 256m^4$ **$(9q^2 + 16m^2)(3q + 4m)(3q - 4m)$**

20. $p^8 - 1$ **$(p^4 + 1)(p^2 + 1)(p + 1)(p - 1)$**

21. $y^{16} - 1$ **$(y^8 + 1)(y^4 + 1)(y^2 + 1)(y + 1)(y - 1)$**

22. $8m^3 - 27n^3$ **$(2m - 3n)(4m^2 + 6mn + 9n^2)$**

23. $125x^3 - 1$ **$(5x - 1)(25x^2 + 5x + 1)$**

24. $x^2 + xy + 5x + 5y$ **$(x + 5)(x + y)$**

25. $x^3 - y^3 + x^2 - y^2$ **$(x - y)(x^2 + xy + y^2 + x + y)$**

26. $8x^2 - 14x^3 + 49x^2 - 16$ **$(7x - 4)(2x + 1)(-x + 4)$**

27. $a^2 + 2ab + b^2 - x^2 - 2xy - y^2$ **$(a + b + x + y)(a + b - x - y)$**

28. $d^2 - 10d + 25 - c^2 + 4c - 4$ **$(d + c - 7)(d - c - 3)$**

29. $(z - w)^2 - 4z^2$ **$(-z - w)(3z - w)$**

30. $9(p - q)^2 - 25q^2$ **$(3p + 2q)(3p - 8q)$**

31. $36(a + 5)^2 - 16a^2$ **$20(a + 3)(a + 15)$**

32. $49(m - 3)^2 - 4a^2$ **$(7m - 21 + 2a)(7m - 21 - 2a)$**

33. $(x + y)^2 + 2(x + y)z - 15z^2$ **$(x + y + 5z)(x + y - 3z)$**

34. $(m + n)^2 + 3(m + n)p - 10p^2$ **$(m + n - 2p)(m + n + 5p)$**

35. $(p + q)^2 - (p - q)^2$ **$4pq$**

36. $(p - q)^2 - (p + q)^2$ **$-4pq$**

37. $(r + 6)^3 - 216$ **$r(r^2 + 18r + 108)$**

38. $(b + 3)^3 - 27$ **$b(b^2 + 9b + 27)$**

39. $27 - (m + 2n)^3$ **$(3 - m - 2n)(9 + 3m + 6n + m^2 + 4mn + 4n^2)$**

40. $125 - (4a - b)^3$ **$(5 - 4a + b)(25 + 20a - 5b + 16a^2 - 8ab + b^2)$**

41. $8u^3 + 27$ **$(2u + 3)(4u^2 - 6u + 9)$**

42. $(3z - 1)^3 + 125$ **$(3z + 4)(9z^2 - 21z + 31)$**

43. $2rp - 6rq - tp + 3tq$ **$(p - 3q)(2r - t)$**

44. $3mx - 6my - nx + 2ny$ **$(3m - n)(x - 2y)$**

45. $n^3 - 3n^2p + 3np^2 - p^3$ **$(n - p)^3$**

46. $z^3 + 3z^2w + 3zw^2 + w^3$ **$(z + w)^3$**

▲ 47. $m^{2n} - 16$ **$(m^n + 4)(m^n - 4)$**

▲ 48. $p^{4n} - 49$ **$(p^{2n} + 7)(p^{2n} - 7)$**

▲ 49. $x^{3n} - y^{6n}$ **$(x^n - y^{2n})(x^{2n} + x^ny^{2n} + y^{4n})$**

▲ 50. $a^{4p} - b^{12p}$ **$(a^{2p} + b^{6p})(a^p + b^{3p})(a^p - b^{3p})$**

▲ 51. $2x^{2n} - 23x^ny^n - 39y^{2n}$ **$(2x^n + 3y^n)(x^n - 13y^n)$**

▲ 52. $3a^{2x} + 7a^xb^x + 2b^{2x}$ **$(3a^x + b^x)(a^x + 2b^x)$**

▲ 53. $25q^{2r} - 30q^rt^p + 9t^{2p}$ **$(5q^r - 3t^p)^2$**

▲ 54. $16m^{2p} + 56m^pn^q + 49n^{2q}$ **$(4m^p + 7n^q)^2$**

▲ **55.** $6(m + p)^{2k} + (m + p)^k - 15$ **$[3(m + p)^k + 5][2(m + p)^k - 3]$**

✍ **56.** When asked to factor $6x^4 - 3x^2 - 3$ completely, a student gave the following result: $6x^4 - 3x^2 - 3 = (2x^2 + 1)(3x^2 - 3)$. Is this answer correct? Explain why or why not.

Factor the variable having the smallest exponent, together with any numerical common factor, from each of the following expressions. For example, factor $9x^{-2} - 6x^{-3}$ *as* $3x^{-3}(3x - 2)$.

57. $m^{-3} + m^{-1}$ **$m^{-3}(1 + m^2)$**

58. $2p^{-2} + p^{-4}$ **$p^{-4}(2p^2 + 1)$**

59. $9k^{-3} + 9k^{-2} + 18k^{-1}$ **$9k^{-3}(1 + k + 2k^2)$**

60. $16b^{-5} + 20b^{-4} - 6b^{-2}$ **$2b^{-5}(8 + 10b - 3b^3)$**

61. $15p^{-5} - 10p^{-1} + 30p^{-3}$ **$5p^{-5}(3 - 2p^4 + 6p^2)$**

62. $48y^{-3} + 32y - 80y^2$ **$16y^{-3}(3 + 2y^4 - 5y^5)$**

63. $4k^{7/4} + k^{3/4}$ **$k^{3/4}(4k + 1)$**

64. $y^{9/2} - 3y^{5/2}$ **$y^{5/2}(y^2 - 3)$**

65. $9z^{-1/2} + 2z^{1/2}$ **$z^{-1/2}(9 + 2z)$**

66. $3m^{2/3} - 4m^{-1/3}$ **$m^{-1/3}(3m - 4)$**

67. $-(2a - 5)^{-3/2} + 3(2a - 5)^{-1/2}$ **$2(3a - 8)(2a - 5)^{-3/2}$**

68. $(3k - 2)^{-1/2} + 4(3k - 2)^{-3/2}$ **$(3k - 2)^{-3/2}(3k + 2)$**

✍ **69.** In Exercises 57–68, why do you think the variable with the smallest exponent was chosen in the common factor? What happens if the variable to one of the other powers is factored out?

▲ *Factor each expression. (The expressions in Exercises 70–77 arise in calculus from techniques called the* product *and* quotient rules.*)*

70. $2(3x - 4)^2 + (x - 5)(2)(3x - 4)(3)$ **$2(3x - 4)(6x - 19)$**

71. $(5 - 2x)(3)(7x - 8)^2(7) + (7x - 8)^3(-2)$ **$(7x - 8)^2(121 - 56x)$**

72. $\dfrac{(p + 1)^{1/2} - p(1/2)(p + 1)^{-1/2}}{p + 1}$ **$(p + 2)/[2(p + 1)^{3/2}]$**

73. $\dfrac{(r - 2)^{2/3} - r(2/3)(r - 2)^{-1/3}}{(r - 2)^{4/3}}$ **$(r - 6)/[3(r - 2)^{5/3}]$**

74. $\dfrac{3(2x^2 + 5)^{1/3} - x(2x^2 + 5)^{-2/3}(4x)}{(2x^2 + 5)^{2/3}}$ **$(2x^2 + 15)/(2x^2 + 5)^{4/3}$**

75. $\dfrac{-(m^3 + m)^{2/3} + m(2/3)(m^3 + m)^{-1/3}(3m^2 + 1)}{(m^3 + m)^{4/3}}$ **$(3m^3 - m)/[3(m^3 + m)^{5/3}]$**

76. $(x^{-1} - 5)^3(2)(2 - x^{-2})(2x^{-3}) + (2 - x^{-2})^2(3)(x^{-1} - 5)^2(-x^{-2})$
$x^{-3}(x^{-1} - 5)^2(2 - x^{-2})(7x^{-1} - 20 - 6x)$

77. $(6 + x^{-4})^2(3)(3x - 2x^{-1})^2(3 + 2x^{-2}) + (3x - 2x^{-1})^3(2)(6 + x^{-4})(-4x^{-5})$
$(6 + x^{-4})(3x - 2x^{-1})^2(54 + 36x^{-2} - 15x^{-4} + 22x^{-6})$

▲ **78.** Factor $4(x^2 + 3)^{-3}(4x + 7)^{-4}$ from $(x^2 + 3)^{-2}(-3)(4x + 7)^{-4}(4) + (4x + 7)^{-3}(-2)(x^2 + 3)^{-3}(2x)$.
$4(x^2 + 3)^{-3}(4x + 7)^{-4}(-7x^2 - 7x - 9)$

▲ **79.** Factor $(7x - 8)^{-6}(3x + 2)^{-3}$ from $(7x - 8)^{-5}(-2)(3x + 2)^{-3}(3) + (3x + 2)^{-2}(-5)(7x - 8)^{-6}(7)$.
$(7x - 8)^{-6}(3x + 2)^{-3}(-147x - 22)$

We can factor $x^4 + 4x^2 + 16$ by grouping as follows.

$$\begin{aligned} x^4 + 4x^2 + 16 &= (x^4 + 8x^2 + 16) - 4x^2 && \textbf{Group; replace } 4x^2 \textbf{ with } 8x^2 - 4x^2. \\ &= (x^2 + 4)^2 - (2x)^2 && \textbf{Factor the perfect squares.} \\ &= (x^2 + 4 + 2x)(x^2 + 4 - 2x) && \textbf{Difference of squares.} \end{aligned}$$

Use this procedure to factor each of the following polynomials.

80. $x^4 + 64$ $\mathbf{(x^2 + 8 + 4x)(x^2 + 8 - 4x)}$

81. $r^4 - 6r^2 + 1$ $\mathbf{(r^2 - 1 + 2r)(r^2 - 1 - 2r)}$

82. $p^4 + 9p^2 + 81$ $\mathbf{(p^2 + 9 + 3p)(p^2 + 9 - 3p)}$

83. $x^4 - 18x^2 + 1$ $\mathbf{(x^2 - 1 + 4x)(x^2 - 1 - 4x)}$

84. $z^4 - 11z^2 + 25$ $\mathbf{(z^2 - 5 + z)(z^2 - 5 - z)}$

85. $m^4 - 22m^2 + 9$ $\mathbf{(m^2 - 3 + 4m)(m^2 - 3 - 4m)}$

1.5 Exercises (text page 41)

Give the restrictions on x for each of the following.

1. $\frac{x - 1}{2x + 3}$ $\mathbf{x \neq -3/2}$

2. $\frac{x + 3}{4x - 1}$ $\mathbf{x \neq 1/4}$

3. $\frac{3x^2}{2x^2 + 11x + 12}$ $\mathbf{x \neq -3/2, x \neq -4}$

4. $\frac{x - 10}{1 + 4x^2}$ **no restrictions**

5. $\frac{8x + 1}{5x^2 + 2}$ **no restrictions**

6. $\frac{5x - 1}{6x^2 + 5x - 4}$ $\mathbf{x \neq -4/3, x \neq 1/2}$

Write each of the following in lowest terms.

7. $\frac{16h^3}{2h^2}$ $\mathbf{8h}$

8. $\frac{27r^5}{9r}$ $\mathbf{3r^4}$

9. $\frac{5t + 25}{3t + 15}$ **5/3**

10. $\frac{32y + 16}{18y + 9}$ **16/9**

11. $\frac{-4(x - 1)}{(x - 1)(x + 2)}$ $\mathbf{-4/(x + 2)}$

12. $\frac{2(4m + 3)}{(4m + 3)(m - 4)}$ $\mathbf{2/(m - 4)}$

13. $\frac{4r^2 + 24}{8r^2}$ $\mathbf{(r^2 + 6)/(2r^2)}$

14. $\frac{12m^2 - 60m}{4m^2}$ $\mathbf{3(m - 5)/m}$

15. $\frac{3y^2 - 10y - 8}{3y^2 + 5y + 2}$ $\mathbf{(y - 4)/(y + 1)}$

16. $\frac{p^2 + 2p - 3}{p^2 + 4p - 5}$ $\mathbf{(p + 3)/(p + 5)}$

17. $\frac{2m^2 + 5m - 25}{4m^2 - 25}$ $\mathbf{(m + 5)/(2m + 5)}$

18. $\frac{6z^2 + 7z + 2}{3z^2 + 14z + 8}$ $\mathbf{(2z + 1)/(z + 4)}$

Perform each operation.

19. $\frac{k(h - 1)}{5} \div \frac{k}{h - 1}$ $\mathbf{(h - 1)^2/5}$

20. $\frac{mn}{m + n} \div \frac{p}{m + n}$ $\mathbf{mn/p}$

21. $\frac{15p^3}{9p^2} \div \frac{6p}{10p^2}$ $\mathbf{25p^2/9}$

22. $\frac{5x^2}{15x^3} \div \frac{25x^4}{10x^2}$ $\mathbf{2/(15x^3)}$

23. $\frac{2k + 8}{6} \div \frac{3k + 12}{2}$ **2/9**

24. $\frac{5m + 25}{10} \cdot \frac{12}{6m + 30}$ **1**

25. $\frac{3p - 18}{8p + 48} \cdot \frac{4p + 24}{9p - 54}$ **1/6**

26. $\frac{12r + 24}{36r - 36} \div \frac{6r + 12}{8r - 8}$ **4/9**

27. $\frac{x^2 + x}{5} \cdot \frac{25}{xy + y}$ $\mathbf{5x/y}$

28. $\frac{a^2 - 9}{a^2 - a - 20} \div \frac{4a + 12}{2a - 10}$ $\mathbf{(a - 3)/[2(a + 4)]}$

29. $\frac{m^2 - 10m + 25}{12m - 60} \cdot \frac{3m - 15}{4m - 20}$ $\mathbf{(m - 5)/16}$

30. $\frac{6r - 18}{9r^2 + 6r - 24} \cdot \frac{12r - 16}{4r - 12}$ $\mathbf{2/(r + 2)}$

31. $\frac{x^2 + 7x + 10}{x^2 + x - 2} \div \frac{x^2 + 2x - 15}{x^2 + 3x + 2}$ $\mathbf{(x + 2)(x + 1)/(x - 1)(x - 3)}$

32. $\frac{x^2 + 2x - 15}{x^2 + 11x + 30} \cdot \frac{x^2 + 2x - 24}{x^2 - 8x + 15}$ $(x - 4)/(x - 5)$

33. $\frac{3y^2 - 7y - 20}{y^2 + 4y - 32} \div \frac{9y^2 - 25}{2y^2 + 13y - 24}$ $(2y - 3)/(3y - 5)$

34. $\frac{n^2 - n - 6}{n^2 - 2n - 8} \div \frac{n^2 - 9}{n^2 + 7n + 12}$ $(n + 4)/(n - 4)$

35. $\left(1 + \frac{1}{x}\right)\left(1 - \frac{1}{x}\right)$ $(x^2 - 1)/x^2$

36. $\left(3 + \frac{2}{y}\right)\left(3 - \frac{2}{y}\right)$ $(9y^2 - 4)/y^2$

37. $\frac{x^3 + y^3}{x^2 - y^2} \cdot \frac{x + y}{x^2 - xy + y^2}$ $(x + y)/(x - y)$

38. $\frac{8y^3 - 125}{4y^2 - 20y + 25} \cdot \frac{2y - 5}{y}$ $(4y^2 + 10y + 25)/y$

39. $\frac{x^3 + y^3}{x^3 - y^3} \cdot \frac{x^2 - y^2}{x^2 + 2xy + y^2}$ $(x^2 - xy + y^2)/(x^2 + xy + y^2)$

40. $\frac{8}{r} + \frac{6}{r} + \frac{r}{2}$ $(28 + r^2)/(2r)$

41. $\frac{3}{y} + \frac{4}{y} - \frac{y}{6}$ $(42 - y^2)/(6y)$

42. $\frac{8}{5p} + \frac{3}{4p} + \frac{2}{5}$ $(47 + 8p)/(20p)$

43. $\frac{2}{3y} - \frac{1}{4y} + \frac{y}{3}$ $(5 + 4y^2)/(12y)$

44. $\frac{6}{11z} - \frac{5}{2z}$ $-43/(22z)$

45. $\frac{a + 1}{2} - \frac{a - 1}{2}$ 1

46. $\frac{y + 6}{5} - \frac{y - 6}{5}$ $12/5$

47. $\frac{2}{y} - \frac{1}{4}$ $(8 - y)/(4y)$

48. $\frac{6}{11} + \frac{3}{a}$ $(6a + 33)/(11a)$

49. $\frac{1}{6m} + \frac{2}{5m} + \frac{4}{m}$ $137/(30m)$

50. $\frac{8}{3p} + \frac{5}{4p} + \frac{9}{2p}$ $101/(12p)$

51. $\frac{1}{y} + \frac{1}{y + 1}$ $(2y + 1)/[y(y + 1)]$

52. $\frac{5}{2(x + 3)} + \frac{7}{3(x + 3)}$ $29/[6(x + 3)]$

53. $\frac{2}{a + b} - \frac{1}{2(a + b)}$ $3/[2(a + b)]$

54. $\frac{3}{m} - \frac{1}{m - 1}$ $(2m - 3)/[m(m - 1)]$

55. $\frac{1}{a + 1} - \frac{1}{a - 1}$ $-2/[(a + 1)(a - 1)]$

56. $\frac{1}{x + z} + \frac{1}{x - z}$ $2x/(x^2 - z^2)$

57. $\frac{m + 1}{m - 1} + \frac{m - 1}{m + 1}$ $(2m^2 + 2)/[(m - 1)(m + 1)]$

58. $\frac{3}{y - 2} + \frac{4}{2 - y}$ $-1/(y - 2)$ or $1/(2 - y)$

59. $\frac{5}{p - 4} - \frac{2}{4 - p}$ $7/(p - 4)$ or $-7/(4 - p)$

60. $\frac{m - 4}{3m - 4} + \frac{3m + 2}{4 - 3m}$ $(-2m - 6)/(3m - 4)$ or $(2m + 6)/(4 - 3m)$

61. $\frac{1}{a^2 - 5a + 6} - \frac{1}{a^2 - 4}$ $5/[(a - 2)(a - 3)(a + 2)]$

62. $\frac{5}{k^2 + 5k + 6} - \frac{-2}{k^2 + 2k - 3}$ $(7k - 1)/[(k + 2)(k + 3)(k - 1)]$

63. $\frac{5x + 2}{x^2 - x - 20} + \frac{2x - 1}{x^2 - 4x - 5}$ $(7x^2 + 14x - 2)/[(x - 5)(x + 4)(x + 1)]$

64. $\frac{3y + 5}{y^2 - 9y + 20} + \frac{2y - 7}{y^2 - 2y - 8}$ $(5y^2 - 6y + 45)/[(y - 5)(y - 4)(y + 2)]$

65. $\left(\frac{3}{p - 1} - \frac{2}{p + 1}\right)\left(\frac{p - 1}{p}\right)$ $(p + 5)/[p(p + 1)]$

66. $\left(\frac{y}{y^2 - 1} - \frac{y}{y^2 - 2y + 1}\right)\left(\frac{y - 1}{y + 1}\right)$ $-2y/[(y + 1)^2(y - 1)]$

67. $\frac{\frac{1}{x + h} - \frac{1}{x}}{h}$ $-1/[x(x + h)]$

68. $\frac{1}{h}\left(\frac{1}{(x + h)^2 + 9} - \frac{1}{x^2 + 9}\right)$ **$-(2x + h)/[(x^2 + 9)((x + h)^2 + 9)]$**

69. $\frac{1 + \frac{1}{x}}{1 - \frac{1}{x}}$ **$(x + 1)/(x - 1)$**

70. $\frac{4 - \frac{1}{z}}{4 + \frac{1}{z}}$ **$(4z - 1)/(4z + 1)$**

71. $\frac{\frac{1}{k - 1} - \frac{1}{k}}{\frac{1}{k}}$ **$1/(k - 1)$**

72. $\frac{\frac{1}{y + 3} - \frac{1}{y}}{\frac{1}{y}}$ **$-3/(y + 3)$**

73. $\frac{1 + \frac{1}{1 - b}}{1 - \frac{1}{1 + b}}$ **$(2 - b)(1 + b)/[b(1 - b)]$**

74. $m - \frac{m}{m + \frac{1}{2}}$ **$m(2m - 1)/(2m + 1)$**

75. In your own words, explain how to find the least common denominator for two fractions.

76. Describe the steps required to add three rational expressions. You may use an example to illustrate.

Perform all indicated operations and write all answers with positive integer exponents. (See Example 5.)

77. $\frac{3^{-1} - 4^{-1}}{4^{-1}}$ **$1/3$**

78. $\frac{6^{-1} + 5^{-1}}{6^{-1}}$ **$11/5$**

79. $\frac{a^{-1} + b^{-1}}{(ab)^{-1}}$ **$a + b$**

80. $\frac{p^{-1} - q^{-1}}{(pq)^{-1}}$ **$q - p$**

81. $\frac{r^{-1} + q^{-1}}{r^{-1} - q^{-1}} \cdot \frac{r - q}{r + q}$ **-1**

82. $\frac{xy^{-1} + yx^{-1}}{x^2 + y^2}$ **$1/(xy)$**

83. $(a + b)^{-1}(a^{-1} + b^{-1})$ **$1/(ab)$**

84. $(m^{-1} + n^{-1})^{-1}$ **$mn/(n + m)$**

Simplify each of the following expressions. (The expressions in Exercises 85–88 arise in calculus from techniques called the chain rule *and* quotient rule *that are used to determine the shape of a curve.)*

85. $\left(\frac{x^3 + 2}{x - 5}\right)^3\left(\frac{(x - 5)(3x^2) - (x^3 + 2)}{(x - 5)^2}\right)$ **$(x^3 + 2)^3(2x^3 - 15x^2 - 2)/(x - 5)^5$**

86. $\left(\frac{2x^2 - 9}{x^2 + 1}\right)^2\left(\frac{(x^2 + 1)(4x) - (2x^2 - 9)}{(x^2 + 1)^2}\right)$ **$(2x^2 - 9)^2(4x^3 - 2x^2 + 4x + 9)/(x^2 + 1)^4$**

87. $5\left(\frac{x + 1}{3x^2 - 4}\right)^4\left(\frac{(3x^2 - 4) - (x + 1)(6x)}{(3x^2 - 4)^2}\right)$ **$[-5(x + 1)^4(3x^2 + 6x + 4)]/(3x^2 - 4)^6$**

88. $7\left(\frac{x^2 + 2}{5x^2 + 3}\right)^6\left(\frac{(5x^2 + 3)(2x) - (x^2 + 2)(10x)}{(5x^2 + 3)^2}\right)$ **$\frac{-98x(x^2 + 2)^6}{(5x^2 + 3)^8}$**

1.6 Exercises (text page 48)

Write each of the following as a radical.

1. $6^{2/3}$ **$\sqrt[3]{36}$** **2.** $5^{1/4}$ **$\sqrt[4]{5}$** **3.** $(3z)^{1/5}$ **$\sqrt[5]{3z}$** **4.** $(4m)^{5/6}$ **$\sqrt[6]{4^5m^5}$** **5.** $2k^{2/3}$ **$2\sqrt[3]{k^2}$** **6.** $7p^{4/5}$ **$7\sqrt[5]{p^4}$**

Simplify each of the following. Assume that all variables represent nonnegative real numbers and that no denominators are zero.

7. $\sqrt{98}$ $\mathbf{7\sqrt{2}}$
8. $\sqrt{27}$ $\mathbf{3\sqrt{3}}$
9. $\sqrt{375}$ $\mathbf{5\sqrt{15}}$
10. $\sqrt{64}$ $\mathbf{8}$
11. $-\sqrt{5/4}$ $\mathbf{-\sqrt{5}/2}$
12. $-\sqrt{3/25}$ $\mathbf{-\sqrt{3}/5}$
13. $4\sqrt{3} - 5\sqrt{12} + 3\sqrt{75}$ $\mathbf{9\sqrt{3}}$
14. $2\sqrt{5} - 3\sqrt{20} + 2\sqrt{45}$ $\mathbf{2\sqrt{5}}$
15. $\sqrt[4]{\dfrac{3}{2}}$ $\mathbf{\sqrt[4]{24}/2}$
16. $\sqrt[4]{\dfrac{32}{81}}$ $\mathbf{2\sqrt[4]{2}/3}$
17. $\sqrt[4]{5\dfrac{1}{16}}$ $\mathbf{3/2}$
18. $\sqrt[3]{3\dfrac{3}{8}}$ $\mathbf{3/2}$
19. $\sqrt[3]{2} - \sqrt[3]{16} + 2\sqrt[3]{54}$ $\mathbf{5\sqrt[3]{2}}$
20. $3\sqrt[3]{4} - 5\sqrt[3]{256} + \sqrt[3]{32}$ $\mathbf{-15\sqrt[3]{4}}$
21. $4/\sqrt{5} - 2/\sqrt{20} + 3\sqrt{5}$ $\mathbf{18\sqrt{5}/5}$
22. $3/\sqrt{3} + 2/\sqrt{27} + 1/\sqrt{12}$ $\mathbf{25\sqrt{3}/18}$
23. $-1/\sqrt[3]{5} + 3/\sqrt[3]{40} - 2/\sqrt[3]{135}$ $\mathbf{-\sqrt[3]{25}/30}$
24. $3/\sqrt[3]{2} + 5/\sqrt[3]{16} - 1/\sqrt[3]{128}$ $\mathbf{21\sqrt[3]{4}/8}$
25. $\sqrt{72p^5q^4r^6}$ $\mathbf{6p^2q^2r^3\sqrt{2p}}$
26. $\sqrt[3]{16z^5x^8y^4}$ $\mathbf{2zx^2y\sqrt[3]{2z^2x^2y}}$
27. $\sqrt[4]{x^8y^6z^{10}}$ $\mathbf{x^2yz^2\sqrt[4]{y^2z^2}}$
28. $\sqrt{a^3b^5} - 2\sqrt{a^7b^3} + \sqrt{a^3b^9}$ $\mathbf{ab\sqrt{ab}(b - 2a^2 + b^3)}$
29. $\sqrt{p^7q^3} - \sqrt{p^5q^9} + \sqrt{p^9q}$ $\mathbf{(pq - q^4 + p^2)p^2\sqrt{pq}}$
30. $(\sqrt{2} + 3)(\sqrt{2} - 3)$ $\mathbf{-7}$
31. $(\sqrt{5} + \sqrt{2})(\sqrt{5} - \sqrt{2})$ $\mathbf{3}$
32. $(\sqrt{2} + \sqrt{7})^2$ $\mathbf{9 + 2\sqrt{14}}$
33. $(\sqrt{5} - 1)^2$ $\mathbf{6 - 2\sqrt{5}}$
34. $(3\sqrt[3]{2} - 4)(3\sqrt[3]{2} + 1)$ $\mathbf{9\sqrt[3]{4} - 9\sqrt[3]{2} - 4}$
35. $(\sqrt[3]{6} + 2)(5\sqrt[3]{6} + 3)$ $\mathbf{5\sqrt[3]{36} + 13\sqrt[3]{6} + 6}$
36. $\sqrt{\dfrac{x^5y^3}{z^2}}$ $\mathbf{x^2y\sqrt{xy}/z}$
37. $\sqrt{\dfrac{g^3h^5}{r^3}}$ $\mathbf{gh^2\sqrt{ghr}/r^2}$
38. $-\sqrt[3]{\dfrac{k^5m^3r^2}{r^8}}$ $\mathbf{-km\sqrt[3]{k^2}/r^2}$
39. $-\sqrt[3]{\dfrac{9x^5y^6}{z^5w^2}}$ $\mathbf{-xy^2\sqrt[3]{9x^2zw}/(z^2w)}$
40. $\sqrt[4]{\dfrac{g^3h^5}{9r^6}}$ $\mathbf{h\sqrt[4]{9g^3hr^2}/(3r^2)}$
41. $\sqrt[4]{\dfrac{32x^6}{y^5}}$ $\mathbf{2x\sqrt[4]{2x^2y^3}/y^2}$
42. $\dfrac{\sqrt[3]{mn} \cdot \sqrt[3]{m^2}}{\sqrt[3]{n^2}}$ $\mathbf{m\sqrt[3]{n^2}/n}$
43. $\dfrac{\sqrt[3]{8m^2n^3} \cdot \sqrt[3]{2m^2}}{\sqrt[3]{32m^4n^3}}$ $\mathbf{\sqrt[3]{4}/2}$
44. $\sqrt[3]{\sqrt{2}}$ $\mathbf{\sqrt[6]{2}}$
45. $\sqrt[3]{\sqrt[4]{5}}$ $\mathbf{\sqrt[12]{5}}$
46. $\sqrt[4]{\sqrt[5]{p}}$ $\mathbf{\sqrt[20]{p}}$
47. $\sqrt[6]{\sqrt[3]{y}}$ $\mathbf{\sqrt[18]{y}}$
48. $\sqrt{2(m + n)^2 + 2(m - n)^2}$ $\mathbf{2\sqrt{m^2 + n^2}}$
49. $\sqrt{2(x + y)^2 - 2(x - y)^2}$ $\mathbf{2\sqrt{2xy}}$
50. $\sqrt{4x^4 + x^4y^4}$ $\mathbf{x^2\sqrt{4 + y^4}}$
51. $\sqrt{-z^2 + (z - 9)^2}$ $\mathbf{3\sqrt{9 - 2z}}$
52. $\dfrac{6}{\sqrt{2} - 1}$ $\mathbf{6(\sqrt{2} + 1)}$
53. $\dfrac{2}{1 + \sqrt{5}}$ $\mathbf{-(1 - \sqrt{5})/2}$
54. $\dfrac{\sqrt{3}}{4 + \sqrt{3}}$ $\mathbf{(4\sqrt{3} - 3)/13}$
55. $\dfrac{2\sqrt{7}}{3 - \sqrt{7}}$ $\mathbf{3\sqrt{7} + 7}$
56. $\dfrac{1}{\sqrt{m} - \sqrt{p}}$ $\mathbf{(\sqrt{m} + \sqrt{p})/(m - p)}$
57. $\dfrac{\sqrt{z}}{-\sqrt{z} + 1}$ $\mathbf{(-z - \sqrt{z})/(z - 1)}$ **or** $\mathbf{(z + \sqrt{z})/(1 - z)}$
▲ 58. $\dfrac{-4}{\sqrt{1 + x} + 3}$ $\mathbf{(-4\sqrt{1 + x} + 12)/(x - 8)}$
▲ 59. $\dfrac{-5}{1 - \sqrt{3 - p}}$ $\mathbf{(-5 - 5\sqrt{3 - p})/(p - 2)}$
▲ 60. $\dfrac{\sqrt{x} + \sqrt{x + 1}}{\sqrt{x} - \sqrt{x + 1}}$ $\mathbf{-2x - 2\sqrt{x(x + 1)} - 1}$
▲ 61. $\dfrac{\sqrt{p} + \sqrt{p^2 - 1}}{\sqrt{p} - \sqrt{p^2 - 1}}$ $\mathbf{(p^2 + p + 2\sqrt{p(p^2 - 1)} - 1)/(-p^2 + p + 1)}$

Writing 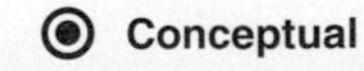 Conceptual ▲ Challenging Connections

▲ 62. $\dfrac{5}{\sqrt[3]{a} + \sqrt[3]{b}}$ **$5(\sqrt[3]{a^2} - \sqrt[3]{ab} + \sqrt[3]{b^2})/(a + b)$**

▲ 63. $\dfrac{1}{\sqrt[3]{m} - \sqrt[3]{n}}$ **$(\sqrt[3]{m^2} + \sqrt[3]{mn} + \sqrt[3]{n^2})/(m - n)$**

▲ 64. $\dfrac{6 - \sqrt{3}}{(5 - \sqrt{2})(3 + \sqrt{5})}$
$(6 - \sqrt{3})(5 + \sqrt{2})(3 - \sqrt{5})/92$

▲ 65. $\dfrac{1 - \sqrt{7}}{(2 + \sqrt{10})(1 - \sqrt{5})}$
$(1 - \sqrt{7})(2 - \sqrt{10})(1 + \sqrt{5})/24$

66. Explain how to rationalize the denominator of $\sqrt[3]{3/2}$.

67. Describe the steps required to multiply $\sqrt[3]{2}$ and $\sqrt{2}$.

▲ *Take the squared terms outside of the following radicals.*

68. $\sqrt{5(3 - \sqrt{10})^2}$ **$(\sqrt{10} - 3)\sqrt{5}$**

69. $\sqrt{(4 - \sqrt{17})^2}$ **$\sqrt{17} - 4$**

▲ *Simplify Exercises 70 and 71, given $x \leq 5$.*

70. $\sqrt{(x - 5)^2}$ **$|x - 5|$ or $5 - x$**

71. $\sqrt{(5 - x)^2(x - 3)^4}$ **$(5 - x)(x - 3)^2$**

72. Use a calculator to find an approximate value for $\sqrt{5 + 2\sqrt{6}}$. **approximately 3.146**

73. Show that $\sqrt{5 + 2\sqrt{6}} = \sqrt{2} + \sqrt{3}$.

1.7 Exercises (text page 55)

Identify each complex number as real or imaginary.

1. -21 **real**
2. -5 **real**
3. $\sqrt{6}$ **real**
4. π **real**
5. $i\sqrt{7}$ **imaginary**
6. $-4i$ **imaginary**
7. $0 - 3i$ **imaginary**
8. $0 + 6i$ **imaginary**

Write each of the following in standard form.

9. $\sqrt{-81}$ **$0 + 9i$**
10. $\sqrt{-14}$ **$0 + i\sqrt{14}$**
11. $-\sqrt{-256}$ **$0 - 16i$**
12. $-\sqrt{-64}$ **$0 - 8i$**
13. $5 + \sqrt{-4}$ **$5 + 2i$**
14. $-7 + \sqrt{-100}$ **$-7 + 10i$**
15. $-6 - \sqrt{-196}$ **$-6 - 14i$**
16. $13 + \sqrt{-16}$ **$13 + 4i$**
17. $\sqrt{-5} \cdot \sqrt{-5}$ **$-5 + 0i$**
18. $\sqrt{-20} \cdot \sqrt{-20}$ **$-20 + 0i$**
19. $\sqrt{-8} \cdot \sqrt{-2}$ **$-4 + 0i$**
20. $\sqrt{-27} \cdot \sqrt{-3}$ **$-9 + 0i$**
21. $\dfrac{\sqrt{-40}}{\sqrt{-10}}$ **$2 + 0i$**
22. $\dfrac{\sqrt{-190}}{\sqrt{-19}}$ **$\sqrt{10} + 0i$**

Add or subract. Write each result in standard form.

23. $(5 - i) + (6 + 2i)$ **$11 + i$**
24. $(7 + 3i) + (5 + 3i)$ **$12 + 6i$**
25. $(-1 + 5i) - (-3 + 2i)$ **$2 + 3i$**
26. $(-3 + 5i) - (-4 + 3i)$ **$1 + 2i$**
27. $(2 - 5i) - (3 + 4i) - (-2 + i)$ **$1 - 10i$**
28. $(-4 - i) - (2 + 3i) + (-4 + 5i)$ **$-10 + i$**

Multiply. Write each result in standard form.

29. $(2 + 4i)(-1 + 3i)$ **$-14 + 2i$**
30. $(1 + 3i)(2 - 5i)$ **$17 + i$**
31. $(-3 + 2i)^2$ **$5 - 12i$**
32. $(2 + i)^2$ **$3 + 4i$**
33. $(2 + 3i)(2 - 3i)$ **$13 + 0i$**
34. $(6 - 4i)(6 + 4i)$ **$52 + 0i$**
35. $(\sqrt{2} + 2i)(\sqrt{2} - 2i)$ **$6 + 0i$**
36. $(\sqrt{3} - 5i)(\sqrt{3} + 5i)$ **$28 + 0i$**
37. $(2 - 7i)(2 + 7i)$ **$53 + 0i$**
38. $i(4 + 9i)(4 - 9i)$ **$0 + 97i$**

Divide. Write each result in standard form.

39. $\dfrac{4 - 3i}{4 + 3i}$ **7/25 − (24/25)*i***

40. $(3 - 4i) \div (2 - 5i)$ **26/29 + (7/29)*i***

41. $(1 - 3i) \div (1 + i)$ **−1 − 2*i***

42. $\dfrac{5 + 6i}{5 - 6i}$ **(−11/61) + (60/61)*i***

43. $\dfrac{2}{i}$ **0 − 2*i***

44. $\dfrac{-7}{3i}$ **0 + (7/3)*i***

45. $\dfrac{1 - \sqrt{-5}}{3 + \sqrt{-4}}$ **$(3 - 2\sqrt{5})/13 + (-2 - 3\sqrt{5})i/13$**

46. $\dfrac{2 + \sqrt{-3}}{1 - \sqrt{-9}}$ **$(2 - 3\sqrt{3})/10 + (6 + \sqrt{3})/10$**

47. Explain why the method of dividing complex numbers (that is, multiplying both the numerator and the denominator by the conjugate of the denominator) works. That is, what property justifies this process?

48. Suppose that your friend, Susan Katz, tells you that she has discovered a method of simplifying a positive power of *i*. "Just divide the exponent by 4," she says, "and then look at the remainder. Then refer to the table of powers of *i* in this section. The large power of *i* is equal to *i* to the power indicated by the remainder. If the remainder is 0, the result is $i^0 = 1$." Explain why Susan's method works.

Find each of the following powers of i.

49. i^{25} ***i***

50. $1/i^9$ **−*i***

51. $1/i^{12}$ **1**

52. i^{-6} **−1**

53. i^{-15} ***i***

54. i^{-49} **−*i***

Perform the indicated operations, and write your answers in standard form.

55. $\dfrac{1 - i}{2 + i} \cdot \dfrac{4 + 3i}{1 + i}$ **2/5 − (11/5)*i***

56. $\dfrac{6 + 2i}{5 - i} \cdot \dfrac{1 - 3i}{2 + 6i}$ **−16/65 − (37/65)*i***

57. $\dfrac{5 - 3i}{1 + 2i} \cdot \dfrac{2 - 4i}{1 + i}$ **−38/5 + (16/5)*i***

58. $\dfrac{4 - 3i}{2 + 5i} + \dfrac{8 - i}{2 + 5i}$ **4/29 − (68/29)*i***

59. $\dfrac{6 + 2i}{1 + 3i} + \dfrac{2 - i}{1 - 3i}$ **17/10 − (11/10)*i***

60. $\dfrac{4 - i}{3 + 4i} - \dfrac{3 + 2i}{3 - 4i}$ **7/25 − (37/25)*i***

Use the definition of equality for complex numbers to solve the following equations for real numbers a and b.

61. $a + 3i = 5 + 3bi + 2a$ **$a = -5$; $b = 1$**

62. $4a - 2bi + 7 = 3i + 3a + 5$ **$a = -2$; $b = -3/2$**

63. $i(2b + 6) - 3 = 4(bi + a)$ **$a = -3/4$; $b = 3$**

64. $3i + 2(a - 1) = 4 + 2i(b + 3)$ **$a = 3$; $b = -3/2$**

Let $z = a + bi$ for real numbers a and b, and let $\bar{z} = a - bi$, the conjugate of z. For example, if $z = 8 - 9i$, then $\bar{z} = 8 + 9i$. Prove each of the following properties of conjugates.

65. $\bar{\bar{z}} = z$

66. $\bar{z} = z$ if and only if $b = 0$.

67. $\overline{-z} = -\bar{z}$

68. $z \cdot \bar{z}$ is a real number.

Evaluate $8z - z^2$ by replacing z with the indicated complex number in Exercises 69 and 70.

69. $2 + i$ **13 + 4*i***

70. $4 - 3i$ **25**

71. Find any restrictions on a and b so that the square $(a + bi)^2$ is real. $\mathbf{a = 0}$ **or** $\mathbf{b = 0}$

72. Find any restrictions on a and b so that the square $(a + bi)^2$ is imaginary. $\mathbf{a = b}$ **or** $\mathbf{a = -b}$

73. Show that $\frac{\sqrt{2}}{2} + \frac{\sqrt{2}}{2} i$ is a square root of i.

74. Show that $\frac{\sqrt{3}}{2} - \frac{1}{2} i$ is a cube root of $-i$.

Chapter 1 Review Exercises (page 56)

Simplify each of the expressions in Exercises 1–8.

1. $14 \div [5 - 4 \cdot (-3)]$ **14/17**

2. $-3 + (-5)(8) + 4$ **−39**

3. $\frac{2 - 3(-7) - 5}{7(-1) + (-5 - 2)}$ **−9/7**

4. $\frac{-6 - (-2)(-3) + 4}{(-9)(-2) + 5(-3)}$ **−8/3**

5. $|3 - \pi|$ $\mathbf{\pi - 3}$

6. $|\sqrt{7} - 2|$ $\mathbf{\sqrt{7} - 2}$

7. $|p - 5|$ if $p < 5$ $\mathbf{5 - p}$

8. $|-2x^2 - 1|$ $\mathbf{2x^2 + 1}$

In each of the following exercises, the coordinates of three points on a number line are given. Find **(a)** *$d(A, B)$ and* **(b)** *$d(A, B) + d(B, C)$.*

9. $A, -3; B, -2; C, -5$ **(a) 1 (b) 4**

10. $A, -8; B, 2; C, -4$ **(a) 10 (b) 16**

Under what conditions are the following statements true?

11. $|x| = x$ **if** $\mathbf{x \geq 0}$

12. $|x| \leq 0$ **if** $\mathbf{x = 0}$

13. $d(A, B) = 0$ **if** $\mathbf{A}$ **and** $\mathbf{B}$ **are the same point**

14. $d(A, B) + d(B, C) = d(A, C)$ **if** $\mathbf{B}$ **is between** $\mathbf{A}$ **and** $\mathbf{C}$ **on a line**

Simplify each of the following. Assume all variables represent nonzero real numbers and variables used as exponents represent rational numbers.

15. $(3x^3 - 9x^2 - 5) - (-4x^3 + 6x^2) + (2x^3 - 9)$ $\mathbf{9x^3 - 15x^2 - 14}$

16. $(2m - 6)(4m - 8)$ $\mathbf{8m^2 - 40m + 48}$

17. $(3r - 2)(r^2 + 4r - 8)$ $\mathbf{3r^3 + 10r^2 - 32r + 16}$

18. $(a - 4b + c)^2$ $\mathbf{a^2 - 8ab + 2ac + 16b^2 - 8bc + c^2}$

19. $(9 - x^y)(9 + x^y)$ $\mathbf{81 - x^{2y}}$

20. $\frac{2x^5t^2 \cdot 4x^3t}{16x^2t^3}$ $\mathbf{x^6/2}$

21. $\frac{(2k^2)^2(3k^3)}{(4k^4)^2}$ $\mathbf{3/(4k)}$

22. Write out the terms of $(3x - y)^4$. $\mathbf{81x^4 - 108x^3y + 54x^2y^2 - 12xy^3 + y^4}$

Factor as completely as possible.

23. $12k^2 - k - 20$ $\mathbf{(3k - 4)(4k + 5)}$

24. $12p^5 - 8p^4 + 20p^3$ $\mathbf{4p^3(3p^2 - 2p + 5)}$

25. $6m^2 - 13m - 5$ $\mathbf{(3m + 1)(2m - 5)}$

26. $10x^2 - 29xy - 21y^2$ $\mathbf{(2x - 7y)(5x + 3y)}$

27. $30m^5 - 35m^4n - 25m^3n^2$ $\mathbf{5m^3(3m - 5n)(2m + n)}$

28. $2x^2p^3 - 8xp^4 + 6p^5$ $\mathbf{2p^3(x - 3p)(x - p)}$

29. $16 - 81y^4$ $\mathbf{(4 + 9y^2)(2 + 3y)(2 - 3y)}$

30. $49m^8 - 9n^2$ $\mathbf{(7m^4 + 3n)(7m^4 - 3n)}$

31. $(x - 1)^2 - 4$ $\mathbf{(x + 1)(x - 3)}$

32. $8z^3 + 27w^3$ $\mathbf{(2z + 3w)(4z^2 - 6zw + 9w^2)}$

33. $r^9 - 8(r^3 - 1)^3$ $\mathbf{(-r^3 + 2)(7r^6 - 10r^3 + 4)}$

34. $(2p - 1)^3 + 27p^3$ $\mathbf{(5p - 1)(7p^2 - p + 1)}$

35. $7(3p - q) + m(3p - q)$ $\mathbf{(3p - q)(7 + m)}$

36. $6(z - 4)^2 + 9(z - 4)^3$ $\mathbf{3(z - 4)^2(3z - 10)}$

37. $2bx - b + 6x - 3$ $\mathbf{(2x - 1)(b + 3)}$

38. $64 - y^{6p}$ $\mathbf{(2 + y^p)(4 - 2y^p + y^{2p})(2 - y^p)(4 + 2y^p + y^{2p})}$

39. Describe the steps needed to find the following sum:

$$\frac{2a + b}{4a^2 - b^2} + \frac{5a}{2a - b}.$$

Perform the indicated operations.

40. $\dfrac{5x^2y}{x + y} \cdot \dfrac{3x + 3y}{30xy^2}$ $\mathbf{x/(2y)}$

41. $\dfrac{2k - 10}{7k} \cdot \dfrac{14(3k - 1)}{8}$. $\mathbf{(k - 5)(3k - 1)/(2k)}$

42. $\dfrac{27m^3 - n^3}{3m - n} \div \dfrac{9m^2 + 3mn + n^2}{9m^2 - n^2}$ $\mathbf{(3m + n)(3m - n)}$

43. $\dfrac{x^2 + 2x - 15}{x^2 + 7x + 12} \div \dfrac{x^2 - 4x + 3}{x^2 + 3x - 4}$ $\mathbf{(x + 5)/(x + 3)}$

44. $\dfrac{m}{4 - m} + \dfrac{3m}{m - 4}$ $\mathbf{2m/(m - 4)}$

45. $\dfrac{3}{8z} - \dfrac{2}{6z} + \dfrac{1}{12}$ $\mathbf{(1 + 2z)/(24z)}$

46. $\left(\dfrac{1}{(x + h)^2 + 16} - \dfrac{1}{x^2 + 16}\right) \div h$ $\mathbf{(-2x - h)/\{[(x + h)^2 + 16](x^2 + 16)\}}$

47. $\dfrac{4a}{a^2 - a - 2} - \dfrac{1}{a^2 - 4}$ $\mathbf{(4a^2 + 7a - 1)/[(a - 2)(a + 1)(a + 2)]}$

48. $\dfrac{x^{-1} - 2y^{-1}}{y^{-1} - x^{-1}}$ $\mathbf{(y - 2x)/(x - y)}$

49. $\dfrac{\dfrac{6}{y + 5}}{\dfrac{3}{y^2 - 25} - 2}$ $\mathbf{6(y - 5)/(53 - 2y^2)}$

Simplify each of the following. Assume all variables represent positive real numbers.

50. $\dfrac{(p^4)(p^{-2})}{p^{-6}}$ $\mathbf{p^8}$

51. $(-6x^2y^{-3}z^2)^{-2}$ $\mathbf{y^6/(36x^4z^4)}$

52. $\dfrac{6^{-1}r^3s^{-2}}{6r^4s^{-3}}$ $\mathbf{s/(36r)}$

53. $\dfrac{(3m^{-2})^{-2}(m^2n^{-4})^3}{9m^{-3}n^{-5}}$ $\mathbf{m^{13}/(81n^7)}$

54. Give some examples of corresponding rules for exponents and radicals, and explain how they are related.

Simplify. Assume that all variables represent positive real numbers.

55. $\sqrt{128}$ $\mathbf{8\sqrt{2}}$

56. $\sqrt[4]{405}$ $\mathbf{3\sqrt[4]{5}}$

57. $\sqrt{\dfrac{2^7y^8}{m^3}}$ $\mathbf{2^3y^4\sqrt{2m}/m^2}$

58. $-\sqrt[3]{\dfrac{r^6m^5}{z^2}}$ $\mathbf{-r^2m\sqrt[3]{m^2z}/z}$

59. $\sqrt[5]{\sqrt[3]{k}}$ $\mathbf{\sqrt[15]{k}}$

60. $(\sqrt[3]{2} + 4)(\sqrt[3]{2^2} - 4\sqrt[3]{2} + 16)$ $\mathbf{66}$

61. $\dfrac{\sqrt[4]{8p^2q^5} \cdot \sqrt[4]{2p^3q}}{\sqrt[4]{p^5q^2}}$ $\mathbf{2q}$

62. $-\sqrt{50x^5} + 3x^2\sqrt{72x} - 2x\sqrt{18x^3}$ $\mathbf{7x^2\sqrt{2x}}$

63. $\sqrt{75y^5} - y^2\sqrt{108y} + 2y\sqrt{27y^3}$ $\mathbf{5y^2\sqrt{3y}}$

64. $\dfrac{-12}{\sqrt[3]{4}}$ $\mathbf{-6\sqrt[3]{2}}$

65. $\dfrac{3m}{\sqrt{m + 2}}$ $\mathbf{3m\sqrt{m + 2}/(m + 2)}$

66. $\dfrac{15}{1 - \sqrt{3}}$ $\mathbf{-15(1 + \sqrt{3})/2}$

67. $\dfrac{4}{\sqrt{7}+3}$ **$-2(\sqrt{7}-3)$**

68. $\dfrac{\sqrt{x}-\sqrt{x-2}}{\sqrt{x}+\sqrt{x-2}}$ **$x-\sqrt{x(x-2)}-1$**

69. Give two ways to evaluate $125^{2/3}$ and then compare them. Which do you prefer? Why?

Simplify each of the following. Assume that all variables represent positive real numbers, and variables used as exponents are rational numbers.

70. $36^{-3/2}$ **$1/216$**

71. $(125m^6)^{-2/3}$ **$1/(25m^4)$**

72. $(8r^{3/4}s^{2/3})(2r^{3/2}s^{5/3})$ **$16r^{9/4}s^{7/3}$**

73. $(7r^{1/2})(2r^{3/4})(-r^{1/6})$ **$-14r^{17/12}$**

74. $\dfrac{p^{-3/4}\cdot p^{5/4}\cdot p^{-1/4}}{p\cdot p^{3/4}}$ **$1/p^{3/2}$**

75. $\left(\dfrac{y^6x^3z^{-2}}{16x^5z^4}\right)^{-1/2}$ **$4xz^3/y^3$**

76. $\dfrac{m^{2+p}\cdot m^{-2}}{m^{3p}}$ **$1/m^{2p}$**

77. $\dfrac{z^{-p+1}\cdot z^{-8p}}{z^{-9p}}$ **z**

Write in standard form.

78. $(-2+i)-(3+6i)$ **$-5-5i$**

79. $(1+3i)-(2-5i)-(1-3i)$ **$-2+11i$**

80. $(7-2i)(3+4i)$ **$29+22i$**

81. $(3+5i)(3-5i)$ **$34+0i$**

82. $(4+7i)^2$ **$-33+56i$**

83. $(5-4i)^2$ **$9-40i$**

84. $(2-i)^3$ **$2-11i$**

85. $(4+i)^4$ **$161+240i$**

86. i^{15} **$0-i$**

87. i^{48} **$1+0i$**

88. $\dfrac{2+7i}{4-3i}$ **$-13/25+(34/25)i$**

89. $\dfrac{3-8i}{2-i}$ **$14/5-(13/5)i$**

90. $\dfrac{1-4i}{1-i}\cdot\dfrac{3+i}{1+i}$ **$7/2-(11/2)i$**

91. $\dfrac{2+5i}{1-3i}\cdot\dfrac{2-5i}{3-i}$ **$0+(29/10)i$**

92. $\sqrt{-27}$ **$0+3i\sqrt{3}$**

93. $\sqrt{-72}$ **$0+6i\sqrt{2}$**

Correct each incorrect statement in Exercises 94–105 by changing the right side of the equation.

94. $x(x^2+5)=x^3+5$ **x^3+5x**

95. $-3^2=9$ **-9**

96. $(m^2)^3=m^5$ **m^6**

97. $(3x)(3y)=3xy$ **$9xy$**

98. $\dfrac{\frac{a}{b}}{2}=\dfrac{2a}{b}$ **$a/(2b)$**

99. $\dfrac{m}{r}\cdot\dfrac{n}{r}=\dfrac{mn}{r}$ **mn/r^2**

100. $\dfrac{1}{\sqrt{a}+\sqrt{b}}=\dfrac{1}{\sqrt{a}}+\dfrac{1}{\sqrt{b}}$ **$\dfrac{\sqrt{a}-\sqrt{b}}{a-b}$**

101. $\dfrac{(2x)^3}{2y}=\dfrac{x^3}{y}$ **$4x^3/y$**

102. $4-(t+1)=4-t+1$ **$4-t-1$**

103. $\dfrac{1}{(-2)^3}=2^{-3}$ **$(-2)^{-3}$**

104. $(-5)^2=-5^2$ **5^2**

105. $\left(\dfrac{8}{7}+\dfrac{a}{b}\right)^{-1}=\dfrac{7}{8}+\dfrac{b}{a}$ **$7b/(8b+7a)$**

106. For which of the following cases does $\sqrt{ab}=\sqrt{a}\cdot\sqrt{b}$?
(a) a and b positive **(b)** a positive, b negative **(c)** a and b negative **(a) and (b)**

107. For what integer values of n does $\sqrt[n]{a^n}=a$? **n odd and positive**

108. For what values of x does $\sqrt{9ax^2}=3x\sqrt{a}$? **$x\geq 0$**

2 Equations and Inequalities

2.1 Exercises (text page 63)

Decide whether each of the following equations is an identity or a conditional equation.

1. $2y - y^2 = y(2 - y)$ **identity**

2. $m^2 - 4 = (m + 2)(m - 2)$ **identity**

3. $2x + 5 = 2(x + 5)$ **conditional**

4. $3x + 8 - x = 3(x + 8) - x$ **conditional**

5. $\dfrac{z - 2}{z} = 1 - \dfrac{2}{z}$ **identity**

6. $\dfrac{y - 1}{y + 3} = -\dfrac{1}{3}$ **conditional**

7. $2(x - 1) = x - 1 + x - 1$ **identity**

8. $4p + 16 = 5(p + 4) - (p + 4)$ **identity**

Decide which of the following pairs of equations are equivalent.

9. $4x - 1 = 10$
$12x - 3 = 30$ **equivalent**

10. $5 = 8 - 2x$
$2x = 3$ **equivalent**

11. $\dfrac{y + 2}{y + 3} = \dfrac{4}{y + 3}$
$y + 2 = 4$ **equivalent**

12. $\dfrac{2x + 5}{9} = \dfrac{4x}{9}$
$5 = 2x$ **equivalent**

13. $\dfrac{x}{x - 2} = \dfrac{2}{x - 2}$
$x = 2$ **not equivalent**

14. $\dfrac{x + 3}{x + 1} = \dfrac{2}{x + 1}$
$x = -1$ **not equivalent**

15. $x = 4$
$x^2 = 16$ **not equivalent**

16. $z^2 = 9$
$z = 3$ **not equivalent**

17. Explain the difference between an identity and a conditional equation.

18. Make a complete list of the steps needed to solve any linear equation. (Some equations will not require every step.)

Solve each of the following equations. Check each solution.

19. $.3x - .7 = .3 + .2x$ **10**

20. $.04x - 2.01 = 3.18x + 4.72$ **−2.14**

21. $(3/4)x - 5 + (2/3) = (5/3) - x$ **24/7**

22. $(-1/2) + (1/4)y + 2 = (3/4)y$ **3**

23. $3r + 2 - 5(r + 1) = 6r + 4$ **−7/8**

24. $5(a + 3) + 4a - 5 = -(2a - 4)$ **−6/11**

25. $2[m - (4 + 2m) + 3] = 2m + 2$ **−1**

26. $4[2p - (3 - p) + 5] = -7p - 2$ **−10/19**

27. $\dfrac{3x - 2}{7} = \dfrac{x + 2}{5}$ **3**

28. $\dfrac{2p + 5}{5} = \dfrac{p + 2}{3}$ **−5**

29. $\dfrac{3k - 1}{4} = \dfrac{5k + 2}{8}$ **4**

30. $\dfrac{9x - 1}{6} = \dfrac{2x + 7}{3}$ **3**

31. $\dfrac{x}{3} - 7 = 6 - \dfrac{3x}{4}$ **12**

32. $\dfrac{y}{3} + 1 = \dfrac{2y}{5} - 4$ **75**

33. $\dfrac{1}{4p} + \dfrac{2}{p} = 3$ **3/4**

34. $\dfrac{2}{t} + 6 = \dfrac{5}{2t}$ **1/12**

35. $\dfrac{m}{2} - \dfrac{1}{m} = \dfrac{6m + 5}{12}$ **−12/5**

36. $\dfrac{-3k}{2} + \dfrac{9k - 5}{6} = \dfrac{11k + 8}{k}$ **−48/71**

37. $\dfrac{2r}{r - 1} = 5 + \dfrac{2}{r - 1}$ **no solution**

38. $\dfrac{3x}{x + 2} = \dfrac{1}{x + 2} - 4$ **−1**

39. $\dfrac{5}{2a + 3} + \dfrac{1}{a - 6} = 0$ **27/7**

Writing 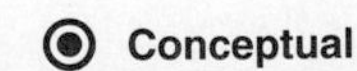 Conceptual  Challenging Connections

40. $\dfrac{2}{x+1} = \dfrac{3}{2x-5}$ **13**

41. $\dfrac{4}{x-3} - \dfrac{8}{2x+5} + \dfrac{3}{x-3} = 0$ **−59/6**

42. $\dfrac{8}{3x+1} + \dfrac{2}{x-1} = \dfrac{5}{3x+1}$ **1/9**

43. $\dfrac{2}{4-3x} - 5 = \dfrac{2}{4-3x}$ **no solution**

44. $\dfrac{5}{2y-3} = 1 - \dfrac{2}{2y-3}$ **5**

45. $\dfrac{3a}{a+1} - 2 = \dfrac{6}{a+1}$ **8**

46. $\dfrac{5p}{2p-1} = \dfrac{15}{2p-1} + 4$ **−11/3**

47. $2(m+1)(m-1) = (2m+3)(m-2)$ **−4**

48. $(2y-1)(3y+2) = 6(y+2)^2$ **−26/23**

49. $(3x-4)^2 - 5 = 3(x+5)(3x+2)$ **−19/75**

50. $(2x+5)^2 = 3x^2 + (x+3)^2$ **−8/7**

Solve each equation in Exercises 51–58 for x.

51. $2(x-a) + b = 3x + a$ **$x = -3a + b$**

52. $5x - (2a+c) = a(x+1)$ **$x = (3a + c)/(5 - a)$**

53. $ax + b = 3(x-a)$ **$x = (3a + b)/(3 - a)$**

54. $4a - ax = 3b + bx$ **$x = (4a - 3b)/(b + a)$**

55. $\dfrac{4x}{2a+1} = ax - 1$ **$x = (2a + 1)/(2a^2 + a - 4)$**

56. $\dfrac{a}{3x+2} + b = 2a$ **$x = (3a - 2b)/(3b - 6a)$**

57. $a^2(2x-3) = 4x$ **$x = 3a^2/(2a^2 - 4)$**

58. $a(x+a) = b(x+b)$ **$x = -(a + b)$**

Graphing Utility Problems

Use a graphing utility to solve the equations in Exercises 59 and 60.

59. **(a)** Solve $\dfrac{2x}{x-1} = 5 + \dfrac{2}{x-1}$. **no solution**

(b) Solve $2x = 5(x-1) + 2$. **1**

(c) Are the equations in (a) and (b) equivalent, identities, or neither? Explain how the graphs illustrate your answer.

60. **(a)** Solve $\dfrac{4x-3}{5} = \dfrac{2x}{5}$. **3/2**

(b) Solve $4x - 3 = 2x$. **3/2**

(c) Are the equations in (a) and (b) equivalent, identities, or neither? Explain how the graphs illustrate your answer.

2.2 Exercises (text page 71)

Solve each of the following equations for y.

1. $3x - 4y = 12$ **$y = (3x - 12)/4$**

2. $6x + 2y = 18$ **$y = 9 - 3x$**

3. $-5x + 3y + 27 = 0$ **$y = (5x - 27)/3$**

4. $-8x - 5y - 35 = 0$ **$y = (-8x - 35)/5$**

5. $x = \dfrac{y}{3} + \dfrac{3}{4}$ **$y = (12x - 9)/4$**

6. $x = \dfrac{4y}{5} - \dfrac{2}{3}$ **$y = (15x + 10)/12$**

Solve each of the following for the variable indicated.

7. $x = m - ax$ for x **$x = m/(1 + a)$**

8. $kr - p = br + c$ for r **$r = (p + c)/(k - b)$**

9. $s = \frac{1}{2}gt^2$ for g **$g = 2s/t^2$**

10. $A = \frac{1}{2}(B + b)h$ for h **$h = 2A/(B + b)$**

11. $A = \frac{1}{2}(B + b)h$ for B **$B = 2A/h - b$**

12. $C = \frac{5}{9}(F - 32)$ for F **$F = 9C/5 + 32$**

13. $S = 2\pi(r_1 + r_2)h$ for r_1 **$r_1 = S/(2\pi h) - r_2$**

14. $A = P\left(1 + \frac{i}{m}\right)$ for m **$m = Pi/(A - P)$**

15. $P = \frac{E^2R}{r + R}$ for R **$R = Pr/(E^2 - P)$**

16. $\frac{1}{R} = \frac{1}{r_1} + \frac{1}{r_2}$ for R **$R = r_1r_2/(r_2 + r_1)$**

17. $m = \frac{Ft}{v_1 - v_2}$ for v_2 **$v_2 = (mv_1 - Ft)/m$**

18. Refer to Example 1. Suppose someone tells you that there is no reason to solve for x, since the right side of the formula is already x. Is this correct? Explain.

19. Suppose two acid solutions are mixed. One is 26% acid and the other is 32% acid. Which of the following concentrations cannot possibly be the concentration of the mixture? Explain. **(a)**

(a) 36% **(b)** 28% **(c)** 30% **(d)** 31%

20. Suppose that a computer that originally sells for x dollars has been discounted 30%. Which of the following expressions does not represent its sale price? **(d)**

(a) $x - .30x$ **(b)** $.70x$ **(c)** $\frac{7}{10}x$ **(d)** $x - .30$

Solve each of the following problems.

21. The length of a rectangular label is 3 cm less than twice the width. The perimeter is 54 cm. Find the width. **10 cm**

22. A puzzle piece in the shape of a triangle has a perimeter of 30 cm. Two sides of the triangle are each twice as long as the shortest side. Find the length of the shortest side. **6 cm**

23. A pharmacist wishes to strengthen a mixture that is 10% alcohol to one that is 30% alcohol. How much pure alcohol should be added to 7 liters of the 10% mixture? **2 liters**

24. A student needs 10% hydrochloric acid for a chemistry experiment. How much 5% acid should be mixed with 60 ml of 20% acid to get a 10% solution? **120 ml**

25. An automobile radiator contains a 10-qt mixture of water and antifreeze that is 40% antifreeze. How much should the owner drain from the radiator and replace with pure antifreeze so that the liquid in the radiator will be 80% antifreeze? **6⅔ qt**

26. In Exercise 24, suppose the student has only pure acid and 5% acid. How much pure acid should be added to the 5% acid to get 12 ml of 10% acid? **12/19 ml**

27. A recycling bin is in the shape of a closed rectangular box. Find the height of the bin if its length is 18 ft, its width is 8 ft, and its surface area is 496 sq ft. **4 ft**

Writing · Conceptual · Challenging · Connections

28. A right circular cylinder has radius 6 inches and volume 144π cu inches. What is its height? **4 inches**

29. On a vacation trip, Le Hong averaged 50 mph traveling from Denver to Minneapolis. Returning by a different route that covered the same number of miles, he averaged 55 mph. What is the distance between the two cities if his total traveling time was 32 hr? **about 838.1 mi**

30. Lindsay left by plane to visit her mother in Hartford, 420 km away. Fifteen minutes later, her mother left to meet her at the airport. She drove the 20 km to the airport at 40 km per hr, arriving just as the plane taxied in. What was the speed of the plane? **560 km per hr**

31. Russ and Janet are running in the Apple Hill Fun Run. Russ runs at 7 mph, Janet at 5 mph. If they leave the starting point at the same time, how long will it be before they are $1/2$ mi apart? **15 min**

32. If the run in Exercise 31 has a staggered start, and Janet starts first, with Russ starting 10 min later, how long will it be before he catches up with her? **25 min**

33. Jon gets to work in 20 min when he drives his car. Riding his bike (by the same route) takes him 45 min. His average driving speed is 14.5 mph greater than his average speed on the bike. How far does he travel to work? **8.7 mi**

34. In the morning, Karen drove to a business appointment at 50 mph. Her average speed on the return trip in the afternoon was 40 mph. The return trip took $1/4$ hr longer because of heavy traffic. How far did she travel to the appointment? **50 mi**

35. Jim Macias drove from Philadelphia to New York, a distance of 100 miles, at an average speed of 45 mph and made the return trip at an average speed of 36 mph.

(a) What was his average speed for the entire trip? **40 mph**

(b) Why was the average speed closer to the lower speed? **He drove for a longer time at the lower speed.**

◉ **(c)** Show that the average speed will be the same no matter what the distance between the two cities.

36. The distance between New York and London is 3469 miles. If an airplane has a cruising speed of 350 mph and a tail wind of 50 mph, how many miles out will it reach the point of no return? (A tail wind blows in the same direction as the plane. The point of no return is the point on the flight where it will take the same amount of time to fly on to the destination as to fly back to the starting point.) **about 1486.7 mi**

37. A town gardener can mow the lawn in a small park in 2 hr. Another gardener can mow the same lawn in 3 hr. How long would it take them to mow the park lawn if they work together? **$1\frac{1}{5}$ hr**

38. Two painters working together can paint an average-sized room in 2.5 hr. One of the painters can do the same job alone in 4 hr. How long would it take the other painter working alone to do the job? **$6\frac{2}{3}$ hr**

39. A sewage treatment plant has two inlet pipes to its settling pond. One can fill the pond in 10 hr, the other in 12 hr. If the first pipe is open for 5 hr and then the second pipe is opened, how long will it take to fill the pond after the second pipe is opened? **$2\frac{8}{11}$ hr**

40. Two chemical plants are polluting a river. If plant A produces a predetermined maximum amount of pollution in half the time as plant B, and together they produce the maximum pollution in 26 hr, how long will it take plant B alone? **78 hr**

41. With both taps open, Mark can fill his kitchen sink in 5 min. When full, the sink drains in 10 min. How long will it take to fill the sink if Mark forgets to put in the stopper? **10 min**

42. If Mark (see Exercise 41) remembers to put in the stopper after 1 min, how much longer will it take to fill the sink? **4.5 min**

43. A VCR is on sale for $245. If the sale price is 30% less than the regular price, what was the regular price? **$350**

44. A jeweler prices his items 60% over their wholesale price. If a watch sells for $152, what is its wholesale price? **$95**

45. Anne Kelly received $52,000 profit from the sale of some land. She invested part at 7.5% interest and the rest at 5.5% interest. She earned a total of $3280 interest during the first year. How much did she invest at each rate? **$31,000 at 5.5%; $21,000 at 7.5%**

46. Jim Marshall invests $20,000 received from an insurance settlement in two ways, some at 5% and some at 6%. Altogether, he makes $1080 the first year in interest. How much is invested at each rate? **$12,000 at 5% and $8000 at 6%**

47. Mary Ellen Heise earned $48,000 from royalties on her book. She paid a 28% income tax on the royalties. She invested $15,000 at one rate and the balance at a rate that was 1% lower, earning $1878 annual interest on the two investments. What was the higher interest rate? **6%**

48. Kevin Conners won $100,000 in a state lottery. He paid income taxes of 33% on the winnings. He invested $40,000 of the balance at one interest rate and the remaining $27,000 at a rate that was 2% lower. The two investments earned $5080 per year in interest. What was the lower rate? **about 6.4%**

49. Diane Gray bought two plots of land for a total of $120,000. On the first plot, she made a profit of 15%. On the second, she lost 10%. Her total profit was $5500. How much did she pay for each piece of land? **$70,000 for land that makes a profit; $50,000 for land that produces a loss**

50. Suppose $10,000 is invested at 6%. How much additional money must be invested at 8% to produce a yield on the entire amount invested of 7.2%? **$15,000**

51. Cathy Wacaser earns take-home pay of $592 a week. If her deductions for taxes, retirement, union dues, and medical plan amount to 26% of her wages, what is her weekly pay before deductions? **$800**

52. Barbara Burnett gives 10% of her net income to her church. This amounts to $80 a month. In addition, her paycheck deductions are 24% of her gross monthly income. What is her gross monthly income? **$1052.63**

53. Adam Bryer wishes to sell a piece of property for $125,000. He wants the money to be paid off in two ways: a short-term note at 10.5% and a long-term note at 9%. Find the amount of each note if the total annual interest on the two notes is $12,600. **$90,000 at 10.5% and $35,000 at 9%**

54. A bank pays 4% interest on passbook accounts and 6% on long-term deposits. Suppose a depositor divides $20,000 among the two types of deposits. Find the amount deposited at each rate if the total annual income from interest is $1060. **$7000 at 4% and $13,000 at 6%**

55. If x represents the number of pennies in a jar, which of the following equations cannot be correct for finding x? (*Hint:* Solve each equation and consider the solution.) **(a) and (b)**
(a) $5x + 3 = 9$ (b) $12x + 3 = -4$
(c) $100x = 50(x + 3)$ (d) $6(x + 4) = x + 24$

56. Which of the following cannot be a correct equation to solve a geometry problem, if x represents the length of a rectangle? (*Hint:* Solve each equation and consider the solution.) **(b) and (c)**
(a) $2x + 2(x - 1) = 14$ (b) $-2x + 7(5 - x) = 62$
(c) $4(x + 2) + 4x = 8$ (d) $2x + 2(x - 3) = 22$

57. Biologists can estimate the number of individual members of a species in an area. Suppose, for example, that 100 animals of the species are caught and marked. A period of time is permitted to elapse, and then b animals are caught. If c of these ($c \leq b$) are marked, find an expression involving b and c that estimates the total number of individuals in the area. **$100b/c$**

58. Suppose B dollars are invested, some at $m\%$ and the rest at $n\%$. If a total of I dollars in interest is earned per year, find the amount invested at each rate.
$(100I - nB)/(m - n)$ at $m\%$; $(Bm - 100I)/(m - n)$ at $n\%$

2.3 Exercises (text page 84)

Solve the following equations by factoring.

1. $p^2 = 5p - 6$ **2, 3**
2. $6x + 9 = -x^2$ **−3**
3. $3q + 4 = q^2$ **−1, 4**
4. $2x^2 - 8x = 0$ **0, 4**
5. $-5x^2 + 30x = 0$ **0, 6**
6. $6y^2 - 5y - 50 = 0$ **−5/2, 10/3**
7. $21p^2 = 10 - 29p$ **−5/3, 2/7**
8. $6r^2 + 7r = 3$ **−3/2, 1/3**

Solve the following equations by the square root property.

9. $x^2 = 20$ **$\pm 2\sqrt{5}$**
10. $y^2 = 48$ **$\pm 4\sqrt{3}$**
11. $(t - 2)^2 = 7$ **$2 \pm \sqrt{7}$**
12. $(4z + 3)^2 = 12$ **$(-3 \pm 2\sqrt{3})/4$**
13. $(5r + 3)^2 = 3$ **$(-3 \pm \sqrt{3})/5$**
14. $(-2w + 5)^2 = 8$ **$(5 \pm 2\sqrt{2})/2$**

Solve the following equations by completing the square.

15. $p^2 - 8p + 15 = 0$ **3, 5**
16. $m^2 + 4m = 1/3$ **$(-6 \pm \sqrt{39})/3$**
17. $9z^2 = 12z - 8$ **$(2 \pm 2i)/3$**
18. $3k^2 - 12k + 12 = 75$ **−3, 7**
19. $2p^2 = -(2p + 1)$ **$(-1 \pm i)/2$**
20. $r^2 + 8r = -13$ **$-4 \pm \sqrt{3}$**

Solve the following equations by the quadratic formula.

21. $m^2 = m + 1$ **$(1 \pm \sqrt{5})/2$**
22. $x^2 + 7 = 6x$ **$3 \pm \sqrt{2}$**
23. $y^2 - 3y = 2$ **$(3 \pm \sqrt{17})/2$**
24. $11p^2 = 7p - 1$ **$(7 \pm \sqrt{5})/2$**
25. $m^2 - \sqrt{2}m - 1 = 0$ **$(\sqrt{2} \pm \sqrt{6})/2$**
26. $z^2 - \sqrt{3}z - 2 = 0$ **$(\sqrt{3} \pm \sqrt{11})/22$**
27. $\sqrt{2}p^2 - 3p + \sqrt{2} = 0$ **$\sqrt{2}/2, \sqrt{2}$**
28. $-\sqrt{6}k^2 - 2k + \sqrt{6} = 0$ **$(-\sqrt{6} \pm \sqrt{42})/6$**

Solve the following equations by any method.

29. $x^2 - x = 6$ **−2, 3**
30. $2s^2 = 10$ **$\pm\sqrt{5}$**
31. $6n^2 = 3n$ **0, 1/2**

32. $9p^2 = 25 + 30p$ **$(5 \pm 5\sqrt{2})/3$**

33. $4 - \frac{11}{x} - \frac{3}{x^2} = 0$ **$-1/4, 3$**

34. $3 = \frac{4}{p} + \frac{2}{p^2}$ **$(2 \pm \sqrt{10})/3$**

35. $\frac{2x - 3}{x + 1} = \frac{x + 1}{x + 2}$ **$(1 \pm \sqrt{29})/2$**

36. $\frac{x}{x - 1} = \frac{3x}{x - 2}$ **$0, 1/2$**

37. Solve $8x^2 + 2x = 5$ by factoring, completing the square, and using the quadratic formula. Compare the three methods, noting advantages and disadvantages of each.

38. In your own words, write a brief explanation of the process of completing the square as you might explain it to another student.

Solve each of the following equations by factoring first and then using the quadratic formula.

39. $x^3 - 64 = 0$ **$4, -2 \pm 2i\sqrt{3}$**

40. $8p^3 + 125 = 0$ **$-5/2, (5 \pm 5i\sqrt{3})/4$**

41. $64r^3 - 343 = 0$ **$7/4, (-7 \pm 7i\sqrt{3})/8$**

Evaluate the discriminant $b^2 - 4ac$ and use it to predict the type of solutions in each of the following. Do not solve the equations.

42. $8y^2 = 14y - 3$ **100; two rational solutions**

43. $3m^2 - 5m + 2 = 0$ **1; two rational solutions**

44. $9k^2 + 11k + 4 = 0$ **-23; two nonreal complex solutions**

45. $4p^2 = 6p + 3$ **84; two irrational solutions**

Find all values of k for which the following equations have exactly one solution.

46. $25m^2 - 10m + k = 0$ **1**

47. $y^2 + 11y + k = 0$ **$121/4$**

48. $kr^2 + (2k + 6)r + 16 = 0$ **1, 9**

49. $ky^2 + 2(k + 4)y + 25 = 0$ **1, 16**

Solve Exercises 50–55 for the indicated variables. Assume that all variables represent positive real numbers.

50. $L = \frac{d^4k}{h^2}$ for h **$h = \sqrt{d^4kL}/L$ or $d^2\sqrt{kL}/L$**

51. $F = \frac{kMv^2}{r}$ for v **$v = \sqrt{Fr/(kM)}$ or $\sqrt{FrkM}/(kM)$**

52. $s = s_0 + gt^2 + k$ for t **$t = \sqrt{(s - s_0 - k)g}/g$**

53. $P = \frac{E^2R}{(r + R)^2}$ for R **$R = [(E^2 - 2Pr) \pm E\sqrt{E^2 - 4Pr}]/(2P)$**

54. $S = 2\pi rh + 2\pi r^2$ for r **$r = (-\pi h \pm \sqrt{\pi^2h^2 + 2\pi S})/(2\pi)$**

55. $pm^2 - 8qm + \frac{1}{r} = 0$ for m **$m = (4qr \pm \sqrt{16q^2r^2 - rp})/(pr)$**

56. Is it possible for the solution of a quadratic equation with real coefficients to consist of a single irrational number? Explain. **yes**

57. Is it possible for the solution of a quadratic equation with real coefficients to consist of one real and one nonreal root? Explain. **no**

▲ *For each of the following, find a quadratic equation that has the given numbers as solutions.*

58. 4, 5 **$x^2 - 9x + 20 = 0$**

59. $-3, 2$ **$x^2 + x - 6 = 0$**

60. $1 + \sqrt{2}, 1 - \sqrt{2}$ **$x^2 - 2x - 1 = 0$**

61. $i, -i$ **$x^2 + 1 = 0$**

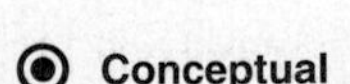
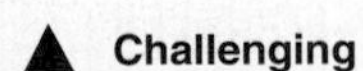

Writing Conceptual Challenging Connections

Solve the following problems.

62. A shopping center has a rectangular area of 40,000 sq yd enclosed on three sides for a parking lot. The length is 200 yd more than twice the width. What are the dimensions of the lot? **100 yd by 400 yd**

63. An ecology center wants to set up an experimental garden. It has 300 m of fencing to enclose a rectangular area of 5000 sq m. Find the dimensions of the rectangle. **50 m by 100 m**

64. A rectangular poster is to have an 18-inch-by-23-inch illustration in the center with equal margins on all four sides. How wide should the margins be if the poster has an area of 594 sq inches? **2 inches**

65. A landscape architect has included a rectangular flower bed measuring 9 ft by 5 ft in her plans for a new building. She wants to use two colors of flowers in the bed, one in the center and the other for a border of the same width on all four sides. If she can get just enough plants to cover 24 sq ft for the border, how wide can the border be? **1 ft**

66. Alfredo went into a frame-it-yourself shop. He wanted a frame 3 cm longer than its width. The frame he chose extends 1.5 cm beyond the picture on each side. Find the outside dimensions of the frame if the area of the unframed picture is 70 sq cm. **10 cm by 13 cm**

67. Juanita wants to buy a rug for a room that is 12 ft wide and 15 ft long. She wants to leave a uniform strip of floor around the rug. She can afford 108 sq ft of carpeting. What dimensions should the rug have? **9 ft by 12 ft**

68. An experienced roofer can do a complete roof in a housing development in half the time that it takes an inexperienced roofer to do the job. If the two work together on a roof, they complete the job in $2\frac{2}{3}$ hr. How long would it take the experienced roofer to do a roof working alone? **4 hr**

69. Two typists are working on a special project. The experienced typist could complete the project in 2 hr less time than the new typist. Together they complete the project in 2.4 hr. How long would it have taken the experienced typist to complete the project working alone? **4 hr**

70. It takes two copy machines 5/6 hr to make the copies for a company newsletter. One copy machine takes 1 hr longer than the other to do the job alone. The slower machine is out of order. How long will it take the faster machine to complete the job alone? **1.3 hr**

71. In 1991 Rick Mears won the (500-mi) Indianapolis 500 race. His speed (rate) was 100 mph (to the nearest mph) faster than that of the 1911 winner, Ray Harroun. Mears completed the race in 3.74 hr less time than Harroun. Find Mears' rate to the nearest whole number. **176 mph**

72. The Branson family traveled 100 mi to a lake for their vacation. On the return trip their average speed was 10 mph faster. The total time for the round trip was $3\frac{2}{3}$ hr. What was the family's average speed on their trip to the lake? **about 50 mph**

Use the Pythagorean theorem, $c^2 = a^2 + b^2$, to solve each of the following.

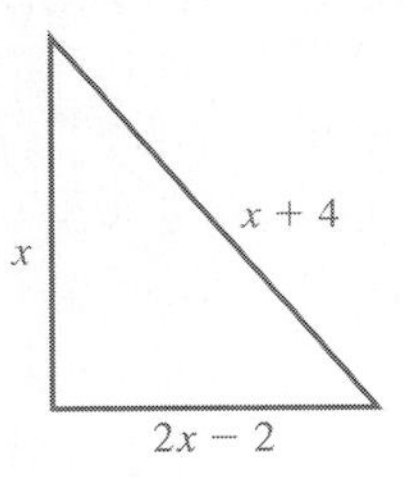

73. To solve for the lengths of the sides of the right triangle shown, which equation is correct? **(d)**
(a) $x^2 = (2x - 2)^2 + (x + 4)^2$
(b) $x^2 + (x + 4)^2 = (2x - 2)^2$
(c) $x^2 = (2x - 2)^2 - (x + 4)^2$
(d) $x^2 + (2x - 2)^2 = (x + 4)^2$

74. If a rectangle is r ft long and s ft wide, which one of the following is the length of its diagonal in terms of r and s? **(c)**
(a) rs **(b)** $r + s$ **(c)** $\sqrt{r^2 + s^2}$ **(d)** $r^2 + s^2$

75. A boat is being pulled into a dock with a rope attached at water level. When the boat is 12 ft from the dock, the length of the rope from the boat to the dock is 3 ft longer than twice the height of the dock above the water. Find the height of the dock above the water. **5 ft**

76. Chris and Josh have received walkie-talkies for Christmas. If they leave from the same point at the same time, Chris walking north at 2.5 mph and Josh walking east at 3 mph, how long will they be able to talk to each other if the range of the walkie-talkies is 4 mi? Round your answer to the nearest minute. **61 min**

Let r_1 and r_2 be the solutions of the quadratic equation $ax^2 + bx + c = 0$. Show that the equations in Exercises 77 and 78 are true.

77. $r_1 + r_2 = -\dfrac{b}{a}$

78. $r_1 r_2 = \dfrac{c}{a}$

79. Suppose one solution of the equation $km^2 + 10m = 8$ is -4. Find the value of k, and the other solution. **$k = 3$; solution is 2/3**

For the equations in Exercises 80 and 81, **(a)** *solve for x in terms of y,* **(b)** *solve for y in terms of x.*

80. $4x^2 - 2xy + 3y^2 = 2$
(a) $x = (y \pm \sqrt{8 - 11y^2})/4$
(b) $y = (x \pm \sqrt{6 - 11x^2})/3$

81. $3y^2 + 4xy - 9x^2 = -1$
(a) $x = (2y \pm \sqrt{31y^2 + 9})/9$
(b) $y = (-2x \pm \sqrt{31x^2 - 3})/3$

Graphing Utility Problems

For each of the following equations, use a grapher to solve the equation. Then solve the equation algebraically and compare the results.

82. $2x^2 = x - 1$ **$(1 \pm i\sqrt{7})/4$**

83. $3x^2 + 2 = x$ **$(1 \pm i\sqrt{23})/6$**

84. What limitation on solving equations with a grapher is suggested by Exercises 82 and 83?

2.4 Exercises (text page 91)

Find all real solutions for each of the following equations.

1. $m^4 + 2m^2 - 15 = 0$ **$\pm\sqrt{3}$**

2. $2x^4 + 5x^2 = 3$ **$\pm\sqrt{2}/2$**

3. $5 = 7r^2 - 2r^4$ **$\pm\sqrt{10}/2, \pm 1$**

4. $3 = 8x^2 - 4x^4$ **$\pm\sqrt{2}/2, \pm\sqrt{6}/2$**

5. $(g - 2)^2 + 8 = 6(g - 2)$ **4, 6**

6. $(p + 2)^2 - 2(p + 2) = 15$ **−5, 3**

7. $4p^{-2} - 9p^{-1} = 5$ **$(-9 \pm \sqrt{161})/10$**

8. $6 + 11x^{-1} - 3x^{-2} = 0$ **$(-11 \pm \sqrt{193})/12$**

9. $(y + 3)^{2/3} - 2(y + 3)^{1/3} = 3$ **−4, 24**

10. $(r - 1)^{2/3} = 12 - (r - 1)^{1/3}$ **−63, 28**

11. $3 + \dfrac{5}{p^2 + 1} = \dfrac{2}{(p^2 + 1)^2}$ **no real solutions**

12. $\dfrac{7}{2y - 3} + \dfrac{3}{(2y - 3)^2} = 6$ **4/3, 9/4**

13. $a^3 - 8a^{3/2} + 7 = 0$ **$1, 7^{2/3}$**

14. $r^3 - 13r^{3/2} + 40 = 0$ **$4, 5^{2/3}$**

15. $5(m^2 + 1)^{-2} = 4(m^2 + 1)^{-1} + 1$ **0**

16. $1 + 3(r^2 - 1)^{-1} = 28(r^2 - 1)^{-2}$ **$\pm\sqrt{5}$**

17. $20(2 - \sqrt{m})^2 + 11(2 - \sqrt{m}) = 3$ **121/16, 81/25**

18. $2(1 + 2\sqrt{x})^2 - (1 + 2\sqrt{x}) = 21$ **25/16**

What is wrong with each solution in Exercises 19 and 20?

19. Solve $4x^4 - 11x^2 - 3 = 0$.
Let $t = x^2$.

$$4t^2 - 11t - 3 = 0$$
$$(4t + 1)(t - 3) = 0$$
$$4t + 1 = 0 \quad \text{or} \quad t - 3 = 0$$
$$t = -1/4 \quad \text{or} \quad t = 3$$

The solutions are $-1/4$ and 3.

20. Solve $x = \sqrt{3x + 4}$.
Square both sides to get

$$x^2 = 3x + 4.$$
$$x^2 - 3x - 4 = 0$$
$$(x - 4)(x + 1) = 0$$
$$x - 4 = 0 \quad \text{or} \quad x + 1 = 0$$
$$x = 4 \quad \text{or} \quad x = -1$$

The solutions are 4 and -1.

Solve the following equations.

21. $\sqrt{6m + 7} - 1 = m + 1$ **−1, 3**

22. $\sqrt{3z + 7} = 3z + 5$ **−1**

23. $\sqrt{4x} - x + 3 = 0$ **9**

24. $\sqrt{2t + 4} = t$ **8**

25. $\sqrt{4k + 5} - 2 = 2k - 7$ **5**

26. $\sqrt[3]{2z} = \sqrt[3]{5z + 2}$ **−2/3**

27. $\sqrt[3]{4n + 3} = \sqrt[3]{2n - 1}$ **−2**

28. $\sqrt[3]{2x^2 - 5x + 4} = \sqrt[3]{2x^2}$ **4/5**

29. $\sqrt[3]{t^2 + 2t - 1} = \sqrt[3]{t^2 + 3}$ **2**

30. $(3m + 7)^{1/3} - (4m + 2)^{1/3} = 0$ **5**

31. $(2r + 5)^{1/3} - (6r - 1)^{1/3} = 0$ **3/2**

32. $\sqrt[4]{q - 15} = 2$ **31**

33. $\sqrt[4]{3x + 1} = 1$ **0**

34. $(3t^2 + 52t)^{1/4} - 4 = 0$ **4, −64/3**

35. $(z^2 + 24z)^{1/4} - 3 = 0$ **−27, 3**

36. $\sqrt{2m} = \sqrt{m + 7} - 1$ **2**

37. $\sqrt{y} = \sqrt{y - 5} + 1$ **9**

38. $\sqrt{r + 2} - 1 = \sqrt{3r + 7}$ **no solution**

39. $\sqrt{2p - 5} - 2 = \sqrt{p - 2}$ **27**

▲ **40.** $\sqrt{x + 4} - \sqrt{x + 3} = \sqrt{3x + 10}$ **−3**

▲ **41.** $\sqrt{5x - 1} + \sqrt{2 - x} = \sqrt{8x + 1}$ **2/9, 1**

▲ **42.** $\sqrt{3\sqrt{2m + 3}} = \sqrt{5m - 6}$ **3**

▲ **43.** $\sqrt{2\sqrt{7x + 2}} = \sqrt{3x + 2}$ **2, −2/9**

▲ **44.** $3 - \sqrt{x} = \sqrt{2\sqrt{x} - 3}$ **4**

▲ **45.** $\sqrt{x} + 2 = \sqrt{4 + 7\sqrt{x}}$ **0, 9**

46. $(z - 3)^{2/5} = (4z)^{1/5}$ **1, 9**

47. $(2r - 1)^{2/3} = r^{1/3}$ **1/4, 1**

48. $(2k - 9)^{-2/3} + 4(2k - 9)^{1/3} = 0$ **35/8**

49. $p(2 + p)^{-1/2} + (2 + p)^{1/2} = 0$ **−1**

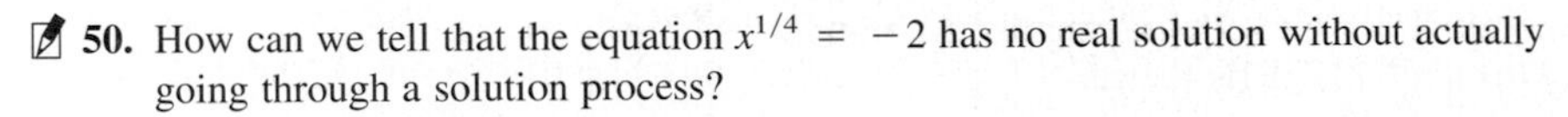

50. How can we tell that the equation $x^{1/4} = -2$ has no real solution without actually going through a solution process?

Solve each equation for the indicated variable.

51. $d = k\sqrt{h}$ for h $\mathbf{h = d^2/k^2}$

52. $v = \dfrac{k}{\sqrt{d}}$ for d $\mathbf{d = k^2/v^2}$

53. $P = 2\sqrt{\dfrac{L}{g}}$ for L $\mathbf{L = P^2g/4}$

54. $c = \sqrt{a^2 + b^2}$ for a $\mathbf{a = \pm\sqrt{c^2 - b^2}}$

55. $x^{2/3} + y^{2/3} = a^{2/3}$ for y $\mathbf{y = \pm(a^{2/3} - x^{2/3})^{3/2}}$

56. $m^{3/4} + n^{3/4} = 1$ for m $\mathbf{m = (1 - n^{3/4})^{4/3}}$

Graphing Utility Problems

Use a grapher to solve each of the following equations. Be careful to use parentheses around the radicand. Check your answers with an algebraic solution.

57. $\sqrt{2x + 1} = 2\sqrt{x}$ **1/2**

58. $\sqrt{5x + 3} = 3\sqrt{x - 1}$ **3**

59. $\sqrt{x + 5} - 2 = \sqrt{x - 1}$ **5/4**

60. $\sqrt{3x - 2} + 1 = \sqrt{3x + 1}$ **1**

2.5 Exercises (text page 95)

Express each of the following as an equation. Use k as the constant of proportionality if none is given.

1. y is proportional to x with a constant of proportionality of 16. $\mathbf{y = 16x}$

2. y varies inversely as x with a constant of variation of 2.6. $\mathbf{y = 2.6/x}$

3. x is inversely proportional to y. $\mathbf{x = k/y}$

4. p varies inversely as y. $\mathbf{p = k/y}$

5. r varies jointly as s and t. $\mathbf{r = kst}$

6. R is proportional to m and p. $\mathbf{R = kmp}$

7. w is proportional to x^2 and inversely proportional to y. $\mathbf{w = kx^2/y}$

8. c varies directly as d and inversely as f^2 and g. $\mathbf{c = kd/(f^2g)}$

Write each of the formulas in Exercises 9–14 as an English phrase using the words varies *or* proportional.

9. $c = 2\pi r$, where c is the circumference of a circle of radius r **The circumference of a circle varies directly as (or is proportional to) the radius.**

10. $d = s/5$, where d is the approximate distance (in miles) from a storm and s is the number of seconds between seeing lightning and hearing thunder **The approximate distance (in miles) from a storm is proportional to (or varies directly as) the number of seconds between seeing lightning and hearing thunder.**

11. $v = d/t$, where v is the average speed when traveling d mi in t hr **The average speed varies directly as the distance and inversely as the time.**

12. $d = 1/(4\pi nr^2)$, where d is the average distance a gas atom of radius r travels between collisions and n is the number of atoms per unit volume **The average distance a gas atom travels varies inversely as the square of its radius and the number of atoms per unit volume.**

13. $s = kx^3$, where s is the strength of a muscle of length x **The strength of a muscle varies directly as (or is proportional to) the cube of its length.**

14. $f = mv^2/r$, where f is the centripetal force of an object of mass m moving along a circle of radius r at velocity v **The centripetal force of an object moving along a circle varies directly as (or is proportional to) its mass and the square of its velocity and varies inversely as the radius of the circle.**

◉ **15.** What happens to y if y varies inversely as x, and x is doubled? **y is half as large as it was before.**

◉ **16.** If y varies directly as x, and x is halved, how is y changed? **y is half as large as it was before.**

◉ **17.** Suppose y is directly proportional to x, and x is replaced by $(1/3)x$. What happens to y? **y is one-third as large as it was before.**

◉ **18.** What happens to y if y is inversely proportional to x, and x is tripled? **y is one-third as large as it was before.**

◉ **19.** Suppose p varies directly as r^3 and inversely as t^2. If r is halved and t is doubled, what happens to p? **p is $1/32$ as large as it was before.**

◉ **20.** If m varies directly as p^2 and q^4, and p doubles while q triples, what happens to m? **m is 324 times as large as it was before.**

21. Simple interest varies jointly as principal, rate, and time. If \$1000 left at interest for 2 yr earned \$110, find the amount of interest earned by \$5000 in 5 yr at the same rate. **\$1375**

22. In electric current flow, the resistance (measured in units called ohms) offered by a fixed length of wire of a given material varies inversely as the square of the diameter of the wire. If a wire .01 inch in diameter has a resistance of .4 ohm, what is the resistance of a wire of the same length and material but .03 inch in diameter? **.044 ohm**

Photographers use the fact that the amount of light required to take a picture varies directly as the square of the F-stop setting and inversely as the ASA number of the film and the shutter speed. For an F-stop of 8, 200 ASA film, and a shutter speed of 1/100 sec, 800 footcandles of light is required. Use this information to solve Exercises 23–25.

23. What F-stop should be used with 200 ASA film and a shutter speed of $1/250$ when 500 footcandles of light is available? **4**

24. What illumination is needed when a photographer is using 400 ASA film, a shutter speed of $1/60$ sec, and an F-stop of 5.6? **117.6 footcandles**

25. If 125 footcandles of light is available and an F-stop of 2 is used with 200 ASA film, what shutter speed should be used? **$1/250$ sec**

26. In a certain manufacturing process, the cost of producing a single item varies inversely as the square of the number of items produced. If 60 items are produced, each costs \$2. Find the cost per item if 400 items are produced. How many items must be produced to reduce the cost to \$1? **\$.045; about 85**

27. Hooke's law for an elastic spring states that the distance a spring stretches varies directly as the force applied. If a force of 15 lb stretches a certain spring 8 inches, how much will a force of 30 lb stretch the spring? See the figure. **16 inches**

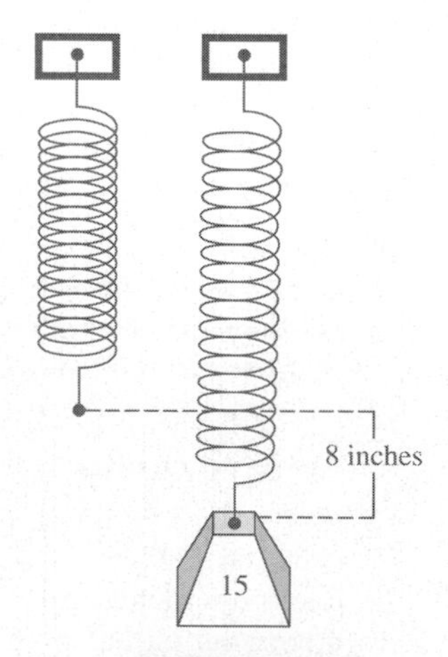

Exercise 27

28. The roof of a new sports arena rests on round concrete pillars. The maximum load a cylindrical column of circular cross-section can hold varies directly as the fourth power of the diameter and inversely as the square of the height. The arena has columns that are 9 m tall, 1 m in diameter, and will support a load of 8 metric tons. How many metric tons will be supported by a column 12 m high and 2/3 m in diameter?
8/9 metric tons

29. The sports arena in Exercise 28 requires a beam 16 m long, 4 cm wide, and 8 cm high. The maximum load of a horizontal beam that is supported at both ends varies directly as the width and the square of the height and inversely as the length between supports. If a beam of the same material 8 m long, 12 cm wide, and 15 cm high can support a maximum of 400 kg, what is the maximum load the beam in the arena will support?
18.96 kg

30. The area of a triangle varies jointly as the lengths of its base and its height. A triangle with a base of 10 ft and a height of 4 ft has an area of 20 square ft. Find the area of a triangle with a base of 3 cm and a height of 8 cm. **12 sq cm**

31. The number of long-distance phone calls between two cities in a certain time period varies directly as the populations p_1 and p_2 of the cities, and inversely as the distance between them. If 10,000 calls are made between two cities 500 mi apart, having populations of 50,000 and 125,000, find the number of calls between two cities 800 mi apart having populations of 20,000 and 80,000. **1600**

32. The distance that a person can see to the horizon from a point above the surface of the earth varies directly as the square root of the height. A person on a hill 121 m high can see 15 km to the horizon. How far can a pilot see from a plane flying at 8100 m?
about 122.7 km

33. The maximum speed possible on a length of railroad track is directly proportional to the cube root of the amount of money spent on maintaining the track. Suppose that a maximum speed of 25 km per hr is possible on a stretch of track for which $450,000 was spent on maintenance. Find the maximum speed if the amount spent on maintenance is increased to $1,750,000. **about 39.3 km/hr**

34. The force needed to keep a car from skidding on a curve varies inversely as the radius of the curve and jointly as the weight of the car and the square of the speed. It takes 3000 lb of force to keep a 2000-lb car from skidding on a curve of radius 500 ft at 30 mph. What force is needed to keep the same car from skidding on a curve of radius 800 ft at 60 mph? **7500 lb**

35. A measure of malnutrition, called the *pelidisi*, varies directly as the cube root of a person's weight in grams and inversely as the person's sitting height in cm. A person with a pelidisi below 100 is considered to be undernourished, while a pelidisi greater than 100 indicates overfeeding. A person who weighs 48,820 g and has a sitting height of 78.7 cm has a pelidisi of 100. Find the pelidisi (to the nearest whole number) of a person whose weight is 54,430 g and whose sitting height is 88.9 cm. Is this individual undernourished or overfed? **92; undernourished**

36. When the brakes of a car are applied, the speed that the car was traveling is proportional to the square root of the distance that the car travels before coming to a stop. Under certain conditions, a car moving at 60 mph will travel 18 m after the brakes are applied. Determine the formula giving speed in terms of the stopping distance.
$d = .005\, s^2$

37. In Exercise 36, how much stopping distance does the car have at 55 mph? **15.125 m**

38. In Exercise 36, what speed would produce a stopping distance of 20 m?
about 63.2 mph

39. Suppose a nuclear bomb is detonated at a certain site. The effects of the bomb will be felt over a distance from the point of detonation that is directly proportional to the cube root of the yield of the bomb. Suppose a 100-kiloton bomb has certain effects to a radius of 3 km from the point of detonation. Find the distance that the effects would be felt for a 1500-kiloton bomb. **about 7.4 km**

40. Under certain conditions, the length of time that it takes for fruit to ripen during the growing season varies inversely as the average maximum temperature during the season. If it takes 25 days for fruit to ripen with an average maximum temperature of 80°F, find the number of days it would take at 75°F. **$26\frac{2}{3}$ days**

Graphing Utility Problems

41. Graph $y = k/x$ for various values of k. Explain how the graph illustrates the idea that as x increases, y decreases.

42. Graph $y = kx$ for various values of k. Explain how the graph illustrates the idea that as x increases, y increases.

43. In Exercises 41 and 42, are negative values of k meaningful? Explain.

2.6 Exercises (text page 104)

Write each of the following in interval notation. Graph each interval.

1. $-3 < x < 2$ **$(-3, 2)$**

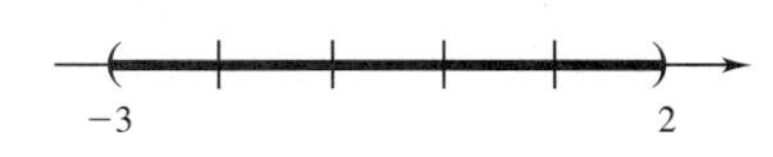

2. $-9 > x$ **$(-\infty, -9)$**

3. $x \leq -4$ **$(-\infty, -4]$**

4. $-4 \geq x \geq -5$ **$[-5, -4]$**

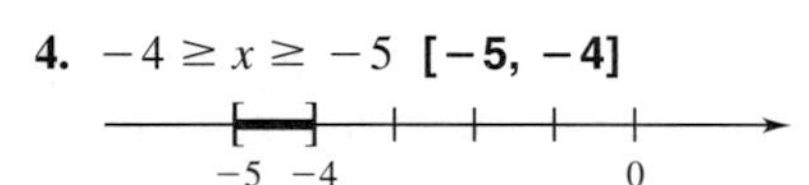

Using the variable x, write each of the following intervals as an inequality.

5. $[-2, 5)$ **$-2 \leq x < 5$**

6. $(-\infty, 3]$ **$x \leq 3$**

7. **$0 < x < 8$**

8. **$-2 \leq x < 6$**

9. Explain how to determine whether a parenthesis or a square bracket should be used when graphing the solution of a linear inequality.

10. The three-part inequality $a < x < b$ means "a is less than x and x is less than b." Which one of the following inequalities is not satisfied by some real number x? **(d)**
(a) $-3 < x < 5$ **(b)** $0 < x < 4$ **(c)** $-3 < x < -2$ **(d)** $-7 < x < -10$

Solve the following inequalities. Write the solutions in interval notation.

11. $4 - 3x \leq 10$ **$[-2, \infty)$**

12. $2y + 7 \geq 12$ **$[5/2, \infty)$**

13. $2(m + 5) - 3m + 1 \geq 5$ **$(-\infty, 6]$**

14. $6m - (2m + 3) \geq 4m - 5$ **$(-\infty, \infty)$**

15. $\dfrac{2x + 4}{5} \leq -2x - 1$ **$(-\infty, -3/4]$**

16. $\dfrac{3z + 7}{-5} \leq 4 - z$ **$(-\infty, 27/2]$**

17. $-2 \leq y - 3 \leq 4$ **$[1, 7]$**

18. $-3 \leq 2t \leq 6$ **$[-3/2, 3]$**

19. $-8 > 2r + 9 > -12$ **$(-21/2, -17/2)$**

20. $5 > 4a - 5 > -2$ **$(3/4, 5/2)$**

21. $-5 < \dfrac{x + 3}{-2} < 4$ **$(-11, 7)$**

22. $-1 < \dfrac{3m - 2}{-3} < 5$ **$(-13/3, 5/3)$**

23. $4 \geq 3 - \dfrac{5}{4}m > -4$ **$[-4/5, 28/5)$**

24. $6 > 2 - \dfrac{3}{2}k > -3$ **$(-8/3, 10/3)$**

25. $x^2 - 6x + 9 < 9$ **$(0, 6)$**

26. $m^2 + 16m + 64 > 64$ **$(-\infty, -16) \cup (0, \infty)$**

27. $r^2 + 4r > -3$ **$(-\infty, -3) \cup (-1, \infty)$**

28. $z^2 + 6z < -8$ **$(-4, -2)$**

29. $4x^2 + 3x + 1 \leq 0$ **no solution**

30. $x^2 + 5x - 2 < 0$ **$((-5 - \sqrt{33})/2, (-5 + \sqrt{33})/2)$**

31. $4m^3 + 7m^2 - 2m > 0$ **$(-2, 0) \cup (1/4, \infty)$**

32. $6p^3 - 11p^2 + 3p > 0$ **$(0, 1/3) \cup (3/2, \infty)$**

33. $\dfrac{m + 2}{m - 4} \leq 0$ **$[-2, 4)$**

34. $\dfrac{r - 3}{r - 1} > 0$ **$(-\infty, 1) \cup (3, \infty)$**

35. $\dfrac{k + 6}{k + 3} > 1$ **$(-3, \infty)$**

36. $\dfrac{a - 2}{a - 5} < -1$ **$(7/2, 5)$**

37. $\dfrac{1}{3k - 5} < \dfrac{1}{3}$ **$(-\infty, 5/3) \cup (8/3, \infty)$**

38. $\dfrac{1}{2m - 7} < \dfrac{5}{4}$ **$(-\infty, 7/2) \cup (39/10, \infty)$**

39. $\dfrac{5}{x - 6} \geq \dfrac{2}{x - 6}$ **$(6, \infty)$**

40. $\dfrac{6}{5 - x} > \dfrac{3}{5 - x}$ **$(-\infty, 5)$**

▲ **41.** $\dfrac{-2}{3r + 2} > \dfrac{1}{r}$ **$(-\infty, -2/3) \cup (-2/5, 0)$**

▲ **42.** $\dfrac{6}{2 - 3h} \geq -\dfrac{2}{h}$ **$(0, 2/3)$**

▲ **43.** $\dfrac{5}{y - 1} \leq \dfrac{2}{y - 2}$ **$(-\infty, 1) \cup (2, 8/3]$**

▲ **44.** $\dfrac{8}{n + 3} < \dfrac{2}{n + 1}$ **$(-\infty, -3) \cup (-1, -1/3)$**

◉ *In each of the following inequalities, the intervals on the number line are $(-\infty, 2)$, $(2, 5)$, and $(5, \infty)$. Without actually solving the inequality, state whether the numbers 2 and 5 will be included in or excluded from the solution of the inequality.*

45. $(x - 2)(x - 5) \geq 0$ **included**

46. $(x - 5)(x - 2) < 0$ **excluded**

47. $\dfrac{x - 2}{x - 5} < 0$ **excluded**

48. $\dfrac{x - 5}{x - 2} \geq 0$ **5 included, 2 excluded**

▲ *Solve the following rational inequalities using methods similar to those used above.*

49. $\dfrac{9x - 8}{4x^2 + 25} < 0$ **$(-\infty, 8/9)$**

50. $\dfrac{(5x - 3)^3}{(8x - 25)^2} \leq 0$ **$(-\infty, 3/5]$**

51. $\dfrac{(9x - 11)(2x + 7)}{(3x - 8)^3} > 0$ **$(-7/2, 11/9) \cup (8/3, \infty)$**

52. Which of the following inequalities have solution set $(-\infty, \infty)$? Explain.

(a) $(x + 3)^2 \geq 0$ **(b)** $(5x - 6)^2 \leq 0$ **(c)** $(6y + 4)^2 > 0$

(d) $(8p - 7)^2 < 0$ **(e)** $\dfrac{x^2 + 7}{2x^2 + 4} \geq 0$ **(f)** $\dfrac{2x^2 + 8}{x^2 + 9} < 0$

53. Which of the inequalities in Exercise 52 has no real solution? Explain.

Use the discriminant to find the values of k for which the following equations have real solutions.

54. $x^2 - kx + 8 = 0$ **$(-\infty, -4\sqrt{2})$ or $(4\sqrt{2}, \infty)$**

55. $x^2 + kx - 5 = 0$ **all real numbers**

56. $x^2 + kx + 2k = 0$ **$(-\infty, 0]$ or $[8, \infty)$**

57. $kx^2 + 4x + k = 0$ **$[-2, 2]$**

In Exercises 58–61, find a quadratic inequality having the given solution.

58. $(-\infty, 2) \cup (5, \infty)$ **$x^2 - 7x + 10 > 0$**

59. $(2, 5)$ **$x^2 - 7x + 10 < 0$**

60. $[-4, 3]$ **$x^2 + x - 12 \leq 0$**

61. $(-\infty, -3] \cup [4, \infty)$ **$x^2 - x - 12 \geq 0$**

In Exercises 62–65, find a rational inequality having the given solution.

62. $(-\infty, -3] \cup (0, \infty)$ **$\dfrac{x + 3}{x} \geq 0$**

63. $[-1, 5)$ **$\dfrac{x + 1}{x - 5} \leq 0$**

64. $(4, 9]$ **$\dfrac{x - 9}{x - 4} \leq 0$**

65. $(-\infty, 4) \cup [9, \infty)$ **$\dfrac{x - 9}{x - 4} \geq 0$**

66. A student attempted to solve the inequality

$$\frac{2x - 1}{x + 2} \leq 0$$

by multiplying both sides by $x + 2$ to get

$$2x - 1 \leq 0$$
$$x \leq \frac{1}{2}.$$

He wrote the solution as $(-\infty, 1/2)$. Is his solution correct? Explain.

67. A student solved the inequality $p^2 \leq 16$ by taking the square root of both sides to get $p \leq 4$. She wrote the solution as $(-\infty, 4]$. Is her solution correct? Explain.

A product will break even or produce a profit only if the revenue from selling the product at least equals the cost of producing it. Find all x-intervals in Exercises 68 and 69 for which the product will at least break even.

68. The cost to produce x underwater cameras is $C = 80x + 10{,}000$; the revenue is $R = 50x$. **The product will never break even.**

69. The cost to produce x chocolate bars is $C = 125x + 5000$; the revenue is $R = 150x$. **$[200, \infty)$**

70. An analyst has found that his company's profits, in hundreds of thousands of dollars, are given by $P = 3x^2 - 35x + 50$, where x is the amount (in hundreds of dollars) spent on advertising. For what values of x does the company make a profit? **$(0, 5/3)$ or $(10, \infty)$**

71. The commodities market is very unstable; money can be made or lost quickly on investments in soybeans, wheat, and so on. Suppose that an investor kept track of her total profit, P, at time t, in months, after she began investing, and found that $P = 4t^2 - 29t + 30$. Find the time intervals in which she has made a profit. (*Hint:* $t > 0$ in this case.) **(0, 5/4) or (6, ∞)**

72. The manager of a large apartment complex has found that the profit is given by $P = -x^2 + 250x - 15{,}000$, where x is the number of apartments rented. For what values of x does the complex produce a profit? **(100, 150)**

73. A projectile is fired from ground level. After t sec its height above the ground is $220t - 16t^2$ ft. For what time period is the projectile at least 624 ft above the ground? **(4, 9.75)**

74. A physicist has found that the velocity (in feet per second) of a moving particle is given by $2t^2 - 5t - 12$, where t is time in seconds since he began his observations. (Here t can be positive or negative; think of t sec before his observations began.) Find the time intervals in which the velocity has been negative. **(−3/2, 4)**

75. Oliver's video club charges an annual fee of \$30 and rents videos for \$2 per day. Stan's video club has no annual fee, but charges \$3 per day to rent videos. Let x be the number of days of rentals during the year.
(a) Express the cost of renting the videos from Oliver in terms of x. **$C = 30 + 2x$**
(b) Express the cost of renting the videos from Stan in terms of x. **$C = 3x$**
(c) Find all x-intervals for which renting from Stan is cheaper. **(0, 30)**

76. Two companies, A and B, offer you a sales position. Both jobs are essentially the same, but Company A pays a straight 7% commission on sales and Company B pays \$100 per week plus 5% commission. Let x be the weekly sales.
(a) Express a week's earnings from Company A in terms of x. **$.07x$**
(b) Express a week's earnings from Company B in terms of x. **$100 + .05x$**
(c) Find all x-intervals for which Company A pays a better salary. **(5000, ∞)**

77. The formula for converting from Celsius to Fahrenheit temperature is $F = 9C/5 + 32$. What temperature range in degrees Fahrenheit corresponds to 0° to 30°C? **32°F to 86°F**

Graphing Utility Problems

Use a grapher to solve the following inequalities.

78. $x^3 - 2x + 3 > 0$ **(−1.89, ∞)**

79. $2x^4 - 5x^3 - 5x - 1 < 0$ **(−1.95, 2.83)**

80. $\dfrac{x}{x^2 - 9} < 0$ **(−∞, −3) ∪ (0, 3)**

2.7 Exercises (text page 112)

Solve each of the following equations.

1. $|x + 3| = 5$ **−8, 2**

2. $|x - 8| = 1$ **7, 9**

3. $|2m + 6| = 4$ **−5, −1**

4. $|3p - 5| = 7$ **−2/3, 4**

5. $|6 - 2x| + 4 = -3$ **no solution**

6. $|-5a + 1| - 2 = -4$ **no solution**

7. $\left|\dfrac{z - 5}{3}\right| = 12$ **−31, 41**

8. $\left|\dfrac{m + 1}{5}\right| = 9$ **−46, 44**

9. $\left|\dfrac{4}{r - 2}\right| = 7$ **10/7, 18/7**

10. $\left|\frac{1}{2h+3}\right| = 5$ **−8/5, −7/5**

11. $\left|\frac{4y+3}{y+2}\right| = 5$ **−7, −13/9**

12. $\left|\frac{3a-4}{2a+3}\right| = 1$ **7, 1/5**

13. $|2k-3| = |5k+4|$ **−7/3, −1/7**

14. $|p+1| = |3p-1|$ **1, 0**

15. $|4-3y| = |7+2y|$ **−3/5, 11**

16. $|2+5a| = |4-6a|$ **2/11, 6**

Solve each of the following inequalities. Give the solution in interval notation.

17. $|m| > 1$ **$(-\infty, -1) \cup (1, \infty)$**

18. $|z| > 4$ **$(-\infty, -4) \cup (4, \infty)$**

19. $|a| < -2$ **no solution**

20. $|b| > -5$ **$(-\infty, \infty)$**

21. $|x| - 3 \le 7$ **$[-10, 10]$**

22. $|r| + 3 \le 10$ **$[-7, 7]$**

23. $|4x-1| < 9$ **$(-2, 5/2)$**

24. $|4-3x| < 3$ **$(1/3, 7/3)$**

25. $|2m-5| > 10$ **$(-\infty, -5/2) \cup (15/2, \infty)$**

26. $|6x-3| > 9$ **$(-\infty, -1) \cup (2, \infty)$**

27. $|4z+3| \ge 7$ **$(-\infty, -5/2] \cup [1, \infty)$**

28. $|6b+1| \ge 19$ **$(-\infty, -10/3] \cup [3, \infty)$**

29. $\left|2x+\frac{1}{4}\right| - 1 < 7$ **$(-33/8, 31/8)$**

30. $\left|3x+\frac{2}{3}\right| + 2 < 4$ **$(-8/9, 4/9)$**

31. $\left|\frac{3x-2}{x}\right| < 2$ **$(2/5, 2)$**

32. $\left|\frac{6-2y}{y}\right| < 4$ **$(-\infty, -3) \cup (1, \infty)$**

33. $\left|\frac{4+y}{y-2}\right| > 3$ **$(1/2, 2) \cup (2, 5)$**

34. $\left|\frac{5-5p}{p+2}\right| > 0$ **$(-\infty, -2) \cup (-2, 1) \cup (1, \infty)$**

35. $\left|\frac{4}{q-1}\right| \le 0$ **no solution**

36. $\left|\frac{3}{t-4}\right| \le 0$ **no solution**

37. Explain why it is incorrect to write the absolute value inequality $|x| > 6$ in any of the following ways: $-6 > x > 6$, $-6 > x < 6$, or $-6 < x > 6$.

38. Without actually going through the solution process, we can say that the equation $|5x-6| = 6x$ cannot have a negative solution. Explain why this is true.

Solve the equations in Exercises 39–46.

39. $|x+1| = 2x$ **1**

40. $|m-3| = 4m$ **3/5**

41. $|2k-1| = k+2$ **3, −1/3**

42. $|3r+5| = r+3$ **−1, −2**

43. $|p| = |p|^2$ **0, ±1**

44. $5|m| = |m|^2$ **0, ±5**

45. $|1-6q|^2 - 4|1-6q| - 45 = 0$ **5/3, −4/3**

46. $|6a-5|^2 + 4|6a-5| - 12 = 0$ **7/6, 1/2**

47. Is $|a-b|^2$ always equal to $(b-a)^2$? **yes**

48. The temperatures on the surface of Mars in degrees Celsius approximately satisfy the inequality $|C+84| \le 56$. What range of temperatures corresponds to this inequality? **$[-140, -28]$**

49. Dr. Tydings has found that, over the years, 95% of the babies that she has delivered have weighed y lb, where $|y-8.0| \le 1.5$. What range of weights corresponds to this inequality? **$[6.5, 9.5]$**

50. The industrial process that is used to convert methanol to gasoline is carried out at a temperature range of 680° F to 780° F. Using F as the variable, write an absolute value inequality that corresponds to this range. **$|F - 730| \le 50$**

51. When a model kite was flown in crosswinds in tests to determine its limits of power extraction, it attained speeds of 98 to 148 ft per sec in winds of 16 to 26 ft per sec. Using x as the variable in each case, write absolute value inequalities that correspond to these ranges. **$|x - 123| \le 25$, $|x - 21| \le 5$**

▲ *Solve each inequality in Exercises 52–55.*

52. $|m^2 + 1| < 3$ $\mathbf{(-\sqrt{2}, \sqrt{2})}$

53. $|z^2 - 5| \geq 4$ $\mathbf{(-\infty, -3] \cup [-1, 1] \cup [3, \infty)}$

54. $|m + 6| \leq |2m - 3|$
(*Hint:* First divide each side by $|2m - 3|$.)
$\mathbf{(-\infty, -1] \cup [9, \infty)}$

55. $|1 - 3x| < 2|x + 5|$ $\mathbf{(-9/5, 11)}$

◉ *Write each statement in Exercises 56–62 using absolute value notation. For example, write "k is at least 4 units from 1" as $|k - 1| \geq 4$.*

56. x is within 4 units of 2. $\mathbf{|x - 2| \leq 4}$

57. z is no less than 2 units from 12. $\mathbf{|z - 12| \geq 2}$

58. p is at least 5 units from 9. $\mathbf{|p - 9| \geq 5}$

59. k is 6 units from 1. $\mathbf{|k - 1| = 6}$

60. r is 5 units from 3. $\mathbf{|r - 3| = 5}$

61. If x is within .0004 units of 2, then y is within .00001 units of 7. **If $|x - 2| \leq .0004$, then $|y - 7| \leq .00001$.**

62. y is within 10^{-6} units of 10 whenever x is within 2×10^{-4} units of 5.
If $|x - 5| \leq 2 \times 10^{-4}$, then $|y - 10| \leq 10^{-6}$.

◉ **63.** If $|x - 2| < 3$, find the values of m and n such that $m < 3x + 5 < n$.
$\mathbf{m = 2, n = 20}$

◉ **64.** If $|x + 8| < 16$, find the values of p and q such that $p < 2x - 1 < q$.
$\mathbf{p = -49, q = 15}$

◉ **65.** If $|x - 1| < 10^{-6}$, show that $|7x - 7| < 10^{-5}$.

Graphing Utility Problems

Use a grapher to solve the following inequalities.

66. $|1 - x| > x + 1$ $\mathbf{(-\infty, 0)}$

67. $|2x + 3| < x - 1$ **no solution**

68. $|3x + 5| < |x - 1|$ $\mathbf{(-3, -1)}$

Chapter 2 Review Exercises (text page 113)

Solve the following equations and check your answers.

1. $2m + 7 = 3m + 1$ **6**

2. $4k - 2(k - 1) = 12$ **5**

3. $\frac{y}{3} - \frac{2y}{5} = 6 + \frac{y}{2}$ **−180/17**

4. $\frac{x + 1}{2} = \frac{2x - 5}{3}$ **13**

5. $\frac{10}{4z - 4} = \frac{1}{1 - z}$ **no solution**

6. $(m + 2)(3m - 1) = 3m^2 + 5m$
no solution

Solve each of the following for the indicated variable.

7. $2x - 5k = 2(kx + 3)$ for x $\mathbf{x = (6 + 5k)/(2 - 2k)}$

8. $F = \frac{9}{5}C + 32$ for C $\mathbf{C = (5/9)(F - 32)}$

9. $A = P + Pi$ for P $\mathbf{P = A/(1 + i)}$

10. $A = I\left(1 - \frac{j}{n}\right)$ for j $\mathbf{j = n - nA/I}$

11. $\frac{1}{k} = \frac{1}{r_1} + \frac{1}{r_2}$ for r_1 $\mathbf{r_1 = kr_2/(r_2 - k)}$

12. $P(r + R)^2 = PE^2R$ for E $\mathbf{E = |r + R|\sqrt{R}/R}$

13. $\frac{xy^2 - 5xy + 4}{3x} = 2p$ for x $\mathbf{x = -4/(y^2 - 5y - 6p)}$

Solve each of the following problems.

14. A computer printer is on sale for 15% off. The sale price is $425. What was the original price? **$500**

15. A realtor borrowed $90,000 to develop some property. He was able to borrow part of the money at 10.5% interest and the rest at 9%. The annual interest on the two loans amounts to $8925. How much was borrowed at each rate? **$55,000 at 10.5% and $35,000 at 9%**

16. An excursion boat travels upriver to a landing and then returns to its starting point. The trip upriver takes 1.2 hr and the trip back takes .9 hr. If the average speed on the return trip is 5 mph faster than on the trip upriver, what is the boat's speed upriver? **15 mph**

17. Wei-jen and Alan Luan are canvassing their neighborhood for their candidate for the school board. Alan can canvass the entire neighborhood alone in 6 hr. Working together, they complete the job in 4 hr. How long would it take Wei-jen to canvass the neighborhood, working alone? **12 hr**

Solve each equation.

18. $(b + 7)^2 = 5$ $\mathbf{-7 \pm \sqrt{5}}$

19. $(3y - 2)^2 = 8$ $\mathbf{(2 \pm 2\sqrt{2})/3}$

20. $2a^2 + a - 15 = 0$ **5/2, −3**

21. $12x^2 = 8x - 1$ **1/2, 1/6**

22. $3x^2 + 2x = -5$ $\mathbf{(-1 \pm i\sqrt{14})/3}$

23. $\sqrt{2}x^2 - 4x + \sqrt{2} = 0$ $\mathbf{\sqrt{2} \pm 1}$

24. $2 - \frac{4}{y} = \frac{21}{y^2}$ $\mathbf{(2 \pm \sqrt{46})/2}$

25. $2 + \frac{4}{a} + \frac{3}{a^2} = 0$ $\mathbf{(-2 \pm i\sqrt{2})/2}$

26. Discuss the method you chose (or would choose) to solve Exercises 19, 21, and 23 and explain why you made that choice.

Evaluate the discriminant for each of the following, and use it to predict the type of solutions for the equation.

27. $4p^2 + 8 = 3p$ **−119; two complex (nonreal) solutions**

28. $5x^2 - 2x = 7$ **144; two rational solutions**

29. $2k^2 + k - 1 = 0$ **9; two rational solutions**

30. $7m^2 - 2m = 5$ **144; two rational solutions**

31. $5x^2 = 8x + 1$ **84; two irrational solutions**

32. $4x^2 + 4x + 5 = 0$ **−64; two complex (nonreal) solutions**

Solve the following problems.

33. Tony Romero has a rectangular-shaped flower box that measures 4 ft by 6 ft. He wants to double the available area by increasing the length and width by the same amount. What should the new dimensions be? **6 ft by 8 ft**

34. Paula Cunningham plans to replace the vinyl floor covering in her 10-by-12-ft kitchen. She wants to have a border of a special material with the same width on each side. She can afford only 21 sq ft of this material. How wide a border can she have? **1/2 ft**

35. It takes two gardeners 3 hr (working together) to mow the lawns in a city park. One gardener could do the entire job in 1 hr less time than the other. How long would it take the slower gardener to complete the work alone? Give the answer to the nearest tenth of an hour. **6.5 hr**

36. In a marathon (a 26-mi run), the winner finished 2/5 hr before a friend. If the difference in their rates was 4/3 mph, what was the winner's average speed in the race? **10 mph**

37. Suppose that one solution of the equation $km^2 - 11m = 3$ is 3. Find the value of k and the other solution. $\mathbf{k = 4, m = -1/4}$

Solve each equation. Check your answers.

38. $3y^4 + 2y^2 = 16$ $\quad \pm\sqrt{2}, \pm 2i\sqrt{6}/3$

39. $2(z - 1)^2 - 11(z - 1) = 21$ $\quad -1/2, 8$

40. $(2n + 3)^{2/3} + (2n + 3)^{1/3} = 6$ $\quad -15, 5/2$

41. $\sqrt{2m - 1} = 3\sqrt{m}$ **no solution**

42. $\sqrt{4y + 3} = \sqrt{2y - 5}$ **no solution**

43. $\sqrt{3x + 4} = 2x + 1$ $\quad 3/4$

44. $x - 2 = \sqrt{3x - 6}$ $\quad 2, 5$

45. $\sqrt{k} = \sqrt{k + 3} - 1$ $\quad 1$

46. $\sqrt{x^2 + 3x} - 2 = 0$ $\quad 1, -4$

47. $\sqrt[3]{2y - 3} = \sqrt[3]{y + 1}$ $\quad 4$

48. $\sqrt[3]{5z} = \sqrt[3]{8z + 4}$ $\quad -4/3$

49. $\sqrt{3 + x} = \sqrt{3x + 7} - 2$ $\quad 6$

50. $\sqrt{4 + 3y} = \sqrt{y + 5} + 1$ $\quad 4$

51. $(x - 2)^{2/3} = x^{1/3}$ $\quad 4, 1$

Write each of the statements below as an equation.

52. Y varies jointly as M and the square of N and inversely as the cube of X. $Y = (kMN^2)/X^3$

53. A varies jointly as the third power of t and the fourth power of s, and inversely as p and the square of h. $A = kt^3s^4/(ph^2)$

Solve each problem below.

54. The power a windmill obtains from the wind varies directly as the cube of the wind velocity. If a wind blowing at 10 km per hr produces 10,000 units of power, how much power is produced by a wind of 15 km per hr? **33,750 units**

55. Hooke's law for an elastic spring states that the distance a spring stretches varies directly as the force applied. If a force of 32 lb stretches a certain spring 48 inches, how much will a force of 24 lb stretch the spring? **36 inches**

56. The weight of an object varies inversely as the square of the distance between the object and the center of the earth. If a man weighs 90 kg on the surface of the earth, how much would he weigh 800 km above the surface? (The radius of the earth is about 6400 km.) **about 71.1 kg**

57. The force of the wind on a sail varies jointly as the area of the sail and the square of the wind velocity. If the force is 8 lb when the velocity is 15 mph and the area is 3 sq ft, find the force when the area is 6 sq ft and the velocity is 22.5 mph. **36 lb**

58. Without actually solving the inequality, explain why 3 cannot be in the solution set of

$$\frac{2x + 5}{x - 3} < 0.$$

59. What is wrong with the following "solution" of $\frac{1}{x-3} \geq 2$?

$$(x-3)\left(\frac{1}{x-3}\right) \geq (x-3)(2) \quad \textbf{Multiply by } x - 3.$$
$$1 \geq 2x - 6 \quad \textbf{Distributive property}$$
$$7 \geq 2x$$
$$\frac{7}{2} \geq x$$

The solution is $(-\infty, 7/2)$.

Solve each inequality. Write solutions in interval notation.

60. $-a + 3 \geq 2(5a + 1)$ **$(-\infty, 1/11]$**

61. $-(2k + 4) < k - 5$ **$(1/3, \infty)$**

62. $2r - 5 > -r + 3(r + 2)$ **no solution**

63. $4m - 3(2m + 2) \leq 6(1 - m)$ **$(-\infty, 3]$**

64. $5 \leq 2x - 3 \leq 7$ **$[4, 5]$**

65. $-8 < 3a - 5 < -1$ **$(-1, 4/3)$**

66. $-5 < \frac{2p - 1}{-3} \leq 2$ **$[-5/2, 8)$**

67. $3 < \frac{6z + 5}{-2} < 7$ **$(-19/6, -11/6)$**

68. $x^2 + 3x - 4 \leq 0$ **$[-4, 1]$**

69. $p^2 + 4p > 21$ **$(-\infty, -7) \cup (3, \infty)$**

70. $z^3 - 16z \leq 0$ **$(-\infty, -4] \cup [0, 4]$**

71. $2r^3 - 3r^2 - 5r < 0$ **$(-\infty, -1) \cup (0, 5/2)$**

72. $\frac{3a - 2}{a} > 4$ **$(-2, 0)$**

73. $\frac{5p + 2}{p} < -1$ **$(-1/3, 0)$**

74. $\frac{3}{r - 1} \leq \frac{5}{r + 3}$ **$(-3, 1) \cup [7, \infty)$**

75. $\frac{3}{x + 2} > \frac{2}{x - 4}$ **$(-2, 4) \cup (16, \infty)$**

76. If $a < b$, on what x-interval is $(x - a)(x - b)$ positive? Negative? Where is the product zero? **$(-\infty, a) \cup (b, \infty)$; (a, b); at $x = a$ and $x = b$**

77. On what x-interval is $(x - a)^2$ positive? Negative? Where is it zero? **$(-\infty, a) \cup (a, \infty)$; at $x = a$**

Work the following problems.

78. A projectile is thrown upward. Its height in feet above the ground after t sec is $320t - 16t^2$.
 (a) After how many seconds in the air will it hit the ground? **20 sec**
 (b) During what time interval is the projectile more than 576 ft above the ground? **for values of t in $(2, 18)$**

79. A company that produces videotapes has found that the revenue from the sale of x units of tapes is $R = 8x$. The cost to produce x units of tapes is $C = 3x + 1500$. In what interval will the company at least break even? **$[300, \infty)$**

Solve each equation.

80. $|a + 4| = 7$ **$-11, 3$**

81. $|-y + 2| = -4$ **no solution**

82. $\left|\frac{7}{2 - 3a}\right| = 9$ **$11/27, 25/27$**

83. $\left|\frac{8r - 1}{3r + 2}\right| = 7$ **$-15/13, -13/29$**

84. $|5r - 1| = |2r + 3|$ **$4/3, -2/7$**

85. $|k + 7| = |k - 8|$ **$1/2$**

Solve each inequality. Write solutions with interval notation.

86. $|m - 3| \le 7$ **[−4, 10]**

87. $|3 - r| < 2$ **(1, 5)**

88. $|z - 2| \ge -1$ **(−∞, ∞)**

89. $|2x + 9| \le 3$ **[−6, −3]**

90. $|5m - 8| \le 2$ **[6/5, 2]**

91. $|7k - 3| > 5$ **(−∞, −2/7) ∪ (8/7, ∞)**

92. $|2p - 1| > 2$ **(−∞, −1/2) ∪ (3/2, ∞)**

93. $|3r + 7| - 5 > 0$ **(−∞, −4) ∪ (−2/3, ∞)**

94. Solve the following equations and inequalities. Compare the solutions.

(a) $\dfrac{x-1}{2x+5} = 4$ (b) $\dfrac{x-1}{2x+5} < 4$ (c) $\dfrac{x-1}{2x+5} > 4$

(d) $\left|\dfrac{x-1}{2x+5}\right| = 4$ (e) $\left|\dfrac{x-1}{2x+5}\right| < 4$ (f) $\left|\dfrac{x-1}{2x+5}\right| > 4$

95. Homing pigeons avoid flying over large bodies of water, preferring to fly around them instead. (One possible explanation is the fact that extra energy is required to fly over water because air pressure drops over water in the daytime.) Assume that a pigeon released from a boat 1 mi from the shore of a lake (point B in the figure) flies first to point P on the shore and then along the straight edge of the lake to reach its home at L. If L is 2 mi from point A, the point on the shore closest to the boat, and if a pigeon needs 4/3 as much energy to fly over water as over land, find an expression for the total energy expended, assuming 1 unit of energy per mi over land.
$(4/3)\sqrt{1 + x^2} + 2 - x$

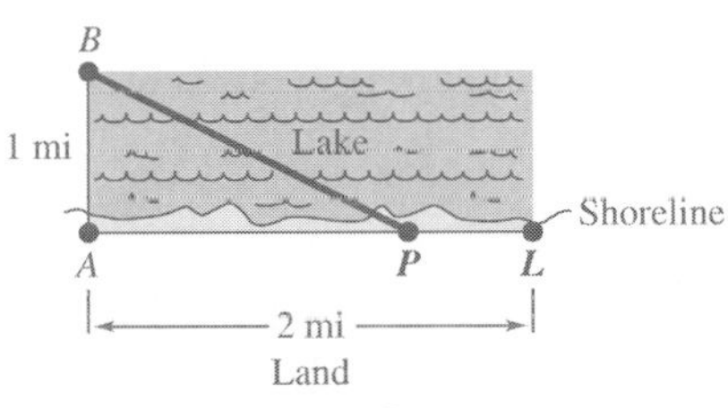

Exercise 95

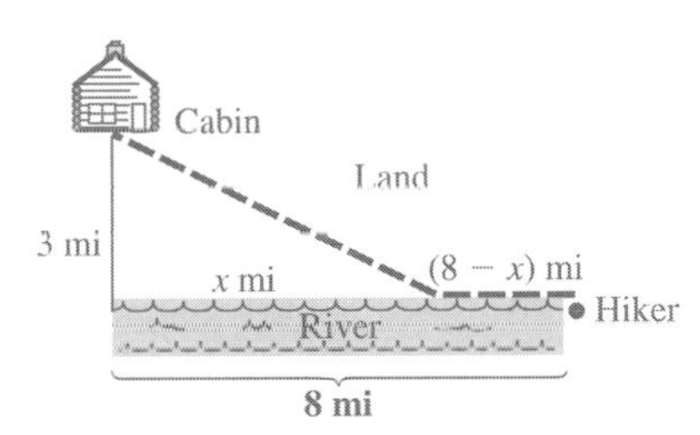

Exercise 96

96. A hiker is at a point on a riverbank. He wants to get to his cabin, located 3 mi north and 8 mi west (see the figure). He can travel 5 mph along the river but only 2 mph on the very rocky ground away from the river. If he travels $8 - x$ mi along the river and then walks in a straight line to the cabin, find an expression for the total time that he travels. **$(8 - x)/5 + \sqrt{9 + x^2}/2$**

*In Exercises 97 and 98, solve for x and express the solution in terms of intervals.**

97. $\dfrac{1}{|x-4|} < \dfrac{1}{|x+7|}$ **(−∞, −7) ∪ (−7, −3/2)**

98. $\dfrac{1}{|x-3|} - \dfrac{1}{|x+4|} \ge 0$ **[−1/2, 3) ∪ (3, ∞)**

99. Find the smallest value of M such that $|1/x| \le M$ for all x in the interval [2, 7].
$M = 1/2$

100. Find the smallest value of M such that $|1/(x + 7)| \le M$ for all x in the interval $(-4, 2)$. **$M = 1/3$**

*Exercises 97–100 from *Calculus with Analytic Geometry* by Howard Anton, pp. 26, 27. Copyright © 1992 by Anton Textbooks, Inc. Reprinted by permission of John Wiley & Sons, Inc.

 **Writing** 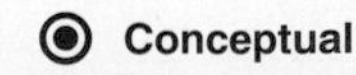 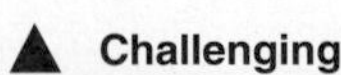 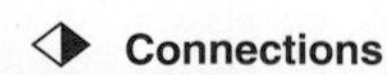

Writing Conceptual Challenging Connections

3 Functions and Graphs

3.1 Exercises (text page 127)

If the point (a, b) is in the third quadrant, in what quadrant is each of the following points?

1. $(-a, b)$ **IV**
2. $(a, -b)$ **II**
3. (b, a) **III**
4. $(-a, -b)$ **I**

In Exercises 5–10, graph the set of all points satisfying the conditions for ordered pairs (x, y).

5. $y = 0$

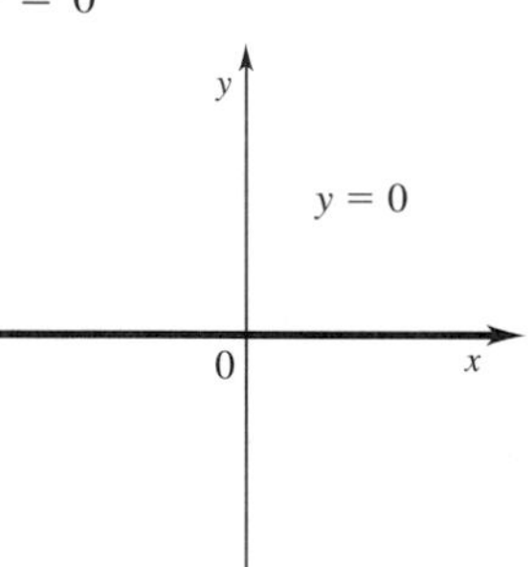

6. $y \le 0$

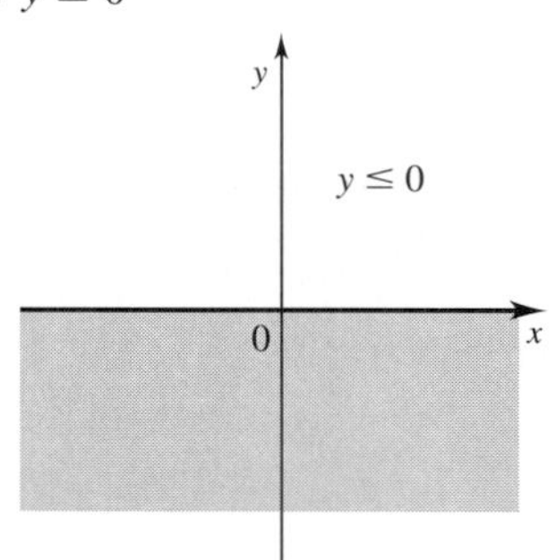

7. $x > 0$

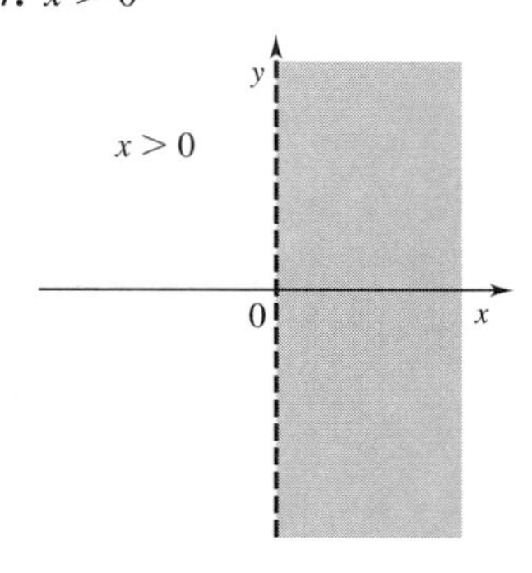

8. $x = 0$

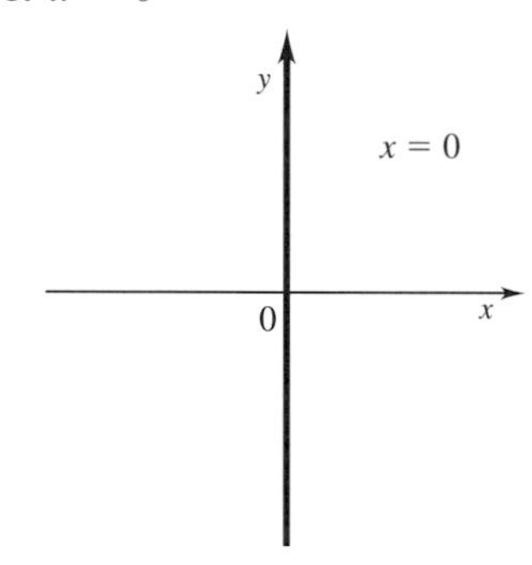

9. $x/y < 0$

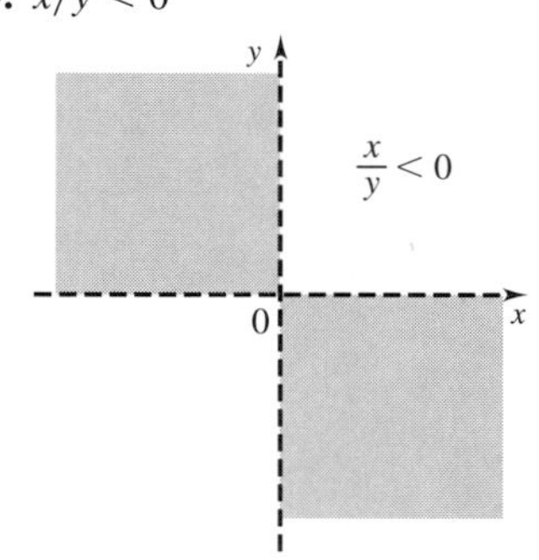

10. $xy > 0$

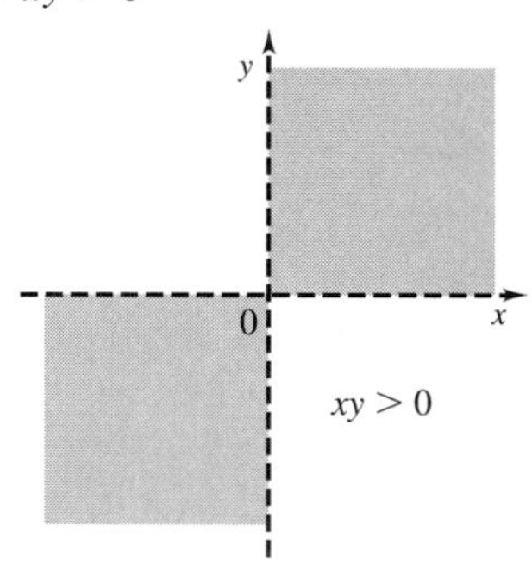

Find the distance $d(P, Q)$ and the coordinates of the midpoint of segment PQ. If necessary, round the distance to the nearest thousandth.

11. $P(4, 8)$, $Q(-1, 3)$ **7.071; (1.5, 5.5)**
12. $P(5, -2)$, $Q(3, -4)$ **2.828; (4, −3)**
13. $P(-7, -5)$, $Q(5, 6)$ **16.279; (−1, 1/2)**
14. $P(-2, -6)$, $Q(-4, -3)$ **3.606; (−3, −9/2)**
15. $P(\sqrt{2}, -\sqrt{5})$, $Q(3\sqrt{2}, 4\sqrt{5})$ **11.533; $(2\sqrt{2}, 3\sqrt{5}/2)$**
16. $P(5\sqrt{7}, -\sqrt{3})$, $Q(-\sqrt{7}, 8\sqrt{3})$ **22.249; $(2\sqrt{7}, 7\sqrt{3}/2)$**
17. $P(3, -7)$, $Q(-5, 19)$ **27.203; (−1, 6)**
18. $P(-9, -2)$, $Q(-1, -15)$ **15.264; (−5, −17/2)**
19. The distance formula equates the distance between two points in a plane with the square root of a quantity. Explain why this formula always produces a nonnegative result.

Find the coordinates of the other endpoint of the segments with endpoints and midpoints having the given coordinates.

20. Endpoint $(-3, 6)$, midpoint $(5, 8)$ **(13, 10)**

21. Endpoint $(2, -8)$, midpoint $(3, -5)$ **(4, −2)**

22. Endpoint $(6, -1)$, midpoint $(-2, 5)$ **(−10, 11)**

23. Endpoint $(-5, 3)$, midpoint $(-7, 6)$ **(−9, 9)**

Decide whether or not the following points are the vertices of a right triangle.

24. $(-2, 5), (1, 5), (1, 9)$ **yes**

25. $(-9, -2), (-1, -2), (-9, 11)$ **yes**

26. $(-4, 0), (1, 3), (-6, -2)$ **no**

27. $(-8, 2), (5, -7), (3, -9)$ **yes**

Use the distance formula to decide whether or not the following points lie on a straight line. (Hint: *The points lie on a straight line if the sum of the two smallest distances equals the largest distance.*)

28. $(0, 7), (3, -5), (-2, 15)$ **yes**

29. $(1, -4), (2, 1), (-1, -14)$ **yes**

30. $(0, -9), (3, 7), (-2, -19)$ **no**

31. $(1, 3), (5, -12), (-1, 11)$ **no**

Find all values of x or y such that the distance between the given points is as indicated.

32. $(3, y)$ and $(-2, 9)$ is 12 **$9 + \sqrt{119}, 9 - \sqrt{119}$**

33. $(x, 11)$ and $(5, -4)$ is 17 **13, −3**

34. (x, x) and $(2x, 0)$ is 4 **$2\sqrt{2}, -2\sqrt{2}$**

35. (y, y) and $(0, 4y)$ is 6 **$3\sqrt{10}/5, -3\sqrt{10}/5$**

36. What does the graph of the equation $(x - h)^2 + (y - k)^2 = A$ look like if $A < 0$? If $A = 0$? Explain why.

Graph each of the following.

37. $x^2 + y^2 = 36$

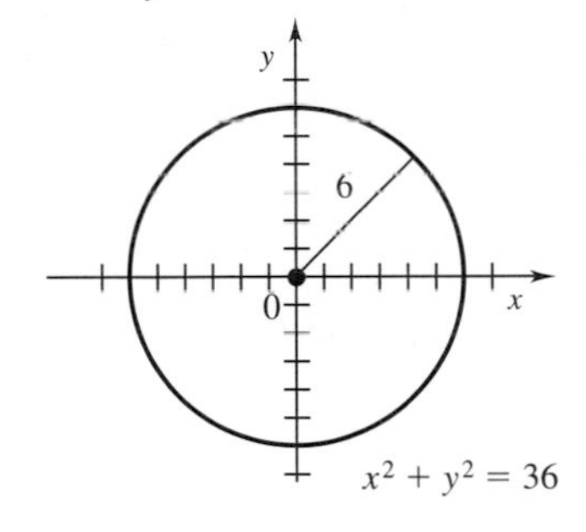

38. $x^2 + y^2 = 81$

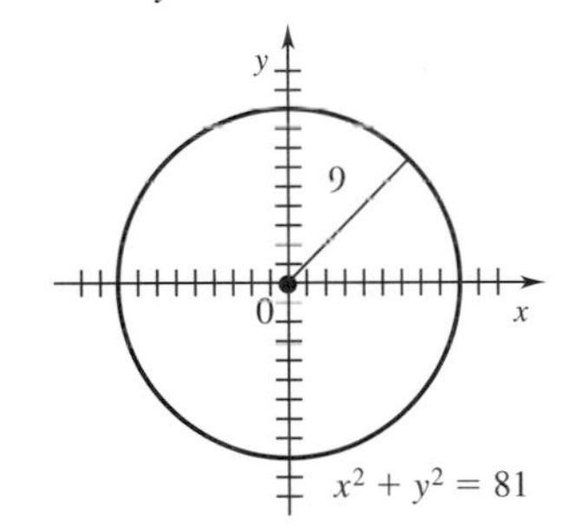

39. $x^2 - 4x + y^2 + 12y + 4 = 0$
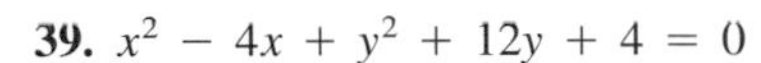

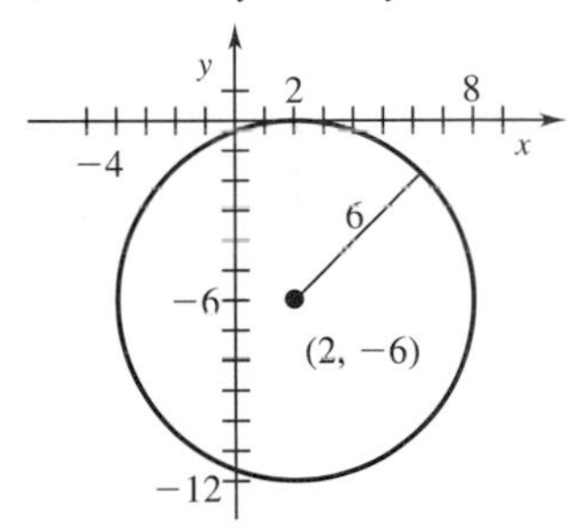

40. $x^2 + 6x + y^2 + 8y = -9$

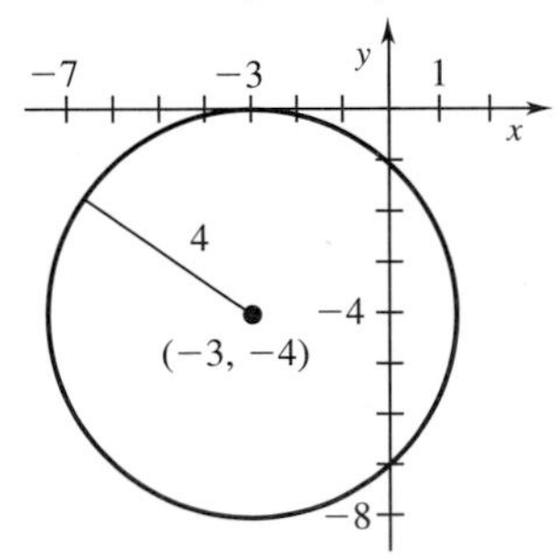

41. $x^2 - 2x + y^2 = -1$
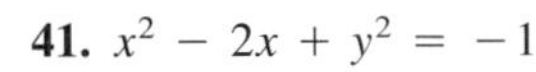

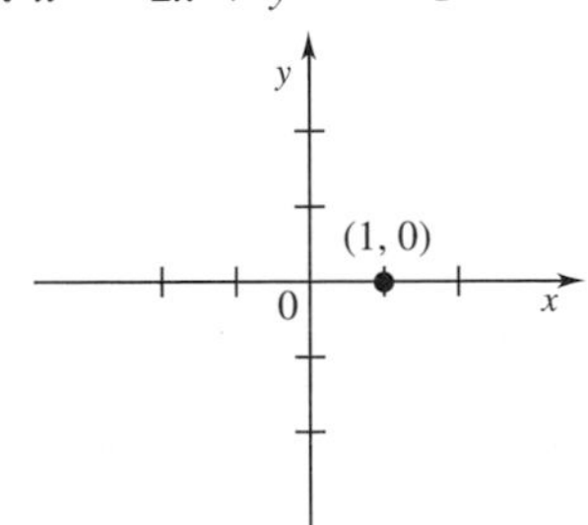

42. $x^2 + 8x + y^2 - 14y + 65 = 0$
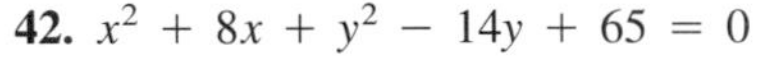

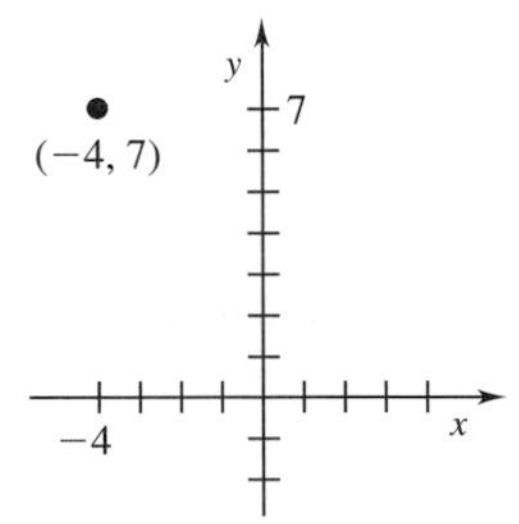

43. $3x^2 + 12x + 3y^2 = 63$

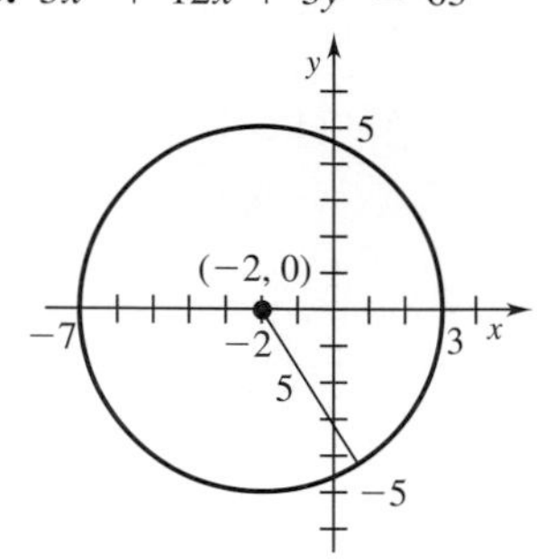

44. $2x^2 + 2y^2 = 4y + 96$

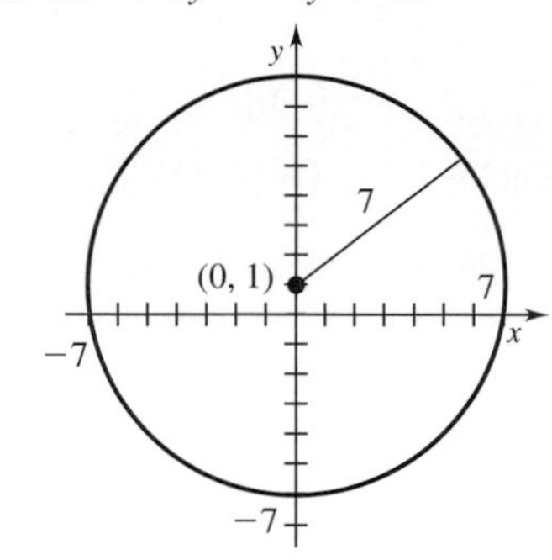

Find equations for each of the following circles.

45. Center (1, 4), radius 3 $\mathbf{(x - 1)^2 + (y - 4)^2 = 9}$

46. Center (−2, 5), radius 4 $\mathbf{(x + 2)^2 + (y - 5)^2 = 16}$

47. Center (−8, 6), radius 5 $\mathbf{(x + 8)^2 + (y - 6)^2 = 25}$

48. Center (3, −2), radius 2 $\mathbf{(x - 3)^2 + (y + 2)^2 = 4}$

49. Center (−1, 2), passing through (2, 6) $\mathbf{(x + 1)^2 + (y - 2)^2 = 25}$

50. Center (2, −7), passing through (−2, −4) $\mathbf{(x - 2)^2 + (y + 7)^2 = 25}$

51. Center (−3, −2), tangent to the x-axis $\mathbf{(x + 3)^2 + (y + 2)^2 = 4}$

52. Center (5, −1), tangent to the y-axis $\mathbf{(x - 5)^2 + (y + 1)^2 = 25}$

53. Endpoints of a diameter at (1, 5) and (7, 8) $\mathbf{(x - 4)^2 + (y - 13/2)^2 = 45/4}$

54. Endpoints of a diameter at (3, −5) and (−7, 2) $\mathbf{(x + 2)^2 + (y + 3/2)^2 = 149/4}$

Use the appropriate formulas to find the circumference and area of each of the circles with equations as follows.

55. $(x - 2)^2 + (y + 4)^2 = 25$ $\mathbf{C = 10\pi, A = 25\pi}$

56. $(x + 3)^2 + (y - 1)^2 = 9$ $\mathbf{C = 6\pi, A = 9\pi}$

The unit circle *is a circle centered at the origin with a radius of 1. In Exercises 57 and 58, show that each point lies on the unit circle.*

57. $(-\sqrt{2}/2, -\sqrt{2}/2)$

58. $(-1/2, \sqrt{3}/2)$

59. Find all points (x, y) with $x = y$ that are 4 units from (1, 3).
$\mathbf{(2 + \sqrt{7}, 2 + \sqrt{7}), (2 - \sqrt{7}, 2 - \sqrt{7})}$

60. Find all points satisfying $x + y = 0$ that are 8 units from (−2, 3).
$$\left(\frac{-5 + \sqrt{127}}{2}, \frac{5 - \sqrt{127}}{2}\right), \left(\frac{-5 - \sqrt{127}}{2}, \frac{5 + \sqrt{127}}{2}\right)$$

61. Decide whether each of the following points is *inside*, *on*, or *outside* the circle with center (1, −4) and radius 6: **(a)** (3, −2) **(b)** (9, 1) **(c)** (7, −4) **(d)** (0, 9).
(a) inside (b) outside (c) on (d) outside

62. A circle is tangent to both axes, has its center in the third quadrant, and has a radius of $\sqrt{2}$. Find an equation for the circle. $\mathbf{(x + \sqrt{2})^2 + (y + \sqrt{2})^2 = 2}$ **or**
$\mathbf{x^2 + 2\sqrt{2}x + y^2 + 2\sqrt{2}y + 2 = 0}$

63. Find the coordinates of a point whose distance from (1, 0) is $\sqrt{10}$ and whose distance from (5, 4) is $\sqrt{10}$. **(2, 3) or (4, 1)**

64. Find the equation of the circle of smallest radius that contains the points (1, 4) and (−3, 2) within or on its boundary. $\mathbf{(x + 1)^2 + (y - 3)^2 = 5}$

65. One circle has center at (3, 4) and radius 5. A second circle has center at (−1, 3) and radius 4. Do the circles cross? **yes**

66. Does the circle with radius 6 and center at (0, 5) cross the circle with radius 4 and center at (−5, −4)? **no**

67. One circle has center (a, b) and radius 4, while another has center (c, d) and radius s. State an inequality that is true if and only if the two circles intersect.
$\mathbf{|4 - s| \le \sqrt{(a - c)^2 + (b - d)^2} \le 4 + s}$

68. Find the coordinates of the points that divide the line segment joining (4, 5) and (10, 14) into three equal parts. **(6, 8) and (8, 11)**

69. Show that the points (−2, 2), (13, 10), (21, −5), and (6, −13) are the vertices of a square.

70. Are the points $A(1, 1)$, $B(5, 2)$, $C(3, 4)$, $D(-1, 3)$ the vertices of a parallelogram (opposite sides equal in length)? Of a rhombus (all sides equal in length)? **yes; no**

71. Find the equation of the circle shown in the figure. $\mathbf{(x - 3)^2 + (y - 3/2)^2 = 45/4}$

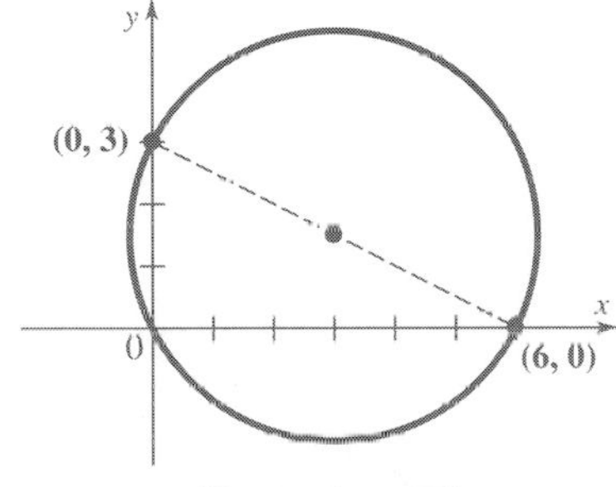

Exercise 71

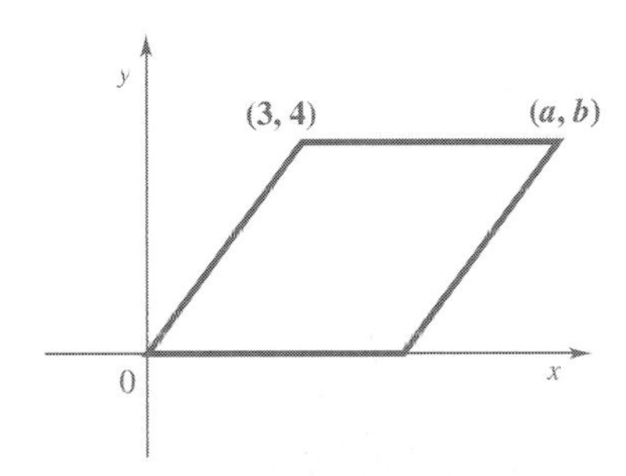

Exercise 72

72. A rhombus is a parallelogram with all sides equal in length. The figure shows a rhombus with one vertex at the origin. Find the values of a and b. $\mathbf{a = 8, b = 4}$

3.2 Exercises (text page 133)

1. Discuss the distinctions among a relation, the equation of a relation, and the graph of a relation.

2. In your own words, write a few sentences describing a relation, its domain, and its range.

In Exercises 3–22, give the domain and range of each relation, then use point plotting or a graphing utility to graph the relation.

3. $y = 3x - 2$
domain: $(-\infty, \infty)$; range: $(-\infty, \infty)$

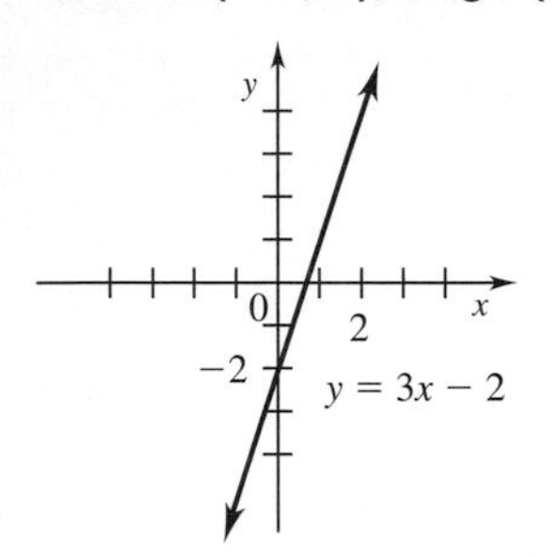

4. $y = -2x + 4$
domain: $(-\infty, \infty)$; range: $(-\infty, \infty)$

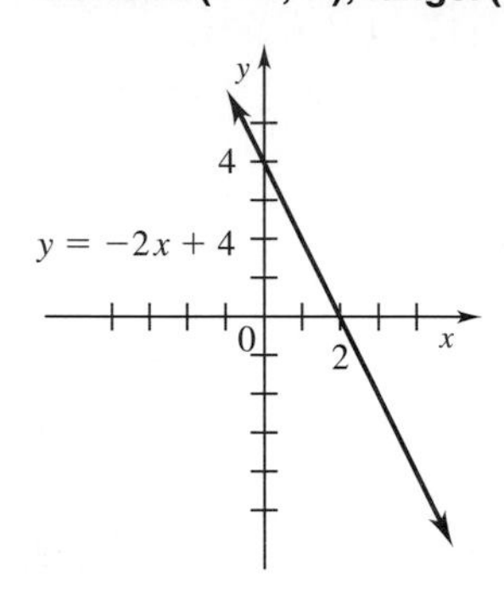

5. $3x = y^2$
domain: $[0, \infty)$; range: $(-\infty, \infty)$

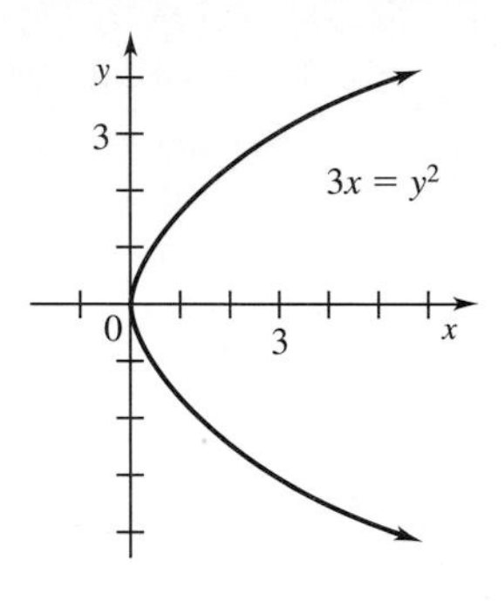

6. $4x = y^2$
domain: $[0, \infty)$; range: $(-\infty, \infty)$

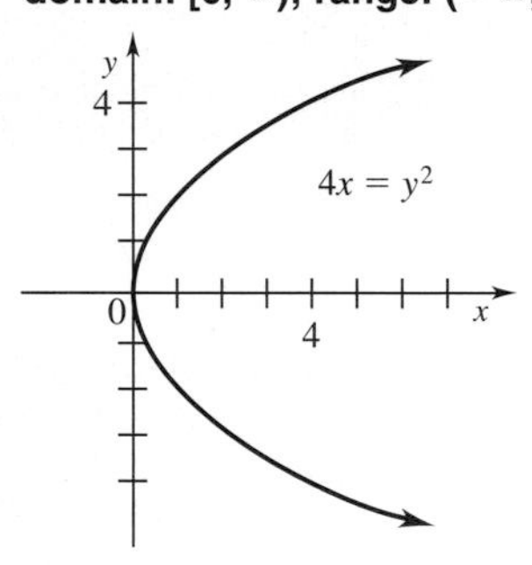

7. $16x^2 = -y$
domain: $(-\infty, \infty)$; range: $(-\infty, 0]$

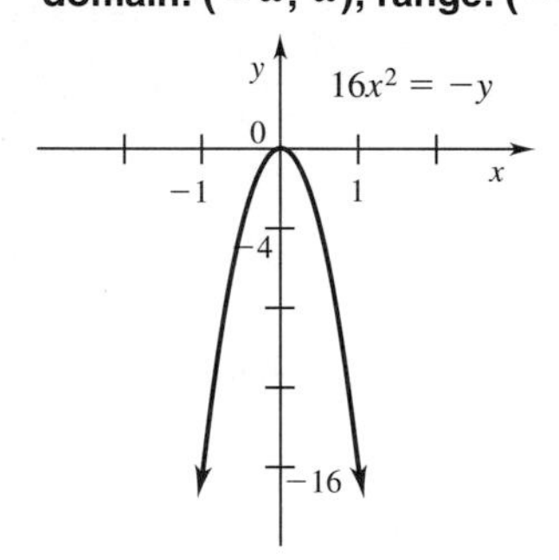

8. $4x^2 = -y$
domain: $(-\infty, \infty)$; range: $(-\infty, 0]$

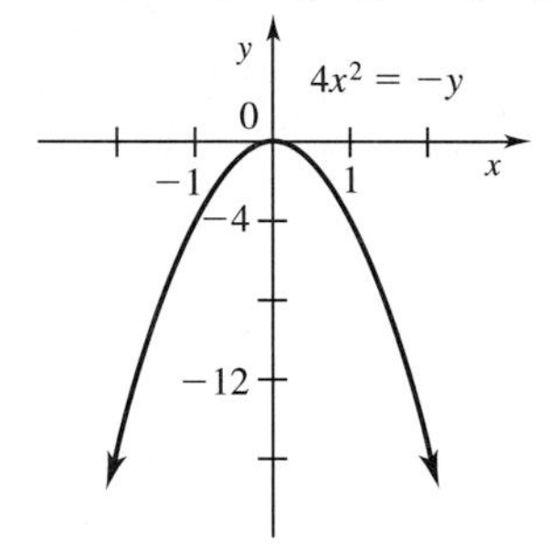

9. $y = |x| + 4$
domain: $(-\infty, \infty)$; range: $[4, \infty)$

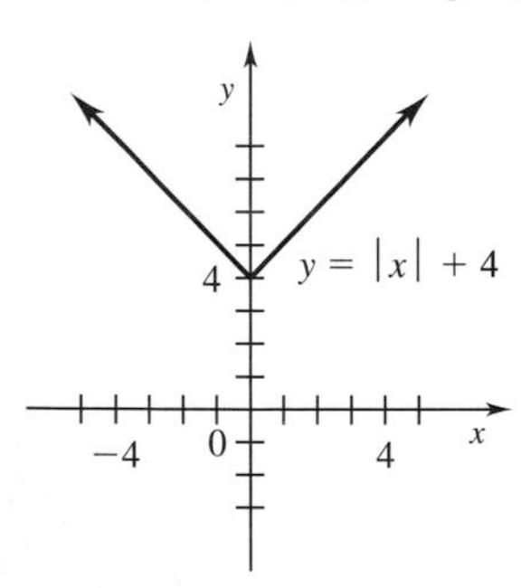

10. $y = |x| - 3$
domain: $(-\infty, \infty)$; range: $[-3, \infty)$

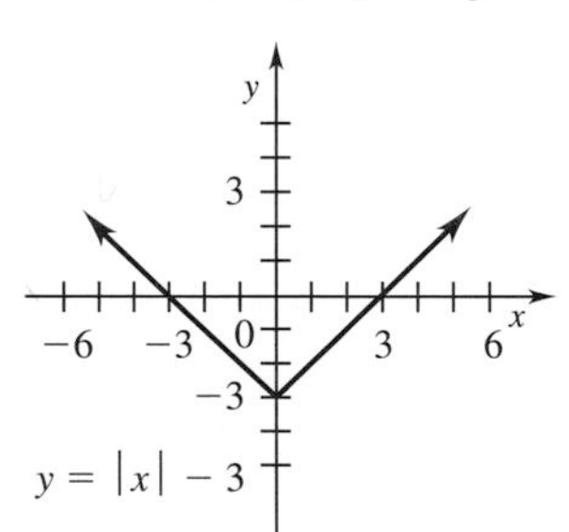

11. $x = |y|$
domain: $[0, \infty)$; range: $(-\infty, \infty)$

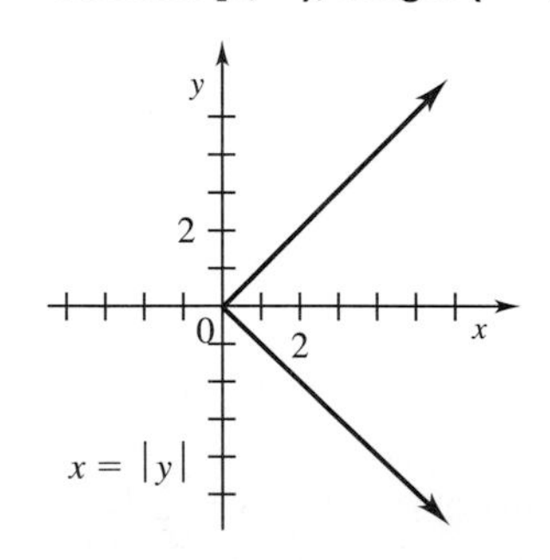

12. $x = |y| - 1$
domain: $[-1, \infty)$; range: $(-\infty, \infty)$

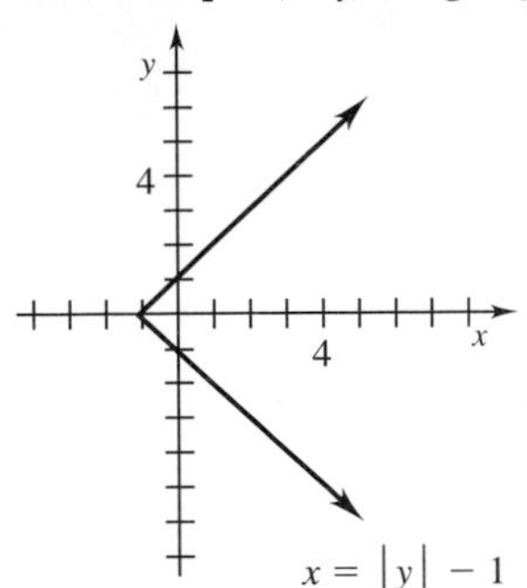

13. $y = -|x + 1|$
domain: $(-\infty, \infty)$; range: $(-\infty, 0]$

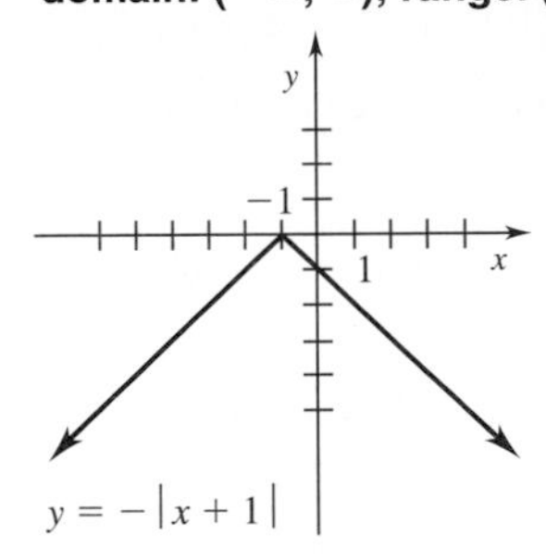

14. $y = -|x - 2|$
domain: $(-\infty, \infty)$; range: $(-\infty, 0]$

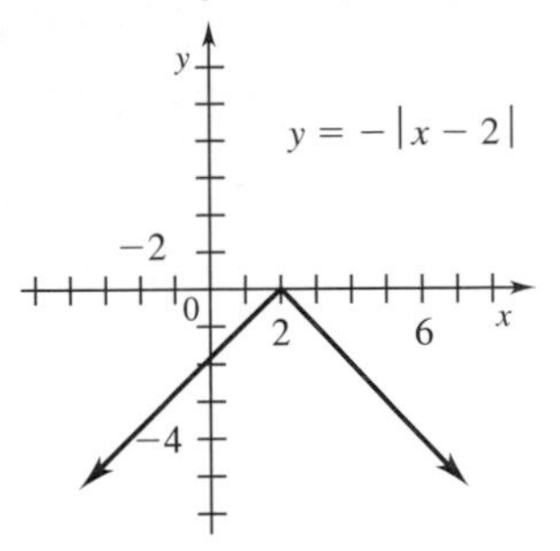

15. $x = \sqrt{y} - 2$
domain: $[-2, \infty)$; range: $[0, \infty)$

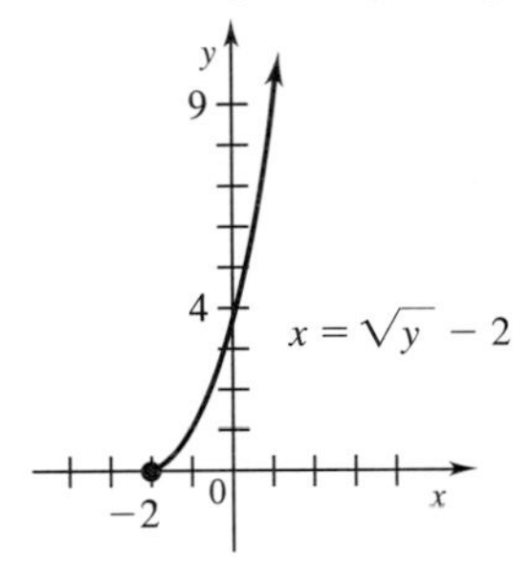

16. $x = -\sqrt{y} + 1$
domain: $(-\infty, 1]$; range: $[0, \infty)$

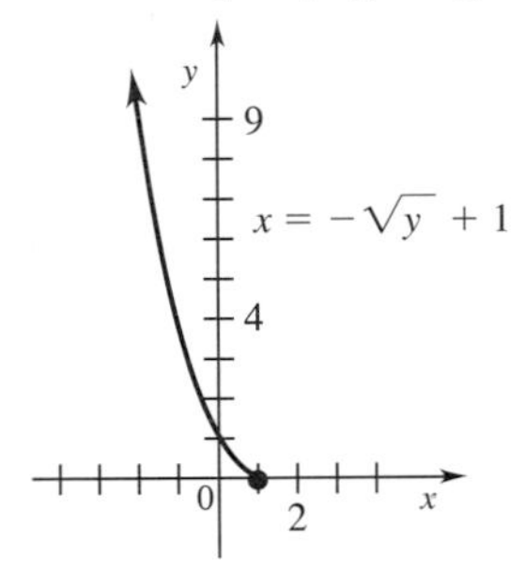

17. $x = -\sqrt{y - 2}$
domain: $(-\infty, 0]$; range: $[2, \infty)$

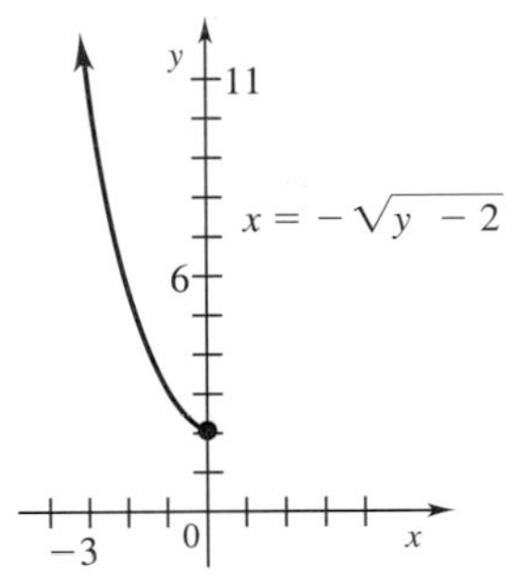

18. $x = \sqrt{y - 4}$
domain: $[0, \infty)$; range: $[4, \infty)$

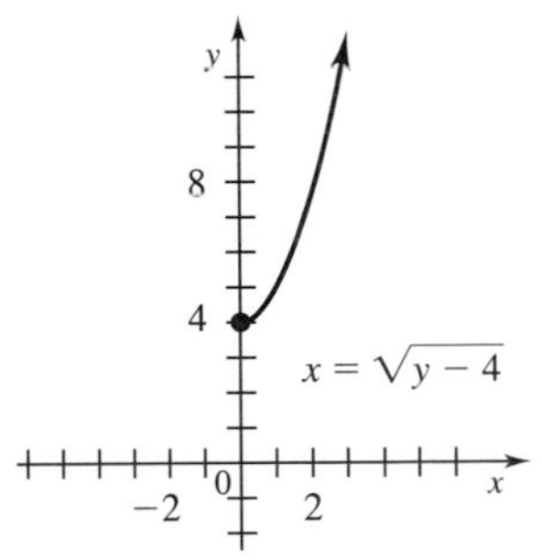

19. $y = \sqrt{2x + 4}$
domain: $[-2, \infty)$; range: $[0, \infty)$

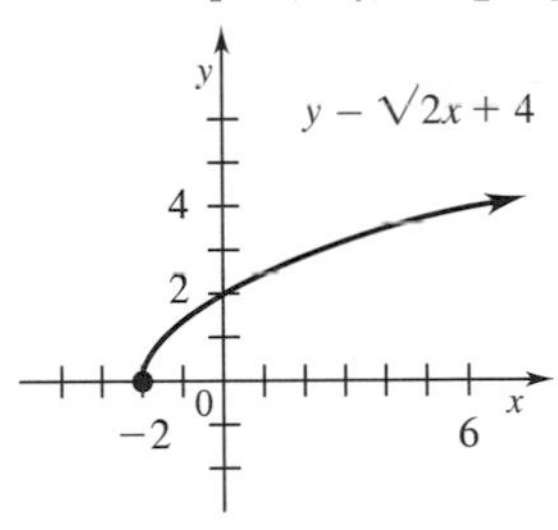

20. $y = \sqrt{3x + 9}$
domain: $[-3, \infty)$; range: $[0, \infty)$

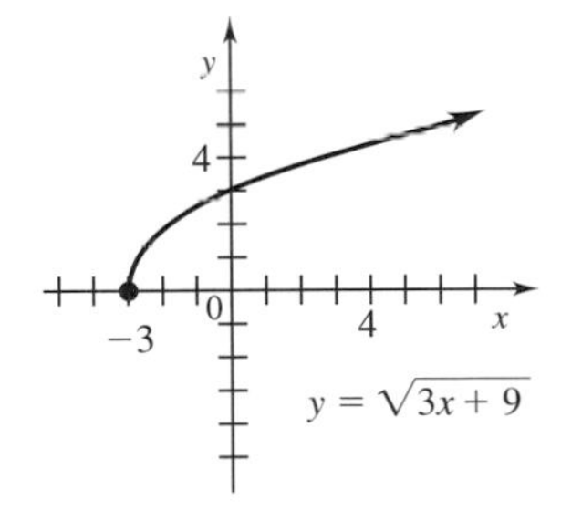

21. $y = -2\sqrt{x}$
domain: $[0, \infty)$; range: $(-\infty, 0]$

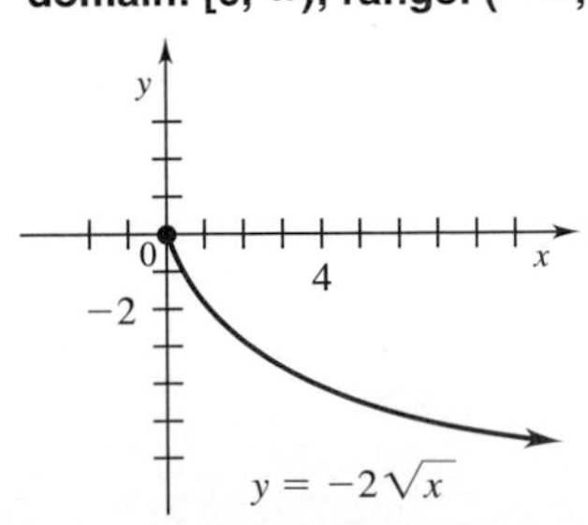

22. $y = -\sqrt{x}$
domain: $[0, \infty)$; range: $(-\infty, 0]$

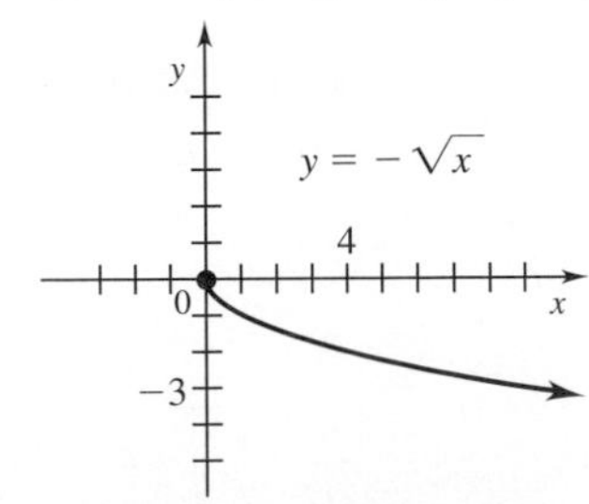

Decide whether the given ordered pair is in the relation defined by the given equation.

23. $(3, -1)$, $2x^2 - y^2 = 17$ **yes**

24. $(-2, 5)$, $3y - x^2 = 2x + 9$ **no**

25. A tool box has a square end with sides of length s and a length 2 inches more than twice the measure of the sides of the square end. Write a relation giving the surface area of the closed box. $\mathbf{A = 10s^2 + 8s}$

26. Write a relation giving the volume of the box in Exercise 25. $\mathbf{V = 2s^2 + 2s^3}$

27. The volume of a cone is found with the formula $V = (1/3)Bh$, where B is the area of the base of the cone and h is the height. Write a relation expressing the volume of a circular cone whose height is three times the radius of the base. $\mathbf{V = \pi r^3}$

28. The formula $A = (1/2)bh$ gives the area of a triangle with base b and height h. Write a relation expressing the area of a triangle in terms of its base if its base is 2/3 its height. $\mathbf{A = (3/4)b^2}$

Graphing Utility Problems

Use a graphing utility to graph each of the following relations. You will have to rewrite each relation as two equations first. Use the graph to determine the domain and range of each relation.

29. $9x^2 - 4y^2 = 36$ $\mathbf{(-\infty, -2] \cup [2, \infty); (-\infty, \infty)}$

30. $9x^2 + 4y^2 = 36$ $\mathbf{[-2, 2]; [-3, 3]}$

31. $x^2 + y^2 - 4x = 32$ $\mathbf{[-4, 8]; [-6, 6]}$

32. $x^2 + y^2 + 6x - 4y = 3$ $\mathbf{[-7, 1], [-2, 6]}$

3.3 Exercises (text page 140)

Decide whether each of the following relations is a function.

1. The $\sqrt{x}$ key on a calculator **yes**

2. The $1/x$ key on a calculator **yes**

3. $\{(x, y) \mid 2x^2 = y + 3\}$ **yes**

4. $\{(x, y) \mid y^2 = 3x + 1\}$ **no**

5. $\{(x, y) \mid 2x - y = 1\}$ **yes**

6. $\{(x, y) \mid y - x = 4\}$ **yes**

7. $\{(x, y) \mid x^2 + y^2 = 4\}$ **no**

8. $\{(x, y) \mid x^2 - y^2 = 4\}$ **no**

9. **no**

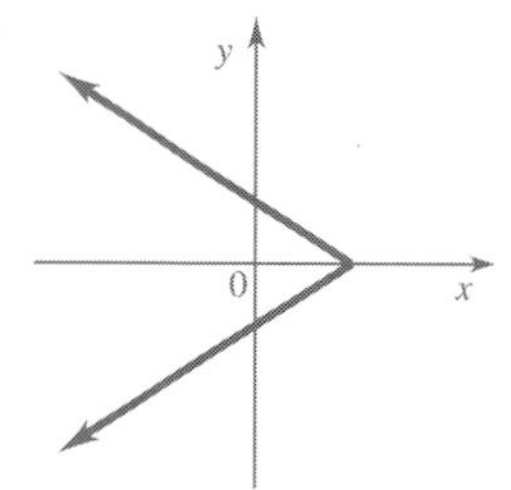

10. **yes**

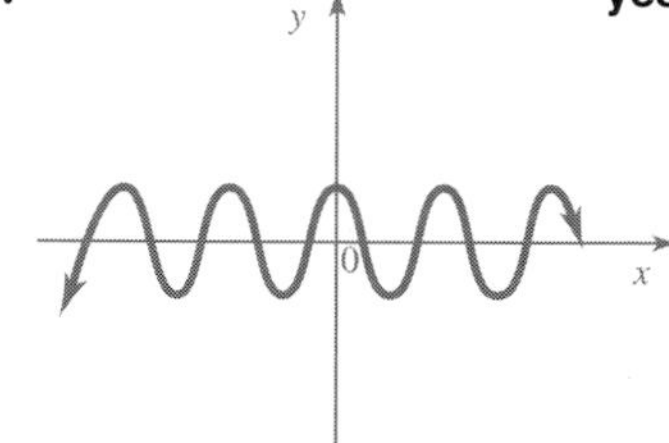

11. **yes**

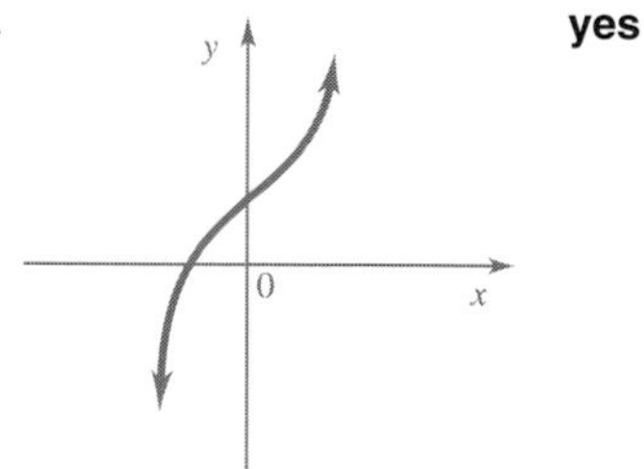

12. Compare relations and functions. How are they alike? How are they different? Be specific; use examples.

For each of the following, **(a)** *find* $f(-2)$, $f(0)$, $f(4)$; **(b)** *find the x-value(s) that correspond(s) to the following function values:* 3, 0, 2.

13.

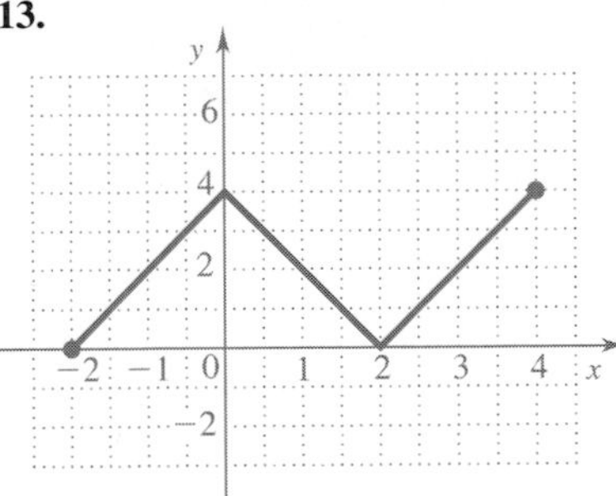

(a) 0, 4, 4 (b) −.5, .5, and 3.5; −2 and 2; −1, 1, and 3

14.

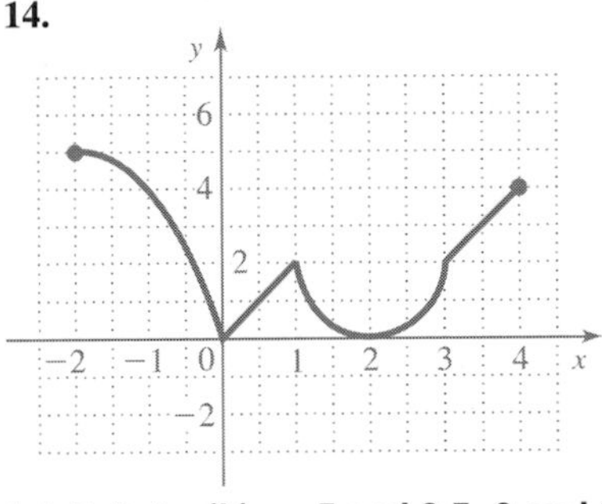

(a) 5, 0, 4 (b) −.5 and 3.5; 0 and 2; about −.3, 1, and 3

15.

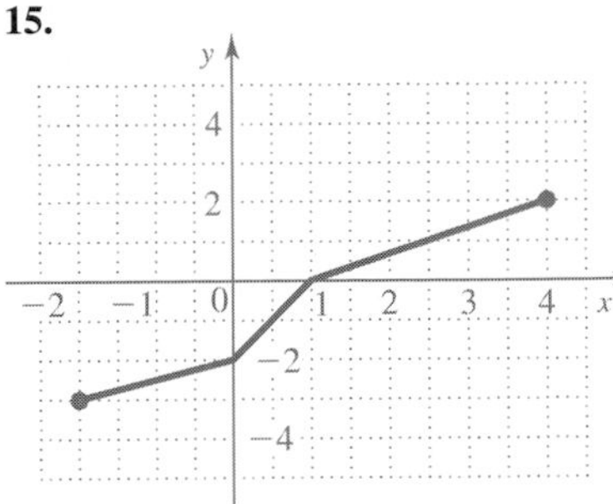

(a) −3, −2, 2 (b) none; 1; 4

Let $f(x) = 3 - 2x^2$, $g(x) = \sqrt{4x - 2}$, *and* $h(x) = x/(2 - x)$. *Find each of the following.*

16. $f(-2)$ $\mathbf{-5}$

17. $h(3)$ $\mathbf{-3}$

18. $g(4)$ $\mathbf{\sqrt{14}}$

19. $g(5)$ $\mathbf{\sqrt{18}}$ **or** $\mathbf{3\sqrt{2}}$

20. $f(-3) + 2$ $\mathbf{-13}$

21. $f(4) - 5$ $\mathbf{-34}$

22. $f(1) + g(1)$ $\mathbf{1 + \sqrt{2}}$

23. $[g(1)]^2$ **2**

24. $[f(-1)]^2$ **1**

25. $f\left(\frac{1}{a}\right)$ $\mathbf{3 - 2/a^2}$ **or** $\mathbf{(3a^2 - 2)/a^2}$

26. $g(3a - 1)$ $\mathbf{\sqrt{12a - 6}, a \geq 1/2}$

27. $\frac{1}{f(a)}$ $\mathbf{1/(3 - 2a^2)}$

28. $f(\sqrt{a})$ $\mathbf{3 - 2a, a \geq 0}$

29. $\frac{f(5a)}{g(a)}$ $\mathbf{(3 - 50a^2)/\sqrt{4a - 2}}$

30. $\sqrt{h(a)}$ $\mathbf{\sqrt{a/(2 - a)}}$ **or** $\mathbf{\sqrt{a(2 - a)}/(2 - a), a \text{ in } [0, 2)}$

Let $f(x) = \begin{cases} 2x^2 & \text{if } x < 0 \\ 3x & \text{if } 0 \leq x \leq 1 \\ |6 - x| & \text{if } x > 1. \end{cases}$ *Find each of the following.*

31. $f(-3)$ **18**

32. $f(1)$ **3**

33. $f(10)$ **4**

34. $f\left(\frac{1}{3}\right) + f(3)$ **4**

35. Figure 18 shows how the range of a function can be determined by looking at the graph of the function. Describe in words how one can tell the range from the graph.

Give the domain and range of the functions defined or graphed in Exercises 36–51. Give only the domain in Exercises 40 and 41.

36. $g(x) = x^2 - 3$ $\mathbf{(-\infty, \infty); [-3, \infty)}$

37. $h(x) = (x - 2)^4$ $\mathbf{(-\infty, \infty); [0, \infty)}$

38. $f(x) = \sqrt{8 + x}$ $\mathbf{[-8, \infty); [0, \infty)}$

39. $f(x) = (3x + 2)^{1/2}$ $\mathbf{[-2/3, \infty); [0, \infty)}$

40. $g(x) = \frac{2}{x^2 - 3x + 2}$ $\mathbf{(-\infty, 1) \cup (1, 2) \cup (2, \infty)}$

41. $h(x) = \frac{-4}{x^2 + 5x + 4}$ $\mathbf{(-\infty, -4) \cup (-4, -1) \cup (-1, \infty)}$

42. $r(x) = -\sqrt{x^2 - 4x - 5}$ $\mathbf{(-\infty, -1] \cup [5, \infty); (-\infty, 0]}$

43. $r(x) = -\sqrt{x^2 + 7x + 10}$ $\mathbf{(-\infty, -5] \cup [-2, \infty); (-\infty, 0]}$

44. $f(x) = |3x - 4| + 2$ $\mathbf{(-\infty, \infty); [2, \infty)}$

45. $k(x) = -|2x - 7| + 5$ $\mathbf{(-\infty, \infty); (-\infty, 5]}$

46.

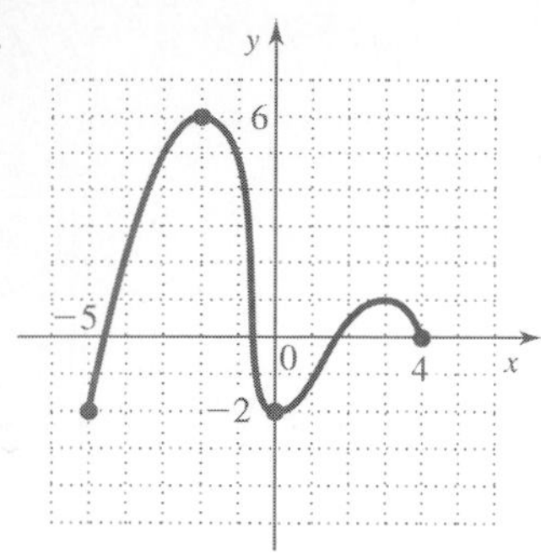

$[-5, 4]$; $[-2, 6]$

47.

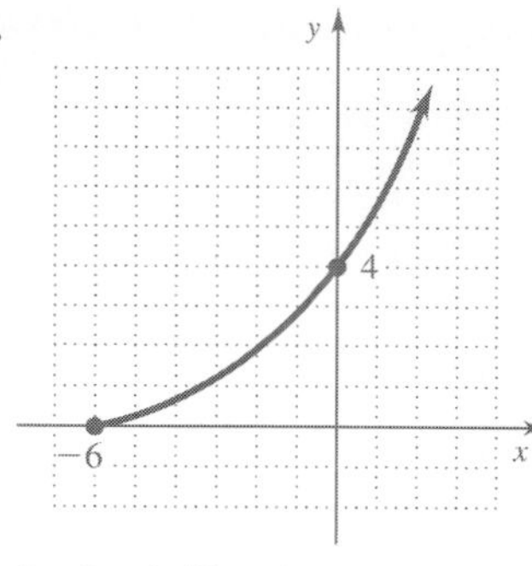

$[-6, \infty)$; $[0, \infty)$

48.

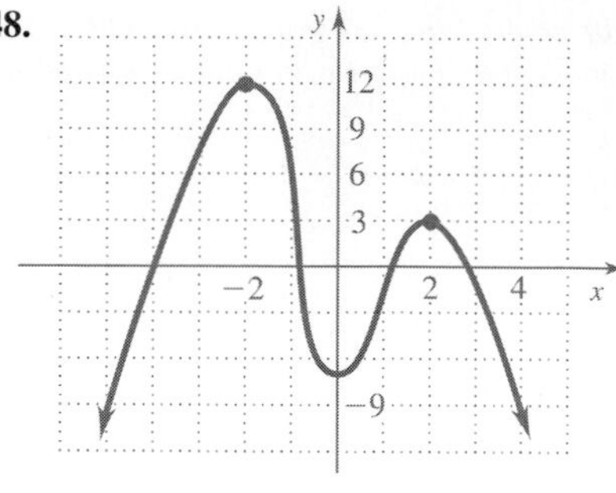

$(-\infty, \infty)$; $(-\infty, 12]$

49. Exercise 11 $(-\infty, \infty)$; $(-\infty, \infty)$ **50.** Exercise 14 **$[-2, 4]$; $[0, 5]$** **51.** Exercise 13 **$[0, 4]$**

52. Identify the maximum and minimum function values of the functions graphed in each of the indicated exercises.

(a) Exercise 46 **(b)** Exercise 47 **(c)** Exercise 48 **(a) 6; −2 (b) none; 0 (c) 12; none**

53. Give the intervals where the function graphed in each of the indicated exercises is increasing and decreasing.

(a) Exercise 46 **(b)** Exercise 47 **(c)** Exercise 48 **(a) $[-5, -2]$, $[0, 3]$; $[-2, 0]$, $[3, 4]$ (b) $[-6, \infty)$; none (c) $(-\infty, -2]$, $[0, 2]$; $[-2, 0]$, $[2, \infty)$**

54. The figure shows the typical relationship between temperature and time of day in April for a town.

(a) Is it the graph of a function? **yes**

(b) What is the temperature at 8 A.M.? At 12 noon? At 6 P.M.? **38°F, 52°F, 66°F**

(c) At what time(s) is the temperature 40°? 65°? **9 A.M. and 3 A.M.; 3 P.M. and 7 P.M.**

(d) What are the maximum and minimum temperatures shown? **68°F, 36°F**

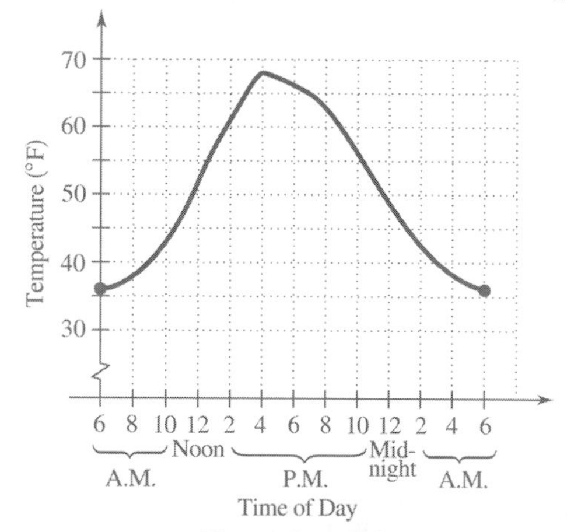

Exercise 54

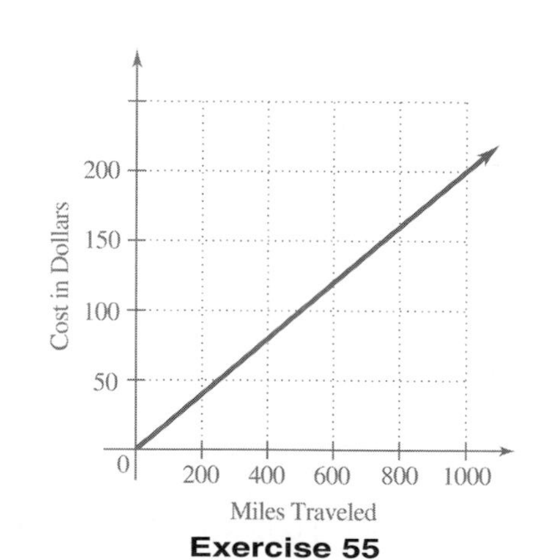

Exercise 55

55. The cost of an automobile trip depends on the distance traveled, as shown in the figure.

(a) Is this the graph of a function? **yes**

(b) What is the cost to travel 100 miles? 250 miles? 500 miles? **$20, $50, $100**

(c) How many miles can be traveled for $100? For $25? **about 500 mi, about 125 mi**

(d) Assume the graph continues in the same way. Does this relation have a maximum or minimum value? **no maximum, a minimum of 0**

56. If $f(x) = x^2/x$ and $g(x) = x$, does $f(x) = g(x)$? Explain why.

For the functions defined in Exercises 57–60, find **(a)** $f(x + h)$, **(b)** $f(x + h) - f(x)$, *and* **(c)** $\frac{f(x + h) - f(x)}{h}$. *(Assume $h \neq 0$).*

57. $f(x) = x^2 - 4$ **(a)** $x^2 + 2xh + h^2 - 4$ **(b)** $2xh + h^2$ **(c)** $2x + h$

58. $f(x) = 8 - 3x^2$ **(a)** $8 - 3x^2 - 6xh - 3h^2$ **(b)** $-6xh - 3h^2$ **(c)** $-6x - 3h$

59. $f(x) = 6x + 2$ **(a)** $6x + 6h + 2$ **(b)** $6h$ **(c)** 6

60. $f(x) = 4x + 11$ **(a)** $4x + 4h + 11$ **(b)** $4h$ **(c)** 4

61. A storage container in the shape of a cube with an edge of length e yd is to be insulated at the cost of $2 per square yard. Find the total cost $C(e)$ for the insulation job as a function of e in yards. $C(e) = 12e^2$

62. A cylindrical water tank has a radius of 6 ft and a height of 9 ft. (See the figure.) The tank is filled with water to a depth of h ft. Express the volume of the water as a function of h. $V(h) = 36\pi h$

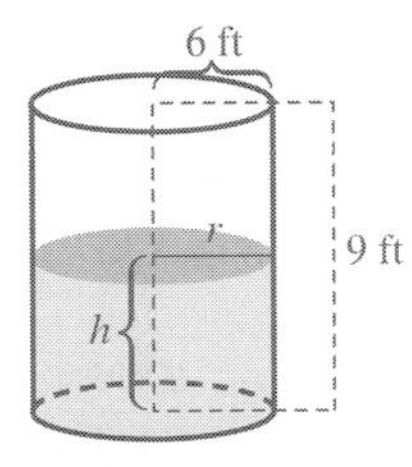

Exercise 62

When a function models a real-life situation, the domain is usually restricted by practical considerations. For instance, in Exercise 62, the value of h must be in the interval [0, 9]. *Determine a meaningful domain for the functions in Exercises 63–65.*

63. A ball is dropped from a 144-ft building. Its height in feet at time t (in seconds) after being dropped is $s(t) = -16t^2 + 144$. **[0, 3]**

64. According to the economic law of demand, the price that should be charged in order to sell x units of a certain commodity is $p(x) = 100 - .05x$ dollars. **(0, 2000)**

65. If a worker earning ten dollars an hour works at least 40 hr a week, with time and a half for overtime, then her weekly salary for h hr of work is $s(h) = 400 + 15(h - 40)$ dollars. **[40, 168)**

66. An architect is designing a building shaped like a rectangle with semicircles of radius r meters attached to each end. See the figure. The perimeter of the building is to be 440 m, and the radius must be at least 5 m.

(a) Show that each straight portion of the building must have length $220 - \pi r$ meters.

(b) Express the area of the entire building as a function of r. $A(r) = 440r - \pi r^2$

(c) Determine a meaningful domain for the function in part (b). $[5, 220/\pi)$

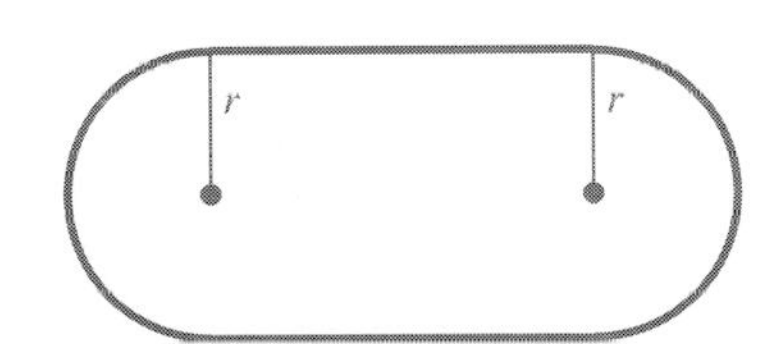

Exercise 66

67. Sixty feet of fencing is used to enclose a rectangular garden in which each side of the garden is at least 12 ft long and one side, of length x, is an extension of a 10-ft stone wall. See the figure.

(a) Show that the length of each side perpendicular to the side with the stone wall is $25 - x$ ft.

(b) Express the area of the garden as a function of x. $A(x) = -x^2 + 15x + 250$

(c) Determine a meaningful domain for the function in part (b). **[2, 13]**

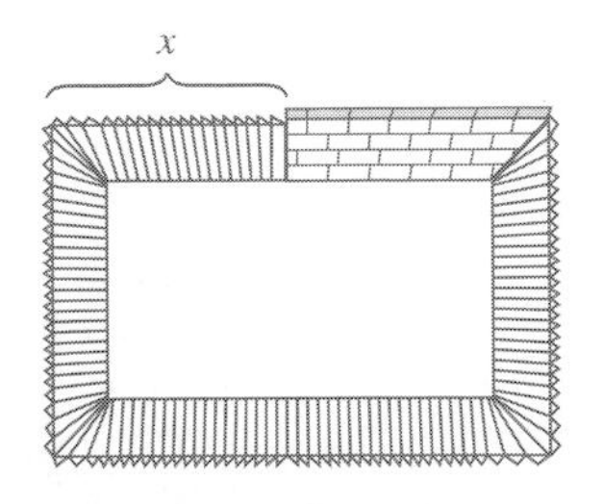

Exercise 67

68. Give the area of a circle as a function of its radius r; also, give the circumference as a function of the radius. $A = \pi r^2$ and $C = 2\pi r$

69. The number of BTUs (British Thermal Units) required to cool a room is 3000 more than 15 times the area of the room in square feet. Let B represent the number of BTUs and A represent the area in square feet. Give A as a function of B. $A = (B - 3000)/15$

70. A rectangle is inscribed in a circle of radius r. Let x represent the length of one side of the rectangle. Give the area of the rectangle as a function of r. $\mathbf{A = 2x\sqrt{r^2 - x^2/4}}$

71. The height of a circular cone is half the radius of the base. Give the volume of the cone as a function of the radius of the base. $\mathbf{V = \pi r^3/6}$

Graphing Utility Problems

Graph each function defined below on the indicated domain. Use the graph to determine the corresponding range. If necessary, approximate the range values to two decimal places.

72. $y = x^2 - 4$, $[-1, 1]$ **[−4, −3]**

73. $y = \sqrt{x - 2}$, $[3, 11]$ **[1, 3]**

74. $y = x^3 - 2x^2 - x$, $[-2, 2]$ **[−14, 0]**

Graph the function whose definition was found in the indicated exercise. Use the graph to approximate the maximum and minimum y-values to two decimal places.

75. Exercise 67 **306.25; 276**

76. Exercise 66 **15406.20; 2121.46**

3.4 Exercises (text page 151)

Graph each of the following.

1. $6y = 3x + 12$

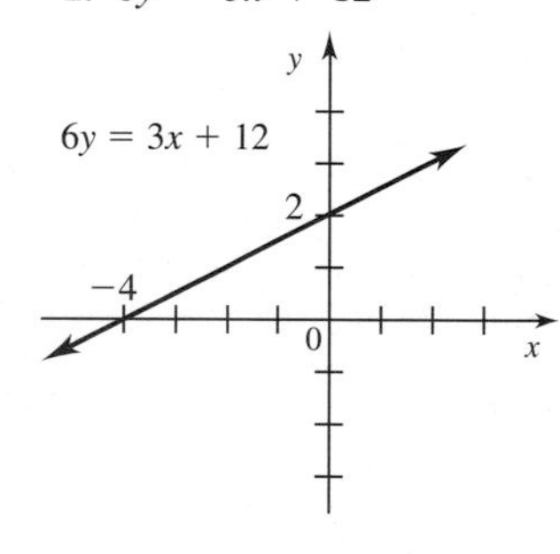

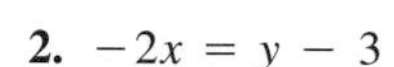
2. $-2x = y - 3$

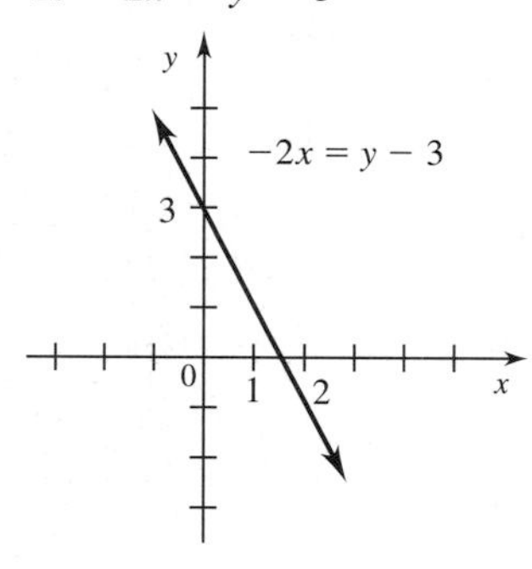

3. $2y = x + 6$

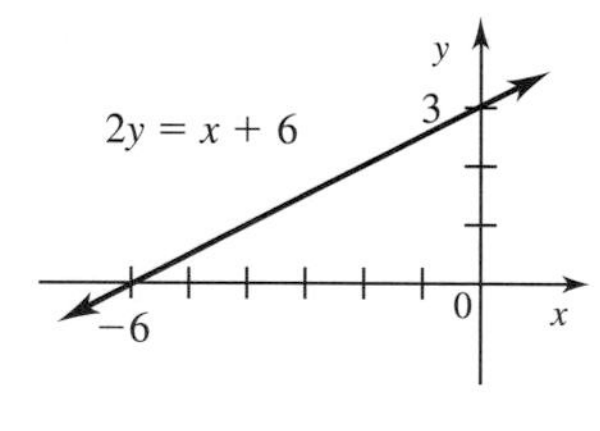

4. $x - 2y = 6$

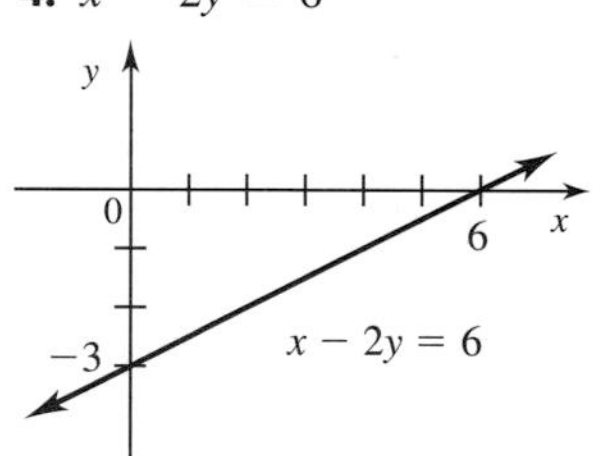

5. $y = -3$

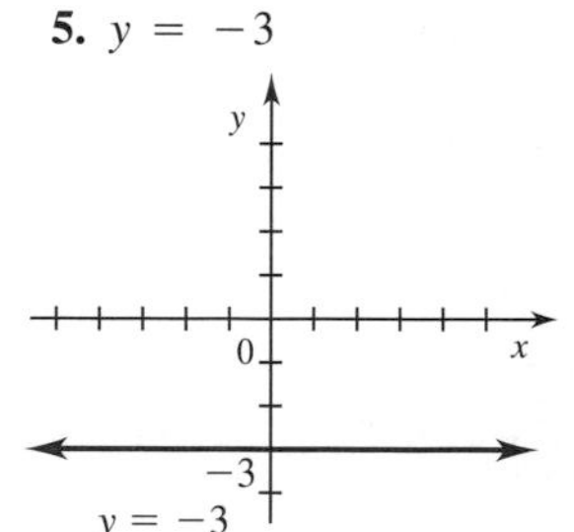

6. $x = 2$

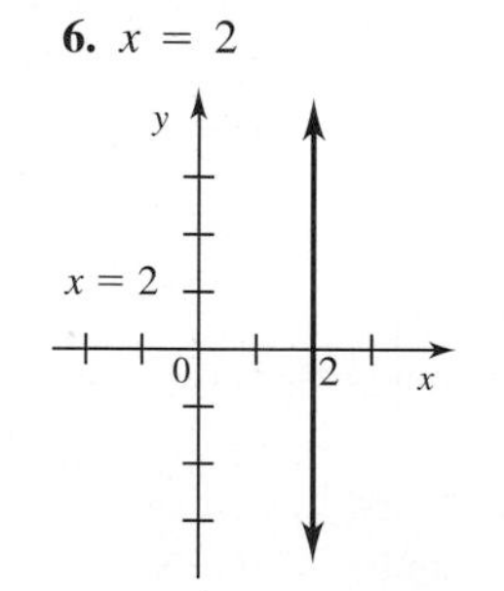

7. $y = 4x$

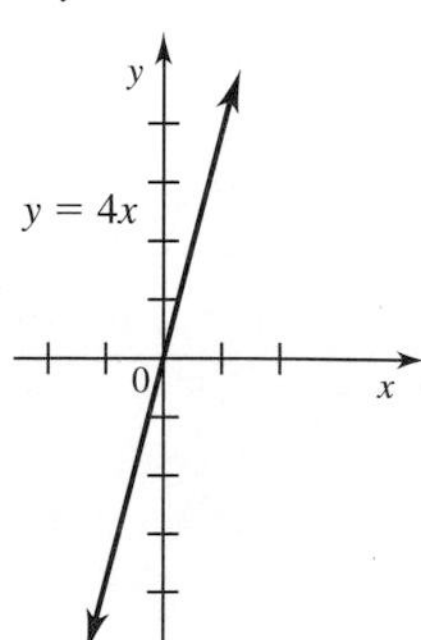

8. $x = -2y$

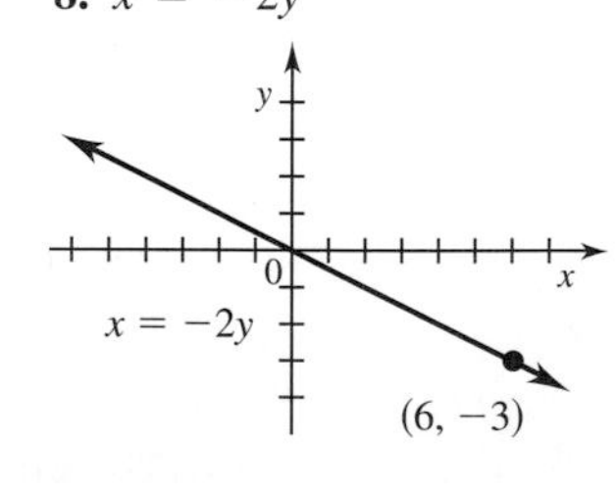

Graph the line passing through the given point and having the indicated slope.

9. Through $(-2, 8)$, $m = -1$

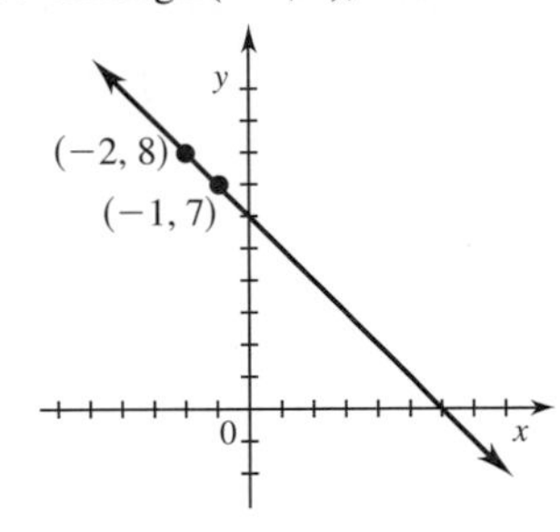

10. Through $(3, -4)$, $m = -1/3$

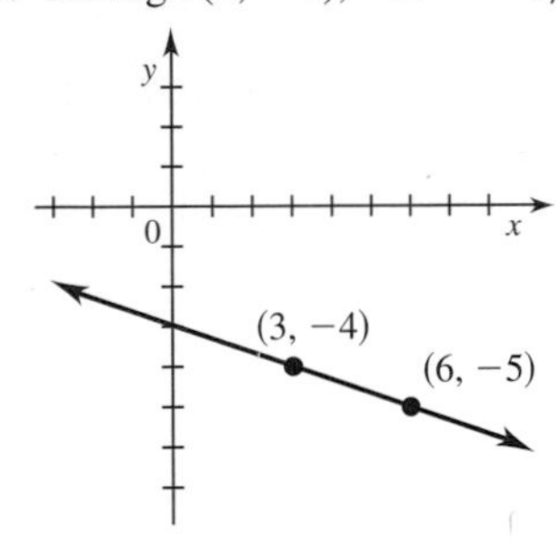

11. Through $(-1, 4)$, $m = 0$

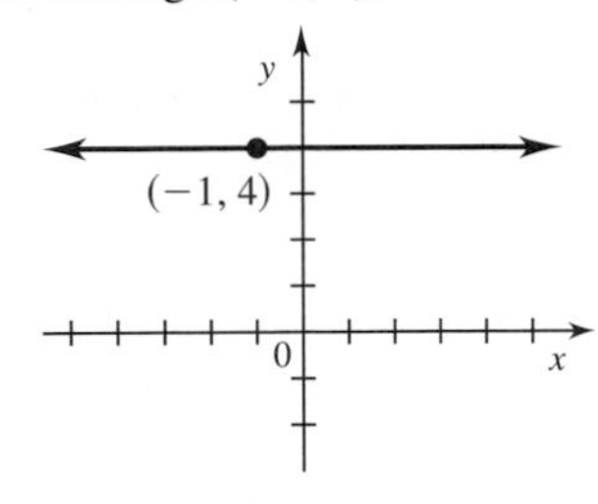

12. Through $(2, -5)$, $m = 0$

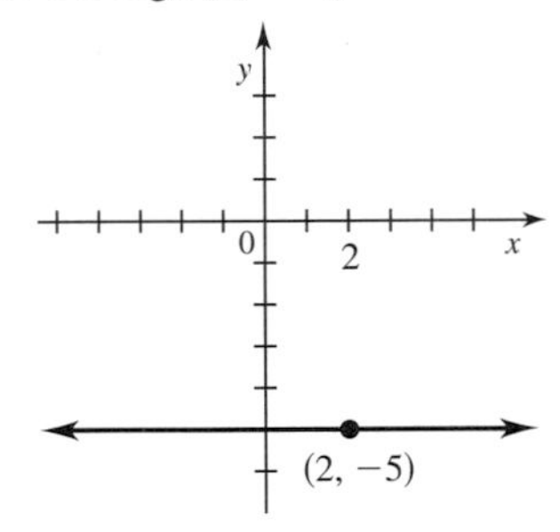

13. Through $(3, 2/3)$, undefined slope

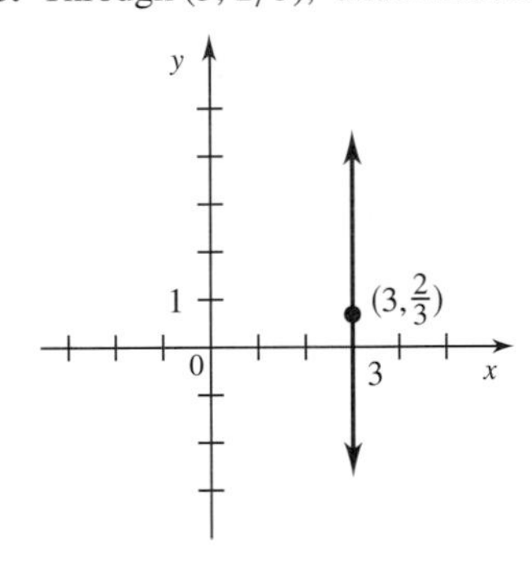

14. Through $(9/4, 2)$, undefined slope

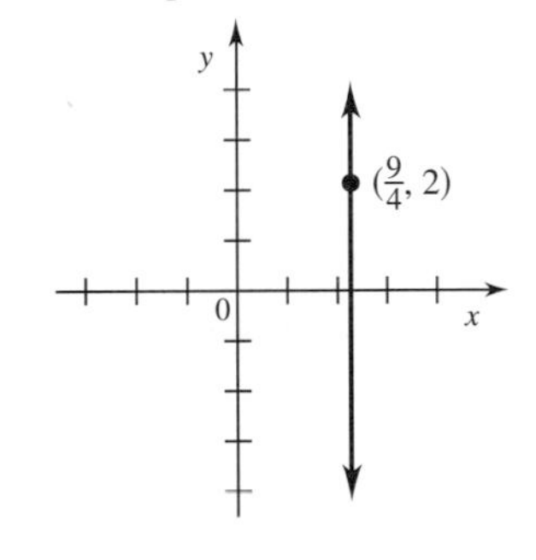

Find the slope of each of the following lines.

15. Through $(8, 4)$ and $(-1, -3)$ **7/9**

16. Through $(-4, -3)$ and $(5, 0)$ **1/3**

17. $-5y = -3x - 11$ **3/5**

18. $2y + 15 = x$ **1/2**

19. $y = 4$ **0**

20. $x = -1$ **undefined**

For each of the following slopes, identify the line below having that slope.

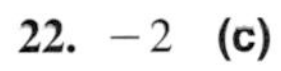

21. $1/2$ **(d)**

22. -2 **(c)**

23. 0 **(a)**

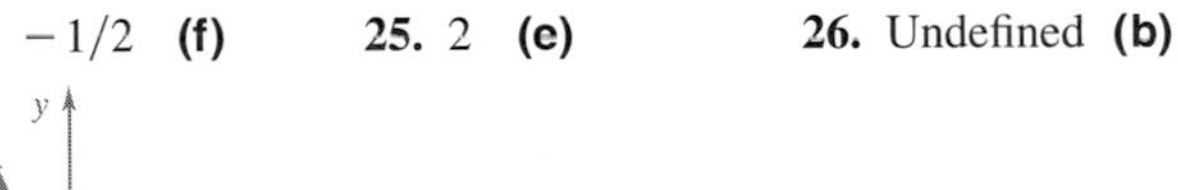

24. $-1/2$ **(f)**

25. 2 **(e)**

26. Undefined **(b)**

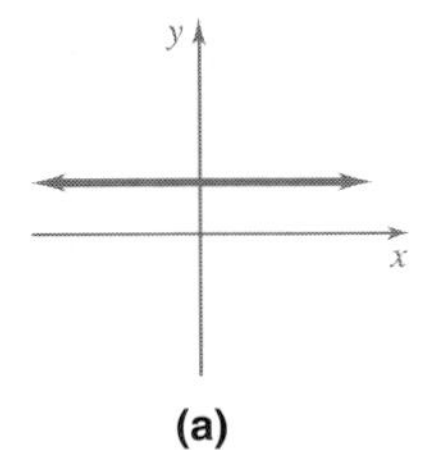

(a)

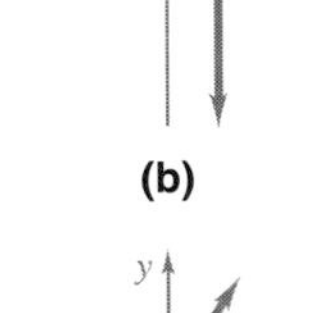

(b)

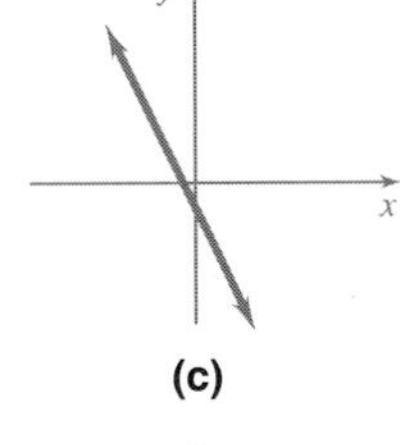

(c)

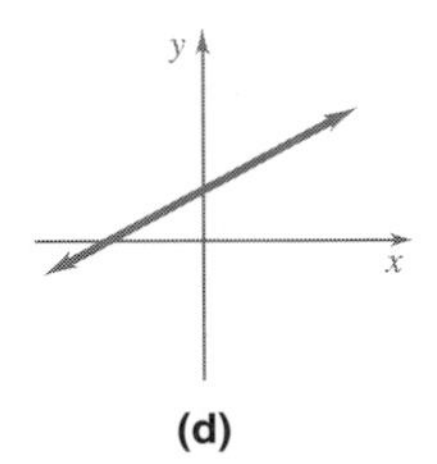

(d)

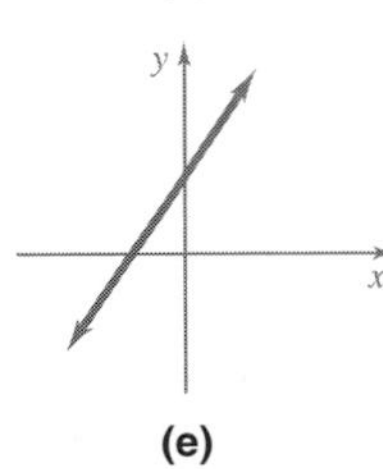

(e)

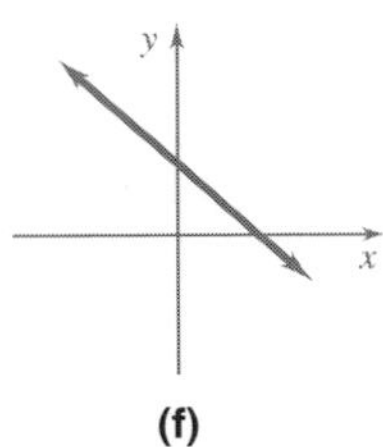

(f)

27. Explain why the equation $y = 5$ defines a linear function, but $x = 5$ does not. What must be true about the slope of a line in order for the line to be the graph of a function?

*Use slopes to decide whether or not the following points lie on a straight line. (*Hint: *Find the slope of the line through the first two points, and then the slope of the line through the last two points.)*

28. $M(4, 3)$, $N(-1, -5)$, $P(2, 0)$ **no**

29. $A(1, 10/3)$, $B(0, 5)$, $C(-3, 10)$ **yes**

Graph the functions defined in Exercises 30–33.

30. $f(x) = \begin{cases} x - 1 & \text{if } x \le 3 \\ 2 & \text{if } x > 3 \end{cases}$

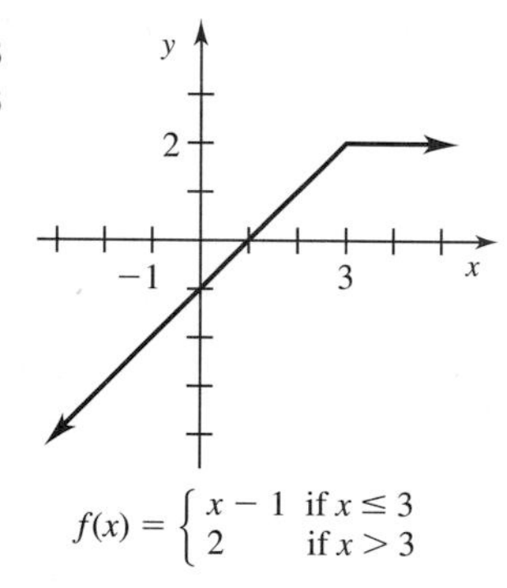

31. $h(x) = \begin{cases} -2 & \text{if } x \ge 1 \\ 2 & \text{if } x < 1 \end{cases}$

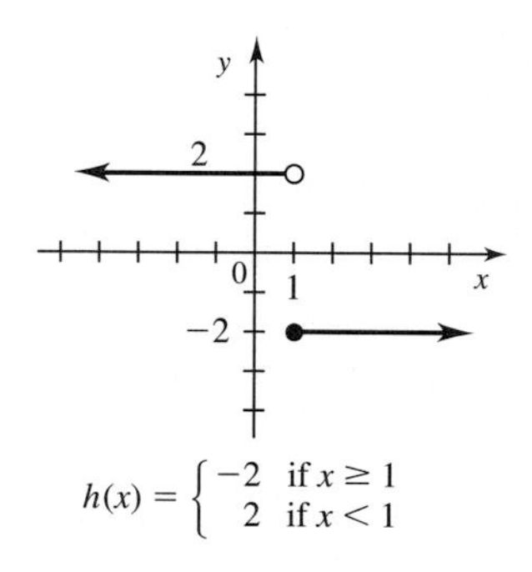

32. $p(x) = \begin{cases} |x| & \text{if } x > -2 \\ x & \text{if } x \le -2 \end{cases}$

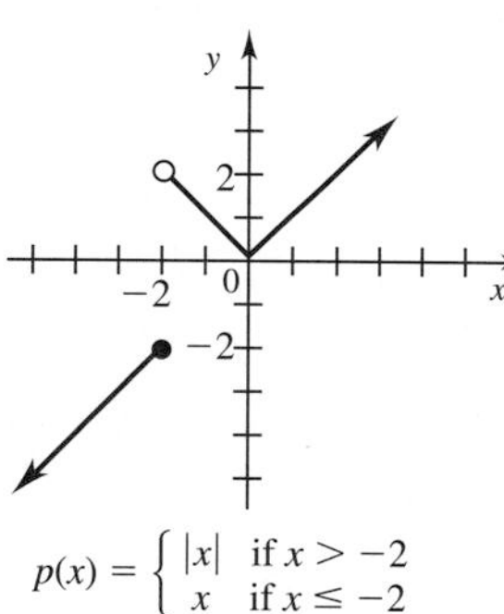

33. $r(x) = \begin{cases} |x| - 1 & \text{if } x > -1 \\ x - 1 & \text{if } x \le -1 \end{cases}$

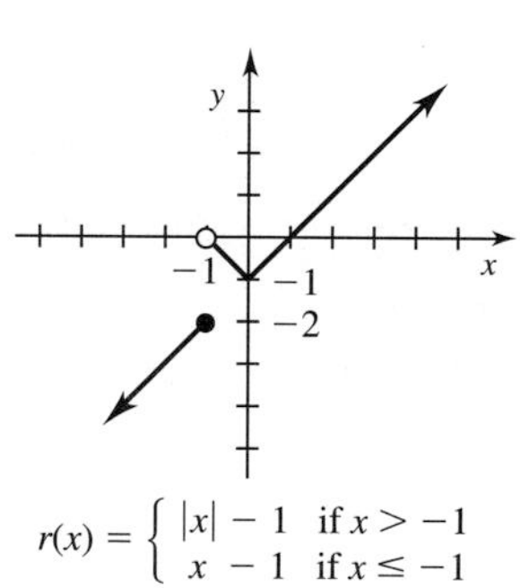

Write an expression similar to those in Exercises 30–33 to define the piecewise functions graphed in Exercises 34–37. Give the domain and range of each function.

34.

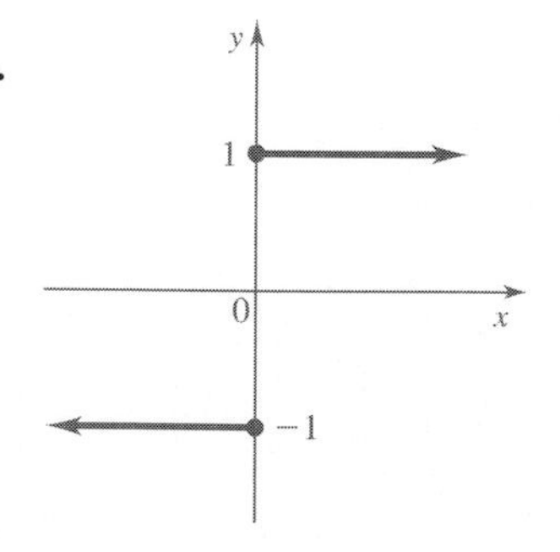

$f(x) = \begin{cases} -1 & \text{if } x \le 0 \\ 1 & \text{if } x > 0 \end{cases}$; **$(-\infty, \infty)$, $\{-1, 1\}$**

35.

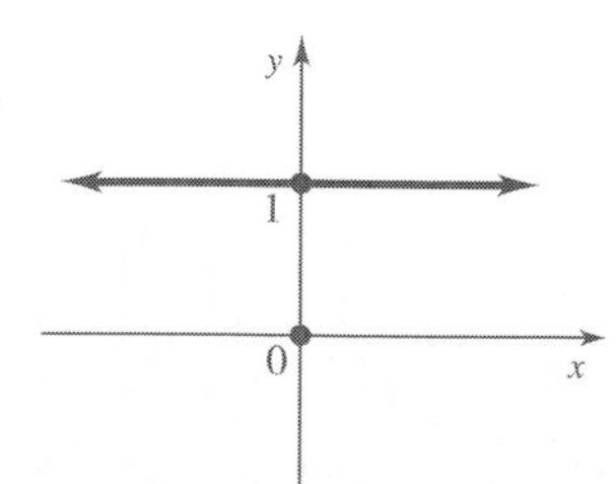

$f(x) = \begin{cases} 1 & \text{if } x \ne 0 \\ 0 & \text{if } x = 0 \end{cases}$; **$(-\infty, \infty)$, $\{0, 1\}$**

36. 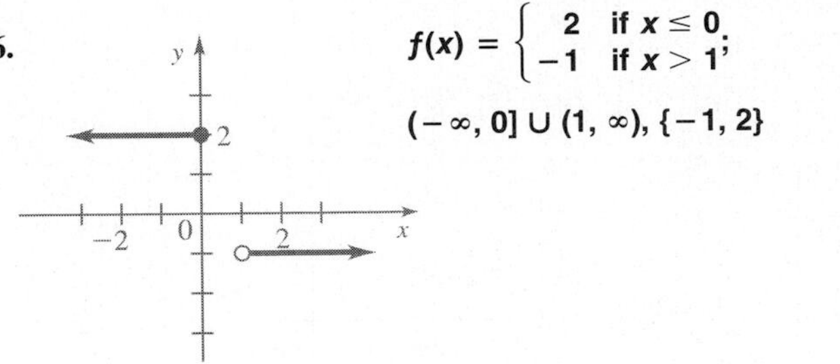

$f(x) = \begin{cases} 2 & \text{if } x \le 0 \\ -1 & \text{if } x > 1 \end{cases}$;

$(-\infty, 0] \cup (1, \infty)$, $\{-1, 2\}$

37. 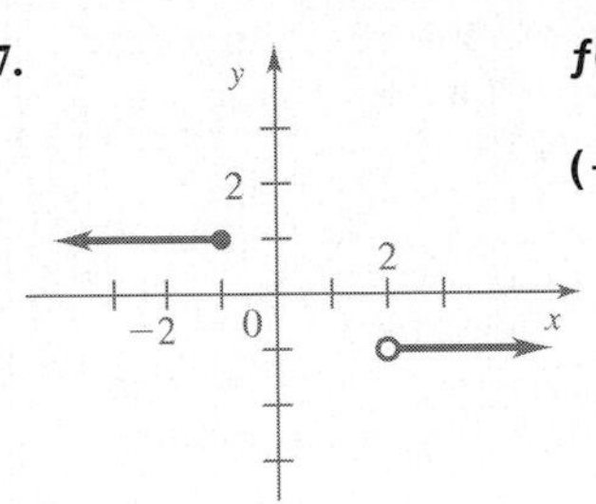

$f(x) = \begin{cases} 1 & \text{if } x \le -1 \\ -1 & \text{if } x > 2 \end{cases}$;

$(-\infty, -1] \cup (2, \infty)$, $\{-1, 1\}$

38. When a diabetic takes long-acting insulin, the insulin reaches its peak effect on the blood sugar level in about 3 hr. This effect remains fairly constant for 5 hr, then declines, and is very low until the next injection. In a typical patient, the level of insulin might be given by the following function.

$$i(t) = \begin{cases} 40t + 100 & \text{if } 0 \le t \le 3 \\ 220 & \text{if } 3 < t \le 8 \\ -80t + 860 & \text{if } 8 < t \le 10 \\ 60 & \text{if } 10 < t \le 24 \end{cases}$$

Here $i(t)$ is the blood sugar level, in appropriate units, at time t measured in hours from the time of the injection. Suppose a patient takes insulin at 6 A.M. Find the blood sugar level at each of the following times: **(a)** 7 A.M. **(b)** 9 A.M. **(c)** 10 A.M. **(d)** noon **(e)** 2 P.M. **(f)** 5 P.M. **(g)** midnight. **(h)** Graph $y = i(t)$.

(a) 140 (b) 220 (c) 220 (d) 220 (e) 220 (f) 60 (g) 60 (h)

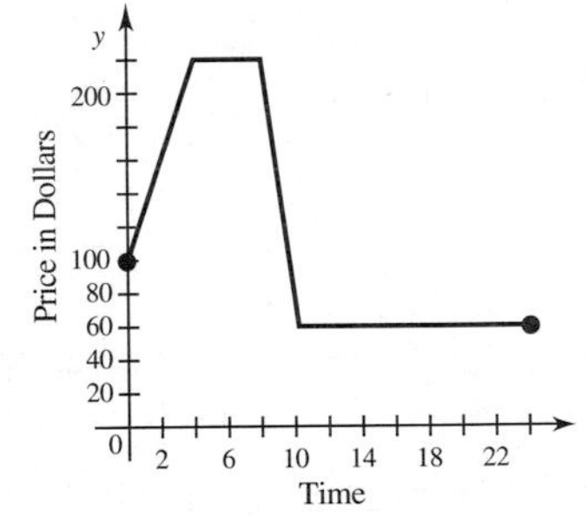

$$i(t) = \begin{cases} 40t + 100 & \text{if } 0 \le t \le 3 \\ 220 & \text{if } 3 < t < 8 \\ -80t + 860 & \text{if } 8 < t \le 10 \\ 60 & \text{if } 10 < t \le 24 \end{cases}$$

39. The snow depth in Michigan's Isle Royale National Park varies throughout the winter. In a typical winter, the snow depth in inches is approximated by the following function.

$$f(x) = \begin{cases} 6.5x & \text{if } 0 \le x \le 4 \\ -5.5x + 48 & \text{if } 4 < x \le 6 \\ -30x + 195 & \text{if } 6 < x \le 6.5 \end{cases}$$

Here, x represents the time in months with $x = 0$ representing the beginning of October, $x = 1$ representing the beginning of November, and so on.

(a) Graph $f(x)$

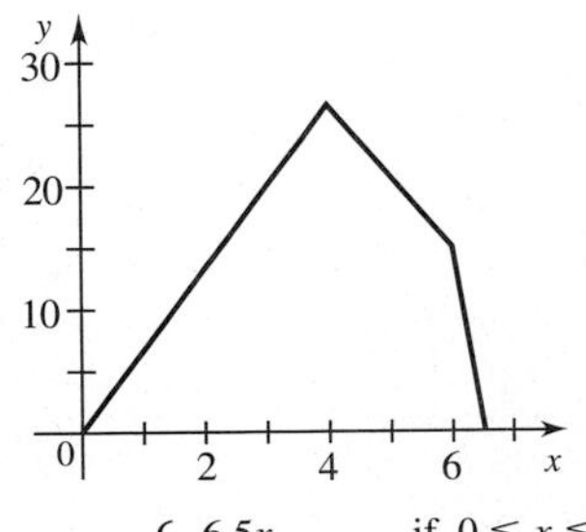

$$f(x) = \begin{cases} 6.5x & \text{if } 0 \le x \le 4 \\ -5.5x + 48 & \text{if } 4 < x \le 6 \\ -30x + 195 & \text{if } 6 < x \le 6.5 \end{cases}$$

(b) In what month is the snow deepest? What is the deepest snow depth?
at the beginning of February; 26 inches

(c) In what months does the snow begin and end?
begins in early October; ends in mid-April

Graph the functions defined as follows.

40. $f(x) = [\![-x]\!]$

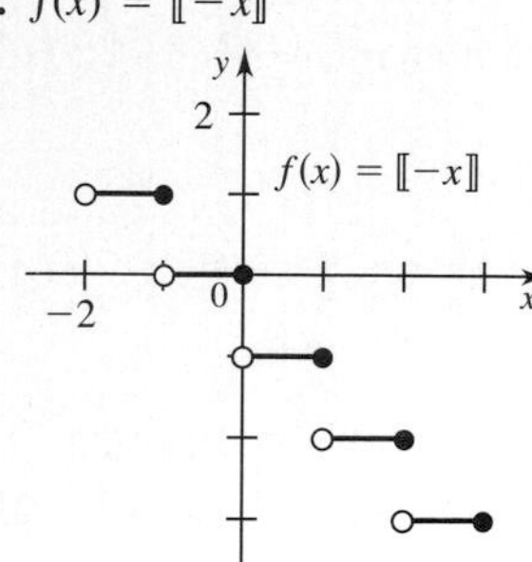

41. $f(x) = [\![2x]\!]$

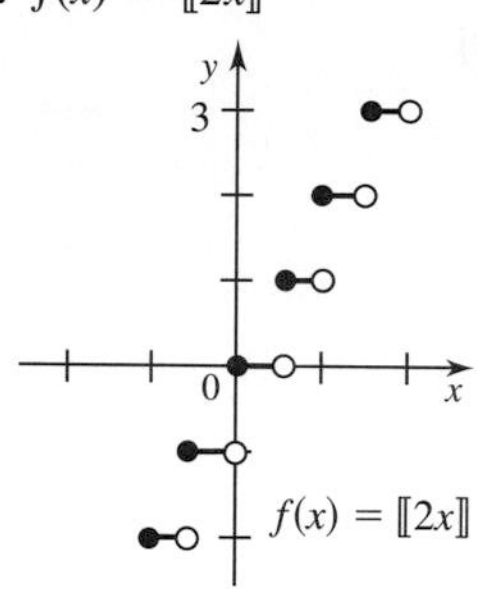

42. $g(x) = [\![2x - 1]\!]$

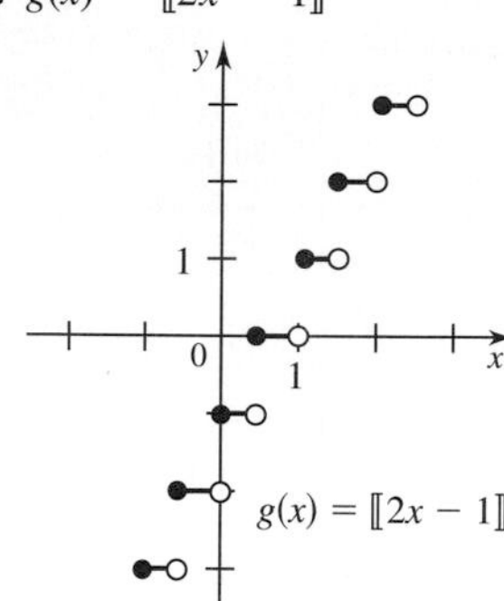

43. $h(x) = [\![3x + 1]\!]$

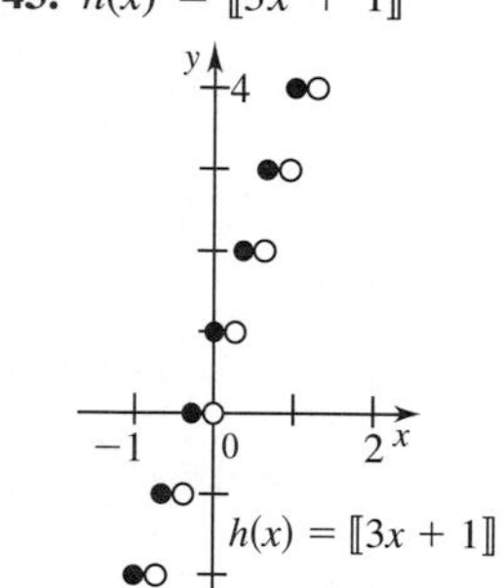

44. $k(x) = [\![3x]\!]$

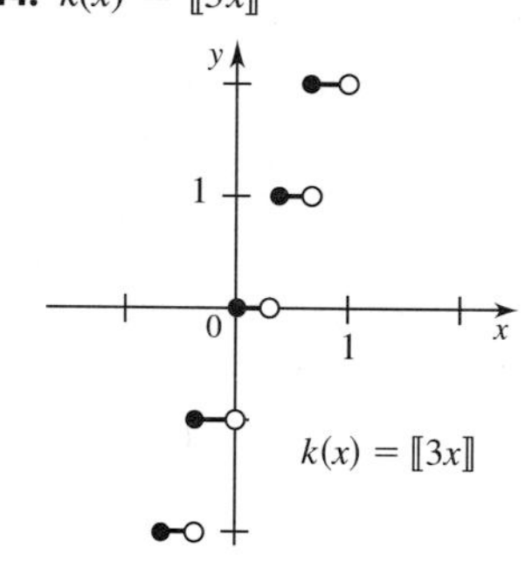

45. $r(x) = [\![3x]\!] + 1$

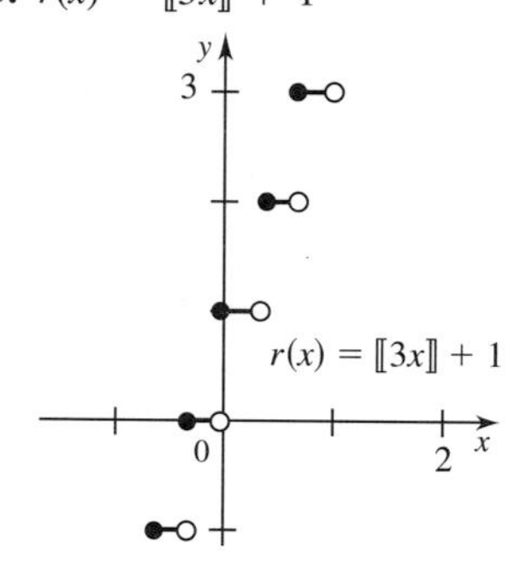

46. Describe how the y-values of the greatest-integer function are determined for negative x-values.

47. At Rick's Rentals, a lift truck can be rented for \$74 per day or fraction of a day plus a fixed charge of \$55. Let $T(x)$ represent the cost to rent a lift truck from Rick for x days.
Find the following: **(a)** $T(1)$ **\$129** **(b)** $T(1.25)$ **\$203** **(c)** $T(3.5)$ **\$351**
(d) Graph $y = T(x)$.

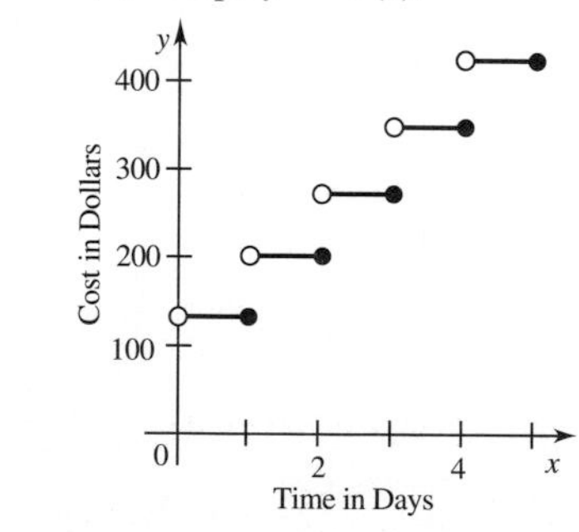

(e) Give the domain and range of T. **$(0, \infty)$; {129, 203, 277, 351, . . .}**

48. A mail-order firm charges 30¢ to mail a package weighing 1 oz or less, and then 27¢ for each additional ounce or fraction of an ounce. Let $M(x)$ be the cost of mailing a package weighing x oz.
Find **(a)** $M(.75)$ **30¢** **(b)** $M(1.6)$ **57¢** **(c)** $M(4)$. **111¢**
(d) Graph $y = M(x)$.

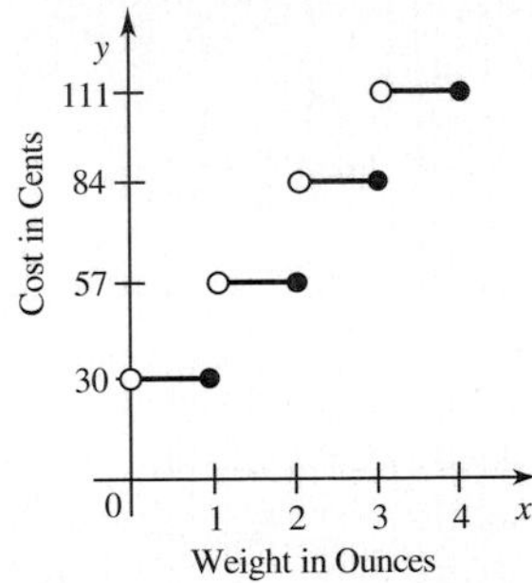

(e) Give the domain and range of M. **(0, ∞); {30, 57, 84, 111, . . .}**

49. Use the greatest-integer function and write an expression for the number of ounces for which postage will be charged on a package weighing x oz (see Exercise 48). $\mathbf{-[\![-x]\!]}$

50. A car rental cost \$37 for one day, which includes 50 free miles. Each additional 25 miles or portion costs \$10. Graph the ordered pairs (miles, cost) for a one-day rental. Use the greatest-integer function to write an expression for the cost of a one-day rental for x miles.

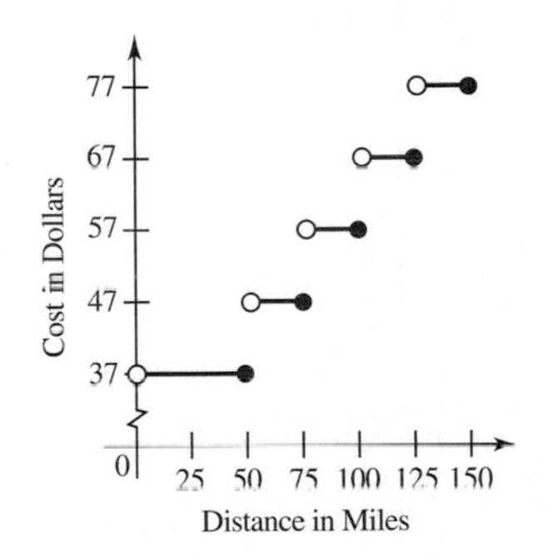

$$C(x) = \begin{cases} 37 & \text{if } 0 < x \le 50 \\ 37 - 10[\![(50 - x)/25]\!] & \text{if } x > 50 \end{cases}$$

51. Suppose that the demand and price for a certain model of electric motor are related by $p = 16 - (5/4)q$, where p is the price in hundreds of dollars and q is the demand in appropriate units. That is, for q units to be sold, the price must be set at $16 - (5/4)q$ hundred dollars.

(a) Find the price at the following demand levels: 0 units, 4 units, 8 units. **\$1600, \$1100, \$600**

(b) Find the demand for the motors at the following prices: \$600, \$1100, \$1600. **8 units, 4 units, 0 units**

(c) According to this function, what is the maximum price that can be charged? What is the maximum demand for the motors? **\$1600, 12.8 units**

(d) Graph $p = 16 - (5/4)q$.

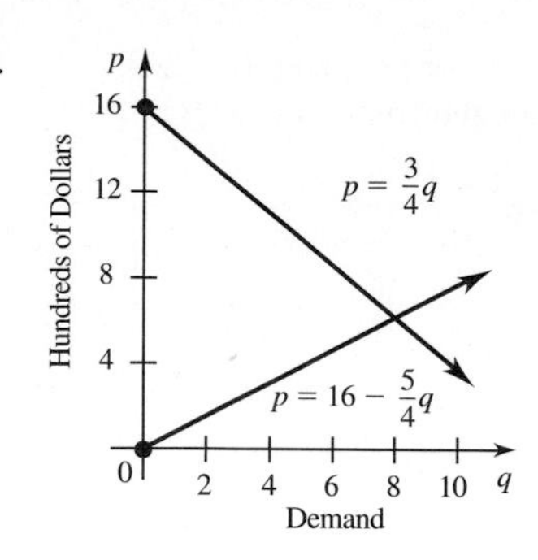

(e) Suppose the price (in hundreds of dollars) and the supply are related by $p = (3/4)q$, where q represents the supply and p the price. Find the supply at the following prices: \$0, \$1000, \$2000. **0, 13 1/3, 26 2/3**

(f) Graph $p = (3/4)q$ on the same axes used for part (d). **See (d).**

(g) Find the equilibrium supply (the supply at the point where the supply and demand graphs cross). **8 units**

(h) Find the equilibrium price (the price at the equilibrium supply). **\$600**

52. Let the supply and demand equations for a new textbook in advanced mathematics be as follows, where x is in thousands and p is in dollars.

$$\text{Supply: } p = (3/2)x \quad \text{and} \quad \text{Demand: } p = 81 - (3/4)x$$

(a) Graph these on the same axes.

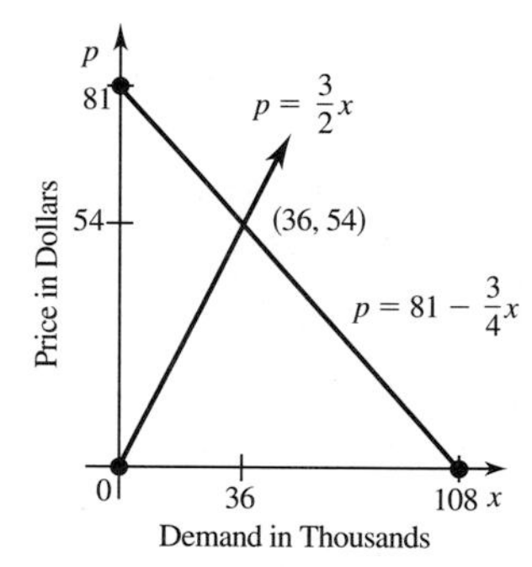

(b) Find the equilibrium demand. (See Exercise 51.) **36,000**

(c) Find the equilibrium price. **\$54**

In Exercises 53–56, decide which of the following numbers gives the slope of the linear function f: **(a)** 3, **(b)** 1, **(c)** 1/3, **(d)** $-1/3$, **(e)** -1, **(f)** -3. *(Hint: Use the slope formula and consider the sign of the differences in the numerator and denominator as well as their comparative sizes.)*

53. 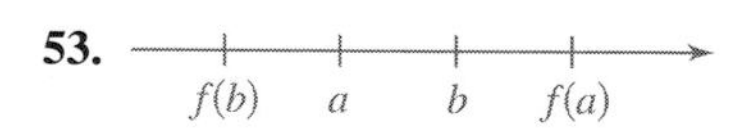

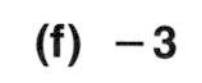
(f) −3

54. 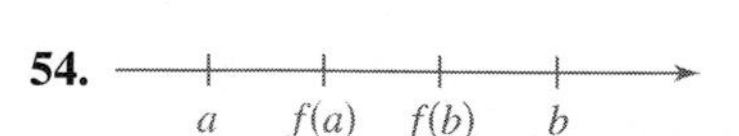

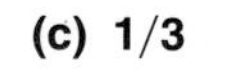
(c) 1/3

55. 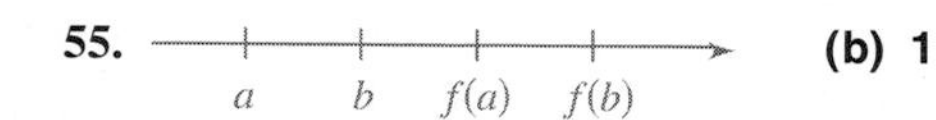

(b) 1

56. 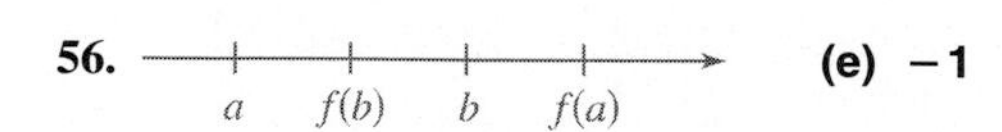

(e) −1

Graphing Utility Problems

Use a graphing utility to graph each of the following lines. You may need to rewrite the equation before graphing it.

57. $5y - 4x = 10$

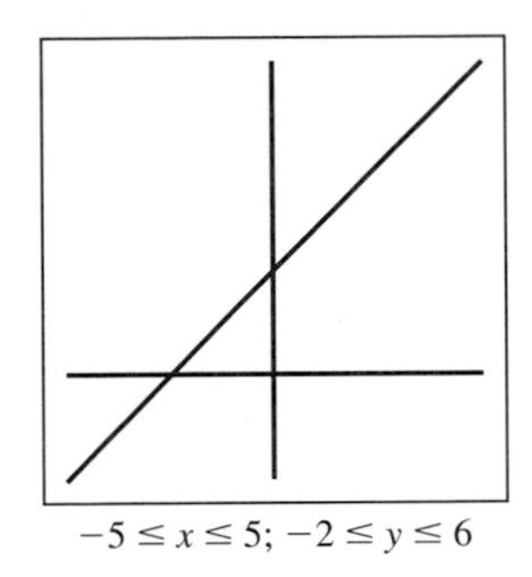

$-5 \le x \le 5; -2 \le y \le 6$

58. $(2y - 2)/(x - 3) = 4$

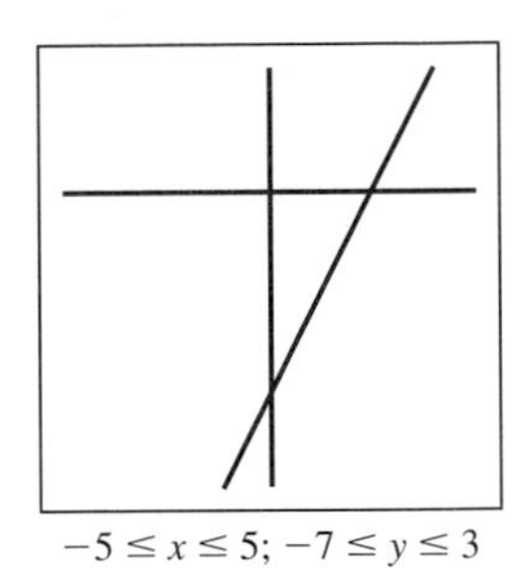

$-5 \le x \le 5; -7 \le y \le 3$

59. Graph $y = x + 4$ and $y = (x^2 + 3x - 4)/(x - 1)$. Do you see a difference between the graphs? Should there be a difference? If so, describe the difference.

60. Graph $y = |x|$. Then graph $y = |x - 2|$, $y = |x| + 2$, and $y = 2|x|$. How are these four graphs related? How do they differ?

3.5 Exercises (text page 160)

Write an equation in standard form for each of the following lines.

1. Through (2, 1), $m = 5$ **$5x - y - 9 = 0$**

2. Through (−3, 2), $m = -4$ **$4x + y + 10 = 0$**

3. Through (−5, 3), $m = -1$ **$x + y + 2 = 0$**

4. Through (0, 0), $m = 2$ **$2x - y = 0$**

5. $f(3) = 1$, $m = 2/3$ **$2x - 3y - 3 = 0$**

6. $f(4) = -1$, $m = .4$ **$.4x - y - 2.6 = 0$**

7. $f(.6) = 2$, $m = 0$ **$y - 2 = 0$**

8. $f(0) = 1/2$, $m = -5$ **$10x + 2y - 1 = 0$**

9. Through (2, 3), vertical **$x - 2 = 0$**

10. Through (3, −1), horizontal **$y + 1 = 0$**

11. Through the origin, horizontal **$y = 0$**

12. Through (−5, 1/2), vertical **$x + 5 = 0$**

13. x-intercept 3, vertical **$x - 3 = 0$**

14. $f(-1) = 2$, horizontal **$y - 2 = 0$**

15. Through the origin, bisects second and fourth quadrants **$x + y = 0$**

16. Through the origin, bisects first and third quadrants **$x - y = 0$**

17. Through (3, 4) and (5, 6) **$x - y + 1 = 0$**

18. Through (−1, 2) and (7, 3) **$x - 8y + 17 = 0$**

19. Through (−4, 2) and (4, 18) **$2x - y + 10 = 0$**

20. Through (.1, .3) and (.3, .1) **$x + y - .4 = 0$**

21. Through (2, 7) and (3, 7) **$y - 7 = 0$**

22. Through (3, 4) and (3, 5) **$x - 3 = 0$**

23. Through (3, −1), parallel to $x - 2y = 4$ **$x - 2y - 5 = 0$**

24. Through (0, 4), parallel to $-2x + 4y = 7$ **$x - 2y + 8 = 0$**

25. Through (−3, 4), perpendicular to $y = 4x + 6$ **$x + 4y - 13 = 0$**

26. Through (5, 6), perpendicular to x-axis **$x - 5 = 0$**

27. $f(-2) = 8$, perpendicular to y-axis **$y - 8 = 0$**

28. $f(4) = 3$, perpendicular to y-axis **$y - 3 = 0$**

29. $f(9) = 0$, parallel to $x = 3y + 4$ **$x - 3y - 9 = 0$**

30. $f(-2) = 5$, parallel to $y = .2x + 3$ **$.2x - y + 5.4 = 0$**

31. x-intercept 3, y-intercept 6 **$2x + y - 6 = 0$**

32. x-intercept −4, y-intercept 2 **$x - 2y + 4 = 0$**

◉ *For each linear equation in Exercises 33–38, identify its graph in (a)–(f) below.*

33. $y = -x - 1$ **(a)**

34. $y = 1 - x$ **(b)**

35. $y = \frac{1}{2}x$ **(c)**

36. $y = 2$ **(e)**

37. $y = 1 - 2x$ **(f)**

38. $y = 2x - 1$ **(d)**

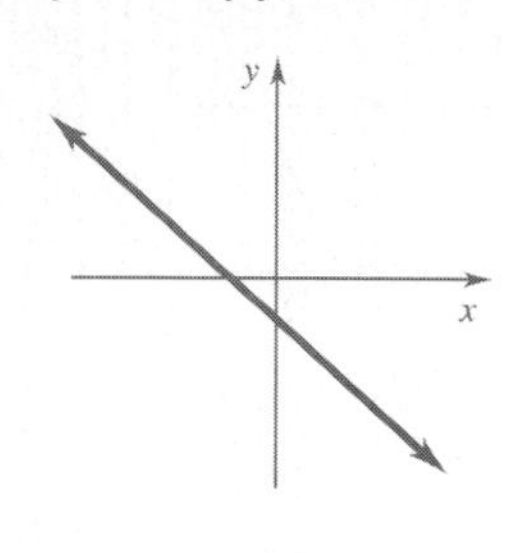

(a)

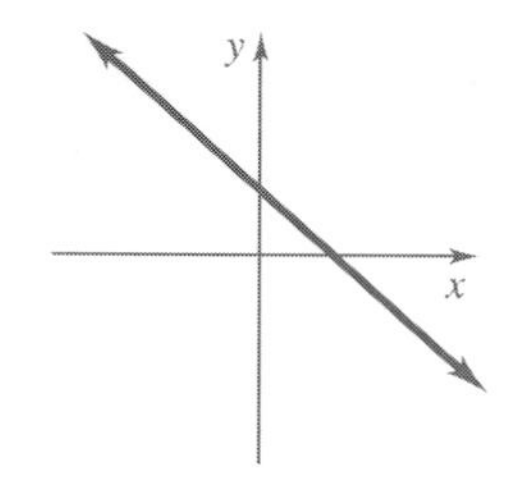

(b)

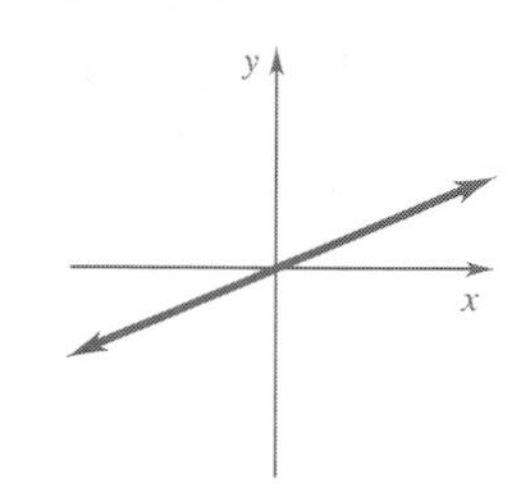

(c)

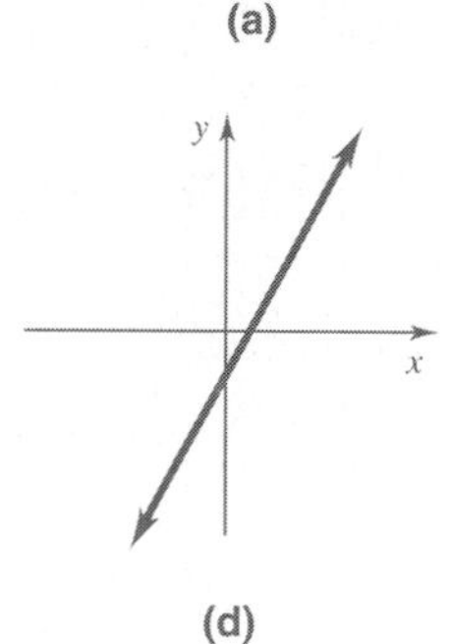

(d)

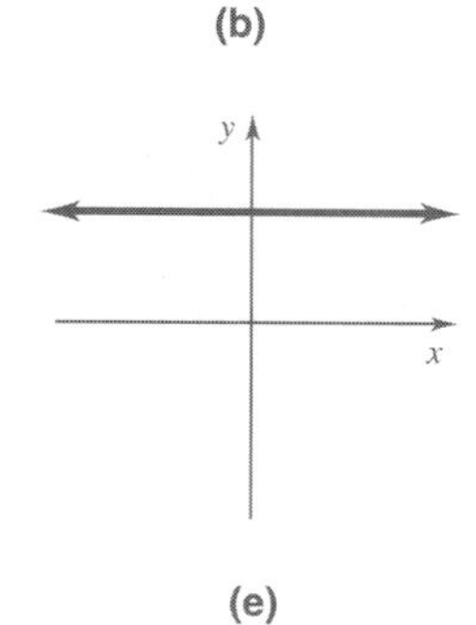

(e)

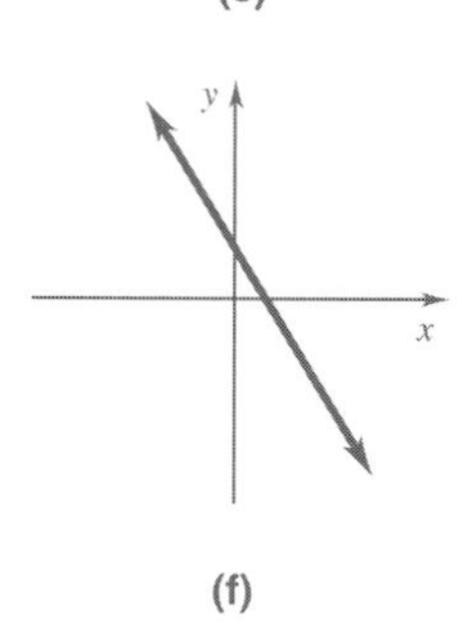

(f)

◉ **39.** If $f(4) = -1$, $f(2) = 0$, and f is a linear function, find $f(x)$ and $f(-4)$. $\mathbf{f(x) = (-1/2)x + 1; f(-4) = 3}$

◉ **40.** A linear function has $f(3) = 4$ and $f(4) = 5$. Find $f(x)$ and $f(8)$. $\mathbf{f(x) = x + 1; f(8) = 9}$

◉ **41.** Find r so that the line through $(2, 6)$ and $(-4, r)$ is as follows.
(a) Parallel to $2x - 3y = 4$ **2**
(b) Perpendicular to $x + 2y = 1$ **−6**

◉ **42.** Find k so that the line through $(4, -1)$ and $(k, 2)$ is as follows.
(a) Parallel to $3y + 2x = 6$ **−1/2**
(b) Perpendicular to $2y - 5x = 1$ **−7/2**

◉ **43.** Two lines intersect at the point $(5, -4)$ and are each perpendicular to one of the axes. What are the equations of the lines? $\mathbf{x = 5 \text{ and } y = -4}$

◉ **44.** The following table gives some points on the line $y = mx + b$. Find m and b. $\mathbf{m = .5, b = 5.5}$

x	2.6	2.8	3	3.2	3.4
y	6.8	6.9	7	7.1	7.2

45. Use slopes to show that the quadrilateral with vertices at $(3/2, 6)$, $(-3/2, 3)$, $(-5/2, 5)$, and $(5/2, 2)$ is not a parallelogram (opposite sides parallel).

46. Use slopes to show that the square with vertices at $(-1, 4)$, $(5, 4)$, $(5, -2)$, and $(-1, -2)$ has diagonals that are perpendicular.

Many real-life situations can be described approximately by a straight-line graph. One way to find the equation of such a straight line is to use two typical data points from the graph and the point-slope form of the equation of a line. In each of Exercises 47–52, assume that the data can be approximated fairly closely by a straight line. Use the given information to find the equation of the line. Then answer the question asked in the problem.*

47. Temperatures of 32° and 212° Fahrenheit correspond to temperatures of 0° and 100° Centigrade. Let y be the Centigrade temperature corresponding to the Fahrenheit temperature x. An increase in temperature of 1° Fahrenheit corresponds to what change in degrees Centigrade? **$y = (5/9)(x - 32)$; 5/9**

48. Suppose a baseball is thrown at 85 mph. The ball will travel 320 ft when hit by a bat swung at 50 mph and will travel 440 ft when hit by a bat swung at 80 mph. Let y be the number of feet traveled by the ball when hit by a bat swung at x mph. (*Note:* This function is valid for $50 \le x \le 90$, where the bat is 35 inches long, weighs 32 oz, and is swung slightly upward to drive the ball at an angle of 35°.†) How much further will a ball travel for each one-mile-per-hour increase in the speed of the bat? **$y = 4x + 120$; 4 ft**

49. The number of farms in the United States declined from 6 million in 1920 to 2 million in 1980. Let y be the number of farms (in millions) x years after 1900. How many farms were there in 1960? **$y = (-1/15)x + 22/3$; about 3.3 million**

50. The average size of a farm in the United States increased from 100 acres in 1920 to 700 acres in 1980. Let y be the average size x years after 1900. In what year was the average size 400 acres? **$y = 10x - 100$; 1950**

51. The worldwide consumption of cigarettes increased from 2.5 trillion in 1960 to 4 trillion in 1980. Let y be the consumption of cigarettes (in trillions) x years after 1940. In what year will the consumption reach 5.5 trillion? **$y = (3/40)\,x + 1$; 2000**

52. The amount of tropical rain forests in Central America decreased from 130,000 sq mi in 1969 to about 80,000 sq mi in 1985. Let y be the amount (in ten-thousands of square miles) x years after 1965. How large will the rain forests be in the year 1997? **$y = (-5/16)\,x + 57/4$; 42,500 sq mi**

53. Use your own words to describe how to find the equation of a line through two given points.

54. Discuss the advantages and disadvantages of each of the three forms of the equation of a line. When would you use each of them and why? Which form is written in function notation?

55. The product of the slopes of two perpendicular lines is -1. Is this true for *any* two perpendicular lines? Explain. (*Hint:* Is slope defined for every line?)

56. Show that the line $y = x$ is the perpendicular bisector of the segment containing (a, b) and (b, a), where $a \neq b$. (*Hint:* Use the midpoint formula and the slope formula.)

57. Let $y = m_1x + b_1$ and $y = m_2x + b_2$ be the equations of two nonvertical lines.
(a) Set $m_1x + b_1$ equal to $m_2x + b_2$ and solve for x to find the x-coordinate of the point of intersection of the lines. **$x = (b_2 - b_1)/(m_1 - m_2)$**
(b) If $m_1 = m_2$ and $b_1 \neq b_2$, the lines must be parallel. Why?

*The information for Exercises 49–52 was taken from *Mathematics and Global Survival*, Second Edition, by Richard H. Schwartz (Ginn Press, 1991).

†Adair, Robert K., *The Physics of Baseball*; (New York: Harper & Row, 1990).

58. Temperature decreases with height above the earth's surface. We can use the following chart to find a function that describes this relationship.

Height (feet)	*Temperature (°F)*
1000	56
5000	41
10,000	23
15,000	5
20,000	−15
30,000	−47
36,100	−69

(a) Let x represent the height in thousands of feet and y represent the temperature in degrees Fahrenheit. Plot the ordered pairs (height, temperature) corresponding to the values in the table. The points should lie in an approximately linear pattern.

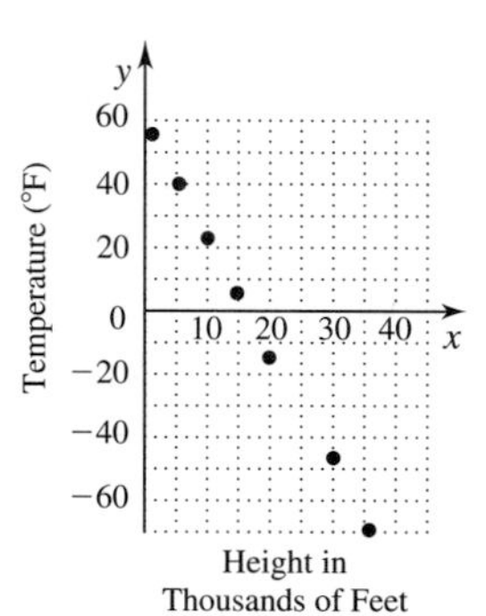

(b) Use the ordered pairs (5, 41) and (30, −47) to find the slope of the line through these two points. Then substitute the values from either ordered pair into the equation $y = mx + b$, which defines a linear function, to find b. Use the values of m and b to write the equation that defines the linear function describing the relationship between height and temperature. $\mathbf{y = -3.52x + 58.6}$

(c) Test the function found in part (b) by substituting other heights from the chart to see if the predicted temperatures are close to the actual ones. What do you find? **The values are reasonably close.**

59. Use the ordered pairs (20, −15) and (10, 23) to repeat Exercise 58(b). Compare the two equations. Are they similar? Do they give equally good predictions?
$\mathbf{y = -3.8x + 61}$

3.6 Exercises (text page 172)

1. Given the graph of $g(x)$ in the figure, sketch the graph of each of the following and explain how it is obtained from the graph of $y = g(x)$.

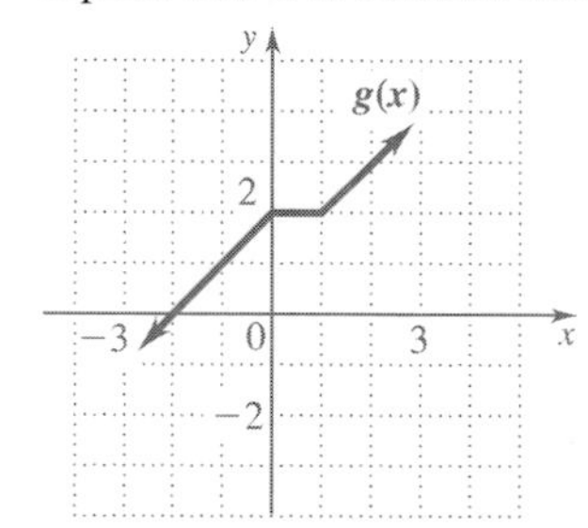

(a) $y = g(-x) + 1$ **(b)** $y = g(x - 2)$
(c) $y = g(x + 1) - 2$ **(d)** $y = -g(x) + 2$

(a) The graph of $g(x)$ is reflected about the y-axis and translated up 1 unit.

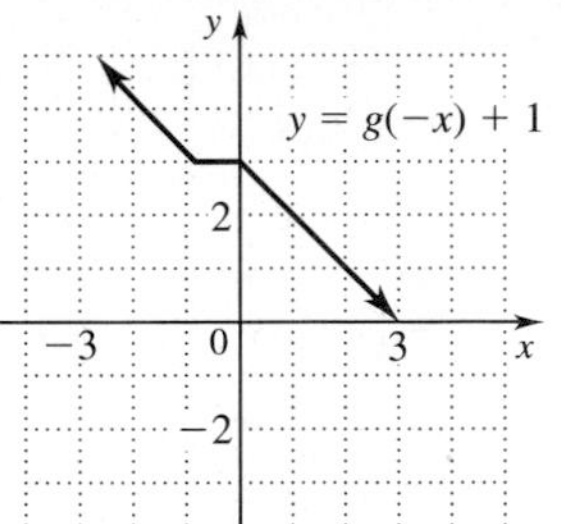

(b) The graph of $g(x)$ is translated to the right 2 units.

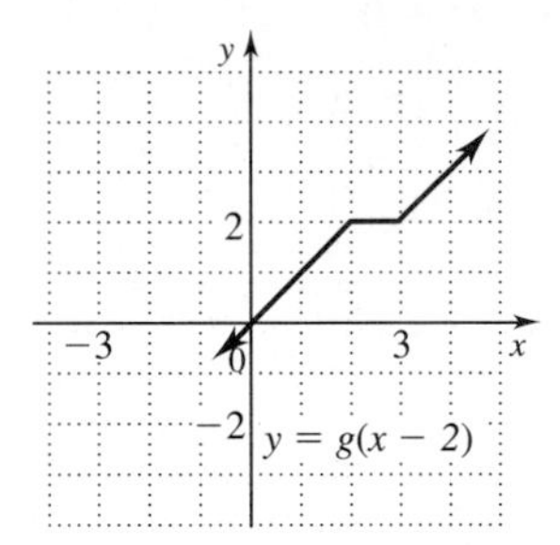

(c) The graph of $g(x)$ is translated to the left 1 unit and down 2 units.

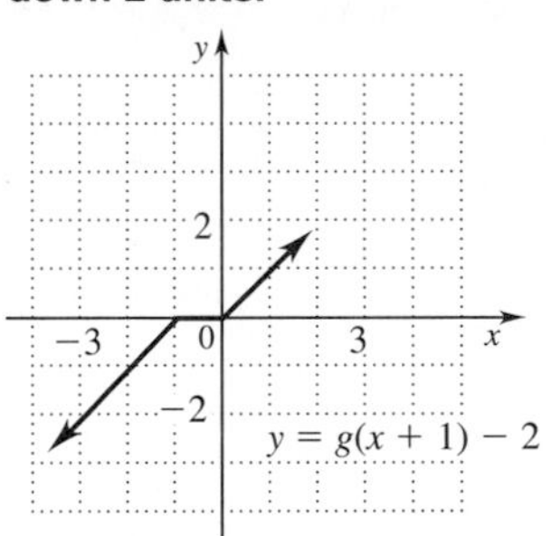

(d) The graph of $g(x)$ is reflected about the x-axis and translated up 2 units.

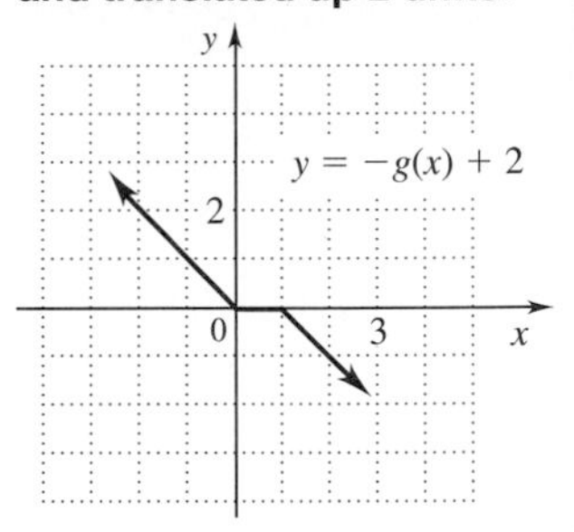

2. Use the graph of $f(x)$ in the figure to obtain the graph of each of the following. Explain how each graph is related to the graph of $f(x)$.

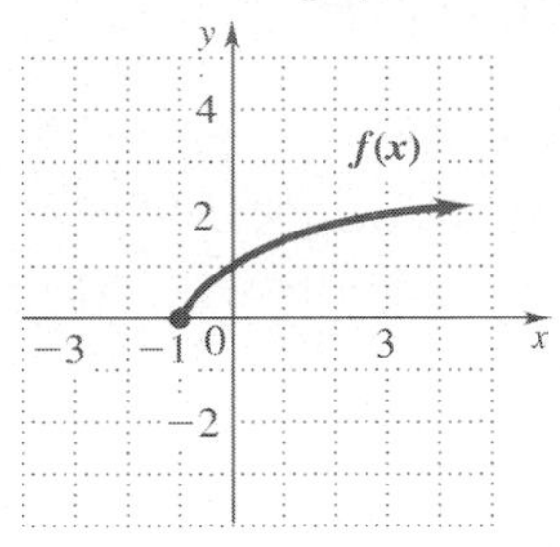

(a) $y = -f(x)$ **(b)** $y = 2f(x)$
(c) $y = f(x - 1) + 3$ **(d)** $y = f(-x)$

(a) The graph of $f(x)$ is reflected about the x-axis.

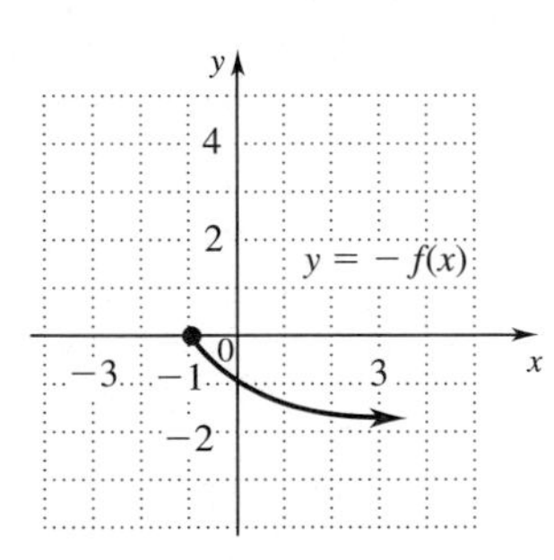

(b) The graph is the same shape as that of $f(x)$, but broader.

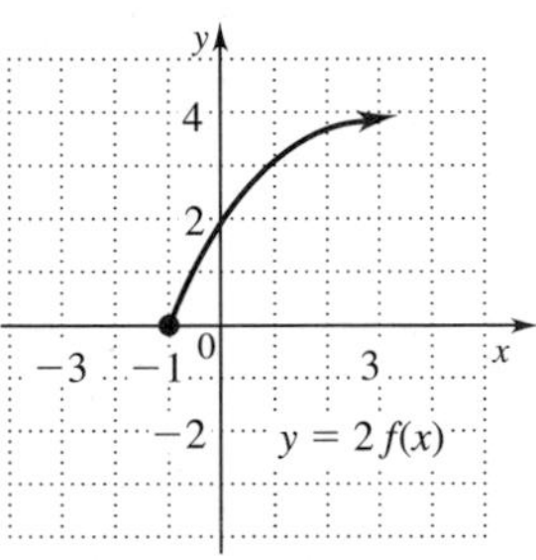

(c) The graph of $f(x)$ is translated 1 unit to the right and 3 units up.

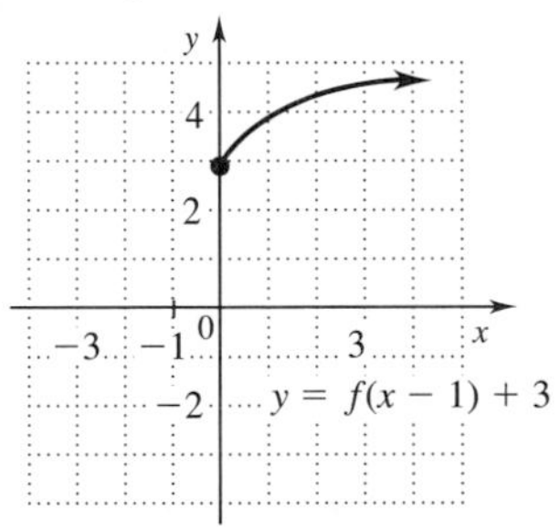

(d) The graph of $f(x)$ is reflected about the y-axis.

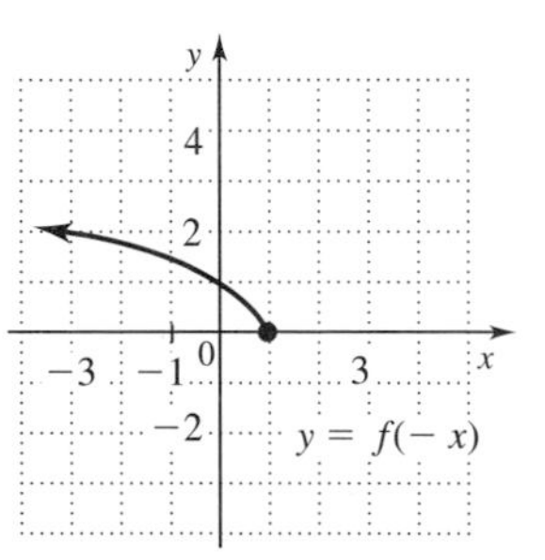

Plot the following points, and then plot the points that are symmetric to the given point with respect to the **(a)** *x-axis,* **(b)** *y-axis,* **(c)** *origin.*

3. $(5, -3)$

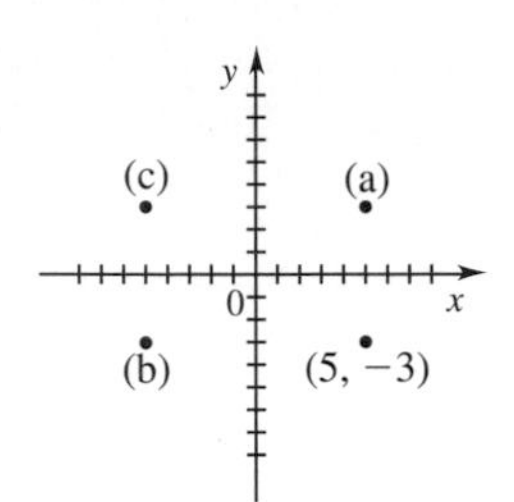

4. $(-6, 1)$

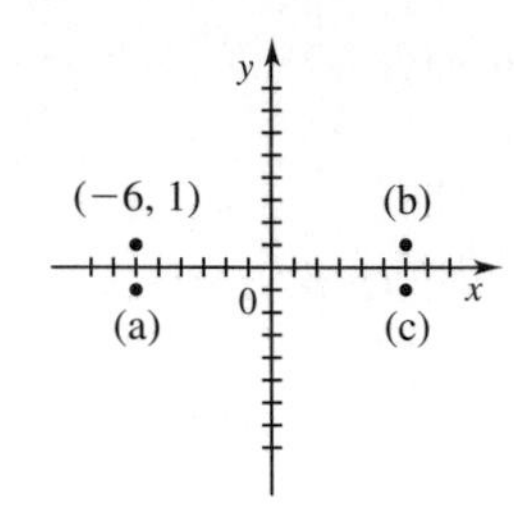

5. $(-4, -2)$

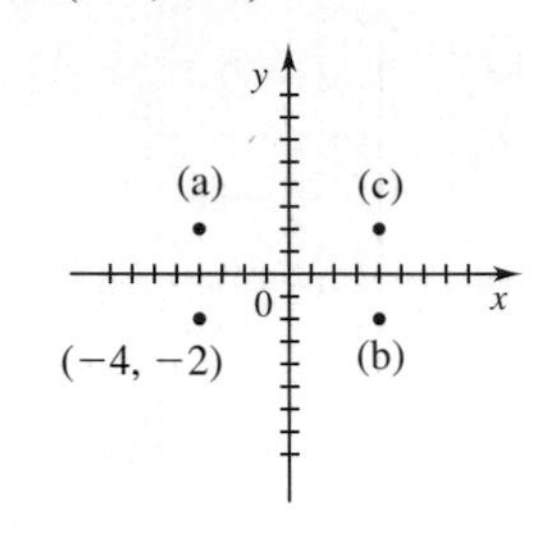

6. $(-8, 0)$

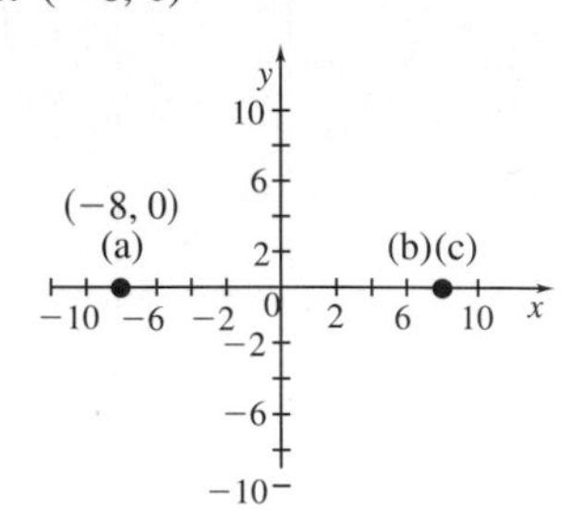

Without graphing, determine whether each equation has a graph that is symmetric with respect to the x-axis, the y-axis, or the origin.

7. $y = (x + 3)(x - 5)$ **none of these**

8. $x^2 + y^2 = 6$ ***x*-axis, *y*-axis, origin**

9. $xy = -3$ **origin**

10. $3x/y = 2$ **origin**

11. $y = -x^3$ **origin**

12. $y + 1 = (x - 3)^3$ **none of these**

13. $x^2 + 4y^2 = 3$ ***x*-axis, *y*-axis, origin**

14. $y = 1/(2 - x^2)$ ***y*-axis**

Use the techniques presented in this section to help graph each of the following.

15. $|x| = y + 1$

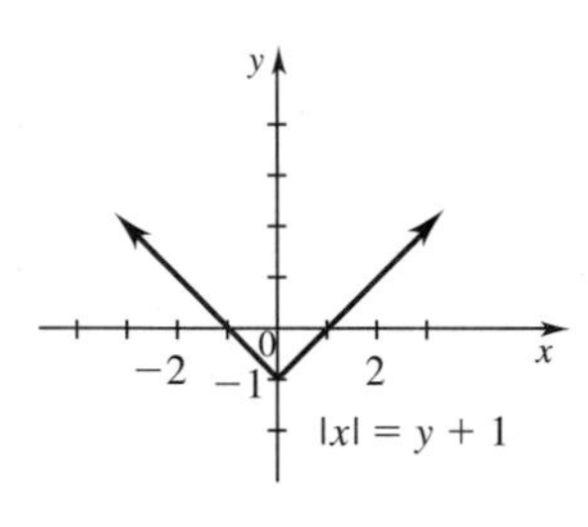

16. 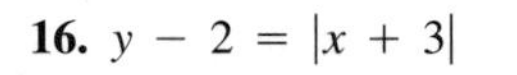 $y - 2 = |x + 3|$

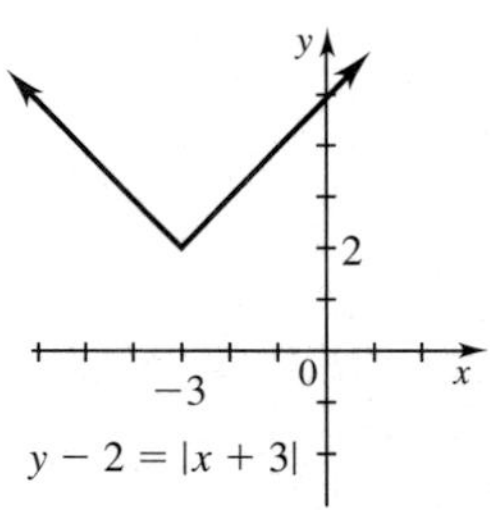

17. $x/y = 2$

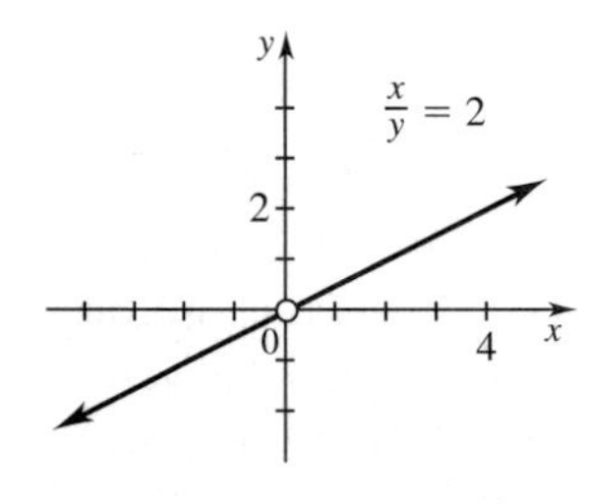

18. 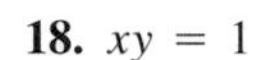 $xy = 1$

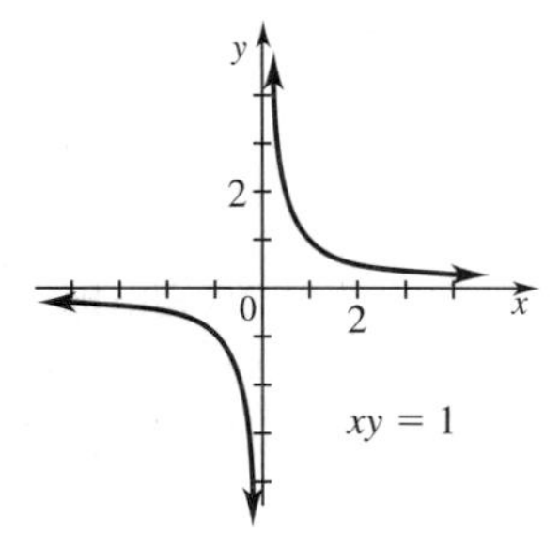

19. $y = -(x + 1)^3$

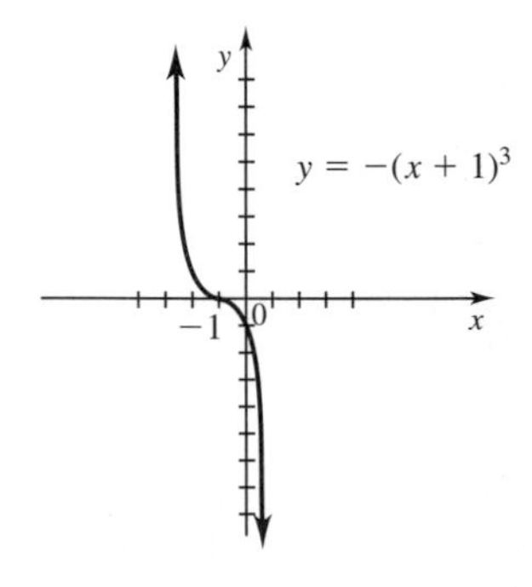

20. $y = (-x + 1)^3$

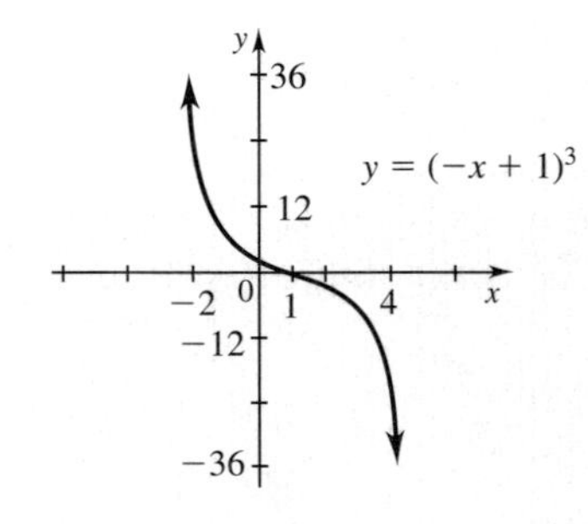

21. $y = 2x^2 - 1$

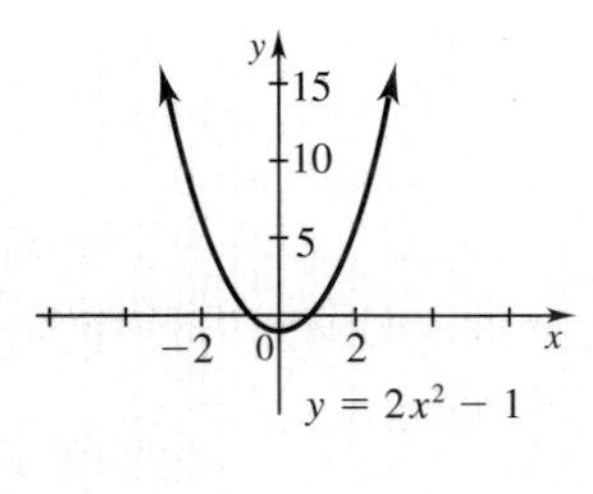

22. 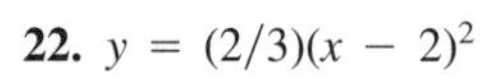 $y = (2/3)(x - 2)^2$

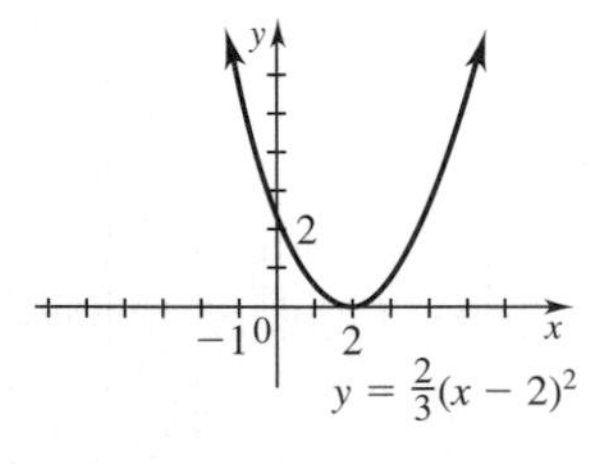

23. $y + 2 = (1 - x)^2$

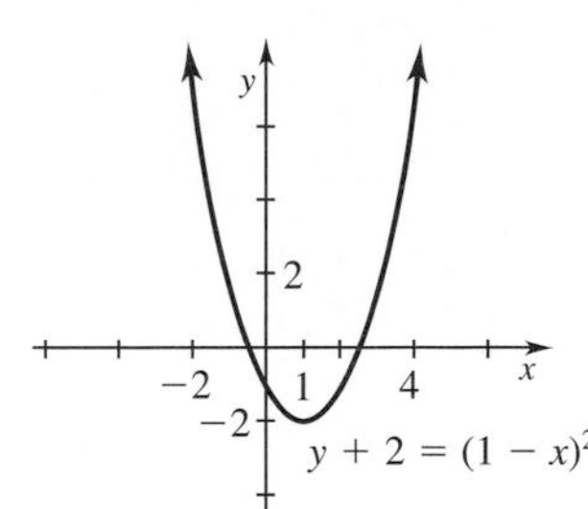

24. $x^2 = y^2$

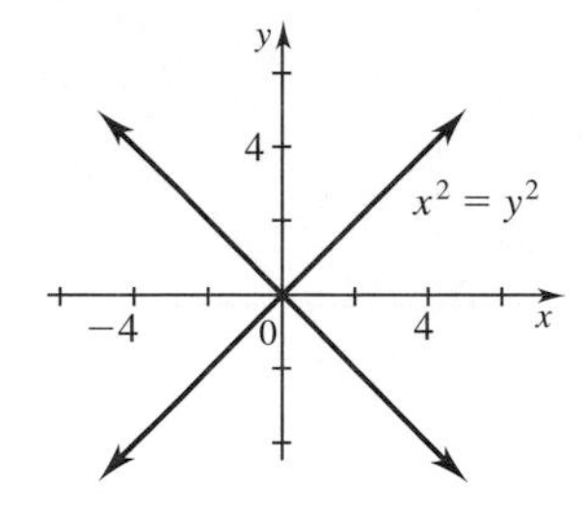

25. $(x - 1)^4 = y - 3$

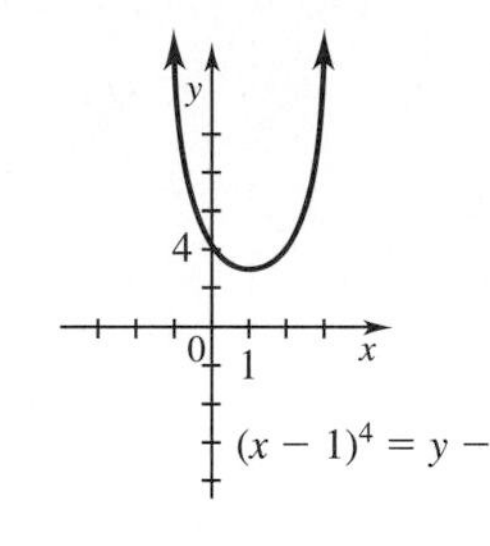

26. $x^2y = 2$

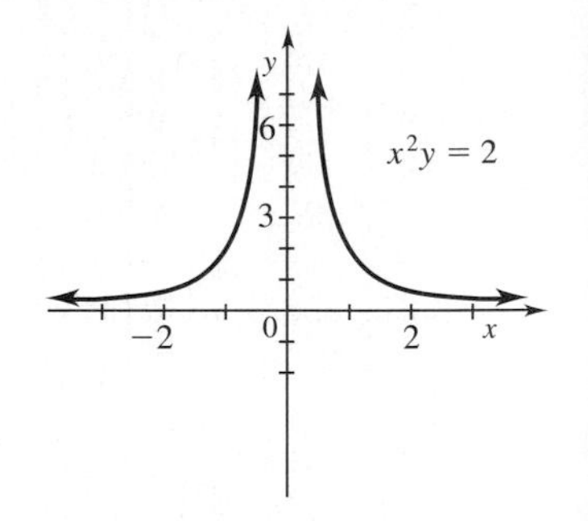

◉ *Suppose that $f(2) = 3$. For each assumption in Exercises 27–29, find another value of the function.*

27. $f(x)$ is symmetric with respect to the origin. $\mathbf{f(-2) = -3}$

28. $f(x)$ is symmetric with respect to the y-axis. $\mathbf{f(-2) = 3}$

29. $f(x)$ is symmetric with respect to the line $x = 6$. $\mathbf{f(10) = 3}$

30. Find the function whose graph can be obtained by translating the graph of $f(x) = 2x + 5$ up 2 units and left 3 units. $\mathbf{g(x) = 2x + 13}$

31. Find the function whose graph can be obtained by translating the graph of $f(x) = 3 - x$ down 2 units and right 3 units. $\mathbf{g(x) = -x + 4}$

✍ *Explain why each of the following statements is true.*

32. The function $f(x) = 0$ is both even and odd.

33. A nonzero function cannot be both even and odd.

34. If (a, b) is on the graph of an even function, then so is $(-a, b)$.

35. If (a, b) is on the graph of an odd function, then so is $(-a, -b)$.

◉ **36.** Complete the left half of the graph of $f(x)$ in the figure for each of the following conditions:
(a) $f(-x) = f(x)$ **(b)** $f(-x) = -f(x)$.

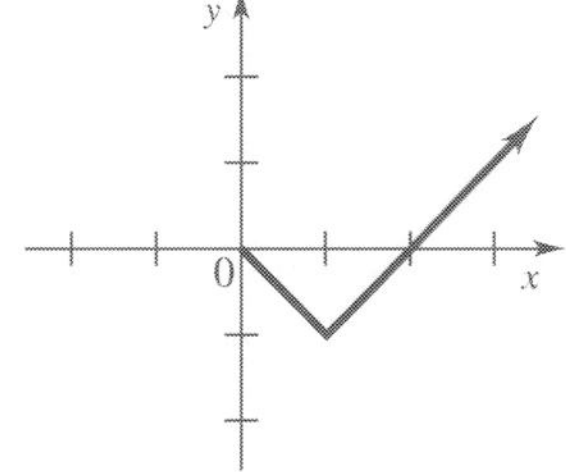

(a)

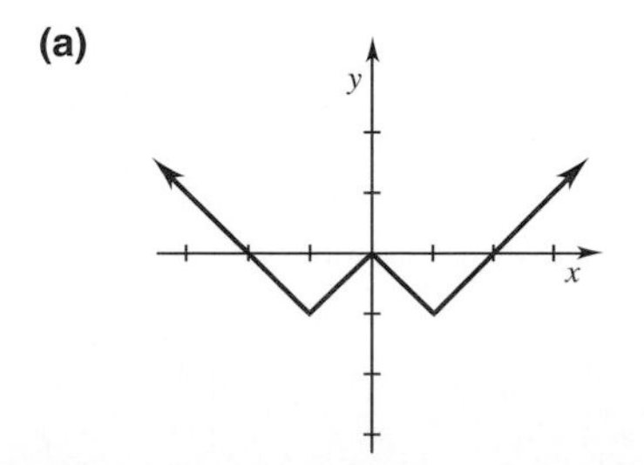

(b)

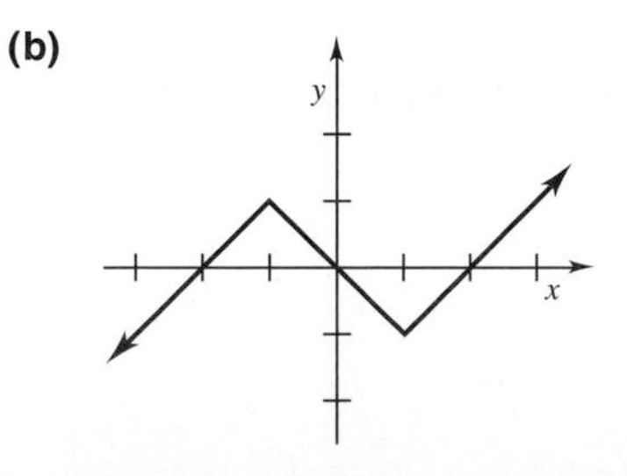

37. Complete the right half of the graph of $f(x)$ in the figure for each of the following conditions: **(a)** $f(x)$ is odd, **(b)** $f(x)$ is even.

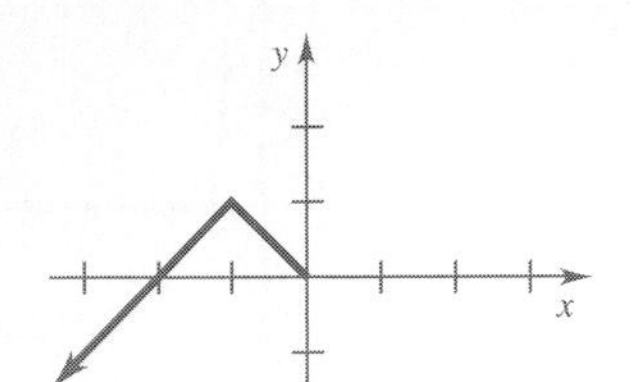

(a)

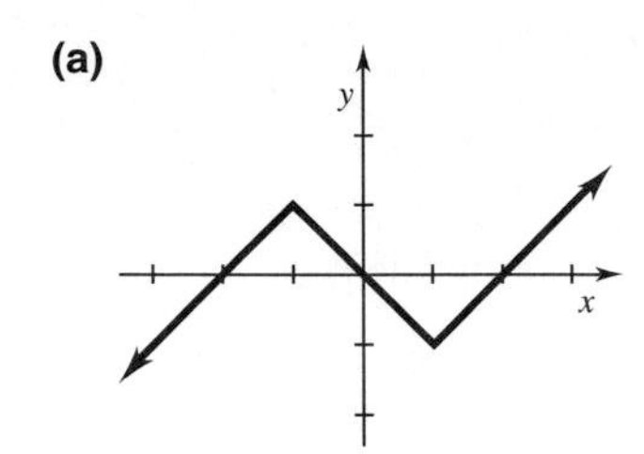

(b)

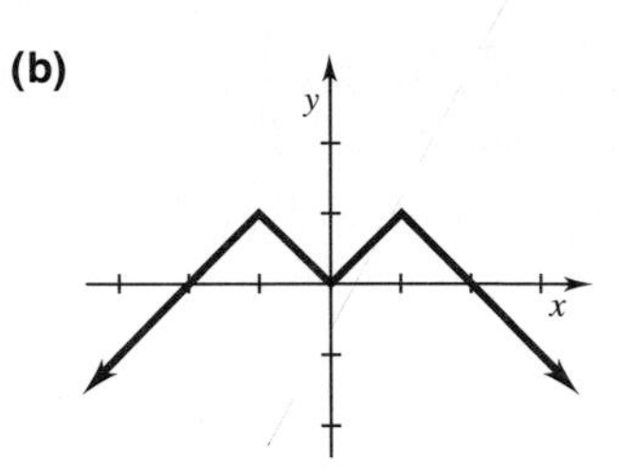

In Exercises 38 and 39, let F be some algebraic expression involving x as the only variable.

38. Suppose the equation $y = F$ is changed to $y = c \cdot F$, for some constant c. What is the effect on the graph of $y = F$? Discuss the effect depending on whether $c > 0$ or $c < 0$, and $|c| > 1$ or $|c| < 1$.

39. Suppose $y = F(x)$ is changed to $y = F(x + h)$. How are the graphs of these equations related? Is the graph of $y = F(x) + h$ the same as the graph of $y = F(x + h)$? If not, how do they differ?

Sketch examples of graphs having the given characteristics.

40. Symmetric with respect to the x-axis but not to the y-axis **Many answers are possible.**

41. Symmetric with respect to the origin but to neither the x-axis nor the y-axis **Many answers are possible.**

Prove each statement below.

42. A graph symmetric with respect to both the x-axis and the y-axis is also symmetric with respect to the origin.

43. A graph possessing two of the three types of symmetry, with respect to the x-axis, y-axis, and origin, must possess the third type of symmetry also.

Answer true *or* false *in Exercises 44–46.*

44. The graph of a nonzero function cannot be symmetric with respect to the x-axis. **true**

45. The graph of an even function is symmetric with respect to the y-axis. **true**

46. The graph of an odd function is symmetric with respect to the x-axis. **false**

Graphing Utility Problems

Let $f(x) = 8x^3 - 12x^2 + 2x + 1$. *Sketch* $f(x)$ *as* y1 *and each function defined as follows as* y2. *Start with x- and y-ranges of* $[-1, 2]$ *and* $[-2, 2]$ *and adjust as necessary. In each case, describe the translation or reflection involved.*

47. $y = f(x - 1)$ **48.** $y = f(x) - 1$ **49.** $y = -f(x)$ **50.** $y = f(-x)$

3.7 Exercises (text page 178)

For each of the pairs of functions defined as follows, find $f + g$, $f - g$, fg, and f/g. Give the domain of each.

1. $f(x) = 3x + 4$, $g(x) = 2x - 5$ **$5x - 1$; $x + 9$; $6x^2 - 7x - 20$; $(3x + 4)/(2x - 5)$; all domains are $(-\infty, \infty)$ except for that of f/g, which is $(-\infty, 5/2) \cup (5/2, \infty)$**

2. $f(x) = 6 - 3x$, $g(x) = -4x + 1$ **$-7x + 7$; $x + 5$; $12x^2 - 27x + 6$; $(6 - 3x)/(-4x + 1)$; all domains are $(-\infty, \infty)$ except for that of f/g, which is $(-\infty, 1/4) \cup (1/4, \infty)$**

3. $f(x) = 2x^2 - 3x$, $g(x) = x^2 - x + 3$
$3x^2 - 4x + 3$; $x^2 - 2x - 3$; $2x^4 - 5x^3 + 9x^2 - 9x$; $(2x^2 - 3x)/(x^2 - x + 3)$; all domains are $(-\infty, \infty)$

4. $f(x) = 4x^2 + 2x - 3$, $g(x) = x^2 - 3x + 2$
$5x^2 - x - 1$; $3x^2 + 5x - 5$; $4x^4 - 10x^3 - x^2 + 13x - 6$; $(4x^2 + 2x - 3)/(x^2 - 3x + 2)$; all domains are $(-\infty, \infty)$ except for that of f/g, which is $(-\infty, 1) \cup (1, 2) \cup (2, \infty)$

5. $f(x) = \sqrt{4x - 1}$, $g(x) = \sqrt{x + 3}$
$\sqrt{4x - 1} + \sqrt{x + 3}$; $\sqrt{4x - 1} - \sqrt{x + 3}$; $\sqrt{(4x - 1)(x + 3)}$; $\sqrt{(4x - 1)/(x + 3)}$; all domains are $[1/4, \infty)$ except for that of f/g, which is $(-\infty, -3) \cup [1/4, \infty)$

6. $f(x) = \sqrt{5x - 4}$, $g(x) = \sqrt{3x - 1}$
$\sqrt{5x - 4} + \sqrt{3x - 1}$; $\sqrt{5x - 4} - \sqrt{3x - 1}$; $\sqrt{(5x - 4)(3x - 1)}$; $\sqrt{(5x - 4)/(3x - 1)}$; all domains are $[4/5, \infty)$

Let $f(x) = 5x^2 - 2x$ and let $g(x) = 6x + 4$. Find each of the following.

7. $(f + g)(3)$ **61**
8. $(f - g)(-5)$ **161**
9. $(fg)(4)$ **2016**
10. $(fg)(-3)$ **-714**
11. $\left(\frac{f}{g}\right)(-1)$ **$-7/2 = -3.5$**
12. $\left(\frac{f}{g}\right)(4)$ **$18/7 \approx 2.5714$**
13. $(f - g)(m)$ **$5m^2 - 8m - 4$**
14. $(f + g)(2k)$ **$20k^2 + 8k + 4$**
15. $(f \circ g)(2)$ **1248**
16. $(f \circ g)(-5)$ **3432**
17. $(g \circ f)(2)$ **100**
18. $(g \circ f)(-5)$ **814**

Find $f \circ g$ and $g \circ f$ for each of the pairs of functions defined as follows.

19. $f(x) = -6x + 9$, $g(x) = 5x + 7$ **$-30x - 33$; $-30x + 52$**
20. $f(x) = 8x + 12$, $g(x) = 3x - 1$ **$24x + 4$; $24x + 35$**
21. $f(x) = 4x^2 + 2x + 8$, $g(x) = x + 5$ **$4x^2 + 42x + 118$; $4x^2 + 2x + 13$**
22. $f(x) = 5x + 3$, $g(x) = -x^2 + 4x + 3$ **$-5x^2 + 20x + 18$; $-25x^2 - 10x + 6$**
23. $f(x) = \frac{2}{x^4}$, $g(x) = 2 - x$ **$2/(2 - x)^4$; $2 - 2/x^4$**
24. $f(x) = \frac{1}{x}$, $g(x) = x^2$ **$1/x^2$; $1/x^2$**
25. $f(x) = 9x^2 - 11x$, $g(x) = 2\sqrt{x + 2}$ **$36x + 72 - 22\sqrt{x + 2}$; $2\sqrt{9x^2 - 11x + 2}$**
26. $f(x) = \sqrt{x + 2}$, $g(x) = 8x^2 - 6$ **$\sqrt{8x^2 - 4}$ or $2\sqrt{2x^2 - 1}$; $8x + 10$**

27. $f(x) = \begin{cases} 2x & \text{if } x < 0 \\ 3x - 5 & \text{if } x \geq 0 \end{cases}$

$g(x) = \begin{cases} x + 3 & \text{if } x \geq 1 \\ 4x & \text{if } x < 1 \end{cases}$

$(f \circ g)(x) = \begin{cases} 8x & \text{if } x < 0 \\ 12x - 5 & \text{if } 0 \leq x \leq 1 \\ 3x + 4 & \text{if } x > 1 \end{cases}$

$(g \circ f)(x) = \begin{cases} 8x & \text{if } x < 0 \\ 12x - 20 & \text{if } 0 \leq x < 1 \\ 3x - 2 & \text{if } x \geq 1 \end{cases}$

28. $f(x) = \begin{cases} 3x - 1 & \text{if } x \leq -1 \\ x - 1 & \text{if } x > -1 \end{cases}$

$g(x) = \begin{cases} x^2 + 1 & \text{if } x < 0 \\ x - 1 & \text{if } x \geq 0 \end{cases}$

$(f \circ g)(x) = \begin{cases} 3x^3 + 2 & \text{if } x \leq -1 \\ x^2 & \text{if } -1 < x < 0 \\ x - 2 & \text{if } x \geq 0 \end{cases}$

$(g \circ f)(x) = \begin{cases} 9x^2 - 6x + 2 & \text{if } x \leq -1 \\ x^2 - 2x + 2 & \text{if } -1 < x < 0 \\ x - 2 & \text{if } x \geq 0 \end{cases}$

29. Describe the steps required to find the composite function $f \circ g$, given $f(x) = 2x - 5$ and $g(x) = x^2 + 3$.

30. Composition is an operation that is unique to functions. Is composition of functions commutative? That is, does $f \circ g = g \circ f$ for all functions f and g? Explain.

The graphs of functions f and g are shown. Use these graphs to find the values in Exercises 31–38.

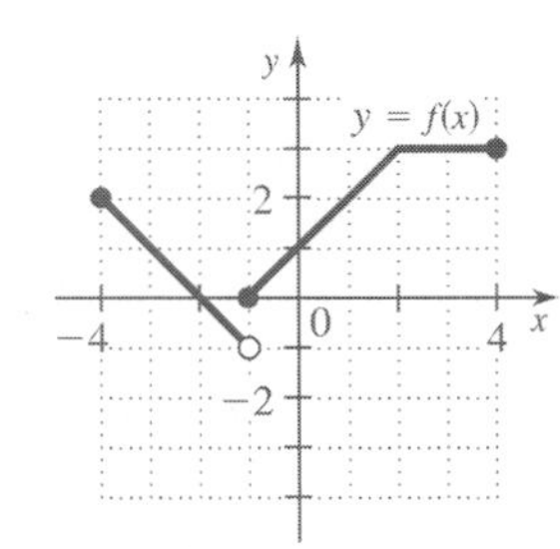

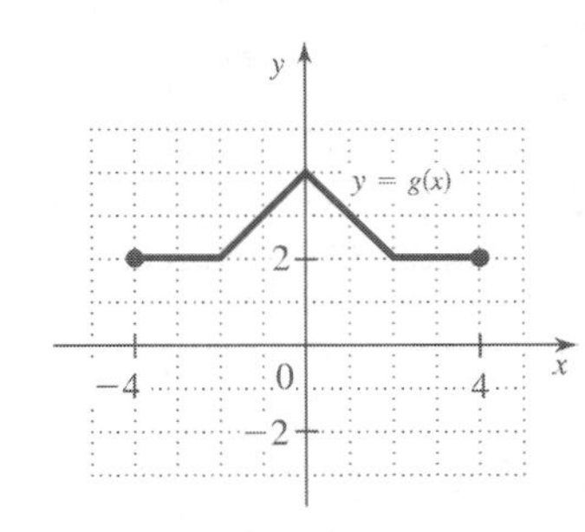

31. $f(1) + g(1)$ **5**

32. $f(4) - g(3)$ **1**

33. $f(-2) \cdot g(4)$ **0**

34. $\dfrac{f(4)}{g(2)}$ **3/2**

35. $(f \circ g)(2)$ **3**

36. $(g \circ f)(2)$ **2**

37. $(g \circ f)(-4)$ **2**

38. $(f \circ g)(-2)$ **3**

39. The graphs of functions f and g are shown. Draw the graph of $f \circ g$.

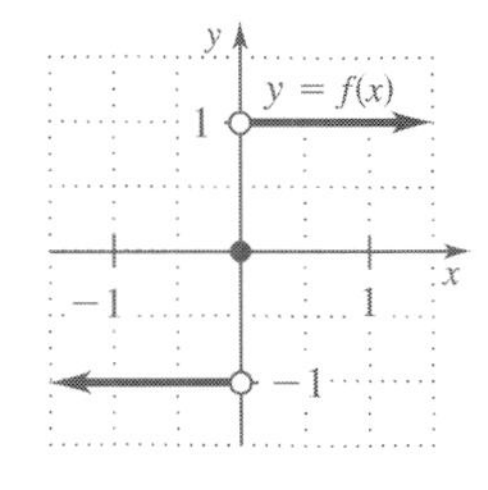

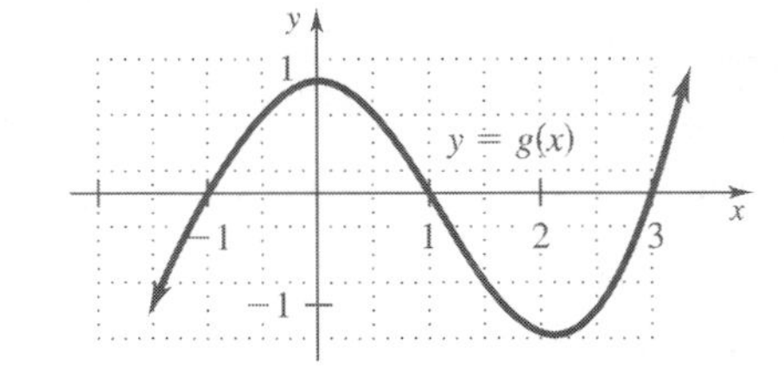

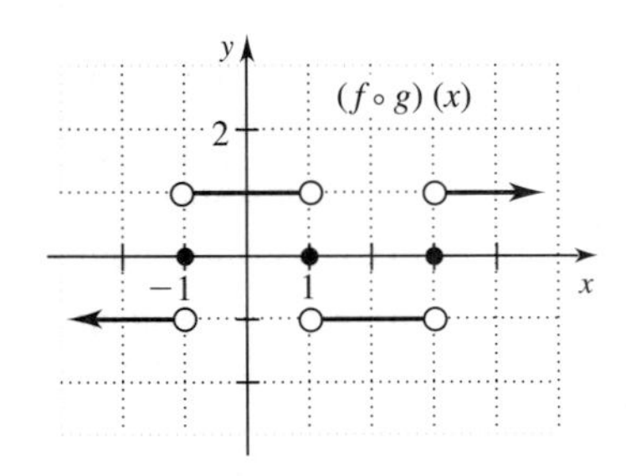

For each of the pairs of functions defined as follows, show that $(f \circ g)(x) = x$ and $(g \circ f)(x) = x$.

40. $f(x) = 2x + 3, \quad g(x) = (x - 3)/2$

41. $f(x) = (x + 5)/3, \quad g(x) = 3x - 5$

42. $f(x) = 2x^3 - 1, \quad g(x) = \sqrt[3]{(x + 1)/2}$

43. $f(x) = x^3 + 4, \quad g(x) = \sqrt[3]{x - 4}$

◉ *In Exercises 44–49, a function h is defined. Find functions f and g such that* $h(x) = (f \circ g)(x)$. Many such pairs of functions exist.

44. $h(x) = (2x - 5)^3$ **Many answers are possible.**

45. $h(x) = (8x^2 - 3x)^4$ **Many answers are possible.**

46. $h(x) = \sqrt{x^2 - 1}$ **Many answers are possible.**

47. $h(x) = \dfrac{1}{x^2 + 2}$ **Many answers are possible.**

48. $h(x) = \dfrac{(x - 2)^2 + 1}{5 - (x - 2)^2}$ **Many answers are possible.**

49. $h(x) = (x + 2)^3 - 3(x + 2)^2$ **Many answers are possible.**

50. The demand for a new camera is given by $D(p) = (-p^2/100) + 500$, where p is the price in dollars. The price, in terms of the cost c to make the camera, is expressed as $p(c) = 2c - 10$. Find the demand in terms of the cost c. $\mathbf{-(2c - 10)^2/100 + 500}$

51. The population P of a certain mammal depends on the number x (in hundreds) of a smaller mammal that serves as its primary food supply. The number x (in hundreds) of the smaller mammal depends upon the amount (in appropriate units) of its food supply, a type of plant. Suppose $P(x) = 2x^2 + 1$ and $x = f(a) = 3a + 2$. Find $(P \circ f)(a)$, the relationship between the population P of the larger mammal and the amount a of plants available to serve as food for the smaller mammal. $\mathbf{18a^2 + 24a + 9}$

52. When a thermal inversion layer is over a city (as happens often in Los Angeles), pollutants cannot rise vertically but are trapped below the layer and must disperse horizontally. Assume that a factory smokestack begins emitting a pollutant at 8 A.M. Assume that the pollutant disperses horizontally, forming a circle. If t represents the time, in hours, since the factory began emitting pollutants ($t = 0$ represents 8 A.M.), assume that the radius of the circle of pollution is $r(t) = 2t$ mi. Let $A(r) = \pi r^2$ represent the area of a circle of radius r. Find and interpret $(A \circ r)(t)$. $\mathbf{4\pi t^2}$ **sq mi**

53. An oil well off the Gulf Coast is leaking, with the leak spreading oil over the surface of the gulf as a circle. At any time t, in minutes, after the beginning of the leak, the radius of the circular oil slick on the water surface is $r(t) = 4t$ ft. Let $A(r) = \pi r^2$ represent the area of a circle of radius r. Find and interpret $(A \circ r)(t)$. $\mathbf{16\pi t^2}$ **sq ft**

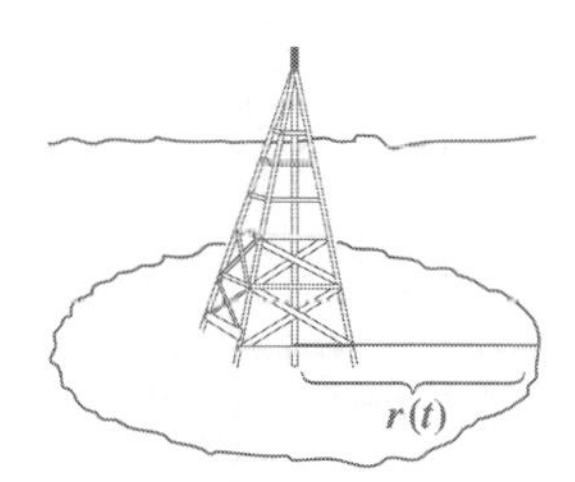

Exercise 53

54. The area of a square is x^2 square inches. If 3 inches is added to one dimension and 1 inch is subtracted from the other dimension, express the area $A(x)$ of the resulting rectangle as a product of two functions. $\mathbf{A(x) = (x + 3)(x - 1)}$

55. A charter flight charges a fare of \$300 per person plus \$20 per person for each unsold seat on the plane. The plane holds 200 passengers. Let x represent the number of unsold seats.

(a) Find an expression in x for the number of people flying. $\mathbf{200 - x}$

(b) Find an expression in x for the price per ticket. $\mathbf{300 + 20x}$

(c) Write an expression for the total revenue $R(x)$. $\mathbf{(200 - x)(300 + 20x)}$

56. The revenue from sales of the camera in Exercise 50 is given by $R(x) = xp$, where x, the number sold, is given by $1.5c$. Find $R(c)$. $\mathbf{R(c) = 3c^2 - 15c}$

✎ **57.** Suppose $g(x) = x - 5$.

(a) For any function f, the graph of $f \circ g$ is a translation of the graph of f. Describe the translation.

(b) For any function f, the graph of $g \circ f$ is a translation of the graph of f. Describe the translation.

58. Suppose $g(x) = -x$. How are the graphs of $f \circ g$ and $g \circ f$ related to the graph of f? (*Hint:* See Exercise 57.)

59. Let $f(x) = x/(x - 1)$ for $x \neq 1$. Show that $f \circ f = x$.

60. If $f(x) = 1 + x$ and $g(x) = 1/(1 + x)$, what is the domain of $f(x) \cdot g(x)$? **$(-\infty, -1) \cup (-1, \infty)$**

Graphing Utility Problems

61. Graph $f(x) = 2x + 5$ and $g(x) = (x - 5)/2$ on the same coordinate system. Add the line $y = x$. What do you notice about the graphs of f and g?

62. Repeat Exercise 61 for $f(x) = .5x^3 - 1$ and $g(x) = \sqrt[3]{2(x + 1)}$.

63. Graph $f(x) = |2x - 1|/(2x - 1)$ using x- and y-ranges of $[-3, 3]$. Separately, graph $g(x) = (2x - 1)/(2x - 1)$. Are the graphs the same? Explain. What is the domain of each function? Do the graphs show the domains correctly?

3.8 Exercises (text page 188)

Which of the functions graphed or defined as follows are one-to-one?

1.

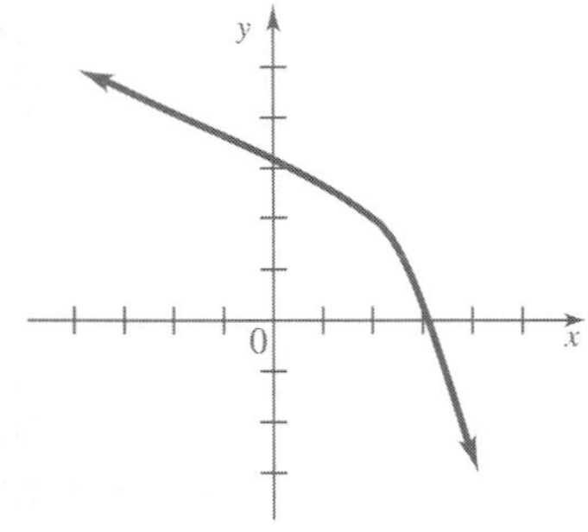

one-to-one

2.

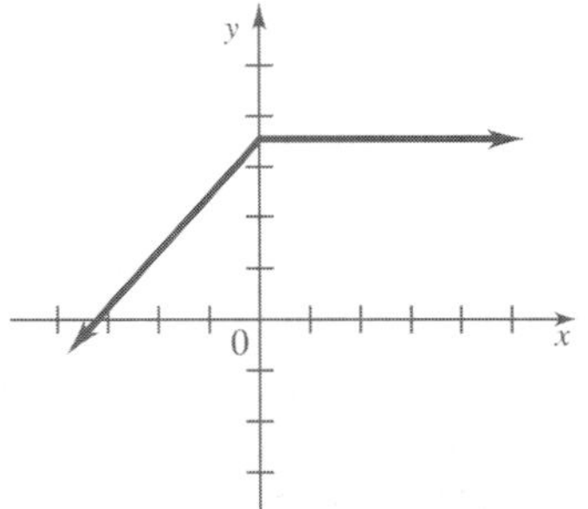

not one-to-one

3.

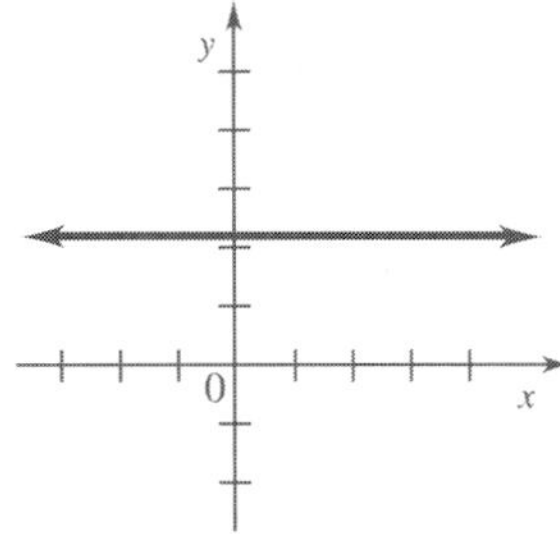

not one-to-one

4.

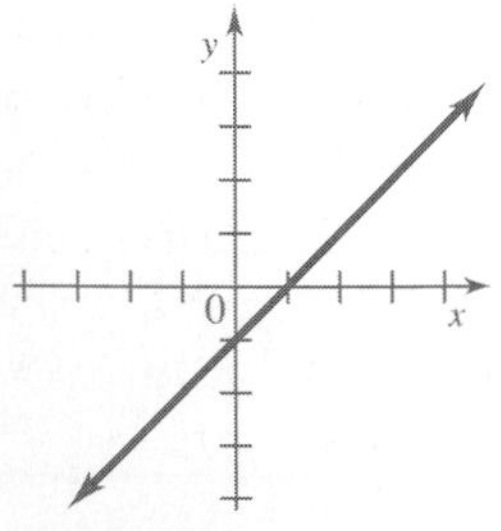

one-to-one

5.

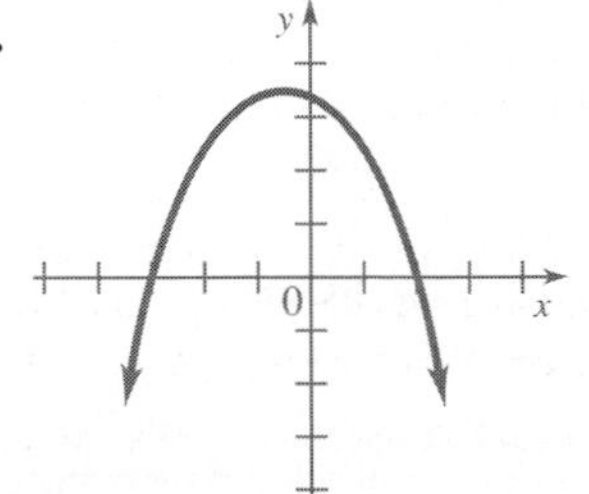

not one-to-one

6.

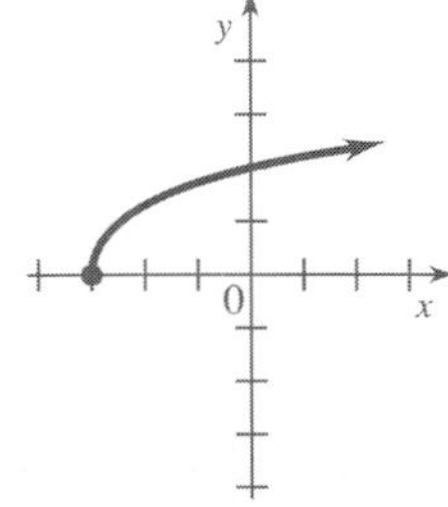

one-to-one

7. $y = 5x - 6$ **one-to-one**

8. $y = 1 - x$ **one-to-one**

9. $y = (1 - x)^2$ **not one-to-one**

10. $y = (x - 5)^3$ **one-to-one**

11. $y = |25 - x^2|$ **not one-to-one**

12. $y = -|16 - x^2|$ **not one-to-one**

13. $y = \sqrt{36 - x^2}$ **not one-to-one**

14. $y = -\sqrt{100 - x^2}$ **not one-to-one**

15. $y = \dfrac{1}{x + 2}$ **one-to-one**

16. $y = \dfrac{-4}{x - 8}$ **one-to-one**

17. $y = x^3 - 1$ **one-to-one**

18. $y = -\sqrt[3]{x + 5}$ **one-to-one**

Which of the pairs of functions graphed or defined as follows are inverses of each other?

19.

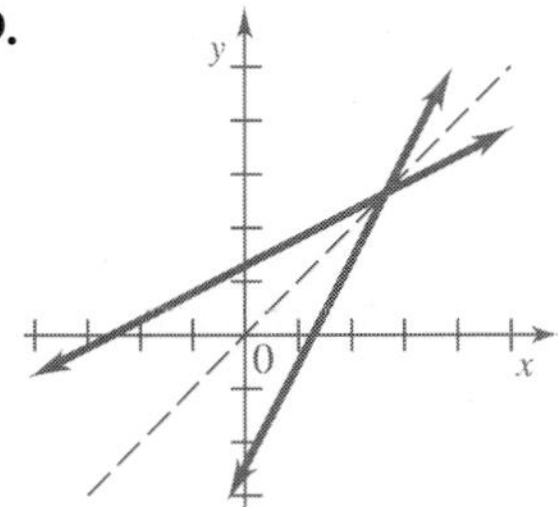

inverses

20.

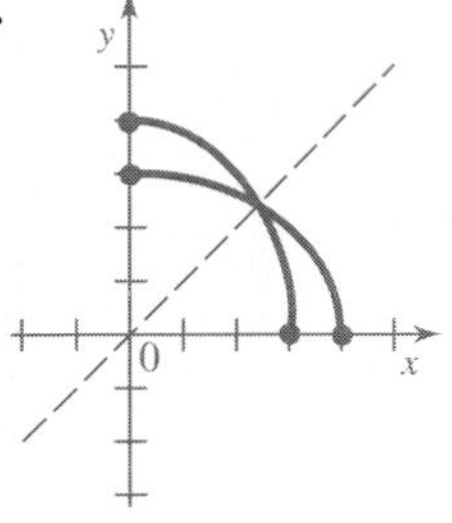

inverses

21.

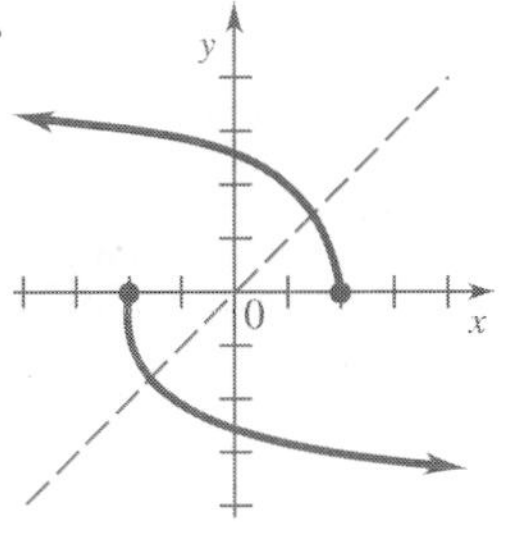

not inverses

22.

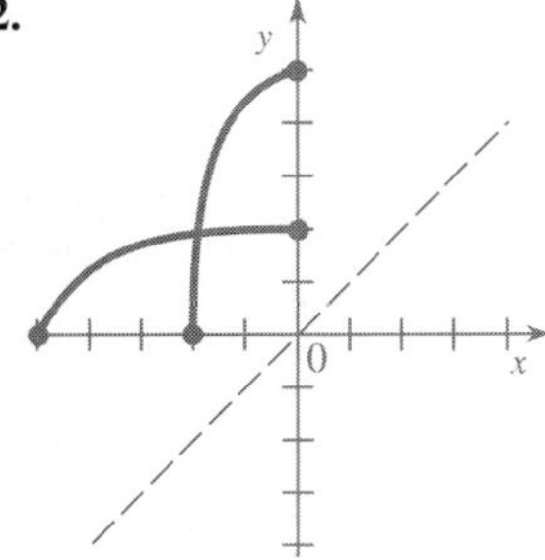

not inverses

23. $f(x) = 2x + 4, \quad g(x) = \dfrac{1}{2}x - 2$ **inverses**

24. $f(x) = 5x - 5, \quad g(x) = \dfrac{1}{5}x + 1$ **inverses**

25. $f(x) = \dfrac{1}{x + 1}, \quad g(x) = \dfrac{1 - x}{x}$ **inverses**

26. $f(x) = \dfrac{2}{x + 6}, \quad g(x) = \dfrac{6x + 2}{x}$ **not inverses**

27. $f(x) = x^2 + 3$, domain $[0, \infty)$, and $g(x) = \sqrt{x - 3}$, domain $[3, \infty)$ **inverses**

28. $f(x) = \sqrt{x + 8}$, domain $[-8, \infty)$ and $g(x) = x^2 - 8$, domain $[0, \infty)$ **inverses**

29. $f(x) = -|x + 5|$, domain $[-5, \infty)$, and $g(x) = |x - 5|$, domain $[5, \infty)$ **not inverses**

30. $f(x) = |x - 1|$, domain $[-1, \infty)$, and $g(x) = |x + 1|$, domain $[1, \infty)$ **not inverses**

Graph the inverse of each one-to-one function.

31.

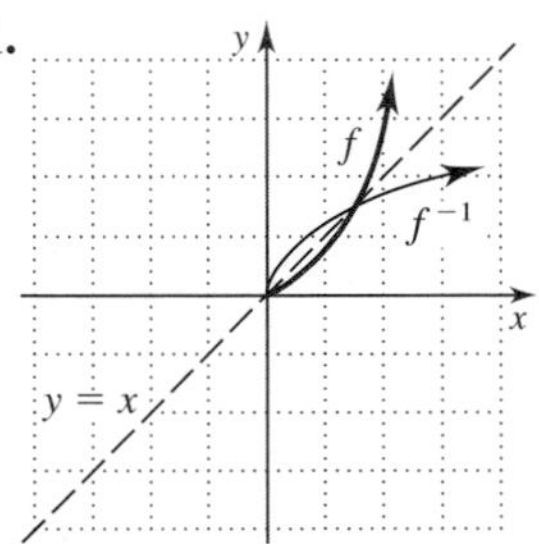

32.

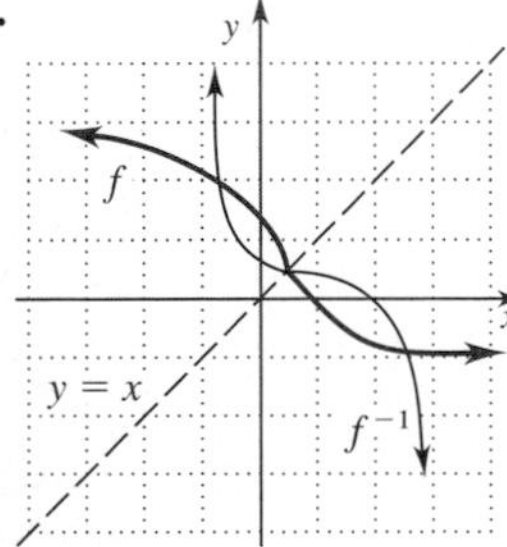

33.

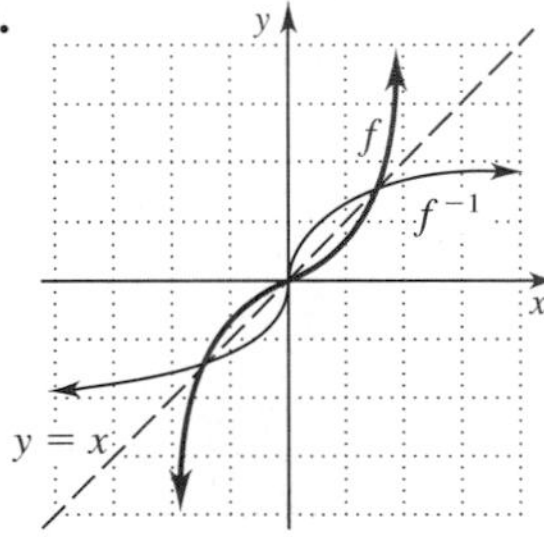

34.

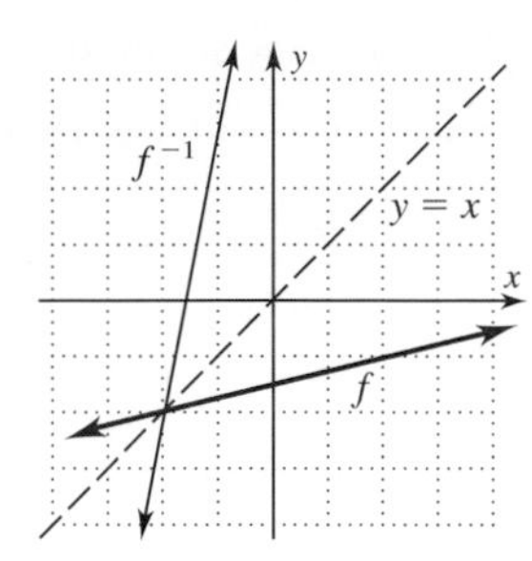

35.

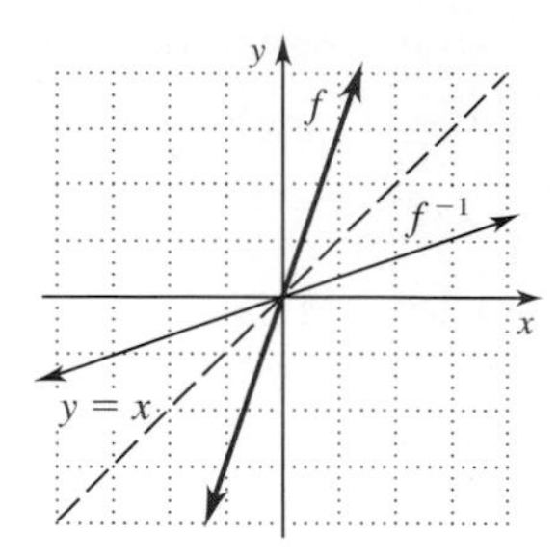

36.

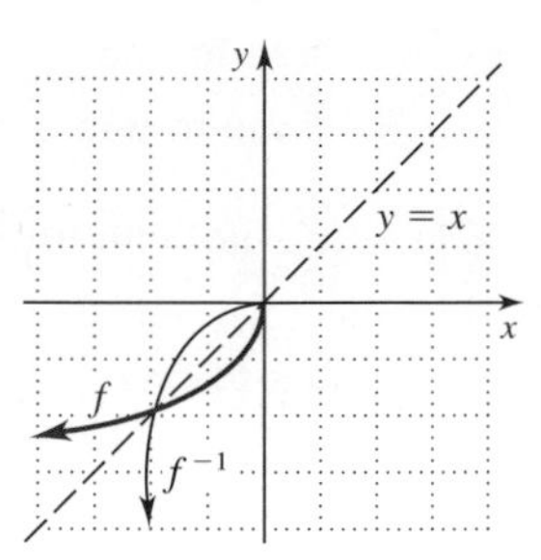

37. As if explaining to another student, describe the steps needed to find the inverse of $f(x) = (x - 3)/x$.

38. As if you were writing a note to a classmate, describe how a function and its inverse are related.

39. Explain why the function defined by $f(x) = x^4$ does not have an inverse. Give examples of ordered pairs to illustrate your explanation.

40. Explain why the function defined by $f(x) = x^5$ is one-to-one.

For each function defined as follows that is one-to-one, write an equation for the inverse function in the form $y = f^{-1}(x)$.

41. $y = 8x - 3$ $\mathbf{f^{-1}(x) = (x + 3)/8}$

42. $y = 2x - 7$ $\mathbf{f^{-1}(x) = (x + 7)/2}$

43. $y = -x^3 - 2$ $\mathbf{f^{-1}(x) = \sqrt[3]{-x - 2}}$

44. $y = -x^3 + 1$ $\mathbf{f^{-1}(x) = \sqrt[3]{1 - x}}$

45. $y = -x^2 + 2$ **not one-to-one**

46. $y = 2x^2 - 3$ **not one-to-one**

47. $y = \dfrac{4}{x}$ $\mathbf{f^{-1}(x) = 4/x}$

48. $y = \dfrac{1}{x}$ $\mathbf{f^{-1}(x) = 1/x}$

49. $y = (x - 3)^2$, domain $(-\infty, 3]$ $\mathbf{f^{-1}(x) = 3 - \sqrt{x}}$

50. $y = \sqrt{2 - x}$ $\mathbf{f^{-1}(x) = 2 - x^2}$**; domain** $\mathbf{[0, \infty)}$

51. $f(x) = \sqrt{6 + x}$ $\mathbf{f^{-1}(x) = x^2 - 6}$**; domain** $\mathbf{[0, \infty)}$

52. $f(x) = (x - 2)^2 + 1$, domain $[2, \infty)$ $\mathbf{f^{-1}(x) = 2 + \sqrt{x - 1}}$

The graph of a function f is shown in the figure. Use the graph to find each of the following values.

53. $f^{-1}(4)$ **4**

54. $f^{-1}(2)$ **3**

55. $f^{-1}(0)$ **2**

56. $f^{-1}(-2)$ **0**

57. $f^{-1}(-3)$ **−2**

58. $f^{-1}(-4)$ **−4**

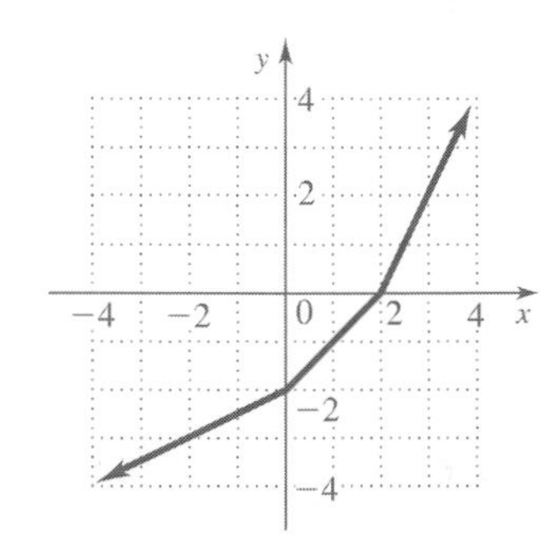

Let $f(x) = x^2 + 5x$ for $x \geq -5/2$. Find the value of the expression in Exercises 59 and 60, rounding to the nearest hundredth.

59. $f^{-1}(7)$ **1.14**

60. $f^{-1}(-3)$ **−.70**

◉ **61.** Suppose that $f(x)$ is the number of cars that can be built for x dollars. What does $f^{-1}(1000)$ represent? **the number of dollars required to build 1000 cars**

◉ **62.** Suppose that $f(r)$ is the volume (in cubic inches) of a sphere of radius r inches. What does $f^{-1}(5)$ represent? **the radius of a sphere with volume 5 cu inches**

63. The brightness B of a star is related to its distance from Earth, d, by the equation $B = k/d$, where k is a constant. Write an equation that expresses d as a function of B. $\mathbf{d = k/B}$

64. Young's rule for determining the correct medicine dosage c for a five-year-old child from the adult dosage d is $c = f(d) = (5/17)d$. Find an expression in terms of c for f^{-1}. $\mathbf{d = f^{-1}(c) = (17/5)c}$

◉ **65.** If a line has slope a, what is the slope of its reflection in the line $y = x$? $\mathbf{1/a}$

◉ **66.** Find $f^{-1}(f(2))$, where $f(2) = 3$. **2**

Let f be a function having an inverse. Prove the statements in Exercises 67 and 68.

67. Every x-intercept of f is a y-intercept of f^{-1}.

68. Every y-intercept of f is an x-intercept of f^{-1}.

Graphing Utility Problems

Graph each of the following using the given x- and y-ranges. Use the graph to decide which functions are one-to-one. If a function is one-to-one, give the equation of its inverse function and graph the inverse function on the same coordinate system.

69. $f(x) = 6x^3 + 11x^2 - x - 6$; $[-3, 2], [-10, 10]$ **not one-to-one**

70. $f(x) = x^4 - 5x^2 + 6$; $[-3, 3], [-1, 8]$ **not one-to-one**

71. $f(x) = (x - 5)/(x + 3)$; $[-8, 8], [-6, 8]$ **one-to-one;** $\mathbf{f^{-1}(x) = (-5 - 3x)/(x - 1)}$

72. $f(x) = -x/(x - 4)$; $[-4, 4], [-4, 4]$ **one-to-one;** $\mathbf{f^{-1}(x) = 4x/(x + 1)}$

Chapter 3 Review Exercises (text page 191)

In Exercises 1 and 2, find $d(P, Q)$ and the midpoint of segment PQ.

1. $P(3, -1)$ and $Q(-4, 5)$ $\mathbf{d(P, Q) = \sqrt{85}; (-1/2, 2)}$

2. $P(-8, 2)$ and $Q(3, -7)$ $\mathbf{d(P, Q) = \sqrt{202}; (-5/2, -5/2)}$

3. Find the other endpoint of a line segment having one end at $(-5, 7)$ and having its midpoint at $(1, -3)$. **(7, −13)**

4. Are the points (5, 7), (3, 9), (6, 8) the vertices of a right triangle? **yes**

5. Find all possible values of k so that $(-1, 2)$, $(-10, 5)$, and $(-4, k)$ are the vertices of a right triangle. **−7, −1, 8, 23**

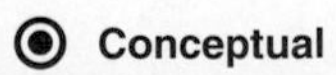 Writing  Conceptual ▲ Challenging  Connections

6. Find all possible values of x so that the distance between $(x, -9)$ and $(3, -5)$ is 6.
$\mathbf{3 \pm 2\sqrt{5}}$

7. Find all points (x, y) with $x = 6$ so that (x, y) is 4 units from $(1, 3)$. **No such points exist.**

8. Find all points (x, y) with $x + y = 0$ so that (x, y) is 6 units from $(-2, 3)$.
$\left(\dfrac{-5 + \sqrt{71}}{2}, \dfrac{5 - \sqrt{71}}{2}\right), \left(\dfrac{-5 - \sqrt{71}}{2}, \dfrac{5 + \sqrt{71}}{2}\right)$

Prove each of the following.

9. The medians to two equal sides of an isosceles triangle are equal in length. (A median is a line segment from a vertex of a triangle to the midpoint of the opposite side.)

10. The line segment connecting midpoints of two sides of a triangle is half as long as the third side.

Find equations for each of the circles below.

11. Center $(5, -2)$, radius 4 $\mathbf{(x - 5)^2 + (y + 2)^2 = 16}$

12. Center $(\sqrt{5}, -\sqrt{7})$, radius $\sqrt{3}$ $\mathbf{(x - \sqrt{5})^2 + (y + \sqrt{7})^2 = 3}$

13. Center $(-8, 1)$, passing through $(0, 16)$ $\mathbf{(x + 8)^2 + (y - 1)^2 = 289}$

14. Center $(-3, 5)$, passing through $(1, 8)$ $\mathbf{(x + 3)^2 + (y - 5)^2 = 25}$

Find the center and radius of each of the following that are circles.

15. $x^2 - 6x + y^2 - 10y + 30 = 0$ **(3, 5); 2**

16. $x^2 + y^2 - 6x + 8y = 11$ **(3, −4); 6**

17. $x^2 + 11x + y^2 - 5y + 46 = 0$ **not a circle**

18. $x^2 + y^2 + x + 5y = 7/2$ $\mathbf{(-1/2, -5/2); \sqrt{10}}$

Tell whether each of the following is true or false.

19. The number a is in the domain of a relation if and only if the vertical line $x = a$ intersects the graph of the relation. **true**

20. The number b is in the range of a relation if and only if the horizontal line $y = b$ intersects the graph of the relation. **true**

Decide whether the following curves are graphs of functions of x. Give the domain and range of each relation.

21.

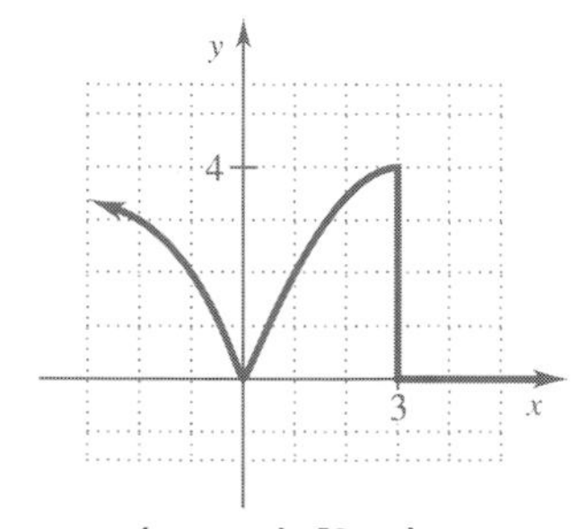

no; $(-\infty, \infty)$; $[0, \infty)$

22.

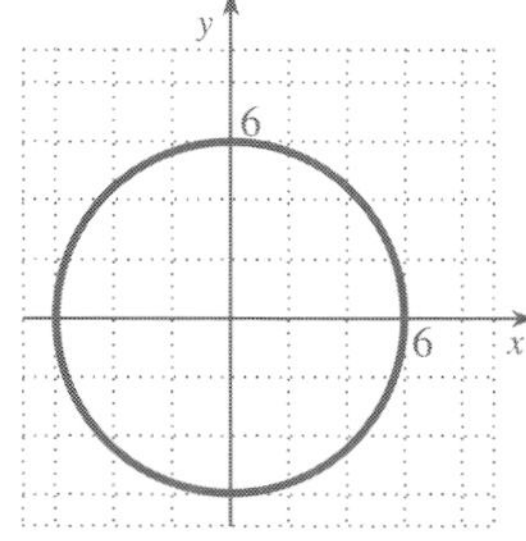

no; $[-6, 6]$; $[-6, 6]$

23.

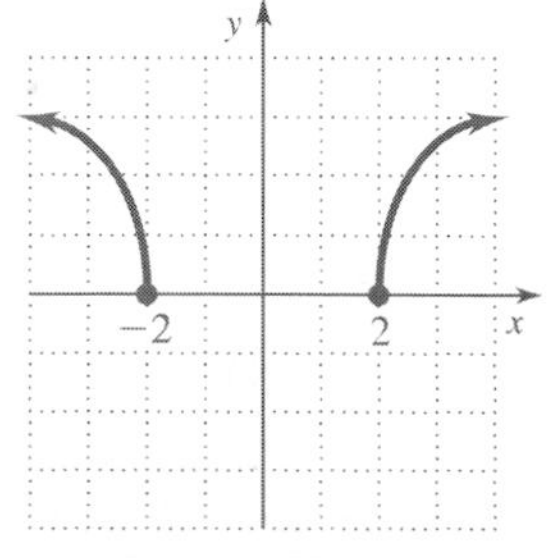

yes; $(-\infty, -2] \cup [2, \infty)$; $[0, \infty)$

24.

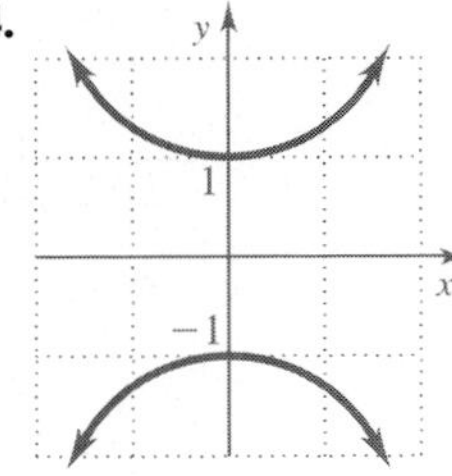

no; $(-\infty, \infty)$; $(-\infty, -1] \cup [1, \infty)$

25.

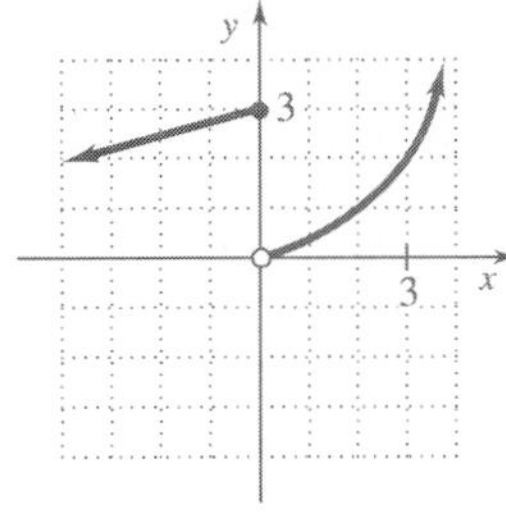

yes; $(-\infty, \infty)$; $(-\infty, \infty)$

26.

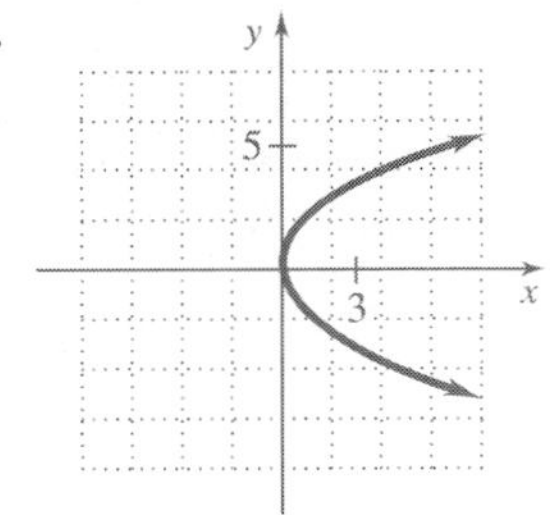

no; $[0, \infty)$; $(-\infty, \infty)$

Identify any equations that define y as a function of x.

27. $x = \frac{1}{2}y^2$ **not a function** **28.** $y = 3 - x^2$ **function** **29.** $y = \frac{-8}{x}$ **function** **30.** $y = \sqrt{x - 7}$ **function**

Give the domain of the functions defined as follows.

31. $y = -4 + |x|$ $(-\infty, \infty)$

32. $y = \frac{8 + x}{8 - x}$ $(-\infty, 8) \cup (8, \infty)$

33. $y = -\sqrt{\frac{5}{x^2 + 9}}$ $(-\infty, \infty)$

34. $y = \sqrt{49 - x^2}$ $[-7, 7]$

Find the slope for each of the following lines.

35. Through (8, 7) and (1/2, −2) **6/5**

36. Through (2, −2), and (3, −4) **−2**

37. $11x + 2y = 3$ **−11/2**

38. $y + 6 = 0$ **0**

Write an equation in standard form for each of the following lines.

39. Through (5, −23) and (−3, 1) $3x + y + 8 = 0$

40. Through (1/4, 5/2) and (−3, −4) $2x - y + 2 = 0$

41. Through (2, 6), with slope 7/3 $7x - 3y + 4 = 0$

42. Through (−4, 8), with slope −4/5 $4x + 5y - 24 = 0$

43. No x-intercept, y-intercept −1 $y + 1 = 0$

44. x-intercept −2, y-intercept 3 $3x - 2y + 6 = 0$

45. Through (−4, 2), perpendicular to $2x - y = 3$ $x + 2y = 0$

46. Through (3, −1), parallel to $3x + 5y = 15$ $3x + 5y - 4 = 0$

Graph each function defined as follows.

47. $3y = x$

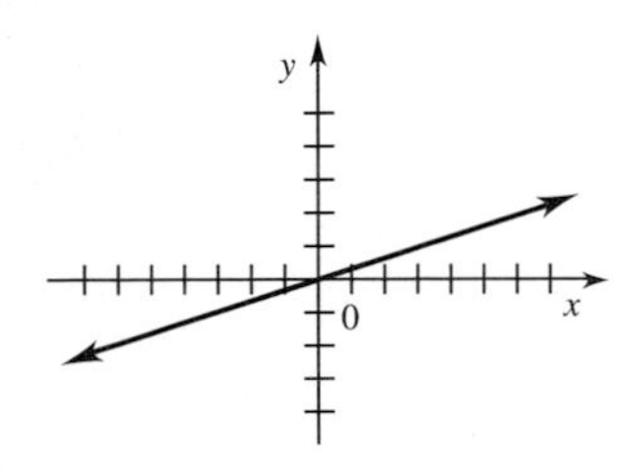

48. $f(x) = [\![x - 1]\!]$

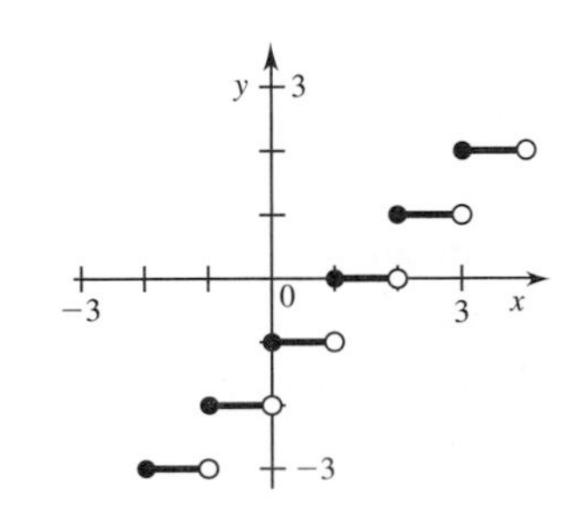

49. $f(x) = \begin{cases} -4x + 2 & \text{if } x \le 1 \\ 3x - 5 & \text{if } x > 1 \end{cases}$

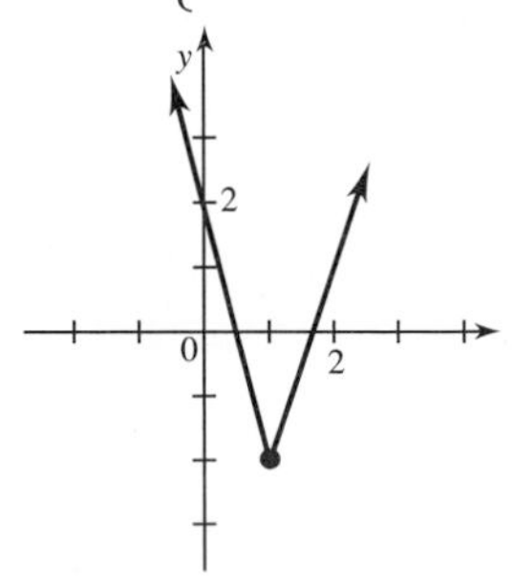

50. $f(x) = x^2 - 4$

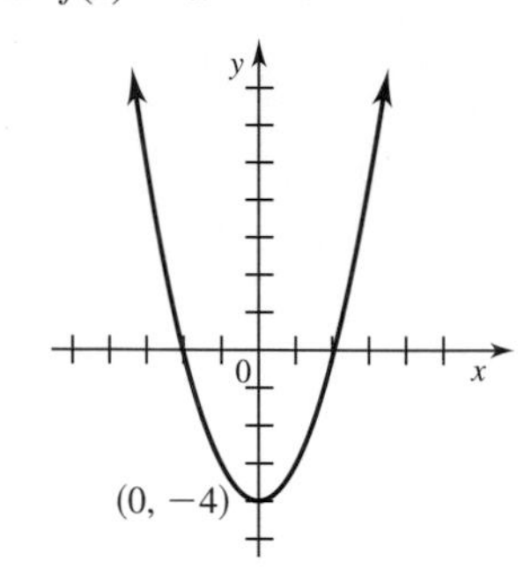

51. The line through $(-3, -2)$, with $m = -1$

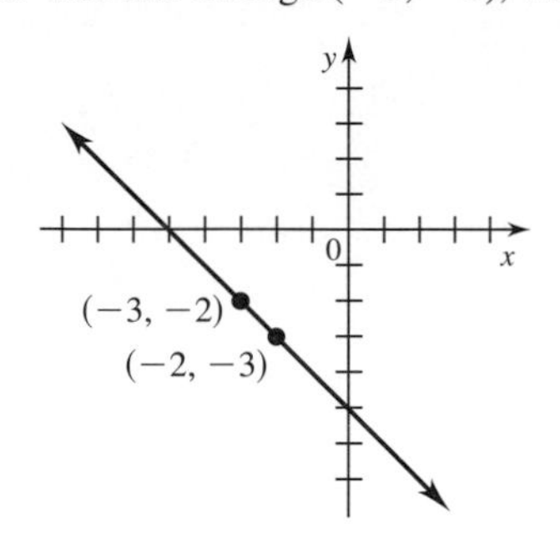

52. $f(x) = \begin{cases} 3x + 1 & \text{if } x < 2 \\ -x + 4 & \text{if } x \geq 2 \end{cases}$

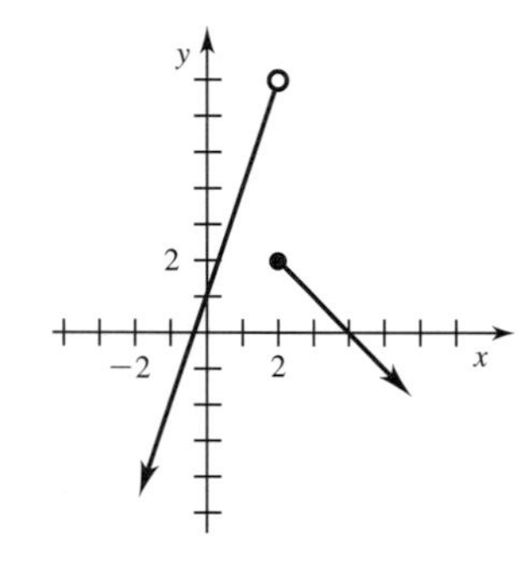

53. $y = 6 - x^2$

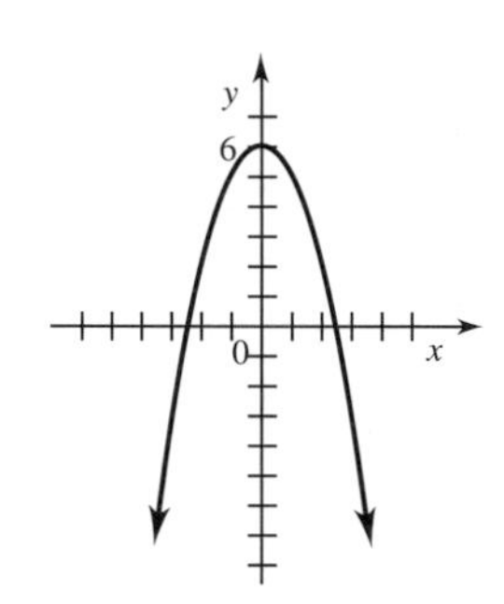

54. The college fieldhouse used for graduation ceremonies has 5000 seats available for family and friends of the graduates. Use the greatest-integer function to write an expression for the number of tickets that can be allocated to each of x graduates. **$y = [\![5000/x]\!]$**

Decide whether the equations in Exercises 55–62 have graphs that are symmetric with respect to the x-axis, the y-axis, the origin, or none of these.

55. $3y^2 - 5x^2 = 15$ **x-axis, y-axis, and origin**

56. $x + y^2 = 8$ **x-axis**

57. $y^3 = x + 1$ **none of these symmetries**

58. $x^2 = y^3$ **y-axis**

59. $|y| = -x$ **x-axis**

60. $|x + 2| = |y - 3|$ **none of these symmetries**

61. $|x| = |y|$ **x-axis, y-axis, and origin**

62. $xy = 8$ **origin**

Describe how the graphs of the following functions can be obtained from the graph of $f(x) = |x|$.

63. $g(x) = -|x|$ **Reflect the graph of $f(x)$ about the x-axis.**

64. $h(x) = |x| - 2$ **Translate the graph of $f(x)$ down 2 units.**

65. $k(x) = 2|x - 4|$ **Double the y-values of $f(x)$ and translate 4 units to the right.**

Let $f(x) = 3x - 4$. *Find an equation for each of the following reflections of the graph of* $f(x)$.

66. About the x-axis **$y = -3x + 4$**

67. About the y-axis **$y = -3x - 4$**

68. About the origin **$y = 3x + 4$**

69. About the line $y = x$ **$y = (1/3)x + 4/3$**

70. The graph of a function f is shown in the figure. Sketch the graph of each function defined as follows.

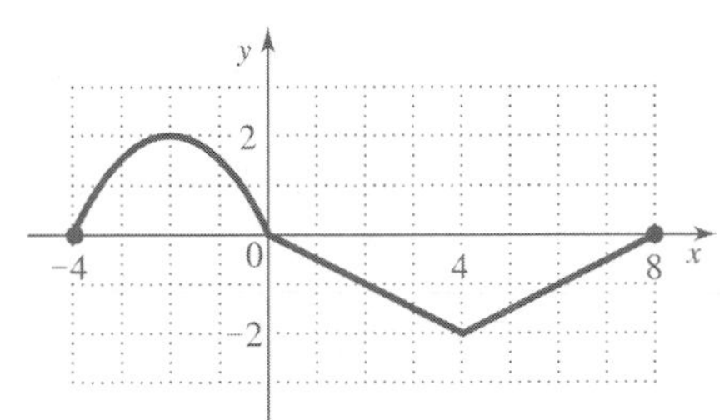

(a) $y = f(x) + 3$

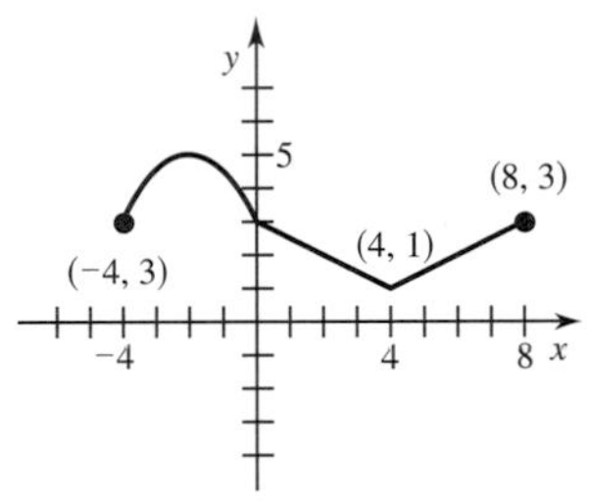

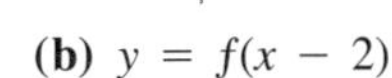

(b) $y = f(x - 2)$

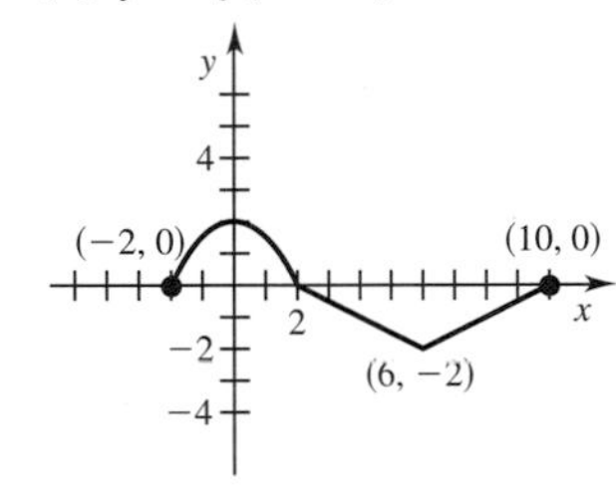

(c) $y = f(x + 3) - 2$

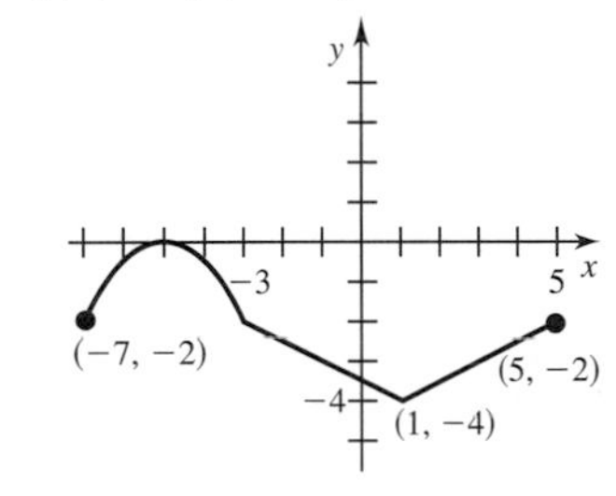

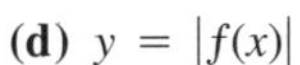

(d) $y = |f(x)|$

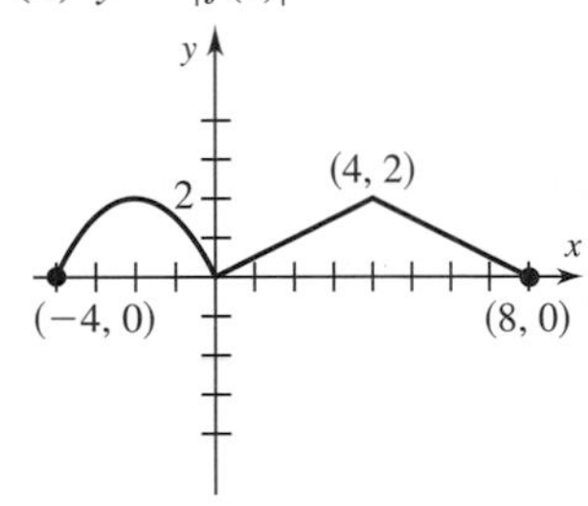

Let $f(x) = 2x + 5$ *and* $g(x) = x^2 - 3x - 4$. *Find each of the following.*

71. $(f + g)(x)$ **$x^2 - x + 1$**

72. $(fg)(x)$ **$2x^3 - x^2 - 23x - 20$**

73. $(f - g)(4)$ **13**

74. $(f + g)(-4)$ **21**

75. $(f + g)(2k)$ **$4k^2 - 2k + 1$**

76. $(fg)(1 + r)$ **$2r^3 + 5r^2 - 19r - 42$**

77. $(f/g)(3)$ **$-11/4$**

78. $(f/g)(-1)$ **undefined**

79. The domain of $(fg)(x)$ **$(-\infty, \infty)$**

80. The domain of $(f/g)(x)$ **$(-\infty, -1) \cup (-1, 4) \cup (4, \infty)$**

Let $f(x) = \sqrt{x - 2}$ *and* $g(x) = x^2$. *Find each of the following.*

81. $(f \circ g)(x)$ **$\sqrt{x^2 - 2}$**

82. $(g \circ f)(x)$ **$x - 2$**

83. $(g \circ f)(3)$ **1**

84. $(f \circ g)(-6)$ **$\sqrt{34}$**

Which of the functions graphed or defined as follows are one-to-one?

85.

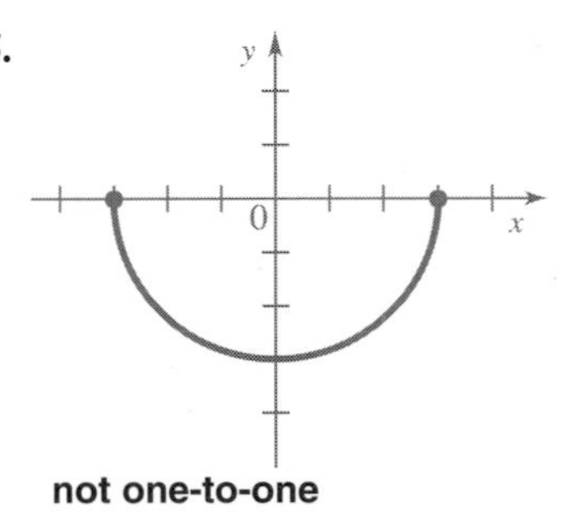

not one-to-one

86.

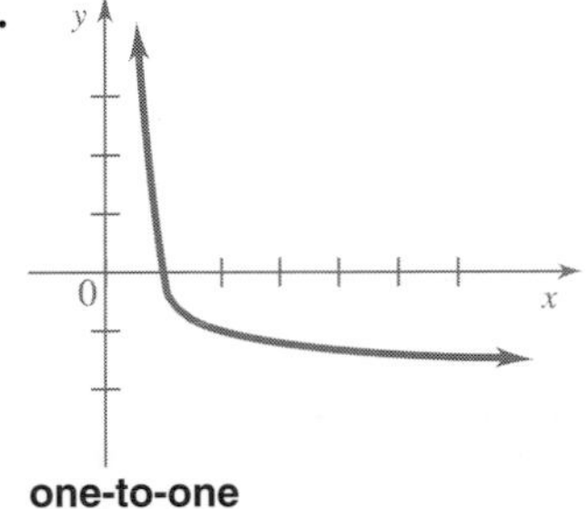

one-to-one

87.

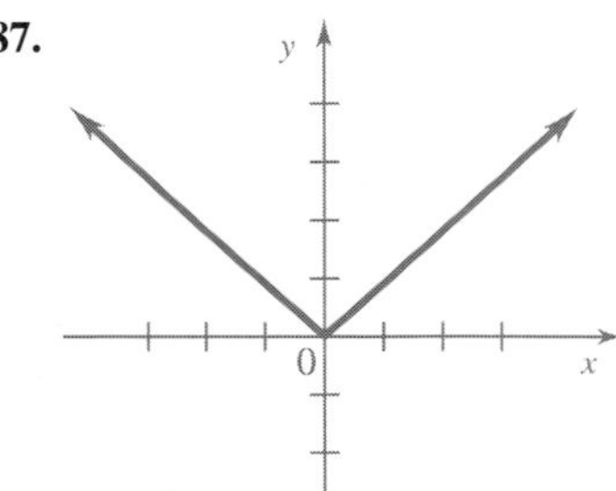

not one-to-one

88. $y = \dfrac{8x - 9}{5}$ **one-to-one**

89. $y = -x^2 + 11$ **not one-to-one**

90. $y = \sqrt{100 - x^2}$ **not one-to-one**

For each of the functions defined in Exercises 91–94, write an equation for the inverse function in the form $y = f^{-1}(x)$ and then graph f and f^{-1}.

91. $f(x) = 12x + 3$ $\boldsymbol{f^{-1}(x) = (x - 3)/12}$

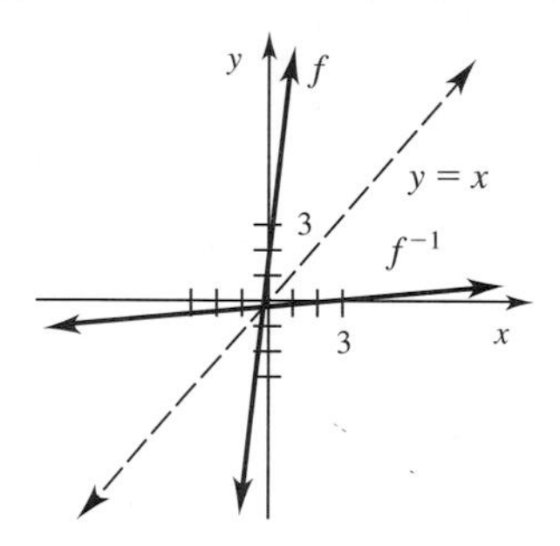

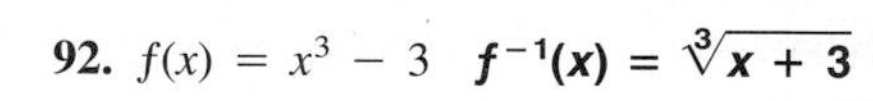

92. $f(x) = x^3 - 3$ $\boldsymbol{f^{-1}(x) = \sqrt[3]{x + 3}}$

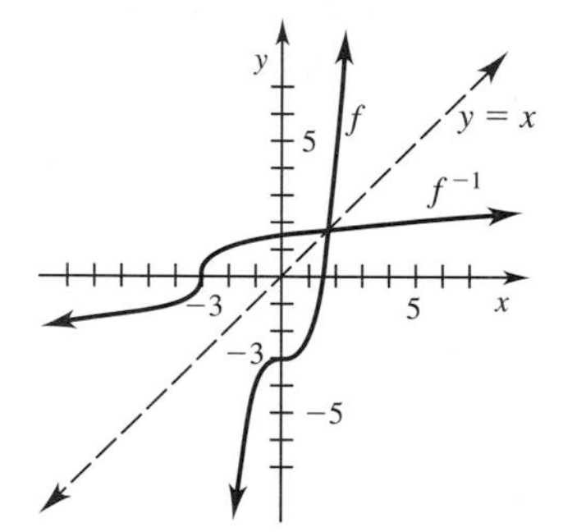

93. $f(x) = x^2 - 6$ **f has no inverse.**

94. $f(x) = \sqrt{25 - x^2}$ **f has no inverse.**

95. Suppose $f(t)$ is the amount an investment will grow to become t years after 1992. What does $f^{-1}(\$50{,}000)$ represent? **the number of years after 1992 required for the investment to reach $50,000**

96. Find the point of intersection of $y = f(x)$ and $y = f^{-1}(x)$ where $f(x) = 3x - 5$. **(5/2, 5/2)**

97. Cylindrical cans make the most efficient use of materials when their height is the same as the diameter of their top.

(a) Express the volume V of such a can as a function of the diameter d of its top. $\boldsymbol{V = \pi d^3/4}$

(b) Express the surface area S of such a can as a function of the diameter d of its top. (*Hint:* The curved side is made from a rectangle whose length is the circumference of the top of the can.) $\boldsymbol{S = 3\pi d^2/2}$

98. A baseball diamond is a square 90 ft long on each side. Casey runs a constant 30 ft per sec whether he hits a ground ball or a home run. Today in his first time at bat, he hit a home run. Write an expression that measures his line-of-sight distance from second base as a function of the time t, in seconds, after he left home plate.*
Let $f(t)$ represent the distance from second base at time t, where $0 \le t \le 12$. Then

$$f(t) = \begin{cases} 30\sqrt{t^2 - 6t + 18} & \text{if } 0 \le t \le 3 \\ 180 - 30t & \text{if } 3 < t \le 6 \\ 30t - 180 & \text{if } 6 < t \le 9 \\ 30\sqrt{t^2 - 18t + 90} & \text{if } 9 < t \le 12. \end{cases}$$

*Exercise 98: From *Calculus*, 4th Edition by Stanley I. Grossman. Copyright © 1988 by Harcourt Brace Jovanovich, Inc. Reprinted with permission of Harcourt Brace Jovanovich, Inc. and the author.

99. Alice, on vacation in Canada, found that her U.S. dollars were increased by 16%. On her return, she expected a 16% decrease when converting her Canadian money back into U.S. dollars. Write an equation for each of these conversion functions. Show that one is not the inverse of the other. What should the conversion factor be for Canadian to U.S. dollars? **For the 16% increase, $y = 1.16x$; the inverse is $y = .8621x$. For the 16% decrease, $y = .84x$; the inverse is $y = 1.1905x$. The Canadian to U.S. dollars conversion factor should be .8621.**

100. The figure shows average prices for domestic crude oil from 1980 to mid-1992.

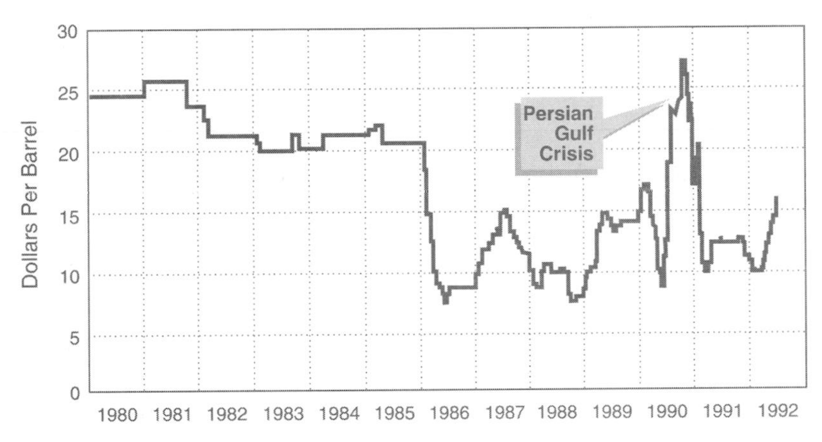

(a) Is this the graph of a function? Of a one-to-one function? **no; no**
(b) In what year were oil prices lowest? Highest? **1986; 1990**
(c) What was the lowest price? The highest price? **about \$7; about \$27.50**
(d) What is the general trend of prices over the given period? **down**
(e) What do the horizontal portions of the graph indicate? **a constant, stable price**

101. The figure shows the number of jobs gained or lost in the Sacramento area from September 1991 to May 1992.

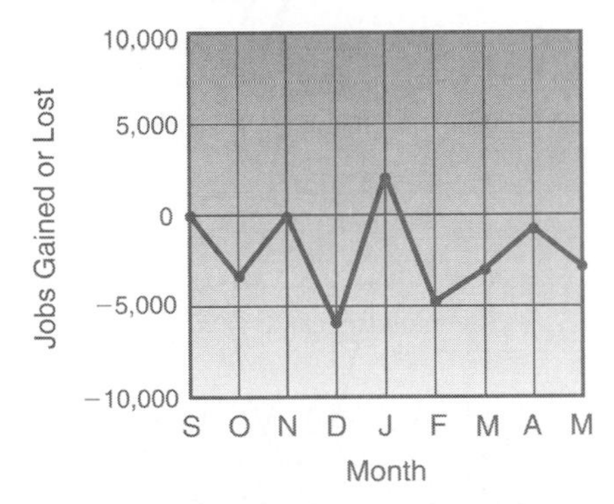

(a) Is this the graph of a function? A one-to-one function? **yes; no**
(b) In what month were the most jobs lost? The most gained? **December; January**
(c) What was the largest number of lost jobs? The most gained? **about 6,000; about 2,000**
(d) Do these data show an upward or downward trend? If so, what is it? **a slight downward trend**

102. Explain the similarities and differences between relations and functions.

103. Discuss what is meant by inverse functions. Give the conditions that are necessary for two functions to be inverses. Describe how a pair of inverse functions are related.

104. Discuss the connections between the following concepts: the distance formula, the equation of a circle, and functions of the form $y = \sqrt{a - x^2}$ or $y = -\sqrt{a - x^2}$.

4 Polynomial and Rational Functions

4.1 Exercises (text page 202)

1. Graph the functions defined as follows on the same coordinate system.

(a) $f(x) = 2x^2$ **(b)** $f(x) = -3x^2$

(c) $f(x) = -\frac{2}{3}x^2$ **(d)** $f(x) = \frac{1}{4}x^2$

(e) How does the coefficient affect the shape of the graph?

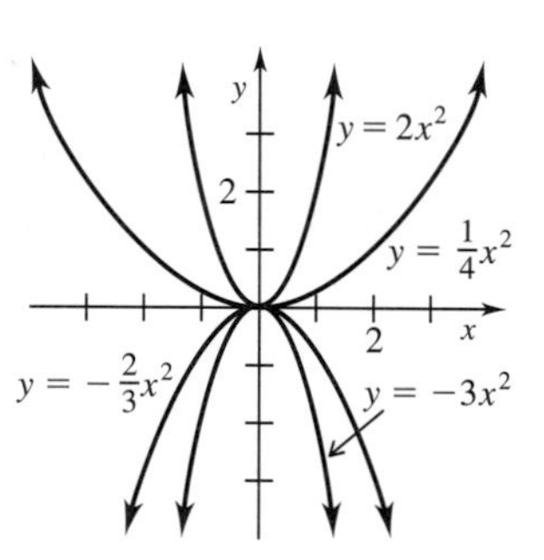

2. Graph the functions defined as follows on the same coordinate system.

(a) $f(x) = x^2 + 3$ **(b)** $f(x) = x^2 - 1$

(c) $f(x) = x^2 + \frac{1}{2}$ **(d)** $f(x) = x^2 - \frac{3}{2}$

(e) How do these graphs differ from the graph of $f(x) = x^2$?

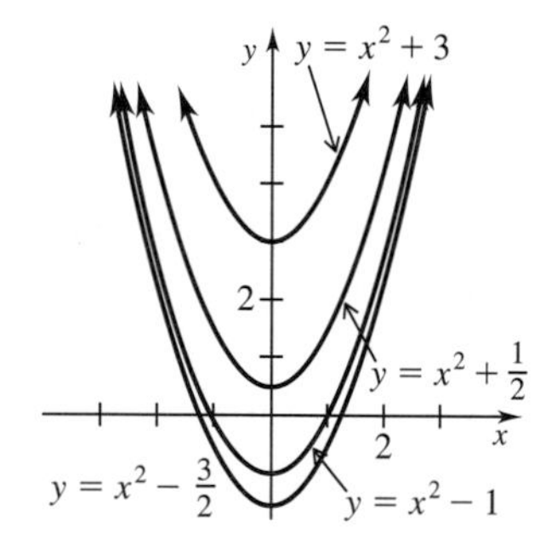

3. Graph the functions defined as follows on the same coordinate system.

(a) $f(x) = (x - 1)^2$ **(b)** $f(x) = (x + 1)^2$

(c) $f(x) = (x + 3)^2$ **(d)** $f(x) = (x - 2)^2$

(e) How do these graphs differ from the graph of $f(x) = x^2$?

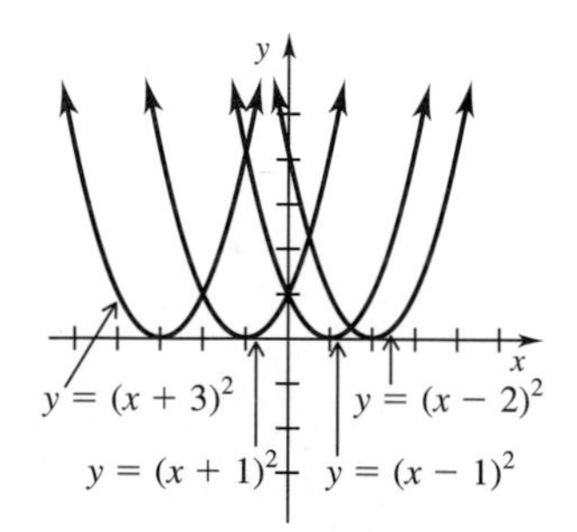

Give the range of each of the functions defined as follows.

4. $f(x) = (x - 3)^2 + 4$ **[4, ∞)**

5. $f(x) = -(x + 3)^2 - 1$ **(−∞, −1]**

6. $g(x) = 2(x + 4)^2 - \frac{1}{2}$ $\left[-\frac{1}{2}, \infty\right)$

Graph the functions defined as follows. Give the vertex, axis, x-intercepts, and y-intercept of each graph.

7. $f(x) = (x + 2)^2$ **vertex: (−2, 0); axis: $x = -2$; x-intercept: −2; y-intercept: 4**

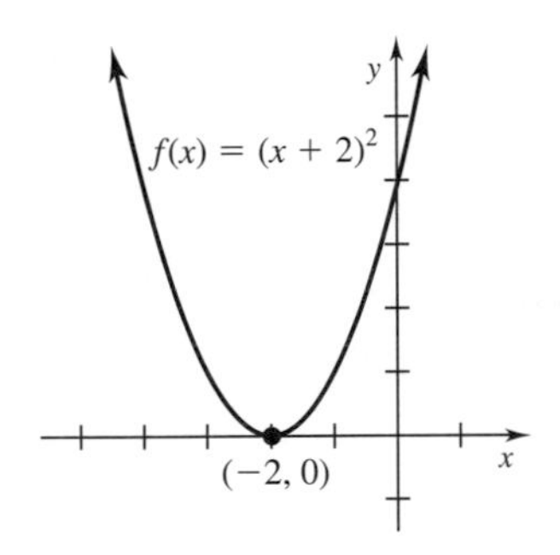

8. $f(x) = (x - 3)^2$ **vertex: (3, 0); axis: $x = 3$; x-intercept: 3; y-intercept: 9**

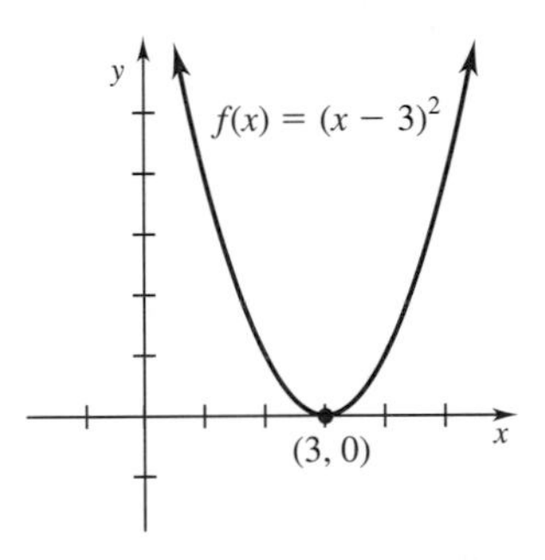

9. $g(x) = 2(x + 3)^2 - 4$ **vertex: (−3, −4); axis: $x = -3$; x-intercepts: $-3 \pm \sqrt{2}$; y-intercept: 14**

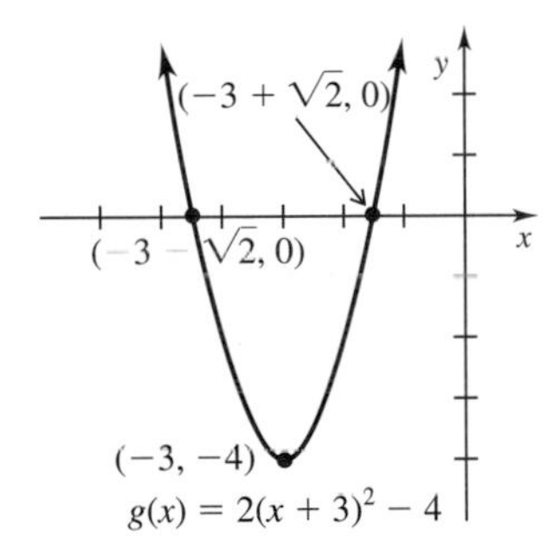

10. $h(x) = (x - 5)^2 - 4$ **vertex: (5, −4); axis: $x = 5$; x-intercepts: 3, 7; y-intercept: 21**

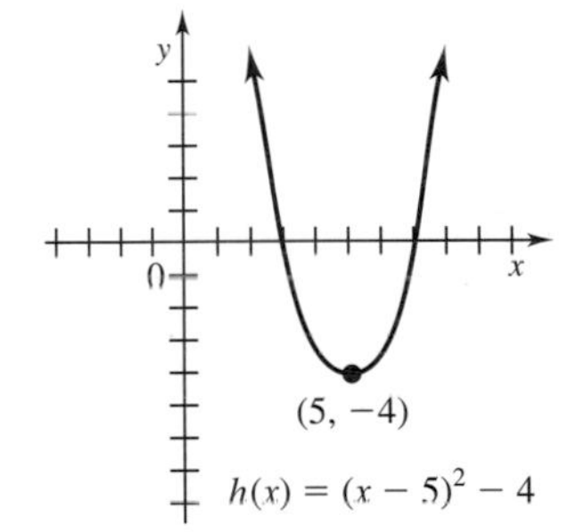

11. $F(x) = -3(x - 2)^2 + 1$ **vertex: (2, 1); axis: $x = 2$; x-intercepts: $(6 \pm \sqrt{3})/3$; y-intercept: −11**

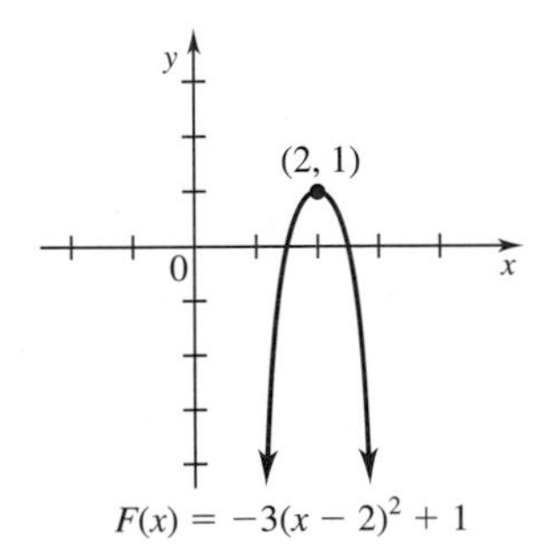

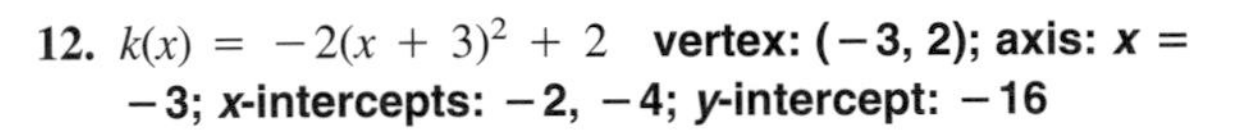

12. $k(x) = -2(x + 3)^2 + 2$ **vertex: (−3, 2); axis: $x = -3$; x-intercepts: −2, −4; y-intercept: −16**

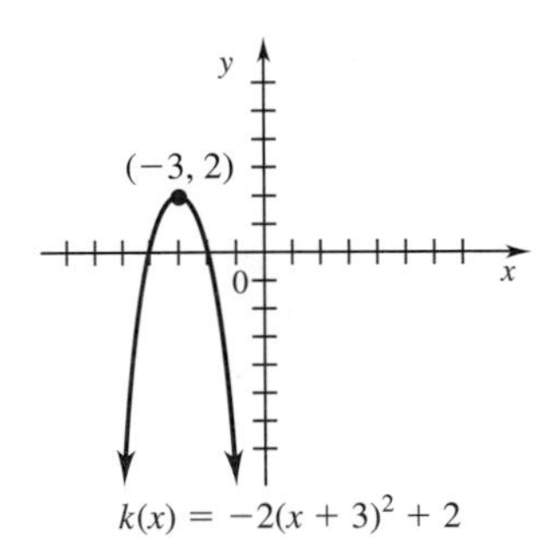

13. $H(x) = \frac{1}{2}(x + 2)^2 - 3$ **vertex: $(-2, -3)$; axis: $x = -2$; x-intercepts: $-2 \pm \sqrt{6}$; y-intercept: -1**

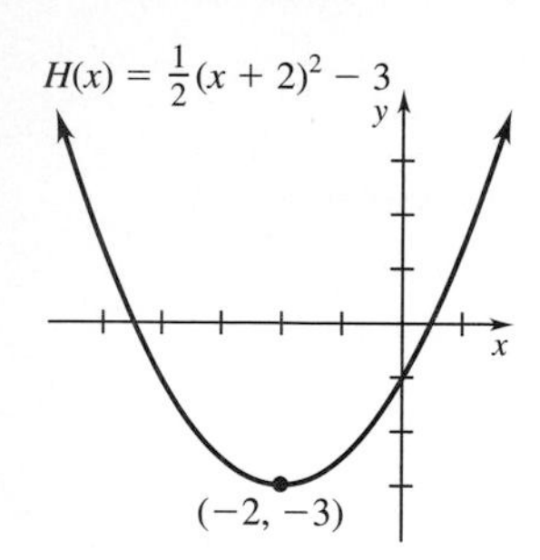

14. $G(x) = -\frac{2}{3}(x - 2)^2 - 1$ **vertex: $(2, -1)$; axis: $x = 2$; x-intercept: none; y-intercept: $-\frac{11}{3}$**

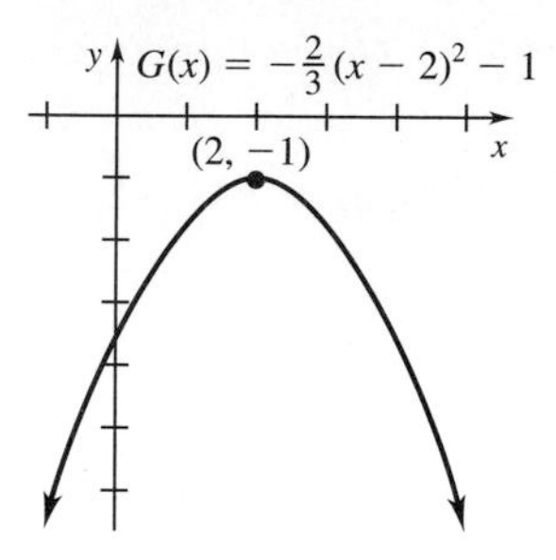

15. $f(x) = x^2 + 4x + 3$ **vertex: $(-2, -1)$; axis: $x = -2$; x-intercepts: $-1, -3$; y-intercept: 3**

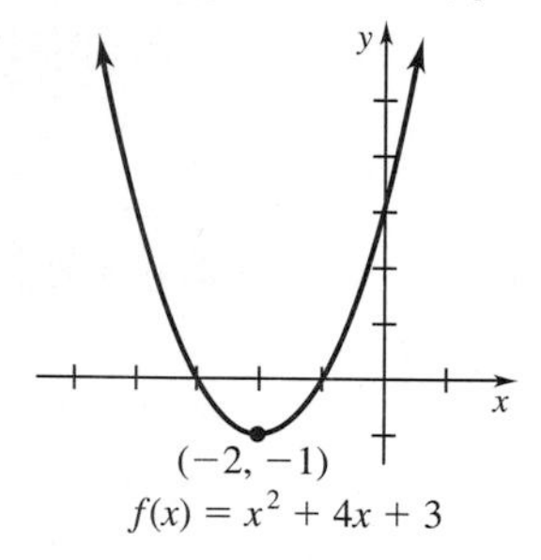

16. $f(x) = x^2 + 6x + 5$ **vertex: $(-3, -4)$; axis: $x = -3$; x-intercepts: $-5, -1$; y-intercept: 5**

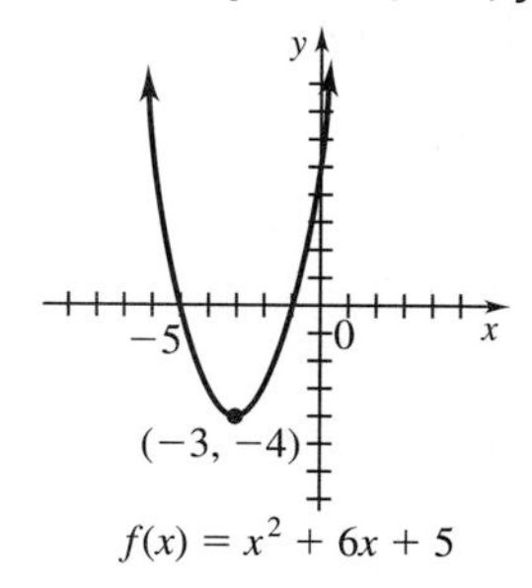

17. $k(x) = -x^2 + 6x - 6$ **vertex: $(3, 3)$; axis: $x = 3$; x-intercepts: $-3 \pm \sqrt{3}$; y-intercept: -6**

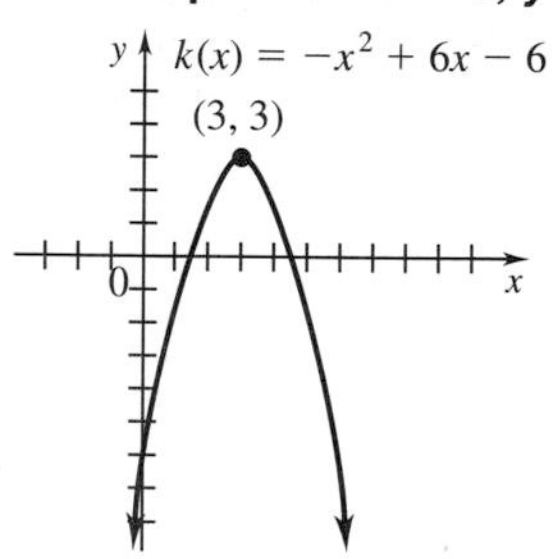

18. $g(x) = -x^2 - 4x + 2$ **vertex: $(-2, 6)$; axis: $x = -2$; x-intercepts: $-2 \pm \sqrt{6}$; y-intercept: 2**

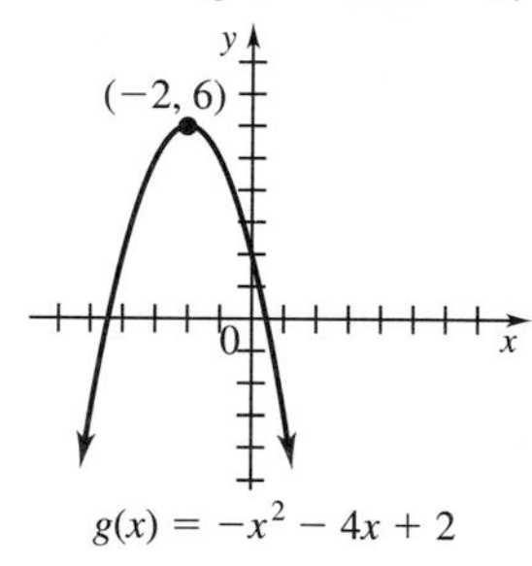

19. $g(x) = -3x^2 + 24x - 46$ **vertex: $(4, 2)$; axis: $x = 4$; x-intercepts: $\frac{12 \pm \sqrt{6}}{3}$, y-intercept: -46**

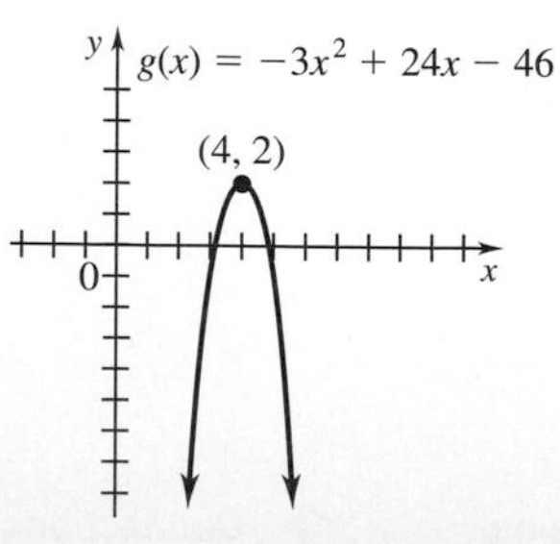

20. $f(x) = 2x^2 - 4x + 5$ **vertex: $(1, 3)$; axis: $x = 1$; x-intercept: none; y-intercept: 5**

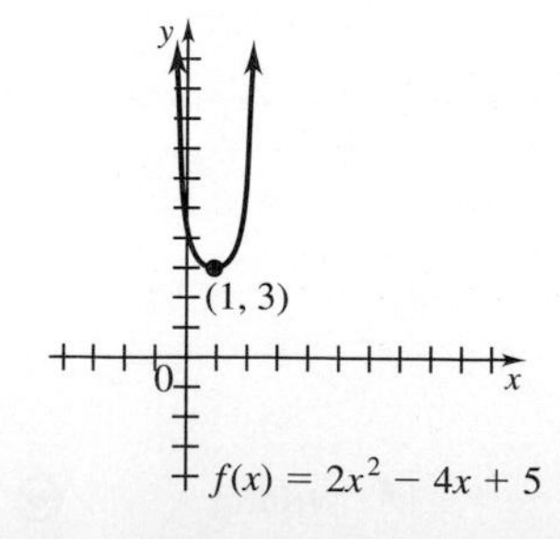

The figure shows the graph of a quadratic function $f(x)$.

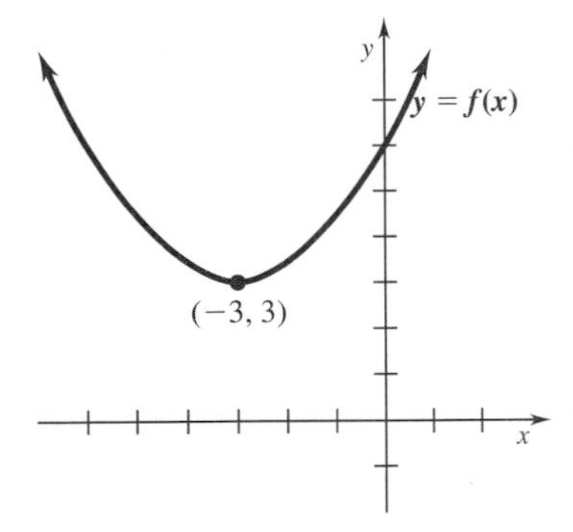

21. What is the minimum value of $f(x)$? **3**

22. For what value of x is $f(x)$ as small as possible? **−3**

23. How many solutions are there to the equation $f(x) = 1$? **0 solutions**

24. How many solutions are there to the equation $f(x) = 4$? **2 solutions**

25. Glenview College wants to construct a rectangular parking lot on land bordered on one side by a highway. It has 320 ft of fencing which it will use to fence off the other three sides. What should be the dimensions of the lot if the enclosed area is to be a maximum? See the figure. (*Hint:* Let x represent the width of the lot and let $320 - 2x$ represent the length. Graph the area parabola, $A = x(320 - 2x)$, and investigate the vertex.) **80 ft wide, 160 ft long**

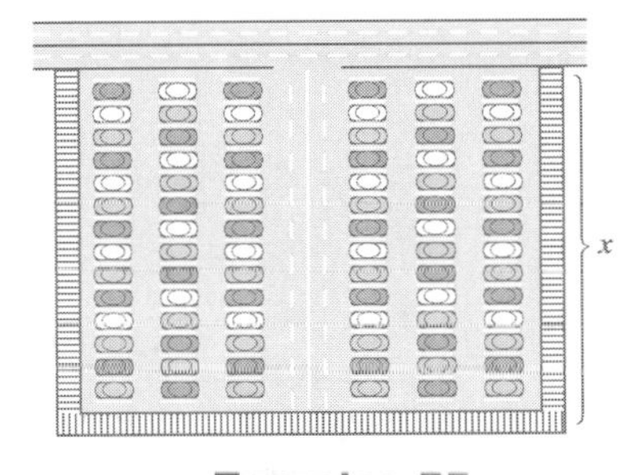

Exercise 25

26. The number of mosquitoes, $M(X)$, in millions, in a certain area of Kentucky depends on the June rainfall, x, in inches, approximately as $M(x) = 10x - x^2$. Find the rainfall that will produce the maximum number of mosquitoes. **5 inches**

27. The revenue and cost of a bus trip depend on the number of unsold seats. If the revenue $R(x)$ is given by $R(x) = 5000 + 40x - x^2$, and the cost $C(x)$ is given by $C(x) = 2000 - 10x$, find the maximum profit. (*Hint:* Revenue minus cost equals profit.) **Maximum profit is $3,625 with 25 unsold seats.**

28. Find the dimensions of a rectangle of maximum area whose perimeter is 8 ft. (*Hint:* Let x be the length and h be the height. Then $2x + 2h = 8$ or $h = 4 - x$.) **2 ft by 2 ft**

29. If an object is thrown upward with an initial velocity of 32 ft per sec, then its height after t sec is given by $h(t) = 32t - 16t^2$. Find the maximum height attained by the object. Find the number of seconds it takes the object to hit the ground. **max. height: 16 ft; 2 sec**

30. Find two numbers whose sum is 20 and whose product is a maximum. (*Hint:* Let x and $20 - x$ be the two numbers, and write an equation for the product.) **10 and 10**

31. Find two numbers whose difference is 30 and whose product is a minimum. **15 and −15**

32. A charter flight charges a fare of $200 per person, with a surcharge of $4 per person for each unsold seat on the plane. If the plane holds 100 passengers, and if x represents the number of unsold seats, find the following.

(a) An expression for the total revenue received for the flight (*Hint:* Multiply the number of people flying, $100 - x$, by the price per ticket.) $R(x) = 20{,}000 + 200x - 4x^2$

(b) The graph for the expression in part (a)

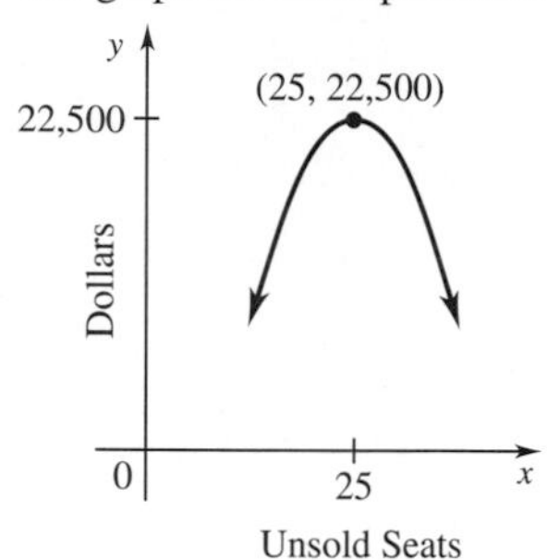

(c) The number of unsold seats that will produce the maximum revenue **25 unsold seats**

(d) The maximum revenue **$22,500**

33. The daily measurement (in particles) of a certain type of pollen during the first 10 days of June is approximated by the function $G(x) = 15 + 24x - 2x^2$, where x is the day in June, with $x = 1$ representing June 1.

(a) Sketch the graph of $G(x)$.

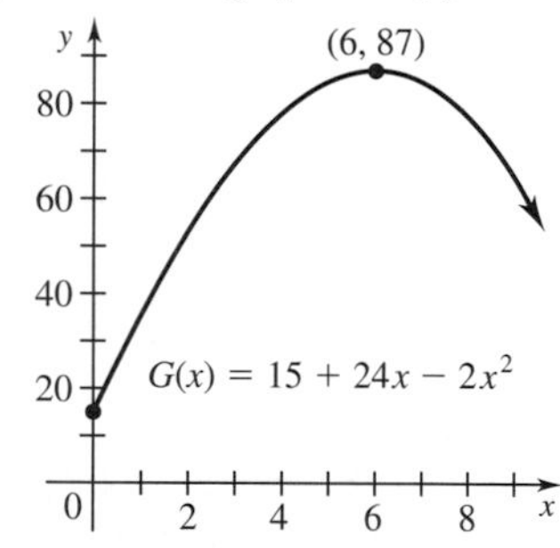

(b) Find the maximum pollen measurement, and determine when it occurs. **87 particles on June 6**

34. The demand for a certain type of cosmetic is given by $p = 500 - x$, where p is the price per unit when x units are demanded.

(a) Find the revenue, $R(x)$, obtained when x units are demanded. (*Hint:* Revenue = Number of units demanded $\times$ Price per unit.) $\boldsymbol{x(500 - x) = 500x - x^2 = R(x)}$

(b) Graph the revenue function defined by $y = R(x)$.

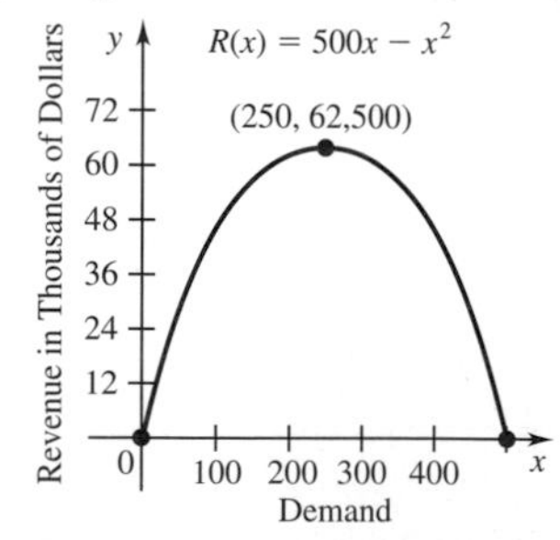

(c) From the graph of the revenue function, estimate the price that will produce the maximum revenue. **$250**

(d) What is the maximum revenue? **$62,500**

35. During the course of a year, the number of volunteers available to run a food bank each month is approximated by $V(x)$, where $V(x) = 2x^2 - 32x + 150$ between the months of January and August. Here x is time in months, with $x = 1$ representing January. From August to December, $V(x)$ is approximated by $V(x) = 31x - 226$. Find the number of volunteers in each of the following months:
(a) January **(b)** May **(c)** August **(d)** October **(e)** December.
(f) Sketch a graph of $y = V(x)$ for January through December. In what month are the fewest volunteers available?
(a) 120 (b) 40 (c) 22 (d) 84 (e) 146 (f) minimum at $x = 8$ (August)

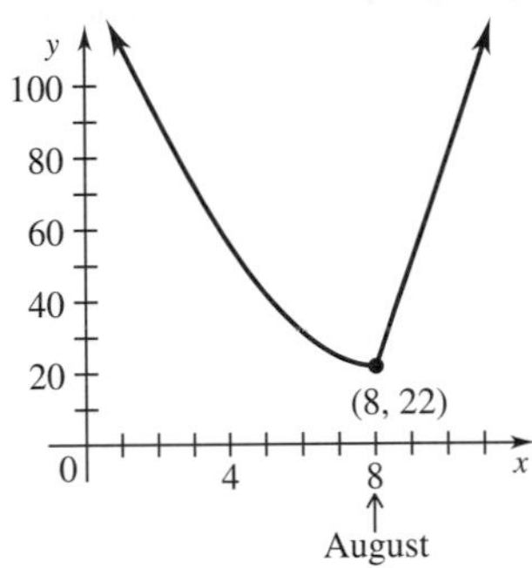

36. Between the months of June and October, the percent of maximum possible chlorophyll production in a leaf is approximated by $C(x) = 10x + 50$. Here x is time in months, with $x = 1$ representing June. From October through December, $C(x)$ is approximated by $C(x) - -20(x - 5)^2 + 100$.
Find the percent of maximum possible chlorophyll production in each of the following months:
(a) June **(b)** July **(c)** September **(d)** October **(e)** November **(f)** December.
(g) Sketch a graph of $C(x)$ from June through December. In what month is chlorophyll production a maximum?
(a) 60 (b) 70 (c) 90 (d) 100 (e) 80 (f) 20 (g) maximum at $x = 5$ (October)

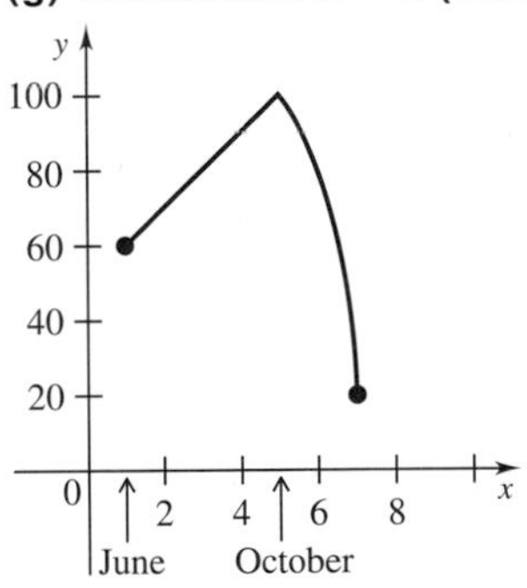

37. The cross section of an irrigation ditch is shaped like a parabola measuring 18 m across the top and 12 m deep. How wide is the ditch 8 m from the top? **$6\sqrt{3}$ m**

38. An arch is shaped like a parabola. It is 30 m wide at the base and 15 m high. How wide is the arch 10 m from the ground? **$10\sqrt{3}$ m $\approx$ 17.32 m**

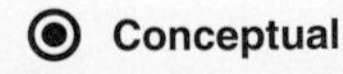

39. Suppose that a quadratic function with $a > 0$ is written in the form $f(x) = a(x - h)^2 + k$. Match each of the items **(a)**, **(b)**, and **(c)** with one of the items **(A)**, **(B)**, or **(C)**.

(a) k is positive.	**(A)** $f(x)$ intersects the x-axis at only one point.
(b) k is negative.	**(B)** $f(x)$ does not intersect the x-axis.
(c) k is zero.	**(C)** $f(x)$ intersects the x-axis twice.

(a) B (b) C (c) A

The figures below show several possible graphs of $f(x) = ax^2 + bx + c$. For the restrictions on a, b, and c given in Exercises 40–45, select the corresponding graph from (a) through (f) below.

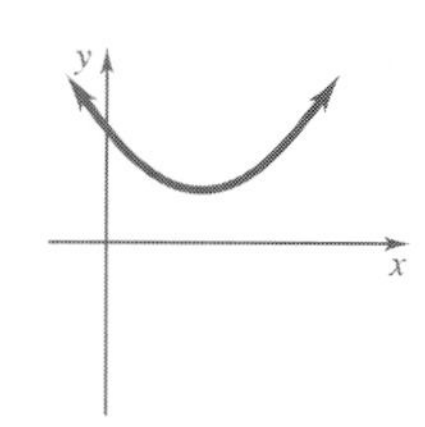

(a)

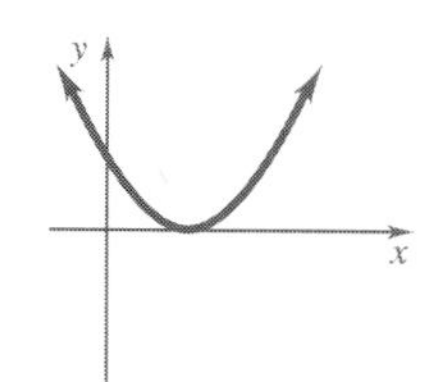

(b)

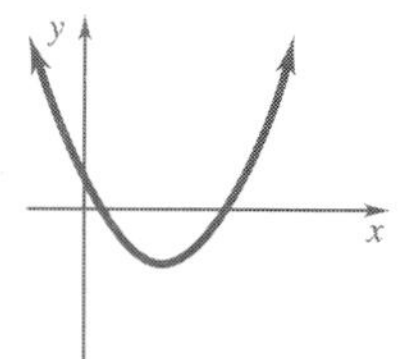

(c)

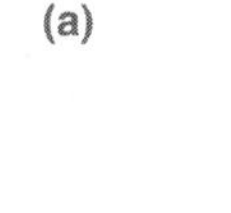

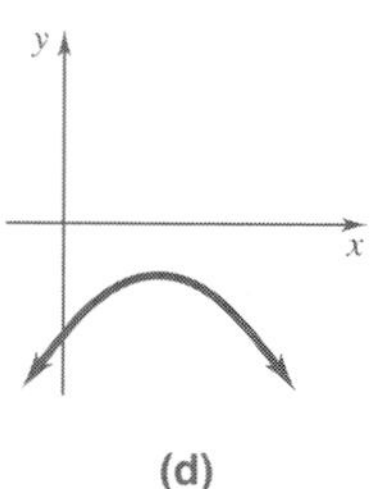

(d)

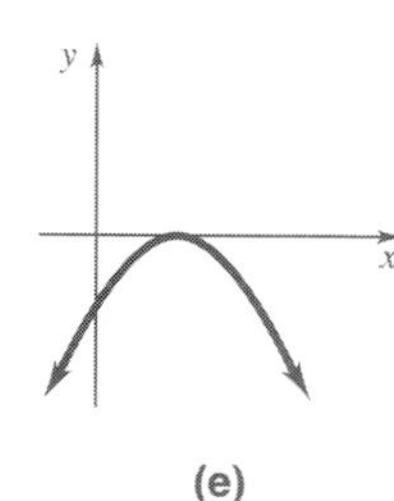

(e)

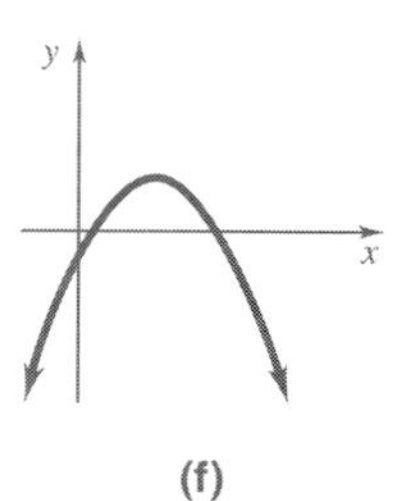

(f)

40. $a < 0,\ b^2 - 4ac = 0$ **(e)**

41. $a > 0,\ b^2 - 4ac < 0$ **(a)**

42. $a < 0,\ b^2 - 4ac < 0$ **(d)**

43. $a < 0,\ b^2 - 4ac > 0$ **(f)**

44. $a > 0,\ b^2 - 4ac > 0$ **(c)**

45. $a > 0,\ b^2 - 4ac = 0$ **(b)**

46. Find a value of c so that $y = x^2 - 10x + c$ has exactly one x-intercept. **$c = 25$**

47. For what values of a does $y = ax^2 - 8x + 4$ have no x-intercepts? **$a > 4$**

48. Find b so that $y = x^2 + bx + 9$ has exactly one x-intercept. **$b = 6$ or -6**

49. Find the quadratic function having x-intercepts 2 and 5, and y-intercept 5. **$f(x) = \frac{x^2}{2} - \frac{7}{2}x + 5$**

50. Find the quadratic function having x-intercepts 1 and -2, and y-intercept 4. **$f(x) = -2x^2 - 2x + 4$**

51. Find the largest possible value of y if $y = -(x - 2)^2 + 9$. Then find the following. **$y = 9$**

(a) The largest possible value of $\sqrt{-(x - 2)^2 + 9}$ **$\sqrt{9} = 3$**

(b) The smallest possible value of $1/[-(x - 2)^2 + 9]$ **1/9**

52. Find the smallest possible value of y if $y = 3 + (x + 5)^2$. Then find the following. **$y = 3$**

(a) The smallest possible value of $\sqrt{3 + (x + 5)^2}$ **$\sqrt{3}$**

(b) The largest possible value of $1/[3 + (x + 5)^2]$ **1/3**

◆ ◉ **53.** From the distance formula in Section 3.1, the distance between the two points $P(x_1, y_1)$ and $R(x_2, y_2)$ is $d(P, R) = \sqrt{(x_1 - x_2)^2 + (y_1 - y_2)^2}$. Using the results of Exercises 51 and 52, find the closest point on the line $y = 2x$ to the point (1, 7). (*Hint:* Every point on $y = 2x$ has the form $(x, 2x)$, and the closest point has the minimum distance.) **(3, 6)**

✎ **54.** Let $f(x) = a(x - h)^2 + k$, and show that $f(h + x) = f(h - x)$. Why does this show that the parabola is symmetric with respect to its axis?

Graphing Utility Problems

55. In Section 3.6 we saw how certain changes to an equation cause the graph of the equation to be stretched, shrunk, reflected about an axis, or translated vertically or horizontally. It is important to notice that the order in which these changes are done affects the final graph. For example, stretching and then shifting vertically produces a graph that differs from the one produced by shifting vertically, then stretching. To see this, use your grapher to graph $y = 3(x^2 - 2)$ and $y = 3x^2 - 2$, and then compare the results. Are the two expressions equivalent algebraically? **no**

In Exercises 56 and 57, find a polynomial function whose graph matches the one in the figure. Then use your grapher to graph the function and verify your result.

56.

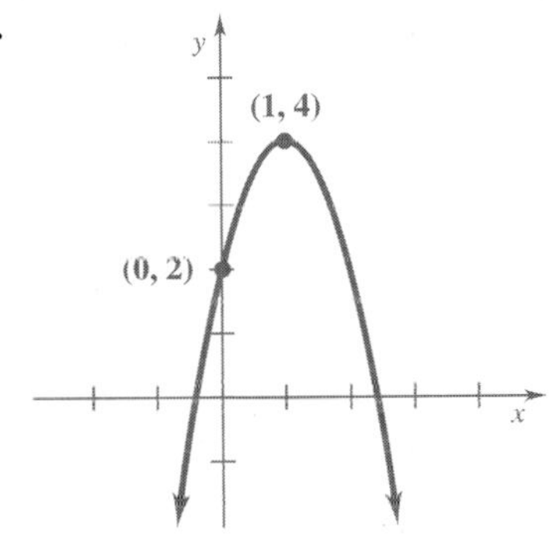

$\boldsymbol{f(x) = -2(x - 1)^2 + 4}$

57.

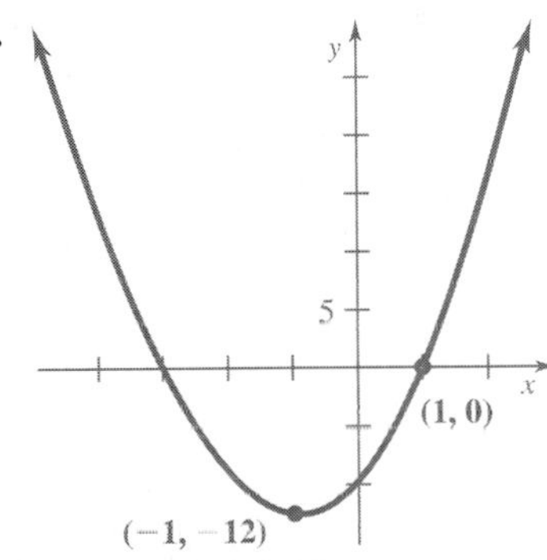

$\boldsymbol{f(x) = 3(x + 1)^2 - 12}$

4.2 Exercises (text page 211)

Use synthetic division to perform each of the following divisions.

1. $\dfrac{x^3 + 4x^2 - 5x + 42}{x + 6}$ $\boldsymbol{x^2 - 2x + 7}$

2. $\dfrac{x^3 + 2x^2 - 8x - 17}{x - 3}$ $\boldsymbol{x^2 + 5x + 7 + 4/(x - 3)}$

3. $\dfrac{4x^3 - 3x - 2}{x + 1}$ $\boldsymbol{4x^2 - 4x + 1 + [-3/(x + 1)]}$

4. $\dfrac{3x^3 - 4x + 2}{x - 1}$ $\boldsymbol{3x^2 + 3x - 1 + 1/(x - 1)}$

5. $\dfrac{x^4 - 3x^3 - 4x^2 + 12x}{x - 3}$ $\boldsymbol{x^3 - 4x}$

6. $\dfrac{x^4 - 3x^3 - 5x^2 + 2x - 16}{x + 2}$ $\boldsymbol{x^3 - 5x^2 + 5x - 8}$

7. $\frac{x^5 + 3x^4 + 2x^3 + 2x^2 + 3x + 1}{x + 2}$ **$x^4 + x^3 + 2x - 1 + 3/(x + 2)$**

8. $\frac{\frac{1}{3}x^3 - \frac{2}{9}x^2 + \frac{1}{27}x + 1}{x - \frac{1}{3}}$ **$\frac{1}{3}x^2 - \frac{1}{9}x + \frac{1}{(x - 1/3)}$**

9. $\frac{x^3 - 1}{x - 1}$ **$x^2 + x + 1$**

10. $\frac{x^5 - 1}{x - 1}$ **$x^4 + x^3 + x^2 + x + 1$**

Express each polynomial in the form $f(x) = (x - k)q(x) + r$ for the given value of k.

11. $f(x) = 2x^3 + x^2 + x - 8;\ k = -1$ **$f(x) = (x + 1)(2x^2 - x + 2) - 10$**
12. $f(x) = 2x^3 + 3x^2 - 16x + 10;\ k = -4$ **$f(x) = (x + 4)(2x^2 - 5x + 4) - 6$**
13. $f(x) = -x^3 + 2x^2 + 4;\ k = -2$ **$f(x) = (x + 2)(-x^2 + 4x - 8) + 20$**
14. $f(x) = -4x^3 + 2x^2 - 3x - 10;\ k = 2$ **$f(x) = (x - 2)(-4x^2 - 6x - 15) - 40$**
15. $f(x) = 4x^4 - 3x^3 - 20x^2 - x;\ k = 3$ **$f(x) = (x - 3)(4x^3 + 9x^2 + 7x + 20) + 60$**
16. $f(x) = 2x^4 + x^3 - 15x^2 + 3x;\ k = -3$ **$f(x) = (x + 3)(2x^3 - 5x^2 + 3) - 9$**

For each of the following polynomials, use the remainder theorem and synthetic division to find $f(k)$.

17. $k = 5;\ f(x) = -x^2 + 2x + 7$ **-8**
18. $k = -3;\ f(x) = 3x^2 + 8x + 5$ **8**
19. $k = 3;\ f(x) = x^2 - 4x + 5$ **2**
20. $k = -2;\ f(x) = x^2 + 5x + 6$ **0**
21. $k = 2;\ f(x) = 2x^2 - 3x - 3$ **-1**
22. $k = 4;\ f(x) = -x^3 + 8x^2 + 63$ **127**
23. $k = -1;\ f(x) = x^3 - 4x^2 + 2x + 1$ **-6**
24. $k = 2;\ f(x) = 2x^3 - 3x^2 - 5x + 4$ **-2**
25. $k = 3;\ f(x) = 2x^5 - 10x^3 - 19x^2 - 45$ **0**
26. $k = 4;\ f(x) = x^4 + 6x^3 + 9x^2 + 3x - 3$ **793**
27. $k = -8;\ f(x) = x^6 + 7x^5 - 5x^4 + 22x^3 - 16x^2 + x + 19$ **11**
28. $k = -\frac{1}{2};\ f(x) = 6x^3 - 31x^2 - 15x$ **-1**
29. $k = 2 + i;\ f(x) = x^2 - 5x + 1$ **$-6 - i$**
30. $k = 3 - 2i;\ f(x) = x^2 - x + 3$ **$5 - 10i$**

Use synthetic division to decide whether or not the given number is a zero of the given polynomial.

31. $3;\ f(x) = 2x^3 - 6x^2 - 9x + 4$ **no**
32. $-6;\ f(x) = 2x^3 + 9x^2 - 16x + 12$ **yes**
33. $-5;\ f(x) = x^3 + 7x^2 + 10x$ **yes**
34. $-2;\ f(x) = 2x^3 - 3x^2 - 5x$ **no**
35. $\frac{2}{5};\ f(x) = 5x^4 + 2x^3 - x + 15$ **no**
36. $\frac{1}{2};\ f(x) = 2x^4 - 3x^2 + 4$ **no**
37. $2 - i;\ f(x) = x^2 + 3x + 4$ **no**
38. $1 - 2i;\ f(x) = x^2 - 3x + 5$ **no**
39. $i;\ f(x) = x^3 + 2ix^2 + 2x + i$ **yes**
40. $-i;\ f(x) = x^3 - ix^2 + 3x + 5i$ **no**
41. Why must the function $r(x)$ in the division algorithm either be zero or have degree less than the degree of $g(x)$? (*Hint:* Think about the division algorithm for whole numbers.)
42. Find the remainder when the polynomial $x^{99} - 2x^{52} + x^2$ is divided by $x + 1$. **-2**
43. Find the remainder when the polynomial $x^{101} + 3x^{20} + x^3$ is divided by $x - i$. **3**
44. Find the value of k so that $(x^2 + 4x + 8) \div (x - k)$ has a remainder of 4. **-2**

4.3 Exercises (text page 218)

Use the factor theorem to decide whether or not the second polynomial is a factor of the first.

1. $4x^2 + 2x + 54$; $x - 4$ **no**

2. $5x^2 - 14x + 10$; $x + 2$ **no**

3. $x^3 + 2x^2 - 3$; $x - 1$ **yes**

4. $2x^3 + x + 2$; $x + 1$ **no**

5. $2x^4 + 5x^3 - 2x^2 + 5x + 6$; $x + 3$ **yes**

6. $5x^4 + 16x^3 - 15x^2 + 8x + 16$; $x + 4$ **yes**

For each of the following, find a polynomial of degree 3 with only real coefficients that satisfies the given conditions.

7. Zeros of -3, 1, and 4; $f(2) = 30$ $\mathbf{f(x) = -3x^3 + 6x^2 + 33x - 36}$

8. Zeros of 1, -1, and 0; $f(2) = 3$ $\mathbf{f(x) = \frac{1}{2}x^3 - \frac{1}{2}x}$

9. Zeros of -2, 1, and 0; $f(-1) = -1$ $\mathbf{f(x) = -\frac{1}{2}x^3 - \frac{1}{2}x^2 + x}$

10. Zeros of 2, -3, and 5; $f(3) = 6$ $\mathbf{f(x) = -\frac{1}{2}x^3 + 2x^2 + \frac{11}{2}x - 15}$

11. Zeros of 5, i, and $-i$; $f(2) = 5$ $\mathbf{f(x) = -\frac{1}{3}x^3 + \frac{5}{3}x^2 - \frac{1}{3}x + \frac{5}{3}}$

12. Zeros of -2, i, and $-i$; $f(-3) = 30$ $\mathbf{f(x) = -3x^3 - 6x^2 - 3x - 6}$

For each of the following polynomial functions, find all zeros and their multiplicities.

13. $f(x) = 7x^3 + x$ $\mathbf{0, \pm\frac{\sqrt{7}}{7}i}$

14. $f(x) = (x + 1)^2(x - 1)^3(x^2 - 10)$ **-1 (multiplicity 2), 1 (multiplicity 3), $\pm\sqrt{10}$**

15. $f(x) = 3(x - 2)(x + 3)(x - 1 + i)$ $\mathbf{2, -3, 1 - i}$

16. $f(x) = 5x^2(x + 1 - \sqrt{2})(2x + 5)$ **0 (multiplicity 2), $-1 + \sqrt{2}$, $-\frac{5}{2}$**

17. $f(x) = (x^2 + x - 2)^5(x - 1 + \sqrt{3})^2$ **-2 (multiplicity 5), 1 (multiplicity 5), $1 - \sqrt{2}$ (multiplicity 2)**

18. $f(x) = (7x - 2)^3(x^2 + 9)^2$ **$\frac{2}{7}$ (multiplicity 3), $3i$ (multiplicity 2), $-3i$ (multiplicity 2)**

For each of the following, find a polynomial of lowest degree with only real coefficients and having the given zeros. **(Other answers are possible.)**

19. $3 + i$ and $3 - i$ $\mathbf{x^2 - 6x + 10}$

20. $7 - 2i$ and $7 + 2i$ $\mathbf{x^2 - 14x + 53}$

21. $1 + \sqrt{2}$, $1 - \sqrt{2}$, and 3 $\mathbf{x^3 - 5x^2 + 5x + 3}$

22. $1 - \sqrt{3}$, $1 + \sqrt{3}$, and 1 $\mathbf{x^3 - 3x^2 + 2}$

23. $-2 + i$, $-2 - i$, 3, and -3 $\mathbf{x^4 + 4x^3 - 4x^2 - 36x - 45}$

24. $3 + 2i$, -1, and 2 $\mathbf{x^4 - 7x^3 + 17x^2 - x - 26}$

25. 2 and $3i$ $\mathbf{x^3 - 2x^2 + 9x - 18}$

26. -1 and $6 - 3i$ $\mathbf{x^3 - 11x^2 + 33x + 45}$

27. $1 + 2i$, 2 (multiplicity 2) $\mathbf{x^4 - 6x^3 + 17x^2 - 28x + 20}$

28. $2 + i$, -3 (multiplicity 2) $\mathbf{x^4 + 2x^3 - 10x^2 - 6x + 45}$

For each of the following polynomials, one zero is given. Find all others.

29. $f(x) = x^3 - x^2 - 4x - 6$; 3 $\mathbf{-1 + i, -1 - i}$

30. $f(x) = x^3 + 4x^2 - 5$; 1 $\mathbf{-\frac{5}{2} \pm \frac{\sqrt{5}}{2}}$

31. $f(x) = 4x^3 + 6x^2 - 2x - 1;\ \frac{1}{2}$ $\mathbf{-1 \pm \frac{\sqrt{2}}{2}}$

32. $f(x) = x^3 - 7x^2 + 17x - 15;\ 2 - i$ $\mathbf{3, 2 + i}$

33. $f(x) = x^4 + 5x^2 + 4;\ -i$ $\mathbf{i, \pm 2i}$

34. $f(x) = x^4 + 10x^3 + 27x^2 + 10x + 26;\ i$ $\mathbf{-5 \pm i, -i}$

Factor $f(x)$ into linear factors given that k is a zero of $f(x)$.

35. $f(x) = 2x^3 - 3x^2 - 17x + 30;\ k = 2$ $\mathbf{f(x) = (x - 2)(2x - 5)(x + 3)}$

36. $f(x) = 2x^3 - 3x^2 - 5x + 6;\ k = 1$ $\mathbf{f(x) = (x - 2)(2x + 3)(x - 1)}$

37. $f(x) = 6x^3 + 13x^2 - 14x + 3;\ k = -3$ $\mathbf{f(x) = (x + 3)(3x - 1)(2x - 1)}$

38. $f(x) = 6x^3 + 17x^2 - 63x + 10;\ k = -5$ $\mathbf{f(x) = (x + 5)(6x - 1)(x - 2)}$

39. $f(x) = x^3 + (7 - 3i)x^2 + (12 - 21i)x - 36i;\ k = 3i$ $\mathbf{f(x) = (x - 3i)(x + 4)(x + 3)}$

40. $f(x) = 2x^3 + (11 - 4i)x^2 + (12 - 22i)x - 24i;\ k = 2i$ $\mathbf{f(x) = (x - 2i)(2x + 3)(x + 4)}$

41. Show that -2 is a zero of multiplicity 2 of $f(x) = x^4 + 2x^3 - 7x^2 - 20x - 12$ and find all other complex zeros. Then write $f(x)$ in factored form.
$\mathbf{-1, 3; f(x) = (x + 2)^2(x + 1)(x - 3)}$

42. Show that -1 is a zero of multiplicity 3 of $f(x) = x^5 - 4x^3 - 2x^2 + 3x + 2$ and find all other complex zeros. Then write $f(x)$ in factored form. **1, 2;**
$\mathbf{f(x) = (x + 1)^3(x - 1)(x - 2)}$

43. How many real zeros (counting multiplicities) are possible for a polynomial with real coefficients of degree 5? Explain your reasoning. **1, 3, or 5**

44. Explain why a polynomial of degree 3 with real coefficients has at least one real zero.

45. Explain why it is not possible for a polynomial of degree 3 with real coefficients to have zeros of 1, 2, and $1 + i$.

46. Show that the zeros of $f(x) = x^3 + ix^2 - (7 - i)x + (6 - 6i)$ are $1 - i$, 2, and -3. Does the conjugate zeros theorem apply? Why?

47. The displacement after time t of a particle moving along a straight line is given by $s(t) = 2t^4 - t^3 + t^2 - 14t$, where t is in seconds and s is in meters. The displacement is 0 after 2 sec. At what other times is the displacement 0? **0, 2 sec; all others are nonreal complex**

48. The cost function (in thousands of dollars) for producing squeeze bottles is given by $C(x) = x^3 - 7x^2 + 20x - 12$ and the revenue function (in thousands of dollars) is given by $R(x) = 4x$, where x is the number of bottles produced (in hundred-thousands). Cost equals revenue if 200,000 bottles are produced ($x = 2$). Find all other break-even points. $\mathbf{x = 3}$ **(300,000)**

If c and d are complex numbers, prove each of the following statements. (Hint: *Let* $c = a + bi$ *and* $d = m + ni$ *and form all the conjugates, the sums, and the products.)*

49. $\overline{c + d} = \bar{c} + \bar{d}$

50. $\overline{cd} = \bar{c} \cdot \bar{d}$

51. $\bar{a} = a$ for any real number a.

52. $\overline{c^n} = (\bar{c})^n$

Graphing Utility Problems

53. Graph $f(x) = x^3 - 9.8x^2 - 2x$ first with the ranges $-1 \leq x \leq 1$, $-3 \leq y \leq 1$, and then with the ranges $-20 \leq x \leq 20$, $-1000 \leq y \leq 1000$. Use TRACE to estimate the three real zeros of $f(x)$. Explain why both sets of ranges were needed to find the zeros.

54. Graph $f(x) = x^4 - 1.9x^2 + .9$ with the ranges $-10 \le x \le 10$, $-1 \le y \le 1000$, and use TRACE to estimate the zeros of $f(x)$. Repeat with the ranges $-2 \le x \le 2$, $-1 \le y \le 10$ and $.9 \le x \le 1.1$ and $-.01 \le y \le .01$. Use TRACE to estimate the three real zeros of $f(x)$. Explain why all three sets of ranges were needed to find the zeros.

In Exercises 55–58, determine the number of distinct real zeros of the polynomial.

55. $f(x) = x^4 + 2x^3 - 4x^2 + 2x - 7$ **2**

56. $f(x) = x^3 - 3x^2 + 4$ **2**

57. $f(x) = 3x^4 - x^2 - 5x - 2$ **2**

58. $f(x) = 10x^4 - 109x^3 + 92x^2 - 20x$ **4**

In Exercises 59–64, each polynomial has one or more rational zeros. Use ZOOM *and* TRACE *to find the rational zero (or zeros) and then factor the polynomial to find the other zeros.*

59. $f(x) = x^3 - x^2 - x - 2$ **2, $(-1 \pm i\sqrt{3})/2$**

60. $f(x) = 2x^3 - 3x^2 - 12x + 18$ **$3/2, \pm\sqrt{6}$**

61. $f(x) = 6x^3 + x^2 - 5x - 2$ **$-2/3, -1/2, 1$**

62. $f(x) = 4x^3 - 7x^2 + 6x + 2$ **$-1/4, 1 \pm i$**

63. $f(x) = x^4 - x^3 + 2x^2 + x - 3$ **$\pm 1, (1 \pm i\sqrt{11})/2$**

64. $f(x) = 2x^4 - 7x^3 + 10x^2 - x - 4$
$1, -1/2, (3 \pm i\sqrt{7})/2$

4.4 Exercises (text page 230)

Graph each of the following polynomials.

1. $f(x) = (x - 1)(x - 3)(x - 5)$

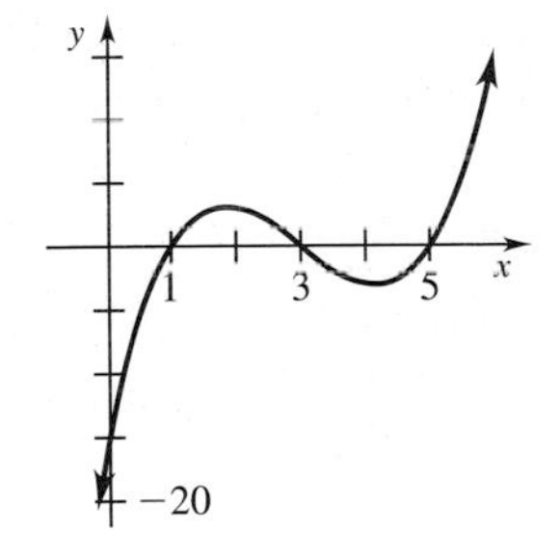

2. $f(x) = -2x(x + 4)(x - 2)$

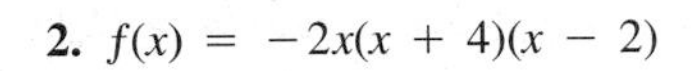

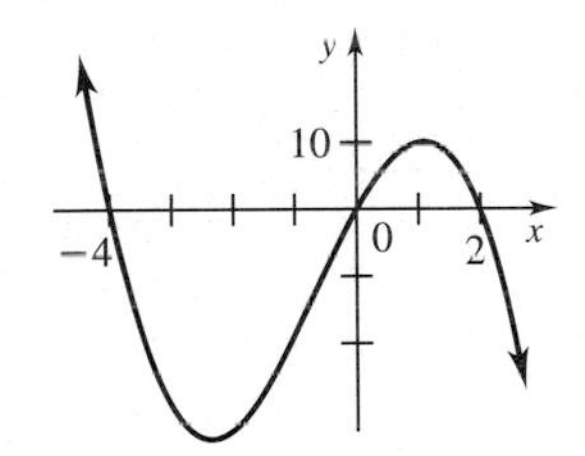

3. $f(x) = -(x + 1)(2x - 3)(2x - 1)$

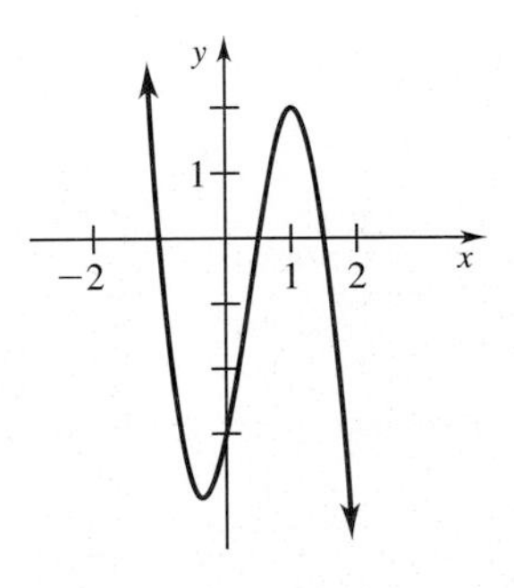

4. $f(x) = x^3 + x^2 - 6x$

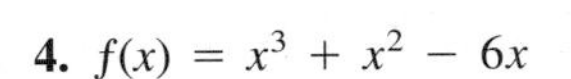

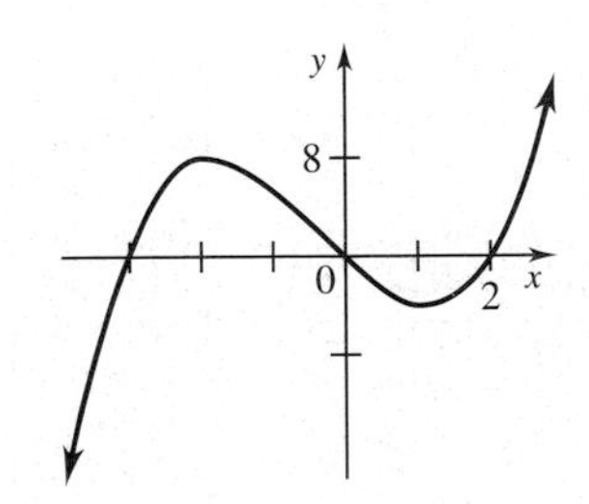

5. $f(x) = x^3 - 4x^2 + 4x$

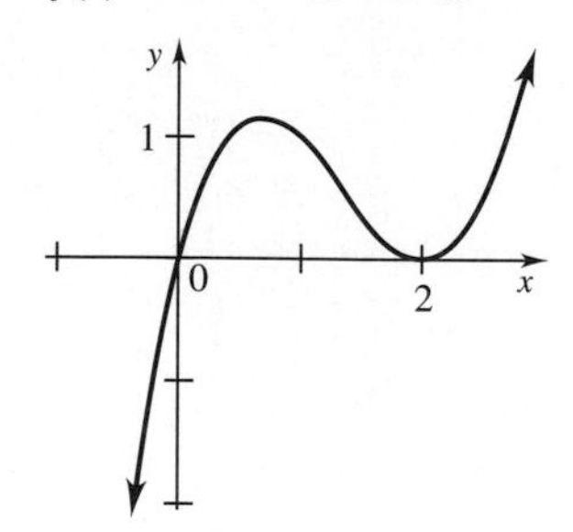

6. $f(x) = \frac{1}{2}(x + 1)(x - 1)^2$

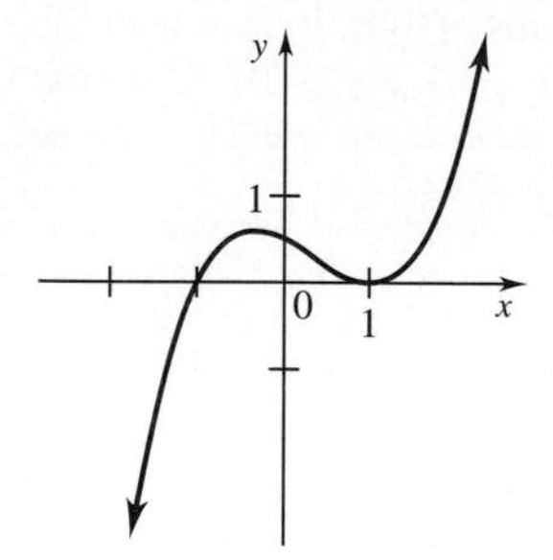

7. $f(x) = -3x(x + 1)^2(x - 1)$

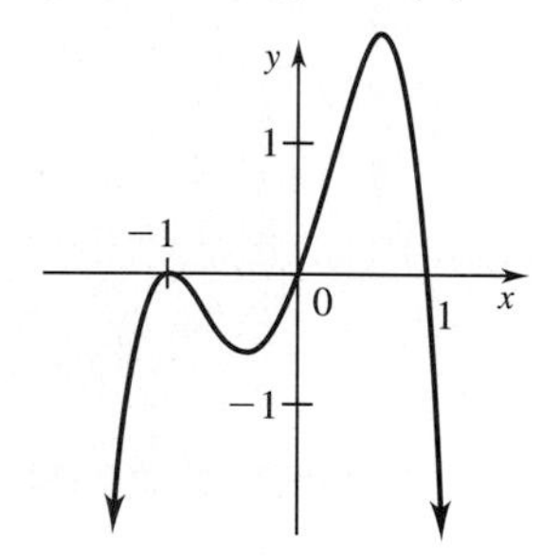

8. $f(x) = 2x^4 - 5x^3$

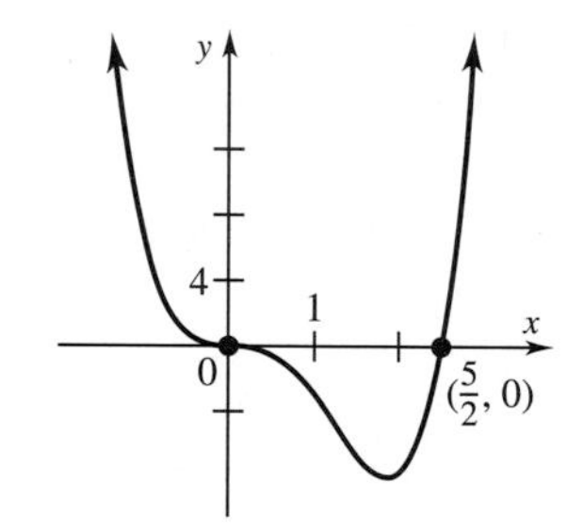

9. $f(x) = x^5 - 2x^3$

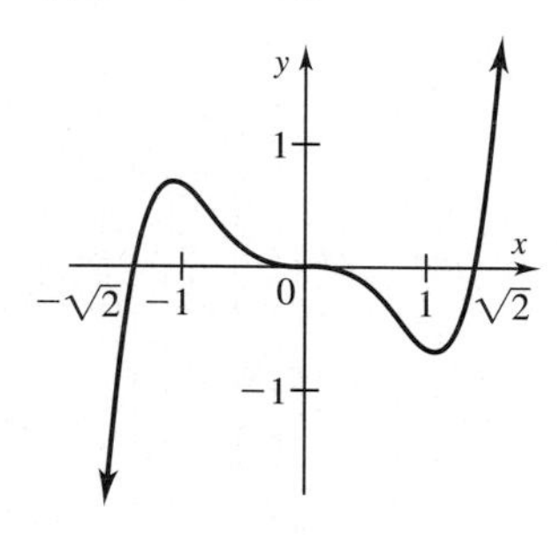

10. $f(x) = x^3 - 3x^2 + 3x - 1$

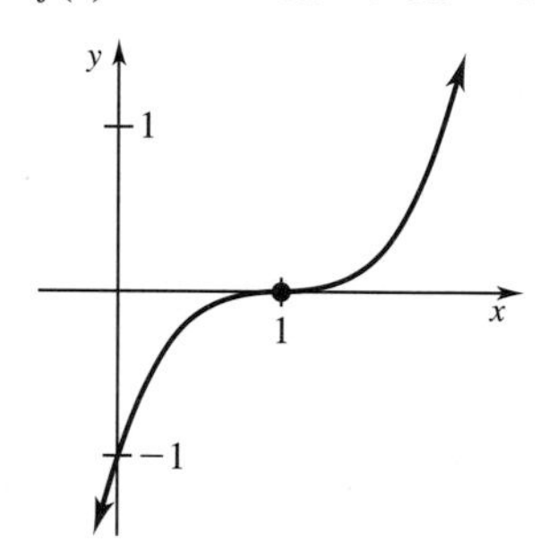

11. $f(x) = (x^2 - 2)(x - 4)^2$

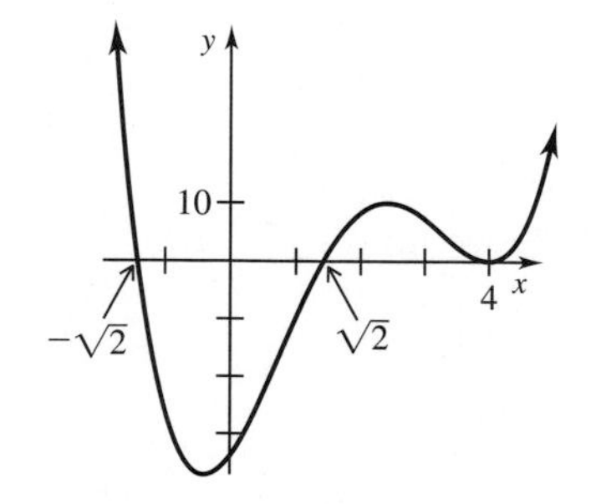

12. $f(x) = -\frac{1}{2}(x + 4)x(x^2 + 5)$

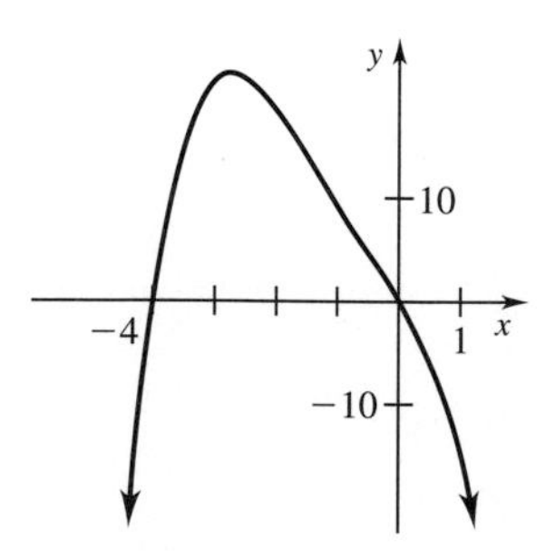

Use the intermediate value theorem for polynomials to show that the following polynomials have a real zero between the numbers given.

13. $f(x) = 2x^2 - 7x + 4$; 2 and 3

14. $f(x) = 3x^2 - x - 4$; 1 and 2

15. $f(x) = 2x^3 - 5x^2 - 5x + 7$; 0 and 1

16. $f(x) = 2x^3 - 9x^2 + x + 20$; 2 and 2.5

17. $f(x) = 2x^4 - 4x^2 + 4x - 8$; 1 and 2

18. $f(x) = x^4 - 4x^3 - x + 3$; 1 and .5

Show that the real zeros of each of the following polynomials satisfy the given conditions.

19. $f(x) = 4x^3 - 3x^2 + 4x + 7$; no real zero greater than 1

20. $f(x) = x^4 - x^3 + 2x^2 - 3x - 5$; no real zero greater than 2

21. $f(x) = x^4 + x^3 - x^2 + 3$; no real zero less than -2

22. $f(x) = x^5 + 2x^3 - 2x^2 + 5x + 5$; no real zero less than -1

Apply Descartes' rule of signs to determine the possible number of positive and negative real zeros of the following polynomials.

23. $f(x) = 5x^3 + 2x^2 - x + 2$ **2 or 0; 1**

24. $f(x) = x^3 + 6x^2 - 4$ **1; 2 or 0**

25. $f(x) = x^4 + 6x^3 - 3x^2 + x - 17$ **3 or 1; 1**

26. $f(x) = x^4 - 2x^3 + x^2 - 1$ **3 or 1; 1**

For each of the following polynomials, approximate each zero as a decimal to the nearest tenth.

27. $f(x) = x^3 + 3x^2 - 2x - 6$ **−3, −1.4, 1.4**

28. $f(x) = x^3 - 3x + 3$ **−2.1**

29. $f(x) = -2x^4 - x^2 + x + 5$ **−1.1, 1.2**

30. $f(x) = -x^4 + 2x^3 + 3x^2 + 6$ **−1.5, 3.1**

The following polynomials have zeros in the given intervals. Approximate these zeros to the nearest hundredth.

31. $f(x) = x^4 + x^3 - 6x^2 - 20x - 16$; [3.2, 3.3] and [−1.4, −1.1] **3.24, −1.24**

32. $f(x) = x^4 - 3x^3 - 2x^2 - 16x + 5$; [.2, .3] and [4.2, 4.3] **.30, 4.28**

Graph each of the functions defined as follows.

33. $f(x) = x^3 - 2x^2 - x + 1$

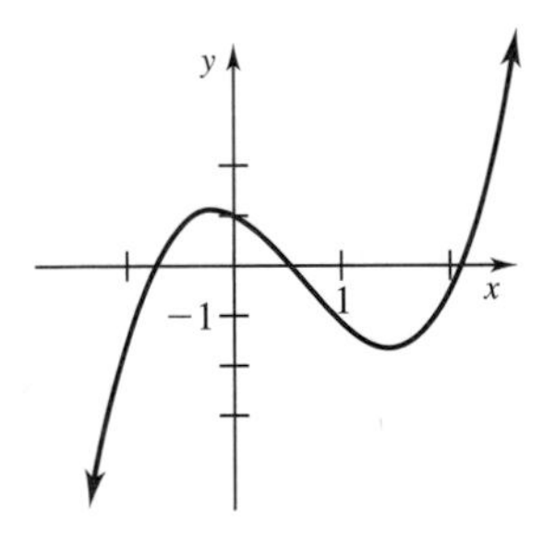

34. $f(x) = -x^3 - x^2 + 2x + 1$

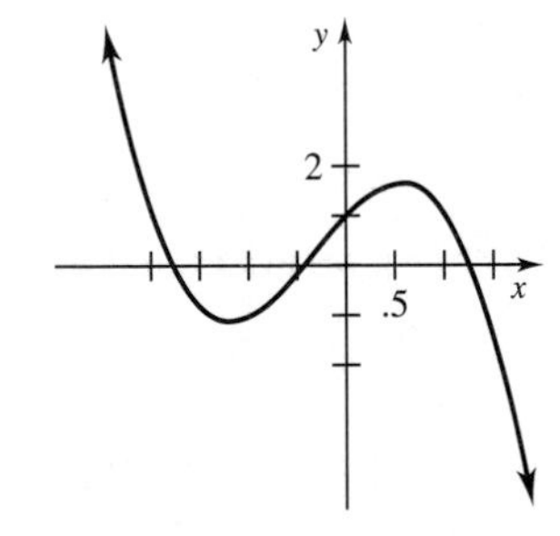

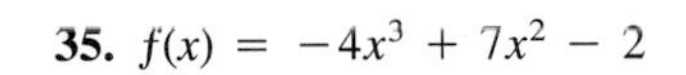
35. $f(x) = -4x^3 + 7x^2 - 2$

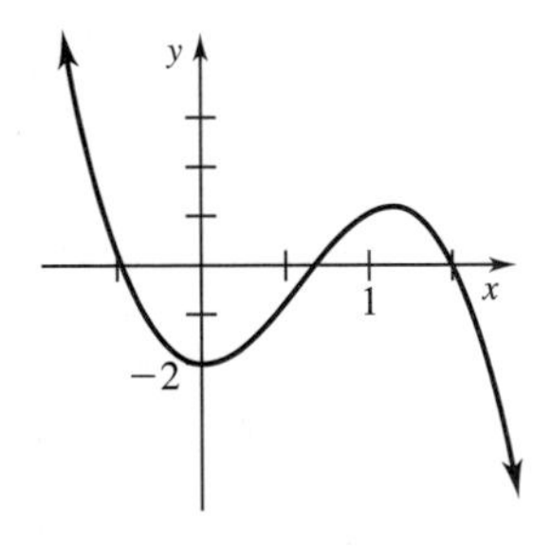

36. $f(x) = 5x^3 - 9x^2 + 1$

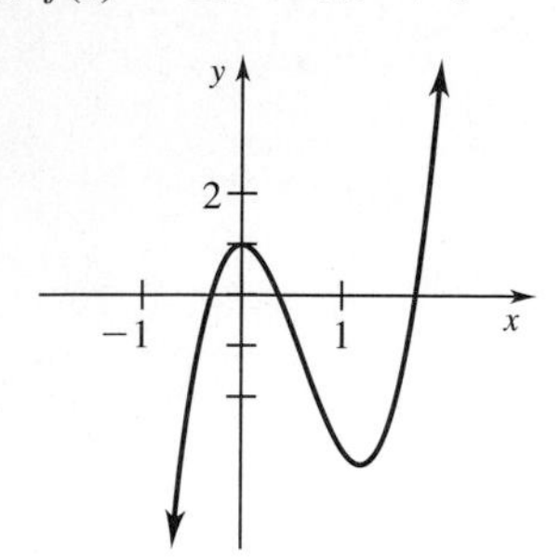

37. $f(x) = x^4 - 5x^2 + 2$

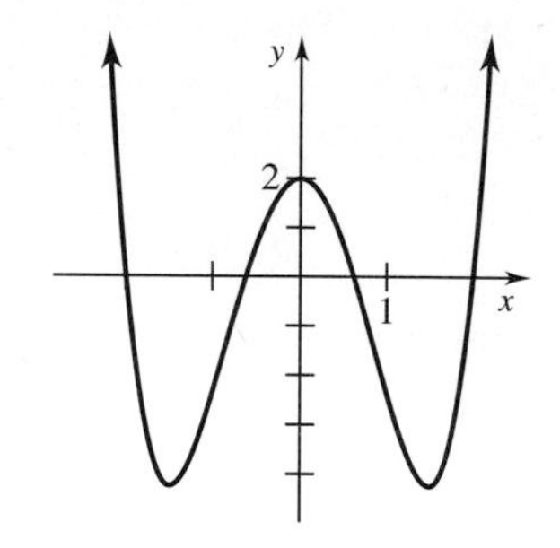

38. $f(x) = 2x^4 - 6x^3 + 7x - 2$

39. Explain why the graph of a polynomial having a as a zero of odd multiplicity crosses the x-axis at $x = a$.

40. Explain why the graph of a polynomial having a as a zero of even multiplicity touches, but does not cross, the x-axis at $x = a$.

In Exercises 41 and 42, find a cubic polynomial having the graph shown.

41.

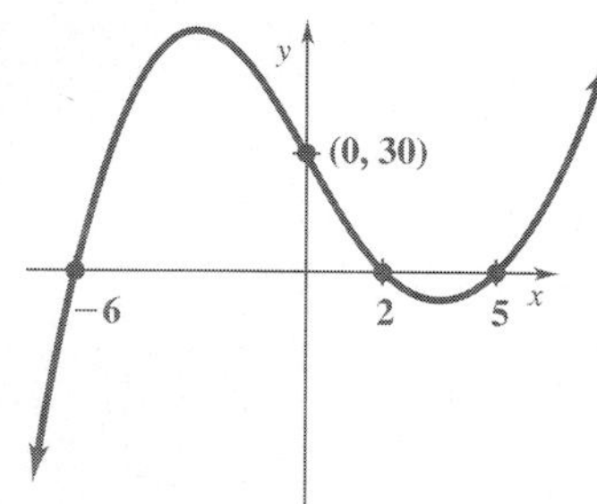

$f(x) = .5(x + 6)(x - 2)(x - 5)$

42.

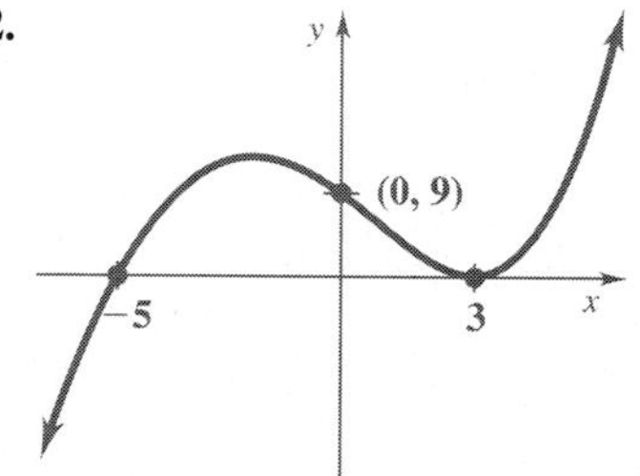

$f(x) = .2(x - 3)^2(x + 5)$

43. Give an example of a polynomial function that is never negative and has -4 and 1 as zeros. **One possible answer is $f(x) = (x + 4)^2(x - 1)^2$.**

44. Give an example of a polynomial function that has -3, 1, and 2 as zeros and is positive only between 1 and 2. **One possible answer is $f(x) = -(x + 3)^2(x - 1)(x - 2)$.**

Determine the domain and range of the functions defined in Exercises 45 and 46.

45. $f(x) = \sqrt{x^3 - x^2 - 6x}$ **domain: $[-2, 0] \cup [3, \infty)$; range: $[0, \infty)$**

46. $f(x) = \sqrt{x^3 - x}$ **domain: $[-1, 0] \cup [1, \infty)$; range: $[0, \infty)$**

47. Summarize the steps used to graph a polynomial.

Graphing Utility Problems

48. The pressure of the oil in a reservoir tends to drop with time. By taking sample pressure readings for a particular oil reservoir, petroleum engineers have found that the change in pressure is given by $P(t) = t^3 - 25t^2 + 200t$, where t is time in years from the date of the first reading.

(a) Graph $P(t)$.

(b) Use the graph from part (a) to decide for what time periods the amount of change in pressure (drop) is increasing or decreasing. **increasing from $t = 0$ to $t = 6\frac{2}{3}$ and from $t = 10$ on; decreasing from $t = 6\frac{2}{3}$ to $t = 10$**

49. During the early part of the twentieth century, the deer population of the Kaibab Plateau in Arizona experienced a rapid increase, because hunters had reduced the number of natural predators and because the deer were protected from hunters. The increase in population depleted the food resources and eventually caused the population to decline. For the period from 1905 to 1930, the deer population was approximated by $D(x) = -.125x^5 + 3.125x^4 + 4000$, where x is time in years from 1905.

(a) Graph $D(x)$.

(b) From the graph, over what period of time (from 1905 to 1930) was the population increasing? Relatively stable? Decreasing? **1910 to 1925; 1905 to 1910; 1925 to 1930**

50. The polynomial function A defined by $A(x) = -.015x^3 + 1.06x$ gives the approximate alcohol concentration (in tenths of a percent) in an average person's blood stream x hours after drinking about 8 oz of 100-proof whiskey. This function is approximately valid for x in the interval $[0, 8.2]$.

(a) Graph $y = A(x)$.

(b) Using the graph from part (a), estimate the time of maximum alcohol concentration. **4.85 hr**

(c) In California a person is legally drunk if the blood alcohol concentration exceeds .08%. Using the graph from part (a), estimate the period in which this average person is legally drunk. **from .76 hr to 8 hr**

51. A survey team measures the concentration (in parts per million) of a particular toxin in a local river. On a normal day, the concentration of the toxin at time x (in hours) after the factory upstream dumps its waste is given by $g(x) = -.006x^4 + .14x^3 - .05x^2 + .02x$, where $0 \leq x \leq 24$.

(a) Graph $g(x)$.

(b) Estimate the time at which the concentration is greatest. **17.3 hr**

(c) A concentration greater than 100 parts per million is considered pollution. Using the graph from part (a), estimate the period during which the river is polluted. **from 11.4 hr to 21.2 hr**

4.5 Exercises (text page 236)

List all possible rational zeros for the following polynomials.

1. $f(x) = 2x^3 + 7x^2 + 12x - 1$ **±1, ±1/2**

2. $f(x) = 2x^3 + 20x^2 + 68x - 40$ **±1, ±20, ±40, ±8, ±4, ±2, ±10, ±1/2, ±5, ±5/2**

3. $f(x) = 8x^3 + 19x^2 - 32x - 2$ **±1, ±2, ±1/2, ±1/4, ±1/8**

4. $f(x) = 15x^3 + 59x^2 + 4x - 1$ **±1, ±1/3, ±1/5, ±1/15**

5. $f(x) = x^4 + 4x^3 + 3x^2 - 10x + 75$ **±1, ±3, ±5, ±15, ±25, ±75**

6. $f(x) = x^4 - 2x^3 + x^2 + 18$ **±1, ±2, ±3, ±6, ±9, ±18**

Find all rational zeros of the following polynomials.

7. $f(x) = x^3 - 3x^2 - 25x - 21$ **−1, −3, 7**

8. $f(x) = x^3 - 2x^2 - 4x + 8$ **−2, 2**

9. $f(x) = x^3 + 9x^2 - 14x - 24$ **no rational zeros**

10. $f(x) = x^3 + 3x^2 - 4x - 12$ **2, −3, −2**

11. $f(x) = x^3 + 6x^2 - x - 30$ **2, −3, −5**

12. $f(x) = x^3 - x^2 - 10x - 8$ **−1, 4, −2**

13. $f(x) = x^4 + 9x^3 + 21x^2 - x - 30$ **1, −2, −3, −5**

14. $f(x) = x^4 + 4x^3 - 7x^2 - 34x - 24$ **−4, −2, −1, 3**

Find the rational zeros of the following polynomials; then write each polynomial in factored form with each factor having only integer coefficients.

15. $f(x) = 6x^3 + x^2 - 5x - 2$ **−1/2, −2/3, 1; $f(x) = (2x + 1)(3x + 2)(x - 1)$**

16. $f(x) = 15x^3 + 61x^2 + 2x - 8$ **−4, −2/5, 1/3; $f(x) = (x + 4)(5x + 2)(3x - 1)$**

17. $f(x) = 2x^3 + 7x^2 + 12x - 8$ **1/2; $f(x) = (2x - 1)(x^2 + 4x + 8)$**

18. $f(x) = 2x^3 + 20x^2 + 68x - 40$ **no rational zeros; cannot be factored**

19. $f(x) = x^4 + 2x^3 - 13x^2 - 38x - 24$ **−1, −2, −3, 4; $f(x) = (x + 1)(x + 2)(x + 3)(x - 4)$**

20. $f(x) = 6x^4 + x^3 - 7x^2 - x + 1$ **1, −1, 1/3, −1/2; $f(x) = (x - 1)(x + 1)(3x - 1)(2x + 1)$**

21. $f(x) = x^4 - 4x^3 - 7x^2 + 34x - 24$ **1, 2, −3, 4; $f(x) = (x - 1)(x - 2)(x + 3)(x - 4)$**

22. $f(x) = x^4 - 2x^3 + x^2 + 18$ **no rational zeros; cannot be factored**

23. $f(x) = x^5 + 3x^4 - 5x^3 - 11x^2 + 12$ **1; $f(x) = (x - 1)(x^4 + 4x^3 - x^2 - 12x - 12)$**

24. $f(x) = 4x^5 + 4x^4 - 37x^3 - 37x^2 + 9x + 9$ **−3, −1, 3, 1/2, −1/2;**
$f(x) = (x + 1)(x - 3)(x + 3)(2x - 1)(2x + 1)$

Find all rational zeros of the following polynomials.

25. $f(x) = x^3 + \frac{1}{2}x^2 - \frac{11}{2}x - 5$ **$-2, -1, \frac{5}{2}$**

26. $f(x) = x^3 - \frac{4}{3}x^2 - \frac{13}{3}x - 2$ **$-\frac{2}{3}, -1, 3$**

27. $f(x) = x^4 + \frac{1}{4}x^3 + \frac{11}{4}x^2 + x - 5$ **$1, -\frac{5}{4}$**

28. $f(x) = \frac{10}{7}x^4 - x^3 - 7x^2 + 5x - \frac{5}{7}$ **$\frac{1}{2}, \frac{1}{5}$**

29. $f(x) = \frac{1}{3}x^5 + x^4 - \frac{5}{3}x^3 - \frac{11}{3}x^2 + 4$ **1**

30. $f(x) = x^5 + x^4 - \frac{37}{4}x^2 + \frac{9}{4}x + \frac{9}{4}$ **no rational zeros**

Find all integer solutions of the following equations.

31. $6x^3 - 31x^2 + 3x + 10 = 0$ **5**

32. $12x^3 - 5x^2 - 9x + 1 = 0$ **none**

33. After a 2-inch slice is cut off the top of a cube, the resulting solid has a volume of 32 cubic inches. Find the dimensions of the original cube. **4 × 4 × 4 inches**

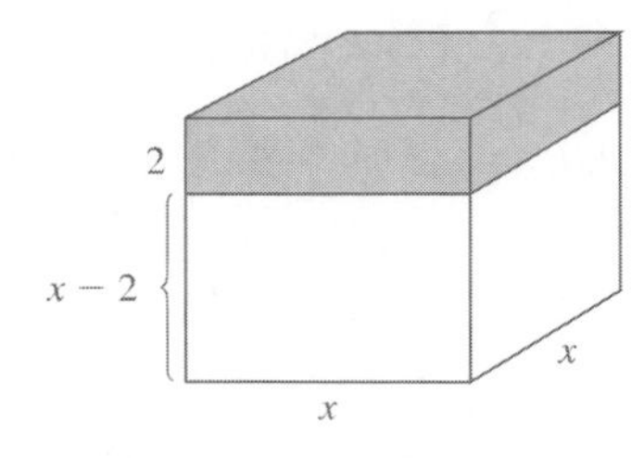

◆ **34.** The width of a rectangular box is three times its height, and its length is 11 inches more than its height. Find the dimensions of the box if its volume is 720 cubic inches. **12 × 4 × 15 inches**

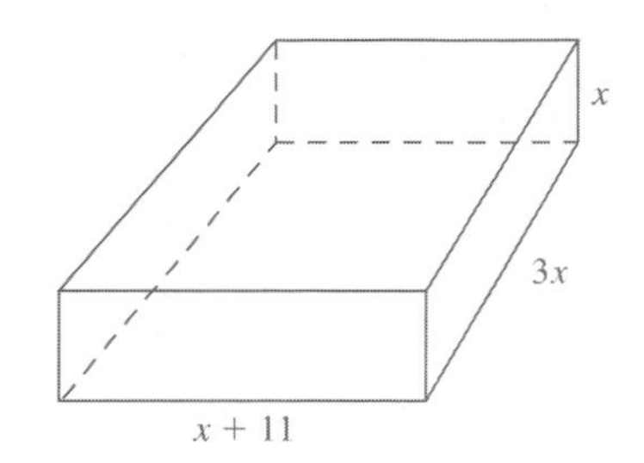

◆ **35.** A rectangle of area 42 square inches has its base on the x-axis and its upper corners on the parabola $y = 16 - x^2$. Find the dimensions of the rectangle. (*Hint:* What are the coordinates of P?) **5 × 7 inches**

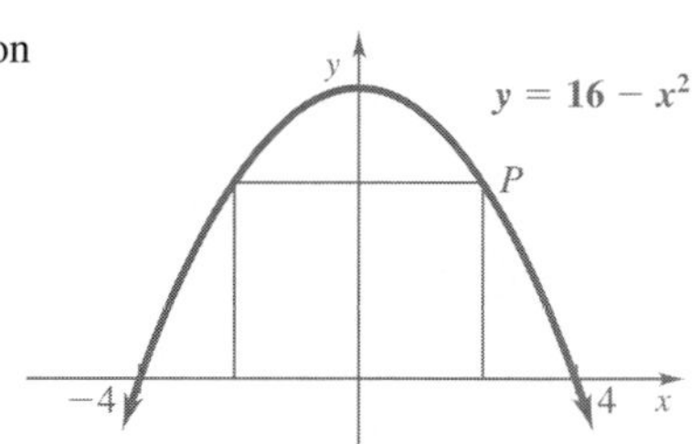

36. Show that $f(x) = x^2 - 7$ has no rational zeros, so $\sqrt{7}$ must be irrational.

37. Show that for any prime p, $\sqrt{p}$ is irrational. (*Hint:* Look at $x^2 - p$.)

38. Show that $f(x) = x^4 + 4x^2 - 13$ has no rational zeros.

▲ **39.** Show that any integer zeros of a polynomial with integer coefficients must be factors of the constant term a_0.

After factoring the polynomial and locating its zeros, sketch the graph of each function.

40. $f(x) = 2x^3 - 7x^2 + 4x + 4$

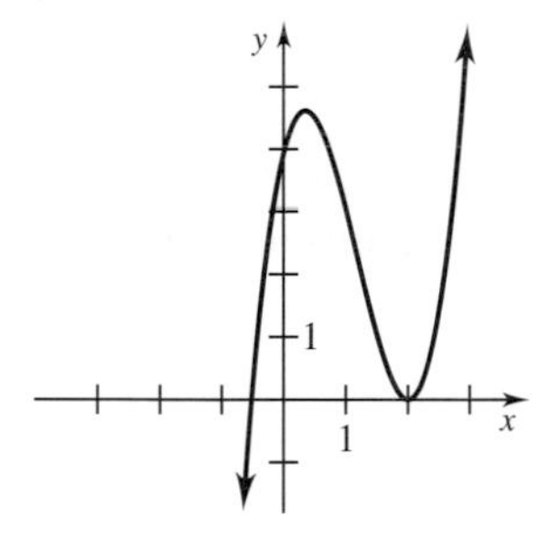

41. $f(x) = 3x^3 + 8x^2 + 3x - 2$

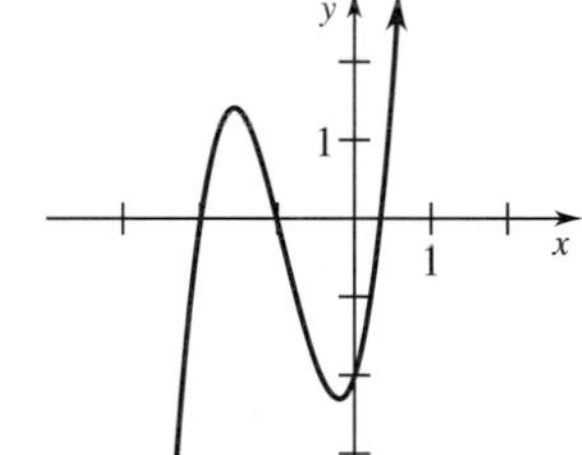
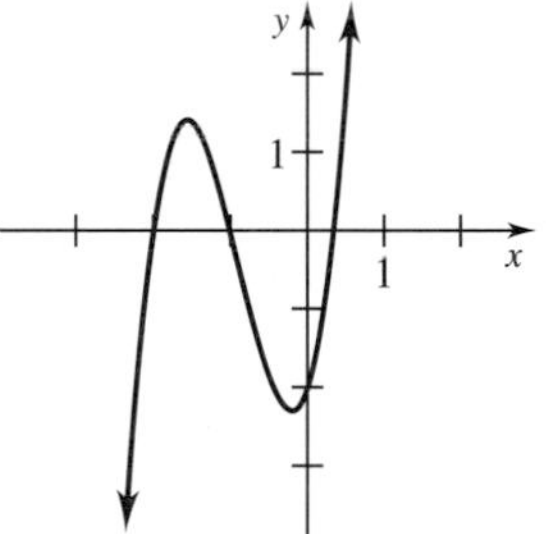

42. $f(x) = 2x^3 + x^2 - x$

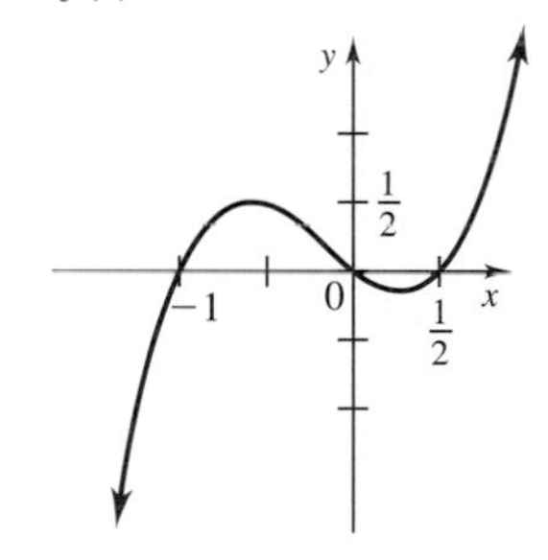

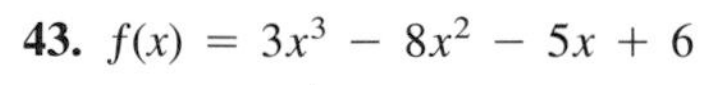

43. $f(x) = 3x^3 - 8x^2 - 5x + 6$

44. $f(x) = x^4 - 3x^2 + 2$

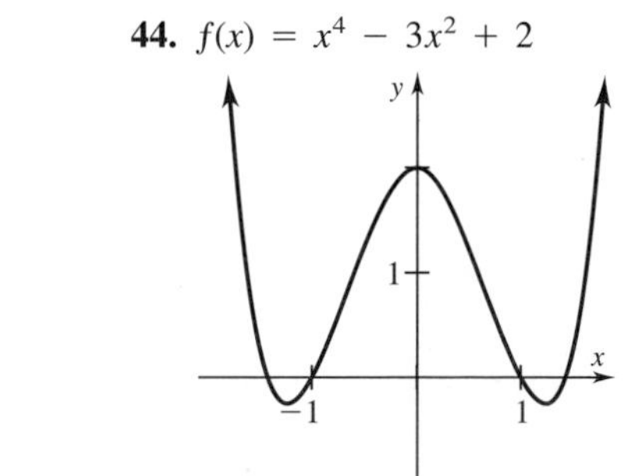

45. $f(x) = 2x^4 + 5x^3 - 5x - 2$

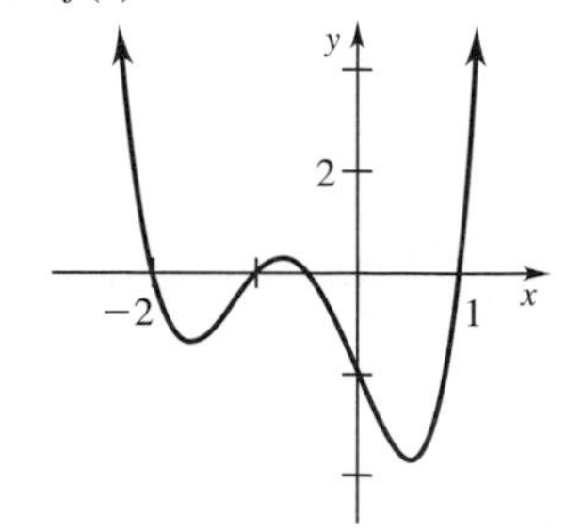

46. Describe some situations in which Descartes' rule of signs would be helpful in searching for rational zeros.

47. Describe some situations in which the boundedness theorem would be helpful in searching for rational zeros.

Graphing Utility Problems

In Exercises 48–53, use the rational zeros theorem to determine the possible rational zeros of the polynomial. Then use the graph of the polynomial to find the actual rational zeros.

48. $f(x) = 2x^3 - 30x^2 + 142x - 210$ **3, 5, 7**

49. $f(x) = 6x^3 - 19x^2 - 65x + 50$ **−5/2, 2/3, 5**

50. $f(x) = -6x^3 + 13x^2 + 99x - 70$ **−7/2, 2/3, 5**

51. $f(x) = 20x^3 + 12x^2 - 128x + 48$ **−3, 2/5, 2**

52. $f(x) = -9x^4 - 33x^3 + 48x^2 + 132x - 48$ **−4, −2, 1/3, 2**

53. $f(x) = x^4 - 17x^3 + 69x^2 + 17x - 70$ **−1, 1, 7, 10**

4.6 Exercises (text page 247)

For Exercises 1–10, find any vertical, horizontal, or oblique asymptotes.

1. $f(x) = \dfrac{1}{x - 3}$ **vertical asymptote: $x = 3$; horizontal asymptote: $y = 0$**

2. $f(x) = \dfrac{-1}{x + 2}$ **vertical asymptote: $x = -2$; horizontal asymptote: $y = 0$**

3. $f(x) = \dfrac{3x + 1}{x - 2}$ **vertical asymptote: $x = 2$; horizontal asymptote: $y = 3$**

4. $f(x) = \dfrac{3x - 5}{2x + 9}$ **vertical asymptote: $x = -9/2$ horizontal asymptote: $y = 3/2$**

5. $f(x) = \dfrac{x^2 - 1}{x + 3}$ **vertical asymptote: $x = -3$; oblique asymptote: $y = x - 3$**

6. $f(x) = \dfrac{2x^2 + 9}{x - 1}$ **vertical asymptote: $x = 1$; oblique asymptote: $y = 2x + 2$**

7. $f(x) = \dfrac{2x^3}{x^2 - 3x - 10}$ **vertical asymptotes: $x = 5$, $x = -2$; oblique asymptote: $y = 2x + 6$**

8. $f(x) = \dfrac{-3x^3}{x^2 - 4x + 3}$ **vertical asymptotes: $x = 3$, $x = 1$; oblique asymptote: $y = -3x - 12$**

9. $f(x) = \dfrac{(x - 3)(x + 1)}{(x + 2)(2x - 5)}$ **vertical asymptotes: $x = -2$, $x = 5/2$; horizontal asymptote: $y = 1/2$**

10. $f(x) = \dfrac{3(x + 2)(x - 4)}{(5x - 1)(x - 5)}$ **vertical asymptotes: $x = 1/5$, $x = 5$; horizontal asymptote: $y = 3/5$**

11. Sketch the following graphs and compare them with the graph of $f(x) = \dfrac{1}{x^2}$.

(a) $f(x) = \dfrac{1}{(x - 3)^2}$

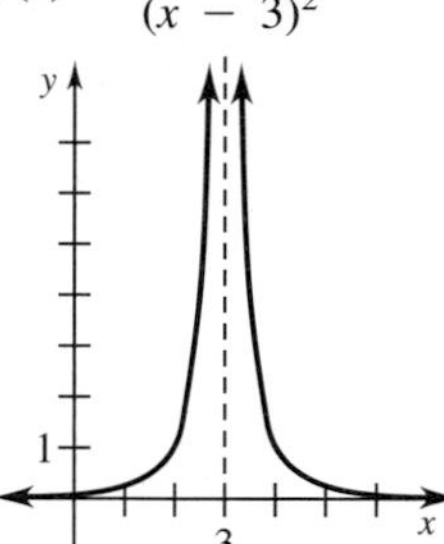

(b) $f(x) = \dfrac{-2}{x^2}$

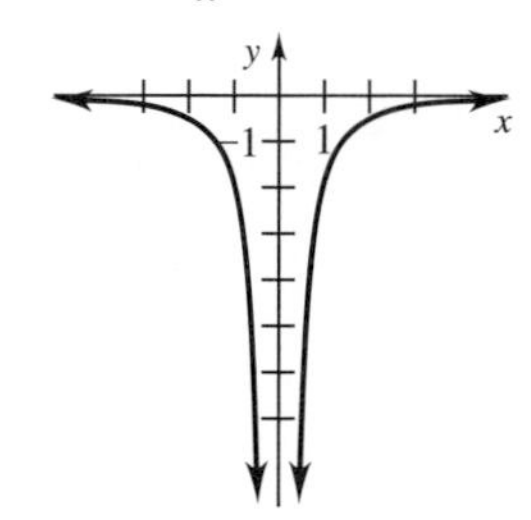

(c) $f(x) = \dfrac{-2}{(x - 3)^2}$

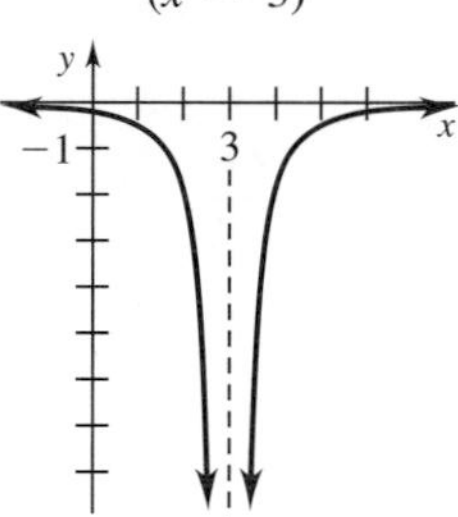

In Exercises 12–17, decide whether the graph has an asymptote or a missing point at $x = a$.

12. $f(x) = \dfrac{1}{x - a}$ **asymptote**

13. $f(x) = \dfrac{ax}{x - a}$ **asymptote**

14. $f(x) = \dfrac{x + a}{x - a}$ **asymptote**

15. $f(x) = \dfrac{x(x - a)}{x - a}$ **missing point**

16. $f(x) = \dfrac{(x + a)(x - a)}{x - a}$ **missing point**

17. $f(x) = \dfrac{x - a}{(x + b)(x - a)}$ **missing point**

Graph each function defined in Exercises 18–41.

18. $f(x) = \dfrac{2}{2 + 5x}$

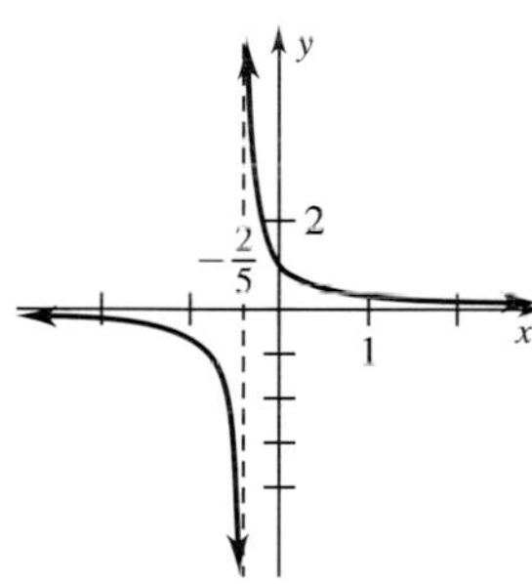

19. $f(x) = \dfrac{3}{(x - 2)^2}$

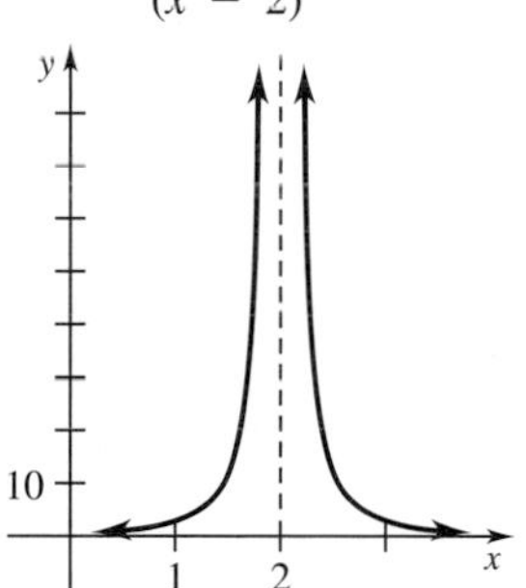

20. $f(x) = \dfrac{-2}{(x - 1)(x + 2)}$

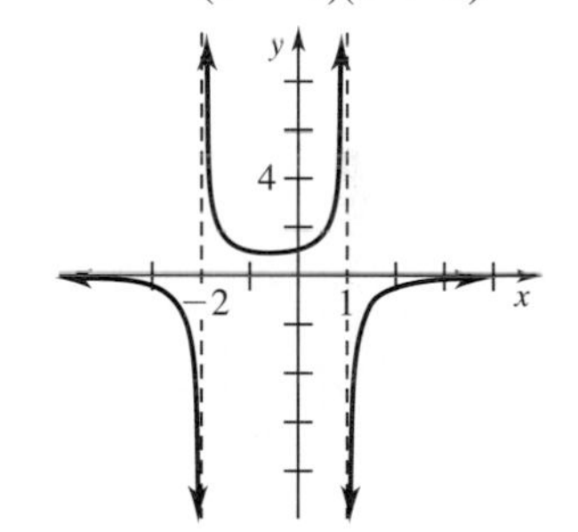

21. $f(x) = \dfrac{3x}{(x + 1)(x - 2)}$

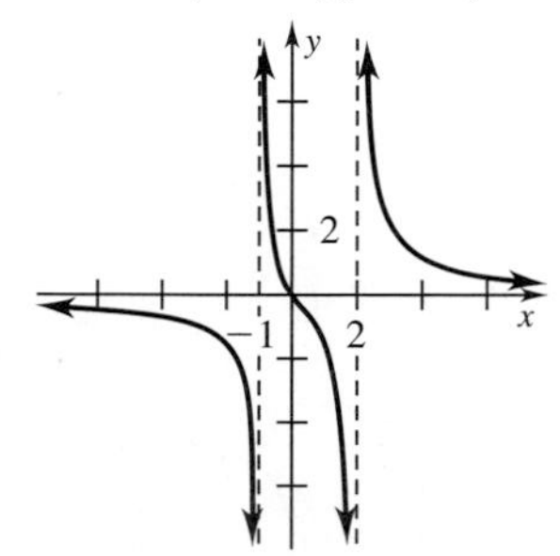

22. $f(x) = \dfrac{2x + 1}{(x + 2)(x + 4)}$

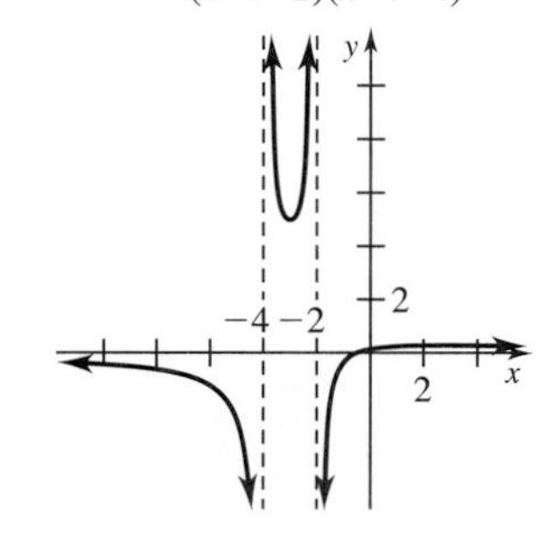

23. $f(x) = \dfrac{-x}{x^2 - 4}$

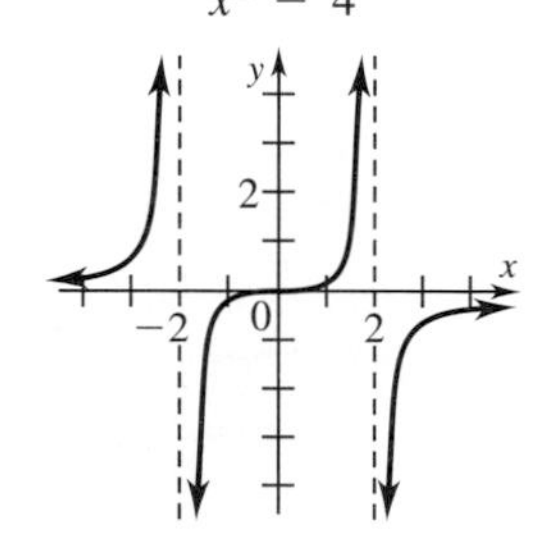

24. $f(x) = \dfrac{3x + 5}{x - 1}$

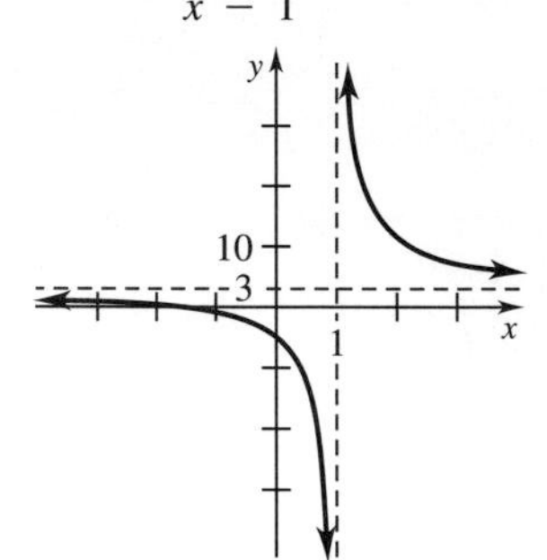

25. $f(x) = \dfrac{x - 2}{x + 2}$

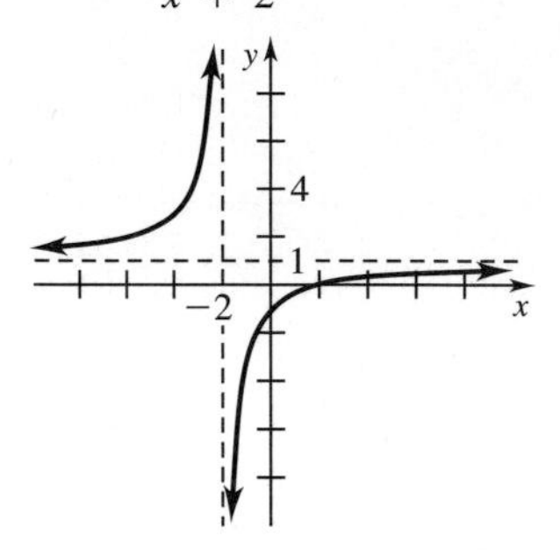

26. $f(x) = \dfrac{3x}{x^2 - 16}$

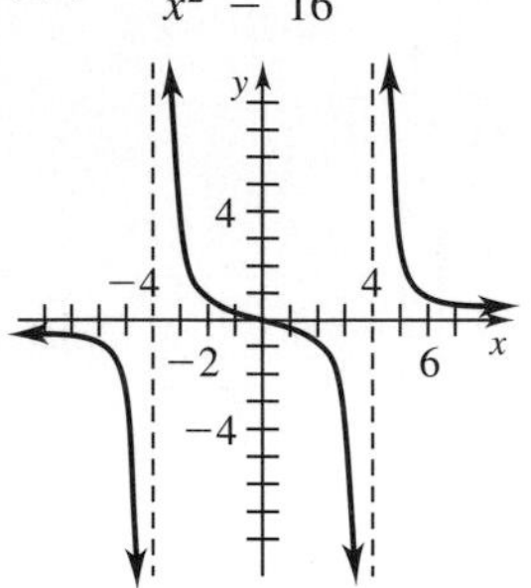

27. $f(x) = \dfrac{x}{x^2 + 6x + 8}$

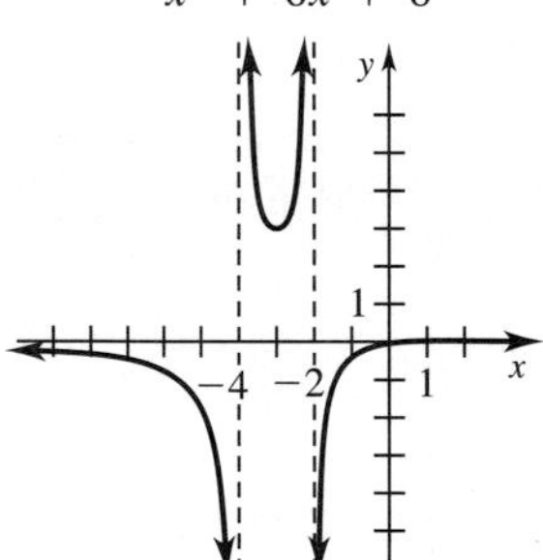

28. $f(x) = \dfrac{x^2 - 5}{x + 2}$

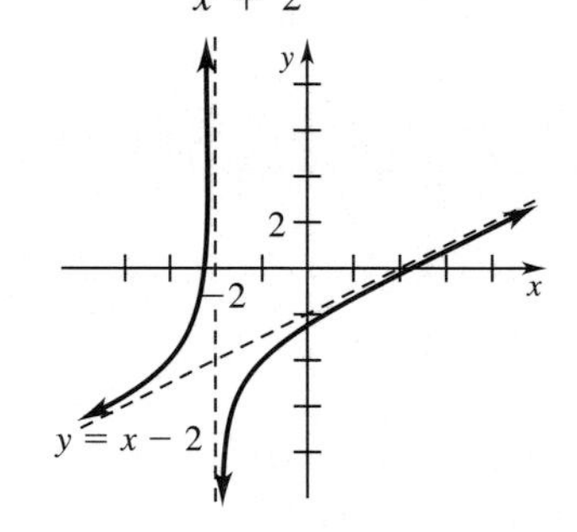

29. $f(x) = \dfrac{x^2 - 2x + 3}{x - 2}$

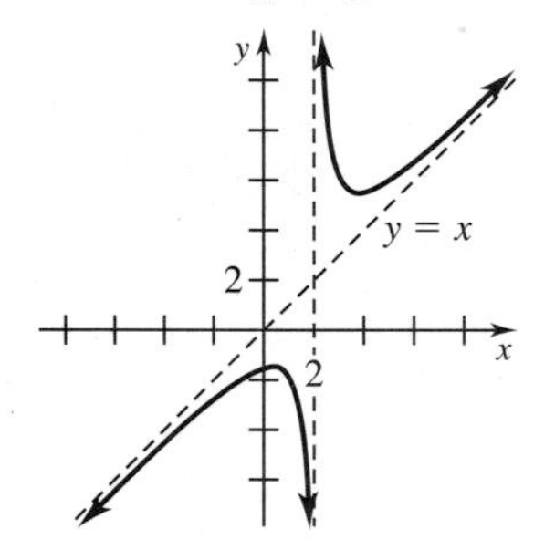

30. $f(x) = \dfrac{x^2 + 1}{x + 3}$

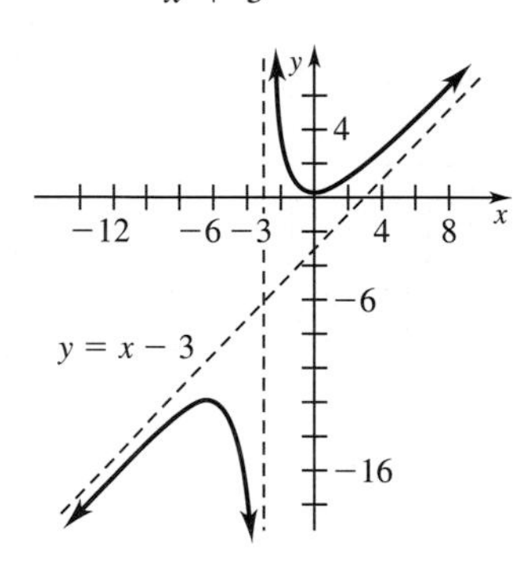

31. $f(x) = \dfrac{x^2 - 6x + 9}{x + 2}$

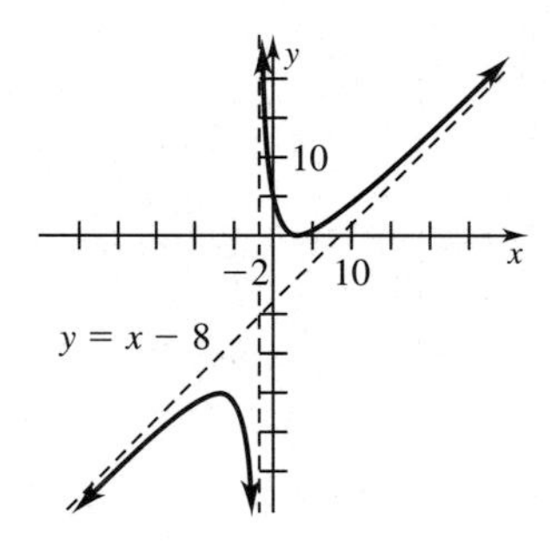

32. $f(x) = \dfrac{3x(2x - 1)}{x^2 + x - 2}$

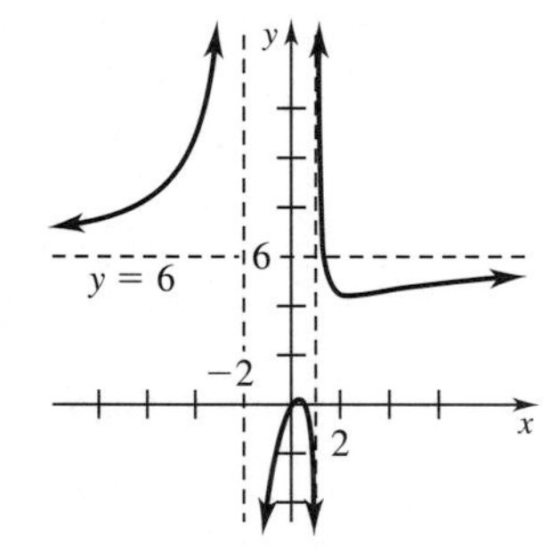

33. $f(x) = \dfrac{(x + 4)(x - 1)}{x^2 + 1}$

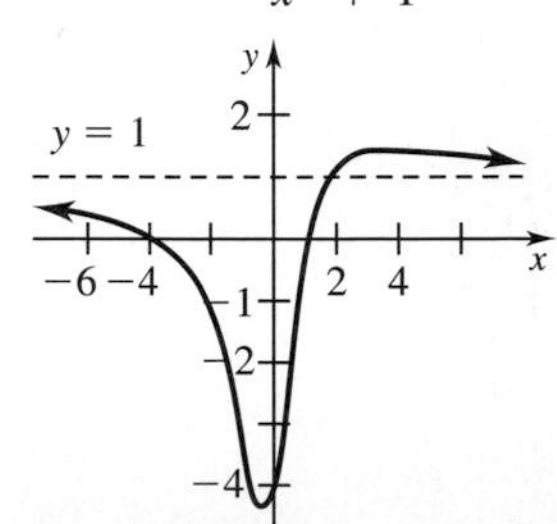

34. $f(x) = \dfrac{1}{x^2 + 1}$

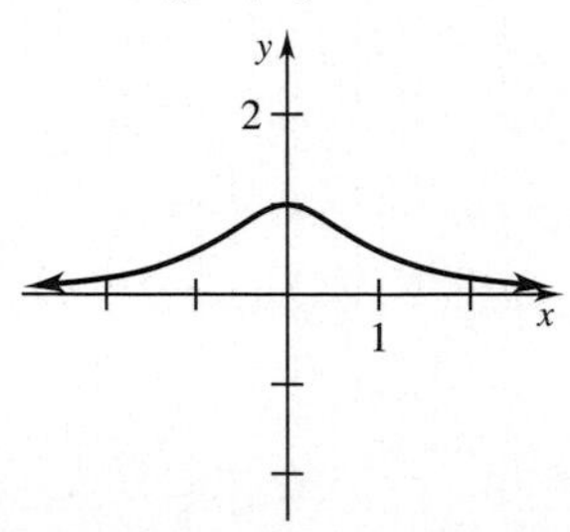

35. $f(x) = \dfrac{-10}{x^2 - 4}$

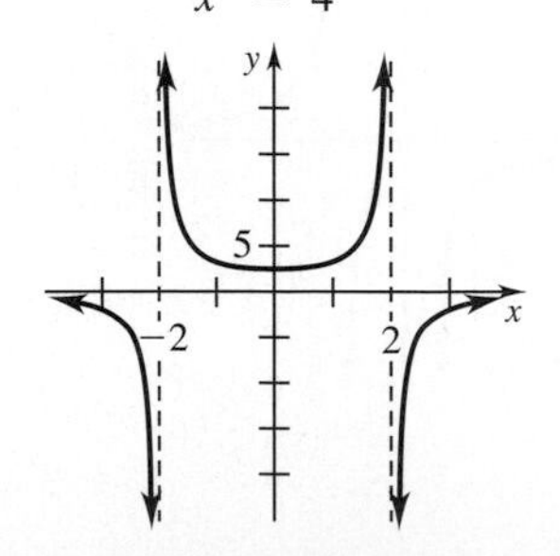

36. $f(x) = \dfrac{x^2 + 3x}{x(x - 2)}$

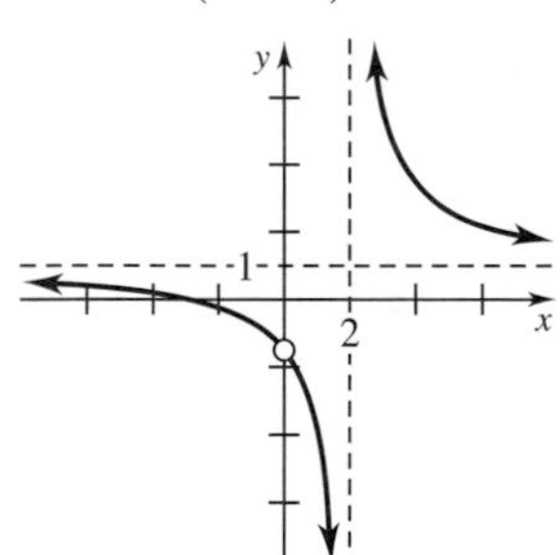

37. $f(x) = \dfrac{(2x - 3)(x - 4)}{x - 4}$

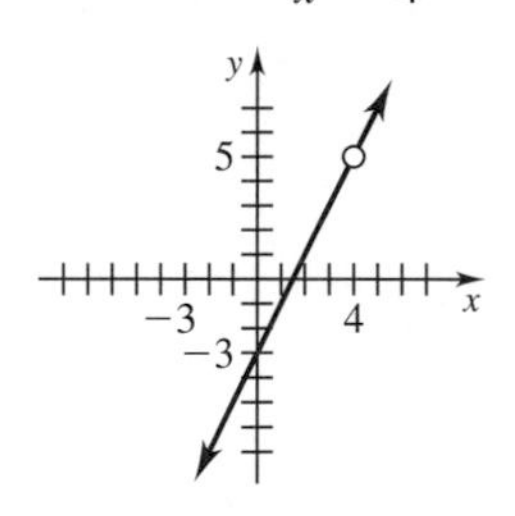

38. $f(x) = \dfrac{(x^2 + 2)(x + 1)}{x + 1}$

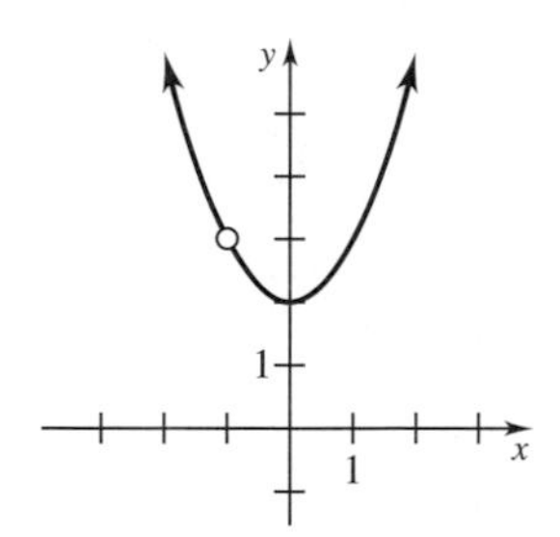

39. $f(x) = \dfrac{(x + 3)(x^2 - 4)}{x + 3}$

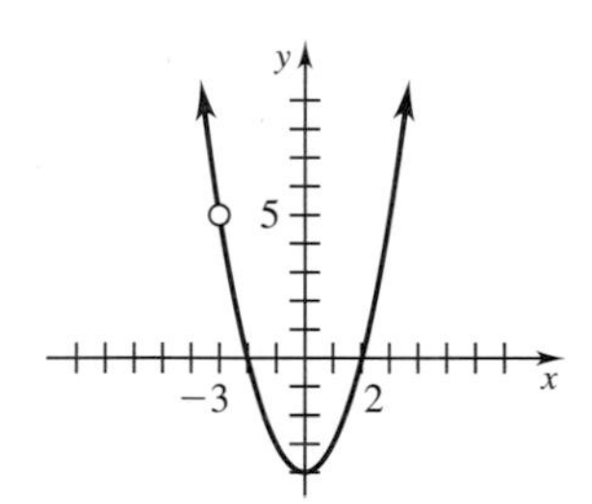

40. $f(x) = \dfrac{x(x^2 + 2x - 15)}{x^3 - x^2 - 5x - 3}$

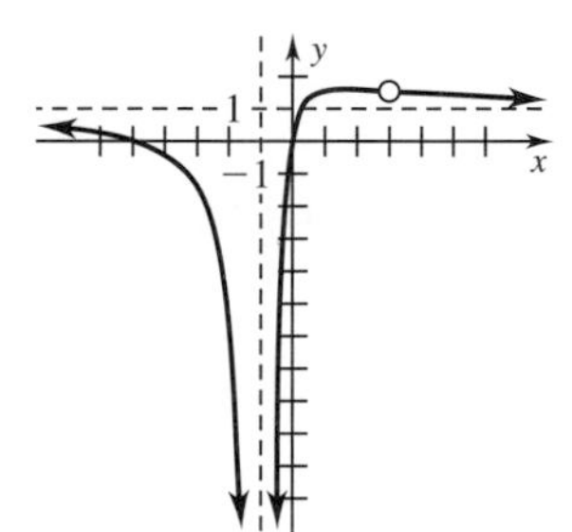

41. $f(x) = \dfrac{2(x - 3)(x - 1)^2}{x^3 - 2x^2 + x}$

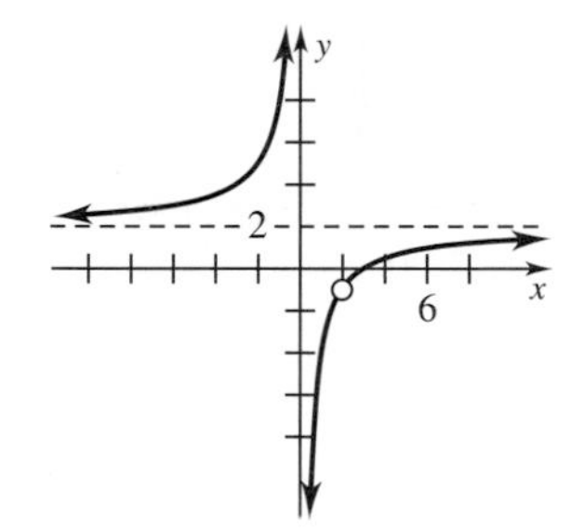

42. The figures below show the four ways that a rational function can approach the vertical line $x = 2$ as an asymptote. Identify the graph of each of the following rational functions.

(a) $f(x) = \dfrac{1}{(x - 2)^2}$ **(b)** $f(x) = \dfrac{1}{x - 2}$ **(c)** $f(x) = \dfrac{-1}{x - 2}$ **(d)** $f(x) = \dfrac{-1}{(x - 2)^2}$

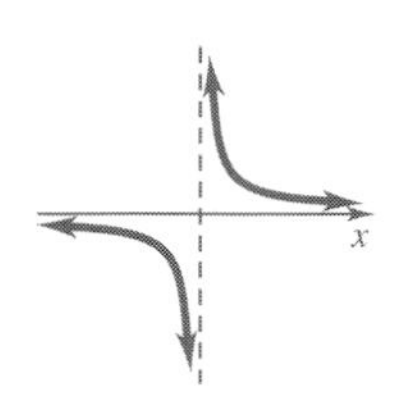

(A)

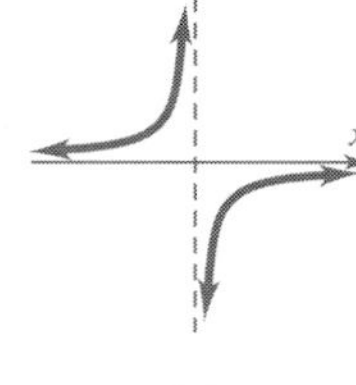

(B)

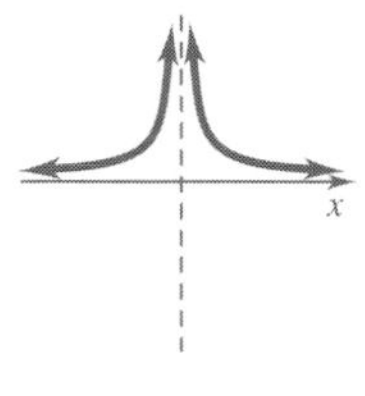

(C)

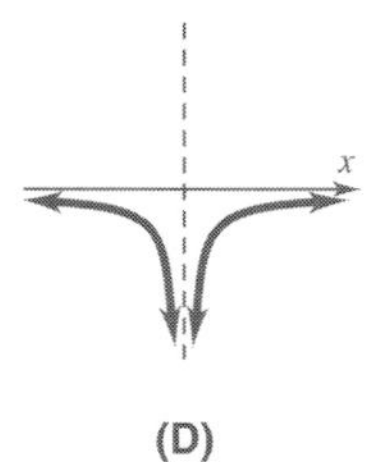

(D)

(a) C (b) A (c) B (d) D

For each of the following graphs, find an equation for a possible corresponding rational function. **Note: In Exercises 43–45, other answers are possible.**

43.

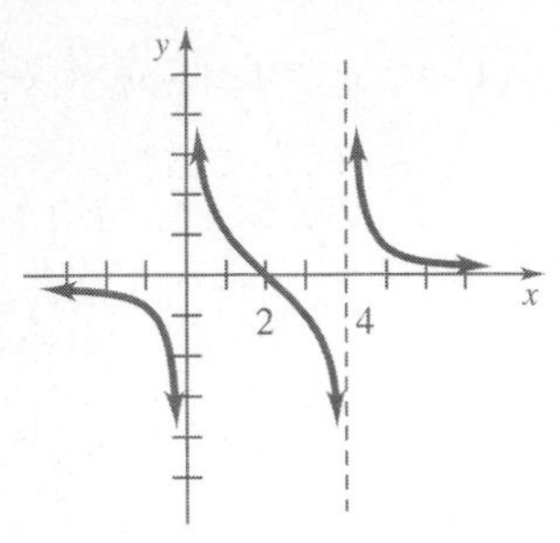

$$f(x) = \frac{x - 2}{x(x - 4)}$$

44.

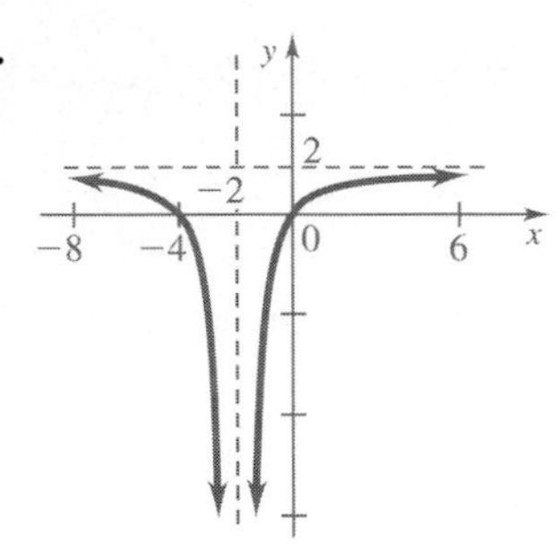

$$f(x) = \frac{2x(x + 4)}{(x + 2)^2}$$

45.

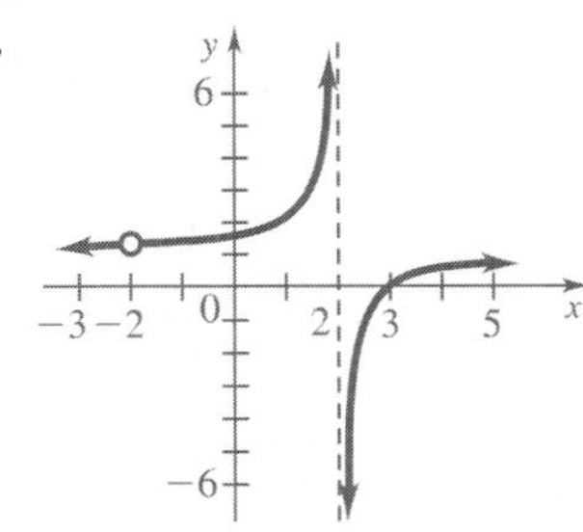

$$f(x) = \frac{(x - 3)(x + 2)}{(x - 2)(x + 2)}$$

46. In situations involving environmental pollution, a cost-benefit model expresses cost as a function of the percentage of pollutant removed from the environment. Suppose a cost-benefit model is expressed as

$$y = \frac{6.7x}{100 - x},$$

where y is the cost in thousands of dollars of removing x percent of a certain pollutant.

(a) Graph the function.

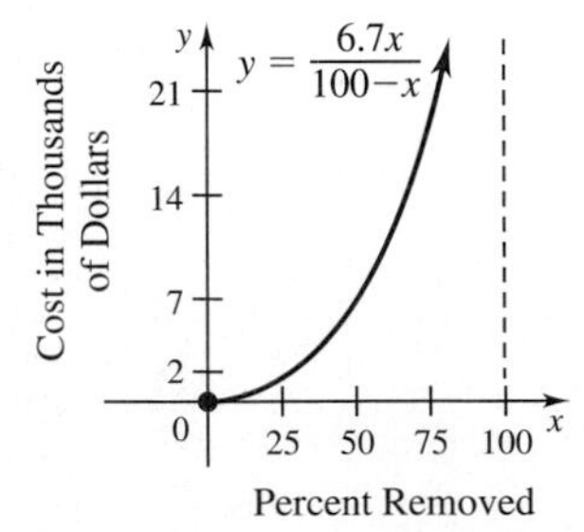

(b) Is it possible, according to this function, to remove all of the pollutant? **no**

47. The quality control department of a parts factory measures the daily percentage of defective parts produced and finds that the function

$$C(x) = \frac{-5x + 500}{x}$$

gives the cost to the factory (in hundred-thousands of dollars) to have x percent defective parts produced.

(a) Graph $C(x)$.

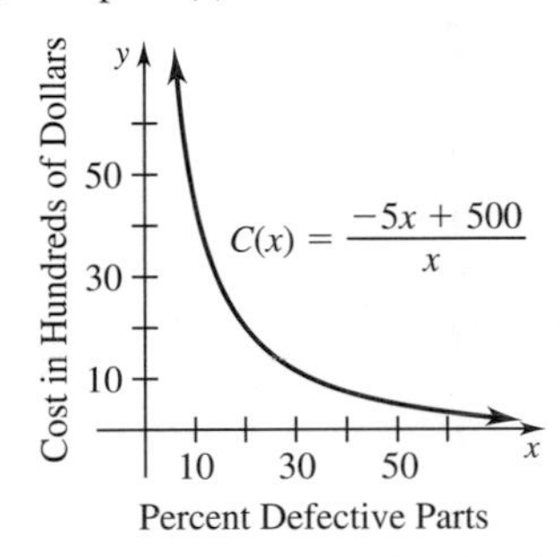

(b) Is it possible, according to this function, to have no defective parts produced? **no**

48. In your own words, describe what happens to the graph of a rational function when the numerator and denominator have a common zero.

49. In your own words, describe how to graph a rational function written in lowest terms.

50. Suppose the graph of a rational function has a horizontal asymptote. How would you determine if the graph crosses the asymptote?

Graphing Utility Problems

An idealized version of the Laffer curve, named for economist Arthur Laffer, is shown here. According to this curve, increasing a tax rate, say from x_1 percent to x_2 percent on the graph, can actually lead to a decrease in government revenue. All economists agree on the endpoints, 0 revenue at tax rates of both 0 percent and 100 percent, but there is much disagreement on the location of the rate x_1 that produces maximum revenue.

51. Suppose an economist studying the Laffer curve produces the rational function

$$y = \frac{10x(100 - x)}{50 + x},$$

where y is government revenue in millions of dollars from a tax rate of x percent. Graph the function and use TRACE to determine the rate that produces maximum revenue. What is the maximum revenue? **36.6%; $268 million**

52. Repeat Exercise 51 for the function

$$y = \frac{10x(100 - x)}{25 + x}.$$ **30.9%; $382 million**

In Exercises 53 and 54, find a rational function with the stated properties. Then graph the function along with its asymptote to confirm that the function satisfies the properties.

53. The denominator is x, the line $y = 2$ is a horizontal asymptote, and the rational function approaches the asymptote from below as x gets larger.

54. The denominator is x^2, the line $y = 2$ is a horizontal asymptote, and the rational function approaches the asymptote from above as x gets larger.

 **Writing** 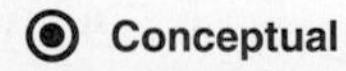 **Conceptual** 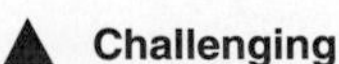 **Challenging**  **Connections**

Chapter 4 Review Exercises (text page 250)

Graph the following quadratic functions. Give the vertex, axis, x-intercepts, and y-intercepts of each graph.

1. $f(x) = 3(x + 4)^2 - 5$ **vertex: $(-4, -5)$; axis: $x = -4$; x-intercepts: $-4 \pm (1/3)\sqrt{15}$; y-Intercept: 43**

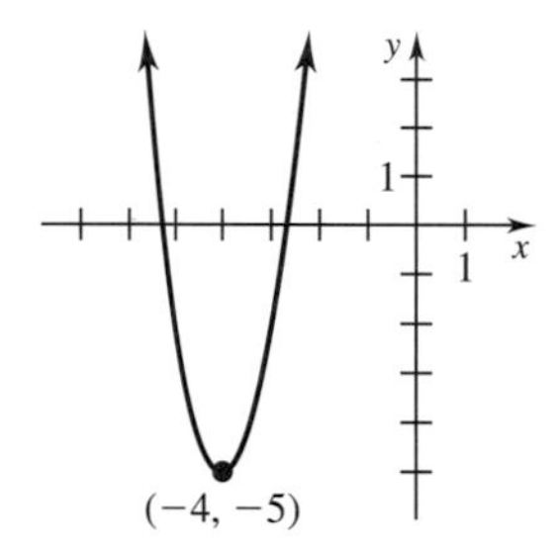

2. $f(x) = -\frac{2}{3}(x - 6)^2 + 7$ **vertex: $(6, 7)$; axis: $x = 6$; x-intercepts: $6 \pm (1/2)\sqrt{42}$; y-intercept: -17**

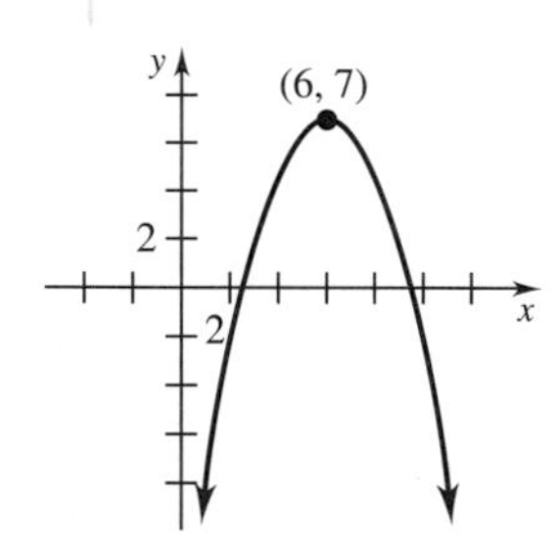

3. $f(x) = -3x^2 - 12x - 1$ **vertex: $(-2, 11)$; axis: $x = -2$; x-intercepts: $-2 \pm \frac{1}{3}\sqrt{33}$; y-intercept: -1**

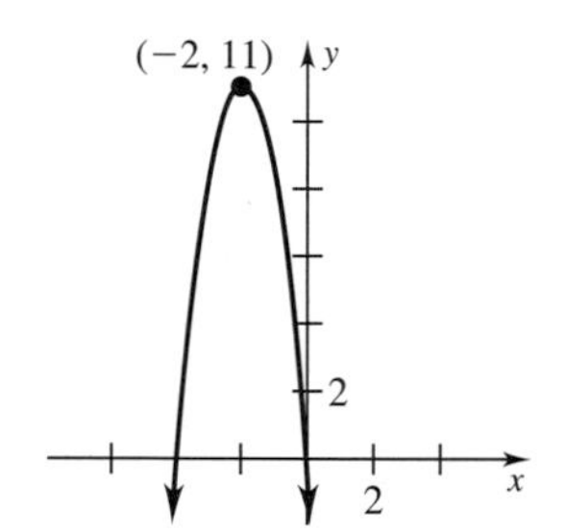

4. $f(x) = 4x^2 - 4x + 3$ **vertex: $(1/2, 2)$; axis: $x = 1/2$; x-intercepts: none; y-intercept: 3**

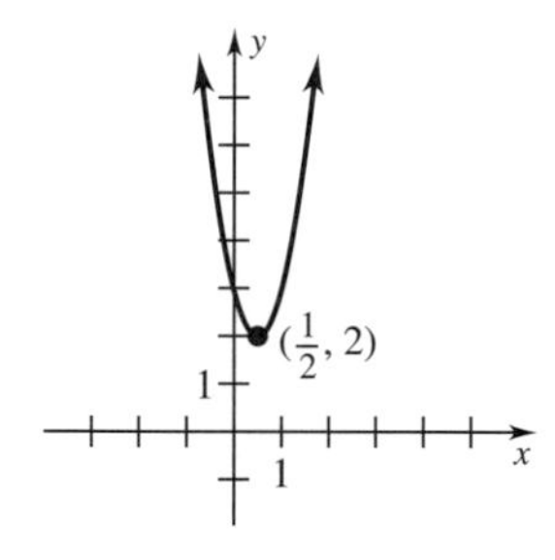

In Exercises 5–8, consider the graph of $f(x) = a(x - h)^2 + k$, with $a > 0$.

5. What is the y-coordinate of the lowest point of the graph? ***k***

6. What is the x-coordinate of the lowest point of the graph? ***h***

7. What is the y-intercept of the graph? $ah^2 + k$

8. Under what conditions, involving the letters a, h, or k, will the graph have one or more x-intercepts? For these conditions, express the x-intercept(s) in terms of a, h, and k.
$\frac{-k}{a} \geq 0$; $\left(h \pm \sqrt{\frac{-k}{a}}, 0\right)$

9. If a is positive, what is the smallest value of $ax^2 + bx + c$? $c - \frac{b^2}{4a}$

10. Use a parabola to find the dimensions of the rectangular region of maximum area that can be enclosed with 180 m of fencing if no fencing is needed along one side of the region. **90 m × 45 m**

Use synthetic division to find $q(x)$ and r for each of the following.

11. $\dfrac{2x^3 + 3x^2 - 4x + 1}{x - 1}$ $\mathbf{2x^2 + 5x + 1; 2}$

12. $\dfrac{4x^3 + 3x^2 + 3x + 5}{x - 5}$ $\mathbf{4x^2 + 23x + 118; 595}$

13. $\dfrac{x^3 + x^2 - 11x - 10}{x - 3}$ $\mathbf{x^2 + 4x + 1; -7}$

14. $\dfrac{3x^3 + 8x^2 + 5x + 10}{x + 2}$ $\mathbf{3x^2 + 2x + 1; 8}$

Use synthetic division to find $f(2)$ for each of the following.

15. $f(x) = -x^3 + 5x^2 - 7x + 1$ $\mathbf{-1}$

16. $f(x) = 2x^3 - 3x^2 + 7x - 12$ $\mathbf{6}$

17. $f(x) = 5x^4 - 12x^2 + 2x - 8$ $\mathbf{28}$

18. $f(x) = x^5 + 4x^2 - 2x - 4$ $\mathbf{40}$

In Exercises 19–22, find a polynomial function of lowest degree having the following zeros.
Note: In Exercises 19–22, other answers are possible.

19. $-1, 4, 7$ $\mathbf{f(x) = x^3 - 10x^2 + 17x + 28}$

20. $8, 2, 3$ $\mathbf{f(x) = x^3 - 13x^2 + 46x - 48}$

21. $\sqrt{3}, -\sqrt{3}, 2, 3$ $\mathbf{f(x) = x^4 - 5x^3 + 3x^2 + 15x - 18}$

22. $-2 + \sqrt{5}, -2 - \sqrt{5}, -2, 1$ $\mathbf{f(x) = x^4 + 5x^3 + x^2 - 9x + 2}$

23. Is -1 a zero of $f(x) = 2x^4 + x^3 - 4x^2 + 3x + 1$? **no**

24. Is -2 a zero of $f(x) = 2x^4 + 7x^3 + 4x^2 - 3x + 4$? **no**

25. Is -2 a zero of $f(x) = 2x^4 + x^3 - 4x^2 + 3x + 1$? **no**

26. Is $x + 1$ a factor of $f(x) = x^3 + 2x^2 + 3x - 1$? **no**

27. Is $x + 1$ a factor of $f(x) = 3x^3 + 2x^2 + 4x + 5$? **yes**

28. Find a polynomial function with real coefficients of degree 4 with 3, 1, and $-1 - 3i$ as zeros, and $f(2) = -36$. $\mathbf{f(x) = 2x^4 - 4x^3 + 10x^2 - 68x + 60}$

29. Find a polynomial function of degree 3 with -2, 1, and 4 as zeros, and $f(2) = 16$. $\mathbf{f(x) = -2x^3 + 6x^2 + 12x - 16}$

30. Find a lowest-degree polynomial function with real coefficients having zeros 2, -2, and $-i$. **Note: In Exercises 30 and 31, other answers are possible.** $\mathbf{f(x) = x^4 - 3x^2 - 4}$

31. Find a lowest-degree polynomial function with real coefficients having zeros 3, -2, and $3i$. $\mathbf{f(x) = x^4 - x^3 + 3x^2 - 9x - 54}$

32. Give an example of a fourth-degree polynomial function having exactly two distinct real zeros, and then sketch its graph. **Any polynomial that can be factored into $a(x - b)^2(x - c)^2$ works.**

33. Give an example of a cubic polynomial function having exactly one real zero, and then sketch its graph. **Any polynomial that can be factored into $a(x - b)^3$ works.**

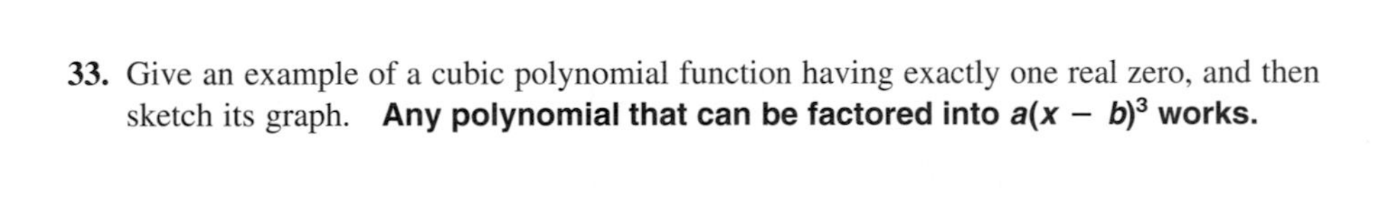

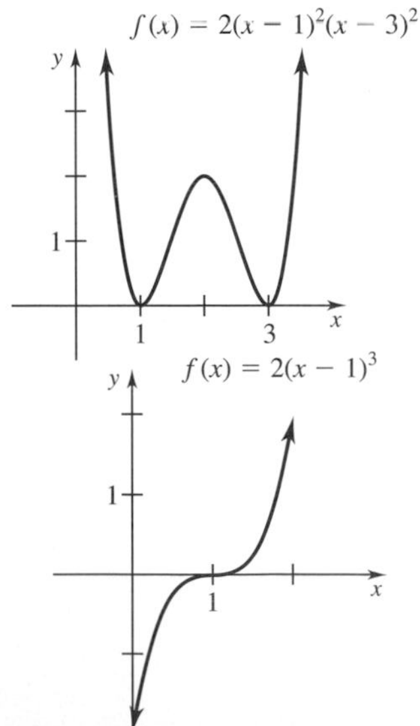

34. Find all zeros of $f(x) = x^4 - 3x^3 - 8x^2 + 22x - 24$, given that $1 - i$ is a zero. **$1 - i, 1 + i, 4, -3$**

35. Find all zeros of $f(x) = 2x^4 - x^3 + 7x^2 - 4x - 4$, given that 1 is a zero. **$1, -1/2, \pm 2i$**

In Exercises 36–38, find all rational zeros.

36. $f(x) = 2x^3 - 9x^2 - 6x + 5$ **$1/2, -1, 5$**

37. $f(x) = 3x^3 - 10x^2 - 27x + 10$ **$-2, 1/3, 5$**

38. $f(x) = 10x^3 - 33x^2 + 18x + 5$ **$-1/5, 1, 5/2$**

39. Show that $\sqrt{5}$ is irrational.

40. Show that $f(x) = x^3 - 9x^2 + 2x - 5$ has no rational zeros.

41. Find all values of s for which the polynomial $f(x) = x^3 + 2x^2 + sx + 3$ has a rational zero. **$-6, 4, -16, -2$**

42. Find a value of s such that $x - 4$ is a factor of $f(x) = x^3 - 2x^2 + sx + 4$. **$-9$**

43. Find a value of s such that when the polynomial $x^3 - 3x^2 + sx - 4$ is divided by $x - 2$, the remainder is 5. **$13/2$**

Show that the polynomials in Exercises 44 and 45 have real zeros satisfying the given conditions.

44. $f(x) = 3x^3 - 8x^2 + x + 2$; zero in $[-1, 0]$ and $[2, 3]$.

45. $f(x) = 6x^4 + 13x^3 - 11x^2 - 3x + 5$; no real zero greater than 1 or less than -3.

46. The function $f(x) = 1/x$ is negative at $x = -1$ and positive at $x = 1$, but has no zero between -1 and 1. Explain why this does not contradict the intermediate value theorem.

Approximate the real zeros of each of the following as a decimal to the nearest tenth. Then graph the function defined by the given expression.

47. $f(x) = x^3 - 8x^2 + 2x + 5$ **$-.7, 1, 7.7$**

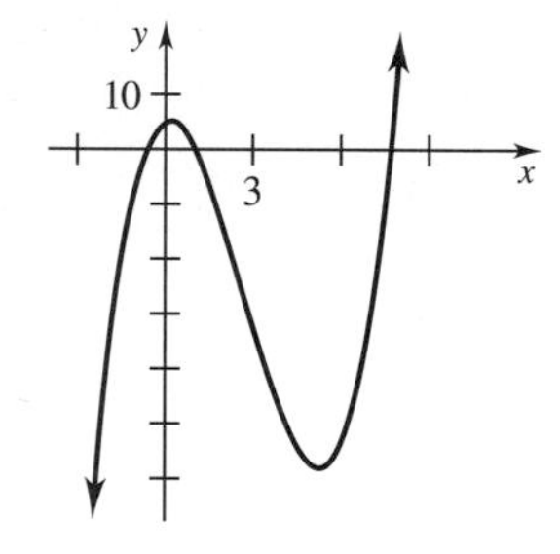

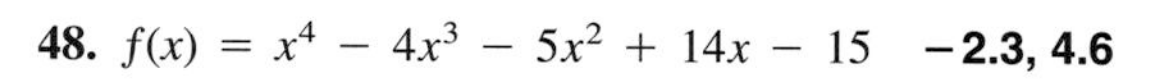

48. $f(x) = x^4 - 4x^3 - 5x^2 + 14x - 15$ **$-2.3, 4.6$**

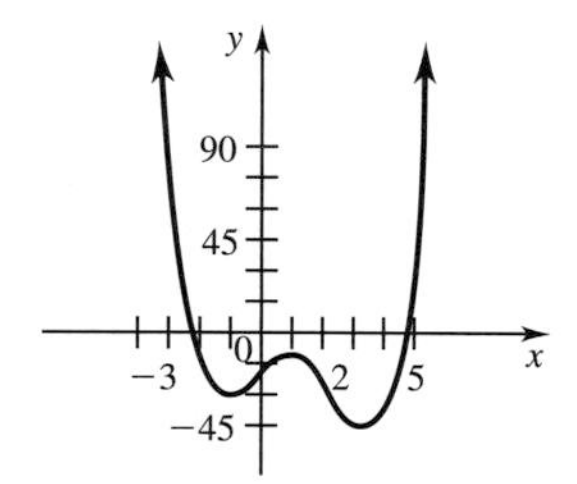

Graph each function defined in Exercises 49 and 50.

49. $f(x) = 3x^4 + 4x^2 + 7x$

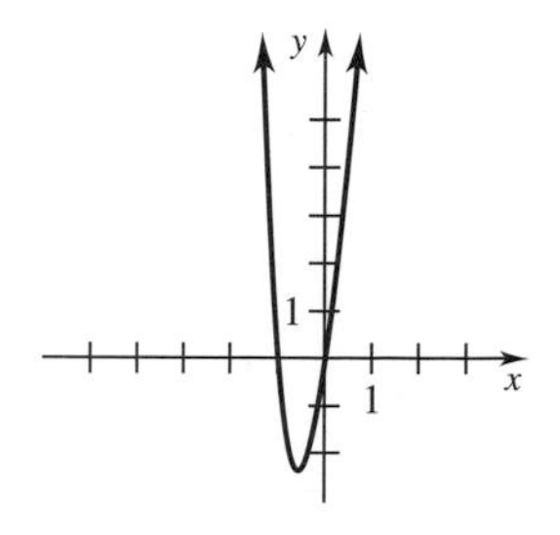

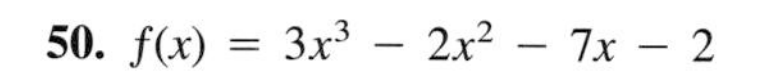

50. $f(x) = 3x^3 - 2x^2 - 7x - 2$

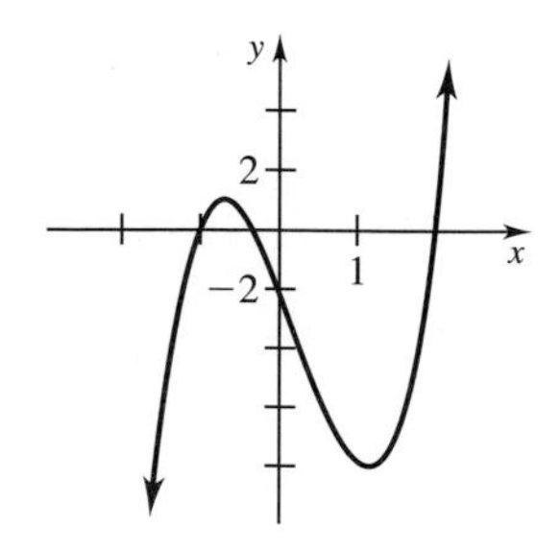

51. (a) Find the number of positive and negative zeros of $f(x) = x^3 + 3x^2 - 4x - 2$.
1 positive; 2 or 0 negative

(b) Show that $f(x)$ has a zero between -4 and -3. Approximate this zero to the nearest tenth. $\boldsymbol{x = -3.9}$

(c) Graph $y = f(x)$.

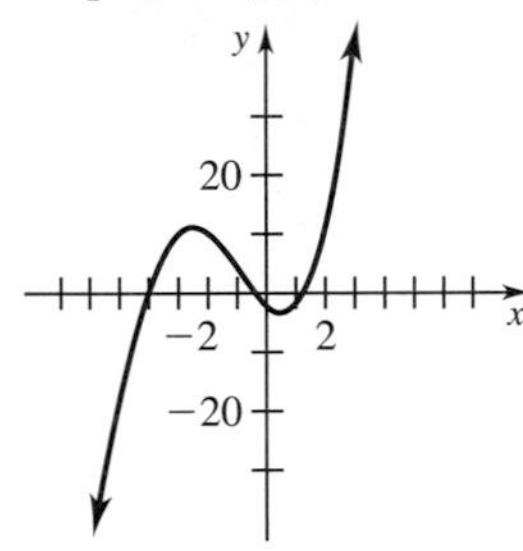

▲ 52. Show without solving that the equation $3x^4 + x^2 + (11/2)x - 1 = 0$ has two real roots and two nonreal complex roots.

Graph each function defined in Exercises 53–58.

53. $f(x) = \dfrac{4}{x - 1}$

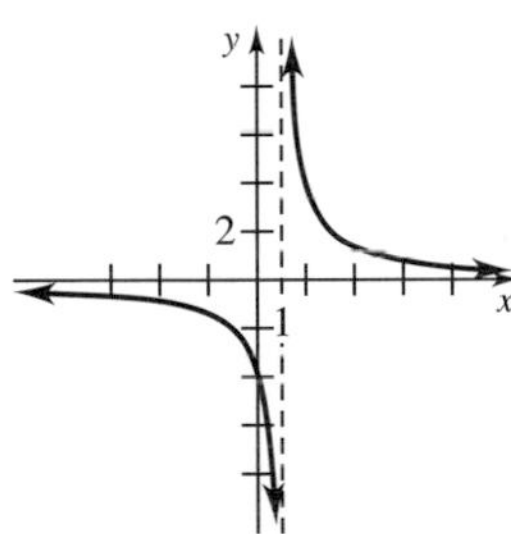

54. $f(x) = \dfrac{4x - 2}{3x + 1}$

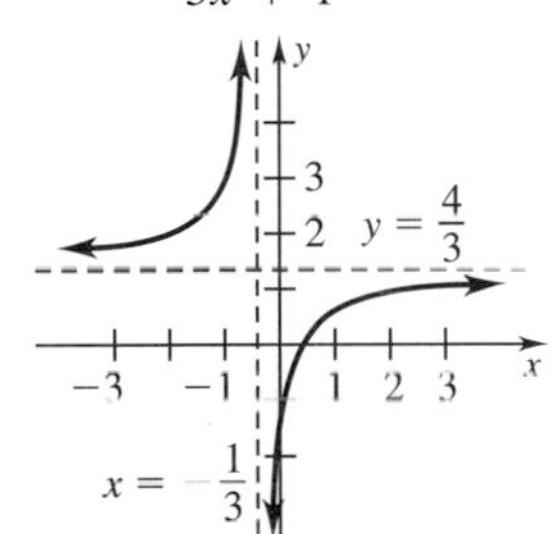

55. $f(x) = \dfrac{6x}{(x + 1)(x + 3)}$

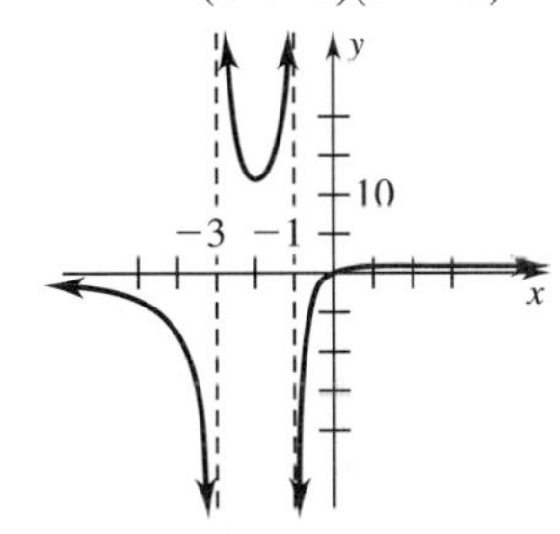

56. $f(x) = \dfrac{2x}{x^2 - 1}$

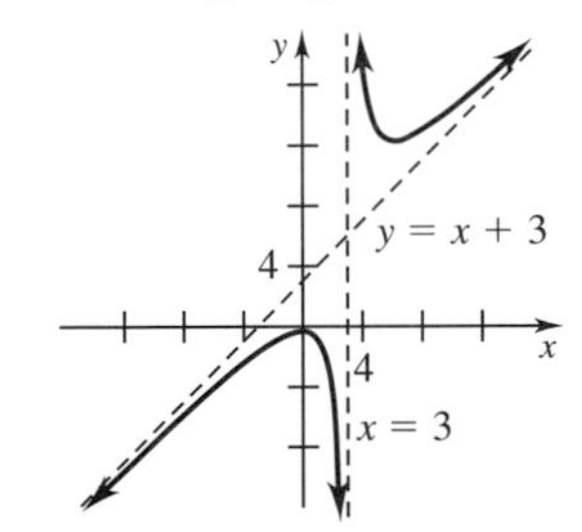

57. $f(x) = \dfrac{x^2 + 1}{x - 3}$

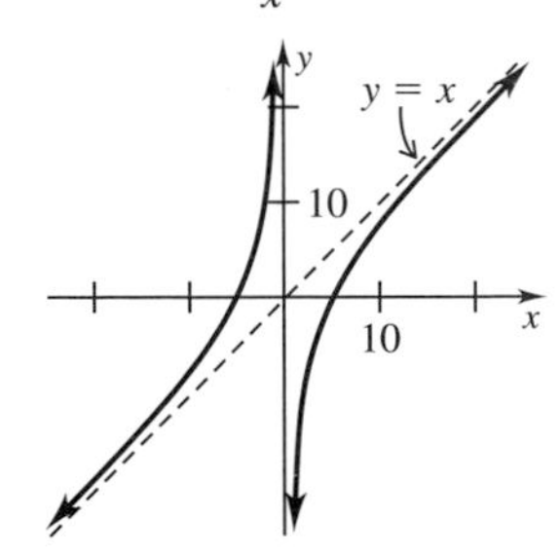

58. $f(x) = \dfrac{x^2 - 1}{x}$

59. **(a)** Sketch the graph of a function that has the line $x = 3$ as a vertical asymptote, the line $y = 1$ as a horizontal asymptote, and x-intercepts 2 and 4.

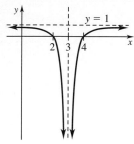

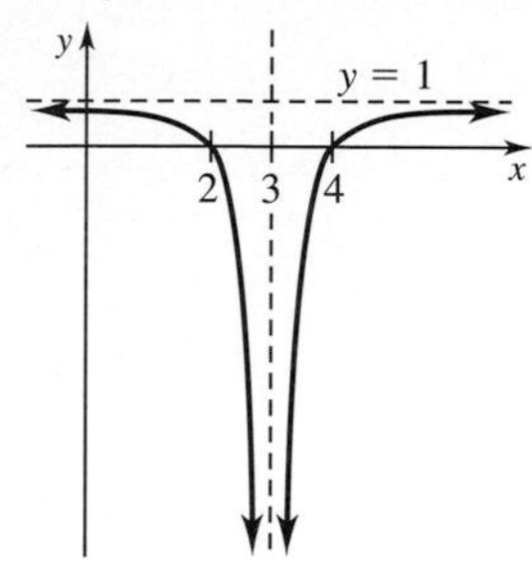

(b) Find an equation for a possible corresponding rational function.

One possibility is $f(x) = \dfrac{(x-2)(x-4)}{(x-3)^2}$.

60. **(a)** Sketch the graph of a function that is never negative and has the lines $x = -1$ and $x = 1$ as vertical asymptotes, the x-axis as a horizontal asymptote, and 0 as an x-intercept.

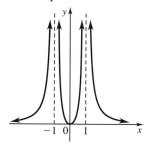

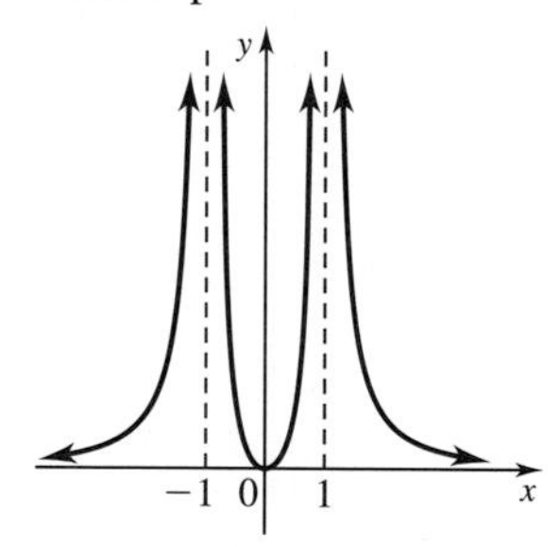

(b) Find an equation for a possible corresponding rational function.

One possibility is $f(x) = \dfrac{x^2}{(x^2-1)^2}$.

61. Suppose the degree of the numerator of a rational function is one more than the degree of the denominator. Explain in your own words how you would find the equation of the oblique asymptote.

5 Exponential and Logarithmic Functions

5.1 Exercises (text page 263)

1. Graph each of the functions defined as follows. Compare the graphs to that of $f(x) = 2^x$.

(a) $f(x) = 2^x - 1/2$ **(b)** $f(x) = 2^x + 3$ **(c)** $f(x) = 2^{x+2}$ **(d)** $f(x) = -2^{x-4}$

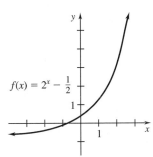

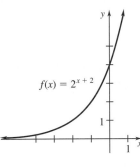

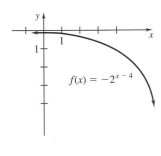

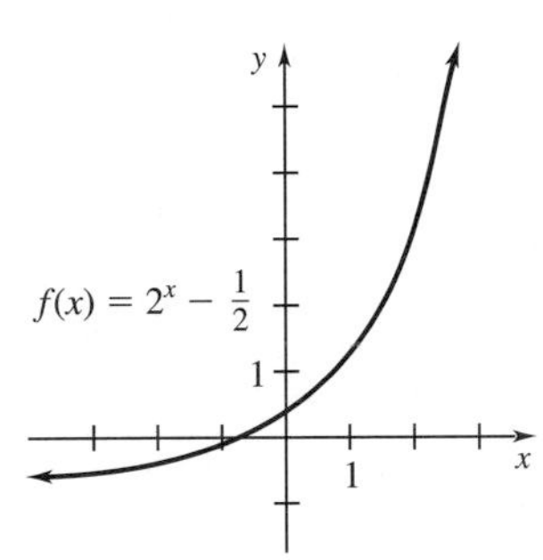

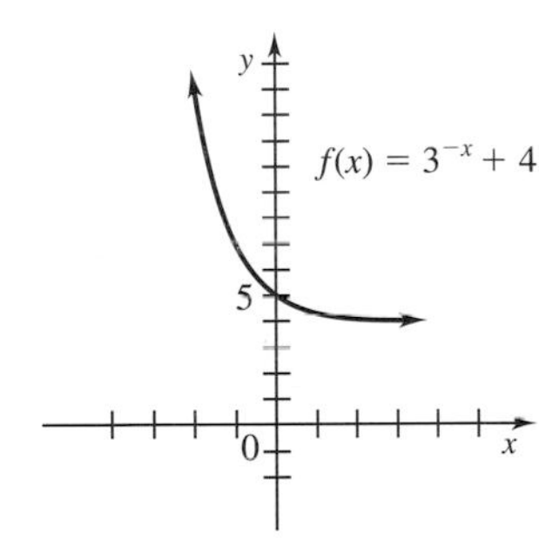

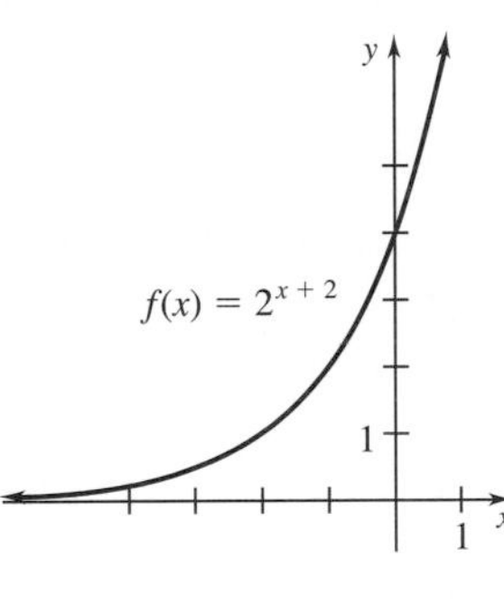

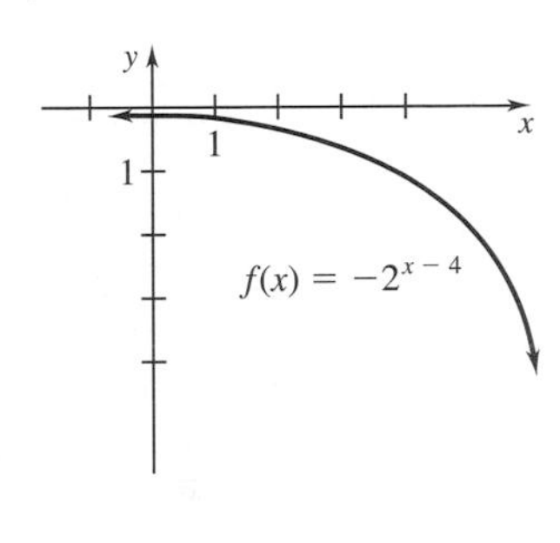

2. Graph each of the functions defined as follows. Compare the graphs to that of $f(x) = 3^{-x}$.

(a) $f(x) = 3^{-x} - 2$ **(b)** $f(x) = 3^{-x} + 4$ **(c)** $f(x) = 3^{-x-2}$ **(d)** $f(x) = 3^{-x+4}$

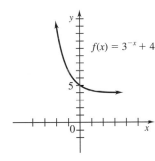

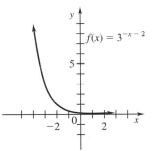

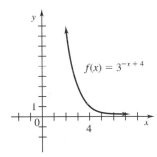

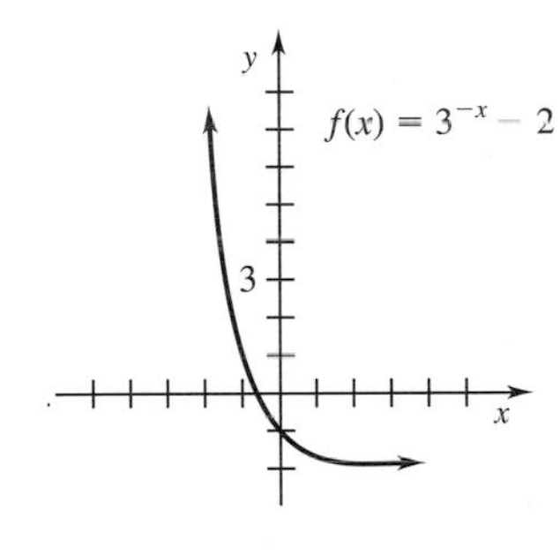

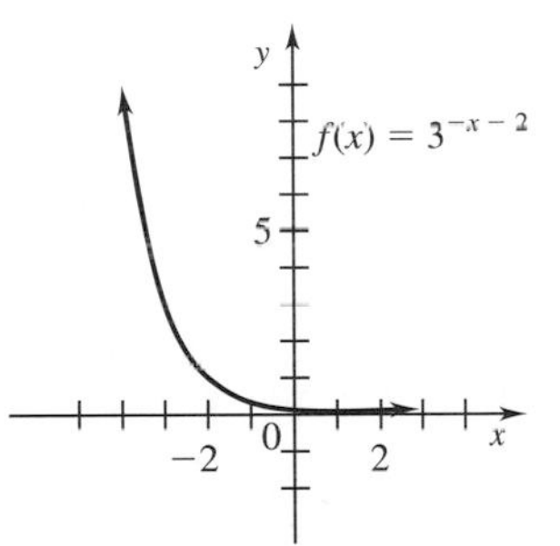

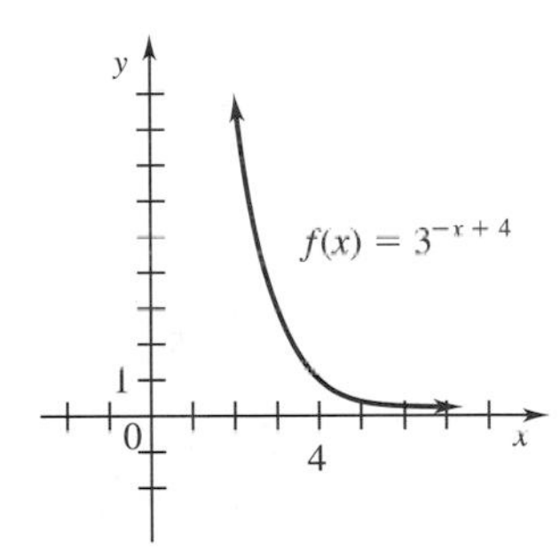

Graph each of the functions defined as follows.

3. $f(x) = 3^x$ **4.** $f(x) = (.5)^x$ **5.** $f(x) = (2/3)^x$ **6.** $f(x) = 4^{-x}$

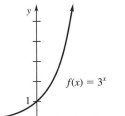

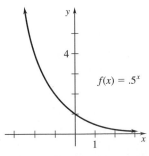

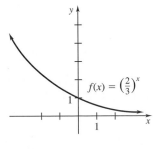

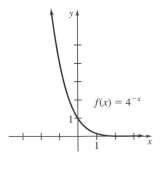

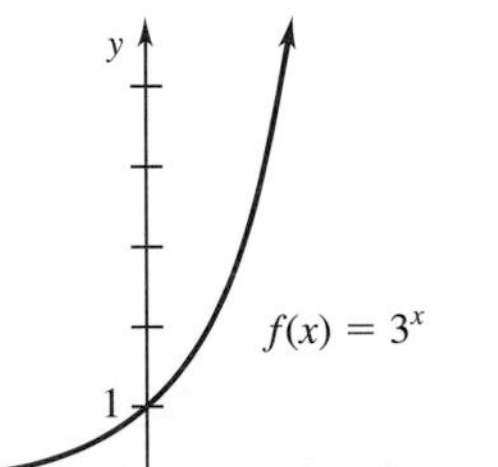

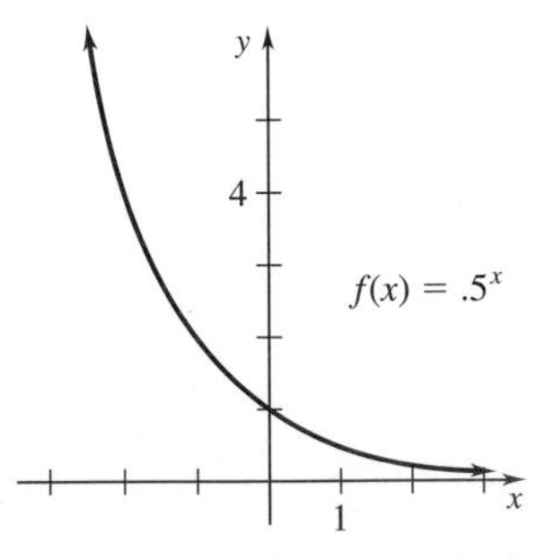

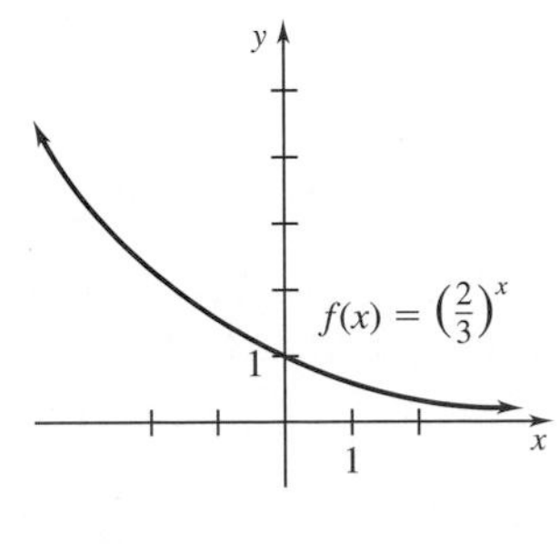

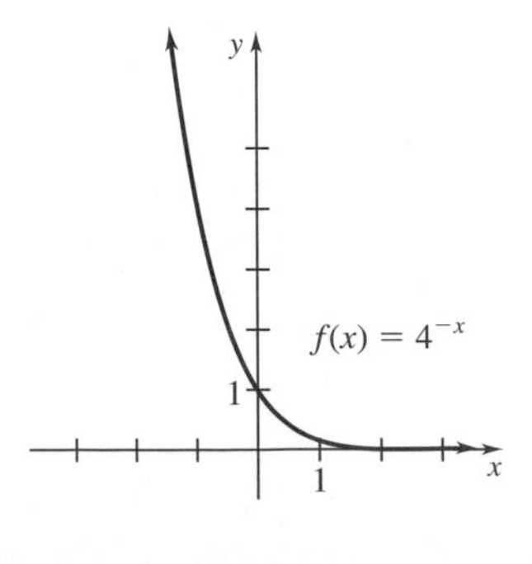

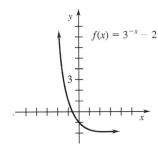

7. $f(x) = e^x$

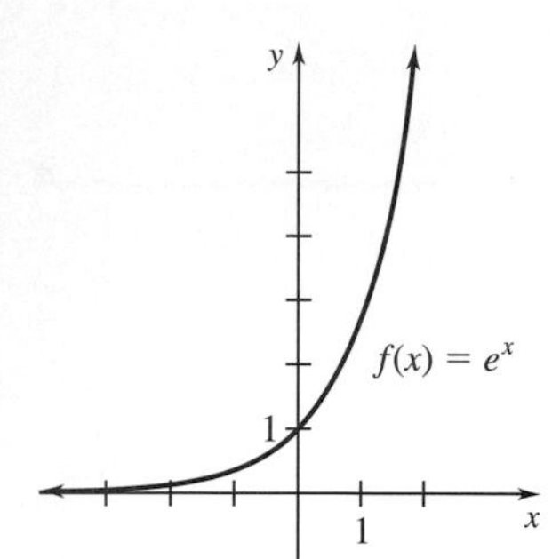

8. $f(x) = e^{1-x}$

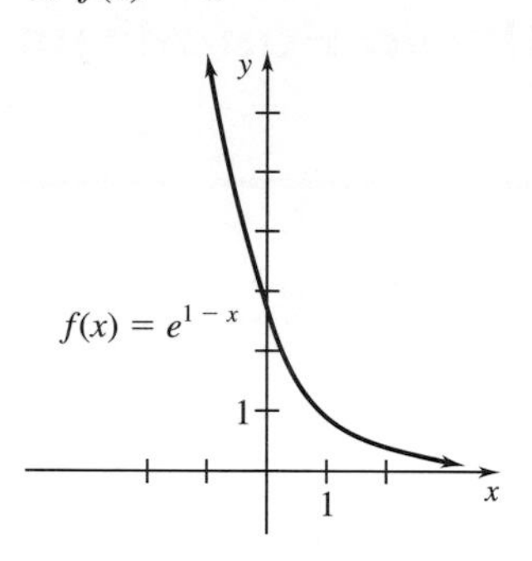

9. $f(x) = 4^x$

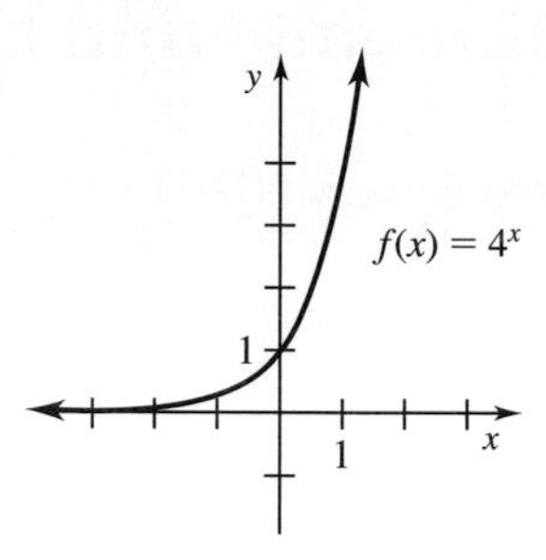

10. $f(x) = 9^x$

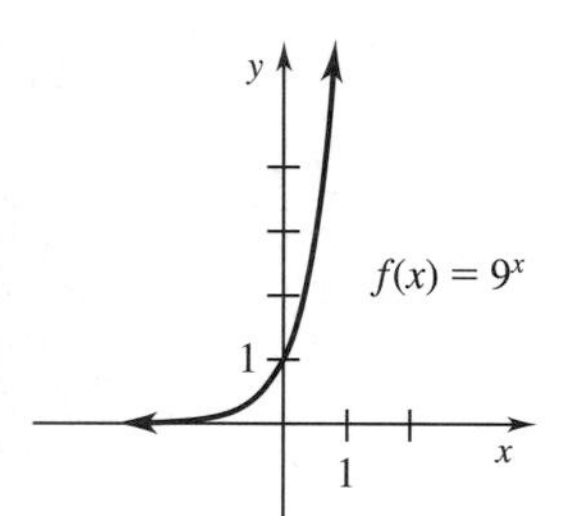

11. $f(x) = e^{|x|}$

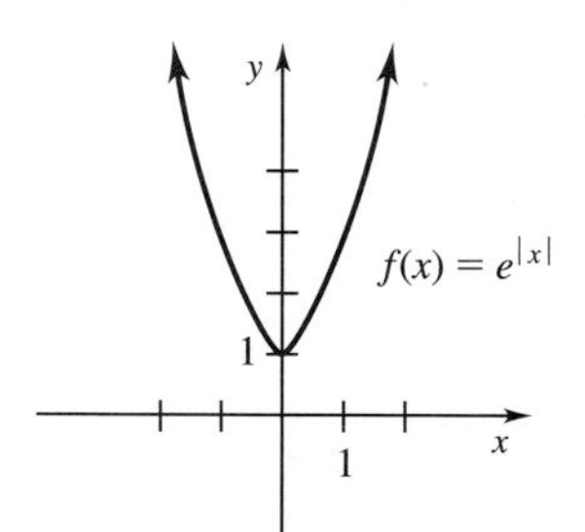

12. $f(x) = 3^{-|x|}$

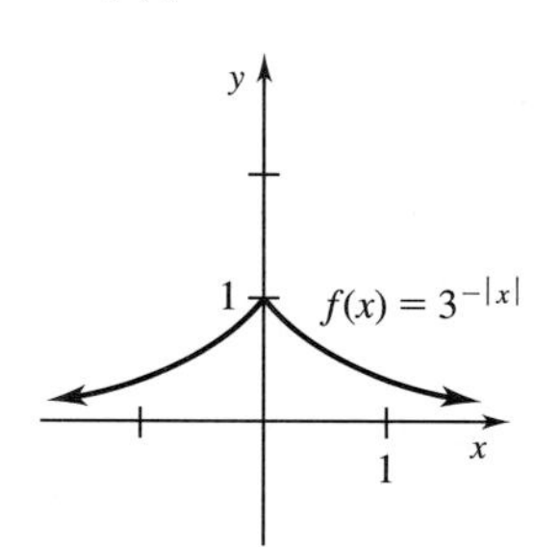

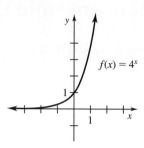

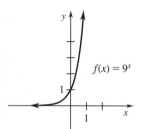

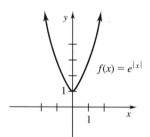

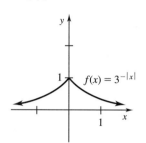

*Solve each equation in Exercises 13–30. (*Hint: *In Exercise 26, note that* $16/81 = (2/3)^4$.)

13. $5^r = 625$ **4**

14. $4^x = 2$ $\mathbf{\frac{1}{2}}$

15. $\left(\frac{1}{2}\right)^k = 8$ **−3**

16. $\left(\frac{3}{5}\right)^x = \frac{125}{27}$ **−3**

17. $3^{2x-5} = \frac{1}{9}$ $\mathbf{\frac{3}{2}}$

18. $5^{2p+1} = 25$ $\mathbf{\frac{1}{2}}$

19. $\frac{1}{27} = b^{-3}$ **3**

20. $\frac{1}{27} = k^{-1/2}$ **729**

21. $4 = r^{2/3}$ **8**

22. $z^{5/2} = 32$ **4**

23. $8^{2x} = 2^{x+3}$ $\mathbf{\frac{3}{5}}$

24. $4^t = 16^{1-t}$ $\mathbf{\frac{2}{3}}$

25. $\left(\frac{1}{2}\right)^{-x} = \left(\frac{1}{4}\right)^{x+1}$ $\mathbf{-\frac{2}{3}}$

26. $\left(\frac{2}{3}\right)^{k-1} = \left(\frac{16}{81}\right)^{k+1}$ $\mathbf{-\frac{5}{3}}$

27. $2^{|x|} = 128$ **±7**

28. $3^{-|x|} = \frac{1}{81}$ **±4**

29. $e^{-5x} = (e^2)^x$ **0**

30. $e^{3(1+x)} = e^{-8x}$ $\mathbf{-\frac{3}{11}}$

31. \$5000 is invested for 4 yr at 8% compound interest. Find the final amount on deposit if the interest is compounded as follows:
(a) annually **(b)** quarterly **(c)** daily (365 days).
(a) \$6,802.44 (b) \$6,863.93 (c) \$6,885.40

32. Find the final amount on deposit if \$5800 is left at interest for 6 yr at 13% and interest is compounded as follows:
(a) annually **(b)** quarterly **(c)** daily (365 days).
(a) \$12,075.32 (b) \$12,496.53 (c) \$12,650.78

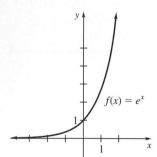

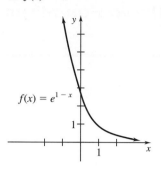

33. Suppose the termite population of a house is given by $P(t)$, where $P(t) = 10{,}000e^{.1t}$, with t representing time in days after some initial day.
Find the following to the nearest thousand.
(a) $P(0)$ **(b)** $P(2)$ **(c)** $P(8)$ **(d)** $P(10)$
(e) Graph $y = P(t)$.
(a) 10,000 (b) 12,000 (c) 22,000 (d) 27,000 (e)

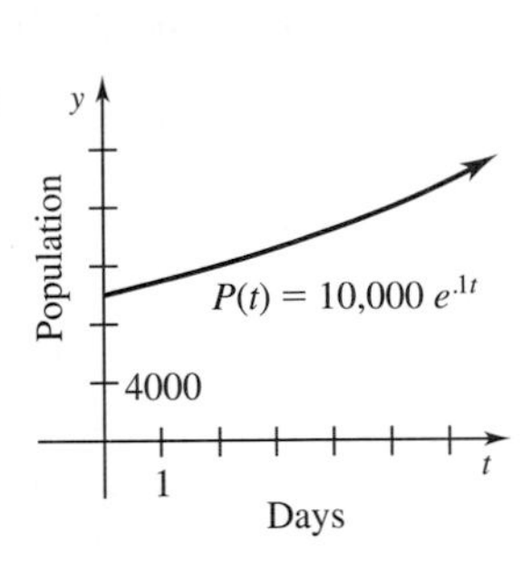

34. Suppose the quantity in grams of a radioactive substance present at time t is $Q(t) = 500e^{-.05t}$. Let t be time measured in days from some initial day.
Find the quantity present (to the nearest 10 g) at each of the following times.
(a) $t = 0$ **(b)** $t = 4$ **(c)** $t = 8$ **(d)** $t = 20$
(e) Graph $y = Q(t)$.
(a) 500 g (b) 410 g (c) 340 g (d) 180 g (e)

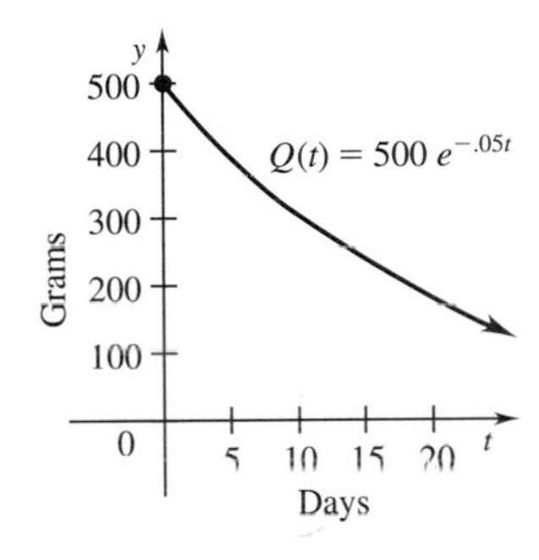

◉ *Give an equation of the form $f(x) = a^x$ to define the exponential function whose graph contains the given point.*

35. $(3, 8)$ $\mathbf{f(x) = 2^x}$

36. $(-3, 64)$ $\mathbf{f(x) = \left(\frac{1}{4}\right)^x}$

37. $(-.5, .4)$ $\mathbf{f(x) = \left(\frac{25}{4}\right)^x}$

38. $(2/3, 4)$ $\mathbf{f(x) = 8^x}$

Use properties of exponents to write each of the following in the form $f(t) = ka^t$, where k is a constant. (Hint: *Recall* $4^{x+y} = 4^x \cdot 4^y$.)

39. $f(t) = 3^{2t+3}$ $\mathbf{f(t) = 27 \cdot 9^t}$

40. $f(t) = 2^{7-t}$ $\mathbf{f(t) = 128 \cdot (1/2)^t}$

41. $f(t) = (1/3)^{1-2t}$ $\mathbf{f(t) = (1/3) \cdot 9^t}$

42. $f(t) = 7^{2t-1}$ $\mathbf{f(t) = (1/7) \cdot (49)^t}$

43. Consider a function of the form $y = Pa^x$. Show that when the value of x is increased by 1, the value of y is multiplied by a.

◉ *Suppose f is an exponential function of the form $f(x) = a^x$ and $f(3) = 4$. Determine the function values in Exercises 44–46.*

44. $f(-3)$ **1/4**

45. $f(0)$ **1**

46. $f(6)$ **16**

47. What two points on the graph of $f(x) = a^x$ can be found without any computation?
(0, 1), (1, *a*)

The pressure of the atmosphere $p(h)$ in pounds per square inch is given by $p(h) = p_0 e^{-kh}$, where h is the height above sea level and p_0 and k are constants. The pressure at sea level is 15 lb/sq in and the pressure is 9 lb/sq in at a height of 12,000 ft. Use this information in Exercises 48 and 49. (The exercises can be answered without solving for k.)

48. Find the pressure at an altitude of 3000 ft. **13.202 lb/sq in**

49. What would be the pressure encountered by a satellite at an altitude of 150,000 ft?
.025 lb/sq in

50. When defining an exponential function, explain why we require $a > 0$.

51. A function of the form x^r, where r is a constant, is called a *power function*. Discuss the difference between an exponential function and a power function.

52. Explain in your own words what the number e is.

53. The techniques of this section can be used to solve $a^x = b$ only in special situations. Describe such a situation.

54. Explain why the graph of $y = a^x$ (where $a > 0$) has a y-intercept but no x-intercept.

Let $f(x) = a^x$ define an exponential function of base a.

55. Is f odd, even, or neither? **neither**

56. Prove that $f(m + n) = f(m) \cdot f(n)$ for any real numbers m and n.

Graphing Utility Problems

Any points where the graphs of functions f and g intersect give solutions of the form $f(x) = g(x)$. Use this idea to estimate the solutions of the following equations.

57. $x = 2^x$ **no solution**

58. $5e^{3x} = 75$ **$x \approx .90$**

59. $6^{-x} = 1 - x$ **$x = 0$; $x \approx .73$**

60. $3x + 2 = 4^x$ **$x = -.5$, $x \approx 1.3$**

61. Graph the function $f(x) = (1 + (1/x))^x$ and the horizontal line $y = 2.71828$ with $1 \le x \le 10{,}000$ and $0 \le y \le 3$. Observe that $f(x)$ gets closer and closer to the line as x gets large.

62. The function e^x grows faster than any power function. Graph the function x^2/e^x for $0 \le x \le 10$ and the function x^{10}/e^x for $0 \le x \le 25$ and observe that the values approach 0 as x gets large. (*Note:* For any n, x^n/e^x approaches 0 as x gets large.)

5.2 Exercises (text page 274)

For each of the following statements, write an equivalent statement in logarithmic form.

1. $10^3 = 1000$ **$\log_{10} 1000 = 3$**

2. $7^2 = 49$ **$\log_7 49 = 2$**

3. $\left(\frac{1}{2}\right)^{-4} = 16$ **$\log_{1/2} 16 = -4$**

4. $e^0 = 1$ **$\ln 1 = 0$**

For each of the following statements, write an equivalent statement in exponential form.

5. $\log_6 36 = 2$ **$6^2 = 36$**

6. $\log .0001 = -4$ **$10^{-4} = .0001$**

7. $\ln 1 = 0$ **$e^0 = 1$**

8. $\log_8 \left(\frac{1}{64}\right) = -2$ **$8^{-2} = \frac{1}{64}$**

Find the value of each of the following. Assume all variables represent positive real numbers.

9. $\log_5 25$ **2**

10. $\log_3 243$ **5**

11. $\log_8 8$ **1**

12. $\log_5 125^2$ **6**

13. $\ln e^5$ **5**

14. $\ln \frac{1}{e}$ **-1**

15. $\ln \sqrt[3]{e}$ **$\frac{1}{3}$**

16. $\log \sqrt{10}$ **$\frac{1}{2}$**

17. $e^{\ln 5}$ **5**

18. $10^{\log 2}$ **2**

19. $5^{\log_5 (x+1)}$ **$x + 1$**

20. $8^{\log_8 2x}$ **$2x$**

21. $e^{\ln 3 + \ln 5}$ **15**

22. $e^{3(\ln 2)}$ **8**

23. $(e^5)^{\ln 2}$ **32**

24. $\ln (6 + 2) - \ln 2$ **$\ln 4$**

Solve each of the following equations.

25. $\log_x 256 = 8$ **2**

26. $\log_x \frac{1}{16} = -2$ **4**

27. $\log_y 12 = \frac{1}{2}$ **144**

28. $\log_r 4 = \frac{1}{3}$ **64**

Write each of the following as a sum or difference of logarithms (or constants times logarithms). Simplify the result if possible. Assume all variables represent positive real numbers.

29. $\log_2 \frac{6x}{y}$ **$\log_2 6 + \log_2 x - \log_2 y$**

30. $\log \frac{p^2}{3q}$ **$2\log p - \log 3q$**

31. $\log_7 \frac{\sqrt{5}}{9}$ **$\frac{1}{2}\log_7 5 - \log_7 9$**

32. $\log_2 \frac{2\sqrt{3}}{5}$ **$1 + \frac{1}{2}\log_2 3 - \log_2 5$**

33. $\log_5 (x + y)$ **not possible**

34. $\log_6 (7m + 3q)$ **not possible**

35. $\log_a \frac{p^2q^3}{r^2}$ **$2 \log_a p + 3 \log_a q - 2 \log_a r$**

36. $\log_z \frac{x^5y^3}{3}$ **$5\log_z x + 3\log_z y - \log_z 3$**

Write each of the expressions in Exercises 37–42 as a single logarithm with a coefficient of 1. Assume that all variables represent positive real numbers.

37. $\log_a x + \log_a y - \log_a m$ **$\log_a \frac{xy}{m}$**

38. $\log_h (4m + 1) + \log_h 2m - \log_h 3$ **$\log_h \frac{2m(4m + 1)}{3}$**

39. $2 \log_m a - 3 \log_m b^2$ **$\log_m \frac{a^2}{b^6}$**

40. $\frac{1}{3}(\log_y p^3 + \log_y q^2) - \frac{1}{2}(\log_y q^{4/3})$ **$\log_y p$**

41. $-\frac{3}{4}\log_x a^6b^8 + \frac{2}{3}\log_x a^9b^3$ **$\log_x \frac{a^{3/2}}{b^4}$**

42. $\log_n (pqr) + 2 \log_n \left(\frac{p}{q}\right)$ **$\log_n \frac{p^3r}{q}$**

43. Show that $\log \left(\frac{1}{2}\sqrt{2}\right)$ simplifies to $-\frac{1}{2}\log 2$.

44. What two points on the graph of $y = \log_a x$, $a > 1$, can be found without computation?
$(a, 1)$ and $(1, 0)$

45. Explain why the natural logarithm function is defined only for $x > 0$.

46. The function $f(x) = \ln|x|$ plays a prominent role in calculus. Give its domain, range, and symmetries. **domain: $(-\infty, 0) \cup (0, \infty)$; range: $(-\infty, \infty)$; symmetric with respect to the *y*-axis**

Use the properties of logarithms to relate the graphs of each of the following functions to the graph of $f(x) = \ln x$.

47. $f(x) = \ln (ex)$ **$f(x) = 1 + \ln x$**

48. $f(x) = \ln (x/e)$ **$f(x) = \ln x - 1$**

For each of the following functions, identify the corresponding graph below.

49. $f(x) = \log_2 x$ **e**

50. $f(x) = \log_2 2x$ **d**

51. $f(x) = \log_2 \frac{1}{x}$ **b**

52. $f(x) = \log_2 \frac{x}{2}$ **c**

53. $f(x) = \log_2 (x - 1)$ ***f***

54. $f(x) = \log_2 (-x)$ **a**

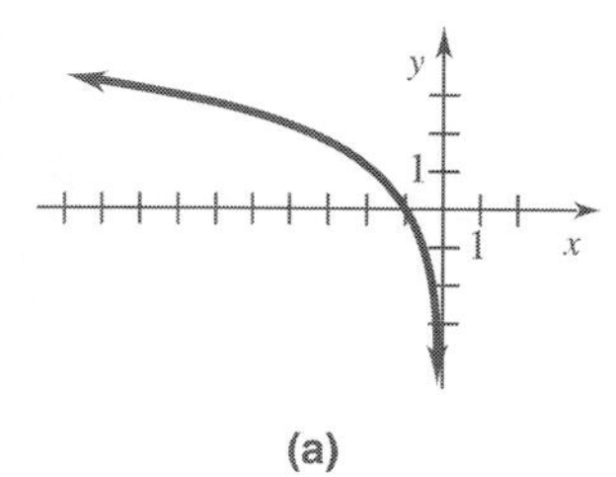

(a)

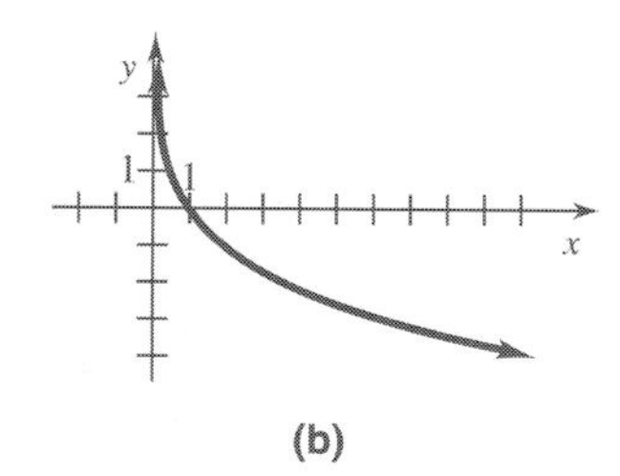

(b)

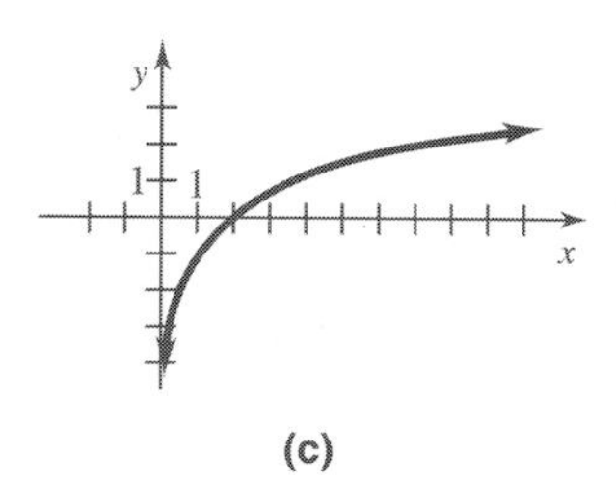

(c)

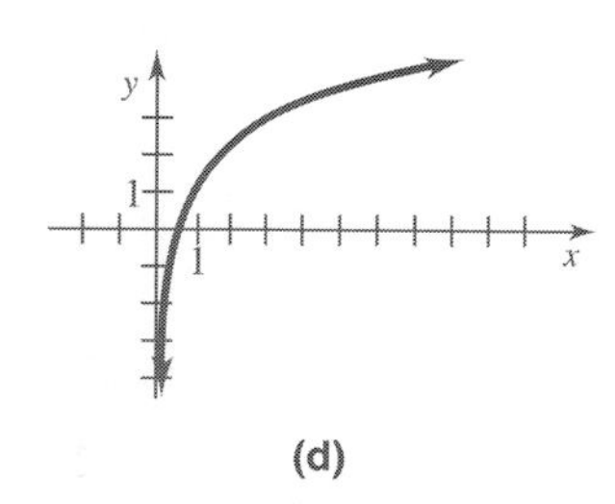

(d)

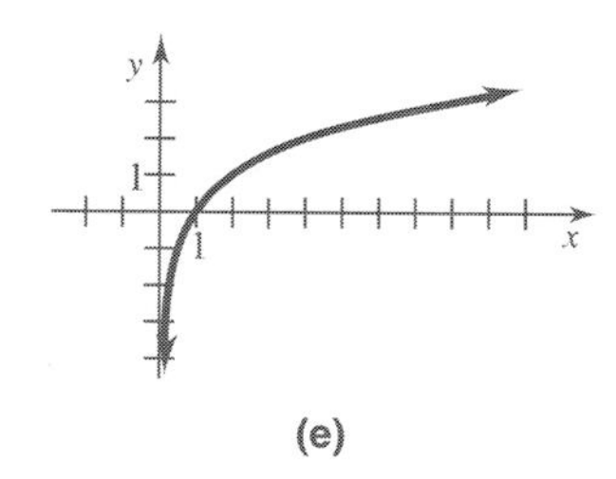

(e)

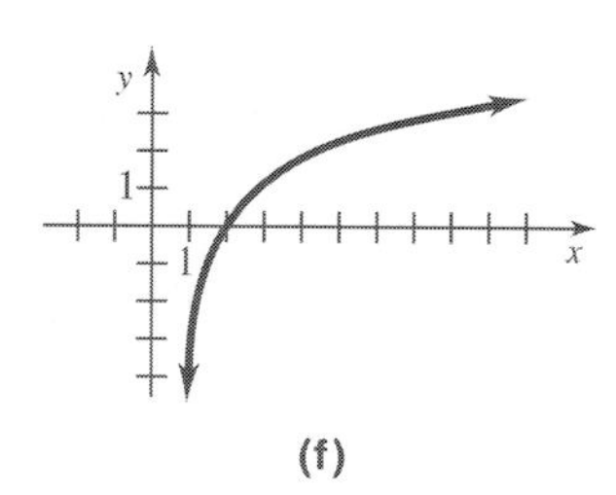

(f)

Graph each of the functions defined as follows. Give any x- and y-intercepts for each graph.

55. $f(x) = \log_5 x$ ***x*-intercept: 1**

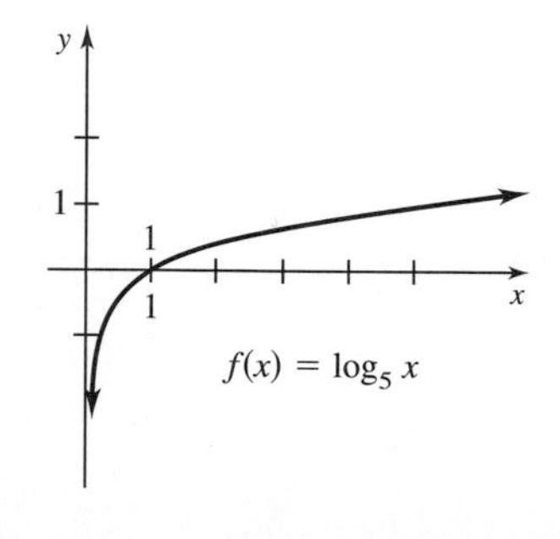

56. $f(x) = \log_3 (2 - x)$
***x*-intercept: 1; *y*-intercept ≈ .63**

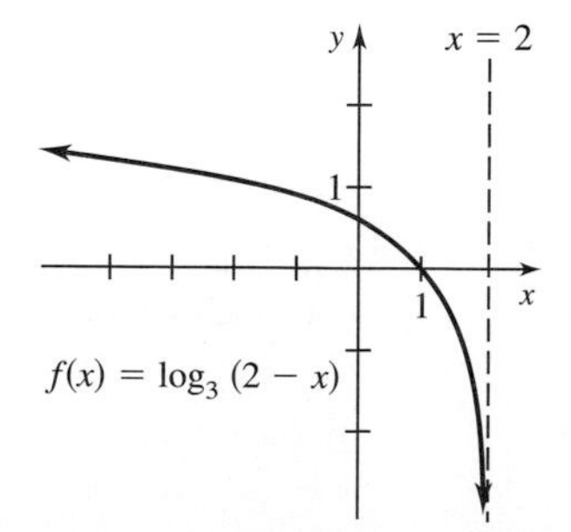

57. $f(x) = \log_3 (x + 3) - 2$
***x*-intercept: 6; *y*-intercept: −1**

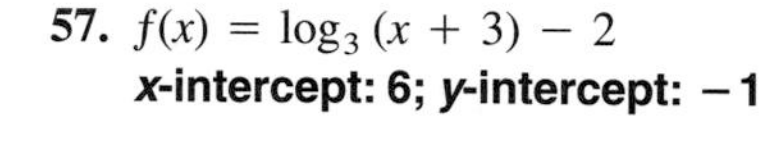

y
x = −3
1
−5
10
25
x
$f(x) = \log_3 (x + 3) - 2$

58. $f(x) = \log_5 (x - 1) + 2$ **x-intercept: 1.04**

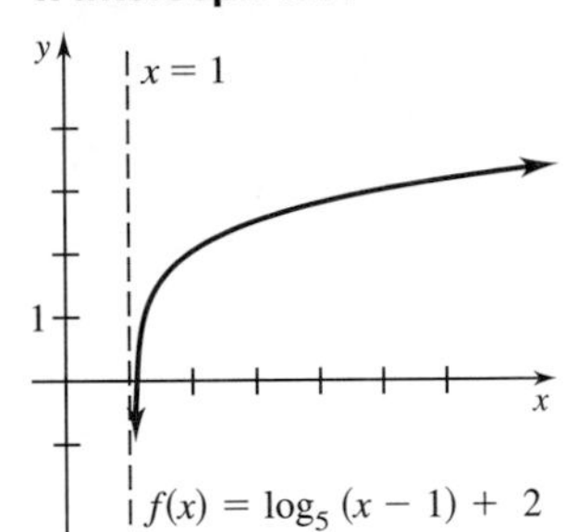

59. $f(x) = \ln (x - 1)$ **x-intercept: 2**

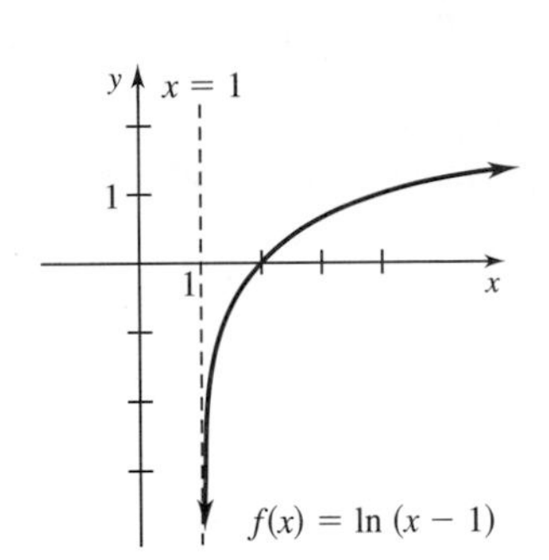

60. $f(x) = 1 - \log x$ **x-intercept: 10**

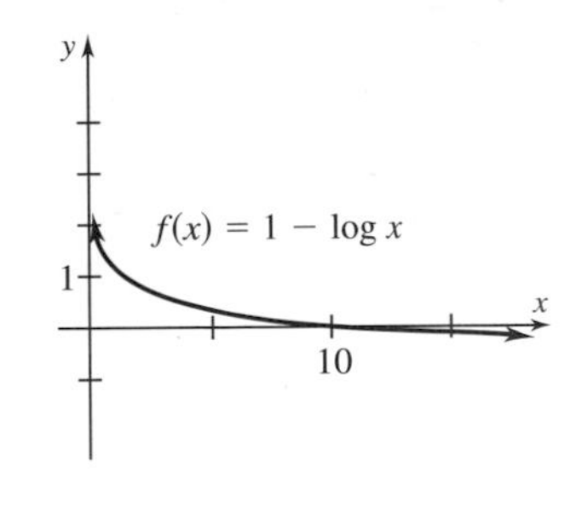

Given $\log 2 = .3010$ *and* $\log 3 = .4771$, *evaluate the logarithms in Exercises 61–68 without using a calculator.*

61. $\log 6$ **.7781**

62. $\log \frac{2}{3}$ **−.1761**

63. $\log 30$ **1.4771**

64. $\log 8$ **.9030**

65. $\log 1.5$ **.1761**

66. $\log 12$ **1.0791**

67. $\log 9$ **.9542**

68. $\log 20$ **1.301**

69. Which is larger, $\log_7 2$ or $\log_6 2$? **$\log_6 2$**

70. Which is larger, $\log_{1/2} 3$ or $\log_{1/3} 3$? **$\log_{1/3} 3$**

Suppose $f(x)$ *is a logarithmic function and* $f(3) = 2$. *Determine the function values in Exercises 71–73.*

71. $f(1/9)$ **−4**

72. $f(1)$ **0**

73. $f(27)$ **6**

74. The population of an animal species that is introduced into a certain area may grow rapidly at first but then grow more slowly as time goes on. A logarithmic function can provide an excellent description of such growth. Suppose that the population of foxes in an area t months after the foxes were introduced there is given by $F(t) = 500 \log(2t + 3)$.
Find the population of foxes at the following times:
(a) When first released into the area (that is, when $t = 0$) **≈ 240**
(b) After 3 months **(c)** After 15 months. **(b) ≈ 480 (c) ≈ 760**
(d) Graph $y = F(t)$.

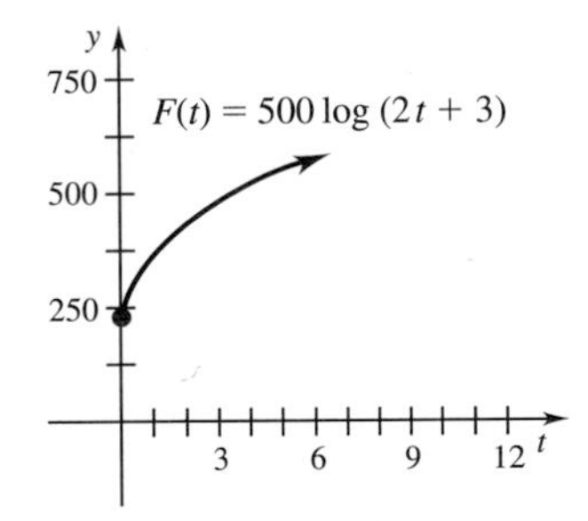

75. A company determines that its monthly sales S are related to its advertising budget x by the function $S(x) = 11{,}000 \log_{10}(3x + 1) + 10{,}000$, where x is in thousands of dollars.
Find the monthly sales for the following advertising budgets:
(a) \$0 **(b)** \$3000 **(c)** \$10,000. **(a) 10,000 (b) 21,000 (c) ≈ 26,000**
(d) Graph $y = S(x)$.

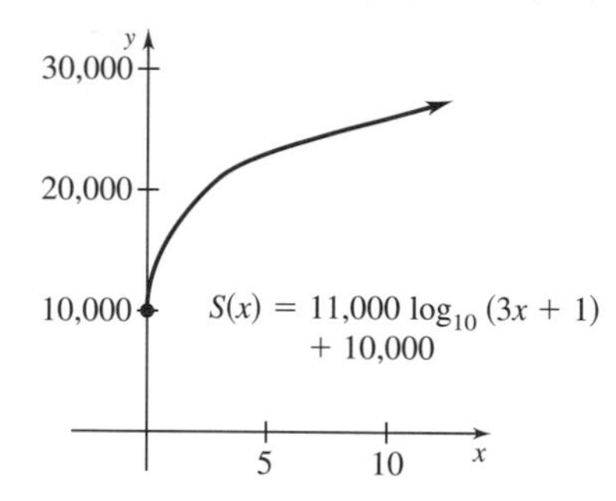

76. The loudness of sounds is measured in a unit called a *decibel.* To measure with this unit we first assign an intensity of I_0 to a very faint sound, called the *threshold sound.* If a particular sound has an intensity I, then the decibel rating d of this louder sound is given by

$$d = 10 \log_{10} \frac{I}{I_0}.$$

Find the decibel ratings of the following sounds, having intensities as given. Round answers to the nearest whole number.
(a) Whisper, $115\ I_0$ **(b)** Rock music, $895{,}000{,}000{,}000\ I_0$ **(a) 21 (b) 120**
(c) Jetliner at takeoff, $109{,}000{,}000{,}000{,}000\ I_0$ **140**
(d) If the intensity of a sound is doubled, by how much is the decibel rating increased? What happens to the decibel rating if the sound intensity is tripled?
≈ 3.0 or 10 · log 2; it increases by approximately 4.8 or 10 · log 3

77. The *Richter scale rating* of an earthquake of intensity I is given by $\log_{10}(I/I_0)$, where I_0 is the intensity of an earthquake of a certain (small) size. Find the Richter scale ratings of earthquakes having the following intensities:
(a) $1000\ I_0$ **(b)** $1{,}000{,}000\ I_0$ **(c)** $100{,}000{,}000\ I_0$.
(a) 3 (b) 6 (c) 8

78. The San Francisco earthquake of 1906 had a Richter scale rating of 8.6. Express the intensity of this earthquake as a multiple of I_0 (see Exercise 77). **398,000,000 I_0**

79. The San Francisco earthquake of 1989 had a Richter scale rating of 7.1. How much more powerful was the 1906 earthquake than the 1989 earthquake?
31.6 times more powerful

80. The number of years, n, since two independently evolving languages split off from a common ancestral language is approximated by $n \approx -7600 \log r$, where r is the ratio of words from the ancestral language common to both languages.
Find n if **(a)** $r = .9$ **(b)** $r = .3$. **(a) ≈ 350 (b) ≈ 4000**
(c) How many years have elapsed since the split if half of the words of the ancestral language are common to both languages? **≈ 2300**

81. Suppose the number of paramecia in a colony is given by $y = y_0 e^{.5t}$, where t is time in hours and y_0 is the population at $t = 0$.
(a) If $y_0 = 1000$, find y when $t = 4$. $\approx$ **7400**
(b) Find y_0 if there are 5400 paramecia when $t = 2$. $\approx$**2000**

82. In the central Sierra Nevada Mountains of California, the percent of moisture that falls as snow rather than rain is approximated reasonably well by $p = 86.3 \ln h - 680$, where p is the percent of snow at an altitude h (in feet). (Assume $h \geq 3000$.) Find the percent of moisture that falls as snow at the following altitudes.
(a) 3000 ft **(b)** 4000 ft **(c)** 7000 ft **(a)** $\approx$ **11** **(b)** $\approx$ **36** **(c)** $\approx$ **84**
(d) Graph p.

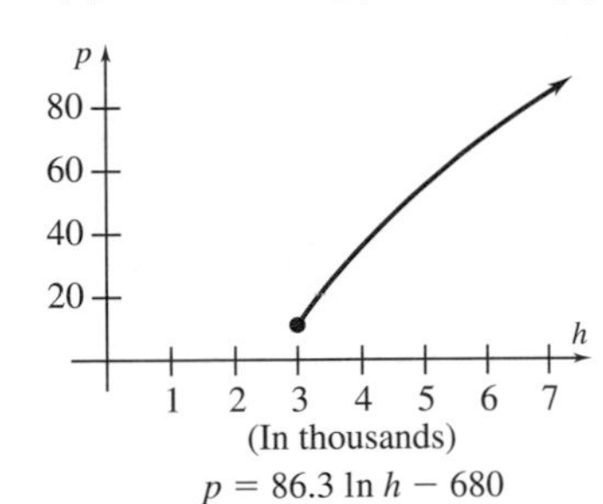

Find each of the following logarithms to the nearest hundredth.

83. $\log_2 10$ **3.32**
84. $\log_8 12$ **1.19**
85. $\log_{1/2} 3$ **−1.58**
86. $\log_{12} 62$ **1.66**
87. $\log_{1/10} 9.2$ **−.96**
88. $\log_{200} 175$ **.97**
89. $\log_{2.9} 7.5$ **1.89**
90. $\log_{6.7} .84$ **−.09**

Use the change-of-base theorem to evaluate the following expressions.

91. $\dfrac{\log_3 25}{\log_3 5}$ **2**
92. $\dfrac{\log_8 32}{\log_8 2}$ **5**

▲ *In Exercises 93 and 94, assume that a and b represent positive numbers other than 1, and x is any real number.*

93. Show that $1/\log_a b = \log_b a$.
94. Show that $a^x = e^{x \ln a}$.

95. Explain the error in the following proof that $2 < 1$.

$$\frac{1}{9} < \frac{1}{3}$$

$$\left(\frac{1}{3}\right)^2 < \frac{1}{3} \quad \text{Rewrite left side.}$$

$$\log\left(\frac{1}{3}\right)^2 < \log\frac{1}{3} \quad \text{Take log of both sides.}$$

$$2 \cdot \log\frac{1}{3} < 1 \cdot \log\frac{1}{3} \quad \text{Use the third property of logarithms.}$$

$$2 < 1 \quad \text{Divide both sides by } \log\frac{1}{3}.$$

96. The first logarithm property can be verbalized as "The logarithm of the product of two numbers is the sum of their logarithms." State the second and third properties in words.

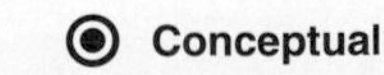

Connections

Graphing Utility Problems

97. Graph $f(x) = 10^x$ and $y = 5$ on the same axes and determine the x-coordinate of their point of intersection. Express this number in terms of a logarithm. **$x \approx .70$, log 5**

98. Use the technique of Exercise 97 to estimate ln 3. **$x \approx 1.1$**

99. Graph the function $f(x) = \log_5 x$.

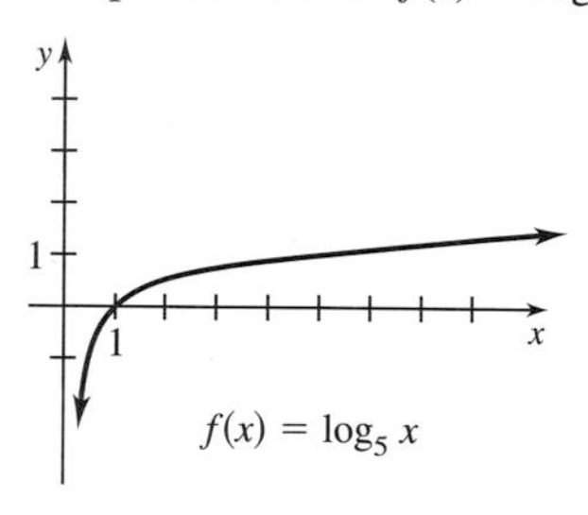

100. Graph the function $f(x) = \log_x 5$.

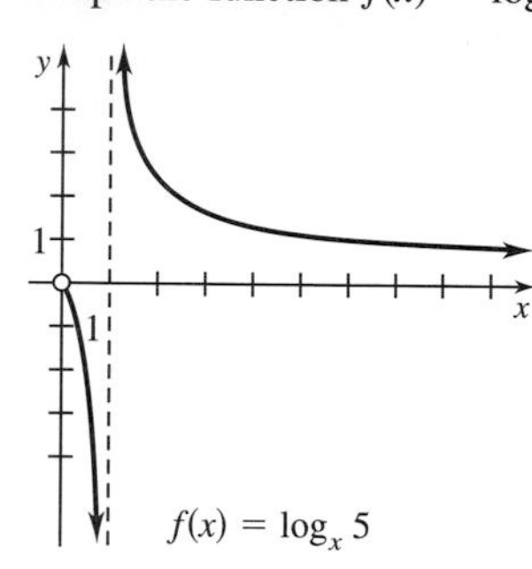

Estimate the solutions of the following equations.

101. $\ln x = x - 2$ **$x \approx .16$, $x \approx 3.1$**

102. $e^{-x} = \ln x$ **$x \approx 1.3$**

103. $\ln x = e^{.3x} - 2$ **$x \approx .42$, $x \approx 4.1$**

104. $\log x = 2 - x/2$ **$x \approx 3.0$**

105. The function $\ln x$ grows so slowly that for any r with $0 < r < 1$, $\ln x/x^r$ approaches 0 as x gets large. Test this statement by graphing $\ln x/\sqrt{x}$ and $\ln x/\sqrt[5]{x}$.

5.3 Exercises (text page 283)

Solve the following equations. Give answers as decimals rounded to the nearest hundredth.

1. $5^x = 7$ **1.21**

2. $4^x = 12$ **1.79**

3. $6^{1-2k} = 8$ **−.08**

4. $3^{k-3} = 11$ **5.18**

5. $4^{3m-1} = 12^{m+2}$ **3.80**

6. $3^{2m-5} = 13^{m-1}$ **−7.96**

7. $e^x = 4$ **1.39**

8. $e^{1-y} = 10$ **−1.30**

9. $3e^{4a+1} = 9$ **.02**

10. $10e^{3z-7} = 5$ **2.10**

11. $2^x = -3$ **no solution**

12. $\left(\frac{1}{5}\right)^q = -3$ **no solution**

13. $100(1 + .02)^{3+n} = 150$ **17.48**

14. $500(1 + .05)^{p/4} = 200$ **−75.12**

15. $2^{x^2-9} = 36$ **± 3.76**

16. $5^{3-x^2} = 6$ **± 1.37**

17. $4(e^x - 1) = 20$ **1.79**

18. $\frac{1}{5}(e^{-x} + 1) = 3$ **−2.64**

19. $\log(t - 1) = 1$ **11**

20. $\log q^{-1} = 1$ **.10**

21. $\log(x + 2) = 2 - \log x$ **9.05**

22. $\log(z - 6) = 2 - \log(z + 15)$ **10**

23. $\ln(y + 1) = \ln(y - 2) + \ln 2$ **5**

24. $\ln p^2 - \ln(p + 2) = \ln 6$ **7.58, −1.58**

25. $\ln(3 + 2y) - \ln(1 + y) = \ln 10$ **$-\frac{7}{8}$**

26. $\log_3(a - 3) = 1 + \log_3(a + 1)$ **no solution**

27. $\log_4(z + 3) + \log_4(z - 3) = 1$ **3.61**

28. $\log 2w + \log(3w - 7) = \log 3$ **2.53**

29. $5^{\log_5(x+1)} = 9$ **8**

30. $\log_2 \sqrt{2y^2} - 1 = \frac{1}{2}$ **± 2** **31.** $\log_x (5x - 6) = 2$ **2, 3** **32.** $\ln e^{x^3 - 2} = 6$ **2** **33.** $7^{2x \log_7 4} = 256$ **2**

34. $\log_3 (\log_3 x) = 1$ **27** **35.** $\log z = \sqrt{\log z}$ **1, 10** **36.** $\log x^2 = (\log x)^2$ **1, 100**

37. The amount of a radioactive specimen present at time t (measured in seconds) is $A(t) = 1000(10)^{-.04t}$, where $A(t)$ is measured in grams. Find the half-life of the specimen—that is, the time it will take until exactly half the specimen remains. **7.53 sec**

A large cloud of radioactive debris from a nuclear explosion has floated over the Pacific Northwest, contaminating much of the hay supply. Consequently, farmers in the area are concerned that the cows who eat the hay will give contaminated milk. The percent of the initial amount of radioactive iodine still present in the hay after t days is approximated by $P(t) = 100e^{-.1t}$, *where t is time measured in days.*

38. Find the half-life of the radioactive iodine in the hay (the time until only 50 percent of the iodine is left). **6.93 days**

39. Some scientists feel that the hay is safe after the level of radioactive iodine has declined to 10 percent of the original amount. Based on this assumption, find the number of days before the hay could be used. Other scientists believe that the hay is not safe until the level of radioactive iodine has declined to 1 percent of the original amount. Find the number of days this would take. **23 days, 46 days**

Solve the following equations for x. (Hint: *In Exercises 42–45 multiply by* e^x.)

40. $2^{2x} - 2^x - 2 = 0$ **1** **41.** $5^{2x} + 2 \cdot 5^x - 3 = 0$ **0** **42.** $2e^x + 1 - 3e^{-x} = 0$ **0**

43. $e^x - 5 + 6e^{-x} = 0$ **ln 2, ln 3** **44.** $\frac{e^x - e^{-x}}{2} = 4$ **$\ln(4 + \sqrt{17})$** **45.** $\frac{e^x + e^{-x}}{2} = 3$ **$\ln(3 \pm 2\sqrt{2})$**

Solve each of the following equations for the indicated variables. Use logarithms to the appropriate bases.

46. $P = 1000e^{t/1000}$ for t **$t = 1000 \ln\left(\frac{P}{1000}\right)$**

47. $I = \frac{E}{R}(1 - e^{-Rt/2})$ for t **$t = (-2/R) \ln (1 - RI/E)$**

▲ **48.** $T = T_0 + (T_1 - T_0)10^{-kt}$ for t **$t = -\frac{1}{k} \log\left(\frac{T - T_0}{T_1 - T_0}\right)$**

▲ **49.** $A = \frac{Pr}{1 - (1 + r)^{-t}}$ for t **$t = \frac{\ln [A/(A - Pr)]}{\ln (1 + r)}$**

▲ **50.** $G = \frac{mH}{H + (m - H)e^{-kmt}}$ for t **$t = \frac{-1}{km} \ln \left[\frac{(m - G)H}{(m - H)G}\right]$**

51. $\log_5 (x + y) = \log_5 (2x - 1)$ for x **$x = y + 1$**

52. $\frac{10^x - 10^{-x}}{2} = y$ for x **$x = \log (y + \sqrt{y^2 + 1})$**

53. $\frac{10^x + 10^{-x}}{2} = y$ for x **$x = \log (y + \sqrt{y^2 - 1})$**

Find the formula, domain, and range of $f^{-1}(x)$.

54. $f(x) = \ln (x - 1)$ **$e^x + 1, (-\infty, \infty), (1, \infty)$**

55. $f(x) = \log (x + 5)$ **$10^x - 5, (-\infty, \infty), (-5, \infty)$**

56. $f(x) = e^{3x+1}$ **$\frac{1}{3}(\ln x - 1), (0, \infty), (-\infty, \infty)$**

57. $f(x) = 3 \cdot 10^x$ **$\log\left(\frac{x}{3}\right), (0, \infty), (-\infty, \infty)$**

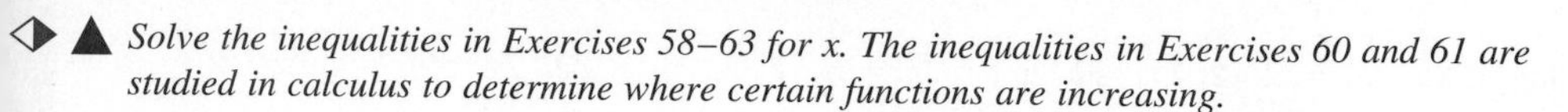

Solve the inequalities in Exercises 58–63 for x. The inequalities in Exercises 60 and 61 are studied in calculus to determine where certain functions are increasing.

58. $\log_2 x < -1$ **(0, 1/2)**

59. $\log_x 64 < 3$ **(0, 1) ∪ (4, ∞)**

60. $\log_3 x > 3$ **(27, ∞)**

61. $\log_x .2 < -.1$ **(1, 5^{10})**

62. $x^2e^x - e^x > 0$ **(−∞, −1) ∪ (1, ∞)**

63. $\dfrac{e^x - xe^x}{e^{2x}} > 0$ **(−∞, 1)**

64. Recall (from Exercise 77 of Section 5.2) the formula for the Richter scale of earthquake intensity, $R = \log_{10}(I/I_0)$. Solve this formula for I. $\mathbf{I = I_0 \cdot 10^R}$

65. Solve the formula for compound interest, $A = P\left(1 + \dfrac{r}{n}\right)^{nt}$, for t, and describe the meaning of t. $\mathbf{t = \dfrac{\ln(A/P)}{n \ln(1 + r/n)}}$

66. Use natural logarithms to solve $A = Pe^{rt}$ for t. $\mathbf{t = (1/r)\ln(A/P)}$

67. Explain why an equation of the form $2^x = a$ does not always have a solution.

68. Without solving, explain why an equation of the form $\log_2 x = a$ must always have a solution for x.

Graphing Utility Problems

69. For each of the functions in Exercises 54–57, graph the function, its inverse, and the line $y = x$. Convince yourself that the two functions are indeed inverses by observing that the two functions are reflections of each other in the line. Also, find the coordinates of a point on one graph, call it (a, b), and check that the point (b, a) is on the other graph.

Solve the following equations by finding the intersection point(s) of the graph of a function and a horizontal line.

70. $e^x + \ln x = 5$ **≈ 1.52**

71. $e^x - \ln(x + 1) = 3$ **≈ −.93, ≈ 1.35**

5.4 Exercises (text page 289)

1. A population that initially contains 40,000 people is growing exponentially with a growth constant $k = .03$.
(a) Give a formula for the size of the population after t yr. $\mathbf{40{,}000\, e^{.03t}}$
(b) How large will the population be after 20 yr? **≈ 73,000**
(c) After how many years will the population reach 60,000? **≈ 13.5 yr**

2. Ten grams of radioactive material with decay constant $k = -.025$ is buried in the ground.
(a) Give a formula for the amount after t yr. $\mathbf{10e^{-.025t}}$
(b) How much will remain after 4 yr? **≈ 9 g**
(c) When will 60 percent of the material have distintegrated? **≈ 36.7 yr**

3. After a bactericide is introduced into a culture of 50,000 bacteria, the number of bacteria decreases exponentially. After 9 hr there are only 20,000 bacteria.
(a) Find the value of the decay constant k. **≈ −.1**
(b) In how many hours will the original population of 50,000 bacteria be reduced by half? **≈ 6.9 hr**

4. The average tuition for public colleges increased exponentially from \$1533 in 1970 to \$8174 in 1990.

(a) Determine the annual rate of increase in tuition from 1970 to 1990. $\approx$ **8.4%**

(b) If tuition continues to increase at the rate found in (a), what will be the average tuition in the year 2000? $\approx$ **\$19,000**

5. A population of bacteria in a culture is increasing exponentially. The original culture of 25,000 bacteria contains 40,000 bacteria after 10 hr. How long will it be until there are 60,000 bacteria in the culture? (*Hint:* This problem must be solved in two steps: first find the growth constant, and then find the time requested.) $\approx$ **19 hr**

6. A radioactive substance is decaying exponentially. The amount of substance is reduced from 800 g to 400 g after 4 days. How much remains after 10 days? (*Hint:* This problem must be solved in two steps: first find the decay constant, and then find the amount requested.) **about 140 g**

7. The amount of a certain radioactive specimen present at time t (in days) decreases exponentially. If 5000 g decreased to 4000 g in 5 days, find the half-life of the specimen. **about 16 days**

8. A population of insects is growing exponentially. After 2 months the population has increased from 100 to 150. How many months will it take for the population to reach 500? **about 8 mo**

9. The amount (in grams) of a substance involved in a chemical reaction increases exponentially as temperature increases (for $0°C \leq t \leq 50°C$). At 0°C, 25 g reacts, and at 20°C, 75 g reacts. At what temperature will 150 g react? **about 33° C**

Exercises 10–12 refer to the carbon dating process of Example 3 in the text.

10. Suppose an Egyptian mummy is discovered in which the amount of carbon 14 is only half of the original amount. About how long ago did the Egyptian die? **about 5600 yr**

11. If an object contains $1/4$ of the amount of carbon 14 that it originally had, how old is the object? How old if the amount is $1/8$? **about 11,200 yr, about 16,800 yr**

12. The Lascaux caves of France contain prehistoric paintings of animals. Charcoal found in these caves contains only about 15 percent of the amount of carbon 14 in living trees. Estimate the age of the paintings. **about 15,300 yr**

Nuclear energy derived from radioactive isotopes can be used to supply power to space vehicles. Suppose that the output of the radioactive power supply for a certain satellite is $P(t) = 30\,e^{-t/250}$ *watts, where t is the time in days.**

13. How much power will be available at the end of one year? (Assume a 365-day year.) **about 6.97 watts**

14. How long will it take for the power to drop to half its original strength? **about 173 days**

15. The equipment aboard the satellite requires 10 watts of power to operate properly. What is the operational life of the satellite? **about 275 days**

*Bernice Kastner, Ph.D., *Spacemathematics.* (NASA, 1985).

Find each of the amounts in Exercises 16–21, assuming continuous compounding.

16. $1,000 at 9% for 1 yr **$1094.17**

17. $2,000 at 8% for 5 yr **$2983.65**

18. $15,000 at 10% for 6 yr **$27,331.78**

19. $135 at 4% for 3 yr **$152.21**

20. $580 at 15% for 6 yr **$1426.57**

21. $10,000 at 12% for 3 yr **$14,333.29**

22. Assuming an inflation rate of 5% compounded continuously, how long will it take for prices to double? **about 13.9 yr**

23. For Exercise 22, how long would it take for prices to double if the inflation rate is 10%? **about 6.9 yr**

24. Solve $A = Pe^{rt}$ for r, and describe the meaning of t. $r = \frac{1}{t} \ln \left(\frac{A}{P}\right)$

25. Solve $2P = Pe^{rt}$ for t, and describe the meaning of t. $t = \frac{\ln 2}{r}$

Newton's law of cooling says that the rate at which a body cools is proportional to the difference in temperature between the body and the environment into which it is introduced. The temperature $f(t)$ of the body at time t in appropriate units after being introduced into an environment having constant temperature T_0 is $f(t) = T_0 + Ce^{-kt}$, where C and k are constants. Use this result in Exercises 26 and 27.

26. Boiling water, at 100°C, is placed in a freezer at 0°C. The temperature of the water is 50°C after 24 min. Find the temperature of the water after 96 min. **6.25°C**

27. A piece of metal is heated to 300°C and then placed in a cooling liquid at 50°C. After 4 min, the metal has cooled to 175°C. Find its temperature after 12 min. **81.25°C**

28. Explain why the half-life of a radioactive material does not depend on the amount of material present initially.

29. The number of bacteria in a jar doubles every minute. If the jar is full at 2:00 P.M., when was it half full? **1:59 P.M.**

30. If interest is compounded continuously and the interest rate is tripled, what effect will this have on the time required for an investment to double? **Time will be divided by 3.**

Graphing Utility Problems

31. Suppose an 80-g sample of radioactive material has decay constant $-.02$. On the same axes, graph the three functions $y = 80e^{-.02t}$, $y = 40$, and $y = 20$. Determine the t-coordinates of the intersection points of the decay curve and the horizontal lines and observe that the time required for the material to decay from 80 g to 40 g is the same as the time required to decay from 40 g to 20 g. Explain why this is true.

32. On the same axes, graph $y = 5\left(1 + \frac{.08}{2}\right)^{2x}$ and $y = 5e^{.08x}$, for $0 \le x \le 25$, $0 \le y \le 75$. Compare the graphs and determine the two y-values when $x = 25$. Repeat with the 2 in the first function replaced by 12.

Many environmental situations place effective limits on the growth of the number of an organism in an area. Many such limited growth situations are described by the logistic function, *defined by*

$$G(t) = \frac{MG_0}{G_0 + (M - G_0)e^{-kMt}}$$

where G_0 is the initial number present, M is the maximum possible size of the population, and k is a positive constant. Assume $G_0 = 100$, $M = 2500$, $k = .0004$, and t is time in decades (10-yr periods).

33. Graph the S-shaped limited growth function using $0 \le t \le 8$, $0 \le y \le 2500$.

34. Estimate the value of $G(2)$ from the graph. Then evaluate $G(2)$ to find the population after 20 yr. **590; 589**

35. Find the x-coordinate of the intersection of the curve with the horizontal line $y = 1000$ to estimate the number of decades required for the population to reach 1000. Then solve $G(t) = 1000$ algebraically to obtain the exact value of t. **2.8; 2.7726**

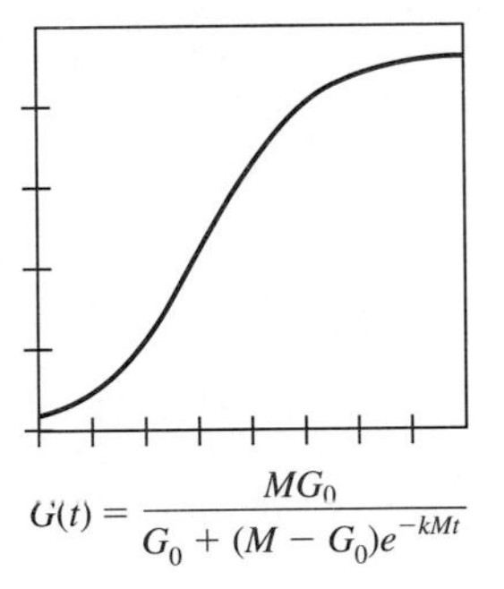

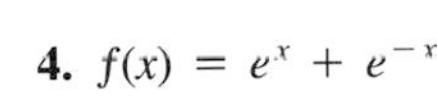

$G(t) = \dfrac{MG_0}{G_0 + (M - G_0)e^{-kMt}}$

Chapter 5 Review Exercises (page 291)

Graph each of the functions defined as follows.

1. $f(x) = 2^x$

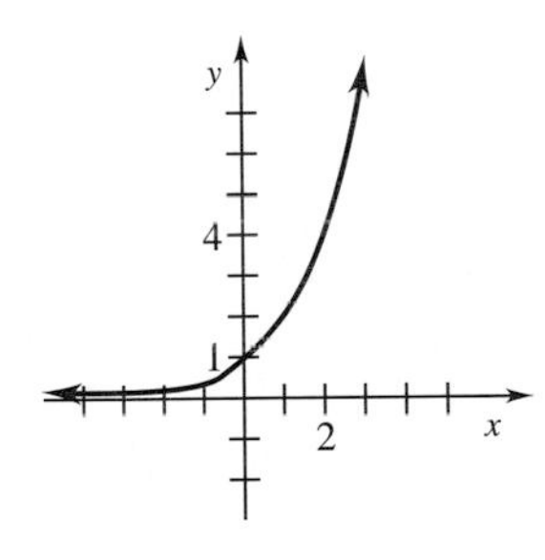

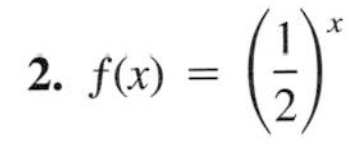

2. $f(x) = \left(\frac{1}{2}\right)^x$

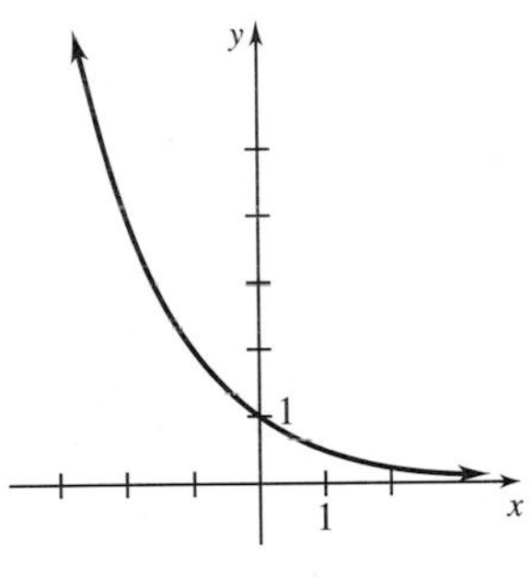

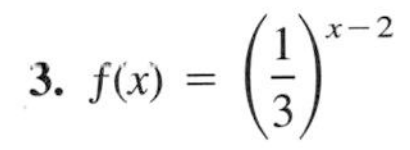

3. $f(x) = \left(\frac{1}{3}\right)^{x-2}$

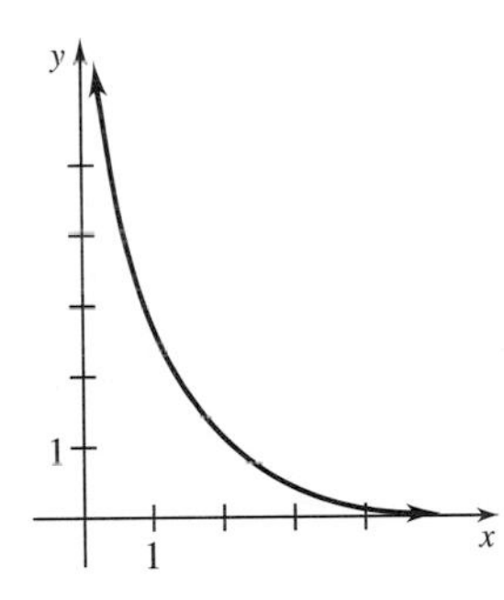

4. $f(x) = e^x + e^{-x}$

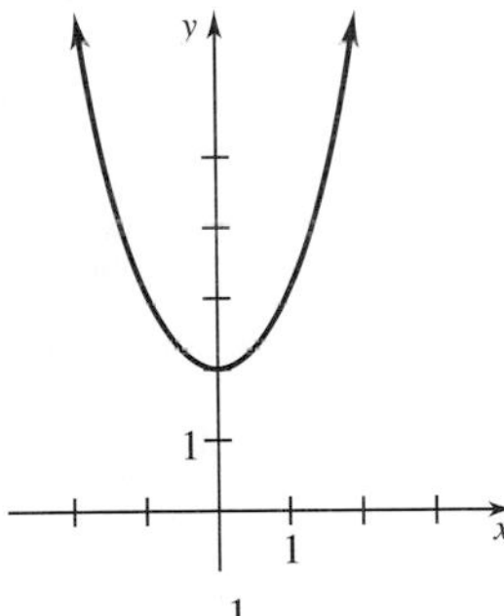

5. $f(x) = (x - 1)e^{-x}$

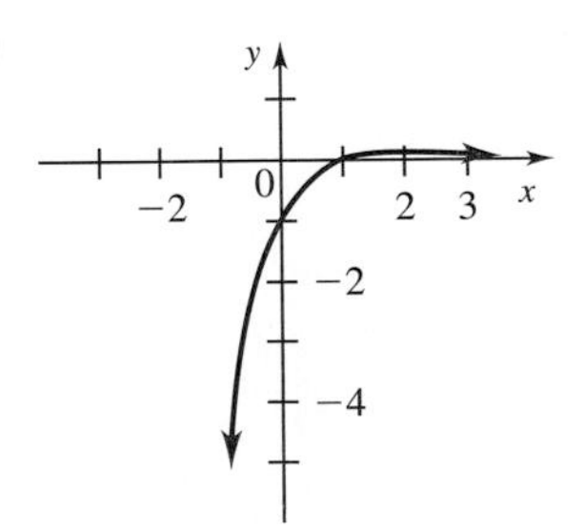

6. $f(x) = \log_3 x$

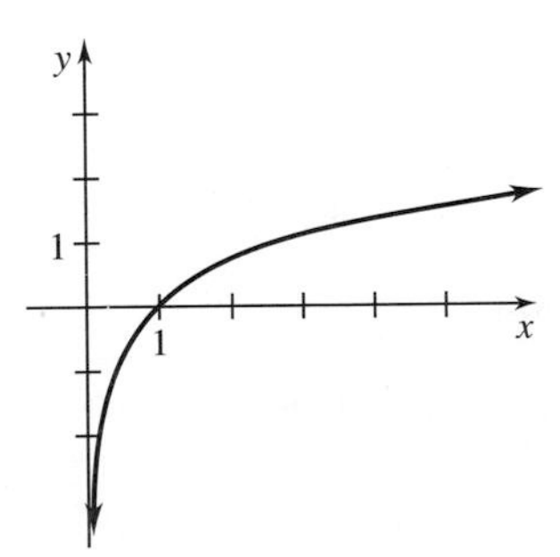

7. $f(x) = \log_2 (x - 1)$

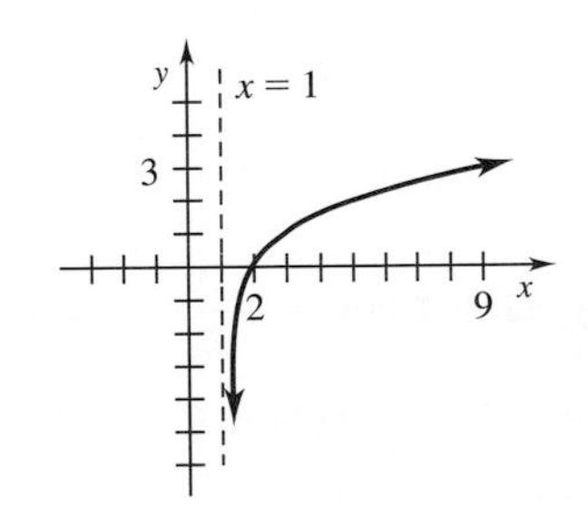

8. $f(x) = \ln \frac{1}{x}$

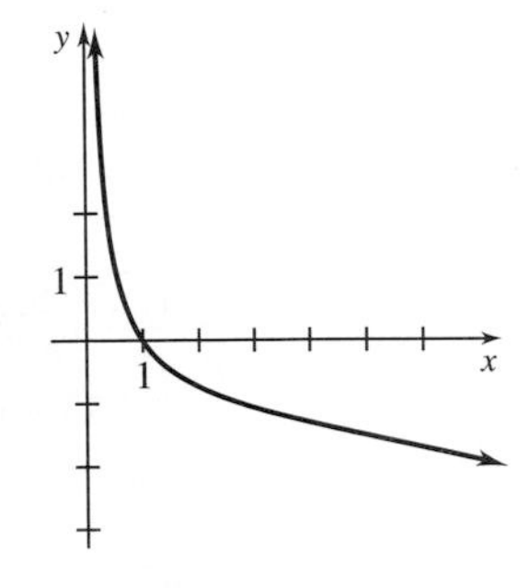

Solve each of the following equations.

9. $16^p = 32$ $\mathbf{\frac{5}{4}}$

10. $9^{y-1} = 27^{2y}$ $\mathbf{-\frac{1}{2}}$

11. $\frac{-8}{125} = b^{-3}$ $\mathbf{-\frac{5}{2}}$

12. $\frac{1}{2} = \left(\frac{b}{4}\right)^{1/4}$ $\mathbf{\frac{1}{4}}$

The amount (in grams) of a certain radioactive material present after t days is given by $A(t) = 800e^{-.04t}$. Find A(t) for the following values of t.

13. $t = 0$ **800 g**

14. $t = 5$ **≈ 650 g**

15. $t = 10$ **≈ 540 g**

How much would $1200 amount to at 10% interest compounded continuously for the following number of years?

16. 1 yr **$1326.21**

17. 2 yr **$1465.68**

18. 10 yr **$3261.94**

19. Historically, the consumption of electricity has increased at a continuous rate of 6% per year. If it continued to increase at this rate, find the number of years before exactly twice as much electricity would be needed. **about 12 yr**

20. Suppose a conservation campaign together with higher rates caused demand for electricity to increase at only 2% per year. (See Exercise 19.) Find the number of years before twice as much electricity would be needed as is needed today. **about 35 yr**

Write each of the following expressions in logarithmic form.

21. $2^8 = 256$ $\mathbf{\log_2 256 = 8}$

22. $100^{1/2} = 10$ $\mathbf{\log_{100} 10 = \frac{1}{2}}$

23. $\left(\frac{1}{16}\right)^{1/4} = \frac{1}{2}$ $\mathbf{\log_{1/16}\left(\frac{1}{2}\right) = \frac{1}{4}}$

24. $\left(\frac{2}{3}\right)^{-2} = \frac{9}{4}$ $\mathbf{\log_{2/3}\frac{9}{4} = -2}$

25. $e^{.1} = 1.1052$ $\mathbf{\ln 1.1052 = .1}$

26. $2^{2.322} = 5$ $\mathbf{\log_2 5 = 2.322}$

27. $10^{.4771} = 3$ $\mathbf{\log 3 = .4771}$

28. $e^{2.4849} = 12$ $\mathbf{\ln 12 = 2.4849}$

In Exercises 29 and 30, write each logarithm in exponential form.

29. $\log_2 \sqrt{32} = 5/2$ $\mathbf{2^{5/2} = \sqrt{32}}$

30. $\ln 45 = 3.806662$ $\mathbf{e^{3.806662} = 45}$

31. What is the base of the logarithmic function whose graph contains the point (81, 4)? **3**

32. What is the base of the exponential function whose graph contains the point $\left(-4, \frac{1}{16}\right)$? **2**

Use properties of logarithms to write each of the following as a sum or difference of logarithms (or constants times logarithms). Assume all variables represent positive real numbers.

33. $\log_3 \frac{mn}{p}$ $\mathbf{\log_3 m + \log_3 n - \log_3 p}$

34. $\log_3 \frac{\sqrt[3]{7}}{4}$ $\mathbf{\frac{1}{3}\log_3 7 - \log_3 4}$

35. $\log_3 (49 \cdot x^2 \sqrt[3]{m^2p})$ $\mathbf{\log_3 49 + 2\log_3 x + \frac{2}{3}\log_3 m + \frac{1}{3}\log_3 p}$

36. $\log_7 (7k + 5r^2)$ **cannot be rewritten using the properties of logarithms**

Find each of the following logarithms to four decimal places.

37. $\ln e^{-e}$ **−2.7183**

38. $\ln e^{.04}$ **.0400**

39. $\log_3 7.1$ **1.7842**

40. $\log_{3.4} 15.8$ **2.2553**

41. $\log_{1/2} 9.45$ **−3.2403**

42. $\log_{11} 520$ **2.6080**

Solve each of the following equations. Round to the nearest thousandth.

43. $5^r = 11$ **1.490**

44. $10^{r+3} = 6$ **−2.222**

45. $2(1 + e^p) = 10$ **1.386**

46. $\left(\frac{1}{2}\right)^{3k+1} = 3$ **−.862**

47. $6^{2-m} = 2^{3m+1}$ **.747**

48. $6(-1 + e^r) = 24$ **1.609**

49. $\log_{50} y = \frac{1}{2}$ **7.071**

50. $\log_3 (q + 1) = 2$ **8**

51. $\ln 6x - \ln (x + 1) = \ln 4$ **2**

52. $\log_{81} \sqrt{x} = \frac{1}{2}$ **81**

53. $\log (3p - 1) = 1 - \log p$ **2**

54. $\log_7 (b + 1)^2 = 3$ **17.520, −19.520**

Solve for the indicated variable.

55. $3 = \frac{5^x - 5^{-x}}{2}$ for x **$x = [\ln (3 + \sqrt{10})]/\ln 5$**

56. $\frac{2^x + 1}{2^x - 1} = -6$ for x **$x = -.485$**

57. $\log_a (x - 1) = 1 + \log_a (x + 1)$ for x **$x = \frac{1 + a}{1 - a}$**

58. $r = r_0 e^{e^{nt}}$ for t **$t = \frac{1}{n} \ln \left| \ln \frac{r}{r_0} \right|$**

59. $N = a + b \ln \frac{c}{d}$ for c **$c = de^{(N-a)/b}$**

60. $P = \frac{k}{1 + e^{-rt}}$ for t **$t = (\ln [P/(k - P)])/r$**

Suppose the gross national product of a small country (in millions of dollars) is approximated by $G(t) = 15 + 2 \log t$ where t is time in years, for $1 \leq t \leq 6$. Find the GNP at the following times.

61. 1 yr **$15 million**

62. 2 yr **$15.6 million**

63. 5 yr **$16.4 million**

64. 6 yr **$16.56 million**

65. The concentration of pollutants, in grams per liter, in the east fork of the Big Weasel River is approximated by $P(x) = .04e^{-4x}$, where x is the number of miles downstream from a paper mill that the measurement is taken.
Find each of the following: **(a)** $P(.5)$ **(b)** $P(1)$. **(a) .0054 g/L (b) .00073 g/L**
(c) the concentration of pollutants 2 mi downstream. **.000013 g/L**
(d) the number of miles downstream where the concentration of pollutants is .002 g per liter. **.75 mi**

66. A person learning certain skills involving repetition tends to learn quickly at first. Then learning tapers off and approaches some upper limit. Suppose the number of symbols per minute a keypunch operator can produce is given by $p(t) = 250 - 120(2.8)^{-.5t}$, where t is the number of months the operator has been in training.
Find each of the following: **(a)** $p(2)$ **(b)** $p(4)$ **(c)** $p(10)$. **(a) 207 (b) 235 (c) 249**
(d) Graph $p(t)$.

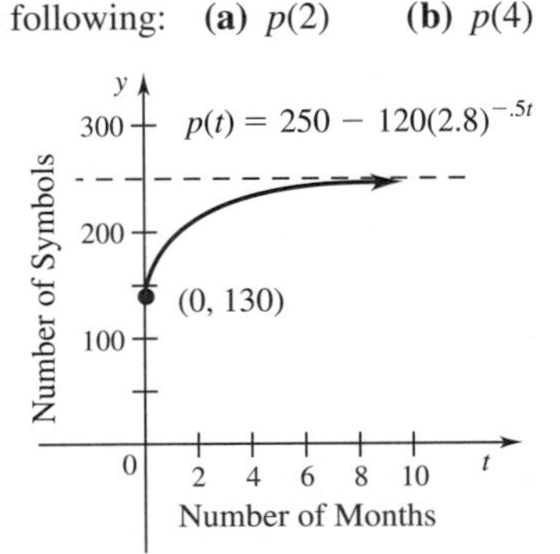

67. A population is increasing according to the growth law $y = 2e^{.02t}$, where y is in millions and t is in years. Match each of the questions (a), (b), (c), and (d) with one of the solutions (A), (B), (C), or (D).

(a) How long will it take for the population to triple? **(B)**
(b) When will the population reach 3 million? **(D)**
(c) How large will the population be in 3 years? **(C)**
(d) How large will the population be in 4 months? **(A)**

(A) Evaluate $2e^{.02(1/3)}$.
(B) Solve $2e^{.02t} = 3 \cdot 2$ for t.
(C) Evaluate $2e^{.02(3)}$.
(D) Solve $2e^{.02t} = 3$ for t.

68. The population of the world is expected to double in the next 44 yr. Without solving for the growth constant k, determine how much the population will increase in 22 yr.
by a factor of $2^{1/2} \approx 1.4$

69. If the world population continues to grow at the current rate, by what factor will the population grow in the next 220 yr. See Exercise 68. (*Hint:* $220 = 5 \cdot 44$.)
by a factor of $2^5 = 32$

70. Find the domain and range of the function $g(x) = \log \left[\frac{x^2 - 3x - 10}{x + 2}\right]$.
domain: $(5, \infty)$; range: $(-\infty, \infty)$

71. Find the domain and range of the function $f(x) = \log_2 (x^2 - x - 2)$. **domain: $(-\infty, -1) \cup (2, \infty)$; range: $(-\infty, \infty)$**

72. A skydiver in free-fall travels at the speed of $v(t) = 176(1 - e^{-.18t})$ ft per sec after t sec. How long will it take for the skydiver to attain the speed of 147 ft. per sec (100 mph)? **10 sec**

73. Give the property that justifies each step of the following derivation. Let a be any number.
(a) (a, e^a) is on the graph of $f(x) = e^x$.
(b) (e^a, a) is on the graph of $g(x) = \ln x$.
(c) $\ln e^a = a$

If R dollars is deposited at the end of each year in an account paying a rate of interest r per year compounded annually, then after t years the account will contain a total of

$$R\left[\frac{(1 + r)^t - 1}{r}\right]$$

dollars. In Exercises 74 and 75, find the final amount on deposit. (Such a sequence of payments is called an annuity.*)*

74. \$800, 12%, 10 yr **\$14,038.99**

75. \$1500, 14%, 7 yr **\$16,095.74**

76. Manuel deposits \$10,000 at the end of each year for 12 yr in an account paying 12% compounded annually. He then puts this total amount on deposit in another account paying 10% compounded semiannually for another 9 yr. Find the total amount on deposit after the entire 21-yr period. **\$580,792.63**

77. Compare the annual yield (interest paid on 1 dollar invested for 1 yr) for accounts paying 7% interest compounded quarterly and 7% compounded continuously. **.0719, .0725**

78. The population of a boom town is increasing exponentially. There were 10,000 people in town when the boom began. Two years later the population had reached 12,000. Assume this growth rate continues.
(a) What will the population be after 5 yr? **about 15,800**
(b) How long will it take for the population to double? **about 7.6 years**

79. Correct the mistakes in the following equation.

$$\log_5 125 - \log_5 25 = \frac{\log_5 125}{\log_5 25} = \log_5 \left(\frac{125}{25}\right) = \log_5 5 = 1$$

The correct statement is $\log_5 125 - \log_5 25 = \log_5 \frac{125}{25} = \log_5 5 = 1$.

Chapter 6 Systems of Equations and Inequalities

6.1 Exercises (text page 303)

Solve each of the following systems by the substitution method.

1. $x - 5y = 8$, $x = 6y$ **(48, 8)**

2. $8x - 10y = -22$, $3x + y = 6$ **(1, 3)**

3. $6x - y = 5$, $y = 11x$ **(−1, −11)**

4. $4x - 5y = -11$, $2x + y = 5$ **(1, 3)**

5. $7x - y = -10$, $3y - x = 10$ **(−1, 3)**

6. $4x + 5y = 7$, $9y = 31 + 2x$ **(−2, 3)**

7. $-2x = 6y + 18$, $-29 = 5y - 3x$ **(3, −4)**

8. $3x - 7y = 15$, $3x + 7y = 15$ **(5, 0)**

9. $3y = 5x + 6$, $x + y = 2$ **(0, 2)**

Solve each system by elimination.

10. $4x + 2y = 6$, $5x - 2y = 12$ **(2, −1)**

11. $2x - y = 8$, $3x + 2y = 5$ **(3, −2)**

12. $3x - y = -4$, $x + 3y = 12$ **(0, 4)**

13. $2x - 3y = -7$, $5x + 4y = 17$ **(1, 3)**

14. $4x + 3y = -1$, $2x + 5y = 3$ **(−1, 1)**

15. $5x + 7y = 6$, $10x - 3y = 46$ **(4, −2)**

16. $12x - 5y = 9$, $3x - 8y = -18$ **(2, 3)**

17. $6x + 7y = -2$, $7x - 6y = 26$ **(2, −2)**

18. $\frac{x}{2} + \frac{y}{3} = 4$, $\frac{3x}{2} + \frac{3y}{2} = 15$ **(4, 6)**

19. $\frac{x}{5} + 3y = 46$, $2x - \frac{y}{5} = 7$ **(5, 15)**

20. $\frac{x}{3} + \frac{2y}{5} = 4$, $x + y = 11$ **(6, 5)**

21. $\frac{3x}{2} - \frac{y}{3} = 5$, $\frac{5x}{2} + \frac{2y}{3} = 12$ **(4, 3)**

22. $\frac{3x}{2} + \frac{y}{2} = -2$, $\frac{x}{2} + \frac{y}{2} = 0$ **(−2, 2)**

23. $\frac{2x - 1}{3} + \frac{y + 2}{4} = 4$, $\frac{x + 3}{2} - \frac{x - y}{3} = 3$ **(5, 2)**

24. $\frac{x + 6}{5} + \frac{2y - x}{10} = 1$, $\frac{x + 2}{4} + \frac{3y + 2}{5} = -3$ **(66, −34)**

25. $-2x + 5y + 3z = -6$
$4x - y - 2z = -6$
$3x + 4y - 2z = -14$
(−2, −2, 0)

26. $-3x + y - z = -13$
$x - y - z = 3$
$x + 2y - z = -3$
(3, −2, 2)

27. $x + y + z = 2$
$2x - y - z = 1$
$3x - y + z = 0$ **(1, 2, −1)**

28. $x + y + z = 2$
$2x + y - z = 5$
$x - y + z = -2$ **(1, 2, −1)**

29. $2x + y + z = 5$
$-x + z = 2$
$3x - y + z = 6$ **(1, 0, 3)**

30. $x + 3y + 4z = 11$
$2x - 3y + 2z = 13$
$3x - z = 3$ **(2, −1, 3)**

31. Consider the linear equation $2x + 3y = 6$. Find a second linear equation for which the system of two linear equations will have the following: **(a)** exactly one solution; **(b)** no solution; **(c)** infinitely many solutions. **Each part of Exercises 31 and 32 has many correct answers. We give one example for each part.**
(a) $x + y = 1$ **(b)** $2x + 3y = 5$ **(c)** $4x + 6y = 12$

32. Consider the linear equation $ax + by = c$, where $a \neq 0$ and $b \neq 0$. Find a second linear equation for which the system of two linear equations will have **(a)** no solution; **(b)** infinitely many solutions. **(a)** $ax + by = d\ (d \neq c)$ **(b)** $2ax + 2by = 2c$

33. Suppose that one of the three allowable transformations of a system of equations has been performed. Is there always an allowable transformation that will convert the new system back to the original system? Explain. **yes**

Write a system of linear equations for each of the following, and then use the system to solve the problem.

34. At the local drugstore, 2 candy bars and 5 lollipops cost \$1.92, while 3 candy bars and 7 lollipops cost \$2.78. Find the cost of a candy bar and the cost of a lollipop. **46¢, 20¢**

35. At the Sharp Ranch, 6 goats and 5 sheep cost \$305, while 2 goats and 9 sheep cost \$285. Find the cost of a goat and the cost of a sheep. **goats cost \$30, sheep cost \$25**

36. The perimeter of a rectangle is 42 cm. The longer side is 7 cm longer than the shorter side. Find the length of the longer side. **14 cm**

37. During summer vacation Hector and Ann earned a total of \$1088. Hector worked 8 days fewer than Ann and earned \$2 per day less. Find the number of days he worked and the daily wage he made if the total number of days worked by both was 72. **32 days at \$14**

38. Chuck Sullivan won \$100,000 in a lottery. He invested part of the money at 5% and part at 6%. His total annual income from the two investments is \$5,500. How much does he have invested at each rate? **\$50,000 at each rate**

39. Ms. Caminiti has some money invested at 4% and three times as much invested at 6%. Her total annual income from the two investments is \$1100. How much is invested at each rate? **\$5000 at 4%, \$15,000 at 6%**

40. A cash drawer contains only fives and twenties. There are eight more fives than twenties. The total value of the money is \$215. How many of each type of bill are there? **15 fives, 7 twenties**

41. Thirty liters of a 50% alcohol solution are to be made by mixing a 70% solution and a 20% solution. How many liters of each solution should be used? **18 liters of 70%, 12 liters of 20%**

42. A merchant wishes to make 100 lb of a coffee blend that can be sold for \$4 per pound. This blend is to be made by mixing coffee worth \$6 per pound with coffee worth \$3 per pound. How many pounds of each will be needed? **33⅓ pounds of the \$6 blend, 66⅔ pounds of the \$3 blend**

43. How many gallons of full-fat milk (4.5% butterfat) must be mixed with 250 gal of skim milk (0% butterfat) to get lowfat milk (2% butterfat)? **200 gal**

44. Two trains leave towns 192 km apart, traveling toward one another. One train travels 40 km/hr faster than the other. They pass one another 2 hr later. What is the speed of each train? **28 km/hr, 68 km/hr**

45. The perimeter of a triangle is 33 cm. The longest side is 3 cm longer than the medium side. The medium side is twice as long as the shortest side. Find the length of each side of the triangle. **15 cm, 12 cm, 6 cm**

46. Carrie O'Day invests $30,000 in lottery winnings in three ways. With part of the money, she buys a mutual fund paying 9% per year. She uses the second part, $2000 more than the first, to buy utility bonds paying 10% per year. She invests the rest in a tax-free 5% bond. The first year her investments bring a return of $2500. How much is invested at each rate? **$10,000 at 9%, $12,000 at 10%, $8000 at 5%**

◆ *The position of a particle moving in a straight line is given by $s = at^2 + bt + c$, where t is time in seconds and a, b, and c are real numbers.*

47. If $s(0) = 5$, $s(1) = 23$, and $s(2) = 37$, find $s(8)$. **37**

48. If $s(0) = -10$, $s(1) = 6$, and $s(2) = 30$, find $s(10)$. **510**

49. The graph of each equation in a system of three linear equations in two variables corresponds to a line. Describe geometrically what happens when the system has a unique solution, no solution, or infinitely many solutions.

Solve each system by the substitution method. In Exercises 50–55, graph the system.

50. $x^2 = y - 1$
$y = 3x + 5$ **(4, 17), (−1, 2)**

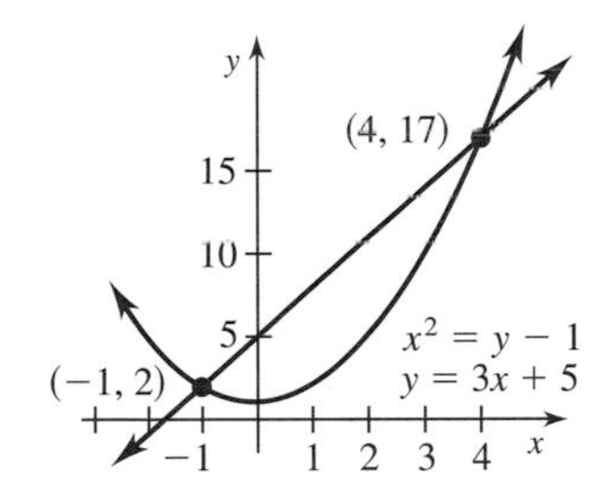

51. $2x^2 = 3y + 23$
$y = 2x - 5$ **(−1, −7), (4, 3)**

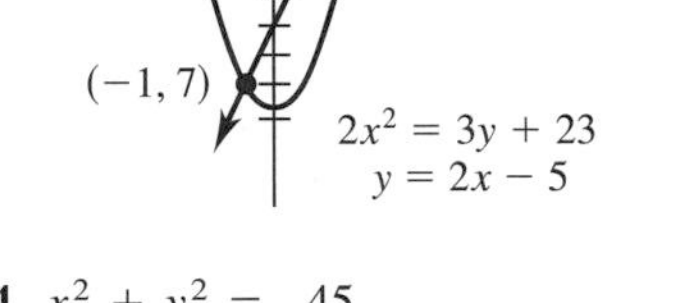

52. $x^2 + y^2 = 5$
$-3x + 4y = 2$
(−2, −1), (38/25, 41/25)

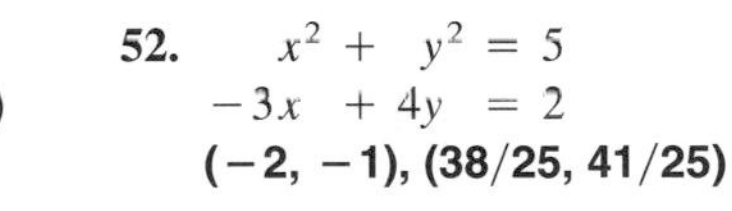

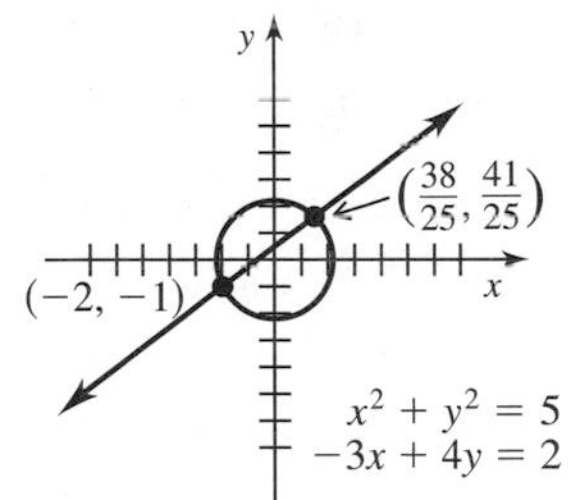

53. $y = x^2$
$x + y = 2$ **(1, 1), (−2, 4)**

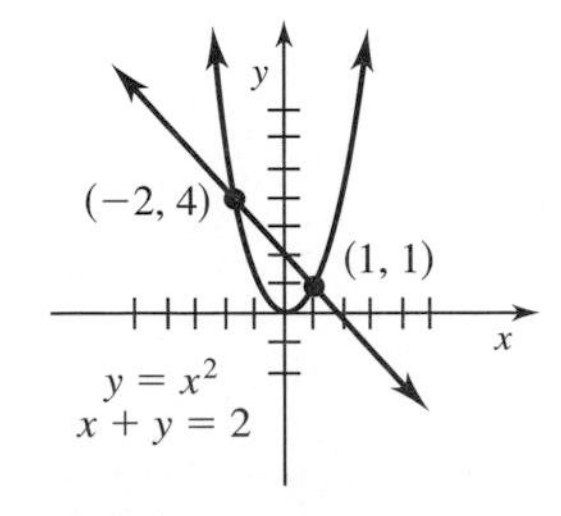

54. $x^2 + y^2 = 45$
$x + y = -3$ **(−6, 3), (3, −6)**

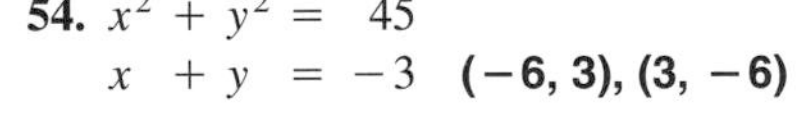

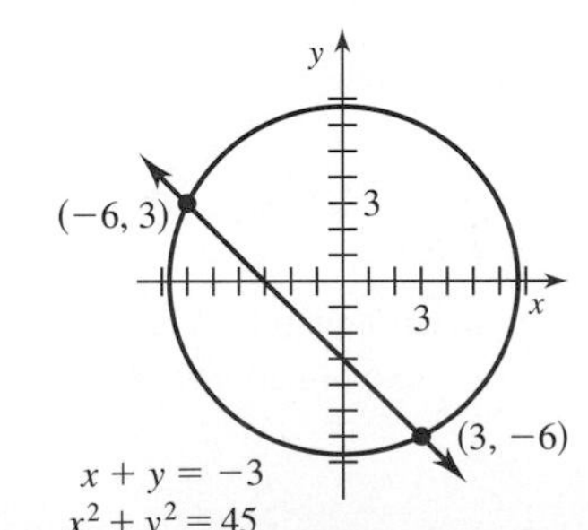

55. $y = x^2 + 6x + 9$
$x + 2y = -2$
(−4, 1), (−5/2, 1/4)

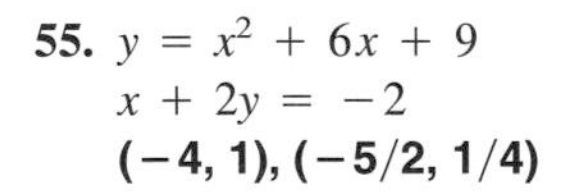

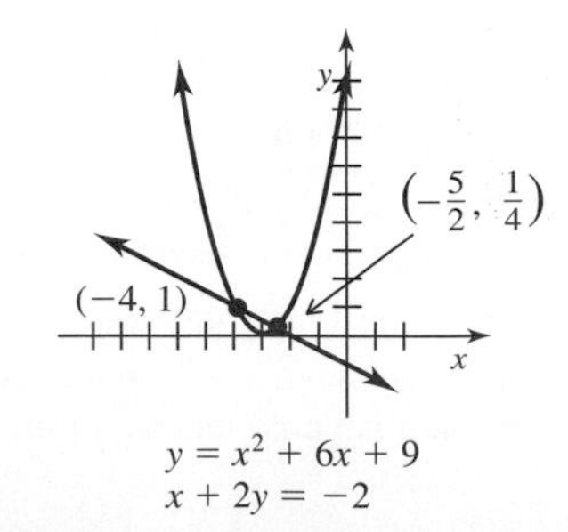

56. $x^2 - y = -1$
$3x = y - 11$ **(−2, 5),(5, 26)**

57. $x = y - 2$
$y^2 = 3x^2 + 4$ **(0, 2), (2, 4)**

58. $y = -x^2 + 2$
$x - y = 0$ **(1, 1), (−2, −2)**

59. $y = x^2 - 2x + 1$
$x - 3y = -1$ **(2, 1), (1/3, 4/9)**

60. $x^2 - 3y^2 = 22$
$x + 3y = 2$ **(−7, 3), (5, −1)**

61. $3x^2 + 2y^2 = 5$
$x - y = -2$ **(−3/5, 7/5), (−1, 1)**

Solve the following systems by any method.

62. $x^2 + y^2 = 32$
$x^2 - y^2 = 0$ **(4, 4), (−4, 4), (4, −4), (−4, −4)**

63. $x^2 + y^2 = 25$
$2x^2 - y^2 = 2$ **(3, 4), (3, −4), (−3, 4), (−3, −4)**

64. $x^2 + y^2 = 4$
$2x^2 - 3y^2 = -12$ **(0, 2), (0, −2)**

65. $5x^2 - y^2 = 0$
$3x^2 + 4y^2 = 0$ **(0, 0)**

66. $3x^2 + 2y^2 = 5$
$-4x^2 + 3y^2 = -1$ **(1, 1), (−1, 1), (1, −1), (−1, −1)**

67. $2x^2 + 2y^2 = 20$
$3x^2 + 3y^2 = 30$ **$\{(x, y) | x^2 + y^2 = 10\}$**

68. $3x^2 + 5y^2 = 17$
$2x^2 - 3y^2 = 5$ **(2, 1), (−2, 1), (2, −1), (−2, −1)**

69. $x^2 + y^2 = 4$
$4x^2 + 4y^2 = 15$ **no solution**

70. $xy = -4$
$2x + y = -7$ **(−4, 1), (1/2, −8)**

71. $xy = 6$
$x + y = 5$ **(2, 3), (3, 2)**

72. $xy = 8$
$3x + 2y = -16$ **(−4, −2), (−4/3, −6)**

73. $x + 3y = -6$
$xy = 3$ **(−3, −1)**

74. $xy = -15$
$4x + 3y = 3$ **(−3, 5), (15/4, −4)**

75. $2xy + 1 = 0$
$x + 16y = 2$ **(4, −1/8), (−2, 1/4)**

76. $-5xy + 2 = 0$
$x - 15y = 5$ **(6, 1/15), (−1, −2/5)**

Use any method to solve the following problems.

77. Find two numbers whose sum is 17 and whose product is 42. **14 and 3**

78. Find two numbers whose sum is 10 and whose squares differ by 20. **6 and 4**

79. Find two numbers whose squares have a sum of 100 and a difference of 28. **8 and 6, 8 and −6, −8 and 6, −8 and −6**

80. The longest side of a right triangle is 13 m in length. One of the other sides is 7 m longer than the shortest side. Find the length of the two shorter sides of the triangle. **5 m and 12 m**

81. For what values of a does the system $x^2 + y^2 = a$ and $x^2 - y^2 = 0$ have no solution? **$a < 0$**

82. Does the straight line $3x - 2y = 9$ intersect the circle $x^2 + y^2 = 25$? (*Hint:* To find out, solve the system made up of these two equations.) **yes**

83. For what values of b will the line $x + 2y = b$ touch the circle $x^2 + y^2 = 9$ at only one point? **$\pm 3\sqrt{5}$**

84. For what values of a do the graphs of $x^2 + y^2 = 25$ and $x^2/a^2 + y^2/25 = 1$ have exactly two points in common? **$a \neq 5, a \neq -5, a \neq 0$**

85. In this section we saw one way that a system of two linear equations could have no solution. If we have a system that is linear in x^2 and y^2, explain another way that this system could have no real solution.

Graphing Utility Problems

In Exercises 86–94, graph each equation and determine the number of solutions (none, exactly one, or infinitely many). If there is exactly one solution, use ZOOM *and* TRACE *to estimate the solution.*

86. $\sqrt{3}x - y = 5$
$100x + y = 9$ **(.138, −4.76)**

87. $\frac{11}{3}x + y = \frac{1}{2}$
$.6x - y = 3$ **(.820, −2.51)**

88. $.2x + \sqrt{2}y = 1$
$\sqrt{5}x + .7y = 1$ **(.236, .674)**

89. $\sqrt{7}x + \sqrt{2}y = 3$
$\sqrt{6}x - y = \sqrt{3}$ **(.892, .453)**

90. $\sqrt{2}x + y = \sqrt{8}$
$2x + \sqrt{2}y = 4$
infinitely many

91. $.2x - y = .1$
$-x + 5y = 2$ **none**

92. $2x - 3y = 4$
$5x + y = 2$
$7x - 4y = 8$ **none**

93. $2x - y = 1$
$6x + 2y = 18$
$-x + 5y = 13$ **(2, 3)**

94. $.4x - .5y = .1$
$-2x + 2.5y = -\frac{1}{2}$
$\frac{6}{5}x - \frac{3}{2}y = .3$ **infinitely many**

6.2 Exercises (text page 313)

Graph the solution of each system of inequalities in Exercises 1–28.

1. $x + y \geq 0$
$2x - y \geq 3$

2. $x + y \leq 4$
$x - 2y \geq 6$

3. $2x + y > 2$
$x - 3y < 6$

4. $4x + 3y < 12$
$y + 4x > -4$

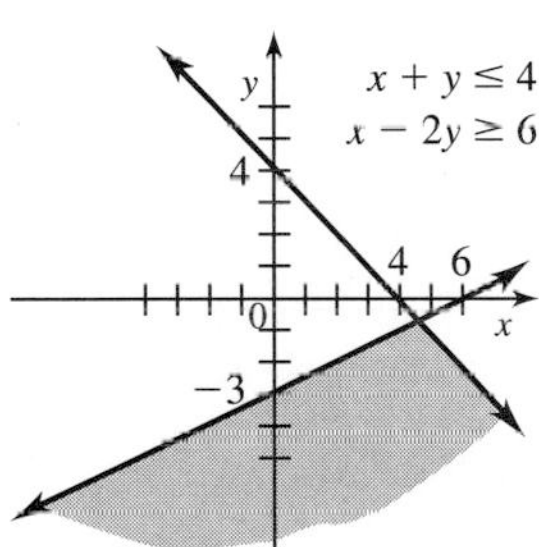

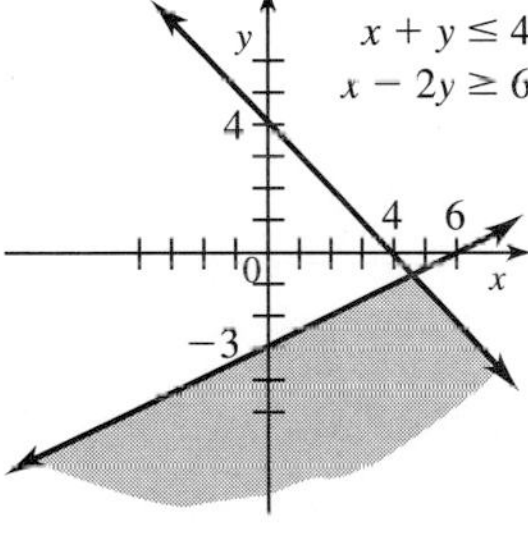

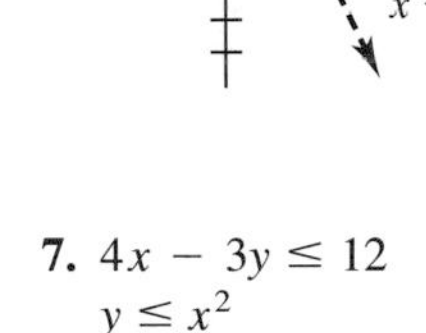

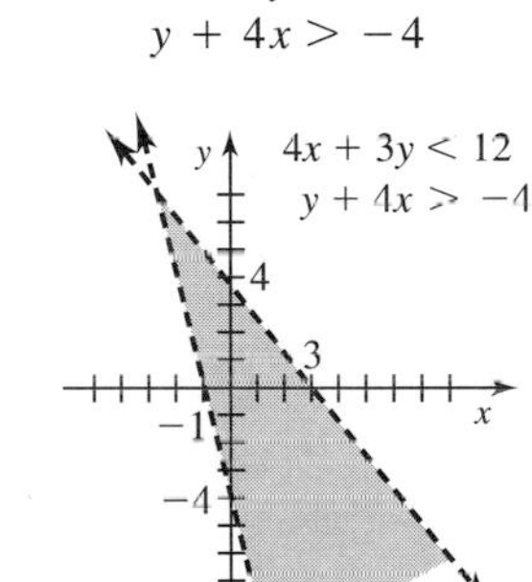

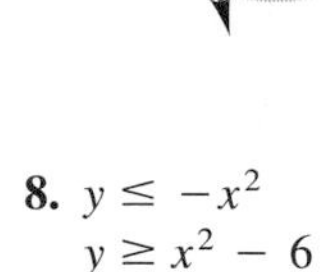

5. $3x + 5y \leq 15$
$x - 3y \geq 9$

6. $y \leq x$
$x^2 + y^2 < 1$

7. $4x - 3y \leq 12$
$y \leq x^2$

8. $y \leq -x^2$
$y \geq x^2 - 6$

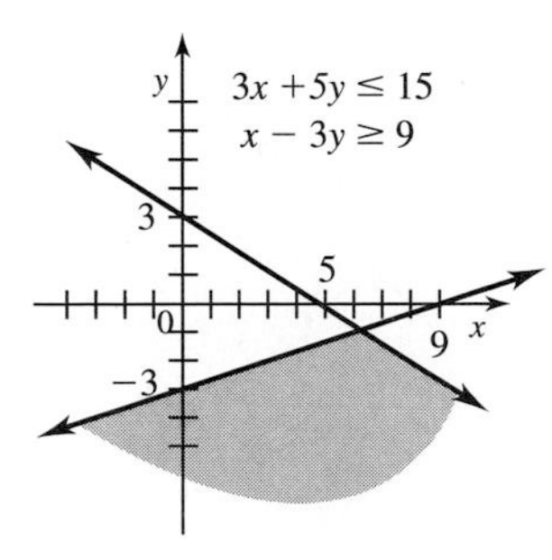

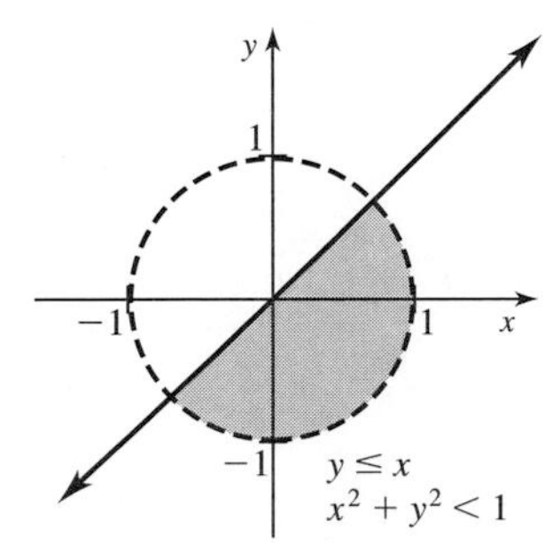

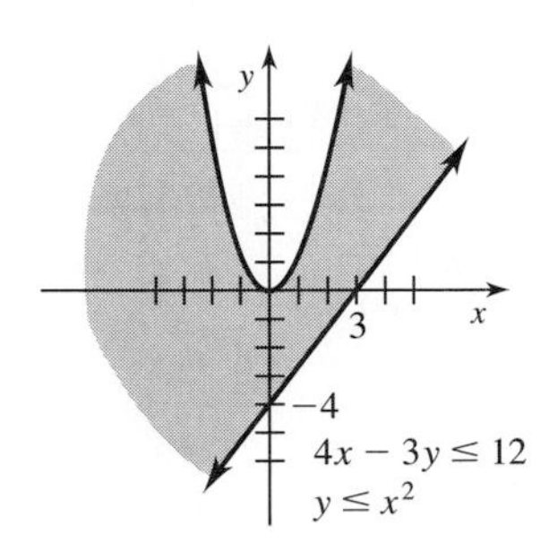

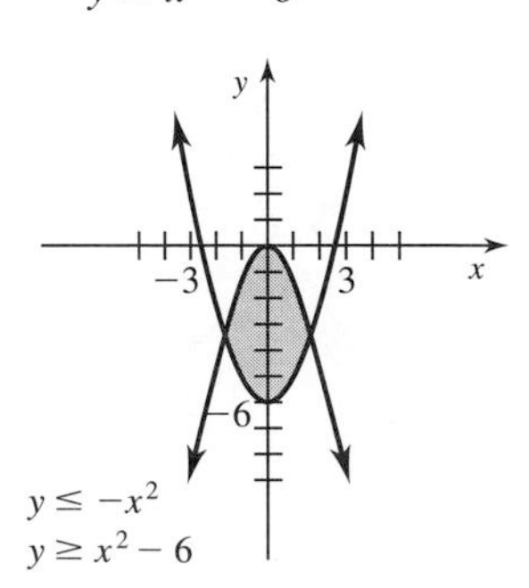

9. $x + y \le 9$
$x \le -y^2$

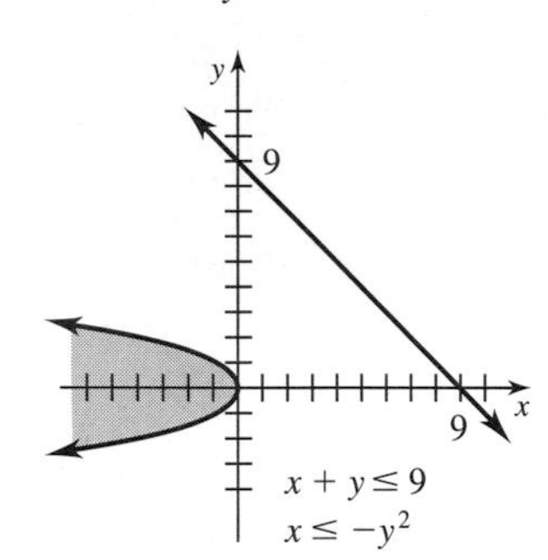

10. $x + 2y \le 4$
$y \ge x^2 - 1$

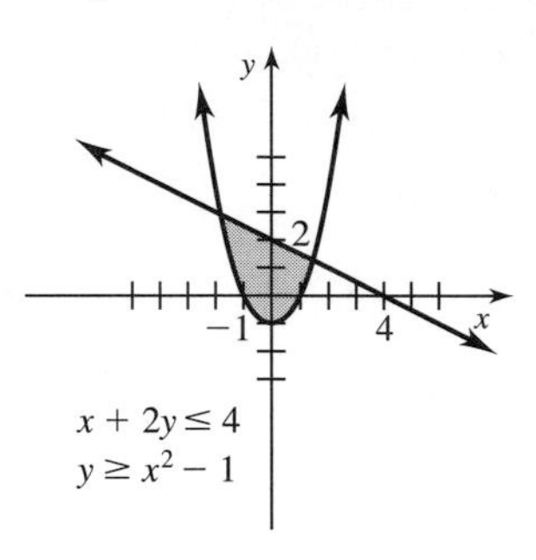

11. $y \le (x + 2)^2$
$y \ge -2x^2$

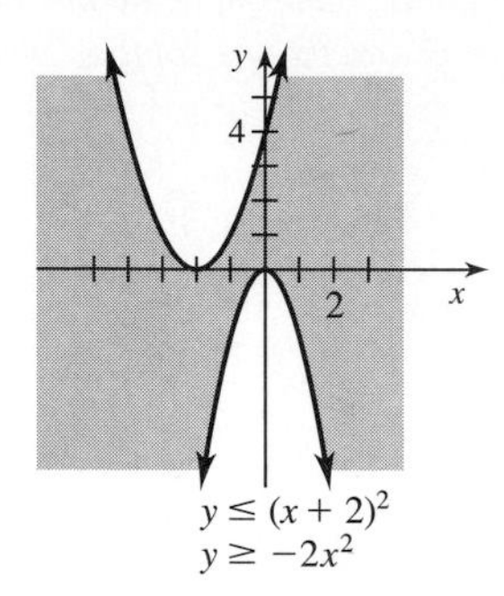

12. $x - y < 1$
$-1 < y < 1$

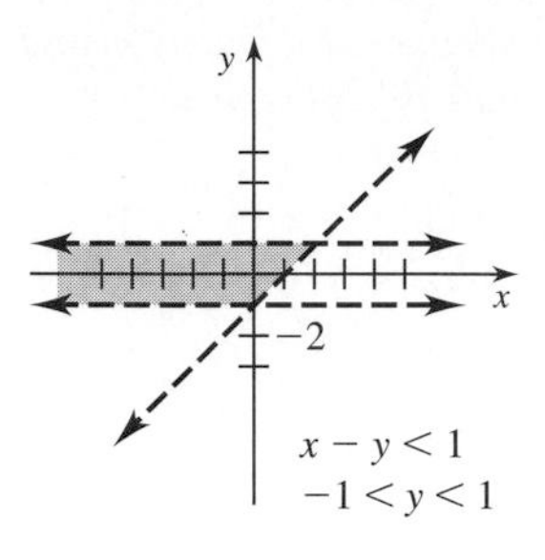

13. $x + y \le 36$
$-4 \le x \le 4$

$x + y \le 36$
$-4 \le x \le 4$

14. $y \ge x^2 + 4x + 4$
$y < -x^2$

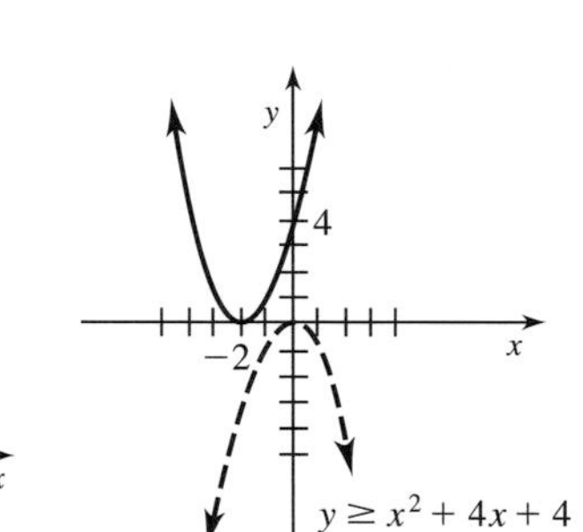

15. $y \ge (x - 2)^2 + 3$
$y \le -(x - 1)^2 + 6$

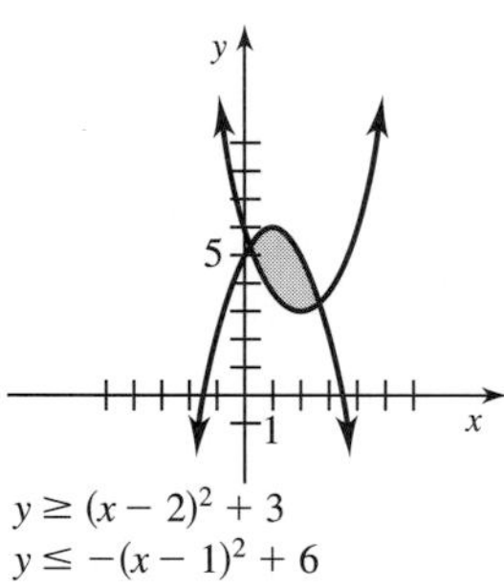

16. $x \ge 0$
$x + y \le 4$
$2x + y \le 5$

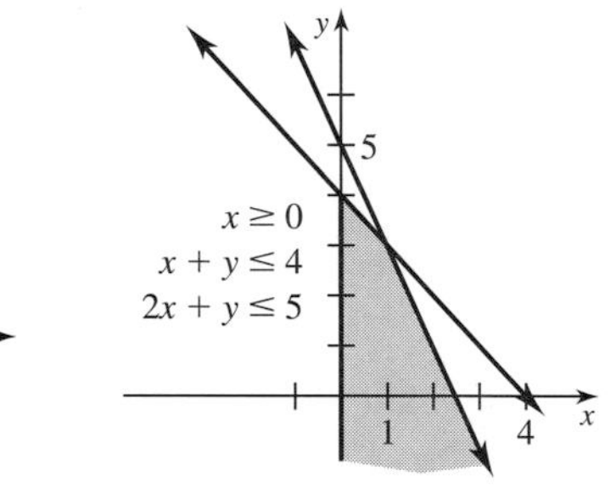

17. $3x - 2y \ge 6$
$x + y \le -5$
$y \le 4$

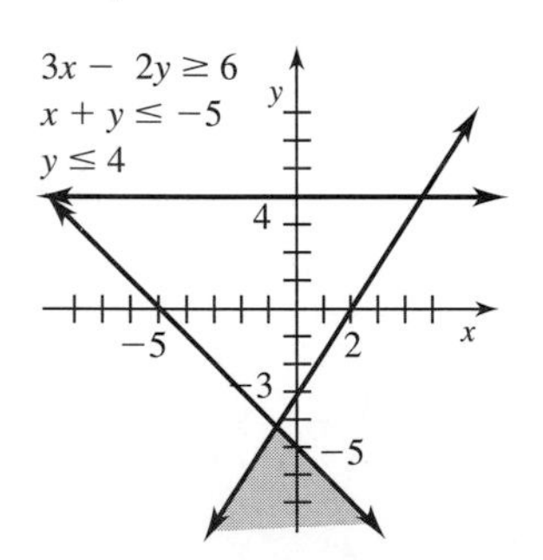

18. $-2 < x < 3$
$-1 \le y \le 5$
$2x + y < 6$

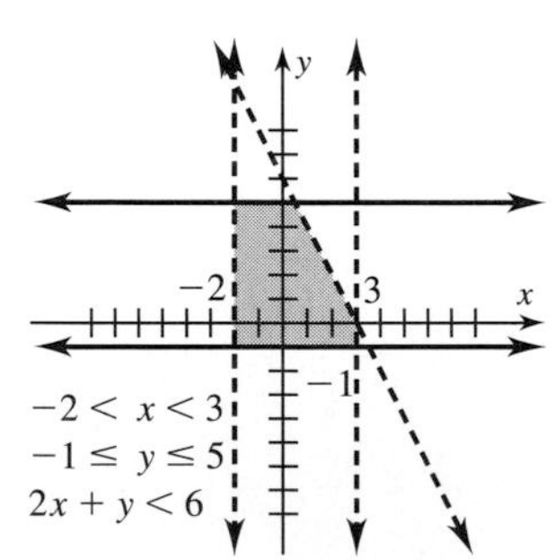

19. $-2 < x < 2$
$y > 1$
$x - y > 0$

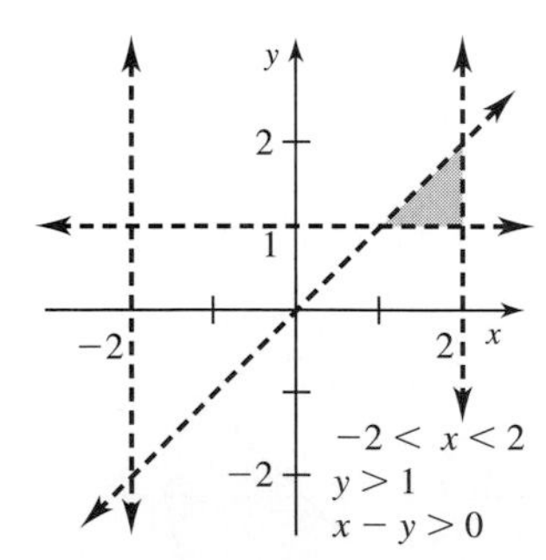

20. $x + y \le 4$
$x - y \le 5$
$4x + y \le -4$

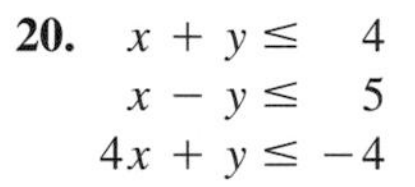

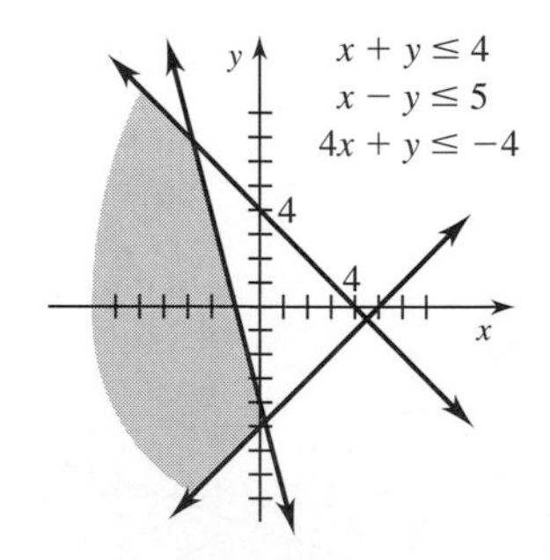

21. $x \le 4$
$x \ge 0$
$y \ge 0$
$x + 2y \ge 2$

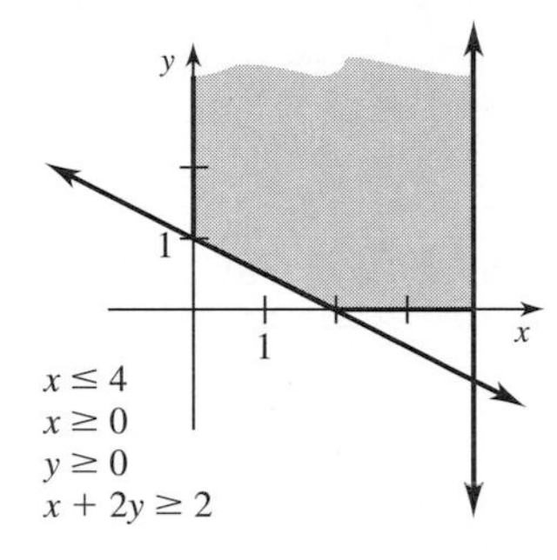

22. $2y + x \ge -5$
$y \le 3 + x$
$x \le 0$
$y \le 0$

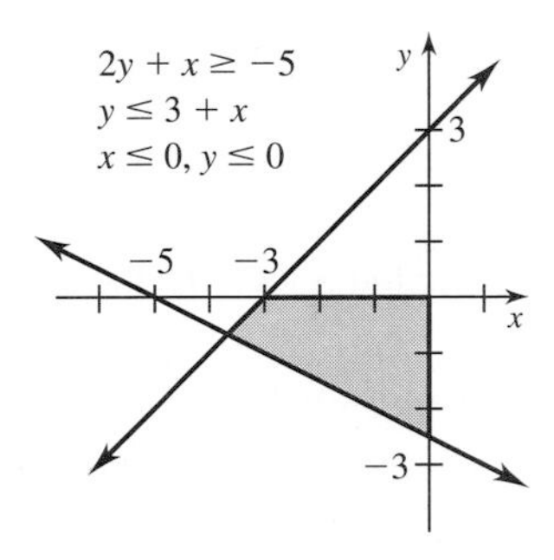

23. $2x + 3y \le 12$
$2x + 3y > -6$
$3x + y < 4$
$x \ge 0$
$y \ge 0$

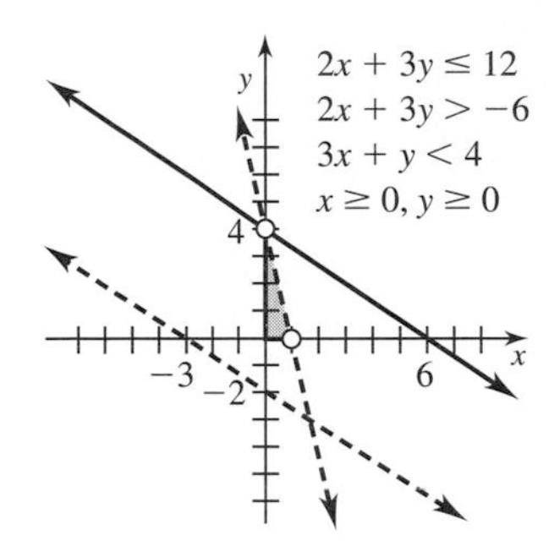

24. $y \ge 3^x$
$y \ge 2$

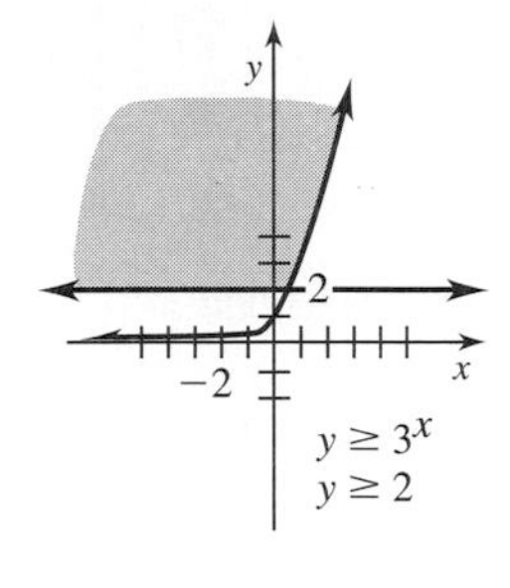

25. $y \le \left(\frac{1}{2}\right)^x$
$y \ge 4$

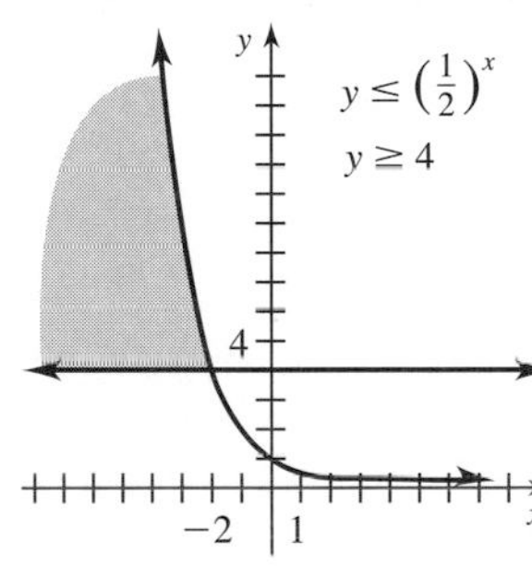

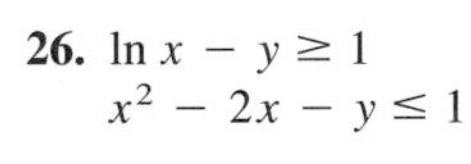

26. $\ln x - y \ge 1$
$x^2 - 2x - y \le 1$

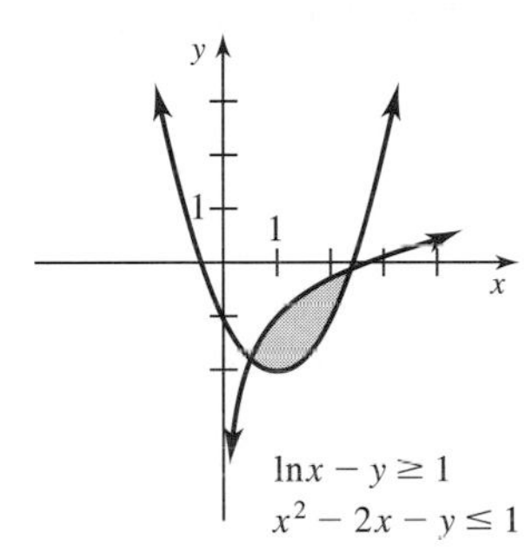

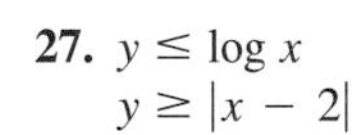

27. $y \le \log x$
$y \ge |x - 2|$

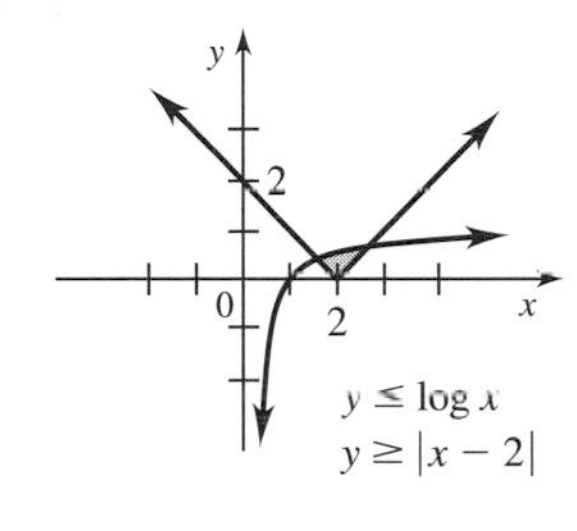

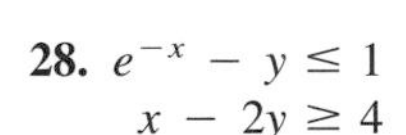

28. $e^{-x} - y \le 1$
$x - 2y \ge 4$

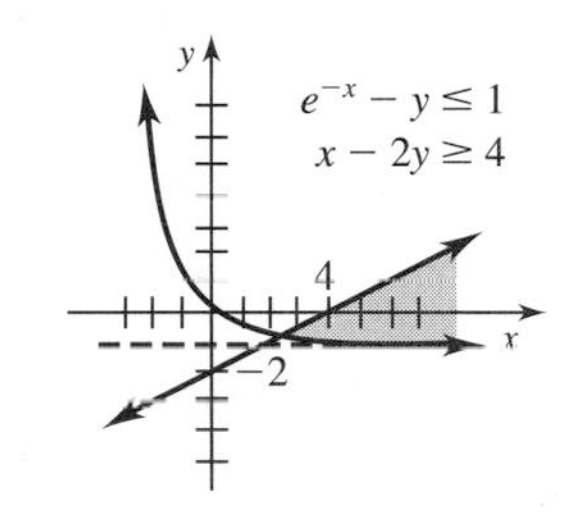

◉ **29.** Find a system of linear inequalities for which the graph is the region in the first quadrant between the pair of lines $x + 2y - 8 = 0$ and $x + 2y = 12$.

$\mathbf{x + 2y - 8 \ge 0}$
$\mathbf{x + 2y \le 12}$
$\mathbf{x \ge 0,\ y \ge 0}$

◉ **30.** Find a linear inequality in two variables whose graph does not intersect the graph of $y \ge 3x + 5$. $\mathbf{y < 3x + 5}$

◆ ◉ **31.** Find an inequality in two variables whose graph does not intersect the graph of $y \le |x|$. $\mathbf{y > |x|}$

The graphs in Exercises 32–35 show regions of feasible solutions. Find the maximum and minimum values of the given expressions.

32. $3x + 5y$

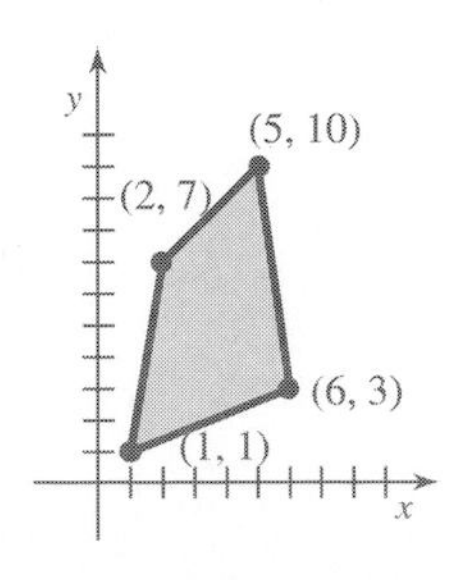

maximum of 65 at (5, 10); minimum of 8 at (1, 1)

33. $6x + y$

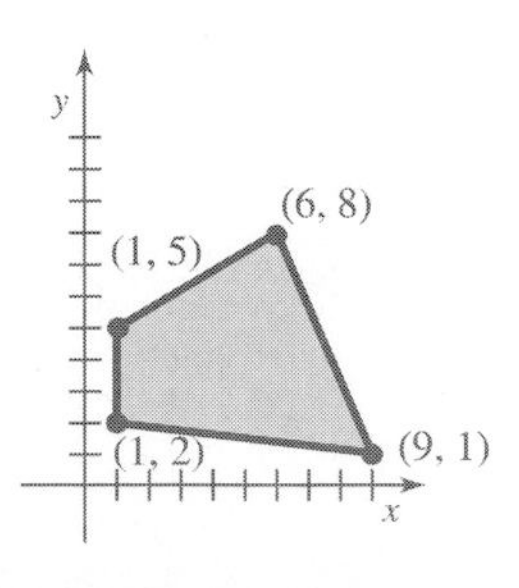

maximum of 55 at (9, 1); minimum of 8 at (1, 2)

34. $40x + 75y$

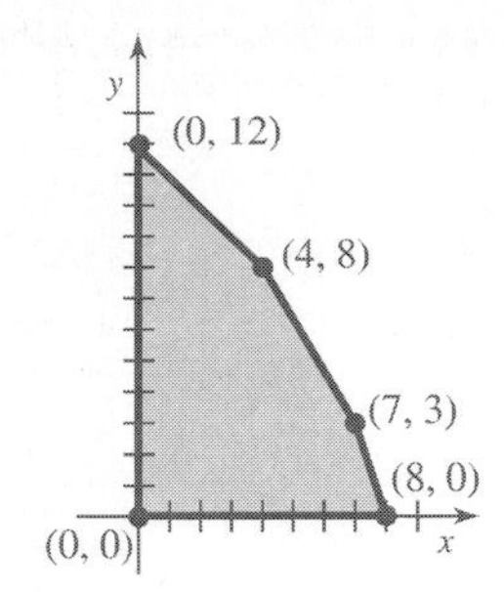

maximum of 900 at (0, 12); minimum of 0 at (0, 0)

35. $35x + 125y$

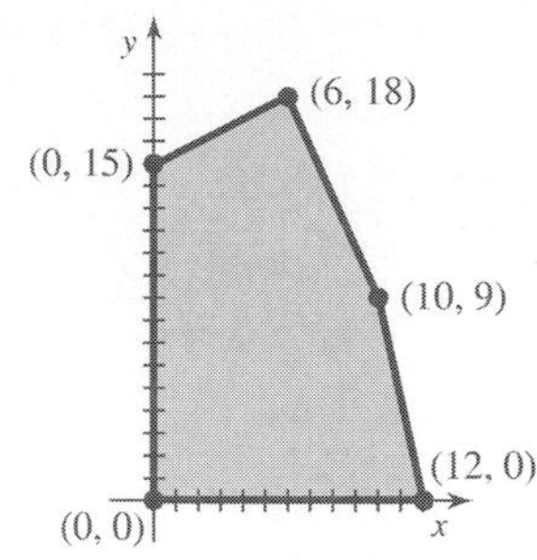

maximum of 2460 at (6, 18); minimum of 0 at (0, 0)

For Exercises 36–39, find the maximum and minimum values of the given expressions over the feasible set shown below.

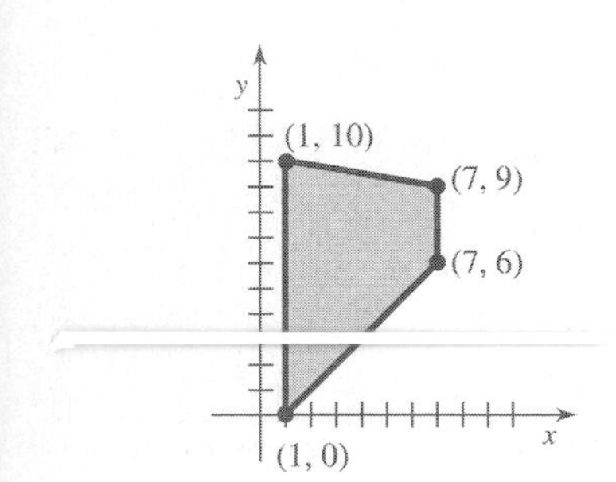

36. $3x + 5y$ **maximum of 66 at (7, 9); minimum of 3 at (1, 0)**

37. $5x + 5y$ **maximum of 80 at (7, 9); minimum of 5 at (1, 0)**

38. $10y$ **maximum of 100 at (1, 10); minimum of 0 at (1, 0)**

39. $3x - y$ **maximum of 15 at (7, 6); minimum of −7 at (1, 10)**

In Exercises 40–43, use graphical methods to solve each problem.

40. Find $x \geq 0$ and $y \geq 0$ such that

$$2x + 3y \leq 6$$
$$4x + y \leq 6,$$

and $5x + 2y$ is maximized.
maximum of 42/5 at (6/5, 6/5)

41. Find $x \geq 0$ and $y \geq 0$ such that

$$x + y \leq 10$$
$$5x + 2y \geq 20$$
$$2y \geq x,$$

and $x + 3y$ is minimized.
minimum of 25/3 at (10/3, 5/3)

42. Find $x \geq 2$ and $y \geq 5$ such that

$$3x - y \geq 12$$
$$x + y \leq 15,$$

and $2x + y$ is minimized. **minimum of 49/3 at (17/3, 5)**

43. Find $x \geq 10$ and $y \geq 20$ such that

$$2x + 3y \leq 100$$
$$5x + 4y \leq 200,$$

and $x + 3y$ is maximized. **maximum of 90 at (10, 80/3)**

44. Why is there no maximum value for the cost function in Example 5? In general, what sort of feasible set has no maximum value for a linear expression $ax + by$ where $a \geq 0$ and $b \geq 0$?

Write a system of inequalities for each of the following problems and then graph the region of feasible solutions of the system.

45. Ms. Oliveras was given the following advice. She should supplement her daily diet with at least 6000 USP units of vitamin A, at least 195 mg of vitamin C, and at least 600 USP units of vitamin D. Ms. Oliveras finds that Mason's Pharmacy carries Brand X and Brand Y vitamins. Each Brand X pill contains 3000 USP units of A, 45 mg of C, and 75 USP units of D, while the Brand Y pills contain 1000 USP units of A, 50 mg of C, and 200 USP units of D. **Let x = number of Brand X pills and y = number of Brand Y pills. Then $3000x + 1000y \geq 6000$, $45x + 50y \geq 195$, $75x + 200y \geq 600$, $x \geq 0$, $y \geq 0$.**

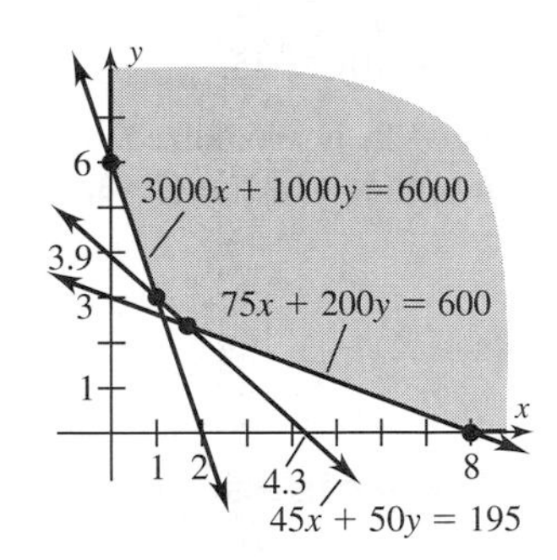

46. The California Almond Growers have 2400 boxes of almonds to be shipped from their plant in Sacramento to Des Moines and San Antonio. The Des Moines market needs at least 1000 boxes, while the San Antonio market must have at least 800 boxes. **Let x = number of boxes to Des Moines and y = number of boxes to San Antonio. Then $x \geq 1000$, $y \geq 800$, $x + y \leq 2400$.**

Solve each of the following linear programming problems.

47. The manufacturing process requires that oil refineries must manufacture at least two barrels of gasoline for every barrel of fuel oil. To meet the winter demand for fuel oil, at least 3 million barrels per day must be produced. The demand for gasoline is no more than 6.4 million barrels per day. If the price of gasoline is \$1.90 and the price of fuel oil is \$1.50 per gallon, how much of each should be produced to maximize revenue? **6.4 million gal of gasoline and 3.2 million gal of fuel oil, for maximum revenue of \$16,960,000**

48. Theo, who is dieting, requires two food supplements, I and II. He can get these supplements from two different products, *A* and *B*, as shown in the following table.

Supplement (g/serving)		*I*	*II*
Product	*A*	3	2
	B	2	4

Theo's physician has recommended that he include at least 15 g of each supplement in his daily diet. If product *A* costs 25¢ per serving and product *B* costs 40¢ per serving, how can he satisfy his requirements most economically? **$3\frac{3}{4}$ servings of A and $1\frac{7}{8}$ servings of B, for minimum cost of \$1.69**

49. Use your own words to describe what a linear programming problem is and how it can be solved graphically.

Graphing Utility Problems

Solve each of the following linear programming problems with a graphing utility. After you graph the region of feasible solutions, use ZOOM *and* TRACE *to estimate the coordinates of the vertices.*

50. Find $x \geq 0$ and $y \geq 0$ such that

$$x + y \leq 8$$
$$3x + y \leq 11,$$

and $6x + 4y$ is maximized.
maximum of 35 at (1.5, 6.5)

51. Find $x \geq 0$ and $y \geq 0$ such that

$$8x + 9y \leq 72$$
$$5x + 2y \leq 26,$$

and $3x + 2y$ is maximized.
maximum of 19.78 at (3.10, 5.24)

 **Writing** ◉ **Conceptual** 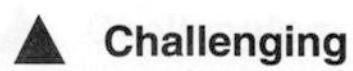 

47. In Exercise 46, suppose we drop the condition that the amount borrowed at 10% is $2000 more than one-half the amount borrowed at 8%. How is the solution changed?

48. Suppose the company in Exercise 46 can borrow only $6000 at 9%. Is a solution possible that still meets the given conditions? Explain.

49. A hospital dietician is planning a special diet for a certain patient. The total amount per meal of food groups A, B, and C must equal 400 g. The diet should include one-third as much of group A as of group B, and the sum of the amounts of group A and group C should equal twice the amount of group B. How many grams of each food group should be included? **44.4 g of A, 133.3 g of B, 222.2 g of C**

50. In Exercise 49, suppose that, in addition to the conditions given there, foods A and B cost 2¢ per gram and food C costs 3¢ per gram, and that a meal must cost $8. Is a solution possible? Explain.

51. At rush hours, substantial traffic congestion is encountered at the traffic intersections shown in the figure. (All streets are one-way.) The city wishes to improve the signals at these corners so as to speed the flow of traffic. The traffic engineers first gather data.

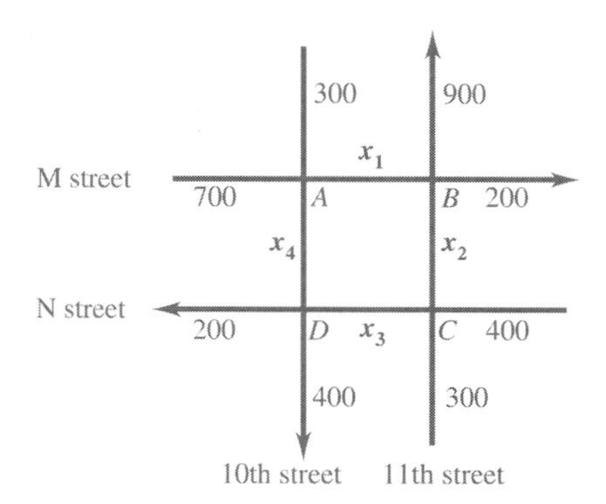

As the figure shows, 700 cars per hour come down M Street to intersection A, and 300 cars per hour come to intersection A on 10th Street. A total of x_1 of these cars leave A on M Street, while x_4 cars leave A on 10th Street. The number of cars entering A must equal the number leaving, so that

$$x_1 + x_4 = 700 + 300$$

or

$$x_1 + x_4 = 1000.$$

For intersection B, x_1 cars enter B on M Street, and x_2 cars enter B on 11th Street. The figure shows that 900 cars leave B on 11th, while 200 leave on M. We have

$$x_1 + x_2 = 900 + 200$$

$$x_1 + x_2 = 1100.$$

At intersection C, 400 cars enter on N Street and 300 on 11th Street, while x_2 leave on 11th Street and x_3 leave on N street. This gives

$$x_2 + x_3 = 400 + 300$$

$$x_2 + x_3 = 700.$$

Finally, intersection D has x_3 cars entering on N and x_4 entering on 10th. There are 400 cars leaving D on 10th and 200 leaving on N.

(a) Set up an equation for intersection D. $\mathbf{x_3 + x_4 = 600}$

(b) Use the four equations to set up an augmented matrix, and then use the Gaussian method to reduce it to triangular form.

$$\left[\begin{array}{cccc|c} 1 & 0 & 0 & 1 & 1000 \\ 1 & 1 & 0 & 0 & 1100 \\ 0 & 1 & 1 & 0 & 700 \\ 0 & 0 & 1 & 1 & 600 \end{array}\right], \left[\begin{array}{cccc|c} 1 & 0 & 0 & 1 & 1000 \\ 0 & 1 & 0 & -1 & 100 \\ 0 & 0 & 1 & 1 & 600 \\ 0 & 0 & 0 & 0 & 0 \end{array}\right]$$

(c) Since you got a row of all zeros, the system of equations does not have a unique solution. Write three equations, corresponding to the three nonzero rows of the matrix. Solve each of the equations for x_4. **$x_4 = 1000 - x_1$, $x_4 = x_2 - 100$, $x_4 = 600 - x_3$**

(d) One of your equations should have been $x_4 = 1000 - x_1$. What is the largest possible value of x_1 so that x_4 is not negative? **1000**

(e) Your second equation should have been $x_4 = x_2 - 100$. Find the smallest possible value of x_2 so that x_4 is not negative. **100**

(f) Find the largest possible values of x_3 and x_4 so that neither variable is negative. **600, 600**

(g) Use the results of (a)–(f) to give a solution for the problem in which all the equations are satisfied and all variables are nonnegative. Is the solution unique?
$x_1 = 1000$, $x_2 = 100$, $x_3 = 600$, $x_4 = 0$; no

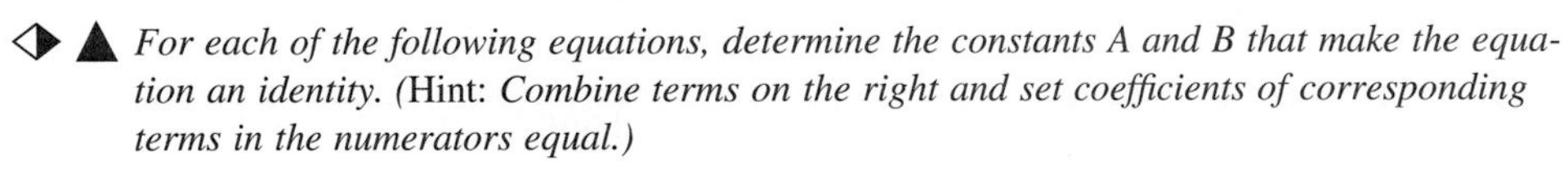

*For each of the following equations, determine the constants A and B that make the equation an identity. (*Hint: *Combine terms on the right and set coefficients of corresponding terms in the numerators equal.)*

52. $\dfrac{1}{(x-1)(x+1)} = \dfrac{A}{x-1} + \dfrac{B}{x+1}$
$A = 1/2$, $B = -1/2$

53. $\dfrac{x+4}{x^2} = \dfrac{A}{x} + \dfrac{B}{x^2}$
$A = 1$, $B = 4$

54. $\dfrac{x}{(x-a)(x+a)} = \dfrac{A}{x-a} + \dfrac{B}{x+a}$
$A = 1/2$, $B = 1/2$

55. $\dfrac{2x}{(x+2)(x-1)} = \dfrac{A}{x+2} + \dfrac{B}{x-1}$
$A = 4/3$, $B = 2/3$

56. Describe the Gaussian reduction method in your own words.

Graphing Utility Problems

If your graphing utility is capable of performing row operations, use the utility to solve the following systems of linear equations by the Gaussian reduction method.

57. $\begin{aligned} \sqrt{2}x - 5y &= .3 \\ -.7x + 3y &= 3/2 \end{aligned}$ **(11.311, 3.139)**

58. $\begin{aligned} 5.1x + \sqrt{3}y &= \sqrt{2} \\ 9x - 2.2y &= 5 \end{aligned}$ **(.439, −.476)**

59. $\begin{aligned} .3x + 2.7y - \sqrt{2}z &= 3 \\ \sqrt{7}x - 20y + 12z &= -2 \\ 4x + \sqrt{3}y - 1.2z &= 3/4 \end{aligned}$ **(.571, 7.041, 11.442)**

60. $\begin{aligned} \sqrt{5}x - 1.2y + z &= -3 \\ (1/2)x - 3y + 4z &= 4/3 \\ 4x + 7y - 9z &= \sqrt{2} \end{aligned}$ **(.407, 9.316, 7.270)**

6.4 Exercises (text page 334)

Find the values of the variables in each of the following.

1. $\begin{bmatrix} w & x \\ y & z \end{bmatrix} = \begin{bmatrix} 3 & 2 \\ -1 & 4 \end{bmatrix}$

$w = 3$, $x = 2$, $y = -1$, $z = 4$

2. $\begin{bmatrix} 0 & 5 & x \\ -1 & 3 & y+2 \\ 4 & 1 & z \end{bmatrix} = \begin{bmatrix} 0 & w+3 & 6 \\ -1 & 3 & 0 \\ 4 & 1 & 8 \end{bmatrix}$

$w = 2$, $x = 6$, $y = -2$, $z = 8$

3. $\begin{bmatrix} 2 & 5 & 6 \\ 1 & m & n \end{bmatrix} = \begin{bmatrix} z & y & w \\ 1 & 8 & -2 \end{bmatrix}$

$m = 8, n = -2, z = 2, y = 5, w = 6$

4. $\begin{bmatrix} -7+z & 4r & 8s \\ 6p & 2 & 5 \end{bmatrix} + \begin{bmatrix} -9 & 8r & 3 \\ 2 & 5 & 4 \end{bmatrix} = \begin{bmatrix} 2 & 36 & 27 \\ 20 & 7 & 12a \end{bmatrix}$

$z = 18, r = 3, s = 3, p = 3, a = 3/4$

5. $\begin{bmatrix} a+2 & 3z+1 & 5m \\ 8k & 0 & 3 \end{bmatrix} + \begin{bmatrix} 3a & 2z & 5m \\ 2k & 5 & 6 \end{bmatrix} = \begin{bmatrix} 10 & -14 & 80 \\ 10 & 5 & 9 \end{bmatrix}$ **$a = 2, z = -3, m = 8, k = 1$**

In each of the following, determine whether the two matrices are equal.

6. $[1 \quad 2], \begin{bmatrix} 1 \\ 2 \end{bmatrix}$ **not equal**

7. $\begin{bmatrix} 1 & 2 \\ 3 & 4 \end{bmatrix}, \begin{bmatrix} 1 & 2 & 0 \\ 3 & 4 & 0 \end{bmatrix}$ **not equal**

Perform each of the operations in Exercises 8–13, whenever possible.

8. $\begin{bmatrix} 6 & -9 & 2 \\ 4 & 1 & 3 \end{bmatrix} - \begin{bmatrix} -8 & 2 & 5 \\ 6 & -3 & 4 \end{bmatrix}$ $\begin{bmatrix} \mathbf{14} & \mathbf{-11} & \mathbf{-3} \\ \mathbf{-2} & \mathbf{4} & \mathbf{-1} \end{bmatrix}$

9. $\begin{bmatrix} 9 & 4 \\ -8 & 2 \end{bmatrix} + \begin{bmatrix} -3 & 2 \\ -4 & 7 \end{bmatrix}$ $\begin{bmatrix} \mathbf{6} & \mathbf{6} \\ \mathbf{-12} & \mathbf{9} \end{bmatrix}$

10. $\begin{bmatrix} -6 & 8 \\ 0 & 0 \end{bmatrix} - \begin{bmatrix} 0 & 0 \\ -4 & -2 \end{bmatrix}$ $\begin{bmatrix} \mathbf{-6} & \mathbf{8} \\ \mathbf{4} & \mathbf{2} \end{bmatrix}$

11. $\begin{bmatrix} 1 & -4 \\ 2 & -3 \\ -8 & 4 \end{bmatrix} - \begin{bmatrix} -6 & 9 \\ -2 & 5 \\ -7 & -12 \end{bmatrix}$ $\begin{bmatrix} \mathbf{7} & \mathbf{-13} \\ \mathbf{4} & \mathbf{-8} \\ \mathbf{-1} & \mathbf{16} \end{bmatrix}$

12. $\begin{bmatrix} 3x+y & x-2y & 2x \\ 5x & 3y & x+y \end{bmatrix} + \begin{bmatrix} 2x & 3y & 5x+y & x-y \\ 3x+2y & x & 2x & 4y \end{bmatrix}$ **Matrices cannot be added.**

13. $\begin{bmatrix} 4k-8y \\ 6z-3x \\ 2k+5a \\ -4m+2n \end{bmatrix} - \begin{bmatrix} 5k+6y \\ 2z+5x \\ 4k+6a \\ 4m-2n \end{bmatrix}$ $\begin{bmatrix} \mathbf{-k-14y} \\ \mathbf{4z-8x} \\ \mathbf{-2k-a} \\ \mathbf{-8m+4n} \end{bmatrix}$

14. A hardware chain does an inventory of a particular size of screw and finds that its Adelphi store has 100 flat-head and 150 round-head screws, its Beltsville store has 125 flat and 50 round, and its College Park store has 175 flat and 200 round. Write this information first as a 3 × 2 matrix and then as a 2 × 3 matrix.
$\begin{bmatrix} \mathbf{100} & \mathbf{150} \\ \mathbf{125} & \mathbf{50} \\ \mathbf{175} & \mathbf{200} \end{bmatrix}; \begin{bmatrix} \mathbf{100} & \mathbf{125} & \mathbf{175} \\ \mathbf{150} & \mathbf{50} & \mathbf{200} \end{bmatrix}$

15. At the grocery store, Miguel bought 4 quarts of milk, 2 loaves of bread, 4 potatoes, and an apple. Mary bought 2 quarts of milk, a loaf of bread, 5 potatoes, and 4 apples. Write this information first as a 2 × 4 matrix and then as a 4 × 2 matrix.
$\begin{bmatrix} \mathbf{4} & \mathbf{2} & \mathbf{4} & \mathbf{1} \\ \mathbf{2} & \mathbf{1} & \mathbf{5} & \mathbf{4} \end{bmatrix}; \begin{bmatrix} \mathbf{4} & \mathbf{2} \\ \mathbf{2} & \mathbf{1} \\ \mathbf{4} & \mathbf{5} \\ \mathbf{1} & \mathbf{4} \end{bmatrix}$

16. For any size matrix A, the zero matrix O of the same size is the *additive identity* such that $A + 0 = A$. For any 2 × 2 matrix A, find the 2 × 2 matrix I that is the *multiplicative identity*, that is, for which $A \cdot I = A$. $\begin{bmatrix} \mathbf{1} & \mathbf{0} \\ \mathbf{0} & \mathbf{1} \end{bmatrix}$

Let $A = \begin{bmatrix} -2 & 4 \\ 0 & 3 \end{bmatrix}$ *and* $B = \begin{bmatrix} -6 & 2 \\ 4 & 0 \end{bmatrix}$. *Find each of the following.*

17. $2A$ $\begin{bmatrix} \mathbf{-4} & \mathbf{8} \\ \mathbf{0} & \mathbf{6} \end{bmatrix}$

18. $-3B$ $\begin{bmatrix} \mathbf{18} & \mathbf{-6} \\ \mathbf{-12} & \mathbf{0} \end{bmatrix}$

19. $2A - B$ $\begin{bmatrix} \mathbf{2} & \mathbf{6} \\ \mathbf{-4} & \mathbf{6} \end{bmatrix}$

20. $-2A + 4B$ $\begin{bmatrix} \mathbf{-20} & \mathbf{0} \\ \mathbf{16} & \mathbf{-6} \end{bmatrix}$

21. $-A + \frac{1}{2}B$ $\begin{bmatrix} \mathbf{-1} & \mathbf{-3} \\ \mathbf{2} & \mathbf{-3} \end{bmatrix}$

22. $\frac{3}{4}A - B$ $\begin{bmatrix} \mathbf{9/2} & \mathbf{1} \\ \mathbf{-4} & \mathbf{9/4} \end{bmatrix}$

Find the matrix products in Exercises 23–34, whenever possible.

23. $\begin{bmatrix} 1 & 2 \\ 3 & 4 \end{bmatrix}\begin{bmatrix} -1 \\ 7 \end{bmatrix}$ $\begin{bmatrix} \mathbf{13} \\ \mathbf{25} \end{bmatrix}$

24. $\begin{bmatrix} 3 & -4 & 1 \\ 5 & 0 & 2 \end{bmatrix}\begin{bmatrix} -1 \\ 4 \\ 2 \end{bmatrix}$ $\begin{bmatrix} \mathbf{-17} \\ \mathbf{-1} \end{bmatrix}$

25. $\begin{bmatrix} 5 & 6 \\ 3 & 4 \end{bmatrix}\begin{bmatrix} 0 & 3 \\ -1 & 2 \end{bmatrix}$ $\begin{bmatrix} \mathbf{-6} & \mathbf{27} \\ \mathbf{-4} & \mathbf{17} \end{bmatrix}$

26. $\begin{bmatrix} -2 & 1 & 3 \\ 7 & 0 & -1 \\ 0 & 2 & 1 \end{bmatrix}\begin{bmatrix} 0 & 1 & 1 \\ -1 & 2 & 0 \\ 3 & 1 & 4 \end{bmatrix}$ $\begin{bmatrix} \mathbf{8} & \mathbf{3} & \mathbf{10} \\ \mathbf{-3} & \mathbf{6} & \mathbf{3} \\ \mathbf{1} & \mathbf{5} & \mathbf{4} \end{bmatrix}$

27. $\begin{bmatrix} -2 & 1 & 4 \\ 0 & 1 & 2 \end{bmatrix}\begin{bmatrix} -2 & 1 & 0 \\ 0 & -2 & 0 \\ 4 & 1 & 2 \end{bmatrix}$ $\begin{bmatrix} \mathbf{20} & \mathbf{0} & \mathbf{8} \\ \mathbf{8} & \mathbf{0} & \mathbf{4} \end{bmatrix}$

28. $\begin{bmatrix} -1 & 0 & 0 \\ 2 & 1 & 4 \end{bmatrix}\begin{bmatrix} 4 & -2 & 5 \\ 0 & 1 & 4 \\ 2 & -9 & 0 \end{bmatrix}$ $\begin{bmatrix} \mathbf{-4} & \mathbf{2} & \mathbf{-5} \\ \mathbf{16} & \mathbf{-39} & \mathbf{14} \end{bmatrix}$

29. $\begin{bmatrix} -3 & 0 & 2 & 1 \\ 4 & 0 & 2 & 6 \end{bmatrix}\begin{bmatrix} -4 & 2 \\ 0 & 1 \end{bmatrix}$
cannot be multiplied

30. $\begin{bmatrix} -1 & 2 & 4 & 1 \\ 0 & 2 & -3 & 5 \end{bmatrix}\begin{bmatrix} 1 & 2 & 4 \\ -2 & 5 & 1 \end{bmatrix}$
cannot be multiplied

31. $\begin{bmatrix} 5 & -1 & 2 \end{bmatrix}\begin{bmatrix} 2 \\ 1 \\ -1 \end{bmatrix}$ $\mathbf{[7]}$

32. $\begin{bmatrix} 7 & 5 & 4 & -6 \end{bmatrix}\begin{bmatrix} 1 \\ 0 \\ 1 \\ 0 \end{bmatrix}$ $\mathbf{[11]}$

33. $\begin{bmatrix} 2 & 1 & -3 \\ 1 & 0 & 4 \end{bmatrix}\begin{bmatrix} 1 \\ 0 \\ 0 \end{bmatrix}$ $\begin{bmatrix} \mathbf{2} \\ \mathbf{1} \end{bmatrix}$

34. $\begin{bmatrix} -6 \\ -1 \\ -2 \end{bmatrix}\begin{bmatrix} 3 & 0 & 1 \end{bmatrix}$ $\begin{bmatrix} \mathbf{-18} & \mathbf{0} & \mathbf{-6} \\ \mathbf{-3} & \mathbf{0} & \mathbf{-1} \\ \mathbf{-6} & \mathbf{0} & \mathbf{-2} \end{bmatrix}$

35. The Bread Box, a small neighborhood bakery, sells four main items: sweet rolls, bread, cake, and pie. The amount of certain major ingredients (measured in cups except for eggs) required to make these items is given in matrix A.

$A =$

Eggs	Flour	Sugar	Shortening	Milk	
1	4	1/4	1/4	1	**Dozen rolls**
0	3	0	1/4	0	**Loaf of bread**
4	3	2	1	1	**Cake (1)**
0	1	0	1/3	0	**Pie (1)**

The cost per cup or per egg (in cents) for each ingredient when purchased in large lots and in small lots is given by matrix B.

Cost

$B =$

Large lot	Small lot
5	5
8	10
10	12
12	15
5	6

(a) Use matrix multiplication to find a matrix representing the comparative costs per item under the two purchase options. $\begin{bmatrix} \mathbf{47.5} & \mathbf{57.75} \\ \mathbf{27} & \mathbf{33.75} \\ \mathbf{81} & \mathbf{95} \\ \mathbf{12} & \mathbf{15} \end{bmatrix}$

Suppose a day's orders consist of 20 dozen sweet rolls, 200 loaves of bread, 50 cakes, and 60 pies.

(b) Represent these orders as a 1×4 matrix and use matrix multiplication to write as a matrix the amount of each ingredient required to fill the day's orders. **[20 200 50 60], [220 890 105 125 70]**

(c) Use matrix multiplication to find a matrix representing the costs under the two purchase options to fill the day's orders. **[11,120 13,555]**

36. In what ways is the matrix I like the real number 1? (See Exercise 16.)

For Exercises 37–51, let

$$A = \begin{bmatrix} a_{11} & a_{12} \\ a_{21} & a_{22} \end{bmatrix}, \quad B = \begin{bmatrix} b_{11} & b_{12} \\ b_{21} & b_{22} \end{bmatrix}, \quad \text{and} \quad C = \begin{bmatrix} c_{11} & c_{12} \\ c_{21} & c_{22} \end{bmatrix}$$

where all the elements are real numbers. Decide which of the following statements are true for these three matrices. If a statement is true, prove that it is true. If it is false, give a numerical example to show it is false.

37. $A + B = B + A$ (commutative property) **true**

38. $A + (B + C) = (A + B) + C$ (associative property) **true**

39. $A + B$ is a 2×2 matrix. (closure property) **true**

40. There exists a matrix O such that $A + O = A$ and $O + A = A$. (identity property) **true**

41. There exists a matrix $-A$ such that $A + (-A) = O$ and $-A + A = O$. (inverse property) **true**

42. $(AB)C = A(BC)$ (associative property) **true**

43. $A(B + C) = AB + AC$ (distributive property) **true**

44. AB is a 2×2 matrix. (closure property) **true**

45. $c(A + B) = cA + cB$ for any real number c. **true**

46. $(c + d)A = cA + dA$ for any real numbers c and d. **true**

47. $c(A)d = cd(A)$ **true**

48. $(cd)A = c(dA)$ **true**

49. $(A + B)(A - B) = A^2 - B^2$ (where $A^2 = AA$) **false**

50. $(A + B)^2 = A^2 + 2AB + B^2$ **false**

51. If $AB = O$, then $A = O$ or $B = O$. **false**

The **transpose**, *A^T, of a matrix A is found by exchanging the rows and columns of A. That is,*

$$\text{if } A = \begin{bmatrix} a & b \\ c & d \end{bmatrix}, \text{ then } A^T = \begin{bmatrix} a & c \\ b & d \end{bmatrix}.$$

Show that each of the following equations are true for matrices A and B, where

$$B = \begin{bmatrix} m & n \\ p & q \end{bmatrix}.$$

52. $(A^T)^T = A$

53. $(A + B)^T = A^T + B^T$

54. $(AB)^T = B^T A^T$

Graphing Utility Problems

In Exercises 55–58, calculate the given expression, where

$$A = \begin{bmatrix} 1.3 & 17 & -.8 \\ 12 & 3/2 & 5.4 \\ .24 & -5 & 9 \end{bmatrix} \quad \text{and} \quad B = \begin{bmatrix} 45 & 7.2 & 4/5 \\ 9/8 & -43 & 2.2 \\ -11 & .33 & 3/4 \end{bmatrix}.$$

55. $A + B$ $\begin{bmatrix} 46.3 & 24.2 & 0 \\ 13.125 & -41.5 & 7.6 \\ -10.76 & -4.67 & 9.75 \end{bmatrix}$

56. $3A$ $\begin{bmatrix} 3.9 & 51 & -2.4 \\ 36 & 4.5 & 16.2 \\ .72 & -15 & 27 \end{bmatrix}$

57. $2A - 3B$ $\begin{bmatrix} -132.4 & 12.4 & -4 \\ 20.625 & 132 & 4.2 \\ 33.48 & -10.99 & 15.75 \end{bmatrix}$

58. AB $\begin{bmatrix} 86.425 & -721.9 & 37.84 \\ 482.288 & 23.682 & 16.95 \\ -93.825 & 219.698 & -4.058 \end{bmatrix}$

A stochastic matrix is a square matrix for which the sum of the entries in each row is 1. An example is $A = \begin{bmatrix} .8 & .2 \\ .4 & .6 \end{bmatrix}$. *Use A in Exercises 59 and 60.*

59. Show that A^2 is also a stochastic matrix. (*Note:* $A^2 = AA$.)

60. Compute A^3, A^4, A^5, and A^6 by successively multiplying by A. (*Note:* The rows of these powers should get closer and closer to [2/3 1/3].) $A^6 = \begin{bmatrix} .6680 & .3320 \\ .6639 & .3361 \end{bmatrix}$

61. Repeat Exercises 59 and 60 for the matrix $\begin{bmatrix} .1 & .6 & .3 \\ .3 & .4 & .3 \\ .7 & .2 & .1 \end{bmatrix}$. What matrix do the rows get closer and closer to? **[1/3 5/12 1/4]**

6.5 Exercises (text page 346)

Find the value of each determinant. All variables represent real numbers.

1. $\begin{vmatrix} 2 & 5 \\ 4 & -7 \end{vmatrix}$ **−34**

2. $\begin{vmatrix} 3 & 4 \\ 5 & -2 \end{vmatrix}$ **−26**

3. $\begin{vmatrix} -9 & 7 \\ 2 & 6 \end{vmatrix}$ **−68**

4. $\begin{vmatrix} 0 & 4 \\ 4 & 0 \end{vmatrix}$ **−16**

5. $\begin{vmatrix} y & 3 \\ -2 & x \end{vmatrix}$ $\mathbf{yx + 6}$

6. $\begin{vmatrix} y & 2 \\ 8 & y \end{vmatrix}$ $\mathbf{y^2 - 16}$

7. $\begin{vmatrix} 3 & 8 \\ m & n \end{vmatrix}$ $\mathbf{3n - 8m}$

8. $\begin{vmatrix} 2m & 8n \\ 8n & 2m \end{vmatrix}$ $\mathbf{4m^2 - 64n^2}$

Find the cofactor of each element in the second row for the following determinants.

9. $\begin{vmatrix} -2 & 0 & 1 \\ 1 & 2 & 0 \\ 4 & 2 & 1 \end{vmatrix}$ **2, −6, 4**

10. $\begin{vmatrix} 1 & -1 & 2 \\ 1 & 0 & 2 \\ 0 & -3 & 1 \end{vmatrix}$ **−5, 1, 3**

11. $\begin{vmatrix} 1 & 2 & -1 \\ 2 & 3 & -2 \\ -1 & 4 & 1 \end{vmatrix}$ **−6, 0, −6**

12. $\begin{vmatrix} 2 & -1 & 4 \\ 3 & 0 & 1 \\ -2 & 1 & 4 \end{vmatrix}$ **8, 16, 0**

Find the value of each determinant. All variables represent real numbers.

13. $\begin{vmatrix} 1 & 0 & 0 \\ 0 & 1 & 0 \\ 0 & 0 & 1 \end{vmatrix}$ **1**

14. $\begin{vmatrix} 1 & 0 & 0 \\ 0 & -1 & 0 \\ 1 & 0 & 1 \end{vmatrix}$ **−1**

15. $\begin{vmatrix} -2 & 0 & 1 \\ 0 & 1 & 0 \\ 0 & 0 & -1 \end{vmatrix}$ **2**

16. $\begin{vmatrix} 1 & -2 & 3 \\ 0 & 0 & 0 \\ 1 & 10 & -12 \end{vmatrix}$ **0**

17. $\begin{vmatrix} 0 & 5 & 2 \\ 0 & 3 & -1 \\ 0 & -4 & 7 \end{vmatrix}$ **0**

18. $\begin{vmatrix} 3 & 3 & -1 \\ 2 & 6 & 0 \\ -6 & -6 & 2 \end{vmatrix}$ **0**

19. $\begin{vmatrix} 0 & 3 & y \\ 0 & 4 & 2 \\ 1 & 0 & 1 \end{vmatrix}$ **$6 - 4y$**

20. $\begin{vmatrix} 3 & 2 & 0 \\ 0 & 1 & x \\ 2 & 0 & 0 \end{vmatrix}$ **$4x$**

21. $\begin{vmatrix} i & j & k \\ 0 & -4 & 2 \\ -1 & 3 & 1 \end{vmatrix}$ **$-10i - 2j - 4k$**

22. $\begin{vmatrix} i & j & k \\ -1 & 2 & 4 \\ 3 & 0 & 5 \end{vmatrix}$ **$10i + 17j - 6k$**

23. $\begin{vmatrix} 2 & 0 & 0 & 1 \\ -2 & 0 & 6 & 0 \\ 2 & 4 & 0 & 1 \\ 2 & 4 & 1 & 2 \end{vmatrix}$ **−40**

24. $\begin{vmatrix} .4 & -.8 & .6 \\ .3 & .9 & .7 \\ 3.1 & 4.1 & -2.8 \end{vmatrix}$ **−5.5**

25. $\begin{vmatrix} -.3 & -.1 & .9 \\ 2.5 & 4.9 & -3.2 \\ -.1 & .4 & .8 \end{vmatrix}$ **−.051**

26. $\begin{vmatrix} -.5 & -.7 & .9 \\ 1.4 & 3.6 & -.2 \\ 1.5 & 2.1 & -2.7 \end{vmatrix}$ **0**

Tell why each determinant has a value of 0. All variables represent real numbers.

27. $\begin{vmatrix} 2 & 3 \\ 2 & 3 \end{vmatrix}$ **Two rows are identical.**

28. $\begin{vmatrix} -5 & -5 \\ 6 & 6 \end{vmatrix}$ **Two columns are identical.**

29. $\begin{vmatrix} 2 & 0 \\ 3 & 0 \end{vmatrix}$ **One column is all zeros.**

30. $\begin{vmatrix} 6 & -8 \\ -3 & 4 \end{vmatrix}$ **If each element of the second row is multiplied by −2, then two rows are identical.**

31. $\begin{vmatrix} 1 & 0 & 0 \\ 1 & 0 & 1 \\ 3 & 0 & 0 \end{vmatrix}$ **One column is all zeros.**

32. $\begin{vmatrix} -1 & 2 & 4 \\ 4 & -8 & -16 \\ 3 & 0 & 5 \end{vmatrix}$ **If each element of the second row is multiplied by −1/4, then two rows are identical.**

33. $\begin{vmatrix} 7z & 8x & 2y \\ z & x & y \\ 7z & 7x & 7y \end{vmatrix}$ **If each element of the second row is multiplied by 7, then two rows are identical.**

34. $\begin{vmatrix} m & 2 & 2m \\ 3n & 1 & 6n \\ 5p & 6 & 10p \end{vmatrix}$ **If each element of the third column is multiplied by 1/2, then two columns are identical.**

Use the appropriate theorems from this section to tell why each statement is true. Do not evaluate the determinants. All variables represent real numbers.

35. $\begin{vmatrix} 4 & -2 \\ 3 & 8 \end{vmatrix} = \begin{vmatrix} 4 & 3 \\ -2 & 8 \end{vmatrix}$ **Rows and columns are interchanged.**

36. $\begin{vmatrix} 2 & 1 & 6 \\ 3 & 0 & 2 \\ 4 & 1 & 8 \end{vmatrix} = \begin{vmatrix} 2 & 3 & 4 \\ 1 & 0 & 1 \\ 6 & 2 & 8 \end{vmatrix}$ **Rows and columns are interchanged.**

37. $\begin{vmatrix} -1 & 8 & 9 \\ 0 & 2 & 1 \\ 3 & 2 & 0 \end{vmatrix} = -\begin{vmatrix} 8 & -1 & 9 \\ 2 & 0 & 1 \\ 2 & 3 & 0 \end{vmatrix}$ **Two columns are interchanged.**

38. $\begin{vmatrix} 2 & 6 \\ 3 & 5 \end{vmatrix} = -\begin{vmatrix} 3 & 5 \\ 2 & 6 \end{vmatrix}$ **Two rows are interchanged.**

39. $-\frac{1}{2}\begin{vmatrix} 5 & -8 & 2 \\ 3 & -6 & 9 \\ 2 & 4 & 4 \end{vmatrix} = \begin{vmatrix} 5 & 4 & 2 \\ 3 & 3 & 9 \\ 2 & -2 & 4 \end{vmatrix}$ **Each element of second column is multiplied by $-1/2$.**

40. $3\begin{vmatrix} 6 & 0 & 2 \\ 4 & 1 & 3 \\ 2 & 8 & 6 \end{vmatrix} = \begin{vmatrix} 6 & 0 & 2 \\ 4 & 3 & 3 \\ 2 & 24 & 6 \end{vmatrix}$ **Each element of second column is multiplied by 3.**

41. $\begin{vmatrix} 3 & -4 \\ 2 & 5 \end{vmatrix} = \begin{vmatrix} 3 & -4 \\ 5 & 1 \end{vmatrix}$ **Elements of first row are multiplied by 1; products are added to elements of second row.**

42. $\begin{vmatrix} -1 & 6 \\ 3 & -5 \end{vmatrix} = \begin{vmatrix} -1 & 5 \\ 3 & -2 \end{vmatrix}$ **Elements of first column are multiplied by 1; products are added to elements of second column.**

43. $\begin{vmatrix} -4 & 2 & 1 \\ 3 & 0 & 5 \\ -1 & 4 & -2 \end{vmatrix} = \begin{vmatrix} -4 & 2 & 1 + (-4)k \\ 3 & 0 & 5 + 3k \\ -1 & 4 & -2 + (-1)k \end{vmatrix}$ **Elements of first column are multiplied by constant *k*; products are added to third column.**

44. $2\begin{vmatrix} 4 & 2 & -1 \\ m & 2n & 3p \\ 5 & 1 & 0 \end{vmatrix} = \begin{vmatrix} 4 & 2 & -1 \\ 2m & 4n & 6p \\ 5 & 1 & 0 \end{vmatrix}$ **Elements of second row are multiplied by 2.**

Use the method of Examples 10 and 11 to find the value of each determinant.

45. $\begin{vmatrix} -5 & 10 \\ 6 & -12 \end{vmatrix}$ **0**

46. $\begin{vmatrix} 2 & 4 \\ 3 & 6 \end{vmatrix}$ **0**

47. $\begin{vmatrix} 6 & 8 & -12 \\ -1 & 0 & 2 \\ 4 & 0 & -8 \end{vmatrix}$ **0**

48. $\begin{vmatrix} 4 & 8 & 0 \\ -1 & -2 & 1 \\ 2 & 4 & 3 \end{vmatrix}$ **0**

49. $\begin{vmatrix} -2 & 2 & 3 \\ 0 & 2 & 1 \\ -1 & 4 & 0 \end{vmatrix}$ **12**

50. $\begin{vmatrix} 3 & 1 & 2 \\ 2 & 0 & 1 \\ 1 & 0 & -2 \end{vmatrix}$ **5**

51. $\begin{vmatrix} -4 & 1 & 4 \\ 2 & 0 & 1 \\ 0 & 2 & 4 \end{vmatrix}$ **16**

52. $\begin{vmatrix} 6 & 3 & 2 \\ 1 & 0 & 2 \\ 5 & 7 & 3 \end{vmatrix}$ **−49**

53. $\begin{vmatrix} 2 & -1 & 1 & 0 \\ 1 & 1 & 0 & 1 \\ 0 & -1 & 1 & 1 \\ 1 & 2 & 1 & 2 \end{vmatrix}$ **−6**

54. $\begin{vmatrix} 1 & 0 & 2 & 2 \\ 2 & 4 & 1 & -1 \\ 1 & -3 & 1 & 0 \\ 1 & 1 & 0 & 1 \end{vmatrix}$ **−32**

Let $A = \begin{bmatrix} a_{11} & a_{12} & a_{13} \\ a_{21} & a_{22} & a_{23} \\ a_{31} & a_{32} & a_{33} \end{bmatrix}$ *for Exercises 55–60.*

55. Find $|A|$ by expansion about row 3 of the matrix. Show that your result is really equal to $|A|$ as given in the definition of the determinant of a 3×3 matrix.

56. Repeat Exercise 55 for column 3.

57. Obtain matrix B by exchanging columns 1 and 3 of matrix A. Show that $|B| = -|A|$.

58. Obtain matrix B by multiplying each element of row 3 of matrix A by the real number k. Show that $|B| = k \cdot |A|$.

59. Obtain matrix B by adding to column 1 of matrix A the result of multiplying each element of column 2 of A by the real number k. Show that $|B| = |A|$.

60. Obtain matrix B by adding to row 1 of matrix A the result of multiplying each element of row 3 of A by the real number k. Show that $|B| = |A|$.

61. Let A and B be any 2×2 matrices. Show that $|AB| = |A| \cdot |B|$, where $|AB|$ is the determinant of matrix AB.

62. Show that $\begin{vmatrix} a_{11}+a & a_{12} & a_{13} \\ a_{21}+b & a_{22} & a_{23} \\ a_{31}+c & a_{32} & a_{33} \end{vmatrix} = \begin{vmatrix} a_{11} & a_{12} & a_{13} \\ a_{21} & a_{22} & a_{23} \\ a_{31} & a_{32} & a_{33} \end{vmatrix} + \begin{vmatrix} a & a_{12} & a_{13} \\ b & a_{22} & a_{23} \\ c & a_{32} & a_{33} \end{vmatrix}$.

Use this fact and Determinant Theorems 4 and 5 to prove Determinant Theorem 6.

◆ *Determinants can be used to find the area of a triangle, given the coordinates of its vertices. Given a triangle PQR with vertices* (x_1, y_1), (x_2, y_2), *and* (x_3, y_3), *as in the figure, it can be shown that the area of the triangle is given by A, where*

$$A = \frac{1}{2}\begin{vmatrix} x_1 & y_1 & 1 \\ x_2 & y_2 & 1 \\ x_3 & y_3 & 1 \end{vmatrix}.$$

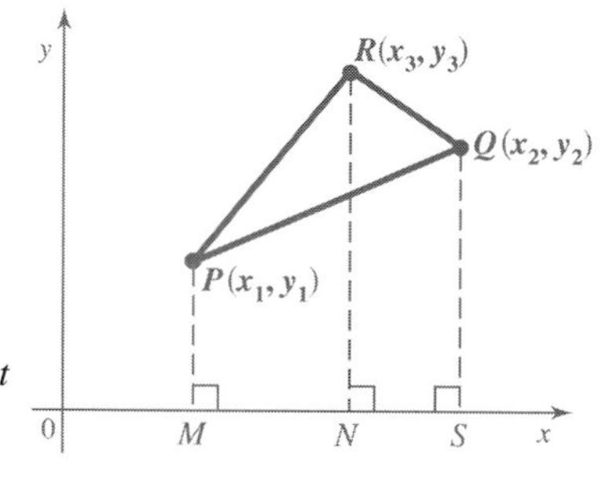

The points (x_1, y_1), (x_2, y_2), (x_3, y_3) *must be taken in counterclockwise order; if this is not done, than A may have the wrong sign. Alternatively, we could define A as the absolute value of 1/2 the determinant shown above. Use the formula given to find the area of the triangles in Exercises 63 and 64.*

63. **(a)** $P(0, 0)$, $Q(0, 2)$, $R(1, 1)$ **1**
(b) $P(0, 1)$, $Q(2, 0)$, $R(1, 3)$ **5/2**
(c) $P(2, 5)$, $Q(-1, 3)$, $R(4, 0)$ **19/2**

64. **(a)** $P(2, -2)$, $Q(0, 0)$, $R(-3, -4)$ **7**
(b) $P(4, 7)$, $Q(5, -2)$, $R(1, 1)$ **33/2**
(c) $P(1, 2)$, $Q(4, 3)$, $R(3, 5)$ **7/2**

◆ **65.** Prove that the straight line through the distinct points (x_1, y_1) and (x_2, y_2) has equation

$$\begin{vmatrix} x & y & 1 \\ x_1 & y_1 & 1 \\ x_2 & y_2 & 1 \end{vmatrix} = 0.$$

◆ **66.** Use the result of Exercise 65 to show that three distinct points (x_1, y_1), (x_2, y_2), and (x_3, y_3) lie on a straight line if

$$\begin{vmatrix} x_1 & y_1 & 1 \\ x_2 & y_2 & 1 \\ x_3 & y_3 & 1 \end{vmatrix} = 0.$$

◆ **67.** Show that the lines $a_1x + b_1y = c_1$ and $a_2x + b_2y = c_2$, when $c_1 \neq c_2$, are parallel if

$$\begin{vmatrix} a_1 & b_1 \\ a_2 & b_2 \end{vmatrix} = 0.$$

68. Prove that $\begin{vmatrix} 1 & 1 & 1 \\ a & b & c \\ a^2 & b^2 & c^2 \end{vmatrix} = (a - b)(b - c)(c - a).$

69. Find $\begin{vmatrix} 1 & 1 & 1 \\ 1 + x & 1 + y & 1 \\ 1 & 1 & 1 \end{vmatrix}$. **0**

70. Prove that $\begin{vmatrix} 1 & 1 & 1 \\ a & b & c \\ bc & ca & ab \end{vmatrix} = (a - b)(a - c)(c - b).$

✎ **71.** In your own words, describe the method for obtaining the determinant of a matrix.

▲ ✎ **72.** Give a method using determinants to decide whether two lines intersect at a single point.

Graphing Utility Problems

For the pairs of matrices in Exercises 73 and 74, show that $|AB| = |A| \cdot |B|$.

73. $A = \begin{bmatrix} 2 & 5 \\ 1 & 7 \end{bmatrix}$, $B = \begin{bmatrix} 9 & 4 \\ 2 & 6 \end{bmatrix}$

74. $A = \begin{bmatrix} 4 & -2 & 11 \\ 5 & 7 & 9 \\ 8 & 6 & 3 \end{bmatrix}$, $B = \begin{bmatrix} 6 & 8 & 7 \\ 3 & 12 & -5 \\ 7 & 1 & 6 \end{bmatrix}$

6.6 Exercises (text page 354)

Use Cramer's rule to solve each of the following systems of linear equations.

1. $2x + 5y = 12$
$x + 3y = 7$ **(1, 2)**

2. $3x + 2y = 8$
$5x + 4y = 14$ **(2, 1)**

3. $-x + 2y = 1$
$4x - 7y = -2$ **(3, 2)**

4. $2x - 5y = 17$
$x + 3y = -8$ **(1, −3)**

5. $4x + 2y = 11$
$3x - y = 2$ **(3/2, 5/2)**

6. $6x - y = -1$
$3x + y = 3$ **(2/9, 7/3)**

7. $5x + 12y = 2$
$-3x + y = 7$ **(−2, 1)**

8. $3x + 5y = 5$
$-4x + 6y = 6$ **(0, 1)**

9. $6x - 15y = 4$
$-2x + 5y = 9$ **D = 0; no solution**

10. $-8x + 6y = -7$
$4x - 3y = 0$ **$D = 0$; no solution**

11. $3x - z = -10$
$y + 4z = 8$
$x + 2z = -1$ **(−3, 4, 1)**

12. $4x - y + 3z = -3$
$3x + y + z = 0$
$2x - y + 4z = 0$ **(−1, 2, 1)**

13. $5x + 2y + z = 15$
$2x - y + z = 9$
$4x + 3y + 2z = 13$ **(3, −1, 2)**

14. $2x - y + 4z = -2$
$3x + 2y - z = -3$
$x + 4y + 2z = 17$ **(−3, 4, 2)**

15. $x + y + z = 4$
$2x - y + 3z = 4$
$4x + 2y - z = -15$ **(−4, 3, 5)**

16. $4x - 3y + z = -1$
$5x + 7y + 2z = -2$
$3x - 5y - z = 1$ **(0, 0, −1)**

17. $2x - 3y + z = 8$
$-x - 5y + z = -4$
$3x - 5y + 2z = 12$ **(4, 0, 0)**

18. $x + 2y + 3z = 4$
$4x + 3y + 2z = 1$
$-x - 2y - 3z = 0$ **$D = 0$; no solution**

19. $2x - y + 3z = 1$
$-2x + y - 3z = 2$
$5x - y + z = 2$ **$D = 0$; no solution**

20. $x - 2y + 3z = 4$
$5x + 7y - z = 2$
$2x + 2y - 5z = 3$ **(197/91, −118/91, −23/91)**

21. $-3x - 2y - z = 4$
$4x + y + z = 5$
$3x - 2y + 2z = 1$ **(5, −4, −11)**

22. $2x + 3y = 13$
$2y - z = 5$
$x + 2z = 4$ **(2, 3, 1)**

23. $3x + 5y = -7$
$2x + 7z = 2$
$4y + 3z = -8$ **(1, −2, 0)**

24. $5x - y = -4$
$3x + 2z = 4$
$4y + 3z = 22$ **(0, 4, 2)**

25. $5x - 2y = 3$
$4y + z = 8$
$x + 2z = 4$ **(24/19, 63/38, 26/19)**

26. $x + 2y = 10$
$3x + 4z = 7$
$-y - z = 1$ **(31/5, 19/10, −29/10)**

27. $x - y + z + w = 6$
$2y - w = -7$
$x - z = -1$
$y + w = 1$ **(0, −2, 1, 3)**

28. $x + z = 0$
$y + 2z + w = 0$
$2x - w = 0$
$x + 2y + 3z = -2$ **(1, 0, −1, 2)**

For the following two exercises, use the system of equations

$$a_1x + b_1y = c_1$$
$$a_2x + b_2y = c_2.$$

29. Assume $D_x = 0$ and $D_y = 0$. Show that if $c_1c_2 \neq 0$, then $D = 0$, and the equations are dependent.

30. Assume $D = 0$, $D_x = 0$, and $b_1b_2 \neq 0$. Show that $D_y = 0$.

31. State Cramer's rule in your own words.

Graphing Utility Problems

Solve each of the following systems of linear equations by Cramer's rule.

32. $17x + .5y = \sqrt{3}$
$9x + .3y = \frac{3}{5}$ **(.3660, −8.981)**

33. $\sqrt{3}x + 6y = 5$
$4.5x - 19y = \sqrt{2}$ **(1.727, .3347)**

34. $3x + 1.4y - z = \sqrt{3}$
$\sqrt{7}x - 6y + .6z = -1$
$x + \sqrt{3}y - 5.1z = 5$ **(.2044, .1685, −.8831)**

35. $9x - 2y + \sqrt{2}z = 1$
$-7x + 3y + \sqrt{3}z = \frac{4}{3}$
$.24x + 17y - z = \sqrt{2}$ **(.0311, .1229, .6828)**

6.7 Exercises (text page 362)

Decide whether or not the given matrices are inverses of each other. (Check to see if their product is the identity matrix I_n.)

1. $\begin{bmatrix} 5 & 7 \\ 2 & 3 \end{bmatrix}$ and $\begin{bmatrix} 3 & -7 \\ -2 & 5 \end{bmatrix}$ **yes**

2. $\begin{bmatrix} 2 & 3 \\ 1 & 1 \end{bmatrix}$ and $\begin{bmatrix} -1 & 3 \\ 1 & -2 \end{bmatrix}$ **yes**

3. $\begin{bmatrix} -1 & 2 \\ 3 & -5 \end{bmatrix}$ and $\begin{bmatrix} -5 & -2 \\ -3 & -1 \end{bmatrix}$ **no**

4. $\begin{bmatrix} 2 & 1 \\ 3 & 2 \end{bmatrix}$ and $\begin{bmatrix} 2 & 1 \\ -3 & 2 \end{bmatrix}$ **no**

5. $\begin{bmatrix} 0 & 1 & 0 \\ 0 & 0 & -2 \\ 1 & -1 & 0 \end{bmatrix}$ and $\begin{bmatrix} 1 & 0 & 1 \\ 1 & 0 & 0 \\ 0 & -1 & 0 \end{bmatrix}$ **no**

6. $\begin{bmatrix} 1 & 2 & 0 \\ 0 & 1 & 0 \\ 0 & 1 & 0 \end{bmatrix}$ and $\begin{bmatrix} 1 & -2 & 0 \\ 0 & 1 & 0 \\ 0 & -1 & 1 \end{bmatrix}$ **no**

7. $\begin{bmatrix} -1 & -1 & -1 \\ 4 & 5 & 0 \\ 0 & 1 & -3 \end{bmatrix}$ and $\begin{bmatrix} 15 & 4 & -5 \\ -12 & -3 & 4 \\ -4 & -1 & 1 \end{bmatrix}$ **yes**

8. $\begin{bmatrix} 1 & 3 & 3 \\ 1 & 4 & 3 \\ 1 & 3 & 4 \end{bmatrix}$ and $\begin{bmatrix} 7 & -3 & -3 \\ -1 & 1 & 0 \\ -1 & 0 & 1 \end{bmatrix}$ **yes**

Find the inverse, if it exists, for each matrix.

9. $\begin{bmatrix} -1 & 2 \\ -2 & -1 \end{bmatrix}$ $\begin{bmatrix} -1/5 & -2/5 \\ 2/5 & -1/5 \end{bmatrix}$

10. $\begin{bmatrix} 1 & -1 \\ 2 & 0 \end{bmatrix}$ $\begin{bmatrix} 0 & 1/2 \\ -1 & 1/2 \end{bmatrix}$

11. $\begin{bmatrix} -1 & -2 \\ 3 & 4 \end{bmatrix}$ $\begin{bmatrix} 2 & 1 \\ -3/2 & -1/2 \end{bmatrix}$

12. $\begin{bmatrix} 3 & -1 \\ -5 & 2 \end{bmatrix}$ $\begin{bmatrix} 2 & 1 \\ 5 & 3 \end{bmatrix}$

13. $\begin{bmatrix} 5 & 10 \\ -3 & -6 \end{bmatrix}$ **no inverse**

14. $\begin{bmatrix} -6 & 4 \\ -3 & 2 \end{bmatrix}$ **no inverse**

15. $\begin{bmatrix} 1 & 0 & 1 \\ 0 & -1 & 0 \\ 2 & 1 & 1 \end{bmatrix}$ $\begin{bmatrix} -1 & 1 & 1 \\ 0 & -1 & 0 \\ 2 & -1 & -1 \end{bmatrix}$

16. $\begin{bmatrix} 1 & 0 & 0 \\ 0 & -1 & 0 \\ 1 & 0 & 1 \end{bmatrix}$ $\begin{bmatrix} 1 & 0 & 0 \\ 0 & -1 & 0 \\ -1 & 0 & 1 \end{bmatrix}$

17. $\begin{bmatrix} 1 & 3 & 3 \\ 1 & 4 & 3 \\ 1 & 3 & 4 \end{bmatrix}$ $\begin{bmatrix} 7 & -3 & -3 \\ -1 & 1 & 0 \\ -1 & 0 & 1 \end{bmatrix}$

18. $\begin{bmatrix} -2 & 2 & 4 \\ -3 & 4 & 5 \\ 1 & 0 & 2 \end{bmatrix}$ $\begin{bmatrix} -4/5 & 2/5 & 3/5 \\ -11/10 & 4/5 & 1/5 \\ 2/5 & -1/5 & 1/5 \end{bmatrix}$

19. $\begin{bmatrix} 2 & 2 & -4 \\ 2 & 6 & 0 \\ -3 & -3 & 5 \end{bmatrix}$ $\begin{bmatrix} -15/4 & -1/4 & -3 \\ 5/4 & 1/4 & 1 \\ -3/2 & 0 & -1 \end{bmatrix}$

20. $\begin{bmatrix} 2 & 4 & 6 \\ -1 & -4 & -3 \\ 0 & 1 & -1 \end{bmatrix}$ $\begin{bmatrix} 7/4 & 5/2 & 3 \\ -1/4 & -1/2 & 0 \\ -1/4 & -1/2 & -1 \end{bmatrix}$

21. $\begin{bmatrix} 1 & 1 & 0 & 2 \\ 2 & -1 & 1 & -1 \\ 3 & 3 & 2 & -2 \\ 1 & 2 & 1 & 0 \end{bmatrix}$ $\begin{bmatrix} 1/2 & 0 & 1/2 & -1 \\ 1/10 & -2/5 & 3/10 & -1/5 \\ -7/10 & 4/5 & -11/10 & 12/5 \\ 1/5 & 1/5 & -2/5 & 3/5 \end{bmatrix}$

22. $\begin{bmatrix} 1 & -2 & 3 & 0 \\ 0 & 1 & -1 & 1 \\ -2 & 2 & -2 & 4 \\ 0 & 2 & -3 & 1 \end{bmatrix}$ $\begin{bmatrix} 1/2 & 1/2 & -1/4 & 1/2 \\ -1 & 4 & -1/2 & -2 \\ -1/2 & 5/2 & -1/4 & -3/2 \\ 1/2 & -1/2 & 1/4 & 1/2 \end{bmatrix}$

Solve each system of equations by using the inverse of the coefficient matrix.

23. $x - 2y = 2$
$3x - 5y = 6$ **(2, 0)**

24. $3x + 7y = 10$
$-x - 9y = -20$ **(−5/2, 5/2)**

25. $-x - 2y = 8$
$3x + 4y = 24$ **(40, −24)**

26. $-x + y = 1$
$2x - y = 1$ **(2, 3)**

27. $3x - 6y = 1$
$-5x + 9y = -1$ **(−1, −2/3)**

28. $3x - 6y = 2$
$-5x + 9y = 1$ **(−8, −13/3)**

Solve each system of equations by using the inverse of the coefficient matrix. The inverses for the first four problems are found in Exercises 17–20 above. Assume b is a constant in Exercises 35–38.

29. $x + 3y + 3z = 1$
$x + 4y + 3z = 0$
$x + 3y + 4z = -1$ **(10, −1, −2)**

30. $-2x + 2y + 4z = 3$
$-3x + 4y + 5z = 1$
$x + 2z = 2$ **(−4/5, −21/10, 7/5)**

31. $2x + 2y - 4z = 12$
$2x + 6y = 16$
$-3x - 3y + 5z = -20$ **(11, −1, 2)**

32. $2x + 4y + 6z = 4$
$-x - 4y - 3z = 8$
$y - z = -4$ **(15, −5, −1)**

33. $x + y - 3z = 4$
$2x + 4y - 4z = 8$
$-x + y + 4z = -3$ **(0, 1, −1)**

34. $x + 2y + 3z = 5$
$2x + 3y + 2z = 2$
$-x - 2y - 4z = -1$ **(−31, 24, −4)**

35. $4x + 2y = 7$
$bx + 5y = 8$ $\left(\frac{19}{20 - 2b}, \frac{-7b + 32}{20 - 2b}\right)$

36. $2x + 3y = 4$
$5x + 6y = b$ $\left(-8 + b, \frac{20 - 2b}{3}\right)$

37. $2x - y = b^2$
$x + y = b$ $\left(\frac{b^2 + b}{3}, \frac{-b^2 + 2b}{3}\right)$

38. $bx + 4y = 1$
$x + by = b$ $\left(\frac{-3b}{b^2 - 4}, \frac{b^2 - 1}{b^2 - 4}\right)$

Solve each system of equations by using the inverse of the coefficient matrix. The inverses were found in Exercises 21 and 22.

39. $x + y + 2w = 3$
$2x - y + z - w = 3$
$3x + 3y + 2z - 2w = 5$
$x + 2y + z = 3$ **(1, 0, 2, 1)**

40. $x - 2y + 3z = 1$
$y - z + w = -1$
$-2x + 2y - 2z + 4w = 2$
$2y - 3z + w = -3$ **(−2, 0, 1, 0)**

Let $A = \begin{bmatrix} a & b \\ c & d \end{bmatrix}$ *and let O be the 2 × 2 matrix of all zeros.*

Show that the statements in Exercises 41–44 are true.

41. $A \cdot O = O \cdot A = O$

42. For square matrices A and B of the same order, if $AB = O$ and if A^{-1} exists, then $B = O$.

43. $A \cdot A^{-1} = A^{-1} \cdot A = I_2$

44. $A^{-1} = \frac{1}{ad - bc} \cdot \begin{bmatrix} d & -b \\ -c & a \end{bmatrix}$

45. Prove that, if it exists, the inverse of a matrix is unique. (*Hint:* Assume there are two inverses B and C for some matrix A, so that $AB = BA = I$ and $AC = CA = I$. Multiply both sides of $AB = I$ on the left by C and then simplify.)

46. The Bread Box Bakery sells three types of cakes, each requiring the amounts of the basic ingredients shown in the following matrix.

	Types of cakes		
	I	II	III
Flour (in cups)	2	4	2
Sugar (in cups)	2	1	2
Eggs	2	1	3

To fill its daily orders for these three kinds of cake, the bakery uses 72 cups of flour, 48 cups of sugar, and 60 eggs.

(a) Write a 3×1 matrix for the amounts used daily. $\begin{bmatrix} 72 \\ 48 \\ 60 \end{bmatrix}$

(b) Let the number of daily orders for cakes be a 3×1 matrix X with entries x_1, x_2, and x_3. Write a matrix equation that you can solve for X, using the given matrix and the matrix from part (a). $\begin{bmatrix} 2 & 4 & 2 \\ 2 & 1 & 2 \\ 2 & 1 & 3 \end{bmatrix}\begin{bmatrix} x_1 \\ x_2 \\ x_3 \end{bmatrix} = \begin{bmatrix} 72 \\ 48 \\ 60 \end{bmatrix}$

(c) Solve the equation you wrote in part (b) to find the number of daily orders for each type of cake. **8, 8, 12**

47. Let $A = \begin{bmatrix} a & b \\ c & d \end{bmatrix}$. Under what conditions on a, b, c, d does A^{-1} exist?

(*Hint:* See Exercise 44.) $ad - bc \neq 0$

48. Give an example of two matrices A and B, where $(AB)^{-1} \neq A^{-1}B^{-1}$. $A = \begin{bmatrix} 1 & 0 \\ 1 & 1 \end{bmatrix}, B = \begin{bmatrix} 1 & 1 \\ 0 & 1 \end{bmatrix}$

49. Suppose A and B are matrices where A^{-1}, B^{-1}, and AB all exist. Show that $(AB)^{-1} = B^{-1}A^{-1}$.

50. Let $A = \begin{bmatrix} a & 0 & 0 \\ 0 & b & 0 \\ 0 & 0 & c \end{bmatrix}$, where a, b, and c are nonzero real numbers. Find A^{-1}. $\begin{bmatrix} 1/a & 0 & 0 \\ 0 & 1/b & 0 \\ 0 & 0 & 1/c \end{bmatrix}$

51. Let $A = \begin{bmatrix} 1 & 0 & 0 \\ 0 & 0 & -1 \\ 0 & 1 & -1 \end{bmatrix}$. Show that $A^3 = I$ and use this result to find the inverse of A. $A^{-1} = A^2 = \begin{bmatrix} 1 & 0 & 0 \\ 0 & -1 & 1 \\ 0 & -1 & 0 \end{bmatrix}$

52. What are the inverses of I, $-A$ (in terms of A), and kA (k a scalar)? I, $-A^{-1}$, $(1/k)A^{-1}$

53. Give two ways to use matrices to solve a system of linear equations. Will they both work in all situations? In which situtations does each method excel?

54. Discuss the similarities and differences between solving the linear equation $ax = b$ and solving the matrix equation $AX = B$.

Graphing Utility Problems

Find the inverses of the following matrices.

55. $\begin{bmatrix} \sqrt{2} & .5 \\ -17 & 1/2 \end{bmatrix}$ $\begin{bmatrix} .0543 & -.0543 \\ 1.8464 & .1536 \end{bmatrix}$

56. $\begin{bmatrix} 2/3 & .7 \\ 22 & \sqrt{3} \end{bmatrix}$ $\begin{bmatrix} -.1216 & .0491 \\ 1.544 & -.0468 \end{bmatrix}$

Use matrix inversion to solve the following systems of linear equations.

57. $x - \sqrt{2}y = 2.6$
$\frac{3}{4}x + y = -7$ **(−3.542, −4.343)**

58. $2.1x + y = \sqrt{5}$
$\sqrt{2}x - 2y = 5$ **(1.687, −1.307)**

6.8 Exercises (text page 372)

Find the partial fraction decomposition for the following rational expressions.

1. $\frac{5}{3x(2x+1)}$ $\mathbf{\frac{5}{3x} + \frac{-10}{3(2x+1)}}$

2. $\frac{3x-1}{x(x+1)}$ $\mathbf{\frac{-1}{x} + \frac{4}{x+1}}$

3. $\frac{4x+2}{(x+2)(2x-1)}$ $\mathbf{\frac{6}{5(x+2)} + \frac{8}{5(2x-1)}}$

4. $\frac{x+2}{(x+1)(x-1)}$ $\mathbf{\frac{-1}{2(x+1)} + \frac{3}{2(x-1)}}$

5. $\frac{x}{x^2+4x-5}$ $\mathbf{\frac{5}{6(x+5)} + \frac{1}{6(x-1)}}$

6. $\frac{5x-3}{(x+1)(x-3)}$ $\mathbf{\frac{2}{x+1} + \frac{3}{x-3}}$

7. $\frac{2x}{(x+1)(x+2)^2}$ $\mathbf{\frac{-2}{x+1} + \frac{2}{x+2} + \frac{4}{(x+2)^2}}$

8. $\frac{2}{x^2(x+3)}$ $\mathbf{\frac{-2}{9x} + \frac{2}{3x^2} + \frac{2}{9(x+3)}}$

9. $\frac{4}{x(1-x)}$ $\mathbf{\frac{4}{x} + \frac{4}{1-x}}$

10. $\frac{4x^2-4x^3}{x^2(1-x)}$ **4**

11. $\frac{4x^2-x-15}{x(x+1)(x-1)}$ $\mathbf{\frac{15}{x} + \frac{-5}{(x+1)} + \frac{-6}{(x-1)}}$

12. $\frac{2x+1}{(x+2)^3}$ $\mathbf{\frac{2}{(x+2)^2} + \frac{-3}{(x+2)^3}}$

13. $\frac{x^2}{x^2+2x+1}$ $\mathbf{1 + \frac{-2}{x+1} + \frac{1}{(x+1)^2}}$

14. $\frac{3}{x^2+4x+3}$ $\mathbf{\frac{3}{2(x+1)} + \frac{-3}{2(x+3)}}$

15. $\frac{2x^5+3x^4-3x^3-2x^2+x}{2x^2+5x+2}$ $\mathbf{x^3 - x^2 + \frac{-1}{3(2x+1)} + \frac{2}{3(x+2)}}$

16. $\frac{6x^5+7x^4-x^2+2x}{3x^2+2x-1}$ $\mathbf{2x^3 + x^2 + \frac{1}{2(3x-1)} + \frac{1}{2(x+1)}}$

17. $\frac{x^3+4}{9x^3-4x}$ $\mathbf{\frac{1}{9} + \frac{-1}{x} + \frac{25}{18(3x+2)} + \frac{29}{18(3x-2)}}$

18. $\frac{x^3+2}{x^3-3x^2+2x}$ $\mathbf{1 + \frac{1}{x} + \frac{5}{x-2} + \frac{-3}{x-1}}$

19. $\frac{-3}{x^2(x^2+5)}$ $\mathbf{\frac{-3}{5x^2} + \frac{3}{5(x^2+5)}}$

20. $\frac{2x+1}{(x+1)(x^2+2)}$ $\mathbf{\frac{-1}{3(x+1)} + \frac{x+5}{3(x^2+2)}}$

21. $\frac{3x-2}{(x+4)(3x^2+1)}$ $\mathbf{\frac{-2}{7(x+4)} + \frac{6x-3}{7(3x^2+1)}}$

22. $\frac{3}{x(x+1)(x^2+1)}$ $\mathbf{\frac{3}{x} + \frac{-3}{2(x+1)} + \frac{-3(x+1)}{2(x^2+1)}}$

23. $\frac{1}{x(2x+1)(3x^2+4)}$ $\mathbf{\frac{1}{4x} + \frac{-8}{19(2x+1)} + \frac{-(9x+24)}{76(3x^2+4)}}$

24. $\frac{x^4+1}{x(x^2+1)^2}$ $\mathbf{\frac{1}{x} + \frac{-2x}{(x^2+1)^2}}$

25. $\frac{3x-1}{x(2x^2+1)^2}$ $\mathbf{\frac{-1}{x} + \frac{2x}{2x^2+1} + \frac{2x+3}{(2x^2+1)^2}}$

26. $\frac{3x^4+x^3+5x^2-x+4}{(x-1)(x^2+1)^2}$ $\mathbf{\frac{3}{x-1} + \frac{1}{x^2+1} + \frac{-2}{(x^2+1)^2}}$

27. $\dfrac{-x^4 - 8x^2 + 3x - 10}{(x + 2)(x^2 + 4)^2}$ $\quad \mathbf{\dfrac{-1}{x + 2} + \dfrac{3}{(x^2 + 4)^2}}$

28. $\dfrac{x^2}{x^4 - 1}$ $\quad \mathbf{\dfrac{-1}{4(x + 1)} + \dfrac{1}{4(x - 1)} + \dfrac{1}{2(x^2 + 1)}}$

29. $\dfrac{5x^5 + 10x^4 - 15x^3 + 4x^2 + 13x - 9}{x^3 + 2x^2 - 3x}$ $\quad \mathbf{5x^2 + \dfrac{3}{x} + \dfrac{-1}{x + 3} + \dfrac{2}{x - 1}}$

30. $\dfrac{3x^6 + 3x^4 + 3x}{x^4 + x^2}$ $\quad \mathbf{3x^2 + \dfrac{3}{x} + \dfrac{-3x}{x^2 + 1}}$

Graphing Utility Problems

Determine whether each of the following partial fraction decompositions are correct by graphing the left side and the right side of the equation on the same coordinate system and observing whether the graphs overlap.

31. $\dfrac{4x^2 - 3x - 4}{x^3 + x^2 - 2x} = \dfrac{2}{x} + \dfrac{-1}{x - 1} + \dfrac{3}{x + 2}$ **correct**

32. $\dfrac{1}{(x - 1)(x + 2)} = \dfrac{1}{x - 1} - \dfrac{1}{x + 2}$ **incorrect**

33. $\dfrac{x^3 - 2x}{(x^2 + 2x + 2)^2} = \dfrac{x - 2}{x^2 + 2x + 2} + \dfrac{2}{(x^2 + 2x + 2)^2}$ **incorrect**

34. $\dfrac{2x + 4}{x^2(x - 2)} = \dfrac{-2}{x} + \dfrac{-2}{x^2} + \dfrac{2}{x - 2}$ **correct**

Chapter 6 Review Exercises (text page 372)

Use the elimination or substitution method to solve each of the following linear systems. Identify any inconsistent systems or systems with dependent equations.

1. $3x - 5y = 7$
   $2x + 3y = 30$ **(9, 4)**

2. $-x + 4y = 3$
   $x + 2y = 9$ **(5, 2)**

3. $6x - 2y = 4$
   $4x + 5y = 9$ **(1, 1)**

4. $\frac{1}{6}x + \frac{1}{3}y = 8$
   $\frac{1}{4}x + \frac{1}{2}y = 12$ **(48 − 2*y*, *y*); dependent equations**

5. $.2x - .3y = 1$
   $.3x + .5y = 11$ **(20, 10)**

6. $3x - 2y = 0$
   $9x + 8y = 7$ **(1/3, 1/2)**

7. $2x - 5y + 3z = -1$
   $x + 4y - 2z = 9$
   $-x + 2y + 4z = 5$ **(3, 2, 1)**

8. $5x - y = 26$
   $4y + 3z = -4$
   $3x + 3z = 15$ **(5, −1, 0)**

Write a system of linear equations for each of the following, and then use the system to solve the problem.

9. Three hundred people attending a club's banquet paid a total of \$4060. Each member paid \$13 and each nonmember paid \$15. How many members and how many nonmembers attended the banquet? **220 members, 80 nonmembers**

10. A cup of uncooked rice contains 15 g of protein and 810 cal. A cup of uncooked soybeans contains 22.5 g of protein and 270 cal. How many cups of each should be used for a meal containing 9.5 g of protein and 324 cal? **1/3 cup of rice, 1/5 cup of soybeans**

11. A company sells 3½″ diskettes for 40¢ each and sells 5¼″ diskettes for 30¢ each. The company receives \$38 for an order of 100 diskettes. However, the customer neglected to specify how many of each size to send. Determine the number of each size of diskette that should be sent. **80 3 1/2″ diskettes, 20 5 1/4″ diskettes**

12. The Waputi Indians make woven blankets, rugs, and skirts. Each blanket requires 24 hr for spinning the yarn, 4 hr for dyeing the yarn, and 15 hr for weaving. Rugs require 30, 5, and 18 hr and skirts 12, 3, and 9 hr, respectively. If there are 306, 59, and 201 hr available for spinning, dyeing, and weaving, respectively, how many of each item can be made? (*Hint:* Simplify the equations you write, if possible, before solving the system.) **5 blankets, 3 rugs, 8 skirts**

Find solutions for the following systems with the specified arbitrary variable.

13. $2x - 6y + 4z = 5$
$5x + y - 3z = 1$; z **$(11/32 + (7/16)z, -23/32 + (13/16)z, z)$**

14. $3x - 4y + z = 2$
$2x + y = 1$; x **$(x, 1 - 2x, 6 - 11x)$**

Solve each system in Exercises 15–18.

15. $x^2 = 2y - 3$
$x + y = 3$ **(1, 2), (−3, 6)**

16. $2x^2 + 3y^2 = 30$
$x^2 + y^2 = 13$ **(3, 2), (3, −2), (−3, 2), (−3, −2)**

17. $xy = -2$
$y - x = 3$ **(−1, 2), (−2, 1)**

▲ **18.** $x^2 + 2xy + y^2 = 4$
$x - 3y = -2$ **(−2, 0), (1, 1)**

19. Find a value of b so that the straight line $3x - y = b$ touches the circle $x^2 + y^2 = 25$ at only one point. **$\pm 5\sqrt{10}$**

20. Do the circle $x^2 + y^2 = 144$ and the line $x + 2y = 8$ have any points in common? If so, what are they? **yes; $\left(\frac{8 - 8\sqrt{41}}{5}, \frac{16 + 4\sqrt{41}}{5}\right), \left(\frac{8 + 8\sqrt{41}}{5}, \frac{16 - 4\sqrt{41}}{5}\right)$**

Graph the solution of each system of inequalities in Exercises 21–23.

21. $x - 3y \geq 6$
$y^2 \leq 16 - x^2$

22. $x + y \leq 6$
$2x - y \geq 3$

23. $x^2 + y^2 \leq 144$
$x^2 + y^2 \geq 16$

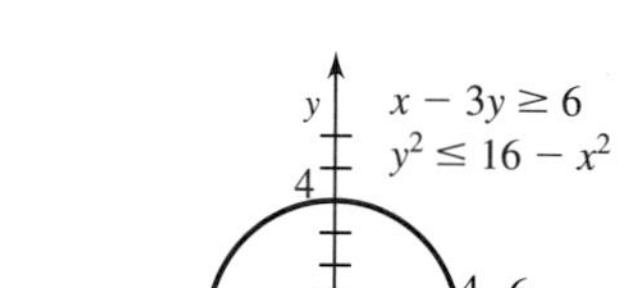

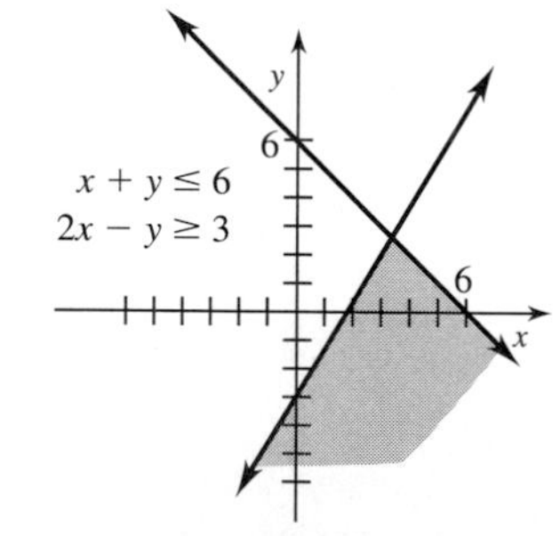

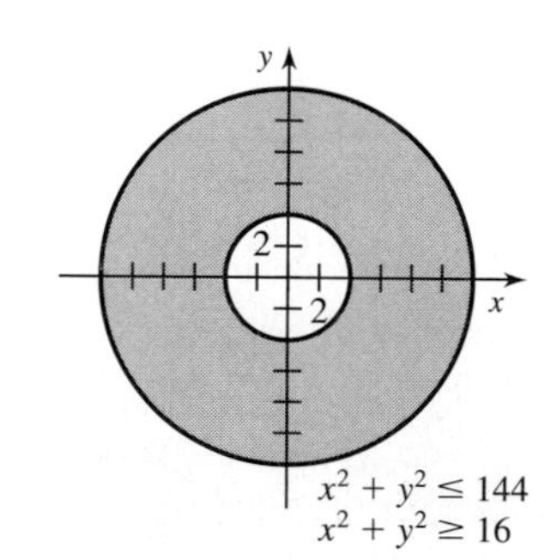

24. A bakery makes both cakes and cookies. Each batch of cakes requires 2 hr in the oven and 3 hr in the decorating room. Each batch of cookies needs $1\frac{1}{2}$ hr in the oven and $\frac{3}{4}$ hr in the decorating room. The oven is available no more than 16 hr a day, while the decorating room can be used no more than 12 hr per day. Set up a system of inequalities expressing this information, and then graph the system. **Let x = number of batches of cakes, y = number of batches of cookies. Then $2x + \frac{3}{2}y \le 16$; $3x + \frac{3}{4}y \le 12$; $x \ge 0$; $y \ge 0$.**

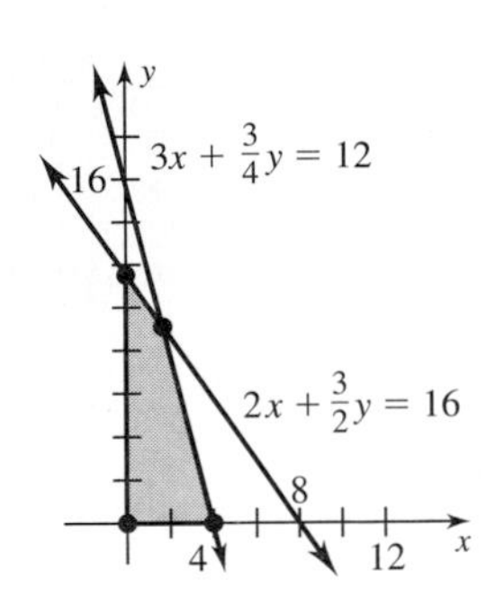

25. A candy company has 100 kg of chocolate-covered nuts and 125 kg of chocolate-covered raisins to be sold as two different mixtures. One mix will contain $\frac{1}{2}$ nuts and $\frac{1}{2}$ raisins, while the other mix will contain $\frac{1}{3}$ nuts and $\frac{2}{3}$ raisins. Set up a system of inequalities expressing this information, and then graph the system. **Let x = number of kg of first mix, y = number of kg of second mix. Then $\frac{1}{2}x + \frac{1}{3}y \le 100$, $\frac{1}{2}x + \frac{2}{3}y \le 125$, $x \ge 0$, $y \ge 0$.**

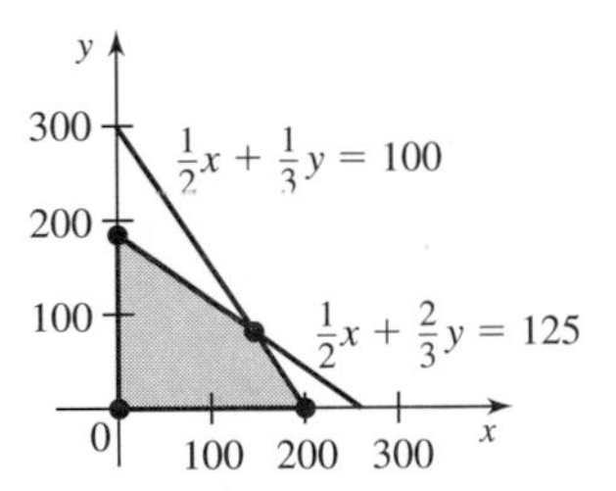

In Exercises 26–29, use graphical methods to find nonnegative values of x and y that meet the constraints.

26. Maximize $2x + 3y$ subject to

$$x + 2y \le 24$$
$$3x + 4y \le 60.$$

There is a maximum of 42 at (12, 6).

27. Maximize $5x + 7y$ subject to

$$2x + 4y \ge 40$$
$$3x + 2y \le 60.$$

There is a maximum of 210 at (0, 30).

28. Minimize $4x + 3y$ subject to

$$x + 2y \le 12$$
$$4x + 3y \ge 12$$
$$y \le 2x.$$

There is a minimum of 12 at (3, 0) or at (6/5, 12/5) or any point between them.

29. Minimize $x + 2y$ subject to

$$5x + 4y \ge 20$$
$$2x + 6y \ge 24$$
$$x + y \le 12.$$

There is a minimum of 92/11 at (12/11, 40/11).

30. In Exercise 24, a batch of cookies produces a profit of \$20; the profit on a batch of cakes is \$30. Find the number of batches of each item which will maximize profit. **The maximum profit is \$220, which occurs when 2 batches of cakes and 8 batches of cookies are made.**

31. In Exercise 25, how much of each mixture should be made to maximize revenue if the first mix sells for \$6.00 per kilogram and the second mix sells for \$4.80 per kilogram? **150 kg of the half-and-half, 75 kg of the other, for a revenue of \$1260**

Use the Gaussian reduction method to solve each of the following.

32. $5x + 2y = -10$
$3x - 5y = -6$ **(−2, 0)**

33. $2x + 3y = 10$
$-3x + y = 18$ **(−4, 6)**

34. $\begin{aligned} 5x - 8y + z &= 1 \\ 3x - 2y + 4z &= 3 \\ 10x - 16y + 2z &= 3 \end{aligned}$ **no solution**

35. $\begin{aligned} 2x - y + 4z &= -1 \\ -3x + 5y - z &= 5 \\ 2x + 3y + 2z &= 3 \end{aligned}$ **(0, 1, 0)**

Find the values of all variables in the following.

36. $\begin{bmatrix} 5 & x+2 \\ -6y & z \end{bmatrix} = \begin{bmatrix} a & 3x-1 \\ 5y & 9 \end{bmatrix}$ **$a = 5, x = 3/2, y = 0, z = 9$**

37. $\begin{bmatrix} -6+k & 2 & a+3 \\ -2+m & 3p & 2r \end{bmatrix} + \begin{bmatrix} 3-2k & 5 & 7 \\ 5 & 8p & 5r \end{bmatrix} = \begin{bmatrix} 5 & y & 6a \\ 2m & 11 & -35 \end{bmatrix}$

$k = -8, y = 7, a = 2, m = 3, p = 1, r = -5$

Perform each of the following operations whenever possible.

38. $\begin{bmatrix} 3 \\ 2 \\ 5 \end{bmatrix} - \begin{bmatrix} 8 \\ -4 \\ 6 \end{bmatrix} + \begin{bmatrix} 1 \\ 0 \\ 2 \end{bmatrix}$ **$\begin{bmatrix} -4 \\ 6 \\ 1 \end{bmatrix}$**

39. $\begin{bmatrix} 3 & -4 & 2 \\ 5 & -1 & 6 \end{bmatrix} + \begin{bmatrix} -3 & 2 & 5 \\ 1 & 0 & 4 \end{bmatrix}$ **$\begin{bmatrix} 0 & -2 & 7 \\ 6 & -1 & 10 \end{bmatrix}$**

40. $\begin{bmatrix} -3 & 4 \\ 2 & 8 \end{bmatrix}\begin{bmatrix} -1 & 0 \\ 2 & 5 \end{bmatrix}$ **$\begin{bmatrix} 11 & 20 \\ 14 & 40 \end{bmatrix}$**

41. $\begin{bmatrix} 2 & 5 & 8 \\ 1 & 9 & 2 \end{bmatrix} - \begin{bmatrix} 3 & 4 \\ 7 & 1 \end{bmatrix}$ **not possible**

42. $\begin{bmatrix} 1 & -2 & 4 & 2 \\ 0 & 1 & -1 & 8 \end{bmatrix}\begin{bmatrix} -1 \\ 2 \\ 0 \\ 1 \end{bmatrix}$ **$\begin{bmatrix} -3 \\ 10 \end{bmatrix}$**

43. $\begin{bmatrix} 3 & 2 & -1 \\ 4 & 0 & 6 \end{bmatrix}\begin{bmatrix} -2 & 0 \\ 0 & 2 \\ 3 & 1 \end{bmatrix}$ **$\begin{bmatrix} -9 & 3 \\ 10 & 6 \end{bmatrix}$**

Find each of the following determinants.

44. $\begin{vmatrix} -2 & 4 \\ 0 & 3 \end{vmatrix}$ **-6**

45. $\begin{vmatrix} -1 & 8 \\ 2 & 9 \end{vmatrix}$ **-25**

46. $\begin{vmatrix} -2 & 4 & 1 \\ 3 & 0 & 2 \\ -1 & 0 & 3 \end{vmatrix}$ **-44**

47. $\begin{vmatrix} -1 & 2 & 3 \\ 4 & 0 & 3 \\ 5 & -1 & 2 \end{vmatrix}$ **-1**

48. $\begin{vmatrix} -1 & 0 & 2 & -3 \\ 0 & 4 & 4 & -1 \\ -6 & 0 & 3 & -5 \\ 0 & -2 & 1 & 0 \end{vmatrix}$ **138**

Explain why each of the following statements is true.

49. $\begin{vmatrix} 4 & 6 \\ 3 & 5 \end{vmatrix} = \begin{vmatrix} 4 & 3 \\ 6 & 5 \end{vmatrix}$ **Rows and columns are interchanged.**

50. $\begin{vmatrix} 8 & 9 & 2 \\ 0 & 0 & 0 \\ 3 & 1 & 4 \end{vmatrix} = 0$ **One row is all zeros.**

51. $\begin{vmatrix} 4 & 6 & 2 \\ -3 & 8 & -5 \\ 4 & 6 & 2 \end{vmatrix} = 0$ **Row 1 and row 3 are identical.**

52. $\begin{vmatrix} 8 & 2 \\ 4 & 3 \end{vmatrix} = 2\begin{vmatrix} 4 & 1 \\ 4 & 3 \end{vmatrix}$ **Elements of first row are multiplied by 2.**

53. $\begin{vmatrix} 8 & 2 & -5 \\ -3 & 1 & 4 \\ 2 & 0 & 5 \end{vmatrix} = -\begin{vmatrix} 8 & -5 & 2 \\ -3 & 4 & 1 \\ 2 & 5 & 0 \end{vmatrix}$ **Columns 2 and 3 are interchanged.**

54. $\begin{vmatrix} 5 & -1 & 2 \\ 3 & -2 & 0 \\ -4 & 1 & 2 \end{vmatrix} = \begin{vmatrix} 5 & -1 & 2 \\ 8 & -3 & 2 \\ -4 & 1 & 2 \end{vmatrix}$ **New row 2 is the sum of old rows 1 and 2.**

Solve each of the following systems by Cramer's rule if possible. Identify any dependent equations or inconsistent systems.

55. $3x + 7y = 2$
$5x - y = -22$ **(−4, 2)**

56. $3x + y = -1$
$5x + 4y = 10$ **(−2, 5)**

57. $5x - 2y - z = 8$
$-5x + 2y + z = -8$
$x - 4y - 2z = 0$ **(16/9, (8 − 9z)/18, z), dependent equations**

58. $3x + 2y + z = 2$
$4x - y + 3z = -16$
$x + 3y - z = 12$ **(−4, 6, 2)**

Find the inverse of each of the following matrices that has an inverse.

59. $\begin{bmatrix} -4 & 2 \\ 0 & 3 \end{bmatrix}$ $\begin{bmatrix} -1/4 & 1/6 \\ 0 & 1/3 \end{bmatrix}$

60. $\begin{bmatrix} 2 & 1 \\ 5 & 3 \end{bmatrix}$ $\begin{bmatrix} 3 & -1 \\ -5 & 2 \end{bmatrix}$

61. $\begin{bmatrix} 2 & 3 & 5 \\ -2 & -3 & -5 \\ 1 & 4 & 2 \end{bmatrix}$ **no inverse**

62. $\begin{bmatrix} 2 & -1 & 0 \\ 1 & 0 & 1 \\ 1 & -2 & 0 \end{bmatrix}$ $\begin{bmatrix} 2/3 & 0 & -1/3 \\ 1/3 & 0 & -2/3 \\ -2/3 & 1 & 1/3 \end{bmatrix}$

Use the method of matrix inverses to solve each of the following.

63. $2x + y = 5$
$3x - 2y = 4$ **(2, 1)**

64. $x + y + z = 1$
$2x - y = -2$
$3y + z = 2$ **(−1, 0, 2)**

65. $x = -3$
$y + z = 6$
$2x - 3z = -9$ **(−3, 5, 1)**

66. $3x - 2y + 4z = 1$
$4x + y - 5z = 2$
$-6x + 4y - 8z = -2$ **((6z + 5)/11, (31z + 2)/11, z)**

◆ *Let $f(x) = ax + b$ and $g(x) = cx + d$. Also, let $\begin{bmatrix} a & b \\ 0 & 1 \end{bmatrix}$ correspond to f and $\begin{bmatrix} c & d \\ 0 & 1 \end{bmatrix}$ correspond to g.**

67. Show that $\begin{bmatrix} a & b \\ 0 & 1 \end{bmatrix}\begin{bmatrix} c & d \\ 0 & 1 \end{bmatrix}$ corresponds to $f \circ g$.

68. Assume that $a \neq 0$ and show that f^{-1} corresponds to $\begin{bmatrix} a & b \\ 0 & 1 \end{bmatrix}^{-1}$.

*Exercises 67 and 68 are from *College Algebra and Trigonometry* by John Schiller and Marie Wurster. Copyright © 1988 by Scott, Foresman and Company.

 Writing 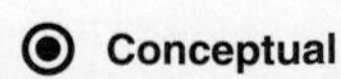 Conceptual 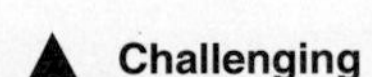 Challenging 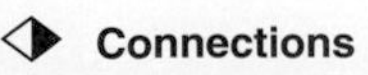 Connections

Find the partial fraction decomposition of the following rational expressions.

69. $\dfrac{x+2}{x^3+2x^2+x}$ $\quad \dfrac{2}{x} + \dfrac{-2}{x+1} + \dfrac{-1}{(x+1)^2}$

70. $\dfrac{x-1}{x^3-x^2+4x}$ $\quad \dfrac{-1}{4x} + \dfrac{x+3}{4(x^2-x+4)}$

71. $\dfrac{2x+1}{(x+1)(x^2-3x+5)}$ $\quad \dfrac{-1}{9(x+1)} + \dfrac{(x+14)}{9(x^2-3x+5)}$

72. $\dfrac{x+3}{x^3+64}$ $\quad \dfrac{-1}{48(x+4)} + \dfrac{x+40}{48(x^2-4x+16)}$

73. Discuss four ways of solving a system of two linear equations in two variables. Discuss the advantages and disadvantages of each method.

Chapter 7 Analytic Geometry

7.1 Exercises (text page 381)

Graph each horizontal parabola.

1. $x = -y^2$

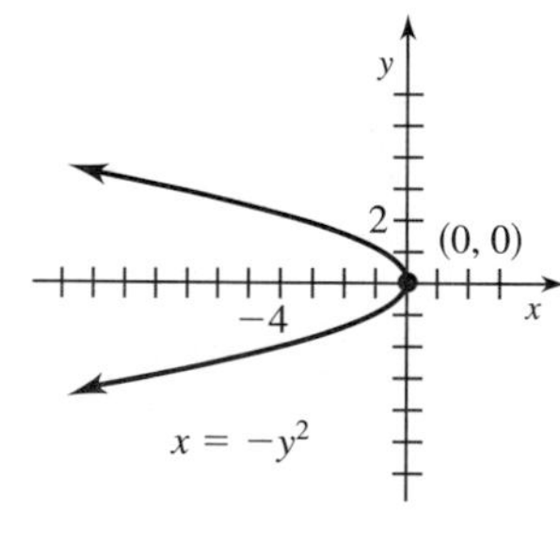

2. $x = y^2 + 2$

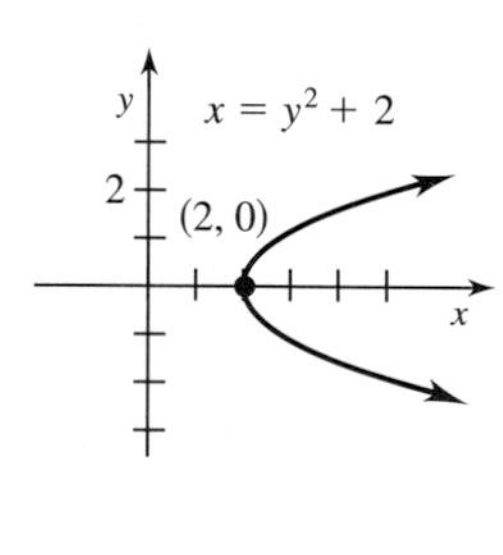

3. $x = (y-3)^2$

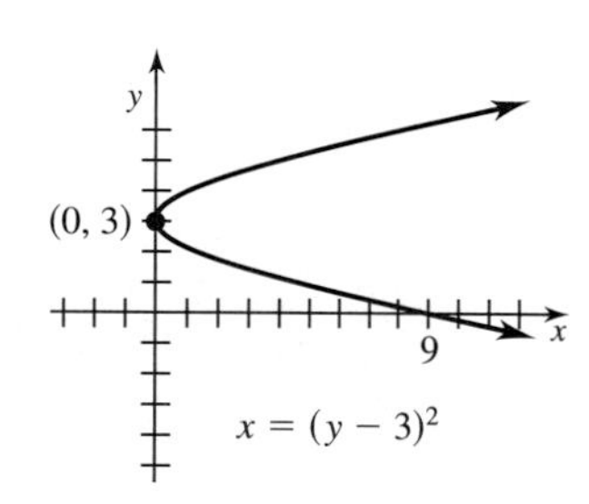

4. $x = (y+1)^2$

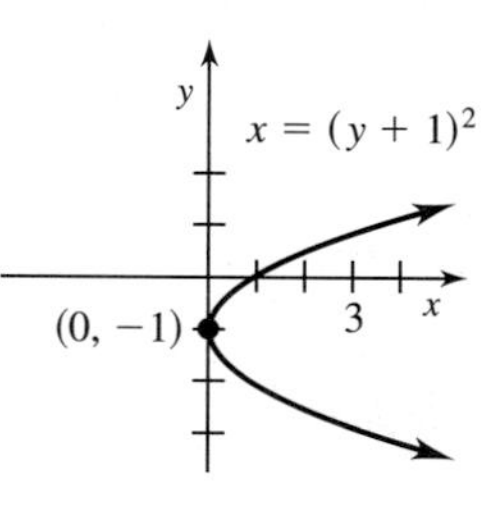

5. $x = (y-4)^2 + 2$

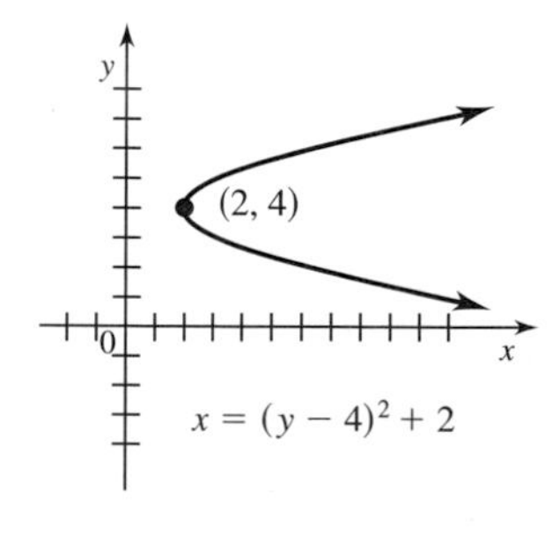

6. $x = (y+2)^2 - 1$

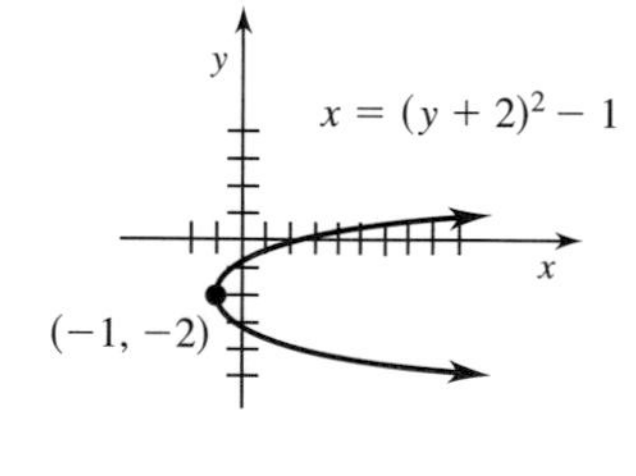

7. $x = -3(y-1)^2 + 2$

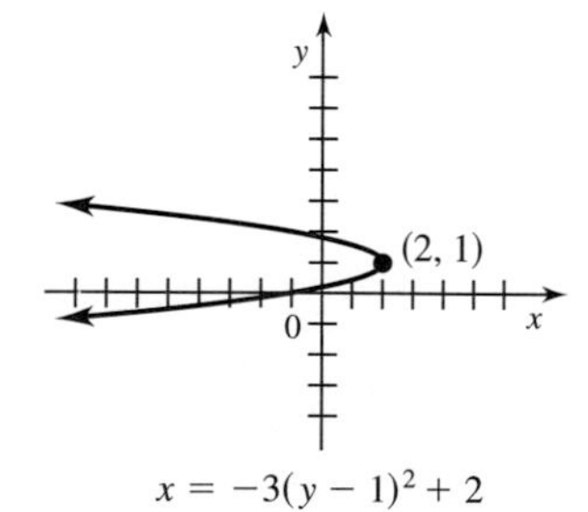

8. $x = -2(y+3)^2$

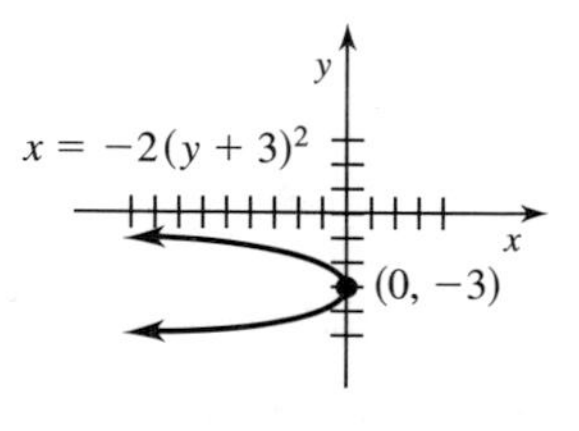

9. $x = \frac{1}{2}(y-1)^2 + 4$

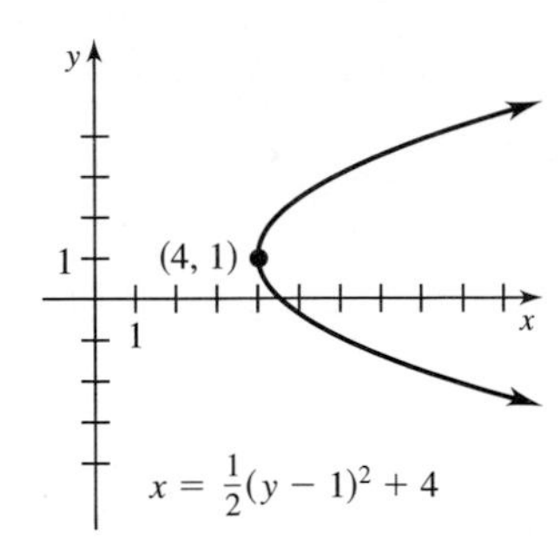

10. $x = -\frac{1}{3}(y-3)^2 + 3$

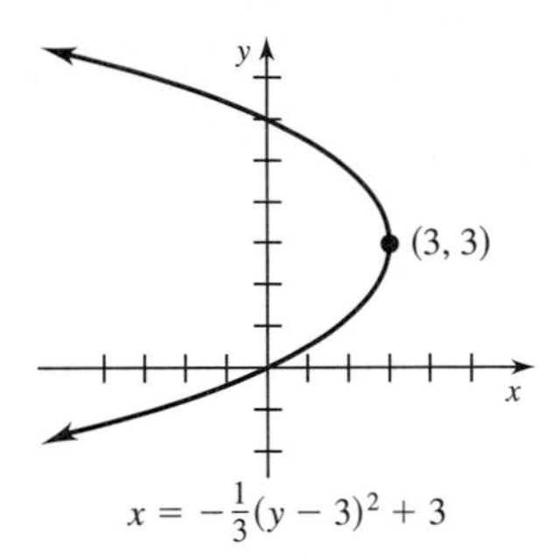

11. $x = y^2 + 4y + 2$

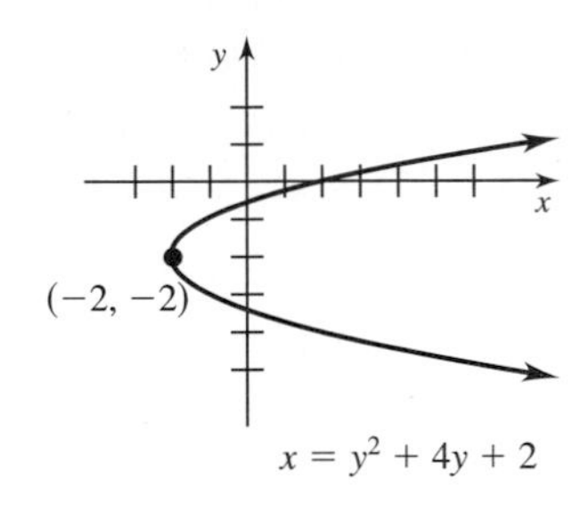

12. $x = 2y^2 - 4y + 6$

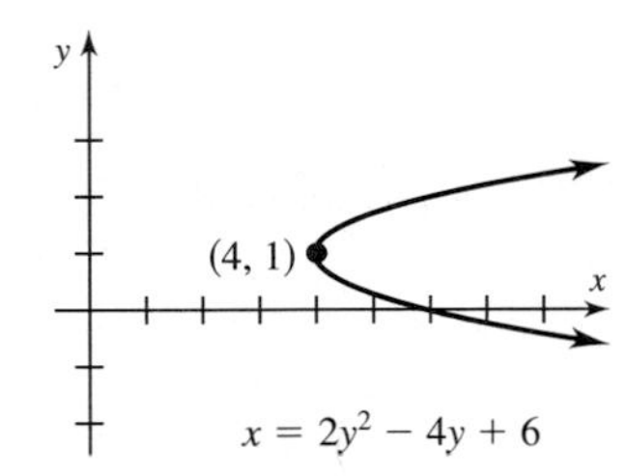

13. $x = -4y^2 - 4y + 3$

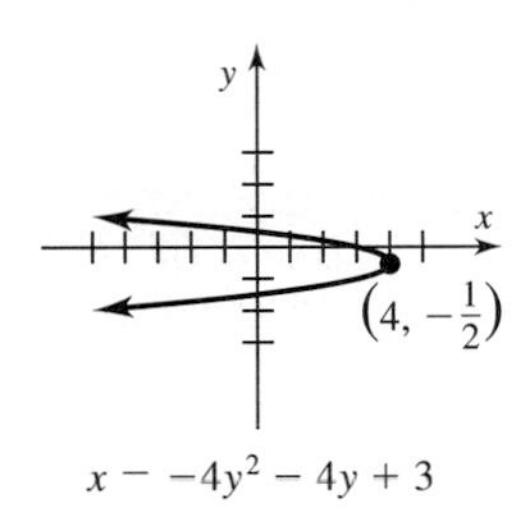

14. $x = -2y^2 + 2y - 3$

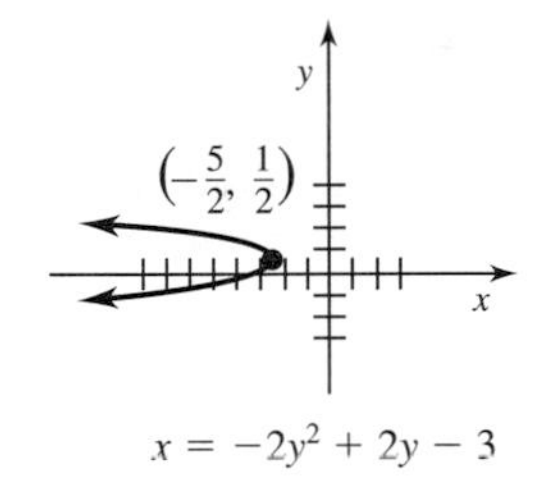

15. $2x = y^2 - 4y + 6$

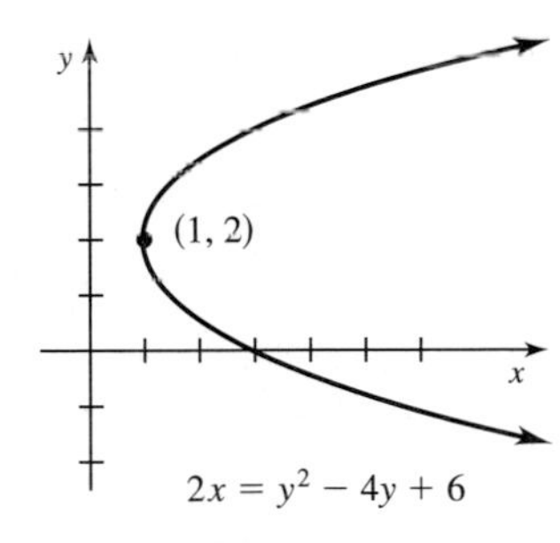

16. $x + 3y^2 + 18y + 22 = 0$

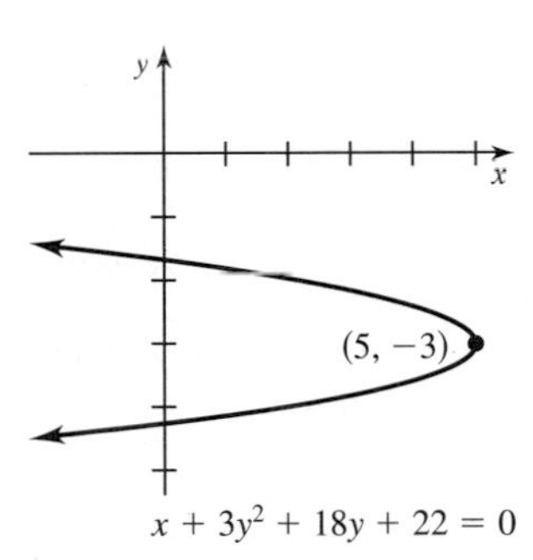

Give the focus, directrix, and axis for each of the following parabolas.

17. $x^2 = 24y$ **(0, 6), $y = -6$, y-axis**

18. $y = 8x^2$ **(0, 1/32), $y = -1/32$, y-axis**

19. $y = -4x^2$ **(0, −1/16), $y = 1/16$, y-axis**

20. $9y = x^2$ **(0, 9/4), $y = -9/4$, y-axis**

21. $x = -32y^2$ **(−1/128, 0), $x = 1/128$, x-axis**

22. $x = 16y^2$ **(1/64, 0), $x = -1/64$, x-axis**

23. $x = (-1/4)y^2$ **(−1, 0), $x = 1$, x-axis**

24. $x = (-1/16)y^2$ **(−4, 0), $x = 4$, x-axis**

25. $(y-3)^2 = 12(x-1)$ **(4, 3), $x = -2$, $y = 3$**

26. $(x+2)^2 = 20y$ **(−2, 5), $y = -5$, $x = -2$**

27. $(x-7)^2 = 16(y+5)$ **(7, −1), $y = -9$, $x = 7$**

28. $(y-2)^2 = 24(x-3)$ **(9, 2), $x = -3$, $y = 2$**

Write an equation for each of the following parabolas with vertex at the origin.

29. Focus (5, 0) $y^2 = 20x$

30. Focus $(-1/2, 0)$ $y^2 = -2x$

31. Focus (0, 1/4) $x^2 = y$

32. Focus $(0, -1/3)$ $x^2 = -(4/3)y$

33. Through $(\sqrt{3}, 3)$, opening upward $x^2 = y$

34. Through $(2, -2\sqrt{2})$, opening to the right $y^2 = 4x$

35. Through (3, 2), symmetric with respect to the x-axis $y^2 = (4/3)x$

36. Through $(2, -4)$, symmetric with respect to the y-axis. $x^2 = -y$

Write an equation for each of the following parabolas.

37. Vertex (4, 3), focus (4, 5) $(x - 4)^2 = 8(y - 3)$

38. Vertex $(-2, 1)$, focus $(-2, -3)$ $(x + 2)^2 = -16(y - 1)$

39. Vertex $(-5, 6)$, focus (2, 6) $(y - 6)^2 = 28(x + 5)$

40. Vertex (1, 2), focus (4, 2) $(y - 2)^2 = 12(x - 1)$

41. The cross-section of a parabolic mirror in a telescope is 8 ft across and 4 inches deep. Find the distance of the focus from the vertex. **12 feet**

42. The cable in the center portion of a bridge is supported as shown in the figure to form a parabola. The center vertical cable is 10 ft high, the supports are 210 ft high, and the distance between the two supports is 400 ft. Find the height of the remaining vertical cables, if the vertical cables are evenly spaced. (Ignore the width of the supports and cables.) **60 ft**

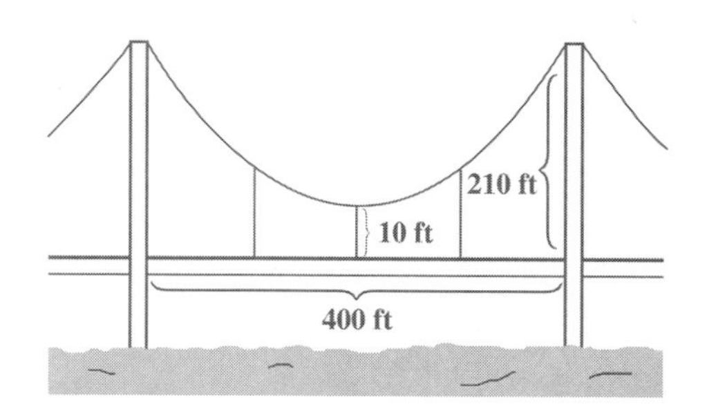

Exercise 42

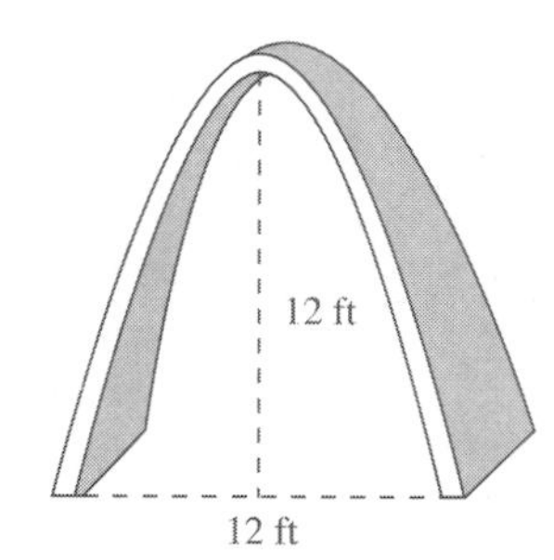

Exercise 43

43. An arch in the shape of a parabola has the dimensions shown in the figure. How wide is the arch 9 ft up? **6 ft**

44. Find the equation of the parabola that has vertex (1, 2), has its axis parallel to the x-axis, and passes through the point (13, 4). $(y - 2)^2 = \frac{1}{3}(x - 1)$

45. Find the equation of the parabola having a horizontal axis and passing through the points $(-3/4, 0)$, (0, 3), and $(0, -1)$. $(y - 1)^2 = 4(x + 1)$

46. A parabolic reflector for a car's headlight is 6 inches across and 2 inches deep. Find the distance of the bulb from the vertex. (*Note:* The bulb is located at the focus.) **1⅛ inches**

47. Explain why the vertex is the point on a parabola that is closest to the focus.

48. Explain why parabolas with horizontal axes were not discussed in the section on quadratic functions.

49. How can you tell by inspecting the equation of a parabola whether the axis is horizontal or vertical?

50. Prove that the parabola with focus $(p, 0)$ and directrix $x = -p$ has equation $y^2 = 4px$.

51. How many (nonfocal) points are needed to determine a parabola? (*Hint:* Look at Exercises 44 and 45.)

Graphing Utility Problems

52. Graph the parabolas $x = \frac{1}{2}y^2$, $x = y^2$, and $x = 2y^2$ using the same scale. How does the coefficient a affect the shape of the graph of $x = ay^2$?

53. The parabola $x = 8y^2$ has $F = (2, 0)$ as focus and the line $x = -2$ as directrix. Graph the parabola and use TRACE to find the coordinates of several points on the graph. Verify that for each point P, the distance of P from the point F is the same as the distance of P from the line L.

Write an equation for each of the following parabolas. Then graph the equation and confirm each of the given properties.

54. Vertex (2, 3), passes through (−18, 1), opens to the left **$(y - 3)^2 = -(1/5)(x - 2)$**

55. Vertex (−1, 5), passes through (2, 4), opens to the right **$(y - 5)^2 = (1/3)(x + 1)$**

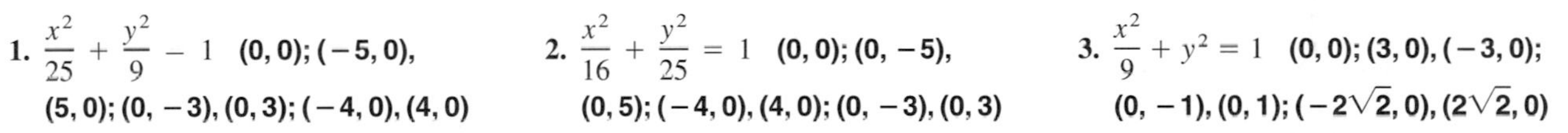

7.2 Exercises (text page 389)

Sketch the graph of each of the following ellipses. Give the center, vertices, endpoints of the minor axis, and the foci for each figure.

1. $\frac{x^2}{25} + \frac{y^2}{9} = 1$ **(0, 0); (−5, 0), (5, 0); (0, −3), (0, 3); (−4, 0), (4, 0)**

2. $\frac{x^2}{16} + \frac{y^2}{25} = 1$ **(0, 0); (0, −5), (0, 5); (−4, 0), (4, 0); (0, −3), (0, 3)**

3. $\frac{x^2}{9} + y^2 = 1$ **(0, 0); (3, 0), (−3, 0); (0, −1), (0, 1); $(-2\sqrt{2}, 0)$, $(2\sqrt{2}, 0)$**

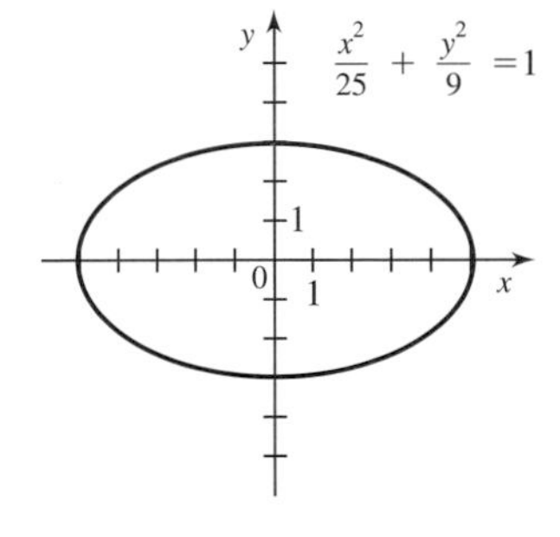

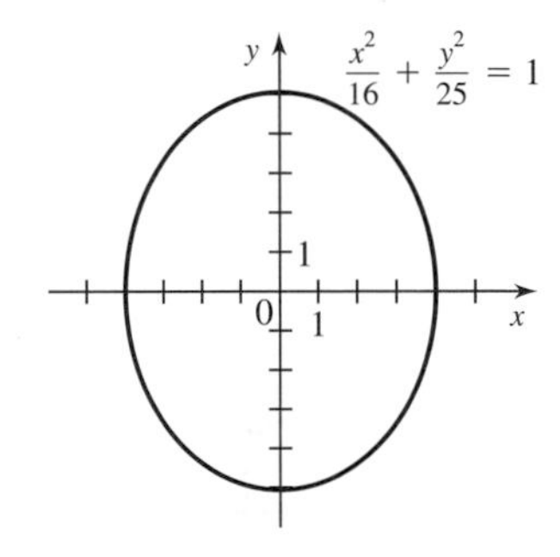

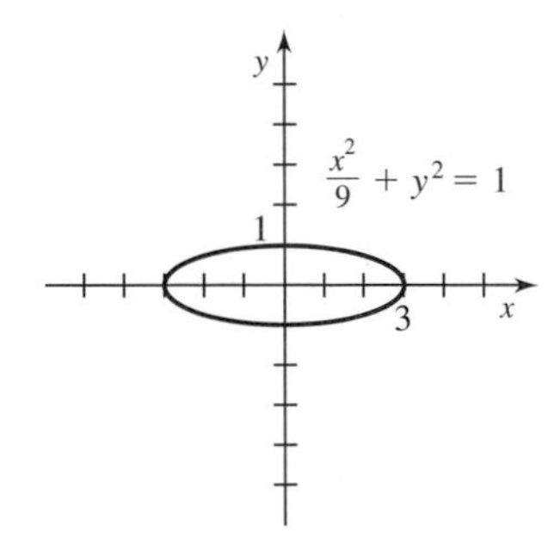

4. $\dfrac{x^2}{36} + \dfrac{y^2}{16} = 1$ **(0, 0); (−6, 0), (6, 0); (0, −4), (0, 4); (−2√5, 0), (2√5, 0)**

5. $9x^2 + y^2 = 81$ **(0, 0); (0, −9), (0, 9); (−3, 0), (3, 0); (0, −6√2), (0, 6√2)**

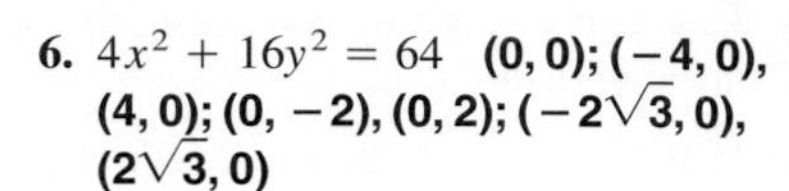
6. $4x^2 + 16y^2 = 64$ **(0, 0); (−4, 0), (4, 0); (0, −2), (0, 2); (−2√3, 0), (2√3, 0)**

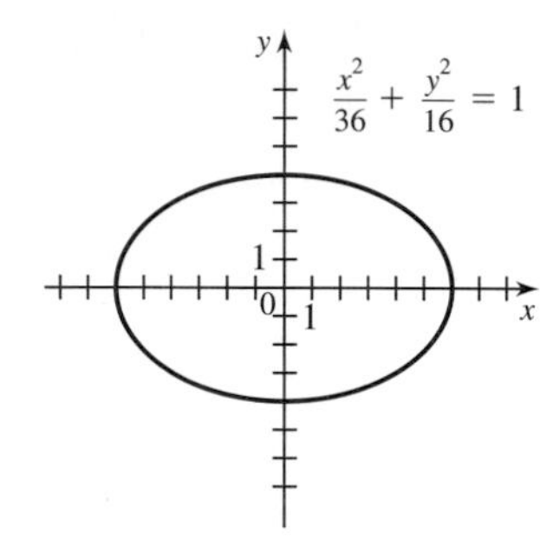

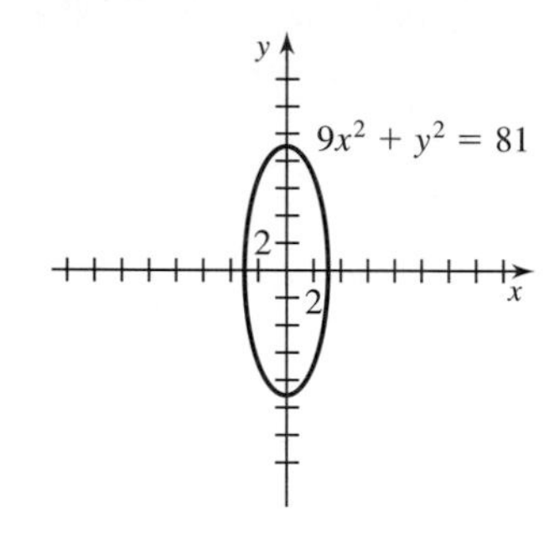

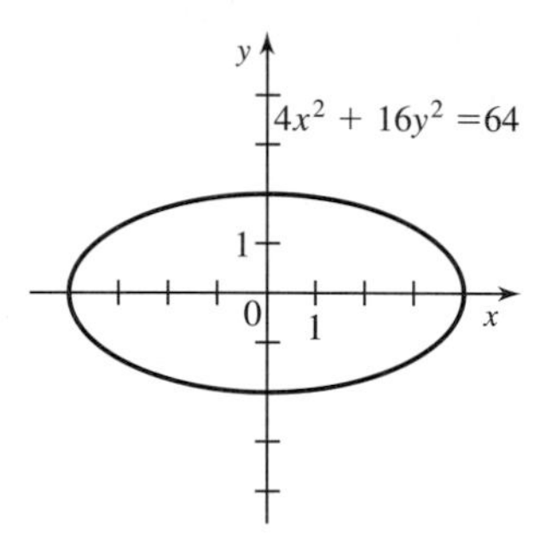

7. $4x^2 + 25y^2 = 100$ **(0, 0); (−5, 0), (5, 0); (0, −2), (0, 2); (−√21, 0), (√21, 0)**

8. $4x^2 + y^2 = 16$ **(0, 0); (0, −4), (0, 4); (−2, 0), (2, 0); (0, −2√3), (0, 2√3)**

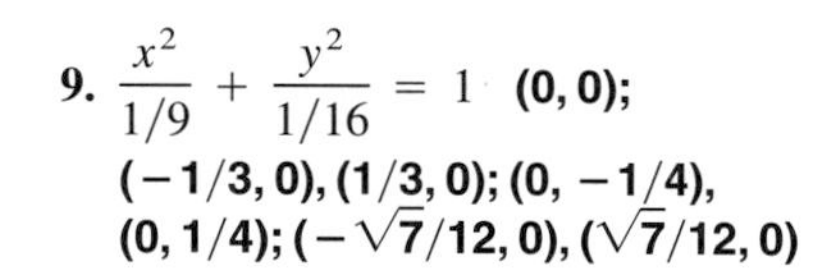
9. $\dfrac{x^2}{1/9} + \dfrac{y^2}{1/16} = 1$ **(0, 0); (−1/3, 0), (1/3, 0); (0, −1/4), (0, 1/4); (−√7/12, 0), (√7/12, 0)**

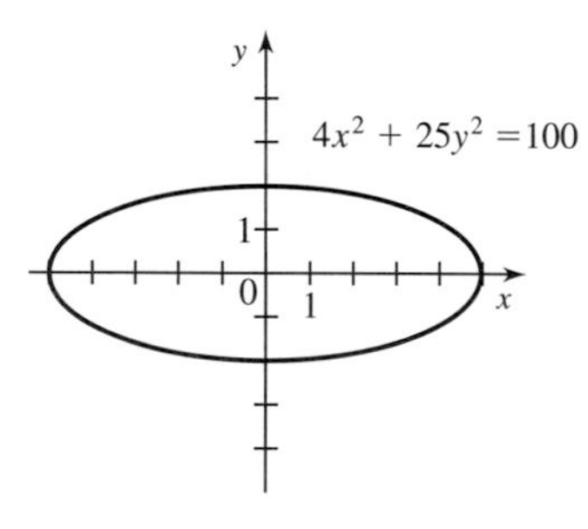

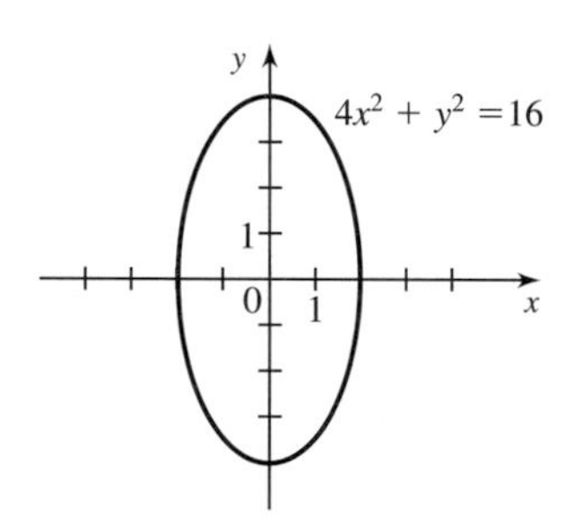

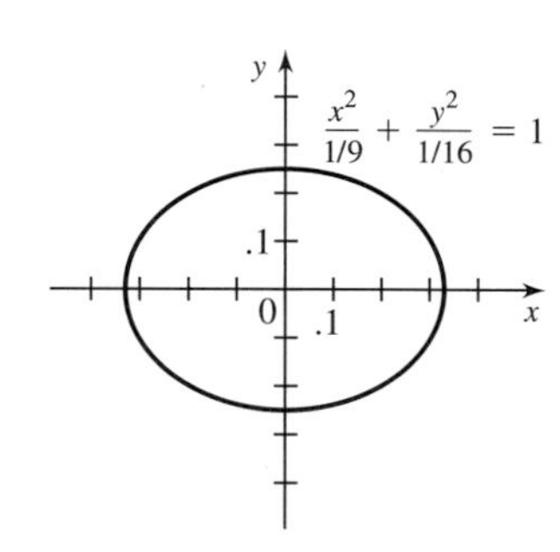

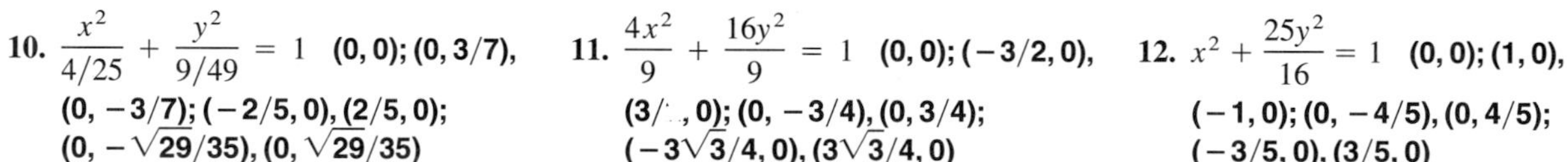
10. $\dfrac{x^2}{4/25} + \dfrac{y^2}{9/49} = 1$ **(0, 0); (0, 3/7), (0, −3/7); (−2/5, 0), (2/5, 0); (0, −√29/35), (0, √29/35)**

11. $\dfrac{4x^2}{9} + \dfrac{16y^2}{9} = 1$ **(0, 0); (−3/2, 0), (3/ , 0); (0, −3/4), (0, 3/4); (−3√3/4, 0), (3√3/4, 0)**

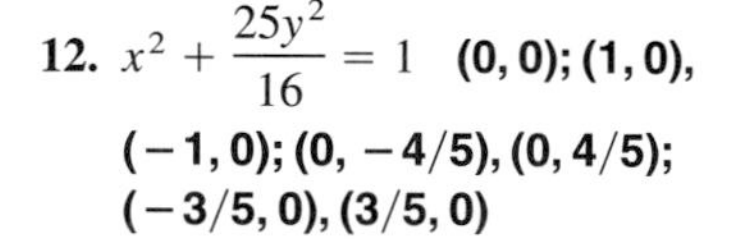
12. $x^2 + \dfrac{25y^2}{16} = 1$ **(0, 0); (1, 0), (−1, 0); (0, −4/5), (0, 4/5); (−3/5, 0), (3/5, 0)**

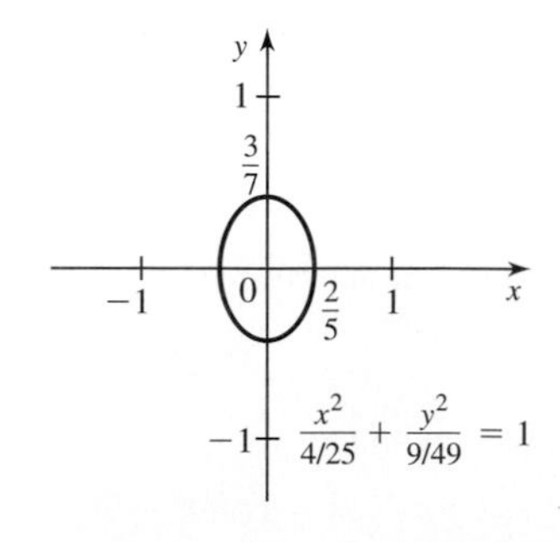

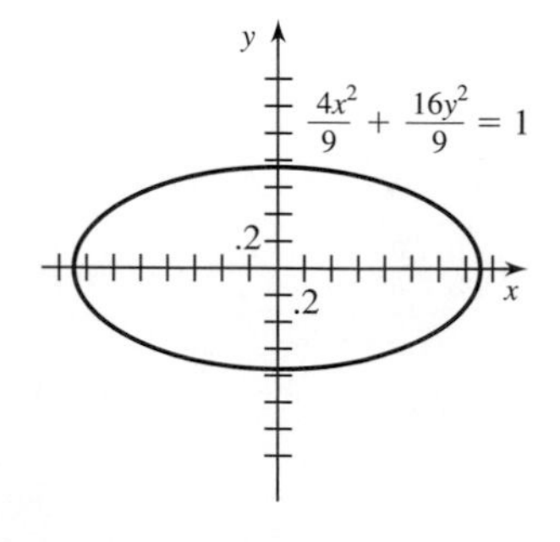

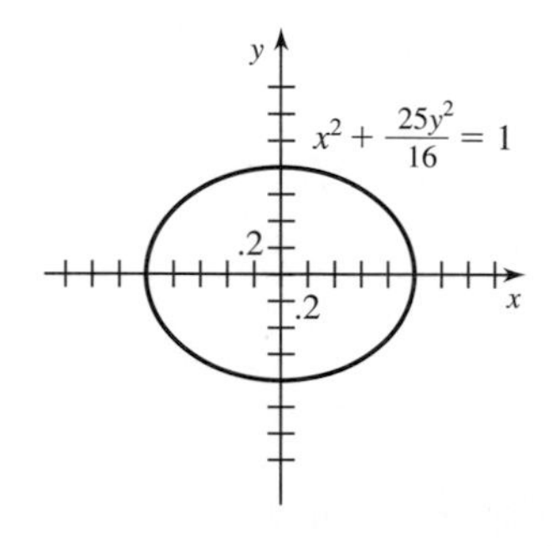

13. $\dfrac{(x-2)^2}{25}+\dfrac{(y-1)^2}{4}=1$ **(2, 1); (−3, 1), (7, 1); (2, −1), (2, 3); $(2-\sqrt{21}, 1), (2+\sqrt{21}, 1)$**

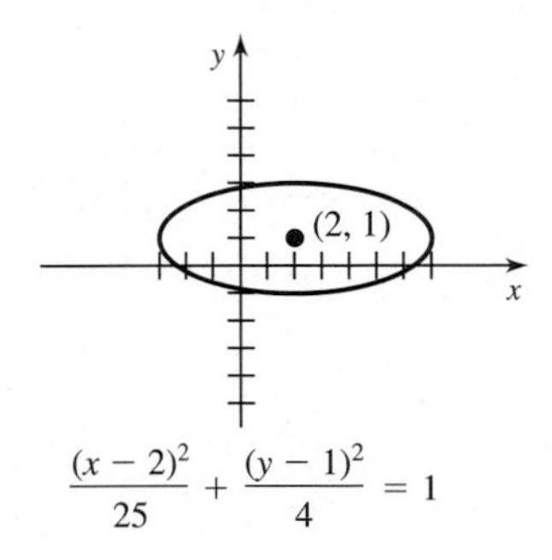

14. $\dfrac{(x+2)^2}{16}+\dfrac{(y+1)^2}{9}=1$ **(−2, −1); (−6, −1), (2, −1); (−2, −4), (−2, 2); $(-2-\sqrt{7}, -1)$, $(-2+\sqrt{7}, -1)$**

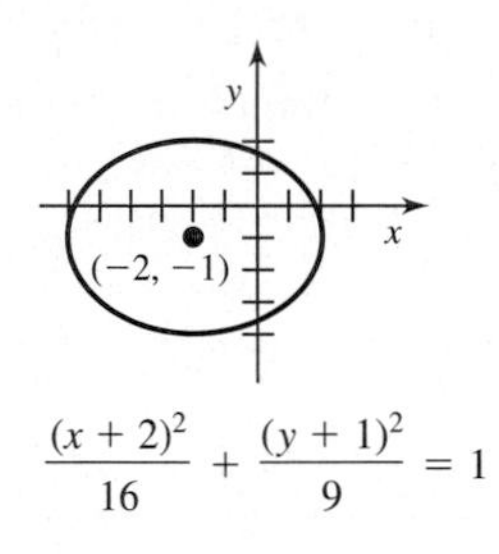

15. $\dfrac{(x+3)^2}{16}+\dfrac{(y-2)^2}{36}=1$ **(−3, 2); (−3, −4), (−3, 8); (−7, 2), (1, 2); $(-3, 2-2\sqrt{5}), (-3, 2+2\sqrt{5})$**

$\dfrac{(x+3)^2}{16}+\dfrac{(y-2)^2}{36}=1$

16. $\dfrac{(x-1)^2}{9}+\dfrac{(y+3)^2}{25}=1$ **(1, −3); (1, −8), (1, 2); (−2, −3), (4, −3); (1, −7), (1, 1)**

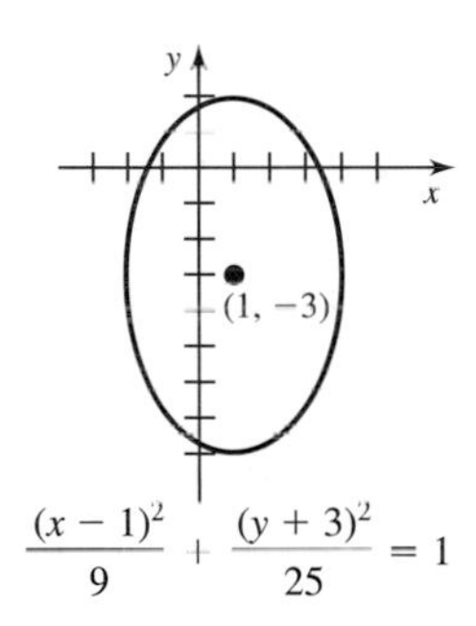

Find equations for each of the following ellipses.

17. x-intercepts ± 5; foci at $(-3, 0)$, $(3, 0)$ $\boldsymbol{x^2/25 + y^2/16 = 1}$

18. y-intercepts ± 4; foci at $(0, -1)$, $(0, 1)$ $\boldsymbol{x^2/15 + y^2/16 = 1}$

19. Major axis with length 6; foci at $(0, 2)$, $(0, -2)$ $\boldsymbol{x^2/5 + y^2/9 = 1}$

20. Minor axis with length 4; foci at $(-5, 0)$, $(5, 0)$ $\boldsymbol{x^2/29 + y^2/4 = 1}$

21. Center at $(5, 2)$; minor axis vertical, with length 8; $c = 3$ $\dfrac{(x-5)^2}{25}+\dfrac{(y-2)^2}{16}=1$

22. Center at $(-3, 6)$; major axis vertical, with length 10; $c = 2$ $\dfrac{(x+3)^2}{21}+\dfrac{(y-6)^2}{25}=1$

23. Vertices at (4, 9), (4, 1); minor axis with length 6 $\dfrac{(x-4)^2}{9}+\dfrac{(y-5)^2}{16}=1$

24. Foci at (−3, −3), (7, −3); (2, 1) on ellipse $\dfrac{(x-2)^2}{41}+\dfrac{(y+3)^2}{16}=1$

25. Foci at (0, −3), (0, 3); (8, 3) on ellipse $x^2/72 + y^2/81 = 1$

26. Foci at (−4, 0), (4, 0); sum of distances from foci to point on ellipse is 9 (*Hint:* Consider one of the vertices.) $4x^2/81 + 4y^2/17 = 1$

27. Foci at (0, 4), (0, −4); sum of distances from foci to point on ellipse is 10 $x^2/9 + y^2/25 = 1$

28. Eccentricity $\frac{1}{2}$; vertices at (−4, 0), (4, 0) $x^2/16 + y^2/12 = 1$

29. Eccentricity $\frac{3}{4}$; foci at (0, −2), (0, 2) $9x^2/28 + 9y^2/64 = 1$

30. Eccentricity $\frac{2}{3}$; foci at (0, −9), (0, 9) $4x^2/405 + 4y^2/729 = 1$

31. Eccentricity $\frac{2}{5}$; vertices at (−20, 0), (20, 0) $x^2/400 + y^2/336 = 1$

Sketch the graph of each of the following. Identify any that are the graphs of functions.

32. $\dfrac{y}{2}=\sqrt{1-\dfrac{x^2}{25}}$ **function**

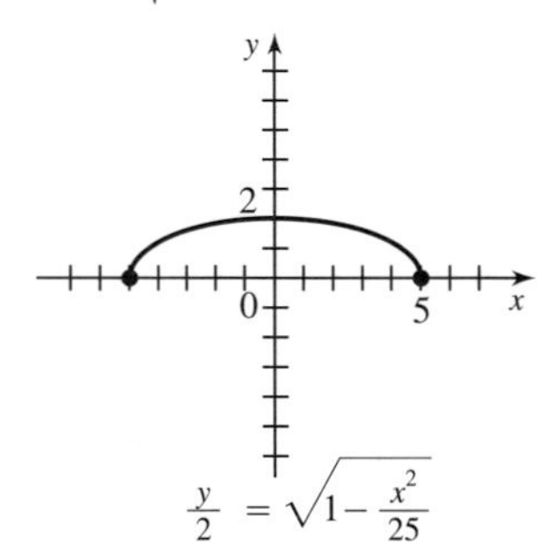

33. $\dfrac{x}{4}=\sqrt{1-\dfrac{y^2}{9}}$

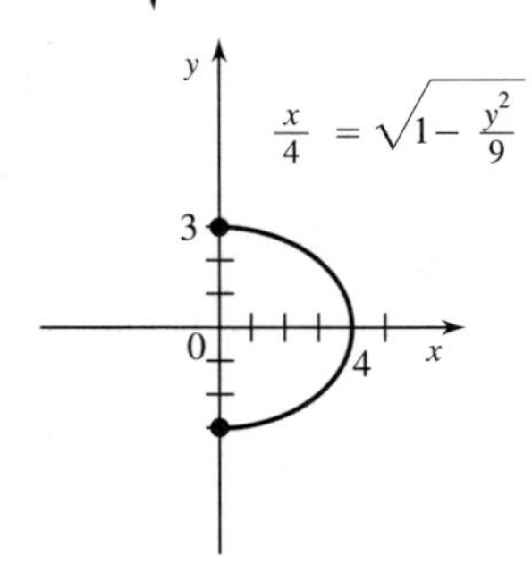

34. $x=-\sqrt{1-\dfrac{y^2}{64}}$

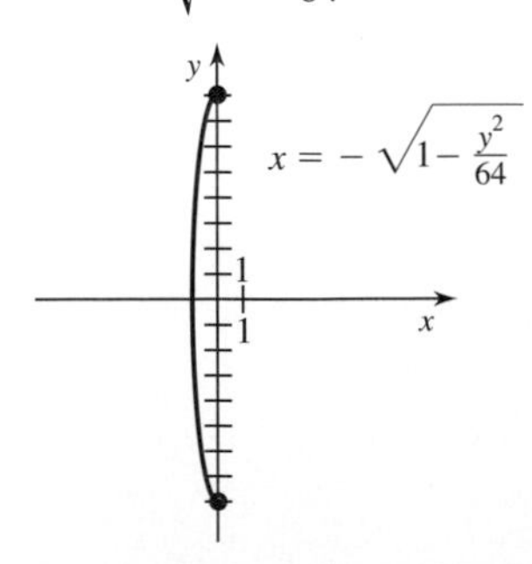

35. $y=-\sqrt{1-\dfrac{x^2}{100}}$ **function**

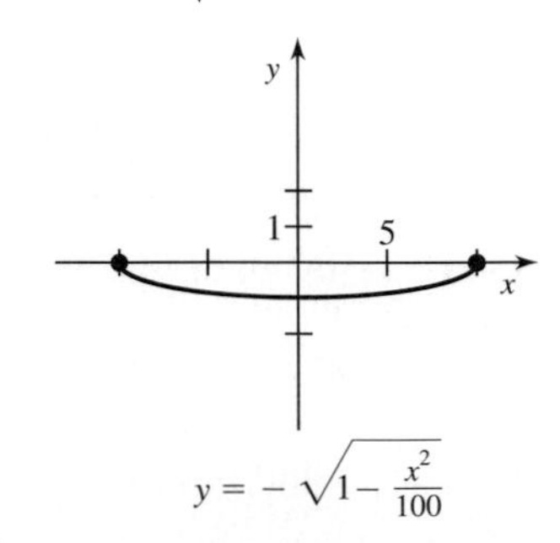

36. The figure shows an ellipse with its foci labeled. Answer the following questions with a method that does not use the equation of the ellipse.
(a) Is the point (2, 4) on the ellipse? **yes**
(b) What is the length of the major axis of the ellipse? **9**

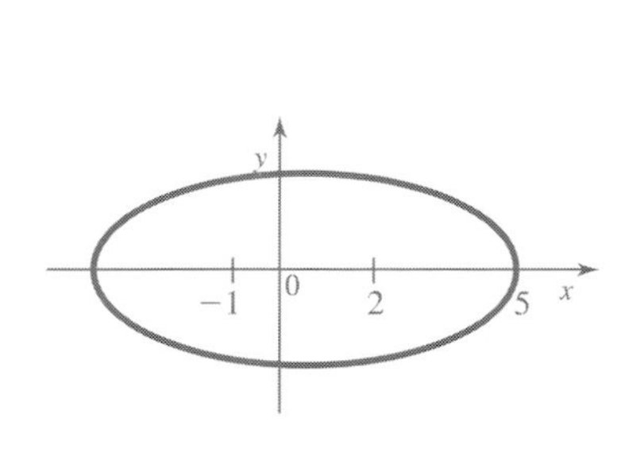

Exercise 36

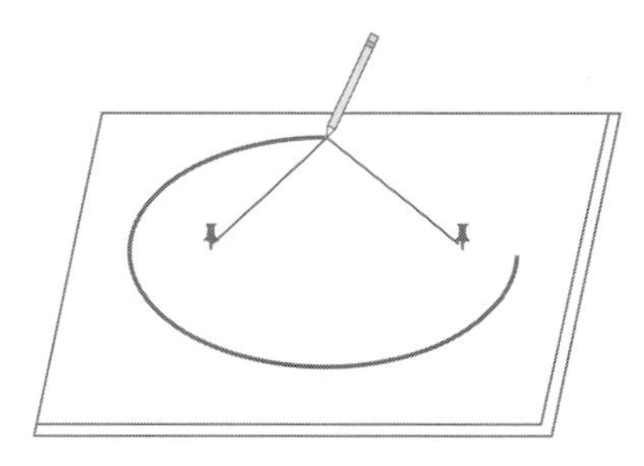

Exercise 37

37. Draftspeople often use the method shown in the sketch to draw an ellipse. Explain why the method works.

38. Suppose you know the foci of an ellipse. What additional piece of information would allow you to graph the ellipse? (Give several examples.) Explain how this information would be used.

39. Halley's Comet has an elliptical orbit of eccentricity .9673 with the sun at one of the foci. The greatest distance of the comet from the sun is 3281 million miles. Find the shortest distance between Halley's Comet and the sun. **≈55 million miles**

40. The orbit of planet Earth is an ellipse with the sun at one focus. The distance between Earth and the sun ranges from 91.4 to 94.6 million miles. Find the eccentricity of Earth's orbit. **≈.0172**

41. An arch of a bridge has the shape of the top half of an ellipse. The arch is 40 ft wide and 12 ft high at the center. Find the equation of the ellipse. Find the height of the arch 10 ft from the center of the bottom. $x^2/400 + y^2/144 = 1$, **≈10.39 ft**

Graphing Utility Problems

In Exercises 42–45, find the equation for the ellipse that has its center at the origin and satisfies the given conditions. Then graph the equation and confirm that the conditions hold.

42. Horizontal major axis with length 6; minor axis with length 4 $x^2/9 + y^2/4 = 1$

43. Vertical major axis with length 10; minor axis with length 2 $x^2 + y^2/25 = 1$

44. x-intercepts ± 1; y-intercepts $\pm\frac{2}{3}$ $x^2 + 9y^2/4 = 1$

45. x-intercepts $\pm\frac{3}{4}$; y-intercepts ± 2 $16x^2/9 + y^2/4 = 1$

46. Graph the ellipse $\dfrac{x^2}{16} + \dfrac{y^2}{12} = 1$. The ellipse has foci $(-2, 0)$ and $(2, 0)$. Use TRACE to find the coordinates of several points on the ellipse. For each of these points P, verify that

$$[\text{Distance of } P \text{ from } (-2, 0)] + [\text{Distance of } P \text{ from } (2, 0)] = 8.$$

47. Find the equation of an ellipse consisting of all points P in the plane for which the sum of the distances of P from $(-4, 0)$ and $(4, 0)$ is 10. Then graph the ellipse and use TRACE to find the coordinates of several points on the graph of the ellipse. For each of these points, verify that the sum of the distances is indeed 10.

7.3 Exercises (text page 397)

Sketch the graph of each of the following hyperbolas. Give the center, vertices, foci, and equations of the asymptotes for each figure.

1. $\dfrac{x^2}{16} - \dfrac{y^2}{9} = 1$ **$(0, 0); (-4, 0), (4, 0); (-5, 0), (5, 0); y = \pm(3/4)x$**

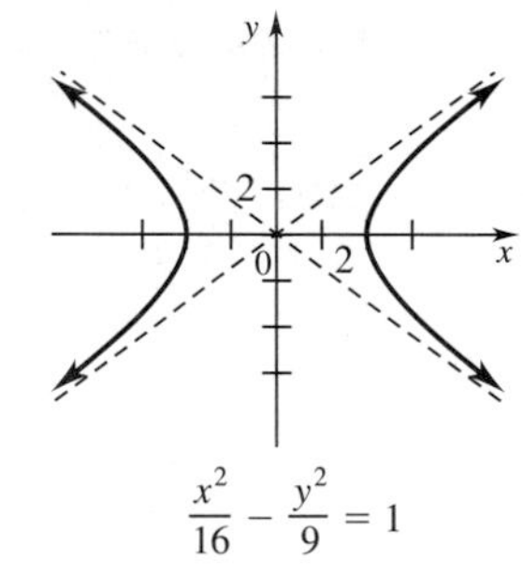

2. $\dfrac{x^2}{25} - \dfrac{y^2}{144} = 1$ **$(0, 0); (-5, 0), (5, 0); (-13, 0), (13, 0); y = \pm(12/5)x$**

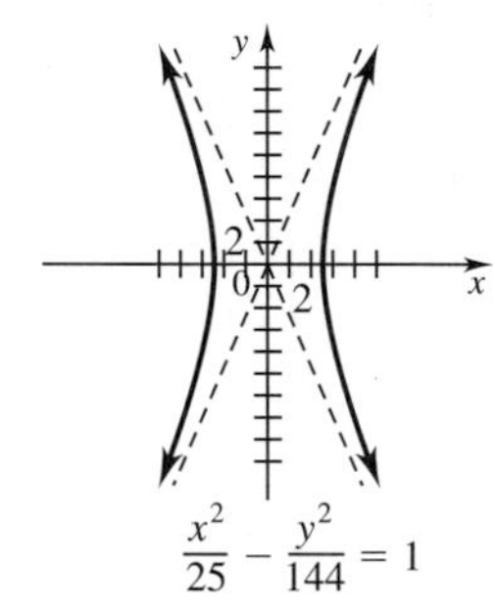

3. $\dfrac{y^2}{25} - \dfrac{x^2}{49} = 1$ **$(0, 0); (0, -5), (0, 5); (0, -\sqrt{74}), (0, \sqrt{74}); y = \pm(5/7)x$**

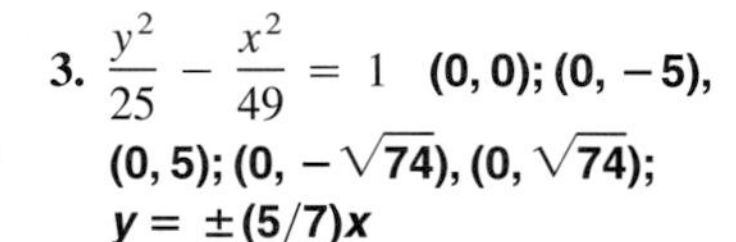

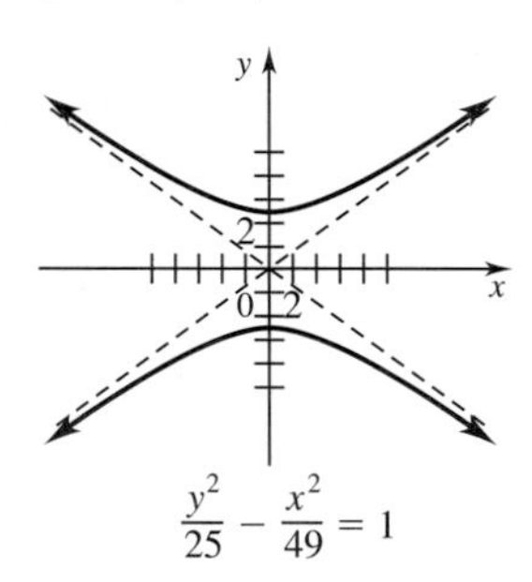

4. $\dfrac{y^2}{64} - \dfrac{x^2}{4} = 1$ **$(0, 0); (0, -8), (0, 8); (0, -2\sqrt{17}), (0, 2\sqrt{17}); y = \pm 4x$**

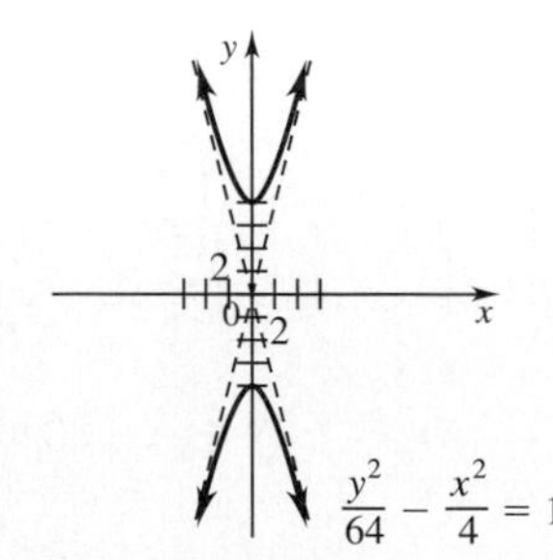

5. $x^2 - y^2 = 9$ **$(0, 0); (-3, 0), (3, 0); (-3\sqrt{2}, 0), (3\sqrt{2}, 0); y = \pm x$**

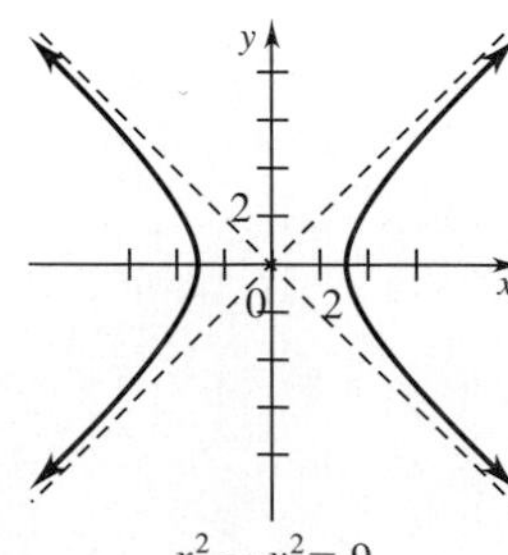
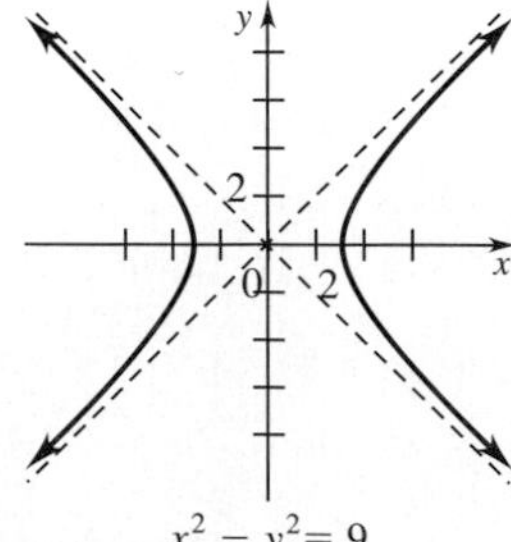

6. $x^2 - 4y^2 = 64$ **$(0, 0); (-8, 0), (8, 0); (-4\sqrt{5}, 0), (4\sqrt{5}, 0); y = \pm(1/2)x$**

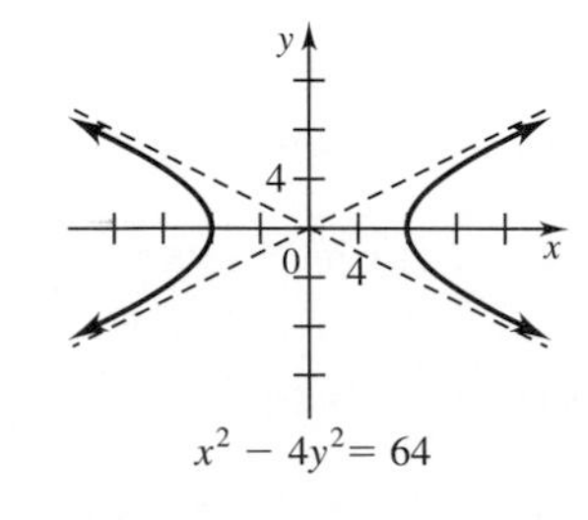

7. $9x^2 - 25y^2 = 225$ **(0, 0); (−5, 0), (5, 0); $(-\sqrt{34}, 0)$, $(\sqrt{34}, 0)$; $y = \pm(3/5)x$**

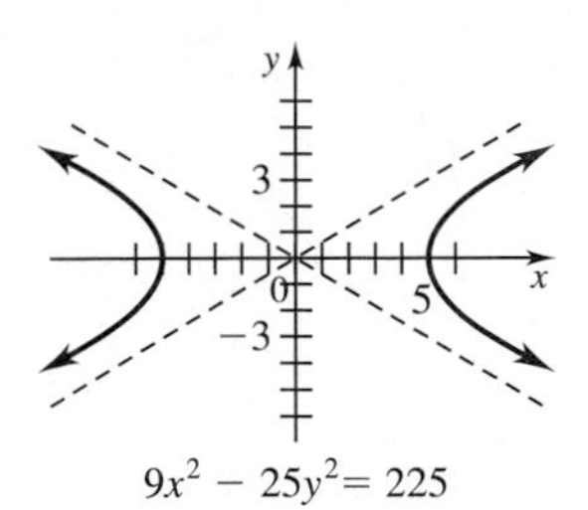

$9x^2 - 25y^2 = 225$

8. $25x^2 - 4y^2 = -100$ **(0, 0); (0, −5), (0, 5); $(0, -\sqrt{29})$, $(0, \sqrt{29})$; $y = \pm(5/2)x$**

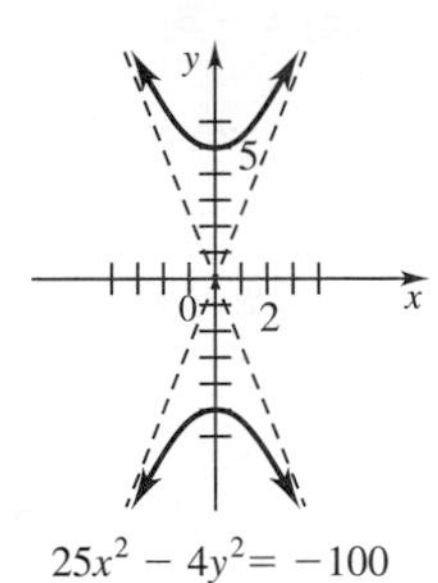

$25x^2 - 4y^2 = -100$

9. $4x^2 - y^2 = -16$ **(0, 0); (0, −4), (0, 4); $(0, -2\sqrt{5})$, $(0, 2\sqrt{5})$; $y = \pm 2x$**

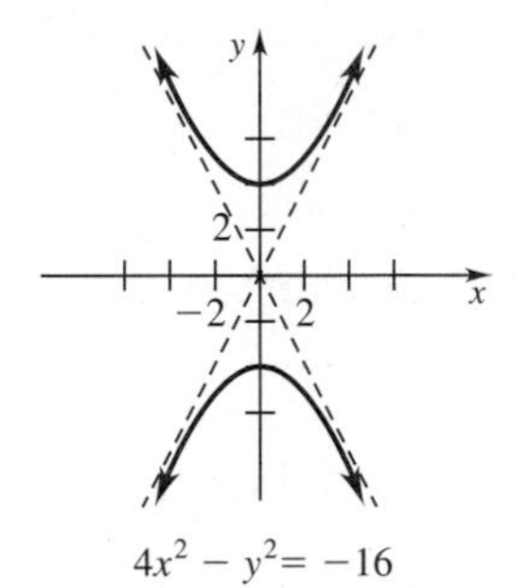

$4x^2 - y^2 = -16$

10. $\frac{x^2}{4} - y^2 = 4$ **(0, 0); (−4, 0), (4, 0); $(-2\sqrt{5}, 0)$, $(2\sqrt{5}, 0)$; $y = \pm(1/2)x$**

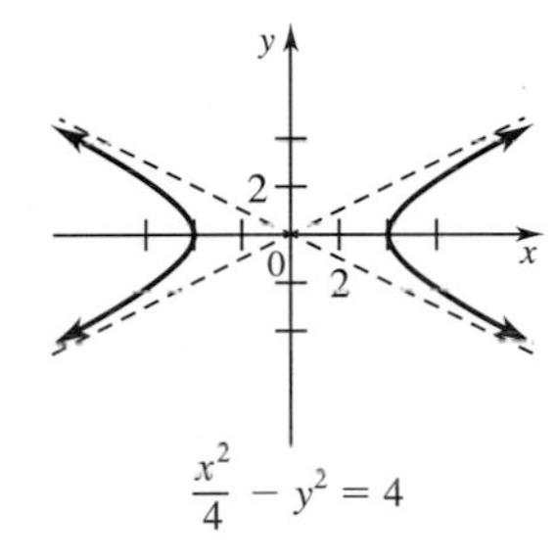

$\frac{x^2}{4} - y^2 = 4$

11. $9x^2 - 4y^2 = 1$ **(0, 0); (−1/3, 0), (1/3, 0); $(-\sqrt{13}/6, 0)$, $(\sqrt{13}/6, 0)$; $y = \pm(3/2)x$**

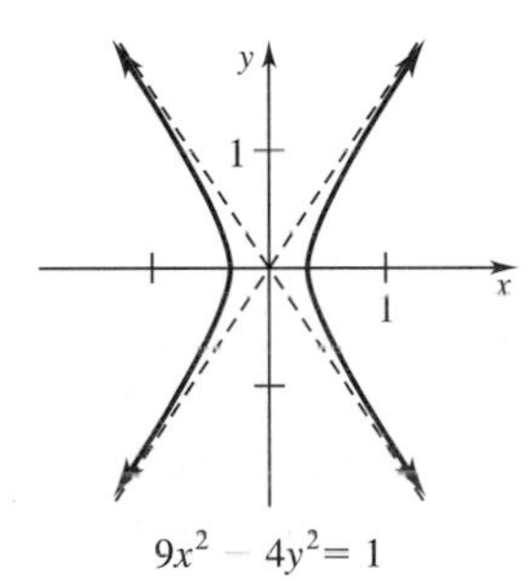

$9x^2 - 4y^2 = 1$

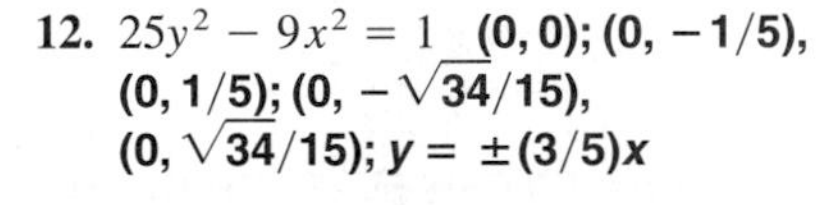
12. $25y^2 - 9x^2 = 1$ **(0, 0); (0, −1/5), (0, 1/5); $(0, -\sqrt{34}/15)$, $(0, \sqrt{34}/15)$; $y = \pm(3/5)x$**

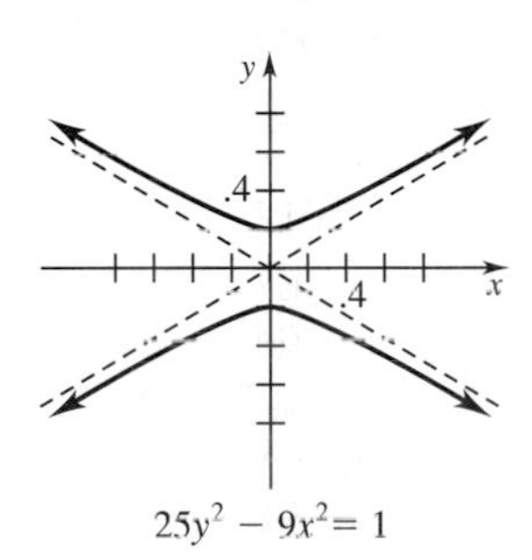

$25y^2 - 9x^2 = 1$

13. $\frac{(y-7)^2}{36} - \frac{(x-4)^2}{64} = 1$ **(4, 7); (4, 1), (4, 13); (4, −3), (4, 17); $y = 7 \pm (3/4)(x - 4)$**

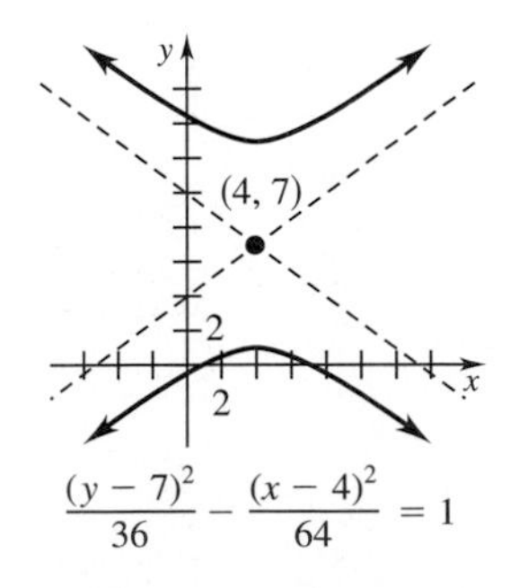

$\frac{(y-7)^2}{36} - \frac{(x-4)^2}{64} = 1$

14. $\frac{(x+6)^2}{144} - \frac{(y+4)^2}{81} = 1$ **(−6, −4); (−18, −4), (6, −4); (−21, −4), (9, −4); $y = -4 \pm (3/4)(x + 6)$**

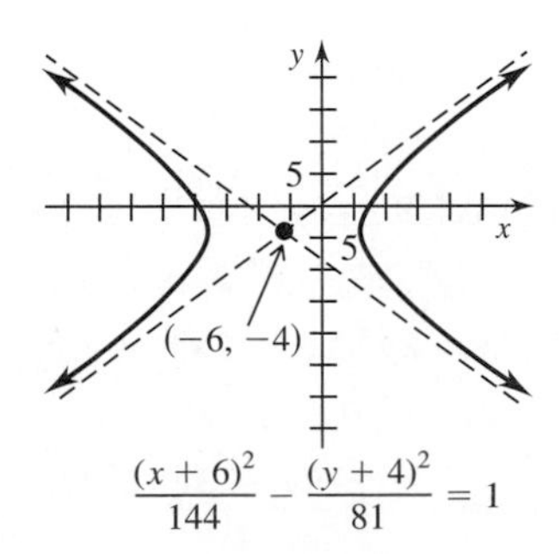

$\frac{(x+6)^2}{144} - \frac{(y+4)^2}{81} = 1$

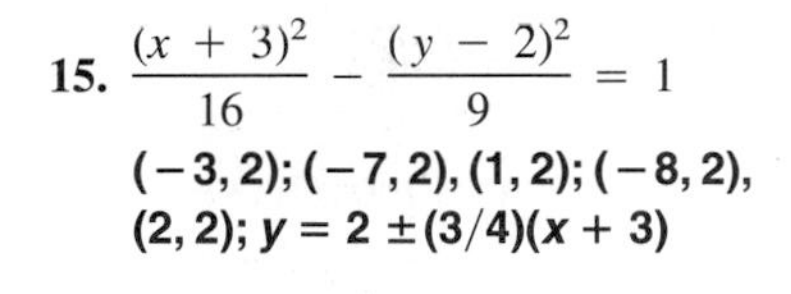
15. $\frac{(x+3)^2}{16} - \frac{(y-2)^2}{9} = 1$ **(−3, 2); (−7, 2), (1, 2); (−8, 2), (2, 2); $y = 2 \pm (3/4)(x + 3)$**

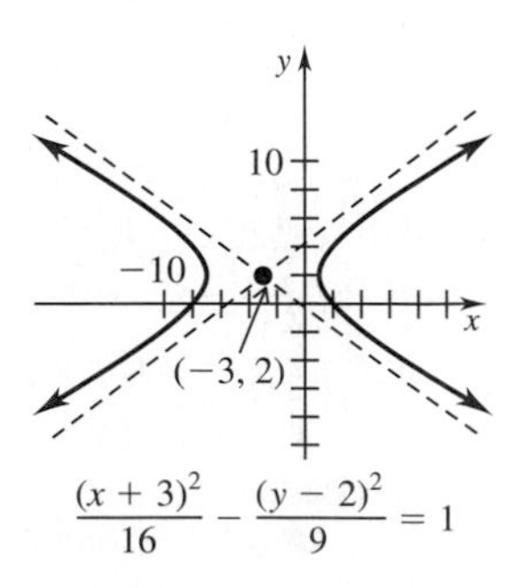

$\frac{(x+3)^2}{16} - \frac{(y-2)^2}{9} = 1$

16. $\dfrac{(y+5)^2}{4} - \dfrac{(x-1)^2}{16} = 1$
(1, −5); (1, −7), (1, −3); (1, −5 − 2√5), (1, −5 + 2√5); y = −5 ± (1/2)(x − 1)

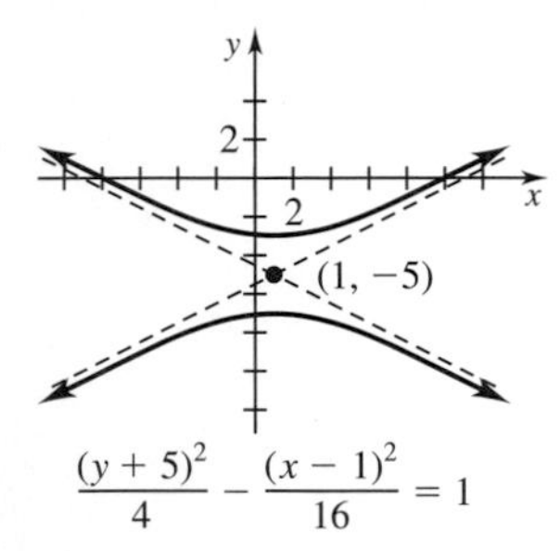

17. $(x-8)^2 - 5(y+7)^2 = 25$
(8, −7); (3, −7), (13, −7); (8 − √30, −7), (8 + √30, −7); y = −7 ± (√5/5)(x − 8)

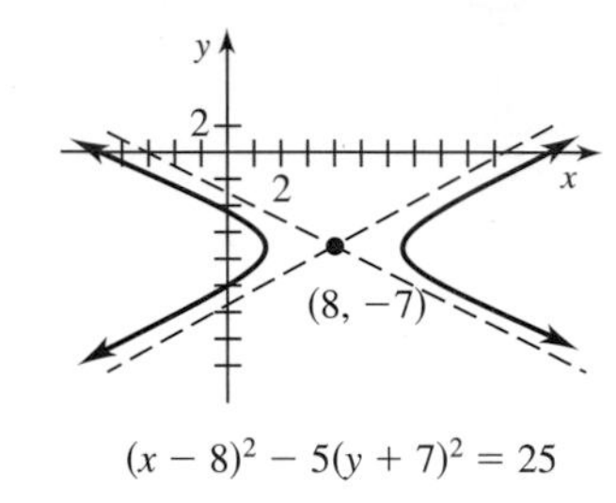

18. $2(x-2)^2 - (y-10)^2 = 8$ **(2, 10); (0, 10), (4, 10); (2 − 2√3, 10), (2 + 2√3, 10); y = 10 ± √2(x − 2)**

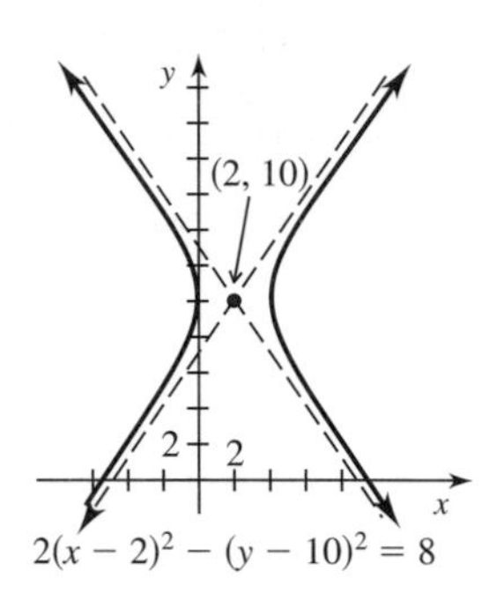

19. $16(x+5)^2 - (y-3)^2 = 1$ **(−5, 3); (−21/4, 3), (−19/4, 3); (−5 − √17/4, 3), (−5 + √17/4, 3); y = 3 ± 4(x + 5)**

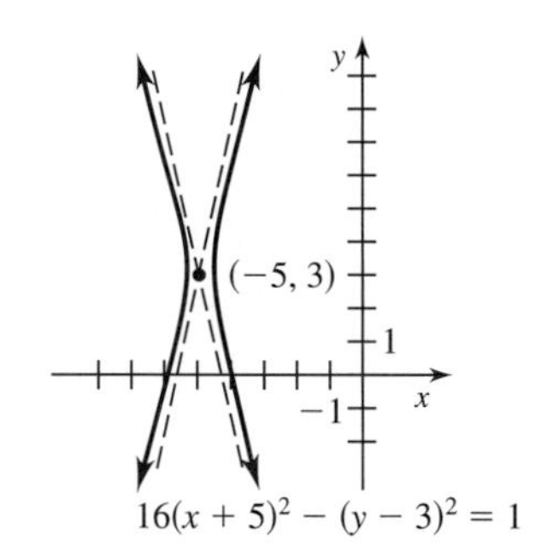

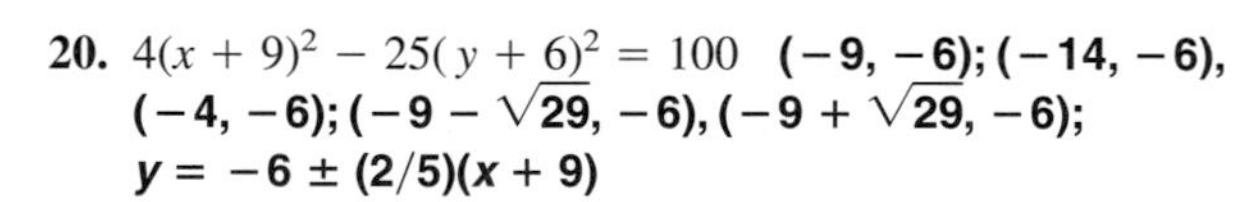

20. $4(x+9)^2 - 25(y+6)^2 = 100$ **(−9, −6); (−14, −6), (−4, −6); (−9 − √29, −6), (−9 + √29, −6); y = −6 ± (2/5)(x + 9)**

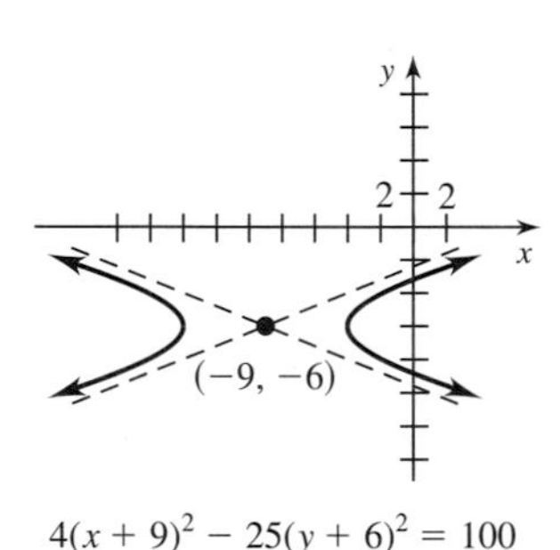

Find equations for each of the following hyperbolas.

21. x-intercepts ± 4; foci at $(-5, 0)$, $(5, 0)$ $\mathbf{x^2/16 - y^2/9 = 1}$

22. y-intercepts ± 9; foci at $(0, -15)$, $(0, 15)$ $\mathbf{y^2/81 - x^2/144 = 1}$

23. Vertices at $(0, 6)$, $(0, -6)$; asymptotes $y = \pm(1/2)x$ $\mathbf{y^2/36 - x^2/144 = 1}$

24. Vertices at $(-10, 0)$, $(10, 0)$; asymptotes $y = \pm 5x$ $\mathbf{x^2/100 - y^2/2500 = 1}$

25. Vertices at $(-3, 0)$, $(3, 0)$; passing through $(6, 1)$ $\mathbf{x^2/9 - 3y^2 = 1}$

26. Vertices at $(0, 5)$, $(0, -5)$; passing through $(3, 10)$ $\mathbf{y^2/25 - x^2/3 = 1}$

27. Foci at $(0, \sqrt{13})$, $(0, -\sqrt{13})$; asymptotes $y = \pm 5x$ $\mathbf{2y^2/25 - 2x^2 = 1}$

28. Foci at $(-\sqrt{45}, 0)$, $(\sqrt{45}, 0)$; asymptotes $y = \pm 2x$ $\mathbf{x^2/9 - y^2/36 = 1}$

29. Vertices at $(4, 5)$, $(4, 1)$; asymptotes $y - 3 = \pm 7(x - 4)$ $\mathbf{\dfrac{(y-3)^2}{4} - \dfrac{49(x-4)^2}{4} = 1}$

30. Vertices at $(5, -2)$, $(1, -2)$; asymptotes $y + 2 = \pm\frac{3}{2}(x - 3)$ $\dfrac{(x-3)^2}{4} - \dfrac{(y+2)^2}{9} = 1$

31. Center at $(1, -2)$; focus at $(4, -2)$; vertex at $(3, -2)$ $\dfrac{(x-1)^2}{4} - \dfrac{(y+2)^2}{5} = 1$

32. Center at $(9, -7)$; focus at $(9, 3)$; vertex at $(9, -1)$ $\dfrac{(y+7)^2}{36} - \dfrac{(x-9)^2}{64} = 1$

33. Eccentricity 3; center at $(0, 0)$; vertex at $(0, 7)$ $y^2/49 - x^2/392 = 1$

34. Center at $(8, 7)$; focus at $(13, 7)$; eccentricity $5/3$ $\dfrac{(x-8)^2}{9} - \dfrac{(y-7)^2}{16} = 1$

35. Vertices at $(-2, 10)$, $(-2, 2)$; eccentricity $5/4$ $\dfrac{(y-6)^2}{16} - \dfrac{(x+2)^2}{9} = 1$

36. Foci at $(9, 2)$, $(-11, 2)$; eccentricity $25/9$ $\dfrac{25(x+1)^2}{324} - \dfrac{25(y-2)^2}{2176} = 1$

37. Vertices at $(5, 5)$, $(5, 11)$; fundamental rectangle of area 18 $\dfrac{(y-8)^2}{9} - \dfrac{4(x-5)^2}{9} = 1$

38. Vertices at $(0, 6)$, $(8, 6)$; fundamental rectangle of area 20 $\dfrac{(x-4)^2}{16} - \dfrac{16(y-6)^2}{25} = 1$

◆ *Sketch the graph of each of the following. Identify any that are the graphs of functions.*

39. $\dfrac{y}{3} = \sqrt{1 + \dfrac{x^2}{16}}$ **function**

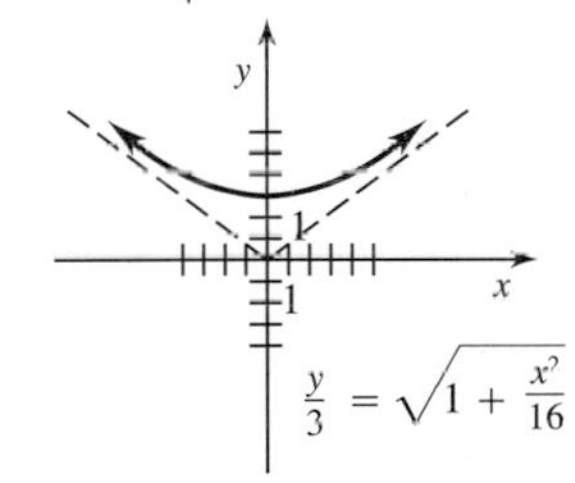

40. $\dfrac{x}{3} = -\sqrt{1 + \dfrac{y^2}{25}}$

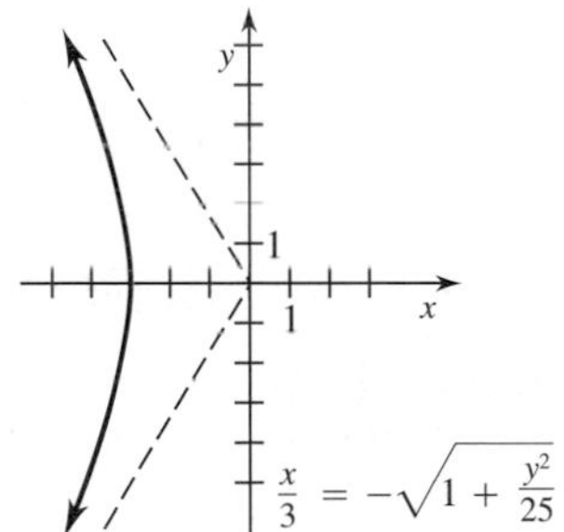

41. $5x = -\sqrt{1 + 4y^2}$

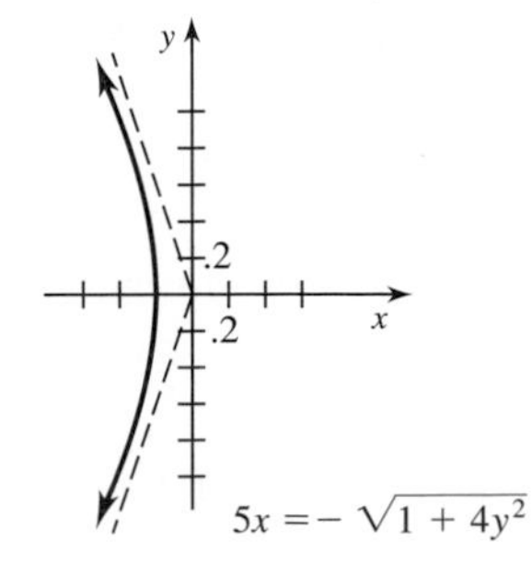

42. $3y = \sqrt{4x^2 - 16}$ **function**

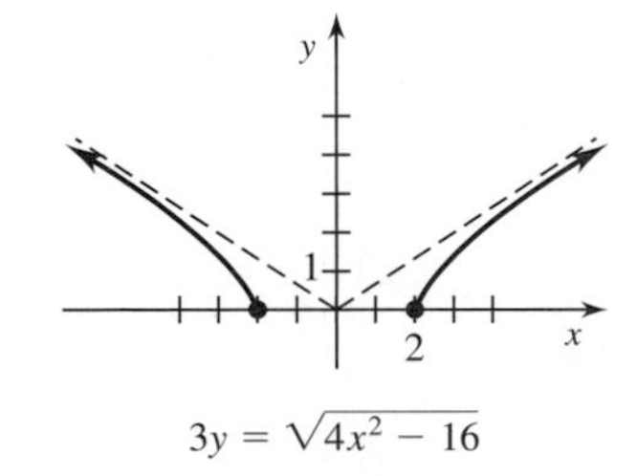

43. Describe the method you would use to determine (without graphing) whether a hyperbola opens left and right or up and down.

44. The figure shows a hyperbola with its foci labeled. Answer the following questions with a method that does not use the equation of the hyperbola.
(a) Is the point (10, 2) on the hyperbola? **no**
(b) What are the x-intercepts of the hyperbola? **0 and 9**

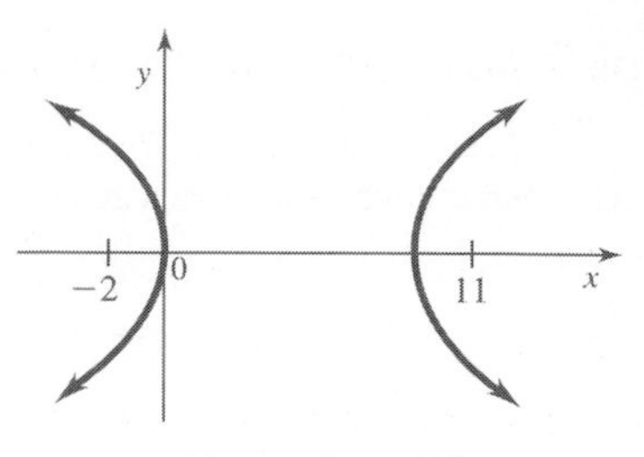

Exercise 44

45. Ships and planes often use a location finding system called LORAN. With this system, a radio transmitter at M on the figure sends out a series of pulses. When each pulse is received at transmitter S, it then sends out a pulse. A ship at P receives pulses from both M and S. A receiver on the ship measures the difference in the arrival times of the pulses. The navigator then consults a special map, showing certain curves according to the differences in arrival times. In this way, the ship can be located as lying on a portion of which curve? (This method requires three transmitters acting as two pairs.)
hyperbola

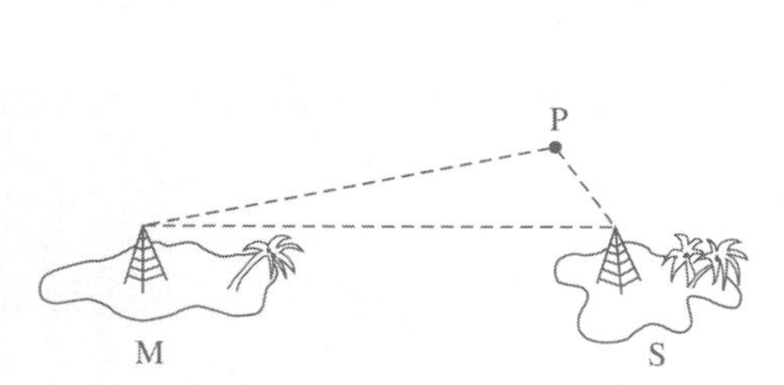

Exercise 45

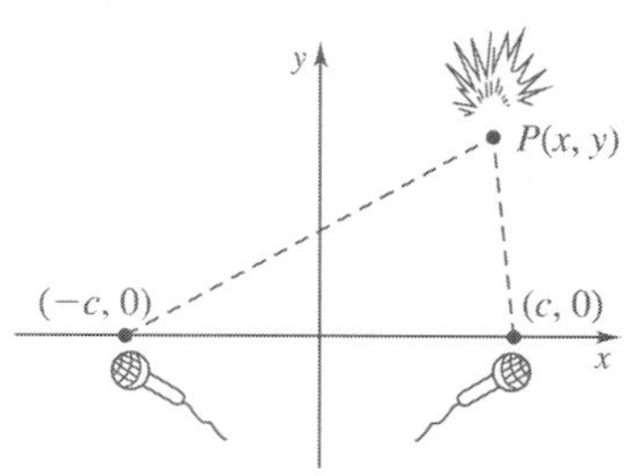

Exercise 46

46. Microphones are placed at points $(-c, 0)$ and $(c, 0)$. An explosion occurs at point $P(x, y)$ having positive x-coordinate. (See the figure.) The sound is detected at the closer microphone t sec before being detected at the farther microphone. Assume that sound travels at a speed of 330 m per sec, and show that P must be on the hyperbola

$$\frac{x^2}{330^2t^2} - \frac{y^2}{4c^2 - 330^2t^2} = \frac{1}{4}.$$

47. Suppose a hyperbola has center at the origin, foci at $F'(-c, 0)$ and $F(c, 0)$, and the value $d(P, F') - d(P, F) = 2a$. Let $b^2 = c^2 - a^2$, and show that an equation of the hyperbola is

$$\frac{x^2}{a^2} - \frac{y^2}{b^2} = 1.$$

Graphing Utility Problems

In Exercises 48–51, find the equations for the asymptotes of the hyperbola. Then graph the hyperbola and one or both (if possible) of the asymptotes to confirm that the equations of the asymptotes are correct.

48. $\frac{y^2}{3} - \frac{x^2}{4} = 1$ $\mathbf{y = \pm\frac{\sqrt{3}}{2}x}$

49. $\frac{x^2}{5} - \frac{y^2}{6} = 1$ $\mathbf{y = \pm\frac{\sqrt{30}}{5}x}$

50. $\frac{(x-4)^2}{5} - \frac{(y-5)^2}{4} = 1$ $\mathbf{y = \pm\frac{2\sqrt{5}}{5}(x - 4) + 5}$

51. $\frac{(x+7)^2}{8} - \frac{(y-6)^2}{2} = 1$ $\mathbf{y = \pm\frac{1}{2}(x + 7) + 6}$

52. Graph the hyperbola $\frac{x^2}{4} - \frac{y^2}{12} = 1$. The hyperbola has foci $(-4, 0)$ and $(4, 0)$. Use TRACE to find the coordinates of several points on the right half of the hyperbola. For each of these points P, verify that

$$[\text{Distance of } P \text{ from } (-4, 0)] - [\text{Distance of } P \text{ from } (4, 0)] = 4.$$

Use TRACE to find the coordinates of several points on the left half of the hyperbola. For each of these points P, verify that

$$[\text{Distance of } P \text{ from } (4, 0)] - [\text{Distance of } P \text{ from } (-4, 0)] = 4.$$

53. Find the equation of a hyperbola consisting of all points P in the plane for which the difference of the distances of P from $(-5, 0)$ and $(5, 0)$ is 8. Then graph the hyperbola and use TRACE to find the coordinates of several points on the graph of the hyperbola. For each of these points, verify that the differences of the distances is indeed 8.

54. Graph the upper half of the hyperbola $\frac{y^2}{4} - \frac{x^2}{25} = 1$ and the asymptote $y = \frac{2}{5}x$. Use ZOOM and TRACE to confirm that the curve approaches the asymptote as x gets larger and larger.

7.4 Exercises (text page 405)

For each of the following equations, identify the corresponding conic section. Draw a graph of each equation that has a graph.

1. $\frac{x^2}{4} - \frac{y^2}{9} = 1$ **hyperbola**

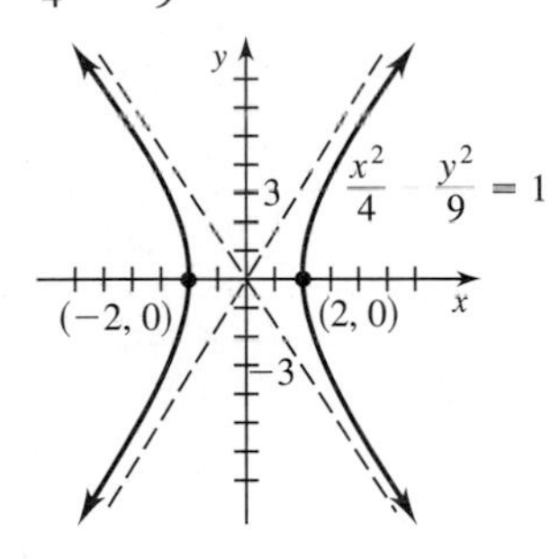

2. $\frac{x^2}{4} + \frac{y^2}{9} = 1$ **ellipse**

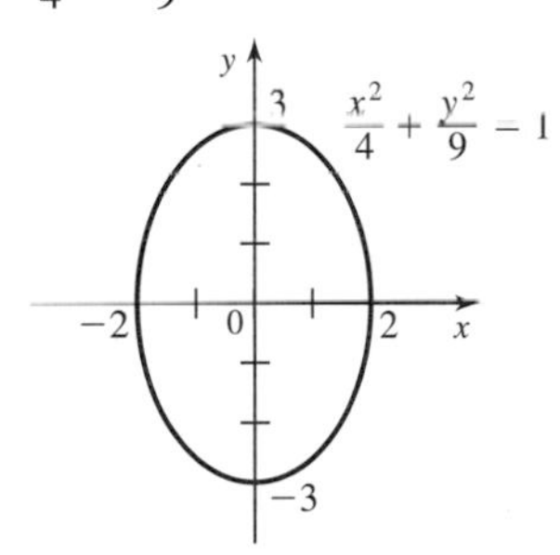

3. $\frac{x^2}{4} + \frac{y^2}{4} = 1$ **circle**

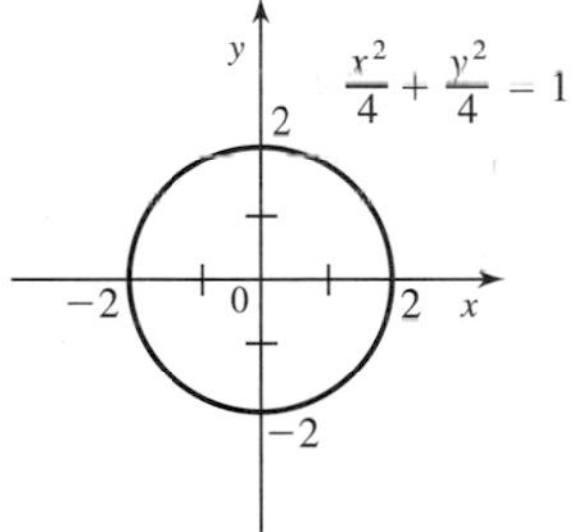

4. $\frac{x^2}{4} - \frac{y^2}{4} = 1$ **hyperbola**

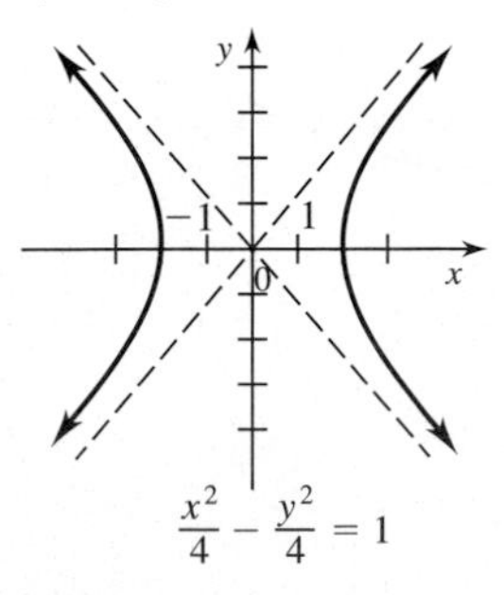

5. $16x^2 + 4y^2 = 64$ **ellipse**

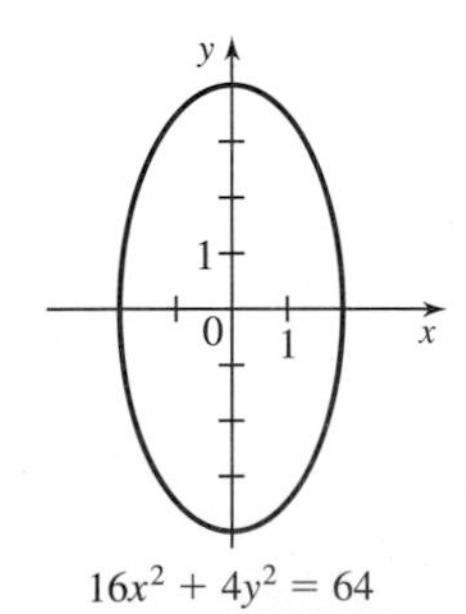

6. $16x^2 + 25y^2 = 1$ **ellipse**

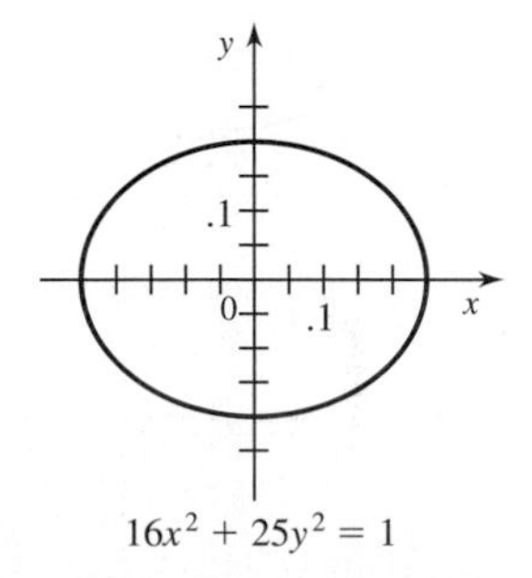

7. $x^2 = y^2 + 36$ **hyperbola**

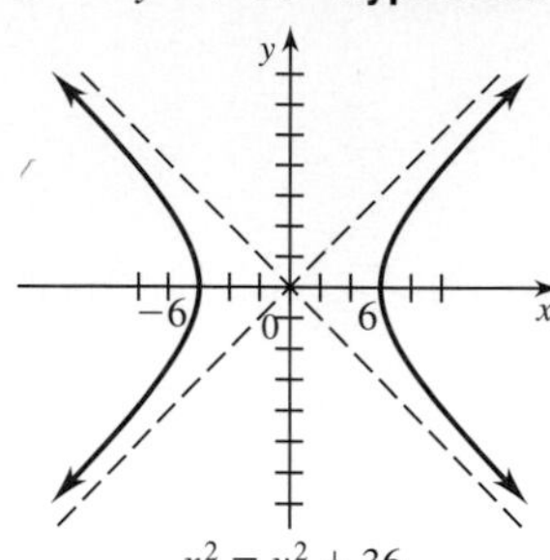

8. $x^2 = y^2 - 36$ **hyperbola**

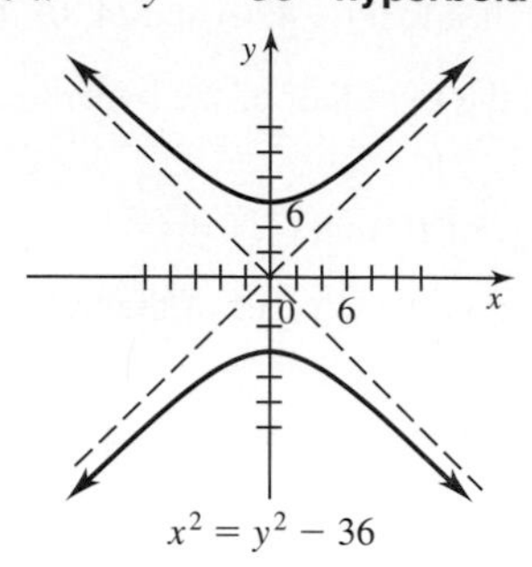

9. $5x^2 + 125y^2 = 5$ **ellipse**

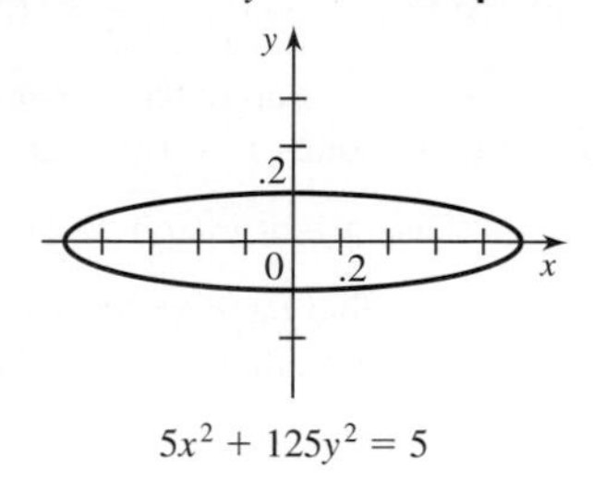

10. $x^2 = 4y - 8$ **parabola**

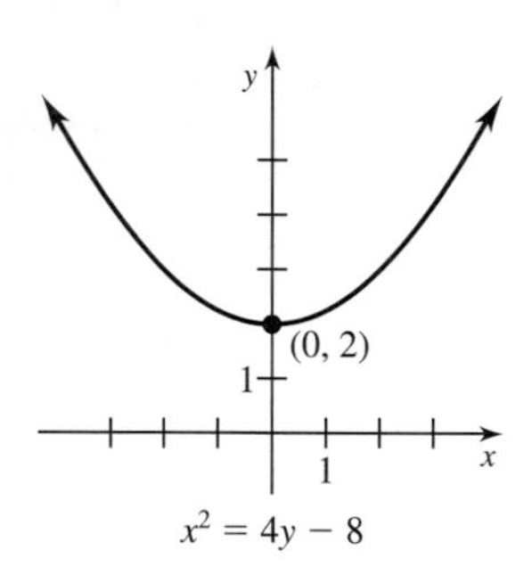

11. $\frac{(x-8)^2}{100} + \frac{(y-5)^2}{49} = 1$
ellipse

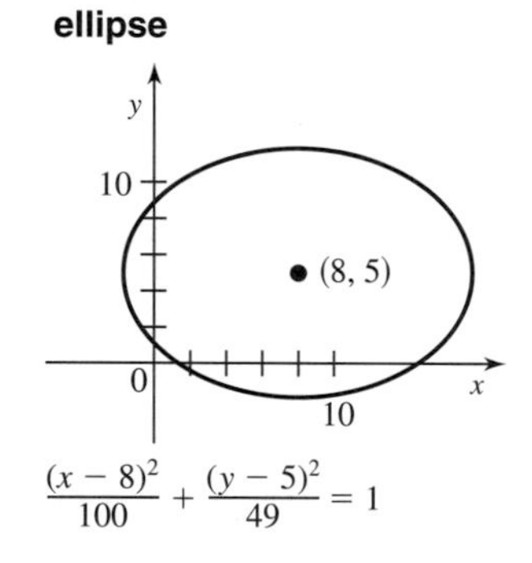

12. $\frac{(x+3)^2}{16} + \frac{(y-2)^2}{16} = 1$ **circle**

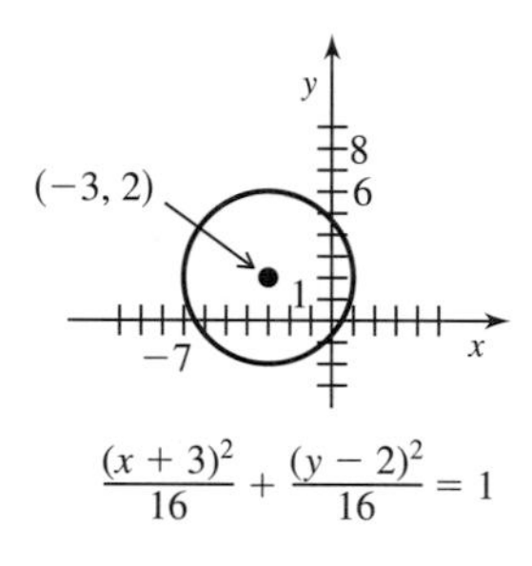

13. $\frac{x+4}{8} - \frac{(y+1)^2}{2} = 1$
parabola

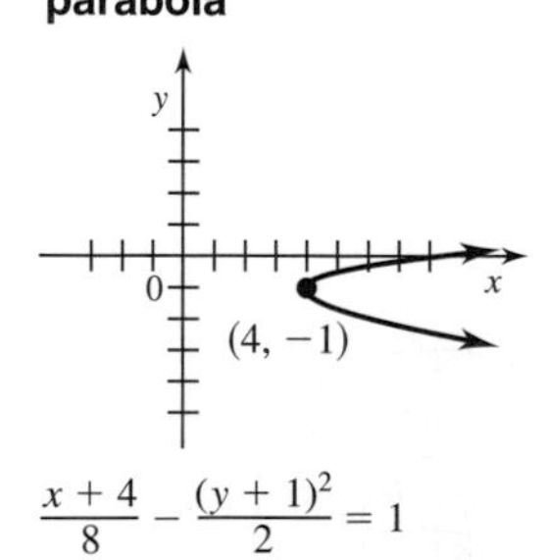

14. $y^2 - 4y = x + 4$ **parabola**

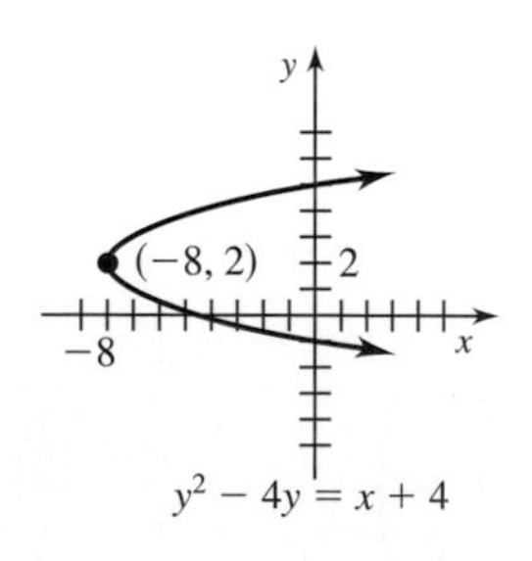

15. $11 - 3x = 2y^2 - 8y$ **parabola**

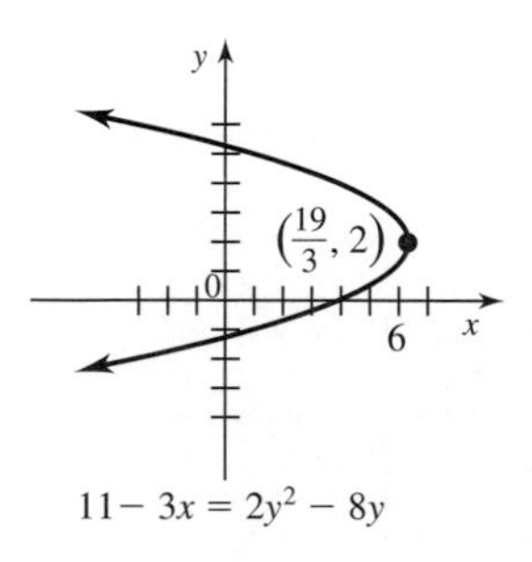

16. $(x + 7)^2 + (y - 5)^2 + 4 = 0$ **no graph**

17. $(x - 3)^2 + (y + 2)^2 = 0$ **point**

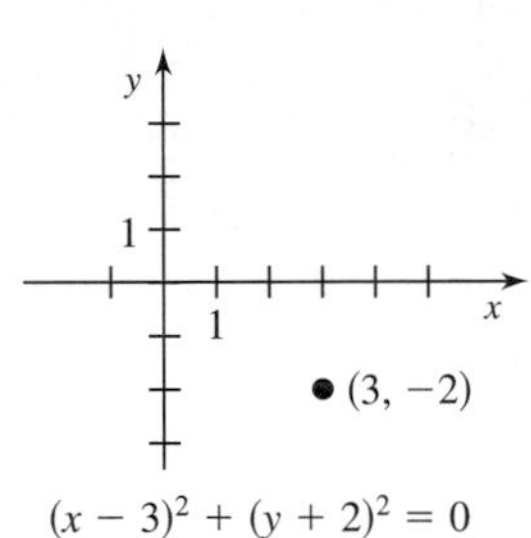

$(x - 3)^2 + (y + 2)^2 = 0$

18. $3x^2 + 6x + 3y^2 + 12y + 15 = 0$ **point**

$3x^2 + 6x + 3y^2 + 12y + 15 = 0$

19. $x^2 - 4x + y^2 + 2y = -6$ **no graph**

20. $x^2 - 6x + y = 0$ **parabola**

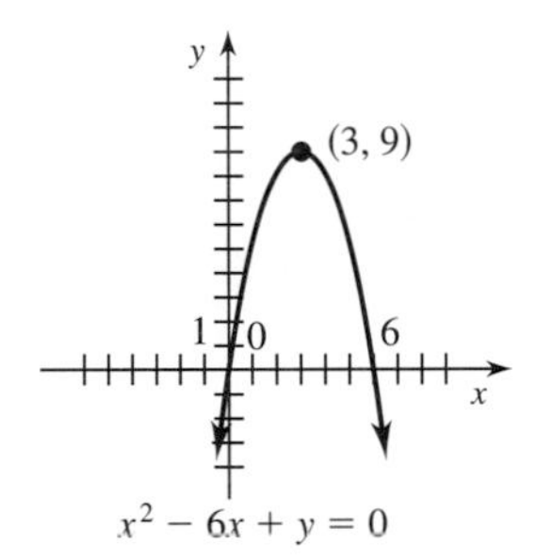

$x^2 - 6x + y = 0$

21. $x - 4y^2 - 8y = 0$ **parabola**

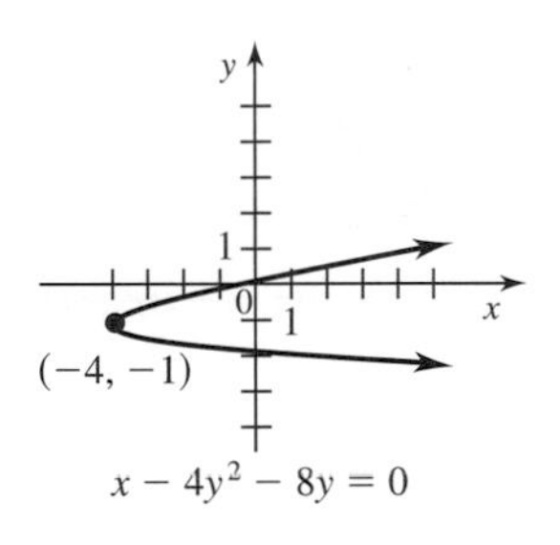

$x - 4y^2 - 8y = 0$

22. $4x^2 - 8x - y^2 - 6y = 6$ **hyperbola**

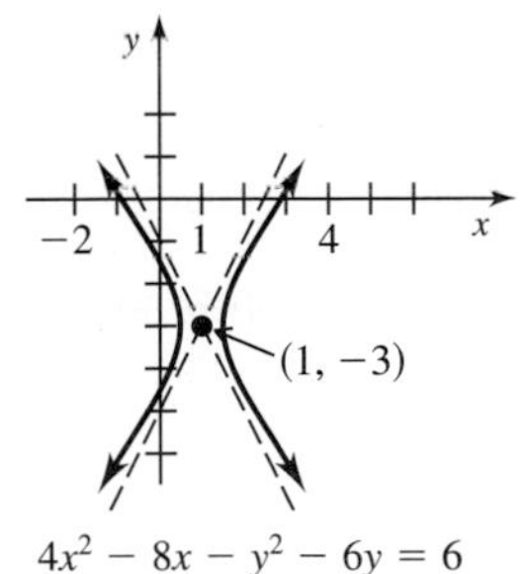

$4x^2 - 8x - y^2 - 6y = 6$

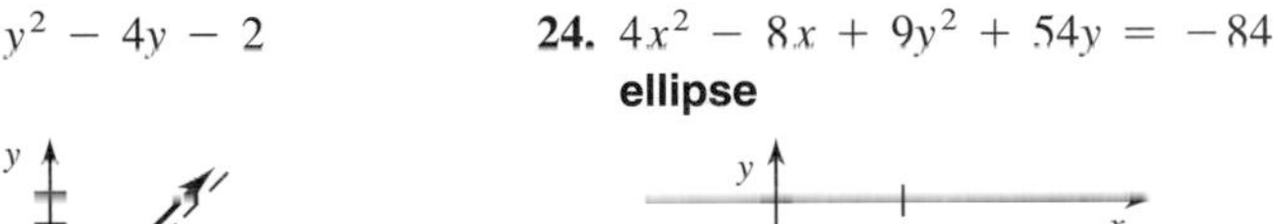

23. $x^2 + 2x = y^2 - 4y - 2$ **hyperbola**

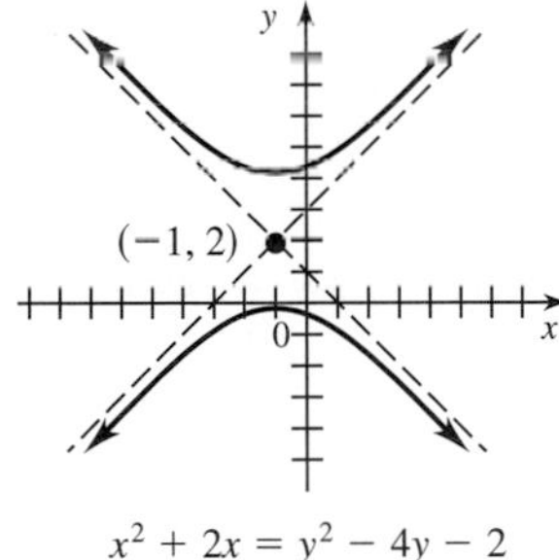

$x^2 + 2x = y^2 - 4y - 2$

24. $4x^2 - 8x + 9y^2 + 54y = -84$ **ellipse**

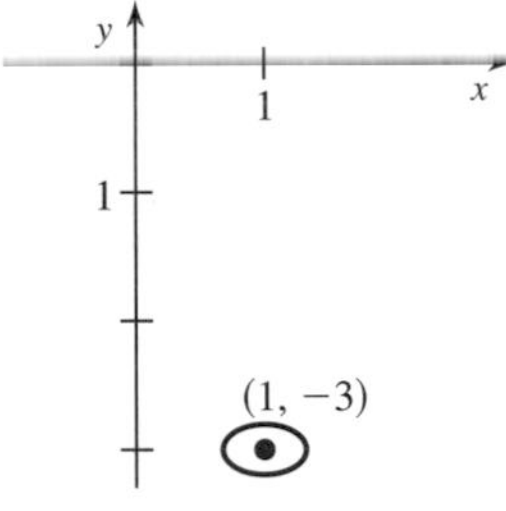

$4x^2 - 8x + 9y^2 + 54y = -84$

25. $3x^2 + 12x + 3y^2 = -11$ **circle**

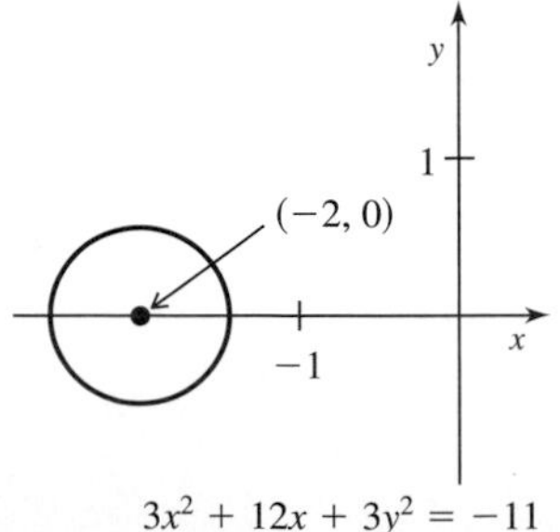

$3x^2 + 12x + 3y^2 = -11$

26. $4x^2 - 12x + 9y^2 - 6y + 12 = 0$ **no graph**

27. $100x^2 + 200x - 81y^2 + 324y - 449 = 0$ **hyperbola**

$100x^2 + 200x - 81y^2 + 324y - 449 = 0$

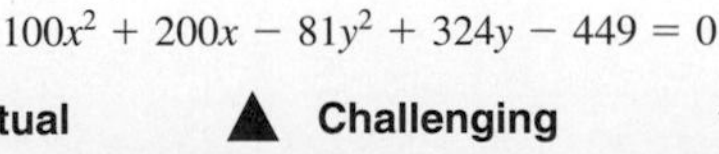

28. Identify the type of conic section consisting of the set of all points in the plane for which the distance from the point (2, 0) is one-third of the distance from the line $x = 10$. **ellipse**

29. Identify the type of conic section consisting of the set of all points in the plane for which the distance from the point (3, 0) is one and one-half times the distance from the line $x = 4/3$. **hyperbola**

In Exercises 30–35, find the eccentricity of the conic section. The point shown on the x-axis is a focus and the line shown is a directrix.

30.

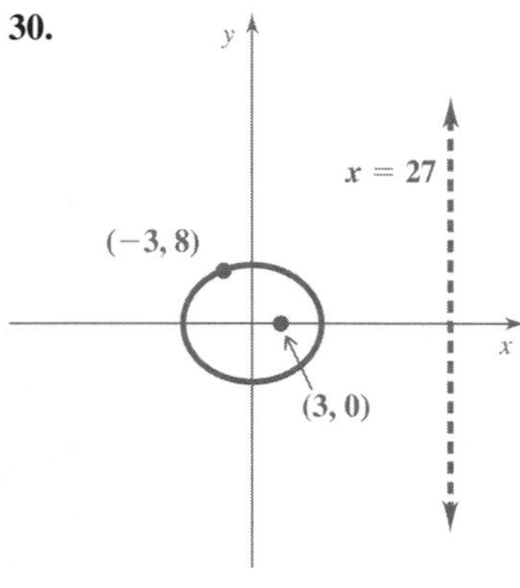

1/3

31.

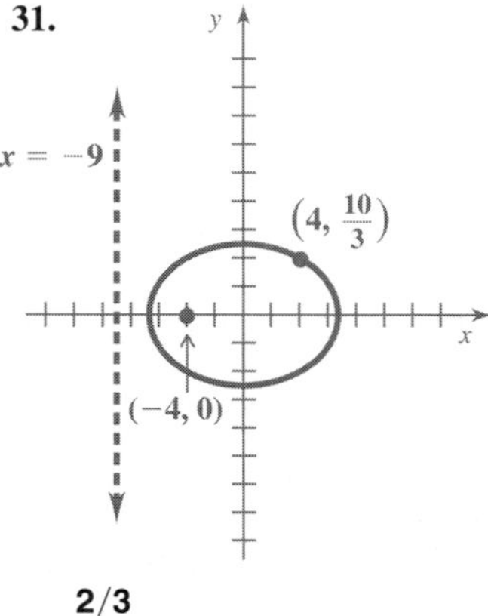

2/3

32.

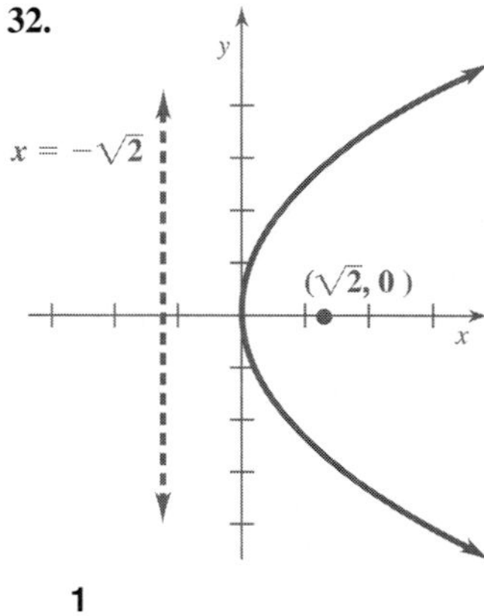

1

33.

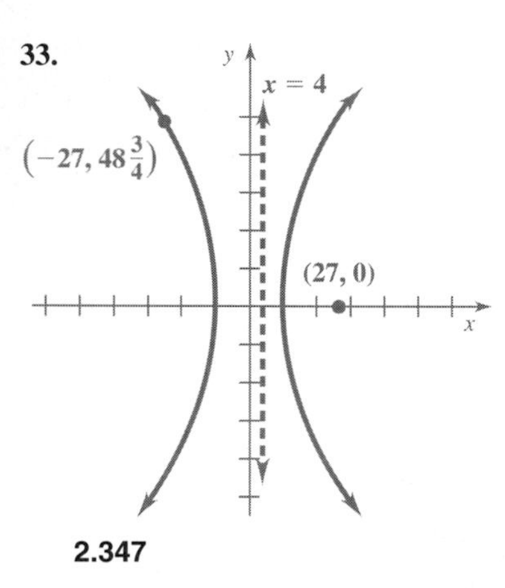

2.347

34.

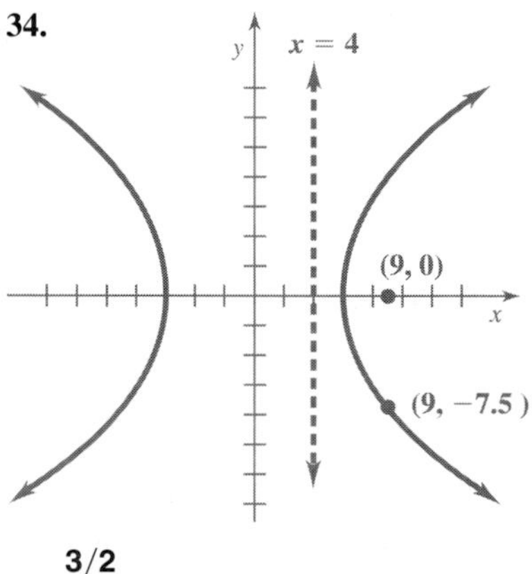

3/2

35.

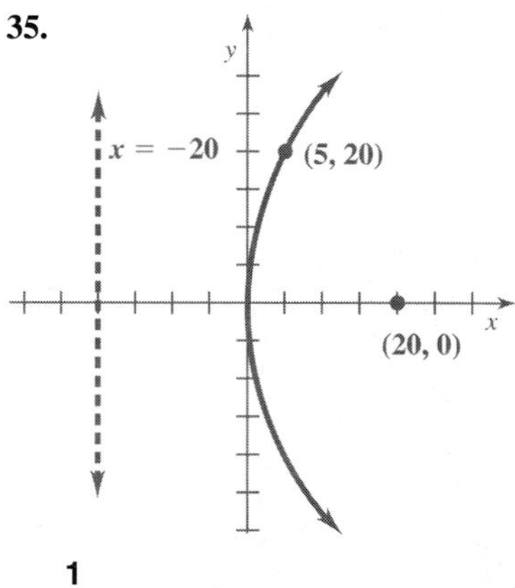

1

36. What is the eccentricity of a circle? (*Hint:* Think of a circle as an ellipse with $a = b$.) **0**

37. If $Ax^2 + Cy^2 + Dx + Ey + F = 0$ is the general equation of an ellipse, find its center point by completing the square. **$(x + D/(2A), y + E/(2C))$**

Graphing Utility Problems

38. Graph the ellipse $\frac{x^2}{16} + \frac{y^2}{12} = 1$. Use TRACE to find the coordinates of several points on the ellipse. For each of these points P, verify that

$$[\text{Distance of } P \text{ from } (2, 0)] = \frac{1}{2}[\text{Distance of } P \text{ from the line } x = 8].$$

39. Graph the hyperbola $\frac{x^2}{4} - \frac{y^2}{12} = 1$. Use TRACE to find the coordinates of several points on the hyperbola. For each of these points P, verify that

$$[\text{Distance of } P \text{ from } (4, 0)] = 2\,[\text{Distance of } P \text{ from the line } x = 1].$$

Chapter 7 Review Exercises (text page 407)

Graph each of the following. Give the vertex and axis of each figure.

1. $x = 4(y - 5)^2 + 2$ **(2, 5); y = 5**

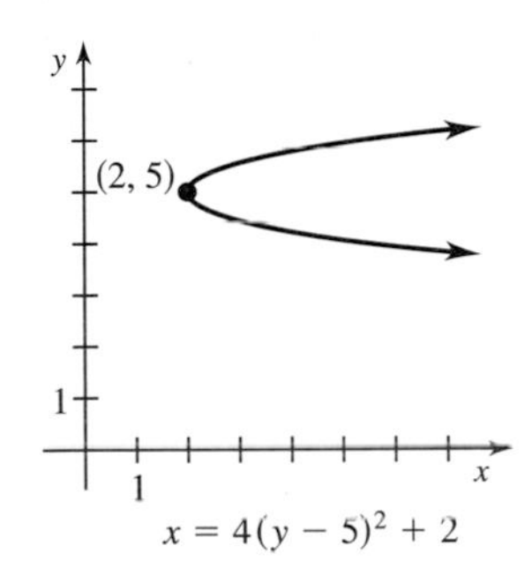

2. $x = -(y + 1)^2 - 7$ **(−7, −1); y = −1**

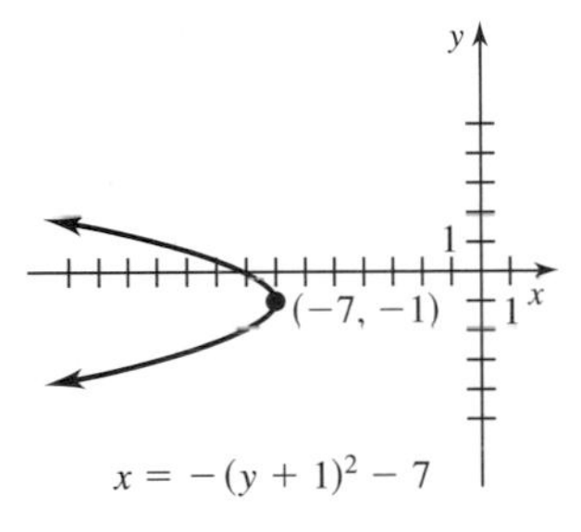

3. $x = 5y^2 - 5y + 3$ **(7/4, 1/2); y = 1/2**

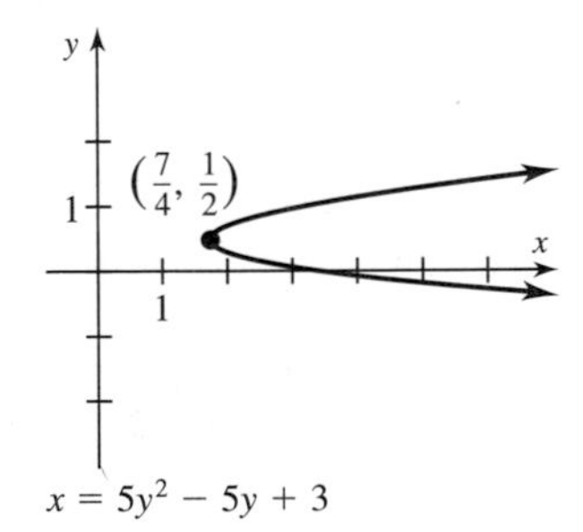

4. $x = 2y^2 - 4y + 1$ **(−1, 1); y = 1**

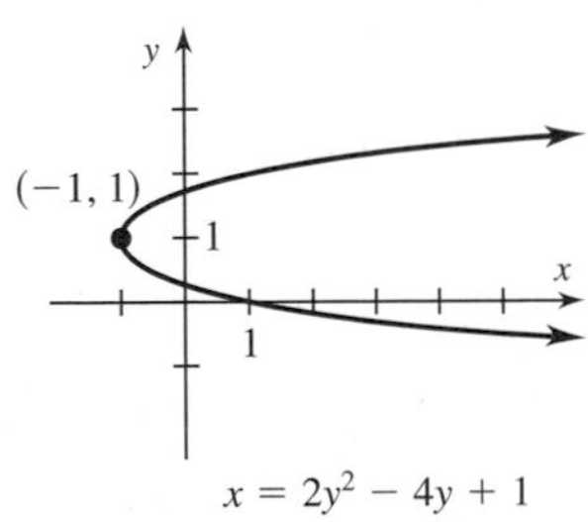

Graph each of the following. Give the focus, directrix, and axis of each figure.

5. $y^2 = -\frac{2}{3}x$ **(−1/6, 0); x = 1/6; y = 0**

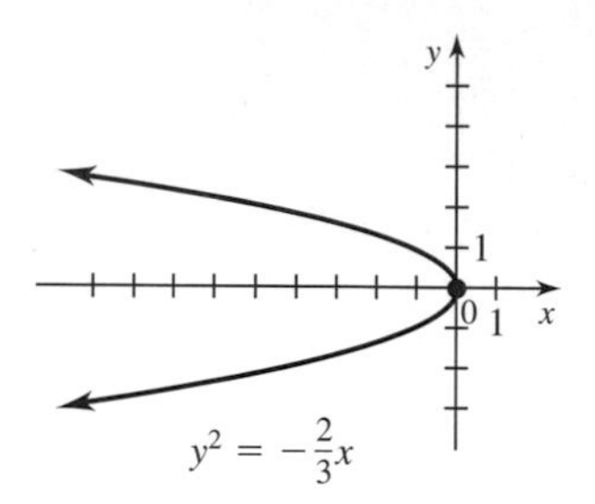

6. $y^2 = 2x$ **(1/2, 0); x = −1/2; y = 0**

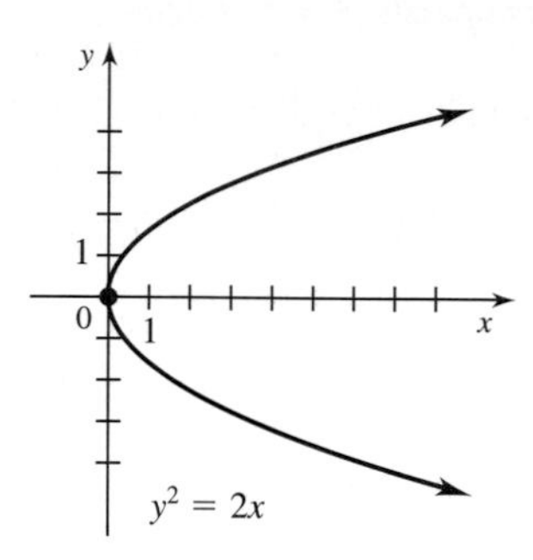

7. $3x^2 = y$ **(0, 1/12); y = −1/12; x = 0**

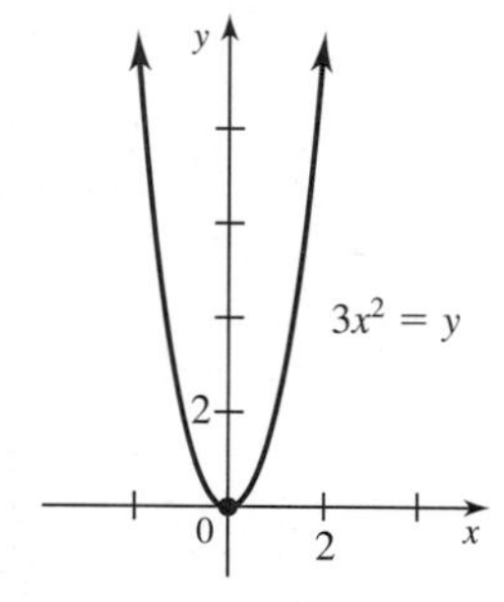

8. $x^2 + 2y = 0$ **(0, −1/2); y = 1/2; x = 0**

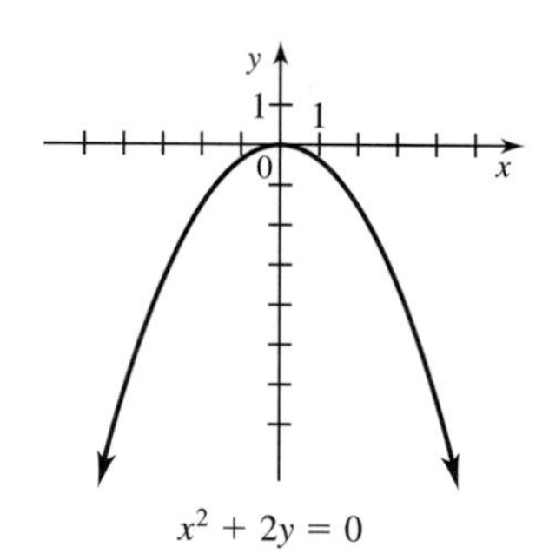

Write an equation for each parabola with vertex at the origin.

9. Focus (4, 0) **$x = y^2/16$**

10. Focus (0, −3) **$y = -x^2/12$**

11. Through (2, 5), opening to the right **$x = (2/25)y^2$**

12. Through (−3, 4), opening upward **$y = (4/9)x^2$**

Graph each of the following and identify each graph. Give the coordinates of the vertices for each ellipse or hyperbola, and give the equations of the asymptotes for each hyperbola.

13. $\frac{x^2}{5} + \frac{y^2}{9} = 1$ **ellipse; (0, −3), (0, 3)**

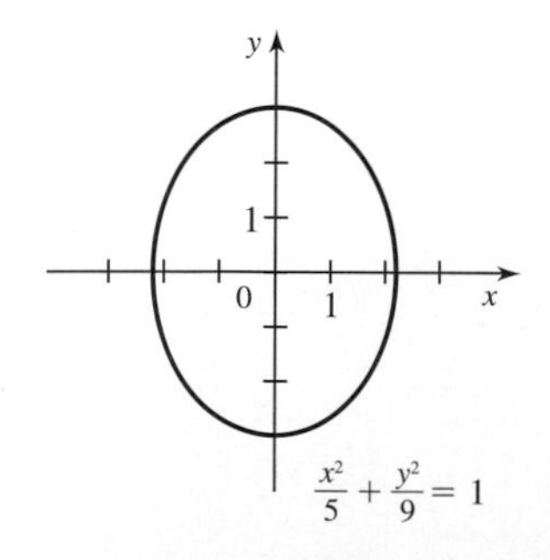

14. $\frac{x^2}{16} + \frac{y^2}{4} = 1$ **ellipse; (−4, 0), (4, 0)**

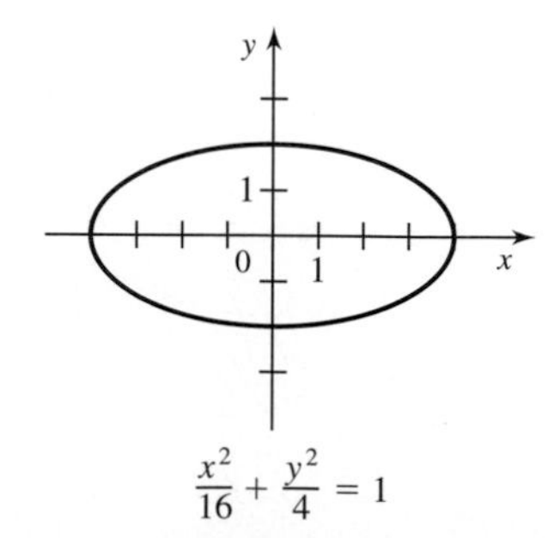

15. $\frac{x^2}{64} - \frac{y^2}{36} = 1$ **hyperbola; $(-8, 0), (8, 0)$; $y = \pm(3/4)x$**

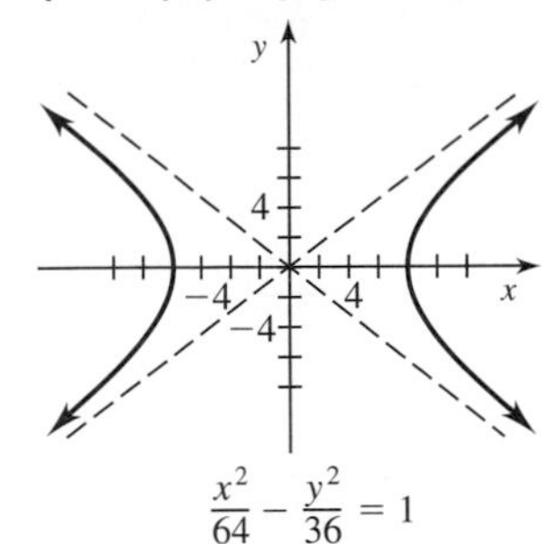

16. $\frac{y^2}{25} - \frac{x^2}{9} = 1$ **hyperbola; $(0, -5), (0, 5)$; $y = \pm(5/3)x$**

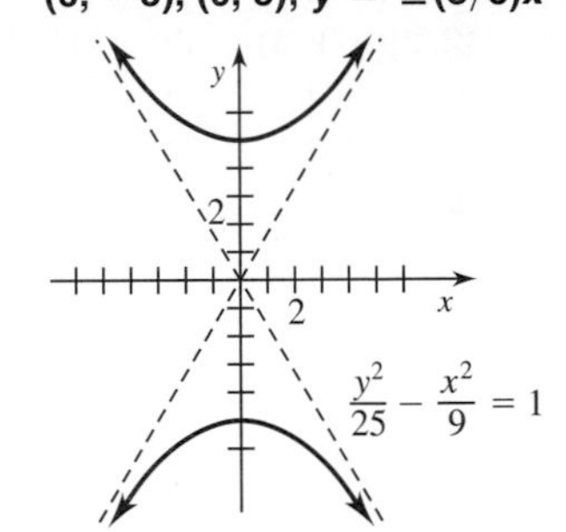

17. $\frac{(x + 1)^2}{16} + \frac{(y - 1)^2}{16} = 1$ **circle**

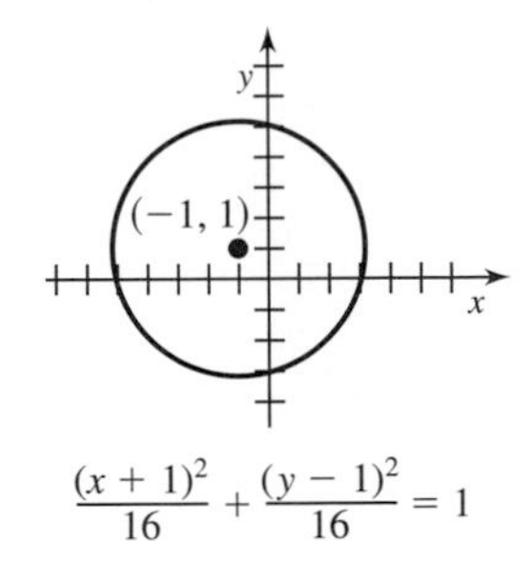

18. $(x - 3)^2 + (y + 2)^2 = 9$ **circle**

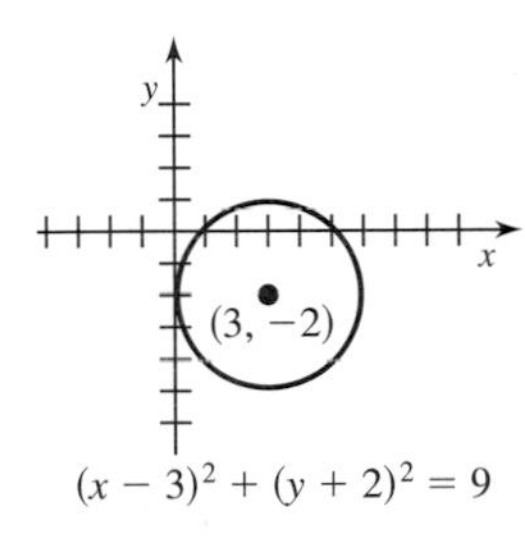

19. $\frac{100x^2}{49} + \frac{9y^2}{16} = 1$ **ellipse; $(0, -4/3), (0, 4/3)$**

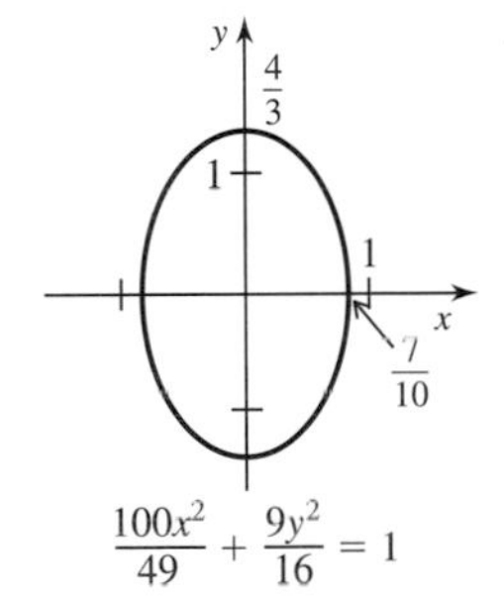

20. $\frac{25x^2}{9} + \frac{4y^2}{25} = 1$ **ellipse; $(0, 5/2), (0, -5/2)$**

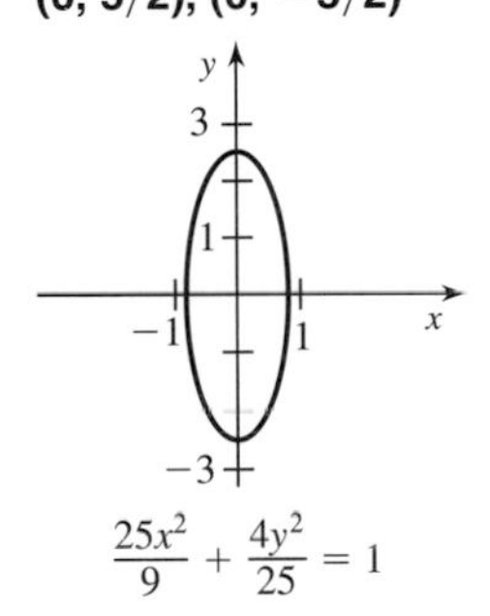

21. $4x^2 + 9y^2 = 36$ **ellipse; $(-3, 0), (3, 0)$**

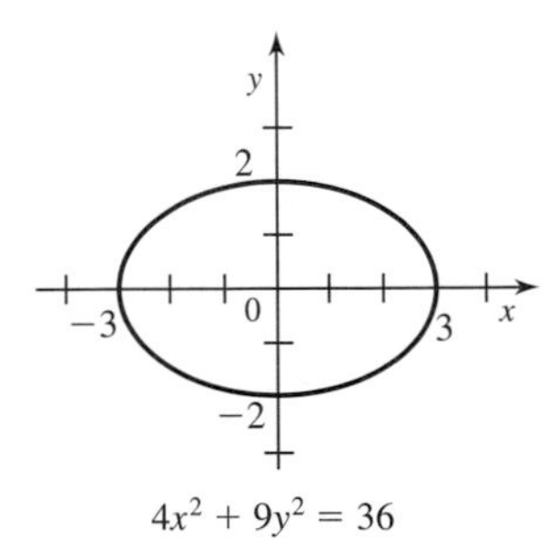

22. $x^2 = 16 + y^2$ **hyperbola; $(-4, 0), (4, 0)$; $y = \pm x$**

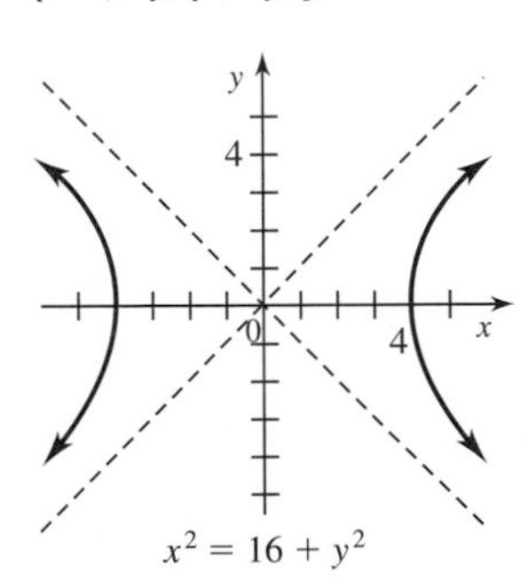

23. $\frac{(x - 3)^2}{4} + (y + 1)^2 = 1$ **ellipse; $(1, -1), (5, -1)$**

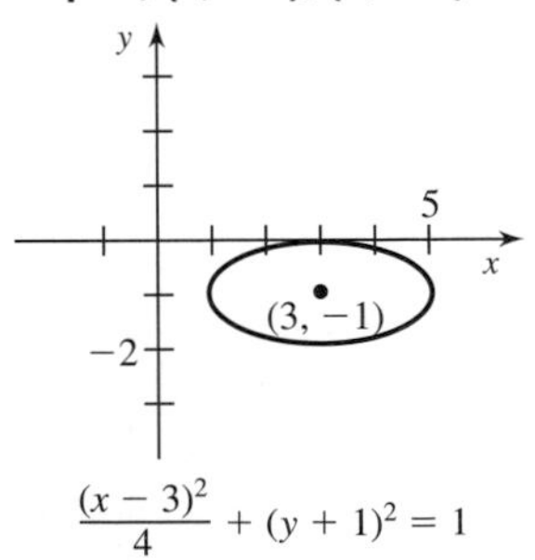

24. $\dfrac{(x-2)^2}{9} + \dfrac{(y+3)^2}{4} = 1$

ellipse; (5, −3), (−1, −3)

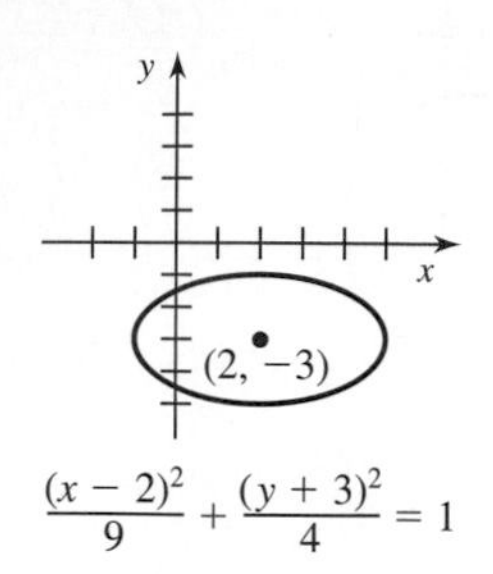

25. $\dfrac{(y+2)^2}{4} - \dfrac{(x+3)^2}{9} = 1$

hyperbola; (−3, 0), (−3, −4);
$y = \pm(2/3)(x+3) - 2$

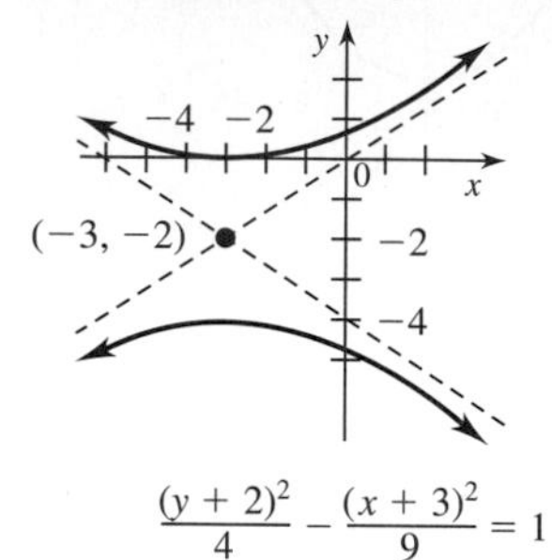

26. $\dfrac{x}{3} = -\sqrt{1 - \dfrac{y^2}{16}}$

semi-ellipse; (0, 4), (0, −4)

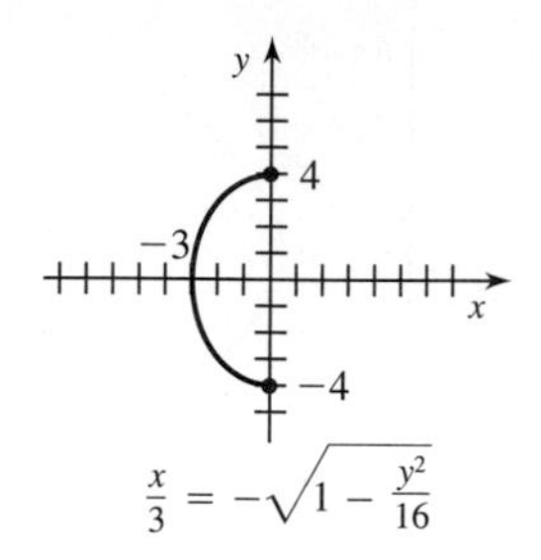

27. $\dfrac{(x+1)^2}{16} - \dfrac{(y-2)^2}{4} = 1$

hyperbola; (3, 2), (−5, 2); $y = \pm(1/2)(x+1) + 2$

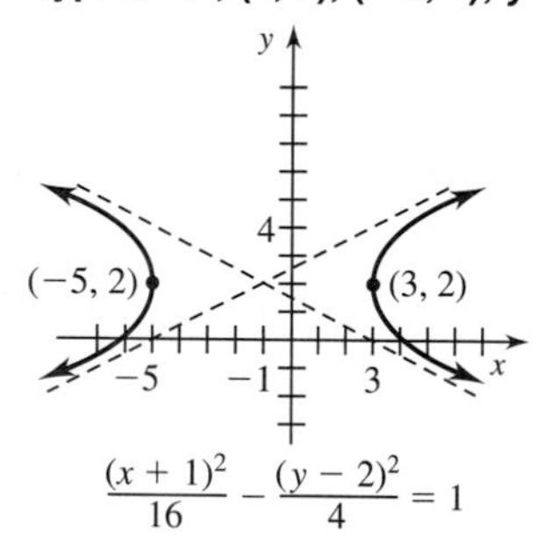

28. $x = -\sqrt{1 - \dfrac{y^2}{36}}$ **semi-ellipse; (0, −6), (0, 6)**

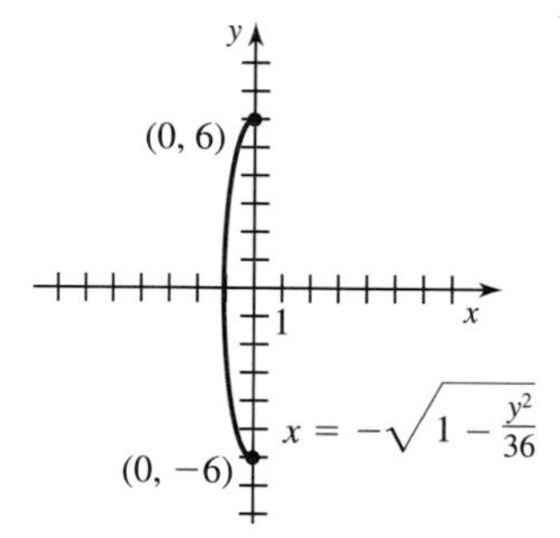

29. $y = -\sqrt{1 + x^2}$ **semi-hyperbola; (0, −1); $y = \pm x$**

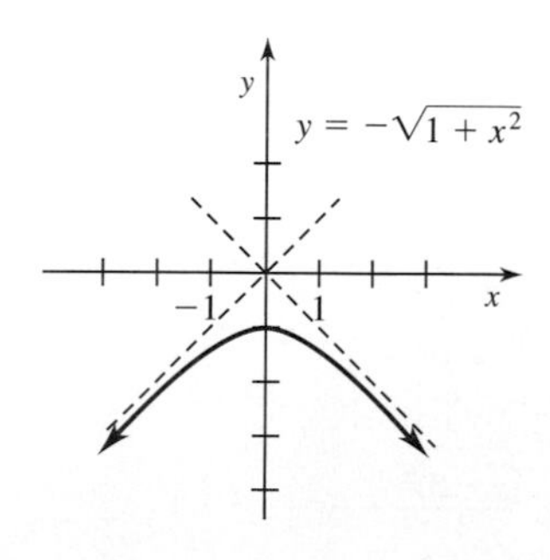

30. $y = -\sqrt{1 - \dfrac{x^2}{25}}$ **semi-ellipse; (−5, 0), (5, 0)**

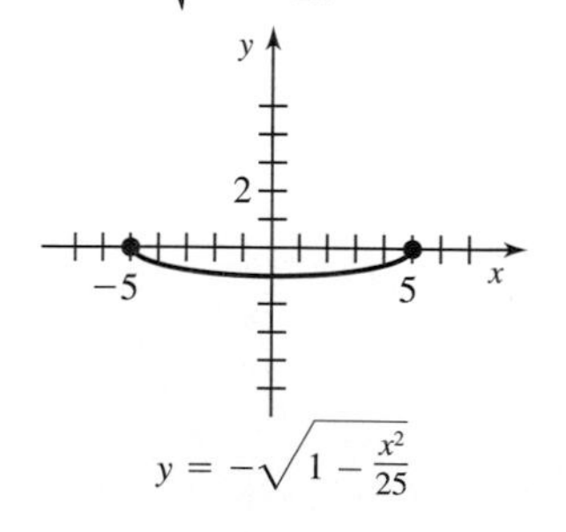

Write an equation for each of the following conic sections (centers at the origin).

31. Ellipse; vertex at (0, 4), focus at (0, 2) $\mathbf{x^2/12 + y^2/16 = 1}$

32. Ellipse; x-intercept 6, focus at $(-2, 0)$ $\mathbf{x^2/36 + y^2/32 = 1}$

33. Hyperbola; focus at $(0, -5)$, transverse axis of length 8 $\mathbf{y^2/16 - x^2/9 = 1}$

34. Hyperbola; y-intercept -2, passing through (2, 3) $\mathbf{y^2/4 - 5x^2/16 = 1}$

For the equations in Exercises 35–42, name the conic section and sketch the graph.

35. $y^2 + 9x^2 = 9$ **ellipse**

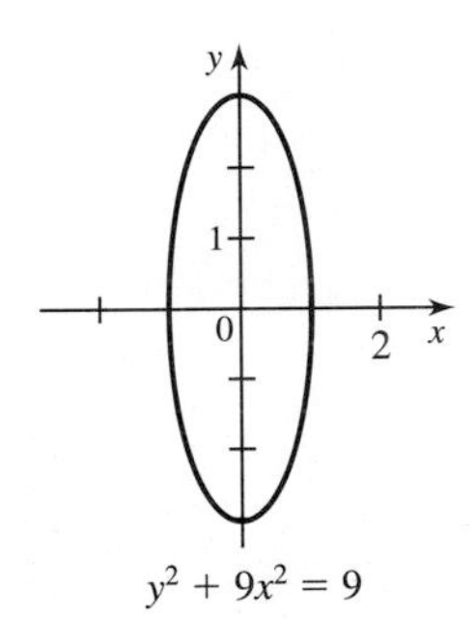

$y^2 + 9x^2 = 9$

36. $9x^2 - 16y^2 = 144$ **hyperbola**

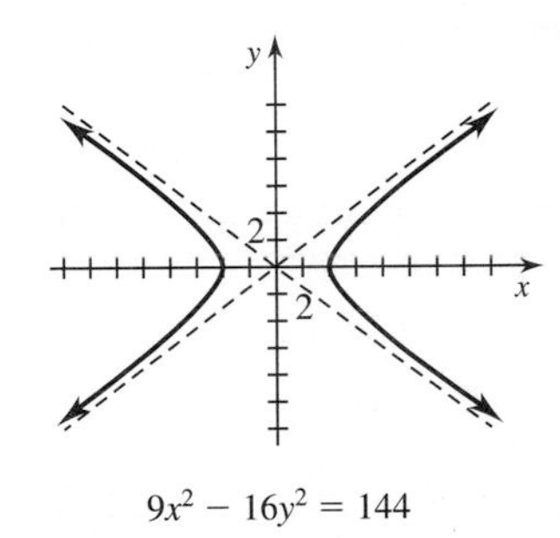

$9x^2 - 16y^2 = 144$

37. $3y^2 - 5x^2 = 30$ **hyperbola**

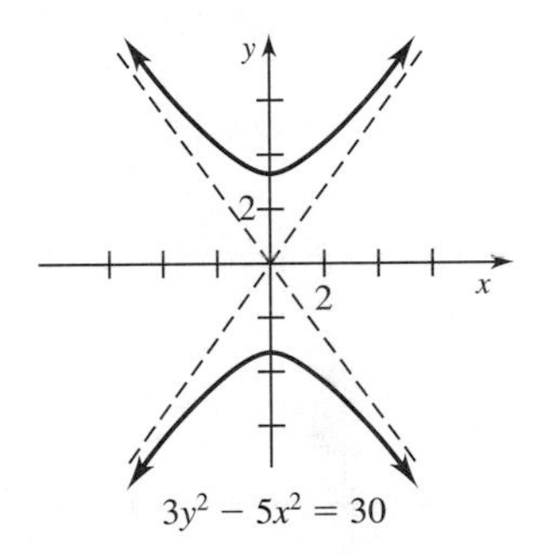

$3y^2 - 5x^2 = 30$

38. $y^2 + x = 4$ **parabola**

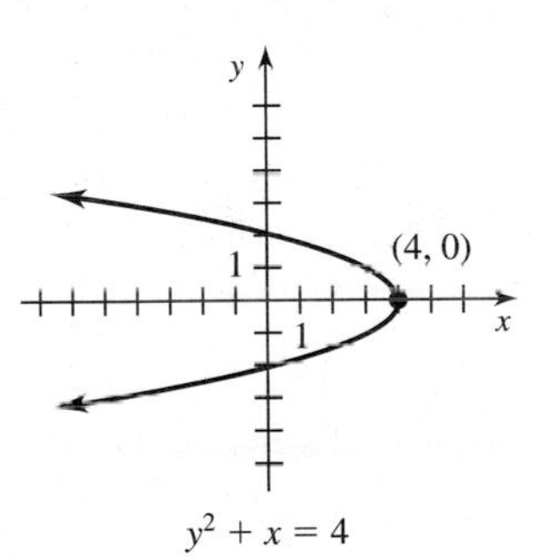

$y^2 + x = 4$

39. $4x^2 - y = 0$ **parabola**

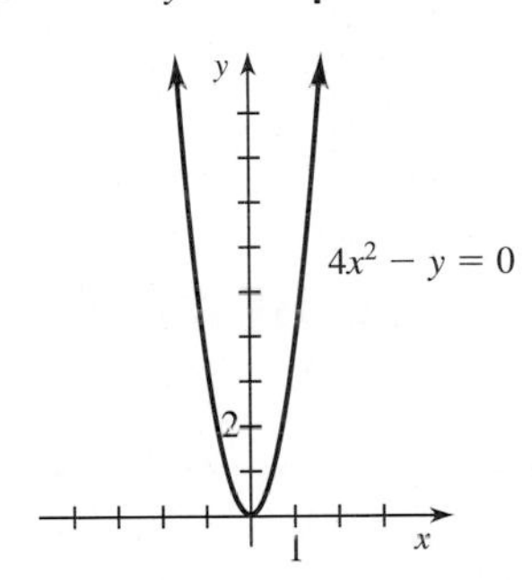

40. $x^2 + y^2 = 25$ **circle**

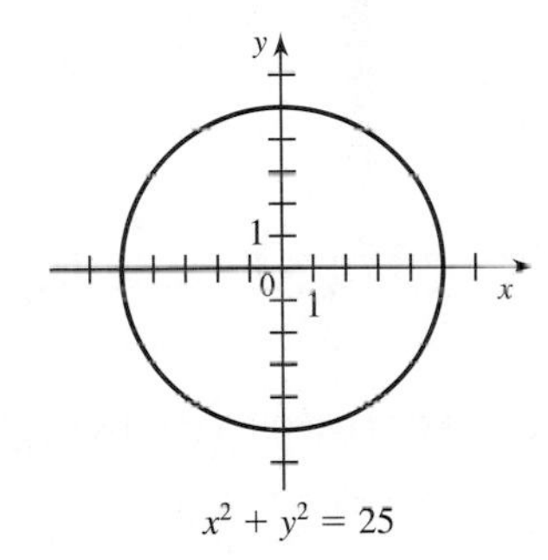

$x^2 + y^2 = 25$

41. $4x^2 - 8x + 9y^2 + 36y = -4$ **ellipse**

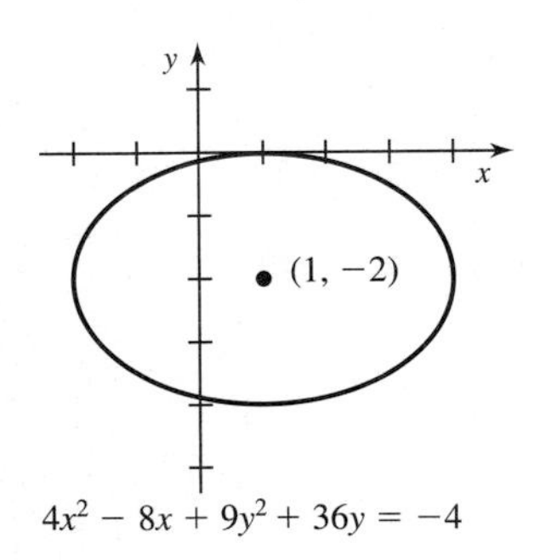

$4x^2 - 8x + 9y^2 + 36y = -4$

42. $25x^2 + 50x + 4y^2 - 24y = 39$ **ellipse**

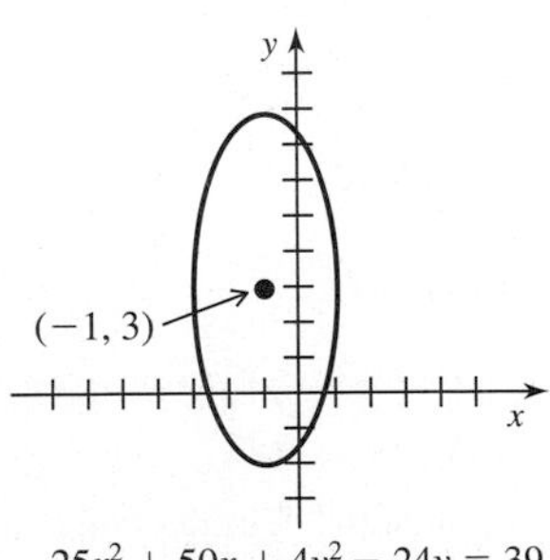

$25x^2 + 50x + 4y^2 - 24y = 39$

43. Find the equation of the ellipse consisting of all points in the plane the sum of whose distances from (0, 0) and (4, 0) is 8. $\mathbf{(x - 2)^2/16 + y^2/12 = 1}$

44. Find the equation of the hyperbola consisting of all points in the plane for which the difference of the distances from (0, 0) and (0, 4) is 2. $\mathbf{(y - 2)^2 - x^2/3 = 1}$

45. The orbit of Venus is an ellipse with the sun at one of the foci. The eccentricity of the orbit is $e = .006775$ and the major axis has length 134.5 million miles. Find the smallest and greatest distances of Venus from the sun. **66.8 and 67.7 million miles**

46. Comet Swift-Tuttle has an elliptical orbit of eccentricity $e = .964$, with the sun at one of the foci. Find the equation of the comet given that the closest it comes to the sun is 89 million miles. $\mathbf{x^2/6{,}111{,}883 + y^2/432{,}135 = 1}$

Graphing Utility Problems

In Exercises 47–52, find the equation of the conic section that satisfies the given conditions. Then graph the equation and confirm that the conditions hold.

47. Parabola with focus at (3, 2) and directrix $x = -3$ $\mathbf{x = (1/12)(y - 2)^2}$

48. Parabola with vertex at $(-3, 2)$ and y-intercepts 5 and -1 $\mathbf{x = (1/3)(y - 2)^2 - 3}$

49. Ellipse with foci at $(-2, 0)$ and $(2, 0)$ and major axis of length 10 $\mathbf{x^2/25 + y^2/21 = 1}$

50. Ellipse with foci at (0, 3) and $(0, -3)$ and vertex at (0, 7) $\mathbf{x^2/40 + y^2/49 = 1}$

51. Hyperbola with x-intercepts ± 3; foci at $(-5, 0)$, $(5, 0)$ $\mathbf{x^2/9 - y^2/16 = 1}$

52. Hyperbola with foci at (0, 12), $(0, -12)$; asymptotes $y = \pm x$ $\mathbf{y^2/72 - x^2/72 = 1}$

Chapter 8 Further Topics in Algebra

8.1 Exercises (text page 415)

Write the first five terms of each of the following sequences.

1. $a_n = 4n + 10$ **14, 18, 22, 26, 30**

2. $a_n = 6n - 3$ **3, 9, 15, 21, 27**

3. $a_n = 2^{n-1}$ **1, 2, 4, 8, 16**

4. $a_n = -3^n$ **−3, −9, −27, −81, −243**

5. $a_n = (1/3)^n(n - 1)$ **0, 1/9, 2/27, 1/27, 4/243**

6. $a_n = (-2)^n(n)$ **−2, 8, −24, 64, −160**

7. $a_n = (-1)^n(2n)$ **−2, 4, −6, 8, −10**

8. $a_n = (-1)^{n-1}(n + 1)$ **2, −3, 4, −5, 6**

9. $a_n = \dfrac{4n - 1}{n^2 + 2}$ **1, 7/6, 1, 5/6, 19/27**

10. $a_n = \dfrac{n^2 - 1}{n^2 + 1}$ **0, 3/5, 4/5, 15/17, 12/13**

11. Your friend does not understand what is meant by the nth term or general term of a sequence. How would you explain this idea?

12. How are sequences related to functions? Discuss some similarities and some differences.

Find the first six terms for the sequences defined as follows.

13. $a_1 = -2, \quad a_n = a_{n-1} + 3, \quad \text{for } n > 1$ **$-2, 1, 4, 7, 10, 13$**

14. $a_1 = -1, \quad a_n = a_{n-1} - 4, \quad \text{for } n > 1$ **$-1, -5, -9, -13, -17, -21$**

15. $a_1 = 1, \quad a_2 = 1, \quad a_n = a_{n-1} + a_{n-2}, \quad \text{for } n \geq 3$ (the Fibonacci sequence) **1, 1, 2, 3, 5, 8**

16. $a_1 = 2, \quad a_n = n \cdot a_{n-1}, \quad \text{for } n > 1$ **2, 4, 12, 48, 240, 1440**

Evaluate the terms for each of the following sums where $x_1 = -2$, $x_2 = -1$, $x_3 = 0$, $x_4 = 1$, $x_5 = 2$.

17. $\sum_{i=1}^{5} (2x_i + 3)$ **$-1 + 1 + 3 + 5 + 7$**

18. $\sum_{i=1}^{4} x_i^2$ **$4 + 1 + 0 + 1$**

19. $\sum_{i=1}^{3} (3x_i - x_i^2)$ **$-10 - 4 + 0$**

20. $\sum_{i=1}^{3} (x_i^2 + 1)$ **$5 + 2 + 1$**

21. $\sum_{i=2}^{5} \frac{x_i + 1}{x_i + 2}$ **$0 + 1/2 + 2/3 + 3/4$**

22. $\sum_{i=1}^{5} \frac{x_i}{x_i + 3}$ **$-2 - 1/2 + 0 + 1/4 + 2/5$**

Evaluate the terms of $\sum_{i=1}^{4} f(x_i)\Delta x$ with $x_1 = 0$, $x_2 = 2$, $x_3 = 4$, $x_4 = 6$, and $\Delta x = .5$ for the functions defined as follows.

23. $f(x) = 4x - 7$ **$-3.5 + .5 + 4.5 + 8.5$**

24. $f(x) = 6 + 2x$ **$3 + 5 + 7 + 9$**

25. $f(x) = 2x^2$ **$0 + 4 + 16 + 36$**

26. $f(x) = x^2 - 1$ **$-.5 + 1.5 + 7.5 + 17.5$**

27. $f(x) = \frac{-2}{x + 1}$ **$-1 - 1/3 - 1/5 - 1/7$**

28. $f(x) = \frac{5}{2x - 1}$ **$-5/2 + 5/6 + 5/14 + 5/22$**

Use summation notation to rewrite each series with the index of summation starting at the indicated number.

29. $\sum_{i=1}^{5} (6 \quad 3i)$; 3 **$\sum_{j=3}^{7} (12 - 3j)$**

30. $\sum_{i=1}^{7} (5i + 2)$; -2 **$\sum_{j=-2}^{4} (5j + 17)$**

31. $\sum_{i=1}^{10} 2(3)^i$; 0 **$\sum_{j=0}^{9} 2(3)^{j+1}$**

32. $\sum_{i=-1}^{6} 5(2)^i$; 3 **$\sum_{j=3}^{10} \frac{5}{16}(2)^j$ or $\sum_{j=3}^{10} 5(2)^{j-4}$**

33. $\sum_{i=-1}^{9} (i^2 - 2i)$; 0 **$\sum_{j=0}^{10} [(j - 1)^2 - 2(j - 1)]$ or $\sum_{j=0}^{10} (j^2 - 4j + 3)$**

34. $\sum_{i=3}^{11} (2i^2 + 1)$; 0 **$\sum_{j=0}^{8} [2(j + 3)^2 + 1]$ or $\sum_{j=0}^{8} (2j^2 + 12j + 19)$**

Use the summation properties to evaluate each series. The following sums may be needed.

$$\sum_{i=1}^{n} i = \frac{n(n + 1)}{2} \qquad \sum_{i=1}^{n} i^2 = \frac{n(n + 1)(2n + 1)}{6} \qquad \sum_{i=1}^{n} i^3 = \frac{n^2(n + 1)^2}{4}$$

35. $\sum_{i=1}^{5} (5i + 3)$ **90**

36. $\sum_{i=1}^{5} (8i - 1)$ **115**

37. $\sum_{i=1}^{5} (4i^2 - 2i + 6)$ **220**

38. $\sum_{i=1}^{6} (2 + i - i^2)$ **-58**

39. $\sum_{i=1}^{4} (3i^3 + 2i - 4)$ **304**

40. $\sum_{i=1}^{6} (i^2 + 2i^3)$ **973**

Use the summation properties to write each of the following without Σ. (Hint: Think of n as a constant.)

41. $\sum_{i=1}^{n}\left[4 + \left(\frac{2}{n}\right)i\right]\frac{2}{n}$ **10 + (2/n)**

42. $\sum_{i=1}^{n}\left[\left(\frac{2}{n}\right)i + 1\right]\frac{2}{n}$ **4 + (2/n)**

43. $\sum_{i=1}^{n}\left[3 + \left(\frac{1}{n}\right)i\right]^2\frac{1}{n}$ **9 + [3(n + 1)]/n + [(n + 1)(2n + 1)]/(6n²)**

44. $\sum_{i=1}^{n}\left[5 + \left(\frac{3}{n}\right)i\right]^2\frac{3}{n}$ **75 + [45(n + 1)]/n + [9(n + 1)(2n + 1)]/2n²**

Graphing Utility Problems

For each sequence defined by a_n, graph the corresponding function defined by $f(x)$. Use the graph to decide whether the sequence converges, and if it does, determine the number to which it converges.

45. $a_n = \frac{n + 2}{2n}$ **converges to 1/2**

46. $a_n = 2e^n$ **diverges**

47. $a_n = n(n + 1)$ **diverges**

48. $a_n = \left(1 + \frac{1}{n}\right)^n$ **converges to $e \approx 2.71828$**

8.2 Exercises (text page 422)

Write the terms of the arithmetic sequences satisfying each of the following conditions.

1. $a_1 = 5,\ d = -2,\ n = 6$ **5, 3, 1, −1, −3, −5**

2. $a_1 = 4,\ d = 3,\ n = 5$ **4, 7, 10, 13, 16**

3. $a_2 = 10,\ d = -3,\ n = 4$ **13, 10, 7, 4**

4. $a_3 = 10,\ d = -2,\ n = 5$ **14, 12, 10, 8, 6**

5. $a_1 = 3 - \sqrt{2},\ a_2 = 3,\ n = 5$ **$3 - \sqrt{2}, 3, 3 + \sqrt{2}, 3 + 2\sqrt{2}, 3 + 3\sqrt{2}$**

6. $a_1 = -5,\ a_2 = -5 + \sqrt{3},\ n = 4$ **$-5, -5 + \sqrt{3}, -5 + 2\sqrt{3}, -5 + 3\sqrt{3}$**

For each of the following sequences, find d and a_n.

7. 18, 15, 12, 9, 6, 3, . . . **−3; 21 − 3n**

8. 5, 11, 17, 23, 29, 35, . . . **6; −1 + 6n**

9. 6, 10, 14, 18, 22, . . . **4; 2 + 4n**

10. 27, 22, 17, 12, . . . **−5; 32 − 5n**

11. $\sqrt{3} + 1,\ 2\sqrt{3} + 1,\ 3\sqrt{3} + 1,\ 4\sqrt{3} + 1, \ldots$ **$\sqrt{3}$; $1 + \sqrt{3}n$**

12. $5 - \sqrt{7},\ 6 - \sqrt{7},\ 7 - \sqrt{7},\ 8 - \sqrt{7}, \ldots$ **1; $4 - \sqrt{7} + n$**

13. $x - 2,\ x - 1,\ x,\ x + 1,\ x + 2, \ldots$ **1; x − 3 + n**

14. $3x + y,\ 3x + 2y,\ 3x + 3y,\ 3x + 4y, \ldots$ **y; 3x + ny**

Find a_{10} and a_n for each of the following sequences.

15. $a_1 = 3,\ d = 5$ **48; −2 + 5n**

16. $a_1 = -4,\ d = 8$ **68; −12 + 8n**

17. $a_1 = -1,\ d = 5$ **44; −6 + 5n**

18. $a_1 = 3,\ d = -4$ **−33; 7 − 4n**

19. $a_1 = 5,\ a_3 = 12$ **73/2; (3 + 7n)/2**

20. $a_2 = 4,\ a_4 = -2$ **−20; 10 − 3n**

21. $a_1 = x,\ a_2 = x + 3$ **x + 27; x + 3n − 3**

22. $a_2 = y + 1,\ d = -5$ **y − 39; y + 11 − 5n**

Find a_1 for each of the following arithmetic sequences.

23. $a_5 = 27,\ a_{15} = 87$ **3**

24. $a_{12} = 60,\ a_{20} = 84$ **27**

25. $S_{16} = -160,\ a_{16} = -25$ **5**

26. $S_{28} = 2926,\ a_{28} = 199$ **10**

27. Which of the following is not an arithmetic sequence?
(a) 4, 6, 8, 10, . . . **(b)** −2, 6, 14, 22, . . .
(c) 1/2, 1, 3/2, 2, . . . **(d)** 5, 10, 20, 40, . . .
For the sequence that is not arithmetic, explain how each term after the first is determined by using the previous term of the sequence.

Find the sum of the first ten terms for each of the following arithmetic sequences.

28. $a_4 = 16,\ a_5 = 19$ **205**

29. $a_2 = 12,\ a_5 = 24$ **260**

30. 4, 12, 20, . . . **400**

31. 20, 15, 10, . . . **−25**

32. $a_1 = 9.428,\ d = -1.723$ **16.745**

33. $a_1 = -3.119,\ d = 2.422$ **77.8**

34. $a_4 = 2.556,\ a_5 = 3.004$ **32.28**

35. $a_7 = 11.192,\ a_9 = 4.812$ **159.77**

Evaluate each of the following sums.

36. $\sum_{i=1}^{8} (5i - 4)$ **148**

37. $\sum_{i=1}^{12} (8i + 17)$ **828**

38. $\sum_{i=1}^{15} (-4i - 1)$ **−495**

39. $\sum_{i=1}^{20} (-2 - 6i)$ **−1300**

40. $\sum_{j=1}^{500} 4j$ **501,000**

41. $\sum_{j=1}^{1200} 2j$ **1,441,200**

42. $\sum_{i=6}^{17} (3i + 7)$ **498**

43. $\sum_{i=8}^{21} (2 - 5i)$ **−987**

In Exercises 44 and 45, find the value of x for which the sequence is arithmetic.

44. 5, x, 19 **12**

45. 2, x, −5 **−3/2**

46. The sum of the first n terms in an arithmetic sequence is given by the formula $S_n = n(3n + 2)$. Find the fourteenth term in the sequence. **83**

47. An arithmetic sequence has $a_{24} - a_{12} = 48$. Find the common difference. **4**

48. Find the sum of the first n positive integers. **$n(n + 1)/2$**

49. Find the sum of the first n odd positive integers. **n^2**

50. If a clock strikes the proper number of tones each hour on the hour, how many tones will it strike in a month of 30 days? **4680**

51. A display of stacked canned goods in a grocery store has 31 cans on the bottom, 25 on the next row, and 1 can on top. Assume the number of cans in the layers form an arithmetic sequence. How many cans are in the display? **96**

52. A skydiver falls 5 m during the first second, 15 m during the second, 25 m during the third, and so on. How many meters will the diver fall during the tenth second? During the first ten seconds? **95 m, 500 m**

53. Ernesto Lopez is hired at a salary of $26,000 a year, with annual increases of $1500. What salary will he earn in his fifth year with this company? **$32,000**

54. Deepwell Drilling Company charges a flat $500 set-up charge, plus $5 for the first foot of well drilled, $6 for the second, $7 for the third, and so on. Find the total charge for a 70-ft well. **$3265**

55. Mei Ling has started on a fitness program. She plans to jog 10 min a day for the first week, then add 10 min a day each week until she is jogging an hour a day. In what week will this occur? What is the total number of minutes she will run during the first four weeks? **the sixth week; 700 min**

56. The sum of four consecutive terms in an arithmetic sequence is 66. The sum of the squares of the terms is 1214. Find the terms. **9, 14, 19, 24**

57. The sum of five consecutive terms of an arithmetic sequence is 5. If the product of the first and second is added to the product of the fourth and fifth, the result is 326. Find the terms. **−17, −8, 1, 10, 19**

58. A super slide of uniform slope is to be built on a level piece of land. There are to be twenty equally spaced supports, with the longest support 15 m in length and the shortest 2 m in length. Find the total length of all the supports. **170 m**

59. How much material would be needed for the rungs of a ladder of 31 rungs, if the rungs taper from 18 inches to 28 inches? Assume that the lengths of the rungs form the terms of an arithmetic sequence. **713 inches**

60. Find all arithmetic sequences $a_1, a_2, a_3, \ldots$, such that $a_1^2, a_2^2, a_3^2 \ldots$, is also an arithmetic sequence. **All terms are the same constant.**

61. Suppose that $a_1, a_2, a_3, \ldots$ and $b_1, b_2, b_3, \ldots$ are each arithmetic sequences. Let $d_n = a_n + c \cdot b_n$, for any real number c and every positive integer n. Show that $d_1, d_2, d_3, \ldots$ is an arithmetic sequence.

62. Suppose that $a_1, a_2, a_3, a_4, a_5, \ldots$ is an arithmetic sequence. Is $a_1, a_3, a_5 \ldots$ an arithmetic sequence? **yes**

63. If $f(x) = mx + b$, show that $f(1), f(2), f(3), \ldots$ is an arithmetic sequence. What is the common difference? (*Note:* This observation shows the connection between linear functions and arithmetic sequences.) ***m***

Explain why each of the following sequences is arithmetic.

64. log 2, log 4, log 8, log 16, log 32, . . .

65. log 12, log 36, log 108, log 324, . . .

8.3 Exercises (text page 431)

Write out the terms of the geometric sequences that satisfy the given conditions.

1. $a_1 = 5/3,\ r = 3,\ n = 4$ **5/3, 5, 15, 45**

2. $a_1 = -3/4,\ r = 2/3,\ n = 4$ **−3/4, −1/2, −1/3, −2/9**

3. $a_4 = 5,\ a_5 = 10,\ n = 5$ **5/8, 5/4, 5/2, 5, 10**

4. $a_3 = 16,\ a_4 = 8,\ n = 5$ **64, 32, 16, 8, 4**

Find a_5 and a_n for each of the following geometric sequences.

5. $a_1 = 5,\ r = -2$ **80; $5(-2)^{n-1}$**

6. $a_1 = 8,\ r = -5$ **5000; $8(-5)^{n-1}$**

7. $a_2 = -4,\ r = 3$ **−108; $(-4/3)(3)^{n-1}$ or $(-4)(3)^{n-2}$**

8. $a_3 = -2,\ r = 4$ **−32; $(-1/8)(4)^{n-1}$**

9. $a_4 = 243,\ r = -3$ **−729; $(-9)(-3)^{n-1}$ or $-(-3)^{n+1}$**

10. $a_4 = 18,\ r = 2$ **36; $(9/4)(2)^{n-1}$ or $9(2)^{n-3}$**

11. −4, −12, −36, −108 . . . **−324; $-4(3)^{n-1}$**

12. −2, 6, −18, 54 . . . **−162; $-2(-3)^{n-1}$**

13. 4/5, 2, 5, 25/2, . . . **125/4; $(4/5)(5/2)^{n-1}$ or $5^{n-2}/2^{n-3}$**

14. 1/2, 2/3, 8/9, 32/27, . . . **128/81; $(1/2)(4/3)^{n-1}$**

15. $a_4 = 36,\ r = 4$ **144; $(9/16)(4)^{n-1}$ or $9(4)^{n-3}$**

16. $a_5 = 12,\ r = 3$ **12; $(4/27)(3)^{n-1}$ or $4(3)^{n-4}$**

◉ *Find the positive value of x for which the sequence is geometric.*

17. 5, x, 3/5 **$\sqrt{3}$**

18. 4/3, x, 16 **$8\sqrt{3}/3$**

Find a_1 and r for each of the following geometric sequences.

19. $a_3 = 5$, $a_8 = 1/625$ **125; 1/5**

20. $a_2 = -6$, $a_7 = -192$ **−3; 2**

21. $a_4 = -1/4$, $a_9 = -1/128$ **−2; 1/2**

22. $a_3 = 50$, $a_7 = .005$ **5000; ±.1**

Use the formula for S_n to find the sum of the first five terms for each of the following geometric sequences.

23. 2, 8, 32, 128, . . . **682**

24. 4, 16, 64, 256, . . . **1364**

25. 18, −9, 9/2, −9/4, . . . **99/8**

26. 12, −4, 4/3, −4/9, . . . **244/27**

27. $a_1 = 8.423$, $r = 2.859$ **860.95**

28. $a_1 = -3.772$, $r = -1.553$ **−14.82**

Find each of the following sums.

29. $\sum_{i=1}^{5} 3^i$ **363**

30. $\sum_{i=1}^{4} (-2)^i$ **10**

31. $\sum_{j=1}^{6} 48(1/2)^j$ **189/4**

32. $\sum_{j=1}^{5} 243(2/3)^j$ **422**

33. $\sum_{k=4}^{10} 2^k$ **2032**

34. $\sum_{k=3}^{9} (-3)^k$ **−14,769**

✍ **35.** Under what conditions does an infinite geometric series converge? Explain why.

Find r for each of the following infinite geometric sequences. Identify any whose sum would not converge.

36. 12, 24, 48, 96, . . . **2; does not converge**

37. 625, 125, 25, 5, . . . **1/5**

38. −48, −24, −12, −6, . . . **1/2**

39. 2, −10, 50, −250, . . . **−5; does not converge**

In Exercises 40–49, find each sum that converges by using the formula from this section where it applies.

40. 16 + 2 + 1/4 + 1/32 + . . . **128/7**

41. 18 + 6 + 2 + 2/3 + . . . **27**

42. 100 + 10 + 1 + . . . **1000/9**

43. 128 + 64 + 32 + . . . **256**

44. 4/3 + 2/3 + 1/3 + . . . **8/3**

45. 1/4 − 1/6 + 1/9 − 2/27 + . . . **3/20**

46. $\sum_{i=1}^{\infty} 3(1/4)^{i-1}$ **4**

47. $\sum_{i=1}^{\infty} 5(-1/4)^{i-1}$ **4**

48. $\sum_{k=1}^{\infty} (.3)^k$ **3/7**

49. $\sum_{k=1}^{\infty} 10^{-k}$ **1/9**

✍ **50.** Explain the difference between an arithmetic sequence and a geometric sequence.

51. The final step in processing a black-and-white photographic print is to immerse the print in a chemical called ''fixer.'' The print is then washed in running water. Under certain conditions, 98% of the fixer in a print will be removed with 15 min of washing. How much of the original fixer would be left after 1 hr of washing? **.000016%**

52. A scientist has a vat containing 100 L of a pure chemical. Twenty liters is drained and replaced with water. After complete mixing, 20 L of the mixture is drained and replaced with water. What will be the strength of the mixture after 9 such drainings? **≈13.4%**

53. The half-life of a radioactive substance is the time it takes for half the substance to decay. Suppose the half-life of a substance is 3 yr, and 10^{15} molecules of the substance are present initially. How many molecules will be present after 15 yr? **$1/32 \times 10^{15}$ or 3.125×10^{13} molecules**

54. Each year a machine loses 20% of the value it had at the beginning of the year. Find the value of the machine at the end of 6 yr if it cost $100,000 new. **$26,214.40**

 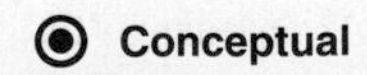 ▲ Challenging 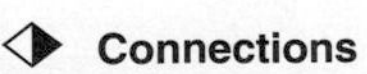

55. A bicycle wheel rotates 400 times in one minute. If the rider removes his or her feet from the pedals, the wheel will start to slow down. Each minute, it will rotate only 3/4 as many times as in the preceding minute. How many times will the wheel rotate in the fifth minute after the rider's feet are removed from the pedals? **about 95 times**

56. A piece of paper is .008 inch thick. Suppose the paper is folded in half, so that its thickness doubles, for 12 times in a row. How thick is the folded paper? **32.768 inches**

57. A sugar factory receives an order for 1000 units of sugar. The production manager thus orders production of 1000 units of sugar. He forgets, however, that the production of sugar requires some sugar (to prime the machines, for example), and so he ends up with only 900 units of sugar. He then orders an additional 100 units, and receives only 90 units. A further order for 10 units produces 9 units. Finally seeing he is wrong, the manager decides to try mathematics. He views the production process as an infinite geometric progression with $a_1 = 1000$ and $r = .1$. Using this, find the number of units of sugar that he should have ordered originally. **10,000/9 units**

58. After a person pedaling a bicycle removes his or her feet from the pedals, the wheel rotates 400 times the first minute. As it continues to slow down, it rotates in each minute only 3/4 as many times as in the previous minute. How many times will the wheel rotate before coming to a complete stop? **1600 rotations**

59. A pendulum bob swings through an arc 40 cm long on its first swing. Each swing thereafter, it swings only 80% as far as on the previous swing. How far will it swing altogether before coming to a complete stop? **200 cm**

60. Mitzi drops a ball from a height of 10 m and notices that on each bounce the ball returns to about 3/4 of its previous height. About how far will the ball travel before it comes to rest? (*Hint:* Consider the sum of two sequences). **70 m**

61. Each person has two parents, four grandparents, eight great-grandparents, and so on. What is the total number of ancestors a person has, going back five generations? Ten generations? **62; 2046**

62. Certain medical conditions are treated with a fixed dose of a drug administered at regular intervals. Suppose that a person is given 2 mg of a drug each day and that during each 24-hr period, the body utilizes 40% of the amount of drug that was present at the beginning of the period.

(a) Show that the amount of the drug present in the body at the end of n days is

$$\sum_{i=1}^{n} 2(.6)^i.$$

(b) What will be the approximate quantity of the drug in the body at the end of each day after the treatment has been administered for a long period of time? **3 mg**

(c) What is the maximum daily dosage that will guarantee that the amount of the drug in the body never exceeds 2 mg? **4/3 mg**

▲ **63.** A sequence of equilateral triangles is constructed. The first triangle has sides 2 m in length. To get the second triangle, midpoints of the sides of the original triangle are connected. What is the length of the side of the eighth such triangle? See the figure below. **1/64 m**

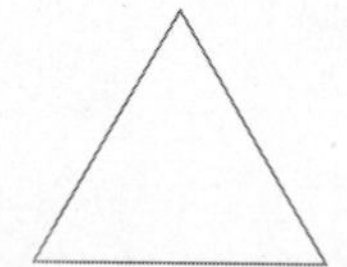 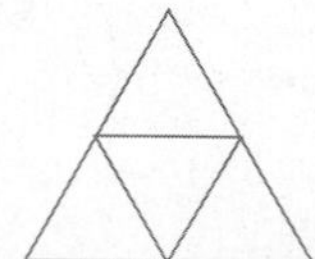 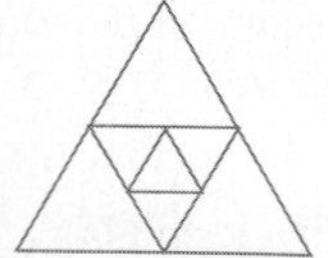 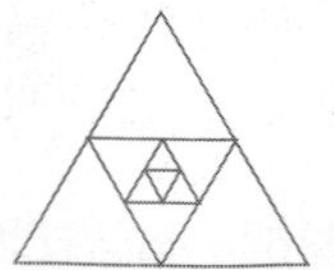

▲ **64.** In Exercise 63, if the process could be continued indefinitely, what would be the total perimeter of all the triangles? What would be the total area of all the triangles, disregarding the overlapping? **12 m; $4\sqrt{3}/3$ sq m**

◉ **65.** Find three numbers x, y, and z that are consecutive terms of both an arithmetic sequence and a geometric sequence. **$x = y = z$, where x is any number**

◉ **66.** Let $a_1, a_2, a_3, \ldots$ and $b_1, b_2, b_3, \ldots$ be geometric sequences. Let $d_n = c \cdot a_n \cdot b_n$ for any real number c and every positive integer n. Show that $d_1, d_2, d_3, \ldots$ is a geometric sequence.

◉ *Suppose that $a_1, a_2, a_3, a_4, a_5, \ldots$ is a geometric sequence with common ratio r. Show that the following sequences are geometric and give their common ratios.*

67. $a_1, a_4, a_7, a_{10}, \ldots$

68. $(a_1)^2, (a_2)^2, (a_3)^2, (a_4)^2, (a_5)^2, \ldots$

69. $\sqrt{a_1}, \sqrt{a_2}, \sqrt{a_3}, \sqrt{a_4}, \sqrt{a_5}, \ldots$

70. $\frac{3}{a_1}, \frac{3}{a_2}, \frac{3}{a_3}, \frac{3}{a_4}, \frac{3}{a_5}, \ldots$

◆ ✎ *Explain why the following sequences are geometric.*

71. log 6, log 36, log 1296, log 1,679,616, . . .

72. log 2, log 4, log 16, log 256, . . .

8.4 Exercises (text page 439)

Evaluate each of the following binomial coefficients.

1. $\binom{10}{6}$ **210** **2.** $\binom{8}{3}$ **56** **3.** $\binom{12}{0}$ **1** **4.** $\binom{20}{17}$ **1140** **5.** $\binom{25}{2}$ **300** **6.** $\binom{28}{28}$ **1**

Write out the binomial expansion for each of the following.

7. $(m^2 + n)^6$ $\mathbf{m^{12} + 6m^{10}n + 15m^8n^2 + 20m^6n^3 + 15m^4n^4 + 6m^2n^5 + n^6}$

8. $(a + b^2)^5$ $\mathbf{a^5 + 5a^4b^2 + 10a^3b^4 + 10a^2b^6 + 5ab^8 + b^{10}}$

9. $(z - 3w)^5$ $\mathbf{z^5 - 15z^4w + 90z^3w^2 - 270z^2w^3 + 405zw^4 - 243w^5}$

10. $(2k - h)^6$ $\mathbf{64k^6 - 192k^5h + 240k^4h^2 - 160k^3h^3 + 60k^2h^4 - 12kh^5 + h^6}$

11. $(r/2 - 3)^5$ $\mathbf{r^5/32 - 15r^4/16 + 45r^3/8 - 135r^2/2 + 405r/2 - 243}$

12. $(5 - x/4)^5$ $\mathbf{3125 - 3125x/4 + 625x^2/8 - 125x^3/32 + 25x^4/256 - x^5/1024}$

13. $(x^{3/5} + 2y^{4/5})^5$ $\mathbf{x^3 + 10x^{12/5}y^{4/5} + 40x^{9/5}y^{8/5} + 80x^{6/5}y^{12/5} + 80x^{3/5}y^{16/5} + 32y^4}$

14. $(3z^{1/3} + w^{2/3})^4$ $\mathbf{81z^{4/3} + 108zw^{2/3} + 54z^{2/3}w^{4/3} + 12z^{1/3}w^2 + w^{8/3}}$

✎ **15.** What is true of the signs of the terms of the expansion of $(x - y)^n$ if $y > 0$? Explain why this is so.

Write only the first four terms in each of the following expansions.

16. $(a + 2b)^{15}$ $\mathbf{a^{15} + 30a^{14}b + 420a^{13}b^2 + 3640a^{12}b^3}$

17. $(3m - n)^{20}$ $\mathbf{3^{20}m^{20} - 20 \cdot 3^{19}m^{19}n + 190 \cdot 3^{18}m^{18}n^2 - 1140 \cdot 3^{17}m^{17}n^3}$

18. $(4m^{-1} + m^{-2})^{12}$ **$4^{12}m^{-12} + 12 \cdot 4^{11}m^{-13} + 66 \cdot 4^{10}m^{-14} + 220 \cdot 4^9m^{-15}$**

19. $(k^{-2} + 3k^2)^9$ **$k^{-18} + 27k^{-14} + 324k^{-10} + 2268k^{-6}$**

In Exercises 20–25 write the indicated term of the binomial expansion.

20. Sixth term of $(4h - j)^8$ **$-3584h^3j^5$**

21. Eighth term of $(2c - 3d)^{14}$ **$-3432 \cdot 6^7c^7d^7$**

22. Fifteenth term of $(a^2 + b)^{22}$ **$319{,}770a^{16}b^{14}$**

23. Twelfth term of $(2x + y^2)^{16}$ **$139{,}776x^5y^{22}$**

24. Fifteenth term of $(x - y^3)^{20}$ **$38{,}760x^6y^{42}$**

25. Tenth term of $(a^3 + 3b)^{11}$ **$55 \cdot 3^9a^6b^9$**

26. Find the middle term of $(3x^7 + 2y^3)^8$. **$90{,}720x^{28}y^{12}$**

27. Find the two middle terms of $(-2m^{-1} + 3n^{-2})^{11}$. **$462(2^6)(3^5)m^{-6}n^{-10}$, $-462(2^5)(3^6)m^{-5}n^{-12}$**

28. Find the value of n for which the coefficients of the fifth and eighth terms in the expansion of $(x + y)^n$ are the same. **11**

29. Find the term in the expansion of $(3 + \sqrt{x})^{11}$ that contains x^4. **$4{,}455x^4$**

30. Find the value of n for which the coefficient of x^2 in the expansion of $(1 + x)^n$ is 21. **7**

In later courses, it is shown that

$$(1 + x)^n = 1 + nx + \frac{n(n-1)}{2!}x^2 + \frac{n(n-1)(n-2)}{3!}x^3 + \ldots$$

for any real number n (not just positive integer values) and any real number x where $|x| < 1$. This result, a generalized binomial theorem, may be used to find approximate values of powers and roots. For example,

$$\begin{aligned}(1.008)^{1/4} &= (1 + .008)^{1/4} \\ &= 1 + \frac{1}{4}(.008) + \frac{1/4(-3/4)}{2!}(.008)^2 + \frac{1/4(-3/4)(-7/4)}{3!}(.008)^3 + \ldots \\ &\approx 1.002.\end{aligned}$$

Use this result to approximate the quantities in Exercises 31–34 to the nearest thousandth.

31. $(1.02)^{-3}$ **.942**

32. $1/(1.04)^5$ **.822**

33. $(1.01)^{3/2}$ **1.015**

34. $(1.03)^{.2}$ **1.006**

35. Let $n = -1$ and expand $(1 + x)^{-1}$. **$1 - x + x^2 - x^3 + x^4 - \ldots$**

36. Use polynomial division to find the first four terms when $1 + x$ is divided into 1. Compare the result with the result of Exercise 35. What do you find? Explain.

37. Find the sum of the first four terms in the expansion of $(1 + 3)^{1/2}$ using $x = 3$ and $n = 1/2$ in the formula above. Is the result close to $(1 + 3)^{1/2} = 4^{1/2} = 2$? Why not? Explain.

38. Use the result above to show that for small values of x, $\sqrt{1 + x} \approx 1 + \frac{1}{2}x$.

39. When $(4x - 5)^7$ is written in the form $a_7x^7 + a_6x^6 + \ldots + a_1x + a_0$, what is the sum of the numbers $a_7, a_6, \ldots, a_1, a_0$? (*Hint:* This question can be answered without determining the values of the coefficients.) **-1**

40. Show that $\binom{n}{0} + \binom{n}{1} + \binom{n}{2} + \ldots + \binom{n}{n} = 2^n$. (*Hint:* Set $x = 1$ in the binomial expansion of $(1 + x)^n$.)

8.5 Exercises (text page 446)

Write out in full and verify each of the statements S_1, S_2, S_3, S_4, and S_5 for each of the following. Then use mathematical induction to prove that each of the given statements is true for every positive integer n.

1. $2 + 4 + 6 + \ldots + 2n = n(n + 1)$ **S_1: $2 = 1(1 + 1)$; S_2: $2 + 4 = 2(2 + 1)$; S_3: $2 + 4 + 6 = 3(3 + 1)$; S_4: $2 + 4 + 6 + 8 = 4(4 + 1)$; S_5: $2 + 4 + 6 + 8 + 10 = 5(5 + 1)$**

2. $1 + 3 + 5 + \ldots + (2n - 1) = n^2$ **S_1: $1 = 1^2$; S_2: $1 + 3 = 2^2$; S_3: $1 + 3 + 5 = 3^2$; S_4: $1 + 3 + 5 + 7 = 4^2$; S_5: $1 + 3 + 5 + 7 + 9 = 5^2$**

Use the method of mathematical induction to prove that each of the following statements is true for every positive integer n.

3. $2 + 4 + 8 + \ldots + 2^n = 2^{n+1} - 2$

4. $1^2 + 2^2 + 3^2 + \ldots + n^2 = \dfrac{n(n + 1)(2n + 1)}{6}$

5. $1^3 + 2^3 + 3^3 + \ldots + n^3 = \dfrac{n^2(n + 1)^2}{4}$

6. $3 + 3^2 + 3^3 + \ldots + 3^n = \dfrac{3(3^n - 1)}{2}$

7. $5 \cdot 6 + 5 \cdot 6^2 + 5 \cdot 6^3 + \ldots + 5 \cdot 6^n = 6(6^n - 1)$

8. $\dfrac{1}{1 \cdot 2} + \dfrac{1}{2 \cdot 3} + \dfrac{1}{3 \cdot 4} + \ldots + \dfrac{1}{n(n + 1)} = \dfrac{n}{n + 1}$

9. $\dfrac{1}{1 \cdot 4} + \dfrac{1}{4 \cdot 7} + \dfrac{1}{7 \cdot 10} + \ldots + \dfrac{1}{(3n - 2)(3n + 1)} = \dfrac{n}{3n + 1}$

10. $\dfrac{1}{2} + \dfrac{1}{2^2} + \dfrac{1}{2^3} + \ldots + \dfrac{1}{2^n} = 1 - \dfrac{1}{2^n}$

In the following statements S_n, find a value of n for which S_n is not true or prove S_n by mathematical induction.

11. $2^n > 2n$ **$n = 1$**

12. $3^n > 2n + 1$ **$n = 1$**

13. $1 \cdot 4 + 2 \cdot 9 + 3 \cdot 16 + \ldots + n(n + 1)^2 = \dfrac{n(n + 1)(n + 2)(3n + 5)}{12}$ **true for all positive integers n**

14. $2^n > n^2$ **$n = 2, 3,$ or 4**

15. $n! > 2n$ **$n = 1, 2,$ or 3**

16. $1 \cdot 2 + 2 \cdot 3 + 3 \cdot 4 + \ldots + n(n + 1) = \dfrac{n(n + 1)(n + 2)}{3}$ **true for all positive integers n**

Prove each result in Exercises 17–26 by mathematical induction.

◆ **17.** $(a^m)^n = a^{mn}$ (Assume a and m are constant.)

◆ **18.** $(ab)^n = a^n b^n$ (Assume a and b are constant.)

19. $2^n > 2n$, if $n \geq 3$

20. $3^n > 2n + 1$, if $n \geq 2$

21. If $a > 1$, then $a^n > 1$

22. If $a > 1$, then $a^n > a^{n-1}$

23. If $0 < a < 1$, then $a^n < a^{n-1}$

▲ **24.** $2^n > n^2$, for $n > 4$

◆ ▲ **25.** If $n \geq 4$, then $n! > 2^n$, where $n! = n(n - 1)(n - 2) \ldots (3)(2)(1)$.

▲ **26.** $4^n > n^4$, for $n \geq 5$

◉ ✎ **27.** Suppose that Step 2 in a proof by mathematical induction can be satisfied, but Step 1 cannot. May we conclude that the proof is complete? Explain.

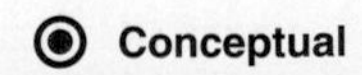
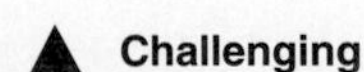

Writing Conceptual Challenging Connections

28. What is wrong with the following "proof" by mathematical induction?
Prove: Any natural number equals the next natural number.
To begin, we assume the statement is true for some natural number k:

$$k = k + 1.$$

We must now show that the statement is true for $n = k + 1$. If we add 1 to both sides, we have

$$k + 1 = k + 1 + 1$$
$$k + 1 = k + 2.$$

Hence, if the statement is true for $n = k$, it is also true for $n = k + 1$. Thus, the theorem is proved.

29. Suppose that n straight lines (with $n \geq 2$) are drawn in a plane, where no two lines are parallel and no three lines pass through the same point. Show that the number of points of intersection of the lines is $(n^2 - n)/2$.

30. The series of sketches at the side starts with an equilateral triangle having sides of length 1. In the following steps, equilateral triangles are constructed on each side of the preceding figure. The lengths of the sides of these new triangles is 1/3 the length of the sides of the preceding triangles. Develop a formula for the number of sides of the nth figure. Use mathematical induction to prove your answer.

31. Find the perimeter of the nth figure in Exercise 30. $\mathbf{3\left(\frac{4}{3}\right)^{n-1}}$ **or** $\mathbf{\frac{4^{n-1}}{3^{n-2}}}$

32. Show that the area of the nth figure in Exercise 30 is $\sqrt{3}\left[\frac{2}{5} - \frac{3}{20}\left(\frac{4}{9}\right)^{n-1}\right]$.

33. A pile of n rings, each smaller than the one below it, is on a peg. Two other pegs are attached to a board with this peg. In the game called the *Tower of Hanoi* puzzle, all the rings must be moved to a different peg, with only one ring moved at a time, and with no ring ever placed on top of a smaller ring. Find the least number of moves that would be required. Prove your result with mathematical induction.

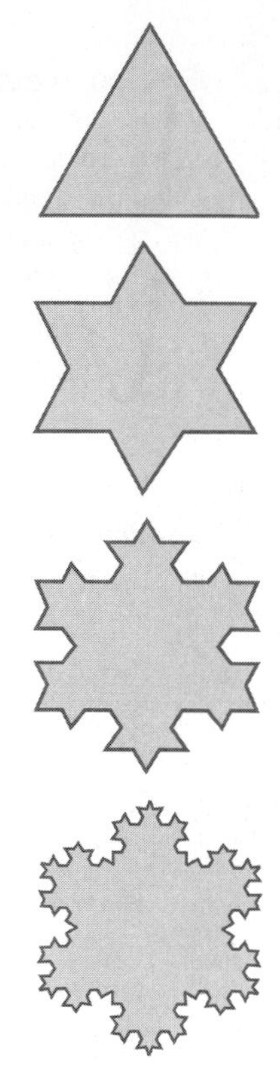

Exercise 30

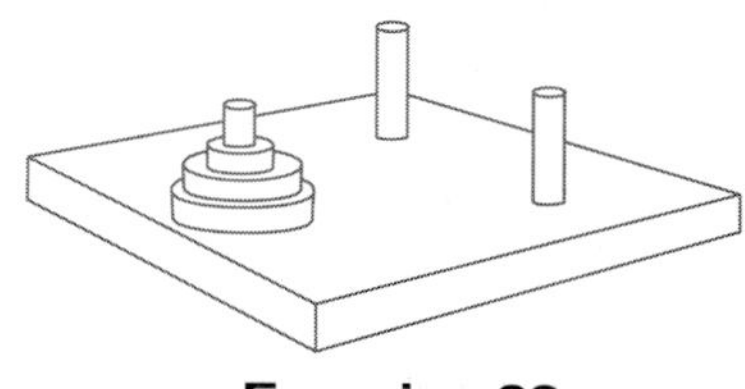

Exercise 33

8.6 Exercises (text page 454)

Evaluate each expression in Exercises 1–12.

1. $P(12, 8)$ **19,958,400**
2. $P(5, 5)$ **120**
3. $P(9, 2)$ **72**
4. $P(10, 9)$ **3,628,800**
5. $P(5, 1)$ **5**
6. $P(6, 0)$ **1**
7. $\binom{4}{2}$ **6**
8. $\binom{9}{3}$ **84**
9. $\binom{6}{0}$ **1**
10. $\binom{8}{1}$ **8**
11. $\binom{12}{4}$ **495**
12. $\binom{16}{3}$ **560**

Use the multiplication principle or permutations to solve the following problems.

13. In an experiment on social interaction 6 people will sit in 6 seats in a row. In how many different ways can the 6 people be seated? **720**

14. In how many ways can 7 of 10 mice be arranged in a row for a genetics experiment? **604,800**

15. For many years, the state of California used three letters followed by three digits on its automobile license plates.
   (a) How many different license plates are possible with this arrangement? **17,576,000**
   (b) When the state ran out of new plates, the order was reversed to three digits followed by three letters. How many additional plates were then possible? **17,576,000**
   (c) Several years ago, the plates described in (b) were also used up. The state then issued plates with one letter followed by three digits and then three letters. How many plates does this scheme provide? **456,976,000**

16. How many 7-digit telephone numbers are possible if the first digit cannot be 0, and
   (a) only odd digits may be used? **78,125**
   (b) the telephone number must be a multiple of 10 (that is, it must end in 0)? **900,000**
   (c) the first three digits must be 456? **10,000**

17. If your college offers 400 courses, 20 of which are in mathematics, and your counselor arranges your schedule of 4 courses by random selection, how many schedules are possible that do not include a math course? **$2.052371412 \times 10^{10}$**

18. In a club with 35 members, how many ways can a slate of 3 officers consisting of president, program chairman, and secretary/treasurer be chosen? **39,270**

19. In how many ways can 5 players be assigned to the 5 positions on a basketball team, assuming that any player can play any position? In how many ways can 10 players be assigned to the 5 positions? **120; 30,240**

20. A softball team has 20 players. How many 9-player batting orders are possible? **$6.09493248 \times 10^{10}$**

21. In how many ways can 6 bank tellers be assigned to 6 different windows? In how many ways can 10 tellers be assigned to the 6 windows? **720; 151,200**

Use combinations to solve the following problems.

22. A homeowners' association has 50 members. If a committee of 6 is to be selected at random, how many different committees are possible? **15,890,700**

23. How many different samples of 4 light bulbs can be selected from a carton of 2 dozen bulbs? **10,626**

24. A group of 5 students is to be selected at random from a class of 30 to participate in an experimental class. In how many ways can this be done? In how many ways can the group that will not participate be selected? **142,506; 142,506**

25. Harry's Hamburger Heaven sells hamburgers with cheese, relish, lettuce, tomato, onion, mustard, or ketchup. How many different hamburgers can be concocted using any 4 of the extras? **35**

26. How many different 5-card poker hands can be dealt from a deck of 52 playing cards? **2,598,960**

27. Seven cards are marked with the numbers 1 through 7 and are shuffled, and then 3 cards are drawn. How many different 3-card combinations are possible? **35**

28. A bag contains 18 marbles. How many samples of 3 can be drawn from it? How many samples of 5 marbles? **816; 8568**

29. In Exercise 28, if the bag contains 5 purple, 4 green, and 9 black marbles, how many samples of 3 can be drawn in which all the marbles are black? How many samples of 3 can be drawn in which exactly 2 marbles are black? **84; 324**

30. In Exercise 23, assume it is known that there are 5 defective light bulbs in the carton. How many samples of 4 can be drawn in which all are defective? How many samples of 4 can be drawn in which there are 2 good bulbs and 2 defective bulbs? **5; 1,710**

31. Explain the difference between a permutation and a combination. What should you look for in a problem to decide which of these is an appropriate method of solution?

32. Is choosing two kittens from a litter of six kittens an example of a permutation or a combination? Explain.

33. Padlocks with digit dials are often referred to as "combination locks." According to the mathematical definition of combination, is this an accurate description? Why or why not?

34. Determine whether each of the following is a permutation or a combination.
(a) Your five-digit postal zip code **permutation**
(b) A particular five-card hand in a game of poker **combination**
(c) A committee of school board members **combination**

Solve each of the following by using either permutations or combinations.

35. From a pool of 7 secretaries, 3 are selected to be assigned to 3 managers, with 1 secretary for each manager. In how many ways can this be done? **210**

36. In a game of musical chairs, 12 children will sit in 11 chairs (1 will be left out). How many seatings are possible? **479,001,600**

37. In an experiment on plant hardiness, a researcher gathers 6 wheat plants, 3 barley plants, and 2 rye plants. Four plants are to be selected at random.
(a) In how many ways can this be done? **330**
(b) In how many ways can this be done if 2 wheat plants must be included? **150**

38. In an office with 8 men and 11 women, how many 5-member groups can be chosen that have each of the following compositions.
(a) All men **(b)** All women **(c)** 3 men and 2 women
(d) No more than 3 women **(a) 56 (b) 462 (c) 3,080 (d) 8,526**

39. From 10 names on a ballot, 4 will be elected to a political party committee. How many different committees are possible? In how many ways can the committee of 4 be formed if each person will have a different responsibility? **210; 5,040**

40. In how many ways can 5 of 9 plants be arranged in a row on a windowsill? **15,120**

41. Velma specializes in making different vegetable soups with carrots, celery, onions, beans, peas, tomatoes, and potatoes. How many different soups can she make using any 4 ingredients? **35**

42. How many 4-letter radio-station call letters can be made if the first letter must be K or W and no letter may be repeated? How many if repeats are allowed? How many of the call letters with no repeats can end in K? **27,600; 35,152; 552**

43. A group of 12 workers decide to send a delegation of 3 to their supervisor to discuss their work assignments.
(a) How many delegations of 3 are possible? **220**
(b) How many if one of the 12, the foreman, must be in the delegation? **55**
(c) If there are 5 women and 7 men in the group, how many possible delegations would include a woman? **105**

44. The Riverdale board of supervisors is composed of 2 liberals and 5 conservatives. Three members are to be selected randomly as delegates to a convention.
(a) How many delegations are possible? **35**
(b) How many delegations could have all liberals? **0**
(c) How many delegations could have 2 conservatives and 1 liberal? **20**
(d) If the supervisor who serves as chairman of the board must be included, how many delegations are possible? **15**

Prove each of the following statements for positive integers n and r, with $r \le n$.

45. $P(n, n-1) = P(n, n)$ **46.** $P(n, 1) = n$ **47.** $P(n, 0) = 1$ **48.** $\binom{n}{n} = 1$

49. $\binom{n}{0} = 1$ **50.** $\binom{n}{n-1} = n$ **51.** $\binom{n}{n-r} = \binom{n}{r}$

8.7 Exercises (text page 462)

Write a sample space with equally likely outcomes for each of the following experiments.

1. A two-headed coin is tossed once. $\boldsymbol{S = \{H\}}$

2. Two ordinary coins are tossed. $\boldsymbol{S = \{HH, HT, TH, TT\}}$

3. Three ordinary coins are tossed. $\boldsymbol{S = \{HHH, HHT, HTH, THH, HTT, THT, TTH, TTT\}}$

4. Five slips of paper marked with the numbers 1, 2, 3, 4, and 5 are placed in a box. After mixing well, two slips are drawn. **S = {(1, 2), (1, 3), (1, 4), (1, 5), (2, 3), (2, 4), (2, 5), (3, 4), (3, 5), (4, 5)}**

5. An unprepared student takes a three-question true/false quiz in which he guesses the answer to all three questions. **Let c = correct, w = wrong. S = {*ccc, ccw, cwc, wcc, wwc, wcw, cww, www*}**

6. A die is rolled and then a coin is tossed. **S = {1*H*, 2*H*, 3*H*, 4*H*, 5*H*, 6*H*, 1*T*, 2*T*, 3*T*, 4*T*, 5*T*, 6*T*}**

Write the events in Exercises 7–10 in set notation and give the probability of each event.

7. In the experiment from Exercise 2:
(a) Both coins show the same face. **{*HH, TT*}, 1/2**
(b) At least one coin turns up heads. **{*HH, HT, TH*}, 3/4**

8. In Exercise 1:
(a) The result of the toss is heads. **{*H*}; 1**
(b) The result of the toss is tails. **∅; 0**

9. In Exercise 4:
(a) Both slips are marked with even numbers.
(b) Both slips are marked with odd numbers.
(c) Both slips are marked with the same number.
(d) One slip is marked with an odd number and the other with an even number.
(a) {(2, 4)}, 1/10 (b) {(1, 3), (1, 5), (3, 5)}, 3/10 (c) ∅, 0 (d) {(1, 2), (1, 4), (2, 3), (2, 5), (3, 4), (4, 5)}, 3/5

10. In Exercise 5:
(a) The student gets all three answers correct.
(b) He gets all three answers wrong.
(c) He gets exactly two answers correct.
(d) He gets at least one answer correct.
(a) {*ccc*}; 1/8 (b) {*www*}; 1/8 (c) {*ccw, cwc, wcc*}; 3/8 (d) {*cww, wcw, wwc, ccw, cwc, wcc, ccc*}; 7/8

11. A student gives the answer to a probability problem as 6/5. Explain why this answer must be incorrect.

12. If the probability of an event is .857, what is the probability that the event will not occur? **.143**

13. A marble is drawn at random from a box containing 3 yellow, 4 white, and 8 blue marbles. Find each probability in (a) – (c).
(a) A yellow marble is drawn. (b) A blue marble is drawn.
(c) A black marble is drawn. **(a) 1/5 (b) 8/15 (c) 0**
(d) What are the odds in favor of drawing a yellow marble? **1 to 4**
(e) What are the odds against drawing a blue marble? **7 to 8**

14. A baseball player with a batting average of .300 comes to bat. What are the odds in favor of his getting a hit? **3 to 7**

15. In Exercise 4, what are the odds that the sum of the numbers on the two slips of paper is 5? **1 to 4**

16. If the odds that it will rain are 4 to 5, what is the probability of rain? **4/9**

17. If the odds that a candidate will win an election are 3 to 2, what is the probability that the candidate will lose? **2/5**

18. Ms. Bezzone invites 10 relatives to a party: her mother, two uncles, three brothers, and four cousins. If the chances of any one guest arriving first are equally likely, find the following probabilities.
(a) The first guest is an uncle or a cousin. **3/5**
(b) The first guest is a brother or a cousin. **7/10**
(c) The first guest is an uncle or her mother. **3/10**

19. A card is drawn from a well-shuffled deck of 52 cards. Find the probability that the card is the following.
(a) A queen (b) Red (c) A black 3 (d) A club or red **(a) 1/13 (b) 1/2 (c) 1/26 (d) 3/4**

20. In Exercise 19, find the probability of the following.
(a) A face card (K, Q, J of any suit) (b) Red or a 3 **(a) 3/13 (b) 7/13**
(c) Less than a four (consider aces as ones) **3/13**

21. Two dice are rolled. Find the probability of the following events.
(a) The sum of the points is at least 10. **1/6**
(b) The sum of the points is either 7 or at least 10. **1/3**
(c) The sum of the points is 3 or the dice both show the same number. **2/9**

22. If a marble is drawn from a bag containing 2 yellow, 5 red, and 3 blue marbles, what are the probabilities of the following results?
(a) The marble is yellow or blue. **1/2**
(b) The marble is yellow or red. **7/10**
(c) The marble is green. **0**

23. The law firm of Alam, Bartolini, Chinn, Dickinson, and Ellsberg has two senior partners, Alam and Bartolini. Two of the attorneys are to be selected to attend a conference. Assuming that all are equally likely to be selected, find the following probabilities.
(a) Chinn is selected. **2/5**
(b) Alam and Dickinson are selected. **1/10**
(c) At least one senior partner is selected. **7/10**

24. The management of a firm wants to survey its workers, who are classified as follows for the purpose of an interview: 30% have worked for the company more than 5 years; 28% are female; 65% contribute to a voluntary retirement plan; half of the female workers contribute to the retirement plan. Find the following probabilities.
(a) A male worker is selected. **.72**
(b) A worker is selected who has been employed by the company for 5 years or less. **.70**
(c) A worker is selected who contributes to the retirement plan or is female. **.79**

25. The table shows the probabilities of a person accumulating specific amounts of credit card charges over a 12-month period.

Find the probabilities that a person's total charges during the period are the following.
(a) \$500–\$999 **(b)** \$5000–\$9999
(c) \$500 to \$2999 **(d)** \$3000 or more
(a) .18 (b) .06 (c) .39 (d) .12

Charges	*Probability*
Under \$100	.31
\$100–\$499	.18
\$500–\$999	.18
\$1000–\$1999	.13
\$2000–\$2999	.08
\$3000–\$4999	.05
\$5000–\$9999	.06
\$10,000 or more	.01

26. In Exercise 25, find the probabilities that a person charges the following amounts.
(a) \$100–\$499 **(b)** \$2000–\$2999
(c) Less than \$2000 **(d)** More than \$499
(a) .18 (b) .08 (c) .8 (d) .51

In most animals and plants, it is very unusual for the number of main parts of the organism (arms, legs, toes, flower petals, etc.) to vary from generation to generation. Some species, however, have meristic variability, *in which the number of certain body parts varies from generation to generation. One researcher studied the front feet of certain guinea pigs and produced the following probabilities.**

$$P(\text{only four toes, all perfect}) = .77$$
$$P(\text{one imperfect toe and four good ones}) = .13$$
$$P(\text{exactly five good toes}) = .10$$

Find the probability of each of the following events.

27. No more than four good toes **.90**

28. Five toes, whether perfect or not **.23**

The probabilities for the outcomes of an experiment having sample space $S = \{s_1, s_2, s_3, s_4, s_5, s_6\}$ *are shown here.*

Outcomes	s_1	s_2	s_3	s_4	s_5	s_6
Probability	.17	.03	.09	.46	.21	.04

Let $E = \{s_1, s_2, s_5\}$, *and let* $F = \{s_4, s_5\}$. *Find each probability in Exercises 29–34.*

29. $P(E)$ **.41**

30. $P(F)$ **.67**

31. $P(E \cap F)$ **.21**

32. $P(E \cup F)$ **.87**

33. $P(E' \cup F')$ **.79**

34. $P(E' \cap F)$ **.46**

*Excerpt from ''Analysis of Variability in Number of Digits in an Inbred Strain of Guinea Pigs'' by S. Wright, in *Genetics*, v. 19 (1934), 506–36. Reprinted by permission of Genetics Society of America.

Chapter 8 Review Exercises (text page 465)

Use summation notation to rewrite each sum with the index of summation starting at the indicated number.

1. $\sum_{i=1}^{8}(3+2i)$; -2 $\quad \sum_{j=-2}^{5}(9+2j)$

2. $\sum_{i=2}^{9}(4-6i)$; 0 $\quad \sum_{j=0}^{7}(-8-6j)$

Use the properties of summation, along with sums given earlier, to evaluate each summation.

3. $\sum_{i=1}^{4}(i^2+2i)$ **50**

4. $\sum_{i=1}^{6}(8+i^3)$ **489**

5. How can you decide whether a given sequence or series is arithmetic or geometric?

Write the first five terms for each sequence in Exercises 6–15.

6. $a_1 = 8, a_2 = 4, a_n = a_{n-1} - a_{n-2}$ for $n > 2$ **8, 4, −4, −8, −4**

7. $b_1 = 5, b_2 = -1, b_n = -2 \cdot b_{n-2}$ if n is odd, and $b_n = 2 \cdot b_{n-2}$ if n is even **5, −1, −10, −2, 20**

8. Arithmetic, $a_2 = 10, d = -2$ **12, 10, 8, 6, 4**

9. Arithmetic, $a_2 = 5, a_3 = 3$ **7, 5, 3, 1, −1**

10. Arithmetic, $a_3 = \pi, a_4 = 1$ $\mathbf{3\pi - 2, 2\pi - 1, \pi, 1, -\pi + 2}$

11. Arithmetic, $a_1 = 4 - \sqrt{3}, a_2 = 3$ $\mathbf{4 - \sqrt{3}, 3, 2 + \sqrt{3}, 1 + 2\sqrt{3}, 3\sqrt{3}}$

12. Geometric, $a_1 = 6, r = 2$ **6, 12, 24, 48, 96**

13. Geometric, $a_3 = 4, r = -1/2$ **16, −8, 4, −2, 1**

14. Geometric, $a_1 = -5, a_2 = -1$ **−5, −1, −1/5, −1/25, −1/125**

15. Geometric, $a_2 = 3, a_5 = 12$ $\mathbf{3\sqrt[3]{2}/2, 3, 3\sqrt[3]{4}, 6\sqrt[3]{2}, 12}$

16. An arithmetic sequence has $a_5 = -3$ and $a_{15} = 17$. Find a_1 and a_n. $\mathbf{-11; -13 + 2n}$

17. A geometric sequence has $a_1 = -8$ and $a_7 = -1/8$. Find a_4 and a_n.
$\mathbf{-1; -8(1/2)^{n-1} = -(1/2)^{n-4}}$ **or** $\mathbf{1; -8(-1/2)^{n-1} = (-1/2)^{n-4}}$

Find a_8 for each of the following arithmetic sequences.

18. $a_1 = 6, d = 2$ **20**

19. $a_1 = -4, d = 3$ **17**

20. $a_1 = 6x - 9, a_2 = 5x + 1$ $\mathbf{-x + 61}$

21. $a_3 = 11m, a_5 = 7m - 4$ $\mathbf{m - 10}$

Find S_{12} for each of the following arithmetic sequences.

22. $a_1 = 2, d = 3$ **222**

23. $a_2 = 6, d = 10$ **612**

24. $a_1 = -4k, d = 2k$ $\mathbf{84k}$

Find a_5 for each of the following geometric sequences.

25. $a_1 = -2, r = 3$ **−162**

26. $a_3 = 4, r = 1/5$ **4/25**

27. $a_1 = 3y, a_2 = y^3$ $\mathbf{y^9/27}$

28. $a_2 = \sqrt{2}, a_4 = 3\sqrt{2}$ $\mathbf{3\sqrt{6}}$ **or** $\mathbf{-3\sqrt{6}}$

Find S_4 for each of the following geometric sequences.

29. $a_1 = 3, r = 2$ **45**

30. $a_1 = -1, r = 3$ **−40**

31. $a_1 = p, r = -2p$ $\mathbf{p(1 - 16p^4)/(1 + 2p)}$

Determine whether each of the following sequences is arithmetic, geometric, or neither. If the sequence is arithmetic, find the common difference. If the sequence is geometric, find the common ratio.

32. $8, -4, 2, -1, \frac{1}{2}, \ldots$ **geometric; $r = -1/2$**

33. $-3, 0, 3, 6, 9, \ldots$ **arithmetic; $d = 3$**

34. $\ln 1, \ln 2, \ln 3, \ln 4, \ldots$ **neither**

35. $\ln 2, \ln 4, \ln 8\ \ln 16, \ldots$ **arithmetic; $d = \ln 2$**

36. Explain the difference between a sequence and a series.

Evaluate each of the following sums that converge.

37. $24 + 8 + 8/3 + 8/9 + \ldots$ **36**

38. $-3/4 + 1/2 - 1/3 + 2/9 - \ldots$ **$-9/20$**

39. $1/12 + 1/6 + 1/3 + 2/3 + \ldots$ **diverges**

40. $.9 + .09 + .009 + .0009 + \ldots$ **1**

Evaluate each of the following sums that exist.

41. $\sum_{i=1}^{7} (-1)^{i-1}$ **1**

42. $\sum_{i=1}^{5} (i^2 + i)$ **70**

43. $\sum_{i=1}^{4} \frac{i+1}{i}$ **73/12**

44. $\sum_{j=1}^{10} (3j - 4)$ **125**

45. $\sum_{j=1}^{2500} j$ **3,126,250**

46. $\sum_{i=1}^{5} 4 \cdot 2^i$ **248**

47. $\sum_{i=1}^{\infty} \left(\frac{4}{7}\right)^i$ **4/3**

48. $\sum_{i=1}^{\infty} -2\left(\frac{6}{5}\right)i$ **diverges**

Evaluate each of the following sums, where $x_1 = 0$, $x_2 = 1$, $x_3 = 2$, $x_4 = 3$, $x_5 = 4$, $x_6 = 5$.

49. $\sum_{i=1}^{4} (x_i^2 - 6)$ **-10**

50. $\sum_{i=1}^{6} f(x_i)\Delta x;\quad f(x) = (x - 2)^3,\ \Delta x = .1$ **2.7**

Write each of the following sums using summation notation.

51. $4 - 1 - 6 - \ldots - 66$ $\boldsymbol{\sum_{i=1}^{15} (9 - 5i)}$

52. $10 + 14 + 18 + \ldots + 86$ $\boldsymbol{\sum_{i=1}^{20} (6 + 4i)}$

53. $4 + 12 + 36 + \ldots + 972$ $\boldsymbol{\sum_{i=1}^{6} 4(3)^{i-1}}$

54. $\frac{5}{6} + \frac{6}{7} + \frac{7}{8} + \ldots + \frac{12}{13}$ $\boldsymbol{\sum_{i=5}^{12} \frac{i}{i+1}}$

55. What is the binomial theorem used for? Give examples.

Use the binomial theorem to expand each of the following.

56. $(x + 2y)^4$ $\boldsymbol{x^4 + 8x^3y + 24x^2y^2 + 32xy^3 + 16y^4}$

57. $(3z - 5w)^3$ $\boldsymbol{27z^3 - 135z^2w + 225zw^2 - 125w^3}$

58. $\left(3\sqrt{x} - \frac{1}{\sqrt{x}}\right)^5$ $\boldsymbol{243x^{5/2} - 405x^{3/2} + 270x^{1/2} - 90x^{-1/2} + 15x^{-3/2} - x^{-5/2}}$

59. $(m^3 - m^{-2})^4$ $\boldsymbol{m^{12} - 4m^7 + 6m^2 - 4m^{-3} + m^{-8}}$

Find the indicated term or terms for each of the following expansions.

60. Sixth term of $(4x - y)^8$ $\boldsymbol{-3{,}584x^3y^5}$

61. Seventh term of $(m - 3n)^{14}$ $\boldsymbol{3{,}003(-3)^6m^8n^6}$

62. First four terms of $(x + 2)^{12}$ $\boldsymbol{x^{12} + 24x^{11} + 264x^{10} + 1{,}760x^9}$

63. Last three terms of $(2a + 5b)^{16}$ $\boldsymbol{480 \cdot 5^{14}a^2b^{14} + 32 \cdot 5^{15}ab^{15} + 5^{16}b^{16}}$

Writing **Conceptual** **Challenging** **Connections**

64. Describe a proof by mathematical induction.

65. What kinds of statements are proved by mathematical induction? Give examples.

Use mathematical induction to prove that each of the following is true for every positive integer n.

66. $1 + 3 + 5 + 7 + \ldots + (2n - 1) = n^2$

67. $2 + 6 + 10 + 14 + \ldots + (4n - 2) = 2n^2$

68. $2 + 2^2 + 2^3 + \ldots + 2^n = 2(2^n - 1)$

69. $1^3 + 3^3 + 5^3 + \ldots + (2n - 1)^3 = n^2(2n^2 - 1)$

70. How do permutations and combinations differ? How are they alike?

Find the value of each expression in Exercises 71–74.

71. $P(9, 2)$ **72**

72. $P(6, 0)$ **1**

73. $\binom{8}{3}$ **56**

74. $\binom{10}{5}$ **252**

75. Four students are to be assigned to 4 different summer jobs. Each student is qualified for all 4 jobs. In how many ways can the jobs be assigned? **24**

76. Nine football teams are competing for first-, second-, and third-place titles in a statewide tournament. In how many ways can the winners be determined? **504**

77. John Jacobs, who is furnishing his apartment, wants to buy a new sofa. He can select from 5 different styles, each available in 3 different fabrics, with 6 color choices. How many different sofas are available? **90**

Write sample spaces for the following.

78. 2 coins are tossed. **{HH, HT, TH, TT}**

79. A card is drawn from a deck containing only the twelve face cards. **{J♥, Q♥, K♥, J♣, Q♣, K♣, J♦, Q♦, K♦, J♠, Q♠, K♠}**

80. A sample of 2 headsets from a box of 5 is tested for defects. **[sets 1 and 2, sets 1 and 3, sets 1 and 4, sets 1 and 5, sets 2 and 3, sets 2 and 4, sets 2 and 5, sets 3 and 4, sets 3 and 5, sets 4 and 5}**

81. The age of one student is determined from a class of college freshmen whose ages range from 17 to 25. **{17, 18, 19, 20, 21, 22, 23, 24, 25}**

A company sells typewriters and copiers. Let E be the event "a customer buys a typewriter," and let F be the event "a customer buys a copier." Write each of the following using ∩, ∪, or ′ as necessary.

82. A customer buys neither. $E' \cap F'$

83. A customer buys at least one. $E \cup F$

Find the odds in favor of a card drawn from an ordinary deck being as follows.

84. A club **1 to 3**

85. A black jack **1 to 25**

86. A red face card or a queen **2 to 11**

87. A sample shipment of five transistors is chosen at random. The probability of exactly 0, 1, 2, 3, 4, or 5 transistors being defective is given in the following table.

Number defective	0	1	2	3	4	5
Probability	.31	.25	.18	.12	.08	.06

Find the probability that at most two are defective. **.74**